CHINESE FOREST INSECTS

(Third Edition)

China Forestry Publishing House

CHINESE FOREST INSECTS

(Third Edition)

Chief Editors Xiao Gangrou, Li Zhenyu

China Forestry Publishing House

图书在版编目（CIP）数据

中国森林昆虫：英文 / 萧刚柔, 李镇宇主编. -- 3版. -- 北京：中国林业出版社, 2020.6
ISBN 978-7-5219-0640-0

Ⅰ.①中… Ⅱ.①萧… ②李… Ⅲ.①森林昆虫学－中国－英文 Ⅳ.①S718.7

中国版本图书馆CIP数据核字（2020）第112379号

Staff Planners: Wang Jiahui, Wen Jin, Liu Jialing
Staff Editors: Liu Jialing, Song Boyang

Copyright China Forestry Publishing House
All Rights Reserved

Library of Congress Cataloging in Publishing Data
Chinese Forest Insects / Chief Editors: Xiao Gangrou, Li Zhenyu. -- Third Edition. -- Beijing: China Forestry Publishing House, 2020.06
ISBN 978-7-5219-0640-0

Ⅰ.①China... Ⅱ.①Xiao ...②Li ... Ⅲ.①Forest Entomology-China-English Ⅳ.①S718.7
China CIP (2020) No. 112379

First Published in the P. R. China in 2020 by
China Forestry Publishing House
No. 7, Liuhaihutong, Xicheng District, Beijing 100009

Printed and bound in Beijing, China

US$ 300.00

Editorial Board

Chief Editors

Xiao Gangrou, Li Zhenyu

Co-editor-in-chief

Wu Jian

Editorial Board Members (*arranged alphabetically*)

Stephen Clarke, Ji Baozhong, Li Chengde, Li Menglou, Luo Youqing, Gary Man, Robert Mangold, Song Yushuang, Sun Jianghua, Wang Changlu, Wen Jin, Wu Chunsheng, Wu Jian, Xu Tiansen, Yang Xiuyuan, Zhang Runzhi, Zhang Zhen, Zhao Wenxia

Translators

Wang Changlu, Liu Houping, Ni Xinzhi, Zhang Yanzhuo, Zhang Qinghe

Preface (Third Edition)

Forestry is an essential part of the ecosystem conservation. It is a basic industry for sustainable economic development and social services. It is also the main battle field for building harmonious relationship between human and the nature. Forestry plays an irreplaceable role in building the ecosystem and protecting our environment. Since China's opening up to the world 40 years ago, Chinese forest coverage significantly improved as a result of afforestation and conservation. Forest protection from pests becomes more important. In the 1980's and 1990's, the first and second edition of Forest Insects of China were published to meet the needs for forest pest research, teaching, and better pest control methods and strategies. As an important reference book, it greatly assisted in the protection of forest ecosystems in China.

It has been nearly 30 years since the second edition was published. China experienced tremendous scientific and economic development during this period. Significant advances occurred in forest entomology research. It becomes necessary to update the second edition of Chinese Forest Insects and to publish an English edition to allow for easy access of the information by international scholars. In the new edition, the following major changes were made.

1. Regrouping the taxa following the newest research findings. For example, Homopter is now Hemiptera. *Dendrolimus punctatus* Walker is now *Dendrolimus punctata punctata* Walker. Chinese common name of *Rammeacris kiangsu* (Tsai) is changed. *Matsucoccus liaoningiensis* Tang *Matsucoccus massonianae* Yang et Hu are considered synonyms of *M. matsumurae* (Kuwana). Some subspecies are raised to species. Synonyms and Chinese common names are added.

2. All species have color photos for identification. Many photos are published for the first time.

3. We added resource insects that only occur in China. About 500 natural enemies were included in pest prevention and treatment section. Some invasive insects were included such as *Phenacoccus solenopsis* Tinsley, *Corythucha ciliata* (Say), *Cydia pomonella* (L.), *Dendroctonus valens* (LeConte), *Carpomya vesuviana* Costa, *Obolodiplosis robiniae* (Haldemann), *Solenopsis invicta* Buren.

4. We deleted insecticides that are no longer suitable for environmental protection and sustainable pest management.

5. We added indexes of host plants and natural enemies for easy search.

6. An English edition is created to help scientific exchange. The English edition is translated by overseas Chinese entomologists who are both fluent in Chinese and English to ensure the accuracy of the translation.

This book introduces nearly 600 species of insects. Each species includes scientific name, taxonomic status, economic importance, distribution, hosts, biology and behavior, prevention and treatment. This book can be used by researchers, students, government agencies, and pest control professionals.

About 450 authors contributed to the contents of this book. Major institutions such as National Forestry and Grassland Administration, Chinese Academy of Science – Institute of Zoology, Chinese Academy of Forestry, Beijing Forestry University, Nanjing Forestry University, Northeast Forestry University, Northwest A & F University, etc. provided assistance to this book.

Professor Li Chengde from Northeast Forestry University reviewed the scientific names and taxonomic status of the species. The prevention and control section was reviewed by Mr. Liang Xiaowen. We thank all the authors, institutions, and reviewers for their assistance. This book received funding from the National Publishing Foundation, US Department of Agriculture Forest Service.

We'd like to pay special homage to Mr. Xiao Gangrou, who is one of the founders of the forest entomology in China. The first and second edition of Chinese Forest Insects he edited played a vital role in the forest entomology research and forest protection in China. He worked tirelessly to prepare for the third edition. With great sadness, he died suddenly on 22 August 2005. His death is a great loss to the forest entomology research in China and to this book. We'd like to dedicate this book to Mr. Xiao Gangrou.

Li Zhenyu
October 2019

Preface (Second Edition)

1. Since the first edition of "Chinese Forest Insects" was published about 10 years ago, there has been much progress in forest insect research in China. It becomes necessary to revise the first edition to meet the needs in forest production, research, and teaching.

2. In the second edition, we added 10 new chapters in the general section, two orders, 44 families, and 402 species. There is a total of 824 species, belong to 13 orders and 141 families in this edition.

3. We deleted five questionable species and 17 species without species names, keys, descriptions to superfamilies and families.

4. Control methods are grouped by families to provide overview of the control methods in each family. Control methods that are unique to species are retained.

5. Species are arranged based upon taxonomy rather than pest status.

6. Distribution of species within China and foreign countries are separated by ";".

7. Except for a few species, the hosts only include those are found in China.

8. Species in Ichneumonidae, Tachinidae, Pentatomidae, Phasmidae, Coleophoridae, Tortricidae, Coccidae, Buprestidae, Cecidomyiidae, and Psyllidae are reviewed by He Junhua, Zhao Jianming, Zheng Leyi, Chen Shuchun, Yang Liming, Bai Jiuwei, Hu Xingping, Peng Zhongliang, Jiang Xingpei, Yang Jikun, and Li Fasheng, respectively. These experts helped to improve the accuracy of the species identification and description.

9. This book is sponsored by Science and Technology Committee of the Ministry of Forestry and Forest Research Institute of the Chinese Academy of Forestry.

10. The first edition of this book is edited by Chinese Academy of Forestry. Editorial board members include Cai Banghua, Xiao Gangrou and the following members (arranged alphabetically): Li Kuansheng, Li Yabai, Li Yajie, Li Zhouzhi, Peng Jianwen, Tian Zejun, Xu Tiansen, Yu Chengming, Zhang Zhizhong.

Xiao Gangrou
July 1991

Preface (First Edition)

Afforestation to increase forest coverage to 30% in China is one of the important goals in our forestry industry. Achieving this goal will allow for greener mountains and clearer water in our country, and more abundant forest resources and crop yield. Yet, as more and more man-made forests increase, forest insect pest management becomes more important. To increase our quality of forest insect pest control research, provide information for teaching, and pest control methods, we organized experts from the country to write a book on forest insects. This book includes important and secondary forest insect pests, natural enemies, resource insects. Each species included distribution, host plants or insects, morphological characteristics, biology, and control methods, and utilization of natural enemies. A total of 444 species of insect pests, 31 natural enemies, 7 resource insects, and 10 pest mites are included.

This is a comprehensive book on forest insects of China. Information on some species are reported for the first time. Contents are from the original publications if possible. Most morphology drawings and host plant damage pictures are based on specimens. Authors are researchers with experiences on the insects.

All insects are ordered following the systems used in Insect Taxonomy by Cai Banghua, and Imm's General Textbook of Entomology.

Many Chinese names used in this book are new and we hope they can be adopted for easy use in the future. The various monographs of Chinese Economic Insects are used when considering the Chinese common names. These monographs are not listed separately in the References section.

Authors and illustrators are listed at the end of each chapter or species. When significant changes were made, then "Editorial board" was added to the author list.

Reviewers of this book included Jiang Shunan, Zhao Yangchang, Liu Youqiao, Zhao Jianming, Yu Peiyu, Yin Huifen, Zhang Youwei, Hou Taoqian, Jiang Shengqiao, Li Hongxing, Pu Fuji, Cai Rongquan, Zhao Zhongling, Huang Xiaoyun, Zhou Shuzhi. Other reviewers participated in reviewing the drafts in Beijing included Wu Cibin, Wang Shufen, Sun Yujia, Zhou Jiaxi, Cheng Liang, Yang Youqian, Yang Jiahuan, Liu Yuanfu, He Jietian, Yang Xiuyuan. We thank their help.

Xiao Gangrou from the Forest Insect Laboratory, Forest Research Institute, Chinese Academy of Forestry is the chief editor of the book. Drafts were first reviewed by the local editorial board members, then they were reviewed again by editorial board members in Beijing. Most of the figures were drawn in Beijing. Xu Tiansen helped editing the figures. There may be errors or omissions due to lack of experience. We welcome readers' comments.

Editorial Board of Chinese Forest Insects
1980

Contents

Preface (Third Edition)

Preface (Second Edition)

Preface (First Edition)

General Section 1

1. Distribution, Fauna and Eco-geographical Regionalization of Forest Insects in China 2
2. Occurrence of Forest Insect Pests and Their Damages in China: an Overview 6
3. Relationships between Occurrences/Outbreaks of Forest Insect Pests and Environmental Factors, Forest Stands and Natural Enemies. 21
4. Historical Development and Technological Advances of Forest Insect Control in China 35

Taxonomic Section 51

Orthoptera

1. *Tarbinskiellus portentosus* (Lichtenstein) 52
2. *Gryllotalpa orientalis* Burmeister 54
3. *Gryllotalpa unispina* Saussure 56
4. *Aularches miliaris* (L.) 58
5. *Ceracris kiangsu* Tsai 59
6. *Ceracris nigricornis* Walker 61
7. *Hieroglyphus tonkinensis* Bolívar 63
8. *Chondracris rosea* (De Geer) 65
9. *Phlaeoba angustidorsis* Bolívar 66

Phasmatodea

10. *Ramulus chongxinense* (Chen et He) 67
11. *Ramulus minutidentatus* (Chen et He) 69
12. *Ramulus pingliense* (Chen et He) 71
13. *Macellina digitata* Chen et He 73
14. *Micadina zhejiangensis* Chen et He 75
15. *Lonchodes bobaiensis* (Chen) 77
16. *Necroscia flavescens* (Chen et Wang) 79
17. *Sinophasma brevipenne* Günther 81
18. *Sinophasma largum* Chen et Chen 82
19. *Sinophasma maculicruralis* Chen 84
20. *Sinophasma mirabile* Günther 86
21. *Sinophasma pseudomirabile* Chen et Chen 88

Blattodea

22. *Cryptotermes domesticus* (Haviland) 90
23. *Hodotermopsis sjostedti* Holmgren 91
24. *Coptotermes formosanus* Shiraki 92
25. *Reticulitermes chinensis* Snyder 94
26. *Reticulitermes flaviceps* (Oshima) 96
27. *Macrotermes barneyi* Light 97
28. *Sinonasutitermes erectinasus* (Tsai et Chen) 99
29. *Nasutitermes parvonasutus* (Nawa) 100
30. *Odontotermes formosanus* (Shiraki) 101

Hemiptera

31. *Lycorma delicatula* (White) 103
32. *Lawana imitata* (Melichar) 105
33. *Aphrophora horizontalis* Kato 107
34. *Omalophora costalis* (Matsumura) 109
35. *Tilophora flavipes* (Uhler) 110
36. *Cryptotympana atrata* (F.) 111
37. *Dundubia hainanensis* (Distant) 114
38. *Macrosemia pieli* (Kato) 116
39. *Hypsauchenia chinensis* Chou 118
40. *Cicadella viridis* (L.) 120

41. *Ricania speculum* (Walker) 124
42. *Agonoscena xanthoceratis* Li..................125
43. *Anomoneura mori* Schwarz..................127
44. *Cyamophila willieti* (Wu)..................130
45. *Colophorina rubinae* (Shinji) 132
46. *Syntomoza homali* (Yang et Li)..................135
47. *Diclidophlebia excetrodendri* (Li et Yang) 136
48. *Carsidara limbata* (Enderlein)..................137
49. *Trioza camphorae* Sasaki 139
50. *Trioza magnisetosa* Longinova..................140
51. *Aleurotrachelus camelliae* (Kuwana) 142
52. *Aphis craccivora* Koch 144
53. *Melanaphis bambusae* (Fullaway) 146
54. *Aphis odinae* (van der Goot) 147
55. *Chaitophorus populialbae* (Boyer de Fonscolombe)......148
56. *Periphyllus koelreuteriae* (Takahashi) 150
57. *Cinara formosana* (Takahashi) 152
58. *Cinara tujafilina* (Del Guercio) 155
59. *Lachnus tropicalis* (van der Goot) 157
60. *Schlechtendalia chinensis* (Bell) 158
61. *Eriosoma dilanuginosum* Zhang 162
62. *Pemphigus immunis* Buckton 164
63. *Takecallis arundinariae* (Essig) 166
64. *Takecallis taiwana* (Takahashi) 167
65. *Kurisakia sinocaryae* Zhang 168
66. *Adelges laricis laricis* Vallot..................170
67. *Drosicha corpulenta* (Kuwana) 173
68. *Icerya purchasi* Maskell..................176
69. *Matsucoccus matsumurae* (Kuwana)..................178
70. *Matsucoccus shennongjiaensis* Young and Liu 180
71. *Matsucoccus yunnanensis* Ferris..................182
72. *Matsucoccus sinensis* Chen..................184
73. *Kerria* spp. 187
74. *Nesticoccus sinensis* Tang..................190
75. *Oracella acuta* (Lobdell) 192
76. *Phenacoccus azaleae* Kuwana 194
77. *Phenacoccus fraxinus* Tang 196
78. *Phenacoccus solenopsis* Tinsley..................197
79. *Asiacornococcus kaki* (Kuwana)..................199
80. *Acanthococcus lagerstroemiae* Kuwana..................201
81. *Kermes nawae* Kuwana..................204
82. *Ceroplastes japonicus* Green 205
83. *Ceroplastes rubens* Maskell..................208
84. *Ericerus pela* (Chavannes)..................209
85. *Eulecanium giganteum* (Shinji) 212

86. *Eulecanium kuwanai* Kanda 214
87. *Metaceronema japonica* (Maskell) 215
88. *Parasaissetia nigra* (Nietner) 217
89. *Parthenolecanium corni* (Bouché)..................219
90. *Cresococcus candidus* Wang..................222
91. *Asterococcus muratae* (Kuwana) 223
92. *Neoasterodiaspis castaneae* (Russell) 224
93. *Bambusaspis hemisphaerica* (Kuwana)..................225
94. *Aulacaspis rosarum* (Borchsenius) 227
95. *Aulacaspis sassafris* Chen, Wu & Su..................229
96. *Hemiberlesia pitysophila* Takagi 230
97. *Lepidosaphes salicina* Borchsenius 233
98. *Lepidosaphes conchiformis* (Gmelin) 235
99. *Prodiaspis tamaricicola* (Malenotti)..................237
100. *Pseudaulacaspis pentagona* (Targioni-Tozzetti)..........238
101. *Diaspidiotus gigas* (Ferris) 240
102. *Comstockaspis perniciosa* (Comstock)..................242
103. *Diaspidiotus slavonicus* (Green)..................244
104. *Shansiaspis sinensis* Tang 246
105. *Unaspis euonymi* (Comstock) 248
106. *Corythucha ciliata* (Say)..................251
107. *Metasalis populi* (Takeya)..................253
108. *Monosteira unicostata* (Mulsant et Rey) 256
109. *Stephanitis illicii* Jing..................258
110. *Stephanitis laudata* Drake et Poor 259
111. *Stephanitis svensoni* Drake..................261
112. *Pirkimerus japonicus* (Hidaka) 262
113. *Sinorsillus piliferus* Usinger..................264
114. *Metacanthus pulchellus* Dallas 265
115. *Yemma signata* (Hsiao)..................266
116. *Notobitus meleagris* (F.)..................267
117. *Notobitus montanus* Hsiao 269
118. *Cyclopelta parva* Distant 270
119. *Aenaria pinchii* Yang..................271
120. *Halyomorpha halys* Stål..................273
121. *Hippotiscus dorsalis* (Stål)..................275
122. *Nezara viridula* (L.) 277
123. *Poecilocoris latus* Dallas 278
124. *Tessaratoma papillosa* (Drury) 281
125. *Urostylis yangi* Maa..................283

Thysanoptera

126. *Scirtothrips dorsalis* Hood..................284
127. *Haplothrips chinensis* Priesner 286

Coleoptera

128. *Anomala corpulenta* Motschulsky 288

129. *Anomala cupripes* (Hope) 291
130. *Proagopertha lucidula* (Faldermann) 293
131. *Nigrotrichia gebleri* (Faldermann).................................295
132. *Eotrichia niponensis* (Lewis)...297
133. *Lepidiota stigma* (F.)..299
134. *Maladera orientalis* (Motschulsky)................................300
135. *Melolontha hippocastani mongolica* Ménétriés............302
136. *Agrilus planipennis* Fairmaire.......................................304
137. *Agrilus moerens* Saunders 306
138. *Agrilus subrobustus* Saunders.......................................308
139. *Agrilus zanthoxylumi* Zhang et Wang 309
140. *Agrilus pekinensis pekinensis* Obenberger...................310
141. *Chalcophora yunnana* Fairmaire 312
142. *Lamprodila limbata* (Gebler) 313
143. *Trachypteris picta picta* (Pallas) 315
144. *Poecilonota variolosa* (Paykull) 317
145. *Agriotes subvittatus* Motschulsky.................................318
146. *Melanotus cribricollis* (Faldermann) 319
147. *Pleonomus canaliculatus* Faldermann 321
148. *Ptilinus fuscus* Geoffroy..323
149. *Sinoxylon japonicum* Lesne 324
150. *Gonocephalum bilineatum* (Walker).............................325
151. *Acalolepta sublusca* (Thomson) 326
152. *Anoplophora chinensis* (Forster)...................................328
153. *Anoplophora glabripennis* (Motschulsky)....................330
154. *Apriona rugicollis* Chevrolat 332
155. *Apriona swainsoni* (Hope) 335
156. *Aristobia approximator* (Thomson) 337
157. *Aristobia hispida* (Saunders)...338
158. *Aromia bungii* (Faldermann)...340
159. *Asemum striatum* (L.)..342
160. *Anoplistes halodendri* (Pallas) 344
161. *Batocera davidis* Deyrolle 346
162. *Batocera lineolata* Chevrolat...348
163. *Batocera rubus* (L.)..351
164. *Callidiellum villosulum* (Fairmaire)..............................354
165. *Clytus validus* Fairmaire 355
166. *Neocerambyx raddei* Blessig 356
167. *Aegosoma sinicum* White 358
168. *Monochamus alternatus* Hope 360
169. *Monochamus urussovii* (Fischer-Waldheim) 362
170. *Philus antennatus* (Gyllenhal) 364
171. *Purpuricenus temminckii* (Guérin-Méneville)...............366
172. *Saperda carcharias* (L.) 368
173. *Saperda populnea* (L.) 370

174. *Semanotus bifasciatus* (Motschulsky)..........................372
175. *Semanotus sinoauster* Gressitt 374
176. *Thylactus simulans* Gahan 376
177. *Trichoferus campestris* (Faldermann)..........................377
178. *Xylotrechus rusticus* (L.)...378
179. *Xystrocera globosa* (Olivier) 381
180. *Acanthoscelides pallidipennis* (Motschulsky) 382
181. *Kytorhinus immixtus* Motschulsky...............................384
182. *Agelastica alni glabra* (Fischer von Waldheim)..........386
183. *Ambrostoma quadriimpressum* (Motschulsky)............388
184. *Basiprionota bisignata* (Boheman)..............................390
185. *Brontispa longissima* (Gestro) 391
186. *Plagiosterna adamsi* (Baly) 393
187. *Chrysomela populi* L. 394
188. *Gastrolina depressa* Baly..396
189. *Gastrolina thoracica* Baly 398
190. *Gastrolina pallipes* Chen 399
191. *Oides leucomelaena* Weise 400
192. *Ophrida xanthospilota* (Baly) 402
193. *Parnops glasunovi* Jacobson...404
194. *Plagiodera versicolora* (Laicharting) 406
195. *Podagricomela shirahatai* (Chûjô)...............................408
196. *Podontia lutea* (Oliver)...409
197. *Pyrrhalta aenescens* (Fairmaire)..................................411
198. *Byctiscus rugosus* (Gelber) 414
199. *Cyllorhynchites ursulus* (Roelofs) 415
200. *Cyrtotrachelus buquetii* Guérin-Méneville..................417
201. *Cyrtotrachelus thompsoni* Alonso-Zarazaga et Lyal 419
202. *Otidognathus davidis* (Fairmaire)................................421
203. *Rhynchophorus ferrugineus* (Olivier)..........................423
204. *Sipalinus gigas* (F.) 425
205. *Apion collare* Schilsky 427
206. *Sternuchopsis juglans* (Chao) 428
207. *Sternu chopsis sauteri* Heller 429
208. *Cryptorhynchus lapathi* (L.) 430
209. *Curculio chinensis* Chevrolat.......................................433
210. *Curculio davidi* Fairmaire 435
211. *Curculio hippophes* Zhang, Chen et Dang...................437
212. *Dyscerus cribripennis* Matsumura et Kono 438
213. *Pimelocerus juglans* (Chao)..440
214. *Eucryptorrhynchus brandti* (Harold) 442
215. *Eucryptorrhynchus scrobiculatus* (Motschulsky).........444
216. *Hylobitelus xiaoi* Zhang..446
217. *Hypomeces squamosus* (F.)...447
218. *Lepyrus japonicus* Roelofs...449

219. *Niphades castanea* Chao 451
220. *Niphades verrucosus* (Voss) 452
221. *Pissodes punctatus* Langor et Zhang 453
222. *Pissodes validirostris* (Sahlberg) 455
223. *Pissodes yunnanensis* Langor et Zhang 457
224. *Rhynchaenus empopulifolis* Chen 459
225. *Scythropus yasumatsui* Kono et Morinoto 461
226. *Shirahoshizo coniferae* Chao 463
227. *Shirahoshizo patruelis* (Voss) 464
228. *Sternochetus frigidus* (F.) 466
229. *Sternochetus mangiferae* (F.) 468
230. *Sternochetus olivieri* (Faust) 470
231. *Sympiezomias velatus* (Chevrolat) 472
232. *Cryphalus tabulaeformis chienzhuangensis* Tsai et Li
     474
233. *Cryphalus tabulaeformis* Tsai et Li 475
234. *Dendroctonus armandi* Tsai et Li 477
235. *Dendroctonus micans* (Kugelann) 480
236. *Dendroctonus valens* LeConte 481
237. *Ips acuminatus* (Gyllenhal) 484
238. *Ips duplicatus* (Sahalberg) 485
239. *Ips hauseri* Reitter 486
240. *Ips nitidus* Eggers 488
241. *Ips sexdentatus* (Börner) 490
242. *Ips subelongatus* (Motschulsky) 492
243. *Ips typographus* (L.) 494
244. *Phloeosinus aubei* (Perris) 496
245. *Phloeosinus sinensis* Schedl 499
246. *Pityogenes chalcographus* (L.) 501
247. *Scolytus schevyrewi* Semenov 503
248. *Tomicus minor* (Hartig) 505
249. *Tomicus piniperda* (L.) 506

Diptera

250. *Dasineura datifolia* Jiang 508
251. *Obolodiplosis robiniae* (Haldemann) 510
252. *Rabdophaga salicis* (Schrank) 513
253. *Chyliza bambusae* Yang et Wang 515
254. *Pegomya phyllostachys* (Fan) 517
255. *Strobilomyia* spp. 519
256. *Carpomya vesuviana* Costa 522

Lepidoptera

257. *Eriocrania semipurpurella alpina* Xu 523
258. *Endoclita signifier* (Walker) 525
259. *Endoclita excrescens* (Butler) 528
260. *Endoclita nodus* (Chu et Wang) 531

261. *Dasyses barbata* (Christoph) 533
262. *Opogona sacchari* (Bojer) 535
263. *Amatissa snelleni* (Heylaerts) 538
264. *Chalioides kondonis* Matsumura 540
265. *Acanthoecia larminati* (Heylaerts) 542
266. *Eumeta minuscula* Butler 543
267. *Dappula tertia* (Templeton) 545
268. *Eumeta variegata* (Snellen) 547
269. *Mahasena colona* Sonan 549
270. *Caloptilia dentata* Liu et Yuan 551
271. *Caloptilia chrysolampra* (Meyrick) 553
272. *Gibbovalva urbana* (Meyrick) 554
273. *Melanocercops ficuvorella* (Yazaki) 556
274. *Leucoptera sinuella* (Reutti) 558
275. *Paranthrene tabaniformis* (Rottemburg) 560
276. *Sesia siningensis* (Hsu) 562
277. *Synanthedon castanevora* Yang et Wang 564
278. *Glyphipterix semiflavana* Issiki 566
279. *Atrijuglans hetaohei* Yang 567
280. *Stathmopoda masinissa* Meyrick 568
281. *Prays alpha* Moriuti 570
282. *Yponomeuta padella* (L.) 571
283. *Atteva fabriciella* Swederus 573
284. *Coleophora obducta* (Meyrick) 575
285. *Coleophora sinensis* Yang 577
286. *Casmara patrona* Meyrick 579
287. *Macrobathra flavidus* Qian et Liu 581
288. *Haplonchrois theae* (Kusnezov) 582
289. *Odites issikii* (Takahashi) 584
290. *Anacampsis populella* (Clerck) 586
291. *Anarsia lineatella* Zeller 588
292. *Anarsia squamerecta* Li et Zheng 590
293. *Dendrophilia sophora* Li et Zheng 592
294. *Dichomeris bimaculatus* Liu et Qian 596
295. *Carposina sasakii* Matsumura 598
296. *Cossus cossus chinensis* Rothschild 600
297. *Cossus orientalis* Gaede 601
298. *Deserticossus arenicola* (Staudinger) 603
299. *Deserticossus artemisiae* (Chou et Hua) 607
300. *Eogystia hippophaecola* (Hua, Chou, Fang et Chen) ... 609
301. *Streltzoviella insularis* (Staudinger) 611
302. *Yakudza vicarius* (Walker) 613
303. *Polyphagozerra coffeae* (Nietner) 615
304. *Xyleutes persona* (Le Guillou) 617
305. *Zeuzera multistrigata* (Moore) 619

306. *Monema flavescens* Walker 621
307. *Phlossa conjuncta* (Walker) 624
308. *Thespea bicolor* (Walker) 626
309. *Parasa consocia* Walker 628
310. *Thosea sinensis* (Walker) 630
311. *Scopelodes testacea* Butler 632
312. *Fuscartona funeralis* (Butler) 634
313. *Histia rhodope* (Cramer) 636
314. *Illiberis ulmivora* (Graeser) 639
315. *Pryeria sinica* Moore 640
316. *Adoxophyes honmai* Yasuda 642
317. *Cerace stipatana* Walker 644
318. *Choristoneura lafauryana* (Ragonot) 645
319. *Cydia pomonella* (L.) 646
320. *Homona issikii* Yasuda 648
321. *Pandemis heparana* (Denis et Schiffermüller) 651
322. *Polylopha cassiicola* Liu et Kawabe 653
323. *Ptycholomoides aeriferanus* (Herrich-Schäffer) 654
324. *Ancylis mitterbacheriana* (Denis et Schiffermüller) 655
325. *Ancylis sativa* Liu 657
326. *Celypha pseudolarixicola* Liu 659
327. *Cryptophlebia ombrodelta* (Lower) 661
328. *Cydia trasias* (Meyrick) 663
329. *Cymolomia hartigiana* (Saxesen) 665
330. *Epinotia rubiginosana* (Herrich-Schäffer) 667
331. *Gatesclakeana idia* Diakonoff 669
332. *Gravitarmata margarotana* (Heinemann) 671
333. *Gypsonoma minutana* (Hübner) 673
334. *Ophiorrhabda mormopa* (Meyrick) 674
335. *Cydia coniferana* (Saxesen) 676
336. *Cydia zebeana* (Ratzeburg) 678
337. *Pammene ginkgoicola* Liu 680
338. *Lobesia cunninghamiacola* (Liu et Pai) 681
339. *Cydia strobilella* (L.) 684
340. *Retinia cristata* (Walsingham) 685
341. *Retinia resinella* (L.) 687
342. *Rhyacionia duplana* (Hübner) 688
343. *Rhyacionia insulariana* Liu 690
344. *Rhyacionia pinicolana* (Doubleday) 692
345. *Strepsicrates coriariae* Oku 693
346. *Zeiraphera griseana* (Hübner) 694
347. *Camptochilus semifasciata* Gaede 696
348. *Rhodoneura sphoraria* (Swinhoe) 698
349. *Striglina bifida* Chu et Wang 700
350. *Asclerobia sinensis* (Caradja et Meyrick) 702
351. *Cryptoblabes lariciana* Matsumura 703
352. *Dioryctria mongolicella* Wang et Sung 705
353. *Dioryctria pryeri* Ragonot 707
354. *Dioryctria rubella* Hampson 710
355. *Etiella zinckenella* (Treitschke) 713
356. *Euzophera alpherakyella* Ragonot 714
357. *Euzophera batangensis* Caradja 716
358. *Hypsipyla robusta* (Moore) 718
359. *Locastra muscosalis* (Walker) 720
360. *Dioryctria yiai* Mutuura et Munroe 723
361. *Propachys nigrivena* Walker 725
362. *Crypsiptya coclesalis* (Walker) 727
363. *Circobotys aurealis* (Leech) 729
364. *Sinibotys evenoralis* (Walker) 731
365. *Demobotys pervulgalis* (Hampson) 733
366. *Eumorphobotys obscuralis* (Caradja) 735
367. *Botyodes diniasalis* (Walker) 738
368. *Cydalima perspectalis* (Walker) 740
369. *Gonogethes punctiferalis* (Guenée) 743
370. *Sinomphisa plagialis* (Wileman) 745
371. *Paliga machoeralis* Walker 747
372. *Cyclidia substigmaria* (Hübner) 749
373. *Apocheima cinerarius* (Erschoff) 751
374. *Ascotis selenaria* (Denis et Schiffermüller) 755
375. *Biston marginata* Shiraki 757
376. *Biston suppressaria* (Guenée) 759
377. *Abraxas suspecta* (Warren) 762
378. *Chihuo zao* Yang 764
379. *Biston panterinaria* (Bremer et Grey) 766
380. *Dilophodes elegans sinica* Wehrli 769
381. *Dysphania militaris* (L.) 772
382. *Erannis ankeraria* (Staudinger) 774
383. *Ectropis excellens* (Butler) 776
384. *Jankowskia fuscaria* (Leech) 778
385. *Meichihuo cihuai* Yang 781
386. *Milionia basalis pryeri* Druce 783
387. *Percnia giraffata* (Guenée) 786
388. *Semiothisa cinerearia* (Bremer et Grey) 788
389. *Thalassodes quadraria* Guenée 790
390. *Zamacra excavate* (Dyar) 792
391. *Xerodes rufescentaria* (Motschulsky) 794
392. *Epicopeia mencia* Moore 796
393. *Epiplema moza* Butler 798
394. *Mirina christophi* (Staudinger) 800
395. *Pyrosis eximia* Oberthür 802

396. *Pyrosis idiota* Graeser 804
397. *Cosmotriche inexperta* (Leech) 806
398. *Kunugia xichangensis* (Tsai et Liu) 808
399. *Dendrolimus grisea* (Moore) 810
400. *Dendrolimus kikuchii kikuchii* Matsumura 813
401. *Dendrolimus punctata punctata* (Walker) 815
402. *Dendrolimus punctata tehchangensis* Tsai et Liu 818
403. *Dendrolimus punctata wenshanensis* Tsai et Liu 820
404. *Dendrolimus spectabilis* Butler 822
405. *Dendrolimus suffuscus suffuscus* Lajonquiere 825
406. *Dendrolimus superans* (Butler) 827
407. *Dendrolimus tabulaeformis* Tsai et Liu 829
408. *Gastropacha populifolia* (Esper) 832
409. *Lebeda nobilis sinina* Lajonquiere 835
410. *Malacosoma dentata* Mell 837
411. *Malacosoma neustria testacea* (Motschulsky) 839
412. *Malacosoma rectifascia* Lajonquière 842
413. *Paralebeda plagifera* (Walker) 844
414. *Suana concolor* (Walker) 846
415. *Trabala vishnou vishnou* Lefebure 848
416. *Trabala vishnou gigantina* Yang 850
417. *Rondotia menciana* Moore 852
418. *Eupterote chinensis* Leech 855
419. *Eupterote sapivora* Yang 857
420. *Actias ningpoana* Felder 859
421. *Antheraea pernyi* (Guèrin-Mèneville) 861
422. *Caligula japonica* (Moore) 863
423. *Eriogyna pyretorum* (Westwood) 866
424. *Samia cynthia* (Drury) 870
425. *Attacus atlas* (L.) 873
426. *Notonagemia analis* (Felder) 875
427. *Callambulyx tatarinovii* (Bremer et Grey) 877
428. *Hyles hippophaes* (Esper) 879
429. *Clanis bilineata bilineata* (Walker) 881
430. *Dolbina tancrei* Staudinger 883
431. *Psilogramma menephron* (Cramer) 885
432. *Smerinthus planus* Walker 887
433. *Amplypterus panopus* (Cramer) 889
434. *Daphnis nerii* (L.) 890
435. *Besaia anaemica* (Leech) 893
436. *Besaia goddrica* (Schaus) 895
437. *Armiana retrofusca* (de Joannis) 897
438. *Cerura menciana* Moore 900
439. *Clostera anachoreta* (Denis & Schiffermüller) 903
440. *Clostera anastomosis* (L.) 906
441. *Euhampsonia cristata* (Butler) 910
442. *Periergos dispar* (Kiriakoff) 912
443. *Micromelalopha sieversi* (Staudinger) 915
444. *Cerura tattakana* Matsumura 917
445. *Phalera bucephala* (L.) 919
446. *Phalerodonta bombycina* (Oberthür) 921
447. *Pterostoma sinica* Moore 923
448. *Syntypistis cyanea* (Leech) 926
449. *Stauropus alternus* Walker 928
450. *Phalera flavescens* (Bremer et Grey) 930
451. *Phalera takasagoensis* Matsumura 932
452. *Agrotis tokionis* Butler 934
453. *Agrotis ipsilon* (Hufnagel) 936
454. *Agrotis segetum* (Denis and Schiffermüller) 938
455. *Apamea apameoides* (Draudt) 940
456. *Kumasia kumaso* (Sugi) 942
457. *Sapporia repetita* (Butler) 944
458. *Bambusiphila vulgaris* (Butler) 945
459. *Hyblaea puera* Cramer 948
460. *Gadirtha inexacta* (Walker) 949
461. *Selepa celtis* Moore 950
462. *Carea angulata* (F.) 952
463. *Earias pudicana* Staudinger 953
464. *Eligma narcissus* (Cramer) 955
465. *Camptoloma interiorata* (Walker) 957
466. *Hypocala subsatura* Guenée 960
467. *Ericeia inangulata* (Guenée) 962
468. *Episparis tortuosalis* Moore 963
469. *Aloa lactinea* (Cramer) 964
470. *Lemyra phasma* (Leech) 965
471. *Hyphantria cunea* (Drury) 967
472. *Calliteara axutha* (Collenette) 972
473. *Arna bipunctapex* (Hampson) 975
474. *Euproctis flava* (Bremer) 978
475. *Arna pseudoconspersa* (Strand) 980
476. *Ivela ochropoda* (Eversmann) 982
477. *Lymantria dispar* (L.) 984
478. *Lymantria mathura* Moore 988
479. *Lymantria monacha* (L.) 990
480. *Lymantria xylina* Swinhoe 992
481. *Orgyia antiguoides* (Hübner) 994
482. *Orgyia antiqua* (L.) 996
483. *Orgyia postica* (Walker) 998
484. *Pantana phyllostachysae* Chao 1000
485. *Pantana sinica* Moore 1004

486. *Parocneria furva* (Leech) 1007
487. *Parocneria orienta* Chao 1010
488. *Somena scintillans* (Walker) 1012
489. *Leucoma candida* (Staudinger) 1015
490. *Leucoma salicis* (L.) 1016
491. *Cifuna locuples* Walker 1019
492. *Perina nuda* (F.) 1021
493. *Polyura athamas* (Drury) 1022
494. *Polyura narcaea* (Hewitson) 1023
495. *Aporia crataegi* (L.) 1025
496. *Papilio elwesi* Leech 1028
497. *Papilio epycides epycides* Hewitson 1030
498. *Papilio slateri* Hewitson 1032
499. *Papilio machaon* L. 1033
500. *Papilio xuthus* L. 1034
501. *Matapa aria* (Moore) 1036

Hymenoptera

502. *Acantholyda erythrocephala* (L.) 1038
503. *Acantholyda flavomarginata* Maa 1040
504. *Acantholyda peiyingaopaoa* Hsiao 1042
505. *Acantholyda posticalis* Matsumura 1043
506. *Cephalcia tienmua* Maa 1045
507. *Cephalcia yanqingensis* Xiao 1047
508. *Chinolyda flagellicornis* (Smith) 1049
509. *Cephalcia kunyushanica* Xiao 1051
510. *Sirex rufiabdominis* Xiao et Wu 1053
511. *Tremex fuscicornis* (F.) 1055
512. *Urocerus gigas taiganus* Benson 1057
513. *Arge captiva* Smith 1059
514. *Agenocimbex ulmusvora* Yang 1061
515. *Cimbex connatus taukushi* Marlatt 1062
516. *Augomonoctenus smithi* Xiao et Wu 1064
517. *Diprion jingyuanensis* Xiao et Zhang 1066
518. *Diprion liuwanensis* Huang et Xiao 1069
519. *Diprion nanhuaensis* Xiao 1071
520. *Gilpinia massoniana* Xiao 1073
521. *Neodiprion dailingensis* Xiao et Zhou 1074
522. *Neodiprion huizeensis* Xiao et Zhou 1076
523. *Neodiprion xiangyunicus* Xiao et Zhou 1078
524. *Nesodiprion huanglongshanicus* Xiao et Huang 1080
525. *Nesodiprion zhejiangensis* Zhou et Xiao 1082
526. *Anafenusa acericola* (Xiao) 1084
527. *Fenusella taianensis* (Xiao et Zhou) 1085
528. *Caliroa annulipes* (Klug) 1087
529. *Dasmithius camellia* (Zhou et Huang) 1089
530. *Eutomostethus longidentus* Wei 1091
531. *Eutomostethus nigritus* Xiao 1093
532. *Eutomostethus deqingensis* Xiao 1095
533. *Hemichroa crocea* (Geoffroy) 1097
534. *Megabeleses liriodendrovorax* Xiao 1099
535. *Moricella rufonota* Rohwer 1101
536. *Nematus prunivorous* Xiao 1103
537. *Pachynematus itoi* Okutani 1104
538. *Pristiphora beijingensis* Zhou et Zhang 1106
539. *Pristiphora conjugata* (Dahlbom) 1109
540. *Pristiphora erichsonii* (Hartig) 1111
541. *Stauronematus compressicornis* (F.) 1113
542. *Dryocosmus kuriphilus* Yasumatsu 1115
543. *Megastigmus cryptomeriae* Yano 1117
544. *Leptocybe invasa* Fisher et La Salle 1118
545. *Quadrastichus erythrinae* Kim 1120
546. *Aiolomorphus rhopaloides* Walker 1121
547. *Eurytoma laricis* Yano 1123
548. *Eurytoma maslovskii* Nikolskaya 1125
549. *Eurytoma plotnikovi* Nikolskaya 1128
550. *Tetramesa phyllostachitis* Gahan 1131
551. *Solenopsis invicta* Buren 1133

References 1135
Host Plant Names 1181
Natural Enemy Names 1193
Scientific Names Index 1200
English Common Names Index 1211

General Section

Forestry is a major component of ecological improvement, one of basic industries for sustainable economic and social development, and also an important public welfare undertaking. As rapid economic growth and continuous improvement of living standards in China has occurred, forestry has gained increasing public attention, and plays an important and ever-increasing role in the modern society.

China has a vast territory with great climatic, geographic and natural resource diversity, including a range of climate zones: north temperate, temperate, warm-temperate, subtropical and tropical zones from north to south that spans 49° of latitude. China also spans 62° of longitude from east to west, starting with the maritime climate forests, followed by semi-humid, and semi-arid forest steppes, a forest-steppe transition zone, before reaching the arid semi-desert and desert area in the west. In general, the land is high in the west and descends to the east coast. Mountains, plateaus and hills account for nearly 70% of the country's land surface, including the world's highest peak, Mt. Everest of the Himalayas. Such diversified terrains and complex climate conditions support a great diversity of flora and fauna; including >6000 species of trees and >2000 species of bushes among the 30,000 species of Spermatophyte (seed plants) recorded. There are also 150,000 species of insects recorded in China, among which 30,000 species are considered as forest insects, and 5000 species as forest insect pests.

Forest insects are an important natural part of the forest ecosystem that has the highest numbers of species and individuals. These species not only have a strong impact on tree growth and reproduction, but also interact closely with other animals and plants in the ecosystem. During the evolutionary process, complex energy flows and food chains among various organisms including forest insects in the forest have formed; they interact (positively or negatively) with each other, support the biodiversity of the forest ecosystem, and maintain an ecological balance. Insect feeding and herbivory are a common natural phenomenon in forests. In a natural ecosystem, insects normally don't cause serious economical and ecological damages; therefore, no or little pest control operations are needed. Some insects might play a positive role in culling out weak and inferior trees, breaking down and recycling litter and nutrients on the forest floor, and act as a food source for birds and other animals. However, certain insects can pose a serious threat to forest health as serious or even outbreak forest insect pests, when their populations reach over the economical and/or ecological thresholds, mostly due to ecological imbalance in the forest ecosystem caused by human or natural disturbances. Some invasive insects are introduced by man from other regions and the populations of the insects can increase unnaturally because of a lack of natural enemies.

1. Distribution, Fauna and Eco-geographical Regionalization of Forest Insects in China

The distribution of forest insects is closely associated with the distribution of various forest vegetations, which are strongly influenced by natural conditions. Forest distribution exhibits a strong spatial distribution both horizontally and vertically that plays an important role in forming the current eco-geographical distribution patterns of forest insects.

The eco-geographical regionalization of forest insect pests in China includes three categories: regions, subregions, and provinces. Each category has its unique climate condition, vegetation type, entomofauna, and its representative species. Within each region, there are widely distributed cosmopolitan species and local endemic species. Insect species in the boundaries between subregions or provinces interpenetrate and intersect, resulting in the so-called edge effect and formation of hybrids or endemic species. There are six entomo-geographical regions in the world [see Ma (1959), Zhu (1987), Fang (1993) and Zhang (1998)]: Palearctic, Oriental, Ethiopian (Afrontropic), Nearctic, Neotropical and Australian. Two of these regions: Palearctic (North China) and Oriental (South China) exist in China.

1.1 Palearctic Insect Fauna

There are two subregions, Northeast Asian subregion and Central Asian subregion, in China's Palearctic region. The Northeast Asian subregion includes two provinces: Northeast China and North China, while the

Central Asian subregion is divided into two provinces: Inner Mongolia-Xinjiang, and Qinghai-Tibet.

1.1.1 Northeast China Province

This province includes the Greater Khingan Range and Lesser Khingan Mountains in the north, Zhangguangcai Mountain in the east, Changbai Mountains in the south and Song-Liao Plateau in the west. Weather in the province is cold and humid, with a relatively long winter ranging from 3 to 7 months depending on the location, and a short summer lasting about 3 months.

In the northern mountain areas (Greater Khingan Range and Lesser Khingan Mountains), forest vegetation is dominated by natural conifer forests of *Larix gmelinii* and *Pinus sylvestris* var. *mongolica*, and secondary forests of *Betula platyphylla*. Common forest insect pests include *Dendrolimus superans*, *Coleophora obducta, Adelges laricis laricis*, *Ips subelongatus*, *Eurytoma laricis*, *Matsucoccus dahuriensis*, *Ips sexdentatus, I. acuminatus, Monochamus sutor, M. urussovii, Rhyacionia pinicolana, Pissodes validirostris*, and *Phigalia djakonvi* among others.

In the Changbai Mountains, the key vegetation includes mixed coniferous (Korean pine dominated) and broad-leaved forests, with *Pinus koraiensis* being the dominant conifer species, followed by *Picea jezoensis* Carr. var. *microsperma, Abies nephrolepis*. and *Larix olgensis*. The major broad-leaved trees include *Ulmus davidiana* var. *japonica, Fraxinus mandshurica, Phellodendron amurense, Juglans mandshurica, Tilia amurensis, Quercus Mongolia*, and *Populus davidiana*. Common forest insect pests in this area include: *Pineus cembrae pinikoreanus, Matsucoccus koraiensis, Pissodes nitidus, Hylobius abietis haroldi, Physokermes inopinatus, Cydia strobilellus, Aphrastasia pectinatae, Lymantria dispar, Malacosoma neustria testacea, Gastrolina thoracica, Prays alpha, Agrilus*, and *Cyllorhynchites ursulus*.

In Song-Liao plateau area, the forest vegetation is mainly consisted of the "Three-North" (Northwest, North China and Northeast) Shelterbelts and plain afforestation plantations, with *Populus* spp., *Ulmus* spp., *Salix* spp., *Pinus sylvestris* var. *mongolica* and *Larix* spp. as major tree species. Common forest insect pests in this area include: *Diaspidiotus gigas, Cicadella viridis, Cerura menciana, Stilpnotia candida, S. salicis, Leucoptera susinella, Paranthrene tabaniformis, Endoclina excrescens, Cryptorhynchus lapathi, Poecilonota variolosa, Saperda populnea, Xylotrechus rusticus, Cimbex connatus taukushi, Ambrostoma quadriimpressum, Holcocerus vicarious, Tuberolachnus salignus*, and *Lepidosaphes salicina*.

The southern end of the Greater Khingan Range is a forest-steppe transition zone, with scattered forests of *Larix gmelinii* (Rupr.) Kuzen, *L. principis-rupprechtii* Mayr, *Pinus tabuliformis* and *Picea mongolica*. Common forest insect pests in this area include *Ips duplicatus, Acantholyda peiyingaopaoa, Cephalcia abietis* and *Gilpinia baiyinaobaoa*.

1.1.2 North China Province

The North China Province lies between Qinghai-Tibet Plateau on the west and the Bohai Gulf and the Yellow Sea on the east. It stretches north from Northeast China province to the north slope of Qinling Mountains and Huai River, which is considered as the boundary line between temperate and subtropical zones.

Little if any original natural vegetation exists in the hilly mountain areas of eastern Liaoning and Shandong. The main forest vegetation is secondary forests of *Quercus wutaishansea*, and plantations of *Robinia pseudoacacia* and *Pinusdensiflora*. Common forest insect pests in this area include: *Matsucoccus matsumurae, Dendrolimus spectabilis, Aphrophora flavipes, Dioryctria sylvestrella*, and *Rhyacionia duplana* plus a serious invasive pest species, *Hyphantria cunea*.

In Huang-Huai Plain (North China Plain), the major forest vegetation types are plain afforestation plantations, shelterbelts and street/roadside trees, with *Populus tomentosa, P. euramericana, P. cathayana, Salix matsudana, Styphnolobium japonicum, Robinia pseudoacacia, Paulownia tomentosa*, and *Platycladus orientalis* as the major tree species. Common forest insect pests in this area include: *Lophoeucaspis japonica, Sesia siningensis, Clostera anachoreta, Anoplophora glabripennis, Apriona rugicollis, Fenusella taianensis, Eulecanium kuwanai, Semiothisa cinerearia, Bruchophagus philorobiniae, Eumta variegate, Drosicha corpulenta, Basiprionota bisignata*, *Eteoneus angulatus*, and *Eupterote chinensis*.

In the Loess Plateau (also known as the Huangtu Plateau) and Yan-Taihang mountain areas, there are several forest vegetation types, including broad-leaved deciduous forests of *Quercus aliena, Q. wutaishansea, Q. variabilis, Tilia amurensis, Betula platyphylla, Populus davidiana, Carpinus turczaninowii, Armeniaca sibirica,* and *Pteroceltis tatarinowii*, evergreen coniferous forests of *Pinus tabuliformis* and *Platycladus orientalis*, and deciduous coniferous forests of *Larixprincipis-rupprechtii*. Common forest insect pests in this area include: *Dendrolimus tabuliformis, Ocnerostoma piniariellum, Rhyscionia pinicolana, Gravitarmata margarotana, Parocneria furva, Semanotus bifasciatus, Phloeosinus aubei, Kermes miyasakii, Erannis dira, Eucryptorrhynchus scrobiculatus, Atrijuglans hetaohei, Stenuchopsis juglans, Gastrolina thoracica, Synanthedon castanevora, Dryocosmus kuriphilus, Asiacornococcus kaki, Ceroplastes japonicas, Chihuo zao, Ancylis sativa* and a serious invasive pest, *Dendroctonus valens*.

1.1.3   Inner Mongolia-Xinjiang Province

This province includes the Northeastern Plateau of Inner Mongolia, Alxa Plateau and Ordos Plateau, Tsaidam Basin (Qinghai), Tarim Basin and Junggar Basin (Xinjiang), with a typical continental climate and a small/limited area of forests.

In the grasslands of the eastern part of the province, the insect fauna consists of typical grass-feeding insects, with *Eodorcadion chinganicum* as the representative species.

In the western desert areas, there are two major vegetation types: the natural scrublands of *Caragana intermedia intermedia, Salix psammophila, Tamarix ramosissima, Haloxylon ammodendron,* and *Nitraria tangutorum*, and the natural poplar forests of *Populus euphratica, P. canescens, P. nigra, P. laurifolia* and *P. alba*. Common forest insect pests in this area include: *Orgyia dubia, Smerinthus kindermanni, Sesia siningensis, Agelastica alni orientalis, Trachypteris picta, Saperda balsamifera, Pristiphora conjugate, Monosteira unicostata, Diorhabda rybakow, Curculio hippophes, Holcocerus hippophaecola, Deserticossus arenicola, Agrilus moerens, Diorhabda elongatadeserticola, Trioza magnisetosa, Apocheima cinerarius, Kytorhinus immixtus,* and *Bruchophagus neocaraganae*.

In the high mountain areas, such as Altay Mountain, Tian Mountain, Qilain Mountain, Helan Mountain, and Yin Mountain, the key forest vegetation is a dark-coniferous forest of *Larix sibirica, Abies sibirica, Picea schrenkiana, Picea crassifolia,* and *Sabina przewalskii*. Common forest insect pests in this area include: *Zeiraphera griseana, Dendrolimus superans, Coleophora sibiricella, Lymantria dispar, Ips hauseri, I. typographus, I. nitidus, Acantholyda piceacola, Cephalcia alashanica, Urocerus gigas taiganus,* and *Megastigmus sabinae*.

1.1.4   Qinghai-Tibet Province

This province covers the entire Qinghai-Tibet Plateau including the northern part of Sichuan, with an average elevation of >4500 m; it has a special climatic environment characterized by extremely severe cold (Alpine) weather, long winter and short summer, and very a short growing season. The dominant vegetation is the alpine desert type.

There are some subalpine coniferous forests of cold-temperate species such as *Abies* spp., *Picea* spp., and *Larix* spp. and temperate species, *Pinus armandii* Franch and *Platycladus orientalis* in Qinghai and southern Tibet. Common forest insect pests in this area include: *Bupalus vestalis, B. mughusaria* and *Eriocrania semipurpurella alpine*.

1.2 Oriental Insect Fauna

The insect fauna in China's Oriental region belongs to the Indo-China subregion, which is further divided into three provinces (areas), Southwest, Central China and South China.

1.2.1   Southwest province

This province is located in Southwestern China, including Himalayas and Hengduan Mountains, with a relatively high insect species richness.

Himalayas, an overall canyon-like landscape, is directly influenced by the Indian Ocean Monsoon. In central parts of Himalayas, the forest vegetation is a montane evergreen broad-leaved forest with *Castanopsis* spp., *Lithocarpus* spp., and the evergreen *Quercus* spp. as major tree species, and several species of Lauraceae, Theaceae and Magnoliaceae, plus conifers as minor species. In the eastern part, the vegetation is dominated by tropical evergreen forests and semi-evergreen rain forests. Common forest insect pests in this area include: *Dendrolimus himalayanus*, *Trabala vishnou vishnou*, *Trichotheca parva*, *Scolytoplatypus* spp. and *Xyleborus* spp.

In Hengduan Mountains, there are several forest vegetation types with the subalpine coniferous forests as the dominant one. This area is also the distribution center of *Pinus yunnanensis*, commonly mixed with *Castanopsis delavayi* Franch., *Cyclobalanopsis* sp., *Quercus franchetii*, *Q. griffithii*, *Lithocarpus dealbatus* (Hook. f. et Thoms.) Rehd., *Keteleeria evelyniana* Mast. and *Pinus armandii*. Common forest insect pests in this area include: *Cosmotriche saxosimilis*, *Kunugia xichangensis*, *Dendrolimus kikuchii kikuchii*, *D. grisea*, *D.punctata tehchangensis*, *D.punctata wenshanensis*, *Matsucoccus yunnanensi*, *Lymantria monacha*, *Dasychira axutha*, *Rhyacionia insulariana*, *Eterusia leptalina*, *Dioryctria rubella*, *Tomicus yunnanensis*, *Cleoporus variabilis*, *Neodiprion xiangyunicus*, and *Cecidomyia yunnanensis*.

1.2.2 Central China Province

This province includes the eastern part of Yun-Gui Plateau, Sichuasn Basin and Yangtze River Basin, belonging to the central and north subtropical zones.

In the western mountain/plateau areas with rugged terrains, high elevations, and dry-cold weather, there is a relatively complex insect fauna. Common forest insect pests in this area include: *Bupalus vestalis* Satudinger, *Pineus sichuananus*, *Gilletteella glandulae*, *Adelges laricis potaninilaricis*, *Matsucoccus shennongjiaensis*, *Dendroctonus armandi*, *Pissodes yunnanensis*, *Shirahoshizo coniferae*, *Baculum Saussure*, *Parocneria orienta*, *Augomonoctenus smithi*, *Chinolyda flagellicornis*, *Plagiosterna adamsi*, and *Hemichroa crocea*.

In the eastern hilly and plain areas, the vegetation is dominated by evergreen coniferous forests with *Cunninghamia lanceolata*, *Pinus massoniana* Lamb. and *Cupressus* spp. as key tree species; and also includes some broad-leaved forests of *Phyllostachys* spp., *Camellia oleifera* and *Vernicia fordii*. Many insect pests often occur in the inter-cropping plantations, such as crop-orchard, crop-forest, and *Cupressus-Vernicia* plantations. Common forest insect pests in this area include: *Dendrolimus punctata punctata*, *Hyssia adusta*, *Dioryctria rubella*, *Monochamus alternates*, *Hylobitelus xiaoi*, *Shirahoshizo patruelis*, *Lobesia cunninghamiacola*, *Callidiellum villosulum*, *Semanotus sinoauster*, *Phloeosinus sinensis*, *Pacaypasoides roesleri*, *Megastigmus cryptomeriae*, *Rammeacris kiangs*, *Hippotiscus dorsalis*, *Algedonia coclesalis*, *Besaia goddric*, *Pantana phyllostachysae*, *Cyrtotrachelus thompsoni*, *Otidognathus davidis*, *Aiolomorphus rhopaloides*, *Trioza camphorae*, *Eriogyna pyretorum*, *Atysa marginata cinnamom*, *Moricella rufonota*, *Arna pseudoconspersa*, *Biston marginata*, *Curculio chinensis*, *Biston suppressaria*, *Aegosoma sinicum*, *Batocera lineolata*, *Podontia lutea*, *Dilophodes elegans sinica*, and *Agrilus zanthoxylumi*.

1.2.3 South China Province

This province includes Yunnan, southern parts of Guangdong and Guangxi, coastal areas of Fujian, Taiwan, Hainan Island and South China Sea Islands, with extremely diversified vegetation, such as tropical rain forests and monsoon forests, plus a great diversity of insects.

In the southern mountain areas of Yunnan, the vegetation is mainly evergreen deciduous rain forests. Common forest insect pests in this area include: *Xyleborus germanus*, *X. interjectus*, *Coptotermes formosanus*, *Macrotermes barneyi*, and *Odontotermes formosanus*.

In the coastal regions of Fujian and Guangdong provinces, ecological forests are the major vegetation in the mountain areas, whereas orchard plantations of litchi, longan, wampee, carambola, mango, olive, and coconut are typical vegetation in the hilly areas. *Casuarina equisetifolia* is the key tree species for the coastal shelter plantations along the shorelines. Common forest insect pests in this area include: *Aularches miliaris*, *Chondracris rosea*, *Icerya purchase*, *Anoplophora chinensis*, *Streblote hai*, *Lymantria xylina*, *Zeuzera multistrigata*, plus several invasive species, such as *Hemiberlesia pitysophila* and *Oracella acuta*.

In Hainan Island, the main forest vegetation consists of summit mossy dwarf forests, tropical montane rain forests, tropical ravine train forests, tropical evergreen monsoon forests and tropical semi-deciduous monsoon forests. Common forest insect pests in this area include: *Aularches miliaris scabiosus*, *Chondracris rosea*, *Tarbinskiellus portentosus*, *Icerya purchase*, *Paracoccus pasaniae*, *Paliga machoeralis*, *Hypsispyla pagodella*, *Stauropus alternus*, *Sraspedonta leayana insulana*, *Aristobia hispida*, *Heterobostrychus aequalis* and *Sinoxylon anale* plus several invasive species, including *Brontispa longissima* and *Quadrastichus erythrinae*.

In Taiwan, there are two major forest vegetation types: subtropical rain forests and tropical rain forests. *Abies kawakamii*, *Picea morrisonicola*, *Tsuga chinensis* var. *formosana* and *Chamaecyparis formosensis* are the major coniferous tree species. Common forest insect pests in this area include: *Dendrolimus punctatus*, *Pissodes nitidus*, *Lymantria dispar*, *L. monacha*, *L. xylina*, *Kunugiaundans fasciatella*, *Porthesia taiwana*, *Trabala vishnou*, and *Polyphagozerra coffeae*.

2. Occurrence of Forest Insect Pests and Their Damages in China: an Overview

Since the 1950s, as the forest plantation areas increase, the insect-infested forest areas and the overall number of major forest pest species also increased continuously. The average total annual forest insect occurrence (infestation) area was estimated in the 1950s at one million ha, mostly caused by several outbreak insect species including *Dendrolimus* spp. (or pine caterpillars), *Rammeacris kiangsu*, *Ambrostoma quadriimpressum* and *Parocneria orienta*. The infestation area was about 1.4 million ha in the 1960s with the following insects as the key outbreak species: *Dendrolimus* spp., *Matsucoccus matsumurae*, *Rammeacris kiangsu*, *Coclebotys coclesalis* and *Ambrostoma quadriimpressum*; the infestation was 3.4 million ha in the 1970s with the key outbreak species: *Dendrolimus* spp., *Matsucoccus matsumurae*, *Polychrosis cunninhamiacola*, *Biston marginata*, plus several species of poplar defoliators and longhorn beetles; in the 1980s 5.8 million ha were infested with the key outbreak species: *Dendrolimus* spp., *Hemiberlesia pitysophila*, *Hyphantria cunea*, poplar longhorn beetles, *Clania variegata*, *Paleacrita vernata*, and *Ambrostoma quadriimpressum*; while in the 1990s about 6.8 million ha was infested with following insects as the key outbreak species: *Dendrolimus* spp., poplar longhorn beetles, *Clania variegata*, *Hyphantria cunea*, *Oracella acuta*, *Matsucoccus matsumurae* and *Hemiberlesia pitysophila*. Infestation area quickly reached up to 10 million ha during the first decade of the 21st century. The average annual infestation areas of the following outbreak insect species exceeded 50,000 ha per species or group: *Dendrolimus* spp., poplar longhorn beetles, poplar defoliators, *Hemiberlesia pitysophila*, bark beetles, *Hylobius xiaoi* Zhang, *Oracella acuta*, *Hyphantria cunea*, *Dendroctonus valens*, *Rammeacris kiangsu*, pine sawflies, *Holcocerus hippophaecolus*, and *Matsucoccus matsumurae*. Thus, the average total annual forest insect infestation areas in China increased >10 times over the past 60 years.

Outbreaks of forest insect pests not only cause direct economic losses by reducing tree growth (timer yield) and timber quality, but also result in a significant reduction in ecological functions (services) of forests. Outbreaks will surely restrain the rapid development of forestry/ecological improvements, and of course, lead to a tremendous waste in man-power, funding and material resources caused by the need for pest control operations. In the "Three-North" region, the great success of the "Three-North Shelterbelt Forest Program" (the so-called "Green Great Wall") over the past 30 years has played a crucial role in maintaining and improving ecological environment, and in safeguarding and accelerating economic growth. However, since the 1980s, continuous outbreaks of wood-boring insects in the shelterbelt forests, especially the Asian longhorn beetle (ALB), *Anoplophora glabripennis*, have killed >100 million trees, resulting in serious sandstorms in some shelterbelt-broken areas, and further weakening this relatively fragile ecological environment. Pines (*Pinus* spp.) are among the most important planting tree species of the timber production forests and ecological forests, playing a very important role in elimination of barren hills, afforestation, windbreak and sand-fixation, water conservation, and ecological enhancement. China's pine forests cover >30 million ha of land area. There are 27 species of the pine caterpillars (*Dendrolimu*s spp.) in China. They have been considered as "historical forest insect pests" and "China's number one forest insect pests" due to their broad distribution ranges, serious infestations and frequent outbreaks. At low or intermediate level of populations, pine caterpillars could significantly reduce the normal growth of pine trees, and could potentially cause the death of entire forests if trees are defoliated over several successive outbreak years. During a severe outbreak, an entire pine forest can be completely defoliated by the pine caterpillars within several days, creating a scene of so called "smokeless forest fire". Serious forest defoliations by these pine moths will not only reduce timber growth and resin production, but also result in outbreaks of secondary bark or wood boring beetles, such as bark beetles (Scolytinae) and longhorn beetles (Cerambycidae), further speeding up the dying process of the infested forests. The fall webworm, *Hyphantria cunea* (Drury) is an invasive pest insect species, with a broad host plant range (extremely polyphagous), a great fecundity (several hundred eggs/female), a strong adaptive capability, and diversified dispersal strategies and pass-ways. It prefers to feed on broad-leaved street trees, roadside trees, shade trees, boulevard trees and park trees; therefore, defoliation of these trees during larval outbreaks can cause significant damage to urban landscapes and ecological environments, and interrupt the daily life of the local residents. It can result in eminent non-economic losses/damages to environmental functions, landscape aesthetic values, productivity and living quality, psychological impact, ecosystem balance, and regional reputation/tourism, as well as direct economic loss. In fact, it is estimated that the non-economic losses are far greater than the direct economic losses.

As shown above, the losses caused by forest insect pests are multi-dimensional. According to the official analytical report entitled "Direct economic losses and ecological service function losses caused by the forest pests of China during 2006-2010" by the "Disaster Loss Assessment Project Team" of the national General Station of Forest Pest Control, State Forestry Administration (SFA), the overall losses were 110.1 billion RMB Yuan, including 24.5 billion Yuan as direct economic losses and 85.6 billion Yuan of ecological service function losses. Forest insect pests are responsible for around 70% of overall direct economic losses.

2.1 A Brief Overview on Forest Insect Pests of China

China is one of the countries in the world with serious forest health issues caused by very severe and frequent outbreaks of forest insect pests. The following insects are considered as China's most serious outbreak/damaging forest pest insect species, including both invasive and native (indigenous) species affecting large outbreak areas.

Invasive species

(1) Japanese pine bast scale, *Matsucoccus matsumurae* (Kuwana)

The Japanese pine bast scale, *M. matsumurae*, is one of the four *Matsucoccus* spp. in China, having the broadest distribution range and the largest occurrence areas, and causing the most serious damages to pine trees (trunks and branches); and is the only invasive species among the four Chinese *Matsucoccus* spp. *M. matsumurae*, native to Japan, was first found at Liaotieshan, Lüshunkou District, Dalian, Liaoning province in 1942, and later in Laoshan, Shandong province in 1950. Since then, it has spread to Jiangsu, Shanghai, Zhejiang, Anhui, and lately to Jilin in 1994. It mainly damages *Pinus densiflora*, *P. tabuliformis* and *P. massoniana* in China, and occasionally *P. hwangshanensis* and *P. thunbergii*. Pine trees infested by the pine bast scale at low or intermediate population level will have their needles wilted, shoots turning yellow, bark cracked, vigor weakened; whereas, at high populations they will cause branches downward-curving and weeping, entire trees turning brownish, attacks by secondary insect pests, and ultimately tree mortalities. The pine bast scale was the most severe outbreak forest pest species in China during the 1970s and 1990s, reaching 150,000 ha annual infestation areas in mid-1980s. Serious outbreaks of *M. matsumurae* were also recorded in Jilin province over the past ten years. The overall national outbreak/infestation area in 2010 was estimated at about 60,000 ha.

Since its invasion, many researches on chemical and biological control measures on this invasive pest have been carried out in Liaoning, Shandong and Zhejiang provinces; and many successful measures from these research projects have been employed in the pest control operations. For instance, in Liaoning province, an IPM has been implemented over years to successfully combat *M. matsumurae* infestations. Their IPM components include an effective detection and monitoring program, an integrated strategy of categorizing infestation areas, corresponding potential and available tools, and suitable/optimal control measures; and well-organized tactics dominated by silvicultural methods, such as "clearing" (removal of infested trees), "closure" (forest fencing or closure to the public or livestock), "replanting", and "adjustment/improvement".

(2) Fall webworm, *Hyphantria cunea* (Drury)

The fall webworm is native to North America and was introduced into many European countries and Japan during World War II via modern transportation tools. The moth now has spread to most European countries (except the Nordics), Korea Peninsula and China, and lately to Central Asia. As an invasive pest, *H. cunea* was first found in Dandong, Liaoning province, China in 1979; since then has spread to several other provinces: Shandong (1982), Shannxi (1984), Hebei (1990), Tianjin (1995), Beijing (2004), Henan (2008), Jilin (2009), Jiangsu (2010), Anhui (2012) and Hubei (2016). Its invasion has caused serious damage to the local forests, agricultural crops and landscaping/ornamental trees, resulting in not only great economical and ecological losses, but also the destruction of landscapes. The fall webworm is an extremely polyphagous species with a great fecundity (several hundred eggs/female) and a quick dispersal capacity. Its larvae are consummate generalists, capable of feeding on >170 species of host plants, mostly broad-leaved trees plus several field crops. They feed in huge nests (silken tents) and are able to completely defoliate trees and shrubs, which would weaken trees, create unripened fruit drops, and even kill young trees after continuous infestations; such serious defoliations can lead to significant destructions to afforestation programs and urban landscaping with ornamental trees, and shelterbelt forests.

In 1998, the State Forestry Administration of China initiated a special pest control operation program against the fall webworm in Beijing (as a preventive area), Tianjin and Hebei (as direct control areas). After three consecutive years of operations, *H. cunea* infestation areas were reduced from 120,000 ha in 1998 down to 70,000 ha. Its occurrence and infestations started bouncing back from 2003, and reached 240,000 ha in 2006, and peaked at 780,000 ha in 2013. The General Office of State Council of China issued a special "notice on further strengthening pest control efforts for combating the infestation of the fall webworm" in 2013. The SFA re-initiated the *H. cunea* control operation program for Beijing-Taijin-Hebei region plus Liaoning province with great efforts and funds after receiving the State Council Notice. After many years of research and operational practices, so far IPM approaches dominated by environmentally sound (pollution or residual free) pest control techniques, have been developed and implemented. The IPM components include monitoring and mass-trapping programs using sex-pheromone baited traps or light traps during adult seasons; manual removal of larval tents (webs), sprays of some bio-pesticides such as *Bacillus thuringiensis*, avermectin and HcNPV, plus bionic pesticides, e.g. chlorbenzuron and tebufenozide against the larvae, and release of the parasitoid (*Chouioia cunea*) as a traditional biological control agent during the late larval instar and pupal stages.

(3) Pine needle hemiberlesian scale, *Hemiberlesia pitysophila* Takagi

The pine needle hemiberlesian scale, native to Japan (mostly Okinawa Islands and Sakishima Islands), was first found in May 1982 in Zhuhai City of Guangdong province, likely introduced into Guangdong, China around 1980. According to the experts, this scale was likely first introduced into Hong Kong and Macau via the imported Christmas trees from Japan; and spread into the neighboring cities, *e.g.* Shengzhen and Zhuhai, through 1st instar larvae by wind. *H. pitysophila* has become one of the major pine tree insect pests in southern China, and has already caused significant damage to pine forest resources, ecological environment, natural landscapes, exports, and social/economical development, and is directly threatening the survival and security of Southern China's forests and ecological environments.

H. pitysophila is oligophagous on pine trees (*Pinus* spp) and is noted for being particularly injurious to over ten species of pines, especially *P. massoniana*, *P. glabra* Walter and *P. caribaea*. It infests mainly the sheaths of pine needles, and then the needles. Heavy infestations can cause premature needle drops and sometimes bending of infested branches, and ultimately lead to tree mortality over large forest areas. There are five over-lapping generations per year in southern China (e.g. coastal areas of southern Guangdong) with no obvious overwintering period. The 1st instar larvae of each generation can be easily dispersed at long distances by wind; both *H. pitysophila* larvae and female adults might be carried in the air by strong winds for up to 6 km. *H. pitysophila* is a serious invasive species with a cryptic feeding behavior, a great fecundity and a rapid dispersal capacity; and is really difficult to control. Currently, it has spread to several provinces in southern China including Guangdong, Fujian (in 2001), Guangxi (2003) and Jiangxi (2006), causing significant damage to the pine forests in these areas, reached an outbreak peak in 2005 with 700,000 ha of heavily or intermediately infested pine forest areas.

Biological control with import and release of a parasitoid, *Coccobius azumai* from the scale's native range, Okinawa Islands of Japan, proved to be very effective, and has been considered as one of best successful examples of "import and release" of foreign natural enemies to control invasive forest insect pests. Lately, enhancements and releases of native parasitoids, such as *Encarsia amicula* Viggiani et Ren, *Marietta carnesi* (Howard), and *Aphytis* sp. also showed good and promising control efficacies.

(4) Loblolly pine mealybug, *Oracella acuta* (Lobdell)

The loblolly pine mealybug, native to North America, was accidentally introduced into the Hongling Pine Seed Orchard, at Taishan, Guangdong Province, China, in 1988 together with the exotic slash pine, *Pinus elliottii*. It is a typical invasion example of an exotic forest pest insect species introduced together with its host plant, *Pinus elliottii*. *Oracella acuta* infests many species of pines, including *P. elliottii*, *P. taeda*, *P. palustris*, *P. clausa*, *P. echinata*, *P. virginiana*, *P. massoniana*, and *P. caribaea*. There are 3-4 generations per year. Young *O. acuta* nymphs can disperse passively through air-movements (wind) for up to 17-22 km; and can also be transported to other regions or areas at long distance via seedlings, seed/cones, and scions. Most of nymphs feed in large numbers in aggregations on pine shoots and suck the sap of pine shoots, resulting in significant reductions in tree growth, shortening current shoots or shoot clumping/curving, needle dropping, and ultimately, great losses in timber and resin production. The mealybug further spread to Guangxi in 2000, Hunan in 2003, and Jiangxi in 2006. Its total annual infestation area in China in 2004 was estimated at about 270,000 ha, and declined to 90,000 ha in 2010.

Introduction and release of the exotic parasitoids, *Acerophagus coccois*, *Zarhopalus debarri* and *Allotropa oracellae* from their native range in U.S.A., seemed to be effective. However, recently it was reported that several native parasitoids might also play a role in controlling this invasive pest insect.

(5) Red turpentine beetle, *Dendroctonus valens* LeConte

The red turpentine beetle (RTB), native to North America and part of Central America has a natural distribution ranging from Northern Canada and Alaska in the north to Mexico, Guatemala and Honduras in the south. The bark beetle was first found in Yangcheng and Bingxin, Shangxi province, China, and now has spread into Henan, Hebei and Shaanxi provinces. RTB primarily colonizes stressed or fire-scorched pine trees or dying pine trees attacked by other primary bark beetles as well as the freshly cut tree stumps; it rarely attacks healthy trees by itself, therefore is considered as a typical secondary pest insect in its native range in North America. However, in China, RTB not only attacks weakened trees, but also colonizes healthy pine trees, especially *P. tabuliformis* trees with >10 cm of DBH. Both RTB adults and larvae feed on phloem and cambium tissues in a mass attack fashion at the lower part of the tree trunk or trunk base, and can kill pine trees in great numbers. The aggressive Chinese populations of RTB have caused great damage to *P. tabuliformis* forests in Shanxi, Henan, Hebai and Shaanxi provinces, and pose a further threat to China's extensive pine forests if its invasive range continues to expand, due to its high fecundity, aggressive attacking behavior, and a capacity of causing significant host tree mortality.

Historical records suggest that *D. valens* was introduced to Shanxi province on untreated logs imported from the western USA during early-mid 1980s. In the beginning, it was not easy to notice its infestations (trunk base and root crown) at low population levels; however, due to the lack of natural enemies and other inhibitory factors in the new range, RTB populations (with excellent dispersal abilities) soon exploded in Shanxi, China, and spread rapidly to the adjacent provinces. Its initial outbreaks in China might be also related to the combination of droughts and consecutive warm winters, a favorable climate condition for many bark beetles. An abundance of susceptible naïve hosts such as large areas of pure *P. tabuliformis* plantations in north China might also have contributed to the success of RTB in China.

Soon after RTB's outbreaks in Shanxi, efforts for introduction and release of a bark beetle predacious natural enemy, *Rhizophagus grandis*, from UK and Belgium as a biological agent, and import and technology transfer of kairomone attractants as a monitoring and mass-trapping tool from Canada and USA were made; these two programs have been implemented in China for many years, and proved to be effective. In 2000, SFA initiated a special national forest pest management program targeting RTB, and its infestation areas were reduced quickly after implementing this national program from 500,000 ha in 2000 down to 50,000 ha in 2010. This program is considered as the most successful example of effective control of an invasive forest pest insect in China; and the experiences obtained from this program could provide invaluable guidance for future prevention and control programs against other invasive insect pests.

(6) Coconut leaf beetle, *Brontispa longissima* (Gestro)

The coconut leaf beetle, *B. longissima*, is native to Indonesia and is mainly distributed in the Pacific Islands. The beetle spreads mostly through the movement of infested palms. In 1992, it was listed as a 2nd tier quarantine pest insect species by the Chinese Plant Quarantine Agency, and was intercepted on the imported palm seedlings multiple times at ports. It was first found in June 2002 from the ornamental palm trees at Fengxiang Road of Haikou City, Hainan province, China. Currently, it has spread into Hainan, Guangdong, Guangxi and Fujian provinces, with a total infestation area reaching up to 20,000 ha. The coconut leaf beetle is one of the most damaging insect pests of coconut and other palms, attacking more than 20 palm species with coconut (*Cocos nucifera*) being the most favored host. Other hosts include Royal palm (*Roystonea* sp.), Alexandra palm (*Archontophoenix alexandrae*), Sago palm (*Metroxylon sagu*), California fan palm (*Washingtonia filifera*), Mexican fan palm (*W. robusta*), Bottle palm (*Hyophorbe lagenicaulis*), Chinese fan palm (*Livistonia chinensis*), Madagascar palm (*Chrysalidocarpus lutescens*) and Areca nut palm (*Areca catechu*). The larvae and adults of *B. longissima* feed on the soft tissues of the youngest leaf in the throat of the palm. Affected leaves dry up, resulting in stunting of the palm and reduced nut production. Prolonged attacks on young palms can lead to their death.

The coconut leaf beetle prefers a warmer climate, with an optimal developmental temperature range of 20-30℃, and a lower threshold temperature of 14.5℃. It has 3-5 generations per year, and is a dangerous invasive pest insect due to its high fecundity and adaptability, therefore, it is really difficult to eradicate once settled in a new range. Currently, chemical control using various insecticides seems to be effective; however, repeated applications may be impractical and uneconomic and cannot be used as a long-term control measure. Two parasitoids of *B. longissima*, *Tetrastichus brontispae* and *Asecodes hispinarum*, have been successfully used in southern China to control the beetle. Use of the entomopathogenic fungus *Metarrhizium anisopliae* is also promising.

(7) Erythrina gall wasp, *Quadrastichus erythrinae* Kim

The Erythrina gall wasp, a likely native to Africa, is a devastating invasive insect species of *Erythrina* (coral) trees, and was first detected in 2005 in Hainan province, China. *Q. erythrinae* has a very short life-cycle (ca. 20 days), a high fecundity, fast dispersal ability, and a cryptic feeding behavior. The Erythrina gall wasp infests the young *Erythrina* leaf and stem tissues via egg laying and larval feeding, which induce the formation of galls in the leaflets and petioles. As the infestation progresses, leaves curl and appear deformed while petioles and shoots become swollen. Heavily galled leaves and stems result in not only a loss of growth and vigor, but also aesthetic value. Severe infestations can cause complete defoliation and tree mortality. Currently, it is distributed in several regions of Hainan, Guangdong and Fujian provinces, attacking several *Erythrina* tree species, including *E. variegate*, *E. variegata* var. *picta*, *E. crista-galli*, and *E. corallodendron*. There are five native and five introduced *Erythrina* tree species in China, mostly distributed in the Southeast, South and Southwest China, and are planted as roadside/street trees, ornamental trees and medical plants.

Q. erythrinae eggs, larvae and pupae are all living inside the galls; can be transported together with their host plants at long distances, therefore, are of great quarantine importance. In 2005, it was listed as an import-port quarantine species and a national quarantine insect species of forest plants in a jointly issued official notice by

Ministry of Agriculture, National General Administration of Quality Supervision, Inspection and Quarantine, and SFA of China. In 2013, China's SFA listed the Erythrina gall wasp as one of the national dangerous forest pest species. Currently, the main control measures include removal of infected branches, leaf sprays of insecticides and stem-base injection of systemic pesticides.

(8) Black locust gall midge, *Obolodiplosis robiniae* (Haldemann)

The black locust gall midge, a native to the eastern part of North America, is a minor pest insect of black locusts, and has never been considered as a serious pest insect in its native range due to its low occurrence and negligible damages to the locust trees. Recently, this gall midge has been introduced into many countries and regions outside of its original distribution area. It was first found in Japan and Korea in 2002; and then in Italy and several other European countries in 2003. Since 2005, *O. robiniae* damages on the black locust trees in China have been observed in Beijing, Hebei (Qinhuangdao, Tangshan, Langfang, Baoding, Zhangjiakou and Chengde), Liaoning (Shenyang, Dalian, Jinzhou, Anshan, Huoludao, Chaoyang) and Shandong (Jinan and Yantai) provinces, reaching a total infestation area of 40,000 ha in 2006.

Black locust (*Robinia pseudoacacia*) was introduced to China from North America >130 years ago. Nowadays, this species, with a great economical and ecological value, is widely planted in many parts of China as one of the key planting trees for various afforestation programs such shelterbelts (wind-break and sand-fixation forests), water and soil conservation forests, barren hill afforestations, plus ornamental and roadside plantations. The margins of *R. pseudoacacia* leaflets are rolled downwards by aggregative feeding of 3-8 *O. robiniae* larvae, forming characteristic leaf galls. Larval feeding inside the galls strongly weakens the black locust trees, can lead to attacks by the secondary insects such as longhorn beetles or flat head beetles, and may ultimately kill the *R. pseudoacacia* trees.

Chemical control using insecticides against larvae and adults seems to be effective. However, enhanced biological control using a parasitoid wasp, *Platygster robiniae*, with a high natural parasitism, looks promising.

(9) Red imported fire ant, *Solenopsis invicta* Buren

The red imported fire ant (RIFA) is a dangerous ant species native to South America (Brazil, Paraguay, and Argentina). *S. invicta* is highly invasive because of its high reproductive capacity, large colony size, wide food range and ability to sting. RIFA has become a major agricultural and urban pest throughout in its new ranges, and caused serious medical and significant ecological/environmental harm. *S. invicta* is one of the most notorious invasive ants and has been nominated for the top 100 of the World's Worst Invaders list compiled by the Invasive Species Specialist Group. In the United States, the RIFA first arrived in the seaport of Mobile, Alabama by cargo ship between 1933 and 1945. *S. invicta* were first discovered in Queensland, Australia in 2001, and then in New Zealand in the same year, in Malaysia in 2003, and in Taibei, Taiyuan and Jiayi of Taiwan in October 2003. RIFA was first detected in mainland China in 2005; since then the ant has spread to Guangdong, Hong Kong, Macau, Hunan and Guangxi, raising a serious threat to China's national economical, environmental and public securities.

In agriculture, RIFA workers feed on plant seeds, fruits, buds, fresh stems and roots, and may cause strong impacts on plant survival and growth. RIFR is also a major public health issue because when disturbed they can attack and sting humans and animals. Most victims experience intense burning and swelling, followed by the formation of sterile pustules which may remain for several days. However, some allergic people may suffer from anaphylaxis which can be fatal if left untreated. RIFR have been reported to attack the irrigation systems, electrical equipment and recreation facilities for unknown reasons and may cause significant damage to the local biodiversity.

Natural dispersal occurs through mating flights, ground migration and by floating in colonies of interlinking ants on water during floods. The long-distance movement of infested articles of commerce such as sod, bales of hay, nursery containers, used soils, agricultural equipment, cabinets and transportation tools have resulted in the spread of *S. invicta* colonies. Analysis of the potential distribution range of *S. invicta* in China indicates that the RIFR can live in most parts of Southeastern China, and may even be able to survive in Shandong, Tianjin, and southern parts of Hebei and Shaanxi provinces to the north.

In order to prevent further spread of *S. invicta*, all the potted plants, flowers, seedlings, turfs and potting soil from the infested areas must be inspected and quarantined before shipping to non-infested areas. Efforts to eradicate newly detected infestations of this species have been made in China; including control measures that are both physical (treated with boiling water, or submerging nests with water) and chemical (broadcast application of granular baits and treatment of individual ant colonies in mounds using contact insecticides).

(10) *Phenacoccus solenopsis* (Tinsley)

Phenacoccus solenopsis is native to North America. It started to cause damage to cotton in 1991 in the U.S. During 2002-2005, it spread to Chile, Argentina, and Brazil. In 2005, it spread to India and Sindh and Punjab provinces of Pakistan. It soon spread to 45 km2 area in this region. This pest caused 12% yield loss in 2006 and 40% yield loss in 2007 in Punjab province alone. The cost of chemical control of this pest within two months period surpassed USD $120 million.

Phenacoccus solenopsis was first recorded in China in 2008. By 2009, it was found on cotton in Hainan, Guangdong, Guangxi, Yunnan, Fujian, Jiangxi, Hunan, Zhejiang, Sichuan and Taiwan. Its hosts include 53 families, 154 species with 64 of them are weeds. Therefore, it spreads quickly. Potential infestation areas include Xinjiang, Gansu, Ningxia, Shaanxi, Shanxi, Hebei, Beijing, Tianjin, Liaoning, and Inner Mongolia. Control methods include quarantine, protect and release Cryptolaemus montrouzieri (Mulsant) and Chrysoperla carnea (Stephen). Insecticide spray containing cyhalothrin, deltamethrin, or neonicotinoid is also effective.

(11) Sycamore lace bug, *Corythucha ciliate* (Say)

Sycamore lace bug is distributed in U.S.A., Canada, Italy, France, Russia, Hungary, Poland, Bulgaria, Greece, Czech Republic, Netherlands, Austria, Croatia, Germany, Serbia, Montenegro, Slovenia, Switzerland, Belgium, Portugal, Slovakia, Spain, Turkey, Chile, Israel, Japan, Korea. Since its discovery in Wuhan, China in 2006, it has invaded Beijing, Hubei, Hunan, Shanghai, Zhejiang, Jiangsu, Anhui, Henan, Chongqing, and Guizhou. Main host plants are in genus Platanus of sycamore. It also causes damage to Broussonetia papyrifera, Carya ovata, Fraxinus spp., Acer pseudoplatanus, Quercus laurifolia, Liquidambar styraciflua. Sycamore lace bug adults and nymphs usually feed on the abaxial side of the leaves on lower layer of the sycamore tree canopy. Initial damage includes yellowish white spots and leaf chlorosis. Severe infestation causes dry and yellow leaves, early senescence of the leaves, slow tree growth, and even death of trees.

Control methods: (1) Scraping loose bark, painting tree trunks with lime, timely collecting and disposing of fallen leaves will reduce the number of overwintering insects; (2) Spraying canopy to control nymphs and newly emerged adults. Insecticides include pyrethroid (lamda-cyhalothrin, fenvalerate), imidacloprid, acetamiprid; (3) Conservation of natural enemies including Nabis pseudoferus, Rhinocoris iracundus, Orius vicinus, Aptus mirmicoides, Aphaenogaster smythiesi, and a spider,Theridion lunatum.

Native species

(12) Pine caterpillars (*Dendrolimus* spp.)

The pine caterpillars, *Dendrolimus* spp., are historic insect pests of China with the first *Dendrolimus* official outbreak (occurred in Zhejiang) record traced back to 1530. A *Dendrolimus* outbreak occurred in 1599 at Changshu county of Jiangsu province with complete defoliation of a pine forest was also recorded in the county's history book. Since the founding of P. R. China in 1949, the pine caterpillars have been considered as the most destructive and influential forest insect pests in China, due to the longest (sustained) outbreak periods, the largest infestation areas and the broadest distribution ranges. *Dendrolimus* spp. are distributed throughout most parts of the China (including 25 provinces/autonomous regions), ranging from the Greater Khingan Range in the north to Hainan Island in the south, and from coastal areas in east to the Altai Mountains, Xinjiang in the west.

Dendrolimus pine moths are among the most destructive pine forest defoliators in China. During a serious outbreak (with several thousands of larvae per tree), an entire pine forest could be completely defoliated by the pine caterpillars within several days, creating a scene of so called "smokeless forest fire". Forest defoliations by these pine caterpillars will not only reduce tree vigor and their timber growth, but also at high population densities could result in significant tree mortality. *Dendrolimus* spp. are indeed the most influential forest insect pests of China due to the fact that outbreaks of the same or different pine caterpillar species occur every year in the same or different regions.

There are 27 species of *Dendrolimus* pine caterpillars in China, with the following six species being the most common in outbreaks: *D. punctatus*, *D. superans*, *D. tabulaeformis*, *D. houi*, *D. spectabilis* and *D. kikuchii*. *D. punctatus* is a major defoliator of *Pinus massoniana*, mainly distributed in Anhui, Henan, Shaanxi, Sichuan, Yunnan, Guizhou, Hunan, Hubai, Jiangxi, Jiangsu, Zhejiang, Fujian, Guangdong and Guangxi provinces. *D. houi* prefers *P. yunnanensis* and *P. kesiya*, as well as some cypresses or cedars; with outbreaks mainly occurring in Yunnan and western Sichuan and occasionally in Guizhou, Hunan, Hubei, Jiangxi, Zhejiang, Fujian and Hainan provinces. *D. tabulaeformis* is a key defoliator of *P. tabuliformis*, with serious outbreaks or infestations occurring mainly in Hebei and Congqing; and occasional outbreaks were also reported in Liaoning, Inner Mongolia, Shandong, Shanxi, and Shaanxi provinces. *D. spectabilis* is a serious defoliator of *P. densiflora* in Shandong province, with occasional outbreaks also occurring in Jiangsu, Liaoning, and Hebei provinces. *D. superans* is a serious defoliator of *Larix* trees, with its outbreaks or infestations commonly occurring in Heilongjiang, Jilin, Liaoning, Inner Mongolia, and Altai Mountains, Xinjiang, and occasional outbreaks reported in Wulingshan, Hebei. *D. kikuchii* is one of the major defoliators of *P. yunnanensis* and *P. kesiya,* distributed mainly in Yunnan and Sichuan provinces.

In China, periodic outbreaks of the pine caterpillars started from mid-1950s, reaching the first outbreak peak during 1956-1958 with a national annual infestation area of 1.2-1.3 million ha; the second outbreak peak during 1964-1965 with a national annual infestation area of 1.3-1.4 million ha; the third outbreak peak during 1972-1973 with a national annual infestation area of 1.9-2.3 million ha. After 1975, this typical 10-year outbreak cycle was broken and serious outbreaks of *Dendrolimus* spp. with an average national annual infestation area of 2.6 million ha continued for the next 15 years. After 1990, the annual outbreak areas dropped and oscillated around 1.5 million ha, but have declined to about 1.0-1.3 million ha after 2000.

The operational pest control progress against the pine caterpillars and their related researches, to a large extent, reflect the overall development of forest pest insect control practices and researches in China. In recent years, new monitoring and detection technologies, such as satellite remote sensing and digital aerial sketch-mapping, dramatically advanced our ability to better monitor population dynamics and predict occurrences/outbreaks of the pine caterpillars. Many new pest control measures, such as aerial application, "biological missiles" enhanced bio-control tactics, bio-pesticides and biomimetic agents, have been employed in control operations against the pine caterpillars. Currently, the IPM program targeting the pine caterpillars, integrating all the available monitoring techniques and environmentally sound control tactics (especially the silvicultural approaches), has proven very effective.

(13) Common defoliators of poplar trees

Poplar defoliators are one of most important forest insect pests in China, causing serious damage to poplar shelterbelt forests and quick-growth poplar timber production forests. Defoliator outbreaks will not only reduce the ecological function of the shelterbelt forests and timber growth, but also might lead to significant tree mortality, especially for young trees at high insect population levels. There are more than a hundred species of poplar defoliators in China, mostly lepidopteran moths from the families of Lymantridae, Noctuidae, Arctiinae, Geometridae, and Notodontidae. Other defoliators include leaf beetles, weevils (Coleoptera) and several species of sawflies (Hymenoptera). They are widely distributed throughout the country, and wherever the poplar trees exist, some poplar defoliators occur. Serious outbreaks of the same or different poplar defoliator species occur every year at the same or different regions, resulting in significant damages to poplar forests, which is one of the major obstacles in the development of poplar plantations in China.

The following species are considered as the major poplar defoliators in China: *Clostera anachoreta*, *Cerura menciana* Moore, *Micromelalopha sieversi*, *Clostera anastomosis*, *Leuoma candida* Staudinger, *Stilpnotia salicis* (L.), *Lymantria dispar*, *Gypsonoma minutana*, *Pyrosis idiota*, *Leucoptera susinella* Herrich-Schaffer, *Paleacrita vernata*, *Chrysomela populi*, and *Stauronematus compressicornis* (F.). *Clostera anachoreta* is a serious defoliator of poplar and willow trees; distributed in most parts of the country (Northeast, North, Northwest,

Central, Southwest and East China); *Cerura menciana* is also a major defoliator of poplar and willow trees, distributed mainly in Northeast, North, Northwest China, Hubei, Hunan, Jiangxi and Jiangsu provinces. *Micromelalopha troglodyte*, infesting both poplar and willow trees, mainly occurs in Heilongjiang, Jilin, Liaoning, Henan, Hebei, Shandong and Anhui provinces. *Clostera anastomosis* is a serious poplar defoliator, occurring in Northeast, North, Northwest China, Jiangsu and Zhejiang provinces. *Stilpnotia salicis* feed on both poplar and willow trees, limiting its distribution range to Northeast and Northwest China, Shandong, Jiangsu and Henan provinces. *Lymantria dispar* occurs in most parts of China, and is a serious defoliator of many broad-leaved trees including poplar and willow trees. *Gypsonoma minutana* is a defoliator of poplar and willow trees in North China. *Pyrosis idiota* is a defoliator of many broad-leaved trees including poplars and elms in Heilongjiang, Jilin, Liaoning and Inner Mongolia. *Leucoptera susinella* infests many species of poplar trees in Northeast and North China. *Paleacrita vernata* is a serious defoliator of many broad-leaved trees, including poplars in Xinjiang, Gansu, Shaanxi, Ningxia, Inner Mongolia and Hebei provinces. *Chrysomela populi* is a serious leaf beetle feeding mainly on poplar leaves in Northwest, North and Northeast China plus Sichuan, Hunan and Hubei provinces. *Stauronematus compressicornis* is a major poplar/willow sawfly defoliator, occurring mainly in Liaoning, Jilin and Heilongjiang provinces.

Over the past decades, poplars have become a favorite tree species for afforestation programs, especially for the fast-growth timber production forests in northern China. The fast growth in the pure poplar plantation areas plus other management and environmental issues, unfortunately, led to more serious occurrences and outbreaks of poplar defoliators in the regions, with a drastic increase (by 1.1 million ha) of the total annual infested areas from 300,000 ha in 1996 to 1.47 million ha in 2013.

Currently, the IPM practices against the poplar defoliators are based on silvicultural approaches, dominated by applications of enhanced biological control, biomimetic agents and botanical insecticides, and supplemented by the coordinated integration of mechanical (manual), physical and chemical tactics. The IPM program has proven to be very effective to reduce the defoliator population densities, their infested areas and damages. In the lighter infestation areas (at low insect population levels), operational focus will be on strengthening monitoring of insect population dynamics, conservation and utilization of natural enemies, and improving the ecological environments. In the mid- to highly infested areas, the IPM components mainly include the biological measures, such as bio-pesticides, NPVs, and biomimetic agents, plus enhancement of natural enemies, in order to keep the defoliators under sustainable controls. However, in the severely infested (at large scales) areas during the outbreaks, more aggressive control measures, such as aerial application of insecticides, should be taken first to quickly knockdown the defoliator's populations; and then other suitable measures can be followed.

(14) Wood-boring insects of poplar trees

Poplar wood-borers are important forest insect pests in China, including beetles from Cerambycidae (longhorn beetles), Buprestidae (flatheaded beetles) and Curculionidae (weevils) as well as moths from Cossidae, Hepialidae and Sesiidae. These insects attack both poplar trunks and branches, and could result in broken branches/stems in heavily mined trees, especially after strong winds, and ultimately tree mortality. The infested trees with hundreds or thousands of wood-borer larval tunnels and emergence holes will have little or no timber use value. The poplar wood-borers, represented by the Asian longhorn beetle (ALB), *Anoplophora glabripennis*, are distributed in >20 provinces/autonomous regions in China, and have caused serious damages to the poplar-based shelterbelt forests in the "Three North" areas, becoming a super biological threat to the development of China's shelterbelt forest systems.

The major wood-boring insects of poplar trees include: *Anoplophora glabripennis*, *Batocera lineolata* Chevrolat, *Apriona germari*, *Saperda populnea*, *Xylotrechus rusticus*, *Megopis sinica*, *Trirachys orientalis*, *Cryptorhynchus lapathi*, *Poecilonota variolosa*, *Melanophila picta*, *Paranthrene tabaniformis*, *Sesia siningensis*, *Phassus excrescens* and *Cossus* spp. These wood-borers are further divided into three species groups based on their feeding locality on the trees, feeding behavior, population size, distribution range, damage features, and economical importance. Group I includes several primary species that are able to attack healthy trees (young or old), and could significantly reduce timber growth and quality grade, and lifespan of polar trees, such as

Anoplophora glabripennis, *Apriona germari*, *Batocera lineolata*, *Saperda populnea*, *Sesia siningensis*, *Phassus excrescens* and *Cossus* spp. Group II represents insect species that only attack branches and young trees, and are considered as major insect pests of forest nurseries and young poplar trees, including *Paranthrene tabaniformis*, *Xylotrechus rusticus* and *Saperda balsamifera*. Group III includes several secondary species that only attack dying and fallen poplar trees with occasional scattered outbreaks mainly in the poplar stands with a great number of old trees, such as *Megopis sinica* and *Trirachys orientalis*.

Outbreaks of polar wood-borers have been a major pest concern for the health of poplar plantations in China. Several factors may have contributed to their outbreaks, including large area of pure (single species or variety) poplar plantations, afforestation under poor site conditions, a lack of timely silvicultural practices, and continuous droughts. The Asian longhorn beetle is distributed throughout China, with serious outbreaks mainly occurring in North and Northwest China, therefore, the species was a destructive pest resulting in serious damage to poplar-based shelterbelt forests planted during Phase I of the "Three North Shelterbelt System" program in Northwest China; and this species nearly destroyed the shelterbelt program. *Apriona germari* is also widely distributed in many parts of China, except Heilongjiang, Inner Mongolia, Ningxia, Qinghai and Xinjiang, and is a serious pest insect of white poplars from *Populus* Section (formerly *Leuce* section). *Saperda populnea* is mainly distributed in Northeast and Northwest China, attacking various species of poplar trees. *Paranthrene tabaniformis* is distributed in northern China and mainly attacks 2-3 year-old poplar seedlings and young poplar trees. *Phassus excrescens* is only distributed in Liaoning, Jilin and Heilongjiang provinces, and is a serious pest of poplar seedlings and young trees.

Populus spp. are major planting tree species for ecological improvement and fast-growth timber forests in the "Three North" region. Currently, there are seven million ha of poplar plantations in China of which many parts are vulnerable to the wood-borers' attacks. The total annual outbreak area of poplar wood-borers was estimated at 360,000 ha in 1996, peaked at 900,000 ha (an all-time high) in 2007, and dropped to 500,000 ha in 2014.

In September 1991, an emergency meeting was held by SFA (formerly Ministry of Forestry) to discuss how to effectively combat the outbreaks of poplar wood-borers in Shaanxi, Gansu, Ningxia, Inner Mongolia and Shanxi provinces; and a national effort targeting the control of the poplar wood-borers was initiated at this meeting. In August 1994, a follow-up meeting on exchanges of pest control operation experiences against the poplar wood-borers in the "Three North" region was held. In 1998, a national pest control program specially targeting the ALB in Inner Mongolia started; in 2000, a similar national pest control model program for combating the poplar longhorn beetles in Shaanxi, Gansu, Ningxia, Qinghai and Heilongjiang provinces was initiated, and then these two programs were consolidated in 2003 and restarted as a national poplar pest (insects and diseases) control program. Over the past ten years, this national program against poplar pests has proven effective, including many successful IPM components such as multiple tree species planting, placement of trap trees/logs, stump grafting technique, topical cutting of tall stem, coppicing, removal of infested branches or galls, conservation and enhancement of woodpeckers, injection of systemic insecticides, inserting mini poison sticks into the larval tunnels and spray of pesticides against adults.

(15) Pine sawflies

There are 32 species of pine sawflies (Diprionidae) in China, of which 11 species are considered as serious insect pests, including *Augomonoctenus smithi* Xiao et Wu, *Diprion liuwanensis*, *D. nanhuaensis*, *D. jingyuanensis*, *Gilpinia massoniana*, *Neodiprion xiangyunicus* Xiao et Zhou, *Nesodiprion zhejiangensis*, *N. huizeensis*, *N. guangxiicus*, *N. fengningensis*, and *N. dailingensis*. All pine sawfly larvae feed on needles of pines, spruces, *Keteleeria* spp., and cypress trees, except for those of *Augomonoctenus smithi* which feed in the developing cones of incense cedar. Infestations by pine saw flies will not only weaken conifer trees, but also can cause serious attacks by secondary insects, and ultimately result in tree mortality. Pine sawfly outbreaks occur periodically (once every 10-12 years) and tend to subside after 2-4 years of heavy defoliation. A distinct feature of pine sawfly population dynamics is that their population increase before the outbreak is very much unpredictable, in most cases abrupt or unexpected; however, population decline after the outbreak is always abrupt.

Pine sawflies occur in 17 provinces/autonomous regions in China, with outbreaks mostly occurring in Shanxi,

Congqing, Sichuan, Shaanxi, Gansu and Ningxia. Occasional outbreaks of *Augomonoctenus smithi* in Sichuan; *Neodiprion zhejiangensis* in Hubei, *Diprion jingyuanensis* in Gansu and Shanxi were reported. The total annual infested area of pine sawflies in China was estimated at 60,000 ha during 2000-2003, and declined after 2004.

Currently, the pine sawfly IPM components include accurate monitoring and prediction of seasonal flight activity and population dynamics using sex-pheromone baited traps; preventive silvilcutural approaches, such as avoidance of irrational or illegal logging; application of environmentally sound insecticides during initial stage of pine sawfly outbreaks; and traditional biological control measures using *Beauveria bassiana*, *Bacillus thuringiensis* and NPVs.

(16) Yellow-spined bamboo locust, *Ceracris kiangsu* (Tsai)

The yellow-spined bamboo locust, *Ceracris kiangsu* (Tsai) (Formerly *Rammeacris kiangsu*), is a serious pest insect of bamboo forests in southern China. Similar to the pine caterpillars, it is also a historic pest insect. *C. kiangsu* is mainly distributed in the provinces south of Yangtze river, Southwest China, Henan, Shaanxi and Taiwan provinces. They prefer to feed on moso bamboo (*Phyllostachys edulis*) and Weaver's bamboo (*Bambusa textilis*), and could often kill entire bamboo forests at large scales during outbreaks. According to the "Yiyang County's History Book", there were several serious *C. kiangsu* outbreaks after 1639 in this county, and each outbreak resulted in complete defoliation of bamboo forests, and great tree mortalities. During the early 1900s, a serious *C. kiangsu* outbreak occurred in 320,000 mu (≈ 21,333 ha) of bamboo forests, killed 70 millions of bamboo trees in Taoyuan county. There were several scientific reports on *C. edulis* outbreaks and their relationships with climatic conditions, even before 1949.

During the early 1950s, severe outbreaks of *C. kiangsu* occurred in some areas. Back then, only primitive control measures, such as manual removal of eggs, handpicking of nymphs/adults, and burning, and application of inorganic pesticides, were available. The national total annual *C. kiangsu* infestation area was estimated at 530,000 ha in 1955, and 620,000 ha in 1957, mostly occurred in Hunan, Jiangxi, Sichuan, and Guangxi provinces. Since early 1960s, chemical control programs using highly toxic organic insecticides began and were able to keep the total annual *C. kiangsu* infested areas under 20,000 ha during 1960s and 1970s. However, after 1980s, the total annual *C. kiangsu* infested areas increased in spurts, and bounced back up to 84,000 ha in 2005, due to the fact that many highly toxic and residual organic insecticides were banned while new and less toxic insecticides were not commercially available yet.

After many years' practices in southern provinces, especially in Hunan province, two effective control strategies against the bamboo locusts were developed and employed, the "positive approaches" and the "passive approaches". The positive control approaches include direct application of insecticides (with ingestion or contact mode of action) to kill the nymphs and mass-trapping (or attract-n-kill) of adults when the pest occurs in relatively small areas. This strategy was proven to be very effective, since it took into account the aggregative behaviors of this locust, especially the aggregative egg-laying behavior, and targeted the most vulnerable newly hatched young nymphs in the focused spots. The "passive" approaches, on the other hand, include aerial or ground applications of insecticides at large scales when serious outbreaks of *C. kiangsu* occur in large areas.

(17) Pine shoot beetles, *Tomicus* spp.

The pine shoot beetles, *Tomicus* spp., are distributed in most parts of China, feeding on >20 species of pines. Their outbreaks occurred periodically in Northeast, North and Southwest China, for instance, serious outbreaks were reported in Hangzhou, Zhejiang during 1955-1956, in Changchun, Jilin during late 1970s to early 1980s, and during the past 20 years in Southwest China, mostly in *P. yunnanensis* and *P. kesiya* forests.

Tomicus spp. in general are secondary insects, mainly infesting weakened, dying and wind-thrown pine trees; however, they might also attack obviously healthy trees at higher population densities. *Tomicus* spp. are boring insects, spending the majority of their lives (eggs, larvae, and adults) beneath the bark of pine trees, and partial adult times in the pine shoots for maturation feedings. Such maturation feeding will kill the current year shoots; weaken the pine trees, which will be followed by the mass attacks on trunks by matured adults for reproductive feedings. Both egg galleries and larval galleries extend in phloem (inner bark) of the infested trunks, cutting off the tree's ability to transport nutrients, ultimately resulting in tree mortality.

The Yunnan pine shoot beetle, *Tomicus yunnanensis* Kirkendall and Faccoli, distributed in Southwest China is a recently described species that was earlier confused with *T. piniperda*, commonly found in Northeast China and Europe due to morphological similarity. Recent studies show there are clear genetic and ecological differences between these two species. Compared with *T. piniperda*, *T. yunnanensis* is much more aggressive and destructive, with an ability to attack and kill healthy pine trees. The following are unique features: aggregative shoot feedings, massive trunk attacks, no overwintering period, overlapping generations and different associated fungi that may have contributed to the success of *T. yunnanensis* as the most aggressive *Tomicus* species. The unusual weather has been considered as the key cause for the serious outbreaks of *T. yunnanensis* that occurred in the pure pine plantations and degraded pine forests in Southwest China, especially the extreme droughts from El Niño phenomena occurred during 1982-1983, 1986-1988, and 1988-1989. The national annual *Tomicus* infested areas were estimated at 160,000 ha in 2010.

The IPM strategy against *Tomicus* spp. includes silvicultural approaches to sustain and enhance the stability of the forest ecosystem, and a combination of biological and chemical control measures to keep the shoot beetle populations under ecological and economical thresholds. Currently, the IPM components consist of monitoring population dynamics of the shoot beetles using attractant-baited traps or logs, timely removal of newly infested pine trees/logs, and application of biological control agent, *Paecilomyces farinosus*. This IPM program has shown a great potential for effective operational control of the aggressive pine shoot pests.

(18) Xiao pine weevil, *Hylobitelus xiaoi* Zhang

The Xiao pine weevil, *Hylobitelus xiaoi*, is currently distributed in Jiangxi, Hunan, Hubei, Guangxi, Guangdong, Guizhou, and Fujian provinces, mainly attacking exotic pines, such as slash pine (*Pinus elliottii*) and loblolly pine (*P. taeda*), and occasionally infesting the native pines, *P. massoniana*, *P. armandii* and *P. hwangshanensis*. *H. xiaoi* is a destructive insect that mostly attacks the base part of the pine trunk (<50 cm above the ground). It bores into the cambium and phloem of the stem, slightly damaging the xylem. The larvae can also bore into the bark of the root. Irregular shaped larval tunnels extend under the bark, resulting in serious resinous exudates and significantly reducing the tree growth and resin production. Serious outbreaks with higher infestation rates often occur in pine stands with high humidity and poor ventilation conditions. In Ji-An, Jiangxi province, *H. xiaoi* has one generation biennially; overwinters in both the adult and larval stages. It is very difficult to spot *H. xiaoi* adults in the field, due to their cryptic lifestyle.

Since it was first described as a new species in 1997, *H. xiaoi* outbreak areas continuously increased, reaching 14,000 ha in 1999, 100,000 ha in 2002, 280,000 ha in 2006, and 160,000 ha in 2010. It became a new major native forest pest insect in China.

Considering the fact that *H. xiaoi* prefers to infest the pine stands with high canopy densities and rich vegetation diversities with more fallen branches and litter, it seems that timely pruning and thinning practices in the middle-age pine stands will increase natural ventilation and light intensity in the stands, therefore, preventing serious *H. xiaoi* attacks. Additionally, injection of systemic insecticides on the infested trees as a direct control measure, and painting the base portion of the tree trunks with whitewashes as a preventative method, seem to be effective.

(19) Yunnan pine weevil, *Pissodes yunnanensis* Langor et Zhang

The Yunnan pine weevil, *Pissodes yunnanensis*, is distributed in Sichuan, Yunnan and Guizhou provinces, mainly attacking *Pinus yunnanensis* and *P. densata*. Young *Pissodes yunnanensis* larvae (1st and 2nd instars) feed in the inner phloem and constructed meandering galleries, and then the 3rd instar larvae leave the phloem and enter the sapwood of tree trunk or the pith of the branch for continued feeding. Serious weevil feeding often results in red wilting of needles, stem forking/crooking, and ultimately tree mortality.

Natural dispersal of *Pissodes yunnanensis* is mainly relying on adult flight activities; however, long distance dispersal can be made via movements and transportations of infested young trees and untreated pine logs. Outbreaks of this weevil are closely related to forest stand conditions and tree vigor.

Since it was first described as a new species in 1999, this weevil has quickly spread, reaching 46,000 ha of infested areas in 2002, and 84,000 ha in 2004. Currently, it is only distributed in the southwest China areas, but has a great potential for further spreading and dispersal into other areas. IPM components against this new pest insect include removal of the weevil-infested pine trees, chemical control during adult maturation feeding period, and application of biological control agents, such as *Paecilomyces farinosus* and *Nomuraea rileyi*.

Another similar weevil species from the same genus, *Pissodes punctatus*, is also distributed in Southwest China, attacking several pine species, especially *P. armandii*.

(20) Sandthorn carpenter worm, *Eogystia hippophaecola* Hua, Chou, Fang et Chen

The sandthorn, *Hippophae rhamnoides*, with a highly developed root crown system, is one of the best planting tree species of afforestation programs for windbreak, sand-fixation, elimination of barren hills, and water-soil conservation; is a great primary tree species for improvement of ecological environments, and a candidate tree species for non-timber-based high value commercial forests in northern China. The sandthorn carpenterworm, *Eogystia hippophaecola*, was first described as a new species in 1990. Since 1999, serious outbreaks occurred in Inner Mongolia, Shanxi, and Liaoning provinces, resulted in vast sandthorn tree mortality and strong negative impacts on the ecological improvement and economical development of the infested regions. Several factors, such as incorrect choice of planting tree species at certain locations, single species/variety monoculture plantations, consecutive years of drought, and poor silvercultural management, may have contributed to their outbreaks. *E. hippophaecola* has one generation every 4-5 years and larval stages occupy most of its life history. The larvae mainly attack the trunk base or roots, with 90.7% located at the trunk base 0-30 cm below the ground, resulting in serious obstruction of water and nutrient transportation of seabuckthorn trees, and ultimately tree mortality. The pest is considered as one of the major threats to the development of seabuckthorn forests and related industries.

Over the past ten years, *E. hippophaecola* infested areas continued to grow, reaching 137,000 ha in 2002, and 160,000 ha in 2006. Current IPM strategy is heavily relying on the silvilcutural approaches, such as regeneration and reformation (or improvement) of seabuckthorn forest stands in order to effectively eliminate the conditions that are suitable for survival and reproduction of *E. hippophaecola*, and thereby significantly reduce its population density down to non-outbreak levels (i.e. below the ecological and economical thresholds).

(21) Oak longhorned beetle, *Massicus raddei* (Blessig)

In the past ten years, the oak longhorned beetle, *Massicus raddei*, has become a serious wood-boring pest insect of natural regeneration forests in Liaoning, Jilin and Inner Mongolia. The beetle mainly attacks *Quercus mongolica* and *Q. liaotungensis* trees, resulting in a significant reduction in timber growth, quality grades or usage values at lower infestation levels, and great mortality of single trees or entire forests during serious outbreaks, causing tremendous economical and ecological losses.

The infestations and damages in natural regeneration forests by *M. raddei* started around late 1980s, and spread quickly in Northeast China, totaling 87,000 ha in area in 1998, 100,000 ha in 2002, and 260,000 ha in 2006.

The oak longhorn beetle has one generation every 3 years, and its life cycle appears to be synchronous with mass adult emergence every 3 years, as recently observed in 1993, 1996, 1999, 2002, 2005 and 2008. During a typical adult-year, pupation starts in mid June, and adults begin to emerge in early July for about one month. The adult emergence period has been considered as the optimal timing for effective pest control. *M. raddei* adults can fly up to 30-80 meters during a single flight, thus, have a great natural dispersal potential.

Currently, IPM components include timely cutting of over-aged forests, mass-trapping of adults with light traps, handpicking of newly emerged beetle adults, as well as the application of biological control agents such as the parasitoid, *Sclerodermus pupariae* and a predacious beetle, *Dastarcus helophoroides*.

(22) *Parocneria orienta* Chao

Parocneria orienta, is also a historical forest pest insect in China. The historical records show its serious outbreaks occurred in Southwest China during the early 1900s, occasional outbreaks appeared periodically during 1950s-1970s, and severe large scale outbreaks happened in the 1980s, and became a serious and abrupt forest defoliator of *Platycladus orientalis* and *Cupressus funebris* in Sichuan and Congqing regions since the 1990s. Therefore, it is now considered as a super threat to the health and survival of many ancient cypress trees in famous historic and cultural sites in the region. It has two generations per year; adults have a great fecundity (300-500 eggs/female); larvae from the second generation overwinter. Young larvae mostly feed on the young/fresh needles and new shoot buds, resulting in strong reductions in growth of needles and branches, whereas the later instar larvae could eat older needles or even fresh branches, causing trees to become withered and yellow, resulting in tree mortality during the outbreaks. The total *P. orienta* infested areas reached 296,000 ha in 2001 and 313,000 ha in 2003. A new *P. orienta* outbreak cycle appeared in 2007, resulted in a total infestation area of >370,000 ha.

Current IPM components consist of silvilcutural approaches: plantation of mixed species forests, fencing/closure of newly planted forest stands, and timely pruning and thinning practices; enhanced biological control tactics with beneficial insects and birds, as well as the microbial agents (NPVs, *Beauveria bassiana* and *Bt*); mass-trapping with light traps, and the aerial application of insecticides as a last resort for combating unprecedentedly huge outbreaks.

(23) Giant globular scale, *Eulecanium giganteum* (Shinji)

The giant globular scale, *Eulecanium gigantean*, is a serious pest insect of many fruit trees, especially the jujubes, with a broadly distributed host range, a highly reproductive and adaptive capability and difficulty to control. It mainly attacks leaves and branches, resulting in both significant defoliations and branch deaths, and is one of the great threats to the development of commercial no-timber forests in China. In recent years, outbreaks of *E. giganteum* occurred in many key jujube commercial production districts in Xinjiang, Ningxia and Shandong provinces, causing dramatic reductions in tree health and yields at lower density levels, and even deaths of jujube branches or entire trees, totaling > tens of millions of RMB Yuans in damage per year.

In addition to the most favorite jujubes (*Ziziphus* spp.), *E. giganteum* can also feed on >40 species of other broad-leaved trees, such as *Amygdalus communis* L., *Robinia pseudoacacia*, *Xanthoceras sorbifolium* Bunge, *Amorpha fruticosa*, *Sorbaria sorbifolia*, *Juglans* spp., *Malus* spp., *Rosa* spp., *Pyrus* spp., *Prunus* spp., *Populus* spp., *Ulmus* spp., and *Acer* spp., with a total infestation area being >10,000 ha. *E. giganteum* is currently distributed mainly in Northwest China, partially in North China. It has one generation per year; females have a high reproductive capability (3,000-6,000 eggs/female). *E. giganteum* are cryptic, have a waxy cover on their bodies, and are strongly adaptive and resistant to various environments. Natural dispersal of this scale mainly relies on the movement of newly hatched young nymphs carried by winds. Long distance dispersals can be made by human activities such as movement of host tree seedlings, scions, rootstocks, infested trees or untreated logs via transportations, or grazing in the forests.

Current IPM strategy focuses on environmentally sound pest control measures, namely, a combination of silvicultural/mechanical tactics and biological control agents. Silvicultural tactics include plantation of insect resistant tree species, reinforcing stand tending with watering and fertilizing, increasing tree growth vigor, and enhancement of tree resistance against *E. giganteum* attacks. Mechanical approaches, including pruning plus manual removals of infested live or dead branches, females and eggs, could quickly reduce the scale populations and their damages. Application of beneficial insects, such as *Blastothrix sericea*, *Chilocorus rubidus* and *C. kuwanae*, has been shown to be effective. If necessary, some environmentally-sound insecticides might be also considered in order to quickly knock down the scale populations.

2.2 Characteristics of Occurrence of Forest Insect Pests in China

According to the results of 9th National Forest Resource Inventory (NFI9: 2014-2018), the total forest area in China is 228 million ha, the relative forest cover is estimated at 22.96%. thus, China now has the largest area of plantations in the world. Most of the plantations are distributed in Guangxi, Guangdong, Hunan, Sichuan, Fujian, Yunnan, Inner Mongolia, Jiangxi, Liaoning, Zhejiang, Shandong and Heilongjiang provinces.

There are five distinct characteristics of occurrence of forest insect pests in China.

2.2.1 Rapid increase in the number of invasive alien forest pest insect species

Currently, the number of invasive alien forest pest species and their infestation areas continue to grow with an increasing threat to the forest health of China. The outbreaks and infestations of earlier arriving invasive alien species, such as *Matsucoccus matsumurae*, *Hyphantria cunea*, *Hemiberlesia pitysophila*, *Oracella acuta* and *Brontispa longissima*, have not been eliminated or under control yet, at the same time more new alien species including *Quadrastichus erythrinae*, *Obolodiplosis robiniae*, *Solenopsis invicta*, *Phenacoccus solenopsis*, and *Opisina arenosella* arrived and have rapidly spread in recent years. Up to date, 25 of the 30 major invasive forest pest species are insects, accounting for 83.3%; six species arrived before 1980, and another 19 species showed up after 1980. The following eight invasive alien insect species were listed on the latest China's national quarantine list for forest pests (revised in 2013): *Dendroctonus valens*, *Hyphantria cunea*, *Heterobostrychus aequalis*, *Cydia pomonella*, *Rhynchophorus ferrugineus*, *Carpomya vesuviana*, *Solenopsis invicta* and *Phenacoccus solenopsis*.

2.2.2   The infestation areas by common outbreak insect pest species remain large

Since the founding of new China in 1949, pest control practices against the historical forest insect pests, such as *Dendrolimus* spp. and *Ceracris kiangsu*, have continued nonstop; however, their outbreaks occurred at relatively high levels in some regions every year, due to their population periodicity and regionality. According to the statistics, there are 27 species of common outbreak forest insect pests, each with an annual total infested area >70,000 ha. For example, the annual total infestation areas caused by pine caterpillars (*Dendrolimus* spp.) and poplar defoliators reached 0.80 and 1.43 million ha, in 2014, respectively.

2.2.3   Abrupt severe outbreaks of insect pests with continuing threat

The reprehensive abrupt forest insect pests are *Massicus raddei*, *Holcocerus hippophaecolus*, *Pissodes pnnetatus*, *Hylobitelus xiaoi*, and many pine sawflies. It is normally very difficult to control them due to the fact in many cases serious damages by the sudden outbreaks of these species have already occurred before they were noticed. For instance, the sudden outbreaks of the destructive oak longhorn beetle, *Massicus raddei* occurred in the oak forests in Jilin and Liaoning provinces in 2006, totaling 200,000 ha in infested area, and resulting in a high mortality of the oak trees, tremendous economic losses, and caused a catastrophic impact on the recovery and rebuilding of secondary forests in the areas. The abrupt and continuous outbreaks of *Eogystia hippophaecolus* in the arid sandy areas in Liaoning, Inner Mongolia, Shanxi and Shaanxi provinces, caused destructive damages to the sea buckthorn (*Hippophae rhamnoides*) plantations, totaling 70,000 ha in the infested area in Jianping county (Liaoning) alone. These outbreaks resulted in both tremendous economical losses to the sea buckthorn-based commercial industry and non-measurable ecological losses to the already fragile arid environment.

2.2.4   Infestations and damages by wood-boring insects continuously worsen

There are many serious wood-boring insect pests in China, including beetles from Cerambycidae (longhorn beetles), Buprestidae (flathead beetles) and Curculionidae (weevils) as well as moths from Cossidae, Hepialidae and Sesiidae. These wood-borers are among the most difficult forest insect pests to control in China, due to the fact that they spend the majority of their lives (eggs, larvae, and adults) beneath bark or inside tree trunks with a minimum negative impact from the surrounding environment (including limited number of natural enemies), and have a high reproductive capability and overlapped generations. For example, the serious infestations and outbreaks of poplar wood-borers in the "Three North" shelterbelt forests continued, with their infested areas doubling within 10 years (from 1996-2006), reaching 800,000 ha infestation area in 2006; recent serious outbreaks of the Yunnan pine shoot beetle, *T. yunnanensis*, occurred in Southwest China, totaling 180,000 ha infested area in Yunnan and Sichuan provinces alone, raising a severe threat to the health of five million ha of pine forests in the Southwestern region.

2.2.5   Rapid increase of the pest infested areas in the commercial non-timber forests

Over the past 20 years, there was a rapid development of various commercial non-timber forests totaling 20 million ha of plantation area. According to incomplete statistics, the insect infestation areas and damages in the commercial non-timber forests increased rapidly, and have continuously worsened in recent years. The major insect pests in commercial non-timber forests are *Eulecanium giganteum*, *Chihuo zao*, *Mecorhis cumulates*, *Atrijuglans hetaohei*, *Biston marginata*, *Cydia pomonella*, *Eurytoma maslovskii*, and *Quadraspidiotus perniciosus*. These insect pests are very difficult to control, due to the fact that commercial non-timber forests involve many tree species, have broadly scattered distributions and there are many restrictions on insecticide safety and application methods. In addition, these pests are able to disperse long distances via transportation of seedlings and fruit-related products.

3. Relationships between Occurrences/Outbreaks of Forest Insect Pests and Environmental Factors, Forest Stands and Natural Enemies

A forest ecosystem is an ecological complex of living organism communities (trees/plants, animals and microorganisms), all interacting among themselves and with the environment (soil, climate, water and light). Forest plants are the essential structural component of the forest ecosystem, whereas the forest animals (mostly forest insects) are important components in the system, and both are affected by many environmental factors. As one of critical natural parts of forest ecosystems, forest insects play an important role during the formation, development and succession processes of forest ecosystems. The classification of an insect as a pest or beneficial is very much subjective, based on its potential damage or benefit to human needs and/or natural habitats and ecosystems. Ever since the beginning of human silvicultural practices, forest benefits related to forest composition, growth rates, production performance, hydrology, grazing, and tourism are the main interests of humans, therefore, forest insects are categorized as forest insect pests or forest beneficial insects if they harm or enhance the human needs or interests.

Growth, development and reproduction of a forest insect are influenced by many environmental factors, forest plants and other forest animals including other forest insect species. At the same time, rapid population growth of certain forest insects to abnormal levels could cause damage directly or indirectly to the growth of forest plants and stand conditions, mostly resulting from the deterioration of forest eco-environments. Forest insect populations survive, develop, decline or disappear, through continuous interactions with all other components in the forest ecosystem.

3.1 Relationship between Occurrences/Outbreaks of Forest Insect Pests and Environmental Factors

Environmental factors include temperature, humidity, precipitation, light, wind and soil. These factors don't act on the insects alone (independently), instead, they affect forest insects interactively. These environmental variables could impact insect growth, development, reproduction, movement and distribution directly or indirectly via their influence on insects' host plants or natural enemies.

3.1.1 Temperature

Ambient temperature controls insect's body temperature, which regulates insect development, metabolic intensity, population density, reproduction, behavior and distribution. The maximum and minimum temperature thresholds for insect survival or development vary among insect species. Some insect species can tolerate a very broad temperature range (known as eurythermal species); for example, *Agrotis ipsilon* is distributed in most parts of world and all the provinces in China, whereas some insect species can only survive in a very narrow temperature range, known as stenothermal insects, e.g. *Hepialus armoricanus*. Some species can tolerate really low temperatures, for instance, *Lymantria monacha* eggs may survive at -40℃, while *Lymantria dispar* adults will die at -4℃ for only 30 min. At the other extreme, *Euproctis chrysorrhoea* larvae start to die at 39℃, but its pupae may still survive at 43℃; *Monochamus alternatus* start dying only when temperature reaches 45℃. These temperature thresholds vary among different development stages, physiological conditions, and seasons even for the same insect species. *Lymantria monacha* larvae can survive at 0℃ to 43℃, but feed mostly at 25-28℃, and become paralyzed when temperature is rapidly increased to 35℃. During the overwintering period, the mortalities of *Dendrolimus punctatus* larvae at constant -7℃ reached 10% after 24 hrs, 20% after 48 hrs, and 40% after 120 hrs. After overwintering, *D. punctatus* larvae feed normally at 20℃; however, dramatic drops in temperature down to 5℃ during this period might result in significant larval mortalities (at 5℃): about 10% after 24 hrs, 40% after 48 hrs and 50% after 72 hrs. Within *D. punctatus*'s distribution range in southern China, winter temperatures are rarely below -7℃, but drastic temperature drops down to 5℃ during the larval feeding period in spring often happen. Thus, compared to the low overwintering larval mortality, a cold spell during the early spring could cause significant larval mortality, due to the fact that *D. punctatus* larvae have a very high metabolic activity and a low cold-resistance during the spring feeding season.

How do forest insect pests adapt to the low temperatures? 1) Insects produce specifically structured eggs or pupae to tolerate the cold weather. Many insects overwinter as eggs or pupae that either have a thick shell or a layer of special secretion cover; for example, the overwintering eggs of *Malacosoma neustria testacea*, or *Lebeda nobilis sinina*. 2) Insects increase bound water in their hemolymph to tolerate the supercooling conditions. The supercooling point and freezing point of *Aporia crataegi* were -9.2℃ and -1.4℃, respectively, when the hemolymph (fluid) index was at 0.55; -8.9℃ and -0.8℃ at a fluid index of 0.61, respectively; and -0.2℃ and 1 to -0.7℃ at a fluid index of 0.62, respectively. 3) Hibernation: many wood-boring insects, such as longhorn beetles or *Cossus* spp., go into hibernation with an increase of cold hardiness and overwinter inside tree trunks.

How do forest insect pests adapt to the high temperatures? From a physiological point of view, 1) insects could reduce body temperatures by water evaporation through their respiratory system; 2) insect fat bodies have a relatively higher boiling point, and therefore could be resistant to high temperatures; 3) summer hibernation should inhibit the increase of insect body temperatures. On the other hand, insects might change their behaviors to adapt to the high temperatures. 1). Moving to the shaded and cool sites: for example, during a typical hot summer day larvae of *Euproctis pseudoconspersa* and *E. bipunctapex* start moving downward from tree crown at 8:00-9:00 and aggregate on the shaded/cool side of tree trunk during daytime, and then move back up to the crown for feeding after 16:00-17:00; also many underground insects such as scarab beetles and mole crickets dig deeper into the ground when soil surface temperature increases. 2) Migration: adults of *Periergos dispar* and *Lymantria dispar* during hot summer often migrate from high temperature areas to the cool areas; the yellow-spined bamboo locusts, *C. kiangsu* often move to shaded and cool areas on the ground during the hot summer daytime, and move back up to bamboo trees for feeding when temperatures start to drop after 16:00 – 17:00.

Over the past one hundred years, climate change and global warming became a reality. China has seen the most significant occurrence of global warming with a national average temperature increase of 0.4-0.5℃ over the past century; such increases are even more significant in northern China, such as Northwest, North and Northeast China. There are several impacts of global warming/climate change on forest insect pests. 1) Increase in effective growing degree days shifting the northern boundaries and upper elevation limits of insect distribution ranges toward higher latitudes (further north) and elevations. For example, *Dendrolimus tabulaeformis* was originally distributed only in Liaoning, Beijing, Hebei, Shaanxi, Shanxi, and Shandong provinces, now the species has extended horizontally and further north/west into Chifeng area in Inner Mongolia, and vertically up to 800 m above sea level in northeast, and between 500 m and 2000 m elevations in the Loess Plateau, Northwest China. In addition, Chinese termites are typical tropical and subtropical insects, but are now spreading toward the north. 2) Global warming/climate change has accommodated insect biology, such as earlier adult emergence or egg hatching and increases in the number of generations per year. It was reported that an average regional temperature increase of 1℃ over the past 30 years in Chaoan county, Guangdong province, which might play a role in the shift of *D. punctatus* overwintering larvae being from the 3rd generation alone previously to now being from both 3rd and 4th (overlapped) generations. *Clostera anachoreta* has only one generation per year within its original range in Northern China; however, at its new range in Hainan province it has 8-9 generations per year without an overwintering (diapause) period. 3) Global warming/climate change might be one of the major causes for current increased numbers of outbreak forest pest insect species, shortened outbreak periodicity (cycles) and larger sustained outbreak/infestation areas.

3.1.2 Humidity and precipitation

Water is a major component of all organisms, and essential for most life processes, including metabolism. Soil moisture and ambient relative humidity (RH) are two key measurements of water contents in the environment; and they have multifold impacts on forest insect pests, such as direct effect on growth, development, metabolism and mortality of insect pests. In general, low humidity may slow down insect metabolism and delay their development; in contrast, high humidity may increase insect metabolism and speed up their development. For example, eggs of the yellow-spined bamboo locusts, *C. kiangsu* developed quickly at 75% RH, reached maximum hatch rates at 85%-95% RH, while low RH (<70%) significantly reduced their hatch rates. The hatch rate of *D. punctatus* eggs reached 90% at 75% RH, and 92%-98% at 90% RH; while it was significantly reduced at <70% RH, and down to 16.4% at 15% RH. The egg period of *Phalerodonta bombycina* lasted only 1-3 days at 50% RH with a 96% hatch rate, whereas it lasted 9-13 days at 30%-40% RH with a 35.8% hatch rate.

Humidity also has a significant impact on insect behavior and activity. Some longhorn beetle species could enter a hibernation stage as larvae inside of tree trunks for several years if the trunk encountered an extremely dry condition. Some underground insect larvae dig deeper into the ground if the soil moisture is low, and tend to stay near ground surface if the soil moisture is high. Subterranean termites build their nests at different vertical levels underground at different topographic/terrain conditions, such as plateau, hills or valleys to adapt to the different precipitation levels. Many insects are hidden on the underside of tree leaves when it rains; and heavy rains may stop or inhibit the dispersal flights of many bark beetles. According to the analysis on a set of 300 years' historical data, outbreaks of *Lymantria monacha* mostly occurred during drought and hot years. Outbreaks of *Lymantria dispar* often appeared after low precipitation (both rains and snows) years; their populations started to increase dramatically during and post drought years. Droughts could also extend *L. dispar* outbreak durations. Egg mortality of *Orgyia antiqua* or *L. dispar* was only 25% at 6% RH. *L. dispar* eggs could tolerate high RH or drought before and during their diapause; but are very vulnerable to the extreme dry or moist conditions after their diapause.

High humidity might be beneficial to some insect species; however, it might also provide a perfect condition for an epidemic of some insect pathogens, such as bacteria, fungi and viruses. For example, in *Dendrolimus punctatus* infested areas, rainy weather during April and May (at 24-28℃ and 90%-100% RH) are optimal for reproduction of insect pathogens (especially the *Beauveria bassiana*), causing *B. bassiana* epidemics, and a quick increase in *D. punctatus* larval mortality. *B. bassiana* spores can't germinate at <70% RH, regardless of the temperature conditions.

There are also interactive effects of relative humidity and temperature on insect egg hatch rate, larval mortality, pupal emergence rate, and adult fecundity (egg production) among others. The optimal condition for *Dendrolimus tabulaeformis* egg hatch is at 20-24 ℃, and 70%-80% RH; its hatch rate will drop significantly at 29%-35% RH, regardless of temperature. A similar pattern is also true for the yellow-spined bamboo locusts which have an optimal condition at 28-30 ℃, and 90%-95% RH for egg hatch, and a low RH (<70%) can result in strong reduction of their egg hatch rate regardless of temperature range. Population density of the pine needle hemiberlesian scale, *Hemiberlesia pitysophila* is negatively correlated to RH when the average monthly precipitation is >100 mm, but has no correlation with RH when the average monthly precipitation is <100 mm.

3.1.3 Light

Light is electromagnetic radiation energy from the sun. Sunlight provides the energy that green plants use to create sugars mostly in the form of starches, which release energy into the living things that eat and digest them. Light property, intensity, and photoperiod all pose a significant impact on insect development. Insect development needs certain amounts of illumination, and light quality and illumination levels affect the speed of insect development. Infestations and outbreaks of *Pineus cembrae* preferably occur in the forest stands with good illumination conditions, such as clear-cuts and forest edges; and are positively correlated with light intensity. After clear-cutting, the original dark and moist pine forest is replaced by a new habitat (clear-cut) with drastic increases in light intensity and temperature, and a decrease in RH, a perfect outbreak site for the highly heliophilous *P. cembrae*. Thus, *P. cembrae* is considered as a dominant pest insect species in the regenerated young pine stands in clear-cuts.

Diurnal insect species, such as locusts and butterflies, prefer to feed during the daytime with plenty of light, whereas nocturnal insects, such as *Agrotis ipsilon*, *Lebeda nobilis* and *Trabala vishnou* feed only during the night, and some of them are also strongly attracted to artificial light sources, especially UV light. In fact, black light traps are currently a common monitoring and mass-trapping tool in the pest control practices in China. Larvae of *Dendrolimus* spp. can feed during both daytime and nighttime, but adult mating and oviposition only occur during the night. Termites are cryptic insects, their workers and soldiers are completely blind as they do not have compound eyes, with a negative phototaxis (repelled by light). In contrast, the reproductive alate (winged) termites have fully developed compound eyes and can fly with a phototaxic behavior toward light.

Radiation heat created by the sunlight sometimes can kill insects directly. For example, the heat energy (transferred from solar radiation) absorbed by tree bark sometimes may kill larvae of bark beetles or longhorn beetles (living under the tree bark), if they are under extremely direct strong solar radiations (might reach 60℃ on the bark surface).

3.1.4. Wind

Wind, especially monsoon, is an important factor affecting precipitation, humidity and temperature. Wind can alter temperature and humidity through its impact on water evaporation and heat dissipation, therefore, influences the water content and energy metabolism of insect pests.

Natural dispersals (migrations) of many insects are totally depending on wind and weather conditions; their flight directions are the same as the dominant wind directions; and their horizontal dispersal ranges are positively correlated with the wind speeds. For instance, natural dispersals of newly hatched *Dendrolimus* spp. larvae are mainly driven by wind, with their dispersal directions and distances being controlled by the wind direction and speeds. Many small sizes of insects, such as leafhoppers, plant hoppers, aphids, whiteflies and small moth species, can be carried by winds up to 170-1,700 m high in the air and as far as several hundred or thousand km away. Wind could blow some insect pests, occurring in the infested foci, to other locations far away from its original sites; therefore, wind is the key force driving the dispersals of some small or less active insects. The newly hatched *Clania vartegata* larvae hang from their host tree on silken threads, and are often carried far away by the wind.

Consistent wind is a type of dispersal vector of insect pests toward certain areas. *Hemiberlesia pitysophila* spread inland from coastal areas since its first entry into China was by the consistent winds; for example, the newly hatched nymphs can be easily carried over 6 km by wind or air current. Natural dispersals of newly hatched nymphs of the loblolly pine mealybug, *Oracella acuta*, are also driven by the air currents, mostly by the southeast monsoon, reaching a distance of 17-22 km. On the other hand, hurricanes or typhoons might have a strong negative impact on pest insect population dynamics or densities.

Wind might also be able to impact insect morphology or behavior. Many insects in islands or high mountains with frequent strong winds are wingless; while many insects living in low elevations or areas without strong winds are mostly winged. Some insect species specially prefer to mate in the air during clear windless days, and will not mate during windy days. Insects can use wind or air current to help them find their foods or avoid dangers from their enemies by smelling the olfactory volatile signals carried by the wind.

3.1.5 Soil

Soil temperatures change hourly, daily and seasonally. Compared to the air temperature, soil temperature changes are relatively small and slow due to the slow soil thermal absorption/desorption process, the deeper into the soil the less the soil temperature changes. During the winter season, soil temperatures are higher than the air temperatures; therefore, many insects overwinter in the soil to avoid the extremely low temperatures in the air. Many underground insects, such as scarab beetles, cutworms and click beetles, adjust their vertical feeding positions based on the temperature changes at different soil layers. They start digging deeper into the soil in the fall and beginning of winter when soil temperature starts to fall, the lower the temperature the deeper they dig into the soil; they move back up to top soil layers next spring when temperature starts to increase.

Soil moisture strongly influences distribution and development of underground insects. For example, the optimal soil moisture contents for *Agriotes subvittatus* and *Agrotis ipsilon* are 50%-60%. Low soil moisture could inhibit or stop pupation of *Curculio chinensis* Chevrolat in the soil. Many parasitoid flies leave their host insects as mature larvae, and enter the ground for pupation; and their pupal survival rates are closely related to the soil moisture. Optimal soil moisture contents for most of them are in the 10%-30% range; thus, too low or too high soil moistures would inhibit their pupation. Flooding water in soil may kill many insects that overwinter or feed underground; thus, a useful pest control approach.

Soil textures (particles) and structures can affect the behavior and movement of underground insects. Many mature larvae prefer to enter more porous and better drained soil to pupate, but if the soil is too loose or excessively drained, it might also inhibit the pupation and be detrimental to the overwintering insects. Jujube trees planted on sandy soil are rarely attacked by *Carposina sasakii*, because the sandy soil is not suitable for *C. sasakii* survival due to the large soil temperature differences between spring and summer, extremely low winter soil temperatures and poor soil thermal insulation.

Insect responses to different soil chemical properties (composition, humus content, pH and base saturation) vary. *Agriotes obscures* prefer to live in the soil with a pH of <6, mostly at a pH of 4-5.2 range, and hardly in soil with a pH of >6. Thus, adjustment of the soil's pH values might prevent attacks from some underground insects, such as adding some calcium oxide (lime) into the soil could reduce the infestations and outbreaks of some click beetles.

3.2 Relationship between Occurrences/Outbreaks of Forest Stand Conditions (Variables)

Outbreaks of forest insect pests don't necessarily occur in any kind of forest stand; outbreaks normally occur only when certain favorable conditions are met due to a significant change of one or more factors in the forest ecosystem. An outbreak starts with a small insect population foci, followed by population increase, expansions and dispersals, and finally reaches the explosive outbreak population level. Thus, forest pest insect outbreaks are influenced by many forest stand factors such as forest stand type, composition and other variables.

3.2.1 Forest stand type

(1) Shelterbelt forests

Shelterbelt is a forest type with a strong and dominant ecological function that is beneficial. The shelterbelt linear forest system (type) mostly consists of a single tree species that has a simple stand structure with plenty of light, a large diurnal temperature range, and a few but dominant natural enemy species. Outbreaks of certain insect pests often occur in shelterbelts due to the simple stand structure and the rich host foods. In addition to frequent infestations of defoliators, outbreaks of wood-boring insects in the shelterbelt forests are often destructive and detrimental to ecological and landscape functions, especially in the older shelterbelts. A typical example is the serious and destructive outbreaks of ALB and other poplar wood-borers in the "Three North Shelterbelt" forest system.

The shelterbelt conditions, affected by silvicultrual practices and management regimes, are intimately affected by the occurrence of insect pests. The infestation rate of *Zeuzera multistrigata* Moore (Cossidae) was extremely low (only at 0.5%-1%) in the healthy costal shelterbelt forests, and very low (at ca. 3.5%) in the well managed young shelterbelt stands, but reached 15% in the inadequately managed young stands. It might even reach >70% in some poorly growing shelterbelt forest stands planted on poor site conditions.

Orientation, landform, and terrain of shelterbelt stands can affect the occurrences and outbreaks of insect pests. Outbreaks or infestations of *Poecilonota uariolosa* in the poplar-based shelterbelt forests occurred more seriously on western sides of the south-north oriented shelterbelts and on southern sides of the east-west oriented shelterbelts than on the northern side, and were the least on the eastern side. The infested rates by *P. uariolosa* in the 1st to 2nd rows of polar trees in the southern (east-west oriented) or western (south-north oriented) side of a shelterbelt might reach as high as 91.6%; decreased gradually when getting deeper into the inner tree rows, and dropped to near 0% in the inner tree rows 20 m away from the most favorable western or southern edge. *P. uariolosa* outbreaks or infestation levels in shelterbelts are, to a certain degree, related to landforms or terrains; the beetle infestation rates could be as high as 76.3% in the poplar shelterbelts planted on high hills or arid areas for windbreaks and/or sand fixation purposes due to poor tree growth, whereas low infested rates (e.g. 22.2%) were recorded in the shelterbelt stands planted on flat areas.

Shelterbelt structures also impact the occurrences and outbreaks of insect pests. The infested rates of *Zeuzera multistrigata* in the mixed *Casuarina equisetifolia* and *Eucalyptus citriodora* (at 1:1 ratio) shelterbelt stands were as low as 1%, while a 15% infested rate was reported in the *Casuarina equisetifolia* monoculture shelterbelt stands. It is very important to choose appropriate tree species for the mixed stands; otherwise a wrong species combination might result in worsening the insect infestations. Mixing the correct non-host and deterrent tree species with host tree species by trees, by rows or even by sections might significantly reduce the occurrences/outbreaks and dispersals of certain pest insect species.

(2) Commercial timber forests

Most commercial timber forests are fast-growing plantations, and are major sources for meeting China's urgent domestic timber needs. On the premise of no land degradation, current commercial timber forests are managed toward a base-oriented, directional, intensive management and large scale industry. Over the past several decades, the commercial timber forests in China are mostly made of following tree species/groups: *Eucalyptus* spp., *Acacia confusa*, *Pinus* spp., and *Poplus* spp.

As quick growth and development of the commercial timber forest industry occurred, forest pest issues, especially related to insects, became more and more obvious. Due to the short growth cycle, insect pests that attack seedlings and young trees have stronger impacts on tree survival rates in the commercial timber forests than in other types of forests. For example, serious occurrences of *Paranthrene tabaniformis* and *Cryptorhynchus lapathi* in the newly planted young poplar stands might result in extensive tree mortality and plantation failure. Serious infestations of termites in a 30 ha young eucalyptus stand in Weidu Forest Farm, Guangxi province, caused >85% tree mortality even after multiple replanting efforts. In commercial timber forests, the timber growth is the most critical and ultimate goal; therefore, outbreaks of forest defoliators, such as *Dendrolimus* spp., and pine sawflies on pines; *Clostera anachoreta*, *Micromelalopha troglodyta*, *Cerura menciana*, *Leuoma candida*, and the poplar leaf beetles on poplars; *Strepsicrates coriariae*, *Buasra suppressaria*, and *Eumeta minuscula* on *Eucalyptus* trees, might cause severe damages to timber growing stocks in the commercial forests. For example, a serious outbreak of *Strepsicrates coriariae* in a 65 ha plantation of eucalyptus with a 98% infestation rate in Hepu county, Guangxi province, resulted in severe damage to the eucalyptus timber growths. Most of commercial timber forests are monoculture plantations with a high tree density, low biodiversity and fewer natural enemies. Therefore, these forests are more susceptible to insect outbreaks and more vulnerable to insect damages.

(3) Commercial non-timber forests (Cash tree crops)

The commercial non-timber forests, in a broader sense, include all the specialized non-timber forests and firewood forests, with significant ecological, economic and social benefits (functions), and are an important part of forest resources in China. In a narrower sense, the commercial non-timber forests or cash tree crops produce the non-timber forest products (NTFPs) with high cash values, such as oils, dry/fresh fruits, industry raw materials, medicines, and other products from tree fruits, seeds, barks, leaves, saps, flowers and fresh buds. These forests play a very important role in the economic improvement of the current forestry industry. By the end of 2013, the total area of commercial non-timber forests in China was estimated at 37.81 million ha.

As the non-timber commercial forestry industry quickly developed in recent years, both the total outbreak areas of forest insect pests and the number of outbreak insect species in these plantations increased significantly, causing a serious threat to the healthy development of China's forestry industry. According to incomplete statistics, the total commercial non-timber forest area infested by forest insect pests reached >1 million ha in 2014. The major insect pest species in these forests are: scale insects, *Dasineura datifolia*, *Carpomya vesuviana* Costa, *Cydia pomonella*, *Atrijuglans hetaohei* Yang, *Agrilus zanthoxylumi* Hou and Feng, *Curculio davidi* Fairmaire, *Pogonopygia pavida*, and *Oides leucomelaena* Weise. In Xinjiang, outbreaks of insect pests in the commercial non-timber forests have become a key destructive factor inhibiting development of the non-timber cash tree (fruits related products) industry in recent years.

The following characteristics regarding the occurrences/outbreaks of forest insect pests in commercial non-timber forests have been recently observed: 1) There is a continuous increase in the number of pest insect species. For instance, 1074 pest insect species were recorded from the commercial non-timber forests (consisting of 28 cash tree species) in Anhui province alone. 2) More new pest insect species appeared consistently. Eighteen species of new insect pests were observed in the commercial non-timber forests of *Castanea mollissima* and *Ginkgo biloba* plus nine other cash tree species in Anhui province. 3) There is an uninterrupted increase in the number of outbreak pest insect species. In Liu-an City, Anhui province, ten species of earlier secondary insect pests including *Cyllorhynchites ursulus* (Roelofs), became serious primary insect pests with continuously growing infestation areas. 4) Many abrupt outbreaks of insect pests frequently occurred. Commercial non-timber forests are mostly monocultures with a single tree species in huge areas, and are susceptible and vulnerable to large scale abrupt outbreaks growing in a short period of time, due to the rich favorable host foods and habitat spaces.

(4) Urban forests

Urban forests play a significant role in sustaining and improving ecological environments in the city, town or suburb, with their environmental service benefits (such as beautification, landscape, tourism and ecology) being considered as the most critical function. Urban forests are highly susceptible and vulnerable to pest insect attacks because these forests have low biodiversity, simple structures and relatively poor habitats, and are usually strongly stressed by water and air pollution, heat island effects, and chemical - dust storms in the urban settings. Firstly, serious infestations of many new forest insect pests have been frequently seen in the urban forest ecosystems in recent years; such as outbreaks of *Hyphantria cunea* in Beijing city, the emerald ash borer (EAB) in Changchun and Tianjin cities, and *Orthaga achatina* on *Cinnamomum camphora* in Shanghai city. Secondly, some of the secondary insect pests have become serious major pests in the urban forests; for example, occurrences of *Chiasmia cinerearia*, *Apriona swainsoni*, *Ambrostoma quadriimpressum*, and *Lymantria dispar* had a serious negative impact on the city's environmental quality and residents' physical and psychological health. Thirdly, outbreaks of some insect pests were triggered by urban environmental pollution. In fact, high levels of sulfur dioxide and hydrogen fluoride in the air are the main cause triggering the outbreaks of aphids, scale insects and whiteflies in the urban forests. Lastly, due to these factors it is not surprising that urban forest ecosystems have a relatively low self-regulatory capacity against pests; therefore, effective control operations against these insect pests are quite difficult. These difficulties are increased because of the close proximity of city residents.

3.2.2 Forest stand composition

Forest stand composition and structure are closely related to the levels of pest insect occurrences and infestations.

(1) Pure and mixed stands

In general, fewer infestations and outbreaks of insect pests occur in the mixed stands than in the pure stands. In the pure stands (single species of monoculture), there are only a few pest insect species with a simple entomofauna and a low level of predation/parasitism by natural enemies; therefore, certain insect populations can increase drastically within a short period of time, and easily cause the insect outbreaks. In the mixed stands, there are many insect species with a complex entomofauna and a high level of predation/parasitism by natural enemies; therefore, pest insect populations often don't increase quickly enough to result in serious outbreaks.

For example, infestation rate of the Chinese fir shoot moth, *Lobesia cunninhamiacola* in the mid-aged fir and *Magnolia macclurei* mixed stands, was around 51.3% (with only 1.4% small branches infested), while it was recorded at 75% (with ca. 4.7% small branches infested) in the pure fir stands. On the other hand, outbreaks of some highly polyphagous insect pests (with a broad host range) might occur more frequently in the mixed stands compared to the pure stands, or sometimes might be more serious in mixed stands than in the pure stands (e.g. *Praps alpha* Moriuiti). The mixed stand types and mixed tree species/ratios in the mixed stands all might have strong impacts on the number of pest insect species, their population densities, and damage levels.

In the mixed stands, existence of more pest insect species offers rich hosts (including intermediate hosts) or prey for their natural enemies (with high species richness and abundance). The overlapped and interlaced development stages of many pest insect species in the mixed stands provide consistently suitable habitats (with plenty of food prey or hosts and reproduction sites) for their natural enemies. For instance, over 110 species of parasitoids (mostly with a relatively high abundance) of *Dendrolimus punctatus* were recorded in the mixed pine-broadleaved stands, whereas in the pure pine stands only about 10 species (mostly with a relatively low abundance) were commonly seen.

(2) Canopy structure

Forest canopy structures are significantly related to community structures of forest insects in the forest ecosystem. A study in three stands with different of canopy structures in Liuyang, Hunan province showed that the populations of forest insects, beneficial insects (especially natural enemies), spiders and birds in the mixed forest stand with multiple layers of canopy structure (i.e. pine-broadleaved tree-understory bush) were twice as high as those in the pure *Pinus massoniana* pine stand (with single canopy layer) or in the pine-broadleaved mixed stand with two layers of canopy structure.

Canopy structure affects not only the number of natural enemy species, but also their population densities. For instance, the population densities of natural enemies, based on their parasitism rates on *Antheraea pernyi* eggs, increased as the number of canopy layers increased (*i.e.* single-layer pure pine (*P. koraiensis*) forests < double-layer pine-broadleaved tree mixed forests < multiple-layer pine-broadleaved tree-understory bush forests).

(3) Nectar plants

Nectar plants in the forests are major food sources for the parasitoid (insect) adults, playing an important role in extending parasitoid life span, enhancing sexual maturation, and increasing the number of eggs laid. There are many species of nectar plants in complex mixed forests; for example, over 100 species of nectar plants were recorded in the temperate forests with flowers consequently blooming throughout the four seasons, such as *Symplocos sumuntia*, *Rhododendro* spp., *Quercus fabri*, *Castanea seguinii*, *Vaccinium bracteatum* Thunb., *Camellia oleifera*, *Camellia cordifolia*, *Lespedeza bicolor* Turcz., *Rosa multiflora* Thunb., and *Lagerstroemia indica*. It was reported that over 120 species of parasitoid insect adults need nectar as their supplementary food; and natural parasitism rates by the egg parasitoids in the pine forests with rich nectar sources were 13.4% higher than those in the pine forests without nectar sources.

3.2.3 Stand condition/characteristics

Forest stand condition is a parameter that reflects the forest health status. Insect outbreaks occur much less in healthy forests than in sub-healthy or unhealthy forests. Many factors, such as age of stand, canopy cover, tree vigor and site sanitary condition affect forest stand health, and they also influence the occurrences of insect pests and their damages.

(1) Stand age and tree vigor

The occurrence and damage levels of forest insect pests are different in stands with different ages and tree vigor. However, some insect species such as *Lymantria dispar*, *Clostera anachoreta* and *Hemiberlesia pitysophila*, can still seriously attack forest stands regardless of the stand ages (young, mid-aged or matured stands) during their outbreaks. Some insect species, including *Polychrosis cunninhamiacola*, *Zeuzera multistrigata*, *Moricella rufonota*, and *Rhyacionia duplana*, only infest young stands of <10 year-old trees, and hardly or don't feed on trees in the >10 year-old stands. *Cossus orientalis*, *Holcocerus vicarious*, and *Agrilus* sp., prefer to infest trees in the >10 year-old forest stands, hardly feeding on trees in the young stands of <10 year-old or on really old trees. *Monochamus alternatus*, *Ips acuminatus*, *Cryphalus massonianus* and *Tomicus piniperda* mainly attack the mature pine forest stands, whereas *Ips subelongatus*, *Acanthocinus griseus* and *Monochamus urussovi* mostly infest weakened, dying/dead trees, tree stumps, or old conifer trees (>100 years old).

Most outbreak insect defoliator species feed on host trees regardless of their ages during outbreaks, but they do prefer to feed on certain ages of host trees when plenty of foods are available. During the outbreaks, *Dendrolimus punctatus* larvae feed on needles of both young and old pine trees. In the mixed aged pine forests, *D. punctatus* egg mass density was the highest on the pine trees of 8-20 years old, and 1-5 m tall. Infested young trees recovered quicker and better than did the infested old trees. *D. punctatus* larval overwintering sites vary by tree ages. *D. punctatus* larvae overwinter in the needle clusters (in canopy) for pine trees of <10 years old; partially in the canopy and partially in bark cracks for trees of 10-20 years age; and mostly in bark cracks and crevices for pine trees >20 years age. Outbreaks of *Haplochrois theae* occurred mostly in *Camellia oleifera* stands of 5-15 years age with an average shoot damage rate of 59.9%, followed by the stands of 15-25 years age with a shoot damage rate of 37.5%, the stands of 25-35 years age with a shoot damage rate of 9.7%, and the stands >35 years age with a shoot damage rate of 6.4%.

Outbreaks/occurrences of wood boring insects are also closely related to stand age and tree vigor. *Zeuzera multistrigata* prefers to attack 3-6 years old young stands with an infested rate of 82%, whereas its infested rate was only 1% for seedlings and <0.5% for stands >10 years age. *Holcocerus vicarious* prefers to infest old elm trees, the older and the heavier. *Cossus cossus orientalis* mostly infests 10-15 years age trees, with an infested rate of 11% on 7-8 year old *Salix matsudana* trees, and of 96.2% on 15-18 year old *S. matsudana* trees; 72% infestation on 13-15 year old *Populus pseudosimonii* trees, and only 20.9% on 15-20 year old *P. pseudosimonii* trees. It seems that after a certain age, infestation levels on *S. matsudana* and *P. pseudosimonii* trees by the *Cossus* moth decrease with the tree age. *Poecilonota uariolosa* prefers to infest weakened trees, especially weakened trees of 15-25 years age, whereas young trees or trees >30 years old are rarely attacked. The larch bark beetle, *Ips subelongatus*, mostly attacks 5-30 year old larch trees, the older and heavier within this age range; the beetle slightly infests larch trees >30 years age, and never attacks young trees (<5 years old) or branches of big trees. Infestation rates of the longhorn beetle, *Batocera davidis*, increase with its host tree age, with 1.24% infestation for 1-5 year old trees, 7.31% for 6-10 year old trees, 17.63% for 11-15 year old trees, 39.25% for 16-20 year old trees, 60.51% for 21-25 year age, and reaches 90% for 30 year old trees. Tree vigor and health status significantly affect the outbreaks and occurrences of *Pissodes punctatus*; therefore, increasing tree vigor and improving stand resistance to insect attacks have been considered as key effective measures to control *P. punctatus*.

(2) Canopy closure

Forest stands with different canopy closures have different microclimates that influence insects' habitat and their development. *Lymantria monacha* often occurs in the dense stands with a canopy closure of >0.5, and is only present in very low numbers in stands with low canopy closures or forest edges. *Cossus cossus orientalis* prefers to infest stands with low canopy closures and high understory illuminations. For example, its infestation rate reached 19.7% in a poplar stand with a canopy closure of 0.3, while no infestations were found in a stand with a canopy closure of 0.9 even though tree height, age, DBH and site conditions were all the same between these two poplar stands. In a 16-17 year old *Salix matsudana* stand, this *Cossus* moth infested 18.8% trees on the forest edges, and only 2.5% inside the forest stand. *Melanophila picta* mostly attacks single standing trees, stands with extremely low canopy closures or at forest edges, and hardly infests mature forests with high canopy closures. The infestation rates of *Poecilonota variolosa* reached 87.7% in the stands with a canopy closure of <0.4 or along forest edges, and were only 25.5% in the stands with a canopy closure of >0.8. *Ips subelongatus* prefers to attack the larch stands with lower canopy closures. *Aiolomorphus rhopaloides* likes to live in a warm well-lighted environment; therefore, no or fewer attacks by *A. rhopaloides* were recorded in the high tree density bamboo stands due to the inhibitory effect of low temperatures and illuminations in the bamboo stands on the development and oviposition activities of *A. rhopaloides*. In contrast, more and heavier infestations by *A. rhopaloides* were found in bamboo forest edges, single standing bamboo trees or bamboo stands on the south-facing slopes.

(3) Sanitary condition

Sanitary condition in the forest stands is closely related to distributions of some forest pest insect species and tree damage levels. Some insects never reach an outbreak level in the forest stands with a good sanitary condition due to the lack of necessary and suitable habitats. For instance, serious outbreaks of many bark beetles mostly occur in the stands with low site class indexes and poor sanitary conditions, and the reverse is true for the stands with high site class indexes and good sanitary conditions. Thus, forest management practices, especially improving stand sanitary conditions, play a very important role in the effective control of bark beetles.

Infestations by wood-boring or shoot-boring insects result in many insect-bearing broken trees, branches and tree stumps; if not removed timely or treated properly, they will become the primary insect population foci for potential future outbreaks. Fruit-boring insects and leaf-miners fall to the forest ground together with infested fruits or dead leaves, and these insect-bearing fruits or leaves on the forest floor are often the insect sources for future outbreaks. Some defoliators pupate and overwinter in the forest litters; thus, the litter layer is not only the pupation and overwintering site but also becomes a shield layer that protects the overwintering insect populations for potential future outbreaks.

Wood-boring insects often attack single standing trees or weakened trees, mostly through tree wounds, knots or cracks. *Poecilonota variolosa* attacks or enters poplar trees mostly via tree knots (36.2%-55.4%) and man-made wounds (26.3%-44.3%). Thus, timely removal of weakened or wounded trees would significantly reduce the attacks by bark beetles and longhorn beetles.

Forest management practices, especially removals of insect-bearing branches and fruits via pruning, could prevent dispersals of some insect pests. For example, *Polychrosis cunninhamiacola* and *Rhyacionia duplana* overwinter in the shoots; thus, removal of insect-bearing shoots before their dispersal flights is a very effective measure to eliminate their population foci. Occurrences of *Poecilonota variolosa* are also affected by the pruning practices. It was reported that infestation rate of *P. variolosa* was only 11% in the stands with a good pruning practice, but reached >70% in the over-pruned stands. Strengthening forest management, such as improving tree growth and health conditions, and healing the wounds, would strongly inhibit oviposition activity and young larval development of many wood-boring insects. According to a survey carried out in Ju County, Shandong province, the infestation rate of the Asian longhorned beetle was only 2% in the commercial fast-growth timber forest stands that were properly fertilized, watered and pruned, and 17% in the non-managed stands; it was about 6.7% in the well-managed healthy non-commercial poplar stands and reached 23.3% in the non-managed poplar stands.

3.2.4 Site condition

Climates, soils and vegetation vary under different hilly/mountain conditions, resulting in various combinations of plant/insect communities, or different phenologies of the same insect species. Thus, mountain/hilly site conditions affect not only the population distribution patterns and development of trees, but also the insects (distributions and survival/development) that feed on these trees.

Temperature and relative humidity in the forests vary significantly under different slope aspect and position conditions, with higher temperatures in the stands on the south-facing slope than on the north-facing slope, and higher Humidities in the stands on the west-facing slope than on the east-facing slope. Insect occurrences also vary as the site slope aspects and positions change. Population densities of *Ips subelongatus* are higher in the larch stands on the south-facing slope than on the north-facing slope, and its infestation level increases as the slope degrees (gradients). The occurrences of *Bambusiphila vulgaris* were more in the stands on south-facing slopes than in the stands on north-facing slopes, but the reverse is true for *Pegomya phyllostachys*. At the same elevations, *Celypha pseudolaricla* mostly prefers to infest stands on the east-facing slope, followed by stands on west-facing and south-facing slopes, whereas the north-facing slope is the least favorite one.

Slope positions also affect insect occurrences and infestations. Infestation levels by *Ips subelongatus* increase gradually from the bottom to the top of mountains/hills. The larch bark beetles normally start their attacks on the trees at the mountain top due to poor soil conditions and existence of more weakened (less resistant) trees, with an infestation rate being 2-3 times as high as the one on the trees at the bottom of the mountain/hill. *Allobremeria plurilineata* mostly occurs at bases of mountains or reentrants, and its infestation levels decrease from the bottom to the mountain top, with a 5-to 10-fold decrease in the population density from the middle section to the top. Occurrences and infestations of *Semanotus sinoauster* are related to the Chinese fir bark textures and tree diameters. On the same slope, more attacks/infestations occur in the bottom section than in the top section, due to the fact that fir trees at mountain bottoms grow faster with a stronger vigor producing larger tree diameters and rougher bark textures. At low elevations, infestations of *Parametriotes theae* Kuznetzov were the most severe at hill tops with an infestation rate of 75.4%, and more serious on the south-facing slope than on north-facing slope. There were no significant differences in the infestation level (rate) between the middle (52.8%) and the bottom (54.2%) sections of the south-facing slope. *Platylomia pieli* mostly occurs in the valleys/saddles of high or deep mountains (accounting for 52.8%), followed by mountain/hill slopes (40%) and hill tops (only 7.2%). Occurrences and infestations of *Aphrodisium sauteri* are more serious on the mountain/hill tops than at mountain/hill bottoms; and more on the south-facing slopes than on the north-facing slopes (at middle section of the slopes); and more on the forest edges than inside the forest stands on the south-facing slopes.

3.3 Relationship between Occurrences/Outbreaks of Forest Insect Pests and Natural Enemies

Natural enemies are a very important part of forest ecosystems, and a natural factor (variable) that regulates the population dynamics of insect pests in the system. Application of natural enemies is one of the most critical IPM measures to combat forest insect pests. In a national survey on forest insects carried out from 1979 to 1983, 1402 species of natural enemies (insects) of forest insect pests were recorded. Since then many more new species or new records of natural enemies of forest insect pests have been documented and published. In addition, a series of studies on the biology, ecology and pest insect-natural enemy interactions of nearly 200 species of key natural enemies have been conducted over the past 30 years.

3.3.1 Parasitoids

There are two major groups of parasitoid insects, parasitoid wasps and parasitoid flies.

(1) Parasitoid wasps

Parasitoid wasps are members of the parasitica group of Hymenoptera, and are an important part of parasitic insects of forest insect pests. There are many (tens of thousands) species of parasitoid wasps with a great diversity in bionomics and ecology. With respect to host stage, there are two types of parasitoids: single host stage parasitoids such as egg parasitoids or larval parasitoids, and cross (multiple: 2-3) host stage parasitoids that complete their development through 2-3 host stages. There are endoparasitoids (living inside their hosts) and ectoparasitoids (living externally on their hosts); solitary parasitoids (one egg or immature per individual host) and gregarious parasitoids (multiple eggs or immatures per individual host). Some parasitoid wasps parasitize another parasitoid species as hyperparasitisms. With respect to other parasitoid species, there are eremoparasitism (single parasitoid wasp species living in the host) and synparasitism (two or more parasitoid species together parasitizing the same host).

A toal of154 species of parasitoid wasps of *Dendrolimus* spp. were reported, including egg parasitoids such as *Trichogramma dendrolimi*, *Telenomus dendrolimusi* and *Pseudanstatus albitarsis*, larval parasitoids: (e.g. *Rogas dendrolimi)* and pupal parasitoids (e.g. *Brachymeria obscurata* and *Hyposoter takagii*). There are 52 species of parasitoid wasps of the gypsy moth recorded, with *Apanteles porthetriae*, *Glyptapanteles liparidis*, *Cotesia melanoscelus* and *Hyposoter takagii* as the key larval parasitoids, and *Coccygomimus disparis* and *Brachymeria intermedia* as the major pupal parasitoids in China. Over ten species of parasitoid wasps were also recorded from *Hyphantria cunea* (16 species), *Malacosoma neustria testacea* (18 species) and *Hemiberlesia pitysophila* (12 species), respectively. Major parasitoids of longhorn beetles, such as poplar longhorn beetles, *Monochamus alternatus* and *Anoplophora chinensis*, include *Sclerodermus guani*, *S. sichuanensis* and *S. hainanica* plus many ichneumonids. Six species of ichneumonids were reported as parasitoid wasps of *Pissodes validirostris* Gyllenhyl, and over 140 species of parasitoid wasps were recorded from 58 species of common bark beetles.

(2) Parasitoid flies

Parasitoid flies are parasitic members of Diptera and are another important part of parasitic insects of forest insect pests. Adults of these parasitoid flies could survive for 1-2 months, depending on temperatures, foods, mating/reproduction status. Adult females of many fly species die soon after mating. Adult flies normally die quicker during the hot and dry season. The reproductive capability of parasitoid flies varies strongly depending on different modes of oviposition and of host invasion. Females of some tachinid species could lay up to a few hundred eggs (or hatching larvae) per female onto the hosts. The larvae of most parasitoid flies develop inside a living host, ultimately killing it within a relative short period of time; however, some species might delay their larval development or extend larval lifespan due to the diapauses of their hosts.

Most parasitoid flies have multiple generations per year, and adult flies are active throughout the year except the winter season. Adults prefer mostly humid environments, and dry summer conditions might result in rapid drops in their population density. Most parasitoid flies are photophilic, and tend to sit on the upper parts of plants or sunny sides of tree trunks in the morning (before 10:00 am). Activity of adults normally increases as the ambient temperature increases; however, they prefer to stay in the shaded areas when temperature reaches 35°C or higher. Female adults of certain fly species like to fly in a less exposed but well ventilated area/habitat. It was reported that the overall parasitic rate of following parasitoid flies, *Nealsomyia quadrimaculata*, *Exorista civilis* and *E. japonica* on *Clania vartegata* larvae could reach 35.8%.

3.3.2   Predatory insects (natural enemies)

There are many (thousands) species of predatory insects, including the commonly known dragonflies, praying mantises, assassin bugs (Reduviidae), minute pirate bugs (Anthocoridae), lacewings, lady beetles, ground beetles, robber flies (Asilidae) and hoverflies. Based on the range of their prey, they are divided into three groups: polyphagous, oligophagous and monophagous predators. Polyphagous predators have a very broad range of prey species that belong to different insect orders; oligophagous predators feed on a narrow range of closely related prey species with similar bionomics, whereas, monophagous predators feed on only one prey species or a very few species from the same genus. Fourteen predatory insect species were recorded associated with pine sawflies. Ants sometimes could kill over 50% of *Monochamus alternatus* eggs. *Dastarcus helophoroides* is a key beneficial predatory insect of longhorn beetles, and has been widely used in IPM practices against several longhorn beetle pests, including *Massicus raddei*. Introductions (from Belgium and UK) and releases of *Rhizophagus grandis* as a biological control agent showed a great success in controlling the invasive the red turpentine beetle, *Dendroctonus valens*; following three predatory rates (in the same year after their releases): 72.5%, 58.6% and 84.2% were recorded from three separated tests.

3.3.3   Other predators

(1) Insect-eating birds

Birds are an important natural enemy group of forest insect pests, with great species diversity and a huge insect intake rate and consumption capability. Beneficial insect-eating birds have been used in the control operations against the forest insect pests as an effective and safe biological control measure. There are >1160 species of birds recorded in China, and many of them are living in the forests and woods, feeding on forest insects. Birds can consume a huge amount of insects, especially during the hatchling periods. For example, *Parus major* could consume 300-450 insect individuals per day during the hatchling period, and a parent pair could kill 10500-15750 insects only during their two incubation periods a year. Most insectivorous bird species are polyphagous, even though some bird species are specialist feeders. *Cuculus* sp. likes to feed on *Dendrolimus* caterpillars, and can eat >300 caterpillars (3rd to 4th instars) per bird per day. *Cyanopica cyana interposita* feed on larvae, pupae and adults of >30 species of forest insects, and could eat 18,000 *Dendrolimus punctatus* 3rd-instar larvae per adult bird per year, keeping *D. punctatus* populations under control within 3.3 ha of pine stands. Insect-eating birds can also effectively control the wood-boring insects; for example, predatory rates of *Dendrocopos major* in a poplar forest with a low Asian longhorn beetle density were 30%-40%, but might reach up to 50%-60% in the heavily infested stands.

(2) Spiders

There are many species of spiders in China, and more than one thousand common species have been reported in the agricultural and forest settings. Spiders are carnivorous animals; feed mostly on live insects, and therefore are important predatory natural enemies of forest insect pests. Within a 1-ha of healthy pine-broadleaved forest, more than 30 species of spiders (with tens of thousands of individuals) were recorded. They feed only the live insects and do not harm trees. Spiders have multiple predation strategies; some of them use webs to capture insect prey and some of them don't build webs at all but actively hunt or ambush prey. *Araneus ventricosus* is a large size of Orb-weaver spider with a high reproduction capability, is a natural enemy of many pest insect species, killing >2400 insect individuals per adult spider. Currently, collections of spider egg-sacs and manually assisted spreading in the forests, and improving stand habitat that is suitable for spiders' growth and reproduction, are critical tactics for the enhanced biological control against forest insect pests using spiders.

Other predators of forest insect pests include bats, shrews, mice, forest frogs and lizards.

3.3.4 *Entomopathogens*

There are many species of entomopathogens, including entomopathogenic fungi, bacteria, viruses and nematodes.

(1) Entomopathogenic fungi

Over 150 species of entomopathogenic fungi have been reported in China; some of them have been applied in pest control practices and operations and as a biological control agent, such as *Beauveria bassiana* (Bals.-Criv.) Vuill., *B. brongniatii* (Sacc.) Petch, *Metarhizium anisopliae* (Metchnikoff) Sorokin, *Paecilomyces farinosus* (Holm. ex Gray) Brown & Smith, *Aschersonia* sp., *Verticillium lecanum*, *Entomophthora* spp., *Hirsutella thompsonii*, and *Nomuraea rileyi*.

Beauveria bassiana is the most commonly used entomopathogenic fungus with the longest application history. *B. bassiana* infested insect larvae usually turn characteristically hard, with cream-colored (white) mycelia on their cuticles. The fungus is distributed throughout China, and found to be able to infect over 100 species of forest insect pests, and has been used to control over 20 species of major pests including *Dendrolimus* spp., *Dasychira argentata* Butler, *Lymantria xylina* Swinhoe and pine sawflies. *B. bassiana* spores germinate and enter an insect's body through cuticle, digestive tract, trachea or wounds. It kills insects by interrupting their metabolism (in cells and tissues), draining the insect of nutrients and by producing toxins. As a biological agent, it could keep the insect problems under control by means of reducing insect population density, killing individual insects and decreasing reproductive capability (such as fecundity) of pest insect populations. In recent years, the national total annual pest control area using *B. bassiana* against forest insect pests reached 500,000 to 800,000 ha.

(2) Entomopathogenic bacteria

There are many species of entomopathogenic bacteria with an extremely fast growing and high reproductive capability, and are an important part of natural enemies of forest insect pests. Currently, most of entomopathogenic bacteria used for pest control operations belong to Bacillaceae, such as *Bacillus thuringiensis* Berliner (*Bt*) and *Paenibacillus popilliae* Dutky (formerly *Bacillus popilliae*). After bacterial infections, insects reduce or even stop feeding and exudates appear at both mouthparts and anus, and then death from septicemia. The dead insect bodies are soft and rotten with a stinky smell.

Bacillus thuringiensis is an entomopathogenic bacterium with a broad spectrum of insecticidal activity against many insect species in different orders, and plays a critically important role in the microbial-based biological control practices. Since it was first introduced to China from overseas in the 1950s as a biological control agent against forest insect pests, it has been very effective for controlling many species of pine caterpillars (*Dendrolimus* spp.) and delaying their outbreak cycles. Currently, *Bt* has been widely used as a biological control agent in IPM operations against more than 20 species of forest insect pests.

(3) Entomopathogenic viruses

Viruses are a very primitive form of life. Entomopathogenic viruses include nuclear polyhedrosis virus (NPV), cytoplasmic polyhedrosis virus (CPV) and granulosis virus (GV). Virus infected larvae exhibit the following symptoms: decreased feeding activity, stopping growth, slow-moving to no movement at all, body softening, liquification and decoloration, no stinky smell, and hanging limply in an inverted "V" position. Currently, *Dendrolimus punctatus* CPV, *D. tabulaeformis* NPV, *D. punctatus tehchangensis* CPV, *D. punctatus wenshanensis* CPV, *Euproctis* sp. NPV, *Lymantria dispar* NPV, *Hyphantria cunea* NPV, *Paleacrita vernata* NPV, *Lymantria xylina* NPV, *Malacosoma neustria testacea* NPV, *Ectropis oblique* NPV and *Clostera anachoreta* GV have either been field tested or applied as a biological control agent against their corresponding target insect pests.

Entomopathogenic viruses are safe to other natural enemies and the environment, have a very short residual time, less probability of insect resistance development, and could cause epizootics in the pest insect populations to achieve the sustainable pest control goal. Therefore, they have become the research (and application) hotspots in recent years.

(4) Entomopathogenic nematodes

Entomopathogenic nematodes are important natural enemies with a great application potential. They are specific in only infecting insects, with a very broad host range (many different types of insects). They actively search and find their hosts, enter insect bodies via natural openings, wounds or inter-segmental membranes, and reproduce rapidly in their hosts. Entomopathogenic nematodes are extremely safe to humans, animals, plants and other beneficial organisms. They are open to mass production and don't require specialized application equipment since they are compatible with standard agrochemical equipment, including various sprayers, and can even be applied in combination with some insecticides. At present, the most common entomopathogenic nematodes used in the biological control of insect pests are the members of Steinernematidae and Heterorhabditidae (Rhabditida). Researches on control of soil-living insects, *Carposina niponensis*, *Alissonotum impressicolle*, and *Rhynchites foreipennis*, and wood-boring insects, *Streltzoviella insularis*, *Zeuzera multistrigata*, *Aristobia testudo* and *Anoplophora chinensis* using entomopathogenic nematodes have been conducted in China. These research results show that entomopathogenic nematodes might have a great potential in future IPM operations against forest insect pests as a biological control agent.

3.3.5 *Tree-pest-natural enemy interaction*

The tritrophic interactions among trees, insect pests and natural enemies, with stable energy flows in the food chains (webs), are results of long-term co-evolutions in the ecosystems. Studying/understanding the tritrophic (tree-pest-natural enemy) interaction is a key foundation for effective pest control using natural enemies.

(1) There are complex dialectical relationships (including both interrestriction and interdependence) between insect pests and their natural enemies. Natural enemies need the insect pests as foods for survival, and the quantity and quality of natural enemies, in turn, strongly affect the population dynamics of insect pests. On the other hand, insect pests might also depend on their natural enemies as a key mortality factor and population regulator. In the absence of natural enemies, the populations of insect pests under normal weather conditions, might increase explosively to such extreme points that their living spaces and food sources are dangerously at risk (both availability and quality), and ultimately become detrimental to the survival of insect pests. The effect of insect pests on natural enemies can also be seen from the "follow-up/tracking" phenomenon of natural enemies to the pest insect populations.

The "follow-up/tracking" phenomenon includes three aspects (timing, population density, and spatial pattern). First, natural enemies normally appear after the establishment of pest insect populations, with a time lag. Second, at the early stage of pest insect outbreaks, populations of natural enemies are very low but increase as the pest insect populations explode; later pest insect populations decrease due to the population increase of natural enemies, and then the natural enemy populations start decreasing due the lack of pest insect food. Third, the spatial distribution patterns (aggregation or dispersal) of natural enemies are also very much shaped by the distributions of insect pests and moving more toward the areas with high pest insect populations. In short, the "follow-up/tracking" phenomenon is a reflection of insect-natural enemy interactions.

(2) With respect to adaptation, natural enemies and insect pests adapt to each other. Each pest insect species has certain species of natural enemies, and each natural enemy species has its own range of host or prey spectrums. Natural enemies aim at finding insect pests for food or reproduction sites, whereas insect pests try really hard to avoid attacks from their natural enemies. Insect pests develop several defensive mechanisms to avoid predation or parasitization by natural enemies, such as by increasing the thickness of egg-shells. Natural enemies on the other hand also develop different strategies to adapt to the diverse insect pests, for instance, the extremely long ovipositors of some parasitoid wasps that bore into trunks to locate wood-boring insects; and synchronization in occurrence with their prey or hosts.

(3) With respect to competition, complex competitions occur between natural enemies and insect pests, and among different species of natural enemies. Migration, dispersal, synparasitism, hyperparasitism, superparasitism and cannibalism are partially resulting from the competitions, and have a strong impact on the control efficacy of natural enemies. *Coccobius azumai* Tachikawa is a parasitic chalcid of *Hemiberlesia pitysophila* in its native range; after the parasitoid was introduced into China, it not only encountered the competitions from several local natural enemies including *Encarsia citrine*, *E. amicula*, *E. fasciata* and *Aphytis vandenboschi*, but also suffered a hyperparasitism from *Marietta carnesi*. *Dolichogenidea* (=*Apanteles*) *lacteicolor* Viereck, is a native natural enemy of the Gypsy moth, and also suffered hyperparasitisms from several species of hyperparasites: *Gelis apantelis*, *G. obscurus*, *Hemiteles aerator* and *Dibrachys cavus*.

4. Historical Development and Technological Advances of Forest Insect Control in China

4.1 Historical Phases of Development

Like many other modern sciences, entomology as a scientific discipline started at the beginning of the 20th century in China. Remarkable progresses on forest entomological researches and forest pest control practices have been made, and universally recognized over the past century through five developmental phases.

4.1.1 Initial phase (1911-1949)

Forest entomology in China started during this phase, with a major focus on field surveys and observations of some major forest insect pests. In 1911, a division of Insect pests and Diseases was established at the China Central Agricultural Experiment Station, and in the same year Mr. Shuwen Zou presented the first entomological talk "on *Ericerus pela*" (as the first Chinese entomology student) at the National Science Conference of America in 1911. Since then, several scientists sequentially published their observational reports on a few forest pests. Anguo Chen (1923) published a paper entitled "Common wood-boring insects" in the journal of "*New Agriculture*" by Beijing Agricultural University (Now China Agricultural University). Pang-hwa Tsai (Bang-Hua Cai) described the major bamboo pest insect (locust) as a new species, *Ceracris kiangsu* Tsai in 1929 (later changed to *Rammeacris kiangsu* Tsai). Jieren Lou (1930) and Zong Yue, Neng Huang, Shi-Bang Quo and Ting-Wei Liu et al. (1935) reported the biology and life cycle of the fir caterpillars in Nanjing area. Jun-Chao Ma (1933) reported the first observation results on *Cyrtotrachelus longimanus* (Fabricius) and *Apamea apameoides* (Draudt). Ming-Xiang Zhou and Zi-Mo Liu (1934) studied *Buasra suppressaria* Guenee for two years in Zhejiang province. More than 470 old cypress/juniper trees (*Platycladus orientalis* and *Juniperus chinensis*) were killed by wood-boring insects in Tiantan Park and Zhingshan Park in Beijing in 1935, and it was later determined to be caused by *Semanotus bifasciatus* according to Xi-tao Yi and Cong-Le Liu. Until the 1930s there were hardly any researchers specializing on forest insects. In early 1930s, the Parasitic Insect Research Lab was established at the Entomological Service (Department) of Zhejiang province, marking the beginning of modern biological control research in China. More than ten research papers related to the parasitoid insects (natural enemies) of

Rondotia menciana and *Dendrolimus* spp. were published during this period, and they became the most important scientific literatures for insect biological control in China during these early days.

In 1947, the first university class on "Forest Protection" was given by Yunnan University, and forest entomology became officially a part of high education discipline system in China.

Overall, however, no or a little pest control operations against forest insect pests were conducted during this period; if any, they were mostly single control measures targeted on the individual pest species.

4.1.2 Second phase dominated by the "eradication" goal (theory) (1950-1974)

Forest entomology in China progressed rapidly since the founding of the new China in 1949 till the mid 1970s. The new China government was strongly concerned about forest pest related issues (outbreaks), and "control of forest insect pests and diseases" was listed in the common program of the Chinese People's Political Consultative Conference (CPPCC), adopted by the First Plenary Session of the CPPCC in 1949. A division of Forest Pest Insect and Disease Control under the Department of Protection of Wildlife Animals and Plants at the Ministry of Forestry (now State Forestry Administration; SFA) was established to administer and oversee nationally the forest pest control programs and operations. In 1955, the Chinese Academy of Forestry, including the Research Institute of Forest Protection, was founded; this institute was responsible for the national basic and applied research works on major forest insect pests in China. In 1965, two Forest Plant Quarantine Stations: the Southern Station in Jiangxi province and the Northern Station in Heilongjiang province, were established by the Ministry of Forestry, focusing on nationwide extensions and technical guidance of forest plant quarantine and forest pest control operations.

During this phase, forest pest insect control strategies changed constantly in China. Due to the lack of understanding of forest pest complexity and pest control difficulty, as well as overwhelming dependence on pesticides, the pest control strategies were dominated by "eradication theory" via pesticide-based chemical control measures, with a general lack of understanding prevention. A pest control strategy called "control timely, continuously and totally" was suggested in 1952. In 1955, people started realizing the potential importance of pest prevention, and suggested that prevention sometimes might be even more critical than direct pest control as a last resort, and a combination of multiple tactics including silvicultural technology, pesticides and others in an integrated pest control (IPC) strategy might be even better. However, due to a strong influence by "the Great Leap Forward" movement in China, a new pest control strategy (guiding principle) was initiated in 1958, *i.e.* "controlling insect pests whenever or wherever they exist; combining domestic and foreign methods; total destruction; and eradication of the key pest species". In 1961, new guiding principles called "control early, control when they are small and control thoroughly" and "prevention first and eradicating actively" were adopted. In May 1965, the State Council of P. R. China issued the "Regulations of Forest Protection", in which (Chapter 5, items 32-36) the following statements related to forest pest control were issued: (a) actively control forest insect pests and diseases, (b) any units in forestry logging operations should pay special attention to silvicultural measures and clear-cut sanitary conditions after logging, and (c) any units in afforestation operations should plant appropriate tree species (plantations, especially mixed species forests) on suitable sites in order to prevent the occurrences/outbreaks and dispersals/spreads of the dangerous forest pests.

After the re-organization of China's higher education systems in 1952, three forestry colleges, Northeast Forestry College, Beijing Forestry College and Nanjing Forestry College, were founded, and many agricultural colleges or universities had a Department of Forestry. Forest entomology class was given to students in the forestry major at various colleges or universities using their own self-complied teaching materials. In 1953, Prof. Jie-Liu Xin published the first book entitled "Forest Entomology" in China, and in 1958 the Chinese translation of the book entitled "Forest Entomology" originally written in Russian by M. N. Rimskii-Korsakov was published in China. C. C. Prozorov, a forest entomology professor from the former USSR, taught forest entomology classes in China during 1957-1958; and his teaching materials were later published in a book entitled "Forest Entomology" in Chinese in 1959. In the same year, the first "Forest Entomology" textbook written by Chinese forest entomologists was published by the China Forestry Publishing House. The above mentioned three forestry colleges started admissions of undergraduate students in forest protection major in 1958 and graduate students in 1959.

Since the 1950s, national-wide field surveys on major forest insects began, and systematic and detailed studies on general biology, ecology and pest control of several important forest insect pests were conducted. Some of the findings were applied to the pest control operations and practices, and dramatically improved the pest control efficacy. During the 1950s, many species of natural enemies, such as *Rodolia cardinalis*, *Cryptolaemus montrouzieri*, *Aphelinus mali*, *Encarsia formosa*, predatory mites, *Bacillus popilliae* Dutky, Bt, microsporidia, nematodes and baculoviruses, were introduced into China from overseas. During the 1960s, a lot of research works on application of natural enemies as biological control agents including *Coccinella septempunctata*, *Anastatus japonicas*, trichogrammatids, pteromalids, Bt, and *Beauveria bassiana*, were carried out.

During this period, there were serious outbreaks of many native forest insect pests, such as *Dendroliumus* spp., *Rammeacris kiangsu*, *Ambrostoma quadriimpressum*, *Parocneria orienta* and *Algedonia coclesalis*. The major control measures against these outbreak pests included mechanical method (handpicking), light trapping, and sprays of pesticides. Aerial sprays of pesticides started from the 1960s. Back then, organochlorines and organophosphates were the most preferred and widely used pesticides since they were highly toxic to insects, resulting in extremely high mortality quickly. However, the national control efficacies ("percent control") against the forest insect pests were <20% in the 1950s, ~30% in the 1960s, and >30% in the mid 1970s. After >20 years of pest control practices in China, it became clear that the good intensions of "eradication theory" based chemical control measures (with pesticides) to solve forest pest insect problems could never be realized.

4.1.3 Onset and rapid development phase of the "IPM Theory and Practice" (1975-1995)

At the jointed conference between the Food and Agriculture Organization of the United Nations (FAO) and the International Organization for Biological Control (IOBC) held in 1967, the "Integrated Pest Control (IPC)" strategy was first proposed. This concept attracted extensive attention worldwide, and was also sequentially accepted by Chinese scientists and colleagues in the plant protection field. At the China National Plant Protection Conference held in 1975, a new guiding principle "Prevention first and IPC" was adopted. Since then the convergence of the concepts of integrated control and pest management, and the ultimate synthesis into "integrated pest management" (IPM), opened a new era in plant protection.

This phase also marks the rapid development period of forest entomology and forest pest insect control practices in China. In 1990, the national General Station of Forest Pest Control, Ministry of Forestry (now State Forestry Administration) was established through a consolidation of the Southern and the Northern Forest Plant Quarantine Stations, in charging of nation-wide extensions and technical guidance. In 1986, the China National Center for Monitoring and Forecasting Forest Insect pests and Diseases was founded. Forest Pest Control and Forest Plant Quarantine Stations were built at more than 90% of the counties (>2500) in China, and were responsible for the local forest pest control/quarantine operations and managements.

Since the resumption of normal university/college admissions after the Cultural Revolution, Beijing Forestry University, Northeast Forestry University and Nanjing Forestry University resumed undergraduate student admissions for the forest pest control major in 1977, and graduate student admissions (Master Degree) in 1978, and started graduate student admissions at PhD level in 1987, indicating that the training and education of forest pest control professionals in China had reached a higher level.

Under the guidance of IPM strategy, the Ministry of Forestry set up several pilot demonstration IPM programs for combating large scale outbreaks of *Dendrolimus* spp. in Chuzhou county (Anhui province), Wuzhou City (Zhejiang province) and Jianping County (Liaoning province); and for controlling serious infestations of the poplar wood-boring insects in Shuo County (Shanxi province) and Yanzhou District (Shandong province) in 1981. These pilot programs succeeded greatly with gaining tremendous experience on both control technologies and operational management. Since 1986, these IPM programs were adopted in >20 provinces at large scales via extension efforts, and achieved significant economic and ecological benefits. The project entitled "Study on

Dendrolimus spp." was listed and funded as one of the Key Scientific Research Projects of China's Sixth Five-Year Plan (1981-1985), focusing on population monitoring/forecasting, IPM, biological control and silvicultural control mechanisms. Great progress and achievements were made from this key research project, and its research quality and depth on IPM of forest insect pests reached the international advanced level. In the "Eighth Five-Year Plan" (1991-1995), two research projects hosted by the Ministry of Forestry, "Directive Breeding Technology of Short Rotation Industrial Timber Forests" and "Technological Systems of Ecological Forestry Engineering" were funded and completed. Within these two large projects, 14 sub-projects involved researches on IPM technologies and operational management tactics of many key forest insect pests, such as *Dendrolimus* spp., ALB, *Apriona germari*, *Batocera horsfieldi*, pine sawflies, pine bark beetles and poplar scale insects among others. Many research outputs and achievements from these projects were soon applied to the IPM operations. For example, synthetic sex pheromones of *Dendrolimus punctatus* and *Paranthrene tabaniformis*, were widely used to monitor the seasonal flights and population dynamics as part of IPM operations. Several new management systems and technologies were adopted in the monitoring and control operations against *D. punctatus*, such as IPM computer system, Expert Decision Support System, and Optimal Management System, plus the "3S" technologies: Remote Sensing (RS), Geographical Information System (GIS) and Global Positioning System (GPS) (or Global Navigation Satellite System (GNSS). Several researches on population dynamics and spatial distribution patterns on *Dendrolimus* spp., leaf beetles, longhorn beetles, notodontid moths and tissock moths (Lymantridae) were conducted, which lay a strong base for accurate pest insect surveys and scientifically sound pest control operations against forest insect pests. Pest control related equipment and technology were also improved during this period, and applications of Ultra-Low Volume (ULV) sprays and aerial sprays of pesticides or other control agents started. Great breakthroughs on silvicultural-based forest pest control measures (e.g. closure and fencing of newly planted forest stands) were achieved, which made a strong positive contribution to environmental protection, maintaining ecological balance and development of sustainable forestry.

A nation-wide survey (inventory) on forest insect pests and diseases was carried out during 1979-1983, with 5020 species of forest insects recorded. Based on the survey, several books, atlases or manuals on local forest insect pests at provincial levels were published in Shandong, Zhejiang, Hunan, Sichuan, Shannxi, Qinghai, Gansu, Inner Mongolia, Heilongjiang, Liaoning, Shanxi, and Hebei provinces. In 1983, the first edition of "Forest Insects of China" edited by Prof. Tsai, Pang-hwa and Prof. Xiao, Gangrou was published [2nd edition (revised and enlarged), edited by Prof. Xiao, was published in 1992]. This book was the most important and influential reference and literature for forest entomology teaching and research in China, presenting the developmental status and the research levels of forest entomology during that period. Several other books, such as "Pine Caterpillars in China" by Prof. Hou, Taoqian (1987), "Integrated Management of Pine Caterpillars in China" by Prof. Chen, Cangjie (1990), "Larch Casebearers, *Coleophora* spp." by Prof. Yang, Liming (1990), "The Cossid fauna of China (Lepidoptera, Cossidae)" by Hua, B. et al (1990), "Small Moths in Jiangxi" by Prof. Liu, Youqiao (1988), "Economic Sawflies Fauna of China" by Xiao, G. et al. (1992), "Natural Enemies of Forest Insect Pests" by Prof. Yan, J. et al. (1989), and "A parasitic Wasp Atlas of Forest Pests" (1990) were consecutively published, which reflected the research status/level of forest entomology in China at the time. The "Decision Concerning Some Questions in Accelerating Agricultural Development" was passed at the 4th Plenary Session of the 11th Central Committee of the Communist Party of China in 1979, in which a special recommendation on "Positively promoting biological control" was made. Since then, the biocontrol-based operations/practices against forest insect pests developed rapidly, including many researches and extension applications of trichogrammatids, bethylids, *Coccobius azumati* Tachikawa, microbial insecticides (*Beauveria bassiana*, Bt, and NPVs), and botanical insecticides. During 1986-1998, a "China-German Cooperation Project for Biological Control of Forest Pest and Disease" with special focus on natural enemies, pathogenic microbes, sex pheromones, and other bionic formulations, was successfully implemented. The 19th International Congress of Entomology was successfully held during 28 June – 4 July 1992 in Beijing, with over 4000 participants from 79 counties, and forest entomology was one of the 18 key sections. During 1994-2001, a China-Japan Special Cooperation Program entitled "Research Project on Forest Protection in Ningxia" was conducted with a special reference to studying control technologies for combating poplar wood-boring insects, especially on the screening of insect-resistant tree species and biological control measures. These cooperative projects and programs strengthened our relationships with foreign colleagues via professional contacts and exchanges, improved our research quality and depth, and some of cooperative research results were quickly applied in the control operations against some serious forest insect pests.

During this period (phase), invasive insect pests became a serious issue, for instance, consecutive invasions and spreads of *Hyphantria cunea*, *Hemiberlesia pitysophila* and *Oracella acuta* among others, resulted in severe damages to the China's forests. Thus, the overall pest control efforts (focuses) started shifting from the native pests towards the invasive pest species. Pest control measures and methods seemed to be more diversified, with many environmentally sound measures, such as biological control agents, low and ultra-low spray techniques, hot (thermal) smoke fogger application, poison "Pen" or "Ring" techniques, being commonly used in the pest control operations. The overall pest control capability against forest insect pests increased continuously over the years, with the average national control efficacy ("percent control") being about 40% in the 1980s, and ~50% in the mid 1990s.

4.1.4 Phase IV: Pest control practices guided by "Sustainable Development Theory" (1996-2013)

The XIII International Plant Protection Congress (with the theme of "Sustainable plant protection to promote the well-being of mankind throughout the world") held in the Netherlands in 1995, determined the overall outlines and strategies for future international plant protection. Since the beginning of China's 9th Five Year Plan (1995-2000), the ideas of sustainable forest management and sustainable disaster reduction were introduced (or incorporated) in the forest pest control operations. Under such circumstances, China's forest protection professionals and experts suggested two concepts: (1) sustainable management and (2) ecological management of forest pests. The sustainable management of forest insect pests is one of the key parts of sustainable forest management; it includes understanding the structures and overall functionalities of forest ecosystems, enhancing bio-control, planting of insect-resistant tree species, monitoring population dynamics of insect pests and their natural enemies, optimal integration and implementation of various ecological control approaches and other control measures. Smart and rational regulations of forest plant-pest-natural enemy interactions in the forest ecosystem might change the "antagonism" into "synergism", or "control" into "regulation"; one should fully exert the roles of various biological resources in the system in odor to achieve the suitable forest development. The ecological management, based on the holism and economical ecology principle, focuses on enhancing and optimizing the structures and functions of the ecosystem via single or combined measures that are safe, healthy, efficient, low energy consumption, stable and sustainable, to keep (maintain) the forest pest insect populations under the economic and ecological thresholds. At the same time, the "forest health" concept (originated from USA) was introduced to China. This concept aims at increasing forest quality and sustainability, realizing the coordinative development of ecological, social and economic benefits of forests, and ultimately achieving the pest management goal with special attention to the prevention. In short, forest entomology research and forest pest insect control practices in China entered an all-round (comprehensive) development phase, guided mainly by the sustainable development theory.

By the end of the 1990s, over 1,000 forest pest monitoring stations at national level, 1,593 stations at provincial level and 22,365 stations at local (county or lower) level were established. These stations are critical parts of China's national forest pest monitoring and forecasting network, headed and overseen by the National Center, and supported by the stations at different national, provincial and local levels. Based on the decision made at the national conference on forest pest control held in 1997, the construction of nationally standardized forest pest control and quarantine stations (including pest control network, quarantine network and monitoring network), funded by the national treasury bonds, started in 2000. By 2003, 1,559 standardized forest pest control and quarantine stations at various levels were established, accounting for 51.3% of the total stations.

A new pest control guiding principle "Prevention first, control scientifically, control according to the law, and promote forest health" was adopted at the national conference on forest pest control held in 2004. "Prevention first" includes strengthening all the preventative measures such as pest insect monitoring, plant quarantine, silvicultural approaches throughout the entire forest production and management processes, and shifting the strategic emphasis from "pest control" to "pest prevention". The "control scientifically" includes thoroughly understanding the biology, ecology, population dynamics and outbreak patterns of the target forest insect pests and implementing the pest control practices using scientifically proven measures and tactics. The "control according to the law" involves the legality related to the forest pest control or forest plant quarantine practices, especially in the efforts of stopping the dispersal and spread of invasive species (or species with quarantine importance) via human activities (e.g. transportation) according to the quarantine laws or regulations. The "promote forest health" focuses on understanding the concept of forest health, and reaching the forest health goal through integrated measures at forest ecosystem level.

The pest control guiding principle suggested in 2004 was slightly adjusted to "Prevention first, control scientifically, administration/inspection according to the law, and strengthening responsibility/accountability" at the national conference on forest pest control held in 2011. This new change reflected the fact that current forest pest control operations in China are also very much involved with the administration, and need the clarification and strengthening of governmental responsibility at different levels.

Scientific and technological supports to forest pest control operations have been continuously intensified over the years. Many important applied research projects were funded and implemented, such as "Sustainable control technologies against poplar wood-boring insects in shelter-belt forests", "Trap log or baiting technologies against longhorn beetles in the fast-growth timber production poplar plantations", "Pest control technologies against the Jingyuan pine sawfly, *Diprion jingyuanensis*", "Pest control technologies against *Chinolyda flagellicornis*", "Pest control technologies against *Oracella acuta*", and "Silviculture-based IPM technologies against pine shoot beetles in Yunnan". The outputs of these applied researches suggest that pest control operations should focus on the prevention of forest pest outbreaks without damaging the forest ecosystems, and incorporate the pest outbreak or disaster control efforts into the forest ecosystem management operations.

Import and transfer of advanced technologies from foreign countries also play an important role in the advancement of forest pest control operations in China. Since 1996, several forest pest control related projects under the "National Plan for Importing International Advanced Agricultural Science and Technology" ("948 Plan") were funded and implemented, including "Importation of *Oracella acuta* natural enemies", "Importation of aerial survey and mapping technology for forest insect pests and diseases", "Importation of synthesis and application technologies of the pine sawfly sex pheromones", "Importation of the predator, *Rhizophagus grandis* Gyllenhaal, for controlling *Dendroctonus valens*", "Importation of *Dendroctonus valens* semiochemicals and their applied technologies", Importation of manufacturing procedures and application protocols of new forest pest control technologies", "Importation of advanced manufacturing procedures and application protocols of *Metarhizium anisopliae*, "Importation of an air-assisted electrostatic spraying system" and "Importation of aerial electrostatic spraying technology". These projects involved in importing new technologies from USA, Sweden, UK, Belgium, Canada, Australia and Germany. Some of the imported natural enemies or technologies were soon digested, absorbed, improved and further transferred into pest control products, processes, applications or services, and this drastically advanced China's pest control technology levels. Implementation of a China-Germany cooperative project entitled "Forest Protection and Sustainable Management in Western China" not only introduced the updated German technologies, but also their advanced management models; established a standardized model and management system for the monitoring and forecasting network of forest pests; strengthened the risk assessments of forest pests in Western China; conducted training programs for forest pest monitoring and forecasting professionals, and undoubtedly enhanced the overall capability (level) of forest pest monitoring and forecasting operations in China. At the same time, several basic research projects such as "Underlying mechanisms for natural control of forest insect pests" and "Taxonomy of forest insects" among others were funded by the National Natural Science Foundation of China and not only enriched our understanding and knowledge about forest insects, but also nurtured the development of forest entomology in China. China positively involved in the pest control efforts against many internationally important forest pests, and made considerable contributions to this endeavor, especially for the Asia-Pacific region. In 2002, invited by the FAO Asia-Pacific Forestry Commission (APFC), Chinese delegates addressed at the 19th APFC Session in detail the successes and challenges of prevention and control of invasive pests in China, and attracted wide attention from delegates of other countrys. In order to strengthen exchange and collaboration among the Asia-Pacific countries on invasion of exotic species, the State of Forestry Administration (China) held an international symposium on prevention and control of invasive pest species in Kunming, Yunnan in 2003. The 15th International Plant Protection Congress was held in Beijing from May 11-16, 2004 with participation of many Chinese and international forest entomologists. At the 20th Session of APFC held in 2004, all the delegates agreed to establish the Asia-Pacific Forest Invasive Species Network (APFISN), as a response to the immense costs and dangers posed by invasive species to the sustainable management of forests in the Asia-Pacific region. APFISN recently opened a branch office in Beijing in 2014.

During this period, rapid developments on technological standardization of forest pest insect control operations were made. Over 40 national or local standards and technical regulations/specifications (codes) on forest pest control or forest plant quarantine were issued, including "Technical Specifications (Codes) for Monitoring and Controlling *Dendrolimus tabulaeformis*, *D. spectabilis* and *D. superans*", "Technical Specifications for Monitoring and Controlling *Dendrolimus punctatus*", "Technical Specifications for Controlling *Ceracris kiangsu*", "Standard of Forest Pest Occurrence and Disaster", "Technical Regulations (Specifications) for Mass Rearing and Applying *Chouioia cunea* Yang (Hymenotera: Chaleidoider, Eulophidae)", "Technical Regulations (Specifications) for Mass rearing and Applying *Scleroderma guani* Xiao et Wu (Hymenotera: Bethylidae)", "*Bacillus thuringiensis* Formulation Specifications", "Classification and Codes for Forestry Resources – Tree Insects", "Technical Regulations (Specifications) for Quarantine of *Hyphantria cunea*", "Technical Regulations (Specifications) for Quarantine of *Cryptorrhynchus lapathi*", "Technical Regulations (Specifications) for Quarantine of *Xylotrechus rusticus*", "Technical Regulation (Specifications) for Quarantine of *Dendroctonus valens*", "Technical Regulations (Specifications) for Controlling *Monochamus alternatus*", "Technical Regulations (Specifications) for Controlling *Hyphantria cunea*" and "Technical Regulations (Specifications) for Controlling *Cydia pomonella*". Adoption of these national or ministry-level (or local) standards (regulations, specifications or codes) surely speeded up the standardization process on forest pest control operations and related administration protocols in China.

The pest risk analysis (PRA) theories and methods have been accepted and adopted in China. During the revision of China's National Forest Plant Quarantine Pest List (in 2003-2004), a PRA working group was set up to carry out the preliminary screening from 178 candidate species recommended by different provinces based on PRA theory and methods; of which 64 species were selected for further detailed assessments by the regulatory committee. During 2011-2013, a new round of pest risk analysis effort on >200 pest species was conducted by the expert assessment panels. Adoption of PRA methods and procedures provides a strong scientific basis for determining and updating Chinas National Forest Plant Quarantine Pest List and the National List for the most dangerous forest pest species in China.

A novel organization of pest control operations was created during this period. China's national "engineering control/management" programs against the key forest insect pests and disease started in 1998. The "engineering control" programs were aimed at the most severe and widely distributed forest pests at appropriate temporal and spatial scales, managing economic and ecological losses caused by these forest pests down to the minimum levels, and achieving a sustainable pest-disaster control by following the "Prevention First and IPM" strategy, and by employing various effective tactics, and engineering project management methods/procedures (with clear focal points, detail plans and operation protocols). The following national "engineering control" programs targeted *Dendrolimus* spp., *Hyphantria cunea*, *Dendroctonus valens*, poplar longhorn beetles, *Hylobitelus xiaoi*, and *Tomicus piniperda*/*T. yunnanensis*, were launched and implemented in succession by SFA, China. These national programs integrated all the proven effective measures, such as satellite (space) remote sensing, aerial photography and videography (monitoring technology), natural enemies, pathogenic microbes, sex pheromones (biotechnology) among others into the control engineering projects for maximum efficacy and sustainability.

After 2000, the average national control efficacy rate against forest insect pests reached >60%; and the percent of pest control operations using sole non-residual or environmentally sound measures exceeded 80% after 2010.

4.1.5 New pest control era guided by the "Eco-civilization Construction" and "Modern Forestry Development" theories (2014 - present)

In May 2014, the General Office of State Council of China issued a special "recommendation on further strengthening and improving forest pest control work", marking the beginning of a new developmental stage for forest pest control operations in China with a special focus on strengthening governmental responsibility/accountability and promoting the prevention. This State Council's "recommendation" is a milestone for accelerating the advancement of China's forest pest control under the "new normality" (or "new normal" theory). In the "recommendation", it was clearly stated that the overall goal is to "reduce losses caused by forest pests and accelerate the development of modern forestry". The key tasks include strengthening pest control capability, perfecting management systems, optimizing the policy making and legislation systems, highlighting the "scientific control", and promoting public awareness of pest control and prevention. With respect to the efficacy of forest pest control operations, one should expect to reach >85% control operations using sole non-residual or environmentally sound measures, 90% pest forecasting accuracy and 100% quarantine rate for tree seedlings, through strengthening of preventative tactics, improvement of control measures for urgent needs ("emergency situations"), and optimizing organizational socialization of pest control practices.

In order to comply with the State Council's "recommendation", SFA launched the 3rd nation-wide survey (lasting for 3 years) on forest pests in 2014, and also officially initiated a special program to build high quality of demonstration pest control/quarantine stations based on the existing national standardized stations.

4.2 Advances in Forest Pest Control Technology in China

Since the founding of P. R. China in 1949, China has experienced at least two major developmental stages, the "rapid development" and the "all-round development". As overall advances in science and technology occur, many new and novel techniques and methods (or measures) have been continuously incorporated into forest pest control operations. The current technology-based forest pest insect control efforts are more efficient and environmentally sound with more accurate monitoring tools (quick and early detections) and more diversified, effective, sustainable control measures (both preventative and direct). The major advances in forest pest insect control technology are described in following eight aspects.

4.2.1 Application of "3S" technologies in monitoring and forecasting of forest insect pests

The "3S" technologies include Remote Sensing (RS), Geographic Information Systems (GIS) and Global Navigation Satellite System (GNSS). Remote sensing technology is used to gather information about the surface of the earth from a distant platform, usually a satellite or airborne sensor. The remotely sensed data used for mapping and spatial analysis is collected as reflected electromagnetic radiation, which is processed into a digital image that can be overlaid with other spatial data. The spectral properties of forest vegetation in different parts of the spectrum can be interpreted to reveal information about the forest types, status of forest stands and forest pest outbreaks. GIS application software enables the storage, management, and analysis of large quantities of spatially distributed data, such as distribution, population dynamics, occurrences, and infestations of forest pests. These data are associated with their respective geographic features, and can be updated, searched, mapped, graphed and outputted. This powerful analytical ability should result in the generation of new information, as patterns and spatial relationships are revealed. GNNS is a satellite-based radio navigation and locational system that provides signals from space that transmit positioning and timing data to the receivers, enabling the user to determine very accurate locations on the surface of the Earth.

Currently, application of the "3S" technologies in forest pest management is leaning toward their integration. Firstly, RS system will obtain the pest outbreak/occurrence related data such as exact locations (geographical coordinates), infestation levels, ranges and areas, through computer image reading and further analysis on the newest raw image data gathered by the RS technology. These data will be stored, managed and statistically analyzed with GIS software with a series of outputs produced by an operational platform, such as pest outbreak/occurrence maps, distribution maps of the various forest pest monitoring stations and survey roadmaps. Such detail, accurate and real-time information together with the historical data on the targeted insect pests stored in the system will surely improve our understanding of the patterns or trends of forest pest outbreaks. The GNSS will guide the pest control professionals on the ground to quickly reach the precise outbreak spots (stands) with all the location, pathway and timing information recorded as part of the database components and ground operation records.

Current data-collection methods mainly include medium resolution satellite remote sensing, airborne video-recording, and digital aerial sketch mapping. On the basis of learning and improving from the aerial surveillance and monitoring technologies originally developed in the USA, Chinese forest pest management professionals recently developed a mid- to long-term monitoring and forecasting system (technology) for the most severe and dangerous forest insect pests by combining the new aerial and satellite data collection technologies with the conventional ground sampling and survey methods. This was applied to demonstration programs of large-scale monitoring, forecasting and early-warning of *Dendrolimus punctatus*. This technology not only increased our early-warning accuracy, but also enhanced our pest control decision-making capability. The development and perfection of the high-resolution satellite remote sensing and unmanned aerial vehicle monitoring technologies in recent years showed a great potential as an important complementary method of the ground monitoring of forest pest outbreaks. In addition, based on the differences in responsibility and functionality among forest pest control stations at different administrative levels (county/province/country), China's C/S (Client-Server) and B/S (Browser-Server) models-based forest pest management information systems (online-networks) were built, and the data and information in the systems can be shared among different administrative levels (Community-County-Province-Country) in order to better serve the decision-making and pest management operations. The "3S" technologies as a new and advanced pest monitoring and forecasting early-warning tool will become a key part of "Digital Forest Pest Control" in the near future.

4.2.2 Application of "3 Trapping" technologies in monitoring and control of forest insect pests

"3 Trapping" technologies include semiochemical-based traps, light traps, and color traps as a part of IPM strategy; are mainly targeting the adults of forest insect pests rather than the conventional larvae-focused control measures. Application of the "3 Trapping" technologies not only enhances pest control efficacy, but also reduces residual insecticide uses and environmental pollutions, resulting in significant economical, ecological and social benefits.

The semiochemical-based trapping technology deploys traps baited with insect pheromone components or plant kairomone components or their combinations for insect monitoring or mass-trapping. Pheromones are chemicals that are produced and secreted to the outside by many insect individuals, and affect the behavior of other insect individuals of the same species; they are intra-specific chemical signals that are beneficial to both the emitters and receivers. In cooperation with colleagues from Czech Republic, USA, Germany and Sweden among others, sex or aggregation pheromones of many Chinese forest insect pests, such as *Paranthrene tabaniformis*, *Sesia siningensis*, *Dendrolimus* spp., *Hyphantria cunea*, *Lymantria dispar*, *Cydia pomonella*, *Cossus orientalis*, *Coleophora* spp., pine sawflies, several species of tortrix moth, geometrid moths and bark beetles (*Ips* spp., *Tomicus* spp.) were identified and synthesized. Traps baited with synthetic pheromone lures were not only applied in population monitoring for all these species, but also successfully employed in the mass-trapping efforts against *C. sinensis*, *H. cunea*, *Tomicus* spp., and *I. duplicatus*. Traps baited with the attractive plant kairomones were also successfully applied for monitoring and mass-trapping of *Monochamus alternatus*, *Semanotus bifasciatus* and *Dendroctonus valens*.

Insect light trap technology was developed to catch flying insects usually at night that are attracted to different wavelengths of light sources, especially UV light, emitted from the traps. Mass-trapping operations with light traps as a control measure have been conducted on *Dendrolimus* spp., *Hyphantria cunea*, *Paleacrita vernata*,

Massicus raddei, and insects from following familes or subfamiles: Lymantriinae, Notodontidae, Arctiinae, Pyralidae and Noctuidae. Although light traps might also capture some natural enemies and non-target insects (as side-effects), it is still considered as a simple, economical, and environmentally sound monitoring and control measure by the local pest control professionals.

The visual (color) traps were developed by integrating the insect positive phototaxis (attraction behavior) to different wavelengths of reflective lights from the colored devices (such as colored plates or boards) with the insect glues to capture the diurnal insect pests. It was shown that khaki colored (earthy yellow) sticky boards were the best visual traps to catch *Paratrioza sinica*, and the yellow sticky board traps were very effective in capturing *Dialeurodes citri* and *Aleurocanthus spiniferus*. Yellow or green colored sticky traps are effective visual traps for capturing the fruit insect pests in the Jujube and citrus orchards, especially when placed at the middle canopy level. Visual trapping technology is simple and safe to fruits, environment, humans and livestock, thus, has a great potential in the IPM operations against the forest insect pests. In recent years, efforts have been to combine the semiochemicals with colored plate traps to increase the efficacy of the visual traps. In the future, efforts to develop insect visual traps with repellent or repulsive features to the natural enemies, in order to reducing side-effects to the natural enemies, are really necessary and beneficial.

4.2.3 Identification technologies of forest insects: towards both micro-scale and distance

Modern molecular biology related technologies, such as esterase isoenzyme electrophoresis (EST), random amplified Polymorphic DNA (RAPD), restriction fragment length polymorphism (RELP), polymerase chain reaction (PCR), DNA sequencing, and DNA hybridization probe, have been tested for identifications and separations of behavioral types or closely related and morphologically similar or inseparable insect species, field collected insect larvae or pupae (no adults available) or intercepted insect larvae during the on-spot quarantine inspections. Currently, these "micro-scale" molecular technologies have been successfully applied to the identification of bark beetles, longhorn beetles, scarab beetles and trichogrammatids.

Distance (or remote) identification/diagnostics is a method of identifying or determining insect species or their damages through digital imaging. Digital pictures of the target insects from the field or under the microscope are taken, and are then submitted (together with other related files) through the Internet to specialists for identification and diagnosis. Distance diagnostics provides rapid identification and diagnosis for many forest pest insect samples without physically sending specimens to the experts. It is extremely important and helpful for the local pest control operators, who don't have the skills to identify insect pests correctly and quickly, and meets the demands for quick and accurate identification and diagnosis during the port quarantine inspections. A Distance Identification/Diagnostics System was recently established and based at Institute of Zoology (Chinese Academy of Sciences; CAS), and is currently being tested in cooperation with several Provincial Forest Pest Control Stations, such as Beijing, Guangdong and Hunnan provinces.

4.2.4 Maturity of quarantine/phytosanitary treatment technologies

Currently, the most common phytosanitary treatment technologies are methyl bromide fumigation and heat treatment. Heat treatment is a method to kill the wood-boring insects by placing the infested logs or timbers in a closed, heated and pressurized room for a certain time period. It is reported that the heat treatment at 65-75°C for 15 hrs, or 157-168°C for 10 min at 9 MPa pressure, can kill 100% of *Monochamus alternatus* and *Bursaphelenchus xylophilus* in the pine lumbers (2.5 cm thick boards or 5 cm x 5 cm square wood columns).

Water treatment is a method of killing wood-boring insects by submerging insect infested logs in the water at certain temperatures. The oxidations of organic compounds in the logs consume oxygen inside the insect tunnels, and suffocate insects in the wood; at the same time, several toxic gases (H_2S, SO_2 and CO) produced during the oxidation might penetrate the insect cuticles and speed up their mortality. Currently, this mature technology has been standardized and widely used in the quarantine treatment operations.

Irradiation treatment is to use ionizing radiation energy of several hundreds of kiloelectronvolts or above, such as γ-rays, x-rays or high-energy particles (electron beams) for killing pests and stopping their normal life cycles, and preventing their dispersals and spreads. It is non-toxic and non-destructive to commodities, has no residual or pollution effects, and can be performed after packaging, thus avoiding recontamination or reinfestation of the product. Currently, irradiation treatment has been successfully applied to kill *Tomicus piniperda* in the pine logs, and several citrus and Chinese chestnut insect pests.

Microwave technology has been widely used in heating, drying, insect killing, sterilization and medical equipment treatments. Breakthroughs of using microwave heating as an effective timber quarantine treatment method for disinfestation of wood-boring insects have been made recently. Several specific commercial microwave products manufactured by Chinese companies for quarantine treatment operations have been applied to kill *Monochamus alternatus* and *Bursaphelenchus xylophilus* in the infested pine logs or timbers, and achieved 100% efficacy at the timber surface temperature of >68℃ for 30 min.

4.2.5 Pesticide application technologies and equipment for forest pest control operations are becoming more diversified, more efficient and safer

Ground chemical control against the forest insect pests on big or tall trees is always a challenging task due to the physical difficulty of reaching the high canopy. Over the past ten years, great efforts were made to develop new pesticide application technologies and equipment with longer effective (spray) range, lower pollution effect and less water consumption.

The long-range pesticide spraying technique (equipment) aimed at combining the air-pressure (air-assisted) and atomization technologies to achieve >20 m vertical spray range with a long liquid jet flow and powerful penetration impact was newly developed. Smoker and fogger technology is a method of using the smoke of burning substrates or from the smoke generators to carry the pesticide particles (in the smoke plumes) to uniformly reach its target insects for killing. Tree trunk injection with systemic insecticides (manually or using equipment) is one of the most effective alternatives to spraying or soil application of pesticides for forest pest insect control with a relatively low cost and a long effective duration. After injected into the trunks, systemic insecticides will be transported to different parts of tree tissues or organs for effective kills; since the pesticide is injected directly into the tree, it has little impact on the environment, natural enemies and the applicator. Trunk injections are effective for many insect pests such as wood-boring insects, especially for areas with less water supplies, aesthetic forests or precious ancient trees. Pesticide "bombs" (bags) dispensing technology is a method or equipment to launch pesticide shells or bombs into air in forest areas (with adjustable launching ranges, angles, explosion heights and areas, pesticide cloud float areas etc.), and the pesticide clouds from the bomb explosion will spread and reach forest canopies to kill the target insects. This technique is probably one of the most effective control approaches for deep mountain forests, dense forests, areas with limited water supply, or where the conventional ground control or aerial control are not feasible. Insecticide-loaded ropes or poison chalk pens were also used in the pest control operations against some forest insect pests.

In addition to the ground applications, aerial pesticide spraying technologies rapidly advanced over the years in China. A new aerial electrostatic spraying system that incorporates various models of fixed-wing aircrafts or rotary-wing aircrafts with electrostatic spraying technology has been widely used in the large scale IPM programs against many serious forest insect pests in recent years. The technology has achieved not only great efficacy with much less environmental pollution (via reduced insecticide dosages), but also significant ecological and economic benefits.

4.2.6 Biological control plays more and more important roles in the environmentally sound IPM operations against forest insect pests

There are over 2000 species of insects recorded as natural enemies of forest insect pests in China; of which, the detailed biology and ecology of >200 species of beneficial insects have been studied. Application of natural enemies as a biological control agent or natural control means has achieved a gratifying success and remarkable progress. Over the past 50 years, surveys and detail studies on natural enemy fauna and key beneficial insect species compositions of many major forest insect pests, such as *Dendrolimus* spp., *Lymantria dispar*, *Clostera anachoreta*, *Cerura menciana*, *Hyphantria cunea*, *Coleophora* spp., *Pyrosis eximia*, *Malacosoma neustria testacea*, *Hemiberlesia pitysophila*, pine swaflies, poplar longhorn beetles, *Monochamus alternates*, *Anoplophora* spp., *Dendroctonus armandi*, *Cryphalus massonianus*, *Agrilus planipennis*, and *Dryocosmus kuriphilus* among others, were conducted with great successes. Biology, ecology and behavior of the following important beneficial insects, *Trichogramma* spp., *Chouioia cunea* Yang, *Coccobius azumai*, *Scleroderma guani* Xiao et Wu, *Dastarcus helophoroides* and *Rhizophagus grandis*, were thoroughly studied; and the technologies on their mass-rearing and field releases were also developed. Applications of *Trichogramma* spp. as a biological control agent for controlling *Dendrolimus* spp., *Scleroderma guani* for controlling several longhorn beetle species, *Coccobius azumai* for controlling *Hemiberlesia pitysophila* and *Chouioia cunea* for controlling *Hyphantria cunea* are some of the successful examples for biological control of forest insect pests in China. China is one of the world's leading countries on mass rearing/production technology of beneficial (entomophagous) insects. Regional collaborations and exchanges through the Asia-Pacific Forest Invasive Species Network (APFISN) surely strengthened the biological control capability against the invasive pests in the region. Some biological control related technologies and equipment were exported to Maldives.

Breakthroughs on laboratory amplification methods for production and field application procedures of *Dendroctonus* spp. NPV and CPV, *Paleacrita vernata* NPV, *Hyphantria cunea* NPV, and the Gypsy moth NPV have been made. These virus preparations are not only effective but also species-specific to the target insects, and have been successfully integrated into the IPM programs as a biological control agent against the corresponding forest insect pests. *Bacillus thuringiensis* formulations have been registered and commercially manufactured in China for many years, and have been commercially used to control many agricultural and forest insect pests with consistent efficacies. The manufacturing techniques and the overall product quality of *Beauveria bassiana* were improved dramatically, and more products and formulations were developed over the years. As a biological control agent, it has been applied to control poplar defoliators, sawflies, *Parocneria orienta*, *Lymantria xylina*, bamboo insect pests, and commercial timber forest insect pests, in addition to the widely routine use in the IPM operations against *Dendrolimus punctatus* in southern China. Recently, the national annual forest pest control area used *Beauveria bassiana* with a satisfactory efficacy in 500,000 ha to 800,000 ha. Significant research progresses on phylogenetic systematics, biology, pathogenesis, toxins, manufacturing procedures of *Metarhizium* spp., were also made, and they have shown a great application potential for controlling the underground insects and wood-boring insects. Entomopathogenic nematodes, *Steinernema* spp. and *Heterorhabditis* spp., have a very broad host range, and powerful host searching capability, and as such are particularly effective on wood-boring or subterranean insects; and have been commercially produced and applied to control operations against several insect pests.

There are many species of birds in China, and many of them are natural enemies of forest insect pests. Setting up of artificial bird nest boxes in forests could attract and enhance bird populations, increases bird community diversity and population densities, and has been considered as an effective biological or natural control measure against forest insect pests. Promising research progresses on attracting the great tits (*Parus major*) with nest boxes in Liaoning and Zhejiang provinces, conservation and application of woodpeckers in Shandong province, and the taming of the azure-winged magpies (*Cyanopica cyanus*) and red-billed blue magpies (*Urocissa erythrorhyncha*) were made. A pair of the great spotted woodpeckers (*Dendrocopos major*) could keep the poplar wood-boring insect populations under control within 10 ha of poplar shelter-belt forests in western Inner Mongolia.

In addition, development and application of botanical pesticides (or natural pesticides) as a part of "bio-control" measures progressed significantly in recent years due to their unique mode of action and the fact that they are environmentally sound and safe to the non-target species. Hundreds of plant species or plant extracts were tested and screened for insecticidal activity, and the active components and their potential mode of actions of the active plant extracts were further investigated. So far, several botanical insecticides (formulations) have been developed and successfully applied in pest control operations with an impressive efficacy against landscape or ornamental insects in an urban environment.

4.2.7 Silvicultural approaches as a critical pest control measure play a more prominent role in improvements of ecological environments and tree resistance to pests

Silvicultural approaches are not only practical measures used throughout the establishment, composition, constitution, and growth of forests, but also critical methods for forest pest insect control, including establishment, maintenance, regeneration, restoration and reconstruction of forest ecosystems. At the end of the 20th century, a China-U.S. cooperation program on "Forest Health and Restoration" was initiated, and the demonstration projects within the program aimed at strengthening tree resistance to insect pests via human (manual) intervention measures. These included optimal selection of tree species for planting, care and regeneration of forests, and irrigation and fertilization management, and were conducted in 14 provinces in China. At the same time, a Sino-Germany cooperative demonstration project on "close-to-nature-forest-management" was also implemented. The major progresses from these two programs are described below. (1) Selection and breeding of pest resistant tree species: it was found that *Pinus taeda* and *P. elliottii* are resistant to *Matsucoccus matsumurae*; *Populus euramericana* cv. I-72/58 and *P. deltoides* cv. I-69/55 are resistant to the Asian longhorned beetles (*Anoplophora glabripennis*) and the Chinaberry trees (*Melia azedarach*) show repellency to the ALB, whereas Mulberry (*Morus alba*) and paper mulberry (*Broussonetia papyrifera*) are attractive to *Apriona germari* (Hope). (2) Planting multiple species mixed forests. The soils in the mixed *Pinus tabuliformis-Robinia pseudoacacia* forests are much richer than ones in the pure *Pinus tabuliformis* forests, which would certainly help the growth of pine trees, increase their tolerance levels to the pine caterpillar attacks, and ultimately are favorable for the natural control against the *Dendrolimus* spp. Adjustment of tree species structures, in particular increasing the proportions (or ratios) of the resistant tree species in the mixed species stands could increase the stand stability and enrich the biodiversity. Considering the limited flight capability of some poplar longhorn beetles, planting non-host repellent tree species together with poplar trees (the host trees) or as a protection (or barrier) row might, to some extent, be able to interrupt or obstruct the natural dispersals of the longhorn beetles. (3) Strengthening regeneration measures, such as regeneration by stump grafting, coppice regeneration and tall stump cutting (similar to pollarding). Stump grafting with resistant poplar species (to the longhorn beetles) such as *Populus tomentosa* would take advantage of the existing stump root systems, and could get the regenerated stands growing up quickly with less afforestation effort and low costs. The "tall stump cutting" (or pollarding) is to cut off (and remove) the section of main trunk that was heavily infested by the wood-boring insects at 1.6-1.8 m height, to reduce the pest populations and infestations, and then to quickly regenerate the forest stands from the insect-free tall stumps. Coppice regeneration takes advantage of the fact that many trees make new growth from the stump or roots if cut down. With proper silvicutural measures on the stools, a new regeneration stand can be quickly restored. Nevertheless, timely removal of the insect-infested trees or logs to eliminate insect sources is one of the most critical silvicultural approaches for controlling the longhorn beetle pests. (4) Strengthening the tending management of young and middle-aged forest stands. Timely thinning of middle-aged or mature forest stands and properly pruning will accelerate tree growth and increase their resistance level to insect pests.

Closure of forest stands with fencing or alike is a simple, easy, economical and effective measure for restoration of thin forests and regeneration forests, for accelerating tree growth in plantations, and for safer pest control against forest insect pests. The "closure" will be able to enrich the stand biodiversity and improve the stand community stability by reducing damages or interruptions caused by human activity or livestock. The forest tending management aims at accelerating the health growth of forest stands, increasing the stand resistance to pests or stress and enhancing their natural self-regulation capability. A combination of the "closure" and "tending management" was tested as a silvilcutural measure for controlling *D. punctatus* in Sanshiliuqu Forest Farm, Qinzhou, Guangxi province. Many years after implementing this program, the numbers of plant species (especially the honey plants), beneficial insects, and birds increased significantly with hardly any occurrences or outbreaks of *D. punctatus* were recorded in the closed forest areas.

4.2.8   Rapid development of environmentally sound chemical insecticides

Along with the public environment protection and ecological consciousness enhancement, and food safety concerns, China's government issued a series of laws, regulations, rules and standards for pesticide manufacturing, registrations and safe uses, and banned or limited the uses of many highly toxic residual pesticides, such as organochlorine pesticides, methamidophos, parathion, methyl parathion, monocrotophos, and phosphamidon. At the same time, strong efforts were made by industries and research institutes for developing effective and low toxic/residual pesticides to meet the urgent demands for environmentally sound pest control operations. In the chemical control practices against forest insect pests, several insect-specific pesticides were favorably selected, such as the chitin synthesis inhibitor, chlorbenzuron, that was widely used to control *Dendrolimus* spp., *Hyphantria cunea, Leuoma candida, Orgyia postica, Lymantria dispar* and *Malacosoma neustria testacea*; the hormonal insect growth regulator, fenoxycarb, for controlling many poplar defoliators, *Dendrolimus* spp. and *Hyphantria cunea*; a new molting hormone compound, tebufenozide, for controlling several species of defoliators. Imidacloprid is a systemic insecticide belonging to a class of chemicals called the neonicotinoids, and was used to effectively control various defoliators, wood-boring insects and cone & seed insects. Emamectin benzoate is an avermectin class insecticide developed for the control of lepidopteron insects, and has shown a great control efficacy for controlling *Hyphantria cunea* even at a very low dosage. The above mentioned insecticides are not only effective and specific, but also relatively safe to environment and natural enemies, therefore they are replacing the old conventional residual pesticides that pose a serious threat to ecological environment and food safety.

In order to reduce or eliminate pesticide-related problems or issues, safer pesticide formulations are leaning towards water-based, granular, slow-release (controlled-release), multi-functional, simple and refined formats. Reducing or limiting the use of emulsified formulations might reduce environmental pollutions caused by the huge amount of aromatic solvents in the formulations. Microcapsule formulation is a very important part of pesticide controlled-release technology, which could prevent the over-release or quick loss of active ingredient(s) (AI), reduce AI dosage, and lessen environment pollution and potential poisoning effects on users, plus have a long-lasting activity. For example, the microcapsule formulations of cypermethrin (a synthetic pyrethroid) and thiacloprid (a neonicotinoid compound) were successfully applied to control several species of longhorn beetles (such as *Anoplophora glabripennis, Batocera lineolata* and *Monochamus alternatus*), and were also tested on *Dendrolinus* spp., and *Hippotiscus dorsalis*. A new controlled-release insecticide bag dispenser loaded with a mixed formulation of acetamiprid and monosultap was developed recently. Hanging these environmentally safe insecticide bags above the young (central heart) leaves in the throat of the infested palm trees has been attempted successfully to control the invasive leaf beetle, *Brontispa longissima*. Monitoring and testing of the long-term toxicity effects of these new insecticides is still necessary as they could have harmful ramifications that may be subtle and affect non-target beneficial insects.

Key References

Bao J-Z, Gu D-X, 1998. Biological control in China[M]. Taiyuan: Shanxi Science & Technology Publishing House.

Editorial Committee of "Forest in China", 1997. Forest in China (Vol. 1)[M].Beijing: China Forestry Publishing House.

Fang S-Y, 1993. The eco-geographical distribution of forest pest-insects in China[M]. Harbin: Northeast Forestry University Publishing House.

He X-Y, 2004. Applied ecology[M]. Beijing: Science Press.

Integrative Research Group on National Natural Disasters, 1996. China natural disasters and disaster reduction strategies[M]. Beijing: Ocean Publishing House.

Li Z-Y, 1992. Overview of forest entomology development in China[J]. Shaanxi Forest Science and Technology, (2): 4-8.

Luo Y-Q, Huang J-F, Li J-G, 2000. Major achievements, problems and prospects of the poplar longhorn beetle research in China[J]. Entomological Knowledge (now Chinese Journal of Applied Entomology), 37 (2): 116-122.

Ma S-J, 1959. Overview of insect eco-geography in China[M]. Beijing: Science Press.

Meng X-Z, 2000. Advances in the research and application of insect pheromones in China[J]. Entomological Knowledge (now Chinese Journal of Applied Entomology), 37 (2): 75-83.

Research Group of China Forestry Sustainable Development Strategy, 2003. Study on China forestry sustainable development strategy[M]. Beijing: China Forestry Publishing House.

Song Y-S, 2000. Historical events of forest pest control operations in China (1949-1999) (Part I)[J]. Forest Pest and Disease, 19 (2): 40-43.

Song Y-S, 2000. Historical events of forest pest control operations in China (1949-1999) (Part II)[J]. Forest Pest and Disease, 19 (3): 45-48.

Song Y-S, 2006. On environmentally sound control operations against forest pests[J]. Forest Pest and Disease, 25 (3): 42-44.

Song Y-S, Su H-J, Yu H-Y, et al, 2011. Evaluation of economic losses caused by forest pest disasters between 2006 and 2010 in China[J]. Forest Pest and Disease, 30 (6): 1-5.

Tsai P -H, 1929. Description of three new species of Acridiides from China with a list of the species hitherto recorded[J]. J. Coll. Agr. Imp. Univ. Tokyo, 10 (2): 139-149.

Wan F-H, Ye Z-C, Guo J-Y, et al, 2000. Advances and prospects of biological control research in China[J]. Entomological Knowledge (now Chinese Journal of Applied Entomology), 37 (2): 65-74.

Wang S-M and Chou Y, 1995. A history of entomology in modern China (1840-1949)[M]. Xi'an: Shaanxi Science & Technology Press.

Xiao G-R, 1992. Recent advances in forest entomology research in China[J]. Forest Pest and Disease, 11 (3): 36-43.

Yang Z-Q, 2004. Advance in bio-control researches of the important forest insect pests with natural enemies in China[J]. Chinese Journal of Biological Control, 20(4): 221-227.

Yan J-J, et al, 1989. Insects as natural enemies of forest insect pests[M]. Beijing:China Forestry Publishing House.

Zhang X-Y and Luo Y-Q, 2003. Major forest biological disasters in China[M]. Beijing:China Forestry Publishing House.

Zhang S-M, 1998. The geographical distribution of agricultural and forest insects in China[M]. China Agriculture Publishing House.

Zhou Z-M and Zhou X-Y, 1984. Introductory ecology: Integrated pest management theory and practice[M]. Congqing: Scientific & Technological Documentation Publishing House.

Zhu H-F, 1987. Fundamental theory of animal taxonomy[M]. Shanghai: Shanghai Science & Technology Press.

Original Chinese texts by Song Yushuang

English translation by Dr. Zhang Qinghe

English proofreading by Dr. John A. Byers

Taxonomic Section

1 *Tarbinskiellus portentosus* (Lichtenstein)

Classification   Orthoptera, Gryllidae
Synonym   *Brachytrupes portentosus* Lichtenstein

Distribution   Zhejiang, Fujian, Jiangxi, Hunan, Guangdong, Guangxi, Hainan, Guizhou, Yunnan, Taiwan; Pakistan, Malaysia, Vietnam.

Hosts   Pinaceae, Taxodiaceae, Lauraceae, *Eucalyptus* spp., *Acacia confuse*, *Citrus reticulate*, *Prunus persica*, *Maninot esculenta*, *Zea mays*, *Arachis hypogaea*, beans and vegetables.

Morphological characteristics   **Adults** Length 40-50 mm. Yellow brown or dark brown. The head is wider than the prothorax. Antennae are filiform, slightly longer than the body. There is a longitudinal streak in the middle of the pronotum, and a transverse coniform-shaped striation on both sides of the pronotum. The hind femur is strong. The hind tibia is thick and armed with two rows of spines, each with 4-5 spines. Longitudinal veins of the male forewings are thick, curved and convex. Those of females are thin, straight and smooth. The ovipositor is tubiform, 5-8mm in length. **Eggs** Cylinder-shaped, light-yellow, and around 4.6mm in length. **Nymphs** Similar to adults; smaller and lighter in color. The length of the 1st -7th instar is 5.4-6.1, 6.5-10.1, 11.1-13.5, 12-21.2, 17-29, 28-33, 35-38 mm, respectively.

Biology and behavior   It has one generation a year. Most of them overwinter in the burrow as the 3rd-5th instar nymphs. Overwintering nymphs start to be active in February to March. Adults appear in May. Nymphs are present until July and up to 260 days or so. Adults reach the peak in June, and copulation peak occurs from July to August. Females oviposit from July to October, and egg stage is up to 20-25 days. Eggs hatch from August to October. Adults die gradually after October and adult stage is around 110 days.

Mating takes place in the burrow. Eggs are laid at the bottom of the burrow. Each egg pile has around 20-50 eggs. Each female can lay more than 200 eggs. The females pile tender host leaves close to eggs for neonates to feed. One adult stays in each burrow. Adults stay in the burrow during daylight and feed at night. They indulge in cannibalism. Only during mating do male and female live together. The depth of the burrow is mainly related to instar, soil texture and temperature. The burrow depth of the first instar nymphs is only 3-7 cm, while the burrow depth of the adults is more than 80 cm. The entrance of the burrows is finger-sized and rounded. Presence of loose soil at entrance indicates an adult is resting inside. Adults use their mandibles to shove away the loose soil at the entrance when they exit the burrow, and they plug the entrance by popping soil with their hind legs after they enter the burrow.

The activity area of *T. portentosus* is around 10 m away from their burrow. They often drag tender shoots, branches, and leaves into their burrows to feed. This insect does not necessarily go out every night, but they do go out more frequently in early spring when it gets warmer and during their mating period during July and August. Most of them forage during 17:00 to 20:00. Peak activity occurs in sunny and muggy, calm, or warm night after a long rain.

T. portentosus mostly occurs at the places where the soil is loose and fertile, is incompletely burned, with sparse vegetation, and is close to dry land crops, at the foot of the mountain and low-lying valley area. Its occurrence and infestation is closely related to the climate. A dry fall and warm winter will favor the nymph overwintering and development. In this case, the damage to the seedlings of the fast-growing eucalyptus will be serious. The adults feed actively during their mating and oviposition period in July and August. They will damage the seedlings that are planted in May and June. *B. portentosus* prefer sour, sweet, fragrant and musty food.

Prevention and treatment

1. Silviculture. (1) Complete burning before afforestation to get rid of the food sources. (2) Destroy nymph

T. portentosus adult (Wang Jijian)

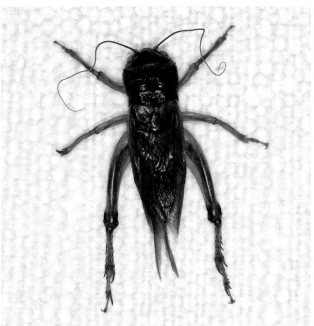

T. portentosus adult male (Li Zhenyu)

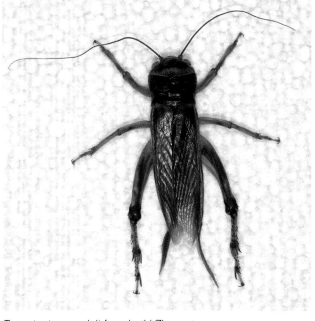

T. portentosus adult female (Li Zhenyu)

burrows and their habitats by whole or strip reclamation. (3) Choose healthy and highly lignified seedlings to plant. During afforestation, apply sufficient fertilizer before planting and re-apply accordingly to increase seedlings growth rate and accelerate seedlings lignification process. These procedures will further reduce pest damage

2. Mechanical control. Monitor pest populations in nurseries and newly planted stands to detect *T. portentosus* and treat early. (1) Catch them at night. People live close to small fields may catch the foraging nymphs and adults after dark using a flashlight and insect net. For large fields, people may set up black light traps to trap adults. (2) Fill burrows with water. During daylight, look for burrows and push aside the pile of loose soil that is used to seal burrows, then place a few drops kerosene and fill the burrow with water using a watering can until water overflows. Then look out the pest that escapes from the burrow. Kill the pest immediately. (3) Weed pile trapping. Take advantage of the pest's habit of preference of living in weed piles. Make piles of weeds in the field. The diameter and thickness of a weed pile is about 25, and 10-20 cm, respectively. The density of the piles can be determined according to the density of the *T. portentosus* and the pattern of the plantings. Generally, weed piles are arranged in lines that are 5 m apart between lines and 3 m between two piles within a line. Collect the pests and kill them the next day. It is more effective if a poison bait is deployed under the weed pile or create burrows using a 3-5 cm diameter stick.

3. Protect natural enemies. Natural enemies of *T. portentosus* include parasitic wasps, scoliids, tachinids, ground beetles, spiders, toads, frogs, lizards and birds. They should be protected.

4. Chemical control. (1) Fill burrows with liquid pesticides. Fill the burrow with cypermethrin 25 g/L at 3,000× or 4,000× of 2.5% deltamethrin, then seal the burrow to fumigate or poison the adults and the nymphs. (2) Attract and kill with pesticides mixed with perfume or food. Mix 47% phoxim, 90% of Dipterex crystal and water at 2:1:10 ratio, then add100 parts of stir-fried wheat bran or rice bran. Add some sugar, vinegar or beer; or add minced pumpkin and greens and mix well. Place 5 g mixtures every 5-10 m in the *T. portentosus* infested field. Apply baits for a few days when no rain is expected. (3) Apply insecticide dust. Mix l,000 mL 47% phoxim solution with 50 kg talcum powder, and then apply the mixture using a duster or by hands in the infested fields at dusk when the surface is dry and there is no rain at night. Apply 7.5-12.5 kg mixture per hectare. The pest will be killed by ingesting the dust when rubbing their mouthparts with their forelegs.

References Zhou Zhongming et al.1983, Beijing Forestry College 1985, Central South Forestry College 1987, Xiao Gangrou 1992, Huang Yanqing 2000, Zhu Tianhui 2002, Hao Wei et al. 2003, Lin Shunde 2003, Pang Zhenghong 2006, Shi Weixiang 2007, Lan Yuguang 2008.

(Jiang Jing; Li Zhenyu; Qin Zebo; Zhao Tingkun)

2 *Gryllotalpa orientalis* Burmeister

Classification Orthoptera, Gryllotalpidae
Common name Oriental mole cricket

Importance Oriental mole cricket is an important underground pest in nursery and farmland. It feeds on sowed seeds and roots of seedlings. It can cut the roots and cause the seedlings to wither and die from water loss. In Yigong tea plantation of Tibet in 1983-1984, the seeded tea plantations suffered 95% mortality in seedlings and almost 20 ha tea plantations were damaged.

Distribution Beijing, Tianjin, Hebei, Shanxi, Inner Mongolia, Liaoning, Jilin, Heilongjiang, Shanghai, Jiangsu, Zhejiang, Anhui, Fujian, Jiangxi, Shandong, Henan, Hubei, Hunan, Guangdong, Guangxi, Hainan, Chongqing, Sichuan, Guizhou, Yunnan, Tibet, Shaanxi, Gansu, Qinghai, Ningxia, Xinjiang, Hong Kong, and Taiwan; Korea Peninsula, Japan, India, Sri Lanka, Central Asia, Southeast Asia, Hawaii, Oceania and Africa.

Hosts Oriental mole crickets are polyphagous. They feed on the seeds and young roots of seedlings of almost all plants. A total of 119 kinds of crops and herbaceous plants were recorded as hosts.

Morphological characteristics **Adults** They are small, 25-35 mm in length, pronotum 7.5-10 mm in length, forewing 8-13 mm in length; cylindrical and grayish brown, with abundant hairs; prothorax well-developed, the center with a dark red heart-shaped marking; front legs are big and fossorial, the tibia with four teeth, the base of tarsus has two teeth; hind tibia with 3-4 teeth; forewing short, hind wings long, fold into a tail that extends beyond the end of the abdomen. The longitudinal veins near the front edge of male forewings curl into a knot called sound file. Sound files on the left and right forewing rub each other to make chirping noise. The sound file tooth number, tooth pitch, and parameters of acoustic pulse are important identification characteristics. Abdomen spindle shape. **Eggs** Oval, ivory white, 2.8×1.5 mm in size; they gradually turn to yellowish brown, and turn to dark purple before hatch. **Nymphs** The shape is similar to adults, but they have wing buds. Length of the 1st to the 9th instars are: 4-6, 8.1, 10.3, 12.6, 14.4, 16.3, 18.8, 22.1, and 24.8 mm, respectively.

Biology and behavior *Gryllotalpa orientalis* occurs one generation a year in South China, and one generation every two years in North China. It overwinters as adults and nymphs. It takes 387-418 days to complete a generation in Zhengzhou. Adults lay eggs between April and October. Each female usually lays eggs 3-4 times with a total of 62-100 eggs. Females construct oval egg chambers of 12-30 cm deep in soil. The chambers are connected. Each egg chamber contains 30-58 eggs. Egg incubation period is 22.4 (15-28) days. The total nymph development period is 130-335 days. The duration for the 1st to the 9th instar is: 17.6, 8.83, 9.2, 10.5, 12.9, 17.2, 32.2, 243.1, and 47.8 days, respectively. Those overwintering nymphs need 198-229 days. Nymphs gradually emerge into adults. Only a small portion of the adults can lay eggs in the same year after emergence. Majority of the adults lay eggs in May and June in the next year. The longevity of adults ranges

Dorsal view of *G. orientalis* adult (Zhang Runzhi)

Ventral view of *G. orientalis* adult (Zhang Runzhi)

from 8 to 12 months. When the soil temperature falls to 11-12°C in early November, the adults and nymphs move to 40-60 cm (maximum 82 cm) depth in soil to overwinter. When the soil temperature rises to 5°C in the next February, mole crickets begin to move upwards. They begin to feed when the temperature rises to over 10°C in early March. The period between early April and late May is when heavy damage occurs. They damage the seedlings from June to August, especially 2-3 days after rain or irrigation. Newly emerged adults and nymphs also infest seedlings in the fall. They overwinter in November in soil. Oriental mole cricket has a high requirement for water content in soil. It needs the water content to be above 22%. Eggs are often laid in soil with high water level adjacent to paddy fields, pools, lakes, rivers and ditches; thus, it often occurs in these areas.

Prevention and treatment

1. Conservation of natural enemies. Frogs and about 60 species of birds are recorded feed on *G. orientalis*.

2. *Gryllotalpa orientalis* is attracted to organic matter. Avoid using organic fertilizer.

3. Using light traps to attract and kill adults during adult period.

4. Apply toxic baits to the soil to kill *G. orientalis*. The bait matrix can be cooked millet hulls or wheat bran. Active ingredient can be 15% chlorpyrifos, phoxim, or 5% carbosulfan.

5. Using treated seeds. Mixing 0.5 kg of 50% imida-

Hind tibia of adult *G. orientalis* (Xu Gongtian)

cloprid and monosultap mixture with 250-500 kg seeds.

References Miao Chunsheng, Cui Jingyue, Zhang Hui 1966; Editorial board of the handbook of forest protection 1971; Wu Juwen 1972; Wu Fuzhen and Gao Zhaoning 1978; Wei Hongjun 1979; Yang Xiuyuan and Wujian 1981; Zhao Xiufu 1981; Cui Guangcheng 1987; Wu Juwen and Chu Huayuan 1987; Wei Hongjun, Zhang Zhiliang, Wang Yinchang 1989; Kang Le 1993; Yin Haisheng and Liu Xianwei 1995; Cui Jingyue, Li Guangwu, Li Zhongxiu 1996; Zhu Shude and Lu Ziqiang 1996; Lei Chaoliang and Zhou Zhibo 1998; Fang Zhigang and Wu Hong 2001.

(Wu Juwen)

Gryllotalpa unispina Saussure

Classification Orthoptera, Gryllotalpidae
Synonym *Gryllotalpa manschurei* Shiraki
Common name Mongolia mole cricket

Importance *Gryllotalpa unispina* is an important underground pest in the nursery and farmland. It feeds on the sowed seeds and seedlings tender roots. It crawls through soil that causes seedlings roots breaking and fibrous roots dehydration, which further leads to seedlings withering. In the early 1960s, its occurrence area is up to 333.33×10^4 ha, and the average pest density is up to 18,510 (4,095-56,610) per hectare. The percentage of damaged seedlings was 10%-20%; in the worst case it could be over 70%.

Distribution Beijing, Tianjin, Hebei, Shanxi, Inner Mongolia, Liaoning, Jilin, Heilongjiang, Jiangsu, Zhejiang, Anhui, Jiangxi, Shandong, Henan, Hubei, Tibet, Shaanxi, Gansu, Qinghai, Ningxia, Xinjiang; Mongolia, Turkey, Russia. It is primarily found in areas north of Huai River in China.

Hosts *Gryllotalpa unispina* is a polyphagous pest, it can feed on almost all the plants in the nursery and farmland, as well as the roots of weeds and seedlings on waste land. Totally, over sixty plant species have been recorded as its hosts, which include: *Larix gmelinii*, *Pinus armandii*, *P. elliottii*, *P. koraiensis*, *P. massoniana*, *P. sylvestris* var. *mongolica*, *P. tabulaeformis*, *Cupressus funebris*, *Ulmus* spp., *Morus alba*, *Cannabis sativa*, *Pterocarya stenoptera*, *Beta vulgaris*, *Spinacia oleracea*, *Camellia sinensis*, *Corchorus* spp., *Gossypium hirsutum*, *Cucumis sativus*, *Populus* spp., *Salix* spp., *Amygdalus persica*, *Malus pumila*, *Pyrus* spp., *Robinia pseudoacacia*, Chenopodiaceae, Cucurbitaceae, Brassicaceae, Fabaceae, *Angelica sinensis*, *Fraxinus mandschurica*, *Sesamu indicum*, Vitaceae, Apiaceae, Solanaceae, Convolvulaceae, Poaceae, and Liliaceae.

Damage Adults and nymphs feed on sowed seeds in soil, especially those seeds just sprouted. They feed on seedlings, tender roots, and rootstocks. They are also active above the ground at night, chewing and cutting stems off. Adults and nymphs crawl through and dig tunnels in the soil, cutting seedlings roots and separating roots and soil, which causes seedlings withering and death due to water loss, and the occurrence of sparse breaks up. When it is serious, the field needs to be replanted.

Morphological characteristics **Adults** Length 38-42 mm (up to 66 mm); pronotum length 12-13 mm, and the forewing length 13-16 mm. The body is cylinder-shaped, brown, and covered with thin hairs. The antennae are short. The prothorax is well-developed. There is a heart-shaped dark red spot in the middle of the pronotum. The forelegs are thick and strong, digging type. There is an indent at the end of the outer ventral side of the front femur, the front tibia is broad and has four dactyls, and there are two dactyls on the base of the tarsi. The dorsal inner side of the hind tibia with or without a spine. The forewings are short, while the hind wings are long and fold into a tail shape that extends beyond the end of the abdomen. The longitudinal veins near the front edge of male forewings curl into a knot called sound file, sound files on the left and right forewing rub each other to make chirping noise. The sound file tooth number, tooth pitch, and parameters of acoustic pulse are important identification characteristics. The abdomen is near cylindrical, it has a pair of long cerci. **Eggs** Oval, yellowish white in color at the beginning and are 1.6-1.8 mm×1.3-1.4 mm in size. They gradually expand and change to yellow-brown color. They are dark grey before hatching and are up to 2.4-3.0 mm×1.5-1.7 mm in size. **Nymphs** They look like adults, but only have wing buds. There are 0-2 spines on the dorsal inner side of the hind tibia. The newly hatched nymph is 3.56 mm in length, and the length of the 1st to the 13rd instars are 4.84, 5.77, 6.68, 7.68, 8.17, 10.9, 14.2, 16.8, 22.1, 26.8, 32.0, 37.4, and 41.2 mm, respectively.

Biology and behavior *Gryllotalpa unispina*

G. unispina adult (Xu Gongtian)

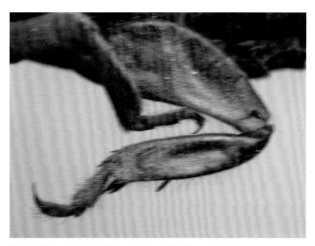

Hind tibiae of adult *G. unispina* (Xu Gongtian)

occurs one generation every three years. Adults and nymphs build chambers and overwinter in soil. The depth of chambers depends on the body size and soil temperature. It is usually 60-70 cm deep (up to 170 cm). Overwintering adults begin to lay eggs in mid- or late June next year. Eggs hatch into nymphs from mid- to late June. Nymphs develop to the 8th-9th instar in the first year and overwinter in deep soil in October or November. They start to be active and feed from early or mid-April in the second year; they continue to develop to the 12nd-13rd instar and overwinter again in the fall. They feed in the third spring and become mature in early or mid-August. After last molt, they become adults and overwinter. In the 4th year, adults mate from May to July, and subsequently between during June and August.

A complete generation takes about 1,131 days in Zhengzhou. The egg duration is 17.1 days (11-23 days), the nymphal stage lasts 735 days (692-817 days). The nymphal duration for the 1st to the 12th instar is: 2, 4, 7, 9, 14, 15, 19, 25, 36, 36, 29, and 38 days, respectively. The overwintering nymphal duration is about 35-283 days. The average life span of the adult is 378 days (278-451 days)

Gryllotalpa unispina prefers to inhabit in loose loam soil or sandy loam that is warm and humid and with rich humus. They are extremely active in Spring and Fall, they rest in soil during the daylight and are active on soil surface at night. Adults have strong phototaxis and flight capacity. *Gryllotalpa unispina* starts climbing up when the soil temperature at 10 cm depth in spring reaches 8°C. When the ground temperature reaches 10-12°C, especially after a spring rain, *Gryllotalpa unispina* is extremely active in the topsoil layer. They dig large numbers of tunnels and loosen up surface soil. The signs of tunnels are visible from the surface.

When the soil temperature reaches 20°C in May, adults come out of ground fly and chirp at night. This is their mating and ovipositing period. After mating males leave the females immediately, otherwise they will be preyed by females. A female builds 2-3 egg chambers that are connected to the tunnels in the depth of 13-33 cm. Females' egg-laying period is more than 50 days. They oviposit 4.7 times on average during their life (the range is 3-9 times). On average one female lays 103-368 (33 to 1,072) eggs. The newly hatched nymphs develop into the 3rd instar within 40-60 days in the egg chamber, then disperse and forage independently. They feed on seedlings sowed in fall. Nymphs begin to dig into deep soil to overwinter when the 10 cm depth soil temperature drops to below 8°C in late fall.

Prevention and treatment

In the 1950s to the 1960s, *Gryllotalpa unispina* occurred rampantly on slight saline-alkali lands in North China Plain area. After the 1970s, the slight saline-alkali lands were transformed into fertile farmlands, which destroyed the pest's suitable habitats. Combined with seed treatment for many years, *Gryllotalpa unispina* infestations are no longer common. The following methods can be adopted when *Gryllotalpa unispina* problem shows up in local areas.

1. Because this insect prefers soil with humus and manure, organic fertilizer should be avoided in its occurrence area.

2. Install light traps.

3. Apply toxic baits to the soil surface. The bait matrix can be cooked millet hulls or wheat bran. Active ingredient can be 15% chlorpyrifos, phoxim, or 5% carbosulfan.

4. In *Gryllotalpa unispina* occurrence area, seeds should be mixed with an insecticide (0.5 kg 50% imidacloprid and monosultap mixture can be mixed with 250-500 kg seeds) before use.

References   Miao Chunsheng, Cui Jingyue, Zhang Hui 1966; Wu Fuzhen and Gao Zhaoning 1978; Wei Hongjun 1979; Yang Xiuyuan and Wujian 1981; Zhang Zhiti, Zhang Lijun, Zhao Changbin et al. 1981; Wu Juwen and Chu Huayuan 1987; Wei Hongjun, Zhang Zhiliang, Wang Yinchang 1989; Kang Le 1993; Yin Haisheng and Liu Xianwei 1995; Cui Jingyue, Li Guangwu, Li Zhongxiu 1996; Lei Chaoliang and Zhou Zhibo 1998; Fang Zhigang and Wu Hong 2001.

(Wu Juwen)

Aularches miliaris (L.)

Classification	Orthoptera, Pyrgomorphidae
Synonym	*Aularches miliaris scabiosus* (F.)
Common name	Spotted locust

Distribution Guangdong, Guangxi, Hainan, Guizhou, Yunnan; India, Pakistan, Sri Lanka.

Hosts *Casuarina eguisetifolia*, *Tectona grandis*, *Cocos nucifera*, *Mangifera indica*, *Saccharum sinense*.

Morphological characteristics **Adults** Female 49-63 mm in length, male 32-57 mm in length; antennae black, dorsal surface of the head black-brown, eyes brownish red. There is a wide yellow stripe under the compound eye. The back of the pronotum brownish black; the tumor-like bulge on the front edge of the pronotum orange; and there is a wide vertical yellow stripe at the lower end of the pleurites. Forewings are yellow, scattered with roundish and reddish orange spots; abdomen black, the posterior border of each abdominal segment orange. **Eggs** Size 7.8 mm long and 2 mm wide. It is orange right after being laid, each egg mass has 40-60 eggs.

Biology and behavior *Aularches miliaris* occurs one generation a year, and it overwinters as eggs in soil. Eggs start hatching in mi-April and reach peak in early May. Nymphs have six instars. Adult period is from late July to late December. Females start laying eggs in October. Adults die in late October. The newly hatched nymphs prefer feeding on short *Casuarina eguisetifolia*. Their mobility is low, and they are gregarious. More than one hundred nymphs can be found in an aggregation. They prefer to hide in shady places when the temperature is high. The older nymphs and adults always bite and cut off in the middle of the twigs when they feed, which causes branches fall to the ground. Adult migration ability is weak. They oviposit at places where it is flat, and the soil is loose.

Prevention and treatment Please refer to the part of the *Ceracris nigricornis* Walker.

References Xiao Gangrou 1992; Zhang Zhizhong 1997.

(Li Zhenyu, Su Xing; Lin Siming; Wu Youchang)

A. miliaris adults
(Li Zhenyu)

5 *Ceracris kiangsu* Tsai

Classification	Orthoptera, Acrididae
Synonym	*Rammeacris kiangsu* (Tsai)
Common name	Yellow-spine bamboo locust

Distribution Jiangsu, Zhejiang, Anhui, Fujian, Jiangxi, Hubei, Hunan, Guangdong, Guangxi, Sichuan, Guizhou, Yunnan.

Hosts They feed on *Phyllostachys* spp., especially *Phyllostachys edulis*, as well as on *Bambusa* spp., primarily on *Bambusa textilis*. They also feed on *Zea mays*, *Oryza sativa*, *Imperata cylindrical*, *Trachycarpus fortune*, and nearly one hundred weed species.

Damage The damage will cause *Phyllostachys edulis* to wither and die. Water will be stored in bamboo stalk and become very stinky. Damaged bamboo will loss all its value and couldn't be burned at all.

Morphological characteristics **Adults** Male 27.5-36.2 mm long, female 29.8-41.4 mm long; green or yellow; gena triangular. There is a yellow vertical line from the top of the frons to the middle of the pronotum, which gradually becomes wider backward. Antennae filiform, dark brown, and the last two segments light-yellow. Compound eyes oval, black. Wings reach beyond the abdomen. Male wings are 24.5-25.6 cm in length, while the female wings are 29.5-34.5 cm in length. The anterior border of the forewing and central part are dark brown; hind femur yellowish green, there are two blackish blue loops near the tibia, there are orderly inverted 'Y' shape brown striations in the middle of the hind femur. The hind tibia is blackish blue; it has two rows of spines; their base is light-yellow and the end is jet black in color. The abdomen has eleven segments, the middle of the tergites is light-yellow and the ventral side is yellow in color. **Eggs** Oval, one end is slightly sharp, the middle is slightly crooked. The long diameter is 6.2-8.5 mm and the short diameter is 1.9-2.6 mm. Brownish yellow in color, with honeycomb shaped marking. Egg sack round-pocket shaped, the lower end is slightly thicker, the long diameter is 18-30 mm and the short diameter is 6-8 mm, it is brown in color. **Nymphs** Nymphs have five instars. The newly hatched nymphs are light-yellow, and then gradually change into linen color that mixed with green, yellow, and brown. The top end of the antennae is light-yellow. The 2nd instar nymph is yellow, and the dorsal middle line of the pronotum and the tergites is more yellowish in color. Most of the 3rd-5th instar nymphs are yellowish black in color, and the dorsal middle line is bright yellow. There is a black longitudinal strip under the dorsal line, below which is yellow. The older nymphs are fresh green before eclosion.

Biology and behavior *Ceracris kiangsu* occurs one generation a year and overwinters as eggs in egg sacs that are 1-2 cm deep in soil. The overwintering eggs start hatching from late May to early June in Yuhang, Zhejiang; from early to mid-May in Leiyang, Hunan; and from mid-April in Guangning, Guangdong. Nymphs have five instars. Adults start mating 10 days after eclosion. They keep on feeding for around 20 days after mating. In Yuhang, females start laying eggs in early August, and peak in mid-August, the egg-laying period can extend to October or November. In Leiyang, females start laying

C. kiangsu nymphs (edited by Zhang Runzhi)

C. kiangsu adults (edited by Zhang Runzhi)

C.kiangsu maiting pair (Xu Tiansen)

eggs in late July, and peak in mid-August. In Guangning, females start laying eggs in late July. The life span of adults is as the following: male is 31-96 days and female is 68-106 days in Yuhang, male is 54-56 days and female is 50-84 days in Leiyang, male is 69-91 days and female is 78-112 days in Guangning.

Prevention and treatment

1. Protect natural enemies. *Ceracris kiangsu* has many natural enemies. The predatory natural enemies of eggs include *Epicauta ruficeps*, those prey on nymphs and adults include *Polyrhachis dives*, *Araneus mitificus*, *Argiope bruennichii*, *Oxyopex lineatipes*, *Pisaura* sp., asilid, katydid, *Isaria cicadae*, *Hierodula patellifera*. The predatory birds include *Cuculus canorus*, swallow, *Garrulax sannio*, thrush, *Garrulax canorus*, *Garrulax perspicillatus*, *Bambusicola thoracica*, crow, *Corvus* sp. Parasitical natural enemies include *Teienomus* spp. which target on eggs, and *Carcelia* sp., *Exorista* sp., *Steinernema glaseri* that target on nymphs and adults; and bacterium *Entomophthora grylli*. *Beauveria bassiana* is effective to control this pest.

2. Dig out eggs by hand. After the occurrence of *Rammeacris kiangsu* in the previous year, look for dead adult bodies as a clue to find female oviposition place, and then dig out eggs before April.

C. kiangsu eggs and egg sacs (Xu Tiansen)

3. Eradicate the newly emerged nymphs. Timely application of 2.5% deltamethrin at 3,000× or cypermethrin at 2,000× will kill most of the young nymphs.

4. Trap and kill pest with human urine. When adults occur, 5 kg human urine that mixed with 50 g of cypermethrin at 25 g/L is enough to treat 0.13-0.2 ha bamboo forest infested with *Ceracris kiangsu*.

References Zhen Zheming and Xia Kaoling 1998; Xu Tiansen and Wan Haojie 2004; Xu Tiansen, Wan Haojie, and Yu Caizhu 2008.

(Xu Tiansen, Wang Haojie)

6 *Ceracris nigricornis* Walker
Classification Orthoptera, Acrididae

Distribution Jiangsu, Zhejiang, Anhui, Fujian, Jiangxi, Hubei, Hunan, Guangdong, Guangxi, Sichuan, Guizhou, Yunnan.

Hosts *Phyllostachys bissetii*, *P. bambusoides f. shouzhu*, *P. flexuosa*, *P. nidular* 'Smoothsheath', *P. makinoi*, *P. sulphurea* 'Viridis', *P. glauca*, *P. iridescens*, *P. nuda*, *Pleioblastus amarus*, *Pleioblastus maculosoides*, *Pleioblastus juxianensis*, *Bambusa textilis*, *Bambusa chungii*, *Bambusa multiplex*, *Dendrocalamus latiflorus*, *Dendrocalamus*, corn, sorghum and other crops and herbaceous plants.

Damage Adults and nymphs feed on bamboo leaves. Young nymphs cause notches on leaves. Large larvae and adults can consume whole leaves and devour all bamboo leaves in an area.

Morphological characteristics **Adults** Males 25.5-32.2 mm long, females 30.8-37.5 mm long; bright green to dark green; top of frons triangular, with coarse punctures; antennae filiform, the last 2-4 segments light-yellow; eyes prominent, oval, dark black; pronotum flat, with coarse punctures, there is a wide longitudinal band from the head to base of the wings. There is not a yellow longitudinal stripe. This is the difference from *Rammeacris kiangsu*. Gena, pleurites of the pronotum, and borders of forewings black. Wings longer than the abdomen; back of the abdomen purplish black, ventral surface yellow; subgenital plate of males short conical shape, the tip obtuse; valvulae of the females thick, short, the tip hook like. **Eggs** Elongated oval, light-yellowish brown, 4.5-6.5 by 1.3-1.9 mm in size; egg mass sack shape, cylindrical, 16-24 by 5-7 mm, dull brown; eggs are separated by spongy material inside egg sacs. **Nymphs** Young nymphs 8-10 mm long; back of thorax and abdomen yellowish white, without black spots on

C. nigricornis adult (edited by Zhang Runzhi)

Damage by *C. nigricornis* (edited by Zhang Runzhi)

C. nigricornis nymphs (edited by Zhang Runzhi)

C. nigricornis adult (edited by Zhang Runzhi)

back of thorax. Length of mature larvae 22-30 mm. Vertex sharp; antennae 20 segmented; back of the body flat and wide, place green.

Biology and behavior *Ceracris nigricornis* occurs one generation a year. It overwinters as eggs in 1-2 mm deep soil. Eggs hatch in mid-April in northern Guangdong, late April in northern Fujian, early and mid-May in Hunan, late May and early June in Zhejiang. Nymphs feed during late April and early August. They mature after 55-65 days. Adults emerge during late June and late July. They lay eggs from early October to early November. Adults disappear in early December. They are not gregarious. They often occur together with *Rammeacris kiangsu*.

Prevention and treatment Protect natural enemies. These include predatory enemies such as: *Epicauta ruficeps*, *Formica japonica*, *Polyrhachis dives*, *Isaria cicadae*, *Hierodula patellifera*; spiders (*Argiope bruennichii*, *Oxyopex lineatipes*, *Pisaura* sp), and birds (*Cissa ergtyrorhyncha*, swallow, *Garrulax canorus*, *Bambusicola thoracica*. Parasitoids include: *Teienomus* spp. (on eggs), *Carcelia sp.*, and *Exorista sp.* (on adults and nymphs). There is an entomopathogen, *Entomophthora grylli*. *Beauveria bassiana* is also effective.

References Xu Tiansen and Wang Haojie 2004.

(Xu Tiansen and Wang Haojie)

7 *Hieroglyphus tonkinensis* Bolívar
Classification　Orthoptera, Acrididae

Distribution　Zhejiang, Fujian, Jiangxi, Hubei, Hunan, Guangdong, Guangxi, Sichuan, Guizhou, Taiwan; India, Vietnam, Thailand.

Hosts　It infests many bamboo species such as *Sinobambusa tootsik*, *Bambusa eutuldoides*, *Bambusa multiplex* 'Fernleaf', *Bambusa textilis*, *B. dissemulator* var. *hispida*, *B. vulgaris* 'Vittata', *B. intermedia*, *B. pachinensis*, *B. chungii*, *Dendrocalamopsis oldhami*, *Phyllostachys heteroclada* var. *pubescens*, *P. sulphurea* 'Viridis', *P. glauca*, *P. dulcis*, *P. edulis* f. *gimmei*, *P. meyeri*, *P. makinoi*, *P. heteroclada*, *P. aureosulcata* f. *pekinensis*, *Pleioblastus amarus*, It also infests crops including *Oryza sativa*, *Zea mays*, *Saccharum sinense*, *Livistona chinensis*, etc.

Damage　Adults and nymphs feed on leaves and can devour all leaves on a plant. Feeding may kill the whole plant and reduce bamboo shoot yield next year and quality of the new bamboo plant.

Morphological characteristics　**Adults** Females 29.5-36.8 mm long, females 39.8-50.5 mm long; grassy green, head bluish green; eyes elliptical, projected, yellowish brown; antennae filiform, 28 segmented, base light-yellow, the last 3-5 segments paler in color; pronotum, pleurites grassy green; pronotum depressed in the middle, saddle shaped from the lateral view; meso- and metathorax yellowish green; wings well developed, reaching or surpassing the end of the abdomen; base of the forewing green, its tip yellowish brown; front and middle legs pale green, femora tip, tibia bluish green; hind legs light-yellowish green, tibiae blue, with two rows of spines, each row with 10 spines. **Eggs** Elongated elliptical, slightly curved; 4.8-6.5 by 1.1-1.3 mm in size;

Feeding by *H. tonkinensis* (edited by Zhang Runzhi)

H. tonkinensis adult male (edited by Zhang Runzhi)

H. tonkinensis adult female (edited by Zhang Runzhi)

Damage by *H. tonkinensis* (edited by Zhang Runzhi)

yellow at beginning, brown later; egg mass elongated elliptical, 14-24 mm long, its bottom 8.6-10.5 mm wide and the top 5.5-6.8 mm wide. **Nymphs** Young nymphs 6-7 mm long, brown; there are 6-7 instars; their length is 6-8, 7-11, 9-12, 13-16, 16-21, 22-26, and 28-34 mm, respectively.

Biology and behavior *Hieroglyphus tonkinensis* occurs one generation a year. It overwinters as eggs 1-2 cm below soil surface. Nymphs hatch in mid- and late May in Zhejiang. The peak is in late May and early June. Eggs hatch between mid-April and early May. Nymphs feed for 50-60 days. Adults emerge in July in Zhejiang, June to August in Guangzhou. Adults feed for 10-20 days. They lay eggs from mid-August in Zhejiang and disappear in early September. In Guangdong, adults lay eggs between July and mid-September.

Prevention and treatment

Predatory enemies include *Camponotus japonicus*, *Polyrhachis dives*, *Hierodula patellifera*, *Cosmolestes* sp., *Argiope bruennichii*, and *Oxyopex lineatipes*. *Beauveria bassiana* is effective to control this pest.

References Xu Tiansen and Wang Haojie 2004.

(Xu Tiansen and Wang Haojie)

8 *Chondracris rosea* (De Geer)

Classification Orthoptera, Acrididae

Synonyms *Chondracris rosea brunneri* Uvarov, *Chondracris rosea rosea* (De Geer), *Gryllus flavicornis* F., *Cyrtacanthacris lutescens* Walker

C. rosea adult female (left) and adult male (right) (edited by Zhang Runzhi)

C. rosea adults (edited by Zhang Runzhi)

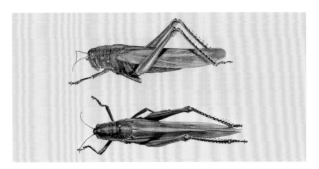

Ventral and dorsal view of C. rosea (Li Zhenyu)

Distribution Beijing, Hebei, Inner Mongolia, Jiangsu, Zhejiang, Fujian, Jiangxi, Shandong, Hubei, Hunan, Guangdong, Guangxi, Hainan, Guizhou, Yunnan, Shaanxi, Taiwan; Myanmar, India, Sri Lanka, Japan, Indonesia, Nepal, Vietnam, Korea Peninsula.

Hosts *Casuarina equisetifolia*, *Terminalia catappa*, *Robinia pseudoacacia*, *Tectona grandis*, *Dalbergia balansae*, *Acacia richii*, *Phyllostachys pubescens*, *Citrus reticulata*, *Trachycarpus fortunei*, *Theobroma cacao*, *Diospyros kaki*, *Vernicia fordii*, *Livistona chinensis*, *Eucalyptous citriodora*, *Mangifera indica*, *Camellia slnensis*, *Dimocarpus longgana*, *Sapium sebiferum*, *Agave sisalana*, *Gossypium hirsutum*, *Zea mays*, *Saccharum sinense*, *Oryza sativa*, *Arachis hypogaea*.

Morphological characteristics **Adults** Females 48-95.3 mm long, males 48-59.3 mm long; bright green and with yellow tint, base of hindwing rose color; antennae filiform, 28 segmented, often pass the posterior border of the pronotum; mid-ridge of the pronotum high, the three transverse grooves obvious and evenly spaced; the ventral projection of prothorax long and conical, reaching mesothorax; wings well developed, reaching the middle of hind tibiae; coxae and femora of front and middle legs green, tibiae and tarsi purplish red; inner side of hind femora in females yellow, tibiae red, with two rows of spines; base of tibiae spines yellow, their tip black.

Biology and behavior *Chondracris rosea* occurs one generation a year. It overwinters as eggs in soil. Eggs hatch from mid-April in Zhanjiang (Guangdong province). The 1st and 2nd nymphs are strongly gregarious. Hundreds or thousands of them may gather on a branch. Adults do not aggregate and do not migrate in groups. They mate multiple times. Eggs are laid in young forests with low canopy coverage, edge of forests, and open land between forests. During October when females are pregnant, they often sunbathe on south facing slopes. They can be easily captured by hands during this period due to low temperature. Eggs are laid in loose soil that is not susceptible to water accumulation.

Prevention and treatment

1. Plant eucalyptus trees and *Pinus elliottii* where the pest is prevalent in southern China.

2. Manual control by catching them with hands.

3. Chemical control. Spray 45% profenofos + phoxim at 1,000-1,500×, or 25 g/L lamda-cyhalothrin at 1,000-1,500× to control nymphs on the ground. Use fogging to control nymphs that on the trees.

References Xiao Gangrou (ed.) 1992, Zhang Zhizhong 1997, You Qijing 2006.

(Liang Hongzhu, Li Zhenyu, Huang Shaobin, Suxing, Yi Xiangdong)

9 *Phlaeoba angustidorsis* Bolívar

Classification Orthoptera, Acrididae

Distribution Jiangsu, Zhejiang, Anhui, Fujian, Jiangxi, Hubei, Hunan, Guangdong, Guangxi, Hainan, Sichuan, Guizhou, Yunnan, Taiwan; India.

Hosts *Phlaeoba angustidorsis* mainly causes damage to *Phyllostachys heteroclada* var. *pubescens*, *P. sulphurea* 'Viridis', *P. glauca*, *P. iridescens*, *P. vivax*, *P. praecox*, *P. bambusoides*, *P. heteroclada*, *Pleioblastus amarus*, and *Pleioblastus* spp. It also feeds on herbaceous plants.

Damage Nymphs and adults feed on bamboo leaves. The insects often consume all bamboo leaves when their population density is high.

Morphological characteristics **Adults** Females 26.5-30.5 mm in length; yellow. Forehead is convex. Antennae are sword-shaped, 20 segmented; the 3rd to the 10th segment slight triangle-shaped and grayish yellow, the others cylinder-shaped and black, the end flesh yellow. The compound eyes are light emerald and have black spots. The median ridge and lateral ridges on pronotum are obvious. The area below the two lateral ridges grayish black. Wings do not cover the end of the abdomen. The last three segments of the abdomen are exposed. There is an oblique line made of a row of black spots on the two laterals of every two segments.

Males 19.5-22.5 mm in length, darker than females; the black spots on the compound eyes darker; eyes with paler emerald color than females. The area below the two lateral ridges of pronotum is black. The junction between hind femora and tibia is black.

Eggs Elongated oval; length 6.5-8.5 mm, width 1.8-2.3 mm; reddish brown. Eggs are enclosed in ovisacs. The ovisac is long cylinder-shaped, slightly curved, and the lower end is blunt round and slightly big, and dark brown in color.

Nymphs The newly hatched nymphs 6.9 mm in length, tawny, gradually becomes darker as they grow. Male nymphs have 4 instars, while female nymphs have 5 instars. Forehead is convex. Antennae are sword shaped. The number of antennae segments of the four instars are 10, 13, 17, 18 and 19, respectively. The 2nd instar has wing buds. Mature nymph forewing buds are 4.8-5.2 mm long and hind wing buds are 4.2-4.6 mm long.

Biology and behavior *Phlaeoba angustidorsis* occurs one generation a year and overwinters in egg stage in 2 cm deep soil. In Zhejiang, the overwintering eggs start hatching in early and mid-May, and finish hatching at the end of May. The average development period of each instar in males are 13.2, 10.7, 11.6 and 11.7 days, respectively. The total male nymph development period is 45-50 days. The average development period of each instar in females are 13.5, 10.7, 11.2, 11.2 and 14.8 days, respectively. The total female nymph development period is 48-63 days. Nymphs are seen from early May to late July. Adults emerge in mid- and late July. Females feed more than males. They mate and lay eggs about 15 days after emergence. After mating and laying eggs, the adults die in the end of October.

Prevention and treatment

Conserve natural enemies. The known predatory natural enemies include birds (*Garrulax canorus*, *Cissa ergtyrorhyncha*, *Bambusicola thoracica*), insects (*Isaria cicadae*, *Hierodula patellifera*, *Camponotus japonicus*, *Formica japonica*, *Haematoloecha nigrorufa*), and spiders (*Dolomedes sulfureus* and *Araneus mitificus*). A bacterium *Entomophthora grylli* can infect *Phlaeoba angustidorsis*. *Beauveria bassiana* is effective to control this pest.

Reference Xu Tiansen and Wang Haojie 2004.

(Xu Tiansen, Wang Haojie)

P. angustidorsis adult (Xu Tiansen)

10 *Ramulus chongxinense* (Chen et He)

Classification Phasmatodea, Phasmatidae
Synonym *Baculum chongxinense* Chen et He

Importance *Ramulus chongxinense* has been an important pest since the 1980's in forests in Chongxin and Huating counties of Gansu province. They affect *Quercus liaotungensis*, *Capinus cordata*, and more than > 20 species of trees. About 700 ha forests were infested in 1988. Each tree had 1,000-5,000 *B. chongxinense*. Outbreaks occurred about every two years since then. In 1998, the infestation area reached 2,667 ha with 2,133 ha suffered significant damage.

Distribution Gansu.

Hosts The most preferred hosts are *C. cordata*, *Q. liaotungensis*, *Tilia tuan*, and *Crataegus pinnatifida*. The medium favorite hosts are: *C. cuneata*, *Prunus pseudocerasus*, *Pyrus berulaefolla*, *Lespedeza bicolor*, *Rosa xanthine*, and *Elaeagnus umbellate*. The least favorite hosts are: *Acer oliverianum*, *Populus davidiana*, and *Elaeagnus pungens*.

Morphological characteristics **Adults** Body stick-shaped, wingless; brown. Two different shades of brown occur among individuals. Darker individuals had white marks on lateral sides of the 1st to 6th abdominal segments. The lighter colored individuals do not have the white marks on the abdominal segments, more individuals belong to this color type. All individuals had dark and pale alternating markings on femora and tibiae.

Females 77 mm in length; mid-sized. Head oval, longer than pronotum. Eyes protruded. A short black tooth occurs between the eyes. The first antennal segment flattened and wide; the 2nd segment short tubular. Pronotum longer than width, with two grooves forming a cross. Mesonotum about 6 times as ong as the pronotum. Metathorax and the median segment together about 6/7 of the length of the mesonotum. The mesothorax, metathorax and the median segment are densely punctured. The median segment slightly wider than long. Front legs long, middle and hind legs shorter. Front femora longer than mesothorax; the exterior side of the femur with 3-4 teeth. The middle and hind femora much shorter than the front femora; with 1-2 small teeth at the interior end. At basal 1/3 of the middle femora, one tooth appears on each of the interior and exterior side. Hind tibiae with 2-3 teeth near the end. Abdomen much longer than the total length of the head and thorax, with dense punctures; the 5th to 7th segments were the longest. The

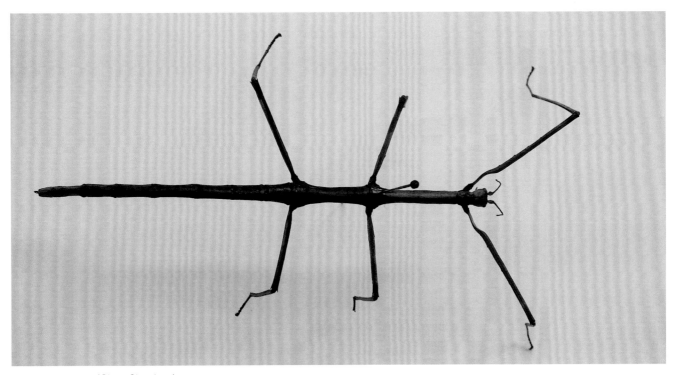

R. chongxinense (Chen Shuchun)

posterior border of the ventral surface of the 7th segment without an obvious medial projection. The anal segment longer than the 9th segment; its posterior border with a triangular notch. Epiproct short; operculum cymbiform. Cerci round.

Nymphs Young nymphs 12 mm in length. Mature nymphs with round heads and 8 segmented antennae. Pronotum with a longitudinal groove in the center; the anterior part had a "v" shaped groove; the middle part had a transverse groove. Mesonotum about 4 times as long as the pronotum and densely punctured. Metathorax shorter than the meso-thorax. The median segment longer than width. Abdomen densely punctured. The 9th dorsal surface with a longitudinal carina. Its posterior border with a triangular notch.

Eggs Brown; flattened; length 3.7 mm, width 2.0 mm. Surface with granules and ridges; the two sides with raised ridges.

Males Unknown.

Biology and behavior *Ramulus chongxinense* occurs one generation every two years. It lays eggs in fallen leaves and soil and overwinters in egg stage. Eggs begin to hatch in mid-April with peak hatching period between late April and early May. Nymphs have 3 instars, each instar lasts for one month. Adult period lasts for 90 days. Adults begin to lay eggs in early August with peak egg laying period occurs in mid-August. The eggs enter diapause the next year and hatch in the 3rd year.

When nymphs emerge from eggs, their heads and abdomen fold to the back. They immediately forage among the weeds and shrubs for food. As they grow, they begin to feed in lower part of the canopy and gradually move upward. This feeding habit renders aerial insecticide application ineffective. Nymphs and adults are gregarious. Food consumption by nymphs accounts for 10.4% and that by adults accounts for 89.6%. When disturbed by wind or vibration, the insects fall off from trees. Nymphs can shed injured legs and regenerate legs. New legs are near the length of the original legs. When at rest, front legs direct to the front resembling the antennae. Nymphs hang themselves on leaves or branches during molting. Newly molted nymphs are milky white, gradually turning to green. They tend to feed their shed skins after molting. Females lay individual eggs from the host trees. Eggs fall to the ground like rain. Each female lays an average of 42 eggs. The 2nd-3rd instars appear in mid- to late July and feed in lower and middle layer of the canopy. This is the most convenient period for control.

Prevention and treatment

1. Aerial ULV spray. This method is most effective for large area suppression of *R. chongxinense* outbreaks. The advantages are that it is efficient and easy to implement. The disadvantage is that it is expensive. Insecticides suitable for ULV application include 25g/L cyhalothrin or 2.5% deltamethrin. Control efficacy can be > 90%.

2. Tank spray of lamda-cyhalothrin, 2-5% deltamethrin at 25 g/L, cypermethrin, 40% chlorpyrifos, phoxim, 8,000 U/mL *B.t.*, and 25% Dimilin. This method is not suitable for treating natural forests where transportation and water source are problematic.

References Chen Shuchun 1994, Chen Peichang and Chen Shuchun 1997, Li Yan and Chen Shuchun 1998, Li Xiushan and Chen Jicai 2002, Chen Shuchun and He Yunheng 2008.

(Chen Shuchun, Xujin)

11 *Ramulus minutidentatus* (Chen et He)

Classification Phasmatodea, Phasmatidae
Synonym *Baculum minutidentatus* Chen et He

Importance Outbreaks of *R. minutidentatum* occurred in Tonghua, Jilin during 1998-2001. It caused mortality of young trees and dead branches of large trees.

Distribution Liaoning, Jilin.

Hosts *Ramulus minutidentatum* infests many broad-leaved trees including: *Quercus mongolica, Betula davurica, Tilia mandschurica, Prunus padus, Ulmus davidiana* var. *japonica, Prunus salicina, Corylus mandshurica, Crataegus pinnatifida* var. *major, Ulmus macrocarpa, Euonymus alatus, Schisandra chinensis, Carpinus cordata, Lonicera maackii, Acer mono, Laspedeza bicolor, Actinidia chinensis, Heracleum barbatum, Malus baccata,* and *Actinidia arguta.* In addition, it also infests young coniferous trees such as *Pinus sylvestris* var. *mongolica, P. koraiensis,* and *Larix gmelini.*

Morphological characteristics **Adults** Females 78 mm in length; large, stick-shaped; covered with dense fine granules. Green to brown. Head broader than pronotum; there are no spines between eyes; eyes small, round, prominent. Antennae about 1/3 length of the front femora. The two pairs of palpi brown. Basal two segments of the maxillary palpi short, the 3rd and 4th segments longer. Labial palpi 3 segmented, with the last segment longest. Pronotum longer than width, the median sulcus not reaching the posterior border; the transverse sulcus situated at the center. Mesonotum is 6 times as long as the pronotum. The posterior end wider, dorsal carina obvious. The metathorax and median segment together is 6/7 of the length of the mesothorax, with a median carina. Median segment length same as width. The legs short. The outer ridge of the abdomen with several small teeth. Median and hind femora shorter than the front femora; base of the median femora with small teeth; the end of the median and hind femora without obvious lobe-shaped structure, but with 3-5 small teeth. Abdomen longer than head and thorax combined, with a dorsal carina; the 3rd-6th segments larger than the other segments; the last 3 segments tectiform; the anal segment slightly longer than the 9th segment, its posterior border emarginated in the middle, dorsal surface with a longitudinal ridge. Epiproct triangular, its posterior border flush with the lateral lobes of the anal segments. Operculum cymbiform, surpassing epiproct, with sharp end, basal 1/3 with a median ridge. Cerci cylindrical, the end narrow, slightly surpassing the

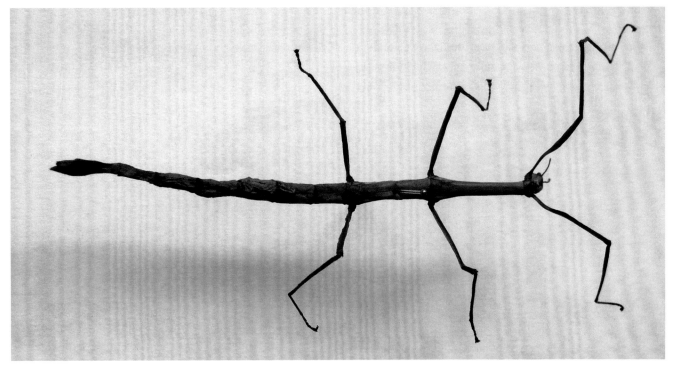

R. minutidentatum (Chen Shuchun)

abdomen. **Eggs** Length 3.1 mm, width 0.8 mm, height 1.83 mm; long and flattened, with dense fine granules; blackish brown. Operculum flat, with granules. Micropylar plate elliptical, edges clear, situated at the middle and lower part of the egg, about 1/3 length of the egg. Micropylar located at the lower border of the micropylar plate. The micropylar cup forming a plate-shaped ridge, a red tubercule located under the micropylar cup. Median line distinct, forming a ridge, reaching the posterior end. Ventral side of the egg convex longitudinally, its sides depressed.

The eggs, nymphs, and adults camouflage in the background very well. The colors of the nymphs vary with the plant leaf colors. The 4th-6th instars and adults have two color types: green and brown. The body color is similar to that of the branches and leaves.

Biology and behavior *Ramulus minutidentatum* occurs one generation in 3 years in Jilin. Eggs pass two winter periods (about 600 days) before hatching. Egg hatching starts in late April and until early June. Peak hatching period is from early May to late May. Nymphal period is about 80-85 days. Adults appear from late July until early September. They lay eggs during early August and mid-October. Those adults emerged in late August and early September die from cold weather before finishing egg laying.

When fed on elm, the 4th-6th instars consumed 0.02, 0.06, and 0.09 g food per day, respectively. Adult consumes 0.15 g per day. The 1st-3rd instars feed on grass and shrubs. From 4th instar, they feed on upper layer of the canopy. They migrate to shrubs and grasses in late September. Un-pregnant females can spread through gliding.

The insect is parthenogenetic. Each female lays 90-120 eggs, it deposits eggs through swinging the abdomen. Eggs can be laid a few centimeter or meters away. Eggs overwinter among dead leaves on ground.

The insect plays death when disturbed. The 1st-5th instar nymphs can regenerate limbs, but with decreasing capability as the instar increases. After two molts, the broken limbs can grow to normal size.

Temperature affects the insect's development and diapause. Eggs require two winters to before hatching. High temperature in winters can increase egg survival. High temperatures in spring and summer increase adult reproduction rate. *Ramulus minutidentatum* prefers moist environment. The insect is more abundant on northern slopes than on southern slopes.

Prevention and treatment

1. Silviculture. Plant mixed forests, using pest resistant varieties, remove dead leaves and branches in winter to reduce the number of eggs.

2. Shake the infested trees and kill the insects that falling to the ground.

3. Conserve natural enemies such as ants, praying mantis, spiders, predatory mites, lizards, and birds. Apply *Beauveria bassiana* during Spring and Summer.

4. Chemical control. During mid- and late June when *B. minutidentatum* was in 1st to 4th instars, spray fenvalerate at 2,000×, 2.5% deltamethrin at 2,000×, or 25 g/L lamda-cyhalothrin 1,000-1,500×. Fogging with emamectin during late June and early July.

References Chen Shuchun et al. 1994; Zhang Heng et al. 1995, 1996; Li Yan and Chen Shuchun 1998, Xia Meiyan and Li Ji 2001; Jiang Yu et al. 2002; Wang Guiqing and Zhou Changhong 2003; Chen Shuchun and He Yunheng 2008.

(Chen Shuchun, Xu Jin)

12 *Ramulus pingliense* (Chen et He)

Classification Phasmatodea, Phasmatidae
Synonym *Baculum intermedium* Chen et Wang

Importance *Ramulus pingliense* is a potential pest of fruit trees and shrubs.

Distribution Guangdong, Guangxi, Hubei, Sichuan, Guizhou, Shaanxi, Gansu.

Hosts *Mimosa sepiaria*, *Quercus* spp., *Broussonetia papyrifera*, *Firmiana platanifolia*, *Leucaena glauca*, *Pterocarya stenoptera*, *Rhus verniciflua*, *Malus pumila*.

Damage *Ramulus pingliense* feeds on leaves and young branches and may completely defoliate the affected branches.

Morphological characteristics **Adults** Females 95-100 mm in length. Head elliptical, without spines between eyes; eyes protruding, with one brown transverse stripe between the eyes. Antennae is 1/3 as long as the front femora, its first segment flat, wide, about 4 times as long as the 2nd segment. Pronotum trapezoid, with two grooves forming a cross; the transverse groove located at the center. Posterior of the mesothorax slightly widened. Metathorax and median segment together is about 4/5 length of the mesothorax; anterior part of the metathorax wider; median segment trapezoid. Front femora with 4-5 black teeth; base of median and hind femora with one tooth; the posterior end with several small teeth. Abdomen longer than the head and thorax together; its 5th segment is the longest and followed by the 4th segment; the anal segment longer than the 9th segment. The posterior border of the anal segment forming a triangular notch. Epiproct slightly surpasses the anal segment. Operculum cymbiform, dorsal surface with longitudinal ridge, slightly passes the epiproct. Cerci cylindrical, not reaching the end of the operculum.

Males 70-88 mm in length; stick shape; yellowish brown to dark brown, meso- and metathorax dark brown; lateral side with yellow longitudinal stripes, dorsal surface with a median ridge. Antennae surpass the front femora by 2/3 of its length; its first two segments paler in color. Legs slender. The posterior end of the 8th abdominal segment and the 9th segment slightly wider. Dorsal plate of the anal segment split into two lobes; the posterior border oblique, sharp at the end; subgenital plate not surpassing the 9th segment, sharp, with a median ridge. Cerci short, slightly curved at center.

Eggs Length 3.84 mm, width 0.59 mm, height 1.44 mm; long and flat; densely covered with granules; yellowish brown. Dorsal and ventral surface flat and straight. The dorsal surface with two longitudinal ridges. Operculum flat, with distinct edges. Below the operculum is a longitudinal ridge. Micropylar plate with a cup that is raised, median line distinct, very short. Ventral surface of the egg with longitudinal ridges.

Nymphs Males with long antennae; legs with colored

R. pingliense male adult (Wang Jijian)

R. pingliense female adult (Wang Jijian)

Damabe from *R. pingliense* (Wang Jijian)

spots. Females with short antennae, bluish green. The 10th abdominal segment of both sexes is black. Newly hatched nymphs greenish. Female nymphs gradually changing to greyish yellow or greyish green; male nymphs changing to dark green, with reddish color, and side of the body being grassy green.

Biology and behavior *Ramulus pingliense* occurs one generation a year in Guangxi. It overwinters as eggs in soil surface among the dead leaves. Eggs hatch during late March and early April. Peak hatching period is in early April. Adults emerge in early June. Peak egg laying period is in mid- and late June. Adults die from late July.

Eggs are scattered on moist soil surface under host trees. They are not easily distinguished from the dead leaves. Egg hatching date is affected by temperature and humidity. They usually hatch in early morning. The nymphs will migrate to host trees the same day after emergence. They rest on underside of branches with the heads oriented in the same direction of the branches. Their abdomen often elevated. Males use their hind legs to attach their body to the branches. They swing sideways. They are active after 19:00. Young nymphs feed on young leaves. Late instars can feed old leaves. When they molt, the nymphs consume most of the skins. Adults feed during the night and occasionally feed during the day. They feed whole young leaves. Each adult consumes 10 leaves per day. Damaged leaves show ladder-like notches. Feces blackish brown, slightly twisted, 5-6 mm in length, 0.8 mm in diameter. Mating occurs during the night, occasionally mating occurs during the day; each mating lasts for > 1 hour. The female to male ratio is 5:8. Deposited eggs fall to the ground similarly as the feces. Each female lays about 100 eggs. During peak egg laying period, a female can lay 15 eggs a day.

Prevention and treatment See "Prevention and treatment of phasmid insects" section (P89).

References Wang Jijian et al. 1993, Chen Shuchun and He Yunheng 2008.

(Wang Jijian, Chen Shuchun)

13 *Macellina digitata* Chen et He

Classification   Phasmatodea, Diapheromeridae

Importance   *Macellina digitata* is the only pest that can cause serious damage to *Baechea frutescens*.

Distribution   Guangdong, Guangxi.

Host   *Baechea frutescens*.

Damage   Young nymphs consume part of the leaves and are difficult to be found. When their density is high, they can completely defoliate *Baechea frutescens*.

Morphological characteristics   **Adults** Males brown; 50.6-54.2 mm in length. Top of head and area behind the eyes with rectangular brown stripes; Center of pronotum brown; lateral sides of the mesothorax and metathorax with dark brown longitudinal stripes. There are brown longitudinal stripes along the median ridge of the thorax and abdomen. The stripes become narrowed from the third segment of abdomen. Body slender; head is distinctly longer than the pronotum, its posterior border almost as wide as the pronotum; with a longitudinal furrow, the furrow becomes shallower towards the back. Antennae 19 segmented, about 1/2 of the length of the front femora. The pronotum is 1.8 times longer than wide; with a longitudinal furrow, center of the furrow deeply depressed. Mesothorax longer than metathorax and median segment combined, the back with a longitudinal ridge. Legs slender, front femora slightly longer than the hind femora; middle femora is the shortest. Median segment is rectangular, longer than wide. The first 5 abdominal segments are equal in length; the posterior border of the anal segment concave; subgenital plate short, the center round, its posterior border concave. There is a finger-like projection on each side under the eighth dorsal plate. Cerci slightly curved inwards, slender toward the end, obtuse, with many small teeth.

Females 76.5-78.0 mm in length, yellowish brown or grass green. Body thick; head smooth, with three shallow vertical grooves on vertex and behind the eyes. Antennae 19 segmented, 1/3 of the length of the front femora. Length of pronotum is 1.5 times of the width, with a distinct median furrow. The two lateral furrows interrupted at the middle. The transverse furrow slightly convex anteriorly. The vertical furrow behind the transverse furrow deeply depressed. There is a vertical ridge on back of the body from mesothorax to the end of the abdomen. Median segment about as wide as long. Among the abdominal segments, the 4th and 5th are the longest. Anal segment forming an arch, its posterior border truncate; epiproct triangular; subgenital plate slightly convex at center, spear shape, sharp, with two distinct borders, not reaching the anal segment. Cerci flat, sharp, directed posteriorly, distinctly surpassing the epiproct.

Eggs   Length 5.10 mm, width 0.78 mm, height 0.78 mm; slender; yellowish brown; irregularly reticulate. Operculum with projections. Micropylar plate long, spindle shape, light-yellow. Micropylar located at center of the micropylar plate; median line surpasses the posterior end. Posterior end concave, edges with projections.

Nymphs   First instar light yellowish green, antennae pale reddish brown, the end segment paler, leg color paler than the body color, eyes brown. Head with 4 whitish vertical stripes. The 2nd instar yellow or grassy green; head and

M. digitate nymph (Wang Jijian)

M. digitate habitat (Wang Jijian)

pronotum with markings similar to bottom of shoes. Body covered with scattered milky white spots. The 3rd instar females grassy green or yellowish brown, covered with white spots. The 3rd instar males light-yellowish brown; head with 4 dark brown stripes at middle of head, and anterior, middle and posterior of the eyes. The 4th instar females with yellowish green stripes at each side; dorsal surface and along the stripes with white spots. The 4th instar males with brown longitudinal stripes along each side from pronotum to the end of the abdomen; the lateral stripes on the mesothorax and metathorax with light-yellow spots.

Biology and behavior *Macellina digitata* occurs one generation per year in Bobai of Guangxi. It overwinters mostly as adults. The adults begin to mate in early June. The peak egg laying period is between mid-June and mid-July. No eggs are laid after July. The egg incubation period is 15-40 days. The peak hatching period is between late July and late August. The development time of the 1st-4th instars are: 12-16, 13-17, 17, and 17 (minimum) days, respectively. Some nymphs may overwinter and develop into adults next year. Male adults' life span is 270 days, female adults' life span is 280-300 days. Adults emerge in early September. Adults from last year die out in early October.

Macellina digitata are usually active during the night. They feed from a leaf tip toward the base. Adults rest in trees higher than the host during the day and move to host trees during the night to feed and mate. A male mates with one female during one night and may mate with several females over a few nights. Copulation lasts for more than one hour. If a male mates with a second female during one night, the mating will last for < 10 minutes. Young nymphs may wave sideways at rest. Nymphs do not play death.

Prevention and treatment See "Prevention and treatment of phasmid insects" section (P89).

References Wang Jijian et al. 1992; Chen Shuchun et al. 1993; Chen Shuchun and He Yunheng 2008.

(Wang Jijian)

14 *Micadina zhejiangensis* Chen et He

Classification Phasmatodea, Lonchodidae

Importance *Micadina zhejiangensis* emerged as a pest since 1985 in Taishun of Zhejiang. It infests evergreen broad-leaved trees. Severe infestations cause complete defoliation, weaken the tree vigor, and cause tree death.

Distribution Zhejiang, Fujian.

Hosts *Castanopsis eyrei*, *C. carlesii*, and other species in Fagaceae.

Morphological characteristics **Adults** Females 48 mm in length, stick-shaped. Green, forewing angular projection black, anal area of the hindwing rosy red. Head broad oval, smooth, broader than the pronotum; occiput elevated, with longitudinal stripes. Eyes oval, protruded. Antennae filiform, much longer than the front legs; its 1st segment flattened; the 2nd segment cylindrical, shorter than the 1st segment. Pronotum nearly rectangular, with two grooves forming a cross; the transverse groove located at 1/3 of the pronotum; the longitudinal groove not reaching the posterior border. Mesonotum slightly cylindrical, dorsal surface granulose; with a medial longitudinal carina and two lateral carinae. Forewing short, nearly rectangular; the median projection at the anterior border obtuse; outer margin truncated. Hindwing short, extends to the posterior border of the 3rd abdominal segment. Legs short; base of the front femora curved; without spines or teeth. The last 4 abdominal segments narrow. The 8th to 10th segments gradually extended. The anal segment longer than the 9th segment and distinctly longer than the 8th segment. The posterior half of the anal segment with a dorsal longitudinal carina; its posterior border with a "v" shaped notch. Epiproct surpasses the lateral lobe of the 10th segment. The operculum surpasses the 9th segment. Ovipositor valves surpass the base of the 10th segment. Cerci long cylindrical, surpass the end of the abdomen.

Males 37-46 mm in length. Dark brown; majority of the antennae, pronotum, forewing projection brown; femora green, their posterior end and the tibiae dull yellow. Head elliptical, Hindwing long, extends to the 6th or 7th segment. The 8th to 10th segments swollen. The 8th and the 10th segments are equal in length. The 10th segment almost vertical with the 9th segment; the posterior border forming a notch; subgenital plate concave, extending to the base of the 10th segment. Cerci long cylindrical, curved inward, the end slender.

Eggs Length 1.82 mm, width 1.33 mm, height 1.40 mm; barrel-shaped; greyish brown; reticulated. Operculum round, narrower than the capsule, its center with a round carina. The micropylar plate bottle shaped. The micropylar located near the bottom of the plate, with a "U" shaped micropylar cup; median line indistinct. Posterior end of the egg flat, with a few radiating ridges.

Nymphs There are 6 instars or occasionally 7 instars. The 1st instar thread shape; light-yellowish green. The

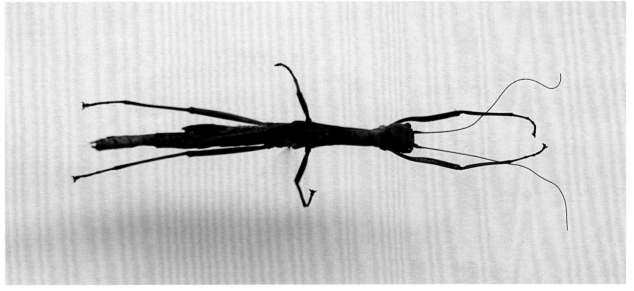

M. zhejiangensis adult (Chen Shuchun)

3rd instars possess hindwing buds and 4th instars possess forewing buds. The 5th to 7th instars gradually change to green color. Other characters except wings and genitalia similar to adults.

Morphological characteristics *Micadina zhejiangensis* occurs one generation a year. It overwinters as eggs in fallen leaves. Eggs hatch from late March and reach peak in early April. Adults mostly emerge in late June. They lay eggs from early July and die after late September. Egg survival best in moist environments. Egg hatching rates of 86.4% and 66.1% were reported under field and laboratory conditions, respectively. Eggs usually hatch during nighttime. Young nymphs reach to canopy a few hours after emergence. They can survive for 3-6 days without feeding. Nymphs do not migrate to ground once on trees. When disturbed, they move away. They usually feed individually with one insect per leaf. Young nymphs feed on leaf edges and cause round holes or notches. When at rest, the body aligns with the main vein on the back of leaves with the head facing the leaf base. They feed 3-6 times a day. Each feeding consumes 1 mm^2 to 127 mm^2 leaf area. The 1st to 4th instars usually feed on leaf edges. From 5th instars, they feed on whole leaves and barks of young branches and causing leaf wilt. The development time for each instar is 9-18 days. They cease feeding one day before molting. During molting, the head faces the ground. The molting time is between 13-22 minutes. Nymphs consume the shed skins after molting and start to feed on leaves about 10 hours after molting. Nymphs can regenerate broken legs. New legs can be regenerated after 3 molting. The nymphal period is between 74-91 days.

Adults are weak fliers; they cannot fly upward. They jump to lower canopy or to the ground when disturbed. Females are parthenogenetic. During outbreaks between 1996-1999, no males were found. Females produced eggs which hatched and developed into the 3rd generation in the laboratory. Females start to lay eggs 9-13 days after emergence. They lay eggs randomly which will fall to the ground. A female lays eggs at 2-hour intervals. Each day a female may lay 0.5-4 eggs. They stop egg-laying about 3-8 days before death. Each female can lay 69-174 eggs (average 118). The adult stage is the most damaging period. Each female consumes 749 mm^2 leaf area or 0.9 leaves a day. Maximum consumption is 2.2 leaves per day. Each female consumes an average of 58.7 leaves during the adult state. During its lifetime, a female *M. zhejiangensis* consumes 76.9 leaves. Adult lives for 46-127 days (average 65.2 days).

Prevention and treatment

1. Chemical control. Spray fenvalerate at 25 g/L lamda-cyhalothrin at 1,000-1,500×, 20% fenvalerate at 2,000×. Fogging with suitable insecticides in areas with gentle slopes. For large trees, inject tree trunk with 20% dinotefuran at 2×, 1 mL for each 1 m diameter. During Spring when eggs begin to hatch, spray 20% chlorantraniliprole suspension to the ground.

References Xiao Gangrou (ed.) 1992, Chen Shuchun 1994, Wu Hong 1995, Bao Qimin et al. 2000, Chen Shuchun and He Yunheng 2008.

(Chen Shuchun, Xujin)

15 *Lonchodes bobaiensis* (Chen)

Classification   Phasmatodea, Lonchodidae.
Synonyms   *Entoria bobaiensis* Chen, *Dixippus bobaiensis* (Chen)

Importance   *Lonchodes bobaiensis* is the only phasmid pest that can cause serious damage to both broad leaved and coniferous trees in China.

Distribution   Guangdong, Guangxi.

Hosts   *Castanopsis carlesii*, *C. hystrix*, *Pinus massoniana*.

Damage   Young nymphs feed on edges of young leaves grown from stumps. The 2nd and 3rd instar consume whole leaves. They can completely defoliate *Castanopsis* spp. stands together with other phasmid pests.

Morphological characteristics   **Adults**   Males 86.5-97.0 mm in length, more slender compared to females. Body bluish green. Head flat on top, with sparse granules. Eyes protruded outward, brown. Antennae filiform, much longer than the front legs. Pronotum slightly longer than wide, rectangular. Front femora with a row of fine teeth on the inner side. Mesothorax long, wider posteriorly. Middle and hind femora with two rows of fine teeth on the inner side. The trochanter, base and end of the femora, and tibiae brown.

Females 113-114 mm in length, yellowish brown. Occiput with three short median grooves. There are black stripes between eyes occipital margin. Antennae surpass the end of the front tibiae. Front femora longer than middle and hind femora, its posterior end with three short teeth on the inner surface. Hind femora longer than middle femora, shorter than the front femora; posterior end of the middle and hind femora with two rows of small teeth on the inner surface; the inner ridge at the base of the middle and hind tibiae lobed.

L. bobaiensis adult female (Wang Jijian)

L. bobaiensis nymph (Wang Jijian)

L. bobaiensis adult (Wang Jijian)

Damage by *L. bobaiensis* (Wang Jijian)

Nymphs Newly hatched nymphs 8-9 mm in length; light-yellow, turning to light green, bluish green, or yellowish green as they grow.

Biology and behavior The insect occurs one generation a year. It overwinters as eggs. Peak hatching period is between late April and early May. Peak mating and egg-laying period is in June and July. Nymphs undergo 5 molts and last for 40-48 days. The sex of the 5th instar can be differentiated. Adults last for 4-5 months. Males appear 2-3 days earlier than females. Males live for 22-74 days and females live for 26-132 days. They lay eggs 2-3 days after mating.

Eggs must hatch in a moisture environment. They hatch in the morning. One insect will reside on each branch. They rest on leaves. Young nymphs feed from tip to base of the *Pinus massoniana* needles, or feed from middle of the leaves and cut off the top part. If the hosts are *Castanopsis* spp., the nymphs would feed on leaf edges. Late instars and adults consume whole leaves. A female adult consumes 0.9 g food a day and 60 g food during its life time. A male adult consumes 1/3 of that of the female adult. A female adult excretes 48-60 fecal pellets a day and a male adult excretes 40 pellets a day. The pest population can be estimated based on the amount of fecal materials on the ground.

Prevention and treatment See "Prevention and treatment of phasmid insects" section (P89).

References Wang Jijian et al. 1990, Chen Shuchun and He Yunheng 2008.

(Wang Jijian)

16 *Necroscia flavescens* (Chen et Wang)

Classification Phasmatodea, Lonchodidae
Synonym *Aruanoidea flavescens* Chen et Wang

Importance *Necroscia flavescens* is an important pest of trees in the family Fagaceae.

Distribution Guangxi.

Hosts *Castanopsis carlesii*, *C. hystrix*.

Damage Older nymphs and adults consume whole leaves of the hosts. They can completely defoliate the *Castanopsis* spp. stands together with other phasmid pests.

Morphological characteristics **Adults** It is the most colorful phasmid insect in China. Body metallic yellow, eyes orange color; the first two antennae segments yellow, other segments brown with 9-10 pale rings; anal area of the hind wing milky white.

Females 75 mm in length. Stick-shaped. Head rectangular, slightly longer than wide, the dorsal surface flat, with longitudinal grooves; occiput with two tubercles. Antennae extend to the end of the abdomen; dorsal view of the 1st antennae segment rectangular, the 2nd segment cylindrical, shorter than the 1st segment. Pronotum rectangular, smooth, with two grooves forming a cross; the transverse groove situated in front of the center. Mesothorax is 2.85 times as long as prothorax, with one longitudinal ridge and two lateral ridges and 6 rows of granules. The front wing oval shape, front margin arched, apex obtuse. Hind wing extends to middle of the 7th abdominal segment. Legs without spines. The posterior border of the anal segment concave and forming two angles laterally. Epiproct with a middle ridge, surpassing the end of the anal segment. Operculum long, narrowed towards the posterior end; its lateral sides slightly directed upward; the posterior border forming a triangular notch. Cerci cylindrical.

Males 55 mm in length, slenderer than females. Antennae surpass the end of the abdomen. Hind wing extends to base of the 6th abdominal segment. Abdomen slender, the 8th and 9th segment widened, anal segment narrow. The 9th segment is the longest among the last three segments. Anal segment with a median ridge, the posterior border concave and forming two thick angles laterally, the inner margin with small teeth. Epiproct with a medi-

N. flavescens adult female (Wang Jijian)

N. flavescens adult male (Wang Jijian)

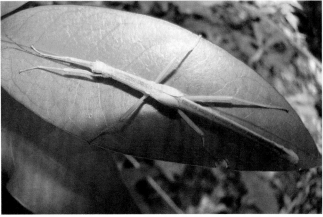

N. flavescens female nymph (Wang Jijian)　　　　N. flavescens female nymph (Wang Jijian)

Damage by N. flavescens (Wang Jijian)　　　　Damage by N. flavescens (Wang Jijian)

an ridge, shorter than the anal segment. Subgenital plate swollen, with sharp tip, extending to the end of the 9th segment. Cerci cylindrical, slightly curved inward, obtuse; distinctly surpassing the end of the abdomen.

Nymphs There are 6 instars. The color changes from light-yellow to yellow as they grow. The length of the 1st to 6th instars is: 12.0, 16.0, 20.0, 25.5, 31.0, and 38.2 mm, respectively.

Biology and behavior *Necroscia flavescens* occurs one generation a year. Eggs overwinter from August to March in the next year. They hatch in early March. Peak nymphal period is in April and May. Adults emerge in early June. Nymphs are absent in mid-June. Peak egg laying period is between June and mid-August. The development time of the 1st-6th instar was 7, 8, 26, 26, 23, 14 days, respectively. Adult live for 33-83 days.

Eggs hatch in the morning before sunrise. Young nymphs feed on leaves from stumps until all young leaves are nearly gone, then they migrate to tall trees. They are easily collected during this period. Their abundance is related to the previous year's egg numbers and distribution. The 1st and 2nd instar feed on very young, light-yellow young leaves. The 3rd instar feed on pale green, young leaves that are already expanded completely. They usually leave the main and side veins un-attacked. The 4th instar consume whole leaves except the main veins. Starting from 5th instar, they can consume whole leaves or move to another leaf before consuming one leaf.

Prevention and treatment See "Prevention and treatment of phasmid insects" section (P89).

References Wang Jijian and Chen Shuchun 1998, Chen Shuchun and He Yunheng 2008.

(Wang Jijian)

17 *Sinophasma brevipenne* Günther
Classification Phasmatodea, Lonchodidae

Importance An outbreak of *S. brevipenne* occurred in Wudang mountain in northern Guizhou in 1983. The middle and top of the mountain had highest populations. Complete defoliation occurred in the infestation center. From there outward, the canopy color shifted from yellow to green. Population density was > 300 per tree in heavily damaged areas with highest density of > 500 per tree. The forest floor had abundant feces. In 1977, a rare simultaneous outbreak of *S. brevipenne* and *S. maculicruralis* occurred in *Castanopsis hystrix* and *C. carlesi* forests in Guangxi. They caused complete defoliation in high population density areas.

Distribution Jiangxi, Guangxi, Guizhou.

Hosts *Castanopsis hystrix*, *C. carlesii*, *Quercus fabri*, *C. fargesii*, *C. seguinii* and other species in Fagaceae.

Morphological characteristics **Adults** Males 56 mm in length. Forewing scale shaped. Hindwing extends to middle of the 7th abdominal segment. The 8th abdominal segment wider posteriorly. The 9th abdominal segment is wider than the 8th abdominal segment and raised to semi-globe shape. Anal segment vertical; its first ¼ nearly quadrilateral; the last ¼ forming a broad petal shape; the anterior border slightly more curved inward than the posterior border. The dorsal surface of the anal segment has a median carina which is depressed near the posterior ¼ of the dorsal surface. Subgenital plate extends to posterior border of the 9th abdominal segment. The posterior border wavy; its two lateral angles obtuse, non-symmetric. Cerci cylindrical, directed down.

Females 69 mm in length, dark green. Body including wings with wide brown longitudinal stripes. Pronotum, mesonotum, apex of both wings, and posterior part of the last abdominal segment with obvious green longitudinal stripes. Hindwings reach the end of the 5th segment. Among the last 4 abdominal segments, the 7th abdominal segment is the longest, the 9th abdominal segment is the shortest. Epiproct surpasses the last abdominal segment. Operculum extends to the end of the anal segment; posterior end obtuse triangular; lateral surface keeled; egg laying apparatus extends to the end of the 9th abdominal segment.

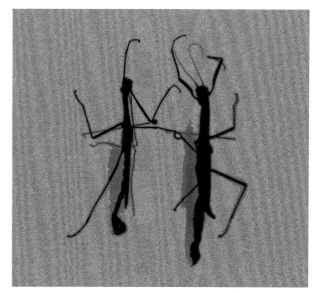

S. brevipenne adult male (left) and female (right) (Chen Shuchun)

Biology and behavior *Sinophasma brevipenne* occurs one generation a year. It overwinters as eggs. The eggs hatch in March in Guizhou. Nymphs have 5 instars. Nymphal period lasts for 1.5 months. Adults appear in May. Peak mating period is between mid-July and early August. Mating occurs at around 11:00. Mating lasts for a few minutes. Newly hatched nymphs do not feed much during the first two days. Feeding starts to increase significantly among 3rd-5th instars. The nymphal period accounts for 10% of their life span. Adults spring backwards when disturbed. They move down the trees at around 17:00. Adults play dead. They die out from late September.

Prevention and treatment Chemical control. Spray 45% profenofos and phoxim mixture, at 1,000-1,500×, or 25 g/L lamda-cyhalothrin at 3,000× to kill 1st to 4th instar nymphs. Fogging with cyhalothrin in forests with > 60% canopy coverage. Also see "Prevention and treatment of phasmid insects" section (P89).

References Chen Shuchun et al. 1985, Chen Shuchun 1986, Chen Shuchun and He Yunheng 2008.

(Chen Shuchun, Xu Jin)

18 *Sinophasma largum* Chen et Chen

Classification Phasmatodea, Lonchodidae

Importance *Sinophasma largum* is a main pest of *Castanopsis* spp.

Distribution Guangdong, Guangxi.

Hosts *Castanopsis carlesii*, *C. hystrix*.

Damage The newly hatched nymphs feed on the edges of new leaves. The 2nd and 3rd instars feed on whole young leaves. Large nymphs and adults may cause complete defoliation of *Castanopsis* trees.

Morphological characteristics **Adults** Female adults 58 mm in length; robust. The 7th abdominal sternite with a median projection at the end. The anal segment roof-shaped, its posterior border curved inward forming a triangular shape. Epiproct slightly pointed; operculum conical; extending to the 9th segment. The end of the egg laying apparatus pointed, extending to the anterior border of the anal segment. Cerci cylindrical, directed posteriorly and laterally.

Male adults 47 mm in length; slender. Head broad oval; convex; with a few longitudinal stripes. Eyes protruded. Antennae filiform, longer than the body length; the 1st segment slightly flattened and wide; the 2nd segment cylindrical; the 3rd segment longer than the 4th and 5th segments combined. Prothorax short, its center with two grooves forming a cross; the transverse groove located at the anterior 1/3 of the pronotum. Mesonotum long; densely punctured; the lateral sides with black markings. Forewings scale-shaped, similar to a square, with short black longitudinal stripes. Hindwings long; extend to the

S. largum adult female (Wang Jijian)

S. largum adult male (Wang Jijian)

S. largum female nymph (Wang Jijian)

S. largum eggs (Wang Jijian)

Damage by *S. largum* (Wang Jijian)

posterior border of the 6th abdominal segment. Among the legs, the hind legs are the longest and the middle legs are the shortest. The abdomen is longer than the head and thorax combined. The 7th dorsal plate wider in the back than in the front. The last three segments swollen. The 8th segment slightly shorter than the 9th segment which is nearly square-shaped and convex. The length of the 9th segment is similar to that of the anal segment and vertical to the anal segment. The anal segment is vertical or oblique; with longitudinal carina and truncated. Subgenital plate convex; with two unsymmetrical sharp teeth. Cerci cylindrical, surpass the anal segment.

Biology and behavior *Sinophasma largum* occurs one generation per year in Bobai of Guangxi. It overwinters as eggs in fallen leaves. The eggs start to hatch in early April; the peak hatching time is in mid-April. Adults emerge in early May. Mating starts in mid-May. Peak mating period is in late May. Peak egg laying period is between early June and mid-August. Adults disappear in late October.

The newly hatched nymphs rest for a short period, then orient themselves to bases of threes and crawl to the branch tips. Young nymphs feed on edges of young leaves and form notches. Later instars feed on whole young leaves, old leaves, and young shoots. Adults play dead; can fly for a short distance and exhibit weak phototaxy. Occasionally they fly to the light during the night. An individual consumes 20.8 g food during its lifetime. The percent consumption of the 1st-4th instar, 5th-6th instar, and adult stages are 3.13%, 21.95, and 74.92%, respectively.

Prevention and treatment See "Prevention and treatment of phasmid insects" section (P89).

References Chen Peichang, Chen Shuchun, Wang Jijian et al. 1998; Chen Shuchun and He Yunheng 2008.

(Wang Jijian)

19 *Sinophasma maculicruralis* Chen

Classification   Phasmatodea, Lonchodidae

Importance   *Sinophasma maculicruralis* feed on leaves of *Castanopsis* spp. and affect tree growth.

Distribution   Guangxi.

Hosts   *Castanopsis carlesii, C. hystrix*.

Damage   Newly hatched nymphs feed on edges of young leaves grown from tree trunks. The 2nd and 3rd instars feed on whole young leaves. Large nymphs and adults feed voraciously and can defoliate all trees.

Morphological characters   **Adults** Males 49.5-55.4 mm long, green. Head broad oval; posterior border convex. Eyes round and protruded outward. Antennae brown. Occiput with a longitudinal groove. Pronotum nearly rectangular, with a shallow groove at the center. Mesonotum distinctly longer than the head and prothorax combined, with a longitudinal carina. Forewings are leathery, very short, and almost square shape. Its front half with a brown obtuse tooth protruding upwards. Hindwings extend to the middle of the 6th abdominal segment. Mesopleuron smooth. The posterior end of the femora and tibiae dark brown. The abdomen slightly compressed; the last three segments swollen and pointed upward; the posterior border of the 8th abdominal dorsal plate widened. The 9th dorsal plate elongated; narrow in the front and wide in the back; lateral view petal shape with the lateral border deeply concave. Subgenital plate trapezoid. Cerci surpass the anal segment.

Females 53.3-61.5 mm long. Pronotum with black longitudinal stripes. The posterior end of the front femora and tibiae black. Hindwings surpass the 5th abdominal segment. The anal segment narrower than the 8th abdominal segment. The two lateral borders of the dorsal plate round. The epiproct and the egg laying apparatus exposed. The later surpasses the 9th abdominal segment. Cerci surpass the last abdominal segment.

Nymphs   There are five instars. The 1st instar light yellow. The 2nd instar light milky yellow. The 3rd instar light yellowish green. The mouthparts, antennae, tibiae and cerci pale bloody red. The dorsal surface with a light-yellow median line; with triangular wing buds. The 4th instar light-yellowish green; eyes blackish brown; tibiae pale reddish brown; wing bugs fan shaped; the males can be distinguished from the females. The 5th instar pale green; claws brown; wing buds feathered fan shaped.

Biology and behavior   *Sinophasma maculicruralis* occurs one generation per year in Bobai of Guangxi. It overwinters as eggs in fallen leaves. The eggs start to hatch in early March; the peak hatching time is in mid-March. Adults emerge in early May. Peak emergence time is in mid-May. Only adults exist in early June. Peak egg laying period is in late June. Adults die after mid-August.

Eggs usually hatch in the morning hours. The newly hatched nymphs usually do not feed until the next day. They move fast once disturbed. The crawling speed of 20-30mm/s was recorded. Nymphs migrate to tip of the branches and rest along the main veins of leaves on the underside of the leaves. They feed on leaf edges and form notches. The egg hatching coincides with the new leaf development of *Castanopsis* spp. The 2nd instar larvae move slowly. Their feeding produces notches or small holes on leaves with veins intact. The 3rd instar larvae do not move when disturbed. Their only leave large veins un-attacked.

S. maculicruralis adult female (Wang Jijian)

S. maculicruralis adult male (Wang Jijian)

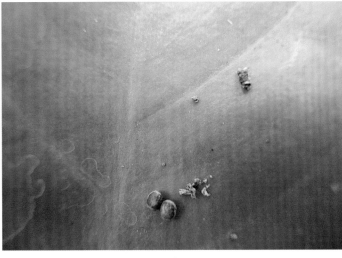
S. maculicruralis eggs (Wang Jijian)

Applying *B. bassiana* to control *S. maculicruralis* (Wang Jijian)

S. maculicruralis being fed by a spider (Wang Jijian)

Damage by *Sinophasma maculicruralis* (Wang Jijian)

S. maculicruralis killed by *B. bassiana* (Wang Jijian)

The antennae and legs extend outside the leaves. They occasionally rest on small branches. The 5th instar larvae move, but do not play dead when disturbed. They feed on both old and young leaves. Young leaves are consumed in whole. Nymphs rest on small branches or between branches and leaves. Adults play dead and drop to the ground when disturbed. They can spring backward and can fly more than 10 cm, but rarely do so. They feed on whole leaves and petiole, young branches, causing leaves fall. After egg laying, their consumption decreases.

Prevention and treatment See "Prevention and treatment of phasmid insects" section (P89).

References Wang Jijian 1988, Xiao Gangrou (ed.) 1992, Chen Shuchun and He Yunheng 2008.

(Wang Jijian)

20 *Sinophasma mirabile* Günther

Classification Phasmatodea, Lonchodidae
Synonym *Sinophasma crassum* Chen et He

Importance *Sinophasma mirabile* became a continuous pest of evergreen broad-leaved forests from 1995. Heavy infestation may lead to complete defoliation and even death of trees.

Distribution Zhejiang, Fujian.

Hosts *Castanopsis eyrei*, *C. carlessi*, *Lithocarpus glaber* and other species in Fagaceae family.

Morphological characteristics **Adults** Males 50-57 mm in length; slender; green to brown. Head oval, brown. Frons with yellowish brown marking. Head with 5 yellowish longitudinal striae. Antennae about 3/4 to 4/5 length of the body. Pronotum small, with a transverse groove near its anterior border. Mesonotum slender, longer than the head and pronotum combined, about the same length as the middle femora, with dense punctures and a central longitudinal carina. Forewing short and small, scale shaped. Its anterior half folds down forming a 90° angle. The folded part is brown color in the front, bright yellow in the back. The posterior half tanned color. The color at the fold is black. Hindwing large, extending to the end of the 6th abdominal segment. Its anterior margin green, other areas brown. The base of the front femora and the middle and hind femora green. The posterior end of the femora and tibiae black. Other areas of the legs yellowish brown. The abdomen longer than head and thorax combined. Abdomen swollen from the 8th segment and arched upward. The 8th abdominal segment is two times as wide as the 7th segment. Its dorsal plate trapezoid. The 9th abdominal segment arched; two times as long as the 8th segment. The anal segment short, curved downward, almost vertical from the 9th segment, with a conical structure on the back. Subgenital plate short. Cerci brown, flattened, pyramid shaped.

Females 60-74 mm in length. Similar to males in color and morphology. The antennae are shorter than males, about 1/3 to 2/3 of the body length. Mesonotum longer than the middle femora and shorter than the hind femora. The folded part of the forewing milky white at the posterior end. The hindwing extends to the end of the 5th abdominal segment. The front legs brown; the base of the femora green; the posterior end of the femora and tibiae black; the middle and hind legs green. The abdomen tapered from front to the rear end. The last three segment

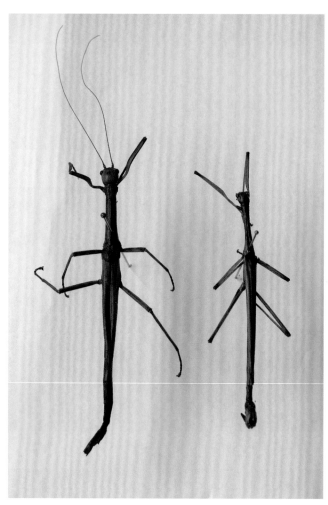

S. mirabile adult female (left) and adult male (right) (Chen Shuchun)

normal size. The 8th segment protrudes into the center of 9th segment from the ventral side, encasing the base of the egg laying apparatus. The poster half of the egg laying apparatus exposed. The posterior border of the last abdominal segment notched and form obtuse angles. Epiproct exposed. Cerci brown.

Eggs Length 1.9 mm, width 1.5 mm; elliptical; laterally compressed. The egg cap black; its margin grayish white. Eggs gray, with brown reticulate ridges. The bottom has a black dot surrounded by an incomplete circular black stripe.

Nymphs There are 6 instars. Occasionally there may be 7th instar. The first instar light brown. Dark brown to tanned from the 2nd instar. The 3rd instar exhibit hindwing buds and the 4th instar start to show forewing buds. The sex can be distinguished from the 4th instar. The 5th and 6th instars are similar to adults except for the undeveloped genitalia.

Biology and behavior *Sinophasma mirabile* occurs one generation a year. It overwinters in fallen leaves. Eggs start to hatch in late March. Peak hatching period is in early April. Adults start to emerge in early June. Peak emergence period is in mid-June. Egg laying period starts in late June until all females die. The development coincides with the plant development. The egg hatching period matches the new shoots development period in Zhejiang. The nymph development matches with the leaf development. Indoor rearing revealed that if hatching or nymph development were slower than young leaf development, the nymphs would die due to the old age of the leaves.

Eggs do not develop the same year after being deposited. They go through a one year diapause period. They require moist conditions to incubate and hatch. They are more tolerant to adverse environmental conditions during the dormant period. Eggs hatch during the night. Nymphs emerge from the egg cap. They crawl to young shoots after about 4 hours of rest and water feeding. One or two nymphs will feed on each leaf. The 1st instar can live for 7 days without feeding. When disturbed, they jump onto other leaves or hide on the underside of the leaves. They feed on leaf edge and form half circles, arches, or notches. They rest on the underside of leaves with the heads toward the petiole. The 1st instar feeds 3-5 times a day, each feeding consumes 1-4 mm^2 leaf area. Nymphs stop feeding 1-2 days before molting. Each molting lasts for 15-25 minutes. After 1-2 hours, the nymph will consume the shed skin. At 10 hours after molting, they start to feed on leaves. The 2nd instar is slower. From the 5th instar, the consumption accounts for 80% of the total consumption of all nymphal instars. Before the 4th instar, nymphs only feed on edges of young leaves. The damaged leaves can continue to grow. From the 5th instar, they will feed on whole leaves including the main veins and petioles and sometimes even the bark of the small branches. The nymphal period lasts for 68-83 days. Males emerge 4-6 days earlier than females. Male to female ratio is 1:1. Adults are weak flyers. They can only fly 1-3 m. They are weakly attracted to lights. From 10 hours after emergence, they start to feed. After 4-5 days, they mate. During mating, males stay on back of the females. The tip of the male abdomen bends downward and reaches the underside of the female abdomen. The mating time is between 13-60 minutes. The female can feed during mating while the male is on the top. Adults may mate multiple times. Females start to lay eggs at 10 days after emergence. They lay eggs randomly and can lay eggs during feeding. They may lay eggs and defecate alternately. After falling onto the ground, eggs overwinter within the leaves and fecal materials. A female lays 3-8 eggs a day until death with less production when getting older. A female can lay a total of 304-614 eggs (average 428). Unfertilized eggs cannot hatch. Adults feed both young and old leaves and are the most damaging stage. Each female consumes an average of 1,082 mm^2 or about 1.3 *Castanopsis eyrei* leaves. The maximum consumption is 2.8 leaves per day. The female consumption accounts for 80% of the consumption of all stages. A female may feed 132.7 leaves during its lifetime. Males consume less than females. Their average life span is 68.4 days compared to 83.7 days in females.

Prevention and treatment See "Prevention and treatment of phasmid insects" section (P89).

References Chen Peichang et al. 1997, Bao Qimin et al. 2000, Chen Shuchun and He Yunheng 2008.

(Chen Shuchun, Xu Jin)

21 *Sinophasma pseudomirabile* Chen et Chen

Classification Phasmatodea, Lonchodidae

Importance *Sinophasma pseudomirabile* is an important pest of *Castanopsis* spp. They, together with other stick insects, may cause complete defoliation of *Castanopsis* spp. and threaten the growth of the trees.

Distribution Guangxi.

Hosts *Castanopsis carlesii* and *C. hystrix*.

Damage First instar nymphs feed on the edge of new leaves grown from stumps on the ground. The second and third instars consume whole young leaves.

Morphological characters **Adults** Males slender, green, 52 mm in length. Head broad oval, with five longitudinal groove on the back. Compound eyes blackish red, round and protruding, with black spots. Antennae are dark brown, filiform, longer than prothoracic legs, and slightly shorter than the body. The end of antennae with alternated three white and four black rings. Prothorax is short, with a transverse groove on the back. Mesonotum with dense granules, longer than the head and prothorax combined, same length as the middle femora. Forewing is short, dorsal view nearly rectangular, with alternately arranged black and yellow short striations. Hindwing is long, extending to the middle of the sixth abdominal segment. Lateral sides reticulate punctured. Legs light green and tarsus gray. The femora of middle legs are the shortest and the femora of the hind legs are the longest. Abdomen is longer than the head and thorax combined. The last three abdominal segments swollen. The 7th dorsal plate gradually widens toward the back. The 9th dorsal plate is the longest, raised, with brown striations; its lateral border arched. The anal segment is nearly vertical to the 9th abdominal segment, slightly concave at center, with a median carina; the posterior angles slightly curved inward. The subgenital plate with three tooth-shaped projections. Cerci red, flattened, tapered at the end.

Females are similar to males but slenderer, 65mm in length. Head flattened, rectangular. The last three abdominal segments are similar to the other abdominal segments. The 8th abdominal plate is short. Ovipositor apparatus exposed. The seventh abdominal plate extends into the eighth abdominal plate and fuse with the operculum. The center of the posterior border of the last segment has a triangular emargination. Epiproct surpasses the last abdominal segment. Cerci not cylindrical, widened at middle, and tapered toward the end, but not as obvious as the males.

Nymphs There are six instars, occasionally there may be seven instars. The 1st and 2nd instars are light yellow.

Damage by *S. pseudomirabile* (Wang Jijian)

S. pseudomirabile adult female (Wang Jijian)

S. pseudomirabile adult male (Wang Jijian)

Wing pads appear in the 3rd and 4th instars. The last three abdominal segments begin to swell in males. Forewings of late instars stick-shaped; hindwing near rectangular, about 5-10mm in length, with many distinct longitudinal ridges. Color is similar to that of adults.

Biology and behavior *Sinophasma pseudomirabile* occurs one generation per year. It overwinters in egg stage. Peak hatching period is in mid- and late March. Peak adult emergence period in mid- and late June. Adults die in mid- and late August.

The first and second instars rest along the mid- vein on the back of leaves. Front legs and antennae extend horizontally. They feed during the night; each feeding lasts for 15-30 minutes. Sometimes they feed during the day. They feed along the edge of the leaves in the same direction. Young nymphs feed on leaf edges, create notches or small holes, and leave veins intact. Late instars cause large holes on leaves, only leave the vein or the petiole intact. The food consumption of the 1st to 4th instars, 5th to 6th (7th) instars, and adults account for 4.5, 31.0, and 64.5% of the total consumption, respectively.

Prevention and treatment See Prevention and treatment of Phasmid insects.

References Chi Xingzhen and Wang Jijian 1998, Chen Shuchun and He Yunheng 2008.

Prevention and treatment of phasmid insects

1. Surveillance. Phasmids have the characteristics of intermittent outbreaks. Their population levels are closely correlated with the egg production and weather in the previous year. Establishing a surveillance program will help predict future population level, emergence period, and distribution patterns.

2. Silviculture. Pruning and thinning will help reduce moisture which is positively correlated with egg survival and development. Removing dead leaves on the forest floor in the fall will expose eggs to sunlight. Turning soil and burying eggs 3-5 cm deep will prevent them from hatching.

3. Mechanical control. Most of the phasmids play death. Thus, adults and nymphs can be killed by shaking the small trees. During the night, using a flashlight to find and kill adults and nymphs while they are active.

4. Conservation of natural enemies. Ants, mantids, spiders, mites, *Calotes versicolor*, and birds can attack nymphs and adults.

5. Biological. Collecting virus infected Phasmids. The diseased insects become lethargic, cease feeding, and turn black. Grind the cadaver and spray the water solution in infested areas or move the cadaver to areas with high numbers of Phasmids. Spray *Beauveria bassiana* at low temperature, after rain, or when dews exist. Apply *B.t.* dust when temperature is high.

6. Chemical control. Phasmids have soft body walls. Good control results can be achieved using stomach insecticides, contact insecticides, or foggers (such as dichlorvos). The following insecticide sprays can provide > 90% population reduction: 2.5% deltamethrin at 5,000×, 25 g/L lamda-cyhalothrin at 800×, fenvalerate at 2,000. cypermethrin at 5,000×. Dusting deltamethrin at 100 mg/kg can reduce pest population by > 85%.

(Wang Jijian)

22 *Cryptotermes domesticus* (Haviland)

Classification Blattodea, Kalotermitidae

Synonyms *Cryptotermes campbelli* Light, *Cryptotermes buxtoni* Hill, *Cryptotermes dentatus* Oshima, *Cryptotermes formosae* Holmgren, *Cryptotermes hermsi* Kirby

Distribution Guangdong, Guangxi, Hainan, Yunnan, Taiwan; India, Thailand, Malaysia, Indonesia, Singapore, New Guinea, Solomon Island, Fiji, Mariana Islands, Marquesas Islands.

Hosts *Ficus* spp., *Litch chinensis*, *Pterocarya stenoptera*, *Castanopsis sclerophylla*, *Lagerstroemia indica*, and wooden structures.

Damage *Cryptotermes domesticus* infests dry hard wood materials. Fecal pellets are pushed out from infested wood.

Morphological characteristics **Soldiers** Head dark brown in the front, reddish-brown in the posterior region; mandibles black; 1st and 2nd antennae segments brown, the other antennae segments light-yellow; thorax and abdomen light-yellow; front part of the pronotum brownish-yellow. Head thick, near rectangular, lateral margins parallel; occipital border round; anterior part of the head truncated, punctuated; borders of the truncated surface elevated; with a notch at center of the upper border.

Alates Head dark yellow; thorax, abdomen, and legs light-yellow; wing scales and veins dark yellow; wing membrane transparent and colorless.

Biology and behavior Nests are built in wooden materials. A colony has dozens to a few hundred individuals. There are no workers in a colony. Larvae are responsible for colony maintenance. The nest shape varies. When larvae are separated from the colony for 7 days, they will start to transform into supplementary reproductives in about 13-20 days. From 24-31 days after separation, supplementary reproductives start to lay eggs. Eggs are laid singly or up to 3 eggs each time. There is a 15-day interval between egg-laying. The minimum incubation period is 51 days, with an average of 53-58 days. Only one pair of supplementary reproductives survive. The others die from cannibalism. Colonies with primary reproductives do not produce supplementary reproductives. Alates are present year around. Swarming usually occurs in afternoon or evening hours between April-June. Peak swarming time is around 19:00. After swarming, alates drop their wings, pair, and dig into wood materials to start a new colony. They prefer decay fungus infested wood and therefore, they are often found congregating in hidden areas. Dealates seal the entrance with saliva.

Prevention and treatment

1. Quarantine of imported wood.

2. Fumigation. Use methyl bromide (CH_3Br) at 35-40g/m^3, chloropicrin (Cl_3NO_2) at 40g/m^3, aluminum phosphide (AlP) at 8-12g/m^3, and sulfuryl fluoride (SO_2F_2) at 20-30g/mm^3.

3. Heat treatment. Infested furniture may be heated to 65 °C for one hour or 60°C for four hours.

References Huang Fusheng et al. 2002, Li Guixiang 2002.

(Tong Xinwang, Dai Xiangguang, Li Guixiang)

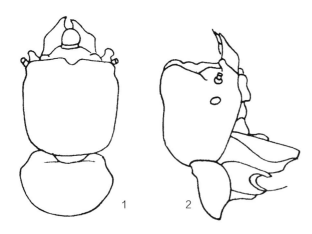

C. domesticus soldier.
1. Dorsal view of head and pronotum, 2. Lateral view of head and pronotum (Chen Ruijin)

23 *Hodotermopsis sjostedti* Holmgren

Classification Blattodea, Hodotermitidae

Synonyms *Hodotermopsis japonicus* Holmgren, *Hodotermopsis lianshanensis* Ping, *Hodotermopsis yui* Li et Ping, *Hodotermopsis orientalis* Li et Ping, *Hodotermopsis dimorphus* Zhu et Huang, *Hodotermopsis fanjingshanensis* Zhu et Wang

Distribution Zhejiang, Fujian, Jiangxi, Hunan, Guangdong, Guangxi, Hainan, Sichuan, Guizhou, Yunnan; Vietnam, Japan.

Hosts *Pinus kwangtungenesis*, *Pinus taiwanensi*, *Fokienia hodginsi*, *Keteleeria davidiana*, *Crytomeria fortune*, *Phoebe zhennan*, *Quercus fabri*, *Schima superba*, *Vaccinium mandarinorum*.

Damage *Hodotermopsis sjostedti* Infest large dying trees. Feed inside wood and may hollow the trees. They leave no obvious signs on tree surfaces.

Morphological characteristics **Soldiers** Head rectangular-oval, black in the front, reddish-black in the posterior region. Mandibles strong, anterior tip sharp; left mandible with 4 irregular large teeth; right mandible with 2 large teeth. Morphology may vary significantly.

Alates Length 12-13.5 cm; body brown, head and pteralia area darker; head almost round. Left mandible with one terminal tooth and three teeth before that. The right mandible with a large terminal tooth, sub-terminal tooth small, first tooth short, 2nd tooth oblique.

Biology and Behavior Nests are built in decaying wood or live trees. All members of a colony reside in wood. It is unclear how long it takes from colony founding by primary reproductives to colony maturation. Eggs hatch within about a month. Cast differentiation starts from the 3rd instar. Members of a colony consist of workers, soldiers, and nymphs. Nymph wing buds appear in November-December. After 6 additional molts and 8 months, alates emerge and swarm in mid- late-July.

Hodotermopsis sjostedti are found along creeks, in primary and secondary forests. Each colony may have 1 king and 1 queen, 2 kings and 2 queens, or 2 kings and 3 queens. Swarming occurs between 17:00-17:30. Swarming usually completes in one day; however, some colonies may swarm in three consecutive days.

Prevention and treatment

1. Silviculture. Burn the stumps and debris before planting trees.

2. Chemical control. Inject 40% Dusban and phoxim mixture at 2,000× or 25 g/L lamda-cyhalothrin at 400× or 350 g/L imidacloprid suspension at 1,500x on top of the infested wood. Apply 4-6 kg solution per colony will achieve 80-98% control.

References Yin Shicai 1982, Huang Fusheng et al. 2002.

(Tong Xinwang, Yin Shicai)

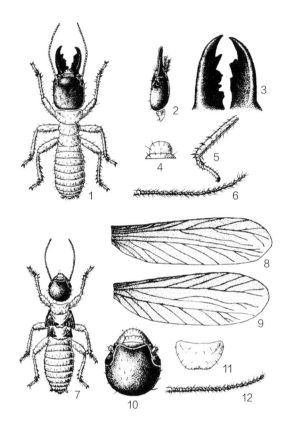

H. sjostedti. 1-6. Soldier, 2. Lateral view of head, 3. Mandible, 4. Clypeus, 5. Hind tibia and tarsus, 6. Antenna. 7-12. Alate, 8-9. Front and hind wing, 10. Head, 11. Pronotum, 12. Antenna (Hou Boxin)

24 *Coptotermes formosanus* Shiraki

Classification	Blattodea, Rhinotermitidae
Synonyms	*Coptotermes hongkongensis* Oshima, *Coptotermes communis* Xia et He, *Coptotermes eucalyptus* Ping, *Coptotermes rectangularis* Ping et Xu
Common name	Formosan subterranean termite

Importance *Coptotermes formosanus* live underground and in wood. They cause serious damage to structures, bridges, communication cables, crops, trees, books, and affect many aspects of people life.

Distribution Jiangsu, Zhejiang, Anhui, Fujian, Jiangxi, Hubei, Hunan, Guangdong, Guangxi, Hainan, Sichuan, Guizhou, Yunnan, Hong Kong, Taiwan; Marshall Islands, U.S.A., Japan, Philippines, Pakistan, Sri Lanka, Myanmar, Thailand, Brazil, South Africa.

Hosts Houses, bridges, communication cables, books, fabric, Chinese medicine, shade trees.

Damage They build mud tubes or covers on wood surface. Brown mud and fecal materials exist near the main nest. Infested wood materials may become hollowed and only the surfaces remain intact.

Morphological characteristics **Soldiers** Head and antennae light-yellow, mandibles blackish-brown, abdomen milky white. Fontanelle large, located on a slightly raised tubular structure, oval, narrower on top and broader at bottom.

Alates Dorsal surface of the head dark yellowish-brown; dorsal surface of the thorax and abdomen brownish-yellow, paler than the head; ventral surface of the abdomen yellow; wings slightly yellowish.

Biology and behavior After alate pairing, the female produces eggs within a week. Eggs hatch after 1 month. Each female lays 1-6 eggs each time. An average of 46 eggs are laid in the first 5 months. Larvae molt 6 times. Cast differentiation starts at the 4th instar. A colony matured (produces swarms) after 8 years in a study. The colony was estimated with 150,000 to 200,000 individuals at 8 years after establishment. Swarming occurs between 16:00-20:00 during April-July. Swarming of a colony may last for 3 months. A colony lives for 30-40 years.

Nests are built underground or above ground. When a colony expands, it forms satellite nests. Foraging distance can be greater than 100m. Large trees are often infested. Commonly infested tree species include *Cinnamomum camphora*, *Liquidambar formosana*, and *Photinia davidisoniae*.

Prevention and treatment *Coptotermes formosanus* control should follow the principle of prevention and integrated pest management based on termite biology, use physical or chemical barrier, colony control, ecological control and biological control.

C. formosanus workers and tunnels (Wang Jijian)

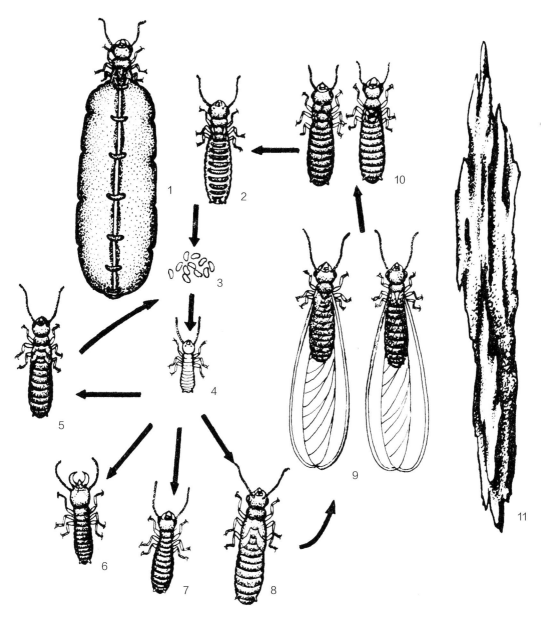

C. formosanus. 1. Queen, 2. King, 3. Eggs, 4. Larva, 5. Supplementary reproductive, 6. Solder, 7. Worker, 8. Brachypterours reproductive, 9. Primary reproductives, 10. Dealates, 11. Wood damage (Yang Kesi)

1. Trapping. Place an open water container filled with water under the light during swarming season to attract and kill alates.

2. Chemical control. Select short residue, low toxicity insecticides to avoid injuries to human and animals. (1) Place bait box to attract and kill termites. Apply direct spray to termites attracted to the bait box. Commonly used insecticides are 40% chlorpyrifos and phoxim mixture. These insecticides can also be directly applied into soil or wood. (2) Place toxic bait to kill the whole colony. Active ingredients include carbosulfan, chrysopyrifos and phoxim mixture. The bait matrices include pine wood, *Cunninghamia lanceolata* saw dust, sugar cane, Paulownia, culture media for *Tremella fuciformis* and *Auricularia auricula-judae*. Feeding stimulants include sugar, pine pollen. Place 4-5 g mixture per paper bag. Hexaflumuron can be impregnated into paper to kill *C. formosanus*. A colony can be killed within 3 months after treatment.

References Huang Fusheng et al. 2000; Tong Xinwang 2004; Liu Zili, Huang Lei, Yi Junji et al. 2005; China Real Estate Management Association-Termite Control Committee, 2008.

(Tong Xinwang, Lu Chuanchuan)

25 *Reticulitermes chinensis* Snyder

Classification Blattodea, Rhinotermitidae
Synonym *Reticulitermes labralis* Hsia et Fan

Distribution Beijing, Hebei, Shanxi, Shanghai, Jiangsu, Zhejiang, Anhui, Fujian, Jiangxi, Shandong, Henan, Hubei, Hunan, Guangxi, Sichuan, Yunnan, Shaanxi, Gansu.

Hosts *Eucalyptus* spp. and many other trees.

Damage *Reticulitermes chinensis* feeds on the roots of eucalyptus seedlings and cause their rapid morality. It is a common structural pest in southern China.

Morphological characteristics **Soldiers** Head and antennae yellow or yellow-brown, mandibles dark brown, abdomen light whitish-yellow. The head covered with sparse hair, the thorax and abdomen with denser hair. Head is long, flat and cylindrical; its occipital border straight; the lateral border is almost parallel. Frontal carinae raised, the area between the carinae depressed. The forehead and post-clypeus forms a near 45° angle. Fontanelle looks like a small dot, located at the first 1/3 of the head. Labrum is less than a half of the mandible length. Mandibles slightly arched, about half of the head length; the middle is almost straight, and the tip curved toward the middle. The left mandible gradually narrows from base to the front; whereas, the right mandible begins to narrow near the tip. The left mandible has one basal tooth and three small teeth beyond the middle point of the mandible; the right mandible is smooth and has no teeth. The front part of the gula is pentagonal, the middle part is thin and long. Antennae have 15-17 segments. The pronotum wider in the front than the back, its anterior border slightly raised and emarginated.

Alates Head and thorax black, abdomen lighter; antennae, femora and wings blackish-brown; tibia and tarsi dark yellow. Body covered with dense hairs. Head ob-

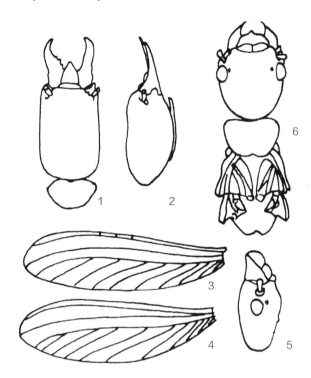

R. chinensis. 1-2. Soldier, dorsal view of head and pronotum, 2. Lateral view of head, 3-6. Alates, 3. Front wing, 4. Hind wing, 5. Lateral view of head, 6. Dorsal view of head and thorax (Chen Ruijin)

R. chinensis in soil (Wang Jijian)

R. chinensis (Wang Jijian)

Damage by *R. chinensis* (Wang Jijian)

long, the occipital border is round, and the two sides are almost parallel. The post-clypeus is slightly lighter than the vertex, convex. Compound eyes small and flat. Ocelli almost round. Fontanelle present, small. Antennae have 18 segments. The pronotum wider in the front than the back; its anterior border is almost flat and straight, emarginated or slightly emarginated. Scales on the forewing are distinctly larger than those of the hindwing.

Workers Body white, evenly covered with short hairs. Head is round and slightly extended around the antennal sockets. Post-clypeus looks like a horizontal strip and slightly raise; its length is less than 1/4 of its width. Occiput flat. Antennae have 16 segments. The front of the pronotum slightly arched; the anterior and posterior borders emarginated.

Biology and behavior *Reticulitermes chinensis* alates appear between mid-April and early May. They swarm around noon time. Eggs hatch in 32-36 days. The development time for the 1st, 2nd, and 3rd instars are 12-14, 12, and 12 days, respectively.

Reticulitermes chinensis often do not build nests. They are found in stumps and under logs. They form small, scattered groups. They may build tunnels under fallen leaves or loosened soil. Multiple tunnels may connect to *Eucalyptus* seedlings. After the seedlings die, the termites will move to nearby areas searching for food. Large trees may also be damaged, but they prefer young trees.

Prevention and treatment

1. Controlled burning before afforestation. Remove dry branches, fallen leaves, and shrubs will reduce termite numbers.

2. Turn up soil to expose termites to natural enemies.

3. Chemical treatment. Treat soil with termiticide one month before planting. Treat seedling roots with 350 g/L imidacloprid 1,500-2,000× or 40.7% chlorpyrifos at 1,200×, or use a mixture of 40.7% chlorpyrifos + 2.5% deltamethrin at 2,000-3,000×.

4. Post-planting treatment. If trees were not treated prior to planting or heavy rainfall occurred after planting, the above-mentioned termiticides may be sprinkled to the roots or using a sprayer to apply 25-50 mL chemicals to the bottom of each seedling.

References Huang Fusheng et al. 1989; Xiao Gangrou 1992; Wei Jiqian, Mo Jianchu, Xuwen et al. 2010.

(Wang Jijian, Dai Zirong, Li Guixiang)

26 *Reticulitermes flaviceps* (Oshima)

Classification Blattodea, Rhinotermitidae
Synonym *Leucotermes speratus* Holmgren

Importance *Reticulitermes flaviceps* damage many different trees.

Distribution Jiangsu, Zhejiang, Fujian, Jiangxi, Hubei, Hunan, Guangdong, Guangxi, Hainan, Hong Kong, Taiwan.

Hosts Logs, dead trees, stumps, barks of live trees, houses, wooden structures.

Damage *Reticulitermes flaviceps* workers build fine mud tubes on surfaces. The mud tubes usually start from bottom of stairs, door frames, etc. and extend upwards. Termites affecting live trees or stumps start their mud tubes from the bark at bottom and then inside the bark and extend further up.

Morphological characteristics **Soldiers** Head yellow-brown, mandibles dark brown; head rectangular, laterally paralleled; labrum pointed at front, with one hair at the tip and two hairs away from the tip. The anterior border of the pronotum slightly emarginated, its posterior border straight.

Dealates Head chestnut-brown, pronotum and gula greyish-yellow, labrum brown, post-labrum yellowish-brown, femur dark yellowish-brown, tibiae yellowish-brown.

Biology and behavior Nests are built in ground or in wood materials. There is no distinct king and queen chamber in a nest. Alates emerge in nests between September-November in Hunan province. Swarming occurs between February-April when temperature is between 17-20°C. Swarming usually happens between 12:00-15:00 on clear days. Based on observations by Shanghai Entomological Research Institute, first batch of eggs are produced 15 days after pairing. A female lays about 20 eggs initially, then again after two months. Eggs hatch after 27-38 days. The 1st instar larvae develop for 12-16 days before molting into 2nd instar. After another 12-16 days, they molt into 3rd instar. Third instar workers have the capacity to convert into apterous supplementary reproductives. When workers are separated from the mother colony for 1.5-2 months, some will transform into supplementary reproductives. In about half a month, they can lay eggs.

Prevention and treatment See *Coptotermes formosanus*.

References Liu Liling et al. 1991, Huang Fu-sheng 2000, Tong Xinwang 2003.

(Tong Xinwang)

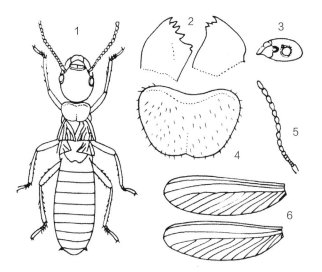

R. flaviceps alate. 1. Adult, 2. Mandibles, 3. Lateral view of head, 4. Pronotum, 5. Antenna, 6. Front and hind wing (Chen Ruijin)

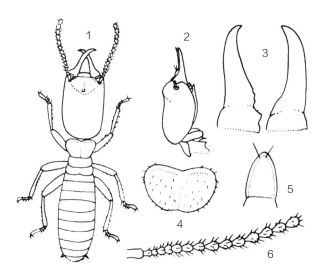

R. flaviceps soldier. 1. Overview, 2. Lateral view of head and pronotum, 3. Mandibles, 4. Pronotum, 5. Clypeus, 6. Antenna (Chen Ruijin)

27 *Macrotermes barneyi* Light
Classification Blattodea, Termitidae

Importance *Macrotermes barneyi* infest trees, crops, communication wires buried underground, and structures. The nests on dikes cause leaks and collapse of dikes. It is only second to *Odontotermes formosanus* in importance.

Distribution Jiangsu, Zhejiang, Anhui, Fujian, Jiangxi, Henan, Hubei, Hunan, Guangdong, Guangxi, Hainan, Sichuan, Guizhou, Yunnan, Hong Kong; Vietnam.

Hosts Crops, herbaceous plants, *Cunninghamia lanceolata*, *Sassafras tzumu*, *Cinnamomum camphora*, *Liquidambar formosana*, *Eucalyptus* spp., etc. Trees belonging to 61 fmilies, 137 genera and 335 species may be damaged.

Damage *Macrotermes barneyi* forms mud tubes or mud covers on tree trunk or other parts of the affected plant. Sometimes the mud covers the whole circumference of the tree trunk. They mainly affect live trees, dying trees, xylem and roots of dead trees.

Morphological characteristics **Major soldiers** Head rectangular, dark yellow; mandibles sickle-shaped, black; labrum tongue-shaped, with a transparent triangular piece in front of it. Pronotum slightly more than half the head width, slightly raised upwards in the front. Both the anterior and posterior borders have a distinct emargination.

Minor soldiers Much smaller than the major solders. Color paler. Head oval, lateral borders curved. Occipital corners round.

Alates Head, thorax, and abdomen dark reddish-brown, legs brown-yellow, wings yellow, post-clypeus reddish-yellow.

Biology and behavior *Macrotermes barneyi* build subterranean nests. The main chamber may be up to 1 m in diameter. Outer surface of the nest consists of multiple thin layers of mud with a total thickness reaching 10-22 cm. The center of the nest is the fungus comb, which is light, sponge-like. The fungus fruit body is large but the numbers of fruit body are small. There are soil plates connecting the fungus combs. The royal chamber is sturdy, made of soil, located at the middle and upper part of the fungus combs. Bottom of the chamber is flat. Numerous termite tunnels are present underneath of the bottom of the nest. The main tunnels from the nests are large, curved. Secondary tunnels are formed from main tunnels far away from the nest. Small chambers are formed on secondary tunnels near ground surface (width 3-5 cm, height 1-2 cm).

Alates swarm the same year after emergence. Swarming occurs between late April and mid-July in Hunan province. It occurs earlier at lower latitude and elevation. When temperature is stable at above 16℃, alates migrate to waiting chambers. Most swarms occur after heavy rain, between 4:00-6:00, with RH > 80% and air pressure of 98.1-99.6 kPa. The number of exit holes range from 30 to 80, with maximum of 239. There are three types of exit holes: flat, concave, and mound-shaped. After flight, alates will drop their wings, pair, dig into soil, and lay

M. barneyi workers (Zhang Peiyi)

M. barneyi alate (Zhang Peiyi)

M. barneyi mud tube (Zhang Peiyi)

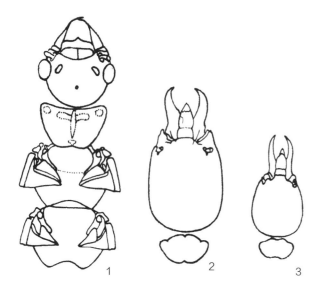

1. *M. barneyi* alate head and thorax, 2. major soldier head and pronotum, 3. minor soldier head and pronotum (Chen Ruijin)

eggs. One male and one female usually start a colony. There are also occasions that one male and several females or vice versa start a colony. A colony matures in 8 years.

Prevention and treatment

1. Silviculture. *Macrotermes barneyi* infestations are most severe in areas with only a single tree species, and lack of shrubs, grasses, leaves and humus. Avoid using controlled fire before planting. Plant mixed trees. Prevent human activities in the forest. Plant resistant species such as *Manglietia fordiana*, *Schima superba*, *Magnolia macclurei*, *M. alba*, *Cornus wilsoniana*.

2. Trapping. Place light traps during swarming period (April-June) each year to attract and kill alates.

3. Nest removal. Identify the nest location and remove the king and queen will cause colony death because no supplementary reproductives are present in the nest. Follow the mud tubes, exit holes, and the symbiotic fungus (*Termitomyces albuminosus*) to find the nest.

4. Chemical control. Spray directly to termites or use toxic baits. Fipronil, chlorpyrifos, and imidacloprid are used to trees or soil in. They can also be applied to soil. A second method is to use termite bait.

References Dai Xiangguang 1987, Huang Fusheng 2000, Tong Xinwang 2004, China Real Estate Management Association - Termite Control Committee 2008.

(Tong Xinwang, Lu Chuanchuan)

28 *Sinonasutitermes erectinasus* (Tsai et Chen)

Classification Blattodea, Termitidae
Synonym *Nasutitermes erectinasus* Tsai et Chen

Distribution Hunan, Hainan, Jiangxi, Zhejiang.

Hosts *Cinnamomum camphora*, *Castanopsis eyrei*, *Litsea cubeba*, *Castanea henryi*, logs, dying trees, dead trees.

Damage *Sinonasutitermes erectinasus* workers build mud tubes on surfaces of trees. Blackish-brown feces are often found near the nests.

Morphological characteristics **Major soldiers** Head and antennae dark reddish-brown, anterior part of the nasus darker in color, thorax and abdomen light-yellow. Head nearly hairless, dorsal and ventral surface with fine hairs, ventral surface has longer hairs. Posterior margin of each segment with a row of erect long hairs. Head oval, with the short end vertical, widest point located after middle of head. Occiput wide, round. Nasus conical, base of nasus wide; the area where nasus meets the head is strongly curved; the angle between labrum and nasus is $> 90°$, tilted. Mandibles with a spine in the lateral-front side. Antennae 13 segmented; the 2nd and 4th segments are similar; the 3rd segment is longer; the 5th segment is longer than the 4th segment. Pronotum forms a $90°$ angle between the front and the rear part. The anterior border of the pronotum slightly emarginated.

Minor soldiers Body color and hair similar to major soldiers but head with sparse hairs; nearly round. Nasus not tilted. Antennae 12 segmented; the 3rd segment is 1-1.5 times as long as the 2nd segment. Anterior border of pronotum not emarginated.

Biology and behavior Nests are built in wood. Both soldiers and workers have two distinct forms. They usually feed on dying trees. The nest usually located in tree trunk from ground level to 1 m above ground. Alates emerge in May. Swarming occur in June and July.

Prevention and treatment

1. Quarantine. Treat infested wood or prevent movement of infested wood.

2. Silviculture. Remove logs, stumps, and hollow live trees.

3. Fumigation. Identify the nest in the tree. Drill a hole below the nest, insert smoker and force smoke into the nest. Apply 0.5 kg fumigant per nest.

4. Liquid treatment. Drill a hole above the nest. Inject liquid pesticide until the nest is saturated with pesticide.

References Xiao Gangrou (ed.) 1992, Huang Fusheng et al. 2000, Wang Yong and Zeng Juping 2013.

(Tong Xinwang, Peng Jianwen, Yin Shicai, Li Zhenyu)

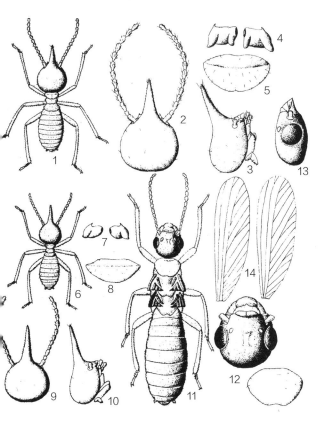

S. erectinasus. 1-5. Major soldier, 2-3. Dorsal and lateral view of head, 4. Mandible, 5. Pronotum, 6-10. Minor soldier, 7. Mandible, 8. Pronotum, 9-10. Dorsal and lateral view of head, 11-14. Alates, 12. Head and pronotum, 13. Lateral view of head, 14. Wings (Hou Boxin)

29 *Nasutitermes parvonasutus* (Nawa)

Classification Blattodea, Termitidae

Distribution Fujian, Jiangxi, Hunan, Guangdong, Guangxi, Hainan, Hong Kong, Taiwan.

Hosts *Cinnamomum* spp., *Magrolia* spp., *Coriaria nepalensis*, *Symplocos* spp., *Castanopsis selerophylla*, *Camellia* spp. and wooden structure.

Morphological characteristics **Soldiers** Head yellow, nasus red with brown tint, abdomen near white. Head without hair or with a few short fine hairs. Head short oval, length slightly larger than width, the widest part is behind the center. Occipital border slightly curved. Nasus tubular, slightly deflected downward. Nasus and vertex form a straight line or a slightly concave line. Lateral tip of the mandibles usually sharp, but not protruded, few may contain a sharp tooth. Antennae 13 segmented; the 4th segment is shortest; the 2nd, 3rd, and 5th segments equal length. If antennae are 12-segmented, then the 4th and 5th segments are not completely separated and forming a long segment. Pronotum as wide anteriorly as posteriorly, the front part vertical or slightly tilt backward, the anterior border of pronotum not emarginated.

Alates Head brown; antennae, labrum, and post-clypeus yellow; pronotum paler than head, with a yellow "T" mark; anterior part of the meso- and metathorax yellow, and posterior part brown; dorsal surface of abdomen brown, ventral surface yellowish. Head wide oval, vertex concave, fontanelle small, slit-like. Antennae 15 segments; the 3rd segment is the smallest; the 2nd segment slightly longer than the 3rd segment; the 4th and 6th segment similar in length.

Biology and behavior Nests are built in live or decaying wood. They are often located at base of live trees belong to Lauraceae, Magnoliaceae, and Fagacea. Nests are blackish-brown and soft, with a honeycomb structure. They are made of feces, saliva, and wood debris. The nest is fragile when moist, but hardened when dry. The king and queen live in a turtle shell-shaped royal chamber. A young nest does not have satellite nests. Mature colonies have 1-2 satellite nests. The nest or infestation site often has blackish-brown mud tubes. Swarms occur in June and July in Hunan (Chenzhou) during humid hot evening. They are attracted to light.

Prevention and treatment Same as *Sinonasutitermes erectinasus*.

References Xiao Gangrou (ed.) 1992, Huang Fusheng 2000.

(Tong Xinwang, Pen Jianwen, Yin Shicai)

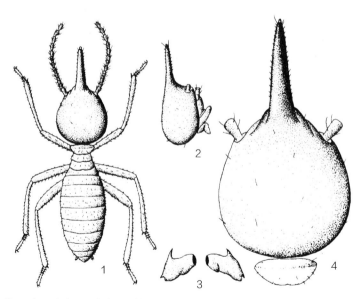

N. parvonasutus soldier. 1. Overview, 2. Lateral view of head, 3. Mandibles, 4. Head and pronotum (Hou Boxin)

30 *Odontotermes formosanus* (Shiraki)

Classification Blattodea, Termitidae
Synonyms *Termes valgaris* Shiraki, *Odontotermes qianyagensis* Lin

Importance *Odontotermes formosanus* cause damage to 70 species trees and also crops, underground communication cables, dikes, and dams. It is a major pest to agriculture and dams.

Distribution Jiangsu, Zhejiang, Anhui, Fujian, Jiangxi, Henan, Hubei, Hunan, Guangdong, Guangxi, Hainan, Chongqing, Sichuan, Guizhou, Yunnan, Shaanxi, Gansu, Hong Kong, Taiwan; Myanmar, Thailand.

Hosts *Cunninghamia lanceolate*, *Platycladus orientalis*, *Cupressus funebris*, *Eucalyptus* spp., *Cinnamomus camphora*, *Schima superba*, *Quercus* spp.

Damage *Odontotermes formosanus* build mud tubes or mud covers on tree trunk or other plant parts. Sometimes the mud covers the whole circumference of the tree trunk. They mainly affect live trees, dying trees, xylem and roots of dead trees.

Morphological characteristics **Soldiers** Head dull yellow, abdomen light-yellow to grey white. Head oval. Mandibles sickle-shaped; tooth of the left mandible located before the mid-point; right mandible with an indistinct tooth. Labrum tongue-shaped, without the small translucent piece in the front. Antennae 16-17 segments; the 2nd segment is about the sum of the 3rd and 4th segments. Anterior part of the pronotum narrow and raised, the posterior part wider. There is a groove on each side of the pronotum running from the top to the bottom. The anterior and posterior borders of the pronotum emarginated.

Alates Dorsal surface of head, thorax, and abdomen blackish-brown, ventral surface brownish-yellow. Anterior half of the labrum orange red; posterior half pale orange in color; a transverse white stripe exists between the two parts. Anterior and lateral margin of the labrum white, transparent. Wings blackish-brown.

Biology and behavior Nests are built underground. The main nest is semi-sphere shape. The long diameter can be as large as 100-140 cm, the short diam-

O. formosanus (edited by Zhang Runzhi)

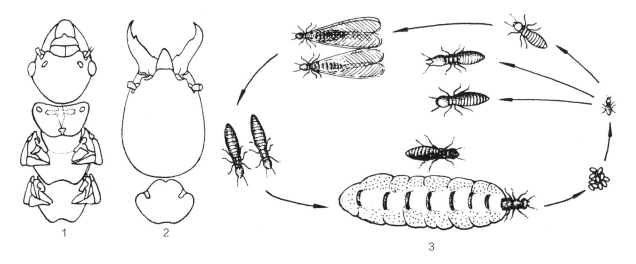

O. formosanus. 1. Alate head and thorax, 2. Head of soldier, 3. Life cycle (Yang Kesi)

O. formosanus worker (SFDP)[1]

O. formosanus queen (SFDP)

O. formosanus alate (SFDP)

O. formosanus mud tube (edited by Zhang Runzhi)

Damage by *O. formosanus* (edited by Zhang Runzhi)

eter is 35-85 cm, height is 30-90 cm. The nest forms a mud shell which separates it from the surrounding soil. The king and the queen live in a royal chamber which is smooth, hard, and oval. The chamber size is about 6-10cm long, 2-3cm wide, and 1-2.5cm high. The fungus garden inside the nest forms layers and is supported by mud columns. Many satellite nests exist within 1 m of the main nest. A few satellite nests may be as far as 8-10 m away. A mature colony's main nest is 1-3 m below the ground. Empty cavities (without fungus comb) are more commonly found at upper part of the main nest and those satellite nests located far away from the main nest.

Alates leave the main nest in March and reside in the waiting chambers. Clumps of soil appear on surface of the nest in April-June. Swarms occur on days with > 20℃, > 85% RH, and during cloudy or rainy evenings (at around 19:00). Swarming often occurs when there is heavy rain. A total of 2,000-9,000 alates may exit from a mature nest. After swarming, alates shed wings, dig a hole on the ground and initiate a new colony. There may be more than two individuals entering the same hole. Therefore, a colony may have more than one king or more than one queen. A newly established nest is merely a small cavity. Fungus comb forms in 3 months. A colony matures in 5-6 years or 7-8 years.

Prevention and treatment Same as *Macrotermes barneyi*.

References Huang Fusheng 2000, Li Guixiang 2002.

(Tong Xinwang, Lu Chuanchuan, Shen Jizeng)

[1] SFDP– "Shanghai Forest Diseases and Pests"

31 *Lycorma delicatula* (White)

Classification Hemiptera, Fulgoridae
Common name Spotted lanternfly

Importance Insect feeding reduces vigor of the trees, in particular, young seedlings and saplings.

Distribution Beijing, Hebei, Liaoning, Jiangsu, Zhejiang, Henan, Shandong, Guangdong, Sichuan, Shaanxi, Gansu, Taiwan; India, Vietnam, Korea Peninsula, U.S.A.

Hosts *Ailanthus altissima*, *A. altissima* 'Qiantouchun', *Toona sinensis*, *Robinia pseudoacacia*, *Melia azedardch*, *Ulmus pumila*, *Populus* spp., *Platanus acerifolia*, *Albizia julibrissin*, *Vitis vinifera*, *Prunus subhirtella*, *P. armeniaca*, *P. salicina*, and *P. persica*.

Damage Both adults and nymphs pierce and suck plant sap from the new growth of the branches and form whitish spots. At the same time, the insects also excrete sticky honeydew, which causes sooty mold that covers branches and trunk. The insect could also cause dwarfed, wilted, or deformed new growth.

Morphological characteristics **Adults** Female length 18 to 22 mm, wingspan 50 to 52 mm; male length 14 to 17 mm, and wingspan 40 to 45 mm. Body protruded upward with acute angle at its head. Antennae are located beneath the compound eyes, bright red, tilted cone in shape, with short cylindrical scape; pedicel enlarged into oval shape. The flagellum segment is extremely slender, and the length is only 1/2 of the pedicel length. Forewings are leathery, oval in shape, and basal 2/3 part is light grayish brown, scattered with more than 20 black spots of varying sizes; the distal 1/3 part is the dark brown with white veins. The hind wings are membranous, and fan-shaped, basal half of the hind wings are in red, and scattered with 7-8 black spots of varying sizes. The middle part of the wings has an inversed white triangular area, and the distal end of the wings, as well as veins, are black in color. **Eggs** The eggs are arranged in egg masses that are covered with gray powdery and loose wax, resembles soil (or mud) in color. Each egg mas has 5 to 10 rows of eggs with 10 to 30 eggs per row. The eggs are oblong, gray, and $3\times1.5\times1.5$ mm in size. On the dorsal surface of the eggs, there are two concaved lines on the sides and a longitudinal ridge in the middle. On the front half of the ridge, there is an elongated-circular operculum. At the front end of the ridge, there is an angled protrusion. The distal end of the eggs is truncated or slightly concave, while the basal end is bluntly circular, and ventral surface is flat. **Nymphs** The nymphs are flat, and the first to the third instars are black, there are white powdery wax spots on the dorsal surface. The dorsal surface of the fourth instar nymphs is red, with alternated black and white spots. The first instar nymphs are 4 mm×2 mm in size, there are white powdery wax spots on the dorsal surface. There are three ridges on the head, and the middle ridge is not as pronounced as the other two. Antennae are black with long pappus, which is 3 times of antennal length. Legs are black in color, there are three white dots on the femur of the front legs, while there is only one white dot in the femur of mid- and hind legs. There are three white dots on the dorsal surface of tibia. The second instar nymphs are 7 mm×3.5 mm. Flagella of antennae are small, the length of pappus is longer than the length of the antennae. The 3rd instar nymphs are 10 mm×4.5 mm in size. The body shape is similar to the second instar; with distinct white spots on the body; head elongated compared to the second instar. Flagella of antennae are small, and the length of pappus is equal to the three segments of the antennae. The fourth

L. delicatula (1) (Zhang Runzhi)

L. delicatula (2) (Zhang Runzhi)

L. delicatula (3) (Zhang Runzhi)

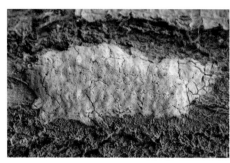

L. delicatula egg mass covered by secretion (Xu Gongtian)

L. delicatula egg mass (unhatched) (Xu Gongtian)

L. delicatula egg mass (hatched) (Xu Gongtian)

L. delicatula nymphs (Xu Gongtian)

L. delicatula 2nd instar (Xu Gongtian)

L. delicatula 4th instar (Xu Gongtian)

instar nymphs are 13×6 mm in size. Their dorsal surface is red. The sharp tip of the head, both lateral sides of the head, and basal areas of compound eyes are all black. Legs are black with white dots. The head is further elongated in comparison with the earlier instars. Wing pads are obvious and extend backwards from the lateral sides of mesothorax and metathorax.

Biology and behavior *Lycorma delicatula* occurs one generation per year. It overwinters by egg masses on tree trunks or adjacent buildings. Eggs hatch in the middle and late April, and there are four instars for the nymphal stage. The peak time for egg hatch is in the early May. Nymphs would jump and escape when they are frightened. The adults emerge from mid-June to early July. Adult feeding causes damage till October. Adults mate and oviposit eggs in the mid-August, eggs are primarily laid on the sunny side of tree trunks or tree forks. Both adults and nymphs prefer to aggregate on either trunk or leaves of the hosts, and move quickly by jumping or flying when frightened. They are good flyers and jumpers. The aggregated nymphs and adults are often found on the young shoots in a straight line.

The natural enemies of *L. delicatula* include an egg parasitoid, *Anastatus disparis* and two parasitoid wasps of nymphs.

Prevention and treatment

1. Physical control. Manually removal of the overwintering egg masses on tree trunks or buildings in winter.

2. Cultural control. *Ailanthus altissima* is the primary host of *Lycorma delicatula*, thus the single species forest with heavy damage could be replaced by planting of mixed tree species.

3. Biological control. Preservation, augmentation, and inundation of natural enemies, such as, parasitoid wasps to suppress the pest population.

4. Chemical control. Application of pesticides on trees to prevent the damage of insects. Pyrethroids, such as 25 g/L lamda-cyhalothrin at 1,000× can be used to control both adults and nymphs.

References Yang Ziqi and Cao Huaguo 2002, Xu Gongtian 2003, Xu Zhihua 2006.

(Qiao Xiurong, Zhou Jiaxi)

32 *Lawana imitata* (Melichar)

Classification Hemiptera, Flatidae

Distribution Fujian, Guangdong, Guangxi, Hainan, Yunnan; Japan.

Hosts *Euphoria longana*, *Litchi chinensis*, *Mangifera indica*, *Prunus dulcis*, *Garcinia tinctoria*, *Clausena lansium*, *Citrus grandis*, *Morus alba*, *Artocarpus heterophyllus*, *Cleidiocarpon caraleriei*, *Prunus* sp., *Olea* sp., *Cinnamomum ovatum*, *Caesalpinia sappan*, *Erythrophloeum fordii*, *Punica granatum* and others with a total of 40 families, 90 species of trees.

Morphological characteristics **Adults** Length 19-25 mm. White or light green covered with powdery wax. Head tapered in cone-shape. Gena area has a ridge. Compound eyes are brown. Antennae are located underneath compound eyes. Prothorax is curved and protruded toward the head. Mesothorax tergite is well developed with three thin ridges. The forewings are nearly triangular, apex angle is acute and anal angle is nearly right. Front margin of the wings is straight, and rear margin of the wings is slightly curved at the base. Radius and the middle of the anal veins yellow. Wax powder at the base of the anal vein forms white dots. Hind wings are white or light green, and translucent. **Eggs** Oblong, yellowish white, surface reticulated. **Nymphs** Length 7-8 mm. Relatively flattened, with wide thorax, well developed wing pads that are truncated at the end. Body white, covered with white fluffy wax.

Biology and behavior In southern Guangxi, there are two generations per year. They overwinter as adults in dense foliage. The oviposition peak period for the first generation occurs in late March to early April. Nymphs appear from April to May. Adults occur in early to mid-June. The peak of the second generation eggs occurs form early July to September. Followed by the emergence of nymphs in early August, it peaks in late August. Adults of the second generation appear in early September and reach peak in October. As the temperature decreases, the adults gradually translocate under dense foliage for overwintering. The overwintering adults resume feeding when weather becomes warm and new growth occurs in the next spring. After a period of post-diapause feeding, the adults mate and lay eggs.

Adults and nymphs suck plant sap from young branches and leaves, in particular young shoots and leaves. The feeding causes abnormal growth of the young shoots and leaf distortion. The feeding on young fruits causes fruit drop. Nymphs are active and can jump. The main feeding

L. imitata adults (Wang Jijian)

L. imitate nymphs (Wang Jijian)

L. imitata nymph (Wang Jijian)

site is on new shoots and young braches. Nymphs move to the abaxial side of the leaves before molting. They return to young shoots for feeding after molting. The wax bundles on the nymphs can extend, resembling the open peacock wing. Adults are often arranged neatly in a straight line. The abundance of this pest is often associated with abundant precipitation in the summer and fall.

Prevention and treatment

1. Manual removal of insect infested branches and leaves during winter pruning.

2. Capture adults with sweep net at the peak of the adult occurrence in the orchard.

3. Protection, augmentation and inundation of natural enemies to reduce *Lawana imitata* population. More than 20 species of natural enemies have been recorded. They are: 13 species in Vespidae (7 in Vespinae and 6 in Eumeninae), one species in Braconidae, five species in Coccinellidae, two species in Reduviidae, one mantis, and one lacewings, respectively. Vespid predators are dominant natural enemies, especially *Vespa velutina nigrihorax*, *V. bicolor*, and *V. tropica leefmansi*. These wasps play an important role in suppressing *Lawana imitata* outbreaks.

4. Using chemical control method by applying insecticides at the peaks of the adult and nymph occurrence, possible insecticides are 20% thiamethoxam, 25 g/L lamda-cyhalothrin or 12% buprofezin and lamda-cyhalothrin mixture at 1,000-1,500×.

References

Department of Plant Protection of South China College of Tropical Plants 1980; Yang Minsheng 1984; Li Tiesheng 1985; Zhou Yao, Lu Jinsheng, Wang Sizheng et al. 1985.

(Yang Minsheng)

33 *Aphrophora horizontalis* Kato

Classification Hemiptera, Aphrophoridae

Distribution Jiangsu, Zhejiang, Anhui, Fujian, Jiangxi, Hunan, Guangdong, Guangxi, and Taiwan.

Hosts Various species of bamboo plants: *Phyllostachys heteroclada* var. *pubescens*, *P. atrovaginata*, *P. aureosulcata*, *P. aureosulcata* f. *aureocaulis*, *P. aureosulcata* f. *pekinensis*, *P. bambusoides*, *P. bambusoides* f. *lacrimadeae*, *P. bissetii*, *P. bambusoides* f. *shouzhu*, *P. nidularia* 'Smoothsheath', *P. flexuosa*, *P. praecox*, *P. iridescens*, *P. sulphurea* 'Viridis', *P. rutile*, *P. glauca*, *P. fimbriligula*, *P. dulcis*, *P. makinoi*, *P. praecox* f. *provernalis*, *P. propinqua*, *P. nigella*, *P. heteroclada*, *P. vivax* and *Pleioblastus amarus*, and *P. juxianensis*.

Damage Both nymphs and adults suck tree sap from young branches, new shoots, and leaf petioles. Light infestations cause leaf chlorosis and wilt and defoliation. Severe infestations cause complete defoliation, death of the shoots, reduced bamboo fiber quality, and also reduce the new shoot development, and lowered bamboo quality for the coming year. All instars are covered under the white foamy bubbles (or spittle) after hatching, and the size of spittle increases as the insect grows. The spittle shape resembles saliva, distributed densely on bamboo branches. After the nymphs leave the feeding sites, the dried spittle forms white residue on bamboo plants with bluish reflection under the sunlight. The residue reduces the aesthetic value of the forest and tourism.

Morphological characteristics **Adults** Length 7.5-9.8 mm, head width 3.8-4.0 mm. Wings surpass the end of the abdomen. Newly emerged adults light-yellow, then change to yellow-brown, with punctures. Gena has a black spot at the center. The compound eyes are grayish black with yellow spots. There are two bright red ocelli. There is a vaguely visible black dot between the compound eyes and the ocelli. Pronotum has four lateral black spots on each side, but the spots on some individuals are vague or absent. At the center of the rear border of the prothorax, there is a large black spot. The forewings yellowish-white, wing base and wing tip grayish black. Sometimes the front margin black at ¼ and ½ from the base, and a yellowish white vertical band appears between them. A black spot exists at the anal part of the forewing. At 2/3 from the apex of the forewings, there is a crescent-moon shaped white spot. Posterior margin of the forewings in light color. **Eggs** Long cylindrical, 1.5 mm×0.4 mm. One end is slightly pointed. Milky white. Eggs are slightly thickened before hatching, the upper end of the egg ruptures, exposing a fusiform-shaped black scar, the egg body gray, tinted with reddish color, and the dark gray compound eyes are visible through the egg shells. **Nymphs** Newly hatched nymphs are 1.4 mm in length, head width 0.6 mm. Pale red in color. Head and thorax become black within half an hour after hatching. Front end of the head spherical. Antennae black, with 9 segments, and 3rd to the 9th segments are fusiform. Compound eyes are protruded and black. Legs are black. Each hind tibia with a row of spines at the tip on its inner side. Abdomen enlarged, especially for the 2nd and the 5th segments. The last segment of the abdomen truncated. The end of the abdomen slightly raised upward. There are five instars. The antennae, compound eyes, pro- and mesothorax, and front wing pads, the dorsal line, as well as the lateral sides of the thorax and abdomen are black in the mature nymphs.

Biology and behavior *Aphrophora horizontalis* occurs one generation per year in Zhejiang. The adults oviposit on dead bamboo branches from mid-October to mid-No-

A. horizontalis adult (Xu Tiansen)

A. horizontalis 3rd instar nymph (Xu Tiansen)

A. horizontalis 1st instar (Xu Tiansen)   A. horizontalis mature nymph (Xu Tiansen)   A. horizontalis nymphs in spittle (Xu Tiansen)

Damage by A. horizontalis (Xu Tiansen)

vember and overwinter as eggs. The eggs hatch in early to mid-April, and end in early May. There are five instars in the nymphal stage. Adults start to emerge in early June. All nymphs develop into adults before mid-June. Newly emerged adults immediately crawl or fly to young bamboo shoots for supplemental feeding. It is difficult to find adults where nymphs existed. The adults mate in mid-September and fly back to the nymphal feeding sites for oviposition. Adults can be seen between early June and mid-November.

Prevention and treatment

1. Practice good bamboo forest management. Remove weak small bamboos. Spittlebug nymphs prefer to feed on the understory weak bamboos. Only adults crawl or fly to tall bamboo plants and cause feeding damage. The adults crawl or fly back to the previously damaged plants for oviposition. Thus, removal of secondary and understory growth of the bamboo plants can effectively reduce spittlebug population.

2. Protection of predators. There is one bird can prey on the nymphs underneath the protected foam. In addition, *Inocellia crassicornis* (Raphidioptera: Inocelliidae) can prey on spittlebug nymphs. Two spiders, *Oxyopex lineatipes* and *Peucetia* sp. can prey on the spittlebug adults. Protection of these predators will help suppress the spittlebug population.

3. Chemical treatment. Nymphs are particularly susceptible to systemic insecticides. When population is high, inject a systemic insecticide such as thiamethoxam mixed with water at 1:1. Inject 1 mL per bamboo stem to kill nymphs.

(Xu Tiansen, Wang Haojie)

34 *Omalophora costalis* (Matsumura)

Classification Hemiptera, Aphrophoridae
Synonym *Aphrophora costalis* Matsumura

Distribution Hebei, Inner Mongolia, Jilin, Heilongjiang; Shaanxi, Gansu, Qinghai, Xinjiang, Korea Peninsula, Japan, Sweden, Russia, Britain, France, Austria, Italy, Germany, Czech Republic, Slovakia, Poland.

Hosts *Salix matsudana, Populus simonii, Ulmus pumila, Hippophae rhamnoides, Medicago sativa, Roegneria kamoji, Artemisia capillaris, Poa sphondylodes*.

Damage Nymphs of *O. costalis* suck sap from branches and produce spittlelike foam. Nymphal feeding reduces host vigor and slow down growth. The feeding location may shrink. Heavy infestation can kill the whole plant.

Morphological characteristics **Adults** Yellowish brown; length 7.6-10.1 mm, width 2.7-3.2 mm.

Eggs Lanceolate; length 1.5-1.8 mm, width 0.4-0.7 mm.

Nymphs There are five instars. The last instar nymphs are 6.4 mm in length, head width 1.7 mm.

Biology and behavior *Omalophora costalis* occurs one generation per year. It overwinters as eggs inside or on the branches. The egg hatching period starts in mid-April, peaks between late April mid-May, and ends in late May. Newly hatched nymphs prefer to aggregate and feed at the base of new shoots. The second instar nymphs not only feed at the base of the new growth, but also feed on the middle and distal part of the new shoots. The 3rd instar and later instars are more mobile and feed on both 1-2 years old branches and 3-5 years old branches. Their feeding increased significantly. The size of the foam increases significantly. The whole insect is covered with foam, and the damaged branches produce water droplets that make the branches wet.

Adults begin to emerge between mid- and late June and reach peak emergence between late June and early July. The newly emerged adults begin to feed on new shoots within 2-4 hours. The adults prefer to feed on 1-2 years old braches at the top and middle parts of the canopy. The infested branches show rings of brown (dead) scars around the branches, which can be easily broken off at the scar. After feeding for 26-40 days, the adults mate. They mate for multiple times during the life time. The females begin to oviposit the day after mating. Most of the eggs are laid on new shoots (with 2-107 eggs per shoot). Eggs may also be deposited on 1-2 years old dead shoots (with 4-44 eggs per shoot). The young shoots with eggs begin to wilt the next day. The lifespan of the females is 28-92 days. Each female can oviposit a maximum of 156 eggs. The lifespan of the males is 36-90 days. Adults exhibit no phototropism. They can fly very fast.

Prevention and treatment

1. Manual control. Manual removal of the wilted shoots with eggs at the end of autumn and the beginning of spring to destroy the overwintering eggs.

2. Chemical control. Spray 2.5% deltamethrin EC at 6,000× during late instar stage or 12% bufrofezin and lamda-cyhalothrin mixture at 1,500-2,000× at the end of May can effectively suppress the population.

References Qinghai Provincial Forestry Bureau 1980-1982, Zhao Yanpeng 1985.

(Du Baoshan, Lü Lujun)

O. costalis adult (Li Zhenyu)

O. costalis adult (Li Zhenyu)

35 *Tilophora flavipes* (Uhler)

Classification Hemiptera, Aphrophoridae
Synonym *Aphrophora flavipes* Uhler
Common name Pine spittle bug

Distribution Beijing, Hebei, Liaoning, Jiangsu, Shandong, Shaanxi; Korea Peninsula, Japan.

Hosts *Pinus thunbergii*, *P. tabulaeformis*, *P. densiflora*, *P. armandii*, *P. bungeana*, and *Larix gmelini*.

Morphological characteristics **Adults** Length 9-10 mm, light brown in color. Head slightly protruded forward. Compound eyes black; two ocelli red; the center of pronotum dark brown, the midline raised; scutellum is nearly triangular, yellowish brown, and darker in the center. Forewing grayish brown. The basal area, transverse band in the middle part, and the marks outside of the band in the front wing brown. The exterior surface of the hind tibia with two distinct spines. **Eggs** Length 1.9 mm, width 0.6 mm. Eggplant shaped or curved lanceolate in shape. Fresh eggs are milky white, become light brown later, with a longitudinal black mark at the distal end. **Nymphs** Mature nymphs dark brown or brown. Compound eyes reddish brown.

Biology and behavior *Tilophora flavipes* occurs one generation per year in Beijing and Shandong. It overwinters as eggs in leaf sheath of pine needles. The eggs hatch in mid- and late April of the following year. Nymphs are gregarious. A cluster often has 3-5 nymphs and could have as high as 30 nymphs. Nymphs suck the plant sap from the base of the new growth. They secrete white foam the abdomen to protect themselves. Nymphs molt four times. The peak of adult emergence is in early July. Adults need a relatively long period of supplemental feeding. They begin to disperse and no longer secrete spittle like foam. Females start to lay eggs in mid-August. Each female lays 28-66 eggs.

Prevention and treatment

1. During the peak of nymphal period, spray 48% chlorpyrifos at 3,500×, or 40% acetamiprid and chlorpyrifos mixture at 1,000×.

2. At the beginning of the egg hatching period, scrape off a thin layer of the bark, brush the tree trunk with 5% cyfluthrin or 45% profenofos and phoxim mixture at 1,000×.

References Zhao Fanggui 1965, Xiao Gangrou (ed.) 1992, Shandong Forest Insect Fauna Committee 1993, Xu Gongtian and Yang Zhihua 2007.

(Xu Gongtian, Li Zhenyu, Zhao Fanggui)

Dorsal view of *T. flavipes* adult (Xu Gongtian)

Lateral view of *T. flavipes* nymph (Xu Gongtian)

T. flavipes adult pronotum (Xu Gongtian)

Lateral view of *T. flavipes* adult (Xu Gongtian)

Dorsal view of *T. flavipes* nymph (Xu Gongtian)

T. flavipes spittle (Xu Gongtian)

36 *Cryptotympana atrata* (F.)

Classification Hemiptera, Cicadidae
Common name Oriental cicada

Distribution Beijing, Hebei, Shanxi, Inner Mongolia, Shanghai, Jiangsu, Zhejiang, Anhui, Fujian, Jiangxi, Shandong, Henan, Hubei, Hunan, Guangdong, Guangxi, Sichuan, Shaanxi, Gansu, U.S.A., Canada, Japan, Indonesia, Malaysia, Philippines.

Hosts There are 144 species belonging to 41 families and 77 genera of plants. The hosts with heavy damage include Poplar (*Populus* spp.), willow (*Salix* spp.), Elm (*Ulmus* spp.), Sycamore (*Platanus hispanica*), toon (*Toona ciliate*), apple (*Malus pumila*), ash (*Fraxinus chinensis*), mulberry (*Morus alba*), the chinaberry (*Melia azedarach*), peach (*Prunus persica*), pear (*Pyrus bretchneideri*), lilac (*Syringa* spp.), pissard plum (*Prunus cerasifera* var. *atropurpurea*), Japanese cherry (*Prunus yedoensis*).

Morphological characteristics **Adults** Head width 10.0-11.7 mm, body length 38-48 mm, wingspan 116-125 mm. Body black, shiny, with dense light-yellow pilosity. There are yellowish brown markings between the compound eyes and antennae. At the center of the mesonotum, there is a pair of diagonal fasciae. Basal 1/3 of forewing is greyish black, covered with short yellowish gray hair, basal cell is black, veins are reddish-brown, and the distal half of the veins are dark brown. Basal 1/3 of hindwing greyish (or smoky) black. Legs black, with irregular yellowish brown spots. The lateral side of the abdominal segments, as well as the rear edge of each abdominal segment is black, except the 8th and 9th segments. The tymbals (sound-producing organs) are located on the 1st and 2nd segments of the male abdomen. The female does not have tymbals; its 9th and 10th abdominal segments yellowish brown, split in the middle, with a spear-shaped ovipositor. **Eggs** Length 3.3-3.7 mm, width 0.5-0.9 mm. White in color. Fusiform-shaped, and slightly curved blunt at one end, tapered at the other. **Nymphs** Head width of the 4th instar 10.0-11.7 mm, body length 24.7-38.6 mm. Brownish grey in color; area in front of the antennae reddish brown, covered with dense short yellowish grey hairs; antennae yellowish brown. There is a yellowish brown longitudinal line between 1/5 and 1/2 of the posterior margin of occiput, which branched to reach the base of the antennae. There is an upside down "M" shaped dark brown mark at the anterior 2/3 part of the pronotum. The anterior part of the wing pads are grayish brown, while the posterior part is dark brown, and the abdomen is also dark brown. Ovipositor is yellowish brown.

Biology and behavior *Cryptotympana atrata* requires five years to complete a generation in middle plain of Shaanxi province. It overwinters as eggs in branches or as nymphs in soil. Mature nymphs emerge from soil and molt into adults between the end of June and early July. The adult population peaks between mid-July and mid-August. Females begin to oviposit in mid-July. Oviposition peaks between late July and mid-August, and ends in mid- and late September. The post-overwintering

Dorsal view of *C. atrata* adult (Xu Gongtian)

Lateral view of *C. atrata* adult (Xu Gongtian)

eggs (from the previous year) hatch between mid-June to early July.

Mature nymphs emerge from the soil at night between 20:00 and 06:00, and 78% of them emerge between 21:00 and 22:00. The emerged nymphs crawl up to nearby weeds, seedlings, shrubs, and tree trunk to molt into adults. About 89.7% adults eclose on tree trunk. Eclosion occcurs at the height of 23 to 800 cm, and 65% of them occur at 250 to 450 cm. The eclosion process lasts for 93-147 minutes. Newly emerged adults with green veins of the wings, and light red body color. The wings gradually unfold and form a roof shape above its body. The body and wing color darkens gradually. At 06:00, the adults gradually crawl upward on the plants. Newly emerged adults are rarely seen in the lower part of the trunk after 06:30 hour.

After their emergence, the adults feed on tree sap for a period of time before mating and oviposition. Females begin to oviposit soon after mating, which is 15 to 20 days after the adult emergence. Fecundity of a female is between 500 to 800 eggs. The females begin their oviposition process by piercing into xylem of the branches to make an oviposition mark (or groove), and then oviposit their eggs in the pith of the branches. The number of eggs per branch can be as high as 634 eggs, with an average of 153 to 358 eggs per branch. A branch could have 1 to 11 oviposition marks with an average of 2.6 per branch. Clusters of eggs (egg masses) per oviposition mark range from 1 to 51, with an average of 11.8 clusters per oviposition mark. Some oviposition marks have no egg clusters, and empty oviposition mark rate is 6.3%. The number of eggs per cluster ranges from 1 to 18, with an average of 6.4 eggs per cluster. The oviposition mark is fusiform in shape, and the clusters of eggs are arranged in two straight rows or a helix. The branch beyond the oviposition site often wilts shortly after oviposition.

Adults can aggregate and massively disperse. They move between 08:00-11:00 from large to small trees and have inverse dispersal between 18:00-20:00. Although the adults have strong flight ability, they usually disperse for only a short distance. The adults can be attracted by light and fire flame when the trees were shaken. The adults do not fly toward lights or fire without being disturbed.

Male adults are well known for their singing. They often sing in chorus (synchronized) during late June and early October. Males start to sing when temperature is above 20℃, and sing in chorus when it is above 26℃. In particular, when temperature is above 30℃, not only did the singing duration is long, but also the singing is more frequent and louder than that under lower temperatures. It has been reported that sound pressure level (SPL) for a single cicada singing is 76 dB, whereas singing by a group of cicadas can reach 87 dB. Based on the ending time of the adult emergence and the ending time of the cicada singing, the estimated adult life span is from 45 to 60 days.

Adult sex ratio is dynamic; at early emergence period, female and male ratio is 1:6-8.5; at the peak of emergence, the sex ratio is 1:1; and at the end is 1:0.067-0.22. The overall sex ratio is 1:1. Humidity greatly influences egg hatch rate. High levels of precipitation and humidity cause early egg hatch with high egg hatch rate, while dry climate conditions lead to the postponed egg hatch, and reduced egg hatch rate. The egg hatch rate is from 75.2% to 92.7%, the egg stage can be from 260 to 345 days.

Nymphs burrow into the soil immediately after hatch, and feed on plant roots by sucking the plant sap. Nymphal stage lasts four years. The nymphs molt once a year between June and September, with a total of 4 instars. The 1st and 2nd instar nymphs attached onto the lateral

C. atrata adult with spread wings (Xu Gongtian)

C. atrata shed skin (Xu Gongtian)

roots and fibrous roots, while the 3rd and 4th instars are attached onto relatively thick roots, and often at the branching points of the roots. Nymphs build oval-shaped chambers for molting and feeding. The chamber is hard and smooth, and very close to the host plant roots, and one nymph per chamber. Nymphs are often located at 0 to 30 cm in depth in soil, and could be up to 80 to 90 cm in the soil.

In Qinling Mountain, *C. atrata* is distributed in the southern slope at an altitude of 1,060 m and in the northern slope at 1,270 m above the sea level. They are common at the altitude of 400 to 600 m. High population density and severe damage is often found near river basins, on the plains away from the towns and villages, or at the base of the mountainous or hilly areas. Young trees suffer more damage than old trees, and low density forests suffer more damage than high density forests. Adults are selective when choosing oviposition sites, therefore, different poplar varieties show different degree of damages due to their differences in the sizes of the 1-2 year-old poplar branches. The thicker the branches, the less is the damage from *C. atrata*. The rainy days with high temperature is conducive to adult emergence and egg hatch, whereas sunny and high temperature day is conducive to adult mating, oviposition, and singing.

The predators of *C. atrata* adults include *Cuculus micropterus*, *Otus scops*, short-eared owl (*Asio flammeus*), *Caprimulgus indicus*, red-footed falcon (*Falco vespertinus*), kestrel (*Falco tinnunculus*), *Apus apus*, *Eurystomus orientalis*, *Anthus hodgsoni*), *Motacilla alba*, *Lanius cristatus*, *Oriolus chinensis diffuses*, *Sturnus cineraceus*, *Pica pica*, *Cyanopica cyana*, *Dicrurus marocercus*, *Passer montanus*, and other birds; and bats, ants, praying mantis, centipedes, spiders and other predators, and pathogenic microbes. There is one parasitic wasp at egg stage. Its parasitization rate is 4.6% to 13.0%. In addition, the multicolored Asian lady beetle (*Harmonia axyridis*) and ants also prey on the eggs. Natural enemies of the nymphs include ants, spiders, earwigs, praying mantis, lady beetles and pathogenic microorganisms.

(Hu Zhonglang, Han Chongxuan)

37 Dundubia hainanensis (Distant)

Classification Hemiptera, Cicadidae

Importance *Dundubia hainanensis* is one of the most important pests within the family. Hundreds or thousands of them can aggregate on trees and suck the tree sap. Their feeding activity reduces yield. The adults produce high level of noise. They frequently excrete while feeding and this phenomenon is sometimes mistakenly considered "trees can rain."

Distribution Guangdong, Guangxi, and Hainan.

Hosts *Pyrus pyrifolia*, *Euphoria longan*, *Litchi chinensis*, *Artocarpus heterophyllus*, *Clausena lansium*, *Acacia confusa*, *Alstonia scholaris*, *Michelia macclurei*, *Michelia alba*, *Melia azedarach*, *Eucalyptus citriodora*, *Paulownia fortunei*, and *Anthocephalus chinensis*.

Damage When high number of nymphs feed on the roots, the vigor and growth of plants are reduced, leaf chlorosis occurs, flowers and fruits fall prematurely. Many emergence holes can be observed during the adult emergence period. The adults often aggregate on the tree branches and sing, and the excretion produced by adults often moisten the ground.

Morphological characteristics **Adults** Female body length 34.2-39.1 mm, forewing length 45-50.8 mm. Male body length 40.2-45.0 mm, forewing length 46-49.2 mm. Head capsule width 12.4-14.1 mm. The top of the head, pronotum and mesonotum are dark green; postclypeus with a brown mark, its posterior border is less than two times of the front border. There is a black spot in the middle of posterior border of the pronotum. There are brown striations in the lateral area of the pronotum. The posterior border of the pronotum is black. The center of front edge of the mesonotum has white thin setae. The ventral surface of its body is castaneous, with long and thin hair and white powdery wax. Femur, trochanter of the front legs and femur and tibia of middle legs have brown spots; tibia and tarsus of front legs and tarsus of middle legs are brown. The dorsal surface of the abdomen is castaneous or ocherous, and the ventral surface is ocherous. The opeculum of male adults are rectangular, and narrow at the base, the distal end is sharp and round, and extends to the seventh abdominal segment or subgenital plate; the opeculum of female adults is triangular, and extends only to the third abdominal segment. Both wings are transparent. **Eggs** White or near ivory-white, elongated, in banana shape and slightly curved, one end is blunt and the other end is sharp, 2.1 mm in length, 0.46 mm in width. **Nymphs** The young nymphs are milky white or yellowish white. The head is yellowish brown, abdomen is expanded to the shape of spiders. The fourth and fifth instar nymphs are yellowish brown with wing pads; their front legs are adapted for digging. The spines on the front legs are dark brown.

Biology and behavior *Dundubia hainanensis* completes one generation in multiple years. Adults emerge between late April and early June. In late April, the mature nymphs emerge from the soil during 19:00-20:30 hours. They climb up to 0.8-1.5 m in height on tree trunks and molt into adults. After about one hour, adults can quickly crawl upward to the shoots. The peak of adult emergence lasts for 9-10 days. After feeding, male adults can sing in groups. The courtship singing is loud and deafening at dawn. The female adults then fly around the tree canopy to choose their mates, and begin to mate after half an hour.

D. hainanensis adult (Wang Jijian)

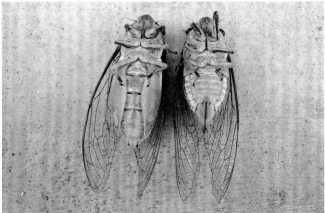

Ventral view of *D. hainanensis* female (right) and male (left) (Wang Jijian)

Adult *D. hainanensis* eclosion in process (1) (Wang Jijian)

Adult *D. hainanensis* eclosion in process (2) (Wang Jijian)

Adult *D. hainanensis* eclosion in process (3) (Wang Jijian)

Adult *D. hainanensis* eclosion in process (4) (Wang Jijian)

The mating lasts for one hour. The males become silent during mating. Each female produces more than 500 eggs. Eggs are mainly deposited on branches of 1-2 years old. After eggs hatch, nymphs fall to the ground and burrow into the soil. They select lateral thin plant roots to anchor and feed. The late instars will translocate to and feed on thick roots until ready to emerge into adults.

Prevention and treatment

1. Preservation of natural enemies. In addition to many birds preying on *D. hainanensis*, praying mantises and *Calotes versicolor* also prey on adults and nymphs. These natural enemies should be protected.

2. Manual control. Based on the narrow window of adult emergence, capture and remove both climbing nymphs and newly emerged adults every night during their emergence is the most simple and effective method for reducing its population.

3. Biological control. *Metarhizium anisopliae* and *Paecilomyces fumosoroseus* show a high level of virulence to the cicada and can be utilized.

4. Chemical control. Based on the size of the tree trunks, drill 2-4 holes per tree, and inject 2-4 mL of 20% carbofuran into the holes, and then seal the holes with sealing compound. During the non-fruit bearing season, 3% carbosulfan granules can be buried into soil at 20-50 g per plant according to the size of trees to kill nymphs underground and adults on the trees.

Reference Wang Jijian 1994.

(Wang Jijian)

38 *Macrosemia pieli* (Kato)

Classification	Hemiptera, Cicadidae
Synonym	*Platylomia pieli* Kato
Common name	Bamboo cicada

Distribution Jiangsu, Zhejiang, Anhui, Fujian, Jiangxi, Hunan, and Sichuan.

Hosts *Phyllostachys heteroclada* var. *pubescens*, *P. edulis* f. *gimmei*, *P. heterocycla* 'Pubescens', *P. sulphurea* 'Viridis', *P. viridis* f. *youngii*, *P. verrucosa*, *P. meyeri*, *P. glauca*, *P. nidularia* 'Smoothsheath', *P. praecox*, *P. tianmuensis*, *P. propinqua*, *P. bambusoides*, *P. vivax*, *P. nuda*, *P. iridescens*, and other species in genus *Phyllostachys* with thick main stem and thick rhizomes. Adults also feed on *Cinnamomum camphora*, *Liquidambar formosana*, *Metasequoia glyptostroboides*, *Schima superba* and other tree species.

Damage All instars of the nymphs build a chamber in the soil near Rhizome, and feed on sap of rhizomes, buds, and shoots resulting in the Rhizome blight, abnormal growth, and cease of growth and decay of the lateral buds. The damaged bamboo plants with less new shoots, thin shoots, and reduced diameter of new plants, and gradually decline of the plant vigor, and the damage is difficult to recover. Adults feed on young bamboo shoot sap, and lead to large number of dead bamboo plants, which become the oviposition sites of the adults for the next 1 or 2 generations.

Morphological characteristics **Adults** Female body length 38.6 to 44.1 mm, male body size 42.9-53.5 mm×16.1-18.7 mm. The newly emerged adults are green, and darken gradually, as well as black and brown embedded markings appear. Antennae are black and think. The compound eyes are protruded in brownish black color. The ocelli are reddish white, shiny and bright as pearls. Enlarged mesothorax, and the markings like facial patterns in the Peking opera, there is a yellow protrude forms an "X"-shaped ridge. There are two depressions in front of the ridge and on both sides. Wings are longer than the abdomen. Femur and tibia of the front legs are relatively thick, and the ventral surface of the tibia had 2 hairs, forming a plier shaped structure. Abdomen is dark brown covered with white powdery wax, scattered with a few short and thin yellow hairs. The end of the female abdomen is cone shaped; ovipositor is hard. The end of the male abdomen is blunt. Male tymbals are well developed, the average size of opeculum for the

A branch with *M. pieli* eggs (Xu Tiansen)   Dorsal view of *M. pieli* adult (Xu Tiansen)   Lateral view of *M. pieli* adult (Xu Tiansen)

M. pieli 1st instar nymphs (Xu Tiansen)   *M. pieli* 3rd instar nymph (Xu Tiansen)   *M. pieli* 5th instar nymph (Xu Tiansen)

tymbal is 23.2 mm.

Eggs Long fusiform, 2.45-2.78 mm×0.51-0.78 mm in size; milky white, shiny as jade.

Nymphs Mature nymphs are orange red. The frons is protruded, and lower sides of the frons have dense brown setae. The compound eyes are protruded, and milky white; antennae are linear with 9 segments. The front part of the pronotum has an inverted triangle, and outside of the hypotenuse of the triangle have a deep groove that ended at the center of the pronotum. Tracheae are powdery white. The femur and tibia of the front legs are enlarged, and modified into a plier like structure, and tibia is triangular with tooth like structure on the top, and dark brown in color. Forewing pads reach the posterior edge of the fourth abdominal segment.

Biology and behavior The life cycle of *Macrosemia pieli* takes 6 years to complete in Zhejiang province. They overwinter as eggs on decaying bamboo shoots, or as nymphs in soil chambers. Adults emerge between late June and late August. The adult stage ranges from late June to mid-September. Adults begin to mate in mid-July, and oviposition starts in late July after supplemental feeding. Oviposition ends at the end of August. The eggs will not hatch until a year later in early July, so the egg stage is 11 to 12 months. There are five instars of the nymphs. They molt once a year. Mature nymphs emerge from soil in the 5th year and eclose to adults.

Prevention and treatment

1. Protection and preservation of natural enemies. Natural enemies of *M. pieli* are abundant. The adult predators include grey magpie (*Cyanopica cyana*), long-tailed blue bird (*Cissa ergtyrorhyncha)*, cuckoo (*Cuculus canorus canorus*), the European cuckoo (*Cuculus canorus*), *Bambusicola thoracica*; insect predators include praying mantis (*Hierodula patellifera*), Chinese sword praying mantis (*Paratenodera sinensis*); egg parasitoid wasps include *Eupelmus* spp.; the predators of nymphs include *Sandalus* spp. There are also fungal pathogens of the nymphs, including *Isaria cicadae* and *Cordyceps sinensis*.

2. Attracting adults for oviposition using dead bamboo shoots. Collection of 2-year-old dead bamboo shoots from the ground in early July and tied in bundles and hung them on bamboo stalks at 2-3 m height to attract adults for oviposition. The dead branches can be burned before egg hatch which occurs next June.

3. Light trapping. Install black light traps in early July to attract and kill emerging adults.

4. Manual removal of emerging nymphs. When the mature nymphs emerge from the soil at night between 19:00 to 21:00, collect and kill them with the aid of a flashlight.

5. Chemical control. When the new shoots start to grow, apply *Beauveria bassiana* around bamboos.

References Xu Tiansen, Wang Haojie 2004; Xu Tiansen, Wang Haojie, Yu Caizhu 2008.

(Xu Tiansen)

39 *Hypsauchenia chinensis* Chou

Classification   Hemiptera, Membracidae

Distribution   Guizhou.

Hosts   Tung oil tree (*Vernicia fordii*), Chinese tallow tree (*Sapium sebiferum*), and *Alnus nepalensis*.

Damage   *Hypsauchenia chinensis* ingests sap from fruits, petioles and twigs of hosts.

Morphological characteristics   **Adults** Length 7-11 mm, wingspan 12-17 mm. Dark brown in color, with dense light brown pubescence and scattered brown bristles. A pair of ocelli are located at the top of the frons; compound eyes are located below the ocelli. Antennae black, 1.5-2.5 mm in length. The metanotum extended to the head and often protruded backward dorsally above the abdomen; length 5-17 mm, arched and compressed; narrowing gradually from its angular base to the end of crown. The end of the crown has a pair of peach shaped structure at opposite positions, on which there are markings resembling turtle shell; dark brown in color. There is a saddle shaped protrusion over meso- and metathorax. Males are smaller, and lighter in color than the females. Other morphological features are similar to females.

Nymphs   Length of the 1st instar 0.7-1.2 mm, with scattered light-yellow bristles. Both head and thorax are light-yellow, abdomen slightly dark color. The middle of the frons is concaved with thorn-like process on both sides. Antennae bristle form, light green, 0.8-1.0 mm in length. Mouthparts have black stylets. Anal plate with 2 segments, reddish brown in color. Cornicles are longer than both sides of the anal plate, cone shaped, and the tip is grey and the base is brown. The 2nd instar 2-3 mm in length, and light brown. The two cornicles of the 1st instar become two spines of 0.5-1.0 mm. The 3rd instar 3.5-4.5 mm long and dark brown. Pronotum with an upward protrusion (1 mm in height). The 4th instar 5-6 mm in length, and dark brown. Both lateral sides of meso- and

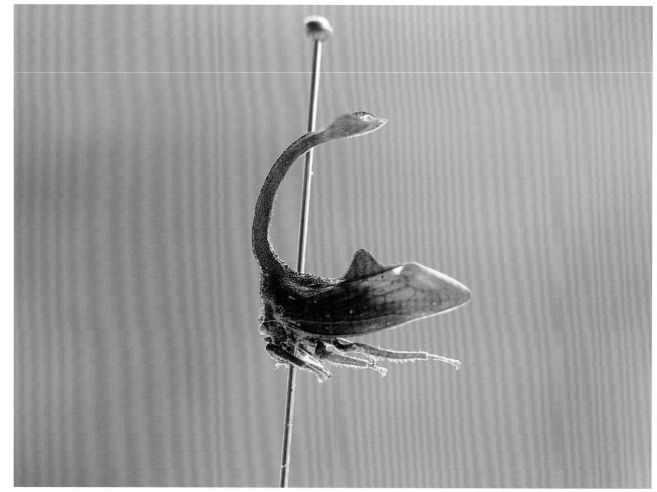

H. chinensis adult (Yu Jinyong)

metanotum extruded posteriorly and attached with the base of the wings.

Biology and behavior There are 4 generations per year. It overwinters as adults on dormant buds, or bark cracks. In late April of the following year, overwintering adults start to feed and mate during the end of the flowering period and when emergence of young buds of the Tung oil tree occurs. In early May, the females oviposit on young shoots, fruits, and leaf petioles. There are 61 to 152 eggs per egg mass and the hatch rate is 70%-90%. In mid-May, the first generation of nymphs appears. Newly hatched nymphs can crawl, and the 1st and the 2nd instar nymphs often aggregate and feed near the egg masses, while the 3rd and 4th instars are dispersed and feed in groups of 3-5 insects or individually. Nymphal stage lasts for 20 to 25 days. Adults of the first generation emerge in early to mid-June. The population of the first generation is low and causes little damage. The second and third generations occur between late June and early September. Due to warm weather conditions, they have high reproduction rate and high population levels and thus cause severe damage. The fourth-generation eggs appear in mid-September, nymphs appear in late September, and adults emerge in October or early November. The adults are agile and ready to jump or fly when they are disturbed. The adult lifespan is 3-6 months. The females can oviposit multiple times. Both nymphs and adults secrete whitish honeydew.

Prevention and treatment

1. Chemical control. The trees can be treated by spraying with 25 g/L lamda-cyhalothrin, 12% thiamethoxam and lamda-cyhalothrin mixture at 1,000-1,500× in May or June to control adults and nymphs.

2. Protection/preservation of praying mantis and yellow ants can reduce population density.

References Forest Bureau of Guizhou Province 1987, Xiao Gangrou (ed.) 1992.

(Yu Jinyong, Feng Junxian)

H. chinensis nymphs (Zhang Peiyi)

H. chinensis adult (Zhang Peiyi)

40 *Cicadella viridis* (L.)

Classification Hemiptera, Cicadellidae
Synonym *Tettigoniella viridis* (L.)
Common name Green leafhopper

Distribution Hebei, Inner Mongolia, Liaoning, Jilin, Heilongjiang, Jiangsu, Zhejiang, Anhui, Fujian, Jiangxi, Shandong, Henan, Hubei, Hunan, Guangdong, Hainan, Sichuan, Guizhou, Shaanxi, Gansu, Qinghai, Ningxia, and Taiwan; Russia, Japan, Korea Peninsula, Malaysia, India, Canada and European countries.

Hosts The hosts of *C. viridis* include more than 160 species in more than 30 families of plants, such as, *Populus* spp., *Salix* spp., *Morus alba*, *Zizyphus jujuba*, *Elaeagnus mooroftii*, *Juglans regia*, *Robinia pseudoacacia*, *Ulmus* spp., *Acer negundo*, *Fraxinus chinensis*, *Punica granatum*, *Syringa valgaris*, and plants from Rosaceae, Asteraceae, Cruciferae, Leguminosae, and Chenopodiaceae.

Damage Nymphs feed on leaves, branches, and trunks in the spring and summer, and causes leaf chlorosis and abnormal growth at the feeding sites. In the autumn, female adults oviposit on young shoots and trunks, and form crescent shaped scars. The damaged area die next spring due to water loss.

Morphological characteristics **Adults** Female length 9.4-10 mm, male 7.2-8.3 mm. Green body with the frons of head is light brown. A pair of black spots appear between the two ocelli. Compound eyes green; pronotum light yellowish green, the posterior half deep green; scutella light yellowish green; forewing turquoise green in females and dark bluish in males; the veins bluish yellow with narrow light black; hind wing greyish black, and semi-transparent; the dorsal surface of the abdomen bluish black; the ventral surface of thorax and abdomen, and legs orange yellow.

Eggs They are milky white, oval, and 1.6 mm in length. One end is slightly thinner than the other and curved in the middle. The egg surface is smooth.

Nymphs Newly hatched nymphs are light yellowish green. Compound eyes are red. After 2-6 days, the nymphs turn light yellow or deep gray. The 2nd instar nymphs are greyish black, and there are two black spots on the vertex. Wing pads appear in the 3rd instar, and reproductive appendage appears in the 4th instar. The 5th instar nymphs are 6-7 mm long. There are two black spots on the vertex, and four brown longitudinal lines on dorsal and lateral sides of the thorax, which extend to the end of the abdomen. The legs are light yellow.

Biology and behavior *Cicadella viridis* has two generations per year in Akesu, Xinjiang. It overwinters in egg stage inside cortex of young shoots and trunks starting in the middle of September. The first and second generations appear from late April to mid-July and from mid-June to early November, respectively. The over-

Dorsal view of *C. viridis* adult (Xu Gongtian)

Lateral view of *C. viridis* adult (Xu Gongtian)

wintering egg stage lasts for more than 5 months. The egg stage in the first generation lasts for 9-15 days; the nymphs in the first generation occur between mid-April and early June and lasts for 44 days. The nymphs of the second generation occur between the late-June and late August and last for 24 days. The peaks of adults for the first and second generations are between mid-May and mid-August, respectively. All adults die in early November.

The newly emerged adults of *C. viridis* are active. They are most active at noon when temperature is high. Adults are weak fliers. They crawl sideways when disturbed. When the adults are frightened, they jump and fly away. Adults show strong phototropism. The sex ratio of females and males is 1.26:1. Adults prefer humid environment and are sensitive to wind, they prefer to inhabit and feed on dense, green and flashy crop and weedy plants in groups and cause damage. Adults begin to mate and oviposit after feeding for more than one month. They usually mate and oviposit during daytime. Females can oviposit one day after mating. They first cut the surface of host plants with serrated ovipositors to form crescent marks, then deposit eggs under the epidermis in rows, and then seal the oviposition wound with a layer of white excretion. Each female adult can deposit 7 (a range of 3-10) egg masses. The number of eggs in each egg mass differs with the hardness of trunks, and generally an egg mass has 11.2 (2-15) eggs. Adults can finish oviposition within days or weeks. In the summer, eggs are mainly deposited in the stalks and leaf sheaths of gramineous plants such as reed, wild oat, *Poa annua*, *Achnatherum splendens* and other grasses, while the overwintering eggs are deposited on young and smooth shoots and trunks of woody plants and fruit trees, and the highest egg densities were found on shoots with the diameter of 1.5-5.0 cm. On 1-2 years old trees, the egg masses are mainly distributed on the main trunks at 0-100 cm in height with higher density near the ground. On 3-4 years old trees, the egg masses are concentrated on trunks and branches that are 1.2-3.0 m above the ground. The density of egg masses is highest at the lower part of the tree canopy. Egg masses are mainly deposited on the north and south sides of the trunks, and more eggs on the north side than on the south side of a tree trunk. If the branches have high density of egg masses, the tree will die in the winter.

Eggs hatch in the morning. Peak hatching time is at 07:00-08:00. The egg masses on the trunks facing south-east side hatch first. Overwintering eggs hatch when pear and willow start flowering, and poplar expands its buds and leaves. The egg hatch peaks when pear flower peaks, and *Elaeagnus angustifolia* and willow unfold leaves. The egg hatch period ends when pear flowering time ends, and mulberry fruit matures, and *E. angustifolia* starts flowering. The egg hatch rate is 72.8% (52%-91%). Newly hatched nymphs aggregate in groups for an hour and then begin to feed. The ability to jump increases within the first 24 hours, and often 10-20 nymphs

C. viridis oviposition scars (Xu Gongtian)

C. viridis oviposition scars (Xu Gongtian)

aggregate and feed in groups on leaves or young shoots. Nymphs are most active at noon when temperature is high, and not active in the morning when temperature is low and the ground is moistened. Within three days, most nymphs move from woody hosts to short herbaceous crop plants. The nymphs usually crawl up on a plant, when they are disturbed, they will move from adaxial to abaxial side of the leaf surface like the adults do.

Prevention and treatment

1. Forest management. Reinforce quarantine and inspection measures. Prevent infested seedlings from being transported. These seedlings should be pruned to the ground. The egg masses on the trees should be removed and destroyed. Plant insect-resistant trees, e.g., white poplar, ash, mulberry, and *Tamarix chinensis* to establish mixed forests. Avoid intercropping other host plants of this pest in the in nursery or orchard. Interplanting of grass or legume plants as trap crops near shelter forests and urban landscape to trap and kill the pest on the trap crops. In mid- and late August, timely cease of irrigation and keep soil surface dry will accelerate lignification process of the young trees and reduce *C. viridis* damage. Between the end of October and mid-November, timely saturated irrigation will reduce death from water loss in next spring.

2. Manual control. Because adults are usually not active on plants in the morning, they can be captured using a sweep net. During the peak oviposition period between mid- and late June, mowing the weedy plants near forest area and crop fields can destroy eggs. Mowing the weedy plants near the forest areas between mid- and late August can eliminate the oviposition habitat and reduce adult population. During winter and early spring when plant is in dormancy, remove lower lateral branches and branches with eggs can reduce the overwintering population. The application of lime salt solution by brushing on the tree trunk below 1.2 m before wintertime can reduce the survival of the overwintering eggs.

3. Physical control. Install black lights in nursery and newly established forest at the peak time to trap adult populations (i.e., between mid-May and mid-June, and between mid-August and mid-September, respectively).

4. Biological control. Apply *Beauveria bassiana* or *Paecilomyces fumosoroseus*. The natural enemies of the eggs include *Allothrombium pulvinum*, *Nabis sinoferus*, and *Pygolampis bidentara*. The natural enemies of the nymphs and adults include *Passer domesticus Formica gagatoides*, *Plagiolepis rothneyi*, etc. Chemical control should be avoided in the spring and summer in nursery and newly established forest to preserve and increase the

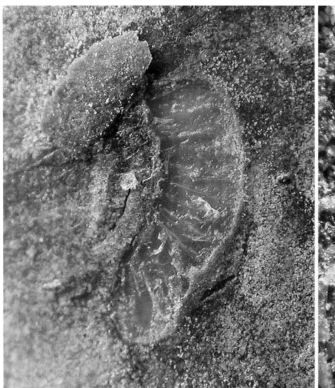

Egg mass of *C. viridis* (Xu Gongtian)

C. viridis eggs (Xu Gongtian)

abundance of natural enemies.

5. Chemical control. The nymphs can be controlled at the peak of egg hatch time with 20% isoprocarb EC or 45% profenofos and phoxim mixture at 1,000-1,200×, 2.5% deltamethrin or 50% pirimicarb ultra-fine WP at 3,000-4,000×. Adults can be controlled using 25 g/L lamda-cyhalothrin at 1,000-1,200×,

References Xiao Gangrou (ed.) 1992; Lu Zhiwei, Li Zhonghuan, and Wang Aijing 1994; Li Zhonghuan et al. 1995; Wang Aijing 1995a, b, 1996; Wang Aijing et al. 1996.

(Wang Aijing, Zou Lijie, Li Zhonghuan)

Dorsal view of *C. viridis* adult head and thorax (Xu Gongtian)

41 *Ricania speculum* (Walker)

Classification Hemiptera, Ricaniidae
Common name Eight-spotted fulgorid

Distribution Jiangsu, Zhejiang, Fujian, Henan, Hubei, Hunan, Guangdong, Guangxi, Hainan, Yunnan, Shaanxi, and Taiwan provinces.

Hosts *Eucalyptus* spp., *Schima superba*, *Juglans regia*, *Nyssa sinensis*, *Rhus verniciflua*, *Camellia oleifera*, *Aleurites fordii*, *Dalbergia hupeana*, *Lagerstroemia indica*, *Toona sinensis*, *Michelia alba*, *Parashorea chinensis*, *Salix babylonica*, *Mallotus apelta*, *Glochidion macrophyllum*.

Damage Nymphs and adults feed on saps of young shoots and reduce growth of plants. Adults oviposit on shoots which become weak and turn yellow or wither.

Morphological characteristics **Adults** Length 6.0-7.5 mm, wingspan 25-27 mm. Head and thorax dark brown; wings, abdomen and legs brown; some individuals yellowish brown. Frons with a median ridge and ambiguous lateral ridges; clypeus with a median ridge. There is also a median ridge on pronotum and an obvious dot on each side. Mesonotum with three longitudinal ridges, the middle ridge is long and straight, the lateral ridges branched forward at the middle. Forewing brown to dark brown. There is a semicircular transparent spot at 2/5 to the end of the costal margin, a large irregular transparent spot in the front but below the semicircular spot, and an elongated transparent spot below and behind the semicircular spot. There is a small elongated transparent spot near the distal corner of the costal margin. Apical margin with two large transparent spots accompanied with several small circular white spots. Hind wing dark brown to blackish brown, semi-transparent; the veins are darker than the base color of the wings. **Eggs** Milky white and spindle shaped. Length 0.8 mm. **Nymphs** There are five instars. Greyish white. The last instar nymph is 4-5 mm long. The body is in shield shape. The end of abdomen is covered with greyish white wax threads.

Biology and behavior *Ricania speculum* occurs one generation per year and overwinters in egg stage inside host plant branches. Nymphs occur between April and May in southern Guangxi, newly hatched nymphs aggregate and feed on new shoots, and the 4th instars disperse and feed individually. The adults emerge between June and July, and then mate and oviposit. A female pierces into xylem of new shoot with its ovipositor to deposit a batch of eggs, and then covers it with white cotton-like wax threads. Each female oviposits 4-5 times. Each time it oviposits from dozens to 150 eggs. Nymphal stage lasts for 40-50 days. Adult life span is 30-60 days. The egg stage lasts for 260-300 days.

Ricania speculum prefers to aggregate in groups. It can quickly jump or flee. The adults have weak flight ability. The upper part of shoots receives more damage than the middle and lower part of the shoots.

Prevention and treatment

1. Manual control. Survey, remove, and burn the infested shoots bearing overwintering eggs in the autumn and winter.

2. Protect and utilize natural enemies. Ants, spiders, praying mantis, birds can prey on the insect pests in forests. These natural enemies should be protected and utilized.

3. Chemical control. In the area with high population density, spray 90% trichlorfon at 800-1,000× combined with 0.2% silicone, thiamethoxam and phoxim mixture at 1,000-1,200×, 40% Dusban EC at 1,000×, or 25 g/L lamda-cyhalothrin at 800-1,000×.

Reference Xi Fusheng, Luo Jitong, Li Yugui et al. 2007.

(Wang Jijian)

Lateral view of *R. speculum* adult
(Wang Jijian)

Dorsal view of *R. speculum* adult
(Wang Jijian)

R. speculum adult
(edited by Zhang Runzhi)

42 *Agonoscena xanthoceratis* Li
Classification Hemiptera, Aphalaridae

Distribution Hebei, Shanxi, Inner Mongolia, Jilin, Liaoning, Heilongjiang, Shandong, Shaanxi, Gansu, Qinghai, Ningxia, and Xinjiang.

Host *Xanthoceras sorbifolia*.

Damage Nymphs and adults feed on sap of buds, leaves and young shoots, which results in leaf rolling, chlorosis and prematurely leaf senescence.

Morphological characteristics **Adults** Male length 1.35-1.50 mm, female length 1.57-1.75 mm. Newly emerged adult body is white, and then changes to pale green, orange yellow, and gradually to greyish brown. Antennae are light yellow with 10 segments, the end of 8th and 9th and 10th segments is black, and there is a pair of bristles at the end. The compound eyes and three ocelli are red. Dorsal surface of thorax has longitudinal brown stripes. Margins of the front wing, dorsal plates, and ventral plates of all abdominal segments have brown spots, and lateral surface is yellow. The tip of the female abdomen is slightly pointed downward, while the end of the abdomen of the males is widely open. **Eggs** Long oval, with egg stalk at the base. Length 0.20-0.23 mm, width 0.09-0.12 mm. Newly deposited eggs are milky white, translucent, and then become light-yellow, and the compound eyes can be seen as orange red spots before egg hatch. **Nymphs** Flat and light green, the size is 1.15-1.20 mm×0.70-0.80 mm. There is a depression at the middle of the front border of its head. Compound eyes are red; antennae are light brown with 7 segments. There are two longitudinal yellow lines from the head to the 4th segment of the abdomen.

Biology and behavior In Inner Mongolia, there are 3 generations per year, with overlapping generations. The adults overwinter under barks at basal part of the

Lateral view of *A. xanthoceratis* (Xu Gongtian)

A. xanthoceratis nymph (Xu Gongtian)

A. xanthoceratis eggs, nymphs, and adults (Xu Gongtian)

A. xanthoceratis adults on a bud (Xu Gongtian)

A. xanthoceratis adult and nymphs on a leaf (Xu Gongtian)

Sooty mold induced by *A. xanthoceratis* infestation (Xu Gongtian)

A. xanthoceratis on *Xanthoceras sorbifolia* leaves (Xu Gongtian)

trunk or under leaves on the ground. In mid-April when new buds grow, the overwintering adults start to mate and oviposit. Adults often aggregate in groups of a few to dozens on leaves or young shoots to mate and feed or jump and fly around the tree canopy. The insect rest on abaxial side of leaves, trunk, or bark in the morning, evening or windy, as well as cloudy and rainy days. The eggs hatch from early to late-May. There are five instars for the nymph stage. The first-generation adults emerge at the end of May, oviposition starts in mid-June; second-generation eggs hatch in late June. The second-generation adults emerge in early July, and the third generation adults emerge in early August. After supplemental feeding, some adults go into summer dormancy. All adults gradually go into overwintering after supplemental feeding in mid-September.

Prevention and treatment

1. Prevent dispersal shipment of the infested plant materials. Timely manage the pest on newly established stands. Improve its canopy density, improve insect resistance; remove and burn branches that are infested with eggs. Clean dead leaves and weeds to eliminate the habitats for overwintering adults.

2. Conservation of natural enemies. The natural enemies include seven spotted lady beetles (*Coccinella septempunctata*), Europe lacewings (*Chrysopa perla*), brown lacewing (*Hemerobius humuli*), and Syrphid larvae, such as *Scaeva pyrastri*.

3. High pest population can be treated with insecticides, such as, spraying of 40% acetamiprid and chlorpyrifos mixture at 800 to 1,000×, buprofezin at 40 g/ha, pyradizinone at 0.05 g/L. In forests of high canopy coverage, fogging with 5% emamectin 2-3 times at the rate of 15 kg/km^2 when overwintering adults become active in mid- and late April.

Reference Xiao Gangrou (ed.) 1992.

(Xie Shouan, Li Menglou, Li Rongbo)

43 *Anomoneura mori* Schwarz

Classification Hemiptera, Psyllidae
Common name Mulberry psyllid

Importance The adults and nymphs can damage mulberry trees by feeding on buds and leaves and remove plant sap. Affected mulberry buds could not develop. The honeydew excreted by the nymphs induces sooty mold, seriously reduces leaf yield and quality, which in turn affects the development of silkworms and silk production relying solely on mulberry leaves as food.

Distribution Beijing, Tianjin, Hebei, Liaoning, Jiangsu, Zhejiang, Anhui, Hubei, Sichuan, Guizhou, Shaanxi, Taiwan; Japan.

Hosts *Morus alba*, *Platycladus orientalis*, *Sabina chinensis*.

Damage Mulberry psyllid nymphs feed on mulberry buds, ingest plant sap, cause reduced growth, unhealthy leaves, and leaf rolling. Severe damage includes tissue necrosis or chlorotic spots. The nymphs excrete wax threads spreading over leaves. Their excretion contaminates both the infested leaves and the leaves at the lower level of the canopy and can induce sooty mold and attract ants.

Morphological characteristics **Adults** Body length is 3.5 mm, wingspan is 8-9 mm. Newly emerged adults are light green, and then become grayish brown. Its body resembles cicadas, compound eyes are hemispherical and in auburn color, and two ocelli are light red. Antennae yellow, distal segment dark brown. The dorsal thorax protruded with several pairs of dark yellow lines. Front wings are translucent with brown markings.

Eggs Fresh eggs are white, and then change to yellow. One end of the eggs is pointed with an angle, while the other end is round with egg stalk. The compound eyes can be seen as a pair of red dots before hatching.

Nymphs The nymphs are yellow green, flat body, and at the end of the abdomen attached with white wax threads.

Biology and behavior *Anomoneura mori* is univoltine and overwinters as adults. The post-overwintering adults mate and oviposit in late March, and a female can oviposit approximately 2,100 eggs. The eggs are deposited on the unfolded leaves of the new

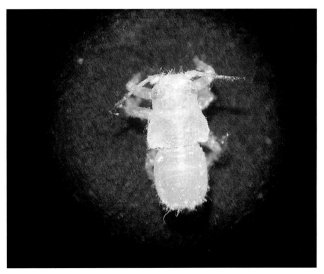
A. mori nymph (Xu Gongtian)

Lateral view of *A. mori* adult (Xu Gongtian)

Dorsal view of *A. mori* adult (Xu Gongtian)

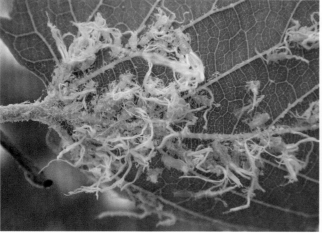
A. mori nymphs (Xu Gongtian)

A. mori nymph and wax threads (Xu Gongtian)

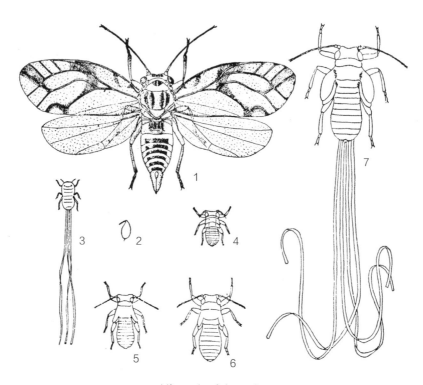

Life cycle of *A. mori*
1. Adult female, 2. Egg, 3-7: 1st to 5th instar nymph (wax thread of the 2nd to 4th instar nymphs were omitted)
(Zhejiang Agricultural University)

buds. The oviposition period lasts for a month, and the egg stage lasts 10-22 days. Eggs hatch in early April. The newly hatched nymphs feed first on the abaxial side of the leaves, and the damaged leaves curl downward from the edge and form a tube- or ear-shaped structure, which then become chlorotic and fall. The nymphs then disperse and feed on other leaves. The abaxial side of the damaged leaves is covered with wax threads excreted from the end of the abdomen of the nymphs, which causes sooty mold. The nymphal stage has five instars. The nymphs emerge into adults in early to mid-May. Adults aggregate and feed on young shoots and the abaxial side of leaves in groups. Once the mulberry branches are cut down in summer, adults migrate to the nearby cypress trees to feed, and migrate back to the mulberry host when new buds appear. *Anomoneura mori* feeds on mulberry trees and cypress trees in the autumn. When the temperature falls from 12 °C to 4.4 °C, the adults start to overwinter underneath the mulberry tree bark, or on cypress tree.

Prevention and treatment

1. Avoid planting mixture stands of mulberry and cypress. Avoid planting any cypress trees near the mulberry trees. The insect will die in 5-10 days without cypress as its alternative host during the summer cutting of mulberry trees.

2. Manual removal of the egg infested leaves. Timely removal of leaves infested with eggs is effective in early spring. After eggs hatch, the aggregated nymphs on the infested leaves can be removed to reduce the psyllid population.

3. Protection of natural enemies. Strengthened protection of the parasitoid *Tetrastichus* sp., lady beetles, lacewings, and other predators can be effective.

4. Chemical control. During the psyllid egg hatch period, spray 40% acetamiprid and chlorpyrifos mixture.

References Zeng Aiguo 1981, Xiao Gangrou (ed.) 1992, Rao Wencong et al. 2006, Xu Gongtian, Yang Zhihua 2007, Forest Disease and Pest Control Station of the Ministry of Forestry 2008.

(Guo Yimei, Wang Zhizheng, Gao Zuxun)

44 Cyamophila willieti (Wu)

Classification Hemiptera, Psyllidae
Common name Japanese pagoda tree psyllid

Importance This pest occurs widely in urban landscape. The insect injury may lead to the occurrence of the locust tree canker, honeydew can easily induce sooty mold, and contaminate the urban landscape and environment and even interfere with the normal activities of the public.

Distribution Beijing, Hebei, Shanxi, Liaoning, Jiangsu, Zhejiang, Shandong, Henan, Hubei, Hunan, Guangdong, Sichuan, Guizhou, Shaanxi, Gansu, Ningxia, and Taiwan.

Hosts *Sophora japonica*, *S. japonica* var. *pendula*.

Damage Both adults and nymphs remove sap from young plant tissues. Nymphal wax and honeydew often induce sooty mold, and reduce photosynthesis, and weaken tree vigor. The adults aggregate on leaves, move toward pedestrians when at close proximity, and therefore becomes a nuisance to the public.

Morphological characteristics **Adults** Male body length is 3.00 to 3.02 mm, the wingspan is 3.86 to 3.96 mm. Head extended downward. Female is 3.03 to 3.50 mm long, and wingspan is 4.25 to 4.70 mm. The winter type adult body and wings are dark brown, and the head and thorax have yellowish brown spots. Summer adult female and male are green to yellowish green. Dorsal thorax has yellow spots. The ocelli are orange. The compound eyes are brown, and the antennae are green, legs yellowish green, and abdomen pale green.

Eggs Oval, 0.4-0.5 mm×0.09 mm in size. Newly deposited eggs are white, and then change to orange. Transparent, red compound eyes can be seen. A hair appears at one end of the egg.

Nymphs Mature nymphs are 2.21 mm×1.34 mm in size. There are five instars. The body is in oval shape, and slightly flat; newly hatched nymphs are yellowish white and change to green. The compound eyes are red, abdomen is light yellow. Head is slightly narrower than the abdomen.

Biology and behavior *Cyamophila willieti* occurs 4 generations per year in Beijing. It overwinters as adults in tree holes, among the weeds under the canopy, or underneath bark seams. Adults become active at the end of March. Eggs are deposited individually on young shoots, leaves, buds, inflorescence, and flower buds. Each female oviposits about 110 eggs. Eggs begin to hatch in mid-April. Nymphs pierce and suck plant sap at young succulent tissues, including abaxial side of leaves, petioles and young branches. The excessive amount of honeydew leads to sooty mold. A large number of adults emerge in May. Severe damage is recorded under drought and high temperature in May and June, whereas psyllid population is reduced in rainy season. The insect population rises again in September and overwinters after October.

Prevention and treatment

1) Protection and utilization of natural enemies. The natural enemies of psyllids include *Harmonia axyridis*, *Coccinella septempunctata*, *Propylaea japonica*, *Chrys-*

Dorsal view of *C. willieti* adult (Xu Gongtian)

Lateral view of *C. willieti* adult (Xu Gongtian)

C. willieti eggs and nymphs (Xu Gongtian)

C. willieti eggs and mature nymphs (Xu Gongtian)

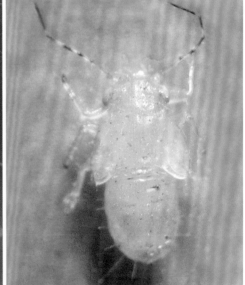

C. willieti nymph (Xu Gongtian)

opa sinica, *C. septempunctata*, *Oxytate parallela*, *Misumenops tricuspidatus* and other natural enemies.

2) Chemical control. In mid-April (the peak of the oviposition by the post-overwintering adults) and early May (the peak of the first-generation nymphs), spray 10% imidacloprid WP at 2,000× dilution, 12% thiamethoxam and lamda-cyhalothrin mixture at 1,000×, or 25% buprofezin WP at 1,000×.

References Yang Youlan, Wang Hongwu, Lü Xiaohu 2002; Qiang Zhonglan and Ma Changlin 2007; Xu Gongtian and Yang Zhihua 2007; Shen Ping, Chang Chengxu, Zhang Yongqiang et al. 2008.

(Zhou Zaibao)

45 *Colophorina robinae* (Shinji)

Classification Hemiptera, Psyllidae
Synonym *Euphalerus robinae* (Shinji)
Common name Honey locust psyllid

Distribution Beijing, Hebei, Liaoning, Shandong, Guizhou, and Shaanxi.

Hosts *Gleditsia sinensis*, *Gleditsia japonica*.

Damage Adults and nymphs of *C. robinae* feed on plant sap of the Chinese honey locust fruit. Infested new leaves form pod-like "insect galls". The damaged leaves join together lengthwise along the main vein and cannot spread. Infested new shoots become deformed, wither or die.

Morphological characteristics **Adults** Female adults 2.1-2.2 mm in length with wingspan of 4.2-4.3 mm; male adults 1.6-2.0 mm in length with wingspan of 3.2-3.3 mm. New emerged adults yellowish white, and gradually turn dark brown. Compound eyes large, oval, purplish red; ocelli brown. Antennae have 10 segments, the distal end of each segment is black, the basal part is yellow, and the two setae at the tip are yellow. Head is yellowish brown dorsally, median suture is brown, with one depressed brown plaque on each side. Prescutum of mesothorax has a pair of brown marks, there are two pairs of brown marks on the prescutum; and marks become gradually inconspicuous as the body color darkens. Forewing transparent at first, and then turns into semi-transparent, the costal, posterior edge and center of the wings have a brown area, and the vein has brown marks. There are scattered small brown spots on the surface of the wing.

Hind wing transparent, the veins at the edge brown. Femora of the legs dark brown and strong; tibia yellow brown with four black spines at the end; metatarsus yellowish brown with two black spines, telotarsus black brown. The end of abdomen of female adults sharp, and the ovipositors are covered with dense white setae; the end of abdomen of male adults is blunt, and genitalia bends dorsally.

Eggs Oval, with a short stalk, 0.28-0.34 mm×0.12-0.19 mm in size. Newly oviposited eggs are ivory white, one end with orange color. Turns purplish brown later, and then turns greyish white before hatch.

Nymphs The fifth instar larvae are 2.10-2.25mm×0.6-0.62 mm in size. They are yellowish green, with dark marks. Compound eyes red brown. Wing pads large.

Biology and behavior *Euphalerus robinae* occurs

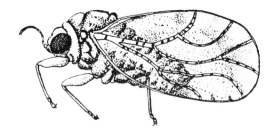

C. robinae adult (Hu Xingping)

Lateral view of *C. robinae* adult (Xu Gongtian)

Dorsal view of *C. robinae* adult (Xu Gongtian)

Overwintering *C. robinae* adult (Xu Gongtian) *C. robinae* nymph (Xu Gongtian)

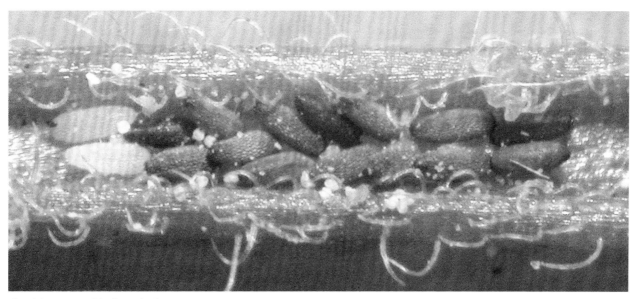

C. robinae eggs (Xu Gongtian)

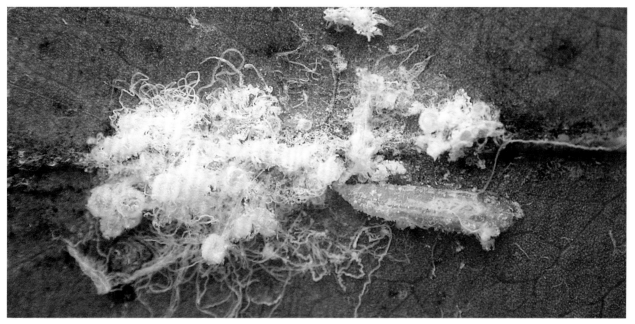

C. robinae on a host (Xu Gongtian)

Deformed host fruit from *C. robinae* infestation (Xu Gongtian)

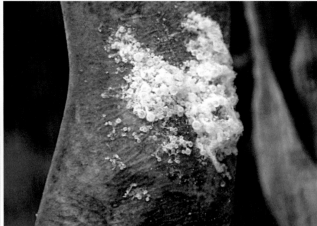
Infested host fruit and *C. robinae* (Xu Gongtian)

C. robinae on a host leaf (Xu Gongtian)

Sooty mold caused by *C. robinae* (Xu Gongtian)

four generations per year in Hebei and overwinters as adults. The post-overwintering adults become active in the early April in the following year and start supplemental feeding. They begin to mate and oviposit in mid-April. Egg stage lasts for 19-20 days. Nymphs are hatched in early May. They have five instars and last for 20-21 days. The peaks of the adults in each generation are in late May, early July, the mid-August and late September, respectively. Adults overwinter in cracks at the base of trunks in October.

Adults emerge mainly during 09:00-12:00. The emergence rate reaches more than 90%. Adults mainly mate at about 06:00 the day after emergence. Mating can be observed any time of the day. Female adults begin to lay eggs two days after mating, primarily in grooves of petioles and near leaf veins, and rarely on leaves. Overwintering adults lay eggs in cracks of the current year branch barks, and the eggs are arranged in rows. Each female adult can oviposit 387-525 eggs. Adults show phototropism and feign death and are often observed jumping.

Nymphs mainly hatch during 08:00-10:00 in the morning. The hatch rate can be more than 95%. Newly hatched nymphs crawl onto the top of small shoots and inhabit among the young leaves. Their feeding of plant sap can make the shoot fail to expand. The development of nymphs is asynchronized. Nymphs of different instars can be observed in the same "gall". Before adult emergence, late instar nymphs often crawl out of the "gall," inhabit on the fork of branches and excrete a large amount of white wax threads to cover their body. These threads are left on petioles after molting.

Prevention and treatment

1. Protect and preserve natural enemies, e.g., parasitic wasps, and lacewings.

2. Spray 40% acetamiprid and chlorpyrifos mixture or profenofos and phoxim mixture at 1,000× dilution to control the newly hatched nymphs and at the peak of adult emergence in each generation.

References Shandong Forest Insects Editorial Board 1993, Xu Gongtian 2003.

(Qiao Xiurong, Sun Lihua, Li Yanjie)

46 *Syntomoza homali* (Yang et Li)

Classification Hemiptera, Liviidae
Synonym *Homalocephala homali* Yang et Li

Distribution Guangxi, Guangdong, Hainan.
Host *Homalium hainanense*.
Damage Nymphs infest plant sap at the terminal bud and abaxial side of the young leaves. The infested terminal buds wither and leaves curl, which reduce the growth of the seedlings.
Morphological characteristics **Adults** Length 1.1 to 1.5 mm. Newly emerged adults light green, and then become yellowish brown or dark brown. Antennae with 10 segments, light yellow in color. The two basal segments are enlarged, the distal end is black with 2 sword-like setae. Compound eyes red or purplish red. The legs are light yellow. The end of the female abdomen is contracted and relatively thin, slightly curved downward, which can be distinguished from the males.
Eggs Size 0.15 mm×0.1 mm, milky white in color, slightly transparent, pear-shaped.
Nymphs The nymphs with wing pads are 1.0-1.2 mm long; oval, slightly flat in shape. Compound eyes are red; antennae are light yellow; body is yellowish brown or green; the end of the abdomen has white waxy threads.
Biology and behavior *Homalocephala homali* occurs 14 to 15 generations per year in Jianfengling, Hainan province. It takes approximately 20 days to complete a generation. There is no overwintering phenomenon observed. The damage on new leaves of the host plants is severe between May and October. In addition, the most severe damage is also observed in the nursery with adequate irrigation during the dry season when seedlings are developing new shoots. As the young forest canopy gradually covers the ground, the insect damage is reduced.

Nymphs aggregate in groups at the terminal bud and abaxial side of the young leaves to ingest plant sap. The infested terminal buds wither and leaves curl, which reduce the growth of the seedlings. The nymphal stage lasts for 12 to 15 days. Adults usually inhabit in the curly young leaves. Adults start to mate and oviposit after supplemental feeding for a few days. Eggs are deposited on young leaves, young shoots, or the abaxial side of the damaged leaves. The eggs are arranged in irregular rows. Egg stage lasts for 4 to 5 days.

When outbreaks occur, lady beetles, syrphids, and spiders are often observed, but their role on pest suppression is limited.

References Chen Zhiqing 1973, Yang Jikun et al. 1986.

(Chen Zhiqing)

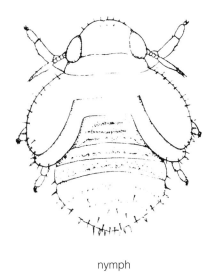

nymph   adult

S. homali (Hu Xingping)

47 *Diclidophlebia excetrodendri* (Li et Yang)

Classification Hemiptera, Liviidae
Synonym *Sinuonemopsylla excetrodendri* Li et Yang

Distribution Guangxi; Vietnam.

Host *Burretiodendron hsienmu*.

Damage Nymphs aggregate and feed in groups on buds and young leaves by ingesting plant sap. Adults often inhabit and feed around the young shoots. Nymphs secrete white waxy substance and cause sooty mold growth. The damaged seedlings or saplings can be identified by wrinkled young leaves, and sagging petioles, which reduces normal bud growth.

Morphological characteristics **Adults** Yellowish green, 2.2-2.6 mm long, Abdomen is flat dorsally, and covered with light-colored long hairs. Head capsule width (including compound eyes) is 0.77-0.79 mm, and compound eyes are dark brown. Antennae are 10 segmented, 1.26-1.27 mm long, yellowish green in color. The distal end of the third to the eighth segments, as well as the ninth and the tenth segments is black or dark brown with two yellow bristles. Forewing length is 1.94-2.25 mm, and width is 1.01-1.21 mm, and oval-shaped, opaque. The costal margin of the forewings has a breaking mark. Forewing is covered with dense brown spots, which form three discontinuous brown bands. Stigma large. The veins are yellowish brown and the end of veins dark brown. The hind wings wide, slightly shorter than the forewings. Legs yellow, and the tibia of the hind legs without basal spine. **Eggs** Elongated pear-shaped with one end narrower than the other, 0.1-0.15 mm in length. Newly deposited eggs are light-yellow, and then change to yellow. The dark spots of the compound eyes can be seen in the eggs before hatching. **Nymphs** The newly hatched larvae are light-yellow in color, and color darkens as they grow. The body is flat. The compound eyes are red. Wing pads are translucent, and greyish brown in color. As nymphs feed, they cover their body with wax threads.

Biology and behavior *Diclidophlebia excetrodendri* has overlapping generations, and no apparent overwintering phenomenon in the southern region of Guangxi province. When the daily average temperature is at 28.8°C, a generation will take approximately 30 days, during which the egg period is 5.24 (5-6 days), nymphs molt four times and lasts for 17 days. Each female lays an average of 212 eggs. Pre-oviposition time is 5 to 6 days. The oviposition period lasts for 8 to 14 days.

Newly emerged adults are covered with wax. They feed on the same shoots where they emerge. The adults mate and oviposit in a few days. Both sexes can mate

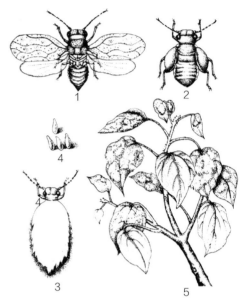

D. excetrodendri. 1. Adult, 2. Nymph, 3. Nymph covered with wax, 4. Eggs, 5. Infested host (Lu Baolong)

multiple times. Adults become more active gradually, including flying, jumping. They often move among the branches, leaves, and terminal bud, seeking new sites to feed and reproduce. Adults fly in short distances, usually 50 to 60 cm. Females oviposit on buds and the concaved spots of the abaxial side of young leaves. Eggs are deposited either individually or in masses. Nymphs often feed aggregately on the terminal bud and young leaves. Nymphs crawl for a short distance if disturbed.

Two population peaks normally occur per year: from March to May, and from September to October, respectively. The 1st peak is particularly large. In southern Guangxi province, temperature increases fast in March. The host plant starts to grow. A rapid increase in insect population follows. In wintertime when daily average temperature reaches 14.4°C, adults can still feed and mate on sunny days. The eggs laid in the winter can normally hatch. There is very little activity when the temperature is below 10°C. Nurseries suffer more damage than young forests. Young forests without complete canopy coverage and with many weeds show greater damage than the forests with good canopy coverage. Plants at the low altitude exhibit more damage than the plants at high altitude.

Natural enemies include *Harmonia dimidiata*, *Menochilus sexmaculate*, and praying mantis.

(Yang Minsheng)

48 *Carsidara limbata* (Enderlein)

Classification	Hemiptera, Carsidaridae.
Synonym	*Thysanogyna limbata* Enderlein
Common name	Parasol psyllid

Distribution Beijing, Hebei, Shanxi, Jiangsu, Zhejiang, Anhui, Fujian, Jiangxi, Shandong, Henan, Guangdong, Guangxi, Shaanxi.

Host *Firmiana simplex*.

Damage *Carsidara limbata* is a monophagous pest, nymphs and adults can ingest plant sap from leaves or young shoots and damage vascular tissues. They cause most damage to young trees. Nymphs excrete white cotton-like wax threads, block stomata of leaves and reduce normal photosynthesis and respiration of plants. The honeydew and wax also induce sooty mold, and negatively affect environmental hygiene. Trees with severe insect damage can lead to early senescence, such as, falling leaves, withered shoots, and rough and fragile bark. These trees can be easily broken by wind.

Morphological characteristics **Adults** Length 5.6-6.9 mm, wingspan about 13 mm. Yellowish green with brown spots. Head width is greater than the length, head crown has a deep groove, frons is exposed, buccae are short and mastoid. Compound eyes reddish brown; ocelli orange. Antennae yellow, 10 segmented, and the last two segments black, the distal end has setae. Pronotum is arched, mesonotum has six vertical striations; metascutum has two small conical prominence. Legs yellow, tibia and tarsus dark brown, claws black. There is one terminal spine pointed outward and the four terminal spines pointed inward. There is a claw-like spine at the outside of the metatarsus. Forewing transparent, veins tea-yellow (yellowish brown). Length of forewing is 2.5× of its width. The dorsal plates of abdomen are light yellow. The front edge of each segment has brown bands. The dorsal plates of the third abdominal segment, the last segment of the male abdomen yellow. The ventral surface of the female abdomen, the end of the abdomen yellow, with a large dorsal valve.

Eggs The eggs are fusiform, transparent and 0.5-0.8 mm in length. Newly deposited eggs are light yellow or yellowish brown, and then change to reddish brown.

Nymphs The body is slightly flat. The last instar is 3.4-4.9 mm in length, tea-yellow and tinted with green in color, almost cylindrical. The body is covered with thick white cotton-like wax threads. The antennae have 10 segments, and the wing pads are obvious, transparent, and light brown in color.

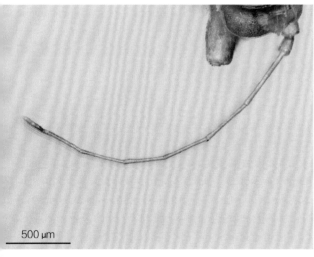

C. limbata antenna (Li Zhenyu)

C. limbata adult (Li Zhenyu)

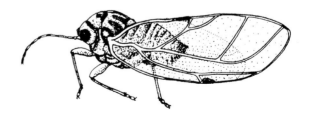

C. limbata adult (Hu Xingping)

C. limbata nymph (Li Zhenyu)

C. limbata nymph (Ji Baozhong)

C. limbata adult and nymph (Li Zhenyu)

Biology and behavior *Carsidara limbata* occurs 2-3 generations per year. The eggs overwinter in cracks of tree bark. Nymphs of the first generation occur between late April and early May and excrete wax and honeydew to contaminate tree trunks and ground. Nymphs have three instars and last for 30 days. The adults of the first generation appear in early June and reach peak in mid-June. The nymphs of the second generation appear at the end of June, and mid-July. Many adults of the second generation emerge in mid-August. Nymphs of the third generation begin to damage plants in early September. The overlapping generations are common and the adults deposit overwintering eggs in late October.

After emergence, adults feed for approximately 10 days before mating. Mating lasts for about two hours. Most adults mate before 08:00 and around 17:00. Females begin to oviposit 2-3 days after mating. Eggs are deposited on the lower side of the main shoots close to the main trunk, or on the lower side (facing ground) of the lateral branches near the main branches, or in the rough parts of tree bark on both main and lateral branches. Some eggs are also deposited on the shaded side (facing away from the sun) on the main trunk, or on abaxial side of the leaves or leaf petioles. Eggs are deposited individually, and egg stage lasts for 10-12 days. The fecundity is approximately 50 eggs per female, and the longevity of adults is about six weeks.

Prevention and treatment

1. Protect and preserve natural enemies, such as parasitic wasps and lacewings.

2. Chemical control. At the beginning of the egg hatch and at the peak of the adult emergence, spray 20% beta-cyhalothrin at 1,000×, or 10% imidacloprid WP at 2,000× can reduce the pest population.

References Editorial Board of the Shandong Forest Insects 1993, Xu Gongtian and Yang Zhihua 2007.

(Qiao Xiurong, Xu Gongtian, Zhou Jiaxi)

49 *Trioza camphorae* Sasaki

Classification Hemiptera, Triozidae
Common name Camphor psyllid

Distribution Zhejiang, Fujian, Jiangxi, Hunan, and Taiwan.

Host *Cinnamomum camphora*.

Damage The nymphs feed on the abaxial side of camphor leaves.

Morphological characteristics **Adults** Length 1.6-2.0 mm, wingspan 4.5-6.0 mm. Yellow or orange. Antennae filiform with 10 segments, the two basal segments are broad and short, the third segment is the longest, while the ninth and the tenth enlarged gradually into club-shaped with two bristles at the end. Compound eyes dark brown, hemispherical. The tibia of all legs has three black spines. **Eggs** Fusiform, 0.3×0.11 mm in size, with egg stalk of 0.06 mm in length. The newly deposited eggs are milky white, transparent, and gradually change to greyish black. The eggs are dark brown and shiny before hatch. **Nymphs** Milky white, the abdomen is egg yolk yellow. Length of the first instar nymphs is 0.3-0.5 mm, and flat oval. Body covered with white waxy secretions. As the insect grows, the number of wax threads also increase. The body color gradually darkens as yellowish green, and mature nymphs are greyish black, and body length increases to 1.6-1.8 mm. The wax threads tightly arranged around the body, especially near the antennae, at the midline where the wing pads attached, and dorsal surface of the abdomen. Compound eyes red. Most of the wax threads fall off prior to adult eclosion.

Biology and behavior The camphor psyllid occurs one generation per year, rarely two generations per year. The psyllid overwinters as nymphs on the abaxial side of leaves. Adult emergence begins in early April of the following year and ends at the end of April. The first generation of nymphs start to hatch in mid-April, a few nymphs emerge in late May. The second-generation nymphs hatch in early June.

Adult eclosion occurs both day and night, but most eclosion occurs between 10:00-14:00. Adult eclosion rate was 75.2%, with 51.2% of them were females. The newly emerged adults aggregate and feed on tender shoots. They become more active next day. They tend to jump and start mating. Each adult mates multiple times. They oviposit soon after mating. Eggs from the overwintering generation are deposited on new shoots and young leaves. The first-generation adults deposit eggs on summer shoots and young leaves. Most eggs are deposited on leaves, and about 7% are deposited on branches. Eggs on leaves are mostly located on tips of the leaves. More eggs are found on the adaxial side than on the abaxial side. The eggs are arranged in rows. Eggs occasionally overlap. Eggs can hatch day and night. The hatch rate is approximately 73%. The newly hatched nymphs feed on the abaxial side of the young leaves. The initial leaf damage symptom is yellowish green oval protrusion on leaves. The damage site becomes purplish red later and the leaf falls prematurely.

The psyllid usually occurs in the middle section of the canopy, followed by the upper section. North side of the canopy often has more insects than the south side. In addition, the forest stand at the lower altitude of the slope usually has higher insect population levels than the stand at the higher altitude of the slope. Regarding the age of forest stand, the 17-year-old forest stand has significantly higher insect population than the 5-year-old young forest stand. Also, mixed forest stands often have less insect infestations than pure stands.

References Zhou Yuemei 1979, Shen Guangpu et al. 1985.

(Shen Guangpu)

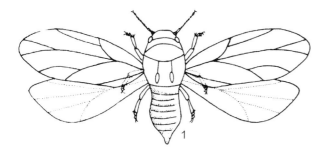
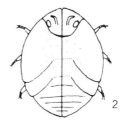

T. camphorae. 1. Adult, 2. Nymph, 3. Eggs (Shen Guangpu)

50 *Trioza magnisetosa* Loginova

Classification Hemiptera, Triozidae
Common name Narrow-leaved oleaster psyllid

Distribution In northern China provinces that are in the arid and semi-arid region, including Inner Mongolia, Shaanxi, Gansu, Ningxia, and Xinjiang.

Hosts The most preferred host is *Elaeagnus angustifolia*, and its general hosts are: *Malus asiatica*, *Pyrus bretschneider*, *Prunus salicina*, *Zizyphus jujube*, *Populus* spp., and *Salix* spp.

Damage The insect feeds on sap of the host. They cause leaf rolling and wither of shoots and branches. Light damage symptoms include deformation of flower buds and reduction of fruit yield. Severe infestations result in leaf wither and early senescence, and even the death of the whole plants.

Morphological characteristics **Adults** Female length 2.6-3.5 mm, male length 2.2-3.0 mm. Yellowish green or yellowish brown, head light yellow; antennae filiform and yellowish brown, its last two segments black, the distal end has two black sward-shaped setae. Compound eyes are large and protruded, and reddish brown. The ocelli are in orange color. Thorax is protruded. The dorsal surface of prothorax is arch-shaped; its anterior and posterior borders are dark brown, with two orange colored vertical striations in the center. The mesonotum is two times wider than its length with four yellow vertical striations. There is a pair of ivory white or relatively dark colored small prominence in the center near the posterior edge of metasternum. Legs are light yellow, claws are black. The ventral surface of abdomen is whitish yellow with brown vertical striations on each segment dorsally. The abdomen end of the female adults contracts sharply. The abdomen of male adults contracts near the end; the terminal segments expand and curve dorsally.

Eggs Length about 0.3 mm; transparent and spindle-shape, the end is tapered with a short thread, the basal part is round; the surface is smooth.

Nymphs Compound eyes are red, the body is flat and sub-orbicular, 2.0-3.4 mm in length. The newly hatched nymphs are white, then turn into light green. Mature nymphs are greyish green; body and wing pads are covered with setae.

Biology and behavior *Trioza magnisetosa* is univoltine in Xinjiang. Adults can overwinter in trunk cracks of *Elaeagnus angustifolia*, under fallen leaves, or among weeds. Overwintering adults become active at the beginning of March when the mean temperature is above 5℃. They feed on young shoots. Large numbers appear in mid-March. Adults mate and oviposit between early April and early June. Egg stage lasts for 8-30 days, which is mainly affected by temperature. Nymphs occur in early May, and their peak time is in mid-May. Nymphs have five instars. Their peak occurrence time of the 1st to 5th instars is in the early and mid-May, mid- and late May, late May and early June, early and mid-June, and mid- and late June, respectively. The development time of the 1st to 5th instars are 7-9, 7-10, 7-10, 8-10, and 13-15 days, respectively. Adults of the first generation emerge in mid-June and reach peak at the end of June and early July. Adults start overwintering (diapause) at the end of October and the beginning of November.

Adults and nymphs feed on plant sap of young shoots and leaves of *E. angustifolia*. Adults usually emerge either in early morning or at dusk, the average of eclosion rate is 90%. Adults begin to feed within half an hour after their emergence. Both post-diapause and newly emerged adults need to have supplemental feeding. They turn greyish brown after 7-10 days. Adults usually do not fly. They can fly briefly for a short distance when they are frightened. Adults usually rest on the abaxial side of leaves during daytime and begin to jump and migrate at nightfall (dusk) to dense forest stands. High adult population is usually found in dense forest stand.

Trioza magnisetosa is more abundant in the middle and lower part of the canopy than in the upper layer. Adults usually feed on the veins on abaxial side of the leaves. When leaves fall, adults move onto shoots. When temperature is low in late autumn, adults usually aggregate and inhabit the weedy ground cover. When temperature rises, adults disperse on the trees to feed. Adults mate primarily in the morning and evening. When the mean temperature falls to below 0℃, they move back into the overwintering sites. The adult lifespan is long, and they can be found year around. Wind is closely related to the migration of adults. Strong wind in spring and autumn is beneficial to the dispersal of the psyllids. The oviposition sites vary with the development of the trees.

Adults oviposit on young developing buds in the early and mid-April. Eggs are tightly arranged. When *E. angustifolia* leaves unfold after May, adults oviposit on the abaxial side of leaves by inserting one end of the eggs into the leaf tissue. The egg-laying period is long. The fecundity is more than 400 eggs.

Newly hatched nymphs are not active, aggregate and feed on the young shoots and both sides of leaves. Affected leaves curl and form a long-tubular structure. Nymphs at this stage are cryptic and often excrete white wax within the rolled leaves. The peak of nymphal stage is in mid-May when the host leaves appear. The newly hatched nymphs can still be seen in mid-June. The damage by nymphs increases when the nymphs are in the third to fourth instars. In addition to leaf rolling, young shoots also begin to deform (curl). Wax deposit and honeydew production also increase, which leads to early senescence and falling of leaves. The ground of heavily damaged forest is white and resembles frost in the autumn. When the rolled leaves gradually turn yellow, rotten and fall, the nymphs in these rolled leaves often move to other rolled leaves. Some nymphs die when moving among the rolled leaves and fall to the ground. The fifth instar nymphs feed much more. Because of lacking optimal feeding sites in the rolled leaves, nymphs move out of the rolled leaves and aggregate neatly on the abaxial side of new leaves and young shoots. This is the period that causes the most severe damage in its entire life cycle. Mature nymphs move from the rolled leaves to inhabit the abaxial side of leaves and shoots.

Prevention and treatment

1. Silviculture. Plant mixed forests and modify tree species composition, for example, interplant *Tamarix chinensis* and *Populus diversifolia*. Planting lure trees in the orchards and shelter forest. In seriously damaged *E. angustifoli* forests, cut the trunk down to 15-20 cm above

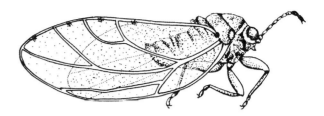

T. magnisetosa adult (Hu Xingping)

the ground, or as high as 1.5-1.7m above ground. Destroy the damaged tree canopy. Plow and irrigate in winter to *destroy* the overwintering adults.

2. Manual control. Clear the withered leaves and weeds in winter and reduce the overwintering adults.

3. Biological control. Apply *Beauveria bassiana* or *Paecilomyces fumosoroseus*. More than 20 species of natural enemies of *T. magnisetosa* were recorded. These natural enemies should be protected and utilized. The dominant species of the parasitoid wasp, *Tetrastichus* sp., can parasitize up to 30% of the nymphs. *Adalia bipunctata*, *Delphastus catalinae*, *Chrysopa formosa*, *C. septempunctata*, and spiders are predators of the nymphs and adults. In addition, there is a fungal pathogen for the adults.

4. Chemical treatment. Spraying of 25% Dimilin or 5.7% emamectin benzoate abamectin at 2,000-3,000× can control the early instar nymphs. During early April and early June, spraying of 40% methidathion EC at 1,000× and 50% fenitrothion EC at 1,000-1,500× can effectively control the adults.

References Xu Zhaoji 1963; Wang Chenggui, Liu Aitu, Xie Zuying 1965; Yang Xiuyuan 1981; Xiao Guangrou 1992; Xi Yong, Renling, Liu Jibao 1996.

(Xi Yong; Wang Aijing, Wang Ximeng)

51 *Aleurotrachelus camelliae* (Kuwana)

Classification Hemiptera, Aleyrodidae
Common name Camellia whitefly

Importance Severe damage induces sooty mold on camellia. Damage on bayberry tree *Myrmica rubra* has been widely reported in recent years.

Distribution Jiangsu, Zhejiang, Shandong; Japan.

Hosts *Catalpa ovata, Camellia sinensis, Prunus persica, Prunus mume, Camellia oleifera, Eriobotrya japonica, Myrica rubra, Castanopsis* sp.

Damage Most nymphs feed on the abaxial side of leaves by piercing-sucking leaf sap. Severe damage of camellia leaves causes yellow wilt, necrosis of shoots, and leads to sooty mold and reduced tree vigor and yield.

Morphological characteristics Adults Female length 1.7-2 mm with wingspan of 3-3.5 mm. Head and thorax are dark brown and shiny; abdomen is orange. There are six yellow wing spots on the forewings, which are located in the costal, exterior, and postal margins of the wings. Among the two spots near the costal margin, one is a relatively narrow, starting at the costal edge and ending where the main vein goes downward. The hind wing is slightly smaller than the forewing, and light brown in color. The males are smaller than the females. Male claspers plier-shape and protrudes from the end of the abdomen. Copulation organ is wedge-shaped.

Eggs Length 0.19-0.21 mm; brown; banana-shaped, slightly curved; smooth, with an egg stalk.

Nymphs The newly hatched nymphs oblong, length 0.25 mm, light-yellow initially, and gradually change to reddish brown. There are 10 long marginal hairs before the thoracic spiracles. There are 10 short marginal hairs at other parts of the body. Anal plate with 4 long bristles. The second instar nymphs are elongated pear-shaped, flat dorsally and ventrally, slightly pointed at the front end, and the posterior end is truncated and slightly concave. The dorsal surface is chitinous and the ventral surface is grey and membranous. Dorsal surface of the nymphs forms a ridge. There is a cluster of white wax threads at the depression of each of the thoracic spiracles. There is also a cluster of white wax threads on the anal plate. The marginal wax glands on the body arranged as an approximately 300 comb tooth-like structure.

Pupae Exarate pupae. Length 1 mm. light-yellow initially, then changes to orange. The compound eyes are black, and wing pads are gray.

Biology and behavior The camellia whiteflies are univoltine and overwinter as the second instar nymphs under the black pupal scale on the abaxial side of the leaves. Nymphs pupate in late March next year. Eclosion begins in early April and reaches peak in mid-April. Eclosion of adults requires daily average temperature of approximately 18°C and relative humidity of greater than 80%. Adult eclosion occurs mostly from 08:00 to 10:00 during the sunny days, whereas the daily eclosion peak is from 12:00 to 14:00 in the rainy days. Adult eclosion duration lasts approximately 20 days, but the emergence peak only lasts for a few days. Adults can mate multiple times and can oviposit soon after mating. Most eggs are deposited on the abaxial side of the leaves or on the margin of the young leaves. Fecundity is from 21 to 44 eggs with an average of 32.6 eggs. Adults move to young shoots after oviposition and are good flyers. No supplemental feeding is observed. Female adult lifespan is 2 to 6 days, and males are 4 to 7 days. Nymphs began to appear in late June. The newly hatched nymphs are mobile, they select a suitable location and pierce and suck leaf sap. In late July to August, most nymphs molt into the

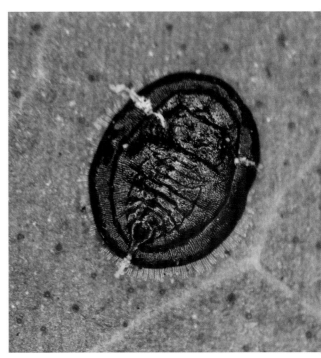

A. camelliae 2nd instar nymph (Ji Baozhong)

A. camelliae 2nd instar nymph (Ji Baozhong, Zhangkai)

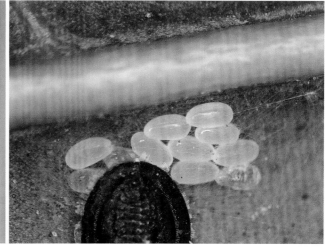
A. camelliae eggs (Ji Baozhong, Zhangkai)

second instar. The legs and antennae of the second instar nymphs degenerate and lose mobility. The dorsal and ventral surfaces of the body are flat, and the formation of black scale occurs. The marginal glands secrete colorless substance that seals the insects onto the leaf tissue, and there is no movement after this period. The nymphal stage has three instars. The second instar lasts approximately 250 days, which is the most damaging instar that not only injures leaves, but also induces sooty mold and reduces tree vigor.

Prevention and treatment

1. Prune and remove the heavily infested foliage can be an effective management strategy to reduce the insect population.

2. From late June and to July, spray of 40% methidathion at 1,500-2,000×, 48% chlorpyrifos at 1,000-1,200×, 25% thiamethoxam EC at 1,500×, 45% profenofos and phoxim mixture at 1,000×, or 25% phosmet EC at 1,500× can effectively reduce the insect populations. The frequency of the spraying is every half month, and for a total of 2 to 3 times.

3. *Aschersonia* sp. is a fungal pathogen of the camellia whitefly. Spray of 3.6×10^7 spore suspension of the fungal pathogen can be effective and results in approximately 80% mortality. The timing of the application can be in late June to July when the nymphs hatch.

References Xiao Gangrou 1992, Chen Weimin 204, 2006, 2008.

(Ji Baozhong, Zhang Kai, Chen Zhuan)

52 *Aphis craccivora* Koch

Classification Hemiptera, Aphididae
Synonym *Aphis robiniae* Macchiati
Common name Black locust aphid

Distribution Beijing, Hebei, Liaoning, Jiangsu, Jiangxi, Shandong, Henan, Hubei, Shaanxi, Xinjiang; Europe and North Africa.

Hosts *Robinia pseudoacacia*, *Amorpha fruticose*, *Sophora japonica* var. *pendula*. There are reports in other countries that the aphid also cause damage to *Colutea* spp., *Coronilla* spp., and *Caragana* spp.

Damage Adults and nymphs feed on new shoots and cause leaf curl, deformation, and shoot wither. The honeydew excreted from the aphid causes sooty mold.

Morphological characteristics Apterous viviparous aphids are parthenogenetic, 2 mm in length, oval, shiny black or dark brown. The dorsal surface of the head, thorax and the first six abdominal segments has reticulated hexagonal marks. The seventh and eighth segments of the abdomen has transverse stripes.

The winged viviparous aphids are oval, shiny black. Wings are gray and transparent.

Biology and behavior The black locust aphid occurs more than 20 generations per year. The apterous viviparous adults, as well as nymphs, overwinter in the sunny habitat on weedy hosts, i.e., wild alfalfa, vetch. The aphids reproduce on the overwintering hosts in the following March and produce winged viviparous aphids in mid- to late April. They migrate to spring peas, locust, and other legumes. This is the first peak of migratory flight. The second peak of migration appears in late May and early June. The aphid population on the locust gradually increases. In mid- to late June, the third peak of adult migration occurs. Severe damage is observed at this time. In mid- and late July, aphid population density significantly decreases due to rain and high temperature, except those hosts in cooler areas. In October, the aphids reproduce and feed on the new shoots of lentils, green beans, and *Amorpha*. Winged adults migrate to the overwintering hosts, reproduce, feed, and overwinter.

Some apterous viviparous aphids start to reproduce when average temperature is -2.6℃. At -0.1℃, 21.85% of the aphids reproduce. The optimum breeding temperature is 19-22℃. Temperature and rainfall are the key factors that regulate the aphid population dynamics. A relative humidity of 60%-75% is suitable for aphid reproduction. Aphid outbreaks frequently occur during April and June. Aphid population reduces significantly in July due to high temperature and high humidity. Stormy weather conditions often cause high aphid mortality. Predators can suppress aphid population to some extent. Common predators are lady beetles, syrphid flies, lacewings, minute pirate bugs, and parasitoid wasps.

Prevention and treatment

1. When aphids migrate to feed on locust trees, remove the heavily infested trunk, branches, and new shoots, or spray water to prevent the spreading of aphids.

2. Protection and preservation of *Coccinella septempunctata*, *Adonia variegata*, *Chrysopa phyllochroma*, *C. septempunctata*, *Orius minutus*, and other predators.

Winged *A. craccivora* (Xu Gongtian)

A. craccivora apterous viviparae (Xu gongtian)

A. craccivora apterous viviparae and nymphs (Xu Gongtian)

A. craccivora being fed by a *Coccinella septempunctata* adult (Xu Gongtian)

3. Chemical treatment. Application of 20% thiamethoxam at 1,500×, or 10% imidacloprid WP at 2,000×, or 50% acetamiprid at 2,000× is effective.

References Zhang Guangxue and Xu Tiesen 1983, Xiao Gangrou (ed.) 1992, Xu Gongtian and Yang Zhihua 2007.

(Guo Yimei, Wang Zhizheng, Sun Yujia, Bao Bailing)

53 *Melanaphis bambusae* (Fullaway)

Classification Hemiptera, Aphididae
Common name Bamboo aphid

Distribution Jiangsu, Zhejiang, Anhui, Fujian, Jiangxi, Hunan, Guangdong, Yunnan and Taiwan; Korea Peninsula, Japan, Malaysia, Indonesia, U.S.A. (Hawaii), Egypt, Russia.

Hosts *Phyllostachys heteroclada* var, *pubescens*, *P. sulphurea* 'viridis', *P. glauca*, *P. praecox*, *P. propinqua*, *P. pubescens*, *P. vivax*, *P. dulcis*, *P. iridescens*, *P. heterolada* var. *funhuaensis*, *P. bambusoides* f. *lacrima*, *P. nigella*, *P. decora*, *P. nuda*, *P. flexuosa*, *P. makinoi*, *P. meyeri*, *P. nigra*, *P. nidularia*, *Pleioblastus*, and *Bambusa*.

Damage The aphids feed on the abaxial side of the bamboo leaves, ingest plant sap, and cause shrinking, and chlorotic, necrotic spots. Honeydew produced by the aphid can induce sooty mold, early senescence of leaves, and the death of plants.

Morphological characteristics **Apterous viviparae** Body length 0.85-1.25 mm; oval; color variable, ranging from black, red brown, yellowish brown, to red; covered with powdery wax. Head smooth, the rostrum short and black. The compound eyes are dark brown with protruded ocular nodule. Antennae have five segments, its length similar to the body length. The distal segment is 4× longer than the basal segment. Legs are slender.

Winged viviparae Body length 1.15-1.40 mm; oval; greenish brown to black; covered with powdery wax. Rostrum short. Compound eyes large, with ocular nodule. Antennae have six segments, its length similar to body length, black. The legs are slender.

Biology and behavior In Yuhang, Zhejiang province, the bamboo aphid occurs 18 to 21 generations per year. The alate viviparae only occur in the overwintering generation and the tenth to the thirteenth generations from July to August. All other generations are apterous. The first generation occurs between mid-March and mid-May. The nymph development needs 17-25 days, with an average of 21 days. They molt four times before turning to adults. The longevity of apterous viviparae is 16 to 33 days (average 21 days).

Development time of nymphs of the other generations is 8.2-15.3 days. They all molt 4 times. The longevity of apterous viviparae 5.2 to 33.4 days. The fourth to thirteenth generations have the shortest life span. The duration of each generation of the bamboo aphids are the first generation- 60 to 70 days, the second generation- 50 to 60 days, the third generation- 40 to 50 days, the fourth to the sixth generations- 30 to 40 days. From late June to the end of August, the seventh to thirteenth generation aphids occur. Their duration is 20 to 30 days. The duration for the fourteenth to twentieth generations is 30 to 60 days. The apterous viviparae produce winged viviparae in early and mid-December and reproduce from mid-December to late January. The aphids develop into apterous viviparous adults. The alate viviparae die between late January and early February. Their life span is 50 days. The apterous viviparae die between the end of February and early March.

M. bambusae (Xu Tiansen)

Prevention and treatment

1. Protection and preservation of natural enemies. Predators include *Lycosa coelestris*, *L. pseudoannulata*, *Oxtopes macilentus*, *Pisaura* sp., *Pardosa tschekiangensis*, *Hyperaspis sinensis*, *Adalia bipunctata*, *Anisolemnia dilatata*, *Propylaea japonica*, *Epistrophe* sp., *Chrysopa formosa*, and *C. sinica*. The parasitoid wasps include *Diaeretiella* sp., and *Aphelinus* sp.

2. Apply *Beauveria bassiana* or *Paecilomyces fumosoroseus*.

3. Chemical control. In early bamboo stand when aphid population density is high, application of systemic insecticides by drilling a hole at the base of the bamboo plant can be effective. The insecticide incudes 5% imidacloprid EC at 2× and 1 mL per hole. Spray 20% thiamethoxam at 2,000×, or 20% fenvalerate at 2,000× is also effective.

References Xu Tiansen and Wang Haojie 2004; Xu Tiansen, Wang Haojie, and Yu Caizhu 2008.

(Xu Tiansen, Wang Haojie)

54 *Aphis odinae* (van der Goot)

Classification　Hemiptera, Aphididae
Common names　Tallow aphid, Mango aphid

Distribution　Hubei, Guangdong, Guangxi and Taiwan; Korea Peninsula, Japan, India, Indonesia.

Hosts　*Eucalyptus* spp., *Sapium sebiferum*.

Damage　Tallow aphids feed on the abaxial side of the young tallow leaves, new shoots, and young branches. Honeydew excreted from the aphid often causes sooty mold and severe damage to young trees.

Morphological characteristics　**Apterous viviparae**　Ovol; about 2.5 mm×1.5 mm in size; brown, reddish brown, dark brown, grayish green or dark green. Body covered with a thin layer of powdery wax. Head, antennae, proboscis and legs are black; cornicles, anal plate, and reproductive plate are black; the end is round. There are more than 20 anal setae at the end of the abdomen.

Alate viviparae　Elongated oval, 2.1 mm×0.96 mm in size. Cornicles are cylindrical, shorter than the anal plate; the anal plate is elongated conical; the end of the anal plate has 14 to 24 setae. All of the other morphological characteristics are the same as the apterous aphids.

Biology and behavior　All tallow tree aphids are viviparae in Hainan in February and March. A large number of winged aphids occur from the end of March to early April. During April and May, both apterous and winged aphids exist. More than 20 generations occur each year. In the northern region, the sexual forms are produced at the end of autumn. Males and females mate and oviposit eggs which will overwinter.

Prevention and treatment

1. Biological control. For example, introduce lady beetles to reduce aphid population.

2. Chemical control. Spray new shoots with insecticides such as 20% fenvalerate at 3,000× and 10% imidacloprid WP at 2,000-4,000×.

(Gu Maobin)

A. odinae apterous viviparous aphids and nymphs feeding on a host (Xu Gongtian)

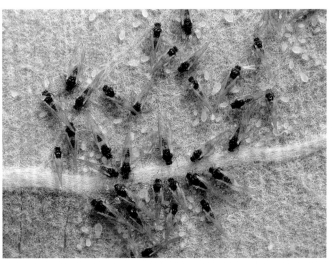

A. odinae winged parthenogenic viviparous females (Xu Gongtian)

55 *Chaitophorus populialbae* (Boyer de Fonscolombe)

Classification	Hemiptera, Aphididae
Synonyms	*Aphis populialbae* Boyer de Fonscolombe, *Chaitophorus albus* Mordvilko, *Chaitophorus inconspicuous* Theobald, *Chaitophorus hickelianae* Mimeur, *Chaitophorus tremulinus* Mamontova
Common name	Poplar leaf aphid

Distribution Beijing, Tianjin, Hebei, Shanxi, Liaoning, Jilin, Shandong, Henan, Shaanxi, Ningxia and other provinces; Europe, Turkey, Siberia, and Central Asia.

Hosts *Populus tomentosa, P. pseudo-simonii, P. simonii, P. tremula.* In the Indian subcontinent, poplar leaf aphid also causes damage on *P. euphratica*.

Damage The poplar leaf aphid excretes a large amount of honeydew, which covers leaves and shoots, causing sooty mold, so that the leaves become black, and cause early defoliation. When the excessive amount of the sooty mold accumulates, the entire branch droops, and seriously reduces the growth and timber quality. Poplar leaf aphid often occurs together with poplar woolen aphid, *Chaitophorus populeti*.

Morphological characteristics **Fundatrices** Approximately 2 mm long, pale green or yellow green. The body size of apterous aphids is 1.9 mm×1 mm, white to pale green in color. The dorsal center of the thorax has two dark green marks. The abdomen has five marks dorsally. The body is covered with dense bristles.

Winged viviparae Body 1.9 mm long, and light green in color. Head is black, and the compound eyes are in

Apterous and winged *C. populialbae* (Xu Gongtian)

C. populialbae winged female (Xu Gongtian)

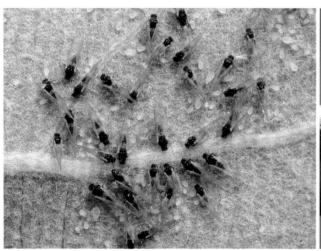

C. populialbae winged females (Xu Gongtian)

C. populialbae apterous viviparae (Xu Gongtian)

auburn color; forewing stigma is brown, and mesothorax and metathorax are black; abdomen is dark green or green, and there are two horizontal black stripes.

Nymphs Initially white, later become green; compound eyes are auburn, and body is white.

Eggs Oblong in shape, and greyish black in color.

Biology and behavior There are more than 10 generations per year in Beijing. They overwinter as eggs in the axillary buds, and underneath bark. When the poplar buds expand in the following spring, the overwintering eggs hatch into fundatrix. Fundatrix often feeds on the abaxial side of in the new leaf surface. The winged viviparae appear in May and June. They occur particularly near and inside the mite galls on the abaxial side of the leaves. They induce sooty mold after June. In October, sexual forms of aphids are reproduced by viviparous aphids. The aphids mate and then deposit overwintering eggs.

C. populialbae on a poplar leaf (Xu Gongtian)

Prevention and treatment

1. Manual control. The spraying of high-pressured water can kill or flush off a lot of aphids. In the spring, when it is under drought condition, spraying water can serve both purposes of physically removing aphids and irrigating plants. It needs to be repeated multiple times during a season.

2. Biological control. Release *Beauveria bassiana* or *Paecilomyces fumosoroseus*. Protect natural enemies of aphids. The natural enemies of the poplar leaf aphid include multicolored Asian lady beetle, *Harmonia axyridis*, seven spotted lady beetle, *Coccinella septempunctata*, *Propylaea japonica*, lacewings, *Chrysopa formosa*, the Chinese lacewing, *C. sinica*, and a parasitoid wasp, *Adialytus salicaphis*.

3. Chemical control. In April to mid-May, the spray of canopy with 10% imidacloprid WP at 2,000×, 50% acetamiprid at 2,000×, 1% matrine at 1,000×, or 3% hypertonic fenoxycarb EC at 3,000× can be effective.

References Zhang Guangxue and Zhong Tiesen 1983, Xiao Gangrou (ed.) 1992, Zhang Guangxue 1999, Xu Gongtian and Yang Zhihua 2007.

(Guo Yimei, Wang Zhizheng, Zhang Shiquan)

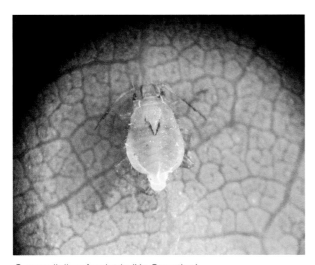

C. populialbae fundatrix (Xu Gongtian)

C. populialbae nymphs (Xu Gongtian)

56 *Periphyllus koelreuteriae* (Takahashi)

Classification Hemiptera, Aphididae
Common name Golden rain tree aphid

Importance *Periphyllus koelreuteriae* is an important insect pest of the *Koelreuteria paniculata*.

Distribution Beijing, Tianjin, Hebei, Liaoning, Shanghai, Jiangsu, Zhejiang, Anhui, Jiangxi, Shandong, Henan, Hubei, Hunan, Taiwan; Japan.

Hosts *Koelreuteria panniculata*, *K. bipinnata* var. *integrifoliola*, and *K. bipinnata*.

Damage *Periphyllus koelreuteriae* feed on young shoots, petioles and leaves, cause rolling of young leaves. They excrete large amount of honeydew which attracts blow flies and ants and induce sooty mold growth. The growth of *K. paniculata* can be affected by the aphid.

Morphological characteristics **Parthenogenetic apterae** Elongated oval, 3.0×1.6 mm, yellowish green with dark brown large stripes dorsally. Antennae 1.8 mm long, ratios of the 1st to the 6th segments are: 13, 11, 100, 51, 47, 22+42. There is a circular primary sensorium on the fifth and sixth segments, respectively. The cornicles are truncated with marginal protrusion, the end with reticulate sculpture.

P. koelreuteriae parthenogenetic apterae and alate (Xu Gongtian)

P. koelreuteriae parthenogenetic alate (Xu Gongtian)

P. koelreuteriae female apterae (Xu Gongtian)

P. koelreuteriae nymphs (Xu Gongtian)

Parthenogenetic alates Size is 3.0×1.6 mm. The head and thorax are black, and the abdomen is yellowish green. The middle and lateral spots on the first to the sixth abdominal segments fuse into a transverse black band. The two pairs of wings overlap resembling the shape of roof. Antennae are 2.0 mm long; the ratios of each segment are: 14, 9, 100, 50, 46, and 20 | 41. There are 33-46, and 0-2 secondary sensoria on the third and the fourth antennal segments, respectively.

Biology and behavior *Periphyllus koelreuteriae* occurs four generations a year in Beijing, and five generations a year in Shandong province. The overwintering eggs hatch in early March next year and become fundatrices. The first generation of apterous parthenogenetic aphids occur in late March; the second generation apterous and winged parthenogenetic aphids occur between in mid-April. Between late April and later August, the third generation of estivation aphids occur. In early September, the aphids develop into sexuparae and produce sexuales. The sexuales mate in late October and deposit eggs in mid- and late November.

Periphyllus koelreuteriae completes its life cycle on the same host. In early spring, the overwintering eggs begin to hatch into fundatrices and intensively damage the developing buds of *K. panniculata*. Each fundatrix produces 50-60 sistentes in 25 days. The sistentes feed for 15 days, then produce 15 second generation sistentes per aphid. The apterae and alates coexist and feed for about 20 days and produce 20 estivation aphids on the abaxial side of the leaves. Both fundatrices and sistentes aggregate and feed on young shoots, young leaves, and petioles. They cause leaf rolling and curl and shorten internodes. The estivation period lasts for approximately 130 days. The aphid growth and development resumes, and mainly cause feeding damage to branches bearing fruits and flowers. Each alate sexupara produces 40 apterous females, and each apterous sexupara produces 20 alate males. The sexuales mate when they are mature. Females can oviposit 20-30 eggs. The eggs are deposited around the buds at the beginning, and later in bark cracks, scars, and the lower part of the branches.

Prevention and treatment

1. Manual control. Prior to November in the autumn, wrap trunks of *K. panniculata* with straw ropes to attract females to oviposit eggs. Then, in the early spring before March, remove and destroy the straw ropes.

Honeydew on tree trunk produced by *P. koelreuteria* (Xu Gongtian)

P. koelreuteriae fundatrices (Xu Gongtian)

2. Chemical control. In late March, spray 40% acetamiprid and chlorpyrifos mixture at 1,000-2,000× can effectively control 95% of the fundatrices. Irrigate the roots using 50% imidacloprid and monosultap mixture at 1,000-1,500×.

References Zhang Guangxue 1983; Wang Nianci, Li Zhaohui, Liu Guilin et al. 1990, 1991; Gu Ping, Zhou Lingqin, and Xu Zhong 2004.

(Yang Chuncai, Tang Yanping)

57 *Cinara formosana* (Takahashi)

Classification Hemiptera, Aphididae
Synonym *Cinara pinitabulaeformis* Zhang et Zhang

Importance This species is one of the important pests of pine trees at lower elevations. In particular, more damage occurs in the sunny stands than in the shaded stands. When population is high, 100% plants show damage symptoms.

Distribution Beijing, Hebei, Inner Mongolia, Shanxi, Liaoning, Shandong, Henan and Shaanxi; Korea Peninsula, Japan, and Europe.

Hosts *Pinus koraiensis*, *P. tabulaeformis*, *P. thunbergii*, *P. densiflora*, *P. sylvestris*, *P. sylvestris* var. *mongolica*, *P. massoniana*.

Damage *Cinara pinitabulaeformis* adults and nymphs feed on sap of young (1 to 2 years old) shoots. The damage symptoms include red needle tip, dry leaves, and yellowish red spots on the needles. Pine needles are covered with honeydew. At high aphid population density, leaves become dry and fall prematurely.

Morphological characteristics **Adults** The winged viviparae length 2.8-3.0 mm; dark brown; body is covered with black bristles, especially on the legs. Abdominal end pointed. The wings are transparent and membranous; the costal edge of the wings are dark brown.

The apterous vivparae are larger than the winged form. The antennae have six segments, and the third segment is the longest. The abdomen is covered with black particles and also with scattered white waxy powder. The end of the abdomen is obtuse.

C. formosana winged vivipara (Xu Gongtian)

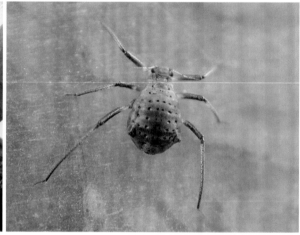

C. formosana apterous female (Xu Gongtian)

C. formosana apterous female depositing an egg (Xu Gongtian)

C. formosana apterous female and eggs (Xu Gongtian)

The males resemble the apterous viviparae but are slightly smaller than the viviparae and the abdomen is more pointed than the viviparae.

Nymphs Similar to the wingless adults, but smaller in size. The viviparous nymphs produced by the fundatrix are light brown in color, length 1 mm, the viviparous nymphs are dark brown after 4 to 5 days.

Eggs The size is 1.8-2.0×1.2 mm, black, oblong.

Biology and behavior *Cinara formosana* overwinters as eggs on pine needles. In Liaoning province. The eggs hatch into nymphs in late April or early May. The fundatrix appears in mid-May. One fundatrix can reproduce parthenogenically more than 30 female nymphs. The nymphs develop into adults and continue their viviparous reproduction. Under optimal conditions, the aphid can complete one life cycle in 3 to 4 days. Winged adults appear between mid-May and early October, they can disperse. From mid-May to early October, all life stages (different instars of nymphs and adults) can be observed. In mid-October, sexual forms (winged male and female adults) appear. The females oviposit eggs after mating. There are two population peaks in Beijing: from May

Freshly laid *C. formosana* eggs (greyish white) (Xu Gongtian)

C. formosana overwintering eggs (dark green) (Xu Gongtian)

C. formosana female placing wax on eggs (Xu Gongtian)

C. formosana apterous viviparae (Xu Gongtian)

A *Harmonia obscurosignata* adult feeding on *C. formosana* eggs (Xu Gongtian)

A syrphid fly lava feeding on *C. formosana* (Xu Gongtian)

C. formosana eggs below a plastic band on a pine tree in the Fall (Xu Gongtian)

C. formosana eggs on a plastic band on a pine tree (Xu Gongtian)

to June, and in October, respectively. They cause severe damage, especially in the autumn. Nymphs have four instars. The eggs hatch in later April in Fuxin, Liaoning. The development duration of the nymphal stage is 19 to 22 days. In May when temperature is high, the developmental duration shortens to 16 to 18 days. Adults can reproduce in 3 to 4 days.

In Beijing and the surrounding area, there is a similar aphid species, *Cinara pinihabitans*. The difference is that the tip of the fifth segment of the antennae of *C. formosana* has two circular sensillae, but *C. pinihabitans* has only one sensilla at the same location.

Prevention and treatment

1) Strengthening the forest management. Timely fertilization and irrigation, removing the weedy host plants, and establishing mixed forest stands can effectively reduce aphid damage. In addition, the removal of aphid-infested pine needles in autumn and winter can reduce the aphid infestation in the following year.

2) Apply *Beauveria bassiana* or *Paecilomyces fumosoroseus*. Protection and utilization of natural enemies. Predators include *Bothrocalvia albolineata*, *Ballia obscurosignata*, *Coccinella septempunctata*, and *Harmonia axyridis*, *Adalia bipunctata*, *Hippodamia tredecimpunctata*.

3) In the early spring before lady beetles and other predators become active, targeted spray of 10% imidacloprid WP or 0.05% avermectin following label directions.

References Zhang Zhizhong et al. 1982; Xiao Gangrou (ed.) 1992; Wang Jijian 1998; Zhang Guangxue 1999; Guan Yongqiang and Zhang Yan 2001; Li Qing 2004; Li Xiaohua and Gao Pei 2005.

(Guo Yimei, Li Zhenyu, Fan Zhongmin)

58 *Cinara tujafilina* (Del Guercio)

Classification Hemiptera, Aphididae
Common name Arborvitae aphid

Importance *Cinara tujafilina* is one of the important pests of the Chinese arborvitae. It significantly reduces the development of young branches. Severe damage causes withered shoots and the bark is covered with a layer of black secretion-honeydew. The growth and ornamental value of arborvitae is reduced by the insect.

Distribution Hebei, Liaoning, Jiangsu, Zhejiang, Jiangxi, Shandong, Yunnan, Shaanxi, Ningxia, and Taiwan; Korea Peninsula, Japan, Turkey, Palestine, Africa, Europe, Oceania, and North America.

Hosts *Platycladus orientalis*, *Sabina chinensis*, *S. virginiana*.

Damage The aphids inhabit at the shady branches in the summer, feed on the sap of the arborvitae branches. The damaged branches show faded green color, poor growth or withering. The feeding sites become softened, depressed. Honeydew on branches leads to sooty mold development.

Morphological characteristics **Adults** Brown; the end of the antennae, compound eyes, and the third to the fifth segments of the rostrum, and the end of femur, tarsi and claws, and abdominal cornicles black. Antenna has 6 segments; the third segment is the longest.

Alate viviparae Length 3.0-3.5 mm, wingspan 7.5-9.0 mm. White setae on legs and dorsal surface of the body dense. There is white pubescence on the wings. There is a "X"-shaped spot on mesonotum. The costal vein of the forewings is dark brown. There are two small dark spots near the apex. The first four segments of the abdomen have two pairs of brown spots arranged neatly on the dorsal surface. The end of the abdomen is pointed.

C. tujafilina apterous viviparae (Xu Gongtian)

C. tujafilina winged viviparae (Xu Gongtian)

C. tujafilina apterous female (Xu Gongtian)

C. tujafilina eggs (Xu Gongtian)

Apterous viviparae Length 3.7-4.0 mm; color slightly lighter; there is an upside down "V" shaped mark on the thorax dorsally. There are six rows of black dots on the abdomen dorsally, each row with 4-6 spots. The abdomen is covered with white wax powder ventrally. The last segment of the abdominal is obtuse.

Male adults Length 3.0 mm; similar to apterous viviparae; the end of the abdomen is pointed.

Nymphs The body shape is similar to the viviparous adult aphids. The new nymphs are reddish orange in color, 1.8 mm long, and then change to brown in 3 days.

Eggs The newly deposited eggs are brownish yellow, and then change to dark brown. The eggs are oblong, approximately 2.0×1.1 mm in size.

Biology and behavior The aphid inhabits the Chinese arborvitae year around. It belongs to non-migratory type of aphid. It occurs multiple generations per year. The aphid overwinters in egg stage. Eggs hatch in late March of the following spring. The fundatrix aggregates on young branches. Apterous viviparae appear in late April. Winged viviparae appear in May and then disperse. W male alates and apterous females occur in October. They mate and oviposit. Eggs start overwintering at the end of November. Very few apterous viviparae overwinter underneath the bark seam and dense branches. Between mid-April and October, different life stages and overlapping generations can be observed.

Prevention and treatment

1. Silviculture. Removal of the branches infested with aphid eggs in winter or scrapping off the overwintering eggs on branches can eliminate the insect source for the following season.

2. Apply *Beauveria bassiana* or *Paecilomyces fumosoroseus*. Protection/preservation and utilization of natural enemies. For example, *Harmonia axyridis*, *Coccinella septempunctata*, lacewings, and syrphid flies can be utilized to reduce aphid population.

3. Spray tree canopy in early spring with 50% imidacloprid and molosultap mixture at 5,000×.

References Wang Xingde and Bi Qiaoling 1988, Xiao Gangrou (ed.) 1992, Guo Zaibin et al. 2000, Xu Gongtian and Yang Zhihua 2007.

(Guo Yimei, Wang Xingde, Bi Qiaoling)

C. tujafilina 1st instar nymph produced by stem mother (Xu Gongtian)

C. tujafilina laying eggs (Xu Gongtian)

A C. tujafilina adult male being fed by a ladybug larva (Xu Gongtian)

59 *Lachnus tropicalis* (van der Goot)

Classification Hemiptera, Aphididae
Synonym *Pterochlorus tropicalis* van der Goot
Common name Chestnut aphid

Distribution Beijing, Hebei, Liaoning, Jilin, Jiangsu, Zhejiang, Fujian, Jiangxi, Shandong, Henan, Hubei, Guangdong, Guangxi, Sichuan, Guizhou, Yunnan, Shaanxi, and Taiwan; Japan, Korea Peninsula, and Malaysia.

Hosts *Castanea mollissima*, *Quercus fabric*, *Q. mongnolica*, and *Q. acutissima*.

Damage Adults and nymphs aggregate and feed on young chestnut shoots, branches and leaves. They can reduce plant vigor by removing sap.

Morphological characteristics **Apterous viviparae** Length 3-5 mm, width 2 mm; shiny black in color. There is fine reticulate sculpture on body surface, which is covered with dense long hairs. Legs are slender, and abdomen is plump. Antennae are 1.6 mm long; the proboscis is long and longer than the hind legs.

Alate viviparae Female length 3-4 mm with the wingspan of about 13 mm. Black, the abdomen light black; veins black. There are two transparent spots from the middle of the costal margin of forewing diagonally to the posterior corner. The costal edge near the top corner has a transparent spot.

Eggs Length 1.5 mm, oblong; newly deposited eggs dark brown, and later become shiny black. Eggs are deposited in one layer and in clusters, which is often found on the shaded side or the base of thick branches.

Nymphs Similar to apterous viviparae, but small in size, pale in color, mostly yellowish brown, and gradually change to black.

Biology and behavior Chestnut aphid occurs 8 to 10 generations per year. After mid-October, the fertilized eggs overwinter under the branch surface, bark cracks, scars, and tree holes in groups. The shady side has more overwintering eggs. In late March to early and mid-April when temperature reaches 10°C, the overwintering eggs begin to hatch. Peak hatching period occurs when the temperature reaches 14 to 16°C. The survival rate is very high when the relative humidity is 65~70%. The cold spill in late spring as well as cold weather greatly reduces the hatch rate. The apterous viviparae aggregate and feed on young shoots, and gradually move to the leaves and lead to the first population peak. In mid- and late May, the alate viviparae appear and move to the whole plant, the inflorescence and other surrounding host plants to reproduce and cause feeding damage. Then, the aphid population declines on the chestnut trees. In late August to early September, when the chestnut fruits are at the rapid expansion stage, the aphid population increases again. The infestation sites are concentrated on the branches, chestnut fruits, and fruit stalk and leads to the second peak of damage. After mid-October, the sexual forms appear, mate and deposit eggs.

Prevention and treatment

1. Manual control. Scraping bark or brushing the chestnut bark to remove/kill the overwintering eggs. When the overwintering egg density is not high, manual removal of the infested branches can effectively reduce the insect population.

2. Apply *Beauveria bassiana* or *Paecilomyces fumosoroseus*. Protection/preservation of natural enemies. Chestnut aphid natural enemies include lady beetles, lacewings and parasitoid wasps. When natural enemies are properly preserved, aphid population can be suppressed by the natural enemies.

3. The spray of insecticides at the canopy of the chestnut trees. Insecticides such as 10% nitenpyram at 2,000×, or 50% pirimicarb WP at 1,500× can be used.

References Zhang Guangxue and Zhong Tiesen 1983, Zhang Guangxue 1999, Liu Jiazhu 2003, Zhao Yanjie 2005.

L. tropicalis (Zhang Peiyi)

(Guo Yimei, Wang Zhizheng, Sun Yujia, Bao Bailing)

60 *Schlechtendalia chinensis* (Bell)

Classification Hemiptera, Aphididae
Common name Chinese sumac aphid

Distribution Jiangsu, Zhejiang, Fujian, Jiangxi, Shandong, Hubei, Hunan, Guangdong, Guangxi, Sichuan, Guizhou, Yunnan, Shaanxi, Gansu, and Taiwan; Japan, Korea Peninsula, Indochina.

Hosts The primary hosts are *Rhus chinensis* and *R. chinensis* var. *roxburghii*. The secondary hosts are *Plagiomnium maximoviczii*, *P. rhyhcnophorum*, *P. succlenlum*, *P. vesicatum*, and *P. acutum*.

Damage *Schlechtendalia chinensis* infests *Rhus* spp. and form aphid galls which is a well-known Chinese herb. The aphids can feed along the central axis of the compound leaves of its primary host *Rhus chinensis* and *R. chinensis* var. *roxburghii*. The mature dry gall is comprised of 60%-65% of soluble tannin. The tannic acid, gallic acid, pyrogallate and other byproducts extracted from the galls are important industrial materials.

Morphological characteristics **Alate parthenogenetic adults (autumn migrant aphids)** Oval; greyish black; 1.93-2.1 mm in length. Thorax dark brown, abdomen pale. Antennae have five segments, 0.62 mm in length, which is shorter than the total length of head and thorax. The ratios of the first to the fifth antennal segment are: 24, 23, 100, 51, 92+13, respectively; there is a small round sensilla at the basal part of the end of the fifth antennal segment. There are 10, 5 and 11 secondary sensilla in the shape of wide-band or open-loop on the surface of the third, fourth and fifth segments of antennae, respectively. Forewing is 2.6 mm in length, and has four diagonal veins; the medial vein does not have a fork; stigma is big, sickle–shaped and extends to the tip of the forewing. Hind wing has two oblique

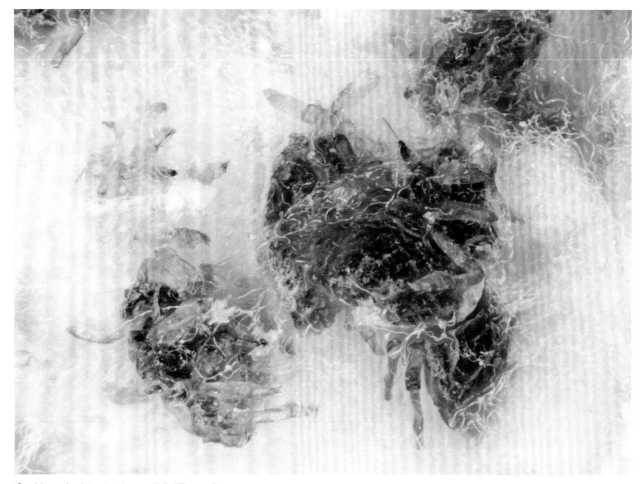

S. chinensis sistentes in a gall (Li Zhenyu)

veins and has no stigma. Hind tibia 0.52 mm in length. The cornicle is absent.

Alate sexupara adults (spring migrant aphids) The body shape and veins are similar to those of alate parthenogenetic adult aphids, but the size is relatively small, 1.4 mm× 0.49 mm. The cornicle is absent.

Sexuparae Apterous. Antennae have four segments. Proboscis is not well developed. The females are in egg- or elongated-egg-shaped and light brown, a size of 0.54 mm×0.25 mm; the overall length of antennae is 0.11 mm; the ratios of each segment are: 15, 17, 22 and 46. The hind tibia length is 0.06 mm. Males are narrow and long, dark-green, and 0.48 mm×0.23 mm in size; the overall length of antennae is 0.13 mm; the hind tibia length is 0.08 mm; copulatory organ is well developed, and dark brown, located ventrally at the end of the abdomen.

Fundatrices Apterous. Oval, shiny black, 0.37×0.17 mm in size. Antennae have four segments, and the total length is 0.11 mm. The distal end of the third segment and the tip of the fourth segment have primary circular sensilla. Proboscis extends to the coxal cavity of middle legs. The tibia of the hind legs is 0.08 mm in length.

Sistentes Apterous. Oval, large, yellow, 0.91×0.54 mm in size. Antennae have five segments, the total length is 0.26 mm. The distal end of the fourth and fifth segments has one circular-shapedprimary sensilla. Proboscis is well developed. The tibia of hind legs is 0.17 mm in length.

Aphid galls They generally grow along the central axis of the compound leaves of *R. coriarioides*. The gall is elongated oval shaped with claw-like branches. The surface has multiple horn-like protrusion. The largest aphid gall is 117 mm×65 mm. There are 45-50 mature fresh aphid galls, and 80-100 dry galls per 500 g of weight.

Biology and behavior *Schlechtendalia chinensis* is a migratory aphid that has two hosts throughout a growing season. Completion of a migration cycle requires 12 months, which spans across two calendar years. Sexupara feed on the stalks of the secondary hosts, excrete wax, form wax balls and overwinter. The aphids develop slowly during the overwintering stage with a total of four instars. The first and second instars last for about 15-20 days, respectively. The third instar lasts for more than 100 days and live through the period with the lowest temperature. Wing pads develop on meso- and meta-thorax in the later third instar; the fourth instar lasts for about 30 days, the wax threads are short and scarce,

S. chinensis sistentes (Li Zhenyu)

S. chinensis galls (Li Zhenyu)

wing pads turn from light yellow to bluish black. The aphids move and establish at the basal part of the stalks of the winter hosts or near the stalks of rizoid, and then molt into alate sexupara adult aphids (spring migrant aphids). The migration period of spring migrant aphids is between mid-April and mid-May in the areas of 1200-1500 m altitude or areas of the further north latitude. The migration period is between early March and early April in areas of below 800 m altitude.

Aphid migration occurs either on sunny or cloudy days with temperature of above 9℃. Optimum temperature is 12-15℃ and optimum relative humidity is less than 80%. No migration was observed on rainy days. The migration time is 10:00-20:00, and the optimal time is 14:00-18:00, which accounts for 98.85% of the total aphid migration per day. They have phototropism, and fly to windows. They migrate to the trunks and branches of the primary host, and inhabit at shady location that is away from the wind. Then the sexupara gradually produce sexuales and die within 1-2 days. Each sexupara produce five sexuales. The ratio of females to males is 3:2 or 2:3. The ratio is influenced by multiple factors such as light, temperature, humidity, and nutrients during the embryo development of the previous two generations (i.e., the autumn migrant aphids). Sexuparae inhabit and develop under tree bark and mosses on tree trunks without feeding. Males have three instars and can mature in five days, while females have four instars and can mature in six days. Males move toward the females to mate. Males can mate multiple times. After mating, females produce fundatrices after 26-27 days. Their life span is 32-33 days.

The newly nymphoposited fundatrices crawl to new shoots within hours, find feeding sites on the adaxial leaf surface. After 1-2 days, the leaf tissue around the insect forms a transparent circle. The leaf tissue shows prolific growth and protrusion after 3-4 days, then the aphid body is encaved. The tissue of the leaf surface overgrows into cone shape and covers the aphid in 5-6 days, the conical structure ceases to grow when the surrounding tissue turns red. Fundatrix molt once at this time. The body color is light green. Their stylets insert into the leaf tissue and stimulate growth of leaf tissue on the abaxial side of the leaves. Globular young galls can be observed after 7-10 days. Young galls are covered with dense soft white hairs, with a diameter of 2-3 mm. They develop into adults after four molts in about 25-30 days. The gall shape change from round to elliptical.

Fundatrices immediately produce two generations of the sistentes. A few galls split from the basal part, and the sistentes occupy both galls. Gall length can be as long as about 20 mm, each with more than 100 first and second generation sistentes. The galls rapidly expand during the third generation. Gall length can reach 70-90 mm, each with thousands of sistentes. When the third generation of the sistentes develop into alate parthenogenetic adults (i.e., autumn migrant aphids), the galls continue to grow and turn from green to yellowish green. The surface facing the sunlight is red, the compound leaves with galls are withered, which signals the maturity and cracking stage of the galls. The duration from the young gall appearance to the cracking stage lasts for 113-177 days. A mature gall usually contains 3,000-5,000 aphids. Large galls can have more than 10,000 aphids. The ratio of alates and the apterous aphids is 9.6:3.4. The gall development vary greatly according to geographical conditions. The young galls appear in late April to late May, and mature and cracking stage is in late September to early October in areas of below 800 m altitude.

Regardless of the altitude, the rapid growth of aphid galls begins from late July when the sistentes of the third generation occur. The fastest growth occurs between mid-August and one week before reaching maturity. At low altitude, the early and late development stage is longer than that at high altitude. It has been observed that when the aphid galls are harvested 30, 20, and 10 days before maturity, the gall weight is 58.1%, 41.3%, 21.6% of the mature galls, respectively. The number of fresh galls per 500 g also decreases from 315 to 113 galls. The water content of the fresh galls reduces from 52.5% to 45%. The mature and cracking stage is when the autumn migrant aphids appear and migrate. The temperature in the forest is around 15-25 °C. Migration occurs between 09:00-17:00. The peak is between 11:00-14:00, which represents 93% of the total emergence.

The optimal flight conditions for the migrant aphids are sunny or cloudy day with breeze (0.1-0.2 m/s), whereas no flight occurs on rainy or windy days when the wind speed is greater than 0.6m/s. Each of the autumn migrant aphid can produce 23 parthenogenic female aphids, with the maximum of 34 aphids. The first instar sexuparae establish their feeding sites, and begin to secrete wax threads in 5-7 days. Then, the large amount of the wax threads covering the second and third instar nymphs form into wax balls, which provide the shelter for the nymphs to survive -10°C temperature in winter. If the temperature is high and sunny before overwintering, some nymphs can develop into adults, but they are extremely weak, die before they can develop into the spring migrant aphids.

The occurrence of gall aphids and production of aphid galls are closely related to the phenology, vigor, and age of *R. coriarioides*. The emergence and migration of the spring migrant aphids occurs during the bud development of *R. coriarioides*. The fundatrices produce galls when there are 5-10 young compound leaves in the new shoots of the host. On vigorous host trees, the rapid growth of compound leaves and excessive thickness are not optimal for the establishment of aphid galls by fundatrices. When the shoots with galls have excessive growth in late season, compound leaves with galls in the lower portion of the branches will senesce and fall down early before the gall maturity. In contrast, when *R. coriarioides* is too weak and the compound leaves are small, there would be no large galls developed. Trees of greater than 10-15 years old with moderate vigor is beneficial to the aphid gall production. The abundance of secondary host is extremely important for the gall production. All secondary hosts are moss of the same genus (*Schlechtendalia*), which is located in shady and cool and humid environment, without direct sunlight, high temperature is below 25°C, and the relative humidity is greater than 80% throughout the year.

The mosses often grow under evergreen broad-leaved forests, on rocks of shady damp areas, on humus soil or on dead wood. The narrow distribution of secondary hosts limit the distribution of gall production forests. In addition, sufficient autumn migrant aphids must migrate to the secondary hosts (mosses) for overwintering so that they can provide enough aphids for gall formation on *Rhus* tree in the next year. Low temperature and rain during migration in the spring often lead to the reduction of gall production.

The leaf-feeding insects on *R. coriarioides* include *Trabala vishnou* and *Mimela lucidula*. Insects on young shoots include lacinid aphids. *Ectatorrhinus adamsi* bores into the trunk. The main natural enemies of *Schlechtendalia chinensis* include ants, spiders, and snails. Syrphid flies, squarrels, and birds also feed on the aphids inside the galls.

References Baker 1917; Takagi 1937; Jiao Qiyuan 1938; Chiao 1939; Jiao Qiyuan et al. 1940; Sun Zhangding 1942; Tao Jiaju 1943, 1948; Yu Dejun 1943; Cai Banghua et al. 1946a, b; Tang Jue 1956, 1976; Tang Jue et al. 1957, 1987; Fang Zhongmin 1962; Xianghe 1980; Wu Faji 1982; Zhang Guangxue et al. 1983; Pan Jianguo et al. 1985; Sichuan Province Wu Bei Zi Research Team 1985; Tian Zejun et al. 1966, 1985, 1986, 1988.

(Tian Zejun)

61 *Eriosoma dilanuginosum* Zhang

Classification Hemiptera, Aphididae
Common name Japanese elm woolly aphid

Distribution Hebei, Liaoning, Jiangsu, Zhejiang, Anhui, Shandong, and Shaanxi.

Hosts *Ulmus pumila, U. parvifolia, U. davidiana* var. *japonica, U. fulva*. The less preferred hosts are: *Pyrrus betulaefolia, P. calleryana, P. communis* var. *satiea, P. pyrifolia, P. pyrifolia* var. *culte*.

Damage Fundatrices, apterous sistentes, and alate sistentes feed and injure young leaves of elm, stimulate prolific growth of the leaves and form false aphid galls. The aphid galls subsequently wither, and small branches lose their ability to produce new shoots. The aphids migrate and cause damage on the roots of pear trees in the summer, and reduce the vigor of the pear tree, and then fruit quality and quantity.

Morphological characteristics **Alate parthenogenetic adults** Oval, 2.00 mm-2.20 mm×0.82-0.97 mm in size. Head, thorax and legs black, abdomen brown. Antennae are thick and short, 0.72 mm in length; the ratios of the first to the sixth antennal segment are: 15, 16, 100, 32, 34, 21+6, respectively. Besides the fifth and the sixth segments having primary circular sensilla, the third to sixth segments have 22-24, 5, 5-7, 1-2 semi-circular secondary sensilla. The medial vein of the forewings has a fork, and the hind wing has two oblique veins. Cornicles are circular, surrounded by 12-16 setae.

Biology and behavior *Eriosoma dilanuginosum* overwinters as eggs. They hatch into fundatrices in the spring when elm begins to sprout new buds (in early March). Affected leaves roll on to the abaxial side. After two generations of apterous sistentes are produced, alate parthenogenetic aphids (alate sistentes) are produced and migrate from elm pear tree to produce apterous parthenogenetic aphids in mid- and late May. In mid- and late November, the alate parthenogenetic aphids return to the trunk of its overwintering host (elm) to produce sexuales. They mate and deposit eggs which will survive through the winter.

In mid- and late March, the fundatrix feeds on the abaxial side of the young leaves for 10-15 days and produces 20-50 first generation of the apterous sistentes; then, after another 10-15 days, each female produces 20-35 second generation alate sistentes. The aphid feeding stimulates the leaves to curl backward and expand, which forms false aphid galls that resembles open fists. Each gall generally contains 200-450 third generation

E. dilanuginosum migratory aphid (Xu Gongtian)

E. dilanuginosum damage (Xu Gongtian)

E. dilanuginosum gall (Xu Gongtian)

alates. The size of the galls is positively correlated to the number of alates, and the diameter of the galls is 3-6 cm. There are scattered white wax masses within the galls. The false aphid galls turn from deep green to brownish green or yellow as the alates occur and migrate. Mature galls harden, turn brown, and die.

In mid- and late May, alate parthenogenetic aphids migrate toward pear trees, crawl into the soil under pear tree canopy, and produce the fourth and fifth generations of apterous parthenogenetic aphids. Each female produces 20-35 nymphs per generation. The migrant aphids continue to disperse. Each adult aphid migrates 4-7 times and produce 5-6 nymphs per site. They continuously excrete white cotton-like wax powder when feeding on the roots. In mid- and late September, the last generation is the alate parthenogenetic aphids (sexuparae). In early and mid-November, the sexuparae return to the branches and trunks of the elm trees. They aggregate either under the rough tree bark or cracks of broken branches. Each sexupara produces 5-10 sexuales with dysfunctional proboscis. After three molts, the sexuales mate, and each female only lays one egg. The dead female covers the surface of the egg. The eggs overwinter in peeled bark and cracks of broken branches.

Prevention and treatment

1. Manual control. Remove the false aphid galls before the early May.

2. Biological control. Spray *Beauveria bassiana* or *Paecilomyces fumosoroseus*.

3. Chemical control. Spray 40% acetamiprid and chlorpyrifos mixture at 1,500× over the canopy of elm trees in early March.

References Zhang Guangxue 1983; Hu Zuodong, Zhang Fuhe, Wang Jianyou et al. 1998; Zhang Guangxue, Qiao Gexia, Zhong Tiesen et al. 1999.

(Tang Yanping, Yang Chuncai)

62 *Pemphigus immunis* Buckton

Classification Hemiptera, Aphididae
Common name Poplar stem gall aphid

Distribution Beijing, Tianjin, Hebei, Inner Mongolia, Liaoning, Jilin, Heilongjiang, Anhui, Shandong, Henan, Gansu, Ningxia, Xinjiang; Egypt, Iran, Iraq, Jordan, Turkey, Russia, and Morocco.

Hosts Primary hosts are *Populus simonii* and *P. cathayana*. The secondary hosts are *Sataria viridis*, *S. faberii*, *S. itelica*, and *S. glauca*.

Damage *Pemphigus immunis* feeds on the middle and lower portions of new shoots, forms pear-shaped galls. Infestations prevent growth of young shoots and thus reduce the vigor of the hosts.

Morphological characteristics **Apterous fundatrices** Oval, 2.8-4.6 mm×2.1-2.6 mm. Antennae have four segments. The body is light green in color and covered with white wax powder.

Alate sistentes Elongated oval, 2-3 mm×0.9-1.02 mm. The body is greyish green and covered with white wax powder. Antennae short and thick; 0.67 mm in length; the first to the sixth segment is weakly imbricated; their ratios are: 26, 29, 100, 43, 57, 67+17. The secondary sensilla horizontal ring-shaped. There are seven sensilla on the third segment; 2-4 sensilla on the fourth segment; one rectangular primary circular sensilla on the fifth segment that covers 2/5 of the whole segment length with an egg-shaped structure inside. The fifth segment also has 1-2 secondary sensilla. The primary circular sensilla on the sixth antennal segment have setae. Forewing has 4 diagonal veins and there is no fork on the medial vein.

Biology and behavior *P. immunis* occurs six to seven generations per year. The aphid overwinters in the egg stage under bark. Eggs hatch into fundatrices in mid-April the following year. Feeding by fundatrices causes damage on new (current year) shoots. In early May, alate sistentes occur. By late May, the alates migrate and burrow around the roots of grassy weed hosts, then produce apterous migrant aphids. They reproduce 4-5 generations. In early or mid-September, the sexuparae return to poplar branches and trunks to produce sexuales. In late-October, the sexuales mate and deposit overwintering eggs on poplar trees.

The fundatrices of *P. immunis* occur in mid-April of the following spring. They crawl to 10-15 cm long new

P. immunis stem mother (Xu Gongtian)

P. immunis gall (old) (Xu Gongtian)

P. immunis gall (fresh) (Xu Gongtian)

shoots, settle down, and feed. The feeding sites gradually form soybean-size green galls. Then, sistentes are produced inside the galls. The reproduction period lasts for 20-25 days. Each fundatrix produces 50-80 sistentes. The size of the galls are positively correlated with the number sistentes. Galls are pear shaped, 1.5-2.8 cm in diameter, thickness of the gall wall is 1.9-2.3 mm; their surface has irregular cracks. Galls are filled with white wax powder. The original opening is facing downward.

The sistentes all develop into alates in early June. They emerge through the primary opening of the galls at 07:00-09:00 and 16:00-19:00 on sunny days. Then they migrate to roots of grassy weed hosts, produce sistentes for 4-5 generations. During this period, they constantly change feeding locations. Each sistentes produces 10-15 nymphs attached with white wax threads. In mid-October, the last generation is the alate sexuparae. They migrate back to poplar branches, trunks and bark cracks, and scars. They produce apterous females and alate males. A female dies after depositing one overwintering egg. The female dies on top of the overwintering egg.

Prevention and treatment

1. Silviculture. Establish fast growing poplar hybrids to prevent damage.

2. Chemical control. In late April, spray 40% acetamiprid and chlorpyrifos mixture at 1,000-2,000×.

References Xiao Gangrou (ed.) 1992; Zhang Guangxue, Qiao Gexia, Zhong Tiesen et al. 1999.

(Yang Chuncai, Tang Yanping)

63 *Takecallis arundinariae* (Essig)

Classification   Hemiptera, Aphididae
Common name   Black-spotted bamboo aphid

Distribution   Beijing, Jiangsu, Zhejiang, Anhui, Fujian, Shandong, Taiwan; Korea Peninsula, Japan, Europe, and North America.

Hosts   *Phyllostachys atrovaginata, P. aureosulcata, P. aureosulcata* f. *aureocaulis, P. bambusoides, P. bambusoides* f. *lacrima-deae, P. bissetii, P. bambusoides* f. *shouzhu, Phyllostachys dulcis, P. flexuosa, P. glauca, P. heteroclada* var. *pubescens, P. iridescens, Bambusa blumeana, P. praecox, P. nigella, P. praecox* f. *provernalis, P. propinqua, P. sulphurea* 'Viridis', *P. gozadakensis, Pleioblastus amarus, Pleioblastus simony, Yushania niitakayamensis, Y. polytricha, Y. qiaojiaensis*, and other bamboo species.

Damage   The damage symptoms include shrink and chlorosis of young leaves. The honeydew on young bamboo leaves causes sooty mold and reduces photosynthesis.

Morphological characteristics   Apterous viviparae   Body length 2.15-2.24 mm, long oval, light-yellow. Head is smooth, with eight long dorsal setae, clypeus with a cystic bulge, proboscis short; compound eyes are big, red, with compound eye wart, and three ocelli. Antennae are light gray with six segments and are about 1.1 times of the body length. Antennal warts are not obvious, Central wart is well developed; legs slender, grey in color.

Alate viviparae   Body length 2.32-2.56 mm, long oval, and light-yellow to yellow in color. Head is smooth, with eight dorsal setae. The central part of the frons is protruded, and cysts on the frons region are extraverted; proboscis is short and smooth. The compound eyes are large with warts, and there as three ocelli. Antennae is slender and light grey with six segments, about 1.6 times of its body length. The first to the seventh abdominal segments has a pair of longitudinal dark brown stripes dorsally. Forewing length 3.42-3.74 mm, the medial vein has two forks. Legs are slender, and light gray in color.

Biology and behavior   In Yuhang, Zhejiang, it occurs 18 to 20 generations per year. The life cycle is similar to *Melanaphis bambusae*, except when temperature is high in June, the population density of this aphid is low in bamboo forests. Its population rebounds after September.

Prevention and treatment

1. Apply *Beauveria bassiana* or *Paecilomyces fumosoroseus*.

2. Protection and preservation of natural enemies. The predators include two spiders: *Clubiona maculata* and *Chiracantium zhejiangensis;* five species of the lady beetles: *Chilocorus rubidus, Harmonia axyridis, Coccinella septempunctata, Hyperaspis sinensis*, and *Ballia obscurosignata;* three species of lacewings: *Chrysopa septempunctata, C. kulingensis*, and *C. sinica*. In addition, *Epistrophe* sp., *Ephedrus* spp., and *Diaeretiella* spp. can be utilized for pest management.

(Xu Tiansen, Wang Haojie)

T. arundinariae winged viviparae (Xu Gongtian)

T. arundinariae nymphs (Xu Gongtian)

Robber fly, a natural enemy of *T. arundinariae* (Xu Gongtian)

64 *Takecallis taiwana* (Takahashi)
Classification Hemiptera, Aphididae

Distribution Jiangsu, Zhejiang, Anhui, Fujian, Shandong, Hunan, Sichuan, Yunnan, and Taiwan; Japan, New Zealand; Europe, North America.

Hosts *Phyllostachys bambusoides, P. dulcis, P. flexuosa, P. glauca, P. heteroclada var. pubescens, P. nuda, P. makinoi, P. iridescens, P. sulphurea* 'Viridis', *P. praecox* and so on.

Damage The aphid prefers to feed on young leaves. Injured bamboo leaves cannot unfold and gradually wither.

Morphological characteristics **Apterous viviparae** Length 2.05-2.14 mm, long oval, pale green or brown. The compound eyes are large and red with warts. There are three ocelli. Antennae have six segments, black, and about 0.65 to 0.75 times of its body length.

Alate viviparae Length 2.35-2.46 mm, long oval, pale green or brown. Head slightly protruding and smooth, with eight dorsal setae. Proboscis is short and thick. The compound eyes are big and red with warts. There are three ocelli. Antennae have 6 segments, black, and are 0.7 to 0.8 times of its body length, and antennal warts are not well-developed. Legs are gray. Forewing length is 2.15-2.24 mm. The medial vein has a fork.

Biology and behavior In Yuhang, Zhejiang province, *Takecallis taiwana* occurs 20 to 23 generations per year. Each generation is 5 to 20 days shorter than *Melanaphis bambusae*. In July and August with high temperature, the completion of a generation only needs 15 days.

Prevention and treatment

Apply *Beauveria bassiana* or *Paecilomyces fumosoroseus*. Protection and preservation of natural enemies can be effective. There are five spiders, i.e., *Dolomedes pallitarsis, Pisaura ancora, Oxtopes* sp., *Lycosa coelestris, Oxyopes sertaus*; and three lady beetles, i.e., *Rodolia rufopilosa, Coccinella trifasciata, Propylaea japonica*. The larvae of syrphids can also feed on both nymphs and adults. In addition, there are parasitoid wasps, e.g., *Aphidius* sp., and *Aphelinus* sp.

(Xu Tiansen, Wang Haojie)

T. taiwana winged viviparae (Xu Gongtian)

T. taiwana winged viviparae (Xu Gongtian)

T. taiwana nymph (Xu Gongtian)

65 *Kurisakia sinocaryae* Zhang
Classification   Hemiptera, Aphididae

Importance   *Kurisakia sinocaryae* causes significant ecological and economic losses in the key pecan production regions.

Distribution   Zhejiang, Anhui.

Host   *Carya cathayensis*.

Damage   The nymphs aggregate and feed on developing buds, young shoots, and young leaves of pecan trees, which leads to wither of young leaves and young shoots, fall off of male flowers, closed female flowers, reduced tree vigor and fruit yield.

Morphological characteristics   The first-generation aphids (fundatrices) The body is reddish brown, wingless, body length is 2-2.5 mm with wrinkles and sarcoma dorsally. The stylets are slender and extend to the last abdominal segment. The antennae are short with four segments and hidden under the abdomen. There are no cornicles. The body shape resembles a tortoise shell.

The second-generation aphids (viviparae) The body is flat, oval-shaped, the abdomen has two green bands dorsally, without well developed or nodular cornicles. Antennae have 5 segments. Compound eyes are red; wingless; body length is about 2 mm.

Third generation aphids (females producing the sexual forms) Apterous aphids. Costal region of the forewing has one stigma; the antennae have 5 segments; the abdomen has two green bands and nodular cornicles. Nymphs are similar to the nymphs of the viviparae, except the last segment of the antennae has a depression laterally.

Fourth generation of aphids (sexual forms) Apter-

K. sinocaryae (Hu Guoliang)

K. sinocaryae (Hu Guoliang)

K. sinocaryae (Hu Guoliang)

K. sinocaryae (Hu Guoliang)

ous aphids without cornicles. Antennae have 4 segments. There are three types: aestivation and sexually differentiated male and female. The aestivation form is flat and resembles a piece of thin paper affixed to the abaxial side of the leaves, and color is yellowish green. The female is yellowish green with black band; at the end of abdomen, that is a circular wax secretion gland on each side. The male aphid body color is darker than the females; deep concave at the front of the head, and there are no wax glands at the end of the abdomen.

Eggs The eggs are oval, 0.6 mm long, newly deposited eggs are white, and change to shiny black and covered with white wax threads later.

Biology and behavior In Zhejiang, there are four generations per year. The eggs hatch in the early and mid-February and become the fundatrix (first genera-

tion). They crawl to the pecan tree and feed on the developing buds and ingest plant sap. The viviparae appear in late March to early April. The third-generation aphid occurs in early and mid-April, and aphid population surge rapidly. Overlapping generations are observed. In late April, the fourth generation (estivation form) occurs. In early May, aphids begin estivation (over-summer) on the abaxial side of the pecan leaves until the beginning of September. In early November, wingless female and winged male aphids occur. The sexual forms of aphids mate and oviposit at the buds, leaf scars and underneath the bark and branch cracks.

Prevention and treatment

1. Protection and preservation of lady beetles, syrphid flies and other natural enemies.

2. Apply *Beauveria bassiana* or *Paecilomyces fumosoroseus*.

3. Chemical control. The application of insecticides in late March to early April can be effective. The insecticides, may include 5% imidacloprid emulsion 1:0 to 1:3 can be injected in the trunk at chest height by drilling a series of holes with hole spacing of 10 cm. The holes can be drilled in the tilt angle of 45° to a depth of more than 1 cm into the xylem tissue. Inject each hole with 2 mL insecticide solution. In early April, the spraying of 5% imidacloprid EC at 1,000 to 1,500× can be effective too.

Reference Hu Guoliang and Yu Caizhu 2005.

(Jiang Ping, Yu Caizhu)

66 *Adelges laricis laricis* Vallot

Classification Hemiptera, Adelgidae
Common names Larch adelgid, Larch wooly aphid

Distribution Beijing, Shanxi, Liaoning, Jilin, Heilongjiang, Qinghai, Ningxia; Japan.

Hosts The first preferred host is spruce *Picea koraiensis*. The second preferred hosts are *Larix gmelini*, *L. kaempferi* and *L. principis-rupprechtii*.

Morphological characteristics **Fundatrices** Orange, surrounded by white cottony secretion. Overwintering nymphs are oblong in shape, approximately 0.5 mm long, dark brown to black in color. The body covered with six rows of glassy, short vertical secretion. Among all of the wax glands, the center one is a large gland with multiple rings, while the surrounding small wax glands with double ridges. There are a total of six wax glands. The cluster of the wax glands is on a sclerotized wax plate. The three rows of the wax glands are diffused, and separated among the midline, which forms two lateral chitinous plates. The three rows of the wax glands can be identified. The plate with ocelli is not connected with wax gland plate, and independently located laterally on both sides of the head. Antennae has 3 segments, and the third segment is the longest, which is ¾ of the total antennal length. The wax gland plates on the lateral edge of the abdomen have diffused into one plate with a pair of the wax glands. Adults are light yellowish green and covered with a dense layer of white wax.

Gall-forming aphids (Gallicolae) Egg color varied from orange, yellow, green, to dark brown before hatch. The first instar nymphs are light-yellow, without wax. White powdery wax secretion starts to show in the second instar, and the color gradually darkens. The fourth instar nymphs are purplish-brown, especially in the winged aphids.

Pseudo-fundatrices The eggs are in orange, and become dark brown before hatch. Overwintering nymphs are dark brown to black. Body surface is not covered with wax, and greatly sclerotized. There are wax glands only on coxae

A. laricis laricis winged migratory aphid (Xu Gongtian)

A. laricis laricis eggs laid on larch tree trunk by pseudo-fundatrices (Xu Gongtian)

A. laricis laricis eggs laid on larch tree needles (Xu Gongtian)

A. laricis laricis nymph (Xu Gongtian)

of the middle and hind legs. The setae on the central tergite still exist. Antennae three segments, and the third segment the longest. Adult dark brown, 1-2 mm long, hemispherical in shape. Only the two last two segments with wax secretion. Dorsal surface with six shiny longitudinal rows of warts.

Sexuparae Some eggs produced by the pseudo-fundatrices become sexuparae. The newly oviposited eggs are orange and covered with a layer of powdery wax, and one end has a filamentous threads. Newly hatched and the second instar nymphs do not produce wax. The nymphs are brown and shiny after the third instar, the sides of thorax are slightly elevated. The fourth instar is pale with wing pads visible. There are six rows of wax glands. Adults are yellowish brown to brown; the wax plates are visible and arranged neatly.

Migrants A portion of the eggs produced by the pseudo-fundatrices become migratory aphids after hatch. The newly hatched nymphs are dark brown, 0.6×0.25 mm in size, and body surface has no wax glands. Head and pronotum form one single plate along the midline; and there is a short seta at the wax gland. Antennae 0.13 mm long, the length of the third segment is 2/3 of the whole antenna length. A portion of the eggs deposited by the pseudo-fundatrices hatches into diapausing nymphs. They are morphologically identical to the overwintering nymphs. Adults appear as a "cotton ball" in size of the mung bean. The body is oblong, brown, 1.4×0.66 mm in size. The wax plate on the head is nearly round, and abdomen is oval, the wax plate is large and less sclerotized. The ocular plate does not merge with wax plate of the head and located on the side of the head. Antennae 0.14 mm long.

Sexuparae Eggs yellowish green. Females orange, males relatively darker than the females. They are active. Antennae and legs are long.

Biology and behavior There are multiple generations per year. The fertilized eggs oviposited by the sexuparae hatch in early August. Nymphs can be observed on the buds of larch trees in early September. They overwinter on the winter buds. They become active in mid- and later April of the following year. The body color changes form black to green, wax secretion expands and curl. Body size increases rapidly after molting. The color changes from dark to pale, and wax secretion increases. Adults appear after another molt. In early June as buds grow, they deform, and aphid galls are formed.

Aphid galls are small and spherical. They start to split in early August and reach peak in mid-August. After gall split, winged form nymphs crawl out of the galls, rest on the nearby needles and molt into winged gall-forming aphid. After wing spread, all of them migrate away from the spruce.

Gall-forming aphids migrate to the larch, and soon start to oviposit parthenogenetically. The eggs are covered under the roof-shaped wings. Both the female body and eggs are covered with white wax threads. The eggs hatch around mid-August. Newly hatched nymphs crawl away from the females after a day. The aphids go into overwintering state in mid-September. In late April of the following year, when the average temperature is approximately 6 ℃, the overwintering pseudo-fundatrices underneath the bark become active, molt, and grow rapidly. The aphids usually molt three times. Adults can be found as early as in early May.

A portion of the nymphs produced by the pseudo-fundatrices (developing type nymphs) ultimately develops into apterous viviparae. There are 4 to 5 generations per year. This is the period that causes serious damage to spruce. The 1st generation of migratory aphid adults occurs? in late May, the second in mid-June, the third in early July, the fourth in late July, and the fifth in mid-August. Eggs hatch in September. They may not be able to enter

A. laricis laricis pseudo-fundatrix and nymph (Xu Gongtian)

A. laricis laricis pseudo-fundatrices on larch tree (Xu Gongtian)

A. laricis laricis galls on spruce (Xu Gongtian)

A. laricis laricis nymphs (Xu Gongtian)

overwintering period when cold weather arrives early.

Among the eggs produced by pseudo-fundatrices, a small portion hatch into diapausing type nymphs resembling the pseudo-fundatrices. A portion of the nymphs hatches from the eggs oviposited by the migratory aphids also belong to this type. This portion increases as the number of generations increase.

Prevention and treatment

1. Silviculture. Maintain proper forest air flow and light penetration. Avoid planting larch and spruce at the same place or in the same nursery. The main protection target is larch. Enhance the control of adelgids in nursery will help prevent the spread of the pest.

2. Manual control. After the formation of adelgid galls and before the winged adult flight, remove and destroy the gall.

3. Biological control. Apply *Beauveria bassiana* or *Paecilomyces fumosoroseus*. Protecting and preservation of natural enemies can be effective. There are seven lady beetle species, i.e., *Harmonia axyridis*, *Coccinella septempunctata*, *Adalia bipunctata*, *Coccinella geminopunctata*, *Coccinella longifaseciata*, *Coccinella transversoguttata*, and *Propylea japonica*.

4. Chemical control. Spray insecticides during the egg hatch time of the first generation of migratory aphids (or the first instar). The effective insecticides include 50% imidacloprid and monosultap mixture at 1,500-2,000× or 25 g/L lamda-cyhalothrin at 1,000×.

References Chinese Academy of Sciences -Institute of Zoology 1980, Xiao Gangrou (ed.) 1992, Zhao Wenjie 1994, Zhang Dehai and Wang Enguang 1998, Gao Zhaoning 1999.

(Wang Shanshan, Li Zhenyu, Li Zhaolin)

67 *Drosicha corpulenta* (Kuwana)

Classification Hemiptera, Monophlebidae
Common name Giant mealybug

Distribution Hebei, Liaoning, Jiangsu, Henan, Hunan, Guangdong, Sichuan, Shaanxi; Japan.

Hosts *Paulownia* spp., *Populus* spp., *Salix* spp., *Platanus* spp., *Melia* spp., *Robinia pseudoacacia*, *Castanea* spp., *Juglans regia*, *Zizyphus jujube*, *Diospyros kaki*, *Pyrus* spp., *Malus pumila*, *Prunus* spp., *P. pseudocerasus*, *Citrus reticulate*, *Litchi chinensis*, *Ficus carica*, *Quercus* spp., and *Morus* spp.

Damage Females and nymphs feed at base of young shoots and buds. The feeding causes buds fail to develop and young leaves become yellow. Continuous infestations of 2-3 years may kill the hosts.

Morphological characteristics Adults Females wingless; length 10 mm; flat and oval; ocherous. Males length 5-6 mm. Front wing span 10 mm, purple-red; hindwing club-like.

Nymphs Similar to female adults, smaller.

Biology and behavior *Drosicha corpulenta* occurs one generation a year. Majority of the insect overwinter as eggs. A small portion overwinter as first instar nymphs. Overwintering eggs start to hatch in early February in Henan province. In mid-February, nymphs start to emerge from soil and migrate onto host trees. In warm winters, eggs may hatch in December. First instar nymphs are not active. They prefer hiding in tree crevices or base of branches and feed around noon time. The nymphs molt during late March and early April. Before molting, the nymphs are covered with white wax powder. After molting, the nymphs become larger and more active and secrete wax materials. Second molting occurs in mid- and late April. Males stop feeding and hide in tree crevices, under bark, or in grass. They secrete wax and pupate. Pupal stage lasts for 10 days. Adults emerge at the end of April and early May. Adults do not feed. They fly around evenings or crawl to trees and search for females to mate. Males die soon after mating. Females continue to feed after mating. They move down from trees to the underneath of rocks or inside cracks of soil during mid- and late June and produce white spongy egg sacs and lay eggs in them. Each female lays 100-180 eggs. The females die after egg laying.

Prevention and treatment

1. Cultural control. Turn soil under the tree canopy and destroy the egg sacs. Destroy young nymphs in early spring as they congregate on tree trunks.

2. Biological control. Apply *Beauveria bassiana* or *Paecilomyces fumosoroseus*. *Rodolia rufopilosa* and *R. limbata* are important natural enemies of *Drosicha corpulenta*.

3. Chemical control. In early spring before nymphs emerge, scrape off the coarse barks and prepare a 30 cm wide smooth ring on the tree trunk. Wrap the ring with plastic film. Cover the plastic film with insect glue (waste engine oil or diesel, grease, 25 g/L lamda-cyhalothrin at

Aggregation of *D. corpulenta* for overwintering or oversummering (Li Zhenyu)

D. corpulenta (Zhang Runzhi)

D. corpulenta eggs under a pine tree (Xu Gongtian)

D. corpulenta shed skins of 1st instar nymphs (Xu Gongtian)

D. corpulenta mature nymphs (Xu Gongtian)

D. corpulenta nymph shedding skin (Xu Gongtian)

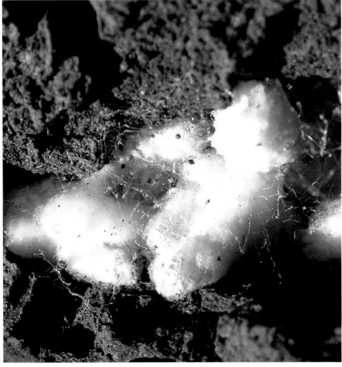
D. corpulenta cocoons (Xu Gongtian)

Rodolia limbata feeding on a *D. corpulenta* adult male (Xu Gongtian)

Rodolia limbata feeding on a *D. corpulenta* adult female (Xu Gongtian)

174 | CHINESE FOREST INSECTS

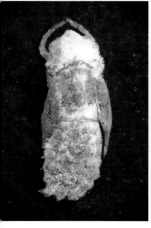
Dorsal view of *D. corpulenta* (Xu Gongtian)

Ventral view of *D. corpulenta* (Xu Gongtian)

D. corpulenta (Zhang Runzhi)

D. corpulenta fertilized female (Xu Gongtian)

D. corpulenta unmated female (Xu Gongtian)

D. corpulenta adult male (Xu Gongtian)

D. corpulenta mating pair (Xu Gongtian)

2:5:0.5 ratio). Nymphs that attempt to cross the ring will be stuck or die from contacting the insecticide. Spray nymphs on trees with 25% malathion at 500-1,000×.

References Yang Youqian et al. 1982, Xiao Gangrou (ed.) 1992, Henan Forest Insect and Disease Control Station 2005.

(Yang Youqian)

68 *Icerya purchasi* Maskell

Classification Hemiptera, Monophlebidae
Common name Cottony cushion scale

Distribution This insect is originated in Australia. It is found in all parts of China except for the northwest. In areas north of Yangtze river, this insect is only found in greenhouses. Foreign distribution includes Sri Lanka, India, Kenya, New Zealand, Palestine, Uganda, Zambia, Portugal, New Mexico, and Pakistan.

Hosts *Casuarina equisetifolia*, *Acacia confuse*, Rutaceae, Rosaceae, Leguminosae, Vitaceae, Oleaceae, Araceae, Pinaceae, and Taxodiaceae.

Damage This insect feeds on underside of leaves, young shoots, and branches. Infestations cause slower growth, premature leaf fall, death of trees, and sooty mold.

Morphological characteristics **Adults** Female length 5-7 mm, width 4-6 mm. Oval. Orange-red or dark orange yellow. The dorsal surface similar to turtle shell; covered with yellowish red wax and silvery white short hairs; lateral hairs long. Males small, length 3 mm, orange red.

Eggs Elongated oval, length 0.6-0.7 mm. Orange-red. Eggs are in egg sacs.

Nymphs There are 3 instars. First instar red; after hatching, the body starts to secrete light yellow wax. The 2nd instar orange yellow; its dorsal surface covered with pieces of yellow wax powder; body with black, slender hairs. The male larva is longer and has a lighter color than the female nymph. The 3rd instars are all females; reddish brown, covered with wax; black whorls of setae well-developed.

Pupae Male pupae length 3.5 mm; flattened, elongated oval; orange yellow. Cocoons white, soft. Pupae visible.

Biology and behavior *Icerya purchasi* occurs 3-4 generations a year in Guangdong, 2-3 generations a year along the Yangtze river. It overwinters as nymphs, female adults, or eggs on trees branches and trunks. First instars feed on the back around the main veins of leaves. From 2nd instar, they feed in groups on shade side of branches. In *Casuarina equisetifolia* forests, it occurs in dense, moist, and dark areas. Nymphs and adults produce honeydew which leads to the growth of sooty mold.

Prevention and treatment

1. Quarantine. Prevent movement of infested seedlings.

2. Silviculture. Create conditions that are not conducive to the occurrence of *Icerya purchasi*. Thin forests to improve light and air movement. Remove infested branches burn them in winter and early spring.

3. Biological control. Apply *Beauveria bassiana* or *Paecilomyces fumosoroseus*. The following predators can be used to control *Icerya purchasi*: *Rodolia cardinalis*, *R. concolor*, *R. limbata*, and *R. pumila*.

4. Chemical control. Spray 1-3 °Bé lime sulfur, 3-5%

I. purchase adult female and egg sac (edited by Zhang Runzhi)

I. purchase adult female after copulation (Xu Gongtian)

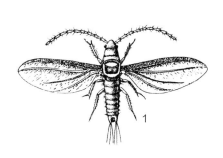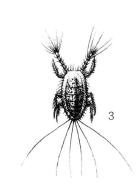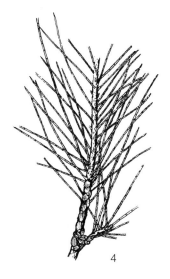

I. purchase. 1. Adult male, 2. Adult female, 3. Nymph, 4. Damage (Yang Kesi)

purchase (edited by Zhang Runzhi)

I. purchase damage (Xu Gongtian)

diesel, resin mixture at 10-15×, or motor oil at 40-50× to kill nymphs and females. To kill first instar nymphs, spray insecticides such as 10% imidacloprid WP, 40% acetamiprid and chlorpyrifos mixture, 40% supracide, 48% Dursban, 20% carbosulfan, etc.

References Xiao Gangrou (ed.) 1992; Gao Jingshun 1995; Hua Fengming, Ni Liangcai, Jin Minxin et al. 1999.

(Huang Shaobin, Yang Youqian, Hu Xingping)

purchase damage (edited by Zhang Runzhi)

Matsucoccus matsumurae (Kuwana)

Classification	Hemiptera, Matsucoccidae
Synonyms	*Matsucoccus liaoningiensis* Tang, *Matsucoccus massonianae* Yang et Hu, *Xylococcus matsumurae* Kuwana
Common name	Japanese pine bast scale

Importance *Matsucoccus matsumurae* infestations reduce pine tree growth, cause yellow leaves, dead shoots, and damage to barks. Severely infested trees may have drooping branches, crooked trunks. Secondary pests may occur as a result of infestation.

Distribution Hebei, Liaoning, Jiangsu, Zhejiang, Anhui, Shandong, Ningxia; Japan, Korea Peninsula.

Hosts *Pinus tabulaeformis*, *P. sylvestris* var. *mongolica*, *P. tabulaeformis*, *P. thunbergii*.

Damage Nymphs feed on tree trunks or branches. Early infestations are often not noticed. They develop fast and may spread quickly and cause mortality of trees in large areas.

Morphological characteristics **Adults** Females elongated oval, orange-brown. Length 2.3-3.3 mm. Body segments not obvious. Antennae 9-segmented, basal two segments thick, other segments moniliform. Mouthparts degenerated, with one pair of black ocelli. Trochanter triangular, with one long seta; femur short, tibia slightly curved. There are two tarsal segments. Claws on tarsi have a pair of globular hairs. Two pairs of spiracles on thorax, large; 7 pairs of spiracles on abdomen, smaller; dorsal surface from 2nd to 7th segments with round 208-384 scars; ventral surface of 8th abdominal segment with 40-78 glands. Both dorsal and ventral surface with glands. **Males** length 1.3-1.5 mm, wingspan 3.5-3.9 mm. Head and thorax blackish brown, abdomen light brown. Antennae filiform, 10-segmented, basal two segments short and thick, the remaining segments slender, each segment with many hairs; eyes large; mouthparts degenerated. Thorax large, legs slender. Forewings developed, translucent, with feather-like stripes; hind wing reduced to halters, its tip with 3-7 hooked hairs. Abdomen brown, 9-segmented, the dorsal side of the 7th segment with a horseshoe shaped hard plate, with a cluster of 10-18 tubular ducts; each secretes silvery white wax threads. The end of the abdomen has a hooked copulation organ, curved downwards.

Eggs Oval; length 0.24 m, width 0.14 mm. Yellowish at beginning, turning to yellowish brown or brownish yellow. Eggs are concealed in a white egg sac. The sac is made of white wax, oval.

Nymphs First instar 0.26-0.34 mm in length; yellowish; elongated oval; antennae 6 segmented, the 1st segment large, the 3rd segment shortest, the 6th segment longest; with one pair of ocelli, purplish black. Mouthparts well developed, proboscis long, rolled inside abdomen. Femora large, tibiae slender, tarsi small, claws strong. Thoracic spiracles 2 pairs, small and indistinct. Abdominal segments distinct, with 7 pairs of spiracles. The end of the abdomen with one pair of short hairs and one pair of long hairs.

The 1st instar parasitic nymphs are 0.42 mm long and 0.23 mm wide; pear or heart shape, orange yellow; dorsal side with rows of wax threads; antennae and legs are visible on ventral side.

The 2nd instar parasitic nymphs do not have appendages. Mouthparts well developed; surrounded by long

M. matsumurae adult female (Xie Yingping)

M. matsumurae adult male (Xie Yingping)

wax threads. Sexual dimorphism distinct; females large, round or flattened, orange yellow; males small, elliptical, brown or blackish brown, attached to the exuviae from the 1st instar.

The 3rd instar male nymphs 1.5 mm long, orange yellow. Mouthparts the legs degenerated; shape similar to female adults; abdomen narrow, without dorsal scar; end of abdomen without emargination.

Male Pupae Male pupae enclosed in white cocoons. Cocoons loose, 1.8 mm long, elliptical. Prepupae thorax raised, with wing bugs. Pupae 1.4-1.5 mm, head and thorax light brown, appendages greyish white; eyes purple brown. Abdomen 9-segmented. Copulation organ conical.

Biology and behavior *Matsucoccus matsumurae* occurs two generations a year. It overwinters or over summers as 1st instar nymphs. In the south, the 2nd instar nymphs appear one month earlier than those in the north. In Zhejiang, adults emerge in from late March to late May. In Shandong, adults emerge during early May and mid-June. Those overwintering 1st instars in the south appear one month later than those in the north. Adults also appear one month later than those in the north. In Shandong, adults appear during late July and mid-October.

In Zhejiang, adults appear during late October and early November. The 1st instars of the overwintering generation start overwintering earlier than those in the south.

In Shandong and Liaoning, the 1st generation appears from April to November. The 2nd generation occurs from October to May in the next year. It overwinters as nymphs. Females hide under barks after mating. They are covered inside sacs made of wax and lay eggs inside the sacs. Males die the same day after mating. Newly emerged nymphs fix themselves to barks and feed. They often occur on shade side where barks peeled. Infested trees grow slowly. Because the sunny side receives less infestation, the tree trunks and branches bend to one side. Nymphs move gradually upward as they develop. After 1st instar, the head and thorax fuse; dorsal surface raised; wax threads are produced. Their body shape changes from spindle to pear form. The insect is small during this period and is referred as "hidden period". The 2nd instars do not have appendages. Males and females can be easily differentiated. The body size increases rapidly. This period is referred as "exposed period" during which the insect causes the most damage. The third instar male larvae pupate on tree branches, under barks, under rocks, in grass, etc. They molt into males. The 2nd instar female nymphs will molt into adult females.

Prevention and treatment

1. Silviculture. Trim and thinning of pine forests that are growing well and with low level of pest populations. Plant broad-leaved trees in forests with heavy infestations.

2. Biological control. Conserve and release natural enemies. They include *Harmonia axyridis*, *H. obscurosignata*, *Exochomus mongol*, *Sospita chinnsis*, *Lestodiposis* sp., *Anystis* sp., *Elatophilus nipponensis* Hiura, *Dufouriellus ater* (Dufour), *Sympherobius weisong* Yang, *Chrysopa kulingensis* Navas, *C. septempunctata* Wesmael, *Inocellia* sp.

3. Chemical control. Spray 1-1.5 °Bé lime sulphur to kill male pupae or 50% sumithion at 200-300× during "exposed period" and egg period. Paint tree trunks with insecticides to kill nymphs during "hidden period" in areas without good transportation and water supply. Drill holes on tree trunks and inject imidacloprid.

References Tang Fangde and He Jingjun 1995; Wang Jianyi, Wu Sanan et al. 2009; Yang Qian, Xie Yingping, Fan Jinhua et al. 2013,

(Xie Yingping, Zhao Fanggui, Wang Liangyan, Jiang Ping, Li Zhenyu)

M. matsumurae 2nd instar female nymph (Xie Yingping)

M. matsumurae cocoons (Xie Yingping)

70 *Matsucoccus shennongjiaensis* Young and Liu
Classification   Hemiptera, Margarodidae

Distribution   Hubei.

Host   *Pinus armandii*.

Damage   The insect is found at 800-240 mm elevation and is most abundant at 1,500-1,800 m elevation. As high as 90% infestation rate and 2,000 nymphs per 100 cm² of bark areas were reported. Mid-aged forests experienced heaviest infestations. The insect occurs at mid- and lower part of tree trunks. Base of large branches have higher population density than mid-portion of the branches. The affected barks break, become rotten, and produce resin. Infested trees grow slowly and die.

Morphological characteristics   **Females** Body elongated oval, the two sides nearly parallel; orange-brown; length 2.5-3.3 mm. The 2nd-3rd abdominal segments widest, width 1.2-1.5 mm. Antennae 0.7 mm long; 9 segmented, its 3rd-9th segments with striae, 5th-9th segments each with a pair of thick sensory spine. Exterior diameter of thoracic spiracle plate 0.02 mm. Thoracic legs 0.6 mm long. The 3rd-6th abdominal segments with 241-327 marks on the dorsal surface. Some individuals have marks starting from the 1st abdominal segment. The marks are nearly round, about 0.009 mm in size. The 8th segment has 56-67 glands. Their exterior diameter 0.012 mm. The central area of each gland has 2 openings surrounded by a ring of 12 smaller openings. The glands arranged in a complete ring on both dorsal and ventral side of the abdomen. The 1st abdominal segment has about 40 bilocular glands. Body with short setae; those on ventral side of the abdomen 0.02 mm long. Male body length 2.0 mm. Eyes developed. Antennae filiform, 10 segmented, basal two segments nearly round, the other segments slender, 4th-5th segments longest; 3rd-10th segments with many setae, 4th-10th segments have thick long setae at the end of each segment, 7th-10th segment with two long setae near the end of each segment. Front wing membrane form, wing length 2.0 mm. Hindwing reduced to halteres, its end with 3-4 hooked setae. The dorsal surface of the 8th abdominal segment with a kidney shaped hard plate. There is a cluster of 18-22 tubular ducts on the plate. Genital sheath wide at base, tapered at end, and curved inward. Anus located on dorsal side of the base of the genital sheath. Aedeagus located at ventral side of the genital sheath, exposed, and curved toward the ventral side.

Eggs   Orange yellow, elongated oval; 0.4 mm long, width 0.2 mm. Egg sac semi-sphere shaped; 3.3 mm long, 3 mm wide.

Nymphs   First instar elongated oval, light-yellow; 0.5 mm long, 0.2 mm wide. Antennae 6 segmented, their length is $2^{nd} < 1^{st} < 3^{rd} < 5^{th}$. The 6th segment has two sensory setae at base and the end. Thoracic legs developed. End of the abdominal segment has one pair of long setae and one pair of short setae. First parasitic nymph 0.41 mm long, 0.22 mm wide. A few days after hatching, the body secretes large amount of wax, the nymph changes from spindle to pear shaped. The 2nd instar does not have appendages. Mouthparts are developed. The body is covered with wax threads. Males and females are easily distinguished. Females are round, purple-brown or dark brown, width 1.7 mm. The 3rd instar 1.5 mm long, similar to female adults, smaller than adults, brown; they pupate in cocoons.

Male pupae   Prepupae similar to nymph but the thorax raised, with wing buds. Pupae naked, 1.4-1.5 mm long; head and thorax pale brown, appendages and wing buds greyish white.

Biology and behavior   *Matsucoccus shennongjiaensis* occurs one generation a year. It overwinters as 2nd instar parasitic nymphs. Females start to emerge in mid- May, peak period occurs in late May; adult emergence may extend to mid-September. Females rest on tree trunks or stumps. Females wait for males to mate. They hide under lichen or crevices on tree trunks around noon and appear around 6:00 in the morning. Adults become sexually mature 40 hours after emergence. Males are weak flyers and may mate with multiple females consecutively. One mating only lasts for a few seconds. Males die the same day after mating. Females hide for a few hours after mating. Each female produces an egg sac made of wax and lays eggs 2-3 days later. Each egg sac has 131-205 eggs. Females die within the egg sacs. Each dead female contains 20-65 eggs inside their body. Egg period lasts for about 60 days under natural environment; those located on sunny side hatch about 15 days earlier than those located on the north side. First instar nymphs congregate in groups of 100-300 under lichen on tree trunks. If little lichen exists, then they congregate in crevices on tree trunks or on smooth surface of tree trunks. They insert their two times body length proboscis into the bark and feed on the sap. The 2nd instars pear

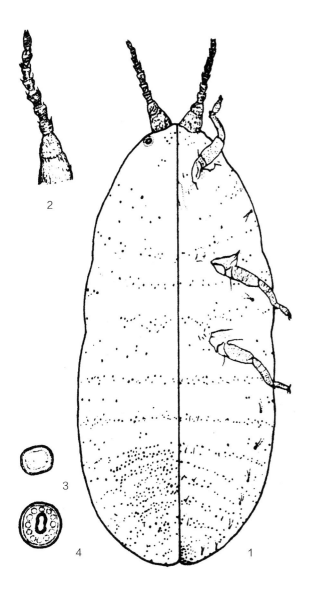

M. shennongjiaensis
1. Adult female, 2. Antennae, 3. Dorsal mark,
4. Multiple-opening glands (Hu Xingping)

shaped at the beginning, then change to elliptical, then change into blackish brown hard-shelled balls. After late April, the insect develops fast. In early May, the insect ruptures the lichen and expose themselves. Some 2nd instars come out from body shell and turn into 3rd instar male larvae. Molting occurs around 14:00-15:00 in sunny days. The male nymphs congregate under lichen, scars of tree trunks, in grass, etc. After about 10 hours, they create cocoon. After 6 days, they pupate. Pupal period is in May. Pupal period is 9 days. Female: male ratio is 40:1.

Egg sac and 1st instar nymphs are easily spread by wind. Its natural speed of spread is several kilometers per year. Rain can wash the egg sacs and 1st instars to the ground and lower part of the forests. Southeastern slopes have higher population than the northwestern slopes. The southeastern side of the trees also has higher population than the northwestern side of the trees. Among the three varieties of *Pinus armandii*, the susceptibility to *M. shennongjiaensis* is "thin bark" < "mosaic bark" < "thick bark". High canopy coverage and pure forests favor higher pest populations.

Prevention and treatment

1. Conservation of natural enemies. Predators include spiders, *Harmonia axyridis*, ants (*Formica* spp. *Paratrechina*, spp. *Prenolepis* spp, *Apahaenogaster* spp.), and larvae of Cecidomyiidae.

References Yang Pinglan et al. 1987, Lü Changren et al. 1989.

(Lü Changren, Zhan Zhongcai, Zhuang Xiaoping)

71 *Matsucoccus yunnanensis* Ferris

Classification   Hemiptera, Matsucoccidae

Distribution   Yunnan.

Hosts   *Pinus yunnanensis*.

Damage   This pest infests mid-aged and mature forests. Annual tree height growth was reduced by an average 5 cm in a study. The average density of *M. yunnanensis* was 91.5 per 10 cm² of bark area on 26-year-old pine trees. Secondary pests often occur after infestation by *M. yunnanensis*. Tree mortality can occur in large areas.

Morphological characteristics   **Females** Elongated elliptical; length 2.5-4.2 mm, width 1.2-2.0 mm; body segments distinct. Antennae moniliform, 9 segmented, 5th-9th segments with striae, 6th-9th segments with a pair of sensory spines on each segment. Eyes black. Mouthparts degenerated. There are two pairs of thoracic spiracles and three pairs of thoracic legs. The legs have striae, trochanter with one long seta. There are seven pairs of abdominal spiracles. Dorsal surface of the 3rd-6th abdominal segments with 50-179 round marks, arranged into transverse bands on each segment. The ventral surface of the 8th abdominal segment has 23-48 glands. The central area of each gland has two openings surrounded by a ring of 12-14 smaller openings. The glands arranged in a complete ring on both dorsal and ventral side of the abdomen. Each gland has only one opening on surface of the body. Female genital opening forms a longitudinal split at the end of the abdomen.

Eggs   Elliptical; length 0.320-0.388 mm, width 0.169-0.219 mm. Pale yellow at beginning and turns to brownish yellow later. Eggs and the female are in a sac made of white wax; size of the sac is 3.0-4.5 mm long, 2.0-3.0 mm wide.

Nymphs   First instar elongated oval; length 0.404-0.489 mm, width 0.185-0.219 mm; light-yellow, eyes black. Antennae 6 segmented. Mouthparts well developed. There are one pair of short and one pair of long setae at the end of the abdomen. The 1st instar grows slowly at the beginning; its shape gradually changes from spindle to olive and pear. Wax production increases significantly as the nymph grows. Each nymph has two pairs of thoracic spiracles, seven pairs of abdominal spiracles, one pair of antennae, one pair of eyes, three pairs of legs, one pair of short and one pair of long cerci. The 2nd instar without appendages. There are two pairs of thoracic spiracles and seven pairs of abdominal spiracles. Thoracic spiracles are larger than abdominal spiracles. Body shape varies with the environment. Solitary nymphs elliptical or round; those in groups take various shapes as a result of crowding conditions. Body color brown.

Biology and behavior   *Matsucoccus yunnanensis* occurs two generations a year. The two generations overlap. In winter, eggs, 1st instar, 1st instar parasitic nymphs, 2nd instar nymphs and female adults exist. There are two egg production peaks each year. They occur during March-May and October-December. Egg period is 18-23 days at 18-23.9°C and 46%-73.3% RH. Egg hatch rate is 100%. More eggs hatch during daytime. Newly emerged nymphs crawl among crevices and peeled barks and settle down at a suitable location approximately one day after hatching. The nymphs either live singly or in groups. They can even develop into adults by inserting their mouthparts into peeled barks. First instar nymphs shed their skin in pieces and remain attached to the 2nd instar, a characteristic of this species. The 2nd instars produce less wax. Although larger, they are not exposed. This is the 2nd distinct characteristics of this species.

When a mature 2nd instar is ready to molt into female

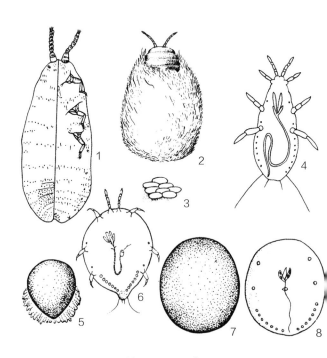

M. yunnanensis
1. Adult female, 2-3. Egg sac and eggs,
4. First instar nymph, 5-6. First instar parasitic nymph,
7-8. Second instar nymph (Li Xichou)

adult, the body becomes shinny and the female adult is visible from the back of the nymph. An adult emerges from the lateral side of a nymph. It takes 10-52 minutes for an adult to emerge from a nymph. Sometimes adults are partially emerged from the nymphs. These females can still produce wax and viable eggs. Adults mostly emerge during 01:00-11:00. No males are seen. Both generations are parthenogenetic. This is the 3rd characteristics of this species.

Newly emerged adults hide under peeled barks. They produce wax threads within 1-5 days after emergence. After 1-2 days, they are enclosed in sacs of wax and start to lay eggs inside. Females continue to produce wax during egg laying and enlarge the sacs. Few females do not lay eggs. Most females lay eggs in consecutive days. Egg laying occurs mostly during 5:00-20:00. Each female lays 1-8 eggs per hour. Females lay an egg every 5-101 minutes; most females lay an egg every 23-28 minutes. Each female lays 1-56 eggs per day. Maximum egg production occurs at 1 day after egg laying starts. Each female lays 27-189 eggs, 44% of females lay 104-189 eggs per female. Egg laying period is 4-12 days. 5.6% of the dead females contain one egg in their body. Egg sacs are rarely arranged in groups. Females live for 10-17 days.

Climate conditions affect growth of the insect. The average daily temperature in Kunming is 15.4°C. In summer and autumn, 1st instars have distinct changes in body shape. This may be related to precipitation. In winter, the body shape change is not obvious and may be related to lower precipitation. Wind is the primary factor of spread. First instars can spread by wind.

Matsucoccus yunnanensis occurs much more often in pure forest than in mixed forests. In *P. yunnanensis*, *Quercus aliena*, and *Q. fabri* mixed forests, infestation rate is only 12%. The occurrence of *Matsucoccus yunnanensis* is also related to host density. Dense forests often suffer more severe infestations. Peeled barks are the primary feeding and resting sites of *M. yunnanensis*. Trees with more peeled barks have higher *M. yunnanensis* populations. The vertical distribution of *M. yunnanensis* on a tree is related to the distribution of peeled barks.

Prevention and treatment

Natural enemies are an important population regulator of *M. yunnanensis*. *Nephus* sp. *Leucopis atratula* Ratzburg, *Lestodiplosis*, predate upon *M. yunnanensis* eggs. When both eggs sacs and these predators were abundant, the pest population diminishes the next year. *Nephus* sp. is most abundant in forests of < 0.5 canopy density. *Leucopis atratula* is more abundant in forests of > 0.6 canopy density.

References

Yang Pinglan et al. 1974, 1976; Qi Jingliang et al. 1981.

(Qi Jingliang, Wang Yuying)

72 *Matsucoccus sinensis* Chen

Classification Hemiptera, Matsucoccidae
Synonym *Sonsaucoccus sinensis* (Chen)

Distribution Jiangsu, Zhejiang, Anhui, Fujian, Jiangxi, Henan, Hunan, Sichuan, Guizhou, Yunnan, Tibet, Shaanxi.

Hosts *Pinus massoniana*, *P. tabuliformis*, *P. thunbergii*, *P. yunnanensis*, and *P. banksiana*.

Damage In Fujian, this insect can infest 90% of the *Pinus massoniana* in a forest. Average population density was 8-10 per needle. The insect sucks the sap from needles, causing yellow leaves and premature leaf fall, and slowed tree growth.

Morphological characteristics **Females** Length 1.5-1.8 mm; upside down oval; orange brown. Body with distinct segments. One pair of black ocelli. Antennae 9 segmented, moniliform. Mouthparts degenerated. Thoracic legs not well developed. Basal part of the legs complete or fused, with transverse striae. Dorsal marks abundant, round, connected to each at the rear end of the abdomen and extend to the anterior part of the abdomen. Body with black leathery shed skins. Reproductive opening small, situated in the depression at the end of the abdomen. Males 1.3-1.8 mm long, wingspan 3.5-4.0 mm. Head and thorax black, abdomen pale brown. Antennae 10 segmented, filiform. Mouthparts degenerated. Eyes purple-brown, large and protruded. Thorax enlarged, legs slender and long. Forewings well developed, semi-transparent, with feathery marking. Hindwings reduced to halteres, each with 3-7 hooked spines. Abdomen 9 segmented, the 8th segment with one cerarius, which produces 10-12 white wax threads. The end of the abdomen with one copulatory hook.

Eggs Oval, tiny. Milky at beginning, turning to light-yellow later. Two black eyes can be seen prior to hatching.

Nymphs First instar nymph oval, tiny, golden yellow. Eyes semi-globular. Mouthparts developed, with long stylets which curled inside proboscis before feeding. Thoracic legs developed. Abdomen 8 segmented with 7 pairs of spiracles; end of the abdomen conical. After fixing to a feeding site, the 1st instar elliptical, 0.35-0.40 mm long, 0.25-0.3 wide. Dark black, covered with white wax. Second instar without appendages; mouthparts developed. Body leathery, black; its end contains the shed skin from the 1st instar. Sexual differentiation is clear. Females upside down oval; 1.4-1.8 mm long, 0.8 mm wide.

Male nymphs small, elliptical; 1.2 mm long, 0.6-0.8 mm wide. Dorsal surface shiny, covered with wax. Third instar elongated elliptical; 1.2-1.5 mm long; orange brown. Mouthparts degenerated, antennae and legs developed. Similar to females, but without marks on the back and without depression at the end of the abdomen.

Male pupae Prepupae and pupae are enclosed in elliptical sacs. Prepupae orange brown. Head and thorax of the pupae light-yellow. Abdomen brown, eyes purple brown, appendages and wing buds greyish white. Abdo-

M. sinensis 3rd instar male nymph (Zhang Gaixiang)

M. sinensis 3rd instar female nymphs on *Pinus tabuliformis* (Zhang Gaixiang)

men 9 segmented, its end conical.

Biology and behavior *Matsucoccus sinensis* occurs one generation a year in Fujian and Shaanxi. It overwinters as 1st instar nymphs. Adults appear during late April to early July; peak period is from mid-May to mid-June. Eggs appear from mid-May to mid-July. First instars emerge from late May to early July; peak period is from early June to mid-July. Damage occurs from early June to early May next year. First instar nymphs undergo diapause from late June to late September. The 1st instar nymphs molt into legless instar from late March to mid-April. Males and females are easily differentiated. Third instar males appear from mid-April to mid-May.

Due to climate differences in Fujian, the emergence time of various stages in the south is about one month earlier than in the north.

After male larvae mature, they move down the tree trunk. They look for crevices, scales of cones, roots near the base of the tree trunk, grass, leaf litter, rocks, etc., and secrete wax and pupate. After another 5-7 days, adults will emerge. Peak emergence period is during 9:00-14:00. Adult males climb up tree trunks or fly for short distances soon after emergence and look for females to mate. Soon after mating, males die. Female adults hide inside the shells of the nymphs during their lifetime. They only expose the end of their abdomen during mating period. After mating, their abdomen retreats into the shell, secretes wax, and lay eggs. One female lays as many as 104 eggs, average egg production per female is 56 eggs. After egg laying, the females die inside the shells. Females live for 5-12 days. Unmated females can live for a maximum of 20 days.

Eggs are laid in sacs. Egg hatch rate is > 95%. Young

M. sinensis adult females on *Pinus tabuliformis* (Zhang Gaixiang)

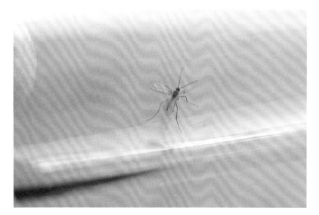
M. sinensis adult males on *Pinus tabuliformis* (Zhang Gaixiang)

M. sinensis male cocoons (Zhang Gaixiang)

M. sinensis damage to *Pinus tabuliformis* (Zhang Gaixiang)

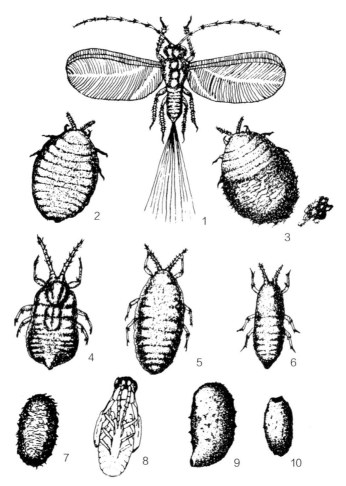

M. sinensis. 1. Male adult, 2. Female adult, 3. Eggs and female adult, 4. Newly hatched nymph, 5. 1st instar parasitic nymph, 6, 3rd instar nymph, 7. Male cocoon, 8. Male pupa, 9. Female nymph shell, 10. Male nymph shell (Kang Yongwu)

nymphs emerge from the end of the shells. They are very active on tree trunks. After 1-2 days, they fix themselves to young leaves of new shoots and become 1st instar parasitic nymphs. The body color turns from light-yellow to dark black. Head and thorax fuse and widen, the dorsal surface raised. They secrete wax. Body shape changes from upside down oval to elliptical. During March and April in the next year, the 1st instar nymphs molt into 2nd instar nymphs and become legless. Their body size increases significantly. This is the most damaging period. Nymphs prefer new shoots and inner side of the needles with their heads facing down. The 2nd instar male larvae molt into 3rd instar male larvae. They molt around noon. After making a cocoon, their thorax becomes raised, antennae and legs stay underside, the dorsal surface shows wing buds and turns into prepupae. Pupal period lasts for 5-7 days.

Spread of *M. sinensis* is mainly assisted by wind, rain, and human activities. At early stage, they are found in spots or blocks, then spread to the whole forest.

Prevention and treatment

1. Apply *Beauveria bassiana* or *Paecilomyces fumosoroseus*

2. Conservation of natural enemies. Important natural enemies include *Chrysopa septempunctata*, *Camponotus japonicas*, *Harmonia axyridis*, and *Chilocorus kuwanae*.

3. Chemical control. Ultra-low volume spray of 40% acetamiprid and chlorpyrifos mixture at 2,000-3,000× to tree trunks and branches to kill 1st instar parasitic nymphs.

References Yang Pinglan 1980; Yang Youqian 1986; Zhang Shimei and Zhao Yongxiang 1996; Shen Fuyong, Zhu Yuxing, Zhang Yujun 2001.

(Li Jiayuan, Chen Wenrong, Wang Shanshan)

73 *Kerria* spp.

Classification Hemiptera, Tachareiidae
Common name Lac insects

Importance There are 19 known species in this genus. Three species are found in China: *Kerria yunnanensis*, *K. ruralise*, and *K. greeni*. *Kerria yunnanensis* produces Chinese lac. *Kerria ruralis* is valuable for breeding. *Kerria greeni* has no lac production value.

Lac insects suck plant sap and secrete scarlet resin – shellac. It has properties of sticks to substrates easily, moisture resistant, insulating, smooth, tolerant to acid, chemically stable, non-toxic, and non-irritating. It is widely used in chemical engineering, electronics, military, medicine, and food industry. Lac insects use more than 200 species of plants as hosts. Some host plants are draught tolerant, fast growing, and can grow on lean soil and germinate easily. The host plants can also help conserve water and soil richness and have important economic and social value.

Distribution Tropical and subtropical region of China; India, Pakistan, Bangladesh, Thailand, Myanmar, Indonesia.

(1) *Kerria yunanensis* Ou et Hong

Distribution Yunnan. It was introduced into Sichuan, Guangxi, and Guangdong.

Hosts Its hosts include more than 100 species of plants. Common hosts are: *Cajanus cajan*, *Dalbergia obtusifolia*, *Dalbergia siemaoensis*, *Moghania macropylla*, *Eriolaena spectabilis*, *Pueraria wallichii*, *Ficus racemosa*, and *Ficus cunia*.

Biology and behavior *Kerria yunnanensis* occurs two generations a year in southern Yunnan. The summer generation occurs from May to October and lasts for approximately 150 days. The winter generation occurs from October to April in the next year and lasts for approximately 210 days. For summer generation, the 1st, 2nd, and 3rd instar female nymph development periods are 20, 15, and 15 days, respectively. The 1st and 2nd instar male larvae development periods are 20 and 18 days, respectively. Pupal development period is 12 days. Adult period is 8 days. For winter generation, the 1st, 2nd, 3rd instar female nymph development periods are 50, 45, and 30 days, respectively. Adult period is 90 days. The 1st and 2nd instar male larval development periods are about 50 and 60 days, respectively. Pupal period is 20 days. Adult period is 15 days. Lac production in summer is higher than that in winter.

(2) *Kerria ruralis* Wang

Distribution Yunnan (Simao, Puwen).

Hosts *Cleidiocapum cavalirii*, *Pueraria tonkinensis*, *Eaphoria longan*, *Litchi chinensis*, *Flemingia macrophylla*, *Cajanus cajan*, and *Pueraria wallichii*.

Biology and behavior *Kerria ruralis* occurs two

K. yunnanensis (Chen Xiaoming)

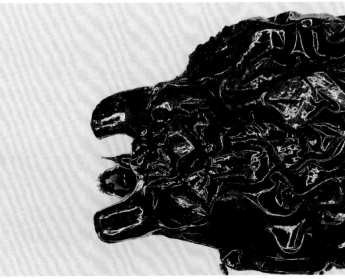

K. yunnanensis female adult (Chen Xiaoming)

generations a year. The summer generation occurs from February-March to July-August. The winter generation occurs from July-August to February-March. The development period of the 1st, 2nd, 3rd, instar female nymphs of the summer generation are 40, 20, and 15 days, respectively. Total nymph period is about 80 days. Adult period is about 90 days. The development period of the 1st and 2nd instar male larvae are 40 and 30 days, respectively. Pupal period is 15 days. The female and male development periods of the summer generation are 170 and 110 days, respectively. For winter generation, the 1st, 2nd, 3rd instar female nymphal periods are 20, 15, and 10 days, respectively. Adult period is 170 days. Both the 1st and 2nd instar male larval period are 20 days. The female and male development periods of the winter generation are 220 and 70 days, respectively. Lac production by *Kerria ruralis* is lower than that by *K. yunnanensis*. However, its shellac has a paler color and higher value than that produced by *K. yunnanensis*.

Common biology and behavior of lac insects

Lac insects occur two generations a year. Males are holometabolous and go through egg, larva, pupa, and adult periods. Females are hemimetabolous and only go through egg, nymph, and adult periods. Lac insects only move once during their lifetime. After eggs hatched, the larvae (nymphs) search for a suitable branch and insert their proboscis into plant tissue and stay there. Male adults with or without wings. They mate with females nearby and die soon after mating. Both females and males can mate multiple times. As the female nymphs develop, they become covered with lac. Females are viviparous.

Lac insects are gregarious. Density of *Kerria yunnanensis* is 160-230/cm^2 of branch surface area and that for *K. ruralis* is 80-100/cm^2. Female: male ratio is 1:2-3. Natural mortality of lac insects is high. The mortality of *K. yunnanensis* is about 90%. Mortality of *K. ruralis* is about 80%. Each female carries 200-1,000 eggs. *Kerria yunnanensis* carries 300-600 eggs, and *K. ruralis* carries 200-500 eggs. The intrinsic rate of increase is 10-15×. When natural enemies and environmental factors are considered, the rate of increase is 5-10×.

Growing lac insects. The lac insect production consists of three elements: the lac insect, host plant, and the environment. First, select a suitable location that favors the growth of lac insect. Second, select and plant host plants according to the insect species. Shrubs can be used 6-8 months after plantation, whereas trees can be used 3 years after plantation. Shrubs are planted at 2 m×2 m grids. Trees are planted at 3 m×4 m or 4 m×4 mgrids.

K. yunnanensis nymphs (Chen Xiaoming)

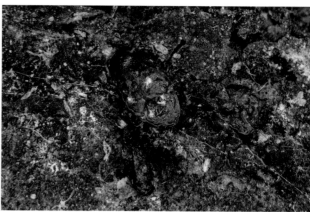

K. yunnanensis nymphs (Chen Xiaoming)

Dorsal view of *K. yunnanensis* female adult (Chen Xiaoming)

K. yunnanensis female adult (Chen Xiaoming)

K. yunnanensis female adult (Chen Xiaoming)

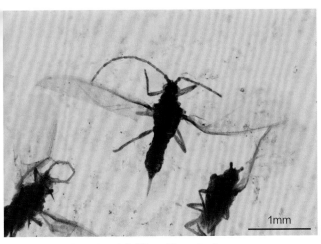

K. yunnanensis male adult (Chen Xiaoming)

Rich and moist soil conditions will yield higher shellac production. Third, culture broodlac that contains eggs ready to hatch. They are used for producing the future generations.

It is relatively simple to cultivate lac insects. Cut sticklac into 10-15 sticks. Make oblique cuts at the two ends to facilitate climbing of the insects. Fix the two ends of the sticklac to 6-month to one-year-old branches. After nymphs emerged, collect the lac. Plant sticklac in May and harvest lac in October each year. When harvesting, the host plant branches are cut down, the harvested sticklac is separated from the branches. Seedlac is produced by removing the impurities.

References   Bose et al. 1963; Information Research Institute of Chinese Academy of Agriculture and Forestry 1974; Varshney 1976, 1984; Wang Ziqing, Yao Defu, Cui Shiying et al. 1982; Ou Bingrong and Hong Guangji 1984, 1990; Chen Xiaoming and Feng Ying 1991; Chen Xiaoming 1998, 2005.

(Chen Xiaoming, Ou Bingrong, Hong Guangji)

Nesticoccus sinensis Tang

Classification Hemiptera, Pseudococcidae
Common name Bamboo nest scale

Importance *Nesticoccus sinensis* is a common serious pest of *Phyllostachys pubescens* and *P. nigra*. Large areas of bamboo forests may die from *N. sinensis* infestations.

Distribution Jiangsu, Zhejiang, Shanghai, Anhui, Shandong, Shaanxi.

Hosts *Phyllostachys pubescens, P. nigra, P. viridis* var. *houzeanana, P. spectabilis, P. glauca, P. propinqua, P. vivax, P. viridis* var. *youngii*, and *P. iridensis*.

Damage Infested bamboo stop growing, cannot generate new shoots, or die. Infested *Phyllostachys nigra* will die after 2-3 years. The insect is small at the beginning of the infestation. Once discovered, the forest may have already suffered significant damage.

Morphological characteristics **Adults** Female pear shaped, reddish brown; length 2.20-3.30 mm, width 1.70-2.60 mm. Antennae globular, two segmented; basal segment ring-like, the apical segment conical, with 6 setae. Mouthparts developed. Stylet ring extends to 3rd and 4th abdominal segment. Ocelli absent. Thoracic legs degenerated. Two pairs of thoracic spiracles developed, cup-like. Spiracles surrounded with groups of trilobular pores. Ventral surface with round discoidal pores and oral-collar ducts which form an elongated elliptical plate on the two sides of the 1st-3rd abdominal segments. Abdomen with 8 visible segments, without dorsal split. Ventral surface of the 3rd-4th segments with a split. Anal ring degenerated to cup-like, without small holes, with two short setae. End of the abdomen with two setae.

Male length 1.25-1.40 mm, wingspan 2.25 mm; orange red, thoracic color darder. Ocelli two pairs; dark reddish brown. Antennae filiform, 10 segmented, with dense pubescence; basal 1st-2nd segment short, enlarged. Forewing white, transparent, with dense pubescence and two longitudinal veins. Hindwing reduced to halteres, the end with one hooked hair. Legs 3 pairs, developed; end of the tibiae with 3 spurs; one pairs of digitules on the claws, without digitules on tarsi. Abdomen with 9 segments, the 1st-6th segments with a few 5-hole glands on the two sides. The 7th segment with a group of 5-hole glands and a tubular gland on the two sides, each tubular gland has one long seta. White wax is attached to the long hair. Copulatory organ hard, conical.

Eggs Oval; light-yellow at the beginning, and brown before hatching; semi-transparent; length 0.30-0.45 mm, width 0.15-0.25 mm.

Nymphs Elongated elliptical; orange yellow at the beginning, and yellowish brown after fixed to host plant; length 0.45-0.50 mm, width 0.15-0.25 mm. Antennae 6 segmented, basal segment enlarged, apical segment equal to the combined length of the 3rd-5th segments; the end with 6 setae. Ocelli two pairs, reddish brown, situated below the antennae. Mouthparts developed. Stylet ring extends to the 5th abdominal segment. Thoracic spiracles 2 pairs, cup-like. Legs developed, 3 pairs. Tibiae slightly shorter than the tarsi, with one pair of tarsal setae and one pair of setae on the claws. Abdomen 8 segmented. Abdominal split round, located at central line of the 3rd-4th segments. Dorsal split one pair, located between the 6th-7th segments. Body scattered with trilobular pores. Anal tube and anal ring developed, with 6 setae. Anal valve slightly obvious, with two setae.

Prepupae Elongated elliptical, orange yellow; length 0.95-1 mm, width 0.45-0.55 mm. Ocelli not obvious. Antennae 6 segmented. Mouthparts degenerated. Legs 3 pairs, tibiae and tarsi not well separated. Abdomen with

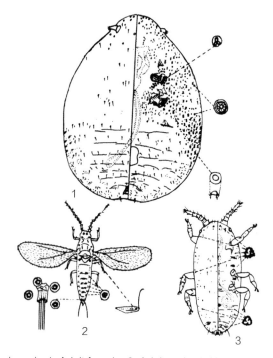

N. sinensis. 1. Adult female, 2. Adult male, 3. Newly hatched nymph (Yan Aojin)

9 visible segments. End of the abdomen conical.

Pupae Elongated; orange yellow at beginning, reddish brown at later stage length 1.00-1.25 mm, width 0.30-0.35 mm. Ocelli two pairs. Dark reddish brown. Antennae filiform, 10 segmented. Mouthparts degenerated. Legs 3 pairs. Wing buds one pair, extending to the 3rd abdominal segment. Abdomen 9 segmented. Copulatory organ conical.

Biology and behavior *Nesticoccus sinensis* occurs one generation a year. Fertilized adults diapause or overwinter within leaf sheets of the new shoots. Females feed from February, become pregnant, enlarge, and form greyish brown, globular wax shells on small branches. The period when females become pregnant vary with temperature. When average temperature in January-February is above 5℃, the females become pregnant in mid-February; when the temperature is below 5℃, females become pregnant in early March. The pregnant period is about two months. The number of eggs produced per female is related to its nutritional status. Those on bamboos grown on plains produce more eggs than those on bamboos grown on hills. Those on *Phyllostachys viridis* var. *youngii* produce more eggs than those on *P. pubescence* and *P. glauca*. Average egg production per female on *P. viridis* var. *youngii*, *P. pubescence*, and *P. glauca* are > 500, > 400, and < 400, respectively. Maximum egg production is 875 per female. Nymph emergence starts from late April and early May, reaches peak during mid-May, and ends during mid-June. Nymphal period is approximately 50 days. Egg hatch rate is 97.5%. Hatched nymphs emerge from a 0.4 mm diameter hole or crack. Emergence period is mostly during 9:00-12:00 at 15-23℃. After 2-3 hours (rarely 12-24 hours) of active movement, they fix themselves within leaf sheath of new shoots. About 90% of them live on branches located at lower half the trees. Nymphs use their proboscis to suck sap from buds or base of young branches. They secrete wax from dorsal and lateral surfaces. Nymphs have 3 instars; 1st instar period is 6-7 days; 2nd instar period is 15-22 days; sexual dimorphism is obvious at this stage. The 3rd instar female nymphs develop for 10 days and molt to adults in early June. Adult emergence reaches peak in mid-June. Male larvae produce white foam and form cocoons in late May. Prepupal period is 2-4 days and after another 3-5 days, pupae are formed. Adults emerge from pupae after 3-6 days. After emergence, they need 3-4 days to leave the cocoons. Males appear from early June, peak in mid-June, and end in early July. Male adults emerge from end of the cocoons. They usually emerge during 5:00-6:00. Emergence rate is 87.5%. Female: male ratio is nearly 1:1. Males are very active and can fly for short distances. During mating period, males search for females with the aid of their antennae. They penetrate the leaf sheath and mate with females. Males live for less than 24 hours. Fertilized females develop slowly, diapause, then overwinter.

Prevention and treatment

1. Quarantine. Prevent movement of infested seedings.

2. Silviculture. Remove infested plants, thinning, and adding fertilizer and improve growth of bamboo plants.

3. Biological control. *Anagrus dactylopii*, *A. nesticoccus*, and *Tetrastichus* sp. are common parasitoids of *Nesticoccus sinensis*. Natural parasitization rate by these wasps is 20-30%. Predators include lady bugs, robber flies, and lacewings. Conserve or propagate and release these natural enemies can suppress the pest populations.

4. Chemical control. During March-April when females are pregnant, inject 20% thiamethoxam, 25% acephate, or 40% acetamiprid and chlorpyrifos mixture at base of infested plants at 3-5 mL per *P. pubescence* plant or 1-2 mL per *P. nigra* plant. Seal the injection holes with mud or tape. During mid-May when nymphs emerge, spray 40% acetamiprid and chlorpyrifos mixture at 1,000-2,000×, 10% imidacloprid WP at 2,000-5,000×, or 2.5% lamda-cyhalothrin at 1,000-2,000×. Re-apply every 7-10 days for a total of 2-3 times will achieve good results.

References Yang Pinglan 1982, Xie Guolin and Yan Aojin 1983, Xu Ji et al. 1983, Xiao Gangrou (ed.) 1992.

(Yan Aojin)

N. sinensis (edited by Zhang Runzhi)

75 *Oracella acuta* (Lobdell)

Classification Hemiptera, Pseudococcidae
Common name Loblolly pine mealybug

Distribution Fujian, Hunan, Guangdong, and Guangxi.

Hosts *Pinus elliottii*, *P. taeda*, *P. echinata*, *P. palustris*, *P. pumila*, *P. clausa* var. *immuginata*, *P. thunbergii*, *P. caribaea*, and *P. massoniana*.

Damage Nymphs suck fluid from young shoots, branches, and cones, causing resin flow, discolor and premature fall of leaves. Severely infested trees may experience defoliation, branch death, shortened branches, sooty mold, and slow growth.

Morphological characteristics **Adults** Male pink red, base of antennae and eyes red. Mesothorax large, yellow. Each side of the 7th abdominal segment with one 0.7 mm long white wax thread. Winged male with one pair of weak white wings; veins simple. Female pale red, pear shaped, encased in wax sac; the end of the abdomen sharp. Eyes distinct, semi-globular. Stylets 1.5x of body length. One large mark spans the median line of the 3rd and 4th abdominal segments.

Nymphs Elliptical to asymmetric elliptical; light-yellow to pink red; 3 pairs of legs. Mid-sized nymphs secrete white wax; end of the abdomen with 3 white wax threads. Large nymphs stay in fixed locations and covered in wax.

Biology and behavior *Oracella acuta* occurs 3-4 generations a year with 3 generations/year being dom-

O. acuta on a young bud (edited by Zhang Runzhi)

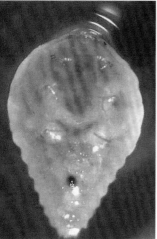
O. acuta nymph (FPC)[1]

O. acuta on a pine shoot (edited by Zhang Runzhi)

tree infested by O. acuta (FPC)

Damage by O. acuta (edited by Zhang Runzhi)

inant. They overwinter as 1st instars within wax sacs or mid-sized nymphs in leaf sheath. They do not have distinct overwintering period. Nymphs grow slowly in winter. Newly emerged nymphs congregate within wax sacs or on young shoots, among needles, or on cones. They spread through wind. They can spread for an average of 17 km and a maximum of 22 km a year. There are two dispersal peaks in a year: from mid-April to mid-May and from mid-September to mid-October. Mid-sized nymphs feed on young shoots. Large nymphs start to secrete wax and form wax sacs. Males have winged and wingless forms and a pupal stage. Females do not have pupal stage. Female adults produce eggs in wax sacs. Egg production period is 20-40 days.

Prevention and treatment

1. Quarantine. Prevent movement of infested timber, seedlings, firewood, and cones. Fumigation of infested wood or seedlings using methane at 20-30 g/m^3.

2. Biological control. Spray *Verticillium lecanii* at 1×10^9 conidia/mL or *Cladosporium cladosporioides* at 2×10^9 conidia/mL. Re-apply every 6 days for one or two times. During late April and early May, release *Cryptolaemus montrouzieri* and *Nephus ryuguus* at 2:5 ratio. The following parasitoids can also be released: *Anagyrus dactylopii*, *Acerophagus coccois*, and *Zarhopalus debarri*.

3. Chemical control. Apply 20% dinotefuran to tree trunk at 5× or spray 20% dinotefuran + 25g/L cypermethrin + water at 1:1:500 to canopy in May and June.

References Liang Chengfeng 2003; Tang Li, Zhao Yumei, Tang Cai et al. 2005; Zhao Yumei, Tang Cai, Lan Cuiyu et al. 2008.

(Yu Zhijun)

[1] FPC – "Forest Pest Control"

76 *Phenacoccus azaleae* Kuwana
Classification Hemiptera, Pseudococcidae

Importance *Phenacoccus azaleae* is a relatively new pest that infests *Zentoxylum bungeanus*. Outbreaks in large areas occurred in recent years.

Distribution Hebei, Henan, Shanxi, Inner Mongolia, Shaanxi; Japan, Korea Peninsula.

Hosts *Rhododendron* spp., *Ulmus* spp., *Prunus salicina*, and *Zentoxylum bungeanus*.

Damage The insect sucks sap from host leaves, weakens the tree growth, causes sooty mold, reduces fruit production, and causes death of branches.

Morphological characteristics **Adults** Females elliptical; length 3.0-3.5 mm; pale brown to dark brown; covered by thick wax. Legs slender; there are teeth under the claws. Cerarius with 18 pairs of pores; each with two setae and a few triocular pores; they appear on dorsal and ventral surfaces. Multi-ocular pores exist on ventral surface of the abdomen. Tubular ducts have two sizes.

Biology and behavior *Phenacoccus azalea* occurs one generation a year. Before overwintering, the 2nd instar move from leaves to base of branches, buds, creases, and cracks. The nymphs form cocoons and molt into 3rd instar. Males and females can be easily differentiated at this stage. Female nymphs and prepupae of males overwinter. In spring, nymphs emerge from cocoons and move to tip of shoots and buds and feed. Their body size increases and wax layer thickens as they grow. In mid-April, female nymphs molt into adults. The adults feed and grow and excrete large amount of honeydew. Males fly or crawl to find females and mate. Males only live for 1-3 days. Mating period is about 20 days. Females continue to feed after mating. They produce white, slender egg sacs and lay eggs inside. The number of eggs inside

P. azalea wax sac and eggs (Xie Yingping)

Overwintering P. azalea wax cocoons (Xie Yingping)

the egg sac increases as the egg sac increases in size. Females stop egg production at the end of May. Female body becomes empty and shinks to dry skins, which located in front of the egg sac. Each female produces one egg sac, each with an average of 446 yellow eggs. They hatch at the end of May. Newly emerged nymphs crawl to leaves and young shoots and feed along veins on the back of leaves. Nymphs light-yellow, without much wax at beginning. After feeding, wax accumulates on dorsal surface and forming a thin wax shell. Nymphs enter into 2nd instar in July and feed until late September to early October. Then they move from leaves to 1-2-year-old branches, make cocoons, and overwinter.

Prevention and treatment

1. Quarantine. Prevent movement of infested seedlings.

2. Silviculture. Cut weeds, loosen soil, and add fertilizer. Remove infested branches, dry branches, and dead trees. Replace weak and old trees.

3. Conserve and release natural enemies.

4. Chemical control. Spray 45% lime sulfur at 200× in mid-April will reduce 80% of the pest population; re-apply after 10-15 days will achieve 95% population reduction. Spray 12% thiamethoxam and lamda-cyhalothrin mixture at 1,500-2,000× for 2-3 times can reduce population by 90%. Spray 35% imidacloprid WP at 1,500× can reduce population by 90%.

References Tang Fangde 1992, Xie Yingping 1998.

(Xie Yingping)

P. azalea damage to young buds (Xie Yingping)

77 *Phenacoccus fraxinus* Tang
Classification Hemiptera, Pseudococcidae

Distribution Beijing, Shanxi, Henan.

Hosts *Fraxinus* spp., *Diospyros kaki*, *Juglans regia*, *Bischofia racemosa*, and *Platanus* spp.

Damage The insect sucks sap from host leaves, causes sooty mold and premature defoliation.

Morphological characteristics **Adults** Female length 4-6 mm, width 2-5 mm; purple brown; elliptical; covered with white wax. Male length 2 mm, wingspan 4-5 mm. Front wing translucent, hindwing reduced to halteres.

Nymphs Pale yellow, elliptical.

Biology and behavior *Phenacoccus fraxinus* occurs one generation a year. It overwinters as nymphs in cracks, under barks, or in buds. They resume activity in early and mid-March. Males and females are differentiated in mid- and late March. Male nymphs secrete wax and pupate inside cocoons. Adults emerge after 3-5 days. Females lay eggs in early April. The eggs hatch in late April to end of May. The nymphs start to overwinter after September.

Male adults fly over tree crown and look for females after emergence. Male adults live for 1-3 days. Female adults feed and excrete honeydew, which promotes growth of sooty mold. After mating, females produce egg sacs, each contains hundreds of eggs. Egg period is approximately 20 days. Nymphs fix themselves to leaves and feed after hatching. They migrate to cracks on tree trunk and other hidden places in autumn before leaf fall.

Prevention and treatment

1. Quarantine. Prevent movement of infested seedlings.

2. Conserve natural enemies. *Nephus ryugus* and *Cheiloneurus claviger* can significantly suppress the pest populations.

3. Chemical control. Spray 0.5 °Bé lime sulfur before buds swell in Spring. Spray 45% profenofos and phoxim mixture at 800-1,000× to kill nymphs.

References Yang Youqian et al 1982, Xiao Gangrou (ed.) 1992, Forest Disease and Pest Control Station of Henan 2005.

(Yang Youqian)

P. fraxinus on back of elm leaves (Xu Gongtian)

P. fraxinus on *Fraxinus* sp. branches (Xu Gongtian)

P. fraxinus female adult and egg sac (Xu Gongtian)

78 *Phenacoccus solenopsis* Tinsley
Classification Hemiptera, Pseudococcidae

Importance When females and nymphs feed on cotton, they can cause dried leaves, premature fall of flower, flower buds. Honeydew accumulation causes sooty mold and death of plants.

Distribution Guangdong, Guangxi, Hainan, Yunnan, Sichuan, Fujian, Zhejiang, Hunan, Jiangxi; U.S.A, Mexico, Pakistan, India, Thailand, Argentina, Brazil, Cuba, Jamaica, Guatemala, Dominica, Ecuador, Panama, Chile, Nigeria, Benin, Cameroom, Australia, New Caledonia.

Hosts About 53 families and 154 species of hosts were recorded in the world. In China, 55 host species were recorded. Main hosts include cotton, Chinese hibiscus, navel orange, watermelon, *Bougainvillea glabra*, *Firmiana simplex*.

Morphological characteristics **Adults** Females light-yellow to orange yellow; a series of black spots often exist on the back, about six pairs of spots from prothorax to mesothorax, mesothorax sometimes without spots; each abdominal segment with a few black spots; thorax and abdomen with linear marks on the side; ventral surface black. Slide specimens measured 3.0-4.2 mm long, 2.0-3.1 mm wide. Anal lobe moderately developed. Antennae 9-segmented (occasionally 8-segmented). Legs well developed. Coxa of hind leg without transparent opening, claws with one small tooth.

Males minute. Red; antennae 10-segmented; legs slender; hind wings reduced to a pair of clubs. End of abdomen with two pairs of wax threads. Slide specimens 1.41 mm long; antennae 2/3 of the length of body; body

P. solenopsis male and female adult (Zhang Runzhi)

P. solenopsis eggs (Zhang Runzhi)

P. solenopsis damage (Wu Ansan)

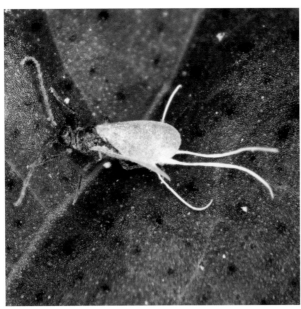
P. solenopsis male adult (Zhang Runzhi)

P. solenopsis female adult (Zhang Runzhi)

covered with sparse setae; antennae and legs with mealy setae.

Biology and behavior *Phenacoccus solenopsis* infests young parts of plants including leaves, flower buds, petioles. Eggs are hatched inside the females. The 1st instars are active. They can feed soon after emerged. One female can lay 400-500 eggs.

Prevention and treatment

1. Quarantine. Prevent movement of infested seedlings to non-infested areas.

2. Conserve and release natural enemies: *Cryptolaemus montrouzieri*, *Chrysoperla carnea*.

3. Chemical control. Spray cyhalothrin, deltamethrin, or neonicotinoid.

References Wang Yanping 2009, Wu Sanan 2009, Zhang Runzhi et al 2010.

(Zhang Runzhi, Li Zhenyu)

P. solenopsis on Chinese hibiscus (Zhang Runzhi)

79 *Asiacornococcus kaki* (Kuwana)

Classification Hemiptera, Eriococcidae
Synonym *Eriococcus kaki* Kuwana

Importance *Asiacornococcus kaki* is an important pest of *Diospyros kaki*. Average fruit infestation rate of 30%-40% and maximum infestation rate of 80%-90% was reported. The pest significantly lowers the fruit quality and threatens the traditional dry fruit industry.

Distribution Hebei, Shanxi, Liaoning, Jilin, Heilongjiang, Zhejiang, Anhui, Shandong, Henan, Guangdong, Guangxi, Sichuan, Guizhou, Shaanxi, Ningxia; Japan, Korea Peninsula.

Host *Diospyros kaki*.

Damage Adults and nymphs suck sap from young branches, leaves, and fruits. Infested young branches do not produce buds or die. Infested leaves deform, become infested with sooty mold, and fall prematurely. The insect prefers feeding between the petal and base of the fruit. The infestation site exhibits yellowish green spots and black marks; fruits turn soft early and fall prematurely.

Morphological characteristics **Adults** Females length 1.5 mm, width 1.0 mm; elliptical; dark purple red. Body segments obvious, dorsal surface with setae, sides with white thin wax threads. Abdomen smooth, setae long. Anal ring with 8 setae. After fertilization, the body secretes white material and forms egg sac. Female shell size is 2.6 mm long and 1.4 mm wide, greyish white, oval or elliptical. Male adult length 1.0-1.2 mm; purple red; antennae slender; without eyes. Male shell 1.1 mm

A. kaki (Zhang Runzhi)

A. kaki (Zhang Runzhi)

A. kaki (Zhang Runzhi)

A. kaki adult female and male cocoons (Xu Gongtian)

A. kaki female sac (Xu Gongtian)

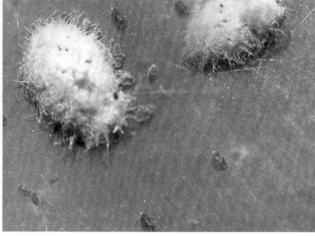

A. kaki females on a *Diospyros kaki* leaf (Xu Gongtian)

long, 0.5 mm wide; white; elliptical.

Eggs Length 0.3 mm; oval; purple red; encased within egg sacs.

Nymphs Bright red at beginning, turning to purple red; oval or elliptical; sides with various sizes of setae; end of the abdomen with a pair of wax threads.

Male pupae Elliptical; purple red; body segments obvious, within elliptical white wax cocoon.

Biology and behavior *Asiacornococcus kaki* occurs 4 generations a year. It overwinters as nymphs in cracks, under thick barks, in buds, etc. They resume activity in April and feed on young buds, new shoots, petioles, and underside of leaves. They produce wax cover on fruit surface and petals. Infestation sites become yellowish brown, concave, hardened, and blackened. Adults mate in mid- and late May. Males die soon after mating. Females form white egg sacs in which they deposit eggs. Each female lays 51-160 eggs. Egg period is 12-21 days. Nymph emerge peak period for the 1st-4th generations are mid-June, mid-July, mid-August, and mid-September, respectively. The first two generations infest leaves and 1-2-year-old branches. The last two generations infest fruits. The 3rd generation causes most severe damage. Nymphs overwinter from mid-October.

Prevention and treatment

1. Quarantine. Prevent movement of infested grafts.

2. Reduce overwintering population. Scrape off old barks. Use steel brush to remove hidden nymphs on barks. Spray 5 °Bé lime sulfur or motor oil at 100× before buds swell.

A. kaki overwintering nymphs (Xu Gongtian)

3. Conservation of natural enemies. The following natural enemies are important predators or parasitoids of *Asiacornococcus kaki*: *Chilocorus rubidus*, *C. kuwanae*, *Chrysopa septempunctata*, and *Aphycus* sp. These natural enemies can reduce population by 50-60%.

4. Chemical control. Spray insecticides before wax shells are formed. The following sprays are effective in suppressing pest populations: 40% acetamiprid and chlorpyrifos mixture at 1,000×, 5.7% emamectin and avermectin mixture at 2,000×, 35% imidacloprid WP at 1,000×, 12% thiamethoxam and lamda-cyhalothrin mixture at 1,500×.

References Xiao Gangrou (ed.) 1992; Zhao Fanggui et al. 1999; Sun Xugen et al. 2001; Wang Jianyi, Wu Sanan, Tang Hua et al. 2001.

(Wang Haiming, Wang Zhuo, Shen Wensheng)

80 *Acanthococcus lagerstroemiae* Kuwana

Classification Hemiptera, Eriococcidae
Synonym *Eriococus lagerstroemiae* Kuwana
Common name Crape myrtle bark scale

Distribution Beijing, Tianjin, Hebei, Shanxi, Liaoning, Jiangsu, Zhejiang, Anhui, Shandong, Guizhou, Ningxia.

Hosts *Lagerstroemia indica*, *Punica granatum*, *Ligustrum lucidum*, *Grewia biloba*, and *Securinega suffruticosa*.

Damage Female adults and nymphs suck sap from host plants, causing premature flower and leaf fall, slow plant growth, and sooty mold development.

Morphological characteristics **Adults** Females length 3 mm; elliptical or elongated oval, the end more tapered than the head; body purple red, with scattered setae and white wax. They form fluffy wax sacs before laying eggs. The body and eggs are enclosed within wax. Antennae 7 segmented; the 3rd segment longest, the 4th segment shorter than the 3rd segment. Adult males 1.2 mm long, purple or brown. Front wing semi-transparent. Antennae 10 segmented. End of the abdomen with a pair of long setae. **Eggs** Length 0.3 mm; elliptical; pale purple. **Nymphs** Length 0.5 mm; elliptical; light-yellow at beginning, then pale purple. Overwintering nymphs 1 mm long; purple red; legs yellow; dorsal surface with some wax threads. **Pupae** Elliptical; purple red; setae on the body forked, encased within dark white cocoon. Cocoon elliptical, slightly compressed dorsal ventrally. Length 1.5-2.0 mm. Posterior half with wax shell. The end with a transverse seam that divides the cocoon into upper and lower layer.

Biology and behavior *Acanthococcus lagerstroemiae* occurs two generations a year in Beijing. It

A. *lagerstroemiae* adults on *Lagerstroemia indica* (Li Zhenyu)

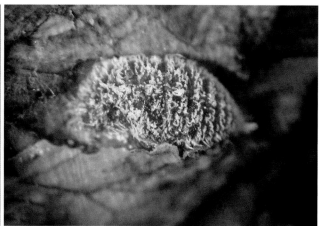

A. *lagerstroemiae* unmated female (Xu Gongtian)

A. *lagerstroemiae* gravid female (Xu Gongtian)

A. *lagerstroemiae* female wax sac (Xu Gongtian)

Dorsal view of *A. lagerstroemiae* pupa (Xu Gongtian)

Ventral view of *A. lagerstroemiae* pupa (Xu Gongtian)

A. lagerstroemiae female wax sac (Xu Gongtian)

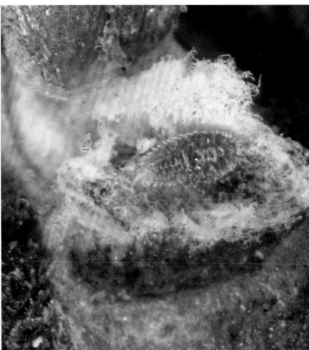
A. lagerstroemiae 2nd instar male nymphs (Xu Gongtian)

overwinters as 2nd instar nymphs on branches, in cracks, under barks, or in empty wax sacs. They resume activity in late March. The males and females differentiated in mid- and late April. Males produce white cocoons and pupate inside the cocoons. Females produce white wax sacs when sexually mature. Males emerge from late April to early May. Females lay eggs in late May and early June. Each female lays 37-124 eggs. They hatch in mid-June. The development time is relatively homogenous. Adults emerge in early and mid-July. Nymphs hatch from mid-August to late September. The emergence time varies a lot. When average temperature drops to 16°C in late October, the 2nd instar will start to overwinter. The insect can reproduce through parthenogenesis.

Prevention and treatment

1. Use non-infested seedlings when planting.

A. *lagerstroemiae* male cocoons (Xu Gongtian)

Overwintering stage of *A. lagerstroemiae* (Xu Gongtian)

Sooty mold developed after *A. lagerstroemiae* infestation (Xu Gongtian)

A. lagerstroemiae adult females (Xu Gongtian)

2. Conservation of natural enemies. The following predators and parasitoids are important in regulating the pest populations: *Chilocorus kuwanae*, *Rodolia limbata*, *Harmonia axyridis*, *Chrysopa sinica*, *Marietta picta*, *Coccophagus yoshidae*, *Metaphycus pulvinariae*, and *Pseudophycus walinus*.

3. Chemical control. Spray 40% acetamiprid and chlorpyrifos mixture at 1,000×, or 35% imidacloprid WP at 2,000× during nymph emergence period.

References Tang Fangde 1977; Xiao Gangrou (ed.) 1992; Xu Gongtian and Yang Zhihua 2007; Wang Jianyi, Wu Sanan, Tang Hua et al. 2009.

(Li Zhenyu, Xu Gongtian, Nan Nan, Zhao Huaiqian, Hu Xingping)

81 *Kermes nawae* Kuwana
Classification   Hemiptera, Kermicoccidae

Importance   *Kermes nawae* is one of the major sucking insect pests of *Castanea* spp.

Distribution   Jiangsu, Zhejiang, Anhui, Jiangxi, Henan, Hubei, Hunan, Sichuan, Guizhou; Japan.

Hosts   *Castanea mollissima*, *C. seguinii*, and *C. henryi*.

Damage   Nymphs and female adults suck sap from host plants, causing delayed budding, reduced fruit production, and dead branches or tree mortality.

Morphological characteristics   **Adults** Females globular or semi-globular, diameter 5 mm. Soft and brittle at beginning, shiny at later state; yellowish brown or brown, with blackish brown irregular round or elliptical marks and several black or dark brown transverse striae; one side has several white wax threads. Male adult 1.49 mm long, wingspan 3.09 mm. Brownish black. Antennae filiform, 10 segmented, each with several hairs. Forewing dirty yellow, transparent. End of the abdomen with a pair of long wax threads.

Eggs   Elongated oval; milky or colorless and transparent at beginning, then purple red.

Nymphs   First instar elongated oval, bloody yellow. First instar parasitic nymphs yellowish brown. Two sides of the thorax with white wax and wax threads. Second instar parasitic nymphs elliptical, bloody red. They often carry the shed skin of the 1st instar.

Pupae   Pupae encased in cocoons. Cocoons compressed elliptical; white; fluffy. Male pupae elongated oval; yellowish brown.

Biology and behavior   *Kermes nawae* occurs one generation a year. It overwinters as 2nd instar nymphs on base of buds, in cracks, and scars. They resume activity

K. nawae adult female (Hu Xingping)

in March when average day temperature is > 10°C. From mid-March, some nymphs turn to adults and continue to feed. This is the main infestation period. The end of the abdomen has a small water bead. The water beads disappear when females are mature with eggs. Some male nymphs migrate to cracks and depressions on branches, scars, under lichen to pupate. Adults emerge in early April. The peak emergence period is in late April. After mating, females lay eggs. The eggs hatch in mid-May; peak hatching period is in late May. Females are viviparous; parthenogenesis occurs. First instar nymphs are active. They turn into 2nd instar in mid- and late June and diapause. The nymphs overwinter from early or mid-November.

Prevention and treatment

1. Select suitable tree species, high quality seedlings according to the field conditions to prevent the occurrence of the pest.

2. Cultural control. Cut down infested branches, slender branches, etc; scrape off diseased scars, rough barks, peeled barks and take them out of the forest.

3. Apply *Pestalotiopsis* fungus. Conservation of natural enemies. Major predator is *Chilocorus rubidus*. Major pathogen is *Cladosporium cladosporioides*.

4. Chemical control. Spray 35% imidacloprid WP at 1,000× in mid- and late March to suppress nymph population.

References   Zhang Fengyi et al. 2000, Chen Jinxiu et al. 2001.

(Fang Minggang, Hu Xingping, Zhao Jinnian)

K. nawae (Ji Baozhong)

82 *Ceroplastes japonicus* Green

Classification   Hemiptera, Coccidae
Common name   Tortoise wax scale

Importance   *Ceroplastes japonicus* reduces fruit quality and quantity and scenic value of the landscape. It is a potential danger to *Zizyphus jujube* production.

Distribution   Beijing, Hebei, Shanxi, Inner Mongolia, Liaoning, Heilongjiang, Jiangsu, Zhejiang, Anhui, Fujian, Shandong, Henan, Hubei, Hunan, Guangdong, Sichuan, Guizhou, Yunnan, Shaanxi, Gansu, Taiwan; Japan, Russia.

Hosts   *Platanus hispanica, Salix* spp., *Ulmus* spp., *Ligustrum lucidum, Koelreuteria aniculate, Acer mono, Rhododendron simsii, Rosa chinensis, Michelia figo, Chaenomeles sinensis, Laurus nobilis, Ficus carica, Malus pumila, Zizyphus jujube, Diospyros kaki, Cutrus reticulate, Punica granatum, Eriobotrya japonica, Prunus salicina, P. mume, P. armeniaca, P. persica, Pyrus bretschneideri, Camellia sinensis, Buxus megistophylla.*

Damage   Nymphs and female adults suck sap from buds and leaves, weaken the hosts, and promote sooty mold growth.

Morphological characteristics   **Adults** Females are covered with thick white wax shells. Elliptical, 4-5 mm long, semi-globular, surface with striae similar to turtle shells. The edge of the wax shells thick, with 8 angular projections; oval; purple red. Males with white wax shell, star-like; the median with an elliptical wax plate which is surrounded by 13 angular projections. Male 1.0-

Dorsal view of *C. japonicus* females (Xu Gongtian)

C. japonicus male (Xu Gongtian)

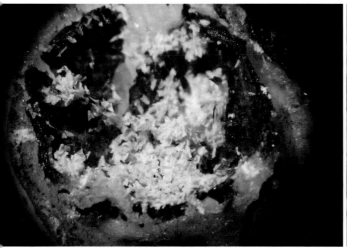
C. japonicus eggshell (Xu Gongtian)

C. japonicus adult and nymph (Xu Gongtian)

1.4 mm long; brownish; eyes black. Antennae filiform. Wings with two thick veins. Legs small. End of the abdomen slightly tapered.

Eggs Elliptical; length 0.2-0.3 mm; orange yellow at beginning, turning to purple red before hatching.

Nymphs New hatched nymphs 0.5 mm long; elliptical, flattened; reddish brown. Antennae and legs developed, grayish white. End of the abdomen with a pair of long setae.

Male pupae Spindle shape; 1 mm long; brown; head, antennae, and wing buds paler in color. Copulatory organ visible.

Biology and behavior *Ceroplastes japonicas* occurs one generation a year. Mated females overwinter on 1-2 year-old branches. They become active in late March and grow rapidly in mid-April. Females deposit eggs under the abdomen from late May or early June. Peak egg production is from early and mid-June. They stop laying eggs in mid-July. Each female lays more than 1,000 eggs or a maximum of 3,000 eggs. Egg period is 20 days. They hatch from mid-June to late July, peak hatching period is in late June to early July. Egg incubation period is 40 days. Males and females differentiate from late July to early August. Male larvae pupate from early August to late September. Pupal period is 15-20 days. Adult males occur from late August to early October. They live for 1-5 days and die soon after mating. Females turn to adults at the end of September after two molts. After mating, they migrate to 1-2-year-old branches and feed and overwinter. Newly hatched nymphs crawl to top of leaves and feed along the main and side veins. After 1-2 days, the dorsal surface secretes two rows of white wax dots. After 3-4 days, the thorax and abdomen form two wax plates and merge into one plate later. The sides of the body secrete13 triangular wax spines. After 12-15 days, the body forms a star shaped wax shell. Male wax shells thicken

C. japonicus young nymph (Xu Gongtian)

C. japonicus female nymph (Xu Gongtian)

C. japonicus wax shell after being parasitized (Xu Gongtian)

C. japonicus wax shell of nymphs (Xu Gongtian)

C. japonicus male nymphs (Xu Gongtian)

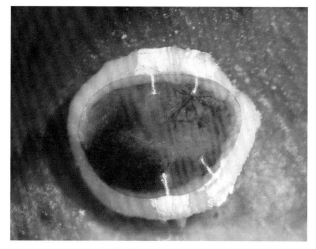

C. japonicus nymph (Xu Gongtian)

over time. Females secrete new wax and form turtle shell-like wax shell. Egg hatch rate can be 100% under high humidity and normal temperature. Dry weather, high temperature will reduce hatch rate and infestation level. Female survival during winter is negatively correlated with low temperature and precipitation.

Prevention and treatment

1. Quarantine. Use treated seedlings and grafts.

2. Cultural control. Remove infested branches or paint the infested branches. When ice accumulates on branches in winter, use wooden sticks to break the ice and overwintering females will fall off the branches.

3. Conserve natural enemies. Major natural enemies include: *Chilocorus kuwanae*, *C. rubidus*, *Scymnus* (*Neopullus*) *babai*, *Microterys clauseni*, and *Chrysopa septempunctata*.

4. Chemical control. After leaf fall and before buds swelling, spray oil suspension at 100× or 5% resin mixture at 8-10×. The following synthetic insecticides can be used: 5.7% avermectin + emamectin benzoate at 3,000×, 40% acetamiprid and chlorpyrifos mixture at 2,000×. Adding 0.2% silicone will increase the efficacy of the insecticides.

References Xiao Gangrou (ed.) 1992, Zhao Fanggui et al. 1999, Sun Xugen et al. 2001.

(Wang Haiming, Sun Jinzhong, Hu Xingping)

83 *Ceroplastes rubens* Maskell

Classification Hemiptera, Coccidae
Common name Red wax scale

Importance *Ceroplastes rubens* is a serious pest of fruit trees, landscape, forests in central, eastern, southern, and even southwestern China. The pest can cause 10%-30% production loss.

Distribution Hebei, Liaoning, Jilin, Heilongjiang, Shanghai, Jiangsu, Zhejiang, Anhui, Fujian, Jiangxi, Hubei, Hunan, Guangdong, Guangxi, Sichuan, Guizhou, Yunnan, Shaanxi, Taiwan; Japan, India, Sri Lanka, Myanmar, Philippine, Indonesia, U.S.A., and pacific countries.

Hosts *Gardenia*, spp., *Osmanthus fragrans*, *Laurus wobilis*, *Camellia japonica*, *Prunus mume*, *Rosa chinensis*, *Aglaia odorata*, *Nandina domestica*, *Punica granatum*, *Magnolia denudate*, *Malus spectabilis*, *Camellia japonica*, *Cedrus deodara*, *Viburnum odoratissimum*, *Magnolia grandiflora*, *Podocarpus macrophyllus*, *Cycas revolute*, *Ilex chinensis*, *Citrus aurantium* var. *amara*, *Citrus medica* L. var. *sarcodactylis*, *Mahonia bealei*, *Trachycarpus fortune*, *Weigela florida*, *Hedara nepalensis* var. *sinensis*, *Cinnamomum Myanmarnnii*, *Michelia figo*, *Fortunella margarita*.

Damage *Ceroplastes rubens* infests young branches, petioles, and leaves. They weaken the plant, kill branches, promote sooty mold growth, and lower the aesthetic value of the plants.

Morphological characteristics **Adults** Female shells reddish brown, semi-globular; diameter of the wax cover is 3.0-5.0 mm; the median area has a round depression with its edge raise like petals; each of the four body corners with a wax thread secreted by the ducts on thoracic spiracles. Female adults elliptical; purple red, depressed at the spiracles. Male wax shells elongated elliptical before pupation, dark purple red. Male adults 1 mm long; dark red; wings semi-transparent; antennae yellow, 10-segmented.

Nymphs Flattened; elliptical; reddish brown; end of the abdomen with two long setae.

Eggs Elliptical; pale purple; the two ends tapered; length 0.1 mm.

Biology and behavior *Ceroplastes rubens* occurs one generation a year. Females overwinter and start to lay eggs in late May. Peak egg laying period is in early and mid-June. Each female lays 150-550 eggs. Eggs are laid under female body. They hatch after a few days. Peak hatching period is in June-July. Nymphs go through 3 instars and a total of 80 days period. The nymphs move to young branches, shoots, and leaves and become sessile in about one hour after hatching. Nymphs start to secrete wax soon after hatching and stop until mature. Adults emerge from late August. After mating, males die and mated females overwinter on branches.

Prevention and treatment

1. Quarantine. Prevent movement of infested seedlings.

2. Physical control. Scrape off overwintering females. Remove infested branches and destroy them.

3. Apply *Pestalotiopsis* fungus. Conserve natural enemies. Major natural enemies include: *Coccophagus yoshidae*, *C. chengtuensis*, *C. hawaiiensis*, *C. ishii*, *C. japonicas*, *Anicetus benificus*, *A. ohgushii*, *Comperiella bifasciata*, *Metaphycus pulvinariae*, *Microterys speciosus*, *M. yunnanensis*, *Chilocorus kuwanae*, and *C. rubidus*.

4. Chemical control. Spray 20% dinotefuran, 45% phofenofos and phoxim mixture or 50% fenitrothion at 1,000× during egg laying period. Spray 12% thiamethoxam and lamda-cyhalothrin mixture at 1,500-2,000× for 1-2 times during egg hatching period.

References Tang Fangde 1991; Xie Yingping, Xue Jiaoliang, and Zheng Leyi 2006; Xu Gongtian and Yang Zhihua 2007.

(Xie Yingping, Li Zhenyu)

C. rubens adult female and hatched nymphs (Xu Gongtian)

84 *Ericerus pela* (Chavannes)

Classification Hemiptera, Coccidae
Common name White wax scale insect

Importance *Ericerus pela* is an economically important beneficial insect. The 2nd instar male larvae produce "white wax" on host plants. White wax is used in chemical engineering, mechanics, precision instruments, medicine, food, and agriculture, etc. Female *Ericerus pela* is a nutritious food source. Mature females contain eggs. Each female contains large number of eggs which are rich in protein, amino acids, micronutrients, etc. Growing *Ericerus pela* requires large areas of host plants which helps improve the ecosystem and generate both economic and social benefits.

Distribution Beijing, Yunnan, Guizhou, Sichuan, Guangxi, Hunan, Shaanxi; Japan, Korea Peninsula.

Hosts *Ericerus pela* uses more than 40 species of plants as hosts. Most preferred hosts are *Ligustrum lucidum* and *Fraxinus chinensis*.

Morphological characteristics **Adults** Female shells reddish brown with dark spots, ventral side yellowish green; antennae slender, 6-segmented; mouthparts needle-like; legs very slender; abdomen 7-segmented. Anal cleft narrow ring-like, anal plate triangular, anal ring with 8 setae; length 1.5 mm, width 1.3 mm. After mating, the body gradually increase in size and becomes globular. During egg production period, females are usually 10 mm in diameter.

Males 2 mm long, wingspan 5 mm; head pale brown to brown; front wing nearly transparent, with iridescence; hind wing club-like, with 3 hooks at distal end; end of abdomen greyish brown, 8-segmented; the 2nd to the last segment with two > 2 mm long wax threads.

Eggs Elongated elliptical, 0.40 mm long, 0.25 mm wide; hidden inside the wax powder; female eggs located at entrance of female shell, male eggs located at bottom of female shell.

Nymphs First instar female nymph near elongated oval; length 0.6 mm, width 0.4 mm; reddish brown, with one pair of ocelli; antennae 6-segmented, the 6th segment with 7 setae; end of abdomen with a pair of wax threads which are about same length as body length; 2nd female nymph oval; length 1 mm, width 0.6 mm; pale yellowish brown; once fixed, body yellowish green; side of body with long and dense wax hairs.

First instar male nymphs similar to female nymphs, but much paler in color; second instar nymphs oval, 0.75 mm long, 0.45 mm wide; pale yellowish brown; antennae 7-segmented.

Pupae Only males develop into pupae. Prepupae pear-shaped, yellowish brown, Length 2 mm, width 1.1 mm; eyes pale reddish brown; antennae short and small; legs short and robust. Pupae 2.4 mm long, 1.1 mm wide; elongated elliptical, antennae 10-segmented.

Biology and behavior *Ericerus pela* occurs one generation a year. Females are hemimetabolous and go through two instars of nymph and adult stages. Males are holometabolous and go through 2 instars of larva, prepupa, pupa, and adult stages. There are two active periods

E. pela dead adult (Xu Gongtian)

E. pela wax cover from male adults (Xu Gongtian)

during its lifetime. The 1st one is when 1st instars search for suitable leaves and become sessile. Based on different requirement of on sunlight, females often fix themselves individually along veins on upper side of leaves. Males fix themselves in groups to the underside of the leaves. The 2nd active period occurs when 1st instars molt into 2nd instars and migrate from leaves to branches. There are no overlapping generations. Females can produce the next generation by parthenogenesis when males are absent.

Each female produces 8,000-12,000 eggs. Female eggs dark red, male eggs yellow. Female: male ratio is 1:2. Mortality during the 1st instar is 70%. Wind and rain are the major mortality factors affecting their survival. After one week of sessile life, the 1st instars molt into 2nd instar and migrate to branches. Females prefer young branches on top of the plants. Males avoid sunlight and prefer 1-2 cm diameter branches. The 2nd instar male larvae secrete wax to protect their body for 2-3 months and stop secreting wax until pupal period. Strands of wax form on the wax cover. Females become sclerotized. Parasitic and predatory enemies are major mortality factors during the 2nd instar period. Mortality is 10%-15%.

About 50% of the nymphs can successfully pupate. Male adults have wings. They live for 8-10 days and die soon after mating. A male can mate with multiple females. Female adults live for 8-9 months. They are round, 0.5-1.5 cm in diameter. They congregate on branches. Mortality is 2%-5%. Parasitic wasps and predators are major mortality factors.

The intrinsic rate of increase is 10-15×. However, 99% of the individuals die within a generation.

Production technique. Wax production normally adopts two-location method where insects are produced at low elevation and wax is produced at high elevation. Females are produced at 1,500-2,000 m elevation in Yunnan province, then transported to Sichuan and Hunan provinces for wax production. The insect is mostly grown in Zhaotongyanshan, Yongshanwanhe, Ludiansuoshan of Yunnan. It is also grown in Xichang and Jinhcko of Sichuan, but the quality is lower than those grown in Yunnan.

1) Seed insect. They are grown on *Ligustrum lucidum*. The insect prefers 1st year-old branches. Prune host plants before buds swell to promote growth of new branches. Old trees can be cut off at 1.5 m height. After 6 months to a year, the new branches can be used for growing wax scale insects. Shorter trees are more convenient for growing wax scale insects.

Place mature female adults into 50-80 mesh nylon pouches to prevent predation. Hang the pouches on branches. Select days when no rain or strong wind will occur. Dozens of females can be placed in a pouch. Place suitable number of pouches on each plant so that maximum production is achieved without plant death. The optimum insect density is when 50%-60% of the branches are utilized by the insect. Leave a small portion of the insect population on the plant during harvesting. The remaining insects will serve as seed population for the next year. This method is simple, convenient, but may suffer from predation and should not be used repeatedly. After harvesting and during transportation, spread the insects to avoid over-heating.

2) Production of wax. Farmers place nylon pouches that contain seed insects onto host branches. Before hanging them onto branches, the insects need to be spread and allow nymphs to hatch for 3-4 days. This method will reduce competition for food sources between males and females. After release, watch for feeding to avoid over-crowding. Optimum insect release density is to utilize 80% of the usable branches.

3) Wax purification. (1) Traditional purification method. Harvest wax before male adults emerge. Place wax into gauze sacs and boil them. After melting and press-

E. pela gravid female (Xu Gongtian)

E. pela nymphs covered by wax (Xu Gongtian)

E. pela nymphs (Xu Gongtian)　　　　　　　　　E. pela nymphs (Xu Gongtian)

ing, remove the supernatant. (2). Mechanical purification. The wax produced from the first method contains large amount of impurities. More pure wax product can be produced through mechanical purification which consisted of multiple steps.

References　Qu Hong 1981; Wu Cibin 1989; Chen Xiaoming, Chen Yong, Ye Shoude et al. 1997; Feng Ying, Chen Xiaoming, Chen Yong et al. 2001; Wang Zili, Chen Xiaoming, Chen Yong et al. 2003; Chen Xiaoming, Wang Zili, Chen Yong et al. 2007a,b, 2008.

(Chen Xiaoming, Wang Fu)

85 *Eulecanium giganteum* (Shinji)

Classification Hemiptera, Coccidae
Synonyms *Lecanium gigantea* Shinji, *Eulecanium diminutum* Borchsenius
Common name Giant globular scale

Distribution Beijing, Hebei, Shanxi, Inner Mongolia, Henan, Jiangsu, Anhui, Shandong, Shaanxi, Gansu, Ningxia, Xinjiang; Japan, Russia.

Hosts Most preferred hosts are: *Ziziphus jujube*, *Z. jujuba* var. *spinosa*, *Prunus dulcis*, *Sophora japonica*, *Acer negundo*, *Xanthoceras sorbifolium*, and *Ulmus pumila*. Preferred hosts are: *Fraxinus chinensis*, *Albizia julibrissin*, *Gleditsia sinensis*, *Zanthoxylum bungeanum*, *Prunus persica*, *Populus* spp., *Salix* spp., *Morus alba*, *Platanus orientalis*, *Punica granatum*, *Ficus carica*, *Juglans regia*, *Vitis vinifera*, *Elaeagnus angustifolia*, *Rosa rugosa*, *Malus pumila*, *Pyrus bretschneideri*, and *Prunus armeniaca*. A total of 45 host species in 20 families are reported.

Damage Nymphs and adults feed on young branches and leaves, causing leaves become yellow and branches die. Infestations will reduce tree growth and fruit production, and even cause death of trees.

Morphological characteristics **Adults** Females semi-globular, smooth, the anterior half raised, the posterior half oblique and narrow, length 10-16 mm, width 8.0 mm, dorsal surface reddish brown, with greyish black markings that are similar to watermelon. There is one median longitudinal band, two bands on the sides, 8 black rhombus markings between the bands; the markings are covered with wax threads. After egg-laying, the females become sclerotized and blackish brown. Male length 1.5-2.1 mm, wingspan 5.0 mm. Head blackish brown, prothorax and abdomen yellowish brown, mesothorax and metathorax reddish brown. Front wing transparent. End of the abdomen with 6 setae and two white anal setae of unequal length.

Eggs Elongated elliptical, length 0.4 mm, milky, turning to reddish brown and covered with white wax powder at later stage.

Nymphs First instar elongated elliptical, length 0.4-0.5 mm, orange yellow. Body segments obvious, covered with white shell. Second instar slightly raised, length 0.6-0.7 mm, width 0.3-0.5 mm, light-yellowish brown. The front and back of the dorsal surface with two ring-like punctures and an orange yellow raised longitudinal band. The 3rd instar flattened, lengh 1.0-1.3 mm, yellowish brown, covered with greyish white semi-transparent wax layer, some wax threads exist on the wax cover. The sides of the body show white setae and white wax plates at late stage. Second instar male larvae are covered with dirty white wax shell.

Pupae Elliptical, 2.3 mm long, pale brown at beginning, dark brown later, covered with dull transparent shell.

Biology and behavior *Eulecanium gigantenum* occurs one generation a year in Xinjiang. It overwinters as 2nd instar on branches and tree trunks. They resume feeding in late March and early April as buds develop.

E. gigantenum gravid female covered in wax (Xu Gongtian)

E. gigantenum dead adult female (Xu Gongtian)

E. gigantenum eggs
(Xu Gongtian)

Male larvae pupate in early April and adults emerge in late April. Pupal period is 12-16 days. Female adults emerge when paulownia and *Robinia pseudoacacia* start to bloom, apple and pear trees in full bloom, and elm trees are at the end of the blooming season. Peak female adult period is in late April and early May. They secrete wax threads and lay eggs. Peak egg production is in early May when blooming season of paulownia and *R. pseudoacacia* ends. Eggs hatch in late May and early June. Egg hatch rate is 95%. Egg period is 20-27 days. Peak nymph period is from mid-July to early August. They start to overwinter from mid- to late October.

Male adults are most active during 10:00-12:00 when weather is clear. They only live for 1-2 days and die soon after mating. Females mate and lay eggs soon after emergence. Their body shrinks as they lay eggs until the abdomen shrinks into a thin shell. Females live for 20-35 days. Female: male ratio is 1:4.3. A female lays 200 to more than 10,000 eggs. Most females lay 3,000-6,000 eggs each. Newly emerged nymphs are very active. They stay for 2-5 days in eggshell and then disperse to leaves. They feed along the main veins of leaves, then the petiole, young branches. The second instar nymphs prefer middle and lower part of the canopy that facing to the northwest. They migrate to branches before defoliation in the fall. Those did not migrate to branches will die after the leaves fall off the trees. Overwintering nymphs are mostly on shade side of those > 1-year-old branches. Female adults are mostly found on lower canopy, shade side, 2 to 5-year-old branches.

Prevention and treatment

1. Quarantine and silviculture. Prevent movement of infested seedlings. Brush lime to tree trunks to prevent diseases and improve tree vigor.

2. Cultural control. Pruning in winter and early spring to remove infested branches, seedlings, and dead trees. Plant bait trees such as locust, elm, *Halimodendron halodendron* and reduce pest population by focusing on the treatment of the bait trees. Hand remove female adults. Use a high-pressure sprayer to wash off overwintering nymphs.

3. Biological control. Apply *Pestalotiopsis* fungus. Conserve the following natural enemies: *Chilocorus geminus*, *Plagiolepis rothneyi*, *Respa germanica*, *Nabis sinoferus*, *Arma chinensis*, *Wesmaelius sufuensis*, *Upupa epop*, and *Passer montanus*. During spring and summer, avoid applying insecticides and utilize *Blastothrix sericae* as natural control.

4. Chemical control. When pest population level reaches $10/10 \text{ cm}^2$ leaf area, spray 5.7% emamectin and avermectin mixture at 3,000×, 12% thiamethoxam and lamda-cyhalothrin mixture at 1,000× during 1st-3rd instar.

References Tang Fangde 1991; Xi Yong et al. 1994, 1996, 2000; Wang Jianyi, Wu Sanan, Tang Ye et al. 2009.

(Xi Yong, Wang Aijing)

86 *Eulecanium kuwanai* Kanda

Classification Hemiptera, Coccidae
Synonym *Lecanium kuwanai* Takahashi

Distribution Beijing, Hebei, Shanxi, Henan, Shandong, Shaanxi, Gansu, Ningxia; Japan.

Hosts *Sophora japonica*, *Albizzia julibrissin*, *Ulmus* spp., *Acer negundo* 'Variegatum', *Lagerstroemia indica*, *Quercus* spp., *Prunus persica*, and *Prunus armeniaca*.

Damage Nymphs feed on branches and cause slow growth of hosts.

Morphological characteristics **Adults** Female semi-globular, reddish brown, length 12.5-18.0 mm. Antennae 7 segmented. Anal cleft shallow, with two triangular anal plates. Anal ring with striae and 8 setae. Before egg-laying, there are 8 greyish black spots between dorsal median band the lateral band, wax cover fluffy. After egg-laying, the female body hardened, yellowish brown to brown. Head of males blackish brown, thorax and abdomen brown, length 1.8-2.0 mm. There are two white wax threads at the end of the abdomen. Antennae 10 segmented, Front wing membranous, Copulatory organ slender. **Eggs** Elongated round, milky or pinkish red. **Nymphs** First instar elliptical, bloody red., length 0.30-0.50 mm. Antennae 6 segmented, Two setae at the end of the abdomen. Anal ring with 6 setae. Sessile nymphs look like paramecium, light-yellowish brown, length 0.60-0.72 mm. Wax cover transparent. **Male pupae** Length 1.7 mm. Segments of antennae and legs are visible. Wing buds and copulatory organ obvious.

Biology and behavior *Eulecanium kuwanai* occurs one generation a year. It overwinters as 2nd instar on

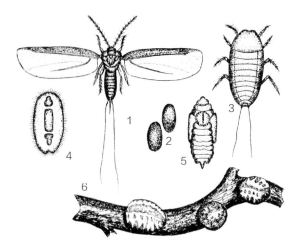

E. kuwanai. 1. Adult male, 2. Eggs, 3. Newly hatched nymph, 4. 2nd instar male nymph, 5, male pupa, 6. Adult female (Liu Qixiong)

1 year-old branches. The males and females differentiate in mid-April in Ningxia. Female nymphs molt into adults. Male nymphs molt into prepupae and pupae. Male adults emerge in early May. They live for 2 days. Females lay eggs in mid- and late May (similar to *Parthenolecaniun corni*). Smooth type females produce 3,241-9,000 eggs per female and their hatch rate is high. The wrinkled females lay 885-3,250 eggs per female and their hatch rate is low. Eggs hatch from late May to mid-June. Newly emerged nymphs migrate to leaves and young branches and suck sap. They migrate to young branches in October and overwinter at base of slender branches. Infestation period is mainly between mid-April and late May.

Prevention and treatment

1. Biological control. Conserve the following natural enemies: *Blastothrix serice* and *Chilocorus rubidus*. Avoid insecticide use when parasitization or predate rate reaches 70%.

2. During late May and June, spray emamectin benzoate + diesel or resin mixture can kill 100% of the population.

References Xiao Gangrou (ed.) 1992; Li Menglou 2002; Wang Fengying, Zhang Chuangling, Li Xuxuan 2007; Liu Yuping, Liu Guifeng, Su Hui et al. 2005; Wang Jianyi, Wu Sanan, Tang Ye et al. 2009.

(Li Menglou, Xue Xianqing, Xie Xiaoxi, Yan Aojin, Wang Ximeng, Wang Jianyi)

E. kuwanai (Li Zhenyu)

87 *Metaceronema japonica* (Maskell)

Classification Hemiptera, Coccidae

Importance *Metaceronema japonica* is a major pest of *Camellia oleifera*. It causes sooty mold growth and may lead to complete loss of tea reduction.

Distribution Zhejiang, Anhui, Jiangxi, Hunan, Sichuan, Guizhou, Yunnan, Taiwan; Japan.

Hosts *Camellia oleifera*, *C. sinensis*, *C. japonica*, and *Eurya japonica*.

Damage Infested branches and leaves turn black. Leaves fall prematurely and the whole plant may die from heavy infestation.

Morphological characteristics **Adults** Length 4-5 mm, width 2-3 mm. Females oval, ventral surface flat, dorsal surface convex, with two while spiral wax threads. Thoracic spiracles two pairs, setae around anal cleft long and thick. Anal plate thick, dorsal surface with two rows of thick and short conical setae, setae on the side of the body tubular or spine-like. Females produce white egg sacs which cover their body. Males orange yellow; legs, antennae, and copulatory organ dark brown. Antennae filiform, 10 segmented. There are three tubercles between the base of the antennae with the middle one being the largest. Wings greyish yellow, with one pair of white setae.

Eggs Length 0.35-0.37 mm, width 0.17-0.18 mm, light-yellow, elliptical. Egg sac length 3.0-6.5 mm.

Nymphs Newly emerged nymphs 0.4 mm long and 0.2 mm wide, light-yellow, upside down oval. Sides of the body notched between the anterior and posterior spiracles. There are two long anal setae. The 2nd instar females with two spiral wax threads on the top. Males secrete wax shells attached with white curly wax threads.

Pupae Length 1.4-1.9 mm, width 0.5-0.6 mm, orange yellow. Ocelli purple brown, Antennae extend to coxa of middle legs. Copulatory organ conical.

Biology and behavior *Metaceronema japonica* occurs one generation a year. It overwinters as mated female adults on branches, leaves or at base of tree trunks. They produce eggs in mid-April and reach peak production in early May. Nymphs emerge in mid-May and peak emergence period is in early June. Males and females differentiate in July. Male larvae pupate in early October. Adults emerge in late October and peak emergence occurs in early November.

Males mate multiple times. They live for 1 day. After mating, females continue to feed and migrate to branches. Each female lays about 800 eggs. Females live for about 6 months. Egg period is 30-35 days. Newly emerged nymphs are active. They aggregate on the underside of leaves and become sessile. Prepupal period is 4-6 days. Pupal period is 8-15 days.

Prevention and treatment

1. Biological control. *Chilocorus rufitarsu* is an important predator of all life stages of *Metaceronema japonica*. Conserve or release this predator is useful.

M. *japonica* female (Zhang Runzhi)

M. *japonica* egg sac and eggs (Zhang Runzhi)

Sooty mold resulted from *M. japonica* infestation (Zhang Runzhi)

Metaceronema japonica male shells (Zhang Runzhi)

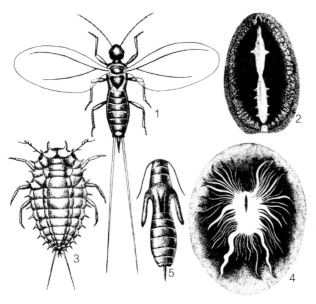
M. japonica. 1. Adult male, 2. Adult female, 3. Newly hatched nymph, 4. Male nymph covered with wax threads, 5. Male pupa (Zhang Peiyi)

Hirsutella sp. and *Aschersonia duplex* are two common entomopathogens.

2. Chemical control. Apply a band of a systematic insecticide (thiamethoxam, imidacloprid, etc.) on tree trunks during clear weather to control heavy infestations.

References Chen Hanlin 1980, 1990; Xiao Gangrou (ed.) 1992.

(Chen Hanlin, Chen Zhuan)

88 *Parasaissetia nigra* (Nietner)

Classification　Hemiptera, Coccidae
Common name　Nigra scale

Importance　Outbreaks of *Parasaissetia nigra* were reported on *Hevea brasiliensis* in Yunnan. It can significantly affect natural rubber production.

Distribution　Fujian, Guangdong, Hainan, Yunnan, Taiwan; Japan, India, Sri Lanka, Malaysia, Philippines, Israel, Egypt, Spain, Australia, U.S.A., Peru, Honduras, South Africa, Pakistan.

Hosts　There are more than 30 host species in China. They include: *Ficus* spp, *Hibiscus* spp., *Annona squamosal*, *Citrus reticulate*, *Coffea arabica*, *Gossypium* spp., *Croton tiglium*, *Psidium guajara*, *Hevea brasiliensis*, *Mangifera indica*, and *Chaenomeles sinensis*.

Damage　*Parasaissetia nigra* feed on phloem or leaves of host plants. They cause yellow leaves, premature leaf fall, fruit fall, or death of branches or the plant. They produce honeydew and lead to sooty mold growth. Infested *Hevea brasiliensis* show yellow and curled leaves, leaf fall, and dead branches. Light infestation will cause stunt growth and reduced rubber production. Heavy infestation may cause complete loss of rubber production or even death of the host tree.

Morphological characteristics　**Adults** Female length 2-5 mm long, elliptical, sometimes asymmetrical. Young adults yellow, some with brown or red marks. When laying eggs, they become shiny, dark brown to purple black. Dead females with a "H" mark on the back. Body with polygonal reticulate pattern sculpture. The edge has one row of wax plates. There are another 5 longitudinal rows of small wax plates. Nymphs flat and yellow.

Biology and behavior　*Parasaissetia nigra* reproduce parthenogenetically. Each female lays 400-1,000 eggs over 20 days period. Eggs are laid inside shells. As the egg numbers increase, the shells become more convex. The females become dry and shrink at the end of the egg laying. First instars wait for 1-2 days under the shells after hatching. They crawl out of the shells when weather conditions are suitable. They quickly spread to surround-

P. nigra female adults (Xie Yingping)

P. nigra female adult (Xie Yingping)

ing leaves and branches, especially the young branches. Their long-distance dispersal is assisted by wind. Once they find a suitable location, they attach their mouthparts to the host plant tissue and become sessile. They only move when feeding site conditions are no longer suitable. The 3rd instars are usually sessile and secrete large amount of honeydew. The honeydew droplets attract ants. There are three reproduction peaks per year. They are in March-April, June-July, and September-October, respectively. The generation during March-April is relatively homogeneous.

Prevention and treatment

1. Quarantine. Prevent movement of infested seedlings and shoots.

2. Silviculture. Maintain vigorous growth of host plants. Remove infested branches and weeds. Collect fallen leaves in winter and destroy them.

3. Apply *Pestalotiopsis* fungus. Conservation of natural enemies. The natural enemies are often able to suppress the pest population below economic injury levels. Parasitic wasps are especially important.

4. During outbreak, use fogging to control 1st instars in rubber production stands. Spray 12% thiamethoxam and lamda-cyhalothrin mixture in young and middle-aged *Hevea brasiliensis* stands.

References Lu Chuanchuan and Wu Huixiong 1991, Wang Ziqing 2001, Zhang Fangping, Fu Yueguan, Peng Zhengqiang et al. 2006.

(Xie Yingping)

89 *Parthenolecanium corni* (Bouché)

Classification Hemiptera, Coccidae
Common names European fruit lecanium scale, Brown elm scale, Fruit lecanium

Distribution Beijing, Hebei, Shanxi, Inner Mongolia, Liaoning, Jilin, Heilongjiang, Jiangsu, Zhejiang, Anhui, Shandong, Henan, Hunan, Hubei, Sichuan, Shaanxi, Gansu, Qinghai, Ningxia, Xinjiang; Western Europe, North Africa, Iran, Korea Peninsula, U.S.A., Canada, Russia.

Hosts More than 100 host plants are recorded. They include: *Acer saccharum, Fraxinus chinensis, F. bungeana, Ulmus pumila, U. densa, Robinia pseudoacacia, Lonicera maachii, Salix alba, Morus alba, Populus lasiocarpa, P. pseudo-simonii, P. alba* var. *pyramidalis, Firmiana platanifolia, Corylus heterophylla, Cassia surattensis, Juglans regia, Xanthoceras sorbifolium, Prunus persica, P. armeniaca, P. salicina, P. mume, Malus pumila, Pyrus bretschneideri, Malus asiatica, Crataegus pinnatifida, Zizyphus jujube, Amorpha fruticosa, Rubus corchorifolius, Albizzia julibrissin, Rosa rugosa, Vitis vinifera, Hibiscus syriacus, Glycine max, Gossypium* spp., *Helianthus annuus*.

Damage *Parthenolecanium corni* produces honeydew as it feeds. Sooty mold may grow on the honeydew, causing blackened areas on leaves and fruit. Heavy infestation causes slow growth and reduced fruit production.

Morphological characteristics **Adults** Female elliptical, domed shape. Length 3.5-6.5 mm, width 3.0-5.5 mm. Body hard, reddish brown. Males 1.2-1.5 mm long, wingspan 3.0-3.5 mm. Body reddish brown, head black.

Eggs Elongated elliptical, 0.2-0.5 mm long, 0.1-0.15mm wide; milky, yellowish brown before hatching.

Nymphs First instar flat, elliptical. Length 0.4-0.6 mm, width 0.25-0.3 mm; light-yellow, eyes black. Second

P. corni mated female and nymphs (Xu Gongtian)

P. corni gravid female (Xu Gongtian)

P. corni dead adult female (Xu Gongtian)

Various *P. corni* stages on a host (Xu Gongtian)

instar 0.8-1.0 mm, width 0.5-0.6 mm. The edges of the dorsal surface with 12 raised wax glands, which secrete radiating long wax threads.

Pupae Length 1.2-1.7 mm. Dark red, with forked copulatory organ at the end.

Biology and behavior *Parthenolecanium corni* occurs two generations a year on *Acer saccharum*, *Robinia pseudoacacia*, and *Vitis vinifera*. It occurs one generation a year on other hosts. In Southern China and Turufan region of Xinjiang, it occurs three generations a year. It overwinters as 2nd instars on young shoots and tree trunks.

In Henan, the 2nd instars become active in mid-March. Females lay eggs from late April to early May. Eggs are

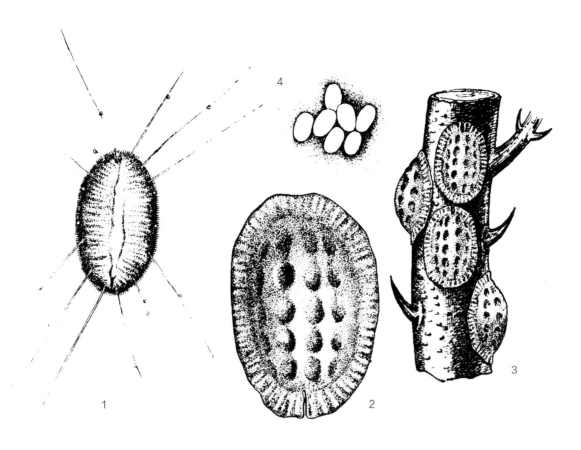

P. corni.
1. 2nd instar nymphs, 2. Adult female, 3. Damage, 4. Eggs (Zhang Peiyi)

laid beneath the female body. Each female produces 867-1,653 eggs (average 1,260 eggs). The eggs hatch after about 20 days. The 2nd instars start to overwinter in October. In Xinjiang, 2nd instars become active in late March and mid-April. Females emerge from late April to early May. Females lay eggs in mid- and late May. Eggs hatch in late May to mid-June. For the one generation per year population, the crawlers migrate to young branches in mid-August. Females lay eggs in late August. Eggs hatch in early and mid-September.

Parthenolecanium corni reproduce parthenogenetically in the central north and northeast. In Xinjiang, a small proportion reproduce sexually, but males only represent 3.5% of the adults. Their body size and egg production vary significantly based on host, season, and location on the host. Egg incubation period is approximately 30 days in April and May when mean monthly temperature is 18°C. The incubation period is approximately 20 days in mid- and late July when mean temperature is 30.5°C. Egg hatch rate is highest at 19.5-23.4°C and 41%-50% RH. The hatch rate at 25.4°C and < 38% RH is 89.3%.

Prevention and treatment

1. Quarantine. Prevent movement of infested seedlings and grafts. Treat infested seedlings before planting.

2. Silviculture. Plant mixed forests; select resistant varies or species; prune; apply lime on tree trunks; irrigate; maintain vigorous growth of plants.

3. Mechanical control. Use brushes to brush off female shells.

4. Biological control. Apply *Pestalotiopsis* fungus. Conserve the following natural enemies: *Chilocorus rubidus*, *C. kuwanae*, *Exochomus mongol*, parasitic wasps, lacewing, and some ants.

5. Chemical control. During spring before buds swell, apply 12% thiamethoxam and lamda-cyhalothrin mixture at 1,000× to trunks or soak the roots at 100×. Spray 45% profenofos and phoxim mixture at 800-1,000×.

References Zhang Zhiguang et al. 1958; Zhou Yajun 1965; Tong Chaoran, Li Xinzheng, Liu Anmin et al. 1985; Sun Dexiang et al. 1989; Xiao Gangrou (ed.) 1992.

(Wen Shouyi, Xu Longjiang, Yang Youqian)

90 *Cresococcus candidus* Wang

Classification Hemiptera, Lecanodiaspididae

Importance *Cresococcus candidus* affects growth and fruit production of *Castanea mollissima*.

Distribution Sichuan, Guizhou, Yunnan.

Hosts *Castanea mollissima*, *Cyclobalanpsis glauca*, *Quercus acutissima*, and *Malus* sp.

Damage *Cresococcus candidus* feed on current year and 2nd year-old branches and leaves. They feed on sap and cause death of branches.

Morphological characteristics **Adults** Females wide oval, hidden in fluffy wax, length 1.92-5.50 mm. Antennae 7 segmented, the 3rd segment the longest. Males 2.10-2.15 mm long, wingspan 2.15-2.25 mm. Body pink red. Antennae 10 segmented.

Eggs Elongated elliptical, length 0.5-0.6 mm, width 0.1-0.2 mm. Newly laid eggs light-yellow, turning to orange yellow later.

Nymphs First instar elliptical, length 0.5-0.7 mm, light-yellow. Antennae 6 segmented, the 3rd segment the longest. The 2nd instar wide oval, 2.7-4.8 mm long. Antennae and legs degenerated. Ocelli absent.

Biology and behavior *Cresococcus candidus* occurs one generation a year in central Yunnan. Females overwinter on branches. They start to lay eggs in mid-May and reach peak egg production in mid-June. The crawlers appear from mid-May to mid-August. The sessile 1st instars appear from mid-June to early September. The 2nd instars appear from late July to mid-November. Females appear from late August to mid-July next year. Male prepupae appear in late August. Male pupae appear in September. Male adults appear in October.

Cresococcus candidus reproduce sexually. Few may reproduce parthenogenetically. Newly hatched nymphs are the main spreading stage. Most plant damage occurs after 2nd instar.

Prevention and treatment

1. Cultural control. Remove infested branches in winter or before early April.

2. Apply *Pestalotiopsis* fungus. Conserve and utilize the following natural enemies: *Anicetus ohgushii*, *A. anulatus*, *Microterys ericeri*, *M. clauseni*, *Microterys* sp., *Tetrastichus ceroplastae*, *T.* sp., *Coccophagus hawaiiensis*, *C. silvestrii*, and *Prococcophagus* sp.

3. Chemical control. During June, spray 20% dinotefuran at 1,500×, 45% profenofos + phoxim and diesel (1:70) at 3,300×, or resin mix or 0.3 °Bé lime sulfur.

References Wang Ziqing 1982; Wang Hailin, Cao Kuiguang, Chen Gang et al. 1998; Tao Mei, Chen Guohua, Yang Benli et al. 2003; Huang Wanbin 2007.

(Li Qiao, Pan Yongzhi, Wang Hailin)

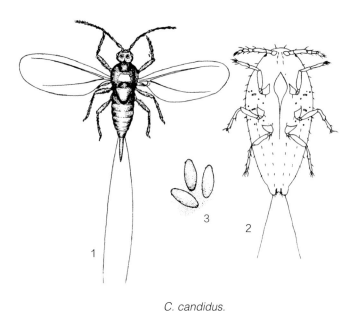

C. candidus.
1. Male adult, 2. Adult female, 3. Eggs
(1 by Hu Xingping; 2-3 by Chen Gang)

91 *Asterococcus muratae* (Kuwana)

Classification Hemiptera, Cerococcidae
Synonym *Cerococcus muratae* Kuwana

Distribution Jiangsu, Zhejiang, Sichuan, Guizhou.

Hosts *Magnolia officinalis* subsp. *biloba*, *Magnolia cylindrical*, *Schima superba*, *Viburnum odoratissimum*, *Vitis* spp., *Camellia sinensis*, *Citrus* spp., *Pyrus* spp., and *Eriobotrya japonica*.

Damage Infestations cause sooty mold growth such as *Capnodium theae*. Infested leaves and branches are covered with black color and the plant photosynthesis is affected.

Morphological characteristics **Adults** Female length 3.0 mm, width 2.5 mm, antennae degenerated, legless, the two thoracic spiracles obvious. Anal vales large, two anal setae long. Anal valves and anal tubes raised dorsally in most individuals. Shells hard; 4 mm long, 3 mm wide; yellowish brown; with 4 radiating white striae; two of them oriented forward in single lines, two of the located in the back as double lines; the radiating white lines converge at middle of the abdomen. The posterior end of the dorsal surface with a projection in the shape of a teapot spout. **Males** Length 1.4 mm, wingspan 1.2 mm. Body brownish yellow. Antennae 10 segmented, each with some thin hairs; the 1st and 2nd segments brown, flagellum yellow. Front wing transparent, with two veins. **Egg** Oval; length 0.35 mm, width 0.2 mm; brownish red.

Nymphs First instar 0.6 mm long, yellow. Antennae 6 segmented, the 3rd segment the longest. Legs developed. Anal valves large, with long setae that are longer than ½ of the body length.

Male pupae Length 1.1 mm, width 0.5 mm, yellowish brown. Appendages and segments obvious, the posterior end sharp.

Biology and behavior *Asterococcus muratae* occurs one generation a year. It overwinters as 2nd instars on branches. Prepupae appear in early April. They are covered with cotton shaped light-yellow secretion. Pupae occur in mid-April. They are enclosed in the secretion and the exuviae of the prepupae. Males emerge in late April. They mate and die soon. Females continue to feed in fixed locations after mating. They lay eggs inside the shells in late August. Each female lays 49-107 eggs (average about 60 eggs). Eggs hatch in late September. The newly hatch nymphs look for suitable places to feed. After a short period of feeding, they molt into 2nd instars. From this stage, they seldom move. Their preferred feeding sites are the tree trunks from bottom to top of the tree. Thick branches may also be infested. On *Magnolia officinalis*, the overwintering nymphs rest mostly on the scars left by the leaf petioles. The insects on tree trunks are similar to the lenticels and are easily missed.

Prevention and treatment

1. Quarantine. Prevent movement of infested seedlings.

2. Conserve natural enemies. The following are important for suppressing *Asterococcus muratae*: *Tetrastichus ceroplastae*, *Microterys postmarginis*, *M. zhaoi*, *Coccophagus yoshidae*, *Anysis saissetiae*, *Chilocorus kuwanae*, *C. hupehanus*, *C. rufitarsus*, *C. rubidus*, *Menochilus sexmaculata*, and *Propylaea japonica*. An outbreak of *Asterococcus muratae* is usually followed by the occurrence of large numbers of natural enemies which can prevent pest outbreaks for many years.

3. Chemical control. When chemical control is necessary, inject 12% thiamethoxam and lamda-cyhalothrin mixture during May-September to tree trunks (for large trees) or paint insecticide on the tree trunks (for small trees).

Reference Chen Hanlin 1995.

(Chen Hanlin)

A. muratae adult wax shell (Xu Gongtian)

92 *Neoasterodiaspis castaneae* (Russell)
Classification Hemiptera, Asterolecaniidae

Importance *Asterolecanium castaneae* affects plant growth and quality and quantity of fruit production of *Castanea mollissima*. It may kill host trees under heavy infestation.

Distribution Jiangsu, Zhejiang, Anhui, Jiangxi, Hubei.

Hosts *Castanea mollissima*.

Damage *N. castaneae* feed on branches and leaves, causing uneven leaves, leaf fall, cracked barks, dead branches or dead trees.

Morphological characteristics **Adults** Female shell almost round, diameter 1 mm, yellowish green or yellowish brown. Shape similar to turtle shell, with 3 longitudinal carinae and several shallow transverse grooves, the edges with pink read brush-like wax threads. Female body brown, 0.5-0.8 mm long.

Male shell elliptical, light-yellow, 1 mm long, 0.5-0.6 mm wide, with one longitudinal carina and several shallow transverse grooves, the edges with pink read brush-like wax threads. Male body length 0.8-0.9 mm, width 1.17-2.10 mm. Wings white, transparent, shinny. Head with one pairs of ocelli on the top and bottom. Antennae filiform, 10

N. castaneae infested branch (Jiang Ping)

segmented, each segment with a cluster of fine hairs. End of the abdomen with a long copulatory organ.

Eggs Milky at beginning, turning to pale red later, dark red before hatching; elliptical, length 0.2-0.3 mm, width 0.15-0.18 mm.

Nymphs White at beginning, turning to pale red later, reddish brown after become sessile; elliptical. Antennae, legs, and mouthparts developed; abdominal segments obvious; abdomen with one pair of long setae.

Pupae Conical, milky, length 0.8-0.9 mm, width 0.4-0.5 mm.

Biology and behavior *Neoasterodiaspis castaneae* occurs two generations a year in Zhejiang. It overwinters as fertilized females on current year branches. They resume activity in late March when temperature reached to 10°C. The 1st generation nymphs hatch in early and mid-May. The 2nd generation nymphs hatch in mid-July.

Prevention and treatment

1. Quarantine. Prevent movement of infested seedlings and grafts.

2. Silviculture. Prune in winter to remove infested branches and weak branches.

3. Chemical control. Spray 40% acetamiprid and chlorpyrifos mixture at 800× or 12% thiamethoxam and lamda-cyhalothrin mixture at 1,500× during peak of the nymph hatching period.

References Yang Yongzhong et al. 1999, Xu Zhihong and Jiang Ping 2001, Liu Yongsheng 2002.

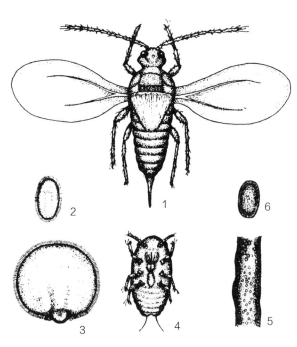

N. castaneae. 1. Male adult, 2. Male shell, 3. Female shell, 4. 1st instar nymph, 5. Damage, 6. Egg (Hu Xingping)

(Jiang Ping, Yuan Changjing)

93 *Bambusaspis hemisphaerica* (Kuwana)

Classification Hemiptera, Asterolecaniidae
Synonym *Asterolecamium hemisphaerica* Kuwana
Common name Bamboo hemisphere scale

Distribution Jiangsu, Zhejiang, Anhui, Jiangxi, Shandong, Guangdong, Shaanxi.

Hosts *Phyllostachys pubescens*, *P. nigra*, *P. praecox*, *P. uiridis*, *P. glauca*, *P. heteroclada*, and *Indocalamus tessellatus*.

Damage *Bambusaspis hemisphaerica* causes stunt growth of young branches, shortened internodes, premature leaf fall, dead branches, and growth of new shoots.

Morphological characteristics **Adults** Female shell 2.5-3.0 mm long, 1.5-2.0 mm wide, domed shape, the front end round, the posterior narrower. The wax shells cover ½ to ¼ of the diameter of the small branches. Shells yellow, smooth, transparent, shinny; the wax threads at the edge white and in broken pieces.

Female body length 2.3-2.7 mm, width 1.4-1.8 mm, semi-globular, the front round, the posterior narrower. Antennae with one segment, tubercle shaped, with two long setae and one short seta. There are 10-20 quinquelocular pores between the antennae and the sides of the body. They are arranged as a band with the width of 3-4 pores. Beside the mouthparts are 10-15 "8"-shaped pores. Spiracles large, round, the openings located in a deep funnel shaped depression. There are quinquelocular pores at the depression and a band of > 100 quinquelocular pores around each spiracle. The "8" shaped pores arranged in two rows along the sides of the body and converge into one row at 1/5-1/4 of the posterior end of the body. There are 10-30 "8" shaped pores at the end of the body. The discoidal pores have similar numbers as the "8" shaped pores on the back and arranged in a row. The row of discoidal pores ends at the anal setae. Multilocular pores with 6-10 pores each. There are 70-100 multilocular pores. They form two complete transverse rows and 6 broken transverse rows on ventral surface of abdomen. Anal ring with 6 setae, surpassing the anal tube. Opening of the anal tube widened. Duct pores scattered on dorsal

B. hemisphaerica adults (Li Zhenyu)

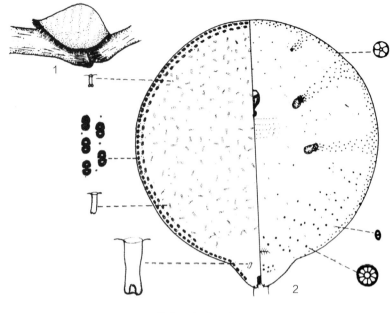

B. hemisphaerica
1. Female wax shell, 2. Female body (dorsal and ventral view) (Hu Xingping)

surface of the body except for the end of the body. There are many small "8" shaped pores and discoidal pores on the dorsal surface and a pair of ducts near the end of the abdomen.

Male length 1 mm, shell elliptical, parallel on the two sides, with sparse wax threads on the side, pale reddish brown, eyes red. Antennae moniliform, 10 segmented. Forewings white, transparent, with two longitudinal veins. Abdomen yellow, slender, copulatory organ needle-like, end of the abdomen with two white wax threads.

Eggs Length 0.4 mm, width 0.2 mm, elliptical, light-yellow.

Nymphs Length 0.40-0.45 mm, width 0.20-0.25 mm, elliptical, pale reddish brown. Antennae 6 segmented. Legs developed. The "8" shaped pores on the sides are obvious. Anal ring well developed, with 6 setae. There are two anal setae at the end of the body.

Biology and behavior The insect occurs one generation a year and occasionally two generations a year in Jiangsu. It occurs two generations a year in Anhui. The fertilized females and 2nd instars overwinter on branches. They resume feeding in February. Females start laying eggs in mid-May. Each female lays an average of 400 eggs (range: 107-553). Egg period is 1-2 days. The first-generation eggs hatch from early May to mid-June. The hatching time during a day is usually between 10:00-15:00. Newly hatched nymphs are active. They crawl at a speed of 3-4 cm per minute and are active for 36-48 hours. They fix to small branches, nodes, buds, and base of petioles. After 3-4 days, they secrete wax power, then forming wax shells. After about 17 days, males and females are differentiated. Male shells elongated elliptical; the sides parallel, wax threads sparse on the sides; they are often on base of petioles of current year leaves. Female nymphs semi-globular, the side with brush-like wax threads; they are mostly found on current year small branches or between nodes.

The 1st generation male larvae pupate in late June. Pupal period is 3-4 days. The males emerge in early July and end in mid-July. About 85% of the pupae hatch into adults. Most females stop development after mating. They resume feeding and lay eggs next spring. A small portion will develop and become pregnant in August. They lay eggs in early and mid-September. The nymphs develop into 2nd instars and overwinter on young branches in late October and early November.

Prevention and treatment Same as *Nesticoccus sinensi*.

References Wu Shijun 1983, Xiao Gangrou (ed.) 1992.

(Yan Aojin)

Aulacaspis rosarum (Borchsenius)

Classification Hemiptera, Diaspididae
Common name Asiatic rose scale

Distribution Jiangsu, Zhejiang, Fujian, Jiangxi, Guangxi, Sichuan, Yunnan.

Hosts Common hosts are: *Rosa multiflora*, *Rosa chinensis*, *R. rugosa*, *R. laevigata*, *Murraya paniculata*, *Cinnamomum camphora*, *Cycas revolute*, and *Sapium sebiferum*.

Damage *Aulacaspis rosarum* infest branches, produce whitish cottony covers, reduce plant growth, and may kill the hosts after continuous infestations.

Morphological characteristics **Shell** Female shell white or greyish white, nearly round. There are two nymph exuviae on the shell; the first one is near the margin, overlapping with the 2nd one, dark brown; the 2nd exuvia is near the center of the shell, blackish brown, shell diameter 2-2.4 mm. Male shell elongated, with two ridges; white; with one exuvia located near the end of the shell; shell length 0.8-1.0 mm, width 0.3 mm.

Adults Female 1.1-1.3 mm long, 0.7-0.9 mm wide, the front end enlarged, the projection on the head obvious. Antennae tubercle shaped, with a long seta. Anterior spiracle with 16-24 pores; posterior spiracle with 8-12 pores. The dorsal surface of the 2nd-6th abdominal segments with groups of duct pores near the center. The dorsal surface of the 3rd-5th abdominal segments with duct pores near the margin. Anal plate margin with 4 groups of pores on each side, they are arranged in 1, 2, 2, 1-2 pattern. Anal plate cerarian setae with 5 groups on each side, they are arranged in 1, 1, 1, 1, 4-7 pattern. Anal lobes with 3 pairs; the median lobes depressed, the basal area congruent; the other pairs of anal lobes divided into two parts each. There are 5 groups of pores around the reproductive opening.

Eggs Purple red, elongated elliptical, length 0.16 mm, width 0.05 mm.

Nymphs Orange red, elliptical. Antennae 5 segmented, the last segment the longest. A pair of long setae at the end of the abdomen.

Biology and behavior *Aulacaspis rosarum* occurs 2-3 generations a year. It overwinters as 2nd instars or rarely as female adults. Male adults emerge from late March to early April. Female adults emerge in mid-March. The 1st generation peak egg production occurs in mid- and late April. Each female lays 57-189 eggs (average: 132) under the shells. Eggs hatch from late April to mid-May. Females emerge in late July. The 2nd generation eggs appear in early and mid-August. Nymphs hatch in mid- and late August. The 3rd generation females emerge in early October. Some females produce eggs and molt into 2nd instars. There are overlapping generations.

Newly hatched nymphs crawl out of the shells after

A. *rosarum* female shell (Xu Gongtian)

A. *rosarum* male shell (Xu Gongtian)

A. rosarum female shell (Xu Gongtian)

A. rosarum adults and host (Xu Gongtian)

A. rosarum male shell (Xu Gongtian)

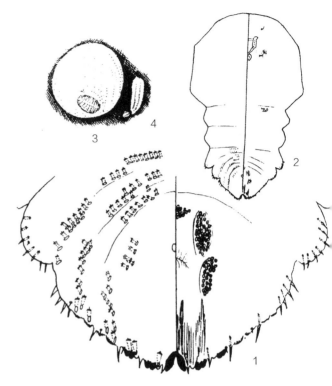

A. rosarum. 1. Adult female pygidium plate, 2. Adult female, 3. Female shell. 4. Male shell (Hu Xingping)

1-2 days. Female nymphs are active. They spread away from the mother. The male nymphs are not as active as the female nymphs. They often stay closer to the mother. Nymphs fix themselves to plant tissues after locating a suitable site. After 3-4 days, they secrete wax. Males and females differentiate during the 2nd instar. Female nymphs produce round wax shells and molt into female adults. Male larvae produce long wax shells. They molt into adults after two pupal periods. Their emergence often occurs during 15:00-17:00. After emergence, they fly to females and mate. They live for less than 24 hours. At about 15 days after fertilization, females produce eggs. Their body shrinks as they lay eggs and eventually die inside the shell. *Aulacaspis rosarum* prefers dark and moist environment. Lower canopy has higher population density, northwestern side of the canopy usually has higher density than southeastern side of the canopy.

Prevention and treatment

Conserve natural enemies. Natural enemies include *Chilocorus chalybeatus*, *Scymnus hoffmanni*, *Delphastus catalinae* earwigs, and two parasitic wasps.

(Xu Peizhen)

95 *Aulacaspis sassafris* Chen, Wu & Su
Classification   Hemiptera, Diaspididae

Importance   *Aulacaspis sassafris* infestations cause death of branches and slowed growth of host trees.

Distribution   Anhui, Jiangxi, Hunan.

Hosts   *Sassafras tzumu* and *Litsea auriculata*.

Damage   *Aulacaspis sassafris* infests young branches and leaves. The branches die from water loss. Infested leaves lose green color, curl, shrink, and fall prematurely.

Morphological characteristics   **Shell** Female shell white, nearly round, slightly raised on the top, diameter 1.20-2.55 mm; there are two yellow or yellowish brown nymph exuviae on the shell, located near the margin or center of the shell. Male shell white, elongated elliptical with the two sides parallel; there is one yellowish brown exuvia located in the front of the shell; three longitudinal ridges on the top, length 0.93-1.09 mm.

Adults Female purple red, length 1.23-1.51 mm. Male spindle shaped, orange red, length 0.57-0.69 mm, wingspan 1.23-1.32 mm.

Biology and behavior   *Aulacaspis sassafris* occurs 3 generations a year. It overwinters as 2nd instars and male pupae on young branches. Female nymphs develop into adults in late March. Male adults emerge in early April. Males are weak flyers. They mate soon after emergence. Females lay eggs under the shell after mating. Their ventral surface raised to allow space for eggs. Peak egg production is in mid-May. Eggs hatch from late May and reach peak in early June. The 2nd generation eggs appear in late June. They hatch in mid- and late July. The

A. sassafris male and female shells (Yang Chuncai, Tang Yanping)

3rd generation eggs appear in late August and reach peak in mid-September. They hatch in mid-October. Nymphs overwinter in late November. They are gregarious. Newly hatched nymphs are active, some feed on young branches or stems, some feed on underside of leaves. They secrete wax and form shells. The overwintering, 1st, and 2nd generation produce 110-210, 30-40, and 15-50 eggs per female, respectively.

Prevention and treatment

1. Silviculture. Maintain health of the trees, avoid dense canopy.

2. Chemical control. Spray 15% imidacloprid at 800-1,000× when nymphs emerge. Re-apply after 7 days.

Reference   Xiao Gangrou (ed.) 1992.

(Yang Chuncai, Tan Yanping)

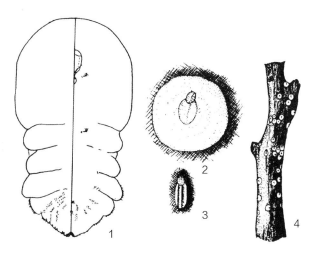

A. sassafris. 1. Female adult, 2. Female shell, 3. Male shell, 4. Damage (Hu Xingping)

Hemiberlesia pitysophila Takagi

Classification Hemiptera, Diaspididae
Common name Pine needle hemiberlesian scale

Importance *Hemiberlesia pitysophila* is a dangerous introduced pest. It was found in Taiwan in the 1950's and in Guangdong in 1982. The pest spreads fast and can kill large areas of *Pinus massoniana* trees, destruct the ecosystem and pose enormous economic loss.

Distribution Fujian, Jiangxi, Guangdong, Guangxi, Hong Kong, Macau, Taiwan; Japan.

Hosts Common hosts are: *Pinus massoniana*, *P. thunbergii*, *P. elliottii*, *P. taeda*, *P. latteri*, *P. caribaea* var. *caribaea*, *P. caribaea* var. *hondurensis*, *P. caribaea*, *P. latteri*, *P. luchuensis*, *P. glabra*, *P. echinata*, *P. kesiya*, *P. rigida* var. *serotina*, *P. patula*, *P. clausa* var. *immuginata*, and *P. oocarpa*.

Damage *Hemiberlesia pitysophila* feed inside the leaf sheath of old leaves. Damaged leaves become brown, black, curled, or leaves fall prematurely. Tree growth is slowed, and trees eventually die after consecutive infestations.

Morphological characteristics **Adults** Female adults near round before pregnant. The shell has three rings and become thickened after the female adult becomes pregnant, pear shaped. Male adult shell is same as the late 2nd instar larvae shell, elongated oval, pale brown, 1.10 mm long, 0.50 mm wide. Male adults light-yellow, 0.8 mm long, 0.22 mm wide. Basal two segments of antennae light-yellow. The other segments blackish brown. Hindwing reduced to halteres. The end of the abdomen with a long copulatory organ.

Nymphs Second instar shell round, the center has orange yellow exuvia, 0.28 by 0.35 mm size. Female shells grow larger and with ringed striae, round, white. Male shells become elongated elliptical, head end slightly raised, pale brown.

Biology and behavior *Hemiberlesia pitysophila* occurs 5 generations a year in Guangdong, 4 generations a year in Fujian. The generations overlap. It can live on all species of the genus *Pinus*, with *P. massoniana* being the preferred host. Females have 3 stages and males have

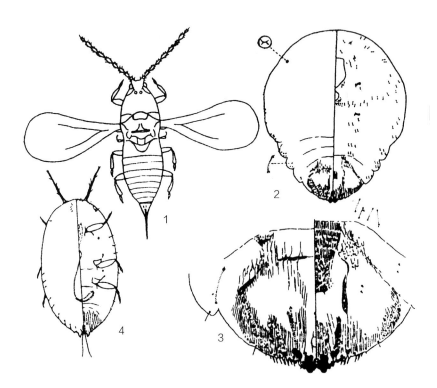

H. pitysophila 1. Adult male, 2. Adult female, 3. Female pygidium plate, 4, Newly hatched crawler (Pan Wuyao)

5 stages. The two additional stages in males are prepupae and pupa. Nymphs are polymorphic. The 1st instars and 2nd instars are distinctly different. The 1st instars are crawlers. The 2nd instars are similar to female adults. There is not an obvious overwintering period. The natural mortality of 1st instars is high.

Eggs are ready to hatch within female adults before being laid. They hatch immediately after being laid. Nymphs look for a suitable location within 10 cm from the female adult after hatching. They insert their proboscis into needles and become sessile. After 5-19 hours of feeding, they secrete wax. After 1-2 days, the wax forms a round shell. The 2nd instars differentiate during middle of the development stage. A portion of the population will lengthen the shells, turn into male prepupae, and then pupae. The rest of the population will have their legs degenerated, proboscis well-developed, and molt into adult females. After mating, females lay eggs over a 33-80 days period. One female lays 1-2 eggs per day and lays approximately 60 eggs during its life time. The 1st and 5th generations are most reproductive. Male pupal stage is 3-5 days. After emergence, they rest in the shells for 1-3 days. One male can mate multiple times. They die a few days after mating. Those feed at base of leaf sheath are usually females. Those feed on top of leaves and cones are usually males.

The highest population density occurs between March-June. It is affected by temperature, relative humidity, wind, and precipitation. Temperature is the dominant factor affecting population densities. High temperature during July-August and low temperature during December-January suppress the population growth. Higher than 23℃ and lower than 18℃ will increase the insect mortality. High canopy coverage will favor the development of *Hemiberlesia pitysophila*.

Prevention and treatment

1. Silviculture. Pruning can improve tree growth and natural defense to *Hemiberlesia pitysophila*.

Trees killed by *H. pitysophila* (Wang Jijian)

H. pitysophila infested branch (edited by Zhang Runzhi)

H. pitysophila infested branches (edited by Zhang Runzhi)

H. pitysophila on needles (Wang Jijian)

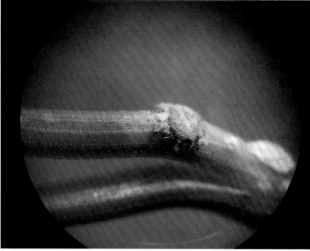

H. pitysophila infected by fungus (Wang Jijian)

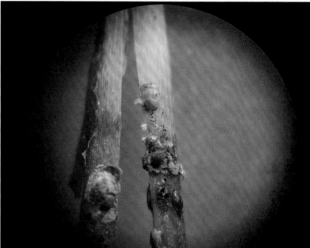

H. pitysophila killed by fungus (Wang Jijian)

2. Biological control. Release *Coccobius azumai* to control *Hemiberlesia pitysophila*. Isolation, propagation, and release of *Pestalotiopsis* sp. can reduce population by > 90%. This is the most easily adopted, effective, and simple biological control method.

3. Chemical control. Spray resin and oil mixture, 12% thiamethoxam and cyhalothrin mixture at 1,000×, 40% acetamiprid and chlorpyrifos mixture at 1,000×, or 12% thiamethoxam and chlorpyrifos mixture at 500× can reduce population by > 90%.

References Xie Guolin, Hu Jinlin, Li Qugan et al. 1984, 1997; Zhou Yao 1985; Pan Wuyao, Tang Ziying, Lian Junhe et al. 1987; Pan Wuyao, Tang Ziying, Xie Guolin et al. 1987, 1993; Tong Guojian, Tang Ziying, Pan Wuyao et al. 1988; Chen Zhiqing 1989; Pan Wuyao, Tang Ziying, Chen Zefan et al. 1989; Wu Jianfen 1990; Xiao Gangrou (ed.) 1992; Chen Yongge and Gu Dexiang 1995, 1997; Zhang Xinyue and Luo Youqing 2003; Chen Shunli, Wu Fuhua, and Hou Qinwen 2004; Wang Zhuhong, Huang Jian, Liang Zhisheng et al. 2004; Huang Zhenyu 2006.

(Wen Xiujun, Huang Shaobin)

97 *Lepidosaphes salicina* Borchsenius

Classification Hemiptera, Diaspididae
Common name Willow oyster scale

Importance *Lepidosaphes salicina* infestations can kill young trees within 3-5 years. Large areas of young trees may die from *Lepidosaphes salicina*.

Distribution North of Yellow River; Northeastern Asia, Far East of Russia, Korea Peninsula, Japan, Mongolia.

Hosts *Populus* spp., *Salix* spp., *Juglans regia*, *Fraxinus chinensis*, *Euonymus alatus*, *Syringa* spp., *Betula* spp., *Ulmus pumila*, and *Rosa* spp.

Damage *Lepidosaphes salicina* feed on branches and tree trunks, causing abnormal growth and death of branches and trees.

Morphological characteristics **Shell** Female length 3.2-4.3 mm, oyster shaped, straight or curved, brown, margin greyish white, covered with grey wax powder. The front sharp, widened in the back, the back convex. Its posterior end rough, with transverse scale-like markings. There are two pale brown markings in the front. Ventral surface of the shell complete, flat and yellowish white; the end is in "∧" shaped. The 1st exuvia elliptical, 6 mm long, its end covers the front of the 2nd exuvia. The 2nd exuvia elliptical, 1 mm long. Male shell similar to female shell, smaller, with one pale brown marking.

Adults Female yellowish white, spindle shaped, narrow in the front and wider in the back. Slide specimen measurements are as the following: length 1.45-1.80 mm, width 0.68-0.88 mm. The 2nd to 4th abdominal segments with lobes on the sides; the 1st-4th abdominal segments with one sharp tooth on each side. The dorsal and lateral margin of the 1st, 2nd-3rd, and 4th abdominal segments with 11-21, 4-6, and 3-4 conical setae around pores, respectively. Anal plate wide round at the end. Anal lobes two pairs, the median lobe large, the distance between lobes shorter than the half width of the lobe, the two transverse margins often with fine teeth. The 2nd pair of anal lobes distinctively smaller and split into

L. salicina (Li Zhenyu)

L. salicina on a branch (Xu Gongtian)

L. salicina dorsal view (Xu Gongtian)

L. salicina female with eggs (Xu Gongtian)

two sub-lobes with the inner one larger than the outer one. The anal plate with 9 pairs of setae around pores. The pair of setae on the 6th abdominal segment are two times long as those on the median anal plate. The 6 pairs of duct pores on anal plate are arranged in 1, 2, 2, 1 pattern. Antennae thick and short, the end serrated, with two long setae. Anterior spiracle with 6-17 pores, posterior spiracle with 2-3 conical setae and arranged transversely. There are many pores on the dorsal surface. The 7th abdominal segment with 4-10 pores on each side or invisible on one side; they are arranged in a band. The 6th abdominal segment with 17-23 pores on each side, forming a longitudinal band parallel with those on the 7th segment. The conical setae on the 1st-2nd abdominal segment are in groups. The 1st-4th abdominal segments usually without dorsal pores at the median area. The pores at dorsal lateral margin and those dorsal pores on the 6th-7th abdominal segments are larger than those below the dorsal lateral margin. There are 5 groups of pores around the reproductive opening.

Males yellowish white, 1 mm long. Head small, eyes black, antennae moniliform and 10 segmented, light-yellow. Mesothorax yellowish brown, scutum pentagon shaped. Wings transparent, 0.7 mm long. Abdomen narrow, copulatory organ 0.3 mm long.

Eggs Length 0.25 mm, elliptical, yellowish white.

Nymphs First instar elliptical, flat, one pair of ocelli on the side. Mouthparts well developed. Antennae 6 segmented, scape short, the end of the flagellum slender and with transverse striae and long setae. The legs are thick and large. Anal plate with two pairs of anal lobes, the median lobes small, the lateral lobes large. The 2nd instars spindle shaped. Abdominal 4th-7th segments with one pore on the lateral margin of each segment. Those pores located dorsally or ventrally are smaller. The anal plate near the lateral and posterior of anus and the two abdominal segments prior to it each has one duct pore. Anal lobes similar to those of adults. Males are usually narrower than females.

Male pupae Length 1 mm, yellowish white, mouthparts invisible.

Biology and behavior *Lepidosaphes salicina* occurs one generation a year in Hebei and Liaoning. It overwinters as eggs in female shells. Eggs hatch in mid-May and peak hatching period is in early June. Egg hatch rate is often 100%. Newly hatched nymphs crawl to trunks, branches and find a suitable location within 1-2 days. They secrete white wax threads which cover their body. Second instars appear in mid-June. The nymphal period is 30-40 days. Female nymphs molt into adults in early July. Male larvae molt into prepupae, then to pupae after 7-10 days. Male adults emerge in early July. Emergence rate is 90%. Male adults crawl on tree trunks looking for females. They usually mate during early evenings. They can mate multiple times. Female: male ratio is 7.3:1. Females start to lay eggs in early August. Females secrete wax threads and form dorsal and ventral wax membrane within the shells. Eggs are laid within the membrane. Female body shrinks as eggs are laid behind the female body. Each female lays 77-137 eggs and dies after finishing egg production. Egg period is 290-300 days. Survival rate of eggs is > 98%.

Population density of *Lepidosaphes salicina* is closely related to environmental conditions. Pure forests suffer more infestation than mixed forests. Poplar trees are more susceptible than other species. Different varieties of poplar trees also vary in their susceptibility: upper portion of trunks are more susceptible than lower part of the trunks, branches are more susceptible than trunks, north side of the slopes has higher pest density than the south side of the slopes. Heavy rain during late May and early June will significantly reduce pest populations.

Prevention and treatment

1. Conserve natural enemies. Common natural enemies include: *Aphytis proclia*, *Exochomus mongol*, and *Propylaea quatuordecimpunctata*.

2. Chemical control. Spray tree trunks or branches with silicone at 50×, 12% thiamethoxam and lamda-cyhalothrin mixture at 1,000× in mid- and late May.

References Cui Wei and Gao Baojia 1995, Xu Zhihua 2006.

(Qiao Xiurong, Xu Gongtian)

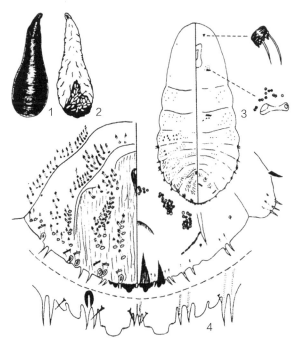

L. salicina. 1-2. Female dorsal and ventral view, 3. Female, 4. Female pygidium plate (Xu Gongtian)

98 *Lepidosaphes conchiformis* (Gmelin)

Classification Hemiptera, Diaspididae
Synonyms *Mytilaspis conchiformis* (Gmelin), *Lepidosaphes conchiformioides* Borchsenius
Common names Fig scalem, Pear oystershell scale

Distribution Hebei, Inner Mongolia, Liaoning, Jilin, Shanghai, Zhejiang, Anhui, Fujian, Shandong, Henan, Hubei, Sichuan, Gansu, Qinghai, Ningxia, Xinjiang; Holarctic, Europe.

Hosts *Prunus mume*, *Malus pumila*, *Prunus avium*, *Pyrus pyrifolia*, *Rosa chinensis*, *Syringa obtata*, and *Elaeagnus angustifolia*.

Damage *Lepidosaphes conchiformis* feed on branches and leaves, causing death of branches and trees.

Morphological characteristics **Shell** Female shell elongated pear shaped, straight or curvy, brown, 1.8-2.5 mm long. Male slender, 1.0 mm long, similar color and texture as the female shell.

Adults Female morphology varies with the host, location on the host, and season. Elongated pear shaped, the posterior half wider. Posterior end of the anal plate pointed. There are no tubercles on the lateral side between abdominal segments. Anterior spiracle with 2-3 pores. Anal lobes two pairs, the 2nd pair has two lobes. Each side of the anal plate has 5 large pores. The dorsal pores on 2nd-6th abdominal segments are off the median and along the border of each segment. Those on mesothorax to 5th abdominal segments are in groups. There are 5 groups of pores surrounding the reproductive opening. Winter type with large median anal lobe; the distance between anal lobes is narrow; the ventral surface of the anal lobe without a sclerotized club; the 2nd anal lobe small and conical, with many dorsal pores; the 2nd-4th abdominal segments with dorsal pores at median region. Summer type with small median anal lobe; the distance between anal lobes is equal to the width of the anal lobe; the ventral surface of the anal lobe with two sclerotized clubs; the 2nd anal

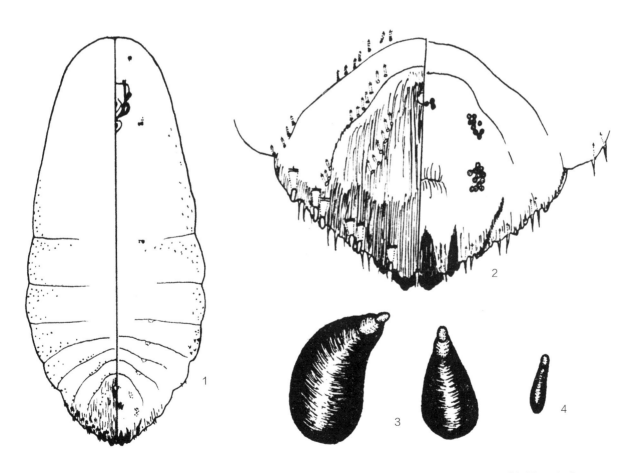

L. conchiformis. 1. Adult female, 2. Female pygidium plate, 3. Female scale, 4. Male scale (Hu Xingping)

lob large and pillar shaped, without dorsal pores; the median region of the abdominal segments without dorsal pores. Some individuals have characteristics between the two types.

Eggs Elliptical, pale purple, slightly transparent, length 0.3 mm.

Nymphs The 1st instars flat, elongated elliptical, dorsal surface with purple markings, pale purple, 0.4 mm long. Mouthparts developed, proboscis long. The end of the abdomen with one pair of anal lobes. The 2nd instar female nymphs elongated spindle shaped, pale purple, antennae and legs degenerated, mouthparts developed, with two pairs of anal lobes. They molt into female adults. The 2nd instar male larvae elongated, pale purple.

Male pupae Elongated, 0.8 mm long, pale purple; with distinct head, thorax, abdomen, antennae, wing buds, legs, and copulatory organ. The copulatory organ needle shaped.

Biology and behavior *Lepidosaphes conchiformis* occurs 2 generations a year. It overwinters as fertilized females on branches or tree trunks. They produce eggs in mid-April. Eggs are laid behind the female body and inside the shell. Each female produces an average of 17.8 eggs. Peak egg production of the 1st and 2nd generation is during early to mid-May and mid- to late July, respectively. The egg hatching period of the 1st and 2nd generation is during late May – early June and late July – early August, respectively. The development periods of the two generations are heterogeneous. Newly hatched nymphs are very active, they crawl upward and fix to smooth and young branches at 2-3 days after hatching. They prefer the base of the branches and northeastern side of the branches. The 1st instars occasionally fix themselves to leaves. After 10 days of sessile stage, they form wax shells which can persist for years. Female nymphs molt into adults after the 2nd instar. The 2nd instar male larvae will molt into prepupae, pupae, then emerge into adults. Male adults mate the same day after emergence. They die soon after mating. Male adults crawl fast but are not seen in flight.

Occurrence of *L. conchiformis* is closely related to environmental conditions. Observations in Yinchuan of Ningxia found infestation rate of *Elaeagnus angustifolia* by *L. conchiformis* was 15% and population density of 5 insects per cm^2 on trees grown on rich and well irrigated soil. However, the infestation rate was 100% and population density was 25 insects per cm^2 on *Elaeagnus angustifolia* grown on poor soil. Mixed forests have lower pest population densities. On each tree, the northeastern side usually has higher *L. conchiformis* density. This is due to the more suitable temperature and relative humidity at this side of the host. The upper part of a tree usually has higher pest density than the middle and lower part of a tree. This is due to the upper part of the tree has smoother, younger branches.

Prevention and treatment

1. Conserve natural enemies. The following are important natural enemies: *Chilocorus kuwanae*, *Adalia bipunctata*, and *Oenopia conglobate*.

References Xiao Gangrou (ed.) 1992; Wang Jianyi, Wu Sanan, Tang Hua et al. 2009.

(Chen Hui, Wang Jianyi)

99 *Prodiaspis tamaricicola* (Malenotti)
Classification Hemiptera, Diaspididae

Distribution Ningxia etc.

Host *Tamarix chinensis*.

Damage *Prodiaspis tamaricicola* feed on branches, causing them turn brown and death of branches and top of the crown.

Morphological characteristics **Shell** Female oval, white, mark on shell orange yellow, off the center.

Adults Female orange yellow, oval, length 0.5-0.6 mm, with clusters of duct pores. Antennae tubercle shaped, with 3 projections. Anal plate is absent. Anal opening situated on the dorsal surface of the 8th abdominal segment. Reproductive opening with a sac valve structure, and without pores surrounding the opening.

Nymphs Pale greyish yellow, elliptical, 0.3 mm long, with groups of duct pores. Antennae 6 segmented. The 2nd instar 0.5-0.6 mm long.

Biology and behavior *Prodiaspis tamaricicola* occurs 2 generations a year. It overwinters as female adults on branches. They produce nymphs from late May to late June. The male nymphs pupate in early June. Adult males emerge from late June to early July. Second generation nymphs appear from mid-July to early September. Male pupae appear in mid-August. Male adults emerge from late August to early September. Males die same day after mating. Females produce nymphs under the shells. The female body shrinks as it produces nymphs. Those nymphs produced first will form shells under the female adults. Those produced later will look for new places to feed. They prefer current year branches.

Prevention and treatment

1. Conserve natural enemies. Avoid using insecticides when natural enemy parasitization rate reaches 50%.

2. When infestation level is high, spray 12% thiamethoxam and lamda-cyhalothrin mixture to control large nymphs, 40% acetamiprid and chlorpyrifos mixture for small nymphs.

References Xiao Gangrou (ed.) 1992; Sun Dexiang, Wang Jianyi, Wu Guangrong et al. 1993.

(Li Menglou, Wang Jianyi)

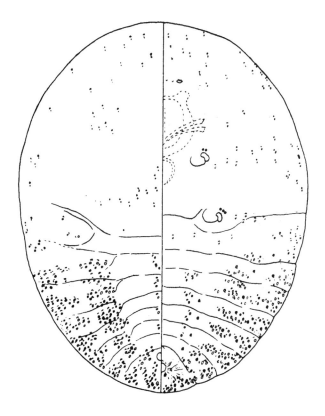

P. tamaricicola female adult (Yang Pinglan)

100 *Pseudaulacaspis pentagona* (Targioni-Tozzetti)

Classification	Hemiptera, Diaspididae
Synonyms	*Diaspis pentagona* Targioni-Tozzetti, *Aulacaspis pentagona* Cockerell, *Sasakiaspis pentagona* Kuwana
Common names	White peach scale, Papaya scale, White mulberry scale, Dadap shield scale

Distribution Hebei, Shanxi, Liaoning, Shandong, Henan, Shaanxi, Gansu.

Hosts *Amygdalus persica, Morus alba, Prunus salicina, P. subhirtella, Armeniaca vulgaris, Sophora japonica, Ailanthus altissima, Hibiscus syriacus, Platanus hispanica, Buxus microphylla, Populus* spp., and *Salix* spp.

Damage *Pseudaulacaspis pentagona* female adults and nymphs feed on branches and occasionally on fruits and leaves. When their population is high, the shells will cover the branches looking like a layer of cotton. Infested trees will have yellow leaves, dead branches, and fallen flowers and leaves.

Morphological characteristics **Adults** Female shell white, yellowish white or greyish white, round or elliptical, dorsal surface raised, the middle with a orange yellow spot, diameter 1.5-2.8 mm. Female body light-yellow or orange yellow, pear shaped, 1.4 mm long. There are five pairs of anal lobes. Median lobe and lateral-inner lobe developed. Outer lobe degenerated. The 3rd-5th lobe conical, median lobe raised as a triangle, not emarginated, its inner and outer margin with 2-3 emarginations, the base connected. The dorsal pores on the 2nd-5th abdominal segments form a line off the median and a line near the margin. The pores on the 6th abdominal segment absent or rare. The 1st abdominal segment with a spot along each side. Anus is near the middle of the anal plate. Base of anal plate with an elongated spot at each side. There are 5 groups of pores around the reproductive opening.

Male adults with wings. Shells white, similar to elliptical cocoons, the front end with a yellow spot, dorsal surface with 3 ridges, 1.2 mm long. Body orange red, head slightly tapered, with a pair of eyes. Front wings membranous, with two veins. Body light-yellow, relatively long, 0.43 mm in length.

Eggs Length 0.25-0.30 mm, elliptical, pale pink red at beginning, soon become light-yellowish brown, apricot red before hatching.

Nymphs Newly hatched nymphs light-yellowish brown, flat and oval. After the 2nd molt, the body shape similar to a pear, light-yellow to dark yellow.

Pupae Elliptical, orange color. Antennae about ½ of the pupa length. Eyes purple black.

Biology and behavior *Pseudaulacaspis pentagona* occurs 2 generations a year in central north and northeastern China, and 5 generations a year in southern China. They overwinter as female adults on branches. They resume activity next year when trees start to have sap flow. Feamles lay eggs under shells. Each female lays a maximum of 150 eggs. The overwintering generation is most reproductive. Males congregate neatly as white and shiny masses. Females overlap with each other in larger groups than the males. Newly hatched nymphs stay in shells for a few hours, then crawl for 1 day, and become sessile. They usually feed on buds, leaf petiole scars. Some are on trunks and base of branches. These individuals usually develop into males.

P. pentagona on black locust (Xu Gongtian)

P. pentagona on apricot (Xu Gongtian)

P. pentagona adult female scales (Xu Gongtian)

P. pentagona adult males and secretion (Xu Gongtian)

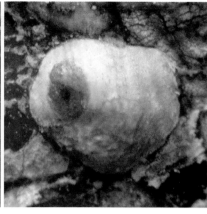
P. pentagona adult female scale (Xu Gongtian)

P. pentagona adult females (Xu Gongtian)

P. pentagona adult male scales (Xu Gongtian)

P. pentagona adult male (Xu Gongtian)

Females go through 2 instars then become adults. Second instar male nymphs form cocoons and go through prepupae and pupae stages before turning into adults. The trunks and base of branches often have accumulation of male cocoons that look like cotton. Male adults crawl more than fly. They live for no more than 1 day and die soon after mating. Their survival is shortened by wind and rain. Newly infested plants have more females. Trees that are infested for a long period have more males with accumulations of cocoons.

Prevention and treatment

1. Cultural control. During winter or early spring, using brushes or cloth to kill overwintering females. Prune off infested branches and destroy them.

2. Biological control. Conserve the following natural enemies: *Coccophagus pulchellus*, *Prospaltella berlesei*, *Encarsia berlesei*, *Aphytis proclia*, *Lestodiplosis pentagone*, *Chilocorus kuwanae*, *C. rubidus*, and *Cybocephalus nipponicus*.

3. Chemical control. During mid-May when 1st generation nymphs are haching, spray 12% thiamethoxam and lamda-cyhalothrin mixture at 1,000-1,500×, or 35% imidacloprid at 1,500-2,000×.

References Yang Ziqi and Cao Guohua 2002; Xu Gongtian 2003; Wang Jianyi, Wu Sanan, Tang Ye et al. 2009.

(Qiao Xiurong, Wang Wenxue, Xu Gongtian, Li Zhenyu)

Diaspidiotus gigas (Ferris)

Classification　Hemiptera, Diaspididae
Synonyms　*Aspidiotus gigas* Thiem et Gerneck, *Aspidiotus multiglandulatus* Borchsenius
Common name　Willow scale

Importance　*Diaspidiotus gigas* is a quarantine pest in China. It was spread to large areas as poplar trees are planted widely in the northwest, north central, and northeast.

Distribution　Hebei, Liaoning, Jilin, Heilongjiang, Inner Mongolia, Shanxi, Shaanxi, Gansu, Qinghai, Ningxia, Xinjiang; Russia, Italy, Spain, former Yugoslavia, Switzerland, Germany, Hungary, Czech Republic, Bulgaria, Turkey, Algeria.

Hosts　*Populus* spp. such as *P. berolinensis*, *P. nigra* var. *thevestine*, *P. nigra* var. *italic*, *P. cathayana*, *P. ussuriensis*. It also infests *Salix matsudana* and *Salix capitata*.

Damage　*Diaspidiotus gigas* feed on tree trunks and thick branches and suck sap from phloem. They form wax shells that cover the body. When pest density is high, shells will cover the breathing holes on the trunks. Infestation sites become concave, yellow, grey, or dark in color. Tree crown become yellow and vigor is reduced. Death of trees may occur. Most infestations occur on 1-15 years old trees. Those 6-15 years old trees suffer the most. Pest density can be 43.5 per cm^2 area.

Morphological characteristics　**Adults** Females upside down pear shaped, green. Spiracles without surrounding pores. Anal plate yellow, with 3 pairs of lobes. Dorsal and ventral pores abundant. There are 5 groups of pores surrounding the reproductive opening. Female shell round and flat, the center raised. There is a spot at or near the center of the shell.

Males slender, orange yellow, 0.71 mm long. Antennae filiform. Forewings milky, hindwings reduced to halteres. Male shell elliptical, shell spot brown, located at one end, the lower part of the shell greyish white.

Nymphs　Near round, 0.12 mm long. Female nymphs light-yellow when crawled out of the shells. Anal plate apricot yellow. Body compressed. Male nymphs slightly larger than the female nymphs, soft, color similar to female nymphs.

Biology and behavior　*Diaspidiotus gigas* occurs one generation a year. It overwinters as 2nd instars. They resume activity in mid-April and molt into adults in mid-May. Females lay eggs in mid-June; egg period 5-6 days; each female lays 64-107 eggs. Nymphs emerge from

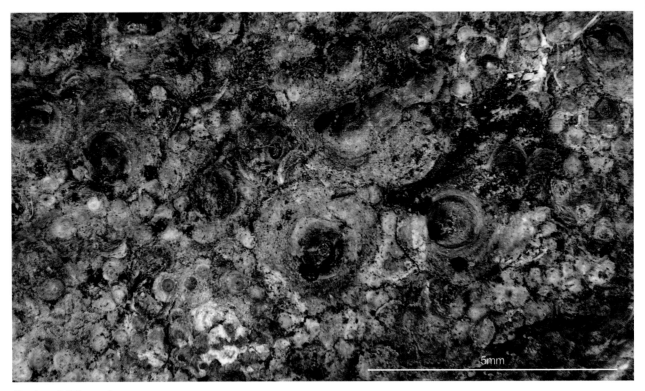

D. gigas (Li Zhenyu)

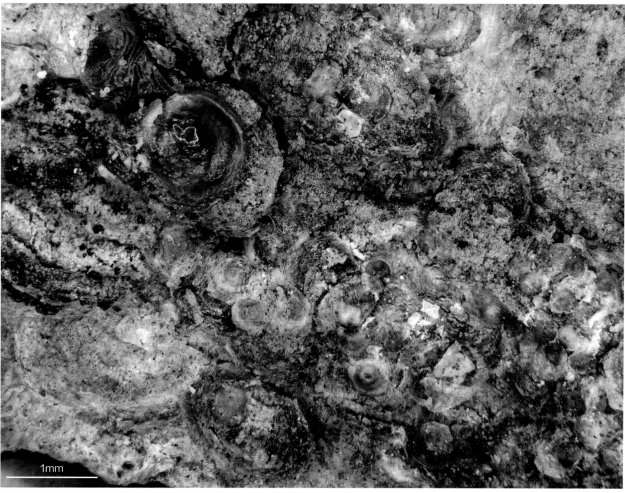

D. gigas (Li Zhenyu)

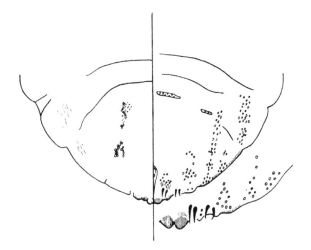

D. gigas female pygidium plate (Zhang Peiyi)

mid-June to mid-July. Egg hatching peak is in late June. Overwintering period stars in September.

Prevention and treatment

1. Quarantine. Prevent movement and use of infested seedlings

2. Silviculture. Proper maintenance of young trees to increase their natural defense.

3. Selection of resistance varieties. Plant varieties that are tolerant to *Quadraspidiotus gigas*, cold, salty soil, and lean soil where the pest is present. Examples are: *Populus simonii* × *P. nigra*, *Populus alba* × *P. beolinen*, *Populus pseudo-simonii* × *P. nigra*.

4. Conserve natural enemies. An adult *Chilocorus kuwanae* can kill 61.3 (average 31.9) *Quadraspidiotus gigas* a day. *Archenomus longicornis* and *Marietta carnesi* are parasitoids of *Q. gigas*.

References   Tang Fangde 1977; Hu Yinyue, Dai Huaguo, Hu Chunxiang et al. 1982; Li Yajie 1983; Zhou Yao 1985; Xiao Gangrou (ed.) 1992; Wang Jianyi, Wu Sanan, Tang Ye et al. 2009.

(Xie Yingping, Li Yabai)

102 *Comstockaspis perniciosa* (Comstock)

Classification	Hemiptera, Diaspididae
Synonyms	*Aspidiotus perniciosus* Comstock, *Diaspidiotus perciciosus* (Comstock), *Aonidiella perciciosus* (Comstock)
Common name	San Jose scale

Importance *Comstockaspis perniciosa* is a dangerous fruit pest. It is listed as a quarantine pest in many countries.

Distribution Beijing, Hebei, Shanxi, Liaoning, Heilongjiang, Jiangsu, Zhejiang, Anhui, Fujian, Jiangxi, Shandong, Henan, Hunan, Guangdong, Sichuan, Yunnan, Shaanxi. Native in East Asia, this pest is now spread to every continent except Antarctic.

Hosts A total of 307 hosts were reported. Some common hosts are: *Pyrus bretschneideri*, *Malus pumila*, *Prunus persica*, *P. mume*, *Vitis vinifera*, *Ziziphus jujube*, *Diospyros kaki*, and *Citrus reticulate*.

Damage *Comstockaspis perniciosa* larvae and female adults feed on trunks, petioles, underside of leaves, and fruits. They reduce plant vigor, delay bud development, or even kill the hosts. It is a major threat to fruit production. Affected fruits often become concave, cracked, and violet-red halo appears around the insect. They significantly reduce the aesthetic value of the fruits.

Morphological characteristics **Female adults** Near circular, raised at center, live shell greyish brown, dead shell greyish white or brown, diameter 1.5-2.0 mm, with 3 concentric rings. Mature females membranous. Antennae tubercle shaped.

Male adults Length 0.6 mm, orange yellow. Eyes dark purple. Antennae 10 segmented, with fine hairs. Thorax brown, the end with Copulatory organ. Shell colors vary with season and generation.

Nymphs First instars elliptical, milky yellow. Antennae 5 segmented. Legs developed. After fixed to the feeding sites, their size increases and greyish white wax shells are formed, length 0.2-0.3 mm. The 2nd instars greyish brown. Males and females are similar. Shell diameter 0.69-0.90 mm. Antennae and legs are degenerated.

Biology and behavior *Comstockaspis perniciosa* occurs 3-4 generations a year. The 2nd instars usually overwinter on branches. They resume feeding after sap begins to flow in spring. After molting to 3rd instars, males and females differentiate. Male adults emerge from mid-April to late April. They fly to female adults to mate. They live for 3-5 days. Females are viviparous. The 1st instars of the 4 generations appear during the following periods: mid-April to early May, early June to mid-July, mid-August to mid-September, early and mid-November, respectively. The young nymphs crawl out of shells soon after being produced. They usually select 2-5 years old branches to feed. At high density, they also feed on fruits and leaves. At 1-2 days after fixing, they produce wax threads and form shells.

Prevention and treatment

1. Quarantine. Prevent movement and use of infested seedlings

2. Silviculture. Avoid plant orchards in close distances. Remove infested branches or trees during early infestation stage.

3. Apply *Pestalotiopsis* fungus. Conserve natural enemies. Dozens of natural enemies are reported. Examples are: *Chilocorus kuwanae*, *C. renipustulatus*, *Sospita chinensis*, *Coelophora saucia*, *Propylaea japonica*, *Ballia obscurosignata*, *Harmonia axyridis*, *Aphytis fisheri*, and *Cybocephalus nipponicus*.

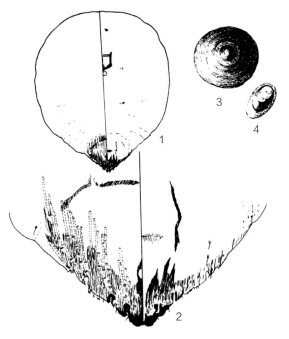

C. perniciosa 1. Adult female, 2. Female pygidium plate, 3. Female scale, 4. Male scale (Hu Xingping)

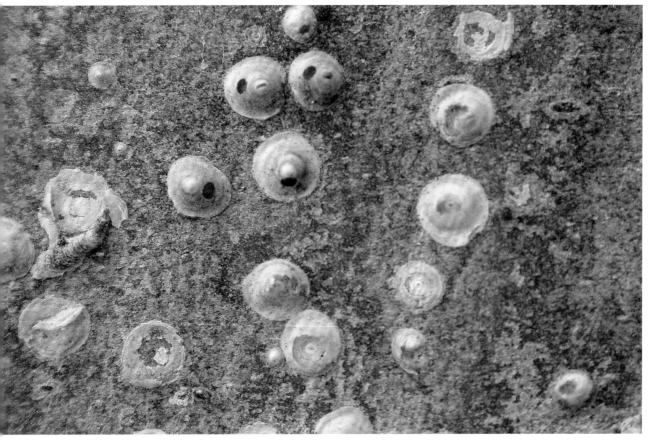

C. perniciosus on poplar tree (Xu Gongtian)

C. perniciosus female scales (Xu Gongtian)

Parasitized *C. perniciosus* scales (Xu Gongtian)

4. Chemical control. When buds start to develop, spray with 0.3%-0.5% pentachlorophenol. For nymphs, control should be concentrated on 1st and 2nd generation nymphs. Insecticides include 12% thiamethoxam and lamda-cyhalothrin mixture at 1,500-2,000×. For adults, spray with 40% acetamiprid and chlorpyrifos mixture at 1,000-1,500× will achieve good results.

References Xu Tiansen 1987; Ministry of Forestry of China 1996; Li Zhanwen, Zhang Aiping, Sun Yaowu et al. 2007; Wang Jianyi, Wu Sanan, Tang Ye et al. 2009.

(Xie Yingping)

103 *Diaspidiotus slavonicus* (Green)

Classification Hemiptera, Diaspididae
Synonym *Quadraspidiotus populi* Bodenheimer

Importance *Diaspidiotus slavonicus* is a dangerous pest of the agriculture protection forests in north of China.

Distribution Inner Mongolia, Shanxi, Shaanxi, Gansu, Qinghai, Ningxia, Xinjiang; Central Asia, Western Asia.

Hosts *Populus* spp. and *Salix* spp.

Damage *Diaspidiotus slavonicus* infest branches and trunks of hosts. When density reaches 60 insects/cm^2, the bark becomes reddish brown, black or rotten. Branches and trunks may die. The 2nd-5th years old trees may die from heavy infestations. Mixed forests usually suffer less damage than pure forests. *Populus nigra* var. *thevestina* is most susceptible. The following trees are less susceptible to the pest: *Populus* × *canadensis*, *P. simonii*, *Populus simonii* × *Populus pyramidalis*, *P. euphratica*, *Salix babylonica*, and *S. matsudana*. *Populus alba* var. *pyramidalis* and *Populus alba* are rarely infested. Pest density is related to tree age, height, coarseness and age of the barks. The pest will migrate down to smoother barks at the upper part of the trees as the trees grow.

Morphological characteristics **Shell** Female shells round, tall, and convex, greyish white, diameter 1.2-2.0 mm. The spot on the shell off the center, orange yellow; shell color white. Male shell shape similar to a shoe bottom; greyish white; 1.0 mm long, 0.7 mm wide; shell spot located at the end; orange yellow.

Adults Color of females change from orange to brown when mature, oval, body hard, anal lobes 3 pairs. Males light-yellow, 0.70-0.86 mm long, antennae 10 segmented, copulatory organ long.

Eggs Elongated elliptical, light-yellow.

Nymphs First instars flattened, elongated, dorsal surface with several symmetrical dark spots, legs developed. Antennae and legs disappear in the 2nd instars. Males and females differentiate during the 2nd instar.

Pupae Pre-pupae light-yellow to yellowish brown. Antennae, legs, and wing buds are rudimentary. Wing buds are obvious. Antennae and legs with visible segments. Copulatory organ conical.

Biology and behavior *Diaspidiotus slavonicus* occurs 1-2 generations a year in Xinjiang. Those occurring one generation a year overwinter as 2nd instars on branches. Those occurring two generations a year overwinter as 1st instar but they will not survive. Nymphs resume activity in March. Males pupate in late April. Adults emerge in early May within a 4-5 days period. The 1st generation nymphs appear in early June. Some nymphs molt into 2nd instar and overwinter. Some nymphs develop into adults in mid-July. The 2nd generation nymphs appear in early August. They develop slowly and cannot molt into 2nd instars. Females reproduce through viviparity or egg production. Each female produces 14-15 eggs. Those hatched early often feed at the rear end of the mother. Those hatched later will crawl to smooth branches and trunks and feed. At 1-2 days after fixing themselves, they produce wax shells. The 1st generation often infests leaves. The 2nd generation only infests branches and trunks.

Prevention and treatment

1. Quarantine. Prevent movement and use of infested seedlings.

2. Silviculture. Select resistant varies when planting new forests.

3. Conservation of natural enemies. The following natural enemies are reported: *Anthemus aspidioti* Marietta

D. slavonicus (Xie Yingping)

sp., *Chilocorus kuwanae*, *Anthocoris pilosus*, and *Cybocephalus nipponicus*.

4. Spray 95% mineral oil at 300× during early June to kill newly hatched nymphs can reduce pest population by 70%.

References Tang Fangde et al. 1980; Zhang Xuezu and Qu Bangxuan 1983; Tang Fangde 1984; Guan Yongqiang 1992.

(Xie Yingping)

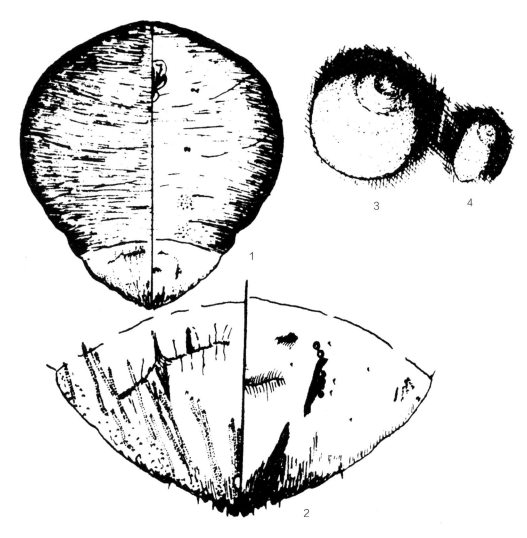

D. slavonicus. 1. Adult female, 2. Female pygidium plate, 3. Female scale, 4. Male scale (Hu Xingping)

104 *Shansiaspis sinensis* Tang
Classification Hemiptera, Diaspididae

Importance The diaspidid scale insect is an important group of pests that threaten *Salix* spp., the main afforestation tree species in the northwest along the sand dunes.

Distribution Shanxi, Inner Mongolia, Shaanxi, Gansu, Qinghai, Ningxia, Xinjiang.

Hosts *Tamarix chinensis*, *Salix matsudana*.

Damage The scale uses its piercing-sucking mouthparts to pierce through tender shoots and leaves and ingest tree sap. They mainly infest young and tender branches. The tip of damaged shoots becomes dry, tree vigor declines, and death of trees may occur.

Morphological characteristics **Scale** Females are nearly pear or triangle shaped, white, shell point protruding toward the tip, yellow, full- length is 2-3 mm. The males are elongated with three longitudinal ridges. The ridges sometimes are not apparent.

Adults Female adult body is fusiform or oval in shape, 1.08 mm×0.48 mm in size, membranous, but anal plate is slightly sclerotized. The lateral protrusion of the body could be apparent or not apparent. Distance between two antennae is wide. The anterior and posterior spiracles exist. There are three pairs of anal lobes, the median lobes are well developed. The gland at the dorsal edge of the anal plate is well developed. Enlarged dorsal tubes are located on the third to sixth abdominal segments. They can be divided into sub-central and sub-lateral groups. The small dorsal tubes are located at the sub-anterior edge, and edge of the body. There are five groups of the genital glands.

S. sinensis (Li Zhenyu)

S. sinensis (Li Zhenyu)

Biology and behavior *Shansiaspis sinensis* occurs two to three generations per year. Generations overlap. It overwinters as nymphs in the trunk, branch, and bud. The overwintering nymphs begin feeding and secreting wax in early April. They molt in mid-April. After two molts, the female nymphs turn to adults. The male nymphs molt once and turn to pre-pupal stage, then become adults. Adult females start egg production in late May and end in mid-June. Each female produces 81 eggs. Females shrink after oviposition and are protected under the scales. Average egg stage is 15 days. The first-generation nymphs begin to appear in mid-June. The nymphs crawl out of the scales and move towards the tip of the trees. After about one day, they enter a stationary period. Nymphal hatching peaks in late June. The first generation of adults emerge in mid-July. They lay eggs in early August. After the end of October, nymphs aggregate and overwinter. The overwintering period is 150 days. The first and second generations of the male adults occur in late April and late July, respectively. The adult life span is only 2 to 3 days. They die soon after mating.

Prevention and treatment

1. Quarantine. Avoid bringing in infested seedlings. Treat the infested seedlings with 52% avermectin granules or 56% aluminum phosphide tablets at $6.6g/m^3$ or 24 to 48 hours can kill more than 99% of the scales.

2. Early removal of the canopy of young trees can eliminate the pest sources.

3. Protection of natural enemies. *Chilocorus kuwanae* is the main predator of the scale. Placement of mulch (bark or grass) underneath the canopy can improve overwintering lady beetle's survival rate up to approximately 26%.

4. Injection of 1.3 mL 20% dinotefuran per square meter of canopy or 40% acetamiprid and chlorpyrifos mixture can reduce pest population by 95%.

References Ma Shirui 1983, Tang Fangde 1986, Liu Mingtang 1989, Shao Qianghua 1990.

(Xie Yingping, Shao Qianghua)

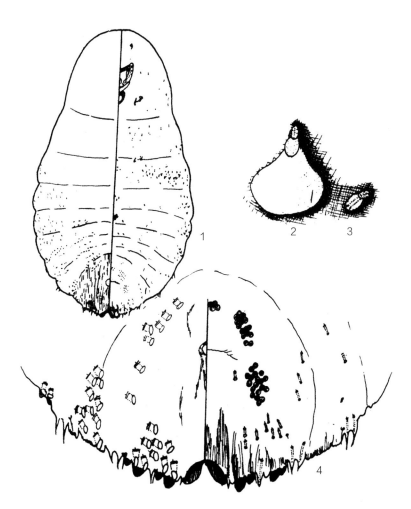

S. sinensis. 1. Adult female, 2-3. Ecological sketch, 4. Female pygidium plate (Hu Xingping)

105 *Unaspis euonymi* (Comstock)

Classification Hemiptera, Diaspididae
Synonyms *Chionaspis euonymi* Comstock, *Unaspis euonymi* Takahashi et Kanda
Common name Euonymus scale

Distribution Inner Mongolia, Shandong, Jiangsu, Shaanxi, Ningxia, Sichuan, Guangdong, Guangxi.

Hosts *Euonymus kiautshovicus*, *E. japonicus*, *E. alatus*, *Hibiscus syriacus*, *Lonicera japonica*, *Ligustrum lucidum*, *Syringa* spp.

Damage Nymphs and female adults pierce and ingest the sap from leaves and young branches, which leads to chlorosis and early senescence of leaves, reduced growth of the infested plants. Severe damage leads to the death of a branch or even a tree. It can also induce sooty mold and reduce the ornamental value of the host plants.

Morphological characteristics **Scale** The size is 1.4-2.0 mm×0.7-1.0 mm. Female scale pear-shaped, often curved with color varies from brown to purple brown. There are two yellowish brown dots on apex, and a longitudinal ridge in the middle of the scale. Male scale flat and elongated, covered with white wax, 0.8-1.1 mm× 0.2-0.3 mm in size. There is a yellow dot on the tip of the scale, and three longitudinal ridges dorsally.

Adults Females fusiform, yellow with yellowish brown anal plates. The antennae are tumor like with a long seta. There are 10-25 anterior spiracle glands and 2-4 posterior glands. The glandular tumors are located on the lateral sides of the first and the second abdominal segments. There are a cluster of small tubular glands on lateral margin of both meso- and meta-thorax and the first abdominal segment. There are three pairs of anal lobes. The middle anal lobe is large and protruded. The basal part is close, but not connected with each other, and the inner margin tilts outward. The second and third anal lobes are divided into two lobes, and the outer lobes are slightly smaller than the inner lobes. There are seven pairs of marginal glands with setae arranged in pairs, more than 60 lines on the back near the end of body, and five groups of glands surrounding the genital area with 3-9 glands per group. Male adult orange, 0.7 mm long, 1.7-1.8 mm in wing spread. It has well-developed thorax, short abdomen and slender copulatory organ.

Eggs The size is 0.2 mm×0.1 mm, elongated oval in shape, and light-yellow in color.

Nymphs The first instar oval, orang. Ocelli orange. Antennae 5 segments; the length of the fifth segment equals to the total length of the basal four segments. Legs are well developed. There are two pairs of anal lobes on the anal plate, the middle anal lobe with three teeth, the second anal lobe has a pair of long hairs at the end. Legs of the 2nd instar nymphs disappear. Antennae are tumor-like. The anal plate is similar to that of the adults, but the glandular spines and the edge glands are all solitary without dorsal abdominal glands.

Pupae Pre-pupae is elongated oval-shaped; orange yellow; mouthparts disappear; antennae, legs, wings and other organ buds present. Organ buds of pupae are more extended.

Biology and behavior *Unaspis euonymi* has three generations per year in the eastern China. There are two generations, and a few of them have 3 generations in Liaoning province. In central Shandong province, the fertilized females overwinter on host branches and leaves in November. The females begin to feed on hosts when the

U. euonymi. 1. Adult female, 2. Female pygidium plate, 3. Female scale, 4. Male scale (Hu Xingping)

U. euonymi adult female scales (Xu Gongtian)

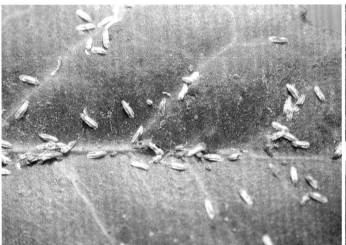

U. euonymi adult male scales (Xu Gongtian)

U. euonymi on *Euonymus japonicus* (Xu Gongtian)

U. euonymi adult female scale (Xu Gongtian)   *U. euonymi* adult male scale (Xu Gongtian)

host trees resume growth in the following spring. They lay eggs in early May, which lasts for about 1 month. The peak for egg hatch of each generation occurs in early and mid-May, early and mid-July and late August to early September, respectively. The overwintering adults begin to oviposit in mid-May in Shenyang, Liaoning Province.

Newly hatched nymphs can be found from mid-May to mid-September with overlapping generations. The two peaks of hatching periods are between mid- and late June, and between early and mid-August, respectively.

Female adults lay eggs under the scales. The fecundity varies in the region and generations with greater fecundity in the south than in the north. The fecundity of the overwintering generation and the first generation is greater than the second generation. The average fecundity of the females in Shandong province is 35-77 eggs.

The eggs hatch within a day, the 1st instar nymphs leave the female scale to crawl and disperse on the branches and leaves. The nymphal dispersal period is very sensitive to temperature, moisture and light. High mortality occurs during storms. The dispersing nymphs settle down and start feeding on the shaded side or fork point of tree branches. One day after their establishment, the nymphs secrete white wax materials on dorsum, and molt after another 2-4 days, then enter into the 2nd instar.

Sexual differentiation can be observed in the 2nd instar nymphs. Male nymphs secrete white cotton-like waxy materials to form the long scale. The nymphs grow under the scale and turn into prepupae after molting, then transform into pupae after another molting. The females, however, continually secrete waxy materials on dorsum as the larvae grow, a dorsal line appears prior to molting. When it molts, the second scale spot is formed by the exuviae and secretion. After eclosion, adults secrete waxy materials to form pear-shaped scales. After emergence, male adults crawl, fly for short distance, and mate.

Prevention and treatment

1. Removal and destroy of the infested twigs and leaves to eliminate insect sources.

2. Apply *Pestalotiopsis* fungus. Protection and utilization of natural enemies. Its natural enemies include predators, such as, *Cybocephalidae*, *Chilocorusrubidus*, *Chilocorus kuwanae*, *Exochomus mongol*, *Harmonia axyridis*, *Chrysopa sinica*, and a parasitoid wasp, *Cheiloneurus claviger*.

3. Spray 3-5% lime sulfur mixture on twigs and branches 15 days before the bud sprouting to control the overwintering nymphs.

4. At the peak of the nymph hatch period, spray 20% thiamethoxam or 12% thiamethoxam and lamda-cyhalothrin mixture at 1,000× to the trunks, especially on the branches and abaxial side of the leaf blades.

References Cui Wei and Gao Baojia 1995; Yang Ziqi and Cao Huaguo 2002; Xu Gongtian 2003; Wang Jianyi, Wu Sanan, Tang Ye et al. 2009.

(Qiao Xiurong, Hu Xingping, Li Zhenyu)

106 *Corythucha ciliata* (Say)

Classification Hemiptera, Tingidae
Common name Sycamore lace bug

Importance The sycamore lace bug is an invasive species that has great potential to cause severe damage. Once it is introduced into a new region, the bug can develop into stable high density populations. It has become an important invasive pest of sycamore. When it was first discovered in 2006 in Wuhan, Hubei provinces, the State Forestry Administration of China added it to the list of dangerous invasive forest pests of China.

Distribution Beijing, Shanghai, Jiangsu, Zhejiang, Anhui, Henan, Hubei, Hunan, Chongqing, and Guizhou; U.S.A., Canada, Italy, France, Russia, Hungary, Poland, Bulgaria, Greece, Czech Republic, Netherlands, Austria, Croatia, Germany, Serbia, Montenegro, Slovenia, Switzerland, Belgium, Portugal, Slovakia, Spain, Turkey, Chile, Israel, Japan, Korea Peninsula.

Hosts Main host plants are in genus *Platanus* of sycamore, including *Platanus occidentalis*, two-ball sycamore *Platanus acerifolia*, three-ball sycamore *Platanus orienfalis*, *It also causes* damage to *Broussonetia papyrifera*, bark hickory *Carya ovata*, *Chamaedaphne* calyculata *Fraxinus* spp., Tong Ye maple *Acer pseudoplatanus*, bay leaf oak *Quercus laurifolia* and rubber sweetgum tree *Liquidambar styraciflua* and other plants.

Damage Sycamore lace bug adults and nymphs usually feed on the abaxial side of the leaves on lower layer of the sycamore tree canopy. Initial damage includes yellowish white spots and leaf chlorosis. Severe infestation causes dry and yellow leaves early senescence of the leaves, interruption of tree growth, and even death of plants. In Europe, sycamore lace bug is a vector of sycamore leaf blight, *Gnomonia platan*, sweet potato disease, *Ceratocystis fimbriata*. Both plant pathogens can reduce the vigor of sycamore trees, and even cause death. In addition, the wound caused by sycamore lace bug leads to the infection of sycamore anthracnose *Gnomonia veneta*, *C. fimbriata*, and sycamore canker *Ceratocystis platani*.

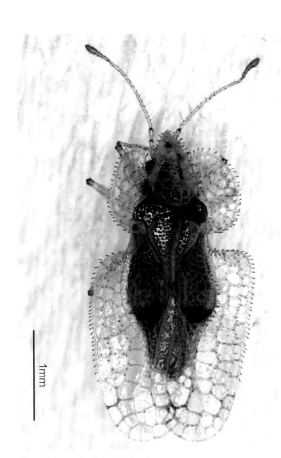

C. ciliate male (Hao Dejun)

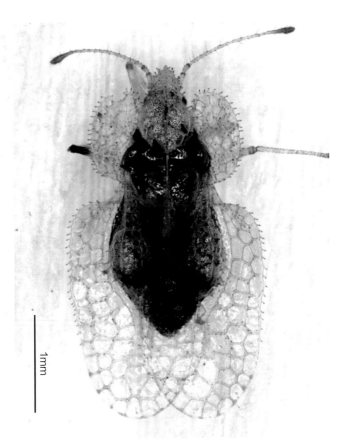

C. ciliate female (Hao Dejun)

C. ciliate 5th instar nymph (Hao Dejun)

Morphological characteristics **Adults** Milky white, 3.2-3.7 mm×2.1-2.3 mm in size. Head and ventral surface of the body are dark brown. Head pocket is developed, helmet shaped. The mesh grid on the head pocket (protrusion) is slightly larger than the lateral panels. From the lateral view, the height of head pocket is higher than the middle longitudinal ridge. There is a brown spot behind the basal protrusion of the two wings. Head pocket, lateral dorsum, middle longitudinal, and the surface of the forewing veins is covered with dense spines. Lateral dorsum and the costal edge have neatly arranged spines. Forewing is significantly longer than the end of the abdomen. The costal base abruptly folds up and protrudes outward. The sub-base is protruded in a right angle, which makes the wings nearly rectangular shape. Femur not thickened, legs and antennae light-yellow. The scent gland on metathorax small.

Eggs Oblong, metallic black. There is a milky white egg cap (or cover).

Nymphs There are a total of five instars. Flat, fusiform in shaped, and wingless. All instars have a spike, except the first instar.

Biology and behavior In Shanghai and Wuhan, the pest has five generations per year. In U.S.A., Italy, Japan, there are 2-3 generations per year, most have three generations. In the autumn, the fifth generation of adults overwinters under the bark, bark cracks, leaf litter, or on plants underneath the canopy of sycamore trees. Overwintering adults become active in early April. The females start to oviposit in late April. They often deposit eggs on the main vein or lateral veins of the abaxial side of the leaves. The eggs are usually in groups of a few dozens of eggs. Each female oviposits 100 to 350 eggs. The eggs hatch in early May, the first to the 3rd instars feed in aggregation. They disperse and feed after the fourth instar. The peak of the first generation nymphs occurs in mid- and late May. They overwinter in late October when the temperature is below 10℃. Low temperature in the winter can significantly reduce the insect population of the following year. Rainfall in spring when adults become active is detrimental to the pest population. Hot and dry weather in the summer and autumn often lead to outbreaks.

Prevention and treatment

1. Physical control. Scraping loose bark, painting tree trunks with lime, timely collection and destroy of fallen leaves will reduce the number of overwintering insects. In the spring when the overwintering adults are active, the treatment of tree trunks in combination with irrigation to wash the tree canopy can also effectively reduce the overwintering insect population.

2. Apply *Pestalotiopsis* fungus. Biological control. Protection and utilization of natural enemies can be effective. The predators include *Nabis pseudoferus*, *Rhinocoris iracundus*, *Deraeocoris nebulosus*, *Orius vicinus*, *Orius insidiosus*, *Aptus mirmicoides*, *Himacerus mirmicoides*, *Cheiracanthium mildei*, *Theridion lunatum*, *Plexippus paykulli*, *Aphaenogaster smythiesi*, and other predators for the lace bug. Entomopathogens include *Beauveria bassiana*, *Verticillium lecanii* and *Paecilomyces farinosus*.

3. Chemical control. The common chemical control methods include spraying canopy or trunk, trunk injection. The canopy spraying is to control nymphs and newly emerged adults. Insecticides include pyrethroid (lamda-cyhalothrin, fenvalerate), imidacloprid, acetamiprid and other botanical pesticides. Early instar nymphs can be controlled using a systemic insecticide for trunk injection. For overwintering adults, the direct spraying of insecticides on the trunk bark cracks can be effective.

References Tremblay and Petriello 1984; Tavella and Arzone 1987; Oszi et al. 2005; Li Chuanren, Xia Wensheng, and Wang Fulian 2007; Xia Wensheng, Liu Chao, Dong Likun et al. 2007; Wang Fulian, Li Chuanren, Liu Wanxue et al. 2008; Ji Rui, Xiao Yutao, Luo Fang et al. 2010; Ju Ruiting and Li Bo 2010; Xiao Yuyu, Wang Feng, Ju Ruiting et al. 2010.

(Hao Dejun)

107 *Metasalis populi* (Takeya)

Classification Hemiptera, Tingidae.
Synonym *Hegesidemus habrus* Darke
Common name Poplar and willow lace bug

Distribution Beijing, Hebei, Shanxi, Jiangsu, Jiangxi, Shandong, Henan, Hubei, Guangdong, Sichuan, Shaanxi, and Gansu.

Hosts *Populus* spp., *Salix* spp.

Damage Nymphs and female adults feed on abaxal side of the host leaves by piercing and ingesting plant sap. The honeydew left on leaves can induce sooty mold growth which reduces photosynthesis, early senescence and defoliation. *M. populi* becomes a key pest to popular trees as the areas of poplar trees grow.

Morphological characteristics **Adults** Females about 3 mm long, males about 2.9 mm long. Head reddish brown, smooth, short, plump; eyes red; with three short yellowish white setae. Antennae light yellow, with short pubescence, the end of the fourth segment is black. Length of the four antennae segments are 0.18, 0.11, 1.03, 0.35 mm, respectively. The head roof-like, with two dark brown spots at the end. The end of the rostrum extends to the middle of meso-sternum. Pronotum light brown to dark brown, covered with fine sculpture. Three longitudinal ridges are grayish yellow. Paranotum narrow, roof-shaped, with a row of small cells. Forewing long oval, longer than abdomen, light yellowish white with many transparent small cells and a dark brown "X"-shaped spot. The hind wing white, ventral surface of the abdomen dark brown, legs yellowish brown.

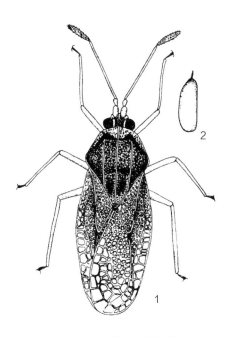

M. populi. 1. Adult, 2. Egg (Liu Yongping)

Eggs Long and oval, slightly curved. The newly deposited eggs are milky white, then change to light yellow. One end (approximately 1/3) becomes light red; after a few days, the other end shows red filaments. Eggs turn to red prior to hatching.

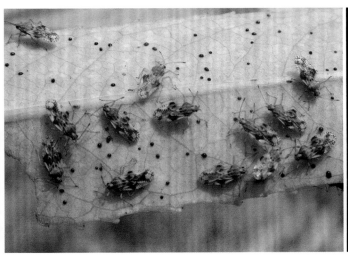

M. populi adults (Xu Gongtian)

M. populi adult (Xu Gongtian)

Nymphs Size 2.17×1.15 mm, head black; wing pads oval, reach to the middle of the abdomen dorsally; the base and end of the wing pads are black; the black spot on the abdomen breaks horizontally and vertically into three areas, which are connected to the cerci.

Biology and behavior *Metasalis populi* occurs three to four generations per year. It overwinters in adult stage under the tree bark cracks, fallen leaves and cracks on ground. During April of the following year when the average temperature is above 12℃, the overwintering adults become active and cause feeding damage. When the average temperature is below 10℃, the adults begin to go down the tree trunk and go back into overwintering stage. Eggs are laid in early May. The eggs are in rows in the leaf tissue near main and lateral veins on the abaxial side of the leaves. One egg per oviposit hole, which is covered with brown sticky secretion. In mid-May, the hatched nymphs pierce and ingest the sap of leaf blades. The damaged leaves can be identified by white chlorotic spots on the abaxial side of the leaves. The second generation of adults emerge in early July, the third generation in early August, and the fourth generation in late August. Adults feed till November then gradually enter into overwintering state. Adults play death and can migrate in large numbers for a short distance. They prefer dark and shaded habitats and usually aggregate on the abaxial side of leaves in the middle and lower parts of the tree canopy. Adult life span is 20-30 days. There are four instars during the nymphal stage. Both adults and nymphs are gregarious.

M. populi adult prothorax (Xu Gongtian)

M. populi adult (Xu Gongtian)

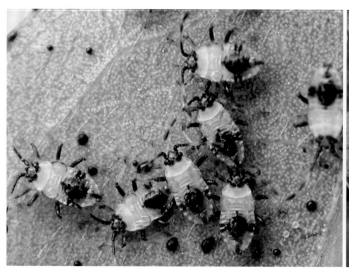
M. populi nymphs (Xu Gongtian)

Willow leaves infested by *M. populi* (Xu Gongtian)

Prevention and treatment

1. Establishing mixed forest stands and strengthening management. Thinning of forest stands with excessive density to improve plant growth and suppress the insect pests. Clean and destroy the withered and fallen leaves on ground by burning or plowing to eliminate the overwintering adults. Paint the host plant trunk with lime.

2. Apply *Pestalotiopsis* fungus. Protecting and utilizing of various natural enemies, such as predatory mite, *Allothrombium pulvinum*.

3. Chemical control. The nymphs and adults can be controlled by the spraying of 50% acetamiprid at 1,500-2,000×, 10% imidacloprid at 1,500-2,000×, or 48% chlorpyrifos at 2,500-3,000×, or 40% acetamiprid and chlorpyrifos mixture 2,000-3,000×.

References Liang Chengjie 1987, Zhao Ling et al. 1989, Zhao Fanggui et al. 1999, Zhao Junfang 2006.

(Wang Haiming, Liang Chengjie)

108 *Monosteira unicostata* (Mulsant et Rey)

Classification Hemiptera, Tingidae
Common name Poplar lace bug

Distribution Inner Mongolia, Gansu, Ningxia, and Xinjiang; Russia, Syria, Morocco, southern Europe, and northern Africa.

Hosts *Populus simonii*, *P. nigra* var. *thevestina*, *P. nigra* var. *italica*, *P. alba*, *P. beijingensis*, *P. canadensis* 'Sacrau', *P. nigra*, and *Salix alba* are preferred hosts. Less preferred hosts are *P. alba* var. *pyramidalis*, *Pyrus bretschneideri*, *Prunus salicina*, *Crataegus pinnatifida*, *Prunus pseudocerasus*, and *Prunus dulcis*.

Morphological characteristics **Adults** Size 1.9-2.3 mm×0.8-1.1 mm, head and thorax greyish black, compound eyes are spherical and reddish black. Antennae rod-shaped with 4-segments. Pronotum is well-developed and elevated upward laterally, with reticulate sculptures. Forewing and scutellum reticulated. The ovipositor of the females is a spherical. The male has a pair of sickle-shaped structure at the end of the abdomen.

Eggs Long oval, 0.2 mm×0.07 mm in size, milky white, surface reticulate.

Nymphs The last instar nymphs is 1.6-1.9 mm×0.9-1.1 mm in size, yellowish grey or light grey. Compound eyes spherical, reddish black. Head with four spines, the posterior two are larger than the front two. Pronotum is light-yellow. The basal and distal ends of the wing pads are black, while the middle is greyish yellow. The edge of the abdomen wavy, each convex part with a thick spin, each concave part has 1-3 fine setae.

Biology and behavior *Monosteira unicostata* occurs five generations per year in Shawan county of Xinjiang province. Adults overwinter in bark cracks and under fallen leaves. During mid-April when the average temperature is above 12.7°C, the overwintering adults become active and cause damage to trees. They mate after feeding for 12-15 days. In early May, the first-generation eggs are deposited in the parenchyma tissue on the abaxial side of the leaves near the primary vein. The eggs hatch in 6-7 days, nymphs develop into adults in 12 days. Adults start to overwinter in late September and early October.

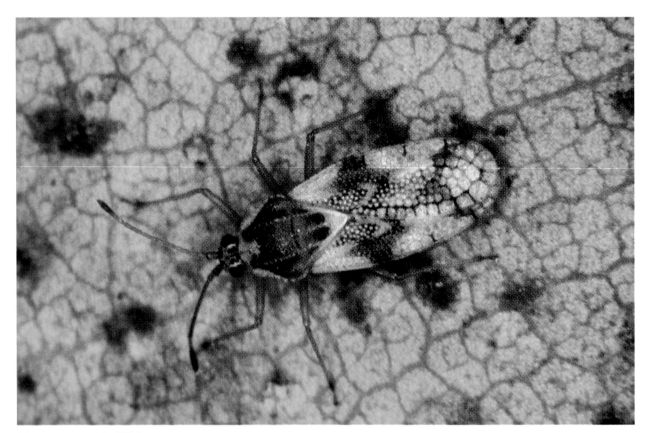

M. unicostate adult (SFDP)

Adults are very sensitive to disturbance and will move immediately. They play death. Adults are weak flyers. It can live in Turpan basin, where is hot and dry, or in the cold Altai Mountains. Adults sometimes bite human and cause pain and itchiness.

There are three instars during the nymphal stage. The 1st and 2nd instar nymphs feed gregariously on the abaxial side of the leaf blades. They disperse and migrate in clusters during the 3rd instar. Nymphs and adults continuously secrete brown honeydew as they feed on the leaves, resulting in leaf discoloration and chlorotic spots. The egg incubation period varies among different generations. The egg stage of the first generation lasts for 7-8 days and the later generations only last for 4-5 days.

Prevention and treatment

1. Establishing mixed forest stands. For example, establishment of mixed forest of poplar, elm, mulberry, *Elaeagnus angustifolia*, *Fraxinus chinensis* and other tree species.

2. Physical control. The overwintering adults can be eliminated by painting the trunk with lime in combination of brushing the bark cracks with metal wire brush, and cleaning up of fallen leaves.

3. Protecting and preserving natural enemies. For example, small grey spider and other natural enemies can be utilized.

4. Chemical control. The spraying of 45% profenofos and phoxim mixture, 50% fenitrothion, 25 g/L lamda-cyhalothrin, 40% chlorpyrifos and phoxim mixture, 20% thiamethoxam at 1,000-1,500× is effective. ULV application of 25% thiamethoxam in medium-aged forest stands can control both nymphs and adults.

References Zhang Shimei et al. 1985, Wen Shouyi and Xu Longjiang 1987, Xiao Gangrou (ed.) 1992.

(Wen Shouyi, Xu Longjiang)

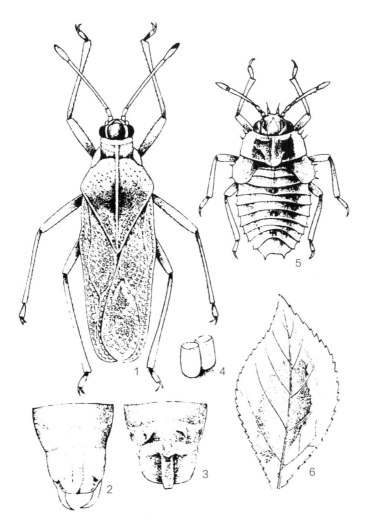

M. unicostate. 1. Adult, 2. Posterior end of the adult male abdomen, 3. Posterior end of the adult female abdomen, 4. Eggs, 5. Nymph, 6. Damage (Zhu Xingcai)

109 *Stephanitis illicii* Jing
Classification Hemiptera, Tingidae

Distribution Yunnan.
Host *Illicium verum*.
Damage Nymphs and adults feed on anise leaves. The damaged leaves are yellowish white, which leads to early senescence of the leaves and young fruits, and even the death of the whole plant.

Morphological characteristics **Adults** Small, 4-6 mm long, flat, and dark brown with white hair. Head and thorax have mesh pattern texture dorsally. Antennae have four segments, the first three segments are yellow, the third segment is particularly long, the fourth segment is covered with dense brown hairs. Forewing has a brown slightly pentagonal mesh patterned texture with a coarse brown spot. The center of the forewing has a brown spot. When at rest, the brown spots form a pitchfork-like marking.

Eggs Milky white, oval, one end slender, semi-translucent.
Nymphs Length 1.5-2.0 mm. The fourth instar nymphs dark brown. There are 12 spiny protrusions on lateral sides of the abdomen. Cercus is apparent.

Biology and behavior *Stephanitis illicii* occurs two generations per year. The insect overwinters as eggs in leaf tissue. In late April to early May of the following year, the overwintering eggs hatch into nymphs. There are four instars for nymphal stage, each instar lasts for 2 to 3 days. A large number of adults occur in mid-May. They mate and oviposit in late June. The egg stage lasts for 60 days, they hatch in late August. In early September, adults of the second generation emerge, and mate and deposit the overwintering eggs in leaf tissue. Overwintering eggs last for approximately 180 days.

Adults can feed in groups or solitary on the abaxial side of the leaves and excrete honeydew at the same time. Adults have weak flight ability and stay inactive in rainy days. They frequently fly in the forest in the evening on sunny days. An adult can damage multiple leaves. Adults mate after approximately 60 days of feeding. Females can mate as they feed. Duration of the mating period is relatively long and can be up to 24 hours. Females can start oviposition in 2 to 3 days after mating. Eggs are mainly deposited on the abaxial side of the leaves in the middle section of the leaf tissue around the main vein. A female can oviposit more than 10 eggs, and up to 40 eggs. Adults die gradually after oviposition. Adult life span is 60 to 90 days. Eggs hatch rate can be over 90%. The newly hatched nymphs are not active. They start to impose damage in the second star. They are gregarious.

Prevention and treatment

1. Removal of dead branches and leaf litter to destroy the overwintering eggs.

2. Chemical control. The spraying of 45% profenofos and phoxim mixture, 50% fenitrothion, 25 g/L lamda-cyhalothrin, 40% chlorpyrifos and phoxim mixture, 20% thiamethoxam at 1,000-1,500× is effective. ULV application of 25% thiamethoxam in medium-aged forest stands can control both nymphs and adults.

References Tea Research Institute of Chinese Academy of Agricultural Sciences 1974, Zhang Shimei et al. 1995.

(Gan Jiasheng)

S. illicii adult (Bu Wenjun)

110 *Stephanitis laudata* Drake et Poor
Classification Hemiptera, Tingidae

Distribution Fujian, Guangdong, Guangxi, and Taiwan.

Host *Cinnamomum camphora*.

Damage The nymphs and adults feed on the abaxial side of the camphor leaves by ingesting sap. The damaged leaves show yellow and white chlorotic spots. Severe infestation can cause large necrotic spots on leaves, and the whole leaf becomes brown. Honeydew discharged on leaves reduces leaf photosynthesis and results in reduced growth and ornamental value of the camphor trees.

Morphological characteristics **Adults** Length 3.12-3.46 mm×1.44-1.80 mm. Head is brown, and the back is covered with white powder. Antennae have four segments and slender; the fourth segment is slightly coarse and covered with white hair. Proboscis reaches sternum of the mesothorax. Pronotum brown, with a uniform engraved point. Head protrusion transparent, with yellowish brown mesh, oval, with tapering tip; paranotum is protruded upward and laterally, posterior margin strongly bent inwards. There are three rows of cells. The central longitudinal ridge is relatively high, and slightly longer than the head protrusion. The two lateral longitudinal ridges are extremely low and short, the length is approximately 1/8 of the central longitudinal ridge. Forewings are narrow and long, basal part of the wing is narrow, distal portion is wide, and the central portion slightly concave. When at rest, the forewings form an apparent "X"-shaped spots. Female abdomen is enlarged with a conical tip, and the ovipositor is obvious. The male abdomen lanky, and the abdominal end truncated, with one pair of claw-like genitalia.

Eggs Size about 0.4 mm×0.2 mm, white, eggplant shaped. The top of the eggs has a cover. The egg cover is grayish brown, oval, and with a central arch protrusion.

Nymphs Size 1.8 mm×0.9 mm. There are five head spines. The three at the front form a triangular shape, which points up and forward. The posterior two spines form a pair and point upward and laterally. Antennae have four segments. The first and the fourth are yellowish

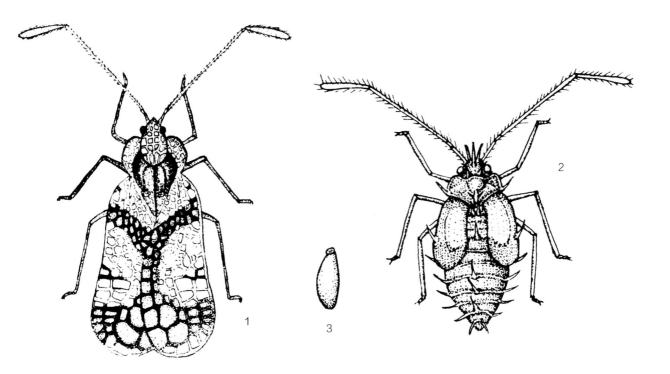

S. laudata. 1. Adult, 2. Nymph, 3. Egg (Li Yougong)

brown. The second and third segments are brown, covered with white hairs. The compound eyes are spherical protrusion in red. Pronotum with head protrusion (pocket), with one pair of spines on the central longitudinal ridge. There is a pair of spines on the scutellum of mesothorax. Wing pads of forewings are brown in both ends, but the middle portion is yellowish white, with a spine at the center of the costal margin. Abdomen is yellowish brown, and there are four spines in center dorsally, and there are six spines on each side laterally.

Biology and behavior There are 5 generations per year. It overwinters in egg stage on the abaxial side of the leaves in the mesophyll tissue around the main veins. In early April, the overwintering eggs begin to hatch. The generations overlap, but the peak of each generation is distinct. The peak of the 1st to 5th generation occurs between May and July, between July and August, between August and September, between September and October, and between October and May of the following year.

The leaves containing overwintering eggs do not fall off in the following March and April when the camphor tree change leaves. Each leaf has 3 to 60 eggs, with an average of 27 eggs. The hatch rate of the overwintering eggs is 67.82%; the egg stage lasts for 170 to 180 days. The egg stage of other generations is 13 to 23 days. Nymphs hatch from eggs by pushing open the top cover. The newly hatched nymphs are transparent, they start to feed in 1 to 2 hours. The abdomen appears green after feeding. The head also gradually becomes yellowish brown. The nymphs prefer to aggregate and feed at the branching point of leaf veins continuously (day and night). The nymphs stop feeding before molting, and the body is enlarged. When molting, the head move upward, the head capsule cracks. The nymphal stage lasts between 14 and 25 days. The overwintering generation is the longest, and the third generation is the shortest. The duration of each instar of nymphs is 3 to 5 days, but the second instar of the overwintering generation is 6 to 7 days, the fifth instar is 7 to 8 days, and in the third generation, each instar has only two days.

The newly emerged adults are transparent. Within an hour of eclosion, the body color gradually darkens. Then, the adults disperse and feed on leaf sap. The adults prefer shaded moist location and are distributed in the middle and lower parts of canopy on the abaxial side of the leaves facing away from the light. Adults have poor flight ability but can crawl very fast. The adults mate in 3 to 5 days after emergence. Female can continue to feed while mating. Adults can mate 1 to 5 times. The females oviposit 2 to 4 days after mating. During the oviposition, the female pierces through mesophyll with its stylets and then moves forward and inserts its ovipositor into the feeding wound to deposit eggs. The fecundity is 3 to 69 eggs per female with an average of 35 eggs. The oviposition period lasts between 2 and 13 days in multiple oviposition events. According to laboratory rearing and numerous random samplings in the forest, the sex ratio is close to 0.5:1 (female: male). Female life span is 8 to 26 days, while the male is 8 to 19 days.

References Xiao Caiyu et al. 1981, Li Yougong et al. 1990.

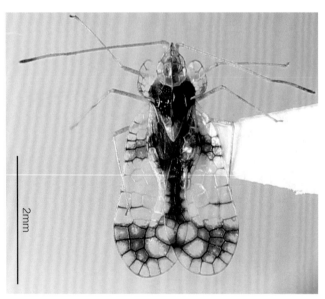

S. laudata adult (Bu Wenjun)

(Li Yougong)

111 *Stephanitis svensoni* Drake
Classification Hemiptera, Tingidae

Importance *Stephanitis svensoni* is a serious pest of *Sassafras* trees, but it also causes damage to trees in the genus *Illicium*.

Distribution Fujian, Hunan, and Guangdong; Japan.

Hosts *Sassafras tzumu*, *Illicium* spp.

Damage Both nymphs and adults of *S. svensoni* ingest plant sap. The affected leaves are covered with greyish or yellowish white chlorotic spots, and then become yellow. Infestations reduce the accumulation of organic matter, weaken the vigor, or even kill the host trees.

Morphological characteristics **Adults** Female size 4.0 mm×1.9 mm, male size 4.0 mm×1.9 mm. Head reddish-brown, antennae light-yellowish brown. Proboscis reaches the posterior edge of the sternum of mesothorax. There is a head pocket, which is spherical on the dorsal view. The front end of the head pocket gradually narrows, reaches above the middle part of the first segment of the antennae, but does not cover the compound eyes. The central longitudinal ridge is approximately the same height as the head pocket. The two lateral ridges are in the shape of leaflet, and relatively low. The length of the lateral ridges is only 1/3 of the central longitudinal ridge. The length of paranotum is greater than the width, and without brown spots. The forewings are elongated, narrow at the base, wide at the tip. The costal edge of the forewings has a slightly wavy curve with two rows of small teeth. The "X"-shaped dark brown pattern is apparent when the forewings are folded. The sternum of the thorax is reddish brown. **Eggs** Size 0.3-0.4 mm×0.1 mm, white, eggplant-shaped. **Nymphs** All nymphal instars are similar in body color and shape, the color is beige to brown, and transparent. The compound eyes are red. Antennae have four segments, and the distal section is in dark color. The body is covered with dark brown spines, those on the side are most obvious. The abdomen is dark brown and shiny.

Biology and behavior *Stephanitis svensoni* occurs 4-5 generations per year in Hunan province and overwinters as adults in leaf litter and tree crevices. The overwintering adults become active in mid-May of the following year. First generation eggs appear in late May, nymphs hatch out at the end of May, and adults emerge in mid-June. The adults of the following generations occur in mid-July, early August, early September, and early October, respectively. Adults start to overwinter when the leaves are beginning to fall in October.

Adult emergence can be observed all day along. The newly emerged adults milky white, and then change to light brown in 48 hours. They mate after supplementary feeding for 2 to 3 days. Mating usually occurs during the day and can be identified by the mating pair forming an acute angle. The eggs are deposited individually on the abaxial side of the leaves. Most of the eggs are located near the main leaf vein. The eggs are covered with black dot-shaped secretion. Often dozens of eggs are laid per leaf. The eggs hatch within a week. Adult life span

S. svensoni adult (Zhou Lijun)

can be as short as one week or as long as more than a month. The overwintering adult life span can be up to seven months. There are 4 instars for the nymphal stage, which lasts for about 15 days. Nymphs feed in aggregation before eclosion. Adults disperse after emergence. The exuviae are often found on the abaxial side of leaves. Both nymphs and adults feed on the abaxial side of leaf surface and cause densely distributed chlorotic spots ranged from greyish to yellowish white spots.

In 1978, a survey was conducted on the *S. svensoni* outbreak in relation to forest composition and slope in Jingzhou County in Hunan province. Results showed that the damage on the pure stand of sassafras is 100%, only 35% in the mixed stand of fir and sassafras. The trees on the slope facing south were 100% infested, the average number of bugs per leaf was 11.6. In contrast, the trees on the slope facing north have an average of 5.6 insect per leaf, although the infestation rate was also 100%. In addition, the population density decreases as altitude increases. The second generation has the greatest number of insects, and the population density reduced in latter generations.

Prevention and treatment

1. Silviculture. Avoid pure forest stand, promote the establishment of mixed forest stands. Select relatively good location in afforestation to enhance plant vigor.

2. Manual control. When the trees are not too tall, remove leaves with aggregated nymphs.

3. Chemical control. Spray 40% chlorpyrifos and phoxim mixture at 1,500-2,500× or 25 g/L lamda-cyhalothrin at 1,000×.

References Xiao Caiyu 1981, Xiao Gangrou (ed.) 1992.

(Zhang Lijun, Zhou Lijun)

112 *Pirkimerus japonicus* (Hidaka)

Classification Hemiptera, Blissidae

Distribution Jiangsu, Zhejiang, Fujian, Jiangxi, Hunan, and Sichuan; Japan, Vietnam.

Hosts *Phyllostachys aurea*, *P. bambusoides* f. *lacrima-deae*, *P. bissetii*, *P. bambusoides* f. *shouzhu*, *P. dulcis*, *P. flexuosa*, *P. fimbriligula*, *P. glauca*, *P. heteroclada* f. *solida*, *P. heteroclada* var. *pubescens*, *P. heterocycla* 'Obliquinoda', *P. iridescens*, *P. sulphurea*, *P. nidularia* 'Smoothsheath', *P. nigra*, *P. nuda*, *P. praecox*, *P. sulphurea* 'Viridis', *P. vivax* f. *huangwenzhu*.

Damage Both nymphs and adults feed in bamboo cavity. The bamboo internodes above the infested internode become wilted and brittle, and bamboo utilization rate declines. Severe infestation can lead to the death of bamboo shoots above the infested internode.

Morphological characteristics **Adults** Body length 7.3-9.5 mm, newly emerged adult body is milky white, and then turns into black and shiny. Head is black. The compound eyes are dark brown, ocelli are brown. Antennae have 4 segments, the distal segment is the longest, fusiform, and dark brown in color. The second and third segments are in equal length. The first segment is the shortest. The third segment is light brown, and the rest are light-yellow. The center of the pronotum is slightly depressed, slightly elevated posteriorly, scattered with punctures of various sizes, and the posterior edge bent forward as an arch. Forewings are black, the base of the forewings has a yellow and white triangular spot. At the middle of forewings of the males, there is a wide horizontal band, while fore wings in females have two yellowish white spots. Abdomen is black, the last segment is truncated and exposed out of the wings at rest. The legs are light yellow, the tibia of all legs are pale black. Hind femur has two rows of spines ventrally, and the number of the spines per row varies. Head, antennae, thorax, legs, wing base and parasternum are covered with long yellow hairs.

P. japonicus adults (Xu Tiansen)

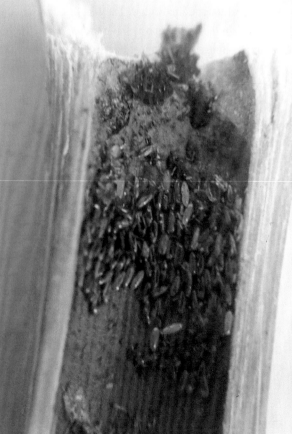

P. japonicus eggs (Xu Tiansen)

Eggs Elongated oval, 1.15-1.40 mm × 0.32-0.45 mm in size. Slightly pointed at both ends, milky white, smooth, shiny. The color gradually changes to creamy yellow. One end of the egg changes to light black before hatching.

Nymphs Length of the newly hatched nymphs is 1.6 mm. The widest body part is at the sixth to eighth abdominal segment, which is 0.6 mm. The body is in long oval shape, milky white in color. Area between head and mesothorax is light-yellowish white, compound eyes are light-yellow and slightly protruded. Antennae have three segments, and the sections have equal length. Length of the first instar can be up to 1.8 mm before molting. The body is covered with light yellow hairs. The 2nd instar length is 3.3-3.8 mm, its width is 0.9-1.2 mm, milky white. Antennae have three segments of equal length; the distal section is fusiform. Length of the 3rd instars is 4.5-5.0 mm, the sides parallel after mesothorax, width is 1.8-2.0 mm. Milky white in color, light-yellow before metathorax, legs light-yellow, compound eyes are bright red, and slightly protruded. Wing pads present, reaching up to the center of the first abdominal segment. The 4th instars are 5.8-6.2 mm in length. Head and thorax are light yellow, abdomen is creamy yellow. The center of the head elevated. Compound eyes are bright red, and antennae have four segments. Wing pads reach the middle of the second abdominal segment. Mature nymphs are 7.5-7.9 mm long, rod shaped. Head and thorax are yellow, abdomen is milky white, the center of the head raised in "Y" shape. Compound eyes are dark red, protruded; two ocelli are distributed below the "Y" protrusion, bright red. Antennae have four segments, the distal segment is the longest, fusiform shape. The middle of the lateral edges of the pronotum has a circular spot on each side. The wing pads reach the front of the third abdominal segment. Femur of the hind legs has two rows of small spines facing inward.

Biology and behavior The number of generations per year reported varies, ranging from 4 generations in Zhejiang, and 2 generations in Jiangxi. Its life history still needs to be further examined. According to incomplete records to date, all life stages (i.e., egg, nymph and adult stages) can be observed in Zhejiang province in winter. The bugs become active in late March of the following year. All developmental stages can be observed throughout the year by dissecting the bamboo cavity. The developmental time for eggs is 15 to 25 days, the developmental time for nymphs is 60 to 80 days. Adult life span is about six months. If overwintering as adults, it is possible that there are three generations per year.

Adults enter the bamboo stalk cavity through damage holes made by bamboo shoot feeding lepidopterous insects, animals, or mechanical injury. The lygaeids oviposit inside the bamboo cavity. Nymphs feed in bamboo cavity until emergence as adults. The adults have supplementary feeding, and then mate and oviposit in the bamboo cavity, and continue their damage on bamboo plants. Some adults crawl out of the original feeding sites, and mate and infest new bamboo internodes. The lygaeid bugs cannot penetrate bamboo stalks, they can only enter and feed on bamboo internode(s) with previous chewing damage by lepidopterous pests.

Prevention and treatment

Protection and preservation of the predators. It has been observed that the adults were preyed by *Sirthenea flavipes*, *Chlaenius bioculatus*, *Camponotus japonicus*, and *Polyrhachis dives*.

Reference Xu Tiansen and Wang Haojie 2004.

(Xu Tiansen, Wang Haojie)

113 *Sinorsillus piliferus* Usinger
Classification Hemiptera, Lygaeidae

Importance This bug is endemic to China, causing damage to fir cones and is one of the key factors that reduces seed production.

Distribution Zhejiang, Fujian, Jiangxi, Hubei, Hunan, Sichuan, Guangdong, Guizhou, Shaanxi.

Host *Cunninghamia lanceolata*.

Damage The adults and nymphs cryptically inhabit in the open fir cone bracts, and feed on the bracts and young seeds by removing sap. The damaged cone bracts become partially or completely reddish brown, and dry and cracked open later. The damaged seeds become shriveled and non-viable. Sometimes the bugs also feed on the tip of new shoots, tender leaves, and inflorescence. Damage symptoms include small brown spots, discoloration, deformation or death of shoots.

Morphological characteristics **Adults** Length is 5.5-9.0 mm, width is 2.5-4.2 mm; pale to dark brown; slightly flattened. The head is reddish or dark brown. The antennae have four segments, and brown in color. Proboscis has four segments, dark brown, it reaches the ninth abdominal segment. Forewings are reddish brown; the membranous area is transparent. Pronotum is trapezoidal in shape, covered with a dense engraved dot. Scutellum is triangular, dark brown.

Nymphs There are five instars. Length of the last instar is 4.0-5.2 mm. The body is elongated oval shaped, and relatively flattened. The body color is brown dorsally, dark brown ventrally, wing pads are exposed.

Biology and behavior *Sinorsillus piliferus* has two generations per year in Fujian province. It overwinters in nymphal stage. Adult eclosion occurs in early April of the following year within the cone fruits. Peak emergence occurs in mid- and late April. The adults cryptically inhabit inside cones during the daytime, and feed on the tender shoots in early morning or after dusk. Peak oviposition occurs in early and mid-May. Peak of the first generation of nymphs occur in late June and early July. The first-generation adults emerge in early August. Most of them lay eggs in late September. In early October, the second generation of nymphs hatch. The nymphs infest and feed on the newly produced cone bracts, which cause discoloration and shriveled non-viable seeds. In late October, the nymphs start overwintering in new fir cone bracts.

Prevention and treatment

1. Utilize new somatic lines with higher insect tolerance in the Chinese fir clones.

2. Manual control. Harvest all the fir cones (including the discolored and small dead cones infested with the bugs) to eliminated overwintering populations. Then, destroy the infested cones.

3. Chemical control. At the peak time of nymphs or adults, applying 20% fenvalerate, or 5% lamda-cyhalothrin at 5,000× spray to the canopy, or fogging of 20% fenvalerate to control both nymphs and adults.

References Qian Fanjun et al. 1992, 1994; Li Kuansheng 1999; Chen Shunli et al. 2004.

(Qian Fanjun)

S. piliferus adult (Qian Fanjun)

114 *Metacanthus pulchellus* Dallas

Classification Hemiptera, Berytidae
Synonym *Gampsocoris pulchellus* (Dallas)

Distribution Hebei, Jiangxi, Shandong, Henan, Hubei, Guangdong, Guangxi, Sichuan, Yunnan, Tibet, Shaanxi.

Hosts *Paulownia* spp., *Populus* spp., *Malus pumila*, *Pyrus* spp., *Prunus persica*.

Damage Feeding on shoots of the host plants by the bug will cause atrophy, stop the dominant growth. Tender leaves shrink as a result of the bug feeding.

Morphological characteristics **Adults** Length 3.5-4.2 mm, narrow, yellowish to greyish brown. The top of the head protruded forward, spherical. Antennae slender, longer than its body length. Pronotum protruded upward, its posterior edge with three conical protrusions. Scutellum appears as an upright thorn. Forewings yellowish white, membranous and transparent. Legs are particularly slender.

Nymphs The last instar nymph is yellowish green, and slender. Wing pads resemble "bubbles." Antennae and legs are slender.

Biology and behavior There are three generations per year in Henan Province. It overwinters as adults in ground cover or leaf litter, inside weeds, or crop stubble fields. The overwintering adults become active at the beginning of April of the following year. In mid-April, adults migrate to the nursery to aggregate, feed, and mate on young shoots. Oviposition begins in early May. Egg hatch in approximately 8 days. The first-generation nymphs reach peak in early June. Nymphs molt four times. The first and the second instars last for 6 days, the third instars for 5 days, and the fourth and fifth instars last for four days. A life cycle in the spring lasts for approximately 33 days. The nymphs of the second-generation peak in early July, and the third-generation peak in late July. The third-generation adults feed until early October, then gradually disperse to their overwintering sites.

Adults are slow moving. They are often seen on leaf surface. Mature nymphs are active. They often crawl on leaves or young shoots, looking for other insects on leaf surface, such as aphids, dipterans, and early instars of lepidopterous larvae and feed on them. After feeding, remains of the dead insect can stay on the plant for a long time. Male and female can mate frequently during both day and night. The pair can stay for more than 24 hours. Sometimes the females pull the males around. When disturbed, the females can drag males and flee. Females oviposit individually. Eggs are scattered inside the glandular hairs on the abaxial side of the leaves. The newly emerged nymphs prefer to feed in aggregation on tender shoots, and young leaves.

Prevention and treatment

1. Manual control. In combination with winter plowing, cleaning of the ground leaves, weeds, can effectively destroy and eliminate the overwintering adults.

2. Silviculture. Selection of the variety without glandular hairs that confer insect resistance to reduce damage. Less eggs and less insect populations and damage will be found on glandular hair-free hosts or varieties, such as *Paulownia fargesii*.

3. Chemical control. At the peak of nymph population, the spraying of 40% chlorpyrifos and phoxim mixture or 50% fenitrothion EC at 1,000× can be effective.

(Yang Youqian)

M. pulchellus adults (Wang Jijian)

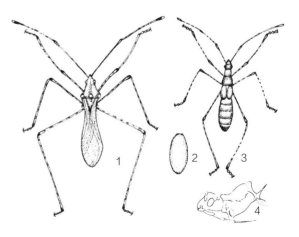

M. pulchellus. 1. Adult, 2. Egg, 3. Nymph, 4. Lateral view of head and thorax (Yang Youqian)

115 *Yemma signata* (Hsiao)

Classification Hemiptera, Berytidae

Distribution Beijing, Henan, Hebei, Hubei, Jiangxi, Zhejiang, Guangdong, Guangxi, Yunnan, Sichuan, Tibet, Shaanxi.

Hosts *Paulownia* spp., *Populus* spp., *Malus pumila*, *Pyrus bretschneideri*, *Prunus persice*.

Damage Infestations often occur on saplings in nurseries. Infested young shoots stop their dominant growth. Young leaves become shriveled and then split.

Morphological characteristics **Adults** Length 6.5-7.5 mm, narrow, pale brown. Scutellum appears as an upright thorn. Forewings narrow, not surpassing the last abdominal segment. Antennae and legs are particularly long, and both surpass the end of the abdomen. **Nymphs** The last instar is 6 mm in length, and yellowish green in color. Wing pads are round resembling bubbles.

Biology and behavior There are two generations per year in Henan province. It overwinters as adults in leaf litter, weeds, or crop stubble fields. The post-overwinter adults migrate to nursery seedlings in the following June. They feed, mate, and oviposit on the saplings. The nymphs of the first-generation peaks in mid-July, and the second generation is in mid-August. The adults of the second-generation feed until October, then disperse to overwintering/hibernation sites.

Adults and nymphs are often active on leaf surface or young shoots. They ingest plant sap, feed on other small insects on the leaf surface, e.g., aphids, flies, and early instar larvae of lepidopterous insects. The feeding remains of other insects are often attached on the leaves and last for a relatively long period of time. Males and females can mate multiple times. Mating occurs both during day and night. Females deposit their eggs individually on the abaxial side of leaves among glandular hairs. After hatching, nymphs aggregate and feed in clusters on young shoots and tender leaves by removing plant sap.

Prevention and treatment

Refer to the methods for *Metacanthus pulchellus*.

References Yang Youqian 1982, Zhang Shimei et al (ed.) 1985, Xiao Gangrou (ed.) 1992, Forest Disease and Pest Prevention and Treatment and Quarantine Station of Henan Province 2005.

(Yang Youqian)

Y. signata nymph (Li Yingchao)

Y. signata mating pair (Li Yingchao)

Y. signata feeding on an insect (Li Yingchao)

Y. signata adult at rest (Li Yingchao)

116 Notobitus meleagris (F.)
Classification Hemiptera, Coreidae

Distribution Zhejiang, Fujian, Jiangxi, Guangdong, Guangxi, Sichuan, Yunnan, Taiwan; also in India, Myanmar, Vietnam, and Singapore.

Hosts *Phyllostachys acuta*, *P. dulcis*, *P. flexuosa*, *P. heteroclada* f. *solida*, *P. heterocycal* 'Tao', *P. nidularia* 'Smoothsheath', *P. sulphurea*, *P. robustiramea*, *Bambusa blumeana*, *B. dissimulator*, *B. flexuosa*, *B. ventricosa*, *B. gibboides*, *B. lenta*, *B. multiplex*, *B. pervariabilis*, *B. textilis*, *B. vulgaris*, *B. surrecta*, *Dendrocalamopsis giganteus*, *D. minor*, *D. latiflorus*, *D. strictus*, *D. oldhami*, *D. daii*, *D. oldhami* f. *revolute*, *D. stenoaurita* and species belonging to genus *Pleioblastus*.

Damage Both nymphs and adults feed on tender bamboo shoots by ingesting plant sap. The damaged bamboo grows slowly. In Zhejiang, the population density of *Notobitus meleagris* on *Pleioblastus* can be as high as hundreds of insects per plant. The damaged plants can become shriveled and even die.

Morphological characteristics **Adults** Size 18-25 mm×6.5-7.0 mm, dark brown to black, covered with short yellowish-brown hairs. Head is short, the ratio of length and width is 2:3. The basal three segments of the antennae have the same length. The basal half of the fourth antennal segment is light in color, and the rest (distal half) is black. Compound eyes are protruded, dark brown. Proboscis dark brown and reaches over the coxae of middle legs. There are densely scattered dots on the pronotum and scutellum. The pronotum has a "collar", which is yellowish brown with superficial transverse wrinkles, the front edge is concaved, and the center of the posterior margin is concaved, and lateral corner is round, not protruded. The leathery forewing portion is dark brown, and membranous portion is brown, and surpasses the last abdominal segment. The outer edge of the lateral margin of the abdomen is bright yellowish brown, area between segments is black; spiracles are black, with light color surrounding them.

Eggs Flat oval, size 1.48-1.64 mm×1.16-1.32 mm. Golden brown and with metallic hue initially, then darkened into

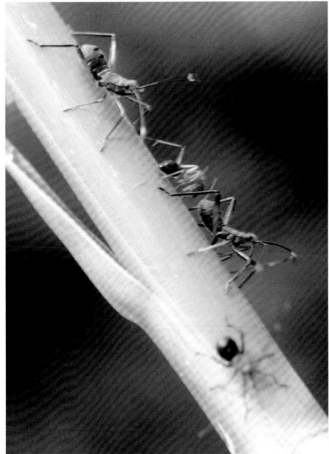

N. meleagris 3rd instar nymphs (Xu Tiansen)

N. meleagris adult (Xu Tiansen)

N. meleagris eggs (Xu Tiansen)

N. meleagris 2nd instar nymphs (Xu Tiansen)

N. meleagris adults (edited by Zhang Runzhi)

dark coppery yellow. **Nymphs** Length of the newly hatched nymphs 3.5 mm, dark brown. Antennae longer than the body, legs slender. The mature nymphs are 19-21 mm long, dark brown or greyish brown, antennae black, and the basal part of the fourth segment of the antennae rusty yellow. The central region of the pronotum, scutellum, and base of the wing pads is dark brown, the scent gland is yellow, and the surrounding area is black. Lateral edges of the abdomen are yellow.

Biology and behavior There are 1-2 generations per year in Zhejiang, and 5 generations per year in Guangdong. It overwinters as adults. In Zhejiang, the overwintering adults become active and start feeding on new shoots in late April to early May of the following year. Females lay eggs between mid-May and early June. Eggs are deposited in two rows with zig-zag arrangement longitudinally on small branches, and abaxial side of the leaves, and other miscellaneous shrubs. Nymphs occur in late May to mid- and late June. Adult emergence starts at the end of June. After supplementary feeding, they overwinter. A small percentage of them oviposit in early July. Nymphs of the second generation appear in mid-July and molt into adults in August.

The overwintering adults in Guangdong become active in late March. They feed and mate in early April and oviposit in mid-April. Nymphs hatch in late April and feed on bamboo. After 30 to 40 days, the first generation of nymphs matures and then adult emerges. The occurrence of future generations is in mid-June to mid-July, mid-July to mid-August, mid-August to mid-September, mid-September to the following mid-April, which is essentially one generation per month. Generations overlap.

Prevention and treatment

Apply *Beauveria bassiana*. Protection and preservation of natural enemies. Predatory birds include *Cuculus canorus canorus*, *Garrulax canorus* and other birds prey on both adults and nymphs. The early instars can also be preyed by *Sirthenea flavipes*, *Dolomedes pallitarsis*, *Oxyopes* sp., and *Pisaura* sp. The nymphs can also be preyed by *Hierodula patellifera*, and *Sphedanolestes gularis*.

References Xu Tiansen and Wang Haojie 2004.

(Xu Tiansen)

117 *Notobitus montanus* Hsiao
Classification Hemiptera, Coreidae

Distribution Zhejiang, Fujian, Jiangxi, Guangdong, Guangxi, Sichuan, Yunnan, Taiwan.

Hosts *Phyllostachys bissetii*, *P. bambusoides* f. *Lacrima-deae*, *P. bambusoides* f. *shouzhu*, *P. dulcis*, *P. flexuosa*, *P. heteroclada* f. *Solida*, *P. heterocycal* 'Tao', *P. heteroclada*, *P. heteroclada* var., *P. iridescens*, *P. nidularia* 'Smoothsheath', *P. nigra*, *P. robustiramea*, *P. sulphurea*, *P. sulphurea* 'Viridis', *P. vivax*, *Bambusa blumeana*, *B. dissimulator*, *B. flexuosa*, *B. ventricosa*, *B. gibboides*, *B. vulgaris*, *B. surrecta*, *Pleioblastus amarus*, and *Pleioblastus juxianensis*. Occasional hosts include *Zea mays* and *Triticum aestivum*.

Morphological characteristics **Adults** Size is 19.6-23.5 mm×5.2-5.8 mm, males slightly smaller than the females; dark brown, covered with greyish yellow hairs. Antennae 10-12 mm long, the first segment shorter than or equal to the head width, the basal half of the fourth segment rusty yellow. Compound eyes are protruded and yellowish brown. Proboscis reaches the front edge of the meso-sternum. Pronotum is trapezoidal, color of the middle and rear parts is lighter than the front part. The femora of the hind legs are enlarged, the distal 2/5 part has a large spine, and there are several small setae in front of and behind the large spine. Hind tibiae are bent slightly inward near the base. The basal half of abdominal dorsum is red. **Eggs** Flat oval, 1.65-1.78 mm×1.1-1.2 mm in size, light brown with bronze hue. **Nymphs** The newly hatched nymphs 2.5 mm long, pink, and change into greyish black later. Thorax small, abdomen large, antennae and legs slender. Mature nymphs 14-17 mm long, soft, wing pads are apparent. Thorax and wing pads are dark brown. The abdomen is relatively large with light color. The scent glands located between the third and fourth, and the fourth and the fifth segment slightly elevated.

Biology and behavior In Zhejiang, the bug is univoltine. In Guangxi, it occurs one to two generations per year and overwinters as adults. The overwintering adults become active in mid- and late April. They lay eggs in late April to early May. The eggs are arranged in the longitudinal strips in each egg mass and forming reversed "V" shape. Eggs are deposited in bamboo twigs and abaxial side of leaf surface, and a few egg masses are also found on bamboo stalks or weeds. There are 18 to 38 eggs per egg mass. Eggs hatch in 12 to 19 days in May. Nymphs can be observed in early and mid-May, and nymphal stage lasts 30 to 45 days. In late June to early July, the mature nymphs molt into adults. In Guangxi, some adults can mate and lay eggs. The second-generation nymphs occur in late September to early October. The first generation of adults start overwintering in late July to early September, while the second-generation adults start overwintering in November.

Prevention and treatment

Apply *Beauveria bassiana*. Protection and preservation of predators. The predators of the early instars of *N. montanus* include *Oxyopes* sp., *Pisaura* sp., *Lycosa coelestris*. Two reduviids, *Sirthenea flavipes*, and *Sphedanolestes gularis*, also prey on *N. montanus*.

Reference Xu Tiansen and Wang Haojie 2004.

(Xu Tiansen)

N. montanus adults (Xu Tiansen)

N. montanus nymphs (Xu Tiansen)

118 *Cyclopelta parva* Distant

Classification Hemiptera, Dinidoridae
Common name Small wrinkled stink bug

Importance *Cyclopelta parva* is widely distributed in our country, mainly on locust and other legumes. When infestations are severe, they can cause great damage to the lumber production.

Distribution Inner Mongolia, Liaoning, Jiangsu, Zhejiang, Fujian, Jiangxi, Hubei, Hunan, Shandong, Guangdong, Guangxi, Sichuan, Yunnan; Myanmar, Bhutan.

Hosts *Robinia pseudoacacia*, *Amorpha fruticosa*, *Lespedeza bicolor*, *Pueraria lobata*, *Phaseolus vulgaris*, *Lablab purpureus*, *Vigna unguiculata*, *Glycine max*, *Citrullus lanatus*, and *Cucurbita moschata*.

Damage Adults and nymphs feed in clusters on 1-3 year-old branches, the base of the young tree trucks and tender parts at branching points. Leaves turn yellow or fall prematurely. The severely infested twigs and young branches become enlarged, cracked, and rotten. Death of whole plants may occur.

Morphological characteristics **Adults** The adults are dark brown, and dull. The size is 12-15 mm×6-10 mm. **Nymphs** The newly hatched nymphs are pale red, the thorax is yellowish brown after molting. The central line has a row of cyst-like protrusions. Compound eyes are dark red.

Biology and behavior There is one generation per year. It overwinters as adults in weeds or under rocks. The overwintering adults become active in mid-March of the following year. At first, the adults crawl out at noon, and crawl back under cryptic habitats before the sunset without feeding. As the ambient temperature rises, the adults gradually move towards the base of the tree trucks. When locust plants are flowering (in late April), the adults crawl upward on the locust trees to feed and cause damage to plants.

Adults begin mating and oviposition in early June. They mate multiple times before oviposition. The females start oviposition the day after mating. The eggs are deposited on branches of 2-7 mm in diameter. They are tightly arranged in a longitudinal row, forming a circle or a semi-circle around the branches. The fecundity is approximately 110 eggs. The peak of the oviposition is in late June to early July. Eggs hatch after 2 to 3 weeks. There are five instars in the nymphal stage. The development time of the nymphal stage is approximately 55 days. The overwintering period starts in late September. In early November, all adults are in overwintering state. Adults and nymphs are gregarious and play death. Feeding damage occurs in August and September. If adults are disturbed, they may spray irritating venom.

Prevention and treatment

1. Manual control. Removal of weedy grass coverage, overturn of the stones can effectively kill the overwintering adults. In late June to early July at oviposition peak time, the removal and destroy (burning) of branches with egg masses can also be effective.

2. Biological control. Under the environment of high humidity, the application of *Beauveria bassiana* can also be effective.

3. Chemical control. At the peak of egg hatching time, the application of diflubenzuron, hexaflumuron can be effective. For nymphs, spray 2% abamectin+ indoxacarb or 5.7% emamectin at 1,000-1,500×.

References Xiao Gangrou (ed.) 1992, Fan Di 1993.

(Li Xianchen, Wang Qinghai, Sun Yujia)

C. parva adult (Xu Gongtian)

C. parva nymph (Xu Gongtian)

119 *Aenaria pinchii* Yang

Classification Hemiptera, Pentatomidae

Distribution Jiangsu, Zhejiang, Anhui, Fujian, Jiangxi, Henan, Hubei, Hunan, Guangdong, Guangxi, Sichuan, Guizhou.

Hosts *Phyllostachys heteroclada* var. *pubescens*, *Phyllostachys vivax* f. *aureocaulis*, *P. bambusoides*, *P. dulcis*, *P. bissetii*, *P. bambusoides* f. *shouzhu*, *P. aureosulcata* f. *pekinensis*, *P. rutila*, *P. nidularia* 'Smoothsheath', *P. makinoi*, *P. flexuosa*, *P. nuda*, *P. praecox*, *P. heteroclada*, *P. glauca*, *P. sulphurea* 'Viridis', *P. praecox* f. *provernalis*, *P. propinqua*, *P. iridescens*, *P. vivax*, *Bambusa pervariabilis*, *B. chungii*, *B. textilis*, *B. eutuldoides*.

Damage Adults and later instar nymphs (older than the third instar) feed on bamboo stalk and nodes of the large branches. The nymphs before the third instar feed on the nodes of small branches and at the branching points to ingest plant sap, causing early senescence and falling of leaves, and subsequently death of the twigs and large branches. The whole plant may die when the *A. pinchii* population density is extremely high.

Morphological characteristics **Adults** Body size 9.5-12.4 mm×5.5-6.2 mm, pale green, covered with dense and uniform black engraved dots. Head is equilateral triangle in shape. Compound eyes are dark brown. Antennae have 5 segments, yellowish brown, and the distal segment is black. The lateral margin of pronotum is white. The lateral margin of the body and wings is white. The opaque portion of the forewing is purplish brown. The area outside radius vein is yellowish white.

Eggs Milky white, barrel shape, diameter 1.2-1.3 mm, the egg cover diameter 1.0 mm, egg length 1.4-1.6 mm. The egg cover has six red dots before hatching; four of them are arranged in a square, the other two arranged in an inverted "V" shape where a black triangle mark exists.

Nymphs Body size 8.8-10.7 mm×5.4-6.2 mm, emerald green. Head is in the shape of an equilateral triangle. Antennae have four segments, brown, the distal segment

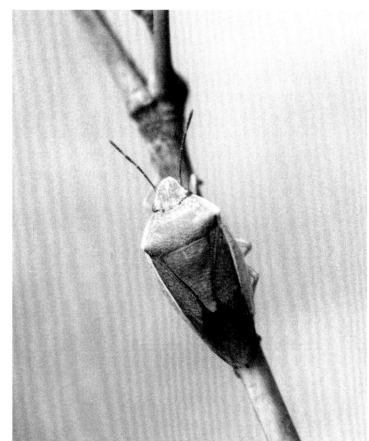

A. pinchii adult (Xu Tiansen)

A. pinchii newly hatched nymphs and empty eggs (Xu Tiansen)

A. pinchii eggs (Xu Tiansen)

A. pinchii nymph (Xu Tiansen)

A. pinchii nymphs (Xu Tiansen)

is black. Compound eyes are dark brown. The lateral margin of the pronotum is white. The margin of the body and wing pads is white. There is a black strip inside the white margin of all abdominal segments.

Biology and behavior In Zhejiang province, *A. pinchii* is univoltine. It overwinters as adults under the fallen twigs, leaf litter, and ground cover. The overwintering adults start feeding between mid-March and early April. They mate in mid- and late April. Each female can produce 1-2 egg masses, and the peak of oviposition is in mid-May. Eggs hatch in 7 to 12 days. Nymphs mature in approximately 50 days. The duration of nymphal stage is 55 to 65 days. The adults are not active in the summer. They start to overwinter on the ground cover from the end of July to November.

Prevention and treatment

Protection and preservation of natural enemies. The predatory birds include *Cuculus canorus canorus*, *Bambusicola thoracica* and others. The spiders include *Araneus mitificus*, *Dolomedes pallitarsis*, *Dolomedes sulfureus*, *Pardosa tschekiangensis* and others. The predatory insects include *Hierodula patellifera*, *Sirthenea flavipes*, *Haematoloecha nigrorufa*, *Chlaenius bioculatus*, *Cicindela* sp., *Formica japonica*, and *Polyrhachis dives*. The egg parasitoid include *Teienomus* spp. The parasitism rate is up to 25%. Both adults and nymphs may be infected with *Beauveria* sp.

Reference Xu Tiansen and Wang Haojie 2004.

(Xu Tiansen)

120 *Halyomorpha halys* Stål

Classification Hemiptera, Pentatomidae
Common name Brown marmorated stink bug

Distribution Beijing, Tianjin, Hebei, Liaoning, Jiangsu, Zhejiang, Anhui, Fujian, Jiangxi, Shandong, Henan, Hubei, Hunan, Guangdong, Sichuan, Guizhou, Yunnan, Shaanxi, Gansu, and Taiwan.

Host *Pyrus* spp., *Malus pumila*, *Malus spectabilis*, *Prunus persica*, *P. salicina*, *P. armeniaca*, *P. pseudocerasus*, *P. mume*, *Crataegus pinnatifida*, *Ficus carica*, *Punica granatum*, *Diospyros kaki*, *Citrus reticulata*, *Ulmus* spp., *Morus* spp., *Syringa* spp.

Damage Nymphs and adults feed on leaves, tender shoots, and fruits of host plants by ingesting plant sap, which causes defoliation, deformation, and lignification of fruits. The damaged young fruits often fall off.

Morphological characteristics **Adults** They are about 15 mm× 8-9 mm in size, flat and oval. The whole body is dark brown.

Eggs The eggs are short and cylindrical, 20 to 30 eggs are deposited side by side in an egg mass, and are shaped like a cup, greyish white in color. The eggs turn into dark brown before hatch.

Nymphs They are similar to adults morphologically, except nymphs are apterous, and with wing pads at later instars.

Biology and behavior *Halyomorpha halys* is univoltine. It overwinters as adults in the corners and the eaves of the vacant houses, under the weeds and ground coverage, in tree holes, under the rocks, and other places. In early May of the following year, the overwintering adults become active gradually, and migrate to fruit trees, forests and crops to feed and cause damage. Oviposition starts in June. Eggs are deposited on the abaxial side of leaf blades. The egg incubation period is 10-15 days. They hatch in early July. Newly hatched nymphs aggregate around the egg masses and then gradually disperse. Adults emerge in mid-August. The adults continue to feed and cause damage on its hosts until September. Then adults seek for appropriate overwintering habitats.

Prevention and treatment

1. Manual prevention and treatment. Manual removal of overwintering adults at its overwintering habitats can be effective. In the areas with frequent severe infestations, the fruits can be protected by installing fruit protective bags. In addition, the removal of egg masses at the oviposition period can also be effective in reducing bug population.

2. Biological control. The release of natural enemies

H. halys adult (edited by Zhang Runzhi)

H. halys nymph (Xu Gongtian)

H. halys adult (Xu Gongtian)

H. halys nymph (edited by Zhang Runzhi)

H. halys nymph (edited by Zhang Runzhi)

can be effective at egg stage. For example, the release of *Trissolcus halyomorphae* can reach parasitic rate of 80% of the eggs, which can effectively reduce the populations and their damage to crops.

3. Chemical control. Because the nymphs often aggregate on branches and the abaxial side of the leaves, it is effective to control the aggregated nymphs by spraying of 4.5% cypermethrin at 1,000× or 40% thiamethoxam or 12% thiamethoxam and lamda-cyhalothrin mixture. at 1,500 to 2,000×.

References Forestry Bureau of Henan Province 1988, Li Zhongxin and Liu Yusheng 2004, Wan Shaoxia and Zhang Lifeng 2004.

(Wan Shaoxia)

121 *Hippotiscus dorsalis* (Stål)

Classification Hemiptera, Pentatomidae

Distribution Shanghai, Jiangsu, Zhejiang, Anhui, Fujian, Jiangxi, Henan, Hunan, Guangxi, Sichuan, Guizhou, Tibet; India.

Hosts *Phyllostachys heteroclada, P. acuta, P. angusta, P. aureosulcata* f. *pekinensis, P. aurita, P. bambusoides, P. bambusoides* f. *shouzhu, P. dulcis, P. flexuosa, P. glauca, P. glauca, P. iridescens, P. sulphurea* 'Viridis', *P. nidularia* 'Smoothsheath', *P. nigella, P. nuda, P. praecox, P. praecox* f. *provernalis, P. propinqua, P. vivax, P. makinoi, P. yunhoensis*.

Damage Nymphs and adults aggregate at or around the nodes of large and small bamboo shoots. They cause leaf fall and death of small branches above the feeding sites. When the bug population density is high, a lot of bamboo branches and twigs die, and the whole bamboo plant may die.

Morphological characteristics **Adults** Size is 13.5-15.5 mm×7.5-8.0 mm, with a relatively high dorsal bulge. The newly emerged adults are creamy yellow and change to shiny greyish blue in 3-4 days. Then, the adults become greyish yellow, greying brown, and greenish brown covered with dense black marker dots and white waxy powder. Head is obtuse triangle in shape. The base of the triangle is shorter than the lateral sides. Compound eyes are dark red, with a smooth area on its interior side. Antennae have five yellowish brown to dark brown segments; the basal half of the distal segment is light yellow. Front margin of the pronotum black, extends slightly outward in an arch shape. At the end of scutellum, there is a yellowish white crescent-moon shaped mark, without engraved dots. Membranous part of the forewing is light black, while the base part of the leathery portion of the front wings is black. The ventral surface of the body is yellow, spiracles are black. Legs are light yellow. **Eggs** Barrel shape, 1.4 mm×1.2 mm in size, with an egg cover in diameter of 1.0 mm; color yellow. Before hatching, a triangle with black margins appears on one side of the egg cover. There is a perpendicular black line dividing the triangle into two parts. **Nymphs** There are 5 instars in the nymphal stage. The first instar is 1.8-2.0 mm×1.4-1.6 mm in size, short oval, yellowish white. Head is triangular. The compound eyes are spherical, and dark red. The antennae have 4 segments, the basal segment and the end of the last segment light black. Pronotum is light grey, and tarsus is light black. Starting from the 2nd instar, the nymphal body length are 2.8-3.5, 4.6-5.3, 7.0-9.1, and 9.5-13.0 mm, respectively; width is 2.0-2.2, 3.2-3.8, 4.5-5.2 mm, respectively. The body of the second instar larvae is greyish yellow with black engraved dots. The lateral lobes of the head is longer than the middle lobe. The fourth instar larvae are brownish yellow in color. Antennae are milky cream in color, and distal segment is light black. The compound eyes are brown. The lateral sides of the meso- and meta-thoraxes are black, extend to the end of the abdomen to form a black "V" shaped mark. Mature nymphs are brownish yellow, with black engraved dots. Antennae have four segments, greyish black. The wing pads are black. Lateral margins of abdomen are light-yellow.

Biology and behavior There is one generation per year in Zhejiang. It overwinters as the second to the fourth instar nymphs at the end of October, or early November. When the daily average temperature falls at approximately 10 ℃, the nymphs fall to the ground, and seek ground cover for overwintering. The majority of the overwintering nymphs are the fourth instar nymphs, which represents approximately 95% of the population. Those overwintering as the second or third instars have high mortality. In the early and mid-April of the following year, the overwintering nymphs become active. Nymphs crawl upon bamboo plants and feed at the bamboo node. If the temperature drops, the nymphs on bamboo plants can fall off to the ground and seek ground cover. When the temperature rises again, the nymphs crawl back to bamboo plants. Adult emergence starts at the end of May to early June, mating starts in mid-June, and peaks in late June. The oviposition period peaks in mid-July, Eggs are deposited in egg masses. Each egg mass contains 14 eggs. They are arranged in two zig-zag rows on the abaxial side of the bamboo leaves. There are occasionally 28 eggs in an egg mass and arranged

Sticky band at base of bamboo to control *H. dorsalis* (Xu Tiansen)

H. dorsalis adult (Xu Tiansen)

H. dorsalis nymphs on bamboo (Xu Tiansen)

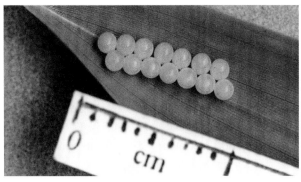
H. dorsalis eggs (Xu Tiansen)

H. dorsalis nymphs caught on glue (Xu Tiansen)

also in two staggered rows. Adults disappear in early October.

Prevention and treatment

1. Protection of natural enemies. Predatory birds include *Cissa ergtyrorhyncha*, *Parus major*, *Cuculus canorus canorus*, *Garrulax canorus*, etc. The spiders including *Pisaura* sp., *Oxyopes* sp., *Oxyopes sertaus*, *Dolomedes pallitarsis*, *Pardosa wuyiensis*, *Chiracantium zhejiangensis*. Other predatory insects include *Isaria cicadae*, *Hierodula patellifera*, *Sirthenea flavipes*, *Lsyndus obscurs*, *Chlaenius bioculatus*, *Cicindela* sp., *Camponotus japonicus*, and *Polyrhachis dives*. There are two egg parasitoid wasps, *Teienomus* spp., and *Ooencyrtus longivenosus* parasitizing eggs. In addition, *Beauveria bassiana* can infect adults and nymphs of the bugs.

2. Oil barrier method. In Zhejiang province, when overwintering nymphs move up to the bamboo stalks in early April, the application of a 15 cm wide oil band containing 1% of any pesticides at the base of bamboo stalks can be effective. Some nymphs are glued on the lubricant band, while most cannot pass the band.

3. Removal of nymphs feeding on bamboo. Using a homemade insect sweeping net with a notch, push against the bamboo stalk and move upward to catch the bugs on bamboo stalk. The nymphs will fall into the net.

4. Chemical control. Before overwintering nymphs crawl to the bamboo stalks, application of an insecticide band (3% thiamethoxam) below 50 cm height on bamboo stalk can be effective. When the population density is high, inject thiamethoxam at 1 mL per bamboo.

References Xu Tiansen and Wang Haojie 2004; Xu Tiansen, Wang Haojie, Yu Caizhu 2008.

(Xu Tiansen)

122 *Nezara viridula* (L.)

Classification Hemiptera, Pentatomidae
Common name Southern green stink bug

Distribution Hebei, Shanxi, Jiangsu, Zhejiang, Anhui, Fujian, Jiangxi, Shandong, Hunan, Guangdong, Guangxi, Sichuan, Guizhou, Yunnan and Taiwan; Russia, Japan, Korea Peninsula, India, Sri Lanka, Myanmar, Malaysia, Vietnam, Indonesia, the Philippines, Australia, New Zealand, Madagascar, South Africa, St. Helena, Cape Verde Islands, Venezuela, Guyana and Cuba.

Hosts *Malus pumila*, *Pyrus bretschneideri*, *Citrus reticulata*, *Oryza sativa*, *Triticum aestivum*, *Zea mays*, soybean, *Glycine max*, *Arachis hypogaea*, *Saccharum sinense*, *Citrus aurantium*.

Damage Both adults and nymphs feed on young and tender shoots and cause branch death and slowed growth.

Morphological characteristics **Adults** Male length is 12.0-14.0 mm, female length is 12.5-15.5 mm. Green, or only the front margin of head is yellow, and the rest of the body is green; the ventral surface of body pale green, compound eyes black, ocelli dark red. At the base of scutellum, there is a row of three small light yellow or green dots. **Eggs** Spherical in shape, and the top with egg cover, egg cover is surrounded by small white spikes. The fresh eggs are yellowish white, and then become reddish brown. **Nymphs** There are total of five instars. Each instar has different markings and color. The fifth instar nymphs are 7.4-10.0 mm in body length. Pronotum has four black dots arranged in a row. The scutellum has four black dots arranged in the shape of trapezoid. There are red spots on dorsal surface of the third and the fourth segments of the abdomen.

Biology and behavior *Nezara viridula* occurs 3 to 4 generations per year in Guangdong province. It overwinters as adults aggregated in loose soil or weed covered ground. The overwintering adults gradually become active in March or early April of the following year. Most adults mate in the daytime and oviposit at nights on abaxial side of the leaves. Eggs are neatly arranged in 2-6 rows, and each row has 30 to 70 eggs. The newly hatched nymphs aggregate near eggshells, and disperse gradually after the second instar.

Prevention and treatment

The bug mainly causes damage to nursery seedlings and newly established saplings. Insecticides can be used for the control of this pest. The insecticides can be used include 25 g/L lamda-cyhalothrin at 1,000×, 20% thiamethoxam or 12% thiamethoxam and lamda-cyhalothrin mixture at 1,000×, 5% lamda-cyhalothrin at 1,500× or 10% imidacloprid WP at 2,000× to 3,000×.

Reference Yang Weiyi 1964.

(Gu Maobin)

N. viridula adult (Xu Gongtian)

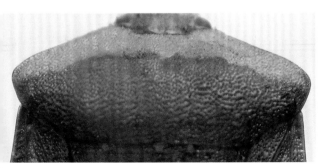

N. viridula adult pronotum (Xu Gongtian)

123 *Poecilocoris latus* Dallas

Classification Hemiptera, Scutelleridae
Common name Tea seed bug

Importance The bug causes important damage on young leaves and fruits, and seriously reduces tea and fruit quality.

Distribution Jiangsu, Zhejiang, Fujian, Jiangxi, Hunan, Guangdong, Guangxi, Sichuan, Guizhou, Yunnan; Vietnam, Laos, Myanmar, India, Malaysia, Indonesia, and Bangladesh.

Hosts *Camellia sinensis*, *Camellia oleifera*.

Damage Newly hatched nymphs ingest leaf sap, causing leaf deformation and reduce shoot growth. Late instar nymphs and adults can also damage fruits.

Morphological characteristics **Adults** Male size 16-19 mm×10.5-12.0 mm, female size 17-20 mm×12-14 mm; wide oval.

Eggs Nearly round, 1.8-2.0 mm in diameter. Fresh eggs are yellowish green. In a few days, two long purple bands appear. The eggs become orange before hatching.

Nymphs Length 3 mm, nearly round, orange color with a metallic hue. There are five instars in the nymphal stage.

Biology and behavior In Jiangxi, Fujian, Guangxi, and Yunnan provinces, there is one generation per year. There are two generations per year in Guangdong province. The insects overwinter as the last instar nymphs under leaves, soil cracks, or on the abaxial side of leaves in low canopy. In warm areas of Guangdong province, there is no overwintering phenomenon. The overwintering nymphs become active in the following March. Adults emerge between late April and the end of June. A small number of adults emerge in July. Adults

P. latus adult (edited by Zhang Runzhi)

are active at daytime, exhibit false death behavior. They prefer to inhabit in closed camellia tree canopy and feed on young and developing fruits. The adult life can be more than three months. They can mate multiple times during the day between 11:00-14:00. The mating can last for two to five days. Eggs are deposited in egg masses on abaxial side of leaves. A female can produce 3 to 9 egg masses, and each has 10 to 15 eggs. A female produces an average of 79.6 eggs. Egg stage lasts for 7 to 15 days. Eggs often hatch between 07:00-10:00 in the morning. All eggs in an egg mass can hatch within one to two hours.

In late July, nymphs aggregate and feed on abaxial side of leaves, ingest leaf sap, causing leaf deformation. Nymphs disperse after the second instar. They feed on young fruits and cause abnormal seed development. The feeding site on fruit often has a star like mark. There are five instars in the nymphal stage; the first to the second instars lasts 12 to 13 days; the second to the third instars lasts 20 to 24 days; the third to the fourth instars last 23 to 26 days, and the fourth to the fifth instars last 35 to 38 days. A total of 90-101 days is needed from the first to the fifth instars. The overwintering period of the fifth instar nymphs lasts from 100 to 120 days. The duration

P. latus 2nd instar nymphs (Zhang Peiyi)

P. latus 3rd instar nymphs (Zhang Peiyi)

P. latus 4th instar nymph (Zhang Peiyi)

P. latus damage to tea tree (Zhang Peiyi)

P. *latus* adults (edited by Zhang Runzhi)

of each instar is shorter under high ambient temperature, and vice versa. Nymphs feed on flower buds in late autumn until late October, then last instar nymphs overwinter. The last instar nymphs exhibit false death behavior. In late November, all nymphs are in overwintering state. Population is low in thinned camellia stands or stands without undergrowth of vegetation (trees or weeds). There is no apparent overwintering phenomenon in Guangdong province. Nymphs become active early in the following year. Adults appear at the end of April and oviposition starts in late June.

Prevention and treatment

1. Manual removal of insects and egg masses. Because the size of the insect is relatively large, with bright color, and aggregation behavior, the manual removal of egg masses, nymphs and adults can be effective in accompany with other tea garden management procedures.

2. Clean camellia stand and remove other tree species.

3. Chemical control. Laboratory efficacy tests showed that 20% fenvalerate emulsion at 6,000× is very effective against both nymphs and adults; 25 g/L lamda-cyhalothrin at 1,000× can cause 88.9% mortality. In addition, 77.5% trichlorfon emulsion at 800× can cause 95.65% mortality of adults and nymphs and can cause 100% mortality of eggs.

References Yang Weiyi 1962, Wang Zongkai 1964, Gan Jiasheng 1982, Jun Xuan 1984, Wei Qiyuan 1985, Hu Keming 1988, Xiao Gangrou (ed.) 1992.

(Wen Xiujun, Huang Shaobin, Su Zhaoqi)

124 *Tessaratoma papillosa* (Drury)

Classification   Hemiptera, Tessaratomidae
Common name   Litchi stink bug

Importance   *Tessaratoma papillosa* is the most important pest in lychee and longan production.

Distribution   Fujian, Jiangxi, Guangdong, Guangxi, Hainan, Guizhou, Yunnan, Taiwan; the Philippines, Vietnam, Myanmar, India, Thailand, Malaysia, Sri Lanka, and Indonesia.

Hosts   *Litchi chinensis*, *Euphoria longan*, *Koelreuteria paniculata*, *Sapindus mukorossi*.

Damage   Flower petioles, surface of fruits, young and tender shoots of the host trees show black spots after feeding by *Tessaratoma papillosa*. A large number flowers and fruits can be found on the ground during heavy infestation. Mature fruits show abnormal growth after feeding.

Morphological characteristics   **Adults** Male 22-26 mm in length, oval, dark or yellowish brown, darker on the dorsal surface. Body covered with white waxy powder except the head, thorax and wings. Antennae have four segments, purplish black. Pronotum is especially enlarged with elevated middle and hind parts, its postal edge extends backward to cover the base of scutellum. Legs are dark brown, and claws are black; there are two spines under the end of femur. **Eggs** Near spherical, with a white line in the middle. Fresh eggs are light yellow, and then turn into reddish black when close to hatch. **Nymphs** There are five instars in the nymphal stage. The newly hatched nymphs are oval and pink, later turn into bluish black. The lateral sides of pronotum are fresh yellow. The second instar nymphs are 8 mm in length, orange, and almost rectangle in shape. The head, antennae, humeral angle of prothorax and dorsal edges of abdomen are dark blue. There are two dark blue lines on the abdomen. The third instar nymphs are 10-12 mm in length. The edge of metathorax is surrounded by mesothorax and the margin of first abdominal segment. The 4th instar nymphs are 14-16 mm in length. The wing pads are apparent, and located at both sides of the mesonotum, which extends to postal edge of metathorax. The last instar nymphs are 18-23 mm in length. Head, thorax and dorsum of abdomen are dark red. Lateral sides of the abdomen are orange in color. The last three segments are black. Wing pads can reach the central part of the third abdominal segment. The abdomen is enlarged and the whole body is covered with white powder before adult eclosion.

Biology and behavior   *Tessaratoma papillosa* occurs one generation per year. It overwinters as sexually immature adults. The overwintering adults become active in early March. After supplementary feeding, the adults mate and oviposit. Eggs can be found from March to October, and nymphs are found from April to October. The nymphs peak during May and July. The new generation adults appear in June. Adult emergence peaks in mid-July. The overwintering adults gradually disappear after June. Eggs hatch in 13-14 days with maximum of 25 days or minimum of 7 days. The duration of the first instar nymphs is 21 days. The second instar is 8 days, the third is 10 days, the fourth is 17 days, and the fifth is 26 days. Nymphs become adults in 82 days. The total nymphal stage is between 58-116 days; the adult life span is 203-371 days with an average of 311 days.

Adults overwinter among leaves, in tree holes, or rock gaps. They can move and feed during overwintering period when temperature is above 15℃. They do not move when temperature is below 10℃. In early March of the following year, when the temperature is about 16℃, the overwintering adults begin supplementary feeding on new growing points and flower buds. They mate and lay eggs when they are sexually mature in March. Peak of

Fresh *T. papillosa* eggs (Wang Jijian)

T. papillosa newly hatched first instar nymphs (Wang Jijian)

T. papillosa mating pair (Wang Jijian)

T. papillosa nymphs on a tree (Wang Jijian)

T. papillosa nymph feeding on a branch (Wang Jijian)

T. papillosa nymph feeding on a fruit (Wang Jijian)

T. papillosa mid-aged and mature eggs (Wang Jijian)

oviposition occurs in April and May. Eggs are usually deposited on the abaxial side of leaves located at lower portion of the canopy. Eggs have also been found on barks, twigs and other part of the trees. Generally, fourteen eggs are deposited per egg mass. Each female can deposit 5-10 (maximum 17) egg masses. Eggs start to hatch in late March. Peak hatching occurs between May and July. All instars inhabit on the abaxial side of leaves with false death behavior. Nymphs fall off when they are disturbed. The third instar nymphs and adults can emit foul odor for self-defense. Nymphs molt into adults beginning in May and June. Adult emergence ends in October.

Prevention and treatment

1. Manual control. Adults can be manually collected and killed during overwintering period by shaking the branches of the plants. Nymphs can be manually removed from trees using a bamboo stick to shake the branches and make nymphs fall to the ground. The egg masses on the low branches can also be manually removed.

2. Protection and utilization of natural enemies. Removing the leaves with eggs and then placing them in a fine mesh metal screen cage with mesh size is smaller than the size of eggs. The screen would allow parasitoid *Anastatus japonicus* to fly out of the cage. Repeated release of *A. japonicus* can be performed to reduce pest population.

3. Biological control. In the spring when the adults become active and during nymphal stage, the application of *Beauveria bassiana* powder over the canopy of the trees can be effective in killing both adults and nymphs.

4. Chemical control. During late March to early April prior to the blossom of lychee and longan and in mid- and late May when it is the peak time of nymphs, one of the following insecticides can be applied: 45% thiamethoxam and lamda-cyhalothrin mixture at 1,500×. The following pyrethroids are also effective: 20% fenvalerate EC, 20% cypermethrin EC, 2.5% lambda-cyhalothrin EC, and 5% esfenvalerate EC at 3,000 - 4,000×.

References Zhang Shimei et al. 1985, Huang Jinyi and Meng Meiqiong 1986.

(Wang Jijian)

125 *Urostylis yangi* Maa
Classification   Hemiptera, Urostylididae

Distribution   Henan, Fujian, Jiangxi, Anhui, Yunnan and Sichuan.

Hosts   *Castanea mollissima* and *C. seguinii*.

Damage   Adults and nymphs feed on tender shoots, which leads to growth cessation of the young twigs. The insect feeding causes new shoots to stop growing, young and tender leaves to curl or wither, or death of the whole sapling.

Morphological characteristics   **Adults** Female size is 11.5-12.5 mm×4.5-5.2 mm, Male size is 8.5-10.5 mm×4.5-5.0 mm. Yellowish green, flat, and oval-shaped. Antennae are extremely long, which is equal to the length of the body. The lateral edge of proxnotum is creamy yellow (or millet yellow), the postal edge is black. The front edge of the leathery portion of the forewing is creamy yellow, the membranous portion is colorless and transparent. **Nymphs** Nymphs have five instars. The last instar is 8.5 mm×4.5 mm in size. Body is fusiform, flat, and yellowish green. Wing pads are visible and bubble shaped.

Biology and behaviour   *Urostylis yangi* has one generation per year in Henan province. It overwinters in egg stage on fallen leaves or under bark crevices. Eggs begin to hatch in late February of the following year. Peak of egg hatch is in early March. Adults emerge in mid- and late May. They mate in late September. Females crawl off the trees to deposit overwintering eggs.

Newly hatched nymphs feed on eggshells and secreted materials on egg masses. As the nymphs turn into the third instar, they leave the egg masses and disperse into the canopy of the chestnut trees. The nymphs prefer to ingest sap from young sprouts and tender leaves. During daytime, the nymphs rest on the tip of young shoots or the abaxial side of leaves. They are active at night and feed in aggregation on new shoots. Adults are rarely in flight. The adults also ingest the sap of tender shoots and leaves. Adults feed for over 4 months. The female and male pair usually stays together during mating and oviposition time. Most of the females crawl downward off the trees to oviposit after mid-October. The eggs are arranged neatly in a single layer. The egg mass is in the elongated strip with 15-40 eggs. They are covered by secreted colloidal material. The life span of adults varies from 120 to 180 days.

Prevention and treatment

1. Forest management. Remove host trees near the chestnut orchard to reduce the bug population. For example, the elimination of wild Chinese chestnut in the vicinity can be very effective in reducing the bug populations for the following year.

2. Manual control. The overwintering eggs can be eradicated by the removal of withered branches and fallen leaves on the ground in chestnut orchard. The collected plant debris with overwintering eggs can be destroyed in a centralized location.

3. Biological control. *Beauveria bassiana* fungus can be applied. The lady beetle, *Rodolia pumila* can prey both eggs and nymphs, which is effective in reducing the pest population. Thus, the protection and preservation of natural enemies should be promoted.

4. Chemical control. In early spring or late February before the third instar nymphs crawl upward to the chestnut tree branches, the trees can be treated. The old tree bark can be scrapped off for a 10 cm band on the tree trunk at 1 m above the ground. The band is covered by a plastic film. Then an adhesive insecticide mixture is applied on the plastic film. The ingredients of the adhesive insecticide mixture include two parts of automotive engine oil, or cotton seed oil sludge, mixed with 0.5 parts of cypermethrin. The adhesive coating on the band can effectively reduce the bug population. If it is possible, the timely spraying of 40% chlorpyrifos and phoxim mixture or 45% acetamiprid and phoxim mixture at 1,000-1,500× during the peak of nymph emergence peak period in the chestnut orchard is also effective.

References   Yang Youqian and Li Xiusheng 1982, Xiao Gangrou (ed.) 1992, Hou Qichang and Yang Youqian 1998.

(Yang Youqian, Hou Qichang)

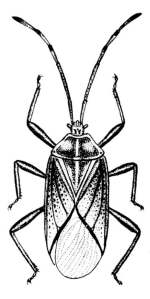

U. yangi adult (Chen Zhongze, Ji Chunfu)

126 *Scirtothrips dorsalis* Hood

Classification Thysanoptera, Thripidae
Common names Yellow tea thrips, Chilli thrips, strawberry thrips

Importance It is an economically important pest of broad hosts and has spread to many countries.

Distribution Zhejiang, Fujian, Guangdong, Guangxi, Hainan, Yunnan, Taiwan; Japan, India, Malaysia, Pakistan.

Hosts *Peltophorum pterocarpum*, *Acacia auriculaeformis*, *A. confuse*, *A. senegal*, *Moringa oleifera*, *Sauropus androgynus*, *Prosopis glandulosa*.

Morphological characteristics Adults Females about 0.9 mm in length, orange yellow. Antennae 8 segmented, dark yellow, the first segment greyish white, the 2nd segment same as body color, the base of the 3rd to 5th segments lighter than body color, the 3rd and 4th segments have forked sense cones, the base of the 4th and 5th segments have one fine ring. The compound eyes dark red; forewings orange yellow, there is a small light-yellow area near the base. Forewings are narrow, and there are 24 setae on the costal margin, 4+3 setae on the costa base, 3 setae distally which the one seta in the middle and two setae at the end, second vein with 2 setae.

There are dark transverse ridges on the 2nd to 8th tergites, the front 1/3 dark brown. There are dark transverse striations on the front of the 4th to 7th sternites. The head width is about two times as the length, but shorter than prothorax; the front of the head extended between the antennae, rear part of the head has thin transverse striations. Gena slightly narrowed after the compound eyes. Setae of head short, there are 2 pairs setae at the front of the anterior ocellus. Maxillary palp has 3 segments. Prothorax wider than its length, covered with fine dense transverse striations, with 4 pairs of posteromarginal setae, one pair of anteromarginal setae, and a pair of setae before the median part of prothorax. One third of the sides of the 2nd to 8th of abdominal tergites are covered with numerous microtrichi, the posteromarginal comb on the 8th segment is complete. Sternites also covered with numerous tiny microtrichi, and long setae of the 2nd to 7th segments are located at the posterior margins and with no affiliated setae.

Eggs Reniform, light-yellow in color.

Nymphs The newly hatched nymphs milky; the 2nd instar nymphs light-yellow, about 0.8 mm in length, its shape is similar to adult but without wings.

Pupae The 4th instar nymphs with ocelli; antennal segments are unclear, directed to the back of head; wing buds obvious.

Biology and behavior *Scirtothrips dorsalis* oc-

S. dorsalis damage to Ginko biloba leaves (Xu Gongtian)

S. dorsalis induced rolled G. biloba leaf (Xu Gongtian)

S. dorsalis adult (Xu Gongtian)

S. dorsalis nymph (Xu Gongtian)

curs throughout the year in Hainan with overlapping generations. It takes more than 10 days to complete a generation. It is capable of reproducing both sexually and parthenogenetically. Females lay eggs on back of leaves along the leaf veins or in mesophyll. A female lays dozens to more than one hundred eggs. Nymphs suck sap from the tender shoots or leaves. The 1st and 2nd instar nymphs and adults cause damage to plants. The 3rd instar nymphs move slowly and no longer feed. They crawl to the ground to pupate. So the 3rd instar is also called pre-pupal stage. The 4th instar is also called pupal stage. Pupae are found in moss, lichen, and leaf litter layer. Adults are active. When disturbed, they jump or fly away. Adults tend to feed and lay eggs on tender leaves. Both adults and nymphs avoid lights and exhibit positive hydrotaxis.

Prevention and treatment

Methods are similar to *Haplothrips chinensis*.

(Gu Maobin, Chen Peizhen)

127 *Haplothrips chinensis* Priesner

Classification Thysanoptera, Phlaeothripidae
Common name Chinese thrip

Importance *Haplothrips chinensis* is the main pest of *Illicium verum*. It reduces the quality and yield of *Illicium verum* fruit. In Guangxi, 38%-50% yield reduction was reported.

Distribution Jiangsu, Zhejiang, Fujian, Henan, Hubei, Hunan, Guangdong, Guangxi, Sichuan, Guizhou, Yunnan, Taiwan.

Hosts *Illicium verum, Diospyros kaki, Prunus persica, Prunus salicina, Fortunella margarita, Camellia vietnamensis, Mangifera indica, Eriobotrya japonica, Antidesma japonicum, Lagerstroemia speciosa, Gardenia jasminoides, Ervatamia divaricate, Michelia alba, Rosa rugosa, Chrysanthemum morifolium*, rice, wheat, barley, millet, and foxtail millet.

Damage The damaged fruit surface has brown patches, become deformed, and small in size.

Morphological characteristics **Adults** Blackish brown to black in color; females 1.9 mm in length. Head width and length 0.18 mm; prothorax 0.19 mm in length, width of mesothorax 0.32 mm; the diameter of the 3rd abdominal segment 0.34 mm; forewings 1 mm in length. Males smaller, 1.4 mm in length; head width 0.15 mm, the diameter of abdomen 0.35 mm. The 3rd to 6th segments of the male antennae are yellow, while the 1st, 2nd, 7th, and 8th segments are blackish brown. The 3rd segment is asymmetric, its exterior surface with a sensory cone. The 7th segment has no handle, and the 8th segment is short and thick. The 3rd to 6th segments of antennae in females are bright yellow. Foreleg tibia is yellowish brown. There are 6-10 long hair-like fringes along the end and posterior margin of the forewing. There are glandular areas on the venter of the 2nd to 7th abdominal segments in males. The end of the abdomen is blunt round.

Eggs Length 0.3 mm, width 0.1 mm; elongated round, the front half is slight bigger; white and semi-transparent, light ivory when near hatching.

Nymphs The 1st instar nymph compound eyes are light red and body is nearly transparent, white in color. The 2nd instar nymph body is greyish white, the 3rd antennae segment is yellow, and the other antennae segments are greyish brown; the end of abdomen with two very thin and long setae and several short and small setae. The 3rd instar nymph antennae scrotiform; compound eyes are small, without ocelli. The 4th instar nymph antennae point to both two of head; wing buds absent. The 5th instar nymph antennae long, ocelli present, wing buds big and dark in color.

Biology and behavior *Haplothrips chinensis* occurs 11-15 generations a year in south of China. The insect starts damage as soon as it enters a flower. It scrapes the epidermis of carpel (tender fruit) and suck the carpel juice with rasping-sucking mouthparts. Adults lay eggs in flowers. Egg development period is around 3 days. All developmental stages often co-occur in one flower. A total of 13-25 individuals may be present in one flower.

H. chinensis eggs (Wang Jijian)

H. chinensis late instar nymph (Wang Jijian)

H. chinensis adult (Wang Jijian)

H. chinensis initial infestation (Wang Jijian)

Light damage from H. chinensis (Wang Jijian)

Chinese thrip has strong ability of locating host flowers. The insect exhibits photophobia. It will move away when the flower heads to the sun. The insect overwinters in 5 cm deep soil when the temperature reduces to 4℃.

Prevention and treatment

1. Silviculture. In high population density areas, plough during the winter to expose overwintering adults.

2. Conservation of nature enemies. Spiders, ants, and birds are important the natural enemies.

3. Chemical control. Mix 2.5% deltamethrin with talcum powder at 1:100 ratio and apply it to canopy using a blower. Spray the following insecticides: 2,500× of 10% imidacloprid wettable powder, 2,000× of 5% lamda-cyhalothrin, 2,000× of 9.5% beta-cypermethrin wettable powder, or 1,500× of 40% profenofos and phoxim mixture.

References Department of Forestry of South China Agricultural University 1985; Wang Jijian et al. 1997, 1998.

(Wang Jijian)

Severe damage from H. chinensis (Wang Jijian)

128 *Anomala corpulenta* Motschulsky

Classification　Coleoptera, Scarabaeidae
Common name　Metallic-green beetle

Importance　*Anomala corpulenta* is a common pest of forest trees and fruit trees. It infests numerous plant species. At high population density, they can cause complete defoliation. Young trees are more susceptible.

Distribution　All provinces except Xinjiang and Tibet; Mongolia, Korea Peninsula, Japan.

Hosts　Trees in the genus *Malus* are most frequently infested. Other hosts include: *Populus* spp., *Salix* spp., *Ulmus* spp., *Morus alba*, *Platanus orientalis*, *Camellia sinensis*, *Cinnamomum camphora*, *Ulmus parvifolia*, *Ligustrum lucidum*, *Toona sinensis*, *Pinus* spp., *Cunninghamia lanceolate*, *Quercus variabilis*, *Vernicia fordii*, *Camellia oleifera*, *Sapium sebiferum*, *Castanea mollissima*, *Juglans regi*, *Cupressus funebris*, *Pterocarya stenoptera*, *Nerium indicum*, *Dimocarpus longan*, *Crataegus pinnatifida*, *Vitis vinifera*, *Syringa* sp., *Pyrus betulaefolia*, *P. bretschneideri*, *Prunus persica*, *P. armeniaca*, *P. pseudocerasus*, *P. salicina P. mume*, *Nicotiana tabacum*, *Diospyros kaki*, *Fragaria ananassa*.

Damage　*Anomala corpulenta* adults feed on buds and leaves. Larvae feed on roots.

Morphological characteristics　**Adults** Length 15-18 mm, width 8-10 mm. Dorsal surface bronze-green color, shiny. Head large, dark bronze-green color. Clypeus brownish green, its anterior border curls upward. Eyes black, large, round. Antennae 9-segment-

A. corpulenta feeding on a willow leaf (Xu Gongtian)

A. corpulenta adult (Xu Gongtian)

A. corpulenta eggs (Xu Gongtian)

A. corpulenta larva (Xu Gongtian)

ed, yellowish brown. Anterior border of the pronotum bends inward, lateral border and posterior border bends outward, notum shiny green, with dense punctures, the two sides with 1 mm wide yellow band. Elytra yellow bronze-green color, shiny. Ventral surface of thorax with yellowish brown fine pubescence. Femur yellowish brown, tibia, tarsi dark brown. Abdomen yellow, shiny; pygidium triangular, with a near triangular black mark. Ventral surface of females milky white, the last segment with brownish yellow transverse band; ventral surface of males brownish yellow.

Eggs Length 1.65-1.94 mm, width 1.30-1.45 mm, white, smooth. Elongated elliptical, turning to round before hatching.

Larvae The 3rd instar length 30-33 mm. Head yellowish brown; vertex with 6-8 setae on each side, arranged in a row. The ventral surface of the last abdominal segment with two rows of yellowish brown long setae, each row with 15-18 setae. The tips of the two rows of setae touch or cross with each other. There are dark yellow hooked setae outside of the two rows of setae.

Pupae Length 18 mm, width 9.5 mm. Elliptical, slightly flattened, yellow soil color, the end round and flat. The dorsal surface has 6 pairs of tymbals. The ventral surface of the last segment of the female pupae with fine rugae, that of the male pupae with a papillate.

Biology and behavior *Anomala corpulenta* occurs one generation a year. It overwinters as 3rd instar or less common 2nd instar in soil. In spring, they migrate upward. Pupae appear in May. Adults appear in early June. Those in the south appear earlier than those in the north. Adults emerge from soil from dusk. Peak adult emergence period is during mid-June to early July. No adults are seen in September. Eggs are found during peak adult period. Larvae appear in August. Adult emergence is associated with precipitation in May and June. Higher precipitation during this period will favor the early appearance of adults. Adults rest in bushes, weeds, or on soil surface during the day. They are active at dusk.

Their optimum development condition is at 25℃ and 70%-80% RH. Adults play death and are phototaxic. Males appear earlier than females. They mate on host trees. The peak activity period is during 20:00-22:00. Adults mate multiple times. Average longevity is 30 days. Eggs are laid in 5-6 cm deep soil under fruit trees or in crop field. Eggs are laid singly. Each female lays 40 eggs. The egg development period is 10 days. Optimum eggs hatch condition is 10%-15% RH and 25℃. Almost all eggs can hatch. Larvae feed on roots. The 1st and 2nd instar do not feed much. From September when most larvae are at 3rd instar, their appetite increases significantly. From October-November, larvae start to overwinter. They feed on roots again in the spring until May. Larvae crawl to soil surface in early morning or dust and feed on main and secondary roots, stems of seedlings. The 1st to 3rd instar development period is 25, 23.1, and 27.9 days, respectively. Mature larvae molt into pupae during late May to early June. They construct soil chambers before pupation. Pre-pupae and pupae development period is 13 and 9 days, respectively.

Prevention and treatment

1. Silviculture. Plant mixed forests. Irrigate and add compost to improve soil condition and root development.

2. Manual control. Shake trees during dusk and collect adults falling to the ground. Loosen soil to kill larvae.

3. Use light traps to kill adults.

4. Attract and kill using 5% lamda-cyhalothrin at 1,500×, or 50% acetamiprid and monosultap mixture

Rose leaved damaged by *A. corpulenta* (Xu Gongtian)

Turf infested by *A. corpulenta* (Xu Gongtian)

Lawn damage from *A. corpulenta* (Xu Gongtian)

at 1,500× treated poplar, willow, or elm branches and leaves. Place bundles of branches every 15 m at evening can protect target tree species.

5. Conservation of natural enemies. Ground beetles, Staphylinid beetles, and birds are important natural enemies.

6. Apply *Beauveria bassiana* to soil to control larvae. Spray canopy with 5% lamda-cyhalothrin at 1,500×, or 50% acetamiprid and monosultap mixture at 1,500×, 50% acetamiprid at 1,500×, 40% chlorpyrifos and phoxim mixture at 4,000× during adult period. Spray insecticides on weeds under canopy and shake trees.

References Beijing Forestry College 1979; Xiao Gangrou (ed.) 1992; Wu Junxiang 1999; Lan Jie, Qinhe, Kang Xiaolong et al. 2001; Run Guixin, Song Fuping, and Zhang Jie 2007; Xu Gongtian and Yang Zhihua 2007; Xue Guishou, Mao Jianping, Pu Guanqin et al. 2007; Zhang Zhixiang 2008.

(Xue Yang, Li Guangwu)

129 *Anomala cupripes* (Hope)

Classification	Coleoptera, Scarabaeidae
Common names	Cupreous chafer red-footed green beetle, Large green chafer beetle, Green flower beetle

Distribution Zhejiang, Fujian, Guangdong, Guangxi, Hainan, Sichuan, Guizhou, Yunnan, Taiwan.

Hosts *Delonix regia, Prunus mume, Michelia alba, Cercis chinensis, Canna indica, Viburnum odoratissimum, Dendranthema morifolium, Cinnamomum camphora, Rose chinensis, Hibiscus rosasinensis, Fokienia hodginsii, Begonia evansiana, Camellia japonica, Camellia oleifera, Aglaia odorata, Rosa rugosa, Bischofia javanica, Canarium album, Acacia mangium, Syzygium jambos, Bischofia polycarpa, Ficus virens, F. microcarpa* var. *pusillifolia, Aleurites fordii, Litchi chinensis, Dimocarpus longan, Averrhoa carambola.*

Damage *Anomala cupripes* adults feed on leaves may consume all leaves on a tree. Larvae feed on roots of trees and crops and may cause death of hosts.

Morphological characteristics **Adults** Length 22 mm. Dorsal surface turquoise green, ventral surface purplish bronze, metallic. Antennae lamellate. Elytra with small round punctures that are arranged into 4-6 indistinctive longitudinal lines. Edges of elytra roll up and with purplish red luster. The posterior end of each elytron with a pointed protuberance. Abdomen with 6 visible segments. Pygidium of the male bends forward and raised, its tip dull. The sternite of the 6th abdominal segment with a blackish brown band. Pygidium of the female slightly pointed.

A. cupripes adult (Li Yingchao)

Eggs Milky, elliptical. Length 2 mm, width 1.5 mm.

Larvae Milky, head yellowish brown, body cylindrical, "C" shape. Ventral surface of the last abdominal segment with yellowish brown setae, arranged into trapezoid shape.

Pupae Length 20-30 mm, width 10-13 mm; elongated elliptical. Yellow, turning to yellowish brown before hatching.

Biology and behavior *Anomala cupripes* occurs one generation a year. It overwinters as 3rd instar. Pupae appear in March-April. Adults emerge in late April. Female: male ratio is 7:3. Adults feed for about one month, then mate and lay eggs. Adults prefer lay eggs singly in compost. A female lays 60-80 eggs a time. They die 4-7 days after finishing egg laying. Egg incubation period is 11-16 days. The 1st-3rd instar development period is 30-40, 40-60, and 20-28 days, respectively. Pupal period is 21-90 days. Adults appear from early May to late July. Adults feed during both day and night. They only rest in forests during very hot days. Adults are phototaxic. They play death after disturbed.

Prevention and treatment

1. Silviculture. Plant mixed forests and resistant species. Reduce humus layer in the forest. Plant trees during rainy season when young trees grow fast and larvae of *A. cupripes* are susceptible to drowning.

2. Biological control. Apply 1×10^{10}/g *Bacillus popilliae* soil at 1,500 g per ha can reduce population by 60%.

3. Physical control. Dig out larvae when their consumption is large. Attract adults with *Hibiscus cannabinus* and kill them. Trap and kill with light traps.

4. Chemical control. During adult period, spray 77.5% trichlorfon at 800-1,000×, 45% profenofos and phoxim mixture at 1,000×, or 50% fenitrothion at 600×.

References Xiao Gangrou (ed.) 1992; Xian Shenghua, Yu Yong, Liang Xueming et al. 2002.

(Wu Jianfen, Jia Yudi, Li Zhenyu)

130 *Proagopertha lucidula* (Faldermann)

Classification Coleoptera, Scarabaeidae

Importance *Proagopertha lucidula* is a widespread pest of many plant species.

Distribution Hebei, Shanxi, Inner Mongolia, Liaoning, Jilin, Heilongjiang, Jiangsu, Zhejiang, Anhui, Jiangxi, Shandong, Henan, Sichuan, Guizhou, Shaanxi, Gansu, Qinghai; Russia.

Hosts *Populus* spp., *Salix* spp., *Ulmus* spp., *Robinia pseudoacacia*, *Morus alba*, *Malus pumila*, *M. spectabilis*, *Pyrus bretschneideri*, *Crataegus pinnatifida*, *Prunus armeniaca*, *P. persica*, *P. serrulata*, *P. mume*, *P. pseudocerasus*, *salicina*, *Quercus mongolica*, *Paeonia lactiflora*, *Acer mono*, *A. truncatum*, *Paeonia suffruticosa*, *P. lactiflora*, *Rosa chinensis*, *Syringa* spp., *Juglans regia*, *Castanea mollissima*, *Zanthoxylum bungeanum*, *Vitis vinifera*, *Celtis bungeana*, *Buxus sinica*, *Acer saccharum*.

Damage *Proagopertha lucidula* adults feed on flowers, buds, young leaves, and immature fruits. They consume a lot and will feed on different plants as plants grow. Larvae feed on fibrous roots, but do not cause significant damage.

Morphological characteristics **Adults** Length 8.0-10.9 mm, width 5.0-6.5 mm. Head, pronotum, scutellum brownish green, with purple luster. Head large, occiput with punctures. Anterior border of clypeus slightly roll up. Eyes black. Antennae 9 segmented, club with 3 segments.

Eggs Length 1.6-1.8 mm, width 1.0-1.2 mm. Elliptical, milky. After 10 days, the size increases to 1.8-2.4 mm×1.3-2.0 mm.

Larvae Length 12-16 mm, head yellowish brown, legs dark yellow, mandible tip blackish brown. Head width of the 1st-3rd instar larvae is 1.3, 2.2, and 3.2-3.5 mm, respectively. There is a row of 7-8 hairs on each side of the anterior part of the vertex and 10-11 hairs on each side of the posterior part of the vertex. There are 5 hairs arranged in an oblique row on each side of the frons. The spines on ventral surface of the anal segment consisted of short and long spines. Each row of short spines has 5-12 spines. Each row of long spines has 5-13 spines.

Pupae Length 14-16 mm. Dark yellow, elongated elliptical. Dorsal median line distinct. The end of the abdomen pointed. The 8th segment widest, trapezoid shape. The 10th segment split at the end, its lateral margin with dense short hairs. The spiracles of the 2nd-3rd segment large, yellowish brown. There is a triangular projection on ventral surface of the 9th abdominal segment in the female.

Biology and behavior *Proagopertha lucidula* occurs one generation a year. It overwinters as adults in loose sandy soil covered with grass or bush and where is sunny. Adults emerge from soil in mid- and late April. It coincides with cherry and peach flower season. About 30% of the adults appear in this season. The 2nd peak emergence is during early May and mid-May. About 65% of the adults emerge during this period. The female: male ratio increases as the time goes by. When ground tem-

P. lucidula adult (lateral view) (Xu Gongtian)

P. lucidula adult feeding on pollen (Xu Gongtian)

perature reaches 12℃ and air temperature reaches 10℃, large numbers of adults often emerge after rain. When air temperature is between 18-23℃, wind is the key factor affecting emergence. On windless days when temperature reaches 20℃, large numbers of adults appear on sunny ground and look for mates. They enter soil after 14:00. Eggs are laid on high, well-drained soil. Adults are not phototaxic. They play death when disturbed. This behavior is most obvious when temperature is below 18℃. Adults prefer flowers, young leaves, and immature fruits. They migrate to different plants as their leaves and flowers emerge. In Liaoning, they first feed on flowers and fruits of *Salix gordejevii*. Once the leaves and flowers of *Salix matsudana*, pear, and *Populus simonii* appear in late April and early May, they migrate to these trees. In mid-May, they migrate to apple trees.

Adult feeding occurs during the day when temperature rises but may rest at noon during hot days. Mating happens before noon. Peak egg laying period is in mid-May. Eggs are laid in loose soil with sparse vegetation. At 18.6-22.5℃, the egg incubation period is 17-35 days. Eggs hatch in mid- and late May. Larvae go through 3 instars and 55-69 days. In August, they migrate to 80-120 cm deep (in northeast of China) or 40-50 cm deep (in Henan, Shandong) under soil and construct pupal chambers. Pupal period is 16-19 days. Eclosion occurs in early September. After eclosion, adults remain in pupal chambers and overwinter.

Proagopertha lucidula is most abundant in grassy land, orchards, and nurseries. In forest stands, leeward side has higher population than the windward side. The insect is tolerant to dry environment.

Prevention and control

1. Silviculture. Plant mixed forests, irrigate promptly, add compost to improve soil condition and root growth. Plant bait plant, *Ricinus communis*, in fields where infestation is heavy. It is best planted early such that when *P. lucidula* adults emerge, the *R. communis* already has 2-3 leaves. The larvae will die from feeding on *R. communis* leaves.

2. Mechanical control. Shake trees during the morning and evening and collect fallen adults.

3. Conservation of natural enemies. Natural enemies include carabid beetle, Staphylinidae, birds, and a parasitic wasp in the family Tiphiidae.

4. Chemical control. Cut flowering elm branches and soak them in 25 g/L lamda-cyhalothrin at 1,500× or 50% imidacloprid monosultap mixture at 1,000×. Then place treated elm branches in infested fields. Adults are attracted to the flowers. Spray canopy with 25 g/L lamda-cyhalothrin at 1,500×, 50% imidacloprid monosultap mixture at 1,000×, 1.2% nicotine and matrine mixture at 1,500×, 48% acetamiprid and chlorpyrifos mixture at 4,000×. Spray weeds under tree canopy and shake trees will also kill adults.

References   Xiao Gangrou (ed.) 1992; Wang Xueshan, Ningbo, Pan Shuqin et al. 1996; Du Xiangge and Zhang Youyan 2003; Xia Xina 2004; Run Guixin, Song Fuping, and Zhang Jie 2007; Xu Gongtian and Yang Zhihua 2007; Xue Guishou, Mao Jianping, Pu Guangqin et al. 2007.

(Xue Yang, Lin Jihui)

131 *Nigrotrichia gebleri* (Faldermann)

Classification Coleoptera, Scarabaeidae
Synonym *Holotrichia diomphalia* Bates, *Holotrichia oblita* (Faldermann)

Importance *Nigrotrichia gebleri* feed on many crops and trees. They cause yield reduction and death of plants.

Distribution Beijing, Hebei, Inner Mongolia, Liaoning, Jilin, Heilongjiang, Gansu; Japan, Mongolia, Russia.

Hosts Adults feed on 94 species of plants belonging to 32 families based on laboratory experiments. Major hosts are: *Malus pumila, Crataegus pinnatifida, Pyrus ussuriensis, Morus alba, Ulmus pumila, Hemiptelea davidii, Corylus heterophylla, Populus simonii, Castanea mollissima, Arachis hypogaea, Pisum sativum, Zea mays, Triticum* spp.*, Sorghum vulgare, Brassica* spp.*, Brassica chinensis, Solanum tuberosum, Solanum melongena, Helianthus annuus, Allium tuberosum, Spinocia oleracea, Beta vulgaris, Ipomoea batatas, Sesamum indicum, Abutilon avicennae.*

Damage Larvae feed on roots and stems. Adults feed on leaves.

Morphological characteristics **Adults** Length 16-21 mm, width 8-11 mm. Elongated elliptical. Blackish brown or black, with luster. Antennae 10 segmented. Pronotum width less than two times the length, with many punctures, the middle of its edge curls outward. Length of elytra about two times the width of the pronotum. The widest part located at the middle of the elytra. Elytra with distinct ridges, there is a distinct ridge where the two elytra meet. Front tibia with three sharp teeth on exterior surface and a spur on inner surface. The middle and hind tibiae with two spurs at the end. End of tarsi have double claws. Middle of the claw has a tooth. Hind tibia with a transverse ridge which has spines. Pygidium raised, the height surpasses the length of the last sternite, short, the apical border with two short and small round projections. Lateral view of the pygidium appears as a round globular surface.

Eggs Milky, oval. Length 2.5 mm, width 1.5 mm. Globular at late incubation stage and the size increases to 2.7 mm by 2.2 mm.

Larvae Milky, horseshoe shape. The 3rd instar length 31 mm, head width 4.7 mm. Each side of the vertex with three hairs, which are arranged in a row; two of the hairs are close and near the coronal suture, the other hair near the middle of the frontal suture. Ventral surface of pygidium scattered with hooked setae.

Pupae Yellow to reddish brown. Length 20 mm, width 8 mm.

Biology and behavior *Nigrotrichia gebleri* requires 2-3 years to complete one generation in Heilongjiang and two years to complete one generation in Liaoning. It overwinters as adults and larvae in alternate years, causing high and low population of adults and larvae in alternate years. For example, adults in 1971, 1973, and 1975 were abundant, whereas high larval populations occurred in even years.

In Liaoning, overwintering adults emerge from soil

N. *gebleri* adult (FPC)

N. *gebleri* larva (FPC)

N. gebleri damage (FPC)

during late April and mid-May. The threshold is average day temperature of 12°C or 10 cm deep soil temperature of 13°C. Adult period lasts until late August.

Adults are active at dusk and reach peak activity during 20:00-21:00. They enter soil and rest around 2:00. All adults enter soil before dawn. They mate multiple times. Adults are phototaxic. Males have stronger phototaxy. They are not phototaxic during the first 15 days after emergence. Eggs are laid in 1.5-17.5 cm deep soil. An average of 102 eggs are laid per female. Egg incubation period is 15-22 days. The egg hatching peak is during mid- and late July. Larvae feed on roots and newly planted seeds. Larvae go through three instars. When temperature at 10 cm deep soil decreases to 12°C, the larvae migrate to 56-149 cm deep and overwinter.

Larvae migrate upward in early April. When the 10 cm deep soil temperature reaches above 10.2°C, all larvae migrate upward and start feeding. This group of larvae cause the most damage to crops in May-June.

Mature larvae migrate down to 20-38 cm soil in late June. Pupal chamber is elongated elliptical. Length is 25.7 mm, width is 15.9 mm. Pupal period is 22 days (female) or 25 days (males). Peak pupal period is from late July to mid-August. After adult emergence, they stay inside soil chambers and overwinter.

This insect is most abundant on unmanaged hills covered with grasses. Soybean, peanut, corn, and sorghum fields are less common. Flat, moist, deep soil fields have higher pest populations than dry, non-fertile or sandy soil. Edges of fields have higher pest population than the middle of the field. Ploughed soil has lower pest populations. Soil with partially decomposed manure has higher pest population.

Prevention and treatment

1. Cultural control. Plough soil. Remove weeds. Avoid adding partially decomposed manure.

2. Use Light traps to kill adults.

3. Conserve natural enemies. *Tiphia* sp. is a parasitoid of *N. gebleri* larvae. *Bacillus popilliae* and *Beauveria bassiana* can be considered as control agents.

4. Chemical control. Mix crop seeds with 45% profenofos and phoxim mixture at 1.0-1.5% by weight. Spray seedlings with 45% profenofos and phoxim mixture at 1,000×.

References   Zhang Zhili 1984, Luo Yizhen 1995.

(Guo Shiying)

132 *Eotrichia niponensis* (Lewis)

Classification Coleoptera, Scarabaeidae
Synonym *Holotrichia titanis* Reitter

Importance *Eotrichia niponensis* is a major pest of crops, forests, and landscape. Larvae feed on roots and stems. They cause death of young seedlings, slow growth or even death of trees.

Distribution Hebei, Shanxi, Liaoning, Jilin, Jiangsu, Zhejiang, Shandong, Henan, Shaanxi; Korea Peninsula, far east of Russia.

Hosts *Zea mays*, *Setaria italic*, *Sorghum vulgare*, *Triticum* spp., *Solanum tuberosum*, Leguminosae, *Gossypium* spp., *Beta vulgaris*, *Malus pumila*, *Pyrus bretschneideri*.

Morphological characteristics **Adults** Length 21.2-25.5 mm, width 11-14 mm. Brownish red, with silk like luster. Head small, clypeus short and wide, its anterior border and the lateral border roll upward, middle of the anterior border emarginated. Antennae 10 segmented, its club very wide, the male antennae club is significantly larger than that of the female. Clypeus, frons, pronotum, elytra with dense punctures. Center of the pronotum with a smooth longitudinal ridge. There is a small barely visible black spot at the projection of the lateral margin. Elytra thin and soft, with 4 longitudinal ridges. Scutellum color darker, about ½ of the scutellum is covered by the yellow hairs on the pronotum. Ventral surface of thorax covered with dense light-yellow pubescence. Exterior surface of the front tibiae with 3 teeth, the inner surface of the front tibiae with a long tooth. Hind tibiae slender and long, its end enlarged into a horn shape, two spurs exist. For the hind leg, the 1st tarsal segment shorter than the 2nd tarsal segment. There are two claws at the end of each tarsus and one tooth under each claw.

Eggs Milky, elliptical. Length 2.8-4.5 mm, width 2.0-2.2 mm.

Larvae Milky. Mature larvae 45-55 mm long, head width 6.1 mm. vertex with two setae on each side. There are two rows of spines on ventral surface of the anal segment, each row with 16-26 spines of various length, additional rows of spines may exist.

Pupae Yellowish white. Length 23.5-25.5 mm, width 12.5-14.5 mm. Two spines at the end of the abdomen, the tip of the spines black. There is a longitudinal ridge from thorax to the end of the abdomen.

Morphological characteristics *Eotrichia niponensis* occurs one generation every two years. It overwinters as adults or larvae alternately. Studies in Wugong (Shaanxi province) found two separate populations coexist. Population A and population B produce adults in different years. Therefore, large number of adults occur each year.

Adults emerge from soil during mid-March and the end of May. Peak egg production is during mid- and late April. Eggs hatch during mid- and late May. Larvae damage seedlings and roots during June-November. When soil is dry and hot in August, the larvae may migrate deeper into the soil. In mid-to late November, the 2nd and 3rd instar overwinter in soil at 30 cm depth. They resume activity near the soil surface in mid-March. Larvae mature in late June. They pupate in soil at 30 cm depth. Pupal period is 32 days. Adults emerge from pupae during

niponensis adult (edited by Zhang Runzhi)

the end of July and early August. They stay inside the pupal chambers until mid-March in the 3rd year.

Adult activity is associated with temperature and light conditions. When average day temperature is above 10.3°C, adults come out from soil at dusk and search for mates. After about 60 minutes of activity, they enter soil. Females are weak flyers. They fly close to the ground surface. Males are strongly attracted to males. Males are better flyers than females. After mating, the female drags the male into soil. It is easy to catch them during this period. Even when pairs of adults are placed in water, they may stay attached. Adults do not spread very far and only appear in a short period. Therefore, a location may suffer heavy infestation in consecutive years. Adults are not phototaxic. Adults do not feed under natural conditions. Females lay eggs singly in 20-30 cm deep soil. Each female lays 20-44 eggs. Egg incubation period is 27 days.

Larval activity is associated with soil temperature. Larval period is 406 days. Egg and larval development are closely associated with soil relative humidity. The optimum relative humidity is 15%-20%. Eggs and larvae cannot survive well below 5% RH or above 30% RH. The 1st and 2nd instar larvae are more susceptible to extreme relative humidity. The most preferred habitat is unmanaged land. The next most preferred habitats are wheat field, clover field, and orchards.

Prevention and treatment

Apply *Beauveria bassiana*. During the 1940's, crows and *Pica pica* are important predators in Shaanxi. They feed on larvae exposed after tillage. They also feed on adults during dusk when *Holotrichia titanis* come out. These birds are much less common after widespread use of insecticides. Conservation of these predators will be helpful.

Other methods are similar to *Eotrichia niponensis*.

References　Wu Dazhang 1951, Luo Yizhen 1955, Pei Jingxian 1980, Zhang Zhili 1984.

(Guo Shiying)

133 *Lepidiota stigma* (F.)
Classification Coleoptera, Scarabaeidae

Importance The insect can cause serious damage to trees under one-year age. Night percent infestation rate was recorded. Hundreds of hectares of newly planted trees can be destroyed.

Distribution Guangdong, Guangxi, Hainan.

Hosts *Manihat esculenta*, *Hevea brasiliensis*, *Eucalyptus* spp., *Arachis hypogaea*, *Saccharum sinense*, *Pachyrhizus erosus*, Fabaceae.

Morphological characteristics **Adults** Elongated oval, head width 6-8 mm, length 36-45 mm; width at middle of the body 18-22 mm. Grayish brown, brown, or pale blackish brown; Head blackish brown; antennae and eyes tanned. Elytra with 3 longitudinal ridges. There are two distinct grayish white elongated elliptical marks near the posterior end of the elytra. The elytra bend downward after the marks. Ventral surface grayish white.

Eggs Elliptical, milky, 3.0-5.6 mm in diameter.

Larvae Mature larvae milky. Length 62-72 mm, width 14-17 mm. There are two rows of spines around the anal segment.

Pupae Elongated elliptical, light-yellowish brown. Length 34-54 mm, width 16-22 mm.

Morphological characteristics *Lepidiota stigma* occurs one generation every two years. It overwinters as 2nd and 3rd instar larvae. Mature larvae pupate in late March. Adults emerge during late April and early May. Adults require supplementary feeding. They mate about 20 days after emergence from soil. Mating occurs between 15:00 and evening. Many males often follow one female. Females lay eggs about 10 days after mating. They lay eggs on sandy soil. Each female lays 40 eggs. Larvae go through three instars. Sandy soil around sea or river is preferred habitat. Previous peanut and *Pachyrhizus erosus* fields often have high populations.

L. stigma adults (Wang Jijian)

Prevention and treatment

1. Set up black light traps during 19:00-21:00.

2. Mechanical control. Collect larvae during tilling. Shake trees and collect fallen adults.

3. Chemical control. Spray 45% profenofos and phoxim mixture or 77.5% trichlorfon at 1,000× to control adults. Trench treatment around young trees using 40% profenofos and phoxim mixture at 1,500-2,000×. Treat 375-450 kg fine soil with 3.7 L of 40% profenofos and phoxim mixture at 10× and then apply the treated soil to one hectare area.

(Gu Maobin)

L. stigma adult male (Wang Jijian)　　*L. stigma* adult female (Wang Jijian)

134 *Maladera orientalis* (Motschulsky)

Classification Coleoptera, Scarabaeidae
Synonym *Serica orientalis* Motschulsky

Importance The insect is an important pest of crops, forests, and landscape plants. Adults eat leaves, buds, and flowers. Larvae cause less damage than adults.

Distribution Beijing, Tianjin, Hebei, Shanxi, Inner Mongolia, Liaoning, Jilin, Heilongjiang, Jiangsu, Zhejiang, Jiangxi, Anhui, Shandong, Henan, Shaanxi, Gansu, Qinghai, Ningxia, Taiwan; Korea Peninsula, Japan, Mongolia, Russia.

Hosts A total of 149 species belonging to 116 genera and 45 families were recorded. Most susceptible hosts include: *Malus pumila, Pyrus bretschneideri, Morus alba, Ulmus pumila, Populus alba, Salix* spp., *Elaeagnus angustifolia, Pisum sativum, Glycine max, Ipomoea batatas, Solanum tuberosum, Melilotus* spp., *Medicago sativa.* Other common hosts include: *Vitis vinifera, Prunus mume, P. persica, P. salicina, P. pseudocerasus, Diospyros kaki, Juglans regia, Crataegus pinnatifida, Cydonia oblonga, Sophora japonica, Robinia pseudoacacia, Dolichos lablab, Arachis hypogaea, Solanum melongena, Lycopersicon esculentum, Brassica* spp., *Daucus carota* var. *sativa, Gossypium* spp., *Helianthus annuus, Beta vulgaris, Sesamum indicum, Sesamum indicum, Nicotiana tabacum, Citrullus lanatus, Ricinus communis, Fragaria* sp., *Oryza sativa, Hordeum* spp., *Triticum* spp., *Sorghum vulgare, Zea mays, Setaria italic, Saccharum* spp.

Damage The insect feeds on leaves and flowers.

Morphological characteristics **Adults** Length 7-8 mm, width 4.5-5.5 mm. Oval, narrow in the front and wide in the back. Males are slightly smaller than females. Blackish brown to black. Body with silk like luster. Clypeus black, shiny, front and later margin curl up, slightly

M. orientalis adult (Li Zhenyu)

emarginated at middle of the front border. Elytra with 9 longitudinal shallow grooves. Punctures small and dense. The lateral margin with setae. Front tibiae with two teeth and one spine. Hind tibia with two spurs.

Eggs Oval. Length 1.2 mm. Milky.

Larvae Milky. The 3rd instar 14-16 mm long, head 2.7 mm wide. Vertex with one seta on each side, frons with one seta on each side. There is a pseudo-ocellus (colored mark) above each antennae socket. Anal segment with 20-23 conical spines arranged in a transverse band with its middle interrupted.

Pupae Length 8 mm; yellowish brown, eyes orange red.

Morphology and behavior

Maladera orientalis occurs one generation a year in China. It overwinters as adults in soil. The adults come out of soil from mid-April. Peak adult period is during late April and early June. They usually emerge in large numbers after rain. Adults are active during 20-25℃. The daily average temperature of above 10℃ and high relative humidity favor the emergence of adults.

Adults are nocturnal and strong flyers. They fly around tree canopy and feed on leaves. During mating, they form a right angle. Peak mating period is mid-May. Females lay eggs in 10-20 cm deep soil. Eggs are laid singly or in groups of > 10 eggs per group. Egg production is related to female adult feeding. Those feeding on elm leaves produce an average 26.1 eggs per female. Egg incubation period is 5-10 days.

Larvae feed on soil humus and some young roots. They do not cause serious damage to crop and seedlings. Larvae go through 3 instars. Their development period is 41, 21, and 18 days, respectively. Mature larvae pupate in 20-30 cm deep soil. Prepupae development period is 7 days. Pupae development period is 11 days. Adults emerge in mid- and late August. Most of them will stay in soil after eclosion and overwinter.

Prevention and treatment

1. Attract and kill. Place poplar and will branches in crop field to attract adults during the night. Spray insecticides on the bait branches. In apple and pear orchards, plant pea, clover, *Melilotus* spp. and spray them with insecticides to control adults.

2. Chemical control. Spray contact insecticides (lamda-cyhalothrin, DDVP, emamectin) in the afternoon will kill adults feeding during the nights.

References Cai Banghua 1963, Zhang Zhili 1984, Luo Yizhen 1995.

(Guo Shiying)

M. orientalis adult (edited by Zhang Runzhi)

135 *Melolontha hippocastani mongolica* Ménétriés

Classification Coleoptera, Scarabaeidae
Common name Oriental brown chafer

Distribution Hebei, Shanxi, Inner Mongolia, Liaoning, Jilin, Heilongjiang, Hunan, Sichuan, Shaanxi, Gansu; Mongolia, Russia.

Hosts Adults infested *Cunninghamia lanceolate*, *Betula* spp., *Populus* spp., *Fructus zanthoxyli*. Larvae infest *Avena nuda*, *Triticum aestivum*, *Pisum sativum*, *Solanum tuberosum*, *Zea mays*, and *Beta vulgaris*.

Damage *Melolontha hippocastani mongolica* is an underground pest to crops, grass, and young trees.

Morphological characteristics **Adults** Length 30 mm, width 13 mm. Antennae 10 segmented. Head covered with dense punctures which have standing pubescence. Pronotum with round punctures, those at the center sparse; the sides with yellowish gray long hairs; its posterior corners form near right angles; the center has one longitudinal line of white pubescence; each of the anterior border and posterior border has two triangular white marks. Elytra have 5 raised ridges and dense punctures and white pubescence between the ridges. The end of pygidium extends to a narrow projection. Abdomen with long and standing yellowish gray pubescence. Front tibiae with 3 teeth on its exterior surface, basal tooth slightly projected. The side of 1st-5th abdominal segments with one triangular white mark per segment.

Eggs Milky. Length 3.5 mm, width 2.5 mm when newly hatched. When fully developed, the length 4.5 mm, width 3.6 mm. Egg shell with irregular sculptures.

Larvae Mature larvae 40-50 mm long, head width 7.8-8.5 mm. Head light castaneous, body yellowish white. Two setae around each side of the frontal suture, one seta on occiput, three setae around each of the frons. Each row of spines on ventral surface of the anal segment has about 30 spines; the first pair of spines near the anus are close to each other; the 2nd and the 3rd pair of spines gradually further apart; the two rows of spines then become parallel to each other. Anus transverse slit shape.

Pupae Golden yellow, turning to blackish brown later. Front wings with 4 longitudinal ridges; only the tip of the hind wing is visible; it reaches to the 3rd abdominal segment. Ventral surface of abdomen has 8 visible segments. Dorsal surface of the abdomen has 9 visible segments. Center of the 1st abdominal segment has one dent; the 2nd to 6th abdominal segments with eye-shaped dentation along the median groove; the posterior border of the 4th and 5th segments with a pair of projections; the posterior end of the 9th segment raised; its tip forked.

Biology and behavior *Melolontha hippocastani mongolica* requires 6 years to complete one generation in Ganzi (Sichuan). Larvae overwinter five times. Adults overwinter once. This insect requires 5 years to finish one generation in Kangding (Sichuan). Overwintering insects migrate to 40 cm deep soil in late October and overwinter. Overwintering larvae resume activity in early

M. hippocastani mongolica adult female (dorsal view)

M. hippocastani mongolica adult female (ventral view)

M. hippocastani mongolica adult male (ventral view)
(Li Zhenyu)

M. hippocastani mongolica adult male (dorsal view)
(Li Zhenyu)

May. The 1st-3rd instar larvae feed on crops with the 3rd instar causing the most damage. Very little damage is done during the last three years period. Adults emerge in mid- and late May. Adults overwinter in soil during the 6th winter. They emerge from soil in May next year, mate, feed, and produce eggs. Females mate 2-3 times. Males mate once. After mating, females lay eggs, then mate and lay eggs again. Each female lays 12-28 eggs (average 21 eggs). Egg incubation period is 30 days. Larvae hatch in July. They overwinter and feed on plants next year.

Adults prefer laying eggs on sandy soil. Adults emerge mostly during 20:00-22:00 on clear days. They left round or elliptical eclosion holes on soil surface. There may be 3-5 eclosion holes per square meter. As many as 20 eclosion holes per square meter were reported. Adults are strongly phototaxic. They feed on top of the shoots. They play death when disturbed. Larvae feed on humus and plant fibrous roots. Therefore, they do not cause much damage to forests, but are very damaging to seedlings. After 4th overwintering, they do not feed much.

Prevention and treatment

1. Set up black light traps.

2. Treat seeds with organophosphate (chlorpyrifos and phoxim mixture, carbosulfan) insecticides or mix soil with organophosphates to kill adults. Spray 3% thiamethoxam at 800×, or 5% lamda-cyhalothrin at 3,000× to kill adults.

3. Dig a 20-30 cm deep, 20 cm wide trench around where larvae occur. Apply a contact insecticide inside the trench to kill dispersing larvae.

References Bu Wangui Yin Chenglong, Zhang Yaorong 1996; Yang Xinyuan, Yang Shirong, Guo Chunhua 2000; Ma Yanfang Xie Zongmou, Zhang Yongqiang et al. 2011.

(Chen Hui, Dai Xiancai)

136 *Agrilus planipennis* Fairmaire

Classification Coleoptera, Buprestidae
Synonym *Agrilus marcopoli* Obenberger
Common name Emerald ash borer

Distribution Beijing, Tianjin, Hebei, Inner Mongolia, Liaoning, Jilin, Heilongjiang, Shandong; Japan, Korea Peninsula, Mongolia.

Hosts *Fraxinus mandshurica, F. chinensis; F. chinensis* var. *rhychophylla.*

Damage *Agrilus planipennis* larvae bore into tree trunks and feed between the phloem and xylem and form S-shaped galleries. These galleries break the nutrient transportation tissue. At low density, local death of bark occurs in areas where larvae exist. At high density, the whole tree will die from infestation. *Fraxinus chinensis* var. *rhychophylla* is the most susceptible species.

Morphological characteristics **Adults** Length 11-14 mm. Body cuneiform, bluish green on the back, light-yellowish green on the ventral surface. Head flat and compressed, the top shield-shape; eyes bronze colored, kidney shape, occupying most part of the head; antennae serrate. Pronotum wider than the head, rectangular, same width as the base of the elytra. Anterior border of the elytra raised to a ridge, elytra with dense punctures, end dull round, the lateral margin with tooth-like projection. Abdomen aeneous color.

Larvae Mature larvae 34-45 mm long, milky, flat and belt-like. Head brown, concealed in the prothorax and only the mouthparts are exposed.

Biology and behavior *Agrilus planipennis* occurs one generation a year in Beijing. It overwinters as mature larvae at the end of the galleries in xylem. Pupation occurs in early April. Adults appear from late April to late June. Egg laying period is from late May to late July.

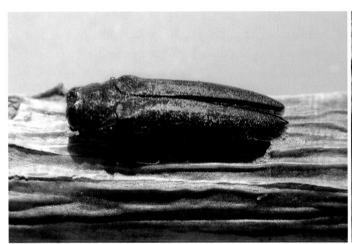
A. planipennis adult (lateral view) (Xu Gongtian)

A. planipennis adult exiting from eclosion hole (Xu Gongtian)

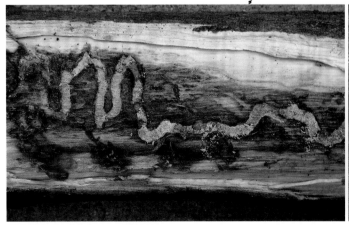
A. planipennis tunnels from larvae (Xu Gongtian)

A. planipennis infested ash tree trunk (Xu Gongtian)

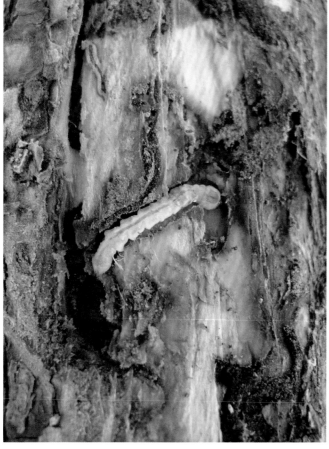

A. planipennis larva (Li Zhenyu)

A. planipennis damage (Li Zhenyu)

Eggs are laid singly. The earliest hatched larvae enter trees in mid-June. They feed on phloem and the surface of xylem. The bark where has larvae shows splits. When larvae grow larger, they form irregular galleries filled with frass. These galleries cause separation of the bark from the xylem. Larvae mature in about 45 days. Mature larvae are seen as early as late July. They overwinter in xylem. Adults show positive phototaxis, prefer warm place, and play death when disturbed. Adults feed on tree leaves for supplementary nutrient.

Prevention and treatment

1. Quarantine. Avoid transporting infested seedlings outside of the quarantine area.

2. Remove heavily infested trees. Shake trees and collect adults that fall to the ground.

3. Conservation of natural enemies. Natural enemies of *A. planipennis* include parasitoids (*Spathius agrili* and *Tetrastichus planipennis*), a mite species (*Pyemotes* sp.), and birds (*Picoides* spp.).

4. During adult emergence period, spray canopy with 50% imidacloprid monosultap mixture at 1,000× or 10% imidacloprid at 3,000× to kill adults.

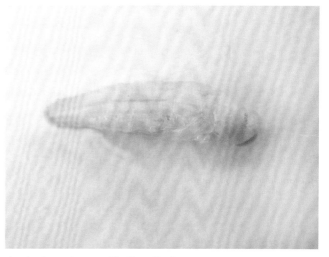

A. planipennis pupa (Xu Gongtian)

References   Xiao Gangrou (ed.) 1992; Gao Ruitong and Zhao Tonghai 2004; Jin Ruozhong, Ruan Qingshu, Yun Lili 2005; Xu Gongtian and Yang Zhihua 2007.

(Qiu Lixin, Cao Chuanjian, Xue Yinshan,
Yu Chengming)

137 *Agrilus moerens* Saunders

Classification Coleoptera, Buprestidae
Synonym *Agrilus ratundicollis* Saunders

Distribution Inner Mongolia, Shaanxi, Ningxia.
Hosts *Salix cheilophila, Caragana korshinskii*.
Damage Adults feed on bark of branches and leaves. Larvae create galleries under bark and cause death of the bark.

Morphological characteristics **Adults** Body aeneous, cuneiform. Length 5.9-7.2 mm, width 1.1-1.5 mm. Body covered with white fine pubescence. Head, pronotum, and elytra with reticulate sculptures. Antennae serrate, 11-segmented; basal segment long, the other segments similar in length. Eyes kidney shape, brown, convex. Elytra narrow and long, with aeneous luster. Abdomen of the female slightly wider than that of the male; the end of the abdomen with a small projection and a depression. The end of the abdomen in males flat, without depression or projection.

Eggs Length 0.8-1.0 mm, width 0.45-0.50 mm. Yellowish white, elongated elliptical.

Larvae Length 9.7-11.8 mm. Pronotum compressed and enlarged, width 0.8-1.38 mm. Mesothorax and metathorax narrower. Mature larvae light-yellow. Head small, only mouthparts are available. The abdominal end with a pair of brown spines.

Pupae Length 5.7-7.1 mm, width 1.66-2.06 mm. Pale yellow at beginning, turning darker later.

Biology and behavior *Agrilus moerens* occurs one generation a year. It overwinters as mature larvae in phloem. They resume activity in April. Pupae appear in mid-June. Adults emerge during mid-July to mid-August. Egg period is 50-60 days. Peak of egg hatching period is in late September. Newly hatched larvae enter bark in early October and start overwintering at the end of October.

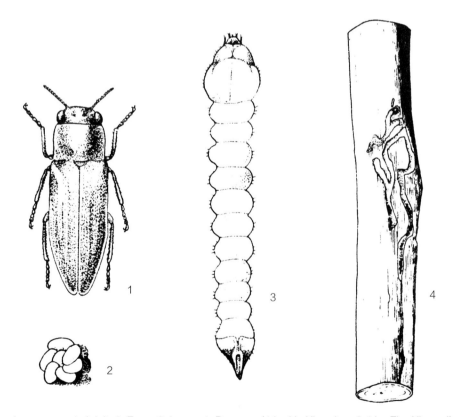

A. moerens. 1. Adult, 2. Eggs, 3. Larva, 4. Damage (1 by Hu Xingping; 2-4 by Zhu Xingcai)

Adults emerge from pupae during 8:00-19:00. They chew open the bark through funnel shaped holes. Emergence galleries are about 14 mm long. Eclosion rate of 85.7% was observed. After about 4 hours, they can fly. Adults feed on leaves and bark of new branches and can cause significant damage. Adults are very active during noon time on clear days. They rest among dense leaves during the early morning and evening hours. They are also not active during cloudy and rainy days. Adults are weak flyers. Each time they can fly 4-5 m. They play death when disturbed. But they will fly away when fallen halfway. Female: male ratio is 1:0.92.

From 2nd day after emergence, they can mate. Mating lasts for approximately two hours. They mate multiple times. Few individuals mate more than 20 times. Females lay eggs 6-8 days after mating. The female scrubs the branch with her abdomen for a few minutes, then lays one egg. A few minutes later, she lays a few more eggs beside the 1st egg. An egg mass contains 2-9 eggs. Eggs were covered with a white glue material. A female lays about 56 eggs during her life time. Non-mated females do not contain eggs in their abdomen. Eggs are usually laid on branches of greater than three years old. Few eggs are laid on 1-2 years old branches. Each branch usually has 3-8 egg masses. Young larvae bore into bark. They create curvy vertical galleries. Later they overwinter between phloem and cambium. A branch often has more than 20 galleries. The bark where larvae occur underneath becomes rotten. Mature larvae build sickle-shaped pupal chambers inside xylem.

Prevention and treatment

1. Silviculture. Cut all hosts in late November in heavily infested areas and destroy the cut-down trees. Two consecutive cutting in two years will significantly reduce the infestations.

2. Conserve the parasitoid *Geniocerus* sp.

3. In mid-July spray trees with 90% trichlorfon, 50% sumithion at 1,000×, or 40% omethoate at 800×. Treat two times will achieve good results.

References Chen Xiaoda, Hu Zhonglang, Yang Pengju et al. 1983.

(Chen Xiaoda)

138 *Agrilus subrobustus* Saunders

Classification Coleoptera, Buprestidae
Synonyms *Agrilus viduus* Kerremans, *Chrysochroa fulminans* F.

Distribution Beijing, Tianjin, Hebei, Shanxi, Liaoning, Jilin, Heilongjiang.

Host *Albizia julibrissin*.

Damage Larvae bore into the bark and xylem. Galleries are filled with frass and sawdust. Once these galleries are connected, nutrient flow is interrupted and the branches or the tree will die.

Morphological characters **Adults** Length 3.5-4 mm. Aeneous, with slightly metallic luster. Occiput flat and straight.

Larvae Mature larvae 5 mm in length. Milky. Head small, dark brown. Thorax wide, abdomen slender. The body shape looks like a nail.

Biology and behavior *Agrilus subrobustus* occurs one generation per year in Beijing and Hebei. They overwinter as larvae in tree trunk. Mature larvae pupate in late May inside the galleries. When flower buds of *A. julibrissin* develop in mid-June, adults begin to emerge. They often crawl on the bark. They feed on leaves to obtain nutrition. Females lay eggs on the trunk and branches. Eggs are laid singly. After hatching, larvae immediately bore into the bark. Black sticky material appears from the infestation site in September-October. Larvae overwinter in November.

Prevention and treatment

1. Quarantine. Avoid transporting infested seedlings out of the quarantine zone.

2. Silviculture. Maintain vigorous growth and smooth trunk. Clear the dead branches and trees. Apply lime slurry to tree trunk to prevent egg laying.

3. Mechanical control. Shake trees and collect adults in early morning. Use a knife to pick out larvae from galleries.

4. Chemical control. Spray 20% avermectin at 1,500-2,000×, 5.7% emamectin to the crown during adult period. Brush trunks with deltamethrin and diesel at 1:1 ratio to kill larvae.

References Xu Gongtian 2003, Xu Zhihua 2006, Wang Xiaoyi 2018.

(Qiao Xiurong)

A. subrobustus tunnels (Xu Gongtian)

139 *Agrilus zanthoxylumi* Zhang et Wang
Classification Coleoptera, Buprestidae

Distribution Shanxi, Shandong, Henan, Shaanxi, Gansu.

Host *Zanthoxylum bungeaum.*

Damage Larvae feed on cambium and xylem layer, causing death of bark, weakened growth, reduced fruit yield, or death of trees.

Morphological characteristics **Adults** Females length 9.0-10.5 mm, male length 8.0-9.0 mm. Body with metallic shine. Vertex with longitudinal depression and dense fine punctures. Eyes brown; antennae blackish brown, serrate, 11-segmeated; pronotum with one depression at center. Elytra greyish yellow, with four pairs of irregular dark spots. The end of the elytra with teeth.

Eggs Elliptical. Length 0.80-0.95 mm, width 0.45-0.65 mm. Milky, semi-transparent.

Larvae Cylindrical. Length 17.0-26.5 mm. Milky, head and caudal projection dark brown; medial groove on the pronotum dark yellow; median groove on abdomen light-yellow. The abdominal end has two pincers.

Pupae Milky, black at late stage. Length 8.0-10.5 mm.

Biology and behavior *Agrilus zanthoxylumi* occurs one generation a year in Shaanxi. It overwinters as larvae at 3-10 mm deep inside branches and trunks. They resume feeding in April. Pupae appear in May. Adults emerge from early June. They lay eggs in mid – and late June. Eggs hatch in early and mid-July. Mature larvae pupate within xylem. Pupal period is 17 days. Adults feed on host leaves for nutrition. They mate same day or next day after emergence. Males and females can mate multiple times. Females lay eggs from 24 hours after mating. Each female lays 11-63 eggs. Eggs are laid in cracks on bark and around old scars. Egg incubation period is 18-19 days. Newly hatched larvae feed between phloem and xylem. They create irregular galleries.

Prevention and treatment

1. Select resistant species. Maintain tree vigor through silviculture, water and fertilizer management. Remove infested trees and branches.

2. Brush mixture of 12% thiamethoxam and lamda-cyhalothrin mixture, diesel, water at 1:1:10 to trunks and branches in April and September. During June-July when larvae start drilling into bark, open the infestation site and brush with lime sulfur.

References Xiao Gangrou (ed.) 1992, Li Menglou 2002, Wu Hai 2006, Gao Huanting and Zhang Guolong 2007.

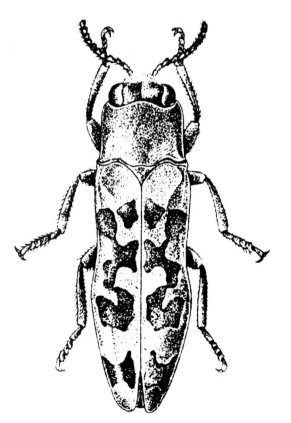

A. zanthoxylumi adult (Zhu Xingcai)

(Tang Guanghui, Li Menglou, Zhang Runke, Wang Tongmou)

Agrilus pekinensis pekinensis Obenberger

140

Classification Coleoptera, Buprestidae

Synonyms *Agrilus nipponigena* Obenberger, *Agrilus cersnii* Obenberger, *Agrilus pekinensis* Obenberger, *Agrilus boreoccidentalis* Obenberger, *Agrilus klapperichi* Obenberger, *Agrilus charbinensis* Thery

Importance Larvae bore into trunks and branches and can cause 90.2% mortality to hosts.

Distribution Beijing, Tianjin, Hebei, Liaoning, Shanghai, Jiangsu, Zhejiang, Hubei, Shandong, Henan, Shaanxi, Gansu, Xinjiang; Republic of Kazakhstan, Mongolia, Korea Peninsula, Russia.

Hosts *Salix* spp.

Damage Larvae feed inside phloem. They create irregular tunnels that are filled with brown frass. Irregular "pest marks" and 2-3 mm diameter scales appear on the bark. The infestation sites form depressions and produce brown fluid. Barks can be easily peeled after dry. Host trees eventually die from heavy infestations.

Morphological characteristics **Adults** Length 5-7 mm, width 2-3 mm. Cuneiform. Blackish brown, with bluish purple sheen, covered with light-yellowish white pubescence.

Eggs Length 1.3-1.5 mm, width 1.0-1.3 mm. Elliptical. Milky.

Larvae Mature larvae 9-12 mm. Flattened. Milky and slightly yellowish. The pronotum with an upside down "^" shaped groove.

Pupae Length 1.2-1.5 mm. Milky.

Biology and behavior *Agrilus pekinensis pekinensis* occurs one generation a year. It overwinters as mature larvae within sapwood. They pupate in early May. Pupal period is 15 days. Adults appear from mid-May to mid-June. Eggs are laid in June. Egg period is 10 days. Larvae appear from early June. Most larval damage occur from mid-June to mid-August. Larvae start to overwinter in late August when temperature is below 25℃.

Agrilus pekinensis pekinensis are active during the day and rest on leaves, branches, or inside cracks during the night. After adult emergence, they feed on host leaves. Adults mate multiple times. Females lay eggs singly. Occasionally, they lay groups of 4-5 eggs. Larvae bore into bark from where they hatch. They stay in phloem layer for 45-65 days. Then they move to xylem layer from the 3rd instar. The larvae form new moon shaped galleries in xylem. The galleries are filled with white sawdust. Only about 1 cm long gallery is used as overwintering space.

Prevention and treatment

1. Silviculture. Maintain good sanitation in forests by removing damaged trees and dead trees. When planting new trees, use trees of at least 10 cm in diameter at breast height.

2. Biological control. Release *Sclerodermus guani* wasps at 1:2 ratio of wasps and pest marks.

3. Chemical control. Before early or mid-May, brush tree trunks below 2.5 m height with lime slurry to reduce number of eggs laid and the hatch rate. Add 40% acetamiprid and chlorpyrifos mixture or 10% imidacloprid to

A. pekinensis pekinensis adult (Xu Gongtian)

A. pekinensis pekinensis tunnels (Xu Gongtian)

A. *pekinensis pekinensis* eggs (Xu Gongtian)

A. *pekinensis pekinensis* exit holes (Xu Gongtian)

A. *pekinensis pekinensis* tunnels (Xu Gongtian)

the lime sulfur to kill larvae.

References   Jiang Sandeng 1990, Xiao Gangrou (ed.) 1992; Yang Zhongqi, Wang Xiaoyi, Cao Liangming et al. 2014

(Huang Pan, Li Zhenyu, Jiang Sandeng)

141 *Chalcophora yunnana* Fairmaire
Classification Coleoptera, Buprestidae

Importance *Chalcophora yunnana* build galleries in sapwood and heartwood. They lower the timber value.

Distribution Henan, Hunan, Guangdong, Guangxi, Yunnan; Japan.

Hosts *Pinus massoniana, P. yunnanensis.*

Damage *Chalcophora yunnana* feed on bark and create galleries in sapwood and heartwood.

Morphological characteristics **Adults** Spindle shape. Length 30-40 mm. Cuprite color or golden copper color. Body covered with yellowish gray powder. Pronotum with blackish longitudinal ridge in the middle, its posterior angles with irregular depression. The elytra with four distinct black longitudinal ridges. Abdominal end round in females, and deeply emarginated in males.

Larvae Yellowish white. Mature larvae 47 mm in length. Pronotum distinctly wider than the 1st abdominal segment. Pronotum and its sternite shield like, with dark coarse punctures.

Biology and behavior *Chalcophora yunnana* requires a few years to finish a generation. It overwinters as larvae in xylem. In Hunan, adults emerge in mid-April. Peak adult emergence period is in May.

Adults are active around noon on clear days. They feed on bark of young branches. After mating, females lay eggs at xylem later of stumps or cut down trees. They can also lay eggs on exposed xylem of physically injured standing trees. Larvae bore into xylem. They create galleries in sapwood and heartwood. Forest fire and storm damage can lead to infestations.

Prevention and treatment

1. Quarantine. Avoid transporting infested materials outside of the quarantine zone. Debark or fumigate infested timber before transporting.

2. Silviculture. Select resistant species; plant mixed forests; remove infested trees; and irrigate adequately. Debark cut down trees before larvae pupate.

3. Conservation of natural enemies.

4. Chemical control. During adult period, spray 90% trichlorfon or 50% fenitrothion at 1,000×, or 45% profenofos and phoxim mixture, or 20% thiamethoxam at 800× to branches. During larvae hatching period, brush 2% abamectin and diesel mixture at 1:40, or 20% dinotefuran at 100× to infested area. Apply every 10 days for three times.

References Huang Fusheng 1987, Xiao Gangrou (ed.) 1992.

(Li Qiao, Pan Yongzhi, Wang Shufen)

C. yunnana adult
(Wang Jijian)

142 *Lamprodila limbata* (Gebler)

Classification	Coleoptera, Buprestidae
Synonyms	*Palmar limbata* Zykor, *Lampra adustella* (Obenberger), *Lampra mongolica* (Obenberger), *Lampra limbata* Gebler, *Poecilonota limbata* Gebler
Common names	Red-sided buprestid, Yellow margined buprestid

Importance *Lamprodila limbata* is a disastrous pest of pear trees. It also causes damage to apple, peach, and apricot trees.

Distribution Beijing, Hebei, Shanxi, Inner Mongolia, Liaoning, Heilongjiang, Jiangsu, Anhui, Jiangxi, Shandong, Henan, Hubei, Shaanxi, Gansu, Qinghai, Ningxia; Mongolia, Russia.

Hosts *Pyrus* spp., *Malus pumila*, *M. asiatica*, *Prunus persica*, *P. pseudocerasus*, *P. armeniaca*, *Crataegus pinnatifida*, *Robinia pseudoacacia*.

Damage *Lamprodila limbata* feed on bark and cambium, causing depression, canker, death of the bark. Infested trees do not grow well and often have small and red leaves. Branches or whole trees may die from infestation.

Morphological characteristics **Adults** Body length 10-18 mm. Bright green, metallic, compressed; pronotum with 5 bluish black longitudinal stripes, the center one more distinct. Elytra have several bluish black longitudinal stripes. Sides of the pronotum and the anterior border of the elytra with a red band.

Eggs Oval, yellowish white.

Larvae Mature larvae 36 mm long, compressed, milky. Head small. Prothorax wide and large; tergites yellowish brown, its center has an upside down "∧" depression. Abdomen slender, each segment rectangular.

Pupae Length 13-18 mm, milky at the beginning, dark brown later.

Biology and behavior *Lamprodila limbata* occurs one generation a year in Jiangxi, one generation in 1-2 years in Hubei, Jiangsu; one generation every two years in Beijing, Shanxi, and Shaanxi; one generation every 2-3 years in Liaoning. It overwinters as larvae in galleries. Overwintering larvae resume activity in spring. Pupation occurs in late March; pupal period is 30 days. Adults emerge in late April but remains in the galleries. Adults come out through an oval opening in mid-May. They are active during the day. They prefer feeding on young leaves. They play death when disturbed. An adult lays eggs over a 10 days period during mid- and late May. Eggs are laid in cracks on bark and injured sites. One to three eggs are laid at each site. Each female lays about 20-60 eggs. The peak egg hatching period is in early June. Larvae bore into bark and feed inside. From the 3rd instar, they form transverse galleries in cambium. Once these galleries are connected around the diameter, the branch or the whole tree will die. The galleries are filled with feces. There are not obvious signs of infestation where the bark is coarse. Larvae overwinter from September.

Prevention and treatment

1. Silviculture. Proper manage the trees and maintain strong growth. Avoid injuries of the trees. Use resistant varieties.

limbate adult (Xu Gongtian)

L. limbate adult male (Li Zhenyu)

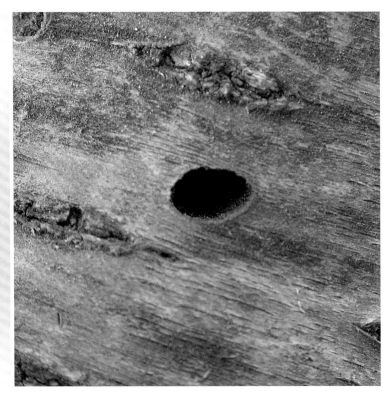
L. limbate exit hole (Xu Gongtian)

2. Mechanical control. 1) Cut down dead trees, remove dead branches and destroy them before adult emergence. 2) During adult period, shake trees and collect them in the early morning. 3) Brush trunks with lime sulfur to deter egg deposition. 4) Use a knife to remove the larvae under the bark. 5) Scrape off the coarse bark on trunks and large branches during pear tree dormancy period can kill some overwintering larvae. 6) Do not plant host species such as *Robinia pseudoacacia* around the orchards.

3. Biological control. Protect natural enemies including birds (*Dendrocopos* spp.), braconid wasps, and *Beauveria bassiana*.

4. Chemical control. Spray 10% imidacloprid on trunks before adult emergence. After adult emergence, spray 1.2% nicotine and matrine mixture at 1,000× or 8% thiamethoxam following the label.

References Luo Junsong et al. 1964; Ye Mengxian 1983; Wang Jinyou and Li Zhixing 1985; Huang Qinghai and Chen Wanfang 1989; Cao Kecheng 1996; Tang Xinpu 1996; Wang Jiangzhu 1997; Fu Ping, Yang Lixin, Gong Xiaoling et al. 2001; Yang Ziqi and Cao Guohua 2002; Zhang Li 2004; Sun Lixin and Yang Lixin 2006; Xu Gongtian and Yang Zhihua 2007.

(Yan Rong)

143 *Trachypteris picta picta* (Pallas)

Classification	Coleoptera, Buprestidae
Synonyms	*Melanophila picta* Pallas, *Buprestis picta* Pallas, *Trachypteris picta* Marseul, *Melanophila picta* Saunders, *Melanophila decostigma* F.
Common name	Ten-spotted buprestid

Distribution Beijing, Shanxi, Inner Mongolia, Heilongjiang, Hunan, Shaanxi, Gansu, Ningxia, Xinjiang; Russia, Turkey, Syria, southern Europe, northern Africa.

Hosts *Populus simonii*, *P. candicans*, *P. nigra* var. *thevestina*, *P. nigra* var. *italica*, *P. alba*, *P. alba* var. *pyramidalis*, *P. nigra*, *P. laurifolia*, *Salix matsudana*, *S. babylonica*.

Damage Larvae bore into branches and trunks of host trees, causing death of bark and secondary infection of diseases.

Morphological characteristics **Adults** Length 11-13 mm, black. Anterior border of the clypeus and frons with yellow fine hairs. Pronotum evenly covered with punctures, those on vertex are finer; with metallic bronze sheen. Antennae serrate. Each elytron with four longitudinal lines and 5-6 yellow spots; most insects with a total of 10 spots on the elytra. There are five visible abdominal segments; the posterior end of the last segment with a spine on each side.

Eggs Oval. Length 1.5 mm, width 0.8 mm. Pale yellow.

Larvae Mature larvae 20-27 mm long. Pale yellow. Head flattened, mouthparts brown. Pronotum yellow; width of pronotum about twice as the width of the abdomen; the area with raised punctures oblate form and its center with a v-shaped mark. The ventral surface of the pronotum has a longitudinal groove.

Pupae Length 11-19 mm. Pale yellow. Head facing downward. Eyes and tip of the mandibles blackish brown. Abdomen with 9 segments and six pairs of spiracles.

Biology and behavior *Trachypteris picta picta* mostly occurs one generation a year. A small portion of the populations undergo diapause and finish one generation within two years. They overwinter as mature larvae in xylem. Pupae appear in mid- and late April. Peak adult period is between mid- and late May. Egg period is between late May and early June. Most eggs hatch during mid-June. Larvae burrow into barks after hatched. They enter xylem in early July and start overwintering in mid- and late October.

T. picta picta adult (Xu Gongtian)

T. picta picta adult (Xu Gongtian)

Adults emerge from elliptical exit holes on the bark. They feed on leaves and young branches before mating. Adults are most active during 10:00-18:00. They may disperse up to 1,500 m distance. Adults live for 7-10 days. Each female lays 22-34 eggs 2-3 days after mating. Eggs are laid in cracks on bark or at base of branches. Egg incubation period is 13-18 days. The newly hatched larvae will burrow into the bark near the eggshell. The entry point slightly changes in color and produces yellowish brown secretion. Feces and sawdust will appear as the larvae grown inside bark. Larvae will live between bark and the sapwood at about 20 days after hatching. Tunnels are often in "L" shape. Mature larvae pupate near the sapwood. Larval period is 180 days. Pupal period is 15 days.

Southwest side of the trunks suffers more damage than other sides due to the adults' behavior toward sunlight. Lower and middle part of the trunk suffers more than the upper part of the trunk. Sparse forest stands suffer more than dense forest stands. Trees at forest edge suffer more damage than those within the forest. Pure forests suffer more damage than mixed forests.

Prevention and control

1. Quarantine. Avoid transporting infested trees outside of the quarantine zone.

2. Silviculture. Select resistant species. Plant mixed forests. Plant trees at high density. Clean out infested trees immediately. Manually collect and kill adults.

3. Biological control. Protect natural enemies.

4. Chemical control. During adult period, apply 50% fenitrothion, or 45% profenofos and phoxim mixture, or 20% thiamethoxam at 800× to infested branches and trunks two times. During larvae hatching period, brush the infestation sites with 2% abamectin and diesel mixture (1:40) three times at 10 days intervals.

References Wen Shouyi and Xu Longjiang 1986; Qu Bangxuan, Chen Hui, Liu Fudai et al. 1991; Xiao Gangrou (ed.) 1992; Li Menglou 2002; Wang Na 2003, Wang Xiaoyi 2018.

(Wen Shouyi, Li Menglou)

144 *Poecilonota variolosa* (Paykull)

Classification Coleoptera, Buprestidae

Synonyms *Buprestis variolosa* Paykull, *Buprestis conspersa* Gyllenhal, *Buprestis rustica* Herbst, *Buprestis tenebrionis* Schaeffer, *Buprestis tenebrionis* (Panzer), *Buprestis plebeia* (Herbst)

Importance *Poecilonota variolosa* is a major pest of the trees used for wind protection in northern China.

Distribution Hebei, Inner Mongolia, Liaoning, Jilin, Heilongjiang, Xinjiang; Russia, Republic of Kazakhstan, Europe, northern Africa.

Hosts *Populus pseudo-simonii*, *P. cathayana*, *P. simonii*, and other species of *Populus* spp., *Salix* spp.

Damage Larvae bore into the phloem and xylem of tree trunks, causing death of bark or trees. Infested trees are easily broken by wind.

Morphological characteristics **Adults** Length 13-19 mm. Flattened, cuneiform. Purplish bronze color, shiny. Each elytron with 10 longitudinal grooves and black short spots and stripes.

Larvae Mature larvae 27-39 mm long. Flattened. Pronotum with an upside down "V" shaped groove, its anterior region had four short longitudinal depressions. There is a longitudinal groove from mesonotum to the end of the abdomen.

Biology and behavior *Poecilonota variolosa* occurs one generation every three years period. It overwinters as larvae in trunks. The larvae start to feed in mid-April and pupate in late April. Adults emerge from early May and reach peak emergence in early June. After feeding for about one week period, adults can mate and lay eggs. Peak egg laying period is in early and mid-July. Eggs start to hatch in early July. After two molts, larvae overwinter. They molt three times during the 2nd and 3rd year. They pupate in the 4th year. Larvae go through 9 instars.

Adults lay eggs in cracks on bark, branches, and injury sites. They lay one egg at each site and lay dozens of eggs. Maximum number of eggs laid by a female can exceed 100. Egg incubation period is 7-10 days. Newly hatched larvae feed on eggshell, then bore into bark. As they grow in size, they bore into phloem, cambium, and xylem. The galleries are flattened and curvy. Larvae start to overwinter in mid-October. The 6th instar and larger larvae can overwinter in xylem layer. Weak trees, sparse forest, edge of the forest, and heavily trimmed trees suffer heavier infestations. *Populus pseudo-simonii* is the most susceptible compared to other host species.

Prevention and treatment

1. Quarantine. Treat infested trees by demarking, heat treatment, or fumigation before moving out of the quarantine zone.

2. Silviculture. Select resistant species. Plant mixed forests. Maintain healthy growth of trees to prevent infestations. Clear infested trees and branches. Debark or utilize the cut-down trees with the pest before pupation.

3. Physical control. During adult period, shake trees in the early morning to collect fallen adults.

4. Biological control. *Dendrocopos major* is a major natural enemy that feed on *P. variolosa* adults and larvae. Place artificial nests to promote the reproduction of the bird.

5. Chemical control. During adult period, spray branches and trunks with 90% trichlorfon, or 45% profenofos and phoxim mixture at 800×, or 50% sumithion at 1,000×, or 20% thiamethoxam at 800× for two consecutive times. During larval period, brush 2% avermectin and diesel mixture (1:40 ratio) at 10 days interval for three times.

References Xiao Gangrou (ed.) 1992. Li Chengde 2004.

(Li Chengde, Qi Mujie, Li Guihe, Zhao Changrun)

P. variolosa adult (Li Chengde)
P. variolosa adult (Xu Gongtian)

145 *Agriotes subvittatus* Motschulsky

Classification	Coleoptera, Elateridae
Synonym	*Agriotes fuscicollis* Miwa
Common names	Narrow-necked click beetle, Barley wireworm

Distribution Beijing, Hebei, Shanxi, Inner Mongolia, Liaoning, Jilin, Heilongjiang, Shandong, Henan, Shaanxi, Gansu, Qinghai, Ningxia; Japan.

Hosts Young buds of tree trees, roots and young stems of seedlings.

Morphological characteristics **Adults** Length 8-9 mm, width 2.5 mm. Slender, flattened, covered with yellow fine pubescence. Head, thorax, blackish brown; elytra, antennae, and legs reddish brown, shinny. The 1st segment of the antennae thick and long, the 2nd segment globular. Pronotum slightly round, each elytron with nine longitudinal grooves consisted of punctures.

Larvae Mature larvae 23 mm long, 1.5 mm wide. Pale yellow and shiny. The terminal segment conical, the tip with a reddish-brown projection, the base with one brown spot and four brown longitudinal stripes.

Pupae Length 8-9 mm. Pale yellow.

Biology and behavior *Agriotes subvittatus* occurs one generation every three years. It overwinters as larvae or adults. In Inner Mongolia, pupae are found in 7-10 cm deep soil in early June. Adults emerge in mid- and late June. Peak egg laying period is from late June to early July. Eggs are laid on soil surface. Adults are attracted to fermenting grass. When average temperature is above 0°C in April, they move up and feed. Most of the damage occurs when soil temperature at 10 cm deep is between 7-13°C. Population is more abundant in moist soil, soil with abundant organic matter, and clay loam soil.

Prevention and treatment
Similar to that for *Pleonomus canaliculatus*.

References Wei Hongjun 1990, Zhang Zhizhong 1997, Xu Gongtian and Yang Zhihua 2007.

(Liang Hongzhu, Li Zhenyu, Wu Yi)

A. subvittatus adult (Li Zhenyu)   *A. subvittatus* adult (Xu Gongtian)   *A. subvittatus* adult (head and thorax) (Xu Gongtian)

146 *Melanotus cribricollis* (Faldermann)

Classification Coleoptera, Elateridae

Distribution Hebei, Jiangsu, Zhejiang, Anhui, Fujian, Jiangxi, Shandong, Guangxi, Sichuan; Japan.

Hosts *Phyllostachys praecox*, *P. propinqua*, *P. tianmuensis*, *P. praecox* f. *viridisulcata*, *P. heteroclada* var. *pubescens*, *P. iridescens*, *P. glauca*, *P. heteroclada*, *P. sulphurea* 'Viridis', *P. kwangsiensis*, *P. heterolada* var. *funhuaensis*, *P. bambusoides*, *P. acuta*, *P. parvifolia*, *P. meyeri*, *P. edulis* f. *gimmei*, *P. atrovaginata*, *P. fimbriligula*, *P. flexuosa*.

Damage Larvae feed on the underground part of the bamboo shoots, causing feeding holes, withering of the above-ground part of the bamboo shoots, and reduction in the bamboo shoot product value.

Morphological characteristics **Adults** Length 9.8-11.6 mm. Width between the two posterior angle of the pronotum 2.6-3.4 mm. Black, ventral surface of abdomen dark, legs brown. Head with dense coarse punctures. Antennae 11 segmented, its 1st segment thick, the 2nd-3rd segment moniliform, the end segment fusiform, other segments serrate. Punctures on pronotum smaller than those on the head. Its posterior angles projected backward about 0.5 mm long and surround the shoulder of the elytra. Elytra more than twice as long as the pronotum, with 9 longitudinal grooves consisted of punctures.

Larvae Mature larvae 27.2-31.5 mm long, front of the pronotum 2.1-2.5 mm wide. Slender, flattened cylindrical shape. Dark red or reddish brown. Head trapezoid, with four longitudinal lines. Mandibles black. The 1st thoracic segment is same length as the meso- and metathorax combined. Each segment has an anterior and posterior border. The last segment conical, its end with three projections.

Pupae Length 10.5-12.8 mm. Pale yellow, becomes greyish black before eclosion. There is a brown seta in front of the antenna. Posterior border of the pronotum with a pair of brown setae. Wing buds reach the posterior border of the 3rd abdominal segment. Hind tarsi reach the posterior border of the 4th abdominal segment.

Soil cocoons Length 22 mm, similar to sunflower seeds.

Biology and behavior The insect occurs one generation every 3-4 years in Zhejiang. It overwinters as adults and larvae. They are active from late April to early July. Adults do not feed or occasionally feed on leaves or outer skin of the bamboo shoots. Adults mate and lay eggs during early and mid-May. Eggs hatch after about 20 days. Small larvae feed on rhizome shoots, bamboo grass roots, or bamboo roots, or even humus. Mature larvae make cocoons in late July and early August. Pupal period is 25 days. Larvae overwinter in soil from November.

Prevention and treatment

Conservation of natural enemies. Adults may be pre-

M. cribricollis adult (Xu Tiansen)

M. cribricollis eggs (Xu Tiansen)

M. cribricollis larva (Xu Tiansen)

M. cribricollis newly molted larva (Xu Tiansen)

dated by *Cuculus canorus canorus*. Larvae may be predated by *Polyrhachis dives* and *Formica japonica*.

References   Xu Tiansen and Wang Haojie 2004.

(Xu Tiansen)

147 *Pleonomus canaliculatus* Faldermann

Classification Coleoptera, Elateridae
Common names Grooved click beetle, Wheat wireworm

Distribution Beijing, Hebei, Shanxi, Inner Mongolia, Liaoning, Jiangsu, Anhui, Shandong, Henan, Hubei, Shaanxi, Gansu, Qinghai.

Hosts Pinaceae, Cupressaceae, *Robinia pseudoacaia*, *Firmiana platanifolia*, *Malus baccata*, *Platanus hispanica*, *Acer truncatum*, *Syringa* spp., *Malus prunifolia*.

Damage *Pleonomus canaliculatus* larvae feed on germinating seeds, roots and stem of the seedlings.

Morphological characteristics **Adults** Female length 14-17 mm, width 4-5 mm. Male length 14-18 mm, width 3.5 mm. Body flattened. Blackish brown covered with golden yellow pubescence. There is a triangular depression on the head. Male antennae 12 segmented, slender, reaching the end of the elytra. Female antennae shorter, 11 segmented, about two times of the length of

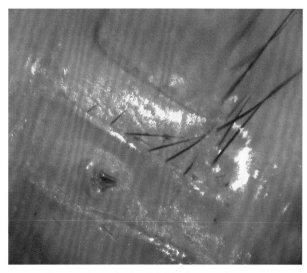

P. canaliculatus larva (spiracle) (Xu Gongtian)

P. canaliculatus adult male (Xu Gongtian)

P. canaliculatus larva (head) (Xu Gongtian)

P. canaliculatus larva (posterior end) (Xu Gongtian)

P. canaliculatus adult female (Xu Gongtian)

the pronotum. Hindwings of the females degenerated.

Eggs Oval. Length 0.7 mm, width 0.6 mm. Milky.

Larvae Mature larvae 20-30 mm long, 4 mm wide. Each body segment wider than long. Body yellow. There is a fine longitudinal groove on dorsal surface of each segment from thorax to the 10th abdominal segment. The anal segment yellowish brown, slightly raised upward, its end forked, with a small tooth on inner side of each of the fork.

Pupae Female length 16-22 mm, width 4.5 mm. Male length 15-19 mm, width 3.5 mm. Yellow, elongated fusiform. The end splits and forms two angular processes which slightly pointed outward, the end of the processes with blackish brown hairs.

Biology and behavior *Pleonomus canaliculatus* occurs one generation in 2-3 years. It overwinters as larvae or adults. In Henan, adults start to be active in late February, reach peak activity between mid-March and mid-April. Adults are inactive during the day. They mate and lay eggs during the night. Each female lays an average 94 eggs. Males are good flyers. They are attracted to lights. Females cannot fly. Mature larvae pupate at 16-20 cm deep in soil in late August. Pupal period is about 16 days. They overwinter in pupal chambers. Larvae become active when soil temperature at 10 cm depth reaches above 6°C. Peak damage period is during March-April. When soil temperature rises to 21-26°C in summer, they move deeper in soil. Very wet soil is unfavorable to larval activity.

Prevention and treatment

1. Set up light traps to catch and kill adult males.

2. Mix 3,000-4,700 mL of 45% profenofos and phoxim mixture with 25-30 kg fine soil. Broadcast treated soil along ridges of the field and then hoe shallowly or apply 40% chlorpyrifos and phoxim mixture at 800× to the trenches.

References Wu Fuzhen and Guan Zhihe 1990, Xiao Gangrou (ed.) 1992, Zhang Zhizhong 1997, Xu Gongtian and Yang Zhihua 2007.

(Liang Hongzhu, Li Zhenyu, Wu Yi)

148 *Ptilinus fuscus* (Geoffroy)

Classification Coleoptera, Anobiidae

Distribution Hebei, Liaoning, Jilin, Shanghai, Zhejiang, Anhui, Shandong, Henan, Hubei, Hunan, Guangdong, Guangxi, Sichuan, Guizhou, Yunnan, Shaanxi, Gansu, Qinghai.

Hosts *Picea asperata*, *Populus* spp., *Salix* spp., leather, paper, fabric, tobacco, Chinese herbs (*Nicotiana* spp., *Glycyrrhiza uralensis*, *Pueraria lobata*)

Damage Larvae bore into dried wood and dead branches on live trees. They create crossed galleries that filled with fine powder. It is a serious pest in Gansu province. Houses using poplar timbers begin to suffer damage within 3-5 years after construction. The wood may break at about 10 years after construction. Among *Picea*, *Populus*, and *Salix*, *Picea* is least preferred.

Morphological characteristics **Adults** Small, black, with dirt yellow pubescence. Female antennae serrate; male antennae plumose. Pronotum round, convex, densely covered with punctures; its center with a fine depression line and small teeth. Elytra with punctures and four inconspicuous longitudinal lines of punctures. Tibiae and tarsi reddish brown; exterior end of the fore- and mid- tibiae each with a tooth.

Larvae Milky, legless.

Biology and behavior *Ptilinus fuscus* occurs one generations two years. After eclosion, adults emerge from holes on wood. They may fly for short distances and are easily caught. Adults mate soon after emergence. After mating, females bore into wood. The entrance holes are round, 2-3 mm in diameter, 5-16 mm deep. Each female produces 40-50 eggs. After egg laying, females die inside wood. Female live for 16-27 days. In Qinghai, adults emerge from late May to early June and reach peak from mid-June to early July.

Prevention and treatment

1. Quarantine. Treat infested wood before use. Remove infested live trees immediately.

2. Mechanical control. Shake trees in the early morning and evening to catch and kill adults.

3. Chemical control. During early June and late July, spray 25 g/L lamda-cyhalothrin at 200-400× at the rate of 3,000-4,000 mL per room, 10% cypermethrin at 1,000×, or 77.5% dichlorophos at 1,000× every 7 days for 5-7 times to control pests in homes.

References Xiao Gangrou (ed.) 1992; Wang Xixin 2000; Wang Xixin, Zhao Minyang, Zhu Zongqi et al. 2001.

(Han Guosheng, Qiu Lixing, Cao Chuanjian, Xu Zhenguo)

P. fuscus damage (FPC)

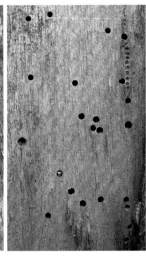
P. fuscus exit holes (FPC)

P. fuscus adult male (FPC)

149 *Sinoxylon japonicum* Lesne

Classification Coleoptera, Bostrychidae

Distribution Beijing, Tianjin, Hebei, Jiangsu, Henan, Guangdong, Guangxi, Hainan, Sichuan, Yunnan, Shaanxi, Ningxia.

Hosts *Sophora japonica*, *Robinia pseudoacacia*, *Diospyros kaki*, *Acacia fornesiana*, *Fraxinus chinensis*, *Malus prunifolia*, *Koelreuteria paniculate*, *Platycladus orientalis*.

Damage Larvae and adults feed under bark in xylem, make galleries, and cause death of branches. Infested trees are easily broken from wind. Seedlings may die from infestations.

Morphological characteristics **Adults** Length 4.2-6.3 mm. Blackish brown. Antennae tanned color, its last three segments pectinate. Pronotum well developed. Head is concealed under the pronotum. Posterior part of the wings oblique, each side armed a sharp tooth. **Larvae** Length 4.2 mm. White, slightly curved. Front legs more developed among the three pairs of thoracic legs.

Biology and behavior *Sinoxylon japonicum* occurs one generation a year in Henan. It overwinters as adults within circular galleries in branches. They start to feed from March-April. Adults come out of galleries in late April, mate, and lay eggs inside galleries. After hatching, larvae feed on xylem. Many holes appear on infested branches and trunks. The galleries are filled with saw dusts and frass. Larvae pupate during May-June. After eclosion, adults stay in xylem layer and feed. Adults come out from galleries in September-October and enter between phloem and xylem and overwinter.

Prevention and treatment

1. Quarantine. Prevent transportation of infested seedlings outside of the quarantine zone.

2. During May-June, cut off infested branches and destroy them along with the pests.

3. Maintain vigorous growth of seedlings and young trees by proper irrigation and fertilization to stimulate natural tolerance to pest infestations.

4. Release *Sclerodermus guani* during larval period.

References Cao Chengyi 1981, Chen Jun and Cheng Huizhen 1997, Yang Youqian 1999.

(Yang Youqian)

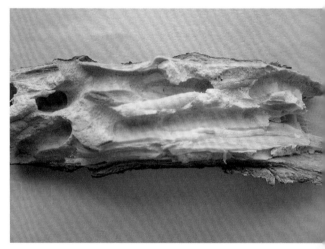

S. japonicum tunnels (Xu Gongtian)

S. japonicum exit hole (Xu Gongtian)

S. japonicum adult (Xu Gongtian)

150 *Gonocephalum bilineatum* (Walker)

Classification Coleoptera, Tenebrionidae
Synonyms *Gonocephalum orarium* (Lewis), *Gonocephalum seriatum* Boisduval

Distribution Fujian, Guangdong, Hainan, Chongqing, Sichuan, Yunnan; Japan, India, Malaysia, Papua New Guinea, Sri Lanka.

Host *Eucalyptus* spp.

Damage Adults feed on bark of young trees near the ground level and kill trees. Large areas of young trees may be killed.

Morphological characteristics **Adults** Elongated elliptical. Length 7.4-8.4 mm. Body and antennae black, dull; with dense granules. Front of head arched, wide; the area in front of the eyes triangular. Pronotum flattened, its posterior border depressed near the side. Punctures on elytra inconspicuous.

Biology and behavior *Gonocephalum bilineatum* occurs one generation every two years. Adults emerge in April-May. After a few days of rest in soil, the adults start to feed on tender bark during the night. They rest in fallen leaves or soil during the day. Adults prefer lay eggs on loose soil that with abundant ground cover. Eggs are laid during June-July. Larvae feed on grass or tree roots. They are active in soil and can have a large feeding area. The mature larvae build pupal chambers in soil next year.

Most damage occurs during mid-May. Population density is related to type of plants in the previous crop. Fields with *Eucalyptus* trees in the previous crop have lower pest density than fields with pineapple or sugar cane in the previous crop. Only trees less than one year old suffer heavy damage. Two-year or older trees suffer much less damage due to their hardened bark.

Prevention and treatment

1. Silviculture. Plant trees in June to avoid damage from adults.

2. Bait and kill. Mix 90% trichlorfon at 100× with rice shells or spray the solution to chopped *Eucalyptus* leaves. Spread them around young trees in the evening.

Reference Zhao Yangchang 1963.

(Gu Maobin)

Gonocephalum bilineatum adult female (NZM)[1]

Gonocephalum bilineatum adult male (NZM)

[1] NZM – "National Zoological Museum of Institute of Zoology, Chinese Academy of Sciences"

151 *Acalolepta sublusca* (Thomson)

Classification   Coleoptera, Cerambycidae

Importance   *Acalolepta sublusca* is a disastrous pest of *Euonymus japonica*.

Distribution   Beijing, Tianjin, Hebei, Shanghai, Jiangsu, Zhejiang, Anhui, Fujian, Jiangxi, Shandong, Hunan, Guangdong, Guangxi, Sichuan, Guizhou; Vietnam, Cambodia, Malaysia, Laos, Singapore, Japan.

Hosts   *Euonymus japonica*, *E. japonica* var. *albo-marginatus*, *E. japonica* var. *aureo-marginatus*, *E. japonica* var. *aureo-variegatus*, *Glochidion* sp.

Damage   Adults feed on bark, causing death of branches and leaves above the feeding site. Larvae feed inside stem of 4-year or older plants and kill whole plants or branches.

Morphological characteristics   **Adults** Male length 13-23 mm, width 5.0-7.5 mm; antennae surpass the body more than one time of its length. Female length 16-24 mm, width 7-9 mm; antennae surpass the body by ½ of its length. Body brown; head, pronotum densely covered with silky, tanned pubescence; elytra densely covered with smooth, shiny, pale grey pubescence and have blackish brown stripes.

Larvae   Length 20-36 mm, cylindrical. Pronotum 4-5.5 mm wide, densely covered with granules; granules coarse at the anterior lateral surface, fine at the posterior lateral surface. Dorsal ambulatory ampulla on abdominal segments with two transverse grooves and four rows of tubercles.

Biology and behavior   *Acalolepta sublusca* occurs one generation a year. It overwinters as larvae in galleries within roots. They feed during early and mid-March, then make pupal chambers at ground level in the roots, and pupate in early May. Adults emerge in mid-

A. sublusca adult (Xu Gongtian)

A. sublusca larva (Xu Gongtian)

A. sublusca adult exit hole (Xu Gongtian)

A. sublusca tunnels (Xu Gongtian)

A. sublusca damage to *Euonymus japonica* (Xu Gongtian)

May. After about 5-7 days after eclosion, they come out from eclosion holes in late May. Adults come out from eclosion holes mostly during 4:00-6:00 and 19:00-21:00.

Adults are weak flyers but can crawl fast. They play death when disturbed. After 3-5 days of feeding on bark or main veins of leaves, they mate at top of trees or branches. After another 1-3 days, they lay eggs on coarse bark about 13 cm above the ground. Egg incubation period is 15 days. A plant that with thin stem only has one larva. Larger plants may have more than two larvae. Each female lays 20-50 eggs. Females live for 20-35 days and males live for 15-25 days. Larvae appear in mid- and late June. Young larvae feed on bark. From the 2nd instar, they move to xylem and then to main roots. During feeding, they create irregular galleries which will lead to death of branches or whole plants. Larvae overwinter in mid-November inside galleries. Larvae period is about 300 days and pupal period is about 20 days.

Prevention and treatment

1. Cultural control. Remove weak, infested and dead plants during September and May and destroy them.

2. Spray *Beauveria bassiana* at root area in early spring.

3. Chemical control. During adult period, spray 3% thiamethoxam or 15% imidacloprid at 3,000×, or 45% profenofos and phoxim mixture at 800× at the crown or root area.

References   Pu Fuji 1980, Jiang Shunan 1989, Yin Chunchu 2003, Xu Gongtian and Yang Zhihua 2007.

(Tang Yanping, Yang Chuncai)

152 Anoplophora chinensis (Forster)
Classification Coleoptera, Cerambycidae

Importance *Anoplophora chinensis* is an extremely damaging wood boring insect pest. In Shandong, it caused death of *Platanus hispanica* trees or branches.

Distribution Beijing, Hebei, Shanxi, Inner Mongolia, Liaoning, Jiangsu, Zhejiang, Fujian, Shandong, Hubei, Hunan, Sichuan, Guizhou, Yunnan, Shaanxi, Gansu, Ningxia, Taiwan; Japan, Korea Peninsula, Myanmar.

Hosts *Populus* spp., *Salix* spp., *Morus alba, Ulmus pumila, Platanus hispanica, Citrus reticulata, Eriobotrya japonica, Ficus carica, Zanthaxylum bungeanum, Malus pumila, Pyrus bretschaneideri, Prunus pseudocerasus, P. armeniaca, P. persica, P. salicina, Juglans regia*.

Damage Larvae feed on phloem and xylem, cause death of branches. Infested branches are easily broken by wind or may suffer disease infection.

Morphological characteristics **Adults** Length 19-39 mm. Black with metallic sheen. Head covered with white pubescence and sparse sculpture; frons and center of vertex with fine longitudinal grooves. Antennae slender, surpassing the posterior end of the body, with white and black alternating color; base of each of the 3rd to 11th segments has a ring of light blue hairs. Center of pronotum with three tubercles, each side of the pronotum has a strong sharp tooth. Basal part of the elytra densely covered with granules, the elytra are dotted with about 20 white hairy spots. Scutellum and tarsi pale bluish green.

Eggs Length 5-6 mm. Elongated elliptical, yellowish brown.

Larvae Mature larvae 45-67 mm long. Cylindrical, slightly flattened. Milky to light-yellow. Head yellowish brown. Mandibles black. Pronotum with bird shaped yellow markings in the front and a yellow large spot at the posterior half. Legs are absent. Ventral surface of mesothorax, dorsal and ventral surface of metathorax and 1st to 7th abdominal segments with elongate ambulatory ampullae.

Pupae Length 30-35 mm. Spindle shape, yellowish brown.

Biology and behavior *Anoplophora chinensis* occurs one generation a year in Fujian and Hunan, one generation every two years in Shandong. They overwinter as larvae in galleries. Larvae pupate in spring. Pupal period is 18-45 days. Adults emerge from early June to August. Adults feed on bark for nutrition. After about half month, they mate and lay eggs on tree trunk below 10 cm height. Females chew "T" shaped oviposition scars that are 2 mm deep and 8 mm long on bark. One egg is laid in each scar. The egg is sealed with a gelatinous material. Each female lays about 32 eggs. Egg incubation period is 9-15 days. Most eggs hatch during mid- and late July. Young larvae bore into bark after hatching. From 2-3 cm deep, they move upward and continue to bore galleries. They create ventilation holes as they extend the galleries. Sawdust and frass are expelled from the ventilation holes.

A. chinensis adult (edited by Zhang Runzhi)

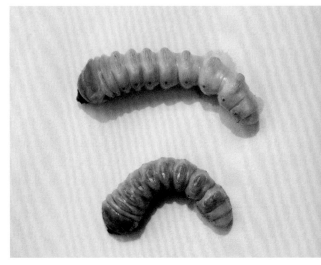

A. chinensis larvae (edited by Zhang Runzhi)

A. chinensis adult feeding on a branch (edited by Zhang Runzhi)

A. chinensis exit hole (Xu Gongtian)

A. chinensis larval damage (Xu Gongtian)

A. chinensis larval damage (edited by Zhang Runzhi)

From late September, larvae move downward, passing the original entrance hole, and continue downward and make new galleries. Larvae overwinter in November.

Prevention and control

1. Silviculture. Plant mixed forests and resistant species. Maintain healthy growth of trees. Cut down infested branches and heavily infested trees.

2. Biological control. Conserve natural enemies. Major natural enemies include *Dastarcus longulus* and *Dendrocopos* spp. In Spring, apply *Beauveria bassiana* around roots.

3. Spray 3% thiamethoxam at 200× to tree trunk during adult emergence period at 15 days intervals for three times. During larval period, inject 2 mL 20% dinotefuran at 5× per hole and then seal the holes. Place 5% cotton ball soaked with cypermethrin at 3× in each hole.

4. Apply lime slurry to tree trunk to prevent adults from laying eggs.

References Chinese Academy of Forestry 1983, Qi Chengjin 1999, Zhao Fanggui et al. 1999.

(Wang Haiming)

153 *Anoplophora glabripennis* (Motschulsky)

Classification	Coleoptera, Cerambycidae
Synonym	*Melanauster glabripennis* Matsumura
Common name	Asian longhorned beetle

Importance *Anoplophora glabripennis* is one of the most important wood boring pests. It often causes serious damage to forests and landscaping and results in tremendous economic losses.

Distribution Beijing, Tianjin, Hebei, Shanxi, Inner Mongolia, Liaoning, Jilin, Heilongjiang, Shanghai, Jiangsu, Zhejiang, Anhui, Fujian, Jiangxi, Shandong, Henan, Hubei, Hunan, Guangdong, Guangxi, Chongqing, Sichuan, Guizhou, Yunnan, Tibet, Shaanxi, Gansu, Qinghai, Ningxia, Xinjiang; Korea Peninsula, Japan, Myanmar, Vietnam, U.S.A., Canada, Austria.

Hosts *Populus* spp., *Salix* spp., *Ulmus punila*, *U. parvifolia*, *Acer mono*, *A. saccharinum*, *Platanus orientalis*, *Elaeagnus angustifalia*, *Aesculus chinensis*, *Betula platyphylla*, *Eucalyptus citridora*.

Damage Larvae feed on phloem and xylem. Average length of galleries is 9.6 cm. Trunks of damaged hosts often show palm-shaped dentation with average dentation area of 166 mm^2. One larva can damage a 12 cm long section, 10 cm diameter wood. When galleries are dense, center of the trunk may become hollowed and the surface forms nodules of 30-70 cm in length. If damaged seriously, the nodules would cut off the nutrient transport, resulting in dead and broken branches. Large areas of dead brown forests look like burned by fire and are described as "smokeless fire disaster".

Morphological characteristics **Adults** Body shiny black. Female length 22-35 mm, width 8-12 mm; antenna is about 1.3 times more than the length of body, the end of the last segment is off white. Males are a little smaller; antenna is about 2.5 times more than the length of

A. glabripennis mating pair (edited by Zhang Runzhi)

A. glabripennis adult on London plane tree (edited by Zhang Runzhi)

A. glabripennis egg (Xu Gongtian)

A. glabripennis larva (Xu Gongtian)

A. glabripennis pupa (Xu Gongtian)　　Sawdust from A. glabripennis (Xu Gongtian)

A. glabripennis exit hole (Xu Gongtian)　　A. glabripennis chew mark for eggs (Xu Gongtian)

body, the end of the last segment is black. Each elytron has about 20 white spots. **Eggs** Milky white, oblong, length 5.5-7 mm. The two ends curl slightly. **Larvae** Young larva milky white, mature larva yellowish. Body length is about 50 mm and head width is about 5 mm. **Pupae** Milky white to yellowish white. Body length 30-37 mm, width 11 mm. Antennae often curl to a circle, located on top of the front and middle legs and the wings.

Biology and behavior *Anoplophora glabripennis* usually occurs one generation every 1-2 years. In some areas, three years may be needed to finish one generation. They can overwinter as eggs, larvae, or adults. Adults emerge from May to October. Peak eclosion period occurs in July. Newly emerged adults feed on leaf veins and young twigs to obtain nutrition. After mating, each female lays about 32 eggs. Before laying eggs, the female makes an incision on the tree, then inserts its ovipositor under the phloem. A female lays one egg into the xylem at each incision. The longest observed adult life-span is 66 days and the shortest being 14 days. Yong larvae feed on phloem after they emerged. Starting from the 3rd instar, larvae began to damage xylem. Mature larvae create pupal rooms at the end of the galleries. Adults come out of round eclosion holes and then fly away, but they are not good flyers.

Prevention and treatment　Natural enemies used for controlling Asian long horned beetles are: Great Spotted Woodpecker (*Dendrocopos major*) and *Dastarcus helophoroides*. In Spring, apply *Beauveria bassiana* around roots.

References　Yan Junjie, Yu Xiulin, and Ren Chaozuo 1989; Xiao Gangrou (ed.) 1992; Haack et al. 1997; Appleby 1999; McManus et al. 1999; Qi Jie, Luo Youqing, Huang Jingfang et al. 1999; Yan Junjie and Yan Huahui 1999; Wang Zhigang, Yan Junjie et al. 2003; Wang Zhigang 2004.

(Yan Junjie, Wang ZhiGang, Qin Xixiang)

154 *Apriona rugicollis* Chevrolat

Classification	Coleoptera, Cerambycidae
Synonym	*Apriona germari* (Hope)
Common name	Mulberry longhorn beetle

Importance Adults and larvae can cause serious damage to trees. As *Populus tomentosa* trees were planted in large areas in the 1980's, outbreaks of this pest occurred. In Guan county of Shandong province, an average of 20% of the *P. tomentosa* trees were infested.

Distribution Beijing, Hebei, Liaoning, Jiangsu, Zhejiang, Shanghai, Anhui, Fujian, Shandong, Henan, Hubei, Hunan, Guangdong, Guangxi, Sichuan, Shaanxi, Taiwan.

Hosts *Morus alba*, *Broussonetia papyrifera*, *Populus tomentosa*, *Populus deltoids* × *Populus nigra*, *Salix* spp., *Ulmus pumila*, *Pterocarya stenoptera*, *Juglans regia*, *Malus pumila*, *M. prunifolia*, *M. asiatica*, *Prunus pseudocerasus*, *Ficus carica*, *Styphnolobium japonicum*, *Artocarpus heterophyllus*.

Damage Larvae bore into branches, affecting tree growth or cause death of trees. Adults feed on tender bark or young branches.

Morphological characteristics **Adults** Length 34-46 mm. Large, black, covered with tawny brown to slightly greenish hairs; hairs on ventral surface brownish yellow. Vertex raised, with a longitudinal groove at the center. Antennae scape black, each of the other segments with anterior half greyish white and posterior half blackish brown in color. Pronotum near square shape, each side with a spine; dorsal surface with transverse rugae. Basal 1/3 of the elytra with up to 150 black shiny tubercles. Legs black, covered with grey short pubescence. In the similar species *A. germari* (Hope), pronotum without transverse rugae or not distinct, tubercles on elytra are located in basal 1/4 of the elytra and the numbers do not surpass 50.

Eggs Elongate, slightly curved. Yellowish white, pale brown before hatching.

Larvae Mature larvae 50 mm long. Cylindrical. Milky. Pronotum square shape, posterior part with dense brown granules.

Pupae Length 50 mm, light-yellow. Spindle shape. Antennae curled. The tail pointed and with a ring of setae.

Biology and behavior It takes 2-3 years to finish a generation. They overwinter as larvae in infested stems or branches. After two winters, the larvae pupate in late June. Adults appear in mid-July. Eggs are laid in late July. Egg-laying period is about 20 days. Egg incubation period is approximately 12 days. After emergence, adults stay in pupal chambers for 5-7 days. Then they fly to mulberry trees and feed on tender branches. The feeding scars are irregular in shape. If cambium layer around the diameter of the branch is destroyed, the branch will die. After 10-15 days of feeding, adults mate and lay eggs. They prefer to lay eggs on 10-15 cm diameter branches. They bite a "U" shaped scar and lay an egg in it and seal it with sticky material. Adults live for about 40 days. Young larvae bore 10 mm upward, then downward. Holes are created along the galleries for expelling frass. As the larva grows, distances between the neighboring holes become longer. The holes are usually lined up vertically. Total distance of the galleries bored by one larva

A. rugicollis adult (Xu Gongtian)

A. rugicollis adult (Wang Jijian)

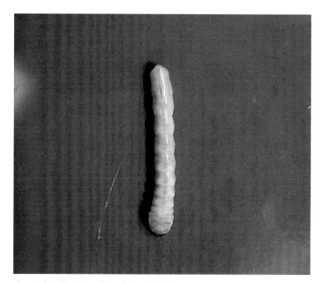

A. rugicollis larva (Xu Gongtian)

is approximately 1.7-2.0 m. Those larvae on *Populus tomentosa* may bore 5-8 m long galleries. Larvae usually feed under the last hole of the gallery. Overwintering larvae may stay some distance above the gallery bottom to avoid water accumulation at the bottom of the galleries. When mature, larvae move upward and pass 1-3 holes. They chew 1-1.6 cm diameter exit holes. Fluid often comes out from the exit holes. Larvae make pupal chambers inside the galleries. It is about 0.7-1.2 m away from the exit holes. Pupal period is 26-29 days.

Prevention and treatment

1. Silviculture. Plant mixed forests. Do not plant Moraceae plants within 1,000 m from *Populus tomentosa*.

2. Cultural control. Use a knife to open the oviposition scars and remove the eggs. Use a metal wire and insert into gallery holes on trunk or branches to destroy the larvae. A knife or screw driver can also be used to open the galleries to kill the larvae.

5. Conservation of natural enemies. Egg parasitoid includes *Aprostocetus fukutai*. *Dendrocopos* spp. eat the larvae. In Spring, apply *Beauveria bassiana* around roots.

Ventilation hole created by A. rugicollis larva (Xu Gongtian)

Sawdust and fecal material expelled by A. rugicollis larva (Xu Gongtian)

A. rugicollis expelled sawdust on the ground (Xu Gongtian)

3. Chemical control. Spray 3% thiamethoxam at 200-300× to trunks and branches to kill adults. Inject gallery holes with 3% thiamethoxam at 3×.

References Chinese Academy of Forestry 1983, Qi Chengjin 1999, Zhao Fanggui et al. 1999, Sun Xugen et al. 2001.

(Li Zhenyu, Ren Lili, Wang Haiming, Zhang Shimei, Shen Rongwu)

155 *Apriona swainsoni* (Hope)
Classification Coleoptera, Cerambycidae

Importance Outbreaks *A. swainsoni* occurred since 1980's in Shandong province. An average 81% of the *Styphnolobium japonica* trees were infested. Each tree had an average 21 *A. swainsoni*.

Distribution Hebei, Jiangsu, Shandong, Henan, Hunan, Sichuan, Guizhou, Yunnan; Vietnam, Laos, India, Myanmar.

Hosts *Styphnolobium japonica*, *Salix* spp., *Caesalpinia sepiaria*, *Dalbergia hupeana*, *Aspidium* spp.

Damage Larvae feed on phloem and xylem. Infested trees show many holes and do not grow well. Severely infested trees may die from disruption of nutrient transportation.

Morphological characteristics **Adults** Male length 28-33 mm, female length 33-39 mm. Rectangular shape, blackish brown, densely covered with rusty short pubescence. Vertex raised, with a central groove. Each side of the pronotum with a sharp and well developed tooth. The dorsal surface with coarse rugae and granules. Base of elytra densely covered with black smooth particles. The rest of the elytra covered with white hair spots and fine sculpture.

Eggs Elongate, 2.0-2.2 mm long, width 0.5-0.6 mm. Yellowish white.

Larvae Mature larvae 42-76 mm long. Cylindrical and slightly flattened. Pale yellowish white. Pronotum yellowish brown, its sclerotized area with small particles.

Pupae Spindle shape, length 35-42 mm, yellowish brown. Wings reach the 2nd abdominal segment. Antennae reach mesothorax, its end curled.

Biology and behavior It takes two years to finish one generation. Larvae overwinter in galleries. They start to feed from early April. After two winters, they pupate

A. swainsoni adult (Xu Gongtian)

A. swainsoni larva (Xu Gongtian)

A. swainsoni adult ready to exit (Xu Gongtian)

A. swainsoni exit holes (Xu Gongtian)

Raised bark from *A. swainsoni* damage (Xu Gongtian)

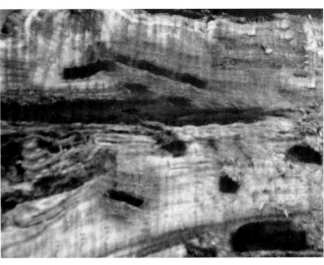
Tunnels inside by *A. swainsoni* (Xu Gongtian)

A. swainsoni damage to tree trunk (Xu Gongtian)

Transverse tunnel made by *A. swainsoni* (Xu Gongtian)

during May. Adults emerge between early June and early September. Peak adult emergence is during mid-June. Adults feed on young shoots after emergence. Adults create oviposition nests on bark before laying eggs. They cover the oviposition nests with green secretion. Egg laying period is about 13 days. Newly emerged larvae bore into phloem and xylem and make "Z" shaped galleries. They overwinter in late October. They continue feeding in early April the next year. From April in the 3rd year, they expel wood threads. When mature in mid- and late April, larvae chew holes on branches and make pupal chambers. Pupal period is approximately 20 days.

Prevention and treatment

1. Cultural control. Adults are weak flyers and fake death when disturbed. Collect adults after their emergence. Physically remove the eggs by looking for the green oviposition nests.

2. Protect natural enemies. *Dastarcus longulus* is a natural enemy of the *A. swainsoni* pupae. Apply *Beauveria bassiana* around roots of host in Spring.

3. Apply lime slurry (lime:sulphur:salt:water at 10:1:0.1:30 ratio) on tree trunks during winter. Add 40% acetamiprid and chlorpyrifos mixture to the slurry will increase the efficacy.

4. Inject insecticides into galleries and seal the injection point. The insecticides include 5.7% emamectin benzoate, 20% fenvalerate, or 5% lamda-cyhalothrin or 45% profenofos and phoxim mixture at 5-10×. Spray canopy with 3% thiamethoxam at 200-300× to control adults during mid-June and mid-July two times at 15 day interval.

References Chinese Bureau of Forestry 1996, Qi Chengjin 1999, Zhao Fanggui et al. 1999, Wang Haiming et al. 2003.

(Wang Haiming)

156 *Aristobia approximator* (Thomson)
Classification Coleoptera, Cerambycidae

Distribution Yunnan; Vietnam, Cambodia, India, Myanmar, Laos, Thailand, Malaysia.

Host *Eucalyptus tereticornis*.

Damage Larvae bore into tree branches and trunk and affect tree growth and quality of timber. Damaged trees are easily broken by wind. Severe infestations may cause tree death. Infestation rate of 60% was reported.

Morphological characteristics **Adults** Female length 25-31 mm, male length 21-24 mm. Black. Head narrow. Thorax with three orange longitudinal bands. Brown elytra with many longitudinally arranged dentations and four rows of orange spots. Scutellum triangular, orange.

Eggs Oval. Length 5 mm. Pale yellow at beginning, greyish later.

Larvae Mature larvae 40-50 mm long. Pale yellow or pale brown. Pronotum light-yellow or blackish brown.

Pupae Pale yellow or brown. Female pupae 26-31 mm long, male pupae 22-25 mm long. Dorsal surface of each of the abdominal segment with a ridge formed by short setae. The posterior end of the abdomen with 10 brown short setae.

Biology and behavior It takes two years to finish one generation. Peak adult emergence period is from early September to early October. They feed for a few days, then mate and lay eggs. Mating occurs around 9:00-11:00. Eggs are laid during 9:00-13:00. Females make 7-10 mm long square shaped oviposition scars and lay eggs. They seal the eggs with sticky material. Egg incubation period is about 15-20 days. Larvae feed on xylem. Mature larvae make pupal chambers and chew round emergence holes above pupal chambers.

Prevention and treatment

1. Quarantine. Prevent transportation of infested seedlings and young trees outside of the quarantine zone.

2. Silviculture. Use resistant species. Plant mixed forests of different species of *Eucalyptus*.

3. Cultural control. Collect adults during their emergence. Use a hammer or metal wire to kill eggs and newly hatched larvae.

4. Chemical control. Inject 25 g/L lamda-cyhalothrin at 40-50×, 40% acetamiprid and chlorpyrifos mixture at 100-200×, or 5% deltamethrin at 400-500× into galleries. Seal holes with cotton or mud after injection.

A. approximator adult (Zhang Peiyi)

(Gu Maobin)

157 *Aristobia hispida* (Saunders)
Classification   Coleoptera, Cerambycidae

Distribution   Beijing, Hebei, Jiangsu, Zhejiang, Anhui, Fujian, Hubei, Hunan, Guangdong, Guangxi, Hainan, Sichuan, Guizhou, Tibet, Shaanxi, Gansu, Taiwan; Vietnam.

Hosts   *Dalbergia odorifera*, *D. balansae*, *D. hainanensis*, *Citrus* spp., *Acacia concinna*.

Damage   Larvae bore into tree trunk and affect tree growth and quality of timber.

Morphological characteristics   **Adults** Female length 26-39 mm, width 10-16 mm. Body covered with brownish red hairs and black and white hair spots. The first a few segments of the antennae brownish red, the other segments become gradually lighter in color; the scape with black spots and erect hairs; the end of each of the 1st-6th flagellum segments black. Female antennae reach the end of the abdomen or slightly shorter than the end of the abdomen; male antennae surpass the end of the abdomen. The spine on each side of the pronotum sharp, slightly bends posteriorly. Pronotum nearly square, its center with a group of tubercles with their bases connected. Scutellum triangular, longer than width. Base of elytra with granules; the posterior end of elytra concave with distinct exterior angle and obtuse interior angle. There are more black spots than white spots on elytra. Ventral surface of the body with a series of large white spots on each side, covered with pubescence and sparse brownish black erect hairs.

Eggs Elongate, 6 mm long, 1.5 mm wide. White, slightly yellowish.

Larvae Mature larvae 70 mm long, 13 mm wide. Milky, slightly yellowish, cylindrical, slightly flat. Head partly concealed within prothorax, yellowish white. Mandibles and surrounding area of the mouthparts brownish black. Labrum with many hairs. Pronotum yellowish brown, its center with a yellowish white mark. There are eight ambulatory ampullae between the metathorax and the 7th abdominal segment. Ventral surface has one more ambulatory ampulla than the dorsal surface.

Pupae Milky, head, mouthparts, and thorax with reddish brown pubescence. Each of the abdominal segment with a transverse band of reddish brown pubescence on the back; those in the front more developed.

Biology and behavior   *Aristobia hispida* occurs one generation a year or finishes one generation in two years. Adults emerge between October and December. They stay in pupal chambers and stay dormant until February and March next year. Peak emergence is in late March. They can still be seen in late August. Adults are active during the night. They feed on tender bark. After about 10 days, they mate and lay eggs. The female:male ratio is about 2:1. Eggs are laid on trunk below 20 cm height. Adults create 9-11 mm long, 1.5-2.5 mm wide new moon shaped oviposition scars. An adult lays one egg a time. A vertical slit is shown at the oviposition site. Egg laying period usually lasts for 47-78 days. The maximum egg laying period observed is 154 days. Each female lays 22-38 eggs. Egg incubation period is 12-18 days. Females live for 4 months. Males live for 3 months.

Young larvae feed on phloem, cambium and occasionally on xylem. They create horizontal galleries under bark. After about 44-55 days, they bore into xylem and

A. hispida adult (dorsal view) (NZM)

A. hispida adult (front view) (NZM) *A. hispida* adult (lateral view) (NZM)

significantly increase their consumption. Larvae mature during July-September in the 2nd year. They pupate in pupal chambers. The gallery length is associated with size of the trees. Galleries in small trees are longer than those in large trees. Total gallery length is 31-50 cm in trees of less than 8 cm diameter. Most galleries are not branched. Some galleries have two or three branches. Each gallery usually has 1-3 holes for expelling frass.

References Liu Yuanfu and Gu Maobin 1977, Fu Guangwen and Zhang Wei 2011.

(Liu Yuanfu)

158 *Aromia bungii* (Faldermann)

Classification	Coleoptera, Cerambycidae
Synonyms	*Cerambyx bungii* Faldermann, *Callichroma bungii* Kolbe
Common name	Red-necked longhorn beetle

Distribution Beijing, Tianjin, Hebei, Shanxi, Inner Mongolia, Liaoning, Jilin, Heilongjiang, Shanghai, Jiangsu, Zhejiang, Fujian, Jiangxi, Shandong, Henan, Hubei, Hunan, Guangdong, Guangxi, Hainan, Sichuan, Guizhou, Yunnan, Shaanxi, Gansu, Ningxia; Korea Peninsula, Russia.

Importance *Aromia bungii* is a major pest of forests, fruit trees, and flower plants. Infestation rate of 90% was reported.

Hosts *Prunus persica, P. armeniaca, P. salicina, P. japonica, P. mume, P. pseudocerasus Malus pumila, M. spectabilis, Pyrus bretschneideri, Punica granatum, Juglans regia, Castanea mollissima, Salix babylonica, Ulmus pumila, Koelreuteria paniculata, Diospyros kaki, Populus* spp., *Zanthoxylum bungeanum, Morus alba.*

Damage Larvae bore into xylem along the base of tree trunk. They create curvy galleries in xylem, causing bark fall off, slow tree growth, sap flow, and death of trees.

Morphological characteristics **Adults** Male 23-28 mm long, female 30-42 mm long. Blackish brown and shiny. Pronotum dark red. The anterior border and posterior border of pronotum each with a pair of protuberance; each side of the pronotum with a large protuberance. Elytra smooth, its base wider than the pronotum; posterior end narrower than the front. Ventral surface of pronotum of the females with many transverse rugae. Antennae surpass the end of the body by two segments.

Ventral surface of pronotum of the males with dense punctures. Antennae 11 segmented, surpassing the body by 5 segments.

Eggs Length 1.2-2.0 mm. Pale green, sometimes milky. Similar to sesame shape.

Larvae Mature larvae 35-45 mm. Yellowish white, flat. Anterior half of the pronotum with a transverse row of 4 yellowish brown spots; each of the spot with a concave anterior border; the two spots on the side are triangular. Non-mature larvae had darker color, anterior

A. bungii adult (lateral view) (Xu Gongtian)

A. bungii adult (dorsal view) (Xu Gongtian)

border of pronotum with an obvious brown spot, mandibles black or blackish brown.

Pupae Length 37 mm. Pale yellowish white. Each side of the pronotum with a protuberance. The center of the anterior border of pronotum also has a protuberance.

Biology and behavior It takes two years to finish one generation in north central China. It takes three years to finish a generation in central Shanxi province. *Aromia bungii* overwinters as larvae in galleries. Adults appear in June-July. They only live for about 10 days. They lay eggs in cracks on bark. Most eggs are laid on tree trunk below 0.3 m height. Eggs hatch after 8-10 days. Young larvae feed under bark. From April next year, they bore into xylem. Galleries are irregular shape. There are many red saw dust and cylindrical frass outside of the bore holes. Most damage occurs during May-June.

Prevention and treatment

1. Trapping with light traps baited with sugar:wine:vinegar at 1:0.5:1.5.

2. Wrap tree trunk with 0.02 mm thick plastic film before egg laying, or apply sulfer lime and 40% acetamiprid and chlorpyrifos mixture slurry to tree trunk to kill eggs.

3. Biological control. Propagate and release *Sclerodermus guani* wasps, pathogeneic nematodes (*Steinernema bibionis*, and *S. feltiae*), *Beauveria bassiana*, and *Dastarcus lingulus*. *Dendrocopos major* and *D. canicapillus* are important predators and should be protected.

4. During adult period, spray 50% fenitrothion at 800× or 25% carbaryl at 500×.

References Jiang Shunan 1989; Liu Zheng and Wang Guilan et al. 1993; Wang Zhicheng 2003; Han Xuejian 2005; Wan Shaoxia and Zhang Lifeng 2005; Yu Guiping and Gao Bangnian 2005; Hu Changxiao, Ding Yonghui, and Sun Ke 2007.

(Zhou Zaibao, Cui Yaqin, Huang Bangkan, Zhao Huaiqian)

159 *Asemum striatum* (L.)

Classification Coleoptera, Cerambycidae
Synonym *Asemum amurense* Kraatz

Importance *Asemum striatum* is a widely distributed pest of many tree species. It is a secondary pest that only infests weak trees. Large areas of *Pinus tabulaeformis* suffered died from *A. amurense* infestations in Fengning county, Hebei province.

Distribution Beijing, Hebei, Shanxi, Inner Mongolia, Liaoning, Jilin, Heilongjiang, Zhejiang, Anhui, Hubei, Shandong, Sichuan, Chongqing, Yunnan, Shaanxi, Gansu, Qinghai, Ningxia, Xinjiang; Korea Peninsula, Russia, Japan, Mongolia.

Hosts *Pinus tabulaeformis*, *P. koraiensis*, *P. densiflora*, *P. armandii*, *P. taiwanensis*, *Larix gmelinii*, *Picea jezoensis*, *Picea asperata*.

Damage Larvae infest stumps and trunk of weak trees. Galleries are elliptical in shape. Some larvae infest tree roots.

Morphological characteristics **Adults** Length 11-20 mm, width 4-6 mm. Blackish brown, densely covered with greyish white pubescence. Ventral surface shiny. Antennae reach to half of the body length; its 5th segment is significantly longer than the 3rd segment. Head densely covered with punctures. Eyes are small. There is a distinct groove between the antennae. Each side of the pronotum with a round protuberance, center of the pronotum slightly concave. Scutellum long, tongue shape, blackish brown. Each of the elytron with two longitudinal ridges; the end of the elytra appears as a "V" notch. Legs short, with dense yellow pubescence.

Larvae Mature larvae 25-30 mm long. Cylindrical, body hairs reddish tanned. Frons convex, with fine rugae and thick setae. Labrum reddish brown, its base with long hairs. Pronotum wide at base, its anterior part with a yellow spot, the side area with reddish tanned hairs. The ambulatory ampullae on dorsal surface of abdominal segments distinct, each with a groove. The dorsal surface of the 9th abdominal segment with dense pubescence. There is a large pair of conical protuberance at the end of the larvae. The tip of the protuberance appears as reddish brown sharp spine.

Pupae Length 18-23 mm. Creamy yellow. Antennae

A. striatum adult male (Xu Gongtian)

A. striatum adult female (Xu Gongtian)

A. striatum larva (Xu Gongtian)

A. striatum larva (Xu Gongtian)

A. striatum pupae (Xu Gongtian)

located on the side of the body, curl back from the 2nd pair of thoracic legs. Wing buds surpass the posterior border of the 3rd abdominal segment.

Biology and behavior *Asemum striatum* occurs one generation a year in Beijing. Adults appear from early June to mid-July. Adults are strongly attracted to lights. Larvae infest newly cut-down trees and weak tree trunk.

Prevention and treatment

1. Quarantine. Avoid transporting infested wood to non-infested areas.

2. Silviculture. Remove infested trees and reduce source population.

3. Protect predatory enemies including *Dendrocopos major*. Apply *Beauveria bassiana* in Spring.

4. Attract and kill adults using baited traps.

5. Spray 3% fenoxycarb at 3,000× during larvae hatching period.

References Chen Shixiang, Xie Wenzhen, and Deng Guofan 1959; Zhao Jinnian et al. 2001; Wang Zhicheng 2003; Wu Haiwei and Luo Youqing 2006; Xu Gongtian and Yang Zhihua 2007; Forest Disease and Pest Control Station of Ministry of Forestry 2008; Shi Shuqing 2012.

(Zhou Zaibao, Cui Yaqi, Li Zhenyu)

160 *Anoplistes halodendri* (Pallas)

Classification	Coleoptera, Cerambycidae
Synonym	*Asias halodendri* (Pallas)
Common name	Red-striped longhorn beetle

Importance *Anoplistes halodendri* is serious wood boring insect pest to forests and landscaping trees.

Distribution Hebei, Shanxi, Inner Mongolia, Liaoning, Jiangsu, Shandong, Henan, Chongqing, Sichuan, Shaanxi, Gansu, Ningxia, Xinjiang; Russia, Mongolia, Korea Peninsula, Japan.

Hosts *Robinia pseudoacacia*, *Caragana sinica*, *Populus* spp., *Salix* spp., *Ulmus punila*, *U. parvifolia*, *Elaeagnus angustifolia*, *Lycium chinense*, *Ziziphus jujube*, *Pyrus* spp., *Malus pumila*, *Hippophae rhamnoides*, *Tetraena mongolica*, *Zanthoxylum bungeanum*.

Damage *Anoplistes halodendri* larvae feed on xylem and create galleries. Once galleries are connected around the diameter of the trunk or branches, the trees will eventually die or are easily broken by wind.

Morphological characteristics **Adults** Length 11.0-19.5 mm, width 3.5-6.0 mm. Head short, antennae slender. Female antennae same length as the body; the 3rd segment of the antennae is the longest. Male antennae is about twice as long as the body; the 11th segment is the longest. Pronotum slightly wider than long; the side protuberance short and obtuse. Elytra narrow and long; its end obtuse round; its base with a pair of dark red spots; the outer margin with a pair of dark red narrow band; elytra covered with blackish brown short hairs. Legs slender; the 1st tarsi of the hind legs longer than the combined length of the 2nd and 3rd tarsi.

Larvae Length 22 cm. Milky, anterior part of the pronotum brown.

Biology and behavior *Anoplistes halodendri*

A. halodendri adult (Li Zhenyu)

A. halodendri larval damage to jujube tree (Xu Gongtian)

A. halodendri adult feeding on jujube tree (Xu Gongtian)

A. halodendri adult (Xu Gongtian)

occurs one generation a year or one generation every two years. It overwinters as larvae deep inside the xylem. They start feeding from March-April. Pupae appear during April-May. Adults emerge during May-October. Adults mate soon after emergence. Egg incubation period is about half month. After hatching, larvae bore into phloem directly from the egg shell. As they grow, they bore into xylem. Adults feed on *Z. jujube* flowers for nutrient and may reduce jujube fruit production.

Prevention and treatment

1. Cultural control. Clear infested branches. Collect adults and kill them.

2. Apply *Beauveria bassiana* in Spring.

3. Brush lime: sulfur: water slurry at 10: 1: 40 ratio plus 40% acetamiprid and chlorpyrifos mixture to tree trunk before adults laying eggs.

4. Spray 8% cypermethrin at 200×, 25 g/L lamda-cyhalothrin at 600× during adult period.

References Zhang Menglin 1977, Xiao Gangrou (ed.) 1992, Fan Renjun et al. 1994, Sun Fenghai et al. 1994, Zhang Yingjuan and Yang Chi 2000, Zong Shixiang et al. 2005.

(Yan Junjie, Huang Dazhuang)

161 *Batocera davidis* Deyrolle
Classification Coleoptera, Cerambycidae

Importance *Batocera davidis* is major pest of *Aleurites fordii*. Large areas of trees died during outbreaks.

Distribution Fujian, Guangdong, Chongqing, Sichuan, Guizhou, Shaanxi, Taiwan; Vietnam, Laos.

Hosts *Aleurites fordii*, *Carya cathayensis*, *Castanea mollissima*, *Malus* spp., *Pyrus* spp., *Melia azedarach*, Chinese tallow.

Damage Young larvae feed on phloem layer. As they develop, the feed on xylem and expel frass and saw dust. Infested trees have small leaves, less branches, and small fruits. Branches or whole trees can die from infestation.

Morphological characteristics **Adults** Length 51-70 mm, width 18-22 mm. Antennae brownish red from the 3rd segment. The first four segments smooth, the other segments covered with grey hairs. Male antennae surpass body by 1/3 of its length. Center of the pronotum has a pair of orange or yellowish spots. The spine on side of the pronotum slender and long; its tip slightly orients toward the rear end. Scutellum covered with dense hairs. Shoulder of elytra with distinct sharp protuberance. Base of elytra with coarse granules; each elytron with 5-6 round orange spots; the exterior rear angle round; the inner rear angle with a short tooth.

Eggs Length 7-8 mm, width 2-3 mm; elongate; milky.

Larvae Pronotum wide, each side with a longitudinal depression, its posterior border concave, with many punctures in the front, rear end with some particles. Each of the ambulatory ampulla on the back with four rows of round protuberance. Those on the ventral surface with two rows of protuberance.

Pupae Length 50-70 mm. Antennae located under thorax. Meso- and metathorax each with one tubercle. Each of the abdominal 1st-6th segment with one black tubercle on the back, densely covered with pubescence.

Biology and behavior It takes two years to finish a generation. It overwinters as larvae in the 1st year and as adults in the 2nd year. Adults emerge from exit holes during late April and early May in the 3rd year. After about 20 days, they mate and lay eggs. Larvae appear in mid-June and overwinter in late October. They continue feeding from late March and pupate in late August next year. Adults emerge in late October and overwinter in pupal chambers. Adults feed on young shoots and tender bark during the day. They lay eggs after 20:00 on tree trunk. Oviposition scars are elliptical and located at base of tree trunk. Young larvae bore into bark and move upward. After seven days, they move downward and feed on phloem. From 2nd instar, they enter xylem. They create both vertical and horizontal galleries and cause disruption of nutrient transportation. Older larvae expel saw dust which accumulates at the base of tree trunk. Bark

B. davidis adult (edited by Zhang Runzhi)

B. davidis adults (edited by Zhang Runzhi)

B. davidis larva (Tang Yanping, Yang Chuncai)

B. davidis pupa
(Tang Yanping, Yang Chuncai)

B. davidis damage
(Tang Yanping, Yang Chuncai)

B. davidis damage to Chinese tallow
(edited by Zhang Runzhi)

B. davidis damage to Chinese tallow
(edited by Zhang Runzhi)

swells as saw dust accumulates under the bark. Mature larvae build pupal chambers in mid- and late August. Pupal period is approximately 45 days.

Prevention and treatment

1. Silviculture. Clear the grass within 1.0-1.5 m diameter from tree trunk to eliminate adult hiding places around tree trunk.

2. Cultural control. Collect and kill adults when they mate at base of tree trunk or rest around tree trunk. Use a hammer to kill eggs and young larvae during June-July.

3. Chemical control. Spray 3% thiamethoxam to canopy and tree trunk during mid-May.

References Pu Fuji 1980, Jiang Shunan 1989, Xiao Gangrou (ed.) 1992.

(Tang Yanping, Yang Chuncai, Wang Wenxue)

162 *Batocera lineolata* Chevrolat

Classification Coleoptera, Cerambycidae
Synonym *Batocera horsfieldi* (Hope)
Common name White-striped longhorn beetle

Importance *Batocera lineolata* is an important boring insect pest of poplar trees along Yangtze River. It also infests walnut, *Fraxinus chinensis*, poplar, and willow trees in other provinces.

Distribution Beijing, Hebei, Jiangsu, Zhejiang, Anhui, Fujian, Jiangxi, Shandong, Henan, Hunan, Guangdong, Guangxi, Sichuan, Guizhou, Yunnan, Shaanxi, Gansu, Xinjiang, Taiwan; Vietnam, India, Japan.

Hosts *Populus* spp., *Salix* spp., *Ulmus pumila*, *Pterocarya stenoptera*, *Quercus* spp., *Castanopsis* spp., *Ficus* spp., *Morus alba*, *Sapium sebiferum*, *Eucalyptus* spp., *Catalpa ovate*, *Castanopsis* spp., *Pyrus bretschneideri*, *Betula* sp., *Eriobotrya japonica*, *Alnus cremastogyne*, *Paulownia* spp., *Casuarina equisetifolia*, *Juglans regia*, *Fraxinus chinensis*, *Ligustrum lucidum*, *Cunninghamia lanceolate*, *Dalbergia* spp., *Schima superba*, *Fagus longipetiolata*, *Ginkgo biloba*, *Patanus hispanica*, *Ailanthus altissima*, *Malus pumila*, *Pinus yunnanensis*, *P. teada*.

Damage *Batocera lineolata* adults eat tender bark and may cause death of branches. Larvae bore into xylem and create galleries. Infested trees grow slowly and are easily broken by wind. Death of trees may also occur. Infested wood loses value.

Morphological characteristics **Adults** Length 32-65 mm, width 9-20 mm. Blackish brown to black. Body covered with grey and greyish brown pubescence. Center of the pronotum with a pair of light-yellow or

B. *lineolate* feeding on wild rose (Qian Fanjun)

B. *lineolate* laying eggs (Qian Fanjun)

B. lineolate larva (Qian Fanjun)

B. lineolata egg and chewing mark (Tian Ling)

B. lineolate mating pair (Tian Ling)

white spots. The tooth on each side of the pronotum large and sharp. Scutellum near half circle shape, covered with white pubescence. Each elytra with irregular light-yellow or white spots arranged into 2-3 longitudinal lines. Base of elytra with black particles. The lateral surface of the body has a white longitudinal band of pubescence.

Eggs Length 6-10 mm. Elongate, slightly curved, one end more slender. White, yellowish white before hatching.

Larvae Length 70-80 mm, yellowish white, plump and with many creases. Pronotum pale brown, with uneven sized brown particles; there are two light-yellow small spots in the front; each spot with one seta on it.

Biology and behavior Batocera lineolata occurs one generation every two years in Hubei and Jiangsu. It overwinters as 5th instar larvae or adults. Eggs appear in early May and reach peak in late May. Peak egg hatching period is in early and mid-June. Larvae feed on xylem. They overwinter from November to next March. Larvae

B. lineolata damage to tree trunk (Qian Fanjun)

resume activity in early April. They pupate from mid-August to mid-September. Adults emerge from mid-September to late October. They overwinter in pupal chambers. Adults exit the pupal chambers starting from mid-April. They feed, mate, and lay eggs until early June.

Adults are good flyers. They feed on tender bark of rose, pear, viburnum, *Betula luminifera*, etc. If these plants are abundant near poplar trees, then the poplar trees will suffer more severe infestations. Thickness of the bark is a major factor affecting the distribution of oviposition scars. Young trees of 2-3 years old have 0.3-0.7 cm thick bark. The oviposition holes are usually located below 1 m height. Four-year-old trees have 0.3-0.9 cm thick bark. Oviposition scars are usually located under 2 m height. The distribution of oviposition scars on 9-year and older trees is: under 2-4 m height 40%, above 4 m height 40%. Infestation rates of poplar trees are associated with habitats. Those in villages are more susceptible than those along roads and irrigation channels. Those in forests suffer least infestation. Trees at edges of forests suffer more than trees away from edges.

Prevention and treatment

1. Cultural control. Collect adults and kill them. Use a hammer to kill eggs and young larvae. Plant bait plants such as roses to attract and kill adults.

2. Biological control. Propagate and release *Dastarcus longulus*.

3. Chemical control. Spray trees with 3% thiamethoxam at 300-400×, or 20% Dimilin EC. Apply 2.5% deltamethrin and diesel at 1: 10-20 to oviposition scars to kill eggs. Drill holes on infested trunk. Place abamectin granules or 10-15 mL 25 g/L lamda-cyhalothrin at 100× into each hole. Then seal the holes with mud to kill larvae.

References Qian Fanjun et al. 1994, 1996, 1997, 1998; Ji Baozhong et al. 1996; Yan Aojin et al. 1997; Wang Wenkai 2000; Xiong Shansong et al. 2005.

(Qian Fanjun, Gong Cai, Liang Chunmei)

163 *Batocera rubus* (L.)

Classification Coleoptera, Cerambycidae
Common name White spotted longhorn beetle

Importance *Batocera rubus* larvae mainly feed on newly transplanted *Ficus microcarpa*, *Ficus virens*, and *Artocarpus nitidus* large seedlings or mature tree trunks. They affect plant growth, cause death of small branches or whole plant death, and cause certain economic loss.

Distribution Guangdong, Hainan, Sichuan, Hong Kong, Macau and Taiwan; Vietnam, Southeast Asia, Korea Peninsula, and India.

Hosts *Ficus* spp., *Mangifera indica*, *Bombax ceiba*, *Bischofia polycarpa*, *Alstonia yunnanensis*, and *Erythrina variegata*.

Damage In the early stage of infestation, *B. rubus* infested tree leaves are not as lustrous as those of healthy trees. As the infestation continues, the affected trees would show signs of water loss, wilt, and early defoliation. Mature larvae move further out from xylem to the phloem under the bark. Their feces and wood debris from feeding cause bulging and cracking on the bark, which can be easily seen.

Morphological characteristics **Adults** Length 30-46 mm, width 10-16 mm; reddish brown or dark red; the head, prothorax and the front femur are darker or near black; body covered with pubescence. Those on the dorsal surface of the body are sparse, gray or brownish gray; those on the ventral side are long and dense, greyish-brown or brown, or slightly yellowish. Both sides have a rather wide white longitudinal stripe. The pronotum has a pair of orange-red spots. The scutellum is densely covered with white pubescence. There are four white spots on each elytron. The fourth from the front is the smallest. The second from the front is the largest and is closer to the middle of the body than other spots. One or two small round spots often join to the upper outer area of the 2nd spot. Male antennae are about 1/3-2/3 longer than the body, with small spines on the inner margin.

B. rubus adult (edited by Zhang Runzhi)

B. rubus adult pair (edited by Zhang Runzhi)

B. rubus larva (edited by Zhang Runzhi)

B. rubus damage and exit hole
(edited by Zhang Runzhi)

B. rubus damage
(edited by Zhang Runzhi)

Starting from the 3rd antenna segment, the end of each segment slightly enlarged with the inner side is more prominent. This is most obvious in the 10th segment, which appears to be triangular and spiny. Female antennae are slightly longer than the body, thinner than the male antennae, with smaller and sparse spines. The end of each segment is not significantly enlarged except the scape. Prothorax has thick lateral spines, the tips slightly bend backwards. Front corners of the elytra have two teeth. The basal area of the elytra is covered with raised granules. They cover 1/3 of the elytra under the shoulder and 1/4 of the elytra at the inner area of the elytra. The end of the elytra is truncated, with an obtuse outer angle and a spine at the inner angle.

Eggs Length 6-8 mm, width 2-3 mm; creamy white, elongate oval, slightly flat.

Larvae Mature larvae 80 mm in length. The prothorax up to 17 mm in width. Similar to *Batocera davidis*, but posterior notum of the prothorax with 5-6 rows tooth like tubercles that are gradually smaller from the front to the back. The 1st row has 22-28 tubercles. Dorsal surface of mandibles conbevex.

Pupae Young pupae yellowish-white, blackish-brown before eclosure; densely covered with pubescence.

Biology and behavior *Batocera rubus* occurs one generation per 1-3 years. They overwinter as mature larvae. Adults appear from late April to early October. They begin to lay eggs in early May. The eggs hatch in 5-8 days (July-August). Newly hatched larvae feed between the phloem and xylem initially, and into the xylem after becoming larger. They stop feeding in mid-December and pupate in March next year. The pupal stage lasts for about 30 days.

In Guangzhou Province, *B. rubus* adults emerge from late April to October. Their life span is 4 to 5 months. Most adults mate during the night. They can mate several times during its life. They feed on tender leaves and green branches to obtain nutrition. They often rest on the tree trunk during the day except when laying eggs. Females prefer lay eggs on larger trunks and less than 2 m from the ground. They first select a suitable site to bite a flat circular groove that can be as deep as to the xylem. Then they insert their ovipositors and lay eggs. One egg is usually laid per groove. It is covered by gelatinous material. *B. rubus* sometimes lay eggs on large branches. Ten to a dozen of eggs can be laid in one tree. Eggs are laid in batches. The egg laying peak is during June – August. No egg-laying was found after late September. Larvae feed between the phloem and xylem after hatching. Their feeding activity form curvy galleries. They bore into the xylem as their size increases. The entrance is round and slightly oval. The feeding galleries are irregular. They cut off the conducting tissues of trees. Mature larvae tend to climb out of the xylem and feed under the bark. The excreted frass and wood debris fill under the bark and cause swelling of the bark. The larger the larvae, the coarser are the wood debris. One tree can be attacked by many *B. rubus* larvae. Mature larvae construct chambers in the xylem where they will pupate. They mostly attack poorly managed large trees after transplanting and rarely attack seedlings. Well established *Ficus microcarpa* trees that grow well are rarely damaged by *M. rubus*.

Prevention and treatment

1. Quarantine. Prevent the movement and planting of trees infested with *B. rubus*.

2. Mechanical control. Wrap the trunks of newly transplanted trees with straw ropes during the egg-laying period. Remove the straw ropes and burn them after the egg-laying period. This method can not only prevent adults from laying eggs on the trunks, but also reduce water evaporation and sun exposure of the trunks of newly transplanted plants. It helps restore vigor of the transplanted trees.

3. Chemical control. (1) Before young larvae entering into the xylem, apply insecticides to the trunk will provide good control results. Common insecticides include: 20% thiamethoxam or 45% profenofos and phoxim mixture at 1,000-1,500×. (2) Inject insecticides into the xylem. Commonly used insecticides include: 50% fenitrothion, 50% malathion EC, 25 g/L lamda-cyhalothrin or 20% dinotefuran at 20-40×. The application rate is 0.5-1 mL of diluted solution per 1 cm trunk diameter. Inject with a syringe or use cotton balls. Plug the holes with cement after application. (3) During the peak adult emergence, spray common stomach or contact insecticides to the foliage and base of the tree trunks.

References Chen Shixiang et al. 1959, Jiang Shunan 1989, Liu Dongming, Gao Zezheng, Xing Fuwu 2003.

(Wu Yousheng; Gao Zezheng; Liu Dongming)

164 *Callidiellum villosulum* (Fairmaire)
Classification   Coleoptera, Cerambycidae

Distribution   Shanghai, Jiangsu, Zhejiang, Fujian, Jiangxi, Henan, Hubei, Hunan, Guangdong, Guangxi, Sichuan, Guizhou.

Hosts   *Cunninghamia lanceolate*, *Cryptomeria fortune*.

Damage   *Callidiellum villosulum* is a common pest. It usually infests 3-5 year-old trees. They feed between phloem and xylem. Their galleries cause disruption in nutrient transportation and eventually cause tree mortality. Mature larvae bore into xylem and lower the quality of wood.

Morphological characteristics   **Adults** Length 6-12 mm. Castaneous. Body covered with sparse long grey hairs. Head and thorax with small punctures. Frons near rectangular; there is a transverse ridge between the antennae. Antennae brownish black. Male antennae longer than the body. Female antennae 2/3 as long as the body; scape with coarse puncture. Pronotum wider than long, the two sides round, without teeth; the dorsum with indistinct particles. Ventral surface of thorax and femora brownish red. Femora much thicker than other segments. Shoulder of the elytra darker than the rest of the elytra; punctures on elytra larger than those on thorax; end of elytra round, obtuse.

Eggs   Length 1 mm. Milky.

Larvae   Mature larvae 10 mm long. Pale yellow. Slightly flattened. Mouthparts blackish brown. Pronotum with a pair of brown spots. Thoracic legs degenerated.

Pupae   Length 7-10 mm. Elongate. Milky. Antennae curled back from the 2nd thoracic legs.

Biology and behavior   *Callidiellum villosulum* occurs one generation a year in Zhejiang. It overwinters as adults in xylem. They emerge from pupal chambers in early March. Peak adult emergence period is from late March to early April. They mate soon after emergence. Mating time is usually during 8:00-10:00. They mate and lay eggs multiple times. Eggs are laid in cracks on bark. Larvae start to appear in late March. They bore into phloem and feed between phloem and xylem. Galleries are flattened, filled with frass and saw dust. From late August, mature larvae enter into xylem; the entrance holes are oblate; the depth of the tunnels in xylem is usually 1 cm. Larvae pupate in late September. Pupal period is 10-15 days. Adults emerge in early October. They plug the entrance holes with sawdust and overwinter within pupal chambers.

Prevention and treatment

1. Promptly remove infested wood and branches and destroy them.

2. Manually kill adults during their mating and egg-laying period.

Reference   Forest Research Institute of Xianju County of Zhejiang 1978.

(Gao Zhaowei)

C. villosulum adult female (SFDP)

C. villosulum adult male (SFDP)

165 *Clytus validus* Fairmaire

Classification   Coleoptera, Cerambycidae

Distribution   Sichuan, Tibet, Gansu.

Host   *Zanthoxylum bungeanum*.

Damage   *Clytus validus* larvae bore into xylem of host plants. Entrance holes are 10-13 mm in diameter. Adults feed on tender shoots and leaves. Infested plants grow slowly, wither, or die.

Morphological characteristics   **Adults** Length 16-20 mm. Black, covered with yellow pubescence. Head with dense punctures; those on the read end coarser. Frons with a longitudinal groove at the center. Antennae 11 segmented, about 1/3 of the body length. Sides of the pronotum arc shape. Punctures on pronotum form winkles. There is a large black spot at center of pronotum. Each elytron with three near round black spots from the base to the end; the spot at the base is largest, the spot at the middle is smallest.

Eggs Elliptical. Length 1.0-1.2 mm. Milky at beginning, yellowish later.

Larvae Young larvae milky, head light-yellow. The 2nd-3rd instar larvae with yellowish brown head. The 4th instar and older larvae with brown head. Mature larvae 23-28 mm long. Pale yellow.

Pupae Yellow. Length 15-20 mm. Pronotum with brown short spines.

Biology and behavior   It takes two years to finish one generation. Few can finish one generation a year. It overwinters as larvae. When air temperature is above 10℃ in early April, larvae come out and feed. They pupate during late May and early July. Adults emerge between June and August. They lay eggs during late June and early August. Larvae appear from late July and early September. Larvae begin to overwinter when temperature is below 10℃.

In areas where it takes two years to finish a generation, adults, eggs, and larvae appear from late May to late July, from late June to mid-August, and from late July to early September, respectively in the 1st year. In the 2nd year, larvae feed from March to October and overwinter as larvae. In the 3rd year, larvae pupate during late April and late July. Adults appear from late May. They mate multiple times. Each female lays eggs one or two times.

C. Validus adult (Li Zhenyu)

They create oviposition scars on base of trunk at 4-50 cm height. Each scar has 5-49 eggs. Adults live for 17-25 days. Egg period is 13-16 days. Pupal period is 18-25 days.

Prevention and treatment

1. Clear old, weak, and heavily infested trees. Trim trees in October. Apply fertilizer in the fall and spring to promote tree growth.

2. Use a tool to knock the oily infestation areas to kill the young larvae.

3. Inject tree trunk with 350 g/L imidacloprid in April when infestation is heavy.

References   Li Menglou, Cao Zhimin, and Wang Peixin 1989; Xiao Gangrou (ed.) 1992; Yang Leifang and Liu Guanghua 2009.

(Wang Hongji, Li Menglou)

166 *Neocerambyx raddei* Blessig

Classification Coleoptera, Cerambycidae
Synonyms *Mallambyx raddei* (*Blessig*), *Massicus raddei* (Blessig)
Common name Deep mountain longhorn beetle

Importance *Neocerambyx raddei* is native to Japan. Outbreaks in oak forests occurred in Jilin, Liaoning, and Inner Mongolia. It spreads fast and often causes tree mortality in large areas.

Distribution Hebei, Shanxi, Inner Mongolia, Liaoning, Jilin, Heilongjiang, Shanghai, Jiangsu, Zhejiang, Anhui, Fujian, Jiangxi, Shandong, Henan, Hubei, Hunan, Hainan, Sichuan, Guizhou, Yunnan, Tibet, Shaanxi, Taiwan; Korea Peninsula, Japan, Russia.

Hosts *Castanea henryi*, *C. mollissima*, *Quercus* spp., *Cyclobalanopsis glauca*, *Carpinus cordata*, *Zelkova serrata*, *Morus alba*, *Ficus carica*, *Cinnamomum cassia*, *Euonymus* sp., *Fraxinus mandschurica*, *Malus pumila*, *Citrus reticulata*, *C. grandis*, *Paulownia fortune*.

Damage *Neocerambyx raddei* young larvae bore 1.5 mm diameter holes into tree trunk or branches. The entrance has accumulation of white, round saw dust and frass. They bore into xylem. As the larvae sizes increase, the galleries become wider and longer. Piles of brown frass and saw dust can be seen under the infested trees.

Morphological characteristics **Adults** Length 40-60 mm. Brownish black, ventral surface and femora brownish red. Body covered with yellow pubescence. Antennae near black.

Larvae Mature larvae 65-70 mm long. Milky. Anterior border of pronotum with two light-yellow spots.

Biology and behavior It takes three years to finish one generation. It overwinters as larvae. Adults emerge from late June. Peak adult eclosion period is in mid-July. They feed on bark. Eggs are laid from early July. Larvae appear from late July. They molt one or two times and then overwinter. They resume activity in early April next year. After another 1-2 molts, they overwinter in early October. The same process repeats in the 3rd year. In the 4th year, larvae pupate in late May.

Adults are gregarious and strong flyers. They are strongly attracted to lights. South slopes, ridges of hills, large trees, and edges of forests often suffer heavier infestation than north slopes, middle of hills, small trees, and center of forests.

Prevention and treatment

1. Quarantine. Treat infested wood before transporting them outside of the quarantine area.

2. Silviculture. Remove infested trees in lightly infested forests. Heavily infested forests may need to be cut down completely.

3. Physical control. Trap adults using black light traps. Manually catch adults during the day.

4. Biological control. Apply *Beauveria bassiana* in Spring. Propagate and release *Sclerodermus guani* and *Dastarcus helophoroides*. Attract predatory birds.

5. Chemical control. To control adults, spray 50% sumithion at 800×, or 50% parathion, 45% profenofos and phoxim mixture, 25 g/L lamda-cyhalothrin, or 90% trichlorfon at 1,000×. To kill eggs, apply 45% profenofos and phoxim mixture: silicon: water at 1:1:20 to oviposition scars. To kill larvae, use a thin metal wire to take out the feces, then plug the holes with 25 g/L lamda-cyhalothrin soaked cotton balls, or 56% avermectin microcapsules 0.15 g per hole and seal with mud. Inject suitable insecticides into trunk at 2×. Effective insecticides include 15% indoxacarb, 20% thiamethoxam, 25% phosmet, 25 g/L lamda-cyhalothrin, and 20% ammonia hydroxide.

6. Fumigation. Infested wood can be fumigated with avermectin microcapsule at 9 g/m^3 for seven days at 15℃

N. raddei adult female (Xu Gongtian)

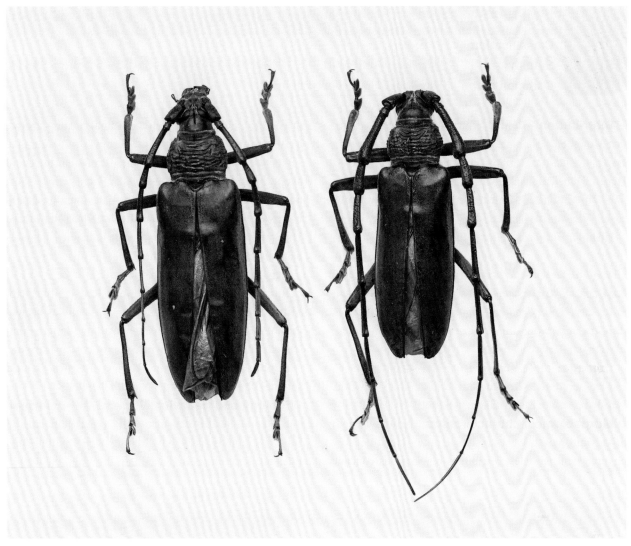

N. raddei adult female (left) and male (right) (Li Chengde)

or higher temperature.

References Sun Xiaoling, Cheng Bin, Gao Changqi et al. 2006; Dang Guojun 2007; Li Guangming, Cheng Ming, Han Chongxuan et al. 2007.

(Li Chengde, Li Xingpeng)

167 *Aegosoma sinicum* White

Classification Coleoptera, Cerambycidae
Synonym *Megopis sinica* (White)
Common name Thin-winged longicorn beetle

Distribution Beijing, Tianjin, Hebei, Shanxi, Inner Mongolia, Liaoning, Jilin, Heilongjiang, Shanghai, Jiangsu, Zhejiang, Anhui, Fujian, Jiangxi, Shandong, Henan, Hubei, Hunan, Guangxi, Sichuan, Guizhou, Yunnan, Tibet, Shaanxi, Gansu, Taiwan; Korea Peninsula, Japan, Vietnam, Myanmar.

Hosts *Firmiana platanifolia*, *Malus spectabilis*, *M. pumila*, *Castanea* spp., *Mallotus apelta*, *Vernicia fordii*, *Hevea brasiliensis*, *Pterocarya stenoptera*, *Populus* spp., *Salix* spp., *Fraxinus chinensis*, *Morus alba*, *Ulmus pumila*, *Quercus* spp., *Crataegus pinnatifida*, *Ziziphus jujuba*, *Zanthoxylum bungeanum*, *Taxus cuspidate* var. *nana*, *Pinus* spp., *Abies nephrolepis*.

Damage *Aegosoma sinicum* is a secondary pest. Larvae bore into xylem and create large holes in wood. Infested trees are easily broken by wind.

Morphological characteristics **Adults** Length 30-52 mm, width 8.5-14.5 mm. Brown. Head densely covered with particles. Mandibles black, frons concave at center. Occiput long, a longitudinal narrow groove exists from occiput to frons. The basal five antennae segments coarse, with dense spines. Pronotum trapezoid shape, with dense particles and greyish yellow short pubescence. Elytra wider than the pronotum, gradually narrower toward the rear end. Each elytron with three longitudinal ridges, the outer one is indistinct. Females have distinct ovipositors at the end of the abdomen.

Eggs Milky, 3-6 mm long. Elongate.

Larvae Mature larvae 50-70 mm long. Cylindrical. Milky. Mouthparts yellowish brown. Mandibles brown. Pronotum light-yellow, its center with one longitudinal smooth line, the two sides with a pair of concave stripes.

Pupae Length 30-60 mm. Yellowish brown. Ventral surface of metathorax with a pair of tubercles.

Biology and behavior It takes 2-3 years to finish one generation. It overwinters as larvae. Adults emerge from late June to early July. Mating occurs in mid-July. Eggs are laid during mid- and late July. Egg incubation

A. sinicum adult female (Xu Gongtian)

A. sinicum adult male (Xu Gongtian)

A. *sinicum* tunnels (Xu Gongtian)   A. *sinicum* exit holes (Xu Gongtian)

A. *sinicum* larva (Xu Gongtian)

period is about one week. Larvae develop into pupae within 2-3 years.

Adults feed on bark and are attracted to lights. Each female lays 200-300 eggs. Females prefer lay eggs on trunk with wound, holes, or decay. They also lay eggs in old tunnels created by other beetles. Adults live for 30-50 days. Tunnels can be 40 cm long. Mature larvae create pupal chambers under the bark. Adult exit holes on bark are elliptical.

Prevention and treatment

1. Silviculture. Maintain healthy growth of trees. Clear dead and weak trees.

2. Cultural control. Collect adults and kill them. Seal holes and wound on trees to prevent egg laying. Apply sulfur, lime, water paste at 1:10:40 ratio to trunk to prevent egg laying. During egg hatching period, scrape off peeled bark to remove eggs and larvae. Remove larvae from holes using tools and seal holes with concrete.

3. Chemical control. Place aluminum phosphate into holes and seal the holes. Attract and kill adults using M99-1 attractant.

References Chen Shixiang, Xie Wenzhen, and Deng Guofan 1959; Wang Zhiming and Ni Hongjin 1991; Xiao Gangrou (ed.) 1992; Zhao Jinnian et al. 2001; Yang Ziqi and Cao Guohua 2002; Wang Zhicheng 2003.

(Yan Rong, Cui Yaqin, Wang Wenxue)

168 *Monochamus alternatus* Hope

Classification	Coleoptera, Cerambycidae
Synonym	*Monochamus tesserula* White
Common name	Japanese pine sawyer beetle

Importance *Monochamus alternatus* is a major wood boring pest of pine trees and vector of pine wilt nematode, *Bursaphelenchus xylophilus*. The nematodes are attached to the body of adult *M. alternatus*.

Distribution Hebei, Jiangsu, Zhejiang, Fujian, Jiangxi, Shandong, Henan, Hunan, Guangdong, Guangxi, Sichuan, Guizhou, Yunnan, Tibet, Shaanxi, Taiwan; Korea Peninsula, Laos, Vietnam, Japan.

Hosts *Pinus massoniana, P. thunbergii, P. densiflora.*

Damage *Monochamus alternatus* larvae bore into bark and feed on xylem. Adults feed on bark. Infested wood have galleries and holes which lead to death of branches.

Morphological characteristics **Adults** Length 15-28 mm, width 4.5-9.5 mm. Orange yellow to reddish brown. Antennae cutaneous. The 1st and 2nd segment, the base of the 3rd segment of the antennae with grey pubescence in the males. Female antennae mostly grey except for the last two or three segments. The last segment darker in color. Male antennae surpass body length by at least one time of the body length. Female antennae surpass the body by 1/3 of the body length. Prothorax wider than long, with wrinkles; the side with large teeth. Pronotum with two broad orange bands. Scutellum densely covered with orange pubescence. Each elytron with five longitudinal stripes; they are composed of black or greyish white hair spots. Ventral surface and legs with greyish white pubescence.

Eggs Length 4 mm; milky; slightly sickle shape.

Larvae Milky; cylindrical; mature larvae length 43 mm. Head blackish brown; pronotum brown, its center has transverse stripes.

Pupae Milky; cylindrical; length 20-26 mm.

Biology and behavior *Monochamus alternatus* occurs one generation a year in Shandong, Jiangsu, Zhejiang, and Anhui. In Guangdong, it mostly occurs two generations a year. It overwinters as mature larvae in xylem. After eclosion, adults exit from 8-10 mm diameter holes. They feed on tender bark and leaves. Peak adult period is during May and June. Adults chew eye shaped scars on trunk to lay eggs. One egg is laid in each scar.

M. alternatus adult (Jiang Ping)

M. alternatus adults (Jiang Ping)

M. alternatus adult inside wood (Zhang Runzhi)

M. alternatus damage (Zhang Runzhi)

M. alternatus larva (Zhang Runzhi)

M. alternatus pupa (Left: ventral view. Right: dorsal view) Zhang Runzhi)

Scars made by egg-laying *M. alternatus* females (Zhang Runzhi)

M. alternatus damage by young larvae (Zhang Runzhi)

After hatching, larvae bore into bark and feed inside the bark. They create galleries between phloem and xylem. In the fall, larvae enter xylem to 3-4 mm depth, then make vertical tunnels. The tunnels are 5-10 cm long. They pupate at the bottom of the tunnels. The tunnels appear as "U" shape.

Sparse forests suffer more damage than dense forests. Infestations tend to start at edge of the forests and spread to center of forests. Cut down trees will easily become infested if left in the forests.

Prevention and treatment

1. Clear the dead trees is the most common method for reducing infestations.

2. Biological control. Release *Sclerodermus guani* and *Beauveria bassiana*.

3. Attract and kill. Use attractant to attract and kill adults.

4. Chemical control. Spray 25% Dimilin at 1,000-5,000× for two times during peak adult emergence period.

References Xu Fuyuan 1994, Zhang Lianqin et al. 1992, Sun Jimei et al. 1997, Chai Ximin and Jiang Ping 2003.

(Jiang Ping, Wang Shufen)

169 *Monochamus urussovii* (Fischer-Waldheim)
Classification Coleoptera, Cerambycidae

Importance *Monochamus urussovii* is a major wood boring pest of forests. It infests weak trees, fire damaged trees, fallen trees, and logs.

Distribution Hebei, Inner Mongolia, Liaoning, Jilin, Heilongjiang, Jiangsu, Shandong, Shaanxi; Russia, Finland, Mongolia, Korea Peninsula, Japan.

Hosts *Picea koraiensis*, *P. jezoensis* var. *microsperma*, *Larix gmelini*, *L. olgensis* var. *changpaiensis*, *Pinus koraiensis*, *Abies nepholepis*, *Betula platyphylla*.

Damage *Monochamus urussovi* larvae bore into xylem and create galleries. Adults cause damage when feeding on bark of small branches and *Picea* leaves.

Morphological characteristics Adults Length 21-33 mm. Black, with dark green or copper sheen. Male antennae about 2.0-3.5 times the body length. Female antennae slightly longer than the body. Pronotum with three indistinct tubercles; each side with a large tooth. Scutellum densely covered with pubescence, yellow. There is a transverse depression at 1/3 of the elytra. Male elytra gradually wider toward the rear end; the two sides of the female elytra near parallel. There are four grey hair spots near the center of the elytra. They may vary a lot.

Eggs Kidney shape, 4.5-5.0 mm long, 1.2-1.5 mm wide; yellowish white.

Larvae Mature larvae 37-50 mm long, Head width 3.0-5.9 mm. Creamy yellow. Head square, the posterior border round; 2/3 of the head hidden in thorax. Pronotum well developed; its length about the combined length of the other two thoracic segments. Pronotum with a reddish-brown spot. Ambulatory ampullae present on both dorsal and ventral surfaces. There are two transverse grooves on each of the dorsal ambulatory ampulla and one transverse groove on each of the ventral ambulatory ampulla.

Pupae Length 25-34 mm, white to creamy yellow. Pronotum with teeth on the sides. There are nine visible abdominal segments.

Biology and behavior *Monochamus urussovii* occurs one generation in two years or one generation a year in Inner Mongolia. It overwinters as larvae. Adults emerge in early June. Eggs are laid during late June and early September. Peak egg laying period is late July and early August. Larvae hatch in early July. They feed on phloem and surface of xylem. After one month, larvae bore into wood and build vertical tunnels in xylem. They overwinter in late September. Larvae feed on phloem in next May and mature in mid-July. They bore into xylem again and create pupal chambers. Larvae go through 6 instars. Mature larvae overwinter in pupal chambers. They pupate in early May in the 3rd year. Adults stay inside the pupal chambers for a week after eclosion. Their exit holes are 8 mm in diameter. They feed before laying eggs. Large numbers of adults can be seen on top of the tree crown. Females lay eggs 10-21 days after exit. They gnaw a straight scar and deposit one or two eggs in each scar. A female lays 14-58 eggs (average 30). The tunnels in xylem can be horseshoe, arc, or straight-line shape. Adults prefer lay eggs on fallen and weak trees.

Prevention and treatment

1. Quarantine. Prevent movement of infested wood.
2. Silviculture. Plant resistant tree species and mixed forests. Place bait wood and attract pests.

M. urussovii adults (Yan Shanchun)

M. urussovii eggs (Yan Shanchun)

M. urussovii larvae (Yan Shanchun)   *M. urussovii* damage (Yan Shanchun)   *M. urussovii* damage (Yan Shanchun)

3. Protect *Dastarcus helophoroides* and *Dendrocopos* spp. and other natural enemies.

4. Chemical control. Fumigate timber or logs with sulfuryl fluoride. Place aluminum phosphate into holes on trees to control larvae. Spray 45% profenofos and phoxim mixture at 40× to newly cut down pine trees to prevent infestation.

References   Meng Gen et al. 1990, Shao Xianzhen et al. 2001.

(Yan Shanchun, Shi Zhenhua)

170 *Philus antennatus* (Gyllenhal)
Classification   Coleoptera, Cerambycidae

Distribution   Hebei, Zhejiang, Fujian, Jiangxi, Hunan, Guangxi, Hainan, Hong Kong; India.

Hosts   *Pinus elliottii, P. massoniana, Fortunella margarita, Morus alba, Thea sinensis.*

Damage   *Philus antennatus* larvae feed on roots. Infested trees become yellow. Their shoots are short and leaves fall prematurely. Trees may eventually die from infestation.

Morphological characteristics   **Adults** Body brown, densely covered with greyish yellow pubescence, ventral surface and legs with rather long and dense pubescence. Females 27.3-31.2 mm long. Pronotum width 4.7-5.9 mm; antennae 15.1-17.0 mm. Head slightly drooped, almost as wide as pronotum; occiput slightly narrower. Antennae reach the middle of elytra. Scape thick, shorter than the 3rd segment. Pronotum trapezoid, posterior half with ridges on the two sides; surface with punctures and hairs. Inner surface of legs with hairs. First tarsal segment of the hind leg is same length as the combined length of the 2nd and 3rd segment.

Males 22.3-24.1 mm long. Pronotum 3.9-4.8 mm wide. Antennae 26.5-31.0 mm long. Antennae surpass the body by 1.3 times of its length. The 3rd-6th segments tooth like; scape thick and short; from the 3rd segment, all are equal in length.

Eggs   Yellowish white. Spindle shape; 2.5-3.5 mm long, 0.8-1.2 mm wide.

Larvae   First instar larvae 4.5-5.5 mm long, pronotum 1.7-3.1 mm wide. Mature larvae 26.6-37.2 mm long, pronotum 6.4-11.5 mm wide. Body plump, near rectangular shape, the last three segments cylindrical and gradually smaller. The 9th abdominal notum surpasses the last segment. Creamy yellow, covered with dense short pubescence. The medial groove of the pronotum milky and wide. Median area with a transverse groove, which forms a "T" stripe with the posterior part of the median groove. Each ambulatory ampulla without transverse groove and tubercles, but with four longitudinal shallow grooves. The claw of each thoracic leg with a thick seta.

Pupae   Yellowish, hairs and terminal spine yellowish brown. Male Pupae 30.0 mm long, 5.6 mm wide. Elongate spindle shape. Female pupae 21.5-22.5 mm long, cylindrical, slightly flattened.

Biology and behavior   *Philus antennatus* occurs

P. antennatus adult (Wang Jijian)

P. antennatus pupa (dorsal view) (Wang Jijian)

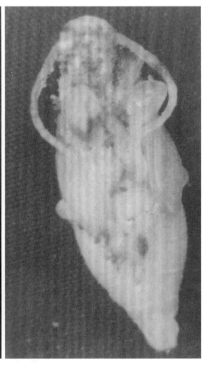

P. antennatus pupa (ventral view) (Wang Jijian)

 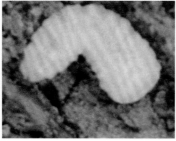

P. antennatus adults (left: male, right: female) (Wang Jijian)

P. antennatus larva (Wang Jijian)

P. antennatus damaged trees (Wang Jijian)

one generation every two years in Guangxi. It overwinters as larvae or adults in soil around. The exit holes are 8-13 mm in diameter. Adults emerge from soil in mid-February. They lay eggs in late March. Larvae hatch in early April. Egg incubation period is 19-34 days. The 1st instar period is the dispersal period, which lasts for 10-15 days. Larvae go through six instars. Mature larvae pupate from mid-September. From mid-October, Adults emerge and overwinter in soil.

Adults are active during the night. Mating duration is about 10 minutes. A few minutes after mating, females look for oviposition sites. They usually lay eggs under 1.5 cm height in bark cracks. The eggs are neatly arranged, covered with gelatinous material and some hairs. The color of the egg mass is slightly lighter than tree bark. Adults rest on tree trunk or on ground. They die about a week after egg laying. At 25-30°C, larvae hatch within 3-5 weeks. New hatched larvae jump onto ground. Within 30 minutes, they find cracks in soil, locate the roots and feed. They feed on root hairs initially. As they develop, they feed on epidermis, phloem, and xylem and infest lateral and primary roots. They feed between 2-150 cm deep. After 2-3 years of infestation, needle leaves become gray and no new shoots stop development. Resin production is also ceased. Trees will eventually die. Infested forests usually have sandy soil. Soil thickness is around 60-100 cm. Forests with hard and sticky soil usually will not suffer much damage. Mature larvae make elliptical pupal chambers in soil.

Prevention and treatment

1. Silviculture. Plant mixed forests. Remove dead trees.

2. Cultural control. Collect adults under the plants. Use a hammer to kill the egg masses.

3. Protection of natural enemies. Birds and ants are natural enemies of young larvae.

4. Chemical control. Spray insecticides to tree trunk under 2 m to kill adults and prevent eggs from hatching.

Reverences Wang Jijian 1989, Xiao Gangrou (ed.) 1992, Yin Xinming 1994.

(Wang Jijian, Huang Jinyi, Meng Meiqiong)

171 *Purpuricenus temminckii* (Guérin-Méneville)

Classification Coleoptera, Cerambycidae

Distribution Hebei, Liaoning, Jiangsu, Zhejiang, Anhui, Fujian, Jiangxi, Shandong, Henan, Hubei, Hunan, Guangdong, Guangxi, Sichuan, Guizhou, Yunnan, Taiwan; Korea Peninsula, Japan.

Hosts *Phyllostachys angusta*, *P. aureosulcata*, *P. aurita*, *P. praecox* f. *provernalis*, *P. aureosulcata* f. *pekinensis*, *P. bambusoides* f. *Lacrima-deae*, *P. bambusoides* f. *shouzhu*, *P. heteroclada* f. *solida*, *P. fimbriligula*, *P. glauca*, *P. heteroclada* var. *pubescens*, *P. heterocycla* 'Obliquinoda', *P. iridescens*, *P. makinoi*, *P. nidularia* 'Smoothsheath', *P. nigra*, *P. prominens*, *P. nuda*, *P. robustiramea*, *P. sulphurea* 'Viridis', *P. sulphurea*, *P. vivax*, *Bambusa blumeana*, *B. multiplex*, *B. pervariabilis*, *B. chungii*, *B. textilis*, *Pleioblastus juxianensis*, *P. amarus*.

Damage *Purpuricenus temminckii* larvae bore into bamboo stem and feed inside bamboo. Infested bamboo plants become yellow. Their leaves fall prematurely. Water accumulates inside the cavity of infested bamboo stem. Infested bamboo loses value.

Morphological characteristics **Adults** Length 11-19 mm. Head black, clypeus yellow. Eyes black. Gena covered with white pubescence. Antennae black, 11 segmented. Male antennae about 1.5 times of the body length. Female antennae reach the posterior border of the elytra. Pronotum red, the side of the prothorax black. There are five black spots on the pronotum; tow in the front, and three in the back. There is a tubercle at the center spot of the back row; the side with tubercles. Scutellum black. Elytra red, its shoulder with longitudinal protuberance. Ventral surface of thorax and abdomen black.

Eggs Length 2.5 mm; elongate; milky.

Larvae The first instar 3 mm long. Milky. Mature larvae 25.5-34.5 mm long, pronotum 5.8-7.5 mm wide. Body light-yellow, head a little darker. Most of the head concealed within prothorax. Mandibles black; pronotum white, 1/3 of it sclerotized, the sclerotized area brownish red; the side of the pronotum also has a sclerotized area.

P. temminckii adult (Xu Tiansen)

P. temminckii pupa (Xu Tiansen)

P. temminckii pupa parasitized by natural enemy (Xu Tiansen)

P. temminckii larva (Xu Tiansen)

There are particles behind the sclerotized area.

Pupae Length 17-22 mm. Flattened, yellowish white. Pronotum covers most of the head. Eyes elongate oval, vertical. Male antennae reach the end of the 4th abdominal segment, then direct back and reach the end of the antennae. Female antennae reach the end of the 2nd abdominal segment and then direct back, the end reach the end of the mid-tarsi. Prothorax upside down triangular when viewed from side; the posterior border with a tubercle at the center and teeth on the sides. The dorsal medial line of the abdomen obvious. Wing buds reach between the 3rd and 4th abdominal segment. Hind femora reach between the 4th and 5th abdominal segment.

Biology and behavior *Purpuricenus temminckii* occurs one generation a year. Few require two years to finish one generation. It overwinters mostly as adults in bamboo. Some overwinter as larvae. In Zhejiang, those live on *Phyllostachys praecox* f. *provernalis* occur one generation every two years. They overwinter as adults and larvae. Adults live for 210-230 days. Adults exit from bamboo in mid-April when mean temperature reaches above 15°C. Adults fly over bamboo plants to find oviposition sites. They lay eggs within 10 cm from bamboo nodes during late April and early May. Eggs hatch within 15-25 (mean 21) days. Larvae appear from mid-May to early June. The larvae immediately bore into stem after hatching. The entrance holes are 2 mm in diameter. Larval period is approximately 270 days. Pupae appear from mid-August. Adults emerge from mid-September. Adult emergence period is about 20-30 days.

Prevention and treatment

Iphiaulax sp. and an ichneumonid wasp are found occasionally parasitizing on *Purpuricenus temminckii*.

Reference Xu Tiansen and Wang Haojie 2004.

(Xu Tiansen, Wu Jianfen)

172 *Saperda carcharias* (L.)

Classification Coleoptera, Cerambycidae
Common name Large poplar borer

Distribution Jilin, Heilongjiang; Russia, Korea Peninsula, Western Europe.

Hosts *Populus davidiana*, *P. tomentosa*, *Salix matsudana*.

Morphological characteristics **Adults** Black, covered with yellow pubescence and sparse setae. Females 28 mm long, 10 mm wide; males 26 mm long, 8 mm wide. Center of frons with a deep longitudinal groove that extends to the occipital border. Eyes black, shiny. Apical part of the antennal segments black, the rest of the antennal segments covered with dense yellow pubescence. Male antennae slightly longer than the body, female antennae not reaching the end of the elytra. Length of the pronotum is same as its maximum width; its center with a longitudinal ridge; covered with greyish yellow pubescence and coarse punctures. Elytra narrow, but wider than the pronotum; shoulders distinct; covered with black shiny coarse punctures, those at base of the elytra large and dense, those toward the tip of the elytra smaller; covered with dense dirty yellow pubescence. Elytra in males constricted toward the back, two sides of the elytra in females almost parallel.

Eggs Elliptical, yellow.

Larvae Length 45 mm; covered with sparse pale pubescence; ambulatory ampullae with small spines. Front border of the head with four depressions, each depression with one seta. Pronotum meat color, with one longitudinal groove at the center and two grooves beside it. Am-

S. carcharias adult (Yan Shanchun)

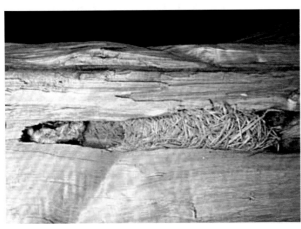

S. carcharias pupal chamber (Yan Shanchun)

S. carcharias excreted material (Yan Shanchun)

 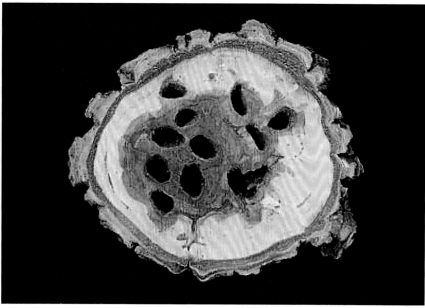

S. carcharias larva (Yan Shanchun)   *S. carcharias* damage (Yan Shanchun)

bulatory ampullae of the abdomen with one transverse groove and two oblique grooves on the side.

Pupae Pale yellowish white, 26 mm long.

Biology and behavior This insect occurs two generations a year. It overwinters as larvae. Larvae feed under the bark. After overwintering, larvae feed in vertical galleries in xylem. The galleries are usually 20-30 cm long. In large trees, the galleries can be 1.0-1.5 m long. After 2nd overwintering, they chew horizontal galleries. They overwinter at the top of the vertical galleries. Adults appear in July-September. They chew holes on leaves or chew the bark of young trees during supplementary feeding.

Prevention and treatment

1. Quarantine.

2. Maintain tree vigor and their natural resistance to pests.

3. Manual control of adults. Cut down the dying or dead trees and burn the trees.

4. Paint the trunks to prevent eggs from hatching.

5. Chemical control. Before adult emergence, place cotton balls soaked with 80% DDVP into the gallery hole of the larvae, then seal the holes. To kill larvae, place aluminum phosphate tablet into gallery holes and seal the holes to kill larvae. Spray canopy with 40% omethoate: DDVP at 800-1,000× during adult emergence peak. During larvae stage, drill 40° angle holes on tree trunks under 30 cm height, 3-5 holes per tree for 20 years or older trees or 1-2 holes for young trees; depth of holes are 8 cm; inject 40% omethoate and 80% DDVP at 100×, then seal the holes.

References Xiao Gangrou (ed.) 1992, Su Xiufeng and Wang Hongmei 2005.

(Yan Shanchun, Liu Kuanyu)

173 *Saperda populnea* (L.)

Classification Coleoptera, Cerambycidae
Synonym *Cerambyx populneus* L.
Common name Small poplar borer

Importance *Saperda populnea* is an important wood-boring insect pest of poplar and willow trees. They can destroy trees in large areas in northeast and northwest of China. They post threats to the safety of pedestrians and the traffic.

Distribution Beijing, Hebei, Inner Mongolia, Liaoning, Jilin, Heilongjiang, Jiangsu, Anhui, Shandong, Henan, Tibet, Shaanxi, Gansu, Qinghai, Xinjiang; Korea Peninsula, Mongolia, Russia, Europe.

Hosts *Populus tomentosa*, *P. alba*, *P. hopeiensis*, *P. canadensis*, *P. simonii*, *P. nigra* var. *thevestina*, *P. cathayana*, *Salix matsudana*, *S. sericocarpa*.

Morphological characteristics **Adults** Length 11-14 mm. Black, densely covered with golden yellow pubescence and mixed with black long pubescence. Eyes black; antennae thread-like, scape large, basal 2/3 of each flagellum grayish white; prothorax without a lateral tooth, each side with a golden yellow longitudinal band; elytra with black coarse punctures and light yellow short pubescence; pubescence on each elytra forms 4-5 golden yellow spots.

Eggs Length about 2.4 mm, width 0.7 mm. Elongate, slightly curved.

Larvae Length 10-15 mm. Milky white at beginning, then became light yellow. Mature larvae dark yellow, head yellowish brown; spiracles brown, dorsal surface with a distinct median line.

Pupae Length 11-15 mm. Brown; abdomen with a median line.

Biology and behavior This insect occurs one generation a year and overwinters as mature larvae in branches. Larvae became active in spring. In Beijing, pupation occurs in late March, pupal period is about 20-34 days. Adults emerge in mid-April. Adults are gregarious.

S. populnea adult male (Li Zhenyu)

S. populnea adult female (Xu Gongtian)

Galls made by *S. populnea* (Xu Gongtian)

S. populnea larva (Xu Gongtian)

They feed on edge of leaves for nutrient. Adults mate 2-5 days after emergence. Eggs appear in mid-May. Eggs are laid in emergence holes on 1-3 year-old trunks or branches. Larvae hatch in May. Larvae feed in branches or trunks. At the beginning, they feed on phloem or sapwood, then around the branches or trunks, forming spindle-shaped galls. The excreted materials are kept in tunnels or sometimes expelled out. Mature larvae create pupal chambers in tunnels in early and mid-October. This insect prefers sunny and warmer environment. Sparely planted and weak trees are more susceptible to *S. populnea*.

Prevention and treatment

1. Quarantine. Prevent movement of infested seedlings to non-infested areas.

2. Silviculture. Trim off the infested branches and burn them.

3. Conserve and release natural enemies such as the larval parasitoid *Schreineria populnea*.

4. Chemical control. Spray 25 g/L lamda-cyhalothrin during adult emergence period.

References Wang Zhicheng 2003, Zhang Xingyao et al. 2003, Sheng Maoling 2005, Xu Gongtian et al. 2007, Forest Disease and Pest Control Station of the Ministry of Forestry 2008, Liu Er et al. 2009.

(Zhou Zaibao, Xu Chonghua)

174 *Semanotus bifasciatus* (Motschulsky)

Classification Coleoptera, Cerambycidae
Common name Juniper bark borer

Distribution Beijing, Hebei, Shanxi, Inner Mongolia, Liaoning, Zhejiang, Anhui, Jiangxi, Shandong, Hubei, Guangdong, Guangxi, Sichuan, Guizhou, Shaanxi, Gansu, Ningxia; Korea Peninsula, Japan.

Hosts *Platycladus orientalis*, *Sabina* spp., *Chamaecyparis*.

Damage Larvae bore into young trees and weak trees. Sawdust appear at the damage site. Needles become yellow after infestation. Oval emergence holes appear on trunks.

Morphological characteristics **Adults** Male length 11.0-17.2 mm, female length 10.6-18.5 mm. Flat and wide. Head black, with dense punctures; mouthparts directed downward. Antennae blackish brown, short; female antennae about 1/2 of the body length, male antennae slightly shorter than the body length. Pronotum black, its two sides arc-shape; with rather long light-yellow pubescence. Notum with five smooth tubercles; the first two round, the last three leaf-like. Meso- and metathorax with yellow pubescence on ventral surface. Elytra with two brownish yellow transverse bands, color dark, with oily appearance; the basal pale transverse band dark, often brownish; the near middle black transverse band becomes paler, often joins the black middle black transverse band. Legs blackish brown, with yellow setae. Abdomen brown, with yellowish brown hairs, the end slightly exposed.

Eggs Length 1.6 mm, long oval, tapered at the rear end, white.

Larvae Mature larvae length 2.2 mm. Pronotum width 4 mm. Slightly flattened, medium-sized, the rear end much narrower. Young larvae pale red, mature larvae milky white. Pronotum with a depression and four yellowish brown marks. Head yellowish brown, near trapezoid, wider near the back. Legless.

Pupae Length about 15 mm. Pale yellow.

Biology and behavior *Semanotus bifasciatus* occurs one generation a year. Adults overwinter in xylem. Few numbers finish a cycle in two years and larvae overwinter in sapwood. Adults emerge from exit holes in March-April. They do not need to feed before laying eggs. Eggs are laid in bark crevices under 2 m height. Egg period is 10-20 days. Young larvae feed on bark for 1-2 days, then bore into bark, causing resin flow. Larvae feed on phloem and sapwood, creating up and down flat galleries. The length can reach 90-120 cm. Sometimes the galleries will surround the tree trunk. The galleries are filled with saw dusts. Larvae build pupal chambers near the sapwood. They pupate in August-October. Pupal period is 20-25 days. Adults emerge from pupae in

S. bifasciatus adult (Xu Gongtian)

S. bifasciatus bait logs (Xu Gongtian)

S. bifasciatus larva (Xu Gongtian)

S. bifasciatus exit hole (Xu Gongtian)

S. bifasciatus excreted material (Xu Gongtian)

S. bifasciatus tunnels (Xu Gongtian)

S. bifasciatus exit holes (Xu Gongtian)

September-November and overwinter in pupal chambers. Population density is higher in pure forests, middle-aged forests, dense forests than that in mixed forests, young forests, and sparse forests. Both healthy and weak trees can be infested, but weak trees often suffer higher damage.

Prevention and treatment

1. Silviculture. Remove sources of infestation by promptly removing dying trees, damaged trees.

2. Attract and kill. Place newly cut 4 cm diameter *Cunninghamia lanceolata* branches at the end of February in forests to attract and kill adults. Oil extracted from *Cunninghamia lanceolata* or essential oils containing alkene can be used to attract adults.

3. Biological control. Release *Sclerodermus guani* or *Pyemotes* sp. to control larvae.

References Yu Lichen et al. 1997, Yang Ziqi and Cao Guohua 2002, Xu Gongtian 2003.

(Qiao Xiurong, Gao Ruitong, Qin Xixiang, Zhou Zhiming)

175 *Semanotus sinoauster* Gressitt

Classification Coleoptera, Cerambycidae
Common name China fir borer

Distribution Jiangsu, Zhejiang, Anhui, Fujian, Jiangxi, Henan, Hubei, Hunan, Guangdong, Guangxi, Sichuan, Guizhou, Taiwan; Korea Peninsula, Japan.

Hosts *Cunninghamia lanceolate*, *Cryptomeria fortune*.

Damage Young larvae feed between xylem and phloem, creating irregular flat galleries. As larvae develop, they bore into xylem. Infested trees grow slowly or die.

Morphological characteristics **Adults** Females length 12-23 mm, width 3.5-6.5 mm; males length 11-21 mm, width 2.9-4.0 mm. Body flat and wide; head black with fine sculptures. Antennae blackish brown, female antennae about half the body length, male antennae about same as the body length. Prothorax black, the sides arc-shaped, with dense light-yellow pubescence. Pronotum with five smooth tubercles arranged in flower shape. Ventral surface of the meso- and metathorax brown, with yellow pubescence. Legs blackish brown, elytra brownish yellow, elytra with two brownish yellow and two black transverse bands. The end of the elytra round; base of elytra with large sculptures, slightly wrinkled; sculptures on other areas of the elytra small. The end of the abdomen in females exposed.

Eggs Length 2-3 mm. Elongated oval, tapered at the rear end. White, turning into light-yellow before hatch, semi-translucent.

Larvae Mature larvae 25-35 mm long. Milky white or light-yellow. Round but slightly flat. Mandibles strong, blackish brown; pronotum wide, the two sides slightly semi-circular; yellowish brown, densely covered with pubescence. Segments after mesothorax gradually narrowed. Spiracles elliptical, brown; the pair on mesothorax is the largest.

Pupae Length 20-25 mm, light-yellow.

Biology and behavior *Semanotus sinoauster* occurs one generation a year or one generation in two years. They overwinter in pupal chambers in galleries. Those finish a cycle in two years will overwinter as larvae in the first year and overwinter as adults in the second year. After eclosion, adults stay in pupal chambers for 30-60 days or 180-200 days if overwintering as adults. Adults emerge from exit holes when temperature is above 10℃. Emergence holes are round, 3-6 mm in diameter. Large numbers of adults emerge between 12:00-16:00. Adults are active during the day. They rest in bark

S. sinoauster adult (dorsal view) (Xu Gongtian)

S. sinoauster adult (lateral view) (Xu Gongtian)

crevices and base of branches, etc. They are weak flyers and play death when disturbed. Males start to mate soon after emergence from trees. Mating duration is 3-10 minutes. They mate multiple times. One to three days after the first mating, females lay eggs in bark crevices. Eggs are laid singly or groups of 2-6 eggs. They are glued in crevices with yellow sticky fluid. Eggs are often located under 2 m height on trunks. Each female lays 50-80 eggs. Adult sex ratio (females: males) is 1.65:1. Longevity after emergence is 7-38 days for females and 4-25 days for males. The egg laying period is 5-20 days. Egg incubation period is 10-20 days. Young larvae feed between phloem and xylem. Galleries are irregular and filled with saw dusts and excretion. As the larvae develop, resin flow increases, larvae move down in xylem, galleries become larger, feces and sawdust remain inside galleries. Mature larvae create pupal chambers at the end of galleries and create elliptical exit holes. Prepupae period is 3-4 days and pupal period is 15-25 days.

Prevention and treatment

1. Silviculture. Create mixed forests. Select resistant varieties. Trim new branches to keep the trunk smooth. Maintain proper tree density. Cut down insect damaged trees, dead trees, and weak trees.

2. Mechanical control. During young larval stage, use a hammer to hit the spot with resin flow to kill the young larvae. Manually collect and kill adults.

3. Biological control. To attract woodpeckers, place 4-5 pieces of wood per 15-20 ha area. The wood should be about 100 m in distance. Maintain them once a year in the fall. Release *Sclerodermus guani*.

4. Spray tree trunks with 40% omethoate at 1,000-1,500× under 2 m height to kill adults.

References Xiao Gangrou (ed.) 1992, Ding Dongsun and Shi Mingqing 1997, Hu Changxiao 2003, Xu Zhizhong 2009.

(Ji Baozhong, Zhang Kai, Zhang Lianqin, Shao Lichao, Wang Shufen)

176 *Thylactus simulans* Gahan
Classification Coleoptera, Cerambycidae

Distribution Hunan, Guangxi, Guizhou, Yunnan; Vietnam, India, Laos.

Hosts *Catalpa bungei, C. duclouxii, C. ovata, Paulownia fortunei*

Damage *Thylactus simulans* larvae bore into branches and trunks, slow down tree growth and kill branches and trunks.

Morphological characteristics **Adults** Length 21-32 mm. Body covered with thick dense greyish yellow to blackish brown pubescence. There is a large upside down blackish brown triangular spot at the base of elytra. There are also broken blackish brown longitudinal bands along the seam of the elytra. The two sides of the 3rd to 5th abdominal segments of the males with hairy pits.

Eggs Length 3.2-3.8 mm. Yellowish white.

Larvae Mature larvae 33-46 mm. Milky white to yellowish white.

Biology and behavior *Thylactus simulans* occurs one generation in two years in Hunan. It overwinters as larvae. Larvae pupate in early May. There is a drainage hole below the pupal chamber and exit hole above the pupal chamber. Pupal period is 18-31 days. Adults appear in late May. They live for 15-33 days. They feed on one-year old small branches after emergence. Mating occurs soon after emergence. About 10 days after mating, they lay eggs on one-year old small branches. A female chews 2-5 rings of marks on a one-year old branch, then lays eggs. The branch above the rings will die soon. One female lays an average of 44 eggs. Eggs hatch in mid-June. Egg incubation period is 9-14 days. After hatch, larvae feed downward. They make 28-193 cm long galleries during the 1st year and 60-123 cm long galleries during the 2nd year. Round holes arranged vertically can be seen on branches. Larval period is 658 days. The distance between the holes made during the 1st year and the 2nd year is 2-10 cm and 10-30 cm, respectively. The diameter of the holes is 3 mm.

Prevention and treatment

1. Silviculture. Maintain vigor of the trees.
2. Manual control. Trim off the dying branches and burn them.
3. Set up light traps to kill adults during adult period.

References Zhang Xiankai and Zuo Yuxiang 1986, Peng Jianwen and Liu Youqiao 1992.

(Tong Xinwang, Zhang Xiankai, Zuo Yuxiang)

T. simulans adult (Li Zhenyu)

177 *Trichoferus campestris* (Faldermann)

Classification Coleoptera, Cerambycidae

Distribution Beijing, Hebei, Shanxi, Liaoning, Shandong, Henan, Sichuan, Guizhou, Yunnan, Shaanxi, Gansu, Qinghai, Xinjiang; Japan, Korea Peninsula, Russia, Mongolia.

Hosts *Robinia pseudoacacia*, *Populus* spp., *Salix* spp., *Ulmus* spp., *Toona sinensis*, *Fraxinus chinensis*, *Betula* spp., *Tectona grandis*, *Pieca asperata*, *Ziziphus jujube*, *Syringa* spp., *Malus pumila*, *Pyrus bretschneideri*.

Damage *Trichoferus campestris* larvae bore into trunks and branches, causing mortality of trees. When infested wood is used for building roof of houses, the roof may collapse.

Morphological characteristics **Adults** Length 13-14 mm; brown; body covered with yellow pubescence; prothorax globular, its dorsal surface with a shallow longitudinal groove near the posterior end.

Larvae Mature larvae 20 mm long; yellowish white.

Biology and behavior *Trichoferus campestris* occurs one generation a year in Henan. It overwinters as larvae in branches. They resume activity in March. Larvae bore flat and wide galleries in xylem and push out sawdust. They pupate during late April and early May. Adult eclosion occurs during late May and early June. Adults are attracted to lights. They prefer laying eggs in bark crevices on branches greater than 3 cm diameter. Newly cut down branches are most preferred. They also lay eggs on timbers that are not fully dried. Eggs are laid singly. The hatch in 10 days. Young larvae feed between phloem and xylem, creating irregular galleries. They start overwintering in October.

Prevention and treatment

1. Manual control. Debark newly cut down timber or submerge timber in water to kill the larvae.

2. Chemical control. Spray 25 g/L lamda-cyhalothrin and kerosene at 1:1 ratio to wood rafter and close the doors and windows. Spray 25 g/L lamda-cyhalothrin and thiamethoxam mixture at 200× to timber, then cover the timber with plastic to kill the larvae through fumigation effect.

References Yang Youqian et al. 1982, Xiao Gangrou (ed.) 1992, Henan Forest Disease and Pest Control and Quarantine 2005.

(Yang Youqian)

T. campestris adult (Li Zhenyu)

178 *Xylotrechus rusticus* (L.)

Classification Coleoptera, Cerambycidae
Common name Gray tiger longicorn beetle

Importance Outbreaks occurred in large areas in northeastern China. Forest belts along farm lands and landscaping trees were killed by this pest.

Distribution Inner Mongolia, Liaoning, Jilin, Heilongjiang, Shanghai, Jiangsu, Xinjiang; Korea Peninsula, Japan, Mongolia, Russia, Iran, Turkey, Europe.

Hosts *Populus* spp., *Salix* spp., *Betula* spp., *Quercus* spp., *Fagus* spp., *Tilia* spp., *Ulmus* spp. *Picea obovate*.

Damage *Xylotrechus rusticus* larvae bore into branches, trunks, and roots, creating irregular galleries. Affected trees grow slowly and are easily damaged by wind.

Morphological characteristics **Adults** Length 11-22 mm, width 3.1-6.2 mm. Black, head and prothorax dark. Frons with two longitudinal ridges forming a

X. rusticus damage (Yan Shanchun)

"V" shape. There is a longitudinal ridge from rear of the head to the vertex. There are two parallel stripes made of yellow pubescence from frons to occiput. Antennae sockets are close to each other. Male antennae reach base of elytra, female antennae only reach posterior border of the pronotum; the 1st and the 4th segment are shorter than the 3rd segment, the last segment is longer than its width. The first five segments without pubescence at the tip. Prothorax globular, wider than length, with irregular wrinkles. Notum with two incomplete light-yellow stripes. Scutellum semi-circular, two sides of the elytra parallel, obtuse at the inner and outer tips. Elytra with fine sculptures and 3-4 yellow opaque stripes; scattered hairs present among the stripes. Base of elytra with wrinkles. Ventral surface with dense yellow pubescence. Legs medium-sized. Front coxal cavity round, not angular. Exterior of the middle coxal cavity open toward the epimerum. Hind femora thick, with two tibia spurs, the 1st tarsal segment is longer than the total length of the other tarsal segments.

Eggs Elongated oval, length 2 mm, width 0.8 mm. Milky white.

Larvae Yellowish white; mature larvae 30-40 mm long, with short hairs. Head light-yellowish brown, hidden in prothorax. Pronotum with yellowish brown stripes. Abdominal segments (except the last one) be-

X. rusticus adult female (Xu Gongtian)

X. rusticus adult male (Xu Gongtian)

come narrower and longer from the 1st segment.

Pupae Yellowish white, 18-32 mm long. Eyes, appendages, and wing buds turn black before eclosion.

Biology and behavior *Xylotrechus rusticus* occurs one generation a year in Shenyang. It overwinters as larvae in xylem in October. Larvae become active in April next year. Galleries are irregular. They pupate near the surface of xylem in late April. Adults appear from late May and reach peak in early June. Exit holes are round, 4-7 mm in diameter.

Adults are active. They can fly for short distances. Soon after emergence, they can mate and lay eggs on trunks and branches. Eggs are laid in groups of a few to dozens in old bark crevices. Each egg laying duration is about one hour. Egg incubation period is 10-12 days. After hatch, larvae feed in groups in bark. Dusts can be seen from outside. After about 7 days, they bore into xylem and do not expel dusts and feces. As they grow, larvae create wider galleries on surface of xylem and scatter. Galleries are 7-10 mm wide and are filled with excretion. Bark become detached from xylem. When larvae reach medium size in July, larvae bore deep into xylem. The galleries from different larvae do not merge. The holes on xylem are elliptical, 10 mm long, 8 mm wide. The insect only infests healthy branches. Young trees are not infested. Edge of the forest suffer more damage than the center of the forest. Single trees suffer more damage than groups of trees. Trees with coarse barks suffer more damage than those with smooth barks. Tree trunk suffer more damage than branches; larger branches suffer more damage than smaller branches.

Prevention and treatment

1. Silviculture. Clear cut, selective cut, and planting mixed forests will help reduce pest populations.

2. Manual control. Collect adults and kill them. Scrape off eggs and young larvae. Paint trunks with lime slurry.

3. Set up light traps to kill adults.

4. Biological control. Attract birds such as *Cuculus* spp., *Pica pica*, *Passer montanus*, *Corvus* spp., and *Upupa epops*.

5. Chemical control. Spray trunks and branches with 20% dinotefuran and 25 g/L lamda-cyhalothrin at 0.125% to kill eggs and young larvae. Inject 0.3% permethrin, or thiamethoxam and diesel at 1:20 ratio into galleries on trunks. Apply 50% thiamethoxam granules, 25 g/L lamda-cyhalothrin, avermectin microcapsules into excretion holes on trunks and seal the holes.

References Jiang Shunan, Pu Fuji, and Hua Lizhong 1985; Meng Xiangzhi 2002, Huang Yonghuai 2004, Li Chengde 2004; Cheng Hong, Yan Shanchun, and Sui Xiang et al. 2006, Jin Gesi and Wuer Lanhan 2006; Yan Shanchun, Li Jinguo, and Wen Aiting et al. 2006; Zhang Yubao, Li Jinguo, and An Kun 2006; Cheng Lichao 2007; Zuo Tongtong, Chi Defu, and Wang Muyuan 2008.

(Yang Shanchun, Xu Gongtian)

X. rusticus adult (Ren Lili)

X. rusticus mating pair (Ren Lili)

X. *rusticus* larva (Yan Shanchun)   X. *rusticus* pupae (Yan Shanchun)

179 *Xystrocera globosa* (Olivier)

Classification Coleoptera, Cerambycidae

Distribution Hebei, Liaoning, Jilin, Heilongjiang, Shanghai, Jiangsu, Zhejiang, Shandong, Guangdong, Guangxi, Sichuan, Taiwan; India, Sri Lanka, Myanmar, Thailand, Malaysia, Indonesia, Korea Peninsula, Japan, the Philippines, Egypt, U.S.A. (Hawaii).

Hosts *Albizzia julibrissin*, *Sophora* spp., *Morus alba*, *Prunus persica*, *Bombax malabaricum*.

Damage *Xystrocera globosa* larvae feed on phloem and xylem. They lower the tree growth and quality of timber. Severely infested trees are easily broken by wind. Branches of 10 cm in diameter are most susceptible.

Morphological characteristics **Adults** Reddish brown to yellowish brown. Center of head with a longitudinal groove. There are metallic blue or green stripes on pronotum and center and edges of elytra. Femora rod-shape.

Larvae Milky white with a tint of greyish yellow. Anterior border of the pronotum with six brown spots which arranged into a band.

Biology and behavior *Xystrocera globose* occurs three generations a year in Shanghai, 1-2 generations a year in Shandong. It overwinters as larvae in galleries. Larvae become active in March and start to create damage in late March. Pupation occurs in early May. Adults appear in late May and reach peak from late June to mid-July. Adults can mate and lay eggs soon after emergence. Egg period is 10-15 days. Adults are attracted to lights.

Prevention and treatment

1. Reduce sources by removing infested wood between fall and spring.

2. Light trapping and manual collection of adults.

3. Spray 25 g/L lamda-cyhalothrin at 1,000-1,500×, 8% cypermethrin at 150-200× to trunks during June-July.

References Zhang Yanqiu and Liu Wei 2002, Ma Xingqiong 2007, Wang Yan 2007, Xu Gongtian and Yang Zhihua 2007.

(Qiu Lixin, Shu Chaoran, Jiang Haiyan)

X. globose eggs (Wang Yan)

X. globose larvae (Wang Yan)

X. globose adult (Wang Yan)

X. globose pupa (Wang Yan)

180 *Acanthoscelides pallidipennis* (Motschulsky)

Classification Coleoptera, Chrysomelidae

Synonyms *Bruchus collusus* Fall., *Bruchus perplexus* Fall., *Acanthoscelides plagiatus* Reiche et Saulcy

Distribution Beijing, Tianjin, Hebei, Inner Mongolia, Liaoning, Jilin, Heilongjiang, Henan, Shaanxi, Ningxia, Xinjiang; Russia, Korea Peninsula, U.S.A., Southeastern Europe.

Host *Amorpha fruticose*.

Damage *Acanthoscelides pallidipennis* infests seeds of *A. fruticose*. Infested seeds cannot germinate.

Morphological characteristics **Adults** Length 2.0-2.6 mm, width 1.1-1.7 mm. Elliptical. Head blackish grey, small, narrower than pronotum. Vertex with dense round sculptures and sparse white hairs. Frons with dense white pubescence. Eyes kidney shape; black. Antennae 11-segmented, serrate; basal 1-4 segments slender, brown; other segments gradually enlarged, each segment with dense white pubescence. Pronotum blackish grey, center slightly convex, with three distinct longitudinal hair bands; the middle hair band covers the whole length of the pronotum, whereas the other two bands are shorter. Scutellum rectangular, with dense white hairs; its posterior angles sharp, posterior border deeply concave. Elytra brown, darker at the seam. Each elytron has 10 brown dotted grooves, 11 bands of white hairs exist between the grooves. The brown color of the elytra appears where hairs are sparse. Base of the 3rd and 4th hair bands each with a dark spot, about 1/4 near the end of the 3rd and 4th hair bands also with a dark spot. There is a large brown

1mm

A. pallidipennis (Li Zhenyu)

A. pallidipennis (Li Zhenyu)

A. pallidipennis (Li Zhenyu)

spot between the 8th and 9th hair bands. There is also a pale brown spot at the end. Dorsal surface of abdomen with seven visible segments, ventral surface with five visible segments. Pygidium bends toward ventral surface in males. Pygidium in females not visible viewed from ventral side.

Eggs Length 0.5 mm, width 0.2 mm. Elongated oval. semi-translucent, Light-yellow, turning into dark before hatch.

Larvae Length 0.4-0.5 mm, light-yellow when hatched. Head and ocelli dark brown. First instar with sparse, pale, long setae. There are three thoracic legs. Starting from the 2nd instar, the thoracic legs degenerated. Mature larvae legless; yellow; 2.4-3 mm long; soft, plump; spiracles round.

Pupae Length 2.7 mm, width 1.6 mm. Milky yellow. Eyes "U" shaped. Pale red at the beginning. Eyes, antennae, femora and end of the tibia become dark brown before eclosion. Spiracles round.

Biology and behavior *Acanthoscelides pallidipennis* occurs two generations a year in Heilongjiang. It overwinters as 2nd to 4th instar larvae in seeds. They pupate during late May and early June. Pupal period is 8-10 days. Newly emerged adults remain inside seeds for 3-5 days before flying out. Exit holes are round, not smooth, diameter 0.8-1.0 mm. Exit holes are usually located at top ¼ of the pods. First generation adults appear in mid-June. Peak emergence is in late June. Adults are strong flyers. They often crawl among flowers and pods on host trees. They fall to ground and play death when disturbed. Adults live for 17-34 days. They feed on nectars, petals, and young skin of pods. Adults can mate soon after emergence. Mating often occurs in the mid-morning. Mating lasts for 8-39 minutes. After another 3 days, they lay eggs. Adults lay eggs in early July. Two to five eggs are laid on each pod. Larvae appear in early July. They bore into seeds after hatch. Pupation occurs between late July and early August. Second generation adults appear in early and mid-August. Larvae appear in late August. They overwinter in September. One larva only infests one seed. Clear cut forests have lower pest populations.

Prevention and treatment

1. Silviculture. Clear cut *Amorpha fruticose* to grow new stands.

2. Chemical control. Fumigate seeds with sulfuryl fluoride at 30-50 g/m^3 for 3-4 days. At low temperature, use aluminum phosphide to fumigate seeds for 3-4 days.

References Tan Juanjie et al. 1980; Zhang Shengfang et al. 1991, 1998.

(Gao Mingchen, Li Zhenyu)

181 *Kytorhinus immixtus* Motschulsky

Classification Coleoptera, Chrysomelidae
Synonym *Mylabris immixta* Baudi

Distribution Inner Mongolia, Heilongjiang, Shaanxi, Gansu, Qinghai, Ningxia, Xinjiang; Russia, Mongolia.

Hosts *Caragana microphylla* and other species in genus *Caragana*.

Damage *Kytorhinus immixtus* larvae infest seeds of the host plants.

Morphological characteristics **Adults** Length 3.5-5.5 mm, width 1.8-2.7 mm. Elongated oval. Black. Antennae, elytra, legs yellowish brown. Head with dense fine sculptures and greyish white pubescence. Antennae 11-segmented, serrate, about 1/2 of the body length. Male antennae pectinate, about same length as the body. Pronotum narrow in the front, with sculptures and greyish white and yellow hairs; its center slightly raised, a longitudinal groove appears near the posterior border. Scutellum rectangular, with a concave posterior border and greyish white hairs. Elytra with 10 longitudinal lines of punctures, the shoulder of elytra distinct, the end of elytra round. Elytra yellowish brown, dark brown at the base, covered with yellow hairs; a cluster of greyish white hairs present at the base. The last two abdominal segments exposed, with punctures and greyish white hairs.

Eggs Length 0.2 mm, width 0.1 mm. Elliptical, light-yellow at the beginning, brown before hatch.

Larvae Mature larvae 4-5 mm long. Head yellowish brown, body light-yellow, "U" shaped.

Pupae Length 4-5 mm, width 3 mm. Pale yellow.

Biology and behavior *Kytorhinus immixtus*

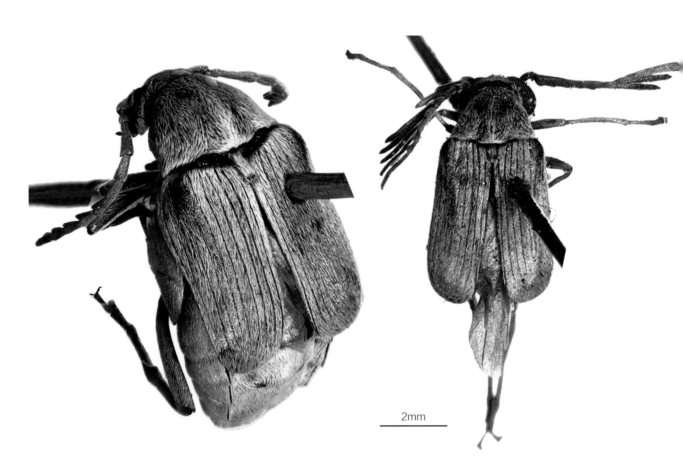

K. immixtus adult female (Li Zhenyu)

K. immixtus adult male (Li Zhenyu)

occurs one generation a year. It overwinters as mature larvae in seeds. Adults emerge between late April and mid-May next year. Larvae appear in late May. They become dormant in August. Some mature larvae undergo diapause for two years.

Adults are strong flyers. They fly away when disturbed. They are active during evening hours. They feed on nectar, petals, and young leaves. Adults mate and lay eggs 2-3 days after eclosion. Male and female life span are 7-8 and 8-12 days respectively. Eggs are laid on pods near calyx. Each pod usually has 3-5 eggs (maximum 13 eggs). Egg incubation period is 11-17 days. Larvae bore into pods after hatch. They feed inside seeds. Larvae have 5 instars. One larva only infests one seed. Infested seeds are blackish brown, with coarse surface, sticky fluid often appears. When harvested, seeds with larvae often jump.

The pest occurs more often in pure forest, sparse forest, and uncut forest. Common natural enemies include *Phanerotomella* sp. and *Entedon* sp. The reported parasitism rates were 7.86-26.93% and 9.79-15.68%, respectively. Others include *Itoplectis* sp., *Chelonus* sp., and *Eupelmus* sp.

Prevention and treatment

1. Quarantine. During harvest, storage, and transportation, follow quarantine regulations.

2. Silviculture. Plant mixed forests. Clear cut *Caragna* spp. and *Amorpha fruticose* to restore new plants. Harvest all pods and treat them.

3. Seed treatment. Place harvested seeds in 0.5%-1.0% saltwater. Remove the floating seeds.

4. Biological control. Two important parasitoids are *Phanerotomella* sp. and *Entedon* sp. Their parasitization rates were reported to reach 7.86-26.93% and 9.79%-15.68%, respectively. Other parasitoids include *Phanerotomella* sp., *Entedon* sp., *Itoplectis* sp., *Chelonus* sp. In addition, *Scleroderma guani* was found to be a promising parasitoid for controlling K. *immixtus*.

5. Chemical control. Spray forests with 45% profenofos and phoxim mixture at 1,000-1,500× to kill adults, or 50% fenitrothion at 500× to kill eggs. Mix seeds with 25% trichlorfon dust at 400:1 (seeds:dust) ratio before storage. Fumigate bags of seeds with 1.5 g avermectin microcapsules per bag and seal for 6 days, or sulfuryl fluoride at 30-35 g/m^3 for 3-4 days.

References Nengnai Zhabu et al. 1980; Tan Juanjie et al. 1980; Zou Lijie et al. 1989; Zhang Shengfang, Liu Yongping, Wu Zengqiang (ed.) 1998.

(Li Kuansheng, Zheng Wenhan, Zou Lijie)

182 *Agelastica alni glabra* (Fischer von Waldheim)

Classification Coleoptera, Chrysomelidae
Synonym *Agelastica orientalis* Baly
Common name Oriental leaf beetle

Distribution Inner Mongolia, Xinjiang; Russia, Europe, North America.

Hosts Most preferred species are *Populus* spp. and *Salix* spp. Less preferred hosts are: *Malus pumile*, *Prunus bretschneideri*, *P. dulcis*, *P. persica*, *Ulmus pumila*, *Vitis vinifera*, and *Morus alba*.

Damage Both larvae and adults feed on leaves, causing leaves to fall or roll. At high population density, most leaves are damaged.

Morphological characteristics Adults Male length 7.0-7.5 mm, female length 7.5-8.0 mm. Head wider than long. Head and pronotum black, with dense small black dots. Antennae black, the 2nd to 4th segment slender, the 5th to 11th segment enlarged. Elytra wider than thorax, dark blue, with lines of punctures.

Eggs Elliptical. Length 2-3 mm. Orange color.

Larvae Mature larvae 11-12 mm long. Body flat. Blackish grey, shiny. Sides of body with two rows of black tubercles that bearing hairs. Ventral surface of abdomen brown.

Pupae Length 6-7 mm. Elliptical. Orange color.

Biology and behavior *Agelastica alni glabra* occurs one generation a year in Xinjiang. It overwinters as unmated adults in 2-4 cm deep soil. They resume activity when average temperature during the day reaches 10°C in March. They feed on young leaves and buds. After 3-4 days of feeding, they mate. Adult mate multiple times, each time lasts about half an hour. They lay eggs 2-3 days after mating. Egg laying period is nearly 2 months. Adults are active for approximately 50 days. Adults play death when disturbed. A female lays 5-6 egg masses, each with about 32 eggs. The average (minimum, maximum) interval between two egg-laying period is 7 (5-11) days. Eggs are laid on leaves. Newly hatched larvae aggregate on tip of leaves. After May, they scatter. The 1st instar feed on leaf tissue and create dents on leaf surface. The 2nd and 3rd instar feed more and cause the leaves net-like. Larvae expose their reversible defensive glands and release dark yellow fluid when disturbed.

Egg incubation period is 7 (5-13) days. Larvae go through three instars. The average (minimum-maximum) development time for the 1st to 3rd instar larvae is 11 (9-

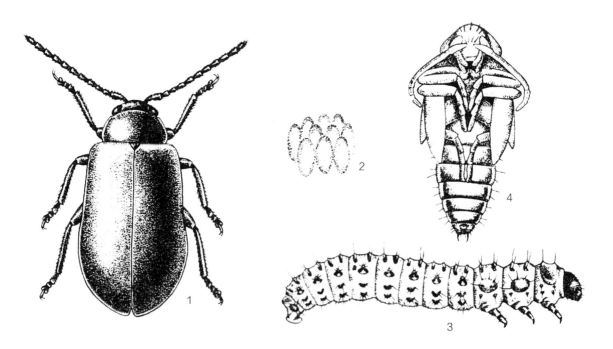

A. alni glabra. 1. Adult, 2. Eggs, 3. Larva, 4. Pupa (Zhang Peiyi)

15), 15 (9-15), 14 (13-20) days, respectively. Total larval period is about 44 days. Mature larvae crawl into loose soil around the trees and build 6-7 mm long soil chambers. After 2-3 days, they become pre-pupae. After 5-7 days of prepupal period, they turn to pupae. The average (minimum-maximum) pupal period is 25 (20-28) days. Adults appear in early July. They stay in soil between mid-July and early September when temperature is above 25°C. They feed on trees between mid-September and late October.

Prevention and treatment

1. Silviculture. Select resistant varies such as *Populus alba* var. *pyramidalis*, *P. alba*. Clear the leaf litter and till soil in late fall and early spring to destroy overwintering adults.

2. Manual control. Manually remove eggs, young larvae, and adults. In late June, cut the grass under trees, till soil in areas of 50 cm diameter around tree trunk.

3. Biological control. Protect *Arma chinensis* and *Plagiolepis rothneyi*. They are predatory enemies of the pest.

4. Chemical control. Spray 5.7% emamectin at 2,000×, or 25% Dimilin at 2,000×, or 2.5% deltamethrin at 2,000-4,000×, to control 1st and 2nd instar larvae. When overwintering adults and newly emerged adults climb up the tree trunks, apply a 10 cm wide insecticidal band on tree trunks can kill more than 90% of the adults. The insecticide band can be 2.5% deltamethrin, rain protectant, water, lime, talcum at 1.5:2,32:40:5 ratio. Irrigate soil with 40% chlorpyrifos and phoxim mixture at 800-1,000× in early Spring and late Fall to kill pupae and adults.

References Yang Xiuyuan 1981; Wang Aijing 1984, 1995; Xiao Gangrou (ed.) 1992.

(Wang Aijing)

183 *Ambrostoma quadriimpressum* (Motschulsky)

Classification Coleoptera, Chrysomelidae
Common name Leaf beetle

Importance *Ambrostoma quadriimpressum* is a major leaf-eating pest to elm. It can consume all elm leaves in early spring each year, seriously affecting the tree growth.

Distribution Beijing, Hebei, Inner Mongolia, Liaoning, Jilin, Heilongjiang; Russia.

Hosts *Ulmus pumila*, *U. macrocarpa*, and *U. propinqua*.

Damage *Ambrostoma quadriimpressum* feeds on elm leaves and cause irregular notches of leaves. When feeding on elm buds, it prevents new leaf development.

Morphological characteristics **Adults** Nearly oval, 10.4-11.0 mm in length. There is purple red alternated gold green luster on pronotum and elytra, especially on elytra. The hind wings are red. The abdomen shows five nodes, the 5th node of the females has a blunt round end.

Eggs Length 1.65-2.3 mm, width 0.8-1.1 mm, with mixed color, light white to brown yellow.

Larvae Gray white to milk yellow, 1.5-10.5 mm in length.

Pupae Milk yellow, slightly flat, nearly oval, 11.5 mm in length.

Biology and behavior *Ambrostoma quadriimpressum* occurs one generation per year in western Jilin Province. It overwinters as adults. They resume activity in mid-April next year and stop laying eggs in mid-August. Egg incubation period is generally 5-17 days. They begin to hatch in early May. Larvae stage lasts about 18 days. Mature larvae pupate in late May to early June. Pupal stage lasts about 9 days. New adults begin to emerge in mid- and late June. They climb down from the trees to overwinter in mid- and late October.

Adults overwinter in 3-12 cm deep soil at base of the trunks. Their lifespan is generally 264-789 days. Adults with longer life can overwinter two times. Adults have wings but do not fly, and can feign death. Adults begin to feed 1 to 2 days after climbing on trees. They feed most before summer aestivation. Adults from the overwintering brood enter aestivation in July-August when the weather is hot. Adults lay eggs on twigs or leaves. Each female lays about 812 eggs in a year. Larval feeding increases as they grow until they enter the soil to pupate. New adults climb up to the crown for nutritional supplements after emergence, but do not lay eggs in the current year.

Prevention and treatment

1. Forest management. Plant mixed forests and max-

A. quadriimpressum adult (Xu Gongtian)

A. quadriimpressum mating pair (Xu Gongtian)

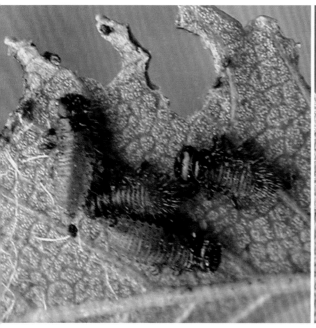

A. quadriimpressum 1st instar larvae (Xu Gongtian)

A. quadriimpressum eggs (Xu Gongtian)

A. quadriimpressum larva (Xu Gongtian)

A. quadriimpressum damage (Xu Gongtian)

imize the role of ecological control. The elm-poplar mixed forest bands are especially beneficial for establishing natural enemies.

2. Biological control. Release *Arma chinensis* which can feed *Ambrostoma quadriimpressum* at 0.4-1.9 larvae/day. Population reduction by *Arma chinensis* is up to 60%-80%. Other important natural enemies include *Macquartia tenebricosa* and *Asynacta ambrostomae*.

3. Chemical control. When overwintering adults and newly emerged adults climb up the tree trunks, apply a 10 cm wide insecticidal band on tree trunks can kill more than 90% of the adults.

References Yu Enyu et al. 1984; Gao Changqi et al. 1987, 1993; Yu Enyu and Gao Changqi 1987; Bi Xianghong 1989; Zhou Yushi et al. 1990; Xiao Gangrou (ed.) 1992; An Ruijun et al. 2005; Wang Xiumei, Zang Liansheng, Zou Yunwei et al. 2012; Zhang Yujun, Zheng Junshan, Pang Xuhong 2013.

(Gao Changqi, Zhang Xiaojun, Li Yajie, Li Zhenyu)

184 *Basiprionota bisignata* (Boheman)
Classification Coleoptera, Chrysomelidae

Distribution Hebei, Jiangxi, Shandong, Henan, Hubei.

Hosts *Paulownia* spp., *Catalpa ovata*, and *C. bungei*.

Damage Adults and larvae feed on leaf lower epidermis and mesophyll and leave the upper epidermis untouched. Damaged leaves look like wicker baskets. Heavily infested canopy becomes greyish yellow, resulting in premature leaf fall.

Morphological characteristics **Adults** Orange, oval; 12 mm in length, 10 mm in width. The first five segments of the antenna are light yellow; the other segments are black. Pronotum extends outward. The elytra convex, extend outward on both sides and form edges. There are two light yellow ridges in the center. There is a large oval black spot at posterior 1/3 of the elytron. **Larvae** Light yellow, spindle shaped, 12 mm in length. There is a light-yellow verruca on each side of each segment. There are two spines on the back of the last two segments. They tilt toward the back and carry shed skins.

Biology and behavior *Basiprionota bisignata* occurs two generations a year in Henan Province. Few go through one generation a year and some may have three generations a year. They overwinter under stones, litter, weed or bushes as adults. They resume activity in early April next year, feed on the new leaves, mate, and lay eggs. Larvae emerge about 9 days after eggs are laid. Larvae pupate in late May. The first-generation adults appear in early June. The second-generation adults appear in August and September. They begin to overwinter at late October.

Adults are active during the day and rest during the night. They are not attracted to lights. They lay eggs on back of the leaves. Dozens of eggs are laid into a mass with the eggs stay vertical within an egg mass. Larvae congregate after hatching and feed on lower epidermis and mesophyll, leaving the upper epidermis. Damaged leaves turn yellow and fall off prematurely. The shed skins are attached to the spines at the end of the body until the larvae pupate. Mature larvae glue their ends to back of the leaves when pupate. Adults feed on leaves after eclosion. Both larvae and adults are present during late May to mid-June, and late July to mid-August. The damage is most serious during these periods.

Prevention and treatment

1. Conservation of natural enemies. The parasitoids (*Podiobina* spp.), ants (*Polyrhachis* spp.), and Russet sparrow (*Passer rutilans rutilans*) play important role in suppressing the pest population.

2. Chemical control. During the adult infestation period in early July, spray 25% Dimilin at 2,500×, 50% phoxim at 1,000×, or 20% fenvalerate at 6,000×.

References Yang Youqian 1982; Wei Xiangdong, Cheng Xuhui, Wang Youfeng 1991; Xiao Gangrou (ed.) 1992; Henan Forest Disease and Pest Control and Quarantine Station 2005.

(Yang Youqian)

B. bisignata adult (Li Zhenyu)

B. bisignata adult (Li Zhenyu)

185 *Brontispa longissima* (Gestro)

Classification	Coleoptera, Chrysomelidae
Synonyms	*Oxycephala longipennis* Gestro, *Brontispa castanea* Lea, *Brontispa javana* Weise, *Brontispa reicherti* Uhmann, *Brontispa selebensis* Gestro, and *Brontispa simmondsi* Maulik
Common names	Coconut leaf beetle, Coconut hispid, Palm leaf beetle, Palm heart leaf miner

Importance *Brontispa longissima* is a disastrous pest of plants belonging to Arecaceae. It is also an internationally quarantined pest. It was introduced into China in the 1990's. It is a serious threat to coconut production and production of ornaments in Arecaceae.

Distribution Guangdong, Guangxi, Hainan, Hong Kong, Taiwan; Indonesia, Malaysia, Vietnam, Maldives, Australia, Peru, Fiji, Papua New Guinea, Vanuatu, Solomon Islands, New Caledonia, Samoa, Micronesia, Polynesia, U.S.A. (Guam).

Hosts *Cocos nucifera*, *Roystonea regia*, *Syagrus schizophylla*, *Livistona chinensis*, *Washingtonia filifera*, *Washingtonia robusta*, *Caryota urens*, *Caryota* sp., *Latania lontaroidea*, *Phoenix daclylifera*, *Metroxylon sagus*, *Arenga pinnala*, *Elaeis guineensis*, *Borassus flabelliformis*, *Phoenix dactylifera*, *Phoenix* sp., *Archontophoenix alexandrae*, *Arecastrum romanzoffianum*, *Chrysalidocarpus lutescens*, *Hyophorbe lagenicaulis*, *Areca catechu*, *Balaka* sp., *Bentinckiopsis* sp., *Calamus* sp.

Damage *Brontispa longissima* infests newly spread or the unopened leaves. Severe infestations cause dead areas, drooped leaf tips, death of leaves, reduced tree vigor, fruit fall, thin trunks, or death of the trees.

Morphological characteristics **Adults** Length 8-10 mm, width 1.9-2.1 mm, body flat, narrow and long. Head, eyes, and antennae blackish brown; pronotum orange yellow; elytra bluish black with metallic sheen, and several rows of punctures; ventral surface blackish brown; legs yellow.

Larvae Mature larvae 8-9 mm long, flattened, yellowish white, head yellowish brown. The end with pincer-like spines.

Biology and behavior *Brontispa longissima* occurs 3-5 generations a year in Hainan. Adults can mate without feeding after eclosion. Egg-laying period is long. Each female lays an average of 119 eggs. Different generations overlap.

Brontispa longissima prefers to infest 4 years or older plants in the family Arecaceae. Adults lay eggs in unopened heart leaves. The eggs are covered with feces or chewed leaves. Young larvae chew on the surfaces of leaves, leaving brown or dark brown longitudinal streaks. During adult mating period, they consume large amount of unopened heart leaves (different from the damage by larvae), causing greater damage than that done by larvae. Adults and larvae often congregate during feeding. Adults avoid lights. They prefer to stay at the base of the unopened heart leaves. Adults can fly for short distances

B. *longissima* damage (Zhang Runzhi)

B. *longissima* adult feeding on leaf (Xu Gongtian)

B. longissima adults (Zhang Runzhi)

B. longissima larvae (Xu Gongtian)

and spread. During the day, they usually crawl slowly. When disturbed, adults play death.

Prevention and treatment

1. Quarantine. Prevent importation of Aracaceae plants from *B. longissima* infested areas. Inspect Arecaceae plants before they are moved out of a region.

2. Biological control. Release *Tetrastichus brontispa*. Parasitization rate of pupae was 72%-92.78%. Apply *Metarhizium anisopliae* var. *anisopliae* achieved 37.0% control at 3 day, 81.1% at 15 day, and 45.1% at 100 day.

3. Chemical control. Place bags of "Yejiaqing" or 50% imidacloprid and monosultap mixture insecticide powder on trees. Inject 14% imidacloprid-dichlorvos into tree trunks at 0.7 mL/cm diameter can achieve 95.1% control in 30 days. Applying 0.1% or 0.05% avermectin can kill 100% of all stages after 120 days and the efficacy lasts for 150 days. Inject carbosulfan + quinalphos at 1:1 ratio into tree trunks at 50 g/tree rate. Spray 50% imidacloprid and monosultap mixture at 600-800× when damage occurred.

4. Low dose of monosultap and *M. anisopliae* mixture has synergistic effect against *B. longissima*. The mixture caused 93.33% mortality compared to 75.56% mortality by *M. anisopliae* alone or 10%-30% by monosultap alone.

References China Plant and Animal Quarantine Bureau 1997; Huang Fayu, Liang Guangqin, Liang Qiongchao et al. 2000; Rong Xuanxiong, Chen Murong, Deng Xianghui et al. 2003; Zeng Ling, Zhou Rong, Cui Zhixin et al. 2003; Zhong Yihai, Liu Kui, Peng Zhengqiang et al. 2003, 2004; Chen Xi, Chen Jiwen, Wang Churong et al. 2004; Chen Yiqun, Huang Honghui, Lin Mingguang et al. 2004; Wu Xiaoying, Zhong Yihai, Li Hong et al. 2004; Zhou Rong, Zengling, Cui Zhixin et al. 2004; Di Fuzhang, Zhang Zehua, Zhang Lisheng et al. 2006; Miao Jiancai, Chi Defu, Chang Guoshan et al. 2006, 2009; Tan Weiquan, Chen Siting, Huang Shanchun et al. 2006; Tang Guanghui, Jiang Zhili, Zhang Wenfeng et al. 2006; Wu Qing, Zengling, Zhang Jingchen et al. 2006; Ding Shaojiang, Li Chaoxu, Liang Minguo et al. 2007; Huang Xiaoqing and Zhang Xianmin 2007; Li Chaoxu, Tan Weiquan, Huang Shanchun et al. 2008; Li Keming, Tan Weiquan, Li Chaoxu et al. 2008; Qin Changsheng, Xu Jinzhu, Xie Penghui et al. 2008; Zhang Zhixiang, Xu Hanhong, Jiang Dingxin et al. 2008.

(Huang Shaobin, Wen Xiujun)

186 *Plagiosterna adamsi* (Baly)

Classification Coleoptera, Chrysomelidae
Synonym *Chrysomela adamsi ornaticollis* Chen

Importance Occasional outbreaks occur. Adults and larvae feed on host leaves and may cause death of branches.

Distribution Anhui, Chongqing, Sichuan, Yunnan.

Host *Alnus cremastogyne*.

Morphological characteristics **Adults** Length 4.5-8 mm. Golden bluish green, with punctures; prothorax yellow, its center with one green spot and one longitudinal depressed line, outer margin with one small green spot at the middle, humeral tubercle large; ventral plate green; first to 7th antennal segments yellowish brown, 8th to 11th antennal segments black.

Eggs Length 1.4 mm, width 0.6 mm; milky white at the beginning, later turns into yellowish white.

Larvae 1st instar 1.2-1.5 mm long, 2nd instar 2.5-5 mm long, 3rd instar 6-11 mm long. Mesothorax and metathorax and abdomen with one black spot on each side; each side of the dorsal line of the 1st to 6th abdominal segment with a rectangular hair spot; there is a tubercle below the spiracular line of each of the 1st to 8th abdominal segment.

Pupae Length 5-6 mm. Light yellowish brown; wing buds and sides of abdomen blackish brown.

Biology and behavior The insect occurs 3 generations per year in Sichuan. It overwinters as adults. Egg stage is 4-6 days. Larval stage is 10 days. Prepupal stage is 1-3 days. Pupal stage is 3-6 days. Adults live for 2-3 months. Larvae molt three times. Adults feed on leaves for nutrient and consume more than larvae. They fake death when disturbed. Each female lays 49-601 eggs.

Prevention and treatment

1. Conserve natural enemies such as lady beetle (*Aiolocaria mirabilis* Motschulsky) and parasitoids.

2. Chemical control. Spray 25 g/L lamda-cyhalothrin or 40% thiamethoxam at 800-1,200×, 45% profenofos and phoxim mixture at 1,000×, or 12% thiamethoxam and lamda-cyhalothrin mixture at 800-1,200× during larval period.

References Xiao Gangrou (ed.) 1992.

(Li Menglou, Wu Cibin)

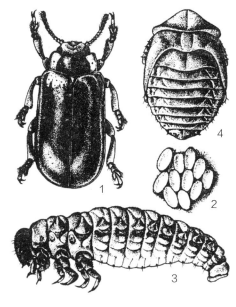

P. adamsi. 1. Adult, 2. Eggs, 3. Larva, 4. Pupa (Zhu Xingcai)

P. adamsi adult (Li Zhenyu)

P. adamsi adult and larva (Li Menglou)

187 *Chrysomela populi* L.

Classification Coleoptera; Chrysomelidae
Common name Poplar leaf beetle

Distribution Beijing, Hebei, Shanxi, Inner Mongolia, Liaoning, Jilin, Heilongjiang, Shandong, Henan, Hubei, Hunan, Sichuan, Guizhou, Shaanxi, Qinghai, Ningxia, Xinjiang; Japan, Korea Peninsula, Russia (Siberia), India, Europe, North America.

Hosts *Populus canadensis*, *P. nigra* var. *italica*, *P. euphratica*, *P. tomentosa*, *P. dakauensis*, *P. canadensis* Moench 'Robusta', *P. laurifolia*, *Salix* spp.

Morphological characteristics **Adults** Female length 12-15 mm, width 8-9 mm; male length 10-11mm, width 6-7 mm. Oval, bluish black, with metallic luster. Antennae short, 11-segmented. Pronotum bluish purple, metallic lustered, with a longitudinal groove on each side, the space between the grooves smooth, with coarse punctures on both sides. Scutellum bluish black, triangular. Elytra orange-red or orange-brown, wider than the prothorax, densely covered with punctures, and with longitudinal ridges along the outer edge. **Eggs** Length 2 mm, width 0.8 mm, oval, newly laid eggs are yellowish, turning to orange or yellow brown before hatching. **Larvae** Mature larvae 15-17 mm in length, near oval; head black; thorax and abdomen dull white; the dorsal surface with black spots on the two sides. There is one pair of large black tubercles on mesothorax and metathorax, and one pair of smaller tubercles on each abdominal segment. When disturbed, these tubercles release milky white odorous liquid. **Pupae** Length 9 mm, gray white, becoming orange when emerging.

Biology and behavior *Chrysomela populi* occurs two generations a year in most areas. It occurs only one generation per year in Bashang of Hebei and Baicheng of Jilin. They overwinter under the leaf litter or in shallow soil. In late April to May next year (Shaanxi), early April (the central and southern Hebei) or late April (Liaoning), adults become active, mate, and lay eggs. Eggs incubation period is 7- 9 days. Larvae have 4 instars; the duration is 12-19 days. Pupal period is 5-8 days. Adult lifespan is 25-30 days. Adults climb down trees to overwinter after mid-September.

Adults begin to lay eggs in late April in Liaoning. Eggs are laid on leaves, arranged in masses. Each egg mass contains a few eggs to 74 eggs. Each female lays up to 695 eggs. The hatch rate is up to 100%. Larvae congregate on the eggshell after hatching and feed on the egg shells. After that, they congregate and feed on leaves. They feed mostly during the night. Mature larvae

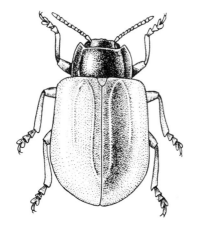

C. populi adult (Zhang Peiyi)

C. populi adult (Xu Gongtian)

C. populi mating pair (Xu Gongtian)

C. populi eggs (Xu Gongtian)

C. populi 1st instar larvae (Xu Gongtian)

C. populi late stage larva (Xu Gongtian)

C. populi larvae (Xu Gongtian)

C. populi pupa (Xu Gongtian)

C. populi damage (Xu Gongtian)

go through 3 to 5 days before entering pre-pupae period. The larvae will stick their end on back of the leaves and begin body contraction when pupating. When the daily average temperature exceeds 25°C. The newly emerged adults will hide in the leaves, under grass, on surface of loose soil and become dormant. They resume feeding in late August. They will play death when disturbed.

Prevention and treatment

1. Plough, irrigate, and clear the litter in winter to destroy the adult overwintering habitat.

2. Shake trunks and kill adults after they fall to the ground.

3. Conservation of natural enemies such as mantis and parasitic wasps.

4. Spray 50% fenitrothion EC at 1,200×, 25 g/L lamda-cyhalothrin or 50% malathion at 1,000×, or 45% profenofos and phoxim mixture at 1,000×, or 12% thiamethoxam and lamda-cyhalothrin mixture at 800-1,200× to kill larvae and adults.

References Chen Yu 1958, Forest Research Academy of Hebei 1959, Li Yajie 1978, Wen Shouyi and Xu Longjiang 1987, Xiao Gangrou (ed.) 1992.

(Wen Shouyi; Zhang Shiquan)

188 *Gastrolina depressa* Baly

Classification Coleoptera, Chrysomelidae
Synonym *Gastrolina depressa depressa* Baly

Importance *Gastrolina depressa* lowers the aesthetic value of the host. Consecutive infestations reduce tree vigor and kill some of the large trees.

Distribution Jiangsu, Zhejiang, Anhui, Fujian, Shandong, Henan, Hubei, Hunan, Guangdong, Guangxi, Sichuan, Guizhou, Shaanxi, Gansu; Korea Peninsula, Japan, Russia.

Hosts *Juglans regia*, *Pterocarya stenoptera*.

Damage Both adults and larvae feed on leaves and may cause complete defoliation.

Morphological characteristics **Adults** Body flat, length 5-7 mm. Antennae short, not reaching the base of elytra, the 3rd segment slender. Base of the pronotum narrower than the elytra, width about 2 times as the length, basal margin with border, front margin deeply concave. Elytra with dense coarse punctures; humerus raised. Front coxa cavity open. Mesosternum surpass front coxae. Legs black. Antennae 6-11 segments filiform. Clypeus not forming a concave area. Base and tip of seminal receptacle wide.

Eggs Length 1.0 mm, width 0.5 mm, long elliptical, yellow, turning black before hatching.

Larvae First instar 1.5-3.0 mm long, 0.7-1.0 mm wide; mature larvae 8-10 mm long, 1.5-2.0 mm wide. Young larvae light-yellow, turning to black later. Dorsal surface with verrucae on the side.

Pupae Length 5.9 mm, width 3.0 mm; milky yellow at beginning, then turning to greyish black; with verrucae.

Biology and behavior *Gastrolina depressa* occurs 2 generations a year in Jiangsu. It overwinters as adults in cracks and crevices on bark, in soil, stone, or

G. depressa adult (lateral view) (Xu Gongtian)

G. depressa adult (dorsal view) (Li Zhenyu)

G. depressa adult (dorsal view) (Xu Gongtian)

G. depressa adult (ventral view) (Li Zhenyu)

G. depressa larvae (Xu Gongtian)

G. depressa empty pupae (Xu Gongtian)

G. depressa damage (Xu Gongtian)

among the fallen leaves. Adults resume activity in early April when *Pterocarya stenoptera* buds develop. After a brief feeding period, they mate and lay eggs on back of leaves. Eggs are laid in masses, each with 10-15 eggs. Egg period is 5-7 days. Mature larvae pupate on back of leaves. Adults emerge in June. Under high temperature, adults aggregate in tree cracks or on trunks and stay dormant. Adults mate in early August and lay eggs until late August. They cause the 2nd damage in the Fall season. Adults overwinter in October.

In Anhui, the insect occurs 4 generations a year and overwinters as adults. In Shandong (Taian), it occurs 2-3 generations a year. Adults resume activity in early April. Egg period is 49.5 days under 20°C. The 1st generation larvae reach peak in May. Larvae go through 3 instars. The 1st instar are gregarious. After 2-3 days feeding at 25°C, they scatter. After another 2-3 days, they molt. Mature larvae feed for 3-4 days. Then they enter prepupal period. After 1-2 days, they become pupae. Adults emerge after another 2-3 days. After a few days feeding, they mate and lay eggs. Adults spend an average of 8.8 days between emergence and laying eggs. A generation requires about 24 days to finish development. The 2nd generation larvae occur in late May. In mid-June, adults become dormant in soil or in fallen leaves. In late July the insect causes the 2nd damage. Adults lay eggs. Peak damage occurs in late August. The 2nd or 3rd generation adults overwinter in late September.

Prevention and treatment

1. Mechanical control. Scrape off curled barks. Collect and burn twigs, fallen barks, dead branches, and leaves to kill adults. Remove infested leaves along with eggs or 1st instar larvae.

2. Conservation of natural enemies. The following predators are helpful to suppress the *G. depressa* populations: *Aiolocaria mirabilis*, *Propylaea japonica*, *Harmonia axyridis*, *Calvia quindecimguttata*, *Aiolocaria nexaspilota*, *Picromerus lewisi*, *Arma chinensis* and Carabidae. Among them, lady bugs may kill significant numbers of *G. depressa* eggs, larvae, pupae, and adults.

3. Chemical control. Shake trees or beat branches to cause adults fall to the ground, then spray the ground with a liquid insecticide. Fogging with 1.2% nicotin and matrine mixture. Spray 8% cypermethrin micro-encapsulated suspension at 200×, 3% fenoxycarb at 3,000×, 2.5% deltamethrin at 8,000-10,000×, or 25% Sevin WP at 200-1,000×, or 45% profenofos and phoxim mixture at 1,000×, or 12% thiamethoxam and lamda-cyhalothrin mixture at 800-1,200×.

References Yu Fangbei 1988; Yu Peiyu and Wang Shuyong 1996; Zhong Xiulin and Fan Li 2001; Ge Siqin, Yang Xingke, Wang Shuyong et al. 2003; Meng Qingying 2007.

(Li Zhenyu, Pan Yanping, Zhang Yan)

189 *Gastrolina thoracica* Baly

Classification Coleoptera, Chrysomelidae
Synonym *Gastrolina depressa thoracica* Baly

Importance Infested walnuts lose yield. *Juglans mandshurica* trees may die from heavy infestations.

Distribution Beijing, Hebei, Liaoning, Jilin, Heilongjiang, Hubei, Sichuan, Gansu; Korea Peninsula, Japan, Russia.

Hosts Walnut, *Juglans mandshurica*.

Morphological characteristics **Adults** Length 6.5-8.3 mm. Median part of pronotum black, sides brownish yellow or brown; distal end of seminal receptacle narrow and long, its base slender; other characteristics similar to *Gastrolina depressa*.

Eggs Length 1.5-2.0 mm. Elongate oval, orange yellow, the end slightly pointed.

Larvae and **Pupae** Similar to *Gastrolina depressa*.

Biology and behavior *Gastrolina thoracica* occurs one generation per year in Liaoning province. It overwinters as adults among dead leaves or under the bark. Adults became active in late April. They feed on young leaves to obtain nutrition, mate and lay eggs. Females mate and lay eggs multiple times. Each female lays 90-120 eggs and up to 167 eggs. Eggs are mostly laid on underside of leaves. Newly emerged adults feed in early morning and evening. They stay dormant in late June and resume activity in late August. Adults are weak flyers. They fake death when disturbed. They are not attracted to lights. Adults live for 320-350 days. Males and females have approximately 1:1 ratio. Newly hatched larvae are gregarious and feed on leaf tissue. From 3rd instar, they disperse and consume much more than early instars. Mature larvae often stay on back of leaves.

Prevention and treatment
Similar to that for *Gastrolina depressa*.

References Wang Weiyi et al. 1988, Xiao Gangrou (ed.) 1992, Xu Shuiwei et al. 2004.

(Li Zhenyu, Yu Siyu, Pan Yanping)

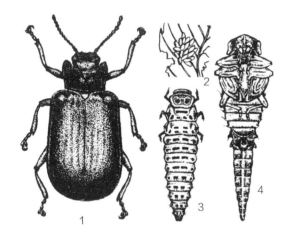

G. thoracica. 1. Adult, 2. Eggs, 3. Larva, 4. Pupa (Yu Enyu)

G. thoracica adult and eggs (Li Kai)

190 *Gastrolina pallipes* Chen
Classification   Coleoptera, Chrysomelidae

Importance   Infested walnut trees lose yield.
Distribution   Yunnan (Yangbi, Weixi, Yunlong).
Host   Walnut.

Morphological characteristics   **Adults** Length 5.3-7.1 mm, width 2.8-3.9 mm. Legs pale brownish yellow; distal end of femur, tibia, and tarsi black; antennae 6-11 segmented, near pectinate; clypeus area forming a triangular depression. Other characteristics similar to *Gastrolina depressa*.

Eggs Length 1.3-1.5 mm, width 0.6 mm. Creamy white at beginning, turning to yellowish later; its top shiny and transparent.

Larvae The 1st to 3rd instars are 1.3-3.2, 3.4-5.8, and 9.0-10.5 mm long, respectively. Head and legs black, ventral side and end of abdomen yellowish white. Dorsal surface of 1st instar gray or dark gray; mature larvae dark brown or grayish brown. Each segment with black pinacula.

Pupae Length 5-6 mm, width 3-4 mm. Dark brown; the back with symmetrical marks.

Biology and behavior   *Gastrolina pallipes* occurs one generation a year. It overwinters as adults under bark. Adults start to feed on walnut young leaves in early and mid-March. Then they mate and produce eggs. Eggs period lasts 20-30 days. Each female lays 383-430 eggs. They are laid on back of leaves as egg masses with each egg mass containing 25-45 eggs or up to 92 eggs. After 5-7 days, eggs hatch. Young larvae feed on leaf tissue on back of leaves. From the 3rd instar, they start to consume whole leaves. Larval stage lasts 19 days. Mature larvae turn to pupate after 1-3 days preputal period. They hang themselves upside down on back of leaves or branches. Pupal period is 5-7 days. Newly emerged adults light-yellow, then turn to copper or bluish colour. After 1-2 days, adults feed and then overwinter. Adult period lasts 350-380 days. They prefer dry environment. Low temperature in winter can cause high adult mortality.

Prevention and treatment
Similar to that for *Gastrolina depressa*.

References   Yang Yuan et al. 1982, Ge Siqin et al. 2003.

(Li Zhenyu, Pan Yanping)

G. *pallipes* adult (dorsal view) (Li Zhenyu)

G. *pallipes* adult (ventral view) (Li Zhenyu)

191 *Oides leucomelaena* Weise

Classification Coleoptera, Chrysomelidae

Distribution Zhejiang, Anhui, Fujian, Hubei, Guangdong, Guangxi, Yunnan.

Hosts *Illicium verum*, *Schisandra* spp.

Damage Adults and larvae feed on leaves and buds. They reduce fruit yield, kill branches or even the trees.

Morphological characteristics **Adults** Female length 12-16 mm, width 8-10 mm. Male length 10-13 mm, width 6-8 mm. Body elliptical, yellowish brown with black marks, metallic. Vertex with a pair of black marks. Eyes black, elliptical, protruding. Antennae filiform, 11-semented, the 1st-7th segment light-yellow, the last four segments black. The last segment of the labial palps and maxillary palps brown. Pronotum with one pair of large and one pair of small black marks. Mesoscutellum triangular, dark reddish brown. Elytra with 10 or 12 large black marks, arranged in three transverse rows. The first two rows with four marks each, the middle four marks are larger. The 3rd row has two or four marks.

Eggs Length 0.8 mm, width 0.4 mm, elliptical, light-yellow. Each egg mass contains dozens of eggs. The egg mass is covered with sticky material, hard, 1 cm wide, firmly attached to branches or base of petioles, greyish brown, similar to the bark of the *Illicium verum*.

Larvae Mature larvae 10-18 mm long, 4-8 mm wide, flat and cylindrical, shiny with bluish black marks. Head bluish black, labrum and clypeus light-yellow. Mesothorax, metathorax, and abdominal 1st-8th segments each with one short spine along the spiracle line. Prothorax and abdominal 1st to 8th segments each with one long spine below the spiracle line. Dorsal surface of the mesothorax, hind thorax, and abdominal 1st to 8th segment with a transverse stripe, which divides each segment into two sub-segments. Each sub-segment with a pair of bluish black marks arranged transversely. There is a round bluish black mark along the subdorsal line of each sub-segment. In addition, there is a small mark between the above marks on each of the 2nd sub-segment of each abdominal segment. Spiracle rings bluish black. The last

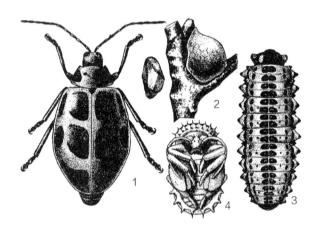

O. leucomelaena. 1. Adult, 2. Egg mass, 3. Larva, 4. Pupa (Zhang Peiyi)

O. leucomelaena adult (Wang Jijian)

O. leucomelaena adult and feeding marks (Wang Jijian)

O. leucomelaena damage (Wang Jijian)

two segments of the thoracic legs bluish black.

Pupae Length 7-12 mm, light-yellow. Frons with two pairs of spines. Pronotum with three rows of spines, the front row with 10 spines, the middle row with two spines, the last row with 8 spines. Dorsal surface of the mesothorax, metathorax, and abdomen with four spines each. There is a large spine under the spiracle line of the abdominal 1st-8th segments. The end of the abdomen with one pair of spines. Pupal chamber elliptical, 20 by 15 mm (length by width), smooth inside, brittle.

Biology and behavior *Oides leucomelaena* occurs one generation a year. It overwinters as eggs. They hatch between late February and late March. Larval period is between March and May, very few larvae exist in June. Pupation occurs from late April to June. Adult eclosion period is from mid-May to July. Eggs are laid between June and August and are rarely laid in September. Egg incubation period is between 245-254 days, larval development period is 27-59 days, pre-pupal period is 10-12 days, and pupal period is 8-16 days. Adult life span is 42-82 days.

Egg hatch rate is as high as 98%. High mortality of the first instars may occur under cold and rainy weather conditions. Larvae molt three times. Young larvae congregate on young shoots and feed on buds. They scatter as they grow. Most damage is caused by the 3rd instar. Feeding mostly occurs during dusk and night. They rest during the day. Once disturbed, they curl up and play death. Mature larvae migrate to the ground around the roots or under the tree canopy. They burrow into 1-2 cm depth and create pupal chambers and pupate. They usually pupate where soil is study. Some pupal chambers are partially exposed.

Adults eclose in the early morning or during dusk. After 2-3 hours, they can fly to trees. They are active after sunrise, during dusk, or after rain. After some period of feeding, they mate. Mating mostly occurs between 7:00-9:00 and 17:00 and 20:00. Adult feeding damage is most serious at 10-15 days prior to egg-laying. Most eggs are laid between 20:00-22:00. Eggs are laid in masses mixed with sticky secretion. The secretion allows eggs to form very hard egg sacs. Each egg mass contains 20-30 eggs. Each female lays 20-60 eggs. Adults are weak flyers. They drop to ground when disturbed. When picked up by hands, the adults secrete yellow liquid material.

Disease epidemics by *Beauveria bassiana* often occurs at high temperature and humidity conditions. Larvae, pupae, and adults all can be infected. Average mortality from *B. bassiana* is 35%, as high as 90% was recorded. Reduviid bugs and ants sometimes predate upon larvae and adults.

References Pingxiang Science and Technology Bureau 1976, Huang Jinyi et al. 1986, Zhang Peikun 1989, Wu Xianxiang 2001.

(Qin Zepo)

192 *Ophrida xanthospilota* (Baly)
Classification　Coleoptera, Chrysomelidae

Distribution　Beijing, Hebei, Shandong, Sichuan.
Host　*Cotinus coggygria*.
Damage　*Ophrida xanthospilota* larvae feed on leaves. Adult feeding creates holes and notches on leaves. Heavy infestations often cause death of branches and affect the growth and aesthetic value of the trees.

Morphological characteristics　**Adults** Female length 7.5-8.5 mm, male length 5.8-7.1 mm, long elliptical, yellowish brown, slightly shiny. End of the abdomen with a round depression. Eyes black. There are two longitudinal grooves between the eyes. Pronotum wide; its anterior border emarginated, forming an arc; its posterior margin slightly convex; the two sides with depressions. Elytra with rows of punctures and yellowish white marks, each elytron has about 70 yellowish white marks. The hind femora thick, the end of the hind tibiae emarginated and with dense stiff setae.

Larvae Mature larvae length 8-13 mm, body with transparent colorless sticky film-like material. Pronotum with a dark reddish-brown mark; the center of the mark with one white fine stripe. The meso- and metanotum with a new moon shaped brown mark. Larvae often carry black feces on their back.

Biology and behavior　*Ophrida xanthospilota* occurs one generation a year in Taishan (Shandong

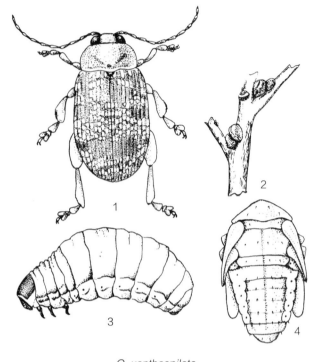

O. xanthospilota.
1. Adult, 2. Egg masses, 3. Larva, 4. Pupa (Hu Xingping)

O. xanthospilota adult (dorsal view) (Xu Gongtian)

O. xanthospilota adult (lateral view) (Xu Gongtian)

O. xanthospilota egg masses (Xu Gongtian)

O. xanthospilota late stage larva (Xu Gongtian)

O. xanthospilota mature larvae (Xu Gongtian)

O. xanthospilota damage to leaves (Xu Gongtian)

province). It overwinters as eggs on branches of the host tree. Larvae hatch in late March when new leaves of *C. coggygria* emerge. Most of the leaves may be consumed in mid-April and only petioles are left. Mature larvae migrate to the ground, make cocoons in 1-3 cm deep soil and pupate in late April. Peak pupation period is in mid-May. Duration of the pupal period is 23-26 days. Adults emerge in late May. After 6-8 days of feeding, they mate and lay eggs. Adults live for more than two months. Egg laying activity lasts until October. The emergence time in Beijing is 10-20 days later than that in Taishan (Shandong province). The insect feeds during day and night. Fecal materials can be seen on the infested leaves.

When all the *C. coggygria* leaves are consumed, *O. xanthospilota* will feed on other trees such as *Robinla pseudoacacia, Ailanthus altissima, Populus tomentosa, Ulmus pumila*, etc. However, the insect cannot survive well on these trees.

It was observed that 60%-65% of the larvae successfully pupated, and 52% of the pupae developed into adults. Adults lay egg masses at the branch collar, base of the petiole, or besides the terminal buds. They cover the eggs with excretion to protect the eggs. Each egg mass contains 12-125 eggs. Each female lays 95-210 eggs.

Prevention and treatment

1. Silviculture. Plant mixed forests to deter the spread of *O. xanthospilota* and reduce pest populations.

2. Conservation of natural enemies. Egg parasitoids include *Trichogramma* sp. and Encyrtidae. *Arma chinensis* can kill 2-4 larvae per day. Ants can feed on *O. xanthospilosa* pupae. Adult natural enemies include *A. chinensis* and an insect in Reduviidae.

References   Zhao Suihua 1985, Bai Jintao 1990.

(Bai Jintao, Chen Chao, Zhao Huaiqian)

193 *Parnops glasunovi* Jacobson

Classification Coleoptera, Chrysomelidae

Importance *Parnops glasunovi* has been increasingly an important pest where *Populus* trees are concentrated. Heavy infestations reduce tree vigor, cause leaf loss or complete defoliation, and reduce survival rate of seedlings especially when spring is dry.

Distribution Beijing, Hebei, Shanxi, Inner Mongolia, Liaoning, Jilin, Heilongjiang, Shandong, Henan, Shaanxi, Gansu, Xinjiang.

Hosts *Populus* spp. and *Salix* spp.

Damage *Parnops glasunovi* adults feed on young shoots and leave petioles. Their feeding causes leaf fall. Complete defoliation may occur when pest density is high.

Morphological characteristics **Adults** Length 6-7.5 mm, elongated elliptical. Head, pronotum and elytra blackish brown; with dense yellow or yellowish green pubescence; the ventral surface with greyish white pubescence.

Eggs Long elliptical, milky white, turning to milky yellow later.

Larvae Mature larvae 7 mm in length, milky yellow, body slightly curved ventrally.

Pupae Length 5 mm, spindle shaped, milky white.

Biology and behavior *Parnops glasunovi* occurs one generation a year. It overwinters as mature larvae. The larvae pupate in April next year. Adults emerge in May. Peak adult activity is during mid-May to early June. Adults lay eggs until mid-August. They feed on petioles and young shoots. They chew off 2/3 of the petiole. The affected leaves fall after 1-2 days. Sometimes the petioles are broken off directly by adult feeding. Heavy infestation may cause complete defoliation. Adults also feed on leaves and create holes or notches. Adults are active during the day and are most active before dark. They play death when disturbed. This behavior is more distinct before 6:00 when temperature is low. After 7:00, disturbed adults will fly away during falling. Eggs are laid in masses between leaves, in cracks of tree bark, weeds, and cracks in soil. Larvae burrow into soil and feed on roots of *Populus* spp., *Salix* spp., or weeds.

Prevention and treatment

1. Till the soil and remove weeds before early April to

P. glasunovi adult (Xu Gongtian)

P. glasunovi adult (Xu Gongtian)

P. glasunovi adult (Xu Gongtian)

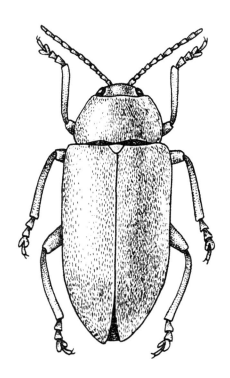

P. glasunovi adult (Zhang Peiyi)

kill pupae.

2. Shake trees and collect adults.

3. Chemical control. Spray 0.3% avermectin + *Bacillus thuringiensis* (3×10^9 spores/g) at 1,000-2,000×, 10% imidacloprid at 1,000-2,000×, 1.8% avermectin EC at 2,000-2,500×, or 3% acetamiprid EC at 1,000-1,500×, or 12% thiamethoxam and lamda-cyhalothrin mixture at 800-1,200×.

References Zhao Fanggui et al. 1999, Sun Xugen et al. 2001.

(Wang Haiming, Qin Xixiang)

194. *Plagiodera versicolora* (Laicharting)

Classification Coleoptera, Chrysomelidae

Importance *Plagiodera versicolora* is a major pest of *Salix* spp. It affects the aesthetic value of the trees and severely slows down the growth of 1-2 year-old seedlings.

Distribution Beijing, Hebei, Inner Mongolia, Liaoning, Jilin, Heilongjiang, Jiangsu, Zhejiang, Anhui, Jiangxi, Shandong, Henan, Hubei, Sichuan, Guizhou, Yunnan, Gansu, Ningxia, Taiwan; Japan, Korea Peninsula, Russia, U.S.A., Canada and Europe.

Hosts *Populus* sp., *Zea mays*, *Glycine max*, *Gossypium* sp., *Morus* sp., and *Salix* spp.

Damage *Plagiodera versicolora* larvae feed on leaves and cause netted appearance.

Morphological characteristics **Adults** Length 3-5 mm, body dark blue, strong metallic. Head wide, its 1st-6th antennal segments small, brown; the 7th-11th segments large, dark brown, with fine pubescence. Eyes blackish brown. Pronotum smooth, wide, the anterior margin emarginate, arc-like. Elytra with rows of punctures. Ventral surface and legs darker in color.

Eggs Length 0.8 mm, elliptical, orange yellow.

Larvae Length 6 mm, flat, greyish yellow, head blackish brown. Pronotum with two large brown marks off the median line. The edge of the meso- and metanotum with large blackish brown verrucae. There are two black spots along the subdorsal line. Each of the 1st-7th segments with a small black verruca along the spiracular line. A verruca is also located along the sub-spiracular line. Two setae are found on each verruca. Ventral surface of each of the abdominal segments has six black spots. One or two setae are located on each black spot. End of the abdomen with a yellow sucking disk.

Pupae Length 4 mm, elliptical, dorsal surface with four rows of black spots.

Biology and behavior *Plagiodera versicolora* occurs six generations a year in Beijing, three generations a year in northeast and Inner Mongolia, 3-4 generations a year in Ningxia, 7-8 generations a year in Shandong, and five generations a year in Hunan. The insect overwinters as adults in fallen leaves, weeds, or in soil. The adults resume activity in spring when buds of *Salix* spp. start to grow. They lay eggs on leaves. Each female lays 1,000-1,500 eggs. Egg incubation period is 7 days. Larvae congregate and feed on mesophyll. The infestation site becomes greyish white, net-like. Larvae go through 4 instars and mature after 5-10 days. They fix them to leaf surface with their ends of the abdomen. Pupal period is 3-5 days. Adults and larvae can be seen from spring to fall. Adults play death when disturbed.

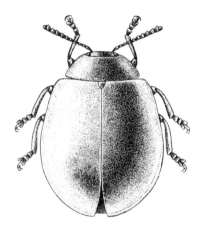

P. versicolora adult (Zhang Peiyi)

P. versicolora adult (Xu Gongtian)

P. versicolora mating pair (Xu Gongtian)

P. versicolora eggs and 1st instar larvae (Xu Gongtian)

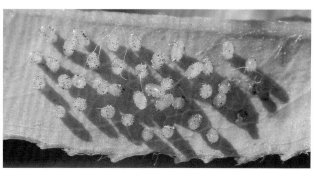
P. versicolora eggs (Xu Gongtian)

P. versicolora larvae (Xu Gongtian)

P. versicolora mature larva (Xu Gongtian)

P. versicolora damage (Xu Gongtian)

Prevention and treatment

1. Silviculture. Remove fallen leaves and weeds. Till the soil.

2. Conserve natural enemies.

3. Mechanical control. Shake trees and collect adults.

4. Chemical control. Spray 1.2% nicotine and matrine mixture at 1,000× or 1% imidacloprid WP at 2,000×, or 45% profenofos and phoxim mixture at 1,000×, or 12% thiamethoxam and lamda-cyhalothrin mixture at 800-1,200×.

References Xia Zhonghui 1990, Xiao Gangrou (ed.) 1992, Yang Zhende et al. 2006, Zhu Yi et al. 2006.

(Chen Chao)

195 Podagricomela shirahatai (Chûjô)

Classification Coleoptera, Chrysomelidae
Synonyms *Clitea shirahatai* (Chûjô), *Podagricomela shirahatai* Chen

Distribution Shanxi, Sichuan, Shaanxi, Gansu.
Host *Zanthoxylum bungeanum*.
Damage Larvae feed inside leaves. Adults feed on young leaves. Infested leaves turn yellow or black and fall. Fruit quality is reduced.

Morphological characteristics **Adults** Elliptical, length 4-5 mm, reddish brown, dull; head, antennae, eyes, and legs black. The grooves on the head complete. Front border of the labrum emarginated. Antennae reach the base of the hind legs. Pronotum with small and dense punctures. Elytra with 11 rows of punctures. The front and mid-femora with sparse pubescence, punctures absent. Hind femora 1.5× as wide as the mid-femora; their posterior part with punctures. Hind tibiae and tarsi with dense pubescence; claws single-toothed.
Eggs Length 0.8-1.0 mm, flat oval, yellowish white. Egg mass greenish brown, turning to blackish brown later. Egg mass is covered with fecal materials and scale-like.
Larvae Mature larvae 5-8 mm long, head and legs black, femora and tibiae, ventral surface light-yellow, pronotum and pygidium with one brown mark.
Pupae Length 4-5 mm, light-yellow, head blackish brown, setae black.

Biology and behavior *Podagricomela shirahatai* occurs two generations a year. It overwinters as adults in soil. They resume activity in early April and feed on the young *Z. bungeanum* leaves. Adults lay eggs from late May to late June. Egg incubation period is 4-7 days. Larvae feed for 14-19 days inside leaves before falling to the ground and pupate in soil in late June. Pupal period is 24-31 days. The 1st generation adults emerge from mid-July to early August. They feed for 8-15 days and

P. shirahatai adult (Zhu Xingcai)

then mate and lay eggs. The 2nd generation adults emerge in late September and overwinter in October.

Adults are good jumpers and fly fast. They feed during the day and hide during the night. Each female produces 2-3 egg masses, each with about 14 eggs. Larvae congregate during the first 2-3 days after hatching. Infested leaves first show translucent patches. When the infested leaves become yellow, the larvae migrate to new leaves. Each leaf is often infested by more than three larvae. Blackish brown feces can be seen expelled from the infested leaves. Larvae go through 4 instars. When their body color turns from white to yellow, they exit from the leaves and pupae in soil. Damage from the larval feeding is similar to fire burn in late June.

Prevention and treatment
1. Silviculture. Remove egg masses and infested leaves during May-September. Remove leaves and weeds on the ground. Till and irrigate in winter.
2. Chemical control. Spray 2.5% deltamethrin or beta-cyfluthrin at 2,000×, or 45% profenofos and phoxim mixture at 1,000×, or 12% thiamethoxam and lamda-cyhalothrin mixture at 800-1,200× in late April in heavily infested forests. Apply insecticide bands on tree trunks during late March and early April. The insecticide mixture consists of 5% imidacloprid, 80% DDVP, and diesel at 8:2:10.

References Yang Yunhan 1986; Ji Weirong, Zhang Liyan, Zhang Yawen 1995; Du Pin, Ren Fang, Mei Liru et al. 1999; Qi Xinhua 2000; Li Menglou 2002; Zhang Bingyan 2006.

P. shirahatai adult (Li Zhenyu)

(Wang Hongzhe, Li Menglou, Dang Xinde)

196 *Podontia lutea* (Oliver)
Classification Coleoptera, Chrysomelidae

Distribution Zhejiang, Anhui, Fujian, Jiangxi, Hubei, Hunan, Guangdong, Guangxi, Hainan, Sichuan, Guizhou, Yunnan, Shaanxi, Hong Kong, Taiwan; Vietnam, Myanmar, Southeast Asia.

Hosts *Toxicodendron verniciflum, T. succedaneum,* and *Pistacia chinensis.*

Damage Adults and larvae feed on buds and leaves. Heavy infestations result in complete defoliation. Consecutive infestations affect growth and production of lacquer.

Morphological characteristics **Adults** Length 14-16 mm, body nearly elliptical, brownish yellow, shiny. Head hidden under the prothorax. Antennae 12-segmented, its first 2-3 segments yellow, the other segments black or blackish brown, with short setae. Eyes black. Elytra with 10 longitudinal rows of punctures. Femora dark yellow; tibiae, tarsi, and claws black. The first tarsal segment of the front and middle tarsi triangular, tarsi inconspicuously 5-segmented. The posterior border of the ventral plate of the last abdominal segment of the females with deep narrow depression at the two sides. The depression is absent in the male.

Eggs Length 1.5-3.0 mm, elliptical, pale grey or greyish white.

Larvae Stout. Dorsal surface with orderly arranged black spots. Base of legs white, other parts black. Young larvae with large heads, club shaped. Milky yellow, 1st instar yellowish green, 2nd instar yellow, 3rd instar yellow or golden yellow.

Pupae Length 10.5-20.0 mm, yellow, head and appendages darker in color. The end of the abdomen with a pair of black spines.

Biology and behavior *Podontia lutea* occurs one generation a year in soil or under rocks in Hubei. It overwinters as adults. The adults resume activity in April. They first feed on leaf tips and back of the leaves, then leaf edges. The peak egg-laying period is in early and mid-May, which is about 15-20 days later than *Ophrida scaphoides*. Egg incubation period is 15-19 days. Larvae appear from early May. Larvae hatch during 5:00-11:00 of the day; most of them hatch between 6:00-8:00. Larval period is 21-28 days. Peak larval period is during May and June. Pupae appear from early August, reach peak at mid- and late August, and end in early September. Mature larvae migrate into soil and pupate in elliptical pupal chambers. Pupal period is 27-31 days. Adult peak emergence period is early and mid-July.

When an adult emerges from the pupal chamber, it takes an hour from breaking the pupa to starting to fly to the trees. They rest on back of leaves. During 7:00-8:00, they can be seen feeding on leaf edges. Adults mate about two weeks after emergence. They mate multiple times and lay eggs next year after overwintering. Sometimes the adults that passed through the winter can still mate. Adults overwinter from early to mid-November in dead leaves, weeds, or cracks on the ground. They lay eggs on back of leaves. A small percentage of them lay eggs on leaf petioles. Each female lays 6-35 eggs a time and a total of 86-400 eggs (average 270 eggs).

Young larvae congregate at the tip and back of the leaves and feed on mesophyll. After 1-2 days, they scatter along leaf edges and feed on the edges. Their consumption increases from the 2nd instar. Each larva consumes 32-49 cm^2 leaf area. When disturbed, they will retrieve their heads and stop feeding. Larvae defecate yellow strips of fecal materials onto the back of their bodies. Adults play death when disturbed. They are weak

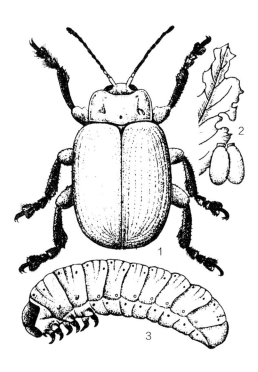

P. lutea. 1. Adult, 2. Eggs, 3. Larva (Hou Boxin)

P. lutea adult (Wang Jijian)

flyers. They disperse through flying or crawling.

Prevention and treatment

1. Silviculture. Deep ploughing between late June and early July to destroy pupal chambers. Clear the leaves and weeds on the forest floor in winter and burn them. Shake trees and collect adults. Collect and destroy egg masses.

2. Chemical control. Spray 4.5% cypermethrin at 1,000-1,500× or fenpropathrin at 1,000×, or 45% profenofos and phoxim mixture at 1,000×, or 12% thiamethoxam and lamda-cyhalothrin mixture at 800-1,200× to kill adults and larvae.

References Chinese Academy of Sciences - Institute of Zoology 1986, Xiao Gangrou (ed.) 1992, Huang Fusheng 2002, Xu Guangyu et al. 2008.

(Chen Jingyuan, Zuo Junjie, Bie Runzhi)

197 *Pyrrhalta aenescens* (Fairmaire)

Classification Coleoptera, Chrysomelidae
Synonym *Galleruca aenescens* Fairmaire

Importance It is a major pest of *Ulmus* spp. Heavy infestations cause complete defoliation. Adults are nuisance pests in homes.

Distribution Beijing, Hebei, Shanxi, Inner Mongolia, Jilin, Heilongjiang, Jiangsu, Anhui, Shandong, Henan, Hunan, Sichuan, Shaanxi, Gansu, Taiwan.

Hosts *Ulmus* spp., *Populus* spp.

Damage Adults and larvae feed on leaves and create holes on leaves or make leaves sieve-like.

Morphological characteristics **Adults** Length 7.5-9 mm, width 3.5-4 mm, body nearly rectangular. Orange yellow to yellowish brown; elytra green, metallic; vertex with one near triangular black spot. The 3rd segment of the antennae longer than the 2nd segment; the 3rd to 5th segments are equal in length. Pronotum twice as wide as long; there is a squash-like black mark with the middle portion deeply constricted. In addition, there is an elliptical black spot on each side of the squash-like mark. Elytra wider than pronotum, its two sides almost parallel; each elytron with irregular longitudinal lines of dense punctures. The posterior border of the ventral plate of the last abdominal segment of the males deeply emar-

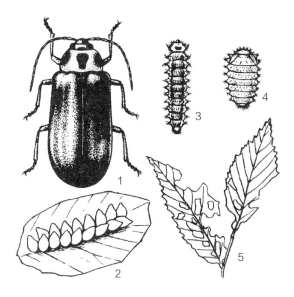

P. aenescens. 1. Adult, 2. Eggs, 3. Larva, 4. Pupa, 5. Damage (Shao Yuhua)

ginate; that of the females only with a small notch.

Eggs Yellow, long elliptical, tip sharp, slender. Length 1.1 mm, width 0.6 mm.

Larvae Mature larvae 11 mm, elongated, slightly flat, deep yellow, the mesothorax, metathorax, and 1st-8th abdominal segments pitch black. Head, thoracic legs, and abdomen with black verrucae. Head small, with sparse, white, long setae. Pronotum with one square black mark. The dorsal surface of the mesothorax and metathorax are divided into two small segments. The first small segment has 4 verrucae, each side has 3 verrucae. Each of the 1st-8th abdominal segment is also divided into two small segments. The first small segment has 4 verrucae, the second small segment has 6 verrucae; each side has 3 verrucae. Pygidium dark yellow, with sparse setae.

Pupae Length 7.5 mm, elliptical, dark brown. The two sides with distinct tubercles.

Biology and behavior The insect occurs 1-2 generations a year in Beijing. It overwinters as adults in cracks of buildings, among bricks, in soil, and among weeds, etc. They resume activity in early April when trees start to bud. Adults feed on buds and leaves. They lay eggs on back of leaves in late April. Each egg mass

P. aenescens adult (Xu Gongtian)

P. aenescens mating pair (Xu Gongtian)

P. aenescens eggs (newly laid) (Xu Gongtian)

P. aenescens eggs (a few days old) (Xu Gongtian)

P. aenescens 1st instar larvae (Xu Gongtian)

P. aenescens mature larva (Xu Gongtian)

P. aenescens mature larva (Xu Gongtian)

P. aenescens larval aggregation (Xu Gongtian) P. aenescens pupae and newly eclosed adult (Xu Gongtian)

P. aenescens damage by young larvae (Xu Gongtian) P. aenescens damage by mature larvae (Xu Gongtian)

has more than 10 eggs, arranged in two rows. Egg incubation period is 7-10 days. Larvae appear in early May and feed on leaves. Mature larvae congregate on scars of branches. The pupal period is 10-15 days. Adults emerge in early July. Large numbers may fly into homes and public areas. Those adults appeared early will lay eggs and produce the 2nd generation. The 2nd generation pupae and adults appear in late August and late September, respectively.

Prevention and treatment

1. Mechanical control. Collect congregating larvae and kill them.

2. Conservation of natural enemies. Coccinellidae and *Arma chinensis* are important predators of *P. aenescens*.

3. Spray 10% imidacloprid WP at 1,000×, or 45% profenofos and phoxim mixture at 1,000×, or 12% thiamethoxam and lamda-cyhalothrin mixture at 800-1,200× to kill larvae. Spray 25% fenoxycarb WP at 300× during adult period,

References Yu Peiyu and Wang Shuyong 1996, Fan Renjun 1999, Xu Gongtian 2007.

(Pan Yanping, Zhang Yan, Li Yajie)

198 *Byctiscus rugosus* (Gelber)

Classification	Coleoptera, Attelabidae
Synonym	*Byctiscus omissus* Voss
Common name	Poplar leaf roller weevil

Importance *Byctiscus rugosus* adults and larvae feed on *Populus davidiana* leaves and young branches. Their feeding causes death of branches and affects normal growth of the host.

Distribution Shaanxi, Gansu, Ningxia, Xinjiang; Mongolia, Russia.

Hosts *Populus davidiana* and other *Populus* spp.

Morphological characteristics **Adults** Length 4.5-7 mm. Green and mixed with shiny copper color; rostrum, femora, and tibiae copper color. Head of males almost equal in length as in width; vertex, gena, and occiput with transverse rugae. Frons with longitudinal rugae. Rostrum pointed to the front and downward, slightly curved, about twice as long as the head. Antennae black, situated at the middle of the rostrum, 11-segmented, with sparse hairs. Pronotum smooth, wider than long; its sides round; the anterior border constricted; with a shallow longitudinal groove; its ventral surface with two spines pointing forward. Scutellum wider than long. Elytra with dense, large punctures that arranged irregularly; humerus slightly raised, the two sides parallel, the posterior end round and narrower with fine punctures and greyish white and greyish brown pubescence. Pronotum of females narrow, its ventral surface lack spines.

Eggs Length 1 mm, wide oval, light-yellow, translucent.

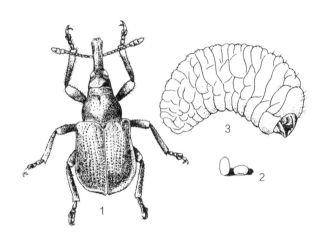

B. rugosus. 1. Adult, 2. Eggs, 3. Larva (Zhu Xingcai)

Larvae Mature larvae 7 mm in length. Milky white, head reddish brown. Body with short sparse pubescence.

Biology and behavior *Byctiscus rugosus* occurs one generation a year. It overwinters as adults in fallen leaves or in soil. The adults emerge in June next year. They feed and mate. Females lay eggs in rolled leaves. They lay 2-4 eggs in each rolled leaf. Larvae feed inside the rolls after hatching. After the leaf rolls dried fell on to the ground, the larvae burrow into soil and build pupal chambers. Adults emerge from pupae in early August and they will enter dormancy until next summer.

Prevention and treatment

1. Mechanical control. Shake trees and collect fallen adults. Remove rolled leaves and burn them to kill larvae.

2. Spray 80% DDVP at 800× or 2.5% deltamethrin EC at 1,500×, or 45% profenofos and phoxim mixture at 1,000×, or 12% thiamethoxam and lamda-cyhalothrin mixture at 800-1,200× when adults are active.

References Shaanxi Forest Research Institute 1984; Chen Yuanqing 1990; Xiao Gangrou (ed.) 1992; Yan Haike, Li Haiqiang, Zhang Yaozeng et al. 2007.

(Li Li, Li Kuansheng, Liu Mantang, Li Menglou)

B. rugosus adult (Ji Baozhong)

199 *Cyllorhynchites ursulus* (Roelofs)

Classification Coleoptera, Attelabidae
Synonyms *Mecorhis ursulus* (Roelofs), *Gyllorhynchites cumulatus* (Voss)

Distribution Hebei, Liaoning, Jilin, Jiangsu, Zhejiang, Anhui, Fujian, Jiangxi, Shandong, Henan, Hubei, Hunan, Sichuan; Japan, Russia.

Hosts *Castanea mollissima* and *C. seguinii*. *Quercus wutaishanica*, *Q. mongolica*, *Q. acutissima*, *Q. variabilis*, *Q. dentata*.

Damage *Cyllorhynchites ursulus* adults break fruit-bearing branches of *Castanea mollissima* and *C. seguinii*. They cause yield loss and future fruit production.

Morphological characteristics **Adults** Length 6.5-8.2 mm, width 3.2-3.8 mm. Bluish black, shiny, densely covered with greyish pubescence with scattered black long hairs. Ventral surface of the abdomen silvery grey. Each elytron with 10 rows of punctures. Rostrum slightly curved, equal length as the elytra. Male antennae located at the distal 1/3 of the rostrum. Female antennae located at the distal ½ of the rostrum. The prothorax with one sharp tooth on each side in the males, but not in the females.

Eggs Elliptical, milky white.

Larvae Milky white at beginning, turning to yellowish white when mature. Length 4.5-8.0 mm, sickle-shaped, mouthparts brown.

Pupae Pale yellow, rostrum pointed to the abdomen, the end of the abdomen with a pair of brown spines.

Biology and behavior

Cyllorhynchites ursulus occurs one generation a year. The larvae overwinter in chambers in the soil. They pupate in early to mid-May next year. Pupal period is about one month. Adults emerge from soil from early June until mid-August. After one week of supplemental feeding, the adults mate and lay eggs. They are active during 9:00-16:00. They play death when disturbed. Females lay eggs in involucres. Before laying eggs, a female breaks a branch at 3-7 cm away from the involucre, causing the broken branch dangling from the tree. It

C. *ursulus* adults (Fang Minggang)   C. *ursulus* adult (Fang Minggang)

C. ursulus egg (Fang Minggang)

C. ursulus damage (Fang Minggang)

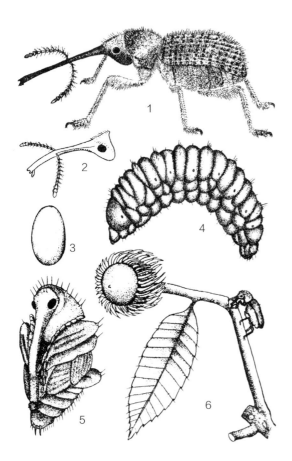

C. ursulus. 1. Adult, 2. Male adult head, 3. Egg, 4. Larva, 5. Pupa, 6. Damage (Xu Tiansen)

chews a dent on one involucre, lays an egg, and seals the egg with crumbs. After finishing laying eggs on all involucres of the broken branch, it chews off the dangling branch which will fall to the ground. Each female lays 25-35 eggs. In heavily infested stands, the ground may be covered with branches and involucres. The eggs hatch in mid-June. The larvae molt two times and mature in 20 days inside the nuts. They chew holes from the nuts and dig into soil at 3-20 cm depth and create oval soil chambers for overwintering. High precipitation and too high or too low moisture will affect the development and survival of larvae.

Prevention and treatment

1. Silviculture. Clean out the *C. seguinii* and other Fagaceae shrubs in *Castanea mollissima* orchards. Fertilize during the fall season to increase natural resistance of the trees.

2. Mechanical control. Collect and destroy the fallen involucres and branches.

3. Chemical control. Spray 4.5% cypermethrin at 2,000× or 20% fenvalerate EC at 2,500×, or 45% profenofos and phoxim mixture at 1,000×, or 12% thiamethoxam and lamda-cyhalothrin mixture at 800-1,200× during adult period (mid-June to mid-July).

References Luo Xizhen 1990, Lu Yingyi et al. 1992, Xiao Gangrou (ed.) 1992, Wang Hailin et al. 1995, Zhao Benzhong et al. 1997.

(Fang Minggang, Liu Zhenlu)

Cyrtotrachelus buquetii Guérin-Méneville

Classification Coleoptera, Dryophthoridae
Common name Bamboo weevil

Distribution Fujian, Guangdong, Guangxi, Sichuan, Guizhou.

Hosts *Bambusa surrecta*, *B. pervariabilis*, *B. blumeana*, *B. chungii*, *B. textilis* var. *gracilis*, *B. cerosissima*, *B. textilis*, *B. textilis* var. *glabra*, *B. eutuldoides*, *B. ventricosa*, *B. vulgaris* 'Wamin', *B. multiplex*, *B. textilis* var. *fasca*, *B. wenchouensis*, *B. tulda*, *B. remotiflora*, *B. textilis* var. *purpurascens*, *Dendrocalamopsis oldhami*, *D. beecheyana*, *D. beecheyana* var. *pubescens*, *D. daii*, *D. oldhami* f. *revolute*, *D. minor*, *D. asper*, *D. latiflorus*, *D. tomentosus*, *D. yunnanicus*, and *Dendrocalamus strictus*.

Damage Adults feed on bamboo shoots of 2 cm diameter or greater. Their feeding reduces quality of bamboos. Infested bamboos are deformed, with broken tips or branches.

Morphological characteristics **Adults** Females 25.5-36.8 mm in length, males 26.5-41.2 mm in length. Body orange yellow or blackish brown, sometimes the whole body black. Head semi-globular. Antennae geniculate, flagellum 7 segmented, the end segment shoe shaped. Female rostrum 9.5-15.5 mm in length, slightly smooth. Male rostrum 8.5-12.5 mm in length, its dorsal surface with a distinct groove, each side of the groove with 7-8 teeth. Pronotum convex; its posterior end has a large black marking; the top of the marking arrow shaped. Elytra yellow or blackish brown. Among the yellow individuals, the anterior border, the posterior border, and middle of the wings with varied sizes of black markings. Its outer margin round, anal angle with a 45° angle tooth. When the two wings are in closed position, the center at the posterior end with a vertical tooth. The female front femur equal or longer than tibia, the inner surface of the tibia with brown, sparse, short hairs. Males with long and large front legs; its femur shorter than tibia, Underside of the tibiae with dense, long, brown hairs.

Eggs Cylindrical, the two ends round; 4.0-5.2 mm long. Milky white, shiny, gradually turns to milky yellow. Surface smooth, without striae. The oviposition hole

C. buquetii adult (Xu Tiansen)

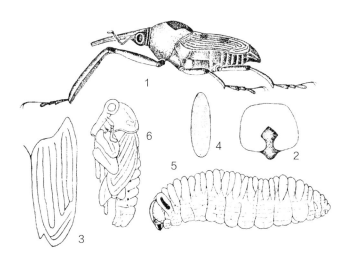

C. buquetii. 1. Adult, 2. Pronotum of adult, 3. Elytra, 4. Egg, 5. Larva, 6. Pupa (Xu Tiansen)

C. buquetii egg (Xu Tiansen)　　*C. buquetii* 4th instar larva (Xu Tiansen)　　*C. buquetii* mature larva (Xu Tiansen)

C. buquetii pupa and cocoon (Xu Tiansen)　　*C. buquetii* pupa and cocoon (Xu Tiansen)　　*C. buquetii* newly eclosed adult (Xu Tiansen)

is 34 mm long, 4 mm wide, with an obvious cavity on the surface of the bamboo shoot. Broken bamboo shoot fibers can be seen around the oviposition holes.

Larvae Young larvae 4.5-5.5 mm long, milky white, turns to milky yellow after feeding. Mature larvae 45-54 mm long. The gena and the area around the coronal suture blackish brown. Pronotum sclerotized, yellowish brown. The dorsal line of the thorax pale, distinct. The dorsal line in the mature larva is indistinct. The end of the abdomen spoon shaped.

Pupae Length 32-50 mm, orange yellow. Pronotum convex, its posterior border arched. Mesonotum upside down triangular shape, the tip covers the metathorax. The metathorax had two stripes arranged in an inversed "∧" shape.

Soil cocoons Long elliptical or kidney shaped. Outer size 55-70 mm, inner size 38-55 mm; the cocoon wall about 10 mm thick. The cocoons are made of grass, mud, and secretion from the larvae. They are hard, smooth inside, and coarse outside.

Biology and behavior *Cyrtotrachelus buqueti* occurs one generation a year in Guangdong and Guangxi. It overwinters as adults in pupal chambers in soil. In Guangdong, adults emerge in mid-June and reach peak in mid- and late August. They feed on bamboo shoots for a few days, then mate and lay eggs. Egg incubation period is 3-4 days. Young larvae move upward following a "Z" pattern as they feed in bamboo shoots. They feed for 12-16 days and may consume the top half of a bamboo shoot. Then they exit the bamboo shoots, drop to the ground, and burrow into soil. Larval occurrence period is from late June to mid-October. They pupate in 10 days after entering the soil. Pupae appear from mid-July to late October. After another 11-15 days, adults emerge from pupae. Adults appear from early August to early November. They will overwinter and lay eggs next year.

Prevention and treatment

Conservation of natural enemies. There is one egg parasitoid. Larvae can be killed by *Beauveria* sp.

(Xu Tiansen)

201 *Cyrtotrachelus thompsoni* Alonso-Zarazaga et Lyal

Classification Coleoptera, Dryophthoridae
Synonyms *Cyrtotrachelus longimanus* Gtyllenhaal, *Curculio longipes* F.

Distribution Fujian, Jiangxi, Hunan, Guangdong, Guangxi, Sichuan, Guizhou, Yunnan, Taiwan; India, Myanmar, Vietnam, Indonesia, Japan, the Philippines.

Hosts *Bambusa dissemulator* var. *hispida*, *B. ramispinosa*, *B. indigena*, *B. albolineata*, *B. gibboides*, *B. intermedia*, *B. lenta*, *B. textilis* var. *purpurascens*, *B. surrecta*, *B. textilis*, *B. chungii*, *B. textilis* var. *glabra*, *B. pervariabilis*, *B. blumeana*, *B. textilis* var. *gracilis*, *B. eutuldoides*, *B. ventricosa*, *B. vulgaris* 'Wamin', *B. multiplex*, *B. textilis* var. *fasca*, *B. wenchouensis*, *B. tulda*, *Dendrocalamopsis oldhami*, *D. beecheyana*, *D. daii*, *D. oldhami* f. *revolute*, *D. minor*, *D. endrocalamopsis asper*, *D. tomentosus*, *D. yunnanicus*, *Pleioblastus amarus*, *P. juxianensis*, *P. hsienchuensis*, *Dendrocalamus strictus*.

Damage Adults feed on smaller bamboos (1-2 cm diameter) or bamboo shoots. They make egg cavities in bamboo shoots, affecting their growth and lower their quality. Larval feeding of bamboo shoots will fill up the galleries with feces, cause tip of the shoots yellow and inhibit the development of the bamboo shoots and eventually kill them.

Morphological characteristics **Adults** Female length 20-32 mm, male length 22-34 mm. Newly emerged adults bright yellow, those coming out from the soil orange yellow; some individuals may be yellowish brown or blackish brown; head black. Antennae geniculate, pedicle 4 mm in length, funiculus 7 segments. Male snout length 7.5-9.5 mm, and female snout 5.4-8.4 mm. There is a black marking at the rear end of the pronotum. The end of the elytra round, without sharp teeth. The center is concave when the two elytra are closed. Front femora and tibiae are the same length as those of the middle and hind legs. Other characteristics are similar to *Cyrtotrachelus buqueti*.

Eggs Cylindrical-shaped, the two ends round; length 3.0-4.1 mm, width 1.2-1.3 mm; milky white; shiny.

Larvae Young larvae 4 mm in length; milky white, becomes milky yellow after feeding; head light-yellowish brown. Mature larvae 38-48 mm, light-yellow, head yellowish brown. There is one dark yellow longitudinal stripe at outside of each coronal suture. Mouthparts black. Pronotum sclerotized. The sides of the prothorax, mesonotum, metanotum, and sides of metathorax with partially sclerotized dark yellow areas. The body with many folds, only the thoracic segments can be differentiated. The dorsal surface is similar to *C. buqueti*. Each segment with many smaller folds. The end of the abdomen spoon-shaped, its margin sclerotized, yellow.

Pupae Length 34-45 mm; milky white, turns to dirty yellow later. Vertex, rostrum around the antennae insertion have a pair of brown hairs. Pronotum with a shallow median groove; its rear border with an area of full of punctures. The rear border of each abdominal segment raised as folds, the middle with one row of brown teeth, each side with 5 teeth; the furthest tooth is more distant from the other 4 teeth and larger.

Cocoons Long elliptical; outer diameter 50-62 mm, inner diameter 38-55mm. the thickness of the wall is about 5 mm; covered with bamboo shoot fibers and soil; sturdy.

Biology and behavior *Cyrtotrachelus thompsoni* occurs one generation a year. It overwinters as adults in soil. They emerge from soil when daily temperature reaches 24-25℃. Peak emergence from soil is when temperature reaches 27-28℃. Most adults emerge from soil during 6:00-9:00 and 16:00-19:00. They feed on bamboo

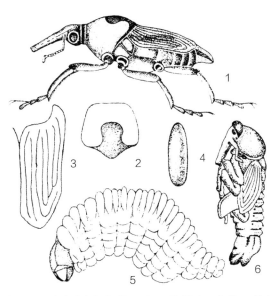

C. thompsoni. 1. Adult, 2. Pronotum, 3. Elytra, 4. Egg, 5. Larva, 6. Pupa (Xu Tiansen)

C. thompsoni adult (Xu Tiansen)

C. thompsoni oviposition site (Xu Tiansen)

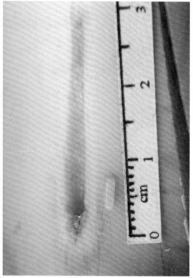
C. thompsoni egg (Xu Tiansen)

C. thompsoni mature larva (Xu Tiansen)

C. thompsoni cocoon and pupa (ventral view) (Xu Tiansen)

C. thompsoni cocoon and pupa (dorsal view) (Xu Tiansen)

shoots after emergence. The peak activity period during each day is among 6:00-7:00, 8:00-10:00 and 15:00-17:00. Adult activity period starts in mid- and late June in Zhejiang, mid-May in Guangdong and ends in late September. Larval period starts in early June in Zhejiang, early May in Guangxi, mid-May in Guangdong and ends in early October. Egg incubation period is 4-5 days in Zhejiang and 2-3 days in Guangdong. Young larvae feed upward to the tip of the bamboo shoots. From the 3rd instar, the larvae feed downward until to 25-30 cm below the oviposition hole.

Larvae go through 5 instars. Larval period lasts for 26-29 days in southern Zhejiang and 12-15 days in Guangdong. Mature larvae move upward in the tunnel after midnight until 13-20 cm below the tip. They chew off the tip, fill the opening with fiber, feces and move downward about 7 cm, then they chew off the shoot. The larva will fall to the ground along with the fallen shoot. The fallen shoot section is 5.7-8.8 cm in length. Sometimes, a larva will fall to the ground along with the 1st cut-off. The larva chews off the tip after the bamboo shoot fell to the ground. The larva will expose its head and thorax and move around along with the cut-off shoot to search for a suitable pupation site. It digs vertically first, then horizontally. The depth is related to soil density, varying from 12-55 cm, with an average of 25 cm. After a larva reaches the suitable depth, it will move to the surface back and force to bring in some bamboo shoot fiber for making pupal chamber. After 10-12 days, the larva molts into pupa. After 12-15 days, the pupa molts into an adult and overwinter in the soil.

Prevention and treatment

Conservation of enemies. An egg parasitic wasp was recorded. Larvae are often found infested with entomopathogenic nematodes and *Beauveria bassiana*.

(Xu Tiansen)

202 *Otidognathus davidis* (Fairmaire)

Classification Coleoptera, Dryophthoridae
Common name Bamboo shoot weevil

Distribution Jiangsu, Zhejiang, Anhui, Fujian, Jiangxi, Henan, Hunan, Hubei, Guangdong, Guangxi, Sichuan, Shaanxi; Vietnam.

Hosts Hosts include 46 bamboo species in the genus of *Phyllostachys*; *Bambusa intermedia*; *Pleioblastus gozadakensis*, *P. hsienchuensis*, and *P. amarus*; *Semiarundinaria lubrica*; *Arundinaria hsienchuensis*; *Sinobambusa tootsik*, *S. rubroligula*, and *S. edulis*; *Indosasa migoi*; and *Pseudosasa guanxianensis*, and *P. amabilis.*

Damage Adults feed on bamboo shoots for supplemental nutrition. New bamboo trees developed from damaged shoots exhibit shortened segments with feeding holes, depressed surface, and stiff wood fibre; resulting in reduced wood value. Furthermore, newly hatched larvae feed inside the shoots at the oviposition site, creating enlarged feeding holes. Some larvae can even feed on and break branches. High pest density leads to bamboo shoot mortality. Moderate damage creates unhealthy bamboo trees with shortened segments containing insects inside. Rigid wood, tip break, branch break, and sparse branches are common among damaged trees. This leads to reduced photosynthesis that affects shoot production and quality in the coming year. Damaged bamboo stands have much lower commercial value.

Morphological characteristics **Adults** Female length 14.5-21.8 mm, milky white immediately after emergence, turn to yellowish gradually. Males 12.4-19.6 mm long, reddish brown; body fusiform; head black, with black and oval compound eyes. Female snout 5.4-8.4 mm long, shiny, slender and smooth, whereas male snout 4.4-7.5 mm long, with spiny granules and a scrobe on top, and one array of teeth on each side. Antennae elbowed; scape 3 mm long, funiculus 7 segments, club rusty yellow, enlarged and boot-like. Pronotum globose, with a fusiform black mark. Posterior edge curved. Each elytron has a black mark in the center and a black mark at at 1/3 of its length. Outside corners of elytra black.

Eggs Oblong oval, slightly curved. Major axis 3.09 mm in length, and minor axis 1.07 mm. Newly laid eggs white, opaque; turning to milky white later; the lower half transparent before hatching.

Larvae Newly hatched larvae milky white, transparent with soft body wall and white dorsal lines, 3.1 mm in length. Body wall hardened from the 3rd instar, yellow. Mature larvae 20.81 mm in length, yellow with red brown head and shiny black mouthparts. Body wall wrinkled with concealed spiracles. Head capsule width of the five larval instars was 1.0-1.2, 1.3-1.5, 2.0-2.3, 2.8-3.3, and 3.8-4.1 mm, respectively; body length was 2.8-3.5, 4.4-6.5, 8.3-9.8, 14.1-17.6, and 20.7-24.8 mm, respectively. Caudal plate slightly bifurcated.

Pupae Milky white initially, turning to light-yellow later; body 16-22 mm in length. Opisthognathous, pronotum conspicuous, snout extends to mid- and hind legs, wing pads reach posterior margin of the 5th sternite.

Cocoons Earthy, oblong oval, 23.5-27.6 mm long, with thick wall that is rugose externally and smooth internally. Mature larvae build strong earthy cocoon inside the soil without fibrous debris from the ground.

Biology and behavior The bamboo shoot weevil undergoes one-year life cycle in small diameter bamboo forests, and 2-year life cycle in biennial forests. It overwinters as adults in the soil. Adults come out the soil in late April and early May. Adults can be seen in bamboo forests until early to mid-June. Adults mate and lay eggs in early to mid-May. Eggs hatch after 3-5 days. Larvae develop for 10-15 days before reaching maturity

O. davidis adult (lower)
(Xu Tiansen)

O. davidis mating pair
(Xu Tiansen)

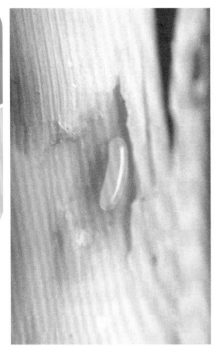
O. davidis egg (Xu Tiansen)

O. davidis soil cocoon and pupa (Xu Tiansen)

O. davidis damage (Xu Tiansen)

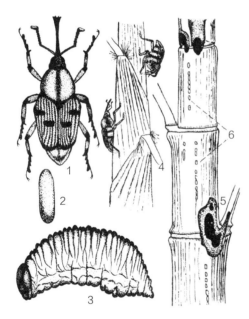
O. davidis. 1. Adult, 2. Egg, 3. Larva, 4. Adults laying eggs, 5. Damage by larva, 6. Damage by adults (Xu Tiansen)

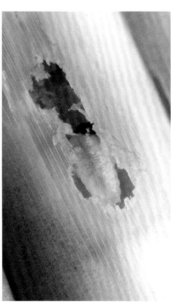
O. davidis 3rd instar larva in bamboo (Xu Tiansen)

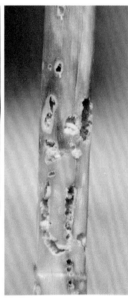
O. davidis damage to bamboo shoot (Xu Tiansen)

and pupate in late May to early June. Adults emerge and overwinter in July. Life cycle in Jiangsu province is 15-20 days behind that in Zhejiang province, whereas 15-20 days ahead in Guangxi province. *Otidognathus davidis* attacks *Phyllostachys heterolada* var. *funhuaensis* in Fenghua area in Zhejiang province. Adults come out from the soil in mid- to late May to coincide with the shooting season, which is usually 15-20 days ahead of other hosts in surrounding areas.

Prevention and treatment

Conservation of natural enemies. *Cuculus canorus canorus*, *Bambusicola thoracica*, and *Cissa ergtyrorhyncha* are natural enemies of *O. davidis* adults. Larvae of *O. davidis* are often killed by entomopathogenic nematodes and *Beauveria* sp.

References   Xu Tiansen and Wang Haojie 2004.

(Xu Tiansen)

203 *Rhynchophorus ferrugineus* (Olivier)

Classification Coleoptera, Dryophthoridae
Common names Red palm weevil, Asian palm weevil, Sago palm weevil

Distribution Widespread in Hainan, isolated areas in Shanghai, Fujian, Guangdong, Guangxi.

Hosts *Cocos nucifera*, *Phoenix dactylifera*, *P. hanceana* var. *formosana*, *P. sylvestris*, *Arenga pinnata*, *Elaeis guineensis*, *Borassus flabellifer*, *Roystonea regia*, *Areca catechu*, *Archontophoenix alexandrae*, *Hyophorbe lagenicaulis*, *Metroxylon sagu*, *Neodypsis decaryi*, and *Saccharum* spp.

Damage Leaves of infested trees turn yellow and break off at the bases later on. Serve damage result in complete leaf loss and death of trees. *Rhynchophorus ferrugineus* is a serious pest of trees in the family of Arecaceae. In Hainan where coconut trees are found everywhere and coconuts and areca nuts account for 80% of the income for local farmers, damage from *Rhynchophorus ferrugineus* has significant impact on local economy. In addition, it is also a pest of various valuable trees species in urban forests.

Morphological characteristics **Adults** Reddish brown, with two rows of black blotches; the first row contains 3 or 5 blotches (larger for the middle one); the second row contains 3 blotches (all relatively large). Elytra short, with six striae on each side.

Larvae Milky white; legless; curved. Head of mature larvae yellowish brown. Pygidium flat, with marginal setae.

Biology and behavior *Rhynchophorus ferrugineus* occurs two to three generations a year with overlapping generations due to unsynchronized development. There are two distinct adult periods in June and November. Females usually lay eggs in tender tissues in the lower crowns of young trees. They may also lay eggs in crevices of leaf stalks, exposed tissues, or places damaged by rhinoceros beetles. Eggs are laid individually with one egg per site. Each female produces 162-350 eggs in her lifetime. Newly hatched larvae bore into soft tissues and feed on the sap, creating interconnected galleries as they bore deeper. Damaged plant fibers are left around the galleries. When *R. ferrugineus* infest mature trees, they occur in damaged locations in the crown, creating localized necrosis. Mature larvae create oblong cocoons using plant tissues and enter prepupal stage before pupation. Pupal stage lasts for 8-20 days. Emerged adults stay inside the cocoons for 4-7 days until sexually matured before exiting the cocoons.

Prevention and treatment

1. Quarantine. Destroy infested trees during quarantine inspection in its native range.

2. Biological control. Inject nematodes *Steinernema* sp., and *Heterorhabditis* sp. through the trunk, or release parasitic mite *Hypoaspis* spp.

3. Trunk injection with 1:50 solution of 25 g/L lamda-cyhalothrin by drilling 1-2 holes on the trunk and slowly apply the insecticide using a hanging bottle between January and March. To control infestations on leaves,

R. ferrugineus adults (FPC)

R. ferrugineus pupa (FPC)

R. ferrugineus infested palm tree base (FPC)

R. ferrugineus infested palm tree tip (FPC)

Damage by *R. ferrugineus* adults (FPC)

R. ferrugineus larva (Xu Gongtian)

R. ferrugineus parasitized pupa (Xu Gongtian)

R. ferrugineus cocoon (FPC)

apply imidacloprid and sand mixture at leaf base. During July and August, apply 25 g/L of lamda-cyhalothrin at 800-1,000×, 40% chlorpyrifos at 1,500×, or avermectins at 1,500×, or 12% thiamethoxam and lamda-cyhalothrin mixture at 1,000×, or 50% imidacloprid and monosultap mixture at 600-800× to tree crown until trunks are wet.

References Zhao Yangchang and Chen Yuanqing 1980; Wu Yousheng, Dong Danlin, Liu Dongming 1998; Wu Kunhong and Yu Fasheng 2001; Liu Kui, Peng Yuqiang, Fu Yueguan 2002; Qin Weiquan, Zhao Hui, Han Chaowen 2002; Zhang Runzhi, Ren Li, Sun Jianghua et al. 2003.

(Zhou Maojian, Li Tao)

Sipalinus gigas (F.)

Classification Coleoptera, Dryophthoridae

Synonyms *Hyposipatus gigas* F., *Sipalus gigas* (F.), *Sipalus hypocrite* Boeheman, *Sipalus misumenus* Boeheman, *Curcuijo gigas* F., *Rhynchophorus gigas* Herbst

Common name Large pine weevil

Importance Larvae bore into weak pine trees, stumps, and fallen trees and lower the timber quality. It is an important pest of log storage facilities.

Distribution Jiangsu, Zhejiang, Anhui, Fujian, Jiangxi, Hunan, Sichuan, Guizhou.

Hosts *Pinus massoniana*, *P. teada*.

Damage The insect bores into xylem of the hosts and creates holes. Yellowish white fecal granules accumulate outside of the boring holes.

Morphological characteristics **Adults** Length 14.2-25.0 mm, black, with greyish brown stripes. Head small hemisphere shaped, snout pointed down; its basal 1/3 thick and opaque; the apical 2/3 smooth, black, and shiny. Pronotum with large tubercles, its center with a longitudinal smooth band. Each elytron with 10 longitudinal lines of punctures, the area between the two adjoining lines slightly raised.

Larvae Length 16.0-27.0 mm, milky white, stout, head yellowish brown, the end of the abdomen with 3 pairs of spines.

Biology and behavior *Sipalinus gigas* occurs one generation a year in Zhejiang. It overwinters as larvae inside galleries in xylem. Pupae appear in late March to late May. Adults appear in late April to early July with peak emergence period in May. Eggs appear between early may and mid-July. Larvae appear in mid-May.

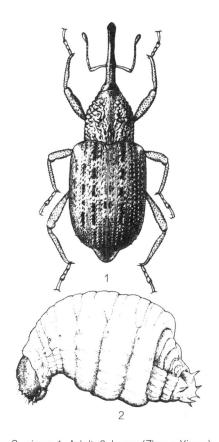

S. gigas. 1. Adult, 2. Larva (Zhang Xiang)

S. gigas adult (edited by Zhang Runzhi)

The adult emergence holes round, located at the base of trunks. Adults are good crawlers and are weakly attracted to lights. They like to congregate at the sap of Fagaceae trees. They prefer lay eggs on large weak pine trees, logs with barks, and new stumps. On stumps of 14.5 cm high and 21.8 cm diameter, the average number of larvae was 14.8 per stump. After hatching, the larvae bore into xylem.

Prevention and treatment

1. Silviculture. Clean out weak trees, infested trees, logs, and stumps in forests.

2. Pheromone trapping. Place pheromone traps in the forest in May to attract and kill adults.

S. gigas larvae (edited by Zhang Runzhi)

S. gigas gallery and a larva (edited by Zhang Runzhi)

S. gigas damage (edited by Zhang Runzhi)

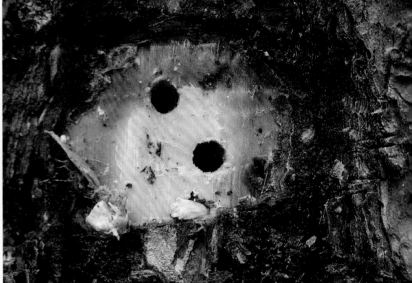
S. gigas exit holes (edited by Zhang Runzhi)

3. Light trapping. In seed plantation or special use forests, place light traps during adult period to attract and kill adults.

4. Apply *Beauveria bassiana* at 3×10^9 conidia/g.

Reference Xiao Gangrou (ed.) 1992.

(Zhao Jinnian, Wang Shufen)

205 *Pseudopiezotrachelus collaris* (Schilsky)

Classification　Coleoptera, Brentidae
Synonym　*Apion collare* Schilsky

Distribution　Guangdong.

Hosts　*Eucalyptus* spp.

Damage　Infested young shoots die back. Young trees may look like a broom. High populations may result in 97% infestation rate. One-year-old young trees are most susceptible. Three-year-old trees suffer the least damage.

Morphological characteristics　**Adults** Length 2.0-3.5 mm. Dark black, shiny. Sides of the pronotum with white pubescence. Rostrum cylindrical, 1.8 mm in length. Antennae 11-segmented, the last three segments forming a club. Eyes black.

Eggs Elliptical, milky white, 1.0-1.2 mm long.

Larvae Mature larvae 3-4.5 mm long, light-yellow, head brown. Body "C" shaped, with sparse short setae.

Pupae Milky white, 2.0-3.0 mm long.

Biology and behavior　*Pseudopiezotrachelus collaris* occurs one generation a year. It overwinters as pupae in soil chamber. Adults emerge from late March to early April. The peak emergence period is in mid-June. They feed on young shoots 3 days after emergence. They feed in the early morning and rest on back of leaves or base of branches in the evening. Adults are weak flyers. They play death when disturbed. After one week of supplementary feeding, adults mate. They lay eggs in young branches. One egg is deposited on each branch. Adults feed on the bark about 3 cm away from the oviposition site and cause the branch to die. Adult period lasts for 5 months. Each adult lays 70-90 eggs, Egg incubation period is 6-9 days. Larvae feed on the soft wood of the dead branches. Larvae molt four times. The last instar larvae migrate out from the branches and enter soil to pupate and overwinter.

Prevention and treatment

1. Silviculture. Plant trees at lower densities to increase vigor, reduce canopy coverage.

2. Conservation of natural enemies. Spiders were observed feeding on *A. collare* adults.

3. Chemical control. Spray 25 g/L lamda-cyhalothrin at 800-1,000×, or 20% fenvalerate EC at 2,000-2,500×, or 45% profenofos and phoxim mixture at 1,000×, or 12% thiamethoxam and lamda-cyhalothrin mixture at 800-1,200×.

References　Xian Shenghua, Kang Shangfu, Yu Yong et al. 1998; Pang Zhenghong 2006.

(Gu Maobin)

P. collaris adult (Zhang Runzhi)

206 *Sternuchopsis juglans* (Chao)

Classification Coleoptera, Curculionidae
Synonym *Alcidodes juglans* Chao

Distribution Henan, Hubei, Guangxi, Chongqing, Sichuan, Yunnan, Shaanxi.

Hosts *Juglans regia*, and *J. sigillata*.

Damage Adults feed on fruits, buds, young branches, and petioles. Infested fruits show 3-4 mm diameter holes and brown juice. Larvae feed on kernels and expel blackish brown feces. Infested fruits have scars on surface and fall prematurely.

Morphological characteristics **Adults** Length 9-12 mm, rostrum 3.4-4.8 mm long. Body pitch black. Sparsely covered with 2-5 forked white scales. Rostrum with dense punctures. Antennae 11-segmented, geniculate, the last 4 segments spindle shaped. Eyes black. Pronotum wider than long, densely covered with tubercles. Scutellum with a median longitudinal groove. Elytra with 10 grooves made of punctures. Femora enlarged, with one tooth, end of the tooth with two small teeth. Tibiae with one hook shaped tooth on the exterior side and two spines on the inner side.

Eggs Long elliptical, 1.2-1.4 mm, milky white or light-yellow, semi-translucent, turning to yellowish brown or brown later.

Larvae Length 9-14 mm, milky white, head yellowish brown or brown, thorax and abdomen sickle shaped.

Pupae Length 12-14 mm, yellowish brown. Thorax and abdomen with scattered small spines; end of abdomen with one pair of spines.

Biology and behavior *Alcidodes juglans* occurs one generation a year in Sichuan and Shaanxi. It overwinters as adults. Adults resume activity in early April and feed on young fruits, buds, young branches, and petioles. They mate multiple times and play death when disturbed and are attracted to lights. Egg laying period is from early May to late August. Eggs are laid on fruits. Adults bite 2-4 mm oval holes and lay one egg on each fruit. Each female lays 105-183 eggs. Adults live for 497-505 days. Egg incubation period is 3-8 days. Young larvae feed on fruit surface for 3-5 days before boring into the fruits. Larval period is 16-26 days (average 21 days). Mature larvae start to pupate from mid-May inside fruits on trees or the fallen fruits. Pupal period is 6-7 days. About 85% of the larvae successfully pated in a study and 80% of the pupae successfully developed into adults. Adult emergence holes are 6-7 mm in diameter. Sex ratio is close to 1:1. Adults feed on trees but do not mate or lay eggs. They overwinter among the cracks on the bark at base of trees.

Prevention and treatment

1. Set up black light traps to attract and kill adults.
2. Scrape off the coarse barks at base of trees. Remove fallen fruits, mix them with CaO and bury in soil.
3. Spray *Beauveria bassiana* at 2×10^9 spores per mL after overwintering adults emerged and until the young larvae appeared. Spray Dimilin to kill larvae.

References Xiao Gangrou (ed.) 1992; Li Menglou 2002; Pu Yonglan, Yang Shizhang, Lin Lin et al. 2003.

(Tang Guanghui, Li Menglou, Jing Heming)

S. juglans adult (Zhang Runzhi)

S. juglans adult (Li Zhenyu)

207 *Sternuchopsis sauteri* Heller

Classification Coleoptera, Curculionidae

Distribution Shanxi, Zhejiang, Fujian, Jiangxi, Hunan, Sichuan, Yunnan, Taiwan; Myanmar, Vietnam.

Host *Zanthoxylum bungeanum.*

Damage Larvae feed on branches of *Z. bungeanum*. Infested branches have rows of holes and are easily broken by winds. Adults feed on young shoots and often cause them to break.

Morphological characteristics **Adults** Length 17-20 mm, width 5-6.4 mm, black; sides of prothorax, humerus, and elytra with forked scales and white powders. Rostrum longer than head and thorax combined; posterior end of the female rostrum has fine rugae on the back. The 4th-6th funicles of the flagellum almost globular. The 7th funicle is equal to the length of the club in the males and two times of the club length in the females. Pronotum densely covered with tubercles, one setae on each tubercle. The posterior 1/4 of the elytra with scattered deep dentations. Femora with one curved tooth and two obtuse teeth at the end.

Eggs Length 2.8-3.2 mm, width 1.6-1.8 mm, long elliptical, milky white.

Larvae Mature larvae 15-17 mm long, white, head and pronotum yellowish brown. Body with many folds and short setae. The tubercles on the sides of abdomen with 2 short setae per tubercle.

Pupae Pale yellow to white, elongated elliptical, 12.5-14.1 mm in length. Dorsal surface with small teeth. End of the abdomen with one pair of spines.

Biology and behavior *Sternuchopsis sauteri* occurs one generation every two years. It overwinters as eggs, larvae, or adults. The overwintering adults emerge in late March when buds are developing. Adults feed on buds. They are most abundant during late April to mid-June. Eggs are laid from early April to late November. A female bites a long groove on the shade side of a branch and creates a row of 1-7 holes in the groove. It lays one egg in each hole and seals the holes with yellowish brown gelatinous materials. Larvae go through 5 instars. They tunnel upward. Long feces can be seen outside of the holes on the branches. One larva consumes about 530 mm^3 wood. Mature larvae create pupal chambers near the exit holes. Adults will exit from the holes 15 days after eclosion. The peak adult emergence period is from early September to late October. Adults can fly about 15 m. They play death when disturbed and look like bird feces. Egg incubation period is 16.1 days. Larval development period is 352.5 days. Pupal development period is 18.7 days. Adults live for 39.5 days. Overwintering adults live for approximately 150 days.

Prevention and treatment

1. Clean out the broken branches, dead branches, and dead trees to reduce pest numbers.

2. For heavy infestations, spray 2.5% deltamethrin EC at 5,000×, 20% fenpropathrin at 3,000×, or 5% esfenvalerate EC at 3,000×, or 45% profenofos and phoxim mixture at 1,000×, or 12% thiamethoxam and lamda-cyhalothrin mixture at 800-1,200× to kill adults. Spray 45% profenofos and phoxim mixture at 30-60×, or 12% thiamethoxam and lamda-cyhalothrin mixture at 30-60× during larval period.

References Zhao Yangchang and Chen Yuanqing 1980; Deng Guofan, Liu Youqiao, Sui Jingzhi et al. 1983, Zhou Mingkuan, Luo Xiuwen, Zhu Yan et al. 1993; Wu Zongxing, Liu Zhifu, Yu Mingzhong et al. 2003.

(Wang Hongzhe, Li Menglou)

S. sauteri adult (Zhang Runzhi)

208 *Cryptorhynchus lapathi* (L.)

Classification Coleoptera, Curculionidae
Synonym *Curculio lapathi* L.
Common name Poplar and willow borer

Importance *Cryptorhynchus lapathi* is a disastrous pest of poplar seedlings and young trees. Larvae feed on cambium along the circumference of the trunk or branch. They cause death of the branches. Mature larvae feed in the wood and reduce wood quality. *Cryptorrhynchus lapathi* is a quarantine pest in China.

Distribution Hebei, Shanxi, Inner Mongolia, Liaoning, Jilin, Heilongjiang, Shaanxi, Gansu, Xinjiang; Japan, Korea Peninsula, Russia, Hungary, Czech Republic, Slovak Republic, Germany, United Kingdom, Italy, Poland, France, Spain, Holland, Canada, U.S.A.

Hosts *Populus suaveolens, P. ×xiaohei, P. ×beijingensis, P. ×berolinensis, P. ×canadensis, P. ×xiaozhuanica* 'Beicheng', *P. ×canadensis* 'Sacrau-79', *P.canadensis* 'I-214', *P. nigra* var. *thevestina, P. simonii, P. guariento, P. tomentosa, P. cathayana, P. pseudo-simonii, P. alba, P. alba* var. *pyramidalis, P. ×canadensis* 'Serotina', *P. ×canadensis* 'Robusta', *P. canadensis×cathagana, Alnus viridis, Salix matsudana, S. viminalis, S. fragilis, S. caprea,* and *Betula pumila*.

Damage *Cryptorhynchus lapathi* infests trunks of seedlings, young trees and sometimes the branches of large trees. Larvae bore galleries along the circumference of the trunk, forming 1-9 round holes on the bark, and expel blackish brown feces from the holes. The bark above the galleries slightly sank and with oil immersion appearance. At later stage, the affected bark often cracks and forms scars. Mature larvae bore into the wood and create 8-20 cm long galleries. They expel white wood crumbs and fluid from the lower end of the galleries. When infesting fast-growing poplar trees, they cause irregular raised or cracked patches where feces and wood crumbs are expelled.

Morphological characteristics **Adults** Length 7-10 mm, elliptical. Blackish brown or brown; rostrum, antennae, and tibia reddish brown. Body covered with dense greyish brown scales and a few irregular transverse white scale bands. Sides of pronotum, the posterior 1/3 of elytra, and femora with dense white scales and scattered clusters of black scales. The numbers of black scale clusters in other areas are base of rostrum – one pair; anterior part of the pronotum – one pair; posterior part of the pronotum – 3; and elytra between 2nd-4th grooves – 6. Rostrum curved, densely punctured, its center with one longitudinal carina. Antennae 9-segmented, geniculate. Eyes round, black. Pronotum wider than long, the two sides round; the front end narrow, with one think median ridge. Elytra wider than pronotum, oblique from posterior 1/3 part and gradually narrows and forming a triangular oblique surface. End of the anal plate round in males and sharp in females. In Shaanxi, there is a color type in which the end of the body is red, the body with pale red

C. lapathi adult (Xu Gongtian)

C. lapathi mature larva (Xu Gongtian)

scales, the end of the elytra more oblique. Sides of the male aedeagus almost parallel, their front not enlarged, similar to bullet shaped, but not raised; there is a "v" suture on the margin of the front end.

Eggs Elliptical; 1.3 mm long, 0.8 mm wide.

Larvae Mature larvae 9-13 mm long. Cylindrical, slightly curved to horse shoe shape. Milky white. Body with many transverse folds and sparse yellow short setae. Head yellowish brown, mandibles blackish brown, maxillary and labial palps yellowish brown. Coronal suture obvious, upside down "Y" shape. Antennae small. Clypeus trapezoid, smooth. Labrum transverse elliptical, the middle of the front border with two pairs of setae; the sides with 3 thick setae on each side; the dorsum with 3 pairs of setae. Each mandible with one obtuse tooth. Maxillary and labial palps are two-segmented. Prothorax with one pair of yellow sclerotized plates. Each of the mesothorax and metathorax consists of two sub-segments. Each of the 1st-7th abdominal segments consist of three sub-segments. The sides of thorax and the sternites convex. Thoracic legs become vestigial. There are several yellow setae located at the leg sites. Spiracles yellowish brown.

Pupae Milky white, length 8-9 mm. The dorsal surface of the abdominal segments with many small spines. The end of the abdomen with a pair of brown hooks that curved inward.

Biology and behavior *Cryptorhynchus lapathi* occur one generation a year. It overwinters as young larvae or eggs (small percentage) in the branches. Occasionally it may overwinter as adults. The overwintering larvae resume activity in late March or early April. The larvae create feeding galleries from the oviposition site. They feed the cambium layer and only leave the topical layer intact. They chew holes on the bark and expel blackish brown fecal materials. They pupate at tips of the galleries in early June. Adults emerge from mid-June. After 6-10 days, the adults emerge from the branches or trunks. Peak adult emergence time from host trees is mid-July. Adults are active during the early morning or late evening. They hide in fallen leaves, under rocks, or inside gaps in soil when resting. Adults prefer to feed on the cambium of the dead or live branches and leave small feeding holes. The feeding sites are attractive to other adults. Adults play death when disturbed. They mate and lay eggs in late July. Eggs are laid on tree trunks under 2 m height. Adults deposit eggs at the leaf scars, branch scars, dormant buds, cracks, and places with thick suberin. They chew a small hole and lay one egg in each hole

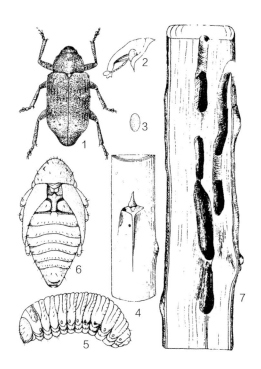

C. lapathi. 1. Adult, 2. Lateral view of head, 3. Egg, 4. Oviposition hole, 5. Larva, 6. Pupa, 7. Damage (Shao Yuhua)

and seal the hole with blackish brown paste. Each female lays an average of 44 eggs. The egg laying period lasts for an average of 36.5 days. Eggs are usually laid on 2-4 years old branches. Sometimes females lay eggs on > 30 years old tree trunks. Adults disperse for long distances through floating on rivers.

Prevention and treatment

1. Area-wide management along the rivers. Control the pest simultaneously along the rivers starting from the upper stream. Eliminate sources of pests before planting poplar trees.

2. Quarantine. Check for oviposition holes on seedlings. Infested seedlings may be soaked in 2.5% deltamethrin ED at 1,000-2,000× for 5 minutes. Infested wood may be debarked, fumigated with methyl bromide or sulfuryl fluoride at 30g/m^3, 20℃, and 24 hours treatment time. Heat treatment of timber for 10 hours at 60% RH and heat the center of the timber to 60℃.

3. Mechanical control. Shake trees and collect adults. Kill larvae with a hammer or knife before they enter the wood.

4. Conservation of natural enemies. Toads, chickens, birds (*Dendrocopos hyperythrus hyperythrus*, *Picoides* spp.), ants (*Formica* sp.) can predate upon adults. Parasitoids include *Dolichomitus* sp. and *Schreineria ceresia*.

C. lapathi mature larva in wood (Xu Gongtian)

C. lapathi damage to trunk (Xu Gongtian)

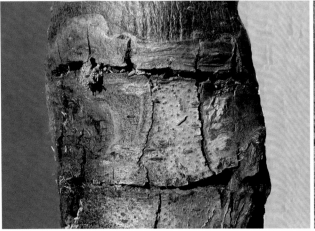
C. lapathi larval damage (Xu Gongtian)

C. lapathi damage (Xu Gongtian)

Beauveria bassiana, *Acremonium strictum*, *Fusarium*, and nematodes may parasitize larvae or pupae.

5. Chemical control. Brush trunks with 45% profenofos and phoxim mixture at 30-60×, or 12% thiamethoxam and lamda-cyhalothrin mixture at 30-60×. Using brushes to apply 10% cypermethrin EC at 20-100×, 2.5% deltamethrin at 50-800×, or 20% fenvalerate at 30-50× at defecation holes to kill larvae before they enter the wood. Alternative chemicals include 2.5% deltamethrin slow release paste and 25% Dimilin oil suspension. Mature larvae and pupae may be controlled by applying 56% aluminum phosphate at 0.05 g per hole. Adults may be controlled by 25% Dimilin spray. Before July 20, apply 5-10 cm wide glue and insecticide mixture bands on tree trunks to prevent adults from oviposition above the bands. The mixture can be insect glue plus 20% fenvalerate EC or 20% fenpropathrin EC at 40-20:1 ratio. Apply insecticides using brushes in spring to kill larvae below the bands.

References Xiao Gangrou (ed.) 1992; Wu Jian and Wang Changlu 1995; Wang Shuying, Zhao Mingguang, Zhu Xinfei et al. 1996; Tang Guangzhong 2005.

(Tang Guanzhong, Hou Aijü, Li Zhenyu)

209 *Curculio chinensis* Chevrolat

Classification Coleoptera, Curculionidae
Common name Camellia weevil

Distribution Jiangsu, Zhejiang, Anhui, Fujian, Jiangxi, Hubei, Hunan, Guangdong, Guangxi, Sichuan, Guizhou, Yunnan.

Hosts *Camellia* spp.

Damage Adults feed on young fruit and leave black marks. They also feed on new shoots, young stems and cause death of branches. Larvae feed on seeds and cause fruit fall or empty fruits.

Morphological characteristics **Adults** Length 7-11 mm (excluding the rostrum). Body elliptical; black or brown; covered with sparse white scales; elytra with grooves formed by punctures and white marks or transverse bands made of scales.

Larvae Length 10-20 mm; mature larvae light-yellow; head reddish brown.

Biology and behavior *Curculio chinensis* occurs one generation every two years in Hunan. In Yunnan, it mostly occurs one generation a year. Only a small percentage occurs one generation every two years. It overwinters as larvae or adults in soil chambers near the roots of *Camellia* spp. The overwintering larvae turn to adults in early August next year. The overwintering adults emerge in late April. They lay eggs on young fruits from early May to late August.

Adults start to mate and lay eggs from 7 days after emerging from the soil. They use their rostrum to bite

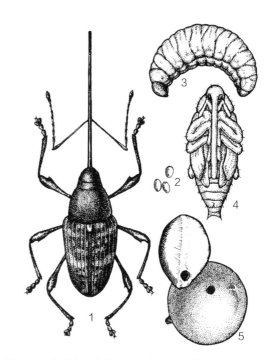

C. chinensis. 1. Adult, 2. Eggs, 3. Larva, 4. Pupa, 5. Damage (Zhang Xiang)

holes on fruits and then lay eggs. One egg is deposited in each hole. Each female lays 51-179 eggs. The egg laying period lasts for 44-54 days. The peak egg laying period is from mid-June to late June. The egg incubation period is 7-15 days. After hatching, larvae feed on seed kernels. A larva infests 2-3 seed kernels during its lifetime. Mature larvae chew round holes on the seed shell and fall out to the ground. They burrow into 3-6 cm depth (maximum 18 cm) in soil. Adults hide in the shade and play death when disturbed.

Population dynamics of *C. chinensis* is closely related to the environment. Adults emerge when temperatures rise to 17-19°C; peak emergence occurs at 23-24°C and die down at 25-28°C. Dense forests suffer more severe damages; middle of the forests usually suffers more severe infestations than forest edges.

Prevention and treatment

1. Silviculture. Trimming, turning soil before March or after June to destroy larvae and adults in soil. Inter-

C. chinensis adult (Zhang Runzhi)

C. chinensis adult (Wang Jijian)

C. chinensis adult (Zhang Runzhi)

C. chinensis adult (Li Zhenyu)

cropping with crops to create less favorable habitat for *C. chinensis*.

2. Mechanical control. Collect and kill adults during 8:00-11:00 and 15:00-18:00 in May and June. Remove fallen fruits every 5 days during July-September to kill the larvae in the fruits.

3. Attract and kill. Placing a sugar-vinegar trap in the evening every 1/3 hectare during May-June to kill adults.

4. Chemical control. Spray 98% carbamic acid at 1,000× or 10% bifenthrin at 6,000×, or 45% profenofos and phoxim mixture at 1,000×, or 12% thiamethoxam and lamda-cyhalothrin mixture at 800-1,200× to the soil during the periods when larvae emerge from fruits or when adults emerge from soil.

References Zhou Shijuan 1981, Xiao Gangrou (ed.) 1992, Tan Jicai 2002.

(Tong Xinwang, Zeng Qingyao, Gan Jiasheng, Chen Sufen)

210 *Curculio davidi* Fairmaire

Classification Coleoptera, Curculionidae

Distribution Beijing, Hebei, Liaoning, Jiangsu, Zhejiang, Anhui, Fujian, Jiangxi, Shandong, Henan, Hubei, Hunan, Guangdong, Yunnan, Shaanxi.

Hosts *Castanea mollissima*, *C. seguinii*, and *C. henryi*.

Damage Larvae feed on the cotyledon of nuts, promote fungal growth, and reduce nut quality.

Morphological characteristics **Adults** Length 6-9 mm, males smaller than females, body black. Elytra with 10 grooves made of punctures. There are white spots or stripes at conjunction between prothorax and head, humerus, and anal angles of the elytra. The spots or stripes are made of white scales.

Eggs Length 1.5 mm, elliptical, transparent when laid, turning to milky color before hatching.

Larvae Mature larvae 8.5-12 mm long, milky white to light-yellow, sickle shaped with many folds and sparse short setae.

Pupae Length 7-11 mm, greyish white.

Biology and behaviour *Curculio davidi* occurs mostly one generation in two years. It was reported occurring one generation a year in Yunnan and one generation in 1-2 years in Anhui. It overwinters as larvae in soil. They pupate in soil June-July. Adults emerge from early July to early October. Peak egg laying period is in September. Larvae live in seeds for about one month and then leave the seeds and enter soil to overwinter.

Adults feed on nectars soon after emergence. Then they feed on seeds, young branch barks of *Castanea mollissim* and *C. seguinii*. They prefer *C. seguinii* in mixed plantations. Adults are active during the day. They play death when disturbed. They mate multiple times. Each mating lasts for 4-7 hours. Males live for 8-16 days (average 10.3 days). Females live for 10-19 days (average 15.8 days). They lay eggs the next day after mating. The females create 1-1.5 mm deep notches on involucres. Each female lays 1-3 eggs a time and lays a maximum of 18 eggs. They usually lay eggs at base of the involucres.

Eggs hatch in 10-15 days. Young larvae feed on surface of cotyledon. Tunnels are 1 mm wide. The feeding tunnels by the 3rd-4th instars are 8 mm wide. The tunnels are filled with greyish white or brown powdery fecal materials. Larvae go through 6 instars. When mature, each larva chews a 2-3 mm diameter hole and burrows into soil and make a 0.5 by 2 cm chamber. They usually stay 10-15 cm deep in soil to overwinter.

Infestation of *C. davidi* is related to harvest time and processing method in plantations in the north. Fallen involucres on the ground, holding of harvested chestnuts near the plantation, and removing the shells and skins will likely to promote large number of *C. davidi* larvae in soil. There may be hundreds to > 1,000 overwintering larvae per square meter. Besides these factors, the *C. davidi* density is also related to the presence of *C. sequinii* surrounding the plantation. A survey in seven counties (cities) in Jiangsu found that 8.26%-68.17% of the 5,628 examined chestnuts (21 varieties) were infested where *C. sequinii* trees were present. In contrast, none of the 4,913 examined chestnuts (25 varieties) were infested where *C. seguinii* trees were absent.

Host susceptibility to *C. davidi* infestation is correlated with the length and density of spines on involucres. *C. davidi* prefers to lay eggs on involucres with short and sparse spines and thin involucres. A survey conducted in a plantation found resistant varies such as Jiaoza, Mici, Zaokui suffered 5.5%-16.98% infestation rates. The susceptible varieties such as Boke, Zhenliangxiang, Zhenzhupu, and Lantouguanyin suffered 31.76%-68.17% infestation rates.

Chestnut mature time is closely correlated with infestation rate. Those matured early suffer less damage. In Jiangsu (Bianyang county), the Chushuhong variety ma-

C. davidi adult (Zhang Peiyi)

tures in late August and early September. It suffered 5% infestation rate. Those mature in early October (Chongyangpu, Chongyanghong) suffered 44.27%-60.36% infestation rates.

Prevention and treatment

1. Quarantine. Treat infested chestnuts with aluminum phosphate before being transported outside of the quarantine area. The treated chestnuts are more easily rotten, though.

2. Harvest involucres as soon as possible in the north to avoid chestnuts falling to the ground. Turn the soil of the chestnut-holding ground in early to mid-June to destroy larvae. In the south, clear the wild chestnuts in mature orchards to reduce source of *C. davidi*.

3. Plant resistant varieties such as Chushuhong and Jiaoza.

4. Chemical control. Spray 50% sumithion EC at 500× two times during adult period. Spray Dimilin at 500× and 1,000×, flufenoxuron at 500× and 1,000× can reduce next generation population by > 90%. ULV application of Dimilin and flufenoxuron during adult period for two times reduced populations by 93%-94%. Spraying mixture of *B.t.* and lannate or *Bt* alone reduced infestation by 95% and 92.3%, respectively.

References Sun Yongchun 1963; Guo Congjian, Liang Zhenhua, Yang Yuzhen et al. 1965; Chinese Academy of Sciences - Institute of Zoology 1986; Chen Jizhong et al. 1987; Wu Yiyou et al. 1999, 2001; Yang Youqian et al. 1995; Sun Shaofang 2003; Zeng Lin et al. 2004; Zhang Haijun et al. 2005; Yin Xiangyu 2006; Qi Xiaoyan 2010, Wu Lifang 2011.

(Sun Yongchun, Xu Fuyuan, Jiang Ping, Li Zhenyu)

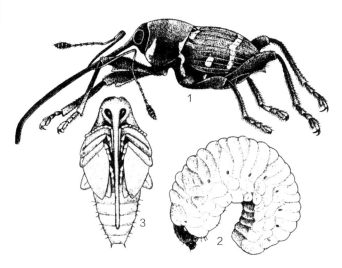

C. davidi. 1. Adult, 2. Larva, 3. Pupa (1-2 by Zhu Xingcai; 3 by Zhang Peiyi)

211 *Curculio hippophes* Zhang, Chen et Dang
Classification Coleoptera, Curculionidae

Distribution Liaoning, Shaanxi, Gansu.

Host *Hippophae rhamnoides*.

Damage Young larvae of *C. hippophes* feed on the cotyledon of *Hippophae rhamnoides* fruit. The large larvae feed on seed kernels.

Morphological characteristics **Adults** Females 3.0 mm long, 1.6 mm wide; beak 1.6-2.0 mm long. Brown, covered with light-yellow scales. Elytra with a transverse band of milky white scales. Flagellum 7-segmented; scape and the first 5 flagellum segments equal in length; club 4-segmented. Pronotum with large and round punctures. The scutellum with dense white scales. There are 2-3 rows of scales between the grooves on elytra. Each femur has a large tooth near the end.

Eggs Milky white, round.

Larvae Mature larvae 3-4 mm; curved; legless. Head yellowish brown; mouthparts blackish brown; thorax and abdomen milky white.

Pupae Length 3-4 mm, width 1.5-2.0 mm. Milky white at the beginning, turning to yellowish brown later; eyes black. The posterior border of each of the 1st-7th abdominal segments with one row of tubercles; each tubercle with one seta. The anal segment with one pair of spines.

Biology and behavior *Curculio hippophes* occurs one generation a year in Shaanxi. It overwinters as mature larvae in soil. They pupate from early June to mid-August. Pupal period is 20-25 days. Adults emerge from late July to mid-September. They play death when disturbed. After a few days of supplementary feeding on *H. rhamnoides*, a female lays one egg on the lateral side of a fruit. One egg is laid on one fruit. A female lays a total of 15-30 eggs. Females live for 19-25 days; males live for 17-19 days. The egg incubation period is 5-8 days. Young larvae feed on surface of the cotyledon in mid-August. Later on, they feed on seed kernels. Larvae go through 4 instars. Their feeding time inside fruits is 15-20 days. Mature larvae fall to ground in mid-October along with the fruits. They enter soil and build soil chambers.

Prevention and treatment

1. In heavily infested forests, cut down and remove all *H. rhamnoides* in winter or in spring before buds expand.

2. Spray *H. rhamnoides* canopy with 2.5% deltamethrin at 4,000×, or 45% profenofos and phoxim mixture at 1,000×, or 12% thiamethoxam and lamda-cyhalothrin mixture at 800-1,200× when adults emerge in early August.

References Chen Xiaoda, Dang Xinde, and Li Feng et al. 1990; Xiao Gangrou (ed.) 1992; Zhang Runzhi, Chen Xiaoda, and Dang Xinde 1992.

(Tang Guanghui, Li Menglou, Chen Xiaoda)

C. hippophes adult (Zhang Runzhi)

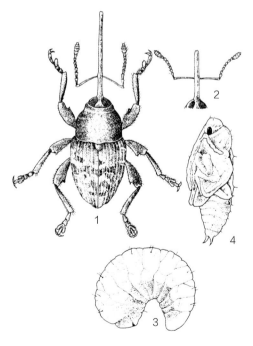

C. hippophes. 1. Male adult, 2. Female adult head, 3. Larva, 4. Pupa (Zhu Xingcai)

212 *Dyscerus cribripennis* Matsumura et Kono

Classification   Coleoptera, Curculionidae

Distribution   Fujian, Shandong, Hunan, Guangxi, Sichuan, Guizhou, Yunnan, Taiwan; Japan.

Hosts   *Olea europaea*, *Melia azedarach*, *Prunus persica*, *Castanea mollissima*, *Toona sinensis*.

Damage   *Dyscerus cribripennis* larvae feed on phloem of trunk, stem-root collar, branches, and may cause tree mortality.

Morphological characteristics   **Adults** Length 13-15 mm; blackish brown, covered with white or golden yellow hair-like scales. Pronotum with round large granules, the front part of the pronotum with a longitudinal ridge at the center. Elytra with lines of large punctures, the distance between the puncture lines wide.

Eggs Elliptical, 1.5 mm in length, pear yellow.

Larvae Curved; mature larvae 17-20 mm in length; head yellowish brown, body milky white.

Pupae Length 16-18 mm, body covered with symmetrical spines.

Biology and behavior   *Dyscerus cribripennis* occurs two generations a year (Type I) or three generations in two years (Type II). The type I population overwinters as adults in soil. The type II populations overwinter as larvae in tree barks. The overwintering adults come out feeding when the temperature in January reaches 6℃. They mate when the temperatures reach to about 18℃. They mate 1-3 times during the one month egg laying period. They lay eggs about 4-5 days after mating. Eggs are laid in phloem on tree trunk, stem-root collar, and large branches. They usually lay two eggs a time, sometimes 4 eggs a time. After hatching, larvae feed inside the phloem. In mid- to late May, larvae make pupal chambers in surface of xylem. Pupal period lasts for about 15 days. The 1st generation adults emerge in late June and July. The 2nd generation adults emerge between late October

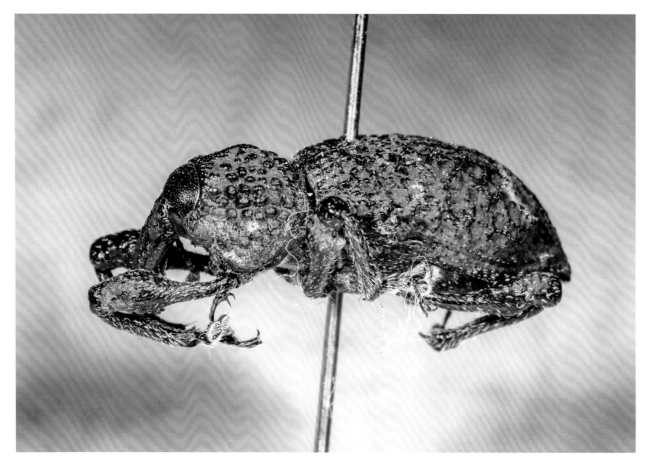

D. cribripennis adult (lateral view) (Zhang Runzhi)

D. cribripennis adult (dorsal view) (Zhang Runzhi)

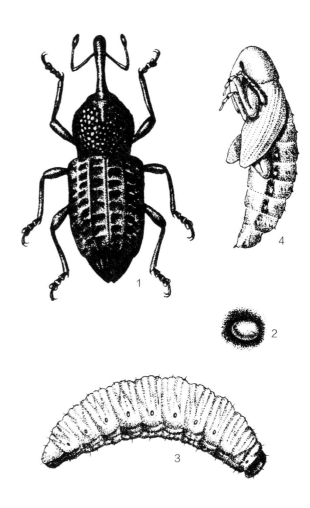

D. cribripennis. 1. Adult, 2. Egg, 3. Larva, 4. Pupa (Zhang Peiyi)

and early November and overwinter in soil around tree roots. Overwintering larvae pupate in late January and early February. Adults emerge in mid-February. They mate in April. Larvae infest hosts during early April and mid-July. Pupation occurs between early July and early August. Adults emerge in mid-August. They lay eggs in October. Eggs hatch into larvae and then overwinter.

There are three peak adult activity periods: February-March, July-August, and October-November. Adults play death when disturbed. They can fly and prefer to stay in the shade area in lower part of the canopy. The insect density is higher in stands with dense grass coverage. They form congregations during the overwintering period.

Prevention and treatment

1. Mechanical control. Collect adults by shaking the trees or searching their overwintering harborages.

2. Chemical control. At damage site, inject 1-2 mL of 20% thiamethoxam or 25 g/L lamda-cyhalothrin at 300-500×, or 45% profenofos and phoxim mixture at 30-60× to kill larvae. Apply 12% thiamethoxam and lamda-cyhalothrin mixture at 30-60× to the tree trunks using brush or sprayer in July to kill the larvae.

Reference Xiao Gangrou (ed.) 1992.

(Tong Xinwang, Zhong Xiaowu)

213 *Pimelocerus juglans* (Chao)

Classification Coleoptera, Curculionidae
Synonym *Dyscerus juglans* Chao

Distribution Hebei, Shanxi, Fujian, Henan, Sichuan, Yunnan, Shaanxi, Gansu.

Hosts *Juglans regia, J. sigillata.*

Damage *Pimelocerus juglans* larvae feed on phloem of the main roots, weaken the trees, reduce yield, or cause tree mortality.

Morphological characteristics **Adults** Length 11-15 mm, width 5.6-6.5 mm; black; covered with yellowish brown needle-like scales. Rostrum densely punctured. There is a groove above the antennae groove. A line of punctures exists in the groove above the antennae groove. Pronotum densely covered with large punctures, with a median longitudinal ridge. Elytra with 10 lines of punctures. There are 10 scattered yellowish-brown pubescence markings. There is a group of orange brown pubescence between the median coxae, there is an arched transverse groove behind the middle and hind coxa fossae.

Eggs Elliptical, 1.6-2.0 mm in length; milky white to yellowish white, gradually change to yellowish or yellowish brown.

Larvae Length 14-18 mm; yellowish white or greyish white, head brown, mouthparts blackish brown. Remnant of front legs with several hairs.

Pupae Length 14-17 mm; yellowish white; the end with two black spines.

Biology and behavior *Pimelocerus juglans* occurs one generation every two years in Sichuan and Shaanxi. It overwinters as adults or larvae. The overwintering 1st instar larvae infest hosts between March and November the next year. The overwintering mature larvae pupate in late April and turn into adults from mid-May to early July. Adults play death and are attracted to lights. They overwinter in pairs or groups in barks at base of main roots. Overwintering adults resume activity in late March, feed on leaves and fruits during early April and early May. They mate multiple times. They chew open a round hole at the stem-root collar in early and mid-June, lay eggs, then seal the holes with crumbs after pushing the eggs into the bottom of the holes. Eggs hatch in early June. Egg incubation period is about 8 days. Young larvae feed between the phloem and the xylem after entering the phloem. The galleries are filled with blackish brown feces and wood crumbs. The larval

P. juglans adult pair ready to mate (the male is located on top) (Zhang Peiyi)

P. juglans adult (dorsal view) (Li Zhenyu)

period lasts for 20-22 months. During heavy infestations, galleries in the phloem of roots between 5-20 cm deep and that of the trunk under 1.4 m will surround the diameter of the trunk or roots.

Prevention and treatment

1. Before May, remove the soil around the stem-root collar. Apply lime slurry ($Ca(OH)_2$) to stem-root collar and cover roots with soil. Remove the coarse barks. These measures can kill 75%-85% of the larvae.

2. Spray 2.5% lamda-cyhalothrin at 4,000×, or 45% profenofos and phoxim mixture at 1,000×, or 12% thiamethoxam and lamda-cyhalothrin mixture at 800-1,200× during June and August to tree canopy to kill adults. During mid- and late June, apply 45% profenofos and phoxim mixture at 39-60×, or 12% thiamethoxam and lamda-cyhalothrin mixture at 30-60× to tree trunks using brush or sprayer to kill larvae.

References Xiao Gangrou (ed.) 1992, Li Menglou 2002.

(Tang Guanghui, Li Menglou, Jing Heming, Han Peiqi)

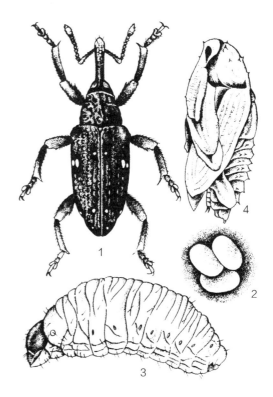

P. juglans
1. Adult, 2. Eggs, 3. Larva, 4. Pupa (Zhu Xingcai)

214 *Eucryptorrhynchus brandti* (Harold)

Classification Coleoptera, Curculionidae

Importance *Eucryptorrhynchus brandti* larvae feed on phloem and xylem of tree trunks and large branches, weaken the host trees, or even cause tree mortality.

Distribution Beijing, Tianjin, Hebei, Shanxi, Liaoning, Jilin, Heilongjiang, Shanghai, Jiangsu, Henan, Sichuan, Shaanxi, Gansu.

Host *Ailanthus altissima*.

Damage Adults feed on young branch tips and leaves. Their feeding cause breaking of branches and damaged leaves and barks. Boring holes are often seen on tree trunks and branches.

Morphological characteristics **Adults** Length 11.5 mm, width 4.6 mm; black or greyish black. Frons much narrower than base of rostrum. There are no grooves beside the middle ridge of the rostrum. Head with small punctures. Pronotum and elytra covered with large punctures, pronotum narrow in the front and wide in the rear. Elytra thick and sturdy, the two elytra fused together. Front thorax, humerus, and the rear end of wings covered with white scales; only a few ochrous scales are present; the scales leaf shaped; the other areas with scattered white dots. The humerus slightly protruded.

Eggs Elongated oval, yellowish white.

Larvae Length 10-15 mm; head yellowish brown, thorax and abdomen milky white.

Pupae Length 10-12 mm, yellowish white.

Biology and behavior *Eucryptorrhynchus brandti* occurs one generation a year. Larvae and adults overwinter in tree trunks or in soil. The overwintering adults resume activity in late April. They are active between late April and mid-May. When laying eggs, females use their rostrum to open the phloem, lay eggs inside, and push the eggs deep into the phloem. White fluid appears at the egg deposition sites afterwards. Eggs hatch in late May. Those overwintering as larvae pupate in May. Adults emerge and lay eggs during June and July. Eggs hatch in late August. Young larvae feed on the bark, cause depression under the think bark. When the size grows larger, they move into the xylem. Adults play death and fall to the ground when disturbed. They feed on young branch tips, leaves for about one month and then lay eggs. Egg incubation period is about 8 days. Young trees and pure stands are more susceptible to damage.

E. brandti adult (Zhang Runzhi)

E. brandti damage (Zhang Runzhi)

E. brandti mating pair (Xu Gongtian)

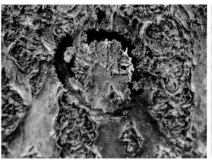
E. brandti exit hole before adult emergence (Xu Gongtian)

E. brandti mature larva (Xu Gongtian)

E. brandti larvae (Xu Gongtian)

E. brandti infested trunk (Xu Gongtian)

E. brandti adult with one elytron raised (Zhang Runzhi)

Prevention and treatment

1. Quarantine. Prevent movement of infested seedlings. Cut down infested seedlings to avoid spread.

2. Mechanical control. Collect adults by shaking the host trees. During larvae period, using a metal hook to take out larvae from the infestation sites. During the adult period, place a plastic barrier at 30 cm above ground on tree trunks. The barrier forms an umbrella shaped around the tree trunk and is covered with a layer of grease to prevent adults from climbing up the tree.

3. Chemical control. During adult period, spray 45% profenofos and phoxim mixture at 1,000×, or brush tree trunks with 12% thiamethoxam and lamda-cyhalothrin mixture at 30-60× to kill newly hatched larvae. Apply 25% carbaryl dust or 2.5% trichlorfon dust at base of trees to kill young larvae.

References Xiao Gangrou (ed.) 1992, Yang Ziqi and Cao Guohua 2002, Xu Gongtian 2003.

(Qiao Xiurong, Wan Shaoxia, Zhou Jiaxi)

E. brandti infested trees (Zhang Runzhi)

215 *Eucryptorrhynchus scrobiculatus* (Motschulsky)

Classification Coleoptera, Curculionidae
Synonym *Eucryptorrhynchus chinensis* (Olivier)

Importance *Eucryptorrhynchus scrobiculatus* is a disastrous pest of *Ailanthus altissima*.

Distribution Beijing, Shanxi, Liaoning, Heilongjiang, Shanghai, Jiangsu, Henan, Sichuan, Shaanxi, Gansu, Ningxia.

Host *Ailanthus altissima*.

Damage Adults feed on young branch tips, leaves and cause defoliation and breaking of branches. Larvae feed on bark and xylem and may cause tree mortality. Infested branches show greyish white resin, feces, and wood dust.

Morphological characteristics **Adults** Length 13.5-18 mm, width 6.6-9.3 mm; long oval; black, shiny. Elytra with milky white, black and ochrous long scales; elytra thick and study, with punctures and scattered white and ochrous color.

Eggs Long oval, yellowish white.

Larvae Length 12-16 mm; head yellowish brown, thorax and abdomen milky white.

Pupae Length 11-13 mm, yellowish white.

Biology and behavior *Eucryptorrhynchus scrobiculatus* occurs one generation a year. The larvae and adults overwinter in 2-20 cm deep soil around the tree trunks or under the bark. The overwintering larvae pupate in May. Peak of adult emergence from pupae is from July to mid-August. The overwintering adults resume activity in late April and reach peak activity period in early and mid-May. Adults play death when disturbed. They feed on young branch tips, leaves for about one month and then lay eggs. Egg incubation period is about 8 days. Young larvae feed on bark, then move into xylem. Ma-

E. scrobiculatus adults feeding on trunk (Xu Gongtian)

E. scrobiculatus adult (Xu Gongtian)

E. scrobiculatus mating pair (Xu Gongtian)

E. scrobiculatus mature larva (Xu Gongtian)

E. scrobiculatus pupa (Xu Gongtian)

ture larvae pupate inside the galleries. The pupal period lasts for about 12 days.

Prevention and treatment

1. Mechanical control. During early to mid-May and late July to mid-August, collect adults by shaking the trees. At 30 cm above ground, place a plastic barrier that forming an umbrella shaped around the tree trunk, apply a layer of grease to prevent adults from climbing up the tree.

2. Chemical control. During adult period, spray 45% profenofos and phoxim mixture at 1,000×. Apply mixture of diesel and 2.5% deltamethrin (1:1) on the bark to kill larvae or apply 45% profenofos and phoxim mixture at 1,000× to the roots.

References Henan Forest Beureau 1988, Xiao Gangrou (ed.) 1992, Zhang Zhizhong 1992.

(Wang Shaoxia, Zhou Jiaxi)

E. scrobiculatus infested tree trunk (Xu Gongtian)

216 *Hylobitelus xiaoi* Zhang

Classification Coleoptera, Curculionidae
Common name Xiao pine weevil

Importance *Hylobitelus xiaoi* is an important wood boring insect affecting pine trees. Since 1988, it caused significant damages in Jiangxi.

Distribution Fujian, Jiangxi, Hubei, Hunan, Guangdong, Guangxi, Guizhou.

Hosts *Pinus elliottii*, *P. teada*, *P. massoniana*, *P. armandii*, and *P. thunbergii*.

Damage When *P. elliottii* is infested, the galleries produce a mixture of purple pulp and white sticky resin. When *P. teada* and *P. massoniana* are infested, the galleries produce white powders or packs of excretion.

Morphological characteristics **Adults** Females 14-16 mm in length, males 12-15 mm in length. Body dark black. Pronotum with rugose punctures, which longitudinally confluent. There are regularly arranged large punctures between the longitudinal lines on elytra.

Larvae White, with yellowish color; slightly "C" shaped; mature larvae 16-21 mm long.

Biology and behavior *Hylobitelus xiaoi* occurs one generation per two years in Jiangxi. It overwinters mostly as 5th to 6th instars in galleries and less commonly as adults in pupal chambers and in soils. The adults emerge in late February. They lay eggs in early May. Larvae hatch in mid-May. They stop feeding in late November. The larvae resume feeding in next March. Pupae appear in mid-August. Adults emerge from early September. Some adults come out from the hosts and overwinter in soil.

Adults fly very little. They are good crawlers and play death when disturbed. They hide inside crevices of barks at the base of tree trunks or in fallen leaves during the day and come out and feed on tender barks during sunset. The favorable temperature range for *H. xiaoi* is between 22-25°C and high relative humidity conditions. Eggs are laid in barks of host trees. After hatching, the larvae enter the phloem and feed there. The 5th instar may feed on cambium and disrupt the nutrient transport. *Hylobitelus xiaoi* is more common in high vegetation coverage and moist forest stands.

Prevention and treatment

1. Silviculture. Clear the shrubs and dead materials around the host trees and the excretion at base of trees.

2. Mechanical control. Open the barks and remove the larvae, pupae, and adults inside the galleries.

3. Biological control. During March and April, place cloth strips treated with *Beauveria bassiana* at 1 m above the ground on tree trunks. This is suitable for stands with > 70% infestation rate.

4. Chemical control. When infestation rate is > 75%, spray 8% cypermethrin at 20× from base of trees to 1.5 m high.

References Zhang Runzhi 1997; Wen Xiaosui, Kuang Yuanyu, Shi Mingqing et al. 2004a, b, 2006; Wen Xiaosui, Shi Mingqing, Kuang Yuanyu 2004, 2005; Peng Longhui, Xu Yongqing, Tang Yanrong et al. 2007; Tang Yanlong, Wen Xiaosui, Xu Yongqing et al. 2007; Wen Xiaosui, Shi Mingqing, Haack et al. 2007; Wen Xiaosui and Wang Hui 2007.

(Wen Xiaosui)

H. xiaoi adult (Zhang Runzhi)

217 *Hypomeces squamosus* (F.)

Classification   Coleoptera, Curculionidae
Common name   Gold dust weevil

Distribution   Jiangsu, Zhejiang, Anhui, Fujian, Jiangxi, Henan, Hubei, Hunan, Guangdong, Guangxi, Sichuan, Guizhou, Yunnan, Taiwan; Vietnam, India, Indonesia, Myanmar, the Philippines.

Hosts   *Camellia sinensis*, *C. oleifera*, *Citrus reticulate*, *Morus alba*, *Pinus massoniana*, *Quercus* spp., *Castanea mollissima* and near 100 species of forest trees, fruit trees, and crops.

Damage   Adults feed on young branches, buds, leaves, and can defoliate all leaves on a tree. The feeding decreases tree vigor and may cause death of trees.

Morphological characteristics   **Adults** Length 15-18 mm; spindle shaped. Overwintering adults purplish brown in soil; after emerging from the soil, their round punctures appear in coppery color; body covered with glittering green scales; some individuals are grey to greyish yellow and covered with yellow powder; some are covered with dense greyish or brownish scales. Dorsal surface of the head flattened, with 5 longitudinal grooves. Antennae short and stout. Eyes prominent. Front thorax wider than long. Dorsal surface with wide and deep median groove and irregular sculptures. Each elytron with 10 lines of punctures. Female scutum with

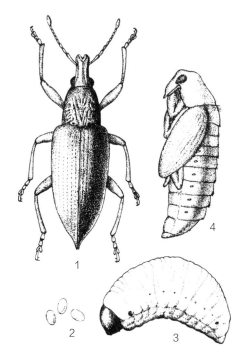

H. squamosus
1. Adult, 2. Eggs, 3. Larva, 4. Pupa (Zhang Xiang)

H. squamosus mating pair (male is on the top) (Li Zhenyu)

H. *squamosus* adult (Li Zhenyu)   H. *squamosus* adult (Wang Jijian)

few pubescence and smooth, the width at wing base wider than the postnotum, abdomen large. Male scutum with abundant pubescence, the width at base of wings as wide as the postnotum, abdomen small.

Eggs Length 1.2-1.5 mm, elliptical, greyish white.

Larvae Mature larvae 10-16 mm in length; milky white to light-yellow, head yellowish brown. Body slightly curved. Spiracles obvious, yellow; those on the pronotum and the 8th abdominal segment are much larger than the others.

Pupae Length 12-16 mm, milky white or light-yellow.

Biology and behavior *Hypomeces squamosus* occurs two generations a year in Fujian, Guangdong, Taiwan, and Guangxi. It occurs one generation a year in other provinces. It overwinters as adults or mature larvae. Adults appear in April to June. They are seen year around in Guangdong. In Zhejiang and Anhui, the insect overwinters mostly as larvae. Adult peak period is in June. They lay eggs in soil in August. In Xishuangbanna of Yunnan province, the peak adult emergence period is in June. In Fuzhou of Fujian province, adults resume activity in mid-April. The number of adults decreases significantly in mid-August. Eggs are seen from late April to mid-October. Pupation occurs in mid-September to mid-October. Those adults emerged in October overwinter in soil chambers.

Adults are active during the day. They chew leaves partially and create notches or only leave the veins intact. During the early morning and evening, they hide in grasses, among fallen leaves, or in soil. They are weak fliers and good crawlers. Individuals congregate and play death when disturbed. Each adult can mate multiple times. Eggs are laid separately on leaves over 80 days period. Each female lays about 80 eggs. Young larvae feed in soil at 10-13 cm depth. Larval period is about 80 days. Those overwintering larvae live about 200 days. Mature larvae pupate at 6-10 cm depth in soil. Pupal period lasts for 17 days.

Prevention and treatment

1. Mechanical control. Collect adults during their peak period by placing a plastic sheet under the infested trees and shaking the trees.

2. Trapping. Place a 10 cm wide sticky band at base of the tree trunk. It lasts for 2 months.

3. Using a repellent. Collect about 50 adults and crush them into pulp, add water, then place under the sun for 48 hours. Once a putrid odor appears, filter out the debris, dilute by 400×, then spray to trees. Adult *H. squamosus* avoid the smell and thus reduce the damage. A 72.33% reduction in adult numbers was reported after treatment.

4. Chemical control. Spray 90% trichlorfon, 50% DDVP at 1,000×, 50% phoxim at 800-1,000×, 44% cypermethrin + profenofos at 2,000×, 2.5% deltamethrin or 50% malathion at 3,000×, or 20% fenvalerate or 10% bifenthrin at 4,000×. Both canopy and the ground need to be sprayed to ensure the adults are directly sprayed.

References Xiao Gangrou (ed.) 1992; Zhang Lixia, Guan Zhibin, Fu Xianhui et al. 2002.

(Yi Xiangdong, Zhu Senhe, Li Zhenyu)

218　*Lepyrus japonicus* Roelofs
Classification　Coleoptera, Curculionidae

Importance　A widely distributed pest. It infests seedlings. Outbreaks of *L. japonicas* occurred in consecutive years in Liaoning, Jilin, and Shanxi.

Distribution　Beijing, Tianjin, Hebei, Shanxi, Inner Mongolia, Liaoning, Jilin, Heilongjiang, Shanghai, Jiangsu, Zhejiang, Anhui, Jiangxi, Shandong, Henan, Hubei, Hunan, Guangdong, Guangxi, Hainan, Sichuan, Guizhou, Yunnan, Tibet, Shaanxi, Gansu, Qinghai, Ningxia, Xinjiang; Japan.

Hosts　*Populus pseudo-simonii*, *P. beijingensis*, *P. canadensis*, *Salix matsudana*.

Damage　Both the larvae and adults cause damage to hosts. Larvae feed on roots of seedlings and kill the seedlings.

Morphological characteristics　**Adults** Length 13 mm; black; covered with greyish scales. Pronotum with one greyish yellow stripe on each side. Rear part of the elytra with a grayish yellow marking. It is located within the number 4-6 grooves.

Eggs　Length 1.5 mm, width 1.2 mm; elongated round; milky white.

Larvae　Mature larvae 10-12 mm; milky white; head brown.

Pupae　Elliptical; white; antennae protruded to the end of the front femora; hind legs covered by elytra.

Biology and behavior　*Lepyrus japonicus* occurs one generation a year. It overwinters as adults and larvae near the roots in soil. Seedlings are most susceptible. The insects resume activity in mid-April in northeastern China and in early April in Beijing. Peak activity period is mid-May in northeastern China and early May in Beijing. Adults lay eggs in early May in northeastern China and in late April in Beijing. Eggs are laid on topsoil. Egg incubation period is 8-10 days. Larvae dig into soil and feed on roots of seedlings. Most damage occur in mid-June in northeastern China and in early June in Beijing. Pupation occurs in mid-July. Adults emerge in early August in northeastern China and in late July in Beijing. Adults crawl fast, can fly for a short distance, and often rest in roots where the main roots meet. They are most active between 9:00-17:00. They are not active in areas with high soil moisture. Therefore, low areas tend to have less numbers. Adults mate during late August and early October. They overwinter in October.

Prevention and treatment

1. Mechanical control. Collect and kill aggregating adults.
2. Dilute *Beauveria bassiana* powder containing 5×10^{10} spores/g by 150×; pour 150-200 g solution to a 10×10 cm^2 pit near the seedling; then cover the pit with soil. An alternative method is to drill two 2-3 cm diame-

L. japonicus adults (Xu Gongtian)

L. japonicus pupal chamber (Xu Gongtian)

L. japonicus damaged roots (Xu Gongtian)

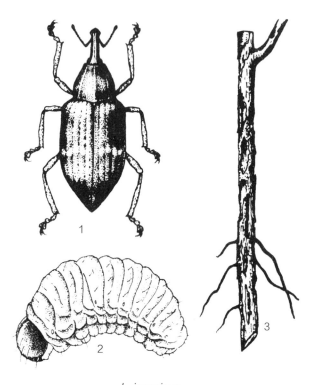
L. japonicus
1. Adult, 2. Larva, 3. Damage (Shao Yuhua, Zhu Xingcai)

ter, 14-15 cm deep holes around the base of the seedling, and inject the solution into the holes; apply 50-100 g solution per seedling.

3. Pour 3% permethrin-fenoxycarb EC at 3,000× or 50% of phoxim EC at 800, or 50% sumithion at 1,00-2,000× to roots of seedlings during the spring. Spray 25 g/L lamda-cyhalothrin at 1,000-2,000×.

References Baichen City Plantation 1975, Wang Runxi 1978, Li Yajie 1983, Xiao Gangrou (ed.) 1992, Xu Gongtian and Yang Zhihua 2007.

(Wang Jinli, Li Yajie, Lin Jihui)

219 *Niphades castanea* Chao
Classification Coleoptera, Curculionidae

Distribution Jiangxi, Henan, Shaanxi, Gansu.
Host *Castanea mollissima.*
Damage *Niphades castanea* larvae bore into the fruit and cause the involucres falling off the trees.
Morphological characteristics **Adults** Length 9-11 mm, width 4.5 mm; brown; each elytra with 10 black longitudinal grooves.
Larvae Length 15 mm; stout, slightly curved.
Biology and behavior *Niphades castanea* occurs one generation a year in Henan. It overwinters as larvae in fallen chestnut fruits inside involucres. Pupae appear in early April. After 20 days, adults emerge from pupae. In late June, adults mate and lay eggs. Eggs start to hatch in early July. The larvae start overwintering from late September to mid-October.

After adult appearance, they bore a hole and fly to trees and feed on barks of branches. They rest on leaf surface during the day and are active in the evening. They fall to the ground and play death once disturbed. Eggs are scattered around the base of the fruits. Usually one egg is laid on one fruit. Each female lays 3-35 eggs. After hatching, the larvae feed on the base of involucres, then migrate to the base of the fruits and feed on the fruits, causing the involucres falling off from the trees. Fallen involucres are most abundant in late August and

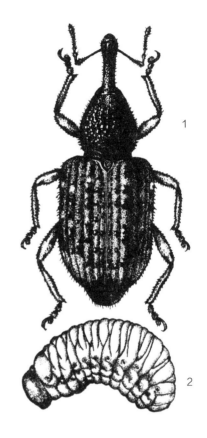

N. castanea. 1. Adult, 2. Larva (Zhang Xiang)

N. castanea adult (Zhang Runzhi)

early September. The larvae feed inside the fruits until late September and overwinter.

Prevention and treatment

1. Mechanical control. Collect and dispose fallen involucres in the winter and spring.

2. Chemical control. Spray 45% profenofos and phoxim mixture at 1,000× when adults are active in the trees.

References Yang Youqian 1982, Xiao Gangrou (ed.) 1992, Forest Disease and Pest Quarantine Station of Henan Province 2005.

(Yang Youqian)

220 *Niphades verrucosus* (Voss)

Classification　　Coleoptera, Curculionidae

Importance　*Niphades verrucosus* is an important wood boring beetle of weak pine trees. Aggregated larval feeding under the bark creates irregular galleries and destroys the phloem, resulting in separation of bark and sapwood that leads to tree mortality from water deficit.

Distribution　Zhejiang, Anhui, Fujian, Jiangxi, Hunan; Japan.

Hosts　*Pinus massoniana*, *P. taiwanensi*, *P. thunbergii*, *P. elliottii*, *P. armandii*, *P. taeda*, and *Pseudolarix kaempferi*.

Damage　Larvae feed on phloem and xylem, creating galleries filled with sawdust and brown fecal pellets. Larval galleries inter-connected when population density is high. Mature larvae create longitudinally arranged pupal chambers in the sapwood and are covered by silk.

Morphological characteristics　**Adults** Length 7.1-10.5 mm. Body dark brown, with conspicuous scattered punctuations on the head. Scutellum with white hairs. Elytra covered with rusty brown and white scales. Scaly white hairs near apex of femora circularly arranged. Scaly white hairs are also found in the abdomen. **Larvae** Length 9.0-15.6 mm. Body yellowish. Head yellowish brown. Mandibles dark brown.

Biology and behavior　*Niphades verrucosus* usually occurs two generations a year in Zhejiang, with a few one-generation populations. It overwinters as mature larvae inside the bark. Pupation occurs between late Mach and mid-June. Adults of the overwinter generation can be found between early April and late October. First generation eggs are laid between early May to late June, with larval stage lasts from mid-May and early August. First generation adults start to emerge in early July and end in early November. Second generation larvae appear from late September to late May the next year.

Adults emerge at any time of the day with peak emergence period between 12:00-16:00. Adults are good crawler and prefer *P. massoniana* flowers and tender bark. They exhibit water feeding and thanatosis behavior. Mating occurs mostly between 20:00-24:00 with the male on top of the female. Each mating lasts for an average of 11 minutes. Both males and females can mate multiple times during their lifetime. Adult longevity ranges from 41 to 126 days, with an average of 111.7 days. Egg stage lasts for 3-4 days. Larvae are mostly found on trunk below 2 m above the ground. Mature larvae build longitudinally arranged pupal chambers in the sapwood, creating audible sound during the process. Pupal chamber 1.1-3.3 cm long. Pupal duration ranged from 9 to 21 days, with an average of 13.8 days.

Niphades verrucosus occurs mostly in high elevation pine stands with moist and fertile soil and dense grass on the floor, or lumberyard with newly felled pine logs. They often co-exist with *Shirahoshizo patruelis*.

Prevention and treatment

1. Bait logs. Place freshly cut pine logs to attract adults. Debark and destroy larvae inside.

2. Mechanical control. Apply insect glue to base of tree trunk to kill larvae.

3. Natural enemy conservation. Suspend chemical treatment during the flight season of *Dolichomitus* sp. (Hymenoptera: Ichneumonidae), a larval and pupal parasitoid of *N. verrucosus* with a parasitism rate up to 32.7%.

4. Pheromone trapping. Trap and kill *N. verrucosus* adults during flight season with wood-boring beetle pheromone (Chinese patent no. ZL03115289.9) produces satisfactory results.

5. Apply *Beauveria bassiana*.

References　Zhao Jinnian 1987, 2001; Xiao Gangrou (ed.) 1992.

(Zhao Jinnian)

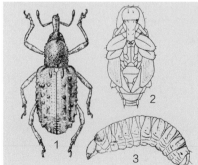

N. verrucosus 1. Adult, 2. Pupa, 3. Larva (Qu Xiaojin)

N. verrucosus adult (Zhang Runzhi)

N. verrucosus damage (Luo Youqing)

221 *Pissodes punctatus* Langor et Zhang

Classification Coleoptera, Curculionidae
Common name Armand pine weevil

Importance *Pissodes punctatus* is a polyphagous species that attacks 17 pine species in China. It is responsible for some of the most serious damages on conifer forests in the country.

Distribution Sichuan, Guizhou, Yunnan, Gansu.

Hosts *Pinus tabulaeformis*, *P. armandii*, *P. yunnanensis*, and *P. massoniana*.

Morphological characteristics **Adults** Female length 4.5-8.5 mm, width 1.6-3.1 mm. No puncture on snout and head. Snout slightly longer than pronotum. Antennae inserted on the snout where it was slightly narrower than the snout tip. Antennal club 1.6 times longer than wide. First abdominal sternite convex. Male length 4.3-8.2 mm, width 1.5-3.0 mm. Black. The area between head and pronotum is reddish brown, with white setae of various shapes and sizes.

Eggs Length 0.6-0.8 mm, width 0.4-0.55 mm. Oval, milky white.

Larvae Length 8.1 mm on average. Crescent, milky white. Legless. Head brownish. Body covered with backwards spinules.

Pupae Length 8.4 mm on average. Milky white, turn to dark brown before emergence.

Biology and behavior *Pissodes punctatus* occurs one generation a year in Sundian area in Yunnan province and overwinters as mature larvae inside pupal chambers. Pupation occurs in April and peaks between late April and early June. Pupal stage lasts for 24-36 days. Adults start to emerge in late June and peak emergence falls between mid-July and early August. Adults mate after about 10 days of supplemental feeding. Females begin to lay eggs 10 days later in late July, with a peak egg- laying period in August. Egg stage lasts for 18-20 days. Eggs start to hatch in early August and peak between mid-August and early September. Mature larvae appear in early October. They overwinter between late December and early January of the next year in pupal chambers. The adults are not very selective toward host plants when it comes to supplemental feeding. They tend to feed on the upper and middle trunks of healthy trees. Eggs are laid mainly on the middle and lower trunks. Pupal chambers and adult exit holes are concentrated on the middle and lower trunks of infested trees. The number of pupal chambers is correlated with the height of the trunk. Aggregate distribution is observed for eggs, pupae, and adult exit holes on *Pinus armandii*. Female lays its eggs at different periods in her lifetime, resulting in overlapping generations. Weak trees are more susceptible to this pest.

Prevention and treatment

Managing *P. punctatus* relies on integrated silviculture measures throughout the entire stand cultivation process to enhance pest resistance of *Pinus armandii*. Semiochemicals can be used to monitor pest populations. The following options can be considered:

1. Silviculture. Use high quality seeds to produce seedlings locally for scientifically designed mixed forest stands. Broad-leaved species such as *Alnus nepalensis*, *Coriaria sinica*, *Ligustrum lucidum*, *Myrica rubra* are fire resistant and therefore should be used as belt or block in mixed forest sands. Thinning may be conducted to increase tree vigor. Prevent forest fire and other forest

P. punctatus adult (FPC)

P. punctatus damage to pine forest (FPC)

P. issodes punctatus damage (FPC)

P. punctatus newly emerged adults (FPC)

Early stage damage by *P. punctatus* (FPC)

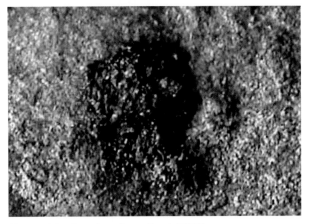

P. punctatus invasion site on trunk (FPC)

pests to promote tree growth and pest resistance.

2. Remove infested trees. Remove infested trees between November and May of the next year during the larval overwintering period. Removed trees need to be debarked or fumigated with aluminum phosphide, and branches and barks be burned to destroy mature larvae inside. Replant in a timely fashion to fill the open space left by tree removal.

3. Biological control. Apply mixture of *B.t.* and avermectins twice within 15 days at 10-15 days interval after peak adult season provides effective control. *Beauveria bassiana* is also proven effective.

4. Chemical control. When pest population exceeds the economic threshold, fumigate with avermectins, or apply 12% thiamethoxam and lamda-cyhalothrin mixture at 30-60 to tree trunks.

References Chai Xiushan and Liang Shangxing 1990; Duan Zhaoyao and Lei Guilin 1998; Langor 1999; Li Shuangcheng, Ma Jin, Gui Junwen et al. 2000; Li Shuangcheng, Li Yonghe, Ma Jin et al. 2001; Li Yonghe, Xie Kaili, and Cao Kuiguang 2002; Lei Guilin, Duan Zhaorao, Feng Zhiwei et al. 2003; Feng Shiming and Situ Yingxian 2004; Liu Jühua, Xiao Yinsong, Luo Zhengfang et al. 2005; Liu Shouli, Yang Shuiqiong, Yang Yuekui et al. 2005; Wang Ge 2006; Ze Sangzi, Yan Zhengliang, Zhang Zhen et al. 2010.

(Zhang Zhen, Wang Hongbin, Kong Xiangbo)

222 *Pissodes validirostris* (Sahlberg)

Classification Coleoptera, Curculionidae
Common name Pine cone weevil

Importance *Pissodes validirostris* is a major cone pest of *Pinus sylvestris* var. *mongolica*. Feeding by larvae and adults on cone scales and seeds leads to premature drop of cones.

Distribution The Greater Khingan Mountains and the Hulunbuir Plateau in Inner Mongolia Autonomous Region, The Greater Khingan Mountains in Heilongjiang province, Shanxi, and the Qilian Mountain forest district in Gansu province in China; Russia, Japan, Korea Peninsula, Turkey, Finland, Poland, Hungary, Germany, France, Spain.

Hosts *Pinus sylvestris* var. *mongolica*, *P. armandii*, *P. tabuliformis*, *P. sylvestris*, *P. pinea*, *P. thunbergii*, *Pseudotsuga menziesii*.

Damage Most damage is caused by larval feeding inside the cones. Initially the area around egg hole becomes dark brown. Galleries constructed along the scales create a dark brown, resin covered serpentine mark on the surface of the infested cone. Larvae feed at the base of the scales and the axis of the cones later, leading to complete consumption of the axis during heavy infestation. Damaged cones fall off early, especially by mid-August. In Jilin, *P. validirostris* is a major pest of *P. sylvestris* var. *mongolica* that infests branches and main trunks of weak trees. Tree mortality occurs due to the disruption of fluid flow inside the tree after infestation. Heavier infestations occurred on single trees compared to stands, on south side compared to north side, in mixed stands of pine and birch or pine and poplar compared to pure stands, and in stands with lower percentage of *P. sylvestris* var. *mongolica*.

Morphological characteristics **Adults** Length 5.5-6.3 mm, width 1.8-2.0 mm. Body dark brown and covered with punctures. Punctures with white or brick red plumose scales. Rostrum black, cylindrical, slightly enlarged apically and extends forward and downward. Head triangular, dark brown. Antennae geniculate, arising laterally from the middle of the rostrum. Scape long, club 8-merous. Pronotum pointed anteriorly, wider from middle toward the posterior margin, arc-like laterally, with a white spot made of plumose scales on either side. Median carina conspicuous. Area connecting head and pronotum brown. Mesosternum connects to metasternum with a band made of yellow scales. There is a white circular blotch between posterior pronotal margin and base of the elytra. Humerus right-angled. Elytra gradually narrowed towards the end and cover the entire abdomen. Elytra of the male extend slightly over the abdomen, whereas elytra of the female do not surpass the abdomen. Each elytron has 11 striae made by punctures and two irregular-shaped transverse bands made by white and yellow scales. The anterior band may be blotch-like while the posterior band wide and conspicuous, covering almost the entire width of the elytron. Legs covered with white scales. Hind tibia with dentate setae externally. Tarsus 3-segmented, with yellow pubescence ventrally.

Eggs Length 0.8-0.9 mm, width 0.5-0.6 mm. Oval, creamy yellow, translucent. The slender end (about 1/4 of the egg length) slightly transparent.

Larvae Milky white, cylindrical, with five instars. First instar length 0.7-1.7 mm, head width 0.3 mm, translucent. Head yellowish, without setae, with undivided body segments. Second instar length 1.7-3.3 mm, head width 0.5 mm. Head yellow, without setae but with obvious body segments. Third instar length 3.2-7.0 mm, head width 0.8 mm. Creamy yellow, with setae on head. Body segments obvious. Both the 4th and 5th instar are white, with brown head and mouthparts, head covered with setae, with well-developed body segments. Forth instar length 5.0-8.0 mm, head width 1.1-1.15 mm, and fifth instar length 7.0-8.7 mm, head width 1.3 mm.

Pupae Length 6.0-6.6 mm. Milky white. Appendages closely attached to the body, with a movable abdomen.

Biology and behavior *Pissodes validirostris* occurs one generation a year and overwinters as adults under the bark on trunk or large branches. Adults become

P. validirostris larvae (Yan Shanchun)

active in mid-May of next year. Females start to lay eggs between late May and early June with a peak egg period in mid- to late June. Egg stage lasts for 10-13 days. Most eggs hatch in early to mid-June. Larval stage lasts for 40-45 days. Pupation occurs in mid- to late July and ends by early September. Pupal stage lasts for 10-15 days. Adults start to emerge in early August with peak emergence occurs between late August and early September. Emergence ends by mid-September.

Adults are phototaxic and do not fly after emergence, instead they move to current year branches or leaf sheaths for supplemental feeding before gradually entering the bark on the trunk or large branches to overwinter. For those emerged between mid- to late September, no supplemental feeding occurred before overwintering. Adults do not migrate after they move up to the host trees. They stay inside deep crevices inside the bark they encountered first until ideal locations are found. Adults generally overwinter individually, rarely with 3-4 individuals together. They overwinter on the trunk and branches from 1.0 m above the ground to 1.2 m from the tree tip. More than 50% of the adults stay on trunks and branches between 5-7 m about the ground, making this section the most frequently used overwintering location. Adults are found mostly on south-faced trunks or right above or below the branch bifurcation. Adults may also overwinter at the tree base of smaller trees. Most adults become active and start to lay eggs next spring. About 20% of the overwintering adults remain in diapause according to observations made between 1962-1964.

Overwintering adults become active only when the average environmental temperature reaches above 10°C and the minimum temperature reaches above -2°C. Supplemental feeding usually occurs between 8:00-12:00 am. They may hide under bark again when the temperature drops suddenly. Most adults become active in early

P. validirostris damage (Yan Shanchun)

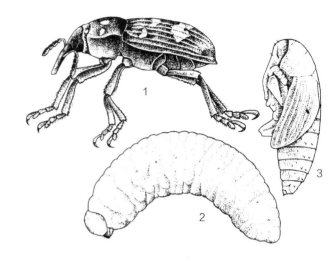

P. validirostris. 1. Adult, 2. Larva, 3. Pupa (Zhu Xingcai)

June when average temperature reaches 18°C. They feed on male flowers and tender cone scales of *P. sylvestris*. Adults spend most of their time on treetops and the south-facing side during the day and hide at base of the cones during the night or on windy days. Adults mate multiple times. Mating occurs mostly between 9:00-15:00 when air temperature reaches 18-20°C. Males initiate the mate search first. A male can mate with the same female for up to four times. Adults feed on young cones for supplemental nutrition during this period. A female chews egg pits on cone scales with its snout before laying eggs. One egg is laid in each pit. About 3-5 eggs are laid on each cone. A maximum of 28 eggs were observed on one cone. Each female lays 30-31 eggs per year. Females enter the bark to overwinter the second time in early July. They mate and lay eggs again next year.

Prevention and treatment

1. Mechanic control. Collect fallen cones from the ground by hand and destroy the emerged weevils.

2. Biological control. Utilize parasitoids from infested cones such as *Scambus brericornis*, *S. eurygenys*, *S. punctatus*, *S. sagax*, *S. sudeticus*.

3. Chemical control. Spray with the mixture of dichlorvos and 40% dimethoate (1:1) at 1,300×, 1% matrine wettable powder at 800×, 0.5% veratridine at 500×, or fumigate with 741 dichlorvos (15 kg/ha) resulted in 90% control in the laboratory and in the field.

References Li Yabai, Li Shuyuan, He Weile 1981; Meng Xiangzhi, Ji Yuhe, Xun Xiufeng 2000; Li Chengde 2003; Deng Xun, Ma Xiaoqian, Wei Xia et al. 2009; Sheng Maoling and Sun Shuping 2010.

(Yan Shanchun)

223 *Pissodes yunnanensis* Langor et Zhang

Classification Coleoptera, Curculionidae
Common name Yunnan pine weevil

Importance *Pissodes yunnanensis* prefers to attack 8-12 years old pine trees. They damage 3-5 years old branches or the top of the main trunk of 4 cm diameter, resulting branch dieback and "headless" trees. Repeated infestation leads to death of young trees in large areas.

Distribution Sichuan, Guizhou, Yunnan.

Hosts *Pinus yunnanensis*, *P. densata*, *P. massoniana*.

Damage Larvae feed inside the pith of young branches or the cambial layer and outer sapwood of the main trunk.

Morphological characteristics **Adults** Length 6.6-7.2 mm, width 2.7 mm; mostly brown. The apical two thirds of the snout, lateral and posterior margins of the pronotum, elytra except the transverse fascia, ventral surface, and legs dark brown. Dorsal and ventral surfaces moderately covered with recumbent, whitish, scale-like setae of various sizes and shapes. Punctures reticulate and similar in size. Snout as long as pronotum, slightly curved and wider apically, with dense and rugose punctures at the base. Head with sparse punctures and scales. Antennae inserted slightly behind the midpoint of the snout; scape shorter than funicle. First funical segment 10% longer than the second and third segments combined. Club oval, 1.5 times longer than wide. Pronotum wider than long, but 15% narrower than the elytra. in dorsal view, sides moderately rounded in basal 80%, width at base slightly narrower than the widest point, deeply constricted anteriorly to form a "collar"; hind angles subrectangular, posterior margin moderately sinuate; in side view, area of disk immediately posterior to collar not elevated, sides, collar, and disk densely and shallowly to moderately puncture, tending to rugose at places on disk; disk lacking mid-dorsal impressions and median carina; moderately covered with oval and elongate scales, with concentrations along midline and side margins. Scutellum small, oval, elevated, and squamous. Elytra sides nearly parallel in anterior two thirds, widening only slightly from humeri to antedeclivity fascia, thereafter narrowing abruptly to posterior apex; anterior margin strongly sinuate; humeri rounded, intervals mostly straight from anterior margin to declivity, intervals 3 and 5 strongly elevated and broader than other intervals, intervals 7 and 9 slightly elevated over posterior half; interval 3 strongly elevated to anterior margin, forming a depression with the humeri. Interval 5 forming a well-developed declivital callus; surface of intervals slightly rugose, striae distinct with shallow to moderate deep, well-separated, round to oval punctures; antedeclivial transverse fascia of whitish oval to elongate scales, best developed between interval 2 and 8; remainder of elytra with sparse, oval to elongate, whitish scales. Venter densely and shallowly punctures. Femora and tibia densely covered with elongate scales, scales not forming a band on the metafemur.

Larvae Length 8.0-9.0 mm; crescent; reddish. Labial sensory organs: 1 pair anteriorly, 3 pairs laterally, one long pair inside the region with a pair of small sensory organs anteriorly and posteriorly. Mandibles with 3 teeth on each side, inner tooth sharp, long and curved; with one seta on the outer surface. Maxillae with 5 equal-sized finger-like structures on the inner side and 2 setae under-

P. yunnanensis adult female (FPC)

P. yunnanensis adult emerging from tree trunk (FPC)

P. yunnanensis larvae (FPC)
P. yunnanensis pupa (FPC)

P. yunnanensis infested forest (FPC)
P. yunnanensis damage to pine shoot (FPC)

side. Spiracles ring-shaped with double openings, with secondary opening surrounded by petal-shaped circular elevation. Cyclorrhapha and secondary spiracle obvious.

Biology and behavior *Pissodes yunnanensis* undergoes one generation per year and overwinters as larvae inside infested branches. Outbreaks occur mostly in Yunnan pine stands in high mountain areas between 2,400-2,800 m of elevation. Aerially seeded stands, north-facing slopes, lowlands, and clay soil favor the *P. yunnanensis* outbreaks. Adults prefer warm weather. They avoid direct sunshine while hiding among the needles during the day, and feed during the night. Supplemental feeding before oviposition lasts for two weeks. Eggs are laid mostly on branch joints two weeks after the mating with 1-2 eggs are laid each time. Larvae feed on pith, cambial region and the outer sapwood. Mature larvae construct pupal chambers and pupate inside. Adults emerge on sunny days in early May. Eggs are laid between late May and mid-August. Adults either die or enter diapause by late September. Eggs hatch in early June. Larvae feed and overwinter inside the branches. Pupation occurs from late March to late April the next year.

Prevention and treatment

1. Remove and destroy infested branches by burning between March and April each year before pupation. Debark infested logs or fumigate logs with aluminum phosphide at 20 g/m^3 for three days.

2. Chemical treatment. Apply 4.5% cypermethrin EC ULV spray when adult exit holes appeared on 30% of the infested branches. Apply 12% thiamethoxam and lamda-cyhalothrin mixture at 30-60× to tree trunks.

References Sichuan Disease and Forest Pest Survey Office 1985, Langor et al. 1999, Zhang Yining 1999, Situ Yingxian et al. 2002.

(Situ Yingxian)

224 *Rhynchaenus empopulifolis* Chen
Classification Coleoptera, Curculionidae

Importance Damage from *Rhynchaenus empopulifolis* weakens poplar trees, resulting in outbreak of canker disease, leaf spot disease, and secondary insect pests such as longhorned beetles that lead to more serious damages to host trees.

Distribution Beijing, Hebei, Shanxi, Inner Mongolia, Liaoning, Jilin, and Shandong.

Hosts *Populus simonii*, *P. cathayana*, *P. beijingensis*, and *P. canadensis*.

Morphological characteristics **Adults** Length 2.3-2.7 mm, width 1.3-1.5 mm; near oval; dark brown to black. Snout, antennae, and most of the legs yellowish brown. Coxae and femora apex reddish brown to dark brown. Pronotum covered with inward-cured yellow recumbent hairs. Elytral interspace contains scattered brownish recumbent short hairs in addition to a line of brown recumbent long hairs. Scutellum covered with white scales. White or brownish hairs also found around the compound eyes, on abdominal sternites, and on legs.

Biology and behavior *Rhynchaenus empopulifolis* occurs one generation a year in Beijing and two generations a year in Shandong. It overwinters as adults. Adults resume activity in late March and feed on buds, young leaves, and the lower epidermis and parenchyma of the mature leaves. Adults start to mate in early April and lay eggs in mid-April. Egg hatching lasts for half a month. Young larvae mine into parenchyma of the leaves. Mature larvae enter prepupal stage and fall onto the ground along with the leaf pods in late April. Prepupal stage ends in early May. Pupal stage lasts for 10 days and adults begin to emerge in early May. Emergence period lasts for 20 days. Adults feed on leaf parenchyma until late October. In late September, adults move down and feed on leaves of the lower crown. Adults overwinter in leaf litters, under the rocks, or inside the soil in late October.

Prevention and treatment

1. Biological control. Natural enemies of *Rhynchaenus empopulifolis* include *Pteromalus miyunensis*, *P. procetus*, *P. semotus*, and *Triaspis* sp.

2. Chemical control. Soil treatment with 25 g/L lamda-cyhalothrin at 1,000×. Trunk injection of 20% Confidor SL (0.4 mL/cm dbh) for control of *Rhynchaenus*

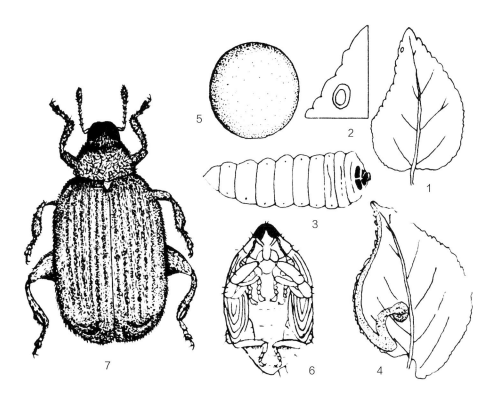

R. empopulifolis. 1. Egg-laying location, 2. eggs, 3. Larva, 4. Feeding gallery inside leaf, 5. Leaf pod, 6. Pupa, 7. Adult (Zhang Lianxiang, Chen Yuanchao)

R. empopulifolis adult (Xu Gongtian)

R. empopulifolis adult (Xu Gongtian)

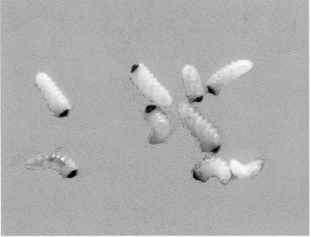
R. empopulifolis young larvae (Xu Gongtian)

R. empopulifolis leaf pods (Xu Gongtian)

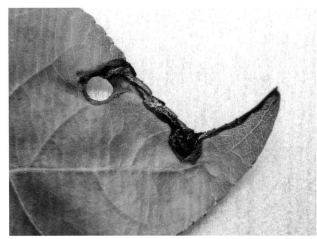
R. empopulifolis larva feeding gallery (Xu Gongtian)

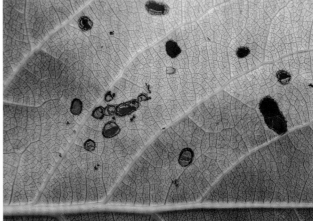
R. empopulifolis adult feeding marks (Xu Gongtian)

empopulifolis and *Stilpnotia candida*. Spray 12% thiamethoxam and lamda-cyhalothrin mixture at 1,000× during larval period.

References Xiao Gangrou (ed.) 1992; Wang Xiaojun, Yang Zhongqi, Wang Xiaoyi 2006; Yao Yanxia and Yang Zhongqi 2008; Hou Yaqin, Wang Xiaojun, Li Jinyu et al. 2009.

(Hou Yaqin, Wang Xiaojun, Li Jingyu, Zhang Lianxiang, Chen Yuanchao)

225 *Scythropus yasumatsui* Kono et Morinoto

Classification Coleoptera, Curculionidae

Distribution Beijing, Hebei, Shanxi, Liaoning, Jiangsu, Henan, Shaanxi, Gansu.

Hosts Mostly on *Ziziphus jujube*, but can also attack *Malus pumila*, *Pyrus bretschnerideri*, *Juglans regia*, *Populus* spp., *Paulownia* sp., *Morus alba*, *Gossypium* spp., and *Glycine max*.

Damage *Scythropus yasumatsui* is an early season defoliator of jujube. Adults feed on young buds. They can destroy all the buds during heavy infestation, resulting in abnormal bud development ("chorista" by jujube farmers) or secondary development. Infestations cause delayed flowering and fruit setting. After leaves are fully spread, adult feeding creates half circles or notches. Larvae can also damage the underground roots.

Morphological characteristics **Adults** Grayish white, males with darker color. Snout strong. Depressed between compound eyes on vertex. Antennae geniculate, brown. Head wide. Snout short, slightly wider than long. Pronotum grayish brown. Elytra arc-shaped, with 10 striae on each side. Striae interspace contains black scaly hairs. Elytra surface with a looming brown halo blotch. Abdominal sterna sliver. Hind wings membranous and are used for flying. Femora toothless, claws connate.

Larvae Fusiform, curved, legless. Pronotum yellowish; abdomen milky white; head brown.

Biology and behavior *Scythropus yasumatsui* occurs one generation a year in the Yellow River basin and overwinters as larvae underground. Larvae pupate in early April. Adults emerge in mid- to late April. They feed in groups on developing buds, resulting discolored and damaged bud tips. After leaves are fully spread, adult feeding creates semi-circles or notches on leaves. Feeding occurs the most during the noon time in May when temperature is relatively low. They are not active in early morning or during the night. They hide in fruit spurs or branch joints and fall to the ground when disturbed. Adults will fly back to the tree in the air during the fall or fly back to the tree after falling on the ground in warm weather. Adults live for about 70 days. Mating and egg laying occur between late April and early May. Eggs are laid on fruiting branches or roots inside the soil. Eggs hatch in mid-May and larvae enter the soil to feed

S. yasumatsui adults (FPC)

S. yasumatsui damage (FPC)

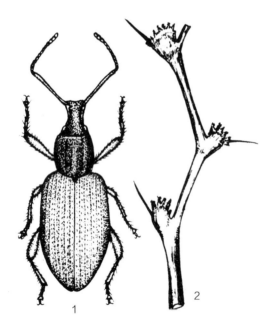

S. yasumatsui. 1. Adult, 2. Damage (Yang Youqian)

S. yasumatsui damage (FPC)

on plant roots.

Prevention and treatment

1. Mechanical control. During the early morning or evening, cover the ground with plastic and shake the trees, collect and kill the adults falling to the ground.

2. Chemical control. Soil application of 1:150-200 solution of 50% phoxim at the 1m diameter area around the base of the tree during adult emergence. Dig 5 cm deep trenches around the tree base and apply Sevin dust to kill emerging adults. Spray 25 g/L lamda-cyhalothrin or 20% dinotefuran at 1,000-1,500× to control adults, Spray 12% thiamethoxam and lamda-cyhalothrin mixture at 1,000×, or 50% imidacloprid and monosultap mixture at 600-800× to the crown during the larval period.

References Yang Youqian et al 1982, Tian Guanghe 1991, Xiao Gangrou (ed.) 1992, Song Jinfeng 2004, Henan Forest Pest Control and Quarantine Station 2005, Hu Weiping and Liang Tingkang 2008.

(Sheng Maoling, Yang Youqian)

226 *Shirahoshizo coniferae* Chao

Classification Coleoptera, Curculionidae

Importance *Shirahoshizo coniferae* is one of the most important seed and cone pests of *Pinus armandii*. Adults feed on young branch tips during supplemental feeding. Larvae feed on seeds and cone scales. They severely hinder the normal growth of trees in naturally regenerated forests or seed plantations.

Distribution Sichuan, Yunnan, Shaanxi.

Hosts *Pinus armandii* and *Keteleeria evelyniana*.

Damage *Shirahoshizo coniferae* adults feed on young branches and cause tip dieback. Larvae feed on cone scales and seeds, resulting fragile, grayish brown cones with shrinking skins. Damaged seeds contain no kernels. Circular holes on seed shells are plopped by filament sawdust.

Morphological characteristics **Adults** Length 5.2-6.5 mm, reddish brown or dark brown with dark brown scales. Pronotum with four small blotches made of white scales. Elytral interspace 4 and 5, as well as stria 3 contains a white scale blotch. Scattered white scales are also found on pronotum and elytra. Dark brown scales concentrated on the frons; anterior portion, besides the median line, and lateral margins of the pronotum; and elytral interspace.

Eggs Length 0.7 mm; oval; yellowish; translucent.

Larvae Length 6.0-8.0 mm at maturity. Body yellowish white, head pale brown.

Pupae Length 7 mm. Exarate; whitish yellow, compound eyes purple.

Biology and behavior *Shirahoshizo coniferae* occurs one generation a year and overwinters as adults in the soil or inside the seeds within the cones or the inner sides of scales. Adults reappear in large numbers in mid-May the next year and start to lay eggs in early June. Eggs are laid in masses

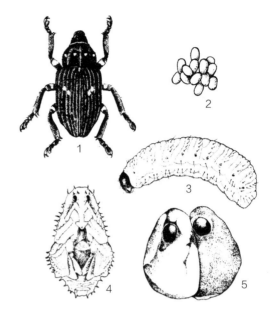

S. coniferae. 1. Adult, 2. Eggs, 3. Larva, 4. Pupa, 5. Damaged seeds (Zhu Xingcai)

under the epidermis on the upper edges of 2nd year cone scales. Scales waxy yellow with resin flow after eggs were laid inside. Each cone usually has 1-3 egg masses. Up to 6 egg masses may be found on a cone. Each egg mass contains an average of 10 eggs, with a maximum of 28. Eggs hatch in late June and larvae feed on epidermal tissues inside the scales by lining up downwards until reaching the base of the scales. They will then feed separately on seeds or the inner side of other scales. Larvae continue feeding until pupation in August.

Prevention and treatment

1. Collect and destroy infested cones.

2. Quarantine. Fumigate infested seeds with aluminum phosphide at 9-30 g/1,000 kg of seeds for 3 days.

3. Aerial application of 1:30-50 solution of 50% sumithion EC or 25 g/L lamda-cyhalothrin at 10×. Apply 12% thiamethoxam and lamda-cyhalothrin mixture at 1,000×, or 50% imidacloprid monosultap mixture at 600-800× to the crown until trunks are wet. Sealing the trunk with plastic film after spray will increase the control efficacy.

References Li Kuansheng et al. 1966, Zhao Yangchang and Chen Yuanqing 1980, Chinese Forest Seeds Coorporation 1988, Wu Chunsheng 1988, Xiao Gangrou (ed.) 1992, Li Kuansheng 1999.

S. coniferae adult (Zhang Runzhi)

(Li Li, Li Kuansheng)

227 *Shirahoshizo patruelis* (Voss)
Classification Coleoptera, Curculionidae

Importance *Shirahoshizo patruelis* larvae feed under the bark of weak Masson pines and other pine trees and lead to tree mortality. Wood of infested trees is susceptible to fungal decay, rendering the timber unusable.

Distribution Jiangsu, Zhejiang, Anhui, Fujian, Jiangxi, Hubei, Hunan, Guangxi, Sichuan, Guizhou, Yunnan, and Taiwan; Japan.

Hosts *Pinus massoniana, P. taiwanensis, P. thunbergii, P. armandii, P. elliottii, P. taeda, P. yunnanensis, P. serotina,* and *Pseudolarix kaempferi*.

Damage *Shirahoshizo patruelis* larvae create cave type galleries and construct pupal chambers inside the outer sapwood longitudinally. Pupal chambers oval and covered with loose fibers. The bark is separated from the wood under high population densities.

Morphological characteristics Adults Length 4.7-6.8 mm; reddish brown or greyish brown. Pronotum with a line of four white blotches of scales in the middle. Elytra have two small white blotches both anteriorly and posteriorly.

Larvae Length 7.0-12.0 mm. Yellowish white, head brownish; slightly curved.

Biology and behavior *Shirahoshizo patruelis* occurs one to four generations a year depends on location. In Kaihua County in Zhejiang province, two generations occur in a year. It overwinters as middle stage instar larvae inside the bark. Larvae mature in mid-March the next year and pupate from mid-March to early June. Adults emerge in mid-May. First generation larvae cause damage from late May to late July. First generation adults appear in late July. Second generation larvae start to emerge in early August. They start to overwinter in late November.

Newly emerged adults stay inside pupal chambers for 4-6 days before chewing circular exit holes with a diameter of 2.0-3.5 mm. Most adults emerge between 14:00-18:00. Adults are good crawlers, phototactic, and play death. Apparent death lasts for 2.0-2.5 seconds with ventral part of the body exposed and legs closely attached to the abdomen. This is very different from the dead adults

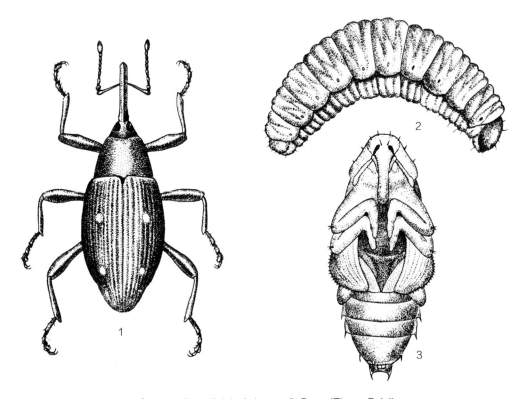

S. patruelis. 1. Adult, 2. Larva, 3. Pupa (Zhang Peiyi)

S. patruelis infested forest (Zhang Runzhi)

S. patruelis damage (Zhang Runzhi)

where legs are fully extended backward. Adults feed on bark of young pine branches and live for 41-62 days. Newly hatched larvae are very active and feed inside the bark, creating meandering galleries. Older larvae feed around original galleries and create cave-type galleries. Pupal chambers 1.1 cm in length, constructed in the outer sapwood, filled with sawdust and frass. Pupal stages last for 11-18 days.

Prevention and treatment

1. Sanitation. Remove weakened, fallen, and damaged trees from winds and snowfall timely to eliminate pest propagation sites.

2. Trapping with trap logs or lures. Place trap logs or traps baited with general woodborer lures during adult season in mid-May to June. Peel barks of trap logs to kill young larvae or collect and kill adults from the traps.

References   Zhao Jinnian 1988.

(Zhao Jinnian)

Sternochetus gravis (F.)

Classification	Coleoptera, Curculionidae
Synonyms	*Cryptorrhynchus gravis* (F.), *Cryptorrhynchus frigidus* (F.), *Rhynchaenus frigidus* (F.), *Curculio frigidus* F.
Common name	Mango pulp weevil

Importance *Sternochetus frigidus* is the most serious pest of mango that could infest 20%-50% of the fruits. Larvae feed on mango pulp and contaminate it with frass, resulting loss of value. It was on the entry quarantine pest list of China as a second category target in the 1990's.

Distribution Guangxi, Yunnan; Thailand, Myanmar, Malaysia, India, Indonesia, Pakistan, Papua New Guinea, and Bangladesh.

Hosts *Mangifera sylvatica* in the tropical forests, *M. siamensis*, various cultivars of *M. indica*.

Damage *Sternochetus gravis* larva feeds and creates irregular gallery inside mongo pulp until fruit maturation. It then makes a larger hole inside the gallery to construct a solid pupal chamber filled with frass and pupates inside. Adult emerges from the fruit by chewing an exit hole in late maturation period.

Morphological characteristics **Adults** Length 4.1-5.2 mm, with a ratio of 5:4 between length and width. Body oval, yellowish brown, covered by bright yellow, brownish, dark brown, or black scales. Antennae rusty red. Head full of punctures, with erect dark brown scales. Snout curved, slightly wider posteriorly, covered by deep punctures. Median carina conspicuous. Antennae arise from one third of the snout posteriorly. Funiculus segment 1 and 2 equal in length, and segment 3 slightly longer than wide. Club oval, two times as long as wide, pilose, and without obvious sutures. No middle depression on frons. Pronotum 1.3 times longer than wide, lateral margins parallel at the posterior half, gradually constricted anteriorly, anterior margin sinuate, convex in the middle. Deeply punctuated and covered by dark brown scales. Scales along the median carina fresh yellow in color. Pronotum contains two brownish scaly blotches on both sides, median carina thin. Scutellum circular, with dark brown scales. Elytra 1.5 times as long as wide. Lateral margins parallel for anterior three fifths. Elytral declivity narrow and flat. Humeri obvious and covered by dark brown scales, with a triangular fresh yellow blotch between humerus and elytral interspace and an incomplete longitudinal band posteriorly. Striae wide. Punctures rectangular. Elytral interspace slightly wider than striae. Stria 3, 5, 7 convex with a few scaly verrucae. Femur contains a tooth, with a groove ventrally. Tibia straight. Abdominal sternites 2-4 with three lines of punctures.

Larvae Length 4.2-5.5 mm. Crescent, milky white. Labial sensory organs: 1 pair anteriorly, 3 pairs laterally

S. gravis adult
(Zhang Runzhi)

S. gravis adult (Zhang Runzhi)

with the external pair slightly longer than the rest, and 3 pairs in the center. Mandibles with 3 teeth on each side, outer tooth obvious, inner tooth curved, and middle tooth shorter than the rest. Maxillae with 4 equal-sized finger-like organs inside, and 2 distant setae.

Pupae Exarate, whitish yellow, compound eyes purplish red.

Biology and behavior *Sternochetus gravis* occurs one generation a year and overwinters inside bark crevices or the break off points on branches. Adults start supplemental feeding in late February and mid-March each year. Mating and egg-laying occurs between March and April. Larvae appear from late March to early June, and pre-pupal period runs from late May to mid-June. Pupation happens in June, and adults emerge between June and July.

Adults prefer warm weather but shun away from strong sunshine. Mating and egg-laying occurs at night. Eggs are laid on fruit surface by sticking with a secretion from females. This secretion dries out and turns dark later and dislodge from the surface as fruit grows. Newly emerged larvae bore directly through epidermis and feed on pulp, creating irregular larval galleries. Mature larva constructs a pupal chamber inside the pulp and covers it with frass and pupates inside. Large quantity of carbon dioxide was generated later when fruit reaches maturity. Adults emerge by chewing a tunnel out of the pulp. Adult spread with the transportation and sales of mongo fruits, as well as seedlings and grafts. Adults enter diapause in late September.

Prevention and treatment

1. Enforce quarantine measures on transportation vehicles and at wholesale locations to prevent its spread in tropical growing regions.

2. Clean fallen fruits after harvest and prohibit eating and storage of mango fruits inside the orchard to prevent accidental introduction of infested fruits.

3. Eliminate adult overwinter sites by sealing branch wounds and breaking points with a mixture of asphalt (1 part) and diesel (2 parts) or fill trunk hollows with lime sludge.

4. Apply 25 g/L lamda-cyhalothrin + 50% padan WP at 1,500× during early flowering or young fruit season to control supplemental feeding or egg laying adults.

References Zhao Yangchang and Chen Yuanqing 1980; Chen Yuanqing 1984; Chinese Academy of Sciences - Institute of Zoology 1986; Huang Yazhi et al. 1986; Situ Yingxian 1992, 1993; Zhou Yousheng et al. 1995; Bureau of Animal and Plant Inspection and Quarantine of the Peoples Republic of China 1997; Situ Yingxian et al. 2000; Zhang Runzhi et al. 2001.

(Situ Yingxian)

229 *Sternochetus mangiferae* (F.)

Classification Coleoptera, Curculionidae
Synonym *Cryptorrhynchus mangiferae* F.
Common names Mango weevil, Mango stone weevil, Mango nut weevil, Mango seed weevil

Importance *Sternochetus mangiferae* is a cosmopolitan mongo pest from India. Larvae feed on seeds inside the stones, resulting massive fruit dropping. Infested seeds will not germinate, which in turn negatively impacts seedling cultivation and field propagation. Adults feed on young leaves and branch tips on seedlings. Adults also feed on mango fruit, apple, peanuts, potato, peach, plum, and beans.

Distribution *Sternochetus mangiferae* occurs in a wide area between the Tropic of Capricorn and the Tropic of Cancer such as India (Chennai), New Zealand, the Philippines, Indonesia (Java Island), Thailand, Vietnam, Japan, Myanmar, Bangladesh, Malaysia, Pakistan, Sri Lanka, Lebanon, Nepal, Bhutan, United Arab Emirates, Oman, South Africa, Madagascar, Mauritius, Tanzania (Zanzibar), Uganda, Kenya, Mozambique, Zambia, Gabon, Fuji, Ghana, Nigeria, U.S.A. (Hawaii), Australia (Queensland), French New Caledonia, Holland, France (Reunion Island, Wallis Island), and Mariana Islands.

Hosts Various cultivars of *Mangifera indica*, especially sweet (ployembryonic) type, or late ripen types such as Neelum and Langra, with an infestation rate of 100% and 98%, respectively. Resin flavored monoembryonic type such as the famous Alphonso suffers with 73% infestation rate, whereas Hawaii cultivars such as Harders, Paris Selection No.1, Wong, and Ah Ping are resistant.

Damage *Sternochetus mangiferae* larva feeds on cotyledon inside the pulp, resulting in blackened cotyledon and decayed seed that loss the ability to germinate. Infestation leads to massive fruit drop and fruit decay for growing or maturing fruits. Adults also feed on young leaves and branch tips.

Morphological characteristics **Adults** (based on Australia specimens) Length 8.0 mm, width 4.2 mm, black, covered by fresh yellow scales, with a ratio of 5:4 between length and width. Body oval, yellowish brown, covered by bright yellow, yellowish brown, dark brown, and black scales. Whitish yellow on observed specimens. Snout slightly curved. Interspace between snout apex and antennae inserting point 10% shorter than that between antennae inserting point and snout base. Apical half of the snout smooth with a few sparse punctures, whereas the posterior half with large punctures covered by yellowish brown scales. Median carina obvious, covered by yellowish brown scales, with an oblique bright yellow blotch in the middle. Antennal funiculus segment 1 slightly longer than segment 2. Frons without central depression. Vertex and space between compound eyes with large punctures and covered by yellowish brown scales. Pronotum longer than wide, covered by yellowish brown scales, posterior one third parallel laterally and extremely constricted toward anterior margin. Anterior margin sinuate with sudden convex in the middle. Pronotal area with dark brown longitudinal band on each side of the median carina and scattered single erect black scales laterally on the posterior half. Scutellum shield-shaped and covered by dense yellowish brown scales. Elytra wider than pronotum, humeri obvious. Inclined and straight bands bright yellow, with inclination ends at interspace 3 on elytra. Interspace 3 and 5 higher than others, with 2-3 short rectangle black blotch interrupting straight bands. Punctures square and narrow. Femur with a tooth on the inner side and deep depression between the tooth and the apex. Groove on metasternum deep and not forked posteriorly. Male aedeagus median apodemes flat and wide, basal sclerite lobe slightly curved in the middle and apically. Basal sclerite with short convex, not hook-like. Female spermatheca toad pod shaped, apical half arc-shaped and gradually constricted with a width of 1/3 of that of the base. Abdominal sternite 8 vase-shaped, constricted collar-like apically.

Biology and behavior In Madras of India: *Sternochetus mangiferae* finishes a lifecycle in 50 days, with an egg period of 1-3 weeks, an egg stage of 7 days, and a pupal stage of 7 days. In Java of Indonesia: 5.5-7 days for egg stage, 30-36 days for larval stage, and 5 days for pupal stage. In Hawaii of US: 7 days for egg stage in March and April when temperature is between 13-17°C; 5-6 days for egg stage, 22 days for larval stage, 7 days for pupal stages, and 40 days for a lifecycle in late May. Others reported an average larval stage of one month or 70 days.

Adults with apparent death and aggregation features. They usually hide inside fence crevices, fruits stones, tree hallows, loose bark on tree trunk. Feeding and mat-

S. mangiferae adult (Zhang Runzhi)　　S. mangiferae adult snout (dorsal view) (Zhang Runzhi)

ing occurs at night and eggs are laid at sunset. Adults can mate and oviposit multiple times. Adult diapause is correlated with photoperiod. Adults can survive for 21 months, but only 140 days without food and water. Eggs are singly laid on fruit surface and protected 50% of the time by a secretion. Each female produces 147-281 eggs with a daily average of 15 eggs. Each fruit contains a maximum of 31 eggs on its surface. Newly hatched larvae bore directly into the seeds. Each fruit contains a maximum of 6 larvae. There are five instars in larval stage. Mature larvae pupate inside cotyledon or in the soil.

Prevention and treatment

1. Orchard management. Seal bark crevices and trunk hallows with kerosene emulsion in August to control adults inside. Collect and destroy fallen fruits twice a week during fruit maturation.

2. Chemical control. Apply chemical insecticides during adult overwintering period, mango flowering and fruit setting season provides good results. Spray of 0.01% fenthion could provide effective control.

3. Post-harvest treatment. Cold storage at 2°C for 5 days or 4°C for 24 days kills all pest stages but injuries the fruit as well (Masan et al. 1934). Methyl Bromide fumigation at 3 g/m^3 at 12°C for 8 hours destroy all life stages but injuries fruits. In addition, residue chemical may cause problem. Radiation treatment at 60 Kr is the proper dosage that delays green fruit maturation and control the pest, prolonging storage to 21 days at 30°C. Steam heat treatment is the safest method for pest management. Steam of 46°C (58% RH) or 48° (98% RH) results in fruit temperature of 43.3°C after 6 hours. Microwave treatment at 2450 MHz and 700 W for 10-15 second with 4-8 cycles leads to an average stone temperature of 51 ± 4°C and pest mortality of 50%.

References　Flecher 1914, Guy and Marshall 1935, Gangolly 1957, Singh 1960, Balock and Kozuma 1964, Hill 1975, Atwals 1976, Shukla and Tandon 1985, Hansen et al. 1989.

(Situ Yingxian)

230 *Sternochetus olivieri* (Faust)

Classification Coleoptera, Curculionidae
Synonyms *Cryptorrhynchus olivieri* Faust, *Cryptorrhynchus olivieri* (Faust)
Common name Mango seed weevil

Importance *Sternochetus olivieri* larvae feed on the seeds inside the mango cotyledon. Feeding leads to loss of ability to germinate, which negatively affects seedling propagation.

Distribution Guangxi, Yunnan; Vietnam, Cambodia.

Hosts *Mangifera siamensis* wild isotypes such as Okrong and Local Minor, *Mangifera indica* cultivars such as Third-Year, Ivory Mans, Maqiesu, Carabao, India Globe, Myanmar No.1, Myanmar No.2, Purple Flower, Coconut Flavored, Eagle Beak, Yuanjiang Ivory.

Damage Laval entrance hole on endocarp will not be healed from normal growth during fruit maturation period, leaving a hole on endocarp. This hole is a typical sign of infestation and becomes the exit hole for adults. Breaking up the endocarp reveals dark powdery damaged seed and incomplete cotyledon. Infested seeds lost ability to germinate.

Morphological characteristics **Adults** Length 7.61 mm width 4.45, with a maximum of 8.20 mm×4.60 mm, width 4.45 mm on average, with a minimumof 5.7 mm×3.42 mm. Body black and covered with bright yellow, rusty red, dark brown, or black scales. After overwintering, bright yellow scales on adults faded to whitish yellow or entirely dark brown. Snout slightly curved, not reaching posterior end of hind coxae. Snout slightly wider apically, with small punctures and sparse scales, median carina obvious. Snout basal half punctures large with dense scales. Antennae arise from the middle of the snout. Antennal groove straight and reaches the base of compound eyes. Funiculus 1 and 2 equal in length. Funiculus 7 as long as wide. Club pubescent, two times as long as wide, acuminate apically, with no significant annulation. Frons with central depression and covered with scales, narrower than snout base. Pronotum 1.5 times as long as wide, posterior margin half arc-shaped, extremely constricted in the middle toward the anterior margin. Anterior margin sinuate with the center suddenly convex. Punctures deep and dense, irregular, surface smooth. Median carina obvious, with one pair of corresponding black scale blotches laterally near anterior margin and the middle. The pair near the middle are more conspicuous. Median carina and middle scale blotches covered by bright yellow circular scales. Scutellum circular with bright yellow scales. Elytra 1.8 times as long as wide, with lateral margins parallel before the basal three fifths portion. Humeri obvious, narrower than the widest portion of the elytra. Elytra constricted apically. Elytral declivity flat and covered by rusty red scales. Oblique bands broad with bright yellow scales, extend from base to one third of the elytra, and from the humeri to interspace 3, occasionally even to interspace 1. Straight bands wide, bright yellow. Interspace 3, 5, and 7 carinate, with near square punctures. Odd number interspaces with scaly verrucae, more significant on interspace 3, 5, and 7. Femur with a tooth. Hind femur with an obvious yellow band. Abdomen contains 7 visible sternites. Sternite 7 with one semi-circular reproductive organ on each lateral margin. Each reproductive organ contains 12-16 longitudinal reproductive files. Male abdominal sternite 7 wrinkled inward apically, whereas in female it merged with pygidium. Both male and female with two lines of punctures on abdominal sternites 2, 3, 4. Male aedeagus sclerite lobe straight and not curved, sclerite hook-like apically.

Larvae First instar crescent, milky white. Dorsal blood vessel purplish brown. Labial sensory organs: 1 pair anteriorly, 2 pairs laterally and 3 pairs in the center. The middle pair in the center with a pair of minute sensory organs inside.

Biology and behavior *Sternochetus olivieri* occurs one generation a year and overwinters inside tree hallows, bark crevices, flanked bark around wounds, airborne roots at tree base, or firewood piles in the orchard. Duration for life stages is 7-8 days for eggs, 40 days for larval, 2 days

S. olivieri adult (lateral view) (Zhang Runzhi)

S. olivieri adult (dorsal view) (Zhang Runzhi)   *S. olivieri* adult (dorsal view) (Bu Wenjun)

for pre-pupal, 9 days for pupal, and 60 days for adults. A few females can survive 121 days without food and water. The average adult longevity is 13-15 months. Adults feed on young branch tips for supplemental nutrition from late February to mid-April. Mating and consequent egg-laying occurs between late March and mid-May. Larval period falls between late May and mid-June, with pre-pupae and pupae found in mid- to late June. Peak adult emergence occurs in late June to mid-July.

Adults prefer moist environment but avoid dampen or extremely dry conditions. The also prefer warm weather but shun away from direct sunshine. They apparent death and usually hidden inside flowers, fruit branches, or the undersides of the leaves. Flying and other activities such as mating occurs after sunset with the aid of aggregation sound produced by stridulating the tympanal organ against elytra. Multiple males may compete for a single female. Activity reduced after midnight. Adults mate multiple times and mating can last as long as 8 hours. Females can lay eggs multiple times as well. Eggs are laid on fruit surface and stuck with a white secretion which dries and hardens later and falls with the fruits. The secretion is not soluble in any organic solvent such as ethyl alcohol. A total 51.8% of the 1,729 fruits observed were found infested, with 2-7 eggs found on 22.4% of the fruits. It is common to found 2-3 larvae inside a fruit. However, only one larva survived to adulthood due to competition. Late hatched larvae that fail to enter the seed by penetrating the membranous endocamp dies when young fruit become 4-6 cm long. Adults emerge when fruits turn yellow with reduced supply of oxygen and increase of ethylene and carbon dioxide. Field rearing results indicate that two days of -1°C results in total mortality for adults inside the fruits. Adults can spread through long-distance movement and sales of infested fruits.

Prevention and treatment

1. Enhance orchard management. Select bud-grafted trees to dwarf fruit trees for easy management. Eliminate overwinter sites by covering the trunk with lime paste, smoothing the cuts, sealing branch wounds with a mixture of asphalt (1 part) and diesel (2 parts), filling trunk hollows with lime sludge, and prohibiting storage of dead branches inside the orchard.

2. Maintain a clean environment to prevent pest from staying. Periodically remove fallen fruits from the orchard during maturation season, not allowing mango eating and stone discarding inside the orchard, prohibit storage of fruits inside the orchard to prevent pest escaping.

3. Trap males by caging sound producing females during mating season.

4. Apply 20% sumicidin at 2,000×, 25 g/L lamda-cyhalothrin at 1,500×, or 12% thiamethoxam and monosultap mixture at 1,000× during early flowering or late fruit setting season to control overwinter adults.

References   Zhao Yangchang and Chen Yuanqing 1980, Chen Yuanqing 1984, Situ Yingxian and Yang Bin 1991, Situ Yingxian 1993, Zhang Runzhi et al. 2001.

(Situ Yingxian)

231 *Sympiezomias velatus* (Chevrolat)

Classification Coleoptera, Curculionidae
Synonym *Sympiezomias lewisi* Roelofs

Distribution Beijing, Hebei, Shanxi, Inner Mongolia, Liaoning, Jilin, Heilongjiang, Shandong, Henan, Hubei, Shaanxi in China; Japan.

Hosts More than 100 species in 41 families and 70 genera, including *Juglans regia*, *Castanea mollissima*, *Ziziphus jujube*, *Robinia pseudoacacia*, *Morus alba*, *Populus* spp., *Amorpha fruticosa*, *Glycine max*, *Beta valgaris*.

Morphological characteristics **Adults** Length about 10 mm, black and densely covered with grey scales. Pronotum dark brown in the middle, covered by rugose convex circles, with a small longitudinal groove in the middle. Pronotum lateral margins and elytra with brown blotches. Head wide. Snout stout and wide, with three surface grooves. Middle groove black, inserted as a triangle anteriorly, with long setae on its margins. Scutellum with a longitudinal groove in the middle. Elytra acuminate apically, with a loop-shaped blotch and 10 striae on each side. Hind winds degenerated.

Eggs Length 1.0 mm, width 0.4 mm, oval, milky white at first, translucent on both ends. Turn to creamy

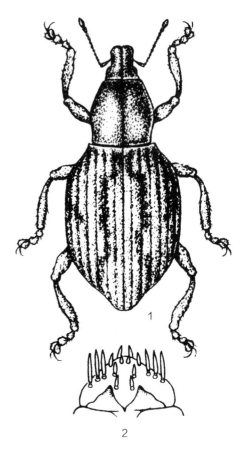

S. velatus. 1. Adult, 2. Inner clypeus of larva (Shao Yuhua)

yellow before hatch.

Larvae Length 14 mm at maturity. Milky white, head whitish yellow. Mandibles with two teeth anteriorly and one obtuse tooth posteriorly. Labium with four pairs of denticles anteriorly and three pairs of denticles in the middle, and two triangular brown lines posteriorly.

Pupae Length 9-10 mm, creamy yellow. Compound eyes brown. Snout extends to prosternum. Mandible large. Elytra tip reaches base of hind tibia. End of the body curved ventrally, with a spine on each lateral margin.

Biology and behavior *Sympiezomias velatus* occurs one generation in two years in Liaoning province and overwinters as larvae or adults. Adults become active in early to mid-April and feed and mate in groups

S. velatus adult (Xu Gongtian)

S. velatus adult (Xu Gongtian)

on seedlings. Eggs are laid on leaves in late May and hatch in early June. Larvae construct earthy cocoons to overwinter in late September. Larvae continue feeding next spring and pupate in early June. Adults emerge in mid-July and overwinter at the same location. It took two years for this weevil to finish a lifecycle.

Adults flightless and hide inside soil crevices or under plant remains during late April when outside temperature is low, rarely come out. Adults can be suffocated by sticking to the soil after the rain. Adults prefer young and juicy seedlings as food but do not feed much individually. However, seedlings at the aggregation sites usually completely devoured. Egg period lasts 19-86 days, with a fecundity of 374-1172 eggs per female.

Newly hatched larva can craw swiftly after dislodging its eggshell. It enters soil through soil crevices or loss surface to feed on humus and minute roots that brings no damage to seedlings.

Prevention and treatment

1. Protect adult predator *Cerceris rufipes evecta*.

2. For nursery and garden lawns, soil drench with *Beauveria bassiana* at the concentration of 2×10^8 conidia/mL twice (late May and early September).

3. Cover spray with 1:1,000 solution of 90% trichlorfon twice (late May and early September).

4. Soil application by mixing 15% chlorpyrifos and phoxim mixture or 50% imidacloprid monosultap mixture with 30-50× fine soil and evenly spread over one square meter of nursery bed to control larvae in the soil.

References Zhang Zhizhong 1959; Lou Shenxiu, Li Zhongxi, Wu Yanru 1991; Yang Ziqi and Cao Huaguo 2002; Xue Zhihua 2006.

(Zhang Zhizhong, Li Zhenyu)

232 *Cryphalus tabulaeformis chienzhuangensis* Tsai et Li

Classification　Coleoptera, Curculionidae

Distribution　Shaanxi.

Hosts　*Pinus tabuliformis*.

Damage　Feeding by females at needle bases on weak trees during colonization attracts males for mating, the consequential egg-laying, as well as the construction of parent galleries inside branches, resulting in tip dieback when infestation is heavy.

Morphological characteristics　**Adults** Length 1.5-1.7 mm. Brown. Male contains a transverse carina on top of frons. No frontal carina in the female. Female with longitudinal striae and verrucae on the gena. Verrucae scattered among the anterior half of the pronotum, with 4-6 along anterior margin. The middle two verrucae larger than the rest. Elytra humeri conspicuous. Striae unpunctured, elytral interspaces flat.

Eggs Oblate oval. Length 0.7-0.8 mm, width 0.2-0.3 mm. White, translucent, and smooth.

Larvae Length 2.0 mm, creamy white and slightly curved. Mouthparts dark brown. Body with sparse setae.

Pupae Length 2.0 mm. Creamy white initially, turns to pale brown later.

Biology and behavior　*Cryphalus tabulaeformis chienzhuangensis* occurs two generations a year in Malan Forest District in Shanxi province. It overwinters as larvae or adults inside the bark of *P. tabulaeformis* branches. Adults become active in early May of the next year and start maturation feeding under the bark. Larvae of the overwinter generation start to pupate in mid- to late May. First year adults begin to emerge in early to mid-July. Female constructs entrance tunnel and lay eggs in mid-July. Eggs hatch in late July. Pupation occurs in early August. Adults of the second generation start to emerge in mid-August, with females laying eggs in late August. Pupation and adult emergence continue to late September. By late October, both adults and larvae of the current generation are ready for overwinter. Each female can produce 15-40 eggs in her lifetime. Egg period lasts about 10 days. Overwintering adult mortality can reach as high as 51.6%, with natural enemy predation accounted for 12.1%. Common nature enemies include woodpeckers, clerids, hister beetles, earwigs, assassin bugs, and ants. This bark beetle damages weak trees and is strongly attracted by fresh cut branches of Chinese red pine, α-pinene, and β-pinene.

Prevention and treatment

1. Remove weak and infested trees through selective thinning. Trap adults with freshly cut pine branches, or α-pinene, and β-pinene for population monitoring.

2. Apply *Paecilomyces* spp. wettable powder at the rate of 15 kg/km^2 in light to moderately infested areas between January and February after the removal of infested trees provides effective control.

3. Chemical control. During larval period, apply 45% profenofos and phoxim mixture at 30-60×, 12% thiamethoxam and lamda-cyhalothrin mixture at 30-60× to tree trunks.

References　Xiao Gangrou (ed.) 1992, Zhou Jiaxi 1994, Zhou Jiaxi, Li Houhun, Sun Qinhang 1997.

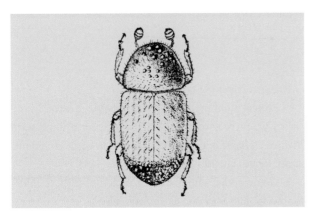

C. tabulaeformis chienzhuangensis adult (Zhou Jiaxi)

(Tang Guanghui, Li Menglou, Zhou Jiaxi)

C. tabulaeformis chienzhuangensis adult (dorsal view) (NZM)

C. tabulaeformis chienzhuangensis adult (lateral view) (NZM)

C. tabulaeformis chienzhuangensis adult (front view) (NZM)

233 *Cryphalus tabulaeformis* Tsai et Li

Classification   Coleoptera, Curculionidae

Distribution   Hebei, Henan, Shaanxi.

Host   *Pinus tabuliformis*.

Damage   Feeding by larvae and adults under the barks creating galleries inside branches, resulting in dieback of branch tips.

Morphological characteristics   **Adults** Length 1.5-2.2 mm. Oval, brown. Males contain a transverse carina on top of the frons and a verruca on the gena. Pronotum anterior margin with 4-5 verrucae, and the middle two are larger than the rest. Elytra punctures small and loosely arranged.

Larvae Length 2.0 mm, white and slightly curved.

Biology and behavior   *Cryphalus tabulaeformis* occurs two generations a year in Henan and overwinters as larvae or adults inside the bark of dead trees. Adults become active between April and May of the next year and start maturation feeding under the bark. Upon emergence, they attack young trees by construct entrance holes on branches and create wide longitudinal parent galleries under the bark into the sapwood. Females lay

C. *tabulaeformis* adult (Xu Gongtian)

C. *tabulaeformis* larva (Xu Gongtian)

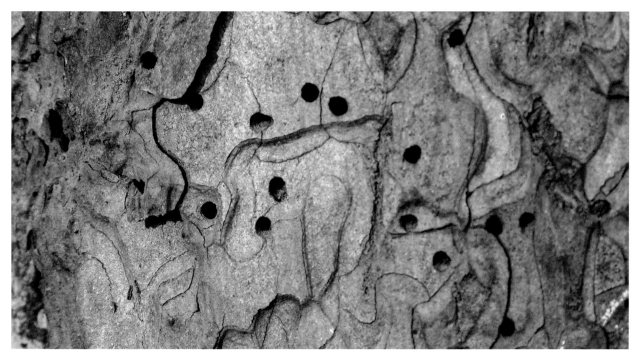

C. *tabulaeformis* exit hole (Xu Gongtian)

C. tabulaeformis infested trees (Xu Gongtian)

eggs on both sides of the parent galleries after mating. Newly hatched larvae tunnel away from parent galleries, creating vertical larval galleries inside the bark. Adults of the first generation enclose, exit the bark and begin to lay eggs between June and July. Wood tissues under the bark are completely devoured due to repeated attach by both the larvae and adults. Branch dieback is common in the same year. Mortality occurs by the end of the summer for severely injured trees. Adults of the second generation emerge between September and October and overwinter inside the original galleries.

Adults prefer branches of young trees for egg-laying, which threats the growth and development of Chinese red pine.

Prevention and treatment

1. Mechanic control. Remove dead branches and trees from the forest in early spring or later fall and destroy beetles inside by burning host materials or submerging them in water.

2. Chemical control. For severe infestation over large area, spray Dimilin through aerial application during peak adult period at the low volume (LV) rate of 600 g/ha,

C. tabulaeformis galleries (Xu Gongtian)

or ultralow volume (ULV) rate of 450 g/ha. During larval period, spray 45% profenofos and phoxim mixture 30-60×, or 12% thiamethoxam and lamda-cyhalothrin mixture at 30-60× to tree trunks.

References Yin Huifen et al. 1984, Xiao Gangrou (ed.) 1992, Yang Youqian 2000.

(Yang Youqian)

234 *Dendroctonus armandi* Tsai et Li

Classification Coleoptera, Curculionidae
Common name Chinese white pine beetle

Importance *Dendroctonus armandi* is a destructive woodboring beetle of Chinese white pine capable of bringing tremendous damage to forests.

Distribution Henan, Hubei, Sichuan, Shaanxi, Gansu.

Hosts *Pinus armandii*, and occasionally *P. tabulaeformis*.

Damage *Dendroctonus armandi* adults and larvae attack healthy living trees gregariously. Dark to pale brown pitch tubes are found at the entrance points along the lower to middle trunk of infested trees. It is a pioneer species capable of weakening apparent healthy trees. Initial infestation by this pest is usually followed by various secondary pest species such as bark beetles (e.g. *Ips sexdentatus*, *I. acuminatus*, *Tomicus piniperda*, *Polygraphus poligraphus*, *Dryocoetes hectographus*, *Hylastes parallelus*, *Cryphalus* spp., etc.), weevils (e.g. *Hylobius abietis haroldi*), and longhorned beetles (e.g. *Asemum amurense*, *Morimospasma paradoxum*, *Acanthocinus griseus*, *Dystomorphus notatu*, etc.). Together they kill pine trees over large areas through feeding under the bark that leads to needle desiccation and bark separation.

Morphological characteristics **Adults** Length 4.4-6.5 mm; oblate oval; dark brown or shiny black, only antennae and tarsi reddish brown. Antennal club oblate, obtuse at the tip. Frons contains a median sulcus, rough with granules and covered by erect long hairs. Pronotum wider than long, narrowed anteriorly, wider toward the posterior end; surface covered with various punctures and hairs and contains an inconspicuous smooth line in the middle; anterior margin concave, posterior margin sinuate with an obtuse angle protruding backward at the center. Elytra parallel-sided, with serrate projections at the base, surface rough with conspicuous punctures that gradually disappear toward both sides and the apex. The space between puncture lines rugose, wider than the strial punctures, contains scattered short pubescence and one line of long hairs. The pubescence is short and inconspicuous on declivity surface. Ventral surface contains dense oppressed pubescence and small punctures ventrally.

Eggs Oval. Length 1.0 mm, width 0.5 mm. Creamy white.

Larvae Length 6.0 mm, creamy white, head yellowish, mouthparts brown.

Pupae Length 4.0-6.0 mm. Creamy white. Each abdominal segment contains a line of transversely arranged spines on dorsal surface. The end has one pair of anal tubercles.

Biology and behavior The number of generations per year varies with elevations. For example, at elevation below 1,700 m in Qinling Forest Region in Shaanxi Province, there are two generations per year, whereas only one generation occurs at above 2,150 m, and three generations in two years occur between 1,700 and 2,150 m. The accumulative degree-day for each generation is 495.5 DD, with a starting point of 9.6°C. It generally overwinters as larvae, although pupae and adults can overwinter as well.

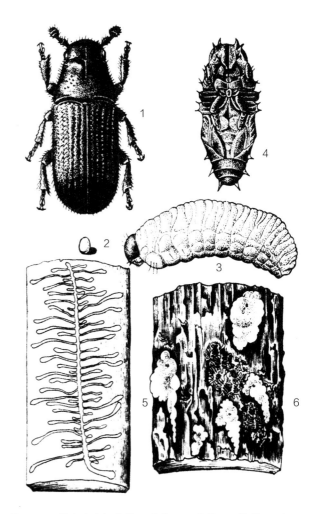

D. armandi. 1. Adult, 2. Egg, 3. Larva, 4. Pupa, 5. Tunnels, 6. Resin at feeding site (Zhu Xingcai)

The primary gallery of *D. armandi* is simple, longitudinal, 10-60 cm long and 2.3 mm wide. Secondary galleries extend laterally for 2-5 cm. Each primary gallery has a pair of male and female. A shoe-shaped mating room is made initially when the pair enters the tree bark. The female lays eggs on each side of the gallery. Each female lays an average of 50 eggs, ranging from 20 to 100. Newly hatched larva feeds the cambium layer. As the larva grows, the secondary galleries enlarge and extend to sapwood and are filled with frass. Mature larva constructs a pupal chamber at the end of the secondary gallery. Newly emerged adult feeds on cambium around the pupal chamber and the secondary gallery to obtain nutrition. Feeding by adults girdles the tree and may completely break the vascular system. Upon a period of feeding, adults chew exit holes and emerge from the host trees and start to attack surrounding healthy trees. The beetle prefers middle-aged and older forest stands.

There are two habitat types of *D. armandi* in the Qinling Forest region – "frequent outbreak habitat" and "dispersal habitat". The "frequent outbreak habitat" type occurs mainly in mature pure stands with high density, older forest stands, poor site conditions, 1,800 to 2,100 m in elevation, and upper part of a hill. Common stands are *P. tabuliformis* -*P. armandi*; on south-faced deep slopes or *P. armandii* stands on top of the mountains. These stands are the old origins of *D. armandi* outbreaks. The "dispersal habitat" occurs mainly in young mixed stands with low density and good site conditions at 1,700 to 2,200 m elevation, as in *P. armandii* istands on north-faced slow slopes and lowlands. Tree habits are usually low land- *P. armandii* and sandy lands- *P. armandii*. They are the new origins of outbreaks.

Prevention and treatment

1. Select appropriate tree species for specific site conditions during the afforestation planning. Maintain tree vigor and pest resistance capability by adopting sound forest management practices such as periodic removal of newly infested, weak, and dead trees.

2. Plant mixed forest stands with Chinese red pine, Armandi pine, oaks, and other broad leaf tree species at low elevations and Armandi pine, aspen, birch and other broad leaf tree species at high elevations.

3. Conserve natural enemies. Natural enemies of *D.*

D. armandi invasion sites and resin (FPC)

D. armandi larvae and galleries (FPC)

armandi include parasitoids such as *Coeloides qinlingensis*, *Eurytoma longicauda*, *Cleonymus pini*, *Roptrocerus qinlingensis*, *Roptrocerus xylophagorum*, *Roptrocerus mirus*, *Dinotiscus armandi*, *Dinotiscus eupterus*, *Rhopalicus guttatus*, *Rhopalicus tutela*, *Tomicobia liaoi*, *Calosota longigasteris*, *Aprostocetus dendroctoni*, and *Tetrastichus armandi*. Predators include ants, rove beetles, ground beetles, clerids, and hister beetles. Other natural enemies include fungi, mites, nematodes, and birds.

4. Disrupt normal communication of the pest by utilizing synthetic pheromones to reduce forest damage.

5. Chemical control. Irrigate the roots with 2×10^8 conidia/mL *Beauveria bassiana*. During laval period, spray 45% profenofos and phoxim mixture 30-60×, or 12% thiamethoxam and lamda-cyhalothrin mixture at 30-60× to tree trunks.

References Cai Banghua and Li Zhaolin 1959; Ren Zuofo et al 1959; Li Kuansheng 1989; Shaanxi Forestry Research Institute 1990; Xiao Gangrou (ed.) 1992; Yang Zhongqi 1996,

(Li Kuansheng, Li Li)

D. armandi adult (FPC)

D. armandi infested forest (FPC)

235 *Dendroctonus micans* (Kugelann)

Classification Coleoptera, Curculionidae
Common name Great spruce bark beetle

Distribution Heilongjiang, Sichuan, Gansu; Japan, Austria, Belgium, Czech Republic, Slovakia, Denmark, Germany, Hungary, Italy, Netherlands, Norway, Poland, Romania, Sweden, Switzerland, Turkey, Great Britain, Russia.

Hosts *Picea jezoensis*, *P. asperata*, *P. crassifolia*, *P. koraiensis*, *Podocarpus macrophylla* var. *maki*.

Damage It attacks mainly spruces at the low to middle bole and can kill healthy trees. In Dahailin of Heilongjiang province, it occurs concurrently with *Ips typographus* and is responsible for large area mortality of Jezo spruce.

Morphological characteristics **Adults** Length 7.2-7.9 mm. Shiny black. Frons flat, coarsely punctured, and with sparse long hairs. Antennae and tarsi reddish brown. Pronotum has a smooth longitudinal line in the middle. Scutellum very small and obsolete. Elytral striae with large and shallow punctures. Elytral interspace wider than elytral striae and contains punctures and verrucae. Punctures and verrucae are only obvious around elytral humeri and scutellum, become sparse and smaller toward the apex and absent in declivity surface. The entire elytra is covered by brown and erect long hairs. Those on declivity surface are longer and denser.

Biology and behavior *D. micans* occurs one generation a year in China and overwinters as adults or larvae. In Qilian Mountain Forest District in Gansu province, adults appear in mid- to late August. Pitch tubes on damaged spruce are good sign of infestation. Primary gallery is short and curved, with an enlarged egg chamber at the end. Eggs scattered inside the egg chamber. Newly hatched larvae feed aggregately, creating cave type common chamber.

Prevention and treatment
1. Remove all logs and peel stumps before summer in the logging area. Remove or peel fallen, wind-broken or snow-damaged trees before May.
2. Conserve parasitic wasp *Calosota qilianshanensis*.

References
Yin Huifen, Huang Fusheng, and Li Zhaolin 1984; Yang Zhongqi 1996.

(Li Zhenyu)

D. micans adult (Bu Wenjun)

236 *Dendroctonus valens* LeConte

Classification Coleoptera, Curculionidae
Common name Red turpentine beetle

Importance *Dendroctonus valens* is an exotic invasive species first discovered in Yangcheng and Qinshui counties in Shanxi province in 1988. It was suggested to be originated from imported timber from the U.S. Different from those in the native range, it attacks both weak and healthy trees, resulting in mass host mortality in affected areas. By the end of 1999, a total of 526,000 ha of forest in Hebei, Henan, and Shanxi provinces were infested, with serve damage found in 130,000 ha of land. Tree mortality among Chinese red pine reached 30% in some areas, with more than six million dead trees.

Distribution Beijing, Hebei, Shanxi, Henan, Shaanxi, Gansu; U. S.A., Canada, Mexico, Guatemala, and Honduras.

Hosts China: *Pinus tabuliformis*, *P. armandii*, *P. pungeana*. North America: more than 40 conifer species in the genera of *Pinus*, *Picea*, *Pseudotsuga*, *Abies*, and *Larix*.

Damage Entrance holes are usually found on tree trunk under 2 m, packed with characteristic pitch tubes and granular frass. Pitch tubes are brown, and moist initially, become greyish brown, hard and dry later. *D. valens* feeds aggregately in unspecialized communal galleries. Gallery is usually linear, cave-type, or irregularly branched.

Morphological characteristics **Adults** Length 5.7-10.0 mm, cylindrical, reddish to dark brown.

Eggs Length 0.9-1.1 mm, width 0.4-0.5 mm, cylindrically oval, opaque white, shiny.

Larvae Body average length 11.8 mm, head average width 1.79 mm when matured. Abdomen contains a small brown patch at the end, with 3 pale-brown hooks dorsally and ventrally for crawling. Dorsal hooks larger than ventral hooks. A row of pale-brown tubercles is evident on both sides of the body, with a seta on each of them.

Pupae Average length 7.82 mm. Wing pads, legs, and antennae closely attached to the body. Milky white at the beginning, becomes yellowish later, with yellow and white stripes on the head and thorax. Wings dirty white, turn to reddish brown to dark brown before emergence.

Biology and behavior The major targets of *D. valens* are weak, large-diameter mature trees. Attacks are most common in stumps and freshly cut trees, with 1-2 generations per year. Overlapping generations are typical and beetles may be found in flight year-around except during the winter. Adult peaks in mid- to late May, and occasionally again in August. The female constructs entrance tunnel through the bark and feeds in the cambium region and outer sapwood. A female tunnels upward when she reaches the cambium region, enlarging its gallery vertically and longitudinally until tree fluid stopped flowing, then downward towards the roots. Males arrive shortly after the females. Characteristic pitch tubes and

D. valens adult (Zhang Runzhi)

D. valens adult (Zhang Runzhi)

D. valens adults (Zhang Runzhi)

D. valens larva (Mai Guoqing)

D. valens tunnel (Xu Gongtian)

D. valens pupa (Xu Gongtian)

D. valens adult in tunnel (Xu Gongtian)

Resin accumulation from *D. valens* (Xu Gongtian)

Fumigation to control *D. valens* (Xu Gongtian)

Trapping *D. valens* using pheromone trap (Xu Gongtian)

granular frass are generally found around entrances holes. It overwinters with all life stages under the bark around the base of the trees.

Prevention and treatment

1. Enforce quarantine to prevent its further spread. Remove dead or dying trees. Fumigate or cover stumps to kill residue beetle population and prevent future oviposition.

2. Spray 45% profenofos and phoxim mixture at 30-60×, 12% thiamethoxam and lamda-cyhalothrin mixture at 30-60× to tree trunks to control larvae. Spray pyrethroid insecticides or fumigate the tree base during adult season to prevent initial infestations. Monitor and control pest population using pheromone traps.

References Smith 1961; Eaton and Lara 1967; Wu Jian et al. 2000; Yin Huifen 2000; Chang Baoshan, Liu Suicun, Zhao Xiaomei et al. 2001; Li Jishun, Chang Guobin, Song Yushuang et al. 2001; Miao Zhenwang, Zhou Weimin, Huo Lüyuan et al. 2001; Zhang Zhen, Wang Hongbin, Kong Xiangbo 2005; Zhang Zhen and Zhang Xudong 2009.

(Zhang Zhen, Wang Hongbin)

237 *Ips acuminatus* (Gyllenhal)

Classification Coleoptera, Curculionidae
Common name Engraver beetle

Importance *Ips acuminatus* is generally regarded as a secondary pest that attacks weakened or wind-damaged trees. When populations build up in weakened or down material, they can attack relatively healthy trees. In some instances, *I. acuminatus* can kill large numbers of trees and cause significant loss of commercial pine volume.

Distribution Beijing, Hebei, Inner Mongolia, Liaoning, Jilin, Heilongjiang, Hubei, Sichuan, Yunnan, Shaanxi, Xinjiang; Mongolia, Korea Peninsula, Japan, Russia, Europe.

Hosts *Pinus koraiensis*, *P. armandii*, *P. densata*, *P. tabuliformis*, *P. massoniana*, *P. yunnanensis*, *P. sylvestris* var. *mongolica*, *P. kesiya* var. *langbianensis*, *Larix gmelini*, *L. olgensis*, and *Picea obovata*.

Damage Primary galleries star-shaped, longitudinally extend to both ends of the trunk. Larval galleries short and sparse, about 5 mm from each other. The number of parent galleries depends on the sex ratio of each "family". Most "families" comprise of 1 male and 6 females, some with 1 male and 5 females. Very few "families" made of 1 male and 4 females, or 1 male and 8 females.

Morphological characteristics Adult. Length 3.8-4.1 mm, cylindrical, reddish brown to dark brown with luster. Compound eyes reniform, covered with un-evenly distributed coarse punctures. There are 2-3 large granules in central frons between the compound eyes. Elytra 1.4 times as long as pronotum, with conspicuous strial punctures. Elytral declivity starts from the posterior third of the elytra and bears 3 ascending spines on each lateral margin. The third spine on the male bifurcate, cylindrical with parallel sides. All three spines of the female acute.

Biology and behavior *Ips acuminatus* occurs one generation a year in Hulunbuir area in Inner Mongolia Autonomous Region and Heilongjiang province and over-

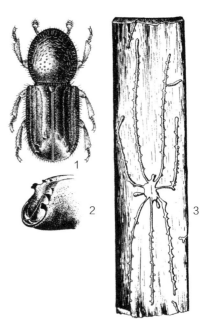

I. acuminatus. 1. Adult, 2. Posterior end of elytra, 3. Tunnel (Zhu Xingcai)

winters as adults in "blind holes" under the bark. Adults of the overwintering generation have a long flight season and oviposition period, enabling constant adult migration and new colonization from May to August. Each female lays an average of 30 eggs. There are 3-4 generations a year in Eshan area in Yunnan province and all life stages are seen year-around under the bark of infested trees.

Prevention and treatment

1. Sanitation measures. Timely removal of small diameter and injured tree after thinning. Pay special attention to fallen trees as *I. acuminatus* disperse to those trees first before spreading to weak and healthy trees.

2. Conserve parasitic wasps such as *Tomicobia ceitneri* and *Ipideurytoma acuminate*.

3. Irrigate soil at base of tree with *Beauveria bassiana* at 2×10^8 conidia/mL. Spray 45% profenofos and phoxim mixture at 30-60×, 12% thiamethoxam and lamda-cyhalothrin mixture at 30-60× to control larvae.

References Yin Huifen, Huang Fusheng, and Li Zhaolin 1984; Xiao Gangrou (ed.) 1992; Yang Zhongqi 1996.

(Li Zhenyu, Yu Chengming, Chen Erhou, Huang Zhongtian)

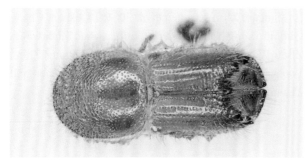

I. acuminatus adult (Li Zhenyu)

Ips duplicatus (Sahalberg)

Classification Coleoptera, Curculionidae
Common name Northern bark beetle

Importance *Ips duplicatus* attacks both weak and healthy trees during outbreaks. High population density severely disrupts the normal physiological functions of the tree and leads to tree mortality in a short period of time.

Distribution Inner Mongolia, Heilongjiang; Russia, Finland, Poland, Czech Republic, Slovakia, Germany, Sweden, and Norway.

Hosts *Picea koraiensis*, *P. mongolica*.

Morphological characteristics **Adult** Length 3.4-4.0 mm, cylindrical, reddish brown to shiny dark brown. Elytral declivity bears 4 teeth on each lateral margin, with the 2nd and 3rd teeth share the same base. The base of these two teeth is closer to the base of the 4th teeth compared to that of the 1st, while the tips of the 2nd, 3rd, and 4th teeth are equally spaced. The interspace between teeth can be expressed as 1-2, 3, and 4. The tips of all teeth are acute except the 4th which is rather obtuse. Elytral declivity is similar between males and females.

Biology and behavior *Ips duplicatus* occurs one generation a year in Inner Mongolia and overwinters as adults in the soil. Adults of overwintering generation resume activity in late May to early June and peaked in early to mid-June. During this time, they mate and lay eggs. Eggs

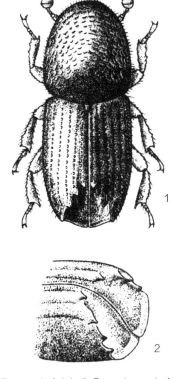

I. duplicatus. 1. Adult, 2. Posterior end of elytra

I. duplicatus tunnel (Li Zhenyu)

hatch in mid- to late June and peaked in late June. Mature larvae start to pupate in late June to early July, and peak pupation period falls in early to mid-July. Adults emerge in early to mid-July with a peak in late July. Adults enter soil to overwinter in early to mid-August after supplemental feeding. *I. duplicatus* sometimes occurs concurrently with other bark beetles such as *I. typographus* and *I. sexdentatus*.

Prevention and treatment

4. Biological control. Parasitoids of *I. duplicatus* include *Tomicobia seitneri*, *Calosota koraiensis*, *Roptrocerus mirus*, and *Calosota conifera*.

5. Trapping with Ipsdienol (Id) + E-myrcenol (EM) + Amitinol (At) (20 mg + 20 mg + 2 mg) provides good results.

References Yin Huifen et al. 1984; Yang Zhongqi 1996; Chen Guofa et al. 2009, Wang Guicheng 1992.

(Nan Nan, Li Zhenyu, Wang Guicheng)

239 *Ips hauseri* Reitter

Classification Coleoptera, Curculionidae
Common name Hauser's engraver

Distribution Xinjiang; Tajikistan, Kyrgyzstan.
Hosts *Picea schrenkiara*, *Larix sibirica*.
Damage Feeding by adults and larvae under the bark create galleries that affect the vascular system of the trees. Heavily infested trees show yellowish or reddish needles, drooping branches, boring holes, and pitch tubes and eventually die.

Morphological characteristics **Adults** Length 4-4.8 mm, cylindrical. Compound eyes reniform. Pronotum dark brown and divided into a verrucose area and a punctate area. The verrucose area contains small and flat verruca, covered by dense long hairs arranged in an inverted "U" curve; the punctate area is hairless with a wide and smooth median line, punctures small and round, concentrated on lateral margins. Elytra brown, with flat strial punctures. Strial interspace wide and flat, with long fine hairs. Elytral declivity bears 4 teeth on each lateral margin. Tooth 2 and 3 arise from the same base but widely separated apically. Male tooth 1 wedge-shaped, tooth 2 short triangular with a tumescent base and acute tip, tooth 3 strong, emarginated at the tip, tooth 4 the smallest of all. All 4 teeth in female conical and spaced equally.

Eggs Oval, milky white when newly laid, changing to dark gray before hatch.
Larvae Length 4-4.8 mm at maturity. Head brown. Newly hatched larvae milky white, changing to reddish brown later.
Pupae Length 4-4.5 mm, milky white, changing to yellowish before emergence.

Biology and behavior *Ips hauseri* occurs one generation a year in Nanshan area in Urumqi of Xinjiang and overwinters as adults. Adults become active in late May and enter the bark to lay eggs in early June. Egg period lasts for 7-15 days. Eggs start to hatch in early June, with a larval period of 20-25 days. Pupation begins in early July and lasts for 4-6 days. Adults emerge in mid-July and overwinter in late August after supplemental feeding for about 40 days. *I. hauseri* is found in the entire tree from roots to top branches with a diameter of > 3 cm, but most abundant in the middle of the tree or places with thin barks.

Ips hauseri adults are active from 9:00 until sunset but more so between 11:00-15:00 on sunny days. They are not active under foggy, windy, or rainy conditions. Adults spend considerable time looking for appropriate entrance locations by moving back and forth on fallen trees. Ide-

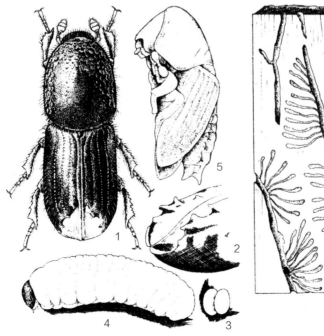
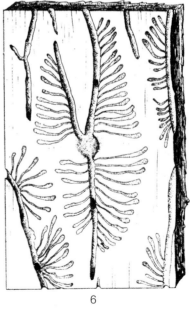

I. hauseri. 1. Adult, 2. Posterior end of elytra, 3. Eggs, 4. Larva, 5. Pupa, 6. Damage (Zhu Xingcai)

al entrance locations include bark crevices, wounds, or branch joints. Normally the male starts the entrance hole and constructs the nuptial chamber before other individuals move in gradually. No more than five individuals will stay in one chamber. Extra individuals will be expelled. Females begin to construct parent galleries after mating. There are usually 3-5 parent galleries with an average length of 7-10 cm and average width of 2.0 mm. Parent galleries are very clean so that adults can move inside freely. Parent galleries are star-shaped initially, become straight afterwards. There are 1-3 openings at the center or both end of each parent gallery. Females start to lay eggs 1-2 days after mating. There are two peak oviposition periods in mid-June and mid-July. Eggs are laid on one side of the parent gallery and spaced un-evenly, with most found near the nuptial chamber. Each female produces an average of 15 eggs in her lifetime, ranging from 8 to 22 eggs.

Newly hatched larvae construct their galleries off the lateral margins of the parent galleries. Larval galleries are narrow at the beginning and widen as the larvae grow. The galleries are filled with sawdust and frass. Larval galleries are fanlike collectively.

Pupal chamber is constructed at the end of the larval gallery, oval, covered with sawdust, and measured 5-7 mm in length. Mature larvae pupate inside the pupal chamber. Newly emerged current generation adults carry out supplemental feeding around pupal chambers, leaving irregular interconnected galleries. Adults start overwintering by late August. Most overwintering adults are found in the soil up to 30 cm in depth and within 60 cm from the base of the tree.

Prevention and treatment

1. Silviculture. Avoid long-term storage of unpeeled logs in the forest. Timely removal of infested, fire damaged, windthrow, and snow-damaged trees through sanitary measures.

Mechanical control. Set up bait logs to attract and destroy beetles.

2. Biological control. Conserve natural enemies such as predatory beetles - *Bledius curricorris*, *Trichodas* sp., and *Carabus* sp.; and woodpeckers - *Dendrocopos leucopterus*, *D. minor kamtschakensis*, and *D. minor tianshanicus*. Irrigate soil at base of trees with 2×10^8 conidia/mL or 1.5×10^{10} conidia/g powder.

3. Chemical control. Apply chemical insecticides on soil surface around infested trees after adults entering overwintering stage or before adult re-emergence in May next year. Spray 45% profenofos and phoxim mixture at 30-60×, 12% thiamethoxam and lamda-cyhalothrin mixture at 30-60× to tree trunks to control larvae.

References Wen Shouyi et al 1959, 1987; Yin Huifen et al. 1984; Xiao Gangrou (ed.) 1992.

(Wen Shouyi)

Holes made by *I. hauseri* (Li Zhenyu)   *I. hauseri* galleries (Li Zhenyu)

240 *Ips nitidus* Eggers
Classification Coleoptera, Curculionidae

Distribution Sichuan, Yunnan, Gansu, Qinghai, Xinjiang; Europe.

Host *Picea crassifolia*.

Damage Larvae feed on branches, stumps, trunks of weak and windthrow trees, and fresh logs.

Morphological characteristics **Adults** Similar to *I. typographus*. Length 4.1-5.5 mm, shiny black, with a large verruca in the middle of the frons. Pronotum with scale-like serrations on lateral margins anteriorly. Elytral declivity surface shiny smooth, bears 4 teeth on each lateral margin. Tooth 3 the largest, constricted in the middle and truncate at the tip. Base of the tooth 2 and 3 separated the furthest compared with other teeth.

Eggs Length 1.0 mm, width 0.5-0.7 mm; oval; milky white.

Larvae Length 6.0 mm; milky white, turns to brownish after feeding.

Pupae Length 5.0 mm; milky white or whitish yellow, turns to yellowish before pupation.

Biology and behavior *Ips nitidus* occurs one generation a year on north-faced slopes in Qilian Mountains in Gansu province. It overwinters as adults in 0-18 cm deep leaf litter and moss layer within one m around the base of the trees, or in galleries at the base of windthrow trees, fresh logs, branches, stumps, and standing trees. Adults emerge from the soil in late April to early May and start to infest new hosts. A male constructs nuptial chambers in the cambium region and releases sex pheromone to attract 1-4 females. Females excavate parent galleries and egg pits to lay eggs after mating. Egg pits symmetrically arranged. Galleries are longitudinal compound type with 1-4 parent galleries, length 5-20 cm, width 2.1-3.3 mm. Larval galleries symmetrically arranged, length 4-6 cm, longer near the nuptial chamber. Each female lays an average of 50 eggs. Egg period lasts for 7-15 days and peaks in mid- to late May. Egg hatches from late May to early June. A larva constructs its gallery on either side of the parent gallery. Larval period lasts for

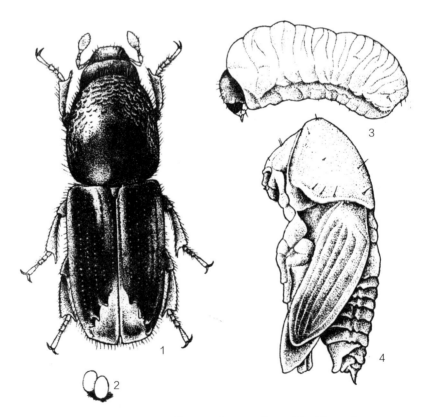

I. nitidus. 1. Adult, 2. Eggs, 3. Lava, 4. Pupa (Zhu Xingcai)

15-25 days. Mature larvae pupate at the end of the larval gallery by mid- to late June. Pupal period lasts for 10-15 days. Adults emerge in late June and start maturation feeding near the pupal chamber for about a month before chewing exit holes and flying out for supplemental feeding. Some of them stay near the pupal chambers and continue feeding until winter. Adults enter the soil for overwintering by mid- to late September when temperatures are below 7°C. Ventilated, sunny, and moist locations are preferred. Infestation rate decreases with the increase in canopy cover and the decrease of artificial cultivation.

Prevention and treatment

Trapping with *Ips typographus*-specific pheromone during adult period provides good results. Periodic removal of heavily infested trees to destroy beetles also helps.

References Xiao Gangrou (ed.) 1992, Li Menglou 2002, Xue Yonggui 2008.

(Li Menglou, Fu Huien)

nitidus damage. 1. Main tunnel, 2. Mating chamber, 3. Secondary tunnel, 4. Pupal chamber (Zhu Xingcai)

I. nitidus adult (Li Zhenyu)

241 *Ips sexdentatus* (Börner)

Classification Coleoptera, Curculionidae

Synonyms *Dermestes sexdentatus* Borner, *Bostrichus pinastri* Bechstein, *Tomicus stenographus* Duftschmidt, *Ips typographus* De Geer

Common name Six-toothed bark beetle

Importance *Ips sexdentatus* is an important secondary pest of conifer forest in northern China. It occurs concurrently with *Ips acuminatus* and other bark beetles on weak trees and cause tree mortality in a short time. It can also infest healthy trees as a pioneer species and causes massive damage to the forest.

Distribution Beijing, Hebei, Inner Mongolia, Liaoning, Jilin, Heilongjiang, Henan, Sichuan, Yunnan, Shaanxi, Gansu, Xinjiang; Korea Peninsula, Mongolia, Thailand, Turkey, Russia, and Europe.

Hosts *Pinus koraiensis*, *P. sylvestris* var. *mongolica*, *P. tabuliformis*, *P. armandii*, *P. kesiya* var. *langbianensis*, *Larix* spp., *Picea koraiensis*, *P. jezoensis*.

Damage *Ips sexdentatus* constructs galleries inside the bark and leaves shallow itching trace on sapwood.

Morphological characteristics **Adults** Length 5.8-7.5 mm; cylindrical; reddish to dark brown; shiny. Elytral declivity bears 6 teeth on each lateral margin. Tooth 4 is the largest, button shaped.

Larvae Length 6.7 mm; cylindrical. Body strong with multiple wrinkles, curved ventrally, U-shaped.

Biology and behavior *Ips sexdentatus* occurs one generation a year in Heilongjiang province and one to two generations a year in Qinba Forest. It overwinters as adults. Overlapping generations are common due to prolonged adult stage and different adult phenotypes. There are two phenotypes in Dailing Forest District in Heilongjiang province. The spring phenotype becomes active and starts to construct galleries and lay eggs in mid- to late May, with adults of the new generation emerge in mid-July and migrate to new hosts for supplemental feeding. The summer phenotype excavates their galleries and lays eggs in early July. The new generation adults emerge in early August and stay inside the original galleries for overwintering by constructing 2-3 cm deep "blind holes" inside the sapwood near the pupal cambers.

Individuals of *I. sexdentatus* are organized in families. Each family is made of 1 male and 2-4 females and lives inside the thick bark on the main trunk and the base of the tree. Longitudinal compound gallery contains 2-4 channels with one on top and two below most of the time. Length 4 cm, width 5 mm. Larval gallery sparse and short with a length of 2.5 - 5 cm. Galleries are filled reddish brown frass which accumulate outside of the tree, forming funnel-shaped frass piles on root flares and at the base of the tree in the mornings or moist days.

I. sexdentatus adult (Li Zhenyu)

I. sexdentatus posterior end of elytra (Li Zhenyu)   *I. sexdentatus* adult (Li Chengde)

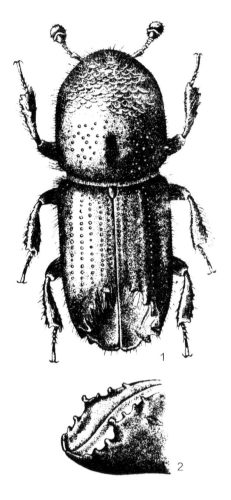

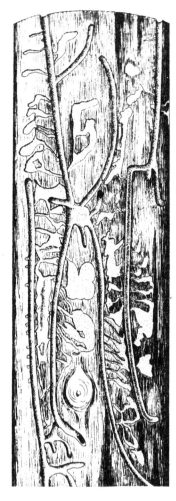

I. sexdentatus. 1. Adult, 2. Posterior end of elytra (Zhu Xingcai)

I. sexdentatus damage (Zhu Xingcai)

Adults are phototactic. Sparse stands, stands on sunny and south-faced slopes, fire-damaged stands, logged off stands, roadside stands, and over matured stands are most susceptible to this pest. In addition, unpeeled logs, newly felled trees and standing dead trees inside the forests promote initial infestation and continuous development of *Ips sexdentatus* populations.

Prevention and treatment

1. Quarantine. Avoid the movement of infested logs. Peel or chemically treat infested logs to prevent spreading the pest.

2. Silviculture. Select resistant tree species. Plant mixed forest with both conifers and broad-leaf trees. Enhance cultivation and limit access to the forest to improve species diversity. Thin and log according to plans. Leave low stumps during logging and peel the barks off. Remove windthrow and windbreak trees and standing dead trees on time. Remove infested trees during sanitation. Remove new logs from the forest or peel the bark off if left in the forest. Place lumberyard away from forest stands.

3. Trapping tree. Set up 1-2 trap trees/800 m^2 before adult flight season when less than 2% of the trees inside the forest are infested. Peel trapping trees to destroy larvae before pupation.

4. Biological control. Conserve parasitoids such as *Pycnetron curculionidis*. Irrigate soil at base of trees using *Beauveria bassiana* at 2×10^8 conidia/mL or 1.5×10^{10} conidia/g powder.

5. Fumigation. Fumigate with methyl bromide at 10-20 g/m^3, aluminum phosphide at 3 g/m^3, or sulfuryl fluoride at 30 g/m^3 by covering the log piles with 0.12 mm agriculture film for 2-3 days will eliminate bark beetles as well as longhorned beetle larvae inside the logs.

6. Pheromone trapping. Set up pheromone traps baited with *I. sexdentatus* adult lures at 1 trap/1 km^2 at 1.5 m above the ground during adult flight season will rapidly reduce mating success rate and population density.

References Xiao Gangrou (ed.) 1992, Li Chengde 2004, FAO 2009.

(Li Chengde, Qi Mujie, Yu Chengming)

242 *Ips subelongatus* (Motschulsky)

Classification Coleoptera, Curculionidae
Common name Larch bark beetle

Importance *I. subelongatus* is an important wood-boring beetle of larch and other pine species. It spends most of its life cycle feeding under the bark, which usually leads to the death of the tree.

Distribution Beijing, Shanxi, Inner Mongolia, Jilin, Heilongjiang, Zhejiang, Shandong, Yunnan, and Xinjiang; Japan, Mongolia, Siberia and the far-east costal area in Russia.

Hosts *Larix* spp., *Pinus koraiensis*, *Pinus densiflora*, *Pinus sylvestris* var. *mongolica*, *Picea koriensis*, and *Picea jezoensis* var. *microsperma*.

Damage Larvae feed on the cambial region, which disrupts the vascular system of the tree, resulting in weakened growth and death of the tree.

Morphological characteristics **Adults** Length 4.4-6.0 mm; milky white at emergence, turns to yellowish, yellow, dark brown or pitchy brown later; shiny.

Eggs Oval; milky white; translucent, shinny.

Larvae Length 4.2-6.5 mm at maturity, milky white, head capsule grayish yellow to yellowish brown.

Pupae Length 4.1-6.0 mm; milky white.

Biology and behavior Generation time for *I. subelongatus* is directly related to temperatures and differs between areas and years. It occurs two and occasionally three generations a year in Siping Area in Jilin province. Adults start to fly in late April and mate in early May. Eggs are laid from the main trunk to large branches. Eggs hatch in mid-May. Adults of the first generation emerge in early July and lay eggs in early August. Larvae of the second generation appear in late August and pupate in late August. Adults of the second generation emerge in early to mid-September and overwinter in mid-October.

It takes 48 days to finish one generation at 18°C and 33 days at 22°C. Overlapping generations are common because of its relatively long egg-laying period. There are three adult flight peaks in early to mid-May, mid-June, and mid-July in Jilin province.

Prevention and treatment

1. Select appropriate trees species for local conditions to boost pest resistance.

2. Enforce quarantine to prevent the movement of infested logs.

3. Sanitation logging. Remove infested trees and treat the logs to prevent beetles from spreading.

4. Trapping with *Ips subelongatus* aggregation pheromone at the density of one trap/2 ha for light infestation, one trap/1 ha for moderate infestations, and one trap/1 ha

I. subelongatus damage (Li Zhenyu)

for heavy infestations after sanitation logging.

5. Trapping logs. Set up trapping logs in open space, forest edges, or in forest with a canopy density less than 0.5 before adult flight season. Debark the logs or treat them with chemicals or submerge them in water to kill eggs, larvae, and pupae.

References Zhang Qinghe, Liu Zhuanfang, Sun Yujian et al. 1990; Xiao Gangrou (ed.) 1992, Gao Changqi, Ren Xiaoguang, Wang Dongsheng et al. 1998; Song Liwen 2005.

(Gao Changqi, Song Liwen, Yu Chengming)

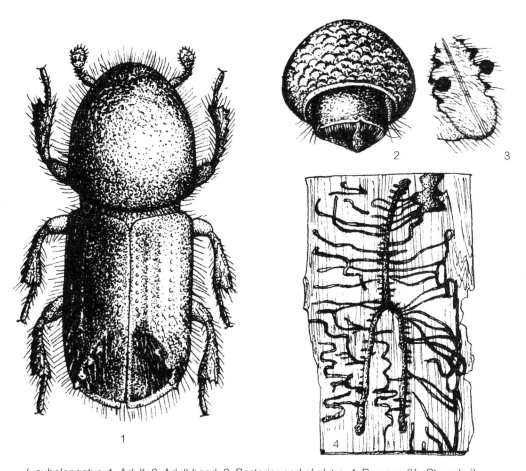

I. subelongatus. 1. Adult, 2. Adult head, 3. Posterior end of elytra, 4. Damage (Yu Changkui)

243 *Ips typographus* (L.)

Classification Coleoptera, Curculionidae
Common name European spruce bark beetle

Importance *Ips typographus* is one of the most dangerous bark beetles in the world, with high survival and reproduction ability. It spends most of its life cycle inside the bark. Feeding under the bark disrupts the vascular system of the trees, resulting in death of host trees over large areas.

Distribution Inner Mongolia, Jilin, Heilongjiang, Sichuan, Gansu, Qinghai, Xinjiang; Japan, Korea Peninsula, Russia, Czech Republic, Denmark, Sweden, Norway, Finland, and France.

Hosts *Picea abies, P. koraiensis, P. jezoensis* var. *microsperona, P. schrenkiana, Pinus thunbergii, P. sylvestris* var. *mongolica*.

Damage Adults of *I. typographus* attack host trees from bark crevices at the middle to lower trunks and construct entrance tunnels between the cambial region and the sapwood. Primary galleries longitudinally compound. Both larvae and adults feed on cambial tissues that destroy vascular system of the tree. Initial No obvious symptoms during the initial attack, however, heavy infestation results in discoloured and fallen needles. Trees are killed over large areas during outbreaks.

Morphological characteristics **Adults** Length 4.2-5.5 mm, reddish brown to dark brown, shiny with brown hairs. Compound eyes reniform. Antennae five segmented, with a club and oval. Frons with evenly distributed verrucae and one large verruca in the middle between lower frons and epistoma. Elytral declivity oblique posteriorly, bears four spines on each lateral margin. Spines conical except spine no. 3, which is button shaped. Spine no. 1 is the smallest among all. Interspace between spines no. 1 and no. 2 is the widest among all.

Larvae Length 5.0 mm, milky white, curved.

Pupae Length 4.0-6.0 mm, milky white.

Biology and behavior *Ips typographus* has one generation a year in high latitude areas in central and northern Europe and two generations a year in southern Europe. In the mountain areas in eastern Jilin Province, it completes one generation a year. Adults start to fly in late May to early June when temperature reaches 20°C, coincides with the budding of elm leaf spirea.

Males initiate attack on host trees and attract females with sex pheromone into the tunnels for mating. There are usually one male and two to three females in each

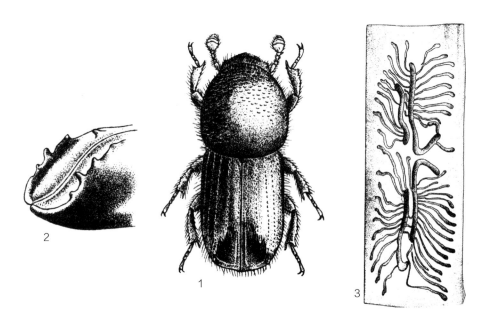

I. typographus. 1. Adult, 2. Posterior end of elytra, 3. Damage (Zhu Xingcai)

typographus damage (Li Zhenyu)

tunnel, occasionally up to five females can be attracted to a single tunnel. Eggs are laid between the cambial region and the sapwood 1-2 days after the invasion. Generation time is 33-55 days from eggs to adults. First generation adults stay inside the bark for an average of 28-30 days before chewing out. Some of them can stay as long as 60 days. Overwintering occurs in mid- to late October when adults complete supplemental feeding, with the majority found in soil under leaf litter around the base of the trees and the minority inside the bark for one-year generation population.

Ips typographus prefers ventilated and light-transmitting environment without direct sun light. Therefore, heavy infestations usually are found on forest edges or in relatively open forests. Outbreak populations are strongly aggregated with clumped or banded distribution that spread diffusively.

Prevention and treatment

Ips typographus spreads through adults and infested logs. Therefore, quarantine enforcement and monitoring are fundamentally important. Stop movement of infested logs. Promptly eliminate source of infestation. Eliminate light pest populations then the heavy infestations.

1. Silviculture. Remove fallen and weak trees and tall stumps to maintain sanitary stand conditions. Remove a small portion (< 20%) of the mature spruces and firs from the stands over long period (8 years) to prevent pest outbreaks.

2. Monitor pest population using lures containing ipsdienol, (S)-cis-verbenal, and 2-methyl-3-buten-1-ol in mid-May. Implement management measures when population levels exceed threshold. Trapping with (S)-cis-verbenal, and 2-methyl-3-buten-1-ol after infested trees are removed at 1-2 trap/ha provides good results.

3. Set trapping logs inside the forest before adult flight season and destroy beetle population inside the logs during pupal stage.

References Xiao Gangrou (ed.) 1992; Sun Xiaoling 2006a, b.

(Gao Changqi, Sun Xiaoling, Huang Xuchang)

244 *Phloeosinus aubei* (Perris)

Classification Coleoptera, Curculionidae
Common name Cypress bark beetle

Importance *Phloeosinus aubei* is an important woodborer of Chinese arborvitae. Feeding activity in the cambial region disrupts the vascular system and leads to the death of the hosts.

Distribution Beijing, Jiangsu, Shandong, Henan, Yunnan, Shaanxi, and Taiwan; Japan, Korea Peninsula, Russia, Germany, France, Italy, Spain, and Bulgaria.

Hosts *Platycladus orientalis*, *Juniperus chinensis*.

Damage *Phloeosinus aubei* infests Chinese arborvitae and juniper selectively. Weak trees and recovering new plantings are easily attacked. Adults feed mostly on small branches with the diameter of 2 mm for supplemental nutrition. This kind of feeding affects the growth and the shape of the tree and creates piles of nipped branches under the tree in heavy infestation. Feeding on the main trunk and branches during the reproduction season results branch dieback and tree mortality.

Morphological characteristics **Adults** Length 2.1-3.0 mm, reddish brown or dark brown. Head small

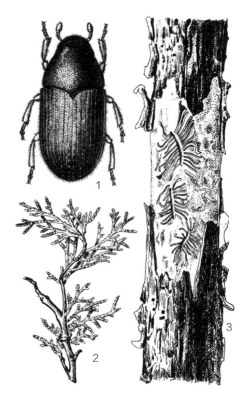

P. aubei. 1. Adult, 2. Damaged branch, 3. Damaged trunk (1 by Zhang Xiang; 2-3 by Zhou Defang)

P. aubei adult (Xu Gongtian)

and concealed under the pronotum. Antennae reddish brown, club tip oval with dense punctures and gray hairs. Elytra with anterior margin convex. There are nine longitudinal striae on each elytron. Elytal declivity depressed, with surface teeth on males.

Eggs White, circular.

Larvae Length 2.5-3.5 mm at maturity. Milky white at hatch, head brownish, body curved.

Pupae Length 2.5-3.0 mm. Milky white.

Biology and behavior *Phloeosinus aubei* occurs one generation a year in Beijing and overwinters as larvae and adults inside the galleries under the bark. There are two damaging peaks in April and June. Overwintering adults become active in April and start to attack weak arborvitae and junipers by initiating entrance tunnels. They construct longitudinal parent galleries on host branches after mating under the bark and start to lay eggs. About 20-30 eggs are laid each time. Newly hatched larvae feed

P. aubei adult (Xu Gongtian)   *P. aubei* larva (Xu Gongtian)

P. aubei tunnel (Xu Gongtian)   *P. aubei* galleries (Xu Gongtian)

between the bark and wood and create star-shaped larval galleries. Larvae mature by mid- to late May and pupate at the end of galleries. Adults emerge in early June and migrate to 2 cm diameter small branches for supplemental nutrition. Adult feeding causes these branches vulnerable to wind damage. By late September adults move back to larger branches to overwinter.

There are 2-3 generations a year for *P. aubei* in Xuzhou area in Jiangsu province. It overwinters as larvae or adults. Overwintering adults become active in early April while overwintering larvae continue their development to pupate and emerge as adults. After supplemental feeding, adults initiate attack on host trees with peak infestation found in mid-May. Infestation from overwintering adults tapered off by mid-June. Eggs of the first generation are laid in early April, with pupation occurs in mid-June. Adult emergence starts in mid-June and peaked in mid-July for the first generation. Eggs and adults of the second generation appear in mid-June and early August, respectively. Adults emerged before early September may lay eggs to start the third generation. Early hatched larvae could further develop to pupae and become adults by the end of October. Third generation adults do not infest new hosts. They overwinter together with larvae and adults of the second generation and larvae of the third generation.

There are 2-3 generations in Taiyuan area in Shanxi province for this pest. It overwinters as adults inside branches or larvae inside the main trunk. Adults fly out from April to May next year and start to lay eggs on host trees. Eggs hatch in late May and larvae pupate in early July. Second generation adults emerge in mid-July. The second generation lasts about two months from egg hatching to adult emergence. The third generation adults emerge in mid-September. Adults will overwinter or lay eggs and overwinter as larvae.

P. aubei exit holes (Xu Gongtian)

P. aubei larva (Xu Gongtian)

There is only one generation a year for this beetle in Jinan and Taian areas in Shandong province. It overwinters as adults in branch tips. Adults become active from March to April. Females construct entrance tunnels under the bark of weak host trees. Males follow the females to the tunnels and start to excavate irregular nuptial chambers for mating. Females start to construct parent gallery upwards after mating and create egg chambers along both sides of the galleries. Eggs are laid inside the egg chambers. At the meantime, males push sawdust out of the parent galleries from the entrance holes. Parent galleries are 15-45 mm long. Each female lays 26-104 eggs in her lifetime. Egg stage lasts for 7 days. Eggs start to hatch in mid-April and larvae create small serpentine galleries mainly inside the cambial region away from the egg chambers. Larval galleries are 30-41 mm long. Larval stage lasts for 45-50 days. Mature larvae pupate inside pupal chambers at the end of the galleries. This chamber is about 4 mm deep and perpendicular to the larval gallery and covered by a semi-transparent film outside. Pupal stage lasts for 10 days. Peak adult emergence occurs in mid- to late June. Newly emerged adults yellowish brown. They crawl upwards along the emergence holes and then fly to the upper crown or outside branches of healthy host trees for supplemental nutrition. Infested branch tips are usually devoured from inside and become vulnerable to wind damage. Adults overwinter in mid-October.

Prevention and treatment

1. Cultivation and management. Water the trees timely and apply fertilizer and tillage when needed. Reinvigorate historical trees to increase their pest resistance capability and prolong their life. Avoid damage to seedlings during transplanting, select raining days for transplanting to avoid egg-laying.

2. Remove damaged trees. Remove damaged trees from infested areas between October of the current year and April of the next year before adult emergence from overwintering. Remove newly killed branches and trees to prevent beetle spread. Debark logs before the end of February to eliminate source populations.

3. Trap logs. Set up 2 mm diameter fresh host branches or log sections to trap and kill adult beetles in early April before adult infestation.

4. Biological control. Conserve natural enemies such as *Tillus notatus*, *Pyemotes trittci*, *Anacallocleonymus gracillis*, *Phleudecatoma platycladi*, *Tetrastichus cupressi*, *Metacolus sinicus*, and *Heydenia scolyti*. Irrigate soil using *Beauveria bassiana* at 2×10^8 conidia/mL or 1.5×10^{10} conidia/g power. Release predatory mite, *Pymeotes scolyti*.

5. Chemical control. Apply chemical insecticides during supplemental feeding for adults. Apply chemicals to overwinter adults provides good results due to life stage uniformity. Fumigate tree trunks with dichlorvos provides good control for larvae and pupae. Spray 45% profenofos and phoxim mixture at 30-60×, 12% thiamethoxam and lamda-cyhalothrin mixture at 30-60× to tree trunks to control larvae.

References Fan Di 1985, Qi Qinglan 1987, Xiao Gangrou (ed.) 1992, Yang Zhongqi 1996, Dong Cunyu 1997, Lü Xiaohong et al. 2000, Yang Yanyan et al. 2004, Xu Gongtian 2007, Zhang Zuoshuang et al. 2008.

(Chen Chao, Fan Di)

245 *Phloeosinus sinensis* Schedl
Classification   Coleoptera, Curculionidae

Importance   *Phloeosinus sinensis* is a common woodborer of Chinese fir in southern China. Feeding by adults and larvae between the cambial region and sapwood creates overlapping galleries under the bark, which disrupts the vascular system, resulting in death of the trees in small spots or over large areas.

Distribution   Zhejiang, Anhui, Fujian, Jiangxi, Henan, Hunan, Chongqing, Shaanxi.

Host   *Cunninghamia lanceolata.*

Damage   Adults of *P. sinensis* feed aggregately on the trunk of Chinese fir that leads to resinosis. Female constructs single longitudinal parent gallery under the bark close to the sapwood, whereas larval galleries are built on both sides of the parent gallery.

Morphological characteristics   **Adults** Length 3.0-3.8 mm. Compound eyes reniform. Pronotum ladder-like. Elytral groove interspace contains dense verrucae, with more than 10 verrucae on rows no. 1 and 3, and 6-7 on row 2.

Larvae Length 5.0 mm, purplish red after feeding, yellowish white at maturity.

Biology and behavior   *Phloeosinus sinensis* occurs one generation a year in Jingde County in Anhui province and overwinters as adults individually inside the trunks of Chinese firs. Overwintering adults become active in late March to late April and aggregately attack

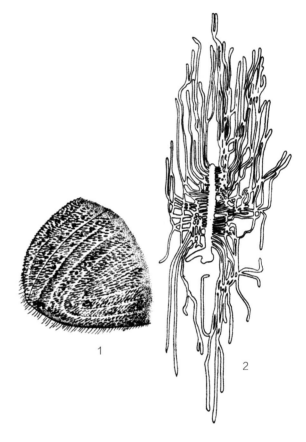

P. sinensis. 1. Posterior end of elytra, 2. Tunnel (Qu Xiaojin)

P. sinensis adult (NZM)

P. sinensis elytra (NZM)

P. sinensis adult (NZM)

P. sinensis adult head (NZM)

5-15 old Chinese fir trees. Eggs are laid between late March and early May, with larvae appear from early April to mid-June. Pupation begins in early May and end by late July. Adults of the first generation emerge from mid-May.

Overwintering adults construct 2.5 mm diameter entrance holes on tree trunk that cause resinosis. Up to 200-300 sites with resinosis can be observed from a single tree with heavy infestation. The female constructs a single longitudinal parent gallery, while the male pushed the sawdust out of the gallery through the entrance hole. Mating occurs between 19:00-21:00 pm. After mating, female creates egg chambers along both sides of the parent gallery. Each chamber contains a single egg and is closed with sawdust after the egg is laid inside. The female lays an average of 48 eggs in her lifetime. Egg stage lasts for 3-5 days. Larval galleries are evenly distributed on both sides of the parent gallery. About 91.8% larvae reach pupal stage. Pupal stage lasts for 8-9 days. Newly emerged adults stay inside the pupal chamber for a while before coming out of the tree. Adults of the first generation emerge in early July for new infestation and create overwintering galleries inside the bark of healthy trees. No sawdust and resin overflow are visible outside overwintering galleries.

Prevention and treatment

1. Bait logs. Remove weak trees early spring and place them inside the forest to attract adults for oviposition. Debark or burn the logs to destroy beetle larvae or adults inside.

2. Conserve natural enemies. *Eurytoma* sp. is the major natural enemy of *P. sinensis*. Stop chemical applications during its adult emergence in June for its protection.

3. Chemical control. Before late March when fresh yellowish sawdust at the base of the tree or overflow resin outside entrance holes along the trunk become evident, apply 75% phoxim EC at 800-1,000× through trunk spray can kill aggregated adults inside the bark. Spray 45% profenofos and phoxim mixture at 30-60×, 12% thiamethoxam and lamda-cyhalothrin mixture at 30-60× to tree trunks to control larvae.

References   Su Shiyou et al. 1898, Zhao Jinnian et al. 1988, Xiao Gangrou (ed.) 1992, Yang Zhongqi 1996.

(Zhao Jinnian)

246 *Pityogenes chalcographus* (L.)

Classification Coleoptera, Curculionidae
Common name Sixtoothed spruce bark beetle

Distribution Beijing, Inner Mongolia, Liaoning, Jilin, Heilongjiang, Sichuan, and Xinjiang; Japan, Korea Peninsula, and Russia.

Hosts Hosts includes almost all conifer species in the native range, *Picea koraiensis*, *Picea jezoensis*, *Pinus koraiensis*, and *Pinus sylvestris* var. *mongolica* in the Greater and Lesser Khingan Mountains, and Changbai Mountain Forest District in northeast China, *Picea schrenkiana* in Nanshan and Hami Forest Districts in Xinjiang Autonomous Region, *Pinus bungeana* in Beijing, and *Picea asperata* in other areas.

Morphological characteristics **Adults** Length 1.4-2.3 mm, shiny brown with sparse hairs. Male frons slightly depressed at the bottom and convex on the top, and occasionally with a verruca in the middle. Female frons slightly convex at the bottom and contains a deep oval depression in the middle. Frontal surface velvet, light yellow to yellowish brown. Pronotum contains large conspicuous punctures and a smooth median line on the posterior half. Elytra 1.7 times as long as wide, and 1.6 times as long as the pronotum. Elytral striae contain small punctures that disappear at the tip. There are no punctures between striae. Spine no. 2 on male elytral declivity slightly closer to spine no. 3. Spine no. 3 contains a small verruca near the suture. Female elytral declivity narrow and shallow, with three pairs of small and inconspicuous spines.

Biology and behavior *Pityogenes chalcographus* co-occurs with *Ips acuminatus*, *Ips sexdentatus*, and *Pityogenes seirindensis* in virgin conifer forests and plantations in the Greater and Lesser Khingan Mountains, and in Mongolian Scots pine forests in Hulunbuir sandy land. It attacks mainly the thin bark, crown, or branches of mature trees and fallen trees at sunny locations.

Pityogenes chalcographus occurs one generation a year in Hulunbuir region in Inner Mongolia and overwinters as adults. Galleries are constructed under the bark of Mongolian Scots pine branches. It took one and a half month for this beetle to develop from eggs to adults. Lifecycle is asynchronized. Adults can be found anytime from early June to early August. Adults feed inside the galleries or under the bark for supplemental nutrition. Most galleries are constructed inside the cambial region, with some on sapwood as well. Large nuptial chambers are conspicuous inside thick bark. Parent gallery is compound and star-shaped, with 3-6 separate tunnels extended from the nuptial chamber. Larval galleries start densely from and perpendicular to the parent gallery. Larval

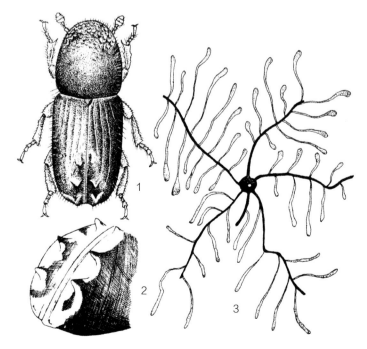

P. chalcographus. 1. Adult, 2. Tip of adult elytra, 3. Gallery (Zhu Xingcai)

P. chalcographus adult (dorsal view) (Li Zhenyu)

P. chalcographus adult (lateral view) (Xu Gongtian)

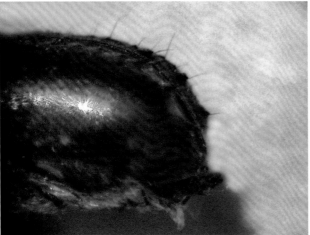
P. chalcographus adult elytra (Xu Gongtian)

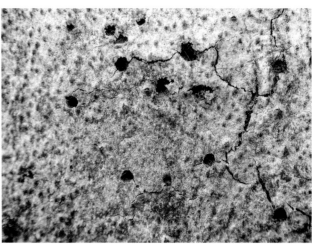
P. chalcographus adult exit holes (Xu Gongtian)

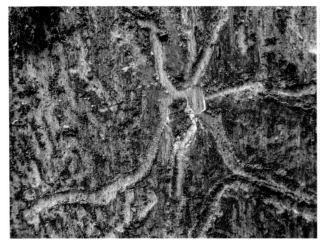
P. chalcographus gallery (Xu Gongtian)

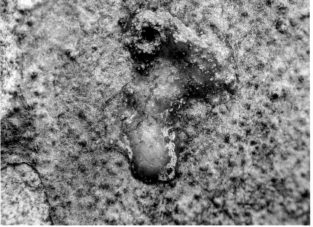
P. chalcographus killed by resin (Xu Gongtian)

galleries start only from the outside margin of parent galleries if two tunnels from the same "family" are close to each other. Pupal chambers are constructed inside the cambial region.

Prevention and treatment

Conserve natural enemies such as *Roptrocerus xylophagorum*, *Roptrocerus mirus*, *Metacolus unifasciatus*, *Eurytoma longicauda*, *Platygerrhus piceae*. Irrigate soil using *Beauveria bassiana* at 2×10^8 conidia/mL or 1.5×10^{10} conidia/g power.

References Cai Banghua et al. 1959, Wen Shouyi et al. 1959, Yu Chengming 1959, Yin Huifen et al. 1984, Jiang Yucai et al. 1989, Xiao Gangrou (ed.) 1992, Yang Zhongqi 1996.

(Li Zhenyu, Yu Chengming)

247 *Scolytus schevyrewi* Semenov

Classification Coleoptera, Curculionidae
Common name Banded elm bark beetle

Distribution Hebei, Heilongjiang, Henan, Shaanxi, Ningxia, and Xinjiang; Turkmenistan, Uzbekistan, Tajikistan, and Kyrgyzstan.

Hosts *Ulmus pumila*, *U. laevis*, *U. davidiana* var. *japonica*, and *Salix* spp.

Morphological characteristics **Adults** 3.19-4.17 mm long with an average of 3.64 mm. Males and females similar in size. Head black, antennae yellow to yellowish brown. Frons of females slightly convex, with carinae running toward clypeus. Frons of males concave, with inward curved yellow hairs on the margins. Pronotum 1.39 mm long and 1.54 mm wide, reddish brown anteriorly and posteriorly, and dark brown in the middle with punctures. Scutellum black. Elytra 1.87 mm long and 1.55 mm wide, reddish brown to dark brown, with a dark-colored transverse band halfway to three quarters toward the end, slightly serrate posteriorly and along the margins. Thorax black laterally. Sterna black. Legs reddish brown with yellowish brown fine hairs. Femur with dark brown blotches at the base and the tip. Sternum 2 constricted extremely upward to form an obtuse angle with sternum 1. Verruca at the center of sternum 2 black with truncate tip. Sternum 7 of males with one pair of long setae.

Eggs 0.82 mm long and 0.56 mm wide, oval, white, translucent when laid, turns to milky white later, and creamy yellow before hatch.

Larvae 4.8-7.5 mm long. Head capsule creamy yellow, hiding inside pronotum posteriorly. Mandibles black, labrum white.

Pupae 3.5-4.8 mm long, milky white. Wing pads reach sternum 5, with circular carinae on surface. Front wing pads shorter than back wing pads and connected at the tips. Abdomen contains a pair of caudal horns. Verrucae large and conspicuous on abdomen segment 3 to 7.

Biology and behavior *Scolytus schevyrewi* occurs two and occasionally three generations a year in Kuitun area in Xinjiang Autonomous Region and overwinters as mature larvae. Larvae pupate in early April when temperatures reach 15°C and peaked in mid-April. Adults start to emerge in late April and peaked in early May. First generation larvae pupate from late May to early June. Adult emergence peaks in early July and ends in late July. A portion of the second generation larvae construct pupal chambers and overwinter inside. The rest develop into the third generation. Generations overlap due to differences in adult longevity, egg periods, and the environmental conditions.

Adults stay inside the pupal chambers for 2-5 days after emergence before chewing out. More than 80% of the adults chew out between 14:00-20:00. Adults then feed on tender bark at the joints of small twigs for supplemental nutrition before migrating to suitable trunks of new hosts. Females enter the trees from bark crevices and start to construct parent galleries. Parent gallery is built above the entrance hole, single longitudinal, 4-6 cm (maximum 9 cm) long. Males search for entrance holes along the trunk and mate with the females after locating the holes. Males continue searching for new females for mating. No nuptial chamber is constructed as mating occurs exclusively at the entrance.

Females lay eggs as they construct the parent galleries. Eggs hatch accordingly and larvae feed in the cambial region on the outer side of the parent gallery, creating larval galleries. Larval gallery is perpendicular to parent

S. schevyrewi adult (dorsal view) (Xu Gongtian)

S. schevyrewi adult (lateral view) (Xu Gongtian)

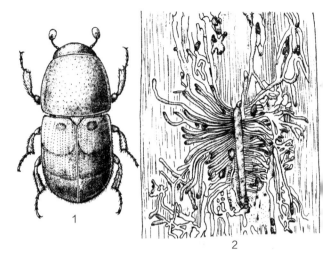

S. schevyrewi. 1. Adult, 2. Damage (Qu Xiaojin)

gallery initially, extends upwards, downwards, serpentine, or interconnected later. Larval galleries become indistinguishable under high density in the sawdust-filled cambial region. There are five instars in the larval stage. Mature larva builds an oblique pupal chamber under the bark at the end of the larval gallery.

Life stage duration is correlated with temperature. At 26°C, egg stage lasts for 3-5 days with an average of 3.8 days. Larval stage lasts for 18-23 days (average 21 days). Pupal stage lasts for 5-7 days (average 6 days). Adult stage lasts for 6-43 days (average 20 days). It took 40-45 days to finish one generation under natural conditions.

Damage on elms from *S. schevyrewi* correlates with tree vigor. Few entrance holes are found on normal healthy trees, with no parent galleries and no adverse impact on tree growth; whereas on weak trees, damage and number of entrance holes on the trees increase as the health conditions of the trees decline. Repeated attacks on the same unhealthy trees further weaken them and eventually kill the trees.

The development and survival of *S. schevyrewi* is directly related to the physiological conditions of the host tree. On weak trees with active sap flow, adult longevity is shortened, and females laid fewer eggs after entry because of sap encroachment. Successfully hatched larvae usually fail to complete their development. The most suitable hosts for this beetle are those trees died recently with no sap flow but intact bark. Adults develop normally inside the bark and lay abundant eggs, followed by successful larval development. They become important sources of new infestation as they harbor large number of adults.

The main hosts of this pest are logs with intact bark, high stumps, trees killed by diseases, frost, and rodents. These sources become the focal points of disperse that lead to more tree damage. Damage to elms is related to the existence of source populations and the distance to the source populations.

Natural enemies of *S. schevyrewi* include one species of ectoparasitic mite on adults, four species of parasitic wasps on larvae, with a combined parasitism of 14%-42.4%.

Prevention and treatment

1. Conserve natural enemies such as *Tetrastichus clavicornis*, *T. thoracicus*, *Dinotiscus aponius*, *Cheiropachus quadrum*, *Dibrachys cavus*, *Eurytoma esuriensi*, *Eurytoma morio*, *Eurytoma ruficornis*, *Eurytoma scolyti*, *Acrocormus wuyingensis*, *Cheiropachus cavicapitis*, *Heydenia scolyti*, *Raphitelus maculatus*, *Eupelmus urozonus*, *Theocolax phlaeosini*, *Oodera pumilae*, *Cleonymus ulmi*, *Entedon ulmi*, *Calosota pumilae*, *Acrocormus ulmi*, *Metapelma zhangi*, and *Callocleonymus ianthinus*.

2. Irrigate soil using *Beauveria bassiana* at 2×10^8 conidia/mL or 1.5×10^{10} conidia/g power.

References Yin Huifen 1984, Li Qingxi et al. 1987, Yang Zhongqi 1996.

(Wang Zhanting, Li Zhenyu)

248 *Tomicus minor* (Hartig)

Classification	Coleoptera, Curculionidae
Synonym	*Blastophagus minor* Hartig
Common name	Lesser pine shoot beetle

Importance Feeding by *T. minor* on pine shoots make them vulnerable to wind damage, with more than 70% branches break off under heavy infestation. Infested trees usually die from wilting.

Distribution Hebei, Jiangxi, Henan, Sichuan, Yunnan, Shaanxi, and Gansu; Southeast Asia, Japan, Russia, Denmark, and France.

Hosts *Pinus massoniana*, *P. tabuliformis*, and *P. yunnanensis*.

Morphological characteristics **Adults** 3.4-4.7 mm long. Elytral interspace with sparse punctures, and a line of setae starts from the middle. Elytral interspace 2 without groove. The second row of setae to the right or the left of the elytral suture complete.

Tomicus piniperda is very similar to *Tomicus minor* but differs in having a groove on elytral interspace 2, and absent of setae on the second row to the right or the left of the elytral suture.

Biology and behavior *Tomicus minor* usually co-occurs with *T. piniperda*. It has one generation a year. Adults become active in late April to late May. They bore into pine shoots in the upper crowns and feed inside until maturation in November before reproducing. In northern China, it overwinters as adults in young shoots or soil. However, no obvious overwintering behavior is observed in southern locations such as the Kunming area in Yunnan province, where adults continue feeding inside the shoots, or move to the trunk and lay eggs in newly created galleries. *T. minor* usually attacks dying trees in weaken stands, and occasionally healthy trees by constructing galleries under the bark at the middle of the trunk. Parent galleries transverse compound, with larval galleries extend longitudinally both upwards and downwards.

Prevention and treatment

1. Biological control. Common parasitoids include *Rhopalicus tutela*, *Roptrocerus mirus*, *Coeloides qinlingensis*, *Metacolus unifasciatus*, *Roptrocerus ipius*, *Eurytoma longicauda*, and *Calosota conifer*. Irrigate soil using *Beauveria bassiana* at 2×10^8 conidia/mL or 1.5×10^{10} conidia/g power.

2. Semiochemicals. Utilizing α-pinene, diepicedrene, and cedrene from Yunnan pine for population monitoring and control of *Tomicus minor* produced satisfactory results.

References Yin Huifen, Huang Fusheng, and Li Zhaolin 1984; Xiao Gangrou (ed.) 1992; Yang Zhongqi 1996; Ye Hui, Lüjun, and Lieutier 2004; Lu Rongchun 2008.

(Nan Nan, Li Zhenyu, Gao Changqi)

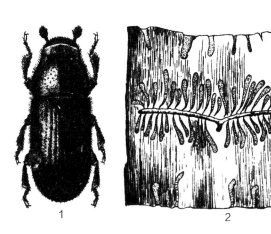

T. minor. 1. Adult, 2. Damage (Zhu Xingcai)

T. minor adult (Li Zhenyu)

T. minor adult (Li Zhenyu)

249 *Tomicus piniperda* (L.)

Classification Coleoptera, Curculionidae
Synonym *Blastophagus piniperda* L.
Common name Pine shoot beetle

Distribution Beijing, Liaoning, Jilin, Jiangsu, Zhejiang, Anhui, Fujian, Shandong, Henan, Hunan, Chongqing, Sichuan, Guizhou, Yunnan, Shaanxi, Gansu; Korea Peninsula, Japan, Russia, Mongolia, Sweden, Finland, and parts of North America.

Hosts *Pinus sylvestris* var. *mongolica*, *P. tabuliformis*, *P. thunbergii*, *P. massoniana*, *P. koraiensis*, *P. densiflora*, *P. kesiya* var. *langbianensis*, *P. taeda*, *P. elliottii*, and *P. serotina*.

Damage *Tomicus piniperda* is a cosmopolitan pest capable of causing severe damage to its hosts. Feeding on shoots disrupts the vascular system that stunts the growth of pine trees, resulting in tree mortality when infestations are heavy. Attack occurs mostly in the middle and lower trunk with single transverse galleries created between the cambial region and outer sapwood. Larvae feed on cambial cells that damages the conducting tissues, while adults attack new shoots during supplemental feeding, resulting in dieback or wind damage to the shoots.

Morphological characteristics **Adults** 3.4-5.0 mm long. Head and pronotum black. Elytra reddish to dark brown, shiny. Compound eyes oblong oval. Antennal club oval, 3-merous. Pronotum 0.8 times as long as wide. Elytra 1.8 times as long as wide, and 2.6 times of pronotal length. Elytral interspace with transverse tubercules and associated erect setae anteriorly. The second row of elytral interspace flat with a groove, and contains no tubercules and setae.

Larvae 5.0-6.0 mm long, milky white, with yellow head and brown mouthparts.
Eggs 0.9 mm long and 0.6 mm wide, oval, and whitish.
Pupae 4.5 mm long, white, with a pair of spiculae at the distal end ventrally.

Biology and behavior *Tomicus piniperda* occurs one generation year in Jilin province and overwinters as adults inside leaf litter or in 5 cm deep soil at the base of the host tree. Adults become active in early to mid-April, with a portion attack fallen and weak trees or stumps and lay eggs after mating, while others bore into pine shoots in upper crowns for supplemental feeding. Adult flight peaks in the end of April and between early to mid-May, with most adults fly out during the first peak. Egg period peaks in early to mid-May with an egg stage of 9-11 days. Most eggs hatch in early June. Larval stage lasts for 5-20 days. Pupation occurs mostly in mid- to late June with a pupal stage of 8-9 days. Adult emergence peaks in mid-July. Newly emerged adults bore into new shoots for supplemental nutrition until mid- to late October before overwintering.

Females initiate attack on recent fallen or weak trees before releasing sex pheromone to attract males for

T. piniperda adult (Xu Gongtian)

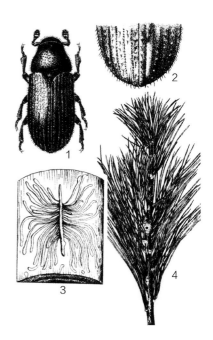

T. piniperda. 1. Adult, 2. Elytra, 3. Gallery, 4. Damage (Zhu Xingcai)

reproduction. The pair then starts to construct a single longitudinal parent gallery in the trunk. Eggs are laid on both sides of the gallery. Each female produces 40-70 eggs. Larvae construct their galleries perpendicular to the parent gallery. Egg period lasts for more than 80 days. The life history of *T. piniperda* is not well synchronized. Newly emerged adults need supplemental feeding on new shoots. Mating occurs the next spring. Late emerged adults may still need supplemental feeding in the spring before reproducing. This beetle prefers sunny and warm forest edges or open space inside the forest.

Prevention and treatment

Tomicus piniperda spreads naturally through adults, or artificially through transporting infested logs. Quarantine enforcement and monitoring is critically important.

1. Prevention. 1) Monitor pest population with semiochemicals for timely removal of infested trees for early eradication; 2) Remove windthrow, dying trees, or high stumps to limit egg laying sites; 3) Enforce quarantine to prevent spread.

2. Silviculture. 1) Enhance silviculture measures to promote healthy growth and species diversity for pest resistance; 2) Limit source populations for further spread through removal of infested trees.

3. Conserve natural enemies: *Rhopalicus tutela*, *R. gutatatus*, *Coloides qinlingensis*, *Rhoptrocerus xylophagorum*, *R. yunnanensis*, *Eurytoma yunnanensis*, *E. longicauda*, *Eupelmus urozonus*, *Dibrachys yunnanensis*.

4. Trap with 95% α-Pinene or 95% α-Pinene + trans-verbenol + nonanal at the density of 1 trap/ha.

5. Chemical control through application of contact insecticides or fumigant to overwintering sites in early spring before adults flying away resulted in 95% control. Spray 45% profenofos and phoxim mixture at 30-60×, 12% thiamethoxam and lamda-cyhalothrin mixture at 30-60× to tree trunks to control larvae.

References Zhao Jinnian et al. 1991, 2004; Xiao Gangrou (ed.) 1992; Yang Zhongqi 1996; Song Liwen, Ren Bingzhong, Sun Shouhui et al. 2005.

(Gao Changqi, Song Liwen, Zhao Jinnian, Li Zhenyu)

T. piniperda adult in tunnel (Xu Gongtian)

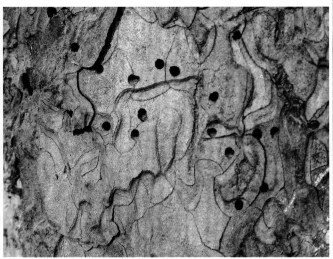

T. piniperda emergence holes (Xu Gongtian)

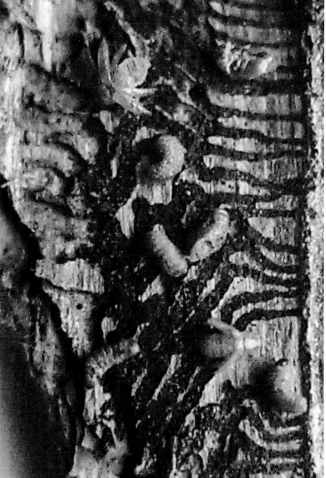

T. piniperda galleries (Xu Gongtian)

250 *Dasineura datifolia* Jiang
Classification   Diptera, Cecidomyiidae

Importance   *Dasineura datifolia* has multiple generations a year. First generation occurs during the budding and leaf expansion period of jujube. Heavy damage from this pest affects the growth, leaf expansion, and flowering of the fruiting branches. Generally young and dwarf trees suffer more than big and tall trees. During the outbreak in the eastern part of Guanzhong area in Shaanxi province in spring 1999, it became a major pest of jujube trees during spring leaf expansion. Buds were damaged from early to mid-April before leaf expansion, with the heaviest damage occurred in this period. Infestation occurs mostly on top buds of the main branches and apical young leaves on secondary branches or fruiting branches. All trees are susceptible, with up to 100% damage rate on apical young leaves. Infestation results in color changing, shrivel, and drop of leaves, which significantly affects the normal growth, flowering, and fruiting of jujube trees.

Distribution   Beijing, Hebei, Shanxi, Shandong, Henan, Sichuan, and Shaanxi.

Host   *Zizyphus jujube*.

Damage   *Dasineura datifolia* mainly damages the tender branch tips and young leaves of jujube trees. Larval feeding on tree sap stimulates leaves and leads to longitudinal rolling of leaf along lateral margins. Infested leaves become thicker and brittle and turn deep red before drop off as dark leaves. Damaged young branches stop growing, which serious weaken the tree and negatively impact the quality and quantity of jujube production.

Morphological characteristics   **Adults** Length 1.4-2.0 mm, like a small size mosquito. Orang or gray, covered by greyish yellow pubescence. Pronotum elevated. Abdomen greyish yellow, yellowish or reddish orange, with slender legs. **Eggs** Length 0.3 mm, oblong oval, slightly narrow at one end, amber colored, shiny, covered by a layer

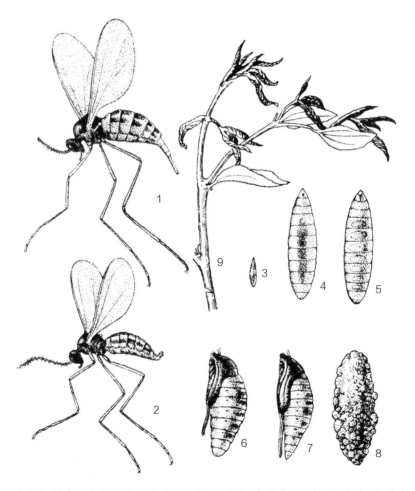

D. datifolia. 1. Female adult, 2. Male adult, 3. Egg, 4. Larva (dorsal view), 5. Larva (ventral view), 6. Female pupa, 7. Male pupa, 8. Cocoon, 9. Damage (Zhou Defang)

D. datifolia damage (SFDP)

of gelatin. **Larvae** Length 2.5-3.0 mm at maturity. Milky white, turn to yellowish before pupation. Maggot-like, body fat and round, tapered at the head and end of the abdomen, with a "Y" shaped and amber colored sternal spatula on the ventral side of the prothorax. **Pupae** Length 1.0-1.4 mm. Milky white at the beginning, turn to yellowish brown later, with a pair frontal bristles on vertex. **Cocoons** Oval, greyish white, soft with small soil granules externally.

Biological and behavior *Dasineura datifolia* occurs 5-6 generations a year in eastern part of the Guanzhong area in Shaanxi province and overwinters as mature larvae inside soil cocoons in shallow soil under the trees. Mature larvae emerge from the soil in early April next year and make new cocoons on soil surface. Peak larval damaging time and period varies each year. There are generally 5-7 larval damaging peaks from late April to mid-May, mid-May to early and mid-June, early and mid-June to mid-July, mid- and late July to early and mid-August, and mid- and late August to early October. Mature larvae begin to enter the soil and construct cocoons for overwintering. Heavy damage occurs mostly between May and June. Adults are weak fliers, prefer shade and shun away from strong sunlight. Each female produces 40 to 100 eggs. Eggs are laid in crevices between unexpanded young leaves. Upon hatching, larvae feed on plant sap, which stimulates leaf tissue and leads to longitudinal rolling of the leaf along lateral margins. Larvae feed inside the leaves. Infested young leaves thicken and turn to reddish or purplish red, become hard and brittle, unable to expand, shrivel, and drop as shriveled and dark leaves. There are usually a few to 10 larvae feeding inside each leaf roll.

Prevention and treatment

1. Elimination of overwinter populations. Bury mature larvae and soil cocoons in late fall, early winter or spring through deep tillage can prevent normal adult emergence from the soil.

2. Protecting parasitic wasp *Nasonia* sp.

3. Chemical control. 1) Shallow tillage after surface application of chlorpyrifos at 200-300×, or 3% phoxim granules at the rate of 45-75 kg/ha for the control of pupating larvae; 2) Cover spray with 1.8% avermectins at 2,500-3,000×, 1% emamectin benzoate at 4,000-5,000×, or 12% thiamethoxam at 1,000× for three times with an interval of 10 days to control overwintering population and larval population of the first generation.

References Xiao Gangrou (ed.) 1992, Zhao Fanggui et al. 1999, Zhao Lingai et al. 1999.

(Wang Haiming, Tong Dequan)

251 *Obolodiplosis robiniae* (Haldemann)

Classification　Diptera, Cecidomyiidae
Common name　Black locust gall midge

Distribution　Beijing, Hebei, Liaoning, and Shandong; U.S.A, Japan, Korea Peninsula, and Italy.

Hosts　*Robinia pseudoacacia* and *R. pseudoacacia*. 'Idaho'.

Morphological characteristics　**Adults** Female length 3.2-3.8 mm. Antennae filiform, 14-segmented. Compound eyes large, occupying most of the vertex. Pronotum elevated, red, with three longitudinal lines extending to the mesonotum, and two lateral black blotches reaching the posterior mesonotal margin. Forewings covered by dense and black pubescence, with three longitudinal veins. Abdomen orange. Male length 2.7-3.0 mm. Antennae 26-segmented. Abdominal terga black, with various light-colored fine hairs. Legs of both male and female slender and covered by scales. Front and middle tibia white.

Eggs Length 0.27 mm, width 0.07 mm, oblong oval, brownish red, translucent.

Larvae Length 2.8-3.6 mm, fusiform to oblong oval, white for early instars, become red at maturity, with a "Y" shaped sternal spatula on the ventral side of the prothorax and a pair of round tubercles on vertex.

Pupae Length 2.6-2.8 mm. Light orange. Abdomen with one row of brown spiculae anteriorly on each sternite for segments 2-8. Head with one erect spine on each lateral side visible from the vertex.

Biology and behavior　*Obolodiplosis robiniae* oc-

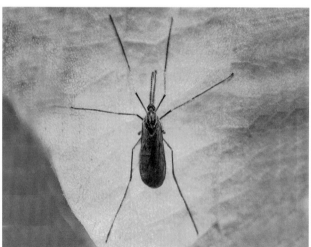

O. robiniae adult (dorsal view) (Xu Gongtian)

O. robiniae adult (with wings spread) (Xu Gongtian)

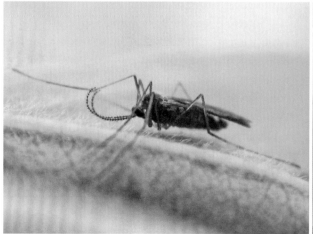

O. robiniae adult (lateral view) (Xu Gongtian)

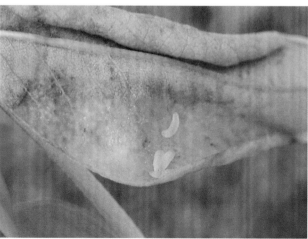

O. robiniae young larvae (Xu Gongtian)

curs five generations a year in Beijing and overwinters as late instar larvae inside the soil. Adults of overwintering generation emerge in mid-April and emergence peaks in mid- to late April. Larvae of the first generation appear in late April and peak in late May. Late instar larvae of the overwintering generation begin to come down from the trees in late September. This process lasts for 4-5 days. Larvae spin cocoons inside the soil for overwintering. A few late instar larvae remain in the galls on fallen leaves. Adults are weak fliers. Sex ratio is 4:7 between females and males. Each female can produce 40-80 eggs. Eggs are mostly laid on either side of the major veins and edges on the leaf. Newly hatched larvae feed aggregately on leaf edges within 3-5 days after migrating from the egg site under the leaves. Larval feeding stimulates abnormal tissue growth on leaves, resulting in longitudinal rolling of leave along leaf margins that forms galls. Larvae feed inside the galls. There are three to eight larvae inside each gall. Up to 32 larvae can feed together on one leaf. Mature larvae fall to the ground with fallen leaves by late fall, with 90% of them enter the soil under leaf litter, 5% on soil surface under the leaf litter, and 5% stay inside the galls.

Prevention and treatment

1. Removing leaf litter on the ground to eliminate habitat for mature larvae leads to significantly lower adult

O. robiniae mature larvae (Xu Gongtian)

O. robiniae larva (Xu Gongtian)

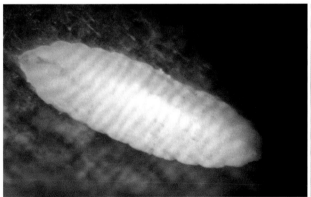

O. robiniae larva (Xu Gongtian)

O. robiniae pupa (lateral view) (Xu Gongtian)

O. robiniae pupa (dorsal view) (Xu Gongtian)

O. robiniae pupa (Xu Gongtian)

O. robiniae infested leaves (Xu Gongtian)

O. robiniae infested leaves (Xu Gongtian)

O. robiniae infested leaves (Xu Gongtian)

O. robiniae infested leaves (Xu Gongtian)

O. robiniae empty pupae (Xu Gongtian)

emergence rate in the following spring.

2. Deep tillage (up to 2 cm deep) to reduce pest population by preventing mature larvae from pupating and adults from emergence.

3. Protecting larval parasitoid *Platygster* sp.

4. Chemical control. Cover spray of 10% imidacloprid at 1,000×, or 4.5% beta-cypermethrin at 1,500×, or 45% profenofos and phoxim mixture at 1,000×, 12% thiamethoxam and lamda-cyhalothrin mixture at 1,000× during adult peak emergence of overwintering generation or peak leaf expansion.

References Yang Zhongqi, Qiao Xiurong, Bu Wenjun et al. 2006; Xu Gongtian and Yang Zhihua 2007; Mu Xifeng, Sun Jingshuang, Lu Wenfeng et al. 2010.

(Mu Xifeng, Sun Jingshuang, Xu Gongtian, Li Zhenyu)

252 *Rabdophaga salicis* (Schrank)

Classification Diptera, Cecidomyiidae
Common name Salix gall midge

Importance *R. salicis* is a regional pest in Linyi, Jining, and Heze with heavy infestations. Infestation trees have reduced timber value and aesthetic value.

Distribution Jiangsu, Anhui, Shandong, Henan, Hubei, Ningxia, and Xinjiang.

Hosts *Salix* sp. and *S. purpurea*.

Damage Larval feeding in the cambial region leads to abnormal tissue growth, resulting in galls on the trunk. Infested areas swell and darken. Diebacks of new branches and branch tips may occur. Chronic infestations lead to tree mortality.

Morphological characteristics **Adults** Female length 3.0-4.0 mm. Deep reddish brown. Head and compound eyes black. Antennae moniliform, grayish yellow, with a whorl of setae on each segment. Forewings membranous and transparent. Hindwings reduced to halters. Mesonotum conspicuous, brown and hairy. Abdomen dark red with the last segment extended to form the pseudovipositor. Male smaller compare to females, deep purplish red, abdominal apex curved upwards.

Eggs Oblong oval, orange red, translucent.

Larvae Length 4.0-5.0 mm at maturity. Oblong oval, milky white at first, orange yellow, with a "Y" shaped sternal spatula on the ventral side of the prothorax.

Pupae Reddish brown, 3-4 mm long.

Biology and behavior *Rabdophaga salicis* occurs one generation a year and overwinters as mature larvae

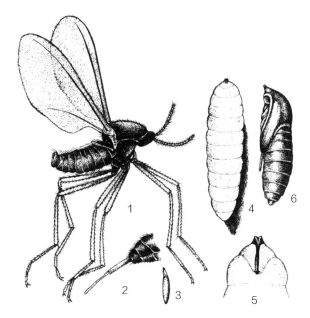

R. salicis. 1. Male adult, 2. End of female adult abdomen, 3. Egg, 4. Larva, 5. Ventral view of larva, 6. Pupa (Zhu Xingcai)

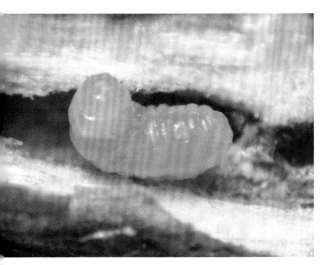

R. salicis overwintering larva (Xu Gongtian)

under the bark of host trees. Overwintering larvae start to pupate in March of the following year. Adults emerge in late March to April and have three emergence peaks in late March, early April, and mid-April depending on temperatures. Mass emergence occurs mostly when temperatures are high, especially during sunny weather right after the rain. Adults mate right after emergence. Egg period lasts for 6-10 days. Larvae feed under the bark until November before overwintering. Mature larvae construct pupal chambers and chew exit holes that are incompletely through the bark before pupation. Pupae expose half of their body on bark surface after pupation by squirming upward from the larval galleries. Mass pupal cases are left at exit holes on the bark after adult emergence, making them easily identifiable. Females lay their eggs mostly between the cambial region and in sapwood inside old exit holes, with a few dozen up to one hundred eggs inside each exit hole. Newly hatched larvae feed on the cambial region after crawling a short distance. "Galled" trunk portion is 1-5 times the size of the normal trunk. The "galling" of the trunk is a gradual process. Eggs are

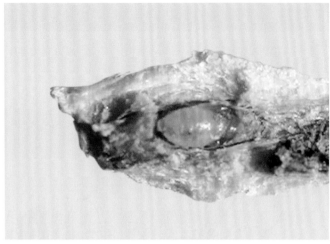

R. salicis pupa (Xu Gongtian)　　　　　　R. salicis galls (Xu Gongtian)

R. salicis exit hole (Xu Gongtian)　　　　A parasitic wasp of R. salicis (Xu Gongtian)

laid inside the bud scars during initial infestation. Larvae enter at the base of tender buds and feed under the bark. Initial infestation has little impact on the branches due to the low pest population density. By the second year, slight swelling is observed on branches with increased pest population. Year after year, tissues around the infestation sites are stimulated to grow abnormally, which in turn increase the size of the gall. Huge galls are formed when abnormal tissue growth covers the circumference of the tree.

Prevention and treatment

1. Enforce quarantine. Avoid direct cutting during planting to prevent pest escaping from the orchards. Stop movement of untreated infested branches.

2. Remove and destroy galls by burning during normal silviculture measures.

3. Utilize characteristic centralized adult emergence period and short oviposition period, cover the gall to prevent emerging adults from leaving, or remove a thin layer of bark on gall surface with a knife to prevent larvae from pupation that lead to larval mortality through desiccation.

4. Chemical control. 1) Remove and destroy bark of the damaged portion or the galls in the winter before the end of March on small trees or during initial infestation; 2) Apply 30% chloramine phosphorus at 2× in late March by brushing the solution on the galls and newly infested sites and cover them with plastic film to kill eggs, larvae, and adults; 3) Apply petroleum oil and used motor oil by brushing them on galls or newly infested sites in the spring before adult emergence to kill mature larvae, pupae, and emerging adults; 4) Trunk injection of 30% chloramine phosphorus at 3-5× at the rate of 1-2 mL per hole for 2-3 holes (0.5-0.8 cm in diameter and 3 cm in depth) between May and June. Seal the holes with mud after injection to prevent chemicals from escaping. Spray 45% profenofos and phoxim mixture at 30-60×, 12% thiamethoxam and lamda-cyhalothrin mixture at 30-60× to tree trunks.

References　Xiao Gangrou (ed.) 1992, Zhao Fanggui et al. 1999, Sun Xugen et al. 2001.

(Wang Haiming, Wang Yongjun)

253 *Chyliza bambusae* Yang et Wang
Classification Diptera, Psilidae

Distribution Shanghai, Jiangsu, Zhejiang, Anhui, Fujian, Jiangxi, Hubei, Hunan, Guangdong, Guangxi, and Sichuan.

Hosts *Phyllostachys heteroclada* var. *pubescens*, *P. praecox*, *P. glauca*, *P. propinqua*, *P. iridescens*, *P. edulis* f. *gimmei*, *P. vivax*, *P. dulcis*, *P. bambusoides*, *P. heterocycal* 'Tao', *P. heteroclada*, and *P. angusta*.

Damage Newly hatched larvae enter young rhizomes at the growing points. Feeding by larvae prevents rhizomes from growing and eventual consumption of their pitch, resulting in reduced rhizomes from their usual length of 80-150 cm to only 10-20 cm. The loss of ability to absorb water and nutrition from the soil by rhizomes leads to weak shoots or shoot mortality. Surviving shoots are expected to produce high tapering grade, low volume ratio bamboo stems later on. Short rhizomes and shallow roots provide little support for bamboo trees to resist lodging.

Morphological characteristics **Adults** Body length 6.0-7.0 mm (females) and 6.0-8.0 mm (males). Wingspan 5.0-6.0 mm for both males and females. Head yellowish brown, frons depressed, with a large black blotch inside and wide black band in the middle. Compound eyes large, slightly depressed at the posterior margin. Three ocelli close to each other. Ocellar

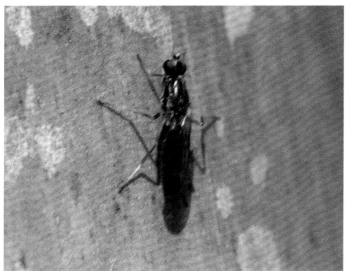
C. bambusae adult (Xu Tiansen)

C. bambusae damaged young rhizomes (Xu Tiansen)

C. bambusae mature lava (Xu Tiansen)

C. bambusae pupa inside rhizomes (Xu Tiansen)

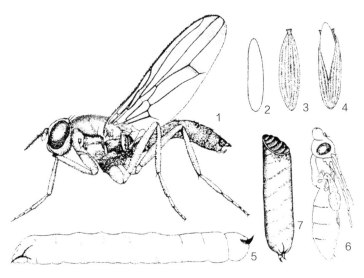

C. bambusae. 1. Adult, 2-4. Egg, 5. Larva. 6-7. Pupa (Xu Tiansen)

bristles present. Arista plumose. Pronotum yellowish brown and with dense pubescent punctures. Notopleuron with one pair of posthumeral bristles and two pairs of notopleural bristles. Transverse suture with two pairs of presutural bristles. Scutellum yellowish brown, with three pairs of scutellar bristles. The posterior pair is relatively large and strong. Fore wings slender, transparent, smoky brown apically, with a notch at the costal third, which extends as a crease line across the wing. Legs slender, yellowish and pubescent. Tarsi significantly longer than tibiae. Abdomen slender, black brown, with pubescence and setae. **Eggs** Length 0.86-0.95 mm for the major axis, and 0.24-0.27 mm for the minor axis. Oblong oval, tapered at one end. Milky white, with creamy yellow ootheca outside of the egg shells. Ootheca with a flask-shaped protrusion at one end and more than 10 ridges on the surface. Length of ootheca major axis is 0.90-0.98 mm. **Larvae** Length 0.75-0.90 mm after hatch. There are three larval instars, with a body length of 0.75-1.21 mm, 3.15-5.08 mm, and 8.5-11.5 mm for the first, second, and third instar, respectively. Larva yellowish, with black mouth hooks which are visible through mesothoracic body wall. Body 12-segmented. Pronotum with a pair of semi-crescent, blackish sclerites and a pair of plumose spiracles below them. Each segment contains a few differently-arranged black spiculae from intersegment between meso- and metanotum to intersegment between abdominal terga 4 and 5, as well as intersegment between metasternum and abdominal sternite 1 to intersegment between sternite 5 and 6. The end of last segment with a convex brown blotch. Caudal plate black. Caudal spiracles arietiform and tilted black. **Pupae** Length 6.2-7.5 mm, milky white right after pupation, turn to creamy yellow before overwintering. Vertex with a cystic protrusion. Compound eyes and ocelli orange red, and mouthparts, antennae, wings, tibia, and some hairs blackish right before adult emergence. Pupal case cylindrical, brownish to reddish brown; length 6.45-8.06 mm on the major axis; oblique, truncate, depressed, black, and contains 5-6 crease lines anteriorly. Pupal case with 10 diagonal segmental marks and arietiform protrusion at the end. Remains of larval intersegment protrusions visible. Pupa located at the underside of apical half of the rhizomes for easy adult emergence.

Biology and behavior *Chyliza bambusae* occurs one generation a year in Zhejiang and overwinters as pupae inside empty rhizomes. A portion of the population undergoes two-year life cycle by having one year diapause in pupa stage in biennial forests. Adults begin to emerge in late March to early April the following year when temperature reaches 10°C. Males emerge two days earlier than females. Peak adult emergence appears when average daily temperature is above 15°C. Emergence completes by late April. Females mate and lay eggs in mid-April and eggs hatch within 4-10 days. Larvae feed from late April to late May. Larval stage lasts for 18-25 days. Mature larvae pupate inside infested rhizomes before the end of May.

Prevention and treatment Conserving natural enemies. Only a few natural enemies attack *C. bambusae*. Adult predators during emergence and egg laying include ants (e.g. *Polyrhachis dives*) and spiders (e.g. *Oxyopes sertaus*). Parasitic wasps such as *Opius* and members in the family of Diapriidae play an important role in pest population regulation. *Opius* sp. is the major parasitoid attacking larvae with an average parasitism of 26.08%. Members of Diapriidae also attack *C. bambusae* larvae, with an average parasitism of 12.18%.

(Xu Tiansen)

254 *Pegomya phyllostachys* (Fan)
Classification Diptera, Anthomyiidae

Distribution Shanghai, Jiangsu, Zhejiang, Anhui, Fujian, Jiangxi, Hubei, Hunan, and Sichuan.

Hosts *Phyllostachys heteroclada* var. *pubescens*, *P. pubescens* f. *gracillis*, *P. glauca*, *P. nuda*, *P. edulis* f. *gimmei*, *P. praecox*, *P. propinqua*, *P. glabrata sinobambusa*, *P. dulcis*, *P. praecox* f. *provernalis*, *P. iridescens*, *P. flexuosa*, *P. heterocycla*, *P. bambusoides*, *P. bambusoides* f. *lacrima-deae*, *P. vivax*, *P. nigra*, *P. sulphurea*, *P. prominens*, *P. fimbriligula*, *P. heteroclada*, *P. nidularia* 'Smoothsheath', and *P. angusta*.

Damage Females lay eggs on weak shoots and larvae feed inside them. In *P. heteroclada* var. *pubescens*, newly hatched larva feeds head-down under inner bamboo sheath, leaving behind hygrophanous feeding trace on both lateral margins. It enters the shoot at the knot where sheath grows out. Larvae feed at random directions after entering the shoots. The hygrophanous trace commonly associated with larval feeding leads to shoot decay and eventually the death of infested shoots. Damaged shoots lose their value as food. Mature larvae bore back to the inner sheath, squirming upward until the edge of the sheath before falling to the ground to pupate inside the soil. Downward feeding by newly hatched larvae on vigorous shoots is not enough to kill the shoots, only results in larval feeding trace on knots of mature bamboo trees. However, for eggs laid on small diameter shoots or small shoot branches, newly hatched larvae can enter the shoots directly and feed inside because of their thin sheath, resulting in branch dieback or shoot mortality from decay.

Morphological characteristics **Adults** Length 7.0-8.0 mm, gray. Frons narrow, brown to dark brown, with 7-8

P. phyllostachys adults (Xu Tiansen)

P. phyllostachys eggs (Xu Tiansen)

P. phyllostachys larva (Xu Tiansen)

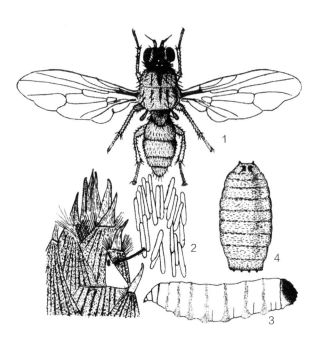

P. phyllostachys. 1. Adult, 2. Oviposition site and enlarged eggs, 3. Larva, 4. Pupa (Xu Tiansen)

Scars on bamboo made by *P. phyllostachys* (Xu Tiansen)

P. phyllostachys damage (Xu Tiansen)

frontal bristles. Antennae long, black. Female compound eyes dichoptic, dirty red. Male compound eyes holoptic, large. Ocelli brownish yellow. Thorax covered dorsally by grayish powder, with a dark median stripe. There are two columns of acrostichal bristles, and only the second pair in front of the transverse suture and one or two pairs before the scutellum relatively big and long, others hair-like. Wings transparent, wing base yellowish. Halters and legs yellow, tarsi black. Abdominal terga covered by gray powder, with clear dark median stripe. **Eggs** Length 1.75-2.00 mm for the major axis, and 0.23-0.24 mm for the minor axis. Cylindrical, milky white. Egg mass irregularly shaped, each with about 50 eggs. **Larvae** Length 1.70-2.01 mm after hatch, milky white. Mature larvae 8.0-11.5 mm in length, yellowish white, cone-shaped, 12-segmeted with slightly enlarged 6th and 9th segments. Head pointed, end of abdomen truncate. Mouth hooks black. Anterior spiracles slightly darker. Intersegment from thorax to the 7th segment slightly protrudes, especially obvious on the ventral side. Oblique and uniform short setae scattered among the protrusion, setae reddish brown to brown. End of abdomen black. There is one pair of spiracles above the center of truncature. Spiracles oval, protrude, with three pairs of brown spiracle clefts. Clefts oblong oval, fan-like arranged, with seven pairs of papillae. **Pupae** Length 4.2-5.3 mm, milky white, stout, with a pair of protrusion above the compound eyes. Tarsi reach last body segment. Pupal case oval, dark brown, 5.8-7.2 mm for the major axis, and 2.5-3.0 mm for the minor axis. Pupa has 10 visible segments. Each segment contains a ring-shaped wrinkle and a pair of elevated wrinkles on the head. Body truncated at the end, with the same numbers of spiracles and papillae at the same locations as the larvae.

Biology and behavior *Pegomya phyllostachys* occurs one generation in one to two years and overwinters as pupae inside the soil. Adults begin to emerge in early to mid-March in Zhejiang. Emergence ends in early April. About 30% of the pupae will diapause in the soil and overwinter the second time before emerging as adults in the third year. Mating starts in mid-March and peaks in late March. Females lay eggs in late March to early April and egg period peaks in mid-April. Eggs are laid at the upper portion between two sheaths. Eggs hatch within 3-5 days under high relative humidity. Hairs on the outer wall of the inside sheath can push eggs to the inner wall of the outside sheath as the shoots grow. This reduces the environmental relative humidity for the eggs and hence stops them from hatching. Larvae mature in about 20 days after feeding inside the shoots and enter the soil for pupation between late April and early May and overwinter together with diapausing pupae of the last generation.

Prevention and treatment Conserving predatory spider *Oxyopex lineatipes*.

(Xu Tiansen)

255 *Strobilomyia* spp.

Classification Diptera, Anthomyiidae
Common name Cone flies

Importance Species in the genus of *Strobilomyia* are important cone and seed pests of larch. Outbreaks damage larch cones and seeds that leads to reduction of seed production.

Distribution Shanxi, Inner Mongolia, Heilongjiang, and Xinjiang; Japan, Russia (Siberia), and Europe (Alps in Austria – the native range). The species group of *Strobilomyia* in the Greater Khingan Mountains includes the following six species: *S. viaria*, *S. baicalensis*, *S. laricicola*, *S. infrequens*, *S. svenssini*, and *S. luteoforceps*. *S. melaniola* is the dominate species.

S. laricicola – Heilongjiang, Inner Mongolia; former USSR (Siberia), and Europe (Alps in Austria – the native range).

S. baicalensis – Heilongjiang, Inner Mongolia; China, and former USSR (Lake Baikal region in Siberia – the native range).

S. svenssini – Heilongjiang; Europe (Austria – the native range), and North America.

S. viaria – Heilongjiang, Liaoning, Inner Mongolia (the native range).

S. infrequens – Heilongjiang, Liaoning, Inner Mongolia, Shanxi; former USSR (Siberia), and UK (the native range).

S. luteoforceps – Heilongjiang (the native range).

Hosts *Larix gmelini* in China; *L. leptolepis*, *L. koreana*, *L. dahurica* var. *japonica* in Japan; *L. dahurica*, *L. sibirica*, *L. subczewii*, *L. olgensis* in former USSR (Siberia); *L. koreana* in northern Korea Peninsula; and *L. deciua* in Europe.

Damage *S. laricicola* feeds inside the cones during the entire larval stage. Newly hatched larva bores directly into the scale at the base and feeds on the young seed. After the seed is completely devoured, it moves to a neighboring seed and continue to feed. Larvae feed exclusively on seeds and can consume seed hulls. About 80% of the seed is consumed if one larva is inside it, whereas 100% of the seed is consumed if there are two larvae inside the seed. No obvious external symptoms is shown on cones at early infestation. By later stage, however, infested scales develop poorly and are smaller than normal scales. They discolor and shrivel prematurely, which leads to bending and malformed cones.

Morphological characteristics **Adults** Length 4.0-5.0 mm. Look like housefly, but smaller. Compound eyes holoptic, dark red, and exposed for males. Eye orbits, mandibles, and gena covered by silver powder. Antennae not reaching anterior margin of the mouthpart. Arista smooth for the basal two fifth, increasing in size to fusiform. There are 3 or 2 rows of the erect oral vibrissae. Middle rostrum short, covered by a thin layer of powder. Thorax black and covered by silver powder. Base of the wing grayish brown. Hairs on the underside of costa disappear before the subcostal. Alulae white. Legs black. Abdomen flat and short, swollen apically, covered by silver powder with a broad longitudinal band in the center. The end of the lateral lobe of the abdominal sternite 5 significantly narrowed; outer margins distinctly curved inward. Anal plate slightly heart-shaped from lateral view. Compound eyes dichoptic, brownish between eyes and darken backwards for females. Fronto-orbital bristles covered by gray powder. Thorax without obvious stripes. Abdomen completely black, somewhat shiny.

Eggs Length about 1.1-1.6 mm for the major axis, and 0.3-0.5 mm for the minor axis. Oblong oval, milky white. One end slightly larger than the other, curved in the middle, with hexagon-shaped reticular marks on eggshell.

Strobilomyia larvae (Yan Shanchun)

Larvae Length 6.0-9.0 mm at maturity, conical, yellowish, opaque. Head pointed, with a pair of black mouth hooks. Prothorax with a pair of flat anterior spiracles laterally. Abdomen with annular ridges and rows of spinules at the lateral margins of each segment. Ridges and spinules are used for locomotion. Abdomen truncated at the end, with 7 pairs of fleshy protrusive papillae on the truncature. Posterior spiracles brown and protrusive.

Pupae Length 3.0-5.5 mm, oblong oval, reddish brown. Mouth hooks sunk in. Anterior spiracles present, forming a pair of anterior protrusions for the coarctate pupa. Posterior spiracles and protrusions visible.

Characteristics used to separate species group of *Strobilomyia*:

1) *S. laricicola*. Length 4.0-5.0 mm. Wing base not yellow. Antennae not reaching the anterior margin of the mouthparts. Arista basal two fifth significantly enlarged. Hairs on the underside of costa rarely extend pass the middle point. Wings almost transparent, only wing base slightly darker. Alula and calypter both white. Middle rostrum short. Abdominal terga with narrow and dark marginal band anteriorly. Anal plate contains no bristles. Abdominal sternite 5 with wide lateral lobe, narrowed at the posterior one quarter. The outer margins of sternite 5 curved inward significantly. The inner margins contain a row of short, dense, uniformed middle-sized bristles, with a row of tiny hairs inside the bristles. Bristles and hairs overlapping only at the base and below the tip.

2) *S. baicalensis*. Length 5.0-5.5 mm. Wing base somewhat yellow, also with yellow veins in this part. Antennae almost reach the anterior margin of the mouthparts. Hairs exit underside of full length of the costa. Abdominal sternite 5 lateral lobe slightly curved inward at the most, with no obvious bristles on the outer margins. Lateral lobe wide and short, resembling a protrusion ventrally. The inner margin of sternite 5 contains dense and short bristles, with one row of hairs inside. There are two rows of oval vibrissae. Anal plate pointed posteriorly from lateral view, with its lateral apex hooked forward.

3) *S. svenssini*. Length 5.5 mm. Abdominal sternite five lateral lobe with obvious long bristles, inner side contains spare, slender hairs, but no bristles, and a row of tiny hairs inside. Lateral lobe semicircular, swelling to resemble the sphere surface, the end circular. Apex of anal plate not hooked laterally.

4) *S. melaniola*. Length 5.0 mm. Abdominal sternite five lateral lobe thin, not concave apically. The inner margin of the lateral lobe smooth and the posterior end round, with one row of long bristles but no pubescence. Bristles at the base dense. Bristles toward the posterior end gradually reduce in size and inclined outward. There is a row of tiny hairs inside the bristles. The inner margin of the inner branch of the anal plate significantly concaved.

5) *S. infrequens*. Length 4.5-5.0 mm. There are two rows of oval vibrissae. Abdominal sternite 5 lateral lobe slender, slightly narrower apically, slightly constricted near the apex. The inner margin of the lateral lobe with hairs, making the lobe looks thicker. Inner bristles sparse and long, consist of about 10 bristles in a row. There are two rows of hairs inside the bristles, with the row close to the bristles bigger. Anal plate contains only one pair of bristles apically. Anula white or yellowish. Calypter yellowish or yellow.

6) *S. luteoforceps*. Length 4.5-5.0 mm. There are two rows of oval vibrissae. Abdominal sternite 5 lateral lobe relatively wide, obtuse triangular, inner margin with a small hairy blotch and a pale membranous portion. Inner marginal bristles mostly concentrated in the middle, become sparse and hair-like posteriorly. There is one row of hair inside the bristles, appears to be three rows in the middle. Anal plate contains two pairs of bristles apically, with the inner pair only reaches the lateral apex of the anal plate. Anula yellowish to yellow. Calypter grayish.

Biology and behavior Most species of *Strobilomyia* occur one generation a year and overwinter as pupae inside leaf litters on soil surface. A few have two-year life cycle due to pupal diapause. Adult emergence starts in early May and ends in late June, with a peak emergence in mid- to late May. Females lay eggs in mid-May. Eggs start to hatch in late May with peak hatching in late May to early June. Larvae mature by mid-June and fall out of the cones to the ground to pupate and overwinter by late June.

There is one generation a year or every two years in the Greater Khingan Mountains region. They overwinter as pupae in leaf litter on soil surface. Adults start to emerge early May. Females lay eggs in mid- to late May. Eggs hatch in late May. Egg stage ends by early to mid-June. Larvae mature in late June to early July and fall out of the cones to the ground to pupate after feeding for 25-30 days. Adults emerge during the day, with >90% of them come out between 6:00-12:00 a.m. Newly emerged adults grayish white, with unexpanded wings, and stay inactive for half an hour on the ground. Adult emergence is closely correlated with temperature. Emergence usually begins when average daily temperature is above 10°C. The higher the temperatures, the more numbers of adults emerge. Adults stay inside bark crevices or leaf litters during the night and in early mornings or evenings when temperatures are low. They become active on grass near the ground after sunrise, and move to low crown by 10:00 a.m. when temperature increases. Adults come back to lower locations from the crown by sunset. Adults are most active during noon, with an active temperature range of 12-26°C. Females start to lay eggs 2-3 days after initial activity. They need supplemental feeding before laying eggs.

Eggs are laid mostly during the noon. Female searches on scale surface with its abdomen before inserting ovipositor and laying an egg between two scales. It usually takes 2-5 minutes to lay one egg. It moves to another cone to lay another egg. Eggs are laid singly on cone scales or host needles at the base of the cones when scales are not open. By the time scales become open, eggs are laid between scales near the seed. Usually 1-2 eggs are found in each cone, with 3-5 eggs per cone sometimes. During years of low seed production, a single cone can contain as many as 18-24 eggs. Female longevity ranged from 6 to 44 days, with an average of 18.6 days; whereas male longevity ranged from 5 to 22 days, with an average of 11 days. Egg stage lasts for 7-10 days.

Mature larvae crawl out of the cones and fall to the ground to pupate in leaf litter or 1-3 cm deep soil. In some locations in Shanxi, mature larvae can also pupate and overwinter inside the cones. The falling of mature larvae to the ground is closely related to precipitation. Most mature larvae fall to the ground to pupate after heavy rain.

Larval distribution is correspondent with forest type, slope direction, canopy density, and position in the crown. Larval density is higher on southern crowns in low canopy density larch-ledum forests on sunny slopes than on northern crowns in high density larch-grass or larch-rhododendron forests on shaded slopes.

Prevention and treatment

1. Silviculture. Destroy overwintering habitats by cleaning leaf litter, shrubs, and dead branches and deep tillage in seed production stands after larvae fell to the ground can reduce pest population significantly. Damage is also mitigated by creating wide isolation belts around seed orchards.

2. Population monitoring. Conduct population monitoring during the outbreaks. Prepare for control action when population density based on individual trees reaches moderate levels and trees have moderate seed production.

3. Mechanic control. 1) Trap adults during emergence based on adult attraction to lights, waves, colors, and tastes, such as frequency vibration light trap at the density of 1 light/2-5.4 ha, trapping with sugar and vinegar mixture in 90% full container (white sugar 40 g, white vinegar 30 mL, white wine 20 mL, water 200 mL; or white wine 1 part, vinegar 4 parts, white sugar 3 parts, water 5 parts, and a few drops of trichlorphon). Fogging with dichlorvos for five consecutive years provided effective control of adults. 2) Blue cup trapping without honey provided the best results. 3) Spray larch crown with semiochemical such as turpentine, camphor, eugenol during oviposition period to change the makeup of plant volatiles from larch cones. This disrupts host locating and host selection for *Strobilomyia* adults, and reduces damage. Turpentine and camphor repel *Strobilomyia* adults.

4. Chemical control. 1) Spray 25 g/L lamda-cyhalothrin at 1,000× or 40% acetamiprid and chlorpyrifos mixture at 1,000×; 2) Trunk injection at the base of the tree with 50% methamidophos or 40% dimethoate emulsifiable suspension before egg hatch provide good control results. Trunk injection is a cost effective and safe chemical control method because chemicals stay inside the trees and are not affected by precipitation, last longer, and requires less chemicals. Current effective chemicals through cover spray include omethoate, decamethrin; 3) Experiments with insect growth regulators such as 5% triflumuron emulsifiable oil, 25% diflubenzuron suspending agent, 20% triflumuron suspending agent, 5% hexaflumuron emulsifiable oil, and 30% phoxim-hexaflumuron emulsifiable oil in the Greater Khingan Mountains resulted in 20-80% reduction in cone damage rate and 35-40% reduction in seed damage rate. Two applications are better than one single application based on analysis on cone and seed damage rates.

References Fan Zide 1988, 1992; Xiao Gangrou (ed.) 1992; Sun Jianghua, Roques, and Fang Sanyang 1996; Yan Shanchun, Hu Yinyue, Zhen Dingfan 1997; Yan Shancun, Sun Xudong, Hu Yinyue 1997; Yan Shanchun, Zhang Xudong, Hu Yinyue et al. 1997; Yan Shanchun, Jiang Haiyan, Li Liqun et al. 1998; Chi Defu, Sun Fan, Qiao Runxi et al. 1999; Yan Shanchun, Hu Yinyue, Sun Jianghua et al. 1999; Qin Xiuyun, Li Fenghua, Zhang Liqing et al. 2000; Yan Shanchun, Sun Jianghua, A. Roques et al. 2000; Zhang Ensheng 2000; Xu Ying, Song Jingyun, Yao Dianjing 2001; Zhang Dandan, Chi Defu, Jiang Haiyan et al. 2001; Yan Shanchun; Jiang Xinglin, Xu Fangling 2002; Zhao Tieliang, Sun Jianghua, Yan Shanchun et al. 2002; Li Chengde 2003; Yan Shanchun, Sun Jianghua, Chi Defu et al. 2003; Yan Shanchun, Chi Defu, Sun Jianghua 2004; Yin Yanbao, Zhao Qikai, Jiang Xinglin 2005; Chen Qinhua, Tian Liming, Lin Shucai 2008.

(Yan Shanchun, Xu Chonghua, Cao Jingqian)

256 *Carpomya vesuviana* Costa

Classification Diptera, Tephritidae
Common name Jujube fruit fly

Distribution It was found in Xinjiang (Turufan region) in 2007. This species is native to India and has been found in approximately 20 countries, mainly in Asia, such as Iran, Oman, and Uzbekistan.

Hosts *Ziziphus* spp.

Damage Larvae bore into fruits, causing decay and early maturity. Infested fruits become deformed.

Morphological characteristics

Adults Yellow, thorax with four white or yellow stripes; scutum yellow or reddish yellow, its center with three fine blackish brown stripes; each side of thorax with four black spots; there are two large black spots after the transverse suture; there is a large black spot at center of thorax near posterior boarder of scutum; wings transparent, each with four yellow or yellowish brown transverse bands, part of the borders of the bands grayish brown.

Larvae Yellow or white, 3rd to 7th abdominal segments with scars in ventral surface; 8th segment with large tubercles; front spiracle with 20-23 finger-like projections, hind spiracle cleft large, 4-5 times as long as wide.

Biology and behavior *Carpomya vesuviana* occurs 6-10 generations a year. It overwinters as pupae in soil around plant roots. It can also overwinter in packaging, warehouses, or inside dry fruits. Adults usually eclose during 9:00-14:00. They mate and lay eggs during the day. Adults rest on trees during night hours. Eggs are laid under bark individually. Each female lays 19-22 eggs. Each fruit may have 1-6 eggs. Larvae feed on fruit. The 1st and 2nd instar generate the most damage.

Prevention and treatment

1. Quarantine. Prevent the movement of infested fruits and plants.

2. Mechanic control. Till soil during January-March and November-December to kill pupae in soil. Cover soil with plastic film to prevent adults from emerging from soil.

3. Sanitation. Collect fallen fruits periodically to remove infested fruits and the pests.

4. Attract and kill. Use traps containing attractants such as sugar-vinegar, methyl eugenol, and a pesticide to attract and kill adults.

5. Biological control. Release braconid wasps or other parasitic wasps.

6. In heavily infested orchards, stop production for two years. Re-grafting the plants.

References Zhang Runzhi et al. 2007, Adil Shattar et al. 2008, Forest Disease and Pest Control Station of the Ministry of Forestry 2008, Wu Jiajiao et al. 2008.

(Zhao Yuxiang)

C. vesuviana adult (Zhang Runzhi)

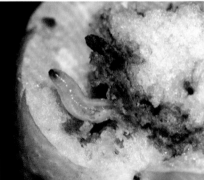

C. vesuviana larva and damage (Zhang Runzhi)

C. vesuviana adult and damaged fruit (Bu Wenjun)

257 *Eriocrania semipurpurella alpina* Xu

Classification Lepidoptera, Eriocraniidae

Importance *Eriocrania semipurpurella alpina* is an important pest of birch in Qinghai province. Larvae burrow into leaves, causing dried leaves. It slows down tree growth and tourism. In heavy infestations, percent of *Betula platyphylla*, *B. albosinensis*, and *B. utilis* leaves damaged were 100, 78, and 35%, respectively.

Distribution Eastern Qinghai province.

Hosts *Betula platyphylla*, *B. albosinensis*, *B. utilis*.

Morphological characteristics Adults Brown with purple luster; vertex and front of prothorax with yellowish brown long hairs. Wings narrow, with scattered golden scales; center with a triangular golden mark; outer margin with brown long hairs. Male wingspan 11-15 mm; female wingspan 9-12 mm.

Eggs Elliptical; length 0.35-0.40 mm, width 0.20-0.25 mm.

Larvae Legless. Length of the 1st to 4th instars are 0.7-1.7, 1.8-3.0, 3.6-5.5, and 7.5-9.0 mm, respectively. First instar head blackish brown, body creamy white. Second instar pronotum and sternum blackish brown, body with black granules. Third instar head brown with reddish color, pronotum and sternum with irregular white marks. Fourth instar pronotum and sternum without blackish brown marks.

Pupae Blackish brown; length 4.0-4.5 mm long.

Cocoons Elongate oval; length 3.4-3.8 mm, width 1.8-2.0 mm.

Biology and behavior *Eriocrania semipurpurella alpine* occurs one generation a year. It overwinters as mature larvae under the trees. Larvae pupate during mid-April to mid-May. Pupal period is 8-14 days. Adults appear from late April to late May. Adults are active during the day. They fake death when disturbed. Mating occurs on tree trunks, it lasts 20-40 minutes. Eggs are laid singly under epidermis on underside of leaves. Egg stage is 7-10 days. Larvae molt 4 times. Larval damage occurs during late May and early July. Mature larvae emerge from leaves in late June, migrate down the trees, pupate, and make cocoons in 0-12 cm deep humus.

Prevention and treatment

1. Conserve natural enemies. Two parasitoids are found in *E. semipurpurella alpine* cocoons: *Grypocentrus basalis* Ruthe, *Lathrolestes eriocraniae* Seyrig. The former had parasitization rate of 28.0%-30.3%. *Pnigalio*

E. semipurpurella alpina adult (Li Tao)

E. semipurpurella alpina egg (Li Tao)

E. semipurpurella alpina larva (Li Tao)

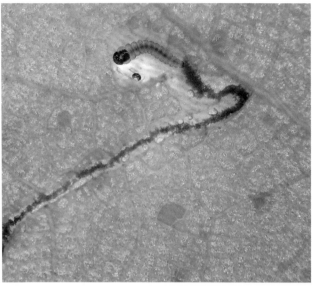

E. semipurpurella alpina 2nd instar larva (Li Tao)   *E. semipurpurella alpina* 3rd instar larva (Li Tao)

E. semipurpurella alpina pupa (Li Tao)   *E. semipurpurella alpina* cocoon (Li Tao)

eriocraniae Li & Yang parasitizes larvae of *E. semipurpurella alpine*. The parasitization rate was 33.8%.

2. Spray 1.8% abamectin or other registered product during adult period.

References Xiao Gangrou (ed.) 1992, Zhang Zenglai et al. 2011, Cai Rang Danzhou et al. 2013, Yu DS et al. 2015, Li Tao and Zeng Hanqing 2016, Zhang Ying et al. 2016, Li Tao and Sun Shuping 2017.

(Li Tao)

258 *Endoclita signifer* (Walker)

Classification Lepidoptera, Hepialidae

Importance *E. signifier* is a serious pest of young and middle-aged fast-growing eucalyptus trees as well as many other species. Infested trees grow slower and more vulnerable to wind damage, affecting the quality and production of the forests.

Distribution Guangdong and Guangxi.

Hosts *Eucalyptus grundis* × *E. urophlla*, Guanglin No. 9, *E. grundis*×*E. tereticomis* DH201-2, *Celtis sinensis*, *Mallotus apelta*, *Clausena lansium*, *Mimosa sepiaria*, *Bridelia tomentosa*, *Mallotus barbatus*, *Trema tomentosa*, *Elaeocarpus sylvestris*, *Clerodendrum serratum*, *Litsea ichangensis*, *Mallotus barbatus*, *Rumex venosus*, *Mallotus barbatus*, *Begonia cathayana*, *Tolypanthus maclurei*.

Damage Larvae bore into the stems and feed inside, pushing sawdust and frass out of the galleries as silken sacs. Silken sacs are made of sticky silk and increase in size as larva grows. The opening of the gallery is relatively large. Gallery is initially constructed transversely to the trunk until reaching the pith, then turns downwards vertically inside the pith, leaving a funnel-shaped opening outside. The silken sac protruding out on the stem is a good indication of infestation.

Morphological characteristics **Adults** Females: Length 50.2-60.8 mm, with an average of 60 mm, grayish brown or brown. Wingspan 80.6-110.4 mm, with an average of 90.9 mm. Males: Length 40.7-50.6 mm, with an average of 50.5 mm, grayish brown or brown. Wingspan 70.9-100.6 mm, with an average of 80.9 mm. Adults have two color types: grayish brown and brown. Hind legs extremely small, significantly retrogressed. The median cell of forewing has a silver transverse mark with slightly wider base at the bottom, and a flat "Y" shaped sliver mark in the middle. Those marks are slightly golden in some adults. Fore- and middle legs well developed with long claws for climbing. Males with dense and brush-like orange hairs on hind femora ventrally. No hair on hind femora for females. Adults dangle like bats or suspended bells when resting.

Eggs Round or nearly round. Length about 0.6 mm on major axis and 0.5 mm on minor axis, white or milky white and slightly metallic at first. Eggshells contain evenly distributed punctures. Eggs become black a few hours before hatch.

Larvae Newly hatched larvae length 1.9-2.3 mm, head capsule width 0.3-0.4 mm. Head brownish. Thorax and abdomen white, translucent. Anal plate brownish with sharp and pointed tip. Abdominal segments contain sparse setae. Setae on the last segment dense and long. Mature larvae cylindrical, 72.4-110.8 mm in length, with a head capsule width of 7.1-11 mm. Head black brown. Body milky white.

Pupae Cylindrical, yellowish brown with a deep brown vertex. Female pupae 55-77 mm long, male pupae 37-55 mm long. Female pupae have a reproduction hole formed as an conspicuous longitudinal groove in the middle on the second to the last abdominal sternite. No reproduction hole in male pupae.

Biology and behavior *Endoclita signifier* occurs one generation a year and overwinters as larvae inside the host in Bobai of Guangxi. Larvae continue to feed during overwintering. Prepupal stage starts from early February to early March. Pupal stage lasts from mid-February to late April and peaks in mid-February. Adult stage lasts from mid-April to mid-May. Egg stage lasts from mid-April to early June. Larvae start to hatch in late April.

E. signifier mating pair (Wang Jijian)

E. signifier adult (Wang Jijian)

E. signifier larval damage (Wang Jijian)

E. signifier infestation (Wang Jijian)

E. signifier egg mass (Wang Jijian)

E. signifier infected with *Beauveria bassiana* (Wang Jijian)

E. signifier larvae and pupae (Wang Jijian)

Mature larva spins 1-2 layers of silk to seal the entrance hole before maturation. Larvae pupate deeper inside the galleries. Pupa squirms upwards to the entrance hole before emergence, by doing so it pushes the pupal case further out of the hole, forming a thumb-sized protrusion. This protrusion will serve as the exit for emerging adult. Emerging adult pushes open the top of the protrude case and expose half of its body. Most adults emerge between 16:00-18:00 p.m. Adult emerges from the pupal case by moving forward along the split from vertex to the median line on pronotum dorsally, and head to the first abdominal sternite ventrally. Newly emerged adult needs 5 minutes to expand its body at 20-25 cm above the pupal case, and an additional 10 more minutes to inflate its wings after the body reaches its normal length to complete the emergence process. They will stay at the same location or crawl away. Adults take flight, search for mates, and mate at night right after emergence. Mating usually lasts for more than three hours with a special posture, during which the female is on top with fore- and middle legs grasp the substance, and the male is dangling upside down clasping to the female with only its genitalia. This posture lasts for the entire mating process. Females can lay eggs with or without mating, but those laid by unmated females will not hatch. Eggs are laid within 2-4 days and scattered on the ground or leaf litter when staying or during flight. Each female

E. signifier pupae (Wang Jijian)　　*E. signifier* larva (Wang Jijian)　　*E. signifier* gall (Wang Jijian)

E. signifier empty pupae (Wang Jijian)　　*E. signifier* mature eggs (Wang Jijian)　　*E. signifier* infested branches (Wang Jijian)

produces a few hundreds to a few thousands eggs. One observed female laid 5,286 eggs. There is strict temperature and humidity requirement for hatching. Less than one third of the eggs laid by mated females will hatch. Newly hatched larvae feed and stay in dampen old silken sacs or leaf litter. They move to the trunk and enter the host later via wounds, branch bases, or leaf axils and forming silken sacs. *Endoclita signifier* mainly infests 1-2 year-old fast growing eucalyptus forests in massive plantations, especially those along the roads below the roadbed. Outbreaks over large areas are also found at source locations and valleys, with up to 85% of the trees are infested under high population density. Over 80% of the silken sacs are found on the trunk below 1 m above the ground. Some can also be found as high as 7 m on the trunk. Generally, 1-3 silken sacs are found on each tree, with a maximum of 17 sacs per tree. Downward larval galleries parallel from each other and are not interconnected. Silken sacs are black due to movement of the larvae inside.

Prevention and treatment

1. Remove and destroy weeds and stumps from the forest before afforestation, and eliminate infested eucalyptus stumps after harvest to destroy larvae inside the roots.

2. Mechanic control. Destroy larvae inside the silken sacs when they are 1-2 cm in diameter by pounding them with wood, or kill larvae inside the sacs by inserting a metal wire and move it up and down.

3. Conserve natural enemies. Bats, lizards, root mice, ants, birds, snakes feed on various life stages of *E. signifier*. Ants are important predators of eggs and larvae. Scrape off silk sacs to allow ants and other natural enemies to attack the larvae inside the tree.

4. Fill sacs with liquid to force larvae out. Inject 10-20 mL of alcohol and water mixture (1 part alcohol and 20 parts water) in the tunnel will force the larva out in 3-5 minutes.

5. Apply *Beauveria bassiana* in vulnerable forest stands during the spring causes repeated infection of pest larvae and pupae.

6. Chemical control. Apply 1-2 mL of 50% dichlorvos emulsifiable suspension through injection or sponge application kills larvae inside the sacs in a day. Spray 45% profenofos and phoxim mixture at 30-60×, 12% thiamethoxam and lamda-cyhalothrin mixture at 30-60× to tree trunks to control larvae.

References　Xi Fusheng et al. 2007, Yang Xiuhao et al. 2013.

(Wang Jijian)

259 Endoclita excrescens (Butler)

Classification	Lepidoptera, Hepialidae
Synonym	*Phassus excrescens* (Butler)
Common names	Swift moth, Japanese swift moth

Importance *Endoclita excrescens* is a domestic quarantine pest attacking various species of trees and bushes. It is a serious threat to the growth and utilization of Manchurian ash forests. It attacks many fruit trees in orchards in Shandong and Hubei provinces, resulting in huge loss to economic forests in China. It is therefore imperative to consider management measures for this pest.

Distribution Hebei, Liaoning, Jilin, Heilongjiang, Zhejiang, Anhui, Shandong, Hubei; Russia (Siberia) and Japan.

Hosts Tree species such as *Fraxinus mandschurica, Fraxinus rhynchophylla, Robinia pseudoacacia, Salix matsudana, Salix viminalis, Acer saccharum, Ailanthus altissima, Firmiana platanifolia, Betula platyphylla, Maackia amurensis* var. *buergeri, Quercus mongolica, Castanea mollissima, Juglans regia, Claoxylon indicum, Xanthoceras sorbifolia, Alangium chinense, Ginkgo biloba, Juglans mandshurica, Euonymus alatus, Metasequoia glyptostroboides, Pterocarya stenoptera, Ulmus pumila, Morus alba, Broussonetia papyrifera, Platanus occidentalis, Fraxinus chinensis, Zanthoxylum bungeanum, Rhamnus davurica, Forsythia suspense, Syringa oblate, Amorpha fruticosa, Sambucus williamsii,* and *Hippophae rhamnoides.* fruit plants such as *Malus pumila, Pyrus bretschneideri, Prunus persica, Cerasus pseudocerasus, Vitis vinifera, Eriobotrya japonica, Actinidia chinensis, Crataegus pinnatifida, Prunus armeniaca, Diospyros kaki,* and *Punica granatum;* crops such as *Humulus lupulus, Zea mays, Triticum aestivum, Solanum melongena;* grass such as *Astragalus membranaceus, Ambrosia artemisiifolia, Artemis iascoparia, Artemisia princeps; Artemisia integrifolia, Artemisia annua, Melilotus suaveolens, Glycyrrhiza pallidiflora, Cirsium souliei, Dictammnus dasycarpus, Leonurus artemisia,* and *Polygonum orientale.* Hosts in other countries include *Populus* spp., *Salix* spp., *Alnus* spp., *Castanea* spp. and *Quercus* spp.

Damage Larvae bore directly into the stems or branches of young trees, while enter through old insect holes or bark crevices on larger trees. They feed on xylem or cambium around the galleries. Most larval galleries extend downwards. Sawdust created through feeding is pushed out the galleries and stuck on silken net at the entrance, creating a sawdust package with silk and sawdust on the outside. Larvae live inside the galleries with entrance holes located at the low trunk, branch joints, or bark cankers, which make them difficult to be found. Larval feeding creates galleries over large areas and reduces timber quality. Infestation to fruit tress leads to reduction of fruit quality and quantity. The damage to young trees is especially serve. Light infestation disrupts translocation of nutrition and water inside the trees, resulting in reduced tree growth, whereas heavy infestation leads to loss of main trunk due to rot caused by pest holes and water damage.

Morphological characteristics **Adults** 30-45 mm long with a wingspan of 60-70 mm, dark brown. Antennae filiform, extremely small. Forewings with seven near circular marks along the anterior margin, and one large dark brown triangular mark in the middle. The obscure parallel arc marks along the outer margin form a wide band which reaches the posterior margin. Hind wings black brown without obvious marks. Fore- and middle legs well developed with long claws, while hind legs small and retrogressed. There are

E. excrescens adult (Xu Gongtian)

E. excrescens adult (Xu Gongtian)

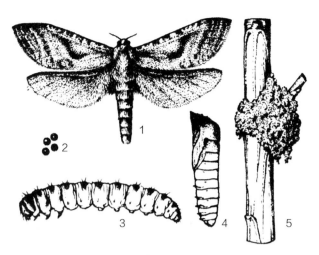

E. excrescens. 1. Adult, 2. Eggs, 3. Larvae, 4. Pupa, 5. Damage. (Shao Yuhua)

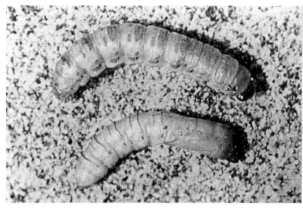

E. excrescens larvae (Yan Shanchun)

E. excrescens pupa (Xu Gongtian)

E. excrescens pupal skin (Xu Gongtian)

dense orange-yellow brush-like long hairs (poisonous scales) on the back of male hind femora. **Eggs** Diameter 0.6-0.7 mm, round. Milky white initially, turn to black later, shiny. **Larvae** Length 44-57 mm at maturity, milky white, cylindrical. Head reddish brown to dark brown. Pronotum large and deep colored. Each notum from mesothorax to 8th abdominal segment contains yellowish brown setae clusters. **Pupae** 29-60 mm long, yellowish brown. Vertex deep reddish brown with protrude median ridges. Wing pads extremely short. There are two rows of hooked spines on the back of abdominal segment 3 to 7. There are also wavy hooked spines on the underside of the abdomen, with one row on segment 4-6, two rows on segment 7, and one row on segment 8.

Biology and behavior *Endoclita excrescens* occurs mostly one generation (some two generations) a year and overwinters as eggs on the ground or as larvae at the base of the tree or inside the pith at breast height of the trunk, or inside the roots under the ground in cut-producing nursery in Liaoning. Eggs start to hatch in mid-May. Larvae migrate to feed inside trunks of poplars, willows, or grass by early June. Pupation begins in early August and ends by late September. Adults start to emerge in late August with a peak emergence in mid-September. Adult emergence ends in mid-October. Adults mate and start to lay eggs right after emergence. The eggs overwinter. A portion of the late hatched larvae or slow developing larvae will overwinter, pupate in late July of the next year, and lay eggs to finish their two-year life cycle.

Most adults emerge between 16:00-18:00. Adults hanging on tree trunks, bushes or grasses under the trees during the day. They start to fly, mate and lay eggs after sunset. Adults show negative phototaxy, with the flight period coincides with sunset. Adults can mate at 2 m height in the air and mate on the same day of emergence. Mating lasts for 14 hours and 20 minutes to 45 hours and 6 minutes. Eggs are laid at various locations right after mating. Most females lay their eggs by flapping their wings a few times, while some can lay eggs while they are mating. Yet many females lay eggs before mating, and some lay their eggs while flying based on field cage observations. Each female produces an average of 2,738 eggs, ranging from 685 to 8,423 eggs. Egg period lasts for 10 days. Females live for 11 (8-13) days while males live for 10 (7-13) days. Eggs are not sticky and usually scattered on the floor or on plant surface. Newly-laid eggs milky white, gradually turn to gray and eventually black within 4-6 hours. Eggs laid by unmated females will not hatch and will dry out within 3 days. Egg stage lasts for 241 (239-243) days. Larvae are

E. excrescens larval damage (Xu Gongtian)   *E. excrescens* larval tunnel (Xu Gongtian)

very active and will back out swiftly or ballooning through silk thread when disturbed. Larvae bore directly into seedlings or branches. Some may also enter the host through axillary buds. Point of entrance is usually 153 (24-246) cm above ground on branches with a diameter of 1.5 (0.8-2.2) cm. On larger trees, larvae normally enter the trees from old insect holes or bark crevices. Most larvae feed downwards after entering the branches. Larval galleries have smooth inner walls. Larvae feed on sapwood around the gallery entry causing a round depression, which makes infested branches vulnerable to wind damage. Larvae usually feed toward the outer side of the branch. Sawdust created through feeding is pushed out the galleries and stuck on silken net at the entrance, creating a sawdust package with silk and sawdust on the outside. Newly hatched larvae feed on humus under leaf litter first. By the 2nd and 3rd instars, they move to feed on 2-year-old seedlings or branches on larger trees, or grasses such as *C. souliei*. Larvae rarely migrate under natural conditions except on small diameter grass until moving to nearby large trees by late July. Larvae develop at different rates on different hosts, resulting in differences in number of molts and development time of each instar. This is extremely true between one-year and two-year generation cycles. Larvae develop faster on *S. matsudana* with larger individuals and earlier adult emergence than larvae on other hosts. Larvae with slower development overwinter in the galleries. Larval stage lasts for 3-4 months. Mature larvae spin white silk to seal the entrance, construct cylindrical cocoons made of white silk and sawdust, and pupate within 2-3 days with heads pointed upward. Pupae can squirm up and down by moving their abdomens utilizing the hooked spines. Pupae are often seen moved near the opening of the galleries at noon. When disturbed, pupae retreat quickly back into the galleries. They turn to dark brown before adult emergence.

Prevention and treatment

1. Enforce quarantine. Enforce quarantine regulations before the shipping out seedlings during cutting, excavating, and counting.

2. Silviculture measures. 1) Select and promote resistant varieties such as *Populus pseudo-simonii* and *Populus × xiaozhuarica* 'Baicheng'; 2) Destroy infested nursery stocks to reduce source populations.

3. Chemical control. 1) Basal spray or cover spray of the ground with beta cypermethrin at 1,000× during the active period of young (2-3rd instars) between mid-May and early June. Spray 2-3 times every 7 days will kill most larvae on the ground. 2) Plug entrance holes with cotton balls containing 10% imidacloprid EC at 10×, or 5.7% emamectin at 10× during the migration of middle-aged larvae between mid-June and mid-July resulted in over 90% mortality. 3) Insert aluminum phosphide stripes into the galleries resulted in 95% control or spray 45% profenofos and phoxim mixture at 30-60×, 12% thiamethoxam and lamda-cyhalothrin mixture at 30-60× to tree trunks to control larvae.

4. Protect natural enemies found in pest larval and pupal stages. Natural enemies are found both in the larval and pupal stages, including *Beauveria bassiana*, *Calathus haleensis*, *Forficula rubusta*, *Grossosmia zelina*, a parasitic fly, and *Dendrocops hyperythrus*.

References   Xiao Gangrou (ed.) 1992; Zhen Zhixian, Chi Defu, Zhang Xiaoyan et al. 2001; Wen Zhenhong 2003; Li Chengde 2004; Li Hongtao, Zhao Bo, Liu Xiaoming 2004.

(Yan Shanchun, Li Yajie, Lin Jihui)

260 *Endoclita nodus* (Chu et Wang)

Classification Lepidoptera, Hepialidae
Synonym *Phassus nodus* Chu et Wang

Importance *Endoclita nodus* is a polyphagous pest with its larvae attacking various species of conifers, broadleaf trees, fruit trees, and rare species of trees. It is an important pest of agriculture, forests, and ornamentals.

Distribution Zhejiang, Anhui, Jiangxi, Henan, and Hunan.

Hosts A total of 51 species in 26 families, including *Cryptomeria fortune*, *Cunninghamia lanceolata*, *Paulownia fortune*, *Castanea mollissima*, *Liriodendron chinense*, *Michelia figo*, *Michelia champaca*, *Nyssa sinensis*, *Toona sinensis*, *Phoebe nanmu*, *Elaeocarpus sylvestris*, and *Pterocarya stenoptera*.

Damage *Endoclita nodus* larvae feed under the bark before entering the heartwood and kill the tree through girdling on conifers, whereas larvae bore into the heartwood from bark surface and create cylindrical galleries on broadleaf trees, leaving annular or pouch-shaped sawdust outside the entrance holes.

Morphological characteristics

Adults Length 28.0-55.0 mm, with a wingspan of 60.0-111.0 mm. Antennae filiform, very short. Forewings yellowish brown, with a nodular elevation in the middle and four marginal blotches made of black and brownish yellow lines anteriorly. The triangular area in the middle of forewings yellowish brown.

Larvae Length 52.0-79.0 mm, yellowish brown, head brownish black. Pronotum strongly sclerotized. Abdomen has a triangle made of one big and two small yellowish brown hairy plates on each segment.

Biology and behavior *Endoclita nodus* occurs one generation in two years and overwinters as eggs in Zhejiang. Eggs hatch in April and larval stage lasts from mid-April to late August. Pupal stage lasts from early August to September, while adults are found between

E. nodus adult (Hu Xingping)

E. nodus adult (SFDP)

E. nodus larva (SFDP)

E. nodus pupa (SFDP)

E. nodus gall (SFDP)

E. nodus formed insect gall (SFDP)

mid-August and early October.

Newly hatched larvae stay and feed inside a ball made of leaves and humus pieces in leaf litter or the humus layer in the forest. Young larvae are very active. On conifer hosts, third instars bore into the trunk and feed around the circumference, kill the host rapidly before moving to another tree. On broadleaf trees, however, they bore directly into the heartwood and stay inside until adult emergence. Larvae usually crawl to entrance holes to feed on the sapwood, creating round and depressed feeding pits. Pupae can move up and down freely inside the tunnels with the help of abdominal spines. Pupal period lasts for 15-20 days. Most adults emerge between 14:00-20:00. Adults hang upside down on bushes or weeds after emergence during the day. By twilight males dash in high speed while females rock their abdomens. Mating mostly occur between 19:00-21:00 and lasts for up to 22 hours. There are no specific egg sites. Females scatter their eggs while flying or resting on branches by swinging their abdomen. Each female can produce 1,604-6,904 eggs with an average of 3,960 eggs.

Prevention and treatment

1. Mechanic control. Nursery is the major source of infestations. The pest spread through transportation of seedlings. Destroy seedlings immediately when boring sawdust is found. Remove garbage and leaf litter from the nursery. Remove weeds such as *Clerodendrum cyrtophyllum* var. *cyrtophyllum* and *Idesia polycarpa* during cultivation in the forest to eliminate feeding and resting sites of the young larvae and overwinter sites of the middle stage larvae.

2. Biological control. Spray *Beauveria bassiana* powder at 1.5×10^{11} conidia/g or suspension at the entrance holes resulted in high infection rate.

3. Chemical control. Inject 50% imidacloprid and monosultap mixture or 40% sumithion at 500× directly into the tunnels to control larvae. Spray 45% profenofos and phoxim mixture at 30-60×, 12% thiamethoxam and lamda-cyhalothrin mixture at 30-60× to tree trunks to control larvae.

References Zhao Jinnian et al. 1988, Zhao Jinnian 1990, Xiao Gangrou (ed.) 1992, Zhu Hongfu et al. 2004.

(Zhao Jinnian)

261 *Dasyses barbata* (Christoph)

Classification Lepidoptera, Tineidae
Synonym *Hapsifera barbata* (Christoph)

Distribution Beijing, Liaoning, Shandong; Russia, Japan.

Hosts *Robinia pseudoacacia*, *Sophora japonica*, *Populus euramericana*, *Salix matsudana*, *Ulmus pumila*, *Castanea mollissima*, *Quercus variabilis*, and *Ziayphus jujube*.

Damage Damaged branches and trunks become swelled, with peeled or cracked barks and decayed tissues under the bark.

Morphological characteristics **Adults** Female length 7-10 mm, with a wingspan of 17-22 mm. Male length 5-8 mm, with a wingspan of 13-17 mm. Body grayish white to dark brown. Forewings grayish white with scattered grayish brown or dark brown scales. The base has a bundle of erect dark brown scales. There are several bundles of inclined erect bundles of dark brown scales between 1/3-2/3 from the wing base and 5-7 small bundles of inclined scales along the subcostal margin.

Eggs Round, white initially, turn to yellowish brown before hatching.

Larvae Length 20 mm. Head reddish brown. Pronotum lip-shaped, the anterior half light brown, the posterior half dark brown. Crochets of prolegs arranged in a circle, while crochets of each anal proleg arranged in a transverse band.

Pupae Length 10 mm, reddish brown, covered by a thin white cocoon. Top of the pupa truncated with circular cap.

Biology and behavior *Dasyses barbata* occurs two generations a year in Beijing and Shandong and overwinters as larvae inside thin cocoons under the bark. Overwintering larvae come out of the cocoons to feed in late March the following year. Pupation starts in mid-May. Adults emerge in late May. Eggs are mostly laid inside wounds. They occasionally lay eggs on branch joints or in bark crevices. The eggs are covered by yellow silk. Each female produces an average of 69 eggs

D. barbata adult (Xu Gongtian)

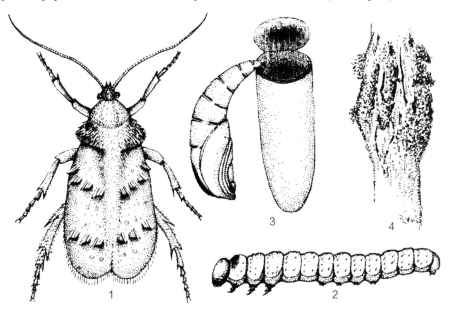

D. barbata. 1. Adult, 2. Larva, 3, empty pupa and cocoon, 4. Damage (Guo Shujia)

D. barbata adult (lateral view) (Xu Gongtian)

D. barbata damage to black locus tree trunk (Xu Gongtian)

(range: 24-146 eggs). Newly hatched larvae bore into old bark near the eggshells and feed in the cambium layer. The entrance holes are covered by frass glued with silk. Larval galleries irregularly shaped, longitudinal, and appear mostly from early August to mid-September. Newly hatched larvae are commonly found in mid-August. Larvae start to overwinter in late October.

Prevention and treatment

1. Apply *Beauveria bassiana* in winter. Spray canopy with 20% diflubenzuron SC at 7,000× during larval infestation period.

2. Spray 25 g/L lamda-cyhalothrin at 200×, or 45% profenofos and phoxim mixture at 30-60×, or 12% thiamethoxam and lamda-cyhalothrin mixture at 30-60× to tree trunks to control the larvae.

References Sun Yujia et al. 1989; Xiao Gangrou (ed.) 1992; Editorial committee of the Atlas of Shandong Forest Pests 1993; Xu Gongtian and Yang Zhihua 2007; Jiang Li, Xuhui, Zhang Qiyu et al. 2009.

(Li Zhenyu, Xu Gongtian, Nan Nan, Sun Yujia)

262 *Opogona sacchari* (Bojer)

Classification Lepidoptera, Tineidae
Synonym *Alucita sacchari* Bojer
Common name Banana moth

Importance *Opogona sacchari* was first introduced to China in the 1990's. It is now found in more than 10 provinces and becomes an important pest of flower and nursery industries, especially on those businesses specializing on the cultivation of *Draceana fragrans* and *Pachira macrocarpa*. It also attacks sugar canes and bananas, threatening the flower industry, tropical agriculture, and sugar industry.

Distribution Beijing, Hebei, Liaoning, Shanghai, Zhejiang, Jiangxi, Shandong, Guangdong, Guangxi, Hainan, and Xinjiang; Japan, India, South Africa, Madagascar, Mauritius, Rwanda, Seychelles, Nigeria, Saint Helena (UK), Cape Verde, France, Germany, Greece, United Kingdom, Belgium, Finland, Sweden, Italy, Denmark, the Netherland, Poland, Portugal, Spain, Switzerland, Bazil, Peru, Venezuela, Barbados, Honduras, U.S.A., and Bermuda (UK).

Hosts *Saccharum sinense*, *Draceana fragrans*, *D. marginata*, *Pachira macrocarpa*, *Yucca elephantipes*, *Cycas revolute*, *Euphorbia pulcherrima*, *Pelargonium hortorum*, *Caryota ochlandra*, *Chysalidocarpus lulescens*, *Roystonea regia*, *Ravenea rivularis*, *Scheffera octophylla*, *Bombax malabaricum*, *Albizia julibrissin*, *Hibiscus syriacus*, *Ficus elastic*, *F. religiosa*, *Broussonetia papyrifera*, *Rhapis excels*, *Musa sapientum*, and *Zea mays*.

Damage *Opogona sacchari* larvae bore under host bark, creating galleries in stems. During heavy infestation, they consume the epidermis. Larvae feeding on cambial region cause wilt and yellowing of leaves and branches, or mortality in *D. fragrans*. On *D. fragrans*, *O. sacchari* bore inside the middle or upper part of the trunk. They feed and completely consume the cambium of the host, leaving only the outer epidermis and xylem intact. Frass and sawdust between outer epidermis and xylem are visible on the surface. Bark is easily separated from xylem after larval feeding. Larvae enter xylem after consuming the cambium, creating serpentine galleries on trunk. Some can even reach and consume the pith. Infested *D. fragrans* do not produce buds or buds cannot develop normally. Affected plants lose their aesthetic value. *Opogona sacchari* feeds on trunk base of *P. macroparca*, creating similar damage as that in *D. fragrans*. Infested *D. fragrans* are susceptible to windbreak. If used in groups of 3-5 plants in bonsai arrangement, the infestation will serious affects its aesthetic value. Among other ornamentals such as *M. sapientum*, damage usually occurs in plant heart tissue and plant growth is affected.

Morphological characteristics **Adults** 8-10 mm long with a wingspan of 18-26 mm, yellowish brown. Head scales large and smooth. Vertex dark colored and flatten backwards while frons bend forwards, with a

O. sacchari larva (Xu Gongtian)

O. sacchari damage to banyan tree trunk (Xu Gongtian)

transverse band of loose hairs in between. Gena flat and inclined, with light small scales. Labial palpi robust and long, slightly tip upwards. Maxillary palpi thin and long, curled. Proboscis extremely small. Antennae small and hair-like, reaching two thirds of the forewings. Flagella relatively large and slightly curved. Scales on thoracic nota large and smooth. Body flat. Wings flat. Forewings dark brown, lanceolate, with a black mark on the top and the later margin of the median cell. Posterior margin of the forewings contains long hairs, which arise to form a cocktail when resting. Forewings in females have a thin black line at the base that reaches half of their length. Hindwings lightly colored, yellowish brown and with long marginal hairs. Hind legs long, extend beyond tip of the hindwings. There are long hairs on hind tibia. Abdomen contains two columns of gray spots. Male genitalia small and specialized. Female ovipositor thin and long, generally extends beyond end of the abdomen. Adults extend their antennae forward when resting, and act like earwigs when crawling with high speed. They can also jump for short distances.

Eggs 0.5-0.7 mm long and 0.3-0.4 mm wide, yellowish, oval.

Larvae 30 mm long and 3 mm wide at maturity, milky white, transparent. Head reddish brown, with four rectangle hair strips (black marks) arranged in two rows for each body segment. There are four hair strips on each pleural plate.

Pupae 10 mm long, brown. Antennae, wing pads, and hind legs stick together and separated from the pupal body. There is a pair of hooked, upward-curved, and strong black anal spines.

Biology and behavior It took about three months for *O. sacchari* to finish a life cycle under 15°C, with 2 days for eggs, 50 days for larvae, 20 days for pupae, and 6 days for adults. The life cycle is shorter at higher temperatures. Up to 8 generations may occur per year.

Opogona sacchari occurs three to four generation a year in Beijing and overwinters as larvae in flower pots. Overwintering larvae resume feeding the next year when temperatures become suitable. Most feed in the cambial region in *D. fragrans* plants of greater than 3-year-old. Young *D. fragrans* is resistant. Larvae bore into the xylem sometimes, leaving light damage. Some of them enter the pith from wounds and crevices and create empty pith. There are seven instars in the larval stage, which lasts for 45 days. Mature larvae construct cocoons with silk and wooden pieces on top of wood post or bark surface during the summer. They construct cocoons in the soil under flowerpots with soil particles attached during the fall. Pupal stage lasts for about 15 days. Adults break silky cocoons and host bark and expose half of the pupal cases. Pupal cases remain on the plants for a long time after adult emergence. Adults are strong crawlers that walks very fast just like earwigs. They can also jump for short distances. Adults need supplemental nutrition and are attracted to sugar. Adults live for about 5 days. Eggs are laid singly or in masses. Egg stage lasts for 4 days. Larvae balloon with silk after hatch and enter host bark quickly. Larvae can also feed on soil, with soil particles commonly visible inside their guts and excreted frass.

Prevention and treatment

1. Enforce quarantine. Enforce internal and external quarantine to prevent outbreak, dispersal, and spread of this pest. Timely inspect imported *D. Fragrans* to remove and destroy infested plants.

2. Population monitoring. Trap adults with sugar water or sex pheromone to prevent this destructive pest from spreading to plants other than flowers and ornamentals.

3. Submerge *D. fragrans* posts in 25 g/L lamda-cyhalothrin at 100× 20% for 5 minutes. Trunk treatment with 5.7% emamectin at 500×.

4. Avoid planting preferred host species such as *D. fragrans* and *P. macrocarpa*. Plant host species in separate locations to avoid pest spreading.

5. Cut treatment. Cuts created during the cultivation are ideal egg laying sites for this pest. Cut treatment is therefore an important control option. Survey results showed that those covered with red and black wax had low infestation rate, whereas those covered by white wax suffered higher damage. Completely sealed cuts are less vulnerable than partially covered cuts. Apply chemical insecticides on wax results in better protection.

6. Winter season control. Control overwintering population in Beijing area by: 1) mixing 90% crystal trichlorfon with soil at 1:200 and apply the mixture under flowerpots every 7-10 days for 21-30 days to control the overwintering larvae; 2) Hang cloth stripes soaked with 25 g/L lamda-cyhalothrin at 100× inside greenhouses at one stripe per 30 m^2. Cloth stripes soaked with dichlorvos every 2 days and hung for 30 days can eliminate this pest.

7. Summer season control. High summer temperatures in Beijing area are conducive for the outbreak of this pest. This is the time to suppress pest population to prevent its spread. 1) Chemical control through spray or drenching on trunks, such as applying 45% profenofos and phoxim mixture at 30-60×, or 12% thiamethoxam

O. sacchari damage to *Dracaena fragrans* (Xu Gongtian)

and lamda-cyhalothrin mixture at 30-60× to tree trunks to control the larvae. 2). Inject *Steinernema carpocapsae* Beijing strain nematodes under the bark at 1,000-2,000 juveniles/mL in the spring or fall. Nematodes are not very active under high or low temperatures.

References Cheng Guifang, 1997, Cheng Guifang and Yang Jikun 1997a, b, Shen Jie et al. 2002, Sun Xuehai 2003, Forest Disease and Pest Control Station of the Ministry of Forestry 2005.

(Guan Ling)

263 *Amatissa snelleni* (Heylaerts)

Classification   Lepidoptera, Psychidae
Synonym   *Kophene snelleni* Heylaerts

Importance   Like other bagworms, *A. snelleni* is a polyphagous defoliator protected in bags. Outbreaks can easily occur.

Distribution   Zhejiang, Jiangxi, Hubei, Hunan, Guangdong, Guangxi, and Yunnan; India.

Hosts   *Eucalyptus* spp., *Cinnamomum camphora*, *Cupressus tunebris*, *Fortunella margarita*, *Litchi chinensis*, *Mangifera indica*, *Malus pumila*, *Prunus salicina*, *Prunus persica*, and *Pyrus pyrifolia*.

Damage   Young larvae feed on leaf tissues and create circular holes, leaving web-like veins or thin upper epidermis. Leaves of the entire trees can be devoured under high population density, spare only the mid veins or a small basal portion of the old leaves. The entire forest canopy may turn yellow due to infestation.

Morphological characteristics   Adults Female length 13-23 mm, width 4-7 mm, yellowish. Head small, with a pair of spines. Thorax slightly curved, with a brown median longitudinal line. Male length 11-15 mm, with a wingspan of 28-33 mm. Forewings with an acute top corner and oblique outer margin, transverse stripes brown, R_3 and R_4 join together at the base, while R_5 is separated from or share only a short portion with R_3+R_4. Body and wings grayish brown to yellowish brown. **Eggs** Length 0.7-0.8 mm, oval, beige colored. **Larvae** Mature larva length 17-25 mm, width 4-6 mm. Head and thoracic nota grayish brown with scattered dark brown dots. Each thoracic notum is divided into two parts, with four black marginal hairs anteriorly along the median line. Pronotal hairs are arranged in a square, whereas those on meso- and metanotum transversely arranged. Abdomen purplish. Anal plate dark brown.

Biology and behavior   *Amatissa snelleni* occurs one generation a year and overwinters as mature larvae

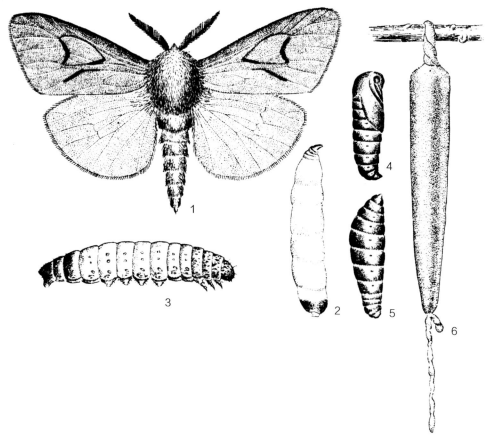

A. snelleni. 1. Male adult, 2. Female adult, 3. Larva, 4-5. Pupa, 6. Bag (Zhang Peiyi)

A. snelleni bag (Wang Jijian)

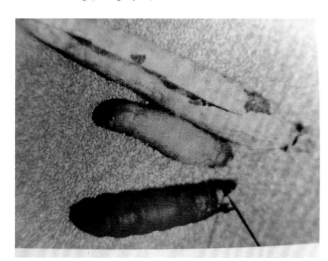

A. snelleni bag, larva, and pupa (Wang Jijian)

inside the bags in Guangxi. Most larvae pupate in mid- to late February. Adult emergence peaks between early to mid-April, and peak egg period falls between mid- and late April. The majority of the eggs hatch from late April to early May. The heaviest damage usually occurs between June and July. Mature larvae overwinter in mid- to late October.

Larva incises leaves and create notches or holes. The damage period lasts from April to October. Mature larvae attach bags to small branches for overwintering using silk bundles. Males are attracted to black lights. Females lay eggs in pupal cases inside the bags.

Prevention and treatment

See general prevention and treatment methods for bagworms.

References Huang Jinyi and Meng Meiqiong 1986, Xiao Gangrou (ed.) 1992.

(Wang Jijian, Chen Lie, Chen Boyao)

264 *Chalioides kondonis* Matsumura

Classification Lepidoptera, Psychidae
Common name Kondo white psychid

Importance *Chalioides kondonis* is a common pest of many trees.

Distribution Shanghai, Jiangsu, Zhejiang, Anhui, Fujian, Jiangxi, Henan, Hubei, Hunan, Guangdong, Guangxi, Sichuan, Guizhou, Yunnan and Taiwan; Japan.

Hosts *Camellia sinensis*, *Salix* spp., *Bambusoideae*, *Cinnamomum camphora*, *Prunus salicina*, *P. armeniaca*, *P. mume*, *P. persica*, *Populus* spp., *Ulmus pumila*, *Morus alba*, *Sophora japonica*, *Quercus* sp., *Castanopsis* spp., *Pyrus bretschneideri*, *Diospyros kaki*, *Zizyphus jujube*, *Vernicia fordii*, *Camellia olefera*, *Cercis chinensis*, *Sapium sebiferum*, *Pterocarya stenoptera*, *Juglans regia*, *Albizzia julibrissin*, *Acacia richii*, *Myrica rubra*, *Lagerstroemia indica*, *Chamaecyparis* spp., *Malus pumila*, *Eriobotrya japonica*, *Punica granatum*, *Ligustrum lucidum*, *Michelia alba*, *Robinia pseudoacacia*, *Citrus reticulate*, *Dalbergia hupeana*, *Amorpha fruticosa*, *Acer buergerianum*, *Bauhinia uariegata*, *Sophora xanthantha*, *Patanus hispanica*, *Bischofia javanica*, *Carya cathayensis*, *Delonix regia*, *Casuarina equisetifolia*, and *Spiraea cantoniensis*.

Damage *Chalioides kondonis* feeds gregariously. Larvae feed on leaves, young branches, or branch bark and fruit skins while carrying the bags, resulting in complete defoliation of portions of the crown.

Morphological characteristics **Adults** Female length 9-16 mm, maggot like. Body yellowish white to yellowish brown with slight purple. Head small, dark yellowish brown. Antennae small, protruding. Compound eyes black. Thoracic notum and the 1st and 2nd abdominal notum shiny with a brown longitudinal line in the middle. Abdominal sternites 1-7 has a purple round dot on each segment. Bundles of pale brownish hairs present on each segment after sternite 3. Abdomen plump, tapered conically towards the end. Male length 8-11 mm, with a wingspan of 18-21 mm, brownish and covered with white long hairs. The end of the body brown. Head brownish. Compound eyes sphere-shaped, dark brown. Antennae plumose, dark brown. Both front and hind wings transparent, slightly darker only at the base and the anal area of the hind wings. Hind wings covered by white long hairs at the base. **Eggs** Length 0.8 mm, oval, beige to fresh yellow. **Larvae** Mature larva length 28 mm. Head orange yellow to brown, with dark brown to black cloudy dots. Thoracic nota yellowish brown with black dots forming three longitudinal lines laterally. Meso- and metathoracic notum divided into two parts. Thorax and abdomen flesh pink, with dark dots laterally on each segment. Abdominal nota 8 and 9 contain large brown blotches. Anal plate brown. Thoracic legs and prolegs present. **Bags** Bags are about 30 mm long for males and 38 mm for females, conical, grayish white, tightly weaved with silk and contains nine longitudinal ridges, smooth externally and free of branches and leaves. **Pupae** Yellowish brown. Male 8-11 mm long, dark brown, fusiform, with a line of spinules on the posterior margin from abdominal segment 3 to 6 and the anterior margin of the segments 8 and 9. Female 12-16 mm long, dark brown, cylindrical, with a line of spinules on the posterior margin of abdominal nota 2 and 5, as well as the anterior margin of notum 7.

Biology and behavior *Chalioides kondonis* occurs one generation a year and overwinters as young larvae in-

C. kondonis adult (Wang Yan)

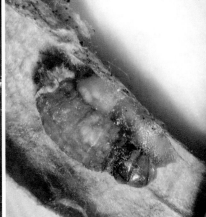
C. kondonis pupa (Wang Yan)

C. kondonis damage (edited by Zhang Runzhi)

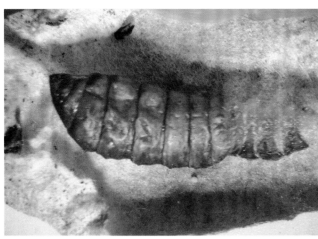
C. kondonis pupa (Wang Yan)

C. kondonis pupal skin exposed from bag (Wang Yan)

C. kondonis bag on *Sterculia lanceolata* (edited by Zhang Runzhi)

C. kondonis bag on *Acacia auriculiformis* (edited by Zhang Runzhi)

C. kondonis bag on *Casuarina equisetifolia* (edited by Zhang Runzhi)

C. kondonis bag on *Acacia mangium* (edited by Zhang Runzhi)

side the bags. Larvae begin to feed when host leaves expand in the next spring. Mature larvae pupate in June. Pupal stage lasts for 15-20 days. Adults emerge between late June and July. Females stay inside the bags and wait for males to come for mating. Eggs are laid inside the bags. Each female can produce up to 1,000 eggs. The egg stage lasts for 12-13 days. Newly hatched larvae feed on eggshells first before moving out of the bags and disperse through ballooning by silk. They then wrap themselves with more silk soon after finding an optimum location. A few larvae usually feed on a leaf. Bags increase in size as the larvae grow. Larvae carry their bags as they move and feed by extending their heads outside the bags. Their heads retract when disturbed. Larvae feed until the fall. They migrate to branches in mid- to later October to overwinter by attaching the bags on branches and closing the entrance holes with silk.

Prevention and treatment

1. Timely removal of bags by hand during pruning. Conserve natural enemies.

2. Attract males with black light traps and sex pheromone.

3. Spray *B.t.* during larval stage.

4. Spray 95% trichlorfon at 800-1,000×, 25 g/L lamda-cyhalothrin at 1,200×, 50% sumithion EC at 1,000×, 50% phoxim EC at 1,500×, or 2.5% deltamethrin EC at 4,000×, 45% profenofos and phoxim mixture at 1,000×, or 2×10^9 PIB/mL nuclear polyhedrosis viruses at 1000-1,500× to control young larvae.

References Xaio Gangrou 1992, Xu Minghui 1993.

(Ji Baozhong, Zhang Kai, Chen Boyao, Chen Lie)

265 *Acanthoecia larminati* (Heylaerts)

Classification Lepidoptera, Psychidae
Synonym *Clania larminati* (Heylaerts)

Importance *Acanthoecia larminati* is a common pest of ornamentals and forests that affects tree growth and landscape aesthetical value during outbreaks.

Distribution Anhui, Fujian, Jiangxi, Hunan, Guangdong, Guangxi, Sichuan, Guizhou, and Yunnan; Japan.

Hosts *Camellia sinensis*, *Morus alba*, *Diospyros kaki*, *Vernicia fordii*, *Platycladus orientalis*, *Dalbergia hupeana*, *Cassia surattensis*, *Cinnamomum Myanmarnnii*, *Mangifera indica*, *Castanea mollissima*, *Citrus reticulate*, *Dimocarpus longan*, *Malus pumila*, *Canarium album*, *Michelia alba*, *Rosa chinensis*, *Camellia japonica*, *Albizzia falcate*, *Patanus hispanica*, *Adenanthera pauonina*, *Delonix regia*, *Eucalyptus citriodora*, *Bauhinia uariegata*, *Duabanga grandiflora*, *Acacia confuse*, *Ormosia pinnata*, and *Elaeocarpus serratus*.

Damage Similar to other bagworms.

Morphological characteristics **Adults** Male length 6-8 mm, with a wingspan of 18-20 mm, head and thorax black, abdomen silver, forewings grayish black with white bases. Hindwings white, grayish brown marginally. Female length 13-20 mm, milky white to yellowish white, cylindrical. **Eggs** Length 0.6-0.7 mm, oval, beige colored. **Larvae** Mature larva length 16-25 mm. Hairs on head, thorax and dorsal surface of the abdominal segments 8-10 grayish black, hairs of the rest areas yellowish white. **Pupae** Males 9-10 mm long, head, thorax, antennae, legs, wings, and abdominal nota dark brown; abdominal sternites and intersegment between abdominal notum grayish brown. There is a row of spinules on the anterior margin of the abdominal segments 4 to 8 and the posterior margin of the abdominal segments 6 and 7. Females 15-23 mm long, cylindrical, smooth. Head, thorax, and end of abdominal nota dark brown, the rest yellowish brown. **Bags** Oblate conical. Female bags 30-50 mm long, male bags 25-35 mm long. Silky, brown in color, free of leaf litter and small branches.

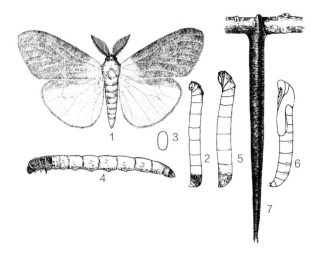

A. larminati. 1. Male adult, 2. Female adult, 3. Egg, 4. Larva, 5. xxx, 6. Pupa, 7. Bag (1 and 7 by Zhang Peiyi, 2-6 by Zhu Xingcai)

Biology and behavior *Acanthoecia larminati* occurs one generation a year in northern Fujian and Nanning region in Guangxi. It overwinters as mature larvae inside the bags. As many as 10 bags are closely arranged along a branch. Larvae can feed during the winter if the weather is warm enough. Larvae pupate in mid- to late February. Adults emerge in early to mid-March. Emergence period lasts for about half a month. Most eggs are laid between late March and early April. Eggs hatch in early May. Damage from larvae mostly observed between June and July. Larvae start to overwinter in mid- to late October. There are seven instars for the males, and eight instars for the females. The egg stage lasts for 30-39 days. The larval stage lasts for 306 days for males and 323 days for females. The pupal stage lasts for 16 days in females and 28 days in males. Adult males live for an average of 3.5 days.

Prevention and treatment See methods for *Acanthopsyche subferalbata*.

References Xiao Gangrou (ed.) 1992, Xu Huiming 1993, Jiang Cuiling and Zhong Jueming 1995.

A. larminati bag

(Ji Baozhong, Zhang Kai, Chen Boyao, Chen Lie)

266 *Eumeta minuscula* Butler

Classification   Lepidoptera, Psychidae
Synonyms   *Cryptothelea minuscula* Butler, *Clania minuscula* Butler
Common name   Tea bagworm

Importance   Like other bagworms, *E. minuscula* is a polyphagous defoliator and often cause serious damages to host.

Distribution   Jiangsu, Zhejiang, Anhui, Fujian, Jiangxi, Hubei, Hunan, Guangdong, Guangxi, Sichuan, Guizhou and Taiwan; Japan.

Hosts   *Eucalyptus* spp., *Pinus massoniana*, *Biota orientalis*, *Casuarina equisetifolia*, *Salix babylonica*, *Bischofia polycarpa*, *Platanus* spp., *Ulmus pumila*, *Dalbergia hupeana*, *Juglans regia*, *Acer fabri*, *Sapium sebiferum*, *Aleurites fordii*, *Punica granatum*, *Pyrus pyrifolia*, *Prunus persica*, *Prunus salicina*, *P. armeniaca*, *Cerasus pseudocerasus*, and *Citrus reticulate*.

Damage   Same as *Amatissa snelleni*.

Morphological characteristics   **Adults** Male length 10-15 mm, with a wingspan of 23-30 mm. Wings dark brown, areas along the veins darker. Forewings with two transparent rectangle blotches between M_2 and Cu_1. Body covered by dense hairy scales. Thorax contains two white longitudinal marks. Female length 15-20 mm, beige colored. Thorax with an obvious yellowish blotch. Abdomen plump, with yellow pubescence on segments 4 to 7. **Eggs** Length 0.8 mm, oval, beige or yellow. **Larvae** Length 16-28 mm at maturity. Head yellowish, scattered with dark reticulated punctures. Each thoracic segment has four dark brown long marks arranged in a line. Abdomen flesh red, each segment with two pairs of slant black pointed protrusions. **Pupae** Females 20 mm long, fusiform, head small. Abdomen contains a row of spinules on posterior margin of notum 3, the anterior margin of nota 4 and 5, and the anterior margin of nota 6 to 8. Spinules on notum relatively bigger. **Bags** Length 25-30 mm, with orderly arranged of small branches externally.

Biology and behavior   *Eumeta minuscula* occurs three generations a year and overwinters as mature larvae in Guangxi. Overwintering larvae pupate in March. Adults emerge and start laying eggs in early April. Most of the 1st generation eggs hatch between mid- and late April and pupate in early June. Most of the 2nd generation

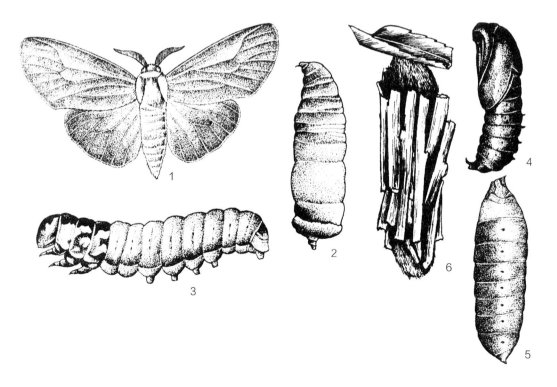

E. minuscula. 1. Male adult, 2. Female adult, 3. Larva, 4. Male pupa, 5. Female pupa, 6. Bag (Chen Boyao, Chen Lie)

E. minuscula adult (Wang Jijian)　　　　E. minuscula bag (Wang Jijian)

E. minuscula female pupa (SFDP)　　　E. minuscula bag (Wang Jijian)

eggs hatch between late June and early July and pupate in early August. Most of the overwintering generation eggs hatch between late August and early September. They start overwintering in mid- to late November.

The first and second instar larvae feed on leaf tissue, leaving translucent yellowish-brown thin films. The third instar larvae consume leaves or branch barks. They create feeding holes, partial leaves, or leaves only with the main veins intact. Heavy infestation affects tree growth or even causes tree mortality. Larvae stop feeding and migrate to branch tips in mid-November and attach the bags on small branches with 10-20 mm long silk bundles before overwintering. There are usually several bags lined along the branches. Each female produces 120-990 eggs. Eggs are laid inside the pupal case in the bag. The egg stage lasts for 15-20 days for the first generation, and 7 days for the second and third generations.

Prevention and treatment

See methods for *Acanthopsyche subferalbata*.

References　Huang Jinyi, Meng Meiqiong 1986, Xiao Gangrou (ed.) 1992.

(Wang Jijian, Chen Boyao, Chen Lie)

267 *Dappula tertia* (Templeton)

Classification Lepidoptera, Psychidae

Importance *Dappula tertia* is the most abundant bagworm during outbreaks in fast growing eucalyptus stands. Similar to other bagworms, *D. tertia* is a polyphagous defoliator and often occurs in huge numbers.

Distribution Zhejiang, Jiangxi, Hubei, Hunan, Guangdong, and Guangxi; Sri Lanka.

Hosts *Eucalyptus* spp., *Aleurites fordi*, *Casuarina equisetifolia*, *Castanea mollissima*, *Mangifera indica*, *Achras zapota*, *Sinopimelodendron kwangsiense*, *Litchi chinensis*, *Euphoria longan*, *Eriobotrya japonica*, *Diospyros kaki*, *Dracontomelon dao*, *Camellia oleifera*, *Cinnamomum cassia*, *Illicium verum*, *Swietenia mahogoni*, *Manglietia glauca*, *Acacoa acuriculaeformis*, *Grevillea robusta*, *Castanopsis hickelii*, *Sapium sebiferum*, *Anthocephalus chinensis*, *Ginkgo biloba*, *Keteleeria fortune*, *Persea Americana*, *Schima superba*, *Citrus reticulate*, *Coffea arabica*, *Cupressus tunebris*, *Cinnamomum camphora*, and *Dalbergia hupeana*.

Morphological characteristics **Adults** Male length 15-18 mm, wingspan of 30-35 mm. Wings grayish black. Forewings with an elongate dark mark on top of the radian cell and a mark on the R vein. Apex with a conspicuous angle. Hindwings grayish dark brown, veins brown. Female length 14-24 mm, yellowish. Head small, thorax elevated dorsally, dark brown. **Eggs** Length 0.7-0.8 mm, oval, beige. **Larvae** Mature larvae length 23-30 mm. Pronotum dark brown. Median line on pronotum and Mesonotum white, with one elongate white mark on each lateral side, forming three parallel lines. Metanotum with a yellowish white mark on each lateral side along the median line. Abdomen black, with numerous transverse wrinkles. **Pupae** Females 12-17 mm long, dark brown to blackish brown. Abdomen contains a row of spinules on the posterior margin of nota 3 and 4, the anterior margin of nota 5 and 6, and the anterior margin of nota 7 to 9. Males 14-25 mm long, dark brown, with a ridge running from thorax to abdominal notum 5. Abdomen contains a row of spinules on the posterior margin of notum 2, the anterior and posterior margins of nota 3, 4, and 5, and the anterior margin of nota 6, 7, and 8.

Bags Length 28-50 mm, brown, elongated conical. Densely weaved and flexible, with broken leaves, half leaves, or entire leaves attached externally.

Biology and behavior *Dappula tertia* occurs one

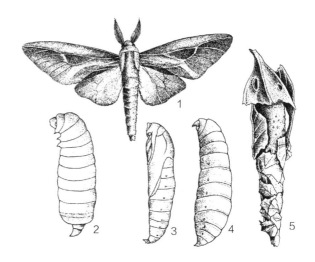

D. tertia. 1. Male adult, 2. Female adult, 3. Male pupa, 4. Female pupa. 5. Bag (Zhu Xingcai)

D. tertia young instar larva bag (Wang Jijian)

D. tertia newly enclosed adult (Wang Jijian)

D. tertia feeding on host (Wang Jijian)

D. tertia larva (Wang Jijian)

D. tertia pupa (Wang Jijian)

D. tertia prepupa (Wang Jijian)

D. tertia bag (Wang Jijian)

generation a year and overwinters as mature larvae in Guangxi. Overwintering larvae pupate mostly in mid-February. Adult emergence peaks in mid- to late March. Most females lay their eggs in late March to early April. Egg stage lasts for 15-20 days. Egg hatching peaks in mid- to late April. Larvae cause most of the damage between June and July. Larvae overwinter in late October.

Larvae feed on leaf tissue and create circular holes or incised leaves. They will continue feed on bark after devouring the leaves. Females lay their eggs inside the protective bags, with an average of 1,500-2,000 eggs per female. Larvae attach the bags to branches with silk bundles before overwintering.

Prevention and treatment   See methods for bagworms.

References   Huang Jinyi and Meng Meiqiong 1986, Xiao Gangrou (ed.) 1992.

(Wang Jijian, Chen Boyao, Chen Lie).

Eumeta variegata (Snellen)

Classification Lepidoptera, Psychidae
Synonyms *Cryptothelea variegata* Snellen, *Clania variegata* (Snellen)
Common name Paulownia bagworm

Importance *Eumeta variegata* is dangerous defoliator with frequent damaging outbreaks that referred as "smokeless forest fire". It is a polyphagous pest with big appetite, high productivity, and easy dispersal pathways. It prefers Paulownia as the host. The large reforestation effort through the creation of continuous Paulownia stands over large areas in agriculture fields in Heze region of Shandong province during the 1980-1990's provided conducive conditions for the outbreak of *E. variegata*. Damage by this pest over large area serious threatens the safety of Paulownia production.

Distribution Hebei, Shanghai, Jiangsu, Zhejiang, Anhui, Fujian, Jiangxi, Shandong, Henan, Hubei, Guangdong, Sichuan, Guizhou, Yunnan, and Taiwan; India, Japan, Sri Lanka, and Malaysia.

Hosts *Paulownia fortunei*, *Robinia pseudoacacia*, *Platanus orientalis*, *Ulmus pumila*, *Ailanthus altissima*, *Quercus acutissima*, *Bischofia polycapa*, *Prunus cerasifera* f. *atropurpurea*, *Cercis chinensis*, *Chimonanthus praecox*, *Rosa* cvs., *Rosa* spp., *Camellia japonica*, *Malus pumila*, *Crataegus pinnatifida*, *Punica granatum*, *Cerasus pseudocerasus*, and *Lagerstroemia indica*.

Damage Young larvae nibble on the lower epidermis and leaf tissues, which creates holes on the leaves.

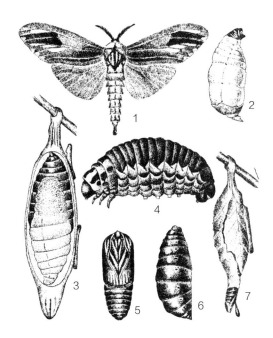

E. variegata. 1. Male adult, 2. Female adult, 3. Female in bag with eggs, 4. Larva, 5-6. Pupa, 7. Male bag (Zhang Peiyi)

E. variegata bag (edited by Zhang Runzhi)

Severely damaged leaves only have veins intact. They can also nibble on bark of the stems.

Morphological characteristics **Adults Female** length 22-30 mm, plump, milky white, body wall transparent which makes eggs inside visible. Wingless, with reduced legs, antennae, mouthparts, and compound eyes. Head small, reddish brown. Thorax with a brown median ridge. Thorax and the first abdominal segment with yellow hairs laterally. Abdominal segment 7 with a band of short yellow hairs along the posterior margin. Abdomen extremely constricted beyond segment 8. Genitalia well developed. Male 15-20 mm long with a wingspan of 35-44 mm, dark brown with light longitudinal marks. Fore wings reddish brown, with 5 transparent blotches apically. Hind wings dark brown, slightly reddish brown. Both wings with obvious branched median veins. **Eggs** Oval, with a diameter of 0.8-1.0 mm, milky yellow at the beginning, turn to yellowish brown, shiny. **Larvae** Oblong round, with smooth surface. There are 3 pairs of thoracic legs. Abdominal legs and anal legs reduced. Mature larvae can be distinguished by sex. The female larvae are black, plump, with an average length of 21.5 mm. Head dark brown, with white sutures. Thorax brown with milky white

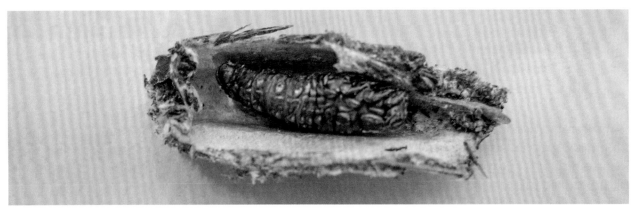

E. variegata pupa (edited by Zhang Runzhi)

blotches, legs dark brown and well developed. Abdomen brownish. Prolegs reduced, with 15-24 crotchets. Male larvae are smaller, yellowish brown, with an average body length of 13.9 mm. **Pupae** Female pupae 25-30 mm long, fusiform, reddish brown. Head small. Thoracic segments fused together, without wind pads, only vestigial of the thoracic legs visible. Male pupae 18-24 mm long, oblong oval, shiny, with a row of transverse spines at the anterior margins of abdominal segment 3-5. Abdomen curved at the tip, with a pair of anal spines. **Bags** Fusiform, with relative large leaf pieces and sometime scattered branches externally. Female bags are large and plump, while male bags are small and slender.

Biology and behavior *Eumeta variegata* occurs one generation a year in most areas or two generations in some areas and overwinters as mature larvae inside the bags. Larvae pupate in early to mid-May and adults emerge in mid- to late May. Males are active at twilight or dawn. Females stay inside their bags after emergence with exposed heads. They produce sex pheromone to attract males. Mating occurs one day after emergence. Eggs are laid inside the pupal cases in the female bags 1-2 days after mating. Each female produces an average of 676 eggs, with the highest of 3,000 eggs per female. Females die after oviposition. Larvae start to feed on eggshells 1-2 days after hatching before moving or floating to nearby leaves and branches. They construct bags with silk and leaf pieces. Bags are finished within 3-4 hours. Except for male adults, all stages stay inside the bags. Larvae start to feed on host plants after the bags are completed with the head and thorax hanging outside. Larvae feed by carrying their bags. Bags increase in size as larvae grow. Bags erect when larvae inside are small and become saggy later. Larvae retreat to the bags when disturbed. Larvae feed mostly in early mornings, later afternoons, and cloudy days, rarely during the noon on sunny days. Most damages are done during July and September. Larvae prefer light and therefore the outer crowns suffer the most. Larvae overwinter by November by attaching the bags with silk on branches, and less commonly on trunk or leaf veins. Newly hatched larvae come out from the bags at the same time after a few days staying inside. They feed aggregately initially. They can disperse by floating with silk in the wind. Upon landing on the leaves, larvae start to construct their own bags. There are five instars in the larval stage. Mature larvae feed voraciously in the fall before entering the bags to overwinter, resulting serious damages to host trees.

Prevention and treatment

1. Mechanic control. Remove bags during fall pruning and destroy them through burning or feeding them to poultry resulted in significant control.

2. Adult trapping. Trap males with frequency vibration-killing lamps or sex pheromone.

3. Conserve and protect natural enemies such as *Nealsomyia rufella*, *Exorista xanthaspis*, *Neophryxe psychidis*, *Chaetexorista klapperichi*, *Exorista sorbillans*, *Coccygomimus aterrima*, *Iseropus kuwanae*, and *Beauveria bassiana*. Up to 50% parasitism was observed for *N. rufella*.

4. Basal trunk injection. Inject 2 mL 30% chloramine phosphorus into each hole at 20 cm intervals on the circumference at the base of the tree during egg hatching to control larvae. Seal injection holes with dirt and cover with leaves.

5. Cover spray. Spray Sendebao (a mixture of 0.18% avermectins and 10^{10} spores/mL *Bacillus thuringiensis*) at 1,000×, 5.7% emamectin at 3,000×, *Bacillus thuringiensis* var. *galleriae* and *Eumeta variegata* nuclear polyhedrosis virus (NPV) mixture at 600×, or *Bacillus thuringiensis* at 200×, 45% profenofos and phoxim mixture at 1,000×, or 2×10^9 PIB/mL nuclear polyhedrosis viruses at 1000-1,500× during larval feeding period.

References Xiao Gangrou (ed.) 1992, Zhao Fanggui et al. 1999, Sun Xugen et al. 2001.

(Wang Haiming, Chen Boyao, Chen Lie)

269 *Mahasena colona* Sonan

Classification Lepidoptera, Psychidae

Importance Similar to other bagworms, *M. colona* is a polyphagous defoliator protected by the bags, with outbreaks going unnoticed until severe damage occurs.

Distribution Jiangsu, Zhejiang, Anhui, Fujian, Guangdong, Guangxi, Sichuan, Guizhou, and Taiwan; India.

Hosts *Eucalyptus* spp., *Camellia oleifera*, *Aleurites fordii*, *Illicium verum*, *Sapium sebiferum*, *Camellia sinensis*, *Cinnamomum camphora*, *Platanus* spp., *Biota orientalis*, *Litchi chinensis*, *Euphoria longan*, *Citrus reticulate*, and *Mangifera indica*.

Damage Young larvae feed on leaf tissues and create circular holes, leaving web-like veins or thin upper epidermis. Leaves of the entire trees can be devoured during adult stage or under high population density, or only the mid veins or a small basal portion of the old leaves are left. Crown of the entire forest stand may turn yellow due to infestation.

Morphological characteristics **Adults** Male about 15 mm long with a wingspan of 24-26 mm. Body brown, no blotch on wing surface. Abdomen metallic. Female about 15 mm long, head small, yellowish, maggot-like, wingless and legless.

Eggs Oval, milky yellowish white.

Larvae Mature larvae 18-25 mm long, head brown, with scattered dark brown blotches. Thoracic nota yellowish, thoracic pleura contain irregular black brown blotches, arranged in two rows laterally.

Pupae Length 20-25 mm, female pupa with three spines apically.

Bags Length 25-40 mm, large and strong, lantern-like, withered brown, loose and silky, with lots of leaf pieces loosely arranged cataphractedly on the outside wall.

Biology and behavior *Mahasena colona* occurs one generation a year and overwinters as old larvae inside bags in Guangxi. It overwinters as young larvae in Hefei area in Anhui. Overwintering larvae resume feeding in March in Guangxi and pupate in mid-June. Pupal stage lasts for 18-21 days. Adults emerge and lay eggs in early to mid-July. Current generation larvae appear in late July. Heaviest damage occurs in August and September. Mature larvae overwinter in mid- to late October.

Larval feeding creates incised leaves or leaves with feeding holes or leaves with only main vein intact. Ma-

M. colona adult (Wang Jijian)

M. colona bag (Wang Jijian)

ture larvae gradually migrate to the end of the branch tops and attach their bags on branches with silk for overwintering. Females lay eggs inside pupal cases in the bags.

Prevention and treatment See general strategies on bagworm prevention and treatment.

References Huang Jinyi and Meng Meiqiong 1986, Xiao Gangrou (ed.) 1992.

(Wang Jijian, Chen Boyao, Chen Lie)

General bagworm prevention and treatment strategies

1. Monitoring and forecasting. Survey regularly for source population since bagworm outbreaks usually start from a relatively centralized location and spread to surrounding areas in all directions. Accurate detection of source population will ensure timely and effective control.

2. Eliminating source populations. Remove bushes and grasses through prescribed burn before afforestation projects can reduce source pest populations. If detected in new forest stands, bags need to be removed and destroyed through burning or mechanic destruction in a timely fashion. Prevent pest from entering new forest stands by removing it from seedlings before shipping.

3. Conserving and protecting natural enemies such as *Beauveria bassiana*, viruses, parasitic wasps in the family of Chalcididae, Ichneumonidae, predatory ladybeetles, ants, spiders, and birds for pest control.

4. Biological control. Apply *Beauveria bassiana*, *Bacillus thuringiensis* var. *galleriae* at the concentration of $1\text{-}2\times10^8$ spores/ml, *Bacillus thuringiensis* (3.5×10^8 spores/g) at 1,000×. Spray Sendebao WP (a mixture of 0.18% avermectins and *Bacillus thuringiensis*) at 15-20 g/ha at 1,000× or Sendebao WP mixed with inert agent at 30-45x. Apply rotenone or pyrethrum at 100× to bags near mulberry gardens.

5. Chemical control. Spray with 90% crystal trichlorfon, 25 g/L lamda-cyhalothrin or 50% fenitrothion at 1,000×, 2.5% decamethrin at 2,000 to 3,000×, 45% profenofos and phoxim mixture at 1,000×, 12% thiamethoxam and lamda-cyhalothrin mixture at 1,000×, or 2×10^9 PIB/mL nuclear polyhedrosis viruses at 1000-1,500× on branches and leaves can result in excellent control.

270 *Caloptilia dentata* Liu et Yuan

Classification   Lepidoptera, Gracillariidae

Distribution   Beijing and Liaoning.

Hosts   *Acer truncatum*, *A. mono*, and *Sorbaria sorbifolia*.

Morphological characteristics   **Adults** have two phenotypes: the summer type and the winter type. The summer type 4-4.6 mm long, antennae longer than body. Thorax dark brown, abdomen grayish brown dorsally and white ventrally. Forewings slender, with yellowish brown long marginal hairs and a golden triangular blotch in the center. Hindwings gray, with relative long marginal hairs. The winter type slightly bigger and darker than the summer type.

Larvae have three pairs of abdominal prolegs with uniordinal penellipse crotchets. Young larvae flat, translucent. Old larvae with well-developed thoracic legs.

Back of pupae yellowish brown, with many dark brown particles, whereas the underside of pupae yellowish green. Antenna longer than body. Compound eyes red.

Biology and behavior   *Caloptilia dentata* occurs 3-4 generations a year in Beijing and overwinters as adults inside grass at the base of the tree. Adults become active in early April when *A. truncatum* start to expand its leaves. They prefer flower nectar for supplemental nutrition. They rest in grass or on underside of the leaves during the day with the body tilted. Females lay eggs near the main veins on leaves, with 1-3 eggs laid on each leaf. The egg stage lasts for about 10 days. Most larvae mine leaves in late April. Later they mine out of the leaves from the tip and feed by rolling the leaves. Leaf-rolling period peaks in early May. Adults are found in May, mid-July, September, and October.

Prevention and treatment

1. Silviculture. Remove grass to reduce overwintering

C. dentata adult (Xu Gongtian)

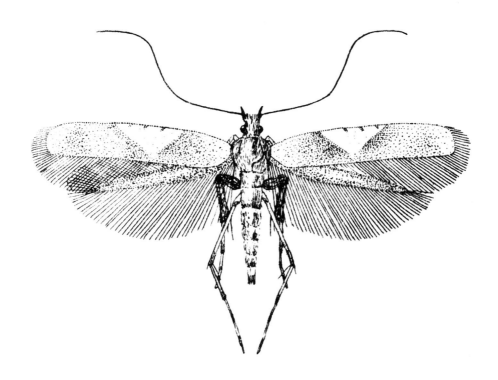

C. dentata adult (Hu Xingping)

populations.

2. Biological control. Protect natural enemies such as Braconids, Aphelinids, Eulophids, ants, and spiders.

3. Chemical control. Cover spay with 1.8% abamectine EC at 2,000×, 5.7% emamectin at 3,000-5,000×, 45% profenofos and phoxim mixture at 1,000×, or 12% thiamethoxam and lamda-cyhalothrin mixture at 1,000×. Avoid treatment during leaf-rolling period for better control results and natural enemy protection.

References Xiao Gangrou (ed.) 1992, Xu Gongtian and Yang Zhihua 2007.

(Xu Gongtian, Nan Nan, Li Zhenyu, Zhang Zuoshuang, Zhou Zhangyi)

C. dentata larva (Xu Gongtian)

C. dentata lava exit hole from leaf (Xu Gongtian)

C. dentata damage to Acer truncatum leaf (Xu Gongtian)

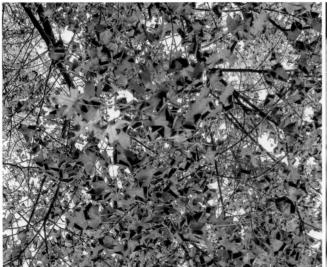

C. dentata damage to Acer truncatum tree (Xu Gongtian)

C. dentata damage to Sorbaria sorbifolia (Xu Gongtian)

271 *Caloptilia chrysolampra* (Meyrick)
Classification Lepidoptera, Gracillariidae

Importance Feeding affects leaf photosynthesis and the aesthetic value of the trees due to large number of insect pods.

Distribution Beijing, Tianjin, Hebei, Shanxi, Inner Mongolia, Liaoning, Jilin, Heilongjiang, Zhejiang, Shandong, Shaanxi, Gansu, Qinghai, Ningxia, and Xinjiang.

Hosts *Salix* spp.

Damage Larvae roll willow leaves from the tip to form triangular pods and feed inside. Affected leaves become reticulate.

Morphological characteristics **Adults** About 4 mm long with a wingspan of about 12 mm. Fore wings yellowish, with a large yellowish white triangular blotch in middle part of the wings extending between anterior and posterior margins. The top corner of this blotch reaches the posterior margin. Posterior margin with a grayish white band between wing base and the triangular blotch. Bands on both wings form a grayish white conical blotch with obtuse anterior portion and pointed posterior portion when at rest. Marginal hairs long, grayish brown. Hairs on apex black or brown with black spots. Forewings with a brown blotch on the surface apically. Antennae extend past the tip of the abdomen. Legs close to body in length, with alternate white and black bands.

Larvae Length 5.3 mm at maturity, cylindrical and slightly flat. Young larvae milky white to yellowish, turn to deep yellow before maturing.

Pupae Fusiform, 4.8 mm long. Thorax yellowish brown. Abdomen lighter in color. Cocoon silky, grayish white, close to fusiform.

Biology and behavior *Caloptilia chrysolampra* occurs several generations a year and overwinters as mature larvae or pupae inside feeding cases in Beijing and Hebei. Early instar larvae are mostly found in the lower crowns in early to mid-June. They roll willow leaves from the tip to form triangular pods and feed inside the pods. Larval feeding creates reticulate leaves. Pupae and adults appear in mid-July. Various stages of larvae but not pupae are found in late July. Most larvae mature by early August. Generations overlap afterwards which results in unsynchronized pest stages. Damage to host trees lasts until September or October.

Prevention and treatment

1. Mechanic control. Remove and destroy infested leaves by hand under low population levels.

2. Biological control. One parasitic wasp species was found rather abundant and should be protected.

3. Chemical control. Spray with 20% thiamethoxam at 1,000×, 1.8% abamectin EC at 3,000×, 45% profenofos and phoxim mixture at 1,000×, or 2×10^9 PIB/mL nuclear polyhedrosis viruses at 1000-1,500× during early instars.

Reference Xu Gongtian 2003.

(Xu Gongtian, Qiao Xiurong)

C. chrysolampra larva (Xu Gongtian)

C. chrysolampra damage to willow leaf (Xu Gongtian)

C. chrysolampra damaged leaves (Xu Gongtian)

272 *Gibbovalva urbana* (Meyrick)

Classification Lepidoptera, Gracillariidae
Synonym *Acrocercops urbana* Meyrick

Importance *Gibbovalva urbana* is a leaf miner with a broad host spectrum. Mining inside host leaves brings damage to important ornamental trees in the Magnoliaceae family, affecting their photosynthesis and aesthetic beauty.

Distribution Beijing, Jiangsu, Zhejiang, Fujian, Jiangxi, Hubei, Guangdong, and Guangxi; Japan, and India.

Hosts Over 70 species in Magnoliaceae family, including *Michelia alba*, *M. figo*, *Magnolia officinalis*, *M. sieboldii*, *M. grandiflora*, *M. soulangeana*, *Liriodendron chinensis*, *Manglietia fordiana*, *Paramichelia baillonii*, and *Tsoongiodendron odorum*.

Damage Young larvae mine inside the leaf and create linear tunnels inside tissue under the epidermis. Older larvae feed on mesophyll and gradually connect all tunnels to form feeding blotches. There are usually frass clumps in the middle of the blotches. Infested leaves wither gradually and fall off.

Morphological characteristics **Adults** Male length 2.94-3.59 mm, with front wing length of 3.15-4.00 mm; female length 3.28-3.74 mm, with a front wing length of 3.47-4.20 mm. Adults with white heads and genae. Vertex scattered with a few gray or brown scales. Labial palpus has four black blotchy rings, curved upwards passing vertex. Maxillary palpus yellowish at the base and black apically. Antennae grayish brown with white base. Fore wing slender, primarily white, almost parallel between anterior and posterior margins, with 4 oblique bands running from the anterior margin to the posterior margin. The band closest to wing base does not reach posterior margin, whereas the band close to the apex is relatively wide and separated by white marks anteriorly. There is a row of three black blotches between the adjacent bands. Wing apex has a brownish black blotch that marked with a white spot in the center. Marginal hairs brownish. The anal area of the fore wing extends outwards to form an acute angle. The margin brown. Hind wing lanceolate, deep gray, with long marginal hairs. Abdomen covered with black scales on the anterior margin of each segment. Legs and spurs robust. Legs with alternate black and white bands. The apical half of the middle tibia has scale bundles arranged in spindle shape on the ventral side. Male genitalia with a fingerlike protrusion on the back of the clasper. **Eggs** Length 0.51-0.56 mm, width 0.33-0.40 mm, oblong oval when newly laid, transparent like liquid drops; turning to yellowish white, lightly elevated with shiny surface later. **Larvae** Mature larvae length 5.05-6.84 mm, width 0.84-1.08 mm. Head width 0.55-0.58 mm. Body orange yellow, become peach red to red before construct cocoons. Head yellowish brown. Abdomen lighter in color ventrally. Early instar flat, yellowish white, translucent. Head flat, near triangular, forming an oval together with enlarged pronotum and mesonotum. Pronotum is the widest, tapering towards the end of the body. Mouthparts well-developed, extended forward, trapezoid shaped, yellowish brown. Abdomen intersegments obviously depressed. The 1st to 8th abdominal segments near oval, while the 9th and 10th segments slender. Body covered by dense and

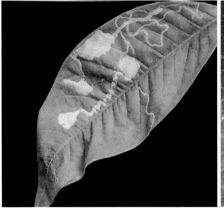

G. urbana damage (edited by Zhang Runzhi)

G. urbana egg (edited by Zhang Runzhi)

G. urbana mature larva (edited by Zhan Runzhi)

semi-transparent white setae. There are two ocelli. Ocular region and ecdysial suture dark brown. There are three thoracic legs, with grayish black blotches. There are three pairs of prolegs, crotchets not well developed, and one pair of backward extended anal legs. The last three segments on abdomen narrower. **Cocoons** Length 7.0-8.77 mm, width 3.64-5.0 mm, yellowish white to yellowish brown, oblong oval, with a few dirty white granules on the surface. **Pupae** Length about 4.77 mm, width about 0.82 mm. Yellow with red compound eyes. Frons contains a red acute projection. Antennae about 0.80 mm long, extending beyond the tip of the abdomen.

Biology and behavior *Gibbovalva urbana* occurs several generations a year in South China Arboretum in Guangzhou with overlapping generations. It starts to feed in March and the first generation causes damage from early to late March. The second generation feeds from late March to mid-April. Adults do not need supplemental feeding and start to lay eggs one day after mating. Eggs are laid singly on the surface of tender leaves. Adults rest on their hind legs and wing tips, with its fore- and middle legs holding up the body to form a 45°angle. Antennae normally vibrate in high speed, slow down only before death. Adults skip for short distance when disturbed. According to laboratory rearing at the average temperature of 23.5°C (ranged from 20.5°C to 28.2°C) conducted during March to April in 2003, egg stage lasted for 2-4 days, larval stage lasted for 7-12 days, pupal stage lasted for 10-12 days. In another experiment at the average rearing temperature of 27.1°C (ranged from 19.8°C to 32.2°C) conducted in May in 1998, duration for egg, larval, pupal stages was 2-3, 8-9, and 8-9 days, respectively. Newly hatched larvae bore directly under the epidermis right after consuming the eggshells and start to construct tunnels. Larval tunnels for first instars linear and narrow, and become wider for the second instars. Head capsules for the first two instars are left inside the linear tunnels. By the third instar, tunnels enlarged and become feeding blotches. Larvae of the 1st to 3d instars feed on leaf tissues under the epidermis, while larvae older than 4th instar start to feed on mesophyll. Head capsules of the 3rd and 4th instar larvae are left inside the blotches near the edges, with the frass stick together to form a mass in the center. There are five instars in the larval stage. A mature larva chews a crescent opening, exits from the edge of the blotch one month later, descends to leaf litter on the ground via a silk thread, forms a cocoon, and pupates inside the cocoon. In the laboratory, cocoons are formed at the corners of the containers. Pupation occurs two days after the cocoon is formed. Adult longevity is 2-4 days.

Prevention and treatment

1. Improve cultivation to promote tree growth and early stand closure. Remove leaf litters, grass, and bushes during the winter.

2. Mechanic control. Use black light to trap and kill adults during their peak emergence.

3. Protect natural enemies. Larval parasitoids include *Litomastix* spp. and *Apanteles* sp. Pupal parasitoids include *Pediobius* spp. Predators include *Polistes rothneyi grahami*, *Hierodula patellifera*, and members of Asilidae. Avoid non-target effect to *G. urbana* natural enemies listed above.

4. Chemical control. Spray with 50% phoxim EC, 40% isocarbophos EC, or 2.5% deltamethrin at 2,500-4,000×, 45% profenofos and phoxim mixture at 1,000×, 12% thiamethoxam and lamda-cyhalothrin mixture at 1,000×, or 2×10^9 PIB/mL nuclear polyhedrosis viruses at 1000-1,500× during larval stages resulted in good control.

References Kuroko 1982; Kumata and Kuroko 1988; Lan Siwen, Wong Xizhao, Huang Jincong et al. 1993; Huang Bangkan 2001; Wu Yousheng and Gao Zezheng 2004.

(Wu Yousheng, Gao Zezheng)

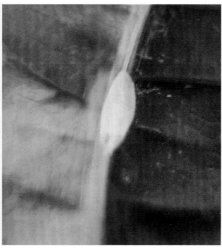
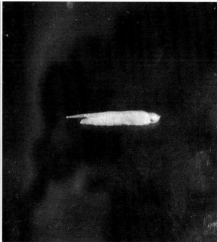

G. urbana adult (edited by Zhang Runzhi)   *G. urbana* cocoon (edited by Zhang Runzhi)   *G. urbana* pupa (edited by Zhang Runzhi)

273 *Melanocercops ficuvorella* (Yazaki)

Classification	Lepidoptera, Gracillariidae
Synonym	*Acrocercops ficuvorella* Yazaki
Common name	Fig leafminer

Importance *Melanocercops ficuvorella* is a leaf miner capable of creating linear or blotch tunnels on host leaves. Although feeding damage is not enough to kill the fig trees, it can affect photosynthesis and aesthetic value of the trees.

Distribution Guangdong; Japan.

Hosts *Ficus microcarpa, F. virens, F. hispida, F. carica, F. pumila*, and *F. erecta*.

Damage Larvae bore into the leaf and feed inside from where the egg is attached, resulting in linear tunnels. Tunnels increase in size as the larvae grow and form white (dirty yellow in late stage) vesicular blotches. Larvae feed and defecate inside the blotches, leaving frass particles scattered inside.

Morphological characteristics **Adults** Females are 2.72-2.95 mm long with front wing length of 2.63-3.10 mm, whereas males are 2.27-2.77 mm long with front wing length of 2.21-2.85 mm. Grayish brown, shiny. Gena light colored. Labial palpi slender and slightly curved upwards. Antennae thin and long, yellowish brown. Compound eyes reddish brown with black marks. Fore wings grayish brown, apical 1/4 lighter colored, with a round black blotch in the middle. Tibia with alternate dark brown and white rings. Hind tibia contains a row of long spiny hairs dorsally and a pair of spurs at 0.5 mm from the base. One of the spurs measured 0.72 mm, about 0.65 times the tibial size. Abdomen with alternating grayish brown and white colors. Scaly hairs on abdomen black apically, truncate in females and mixed with yellowish white scaly hairs in males. Male claspers wide at the middle, narrow and obtuse apically, with several downward curved hairy spines. There is one row of approximately 20 long setae on dorsal surface of the basal one third of the claspers. Setae longer than the end of claspers. **Eggs** 0.37-0.41 mm long and 0.31-0.33 mm wide, oval, yellowish white first, near transparent, with reticulate punctures. **Larvae** Young larvae yellowish white, translucent, narrow anteriorly and wide pos-

M. ficuvorella damage (edited by Zhang Runzhi)

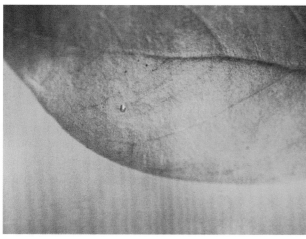
M. ficuvorella egg (edited by Zhang Runzhi)

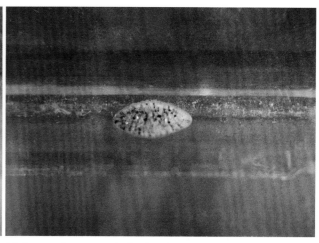
M. ficuvorella cocoon (edited by Zhang Runzhi)

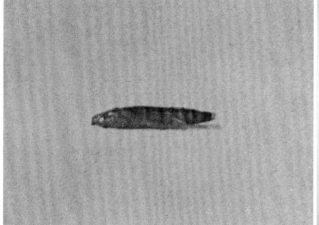
M. ficuvorella pupa (edited by Zhang Runzhi)

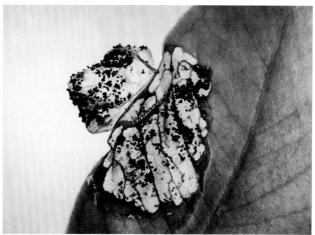
M. ficuvorella larva (edited by Zhang Runzhi)

teriorly. Head near triangular, brown. Mouthparts oval, extend forwards. Head capsule semi-circularly depressed posteriorly. Mature larvae 4.40-4.56 mm long and 0.53-0.58 mm wide, yellowish brown with slight reddish shade. Body shrinks before pupation and turns to peach red. Majority part of the thoracic segments yellow posteriorly. Head yellowish brown, 0.37-0.45 mm in width. There are three pairs of thoracic legs. Prolegs are found on abdominal segment 3-5. There is one pair of anal legs. **Cocoons** 5.87-7.01 mm long and 2.64-3.25 mm wide. Dirty white to yellowish brown, oval, with a few dirty white and black particles on the surface. **Pupae** 3.28-3.53 mm long and 0.56-0.65 mm wide. Yellowish, with wide peach red band on each abdominal segment, Compound eyes peach red or dark colored.

Biology and behavior Larvae of *M. ficuvorella* start to feed on young leaves in late April in Guangzhou area, with two damage peaks in late August and late October to November. Females prefer to lay eggs on the surface of young leaves before they turn to dark green. Eggs are laid singly, with 1-3 eggs on each leaf. Newly hatched larvae bore directly into the leaf from the bottom of the egg after consuming the eggshell, creating linear tunnels initially. Tunnels increase in size as the larvae grow and form white (dirty yellow in late stage) vesicular blotches. Larvae feed and defecate inside the blotches, leaving frass particles scattered inside. There are five larval instars. Mature larvae chew arc-like holes at the edge of the blotches before coming out and fall to the ground. They produce cocoons in leaf litters or crevices. In the laboratory, cocoons are usually found at corners of the rearing containers. No supplemental feeding is needed for adults after emergence and before mating. Adults are fast fliers. They rest on their hind legs and wing tips, with their front and middle legs holding up the body to form a 45° angle. Antennae normally vibrate in high speed. Larval parasitoid includes one species in Eulophidae.

Prevention and treatment See entry for *G. urbana*.
Reference Wu Yousheng and Gao Zezheng 2004.

(Wu Yousheng, Gao Zezheng)

274 *Leucoptera sinuella* (Reutti)

Classification Lepidoptera, Lyonetiidae
Synonym *Leucoptera susinella* (Reutti)

Importance *Leucoptera sinuella* is a major pest of poplar trees. Leaves of infested trees turn black and dry from mining by this pest, resulting in premature falling of the leaves during heavy infestation, and subsequently negative impact on the growth of the tree.

Distribution Beijing, Hebei, Inner Mongolia, Liaoning, Jilin, Heilongjiang, Shandong, and Henan; Japan, Russia, Western Europe.

Hosts *Populus tomentosa*, *Populus* sp., *Salix* sp.

Damage Larval mining inside the leaf creates interconnected feeding blotches during outbreaks, which lead to large areas of brown canopy.

Morphological characteristics **Adults** Length 3-4 mm with a wingspan of 8-10 mm. Vertex white with slight milky yellow on top, with a bundle of white long hairs. Base of antennae wide, contains dense white long hairs that usually cover the compound eyes. Proboscis short, reaches only a third of the fore femur. Thorax white. Forewings white and shiny, with four brown marks near the top. Anal angle contains a near triangular blotch with black top and bottom, while the rest of the blotch silver colored. Hindwings white, narrowly lanceolate with long marginal hairs. Legs grayish white. Mid-tibia contains a pair of apical spurs with one longer than the other. Hind tibia contains backward-inclining long hairs and one pair of median spurs and one pair of apical spurs. Abdomen white. **Eggs** Length about 0.3 mm; oblong oval; surface reticulated. **Larvae** Length about 6.5 mm at maturity, milky white. Body flat, head relatively small, and

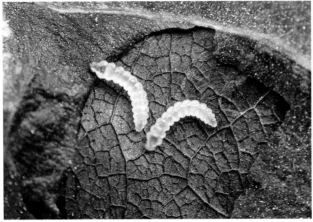
L. sinuella larvae (Xu Gongtian)

L. sinuella early stage damage to leaf (Xu Gongtian)

L. sinuella late stage damage to leaf (Xu Gongtian)

L. sinuella cocoon (Xu Gongtian)

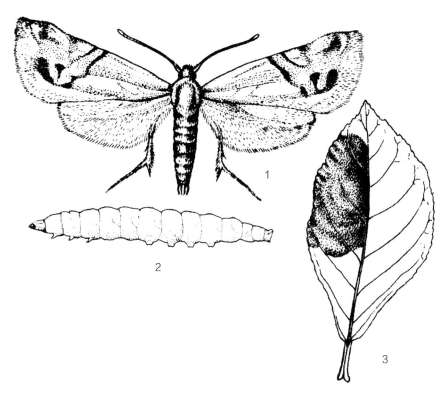

L. sinuella. 1. Adult, 2. Larva, 3. Damage (1, 3 by Shao Yuhua; 2 by Zhu Xingcai)

mouthparts brown and extends forward. Thoracic nota milky white. Thoracic legs relatively small. Locomotion depends mainly on the expansion and contraction of body segments. There are three long hairs on each side of the head and each body segment. **Pupae** Length about 3 mm, fusiform, yellowish, and hiding inside H-shaped white cocoons.

Biology and behavior *Leucoptera sinuella* occurs three generations a year and overwinters as pupae inside infested leaves or bark crevices in Beijing. Adults of various generations appear in early May, late July, and late August. Adults prefer to stay on the glands at the base of poplar leaves and are attracted to light. They mate and lay eggs right after emergence on the same day, with a peak mating time between 11:00-16:00. Females stay for about half an hour before seeking suitable egg sites by moving back and forth on leaf surface. Eggs are generally laid in masses of 3-5 eggs/mass on the upper surface of the middle-aged leaves, aligned with the major vein or side veins. Each female produces an average of 40 eggs. Hatching rate is high with eggs from the same mass hatch during the same day. Newly hatched larva bores directly from the bottom of the eggshell into the mesophyll. Mature larvae can feed through side veins although they cannot pass the major vein. Feeding blotches are filled with frass and appear black. Interconnected small feeding blotches form a large brownish black dead spot on the leaf, which could lead to the death and consequent drop of the entire leaf. Mature larva chews out from the upper surface and makes an H-shaped white cocoon on the underside of the leaf and pupate inside during the growing season, while cocoons of the overwintering generation are usually found on the upper surface of the leaves or inside tree crevices. Mature larvae of the last generation construct cocoons and pupate in tree crevices, under thick bark, or on leaf litters.

Prevention and treatment

1. Remove fallen leaves and burn them or dump them in compost pits during the winter and spring season to destroy overwintering pupae and reduce source population.

2. Trap adults with black light traps during adult period in nurseries, fast-growing stands, and forest farms.

3. Spray emamectin benzoate at 1,500-2,000×, avermectin at 2,000-2,500×, hexaflumuron at 2,000-3,000×, Dimilin at 600-800×, or a mixture of 0.18% avermectin and 10^{10} spores/mL *B.t.* at 1,000-2,000×, 45% profenofos and phoxim mixture at 1,000×, 12% thiamethoxam and lamda-cyhalothrin mixture at 1,000×, or 2×10^9 PIB/mL nuclear polyhedrosis viruses at 1000-1,500×.

References Xiao Gangrou (ed.) 1992, Zhao Fanggui et al. 1999, Sun Shuigen et al. 2001.

(Wang Haiming, Zhang Shiquan)

275 *Paranthrene tabaniformis* (Rottemburg)

Classification Lepidoptera, Sesiidae
Common name Dusky clearwing

Distribution Beijing, Hebei, Shanxi, Inner Mongolia, Liaoning, Jilin, Zhejiang, Henan, Shaanxi, and Xinjiang in China; Russia.

Hosts *Populus* spp. and *Salix* spp.

Damage Mostly occur on seedlings in nurseries. Larval boring inside seedlings creates galls that make seedlings vulnerable to wind damage.

Morphological characteristics **Adults** Length 11-20 mm with a wingspan of 22-38 mm. Forewings slender, dark brown, with a slightly transparent median cell and posterior margin. Hindwings completely transparent. Abdomen greenish black, with five orange yellow rings. **Larvae** Length 30-33 mm, yellowish white, with two dark brown anal spines pointed slightly upward.

Biology and behavior *Paranthrene tabaniformis* occurs one generation (occasionally two generations) a year and overwinters as larvae inside the galleries in Henan and Beijing. Overwintering larvae resume feeding in April and start to pupate in late May and early June. Adults begin to emerge in early June (or in early to mid-May in Beijing) with a peak emergence between late June to early July. About one third of the pupa length extends outside the exit hole during adult emergence. Pupal cases remain on the exit holes for a long period. Adults fly, mate and lay eggs during the day and rest at night. Eggs are laid mostly at the base of leaf stalks or in branch crevices. Egg stage lasts for 10 days. Newly hatched larvae bore directly into the bark or enter the seedlings from the wounds. They feed between the cambium and sapwood. After circling around the trunk, the larva tunnels upwards along the trunk and pushes frass and sawdust out of the entrance hole. Larval gallery is 2-10 cm long. Larvae rarely migrate after entering the trunk. The infested area gradually grows into a gall, which makes the trunk vulnerable to wind damage. If the trunk is broken by wind, the larva migrates to another suitable location for secondary invasion. Larva feeds until the fall when it constructs a thin cocoon at the end of the gallery for overwintering.

Prevention and treatment

1. Quarantine. Thoroughly inspect seedlings and cut branches before exportation and remove galls to prevent spread.

2. Mechanic control. Remove galls with a pocketknife during larval invasion by searching for sawdust and

P. tabaniformis adult (Li Zhenyu)

P. tabaniformis larva (SFDP)

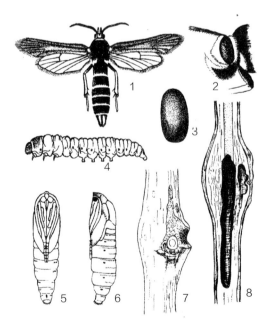

P. tabaniformis. 1. Adult, 2. Lateral view of adult head, 3. Egg, 4. Larva, 5. Pupa, 6. Lateral view of pupa, 7-8. Damage (Xu Tiansen)

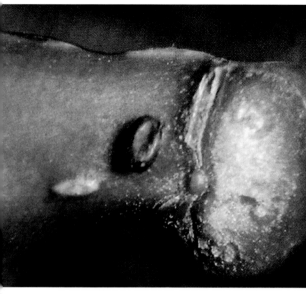
P. tabaniformis egg (SFDP)

P. tabaniformis damage (SFDP)

P. tabaniformis pupa (SFDP)

P. tabaniformis empty pupa (SFDP)

P. tabaniformis larval damage (SFDP)

small galls along the trunk. Cut off branches from below the galls during heavy infestation to promote sprouts.

3. Pheromone trapping. Trapping with sticky trap baited with poplar clearwing sex pheromone – (E, Z)-3, 13-octadecadien at 200 mg / lure rate during adult stage. Place traps at 1 m above the ground inside the forest or along forest edges effectively attracted males within 100-150 m.

4. Chemical control. Inject 25 g/L lamda-cyhalothrin at 500× to trunks during larval invasion to control larvae, or apply 45% profenofos and phoxim mixture at 39-60×, 12% thiamethoxam and lamda-cyhalothrin mixture at 30-60× to tree trunks.

References Yang Youqian et al. 1982, Xiao Gangrou (ed.) 1992, Zhang Xingyao and Luo Youqing 2003, Forest Insect and Disease Control and Quarantine Station of Henan Province 2005.

(Yang Youqian, Li Zhenyu)

276 *Sesia siningensis* (Hsu)

Classification Lepidoptera, Sesiidae
Synonyms *Sphecia siningensis* Hsu, *Aegeria apiformis* Clerck
Common names Poplar bole clearwing moth, Hornet moth

Distribution Shanxi, Inner Mongolia, Shandong, Yunnan, Tibet, Shaanxi, Gansu, Qinghai, and Ningxia.

Hosts *Populus cathayana, P. nigra* 'Thevestina', *Populus × canadensis, P. hopeiensis, Populus× xiaozhuanica* 'Opera', *P. simonii, Populus × Euranericana, P. alba* var. *bolleama, Salix matsudana.*

Morphological characteristics **Adults** Female length 25-30 mm, wingspan of 45-55 mm. Male length 20-25 mm, wingspan of 45-50 mm. Very similar to hornets in appearance. Wings transparent. Forewings slender. Hindwings fan-like with dark brown marginal hairs. M_3 and Cu of hindwings fused at the base. Abdomen with five alternate yellow and brown rings. Front coxa yellowish red. Hind tibia with a white blotch externally. Female antennae brown, club-like; the tip sharp and slightly pointed backward, with a yellowish brown hair bundle. The end of the abdomen pointed. Male antennae pectinate, straight with pubescence on the outer margin of each segment. Abdomen narrow and small, with a bundle of brown scaly hairs apically. **Eggs** Length 1.2-1.4 mm, width 0.6-0.8 mm. Oblong oval, relatively flat and slightly depressed. Burgundy at first, turn to yellowish brown before hatching. **Larvae** Length 40-45 mm at maturity, cylindrical. Head of newly hatched larva black, body grayish white. Head of mature larva dark brown, body milky yellow. Pronotum with a brown longitudinal groove on each side and two parallel brown blotches near the anterior margin around the median line. Abdomen with a tiny brown spine on the back apically. **Pupae** Length 24-32 mm, fusiform, brown. Abdomen with two rows of tiny spines on each of the segments 2-6 and 10 robust anal spines.

Biology and behavior *Sesia siningensis* occurs one generation every two years. It overwinters as larvae under the bark in the first calendar year. Overwintering larvae become active in late March the next year and feed inside the wood. By early October, larvae overwinter again inside the wood. Larvae pupate in early August of the third year with a peak pupation period of mid-August. Adults emerge in early to mid-August. Peak adult emergence falls in early September in Taigu area of Shanxi, with a larvae period of 22 months. Emergence peak varies among different areas. For example, it falls between late August and early September in Yulin of Shanxi, but with two peaks in

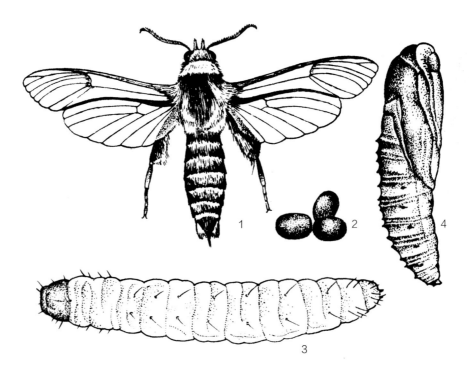

S. siningensis. 1. Adult, 2. Eggs, 3. Larva, 4. Pupa (Zhu Xingcai)

early June and mid-August in Xining of Qinghai.

Adults are active during the day and are not attracted to light at night. Emergence occurs mostly between 9:00-10:00 am. The pupa moves its body before emergence by rubbing the body spines against the wall of the exit hole. This enables the pupa breaks open the exit hole. About 1/3-1/2 of the pupa is exposed outside of the exit hole. The pupal case remains in the exit hole after adult emergence. Eggs are scattered in bark crevices at the base of the trees. Newly hatched larvae bore into the bark through tender tissues in bark crevices of existing wounds, reaching the area between cambium and sapwood by late October for overwintering. They bore deeper into the heartwood by next spring and create vertical galleries align with the trunk. Mature larva feeds transversely back to the bark, chews an oval exit hole in the bark with only intact upper epidermis, and constructs a pupal chamber for pupation. Most larvae feed at the base of the tree along the trunk 30 cm above the ground, while a few can feed on existing wounds and branch joints. Life stages are not well synchronized due to different invasion times.

Prevention and treatment

1. Quarantine. Ban the movement of poplar timber or transplanting of large diameter poplar trees from the infested areas to prevent artificial spread.

2. Trapping adults with sex pheromone such as Z3, Z13-18:OH. Place one trap (800 mg/lure) at the lower crown every 100 m can attract males within 2,000 m. This reduces the average mating rate by 42.8%-47.6%. Continuous trapping for three years resulted in a population reduction of 38.2%.

3. Protect and conserve natural enemies. Apply *Beauveria bassiana* at 20× using cotton balls to larval galleries and seal with mud resulted in 66% infection rate of the larvae.

4. Apply aluminum phosphide at 0.4 g/hole and seal with mud, or insert zinc phosphide sticks in larval galleries resulted in 96% and 89% mortality for larvae and pupae, respectively.

References Xu Zhenguo 1981; Zhang Zhiyong 1983; Li Zhenyu et al. 1989; Xu Shouzhen, Wu Jiangong, Meng Changxiao et al. 1996; Feng Shiming, Zeng Shusheng, Yang Lingxuan et al. 1999; Li Zhenyu 2003; Huang Dazhuang, Li Liangming, Tang Xiaozhen et al. 2004.

(Li Zhenyu, Tao Jing, Xu Zhenguo, Qu Qiuyun, Wang Xinmin)

S. siningensis damaged trees (Li Zhenyu)

S. siningensis pheromone trap (Li Zhenyu)

277 Synanthedon castanevora Yang et Wang

Classification Lepidoptera, Sesiidae

Distribution Beijing, Hebei, and Shandong.

Host *Castanea mollissima.*

Morphological characteristics **Adults** Female length 10 mm, wingspan 19 mm; male length 9 mm, wingspan 16 mm. Black with shiny bluish purple. Head with white scaly hairs at the base and dense white scales along the inner margin of the compound eyes. Wings transparent. Forewings with a black transverse band at the end of the median cell. Females have 6 abdominal segments, with yellow transverse bands on the posterior margin of the 2nd, 4th, and 6th segment. The transverse band on the 4th abdominal segment is wider than the rest. Males have 7 abdominal segments, with transverse yellow bands on the 2nd, 4th, and 6th segment. Some individuals have a small yellow ring of scales on the 7th segment. The end of the abdomen with well-developed fan-shaped black scales. Those scales on the two lateral sides are white. **Eggs** Length about 0.4 mm and width about 0.3 mm, oval and slightly flat, depressed in the middle, dark brown, surface white reticulated. **Larvae** Newly hatched larvae about 0.9 mm in length; white, semi-transparent; head yellowish brown, ocular region reddish brown. Body covered with thin and long setae. Pronotum contains an inverted V-shaped brownish mark from the third instar. Mature larvae 12-15 mm long, milky white with a reddish brown head. There are 1-5 ocelli on the head, which are surrounded by black blotches. Thoracic legs well-developed, yellowish brown. There are four pairs of prolegs with uniordinal biserial band, and one pair of anal prolegs with uniordinal uniserial band. Anal plate the same color as the body and covered by brown marks. **Pupae** Length about 10 mm and width about 2 mm. Yellowish brown originally, turn to dark brown before adult emergence. **Cocoons** Length about 11.5 mm and width about 3.6 mm. Fusiform, covered by brown frass particles, with a round cap on top.

Biology and behavior *Synanthedon castanevora* occurs two generations a year and overwinters as 3rd to 5th instar larvae in Hebei. The overwintering larvae re-

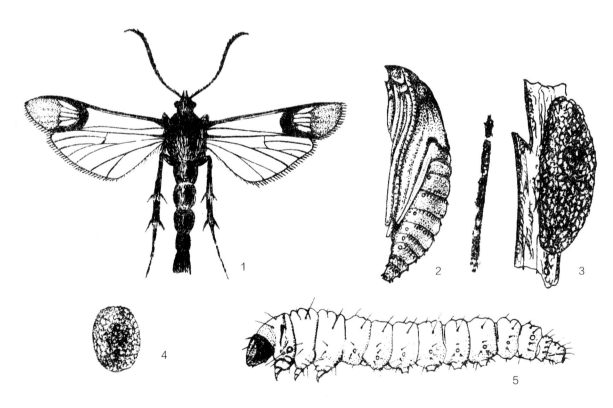

S. castanevora. 1. Adult, 2. Pupa, 3. Cocoon, 4. Egg, 5. Larva (Hu Xingping)

S. castanevora adult (Li Zhenyu)

sume activity in April and pupate in mid-April. Adults begin to emerge in early May. Adult emergence peaks in mid- to late May and ends in late June. The first generation larvae start to hatch from late May to early June and peak in mid-June. They begin to pupate in mid-July and pupation peaks between late July and early August. Peak adult emergence falls between the end of July and mid-August. Most of the overwintering generation larvae hatch between mid- to late August. They feed until early November and then chew deep pits at the end of the galleries and weave oval cocoons to overwinter.

Adults emerge from pupae and are active during the day and rest at night. They are not attracted to light. Females start to lay eggs one day after emergence. Eggs are laid inside bark crevices of the main and side branches, mostly inside feeding sites and existing wounds. Larvae hatch during the day by breaking eggshells from the top. They search rapidly for suitable feeding sites to bore and feed in the cambial region. They seal the crevice with silk or frass. The larvae bore deeper gradually and feed over large areas. Larval feeding leads to bark swelling. The affected bark turns swelled and turns red, with tiny brown frass inside crevices. The swelling increases in size and cracks longitudinally as the larva grows. By this time, silk-linked brown frass fills the area between the bark and the sapwood. The most severe damage is caused by the overwintering larval feeding and feeding of the first generation larvae before maturity. Tree mortality usually occurs in May or July.

Prevention and treatment

1. Protect and utilize natural enemies such as *Apanteles* sp. (with a parasitism of 17%), *Leskia aurea* (with a parasitism between 10%-12%), and *Goniozus sinicus* (with a parasitism of 2%).

2. Trap Adults with sex pheromone such as Z3, Z13-18:OH can result in moderate control.

3. Spray 10% cypermethrin EC at 2,000×, 45% profenofos and phoxim mixture at 1,000×, or 12% thiamethoxam and lamda-cyhalothrin mixture at 1,00× during the egg hatching period to control larvae.

References Wu Peihuan et al. 1988; Liu Huiying, Zhou Qingjiu, and Wu Dianyi et al. 1989; Yang Jikun and Wang Yin 1989; Liu Huiying and Zhou Qingjiu 1992, 1994.

(Li Zhenyu, Tao Jing, Liu Huiying, Zhou Qingjiu)

278 *Glyphipterix semiflavana* Issiki
Classification Lepidoptera, Glyphipterigidae

Distribution Hebei, Liaoning, Jilin, Zhejiang, Fujian, Jiangxi, Henan, and Hunan; Japan.

Hosts *Phyllostachys pubescens* and *Phyllostachys* spp.

Damage Larvae feed on the new leaves of current-year bamboo or new leaf sheaths of the old bamboo, resulting in wilt of the central leaves.

Morphological characteristics **Adults** Length 4-5 mm, wingspan about 14 mm. Grayish black. Forewings dark grayish black, with six short white stripes along the anterior margin between 2/5 from the base and the apical angle. There is a white blotch below the second and third stripe, respectively. Anal angle contains five silvery white spots. **Eggs** Oval, milky white, with longitudinal stripes on the surface. **Larvae** Length 7.2-10.5 mm, purplish green, semi-transparent, with visible white internal tissues. **Pupae** Length 6 mm, with a pair of spines on both sides of the pronotum.

Biology and behavior *Glyphipterix semiflavana* occurs one generation a year and overwinters as mature larvae inside infested sheath in Zhejiang and Jiangxi. Larvae become active in late March and early April when average daily temperature is above 10°C for several consecutive days. They come out from the entrance holes and construct reticular cocoons on leaf sheath, small branches, or bamboo stems; or other suitable locations after searching through crawling or ballooning by silk threads. Larva pupates inside the cocoon. Pupal stage lasts for about 25 days. Adults emerge in mid-April to mid-May when average daily temperature is about 20°C. Females lay their eggs on top of the sheath or leaf ears. Newly hatched larva crawls on the surface and enters the sheath from its base and feeds inside. Entrance hole is very small with a visible dark brown spot 1-2 days after initial invasion. Larva matures in later June and aestivates and overwinters in the host.

Prevention and treatment

1. Chemical control. Cover spray with methamidophos at 2,000× during larvae feeding period resulted in 62.5% and 84.0% mortality after 24 and 48 hours, respectively.

2. Protect natural enemies. The spiders, *Araneus mitificus*, *Oxyopes* sp, and *Pisaura* sp. predate upon larvae when they come out of leaf stalks to make cocoons.

References Lü Ruoqing 1988, Xu Tiansen and Wang Haojie 2004.

(Tong Xinwang)

G. semiflavana damage (Xu Tiansen)

G. semiflavana adult (Xu Tiansen)   *G. semiflavana* mature larva (Xu Tiansen) *G. semiflavana* cocoon (Xu Tiansen)

279 *Atrijuglans hetaohei* Yang

Classification Lepidoptera, Heliodinidae

Distribution Beijing, Hebei, Shanxi, Shandong, Henan, Sichuan, Guizhou, and Shaanxi.

Hosts *Juglans regia* and *J. mandshurica*.

Morphological characteristics **Adults** Length about 6 mm, wingspan about 14 mm; shiny black. Antennae filiform, brown and covered by dense white hairs. Head brown, covered by large silver gray scales. Labial palpi silvery white, slender, and curved passing the vertex. Compound eyes red. Forewings black, with a near oval white blotch at 1/3 length from the wing base. Abdomen silvery white ventrally. Hind legs stout, with three bundles of circular white and black setae.

Eggs Oval, 0.3-0.4 mm long, milky white, turning to reddish yellow before hatching.

Larvae Length 9.5-12 mm at maturity, head yellowish brown, body yellowish with a purplish red blotch in the middle. Prolegs with crochets arranged in a single complete ellipse.

Pupae Length 5-6 mm, width about 2.5 mm, yellowish white originally, turning to black brown before emergence with visible red compound eyes. Antennae reach the end of the wing tips and flush with hind legs. Spiracles protruding and conspicuous.

Cocoons Length 6-9 mm, width 3-6.5 mm, flat oval, brownish with fine soil particles attached externally, earthy colored. There is an obvious reddish or grayish white suture along the widest portion, which often exposed above soil surface.

Biology and behavior *Atrijuglans hetaohei* occurs one to two (Beijing, Sichuan and Shaanxi) or two (Henan) generations a year and overwinters as mature larvae inside cocoons. Adults first appear in Beijing area in early May and peak in late May and early June. A small portion of the larvae bore into the fruit in mid-May and feed inside the cotyledons, leaving no visible damage externally. Larvae can enter at this time as developing fruits contain soft skins. There are usually 1-2 larvae per fruit. Massive fruit dropping occurs from mid-June to early July. Larvae of the overwintering generation cause most damage in mid-August. However, they can only feed on mesocarp as the endocarp hardens by this time. Massive fruit dropping in the first generation reduces number of fruits on the tree, which leads to higher population density in the infested fruits. The average density is five larvae per fruit, with the highest of 25 larvae. Infestation leads to the blackening of meso- and endocarp caused by tannin oxidation. The symptom is commonly referred as "black walnuts".

Atrijuglans hetaohei occurs less on flat plains and sunny slopes. It is more prevalent on shady slopes.

Prevention and treatment

1. Deep tillage in dry and mountainous areas after walnut harvest in the fall will prevent cocoons from developing into adults.

2. Remove fallen fruits from the ground in mid-June can reduce damage in the next generation. Blackened fruits should be removed promptly.

3. Apply 20% fenvalerate or 2.5% deltamethrin at 3,000-4,000× twice during adult stage at 10-15 days interval.

4. Apply nematode *Steinernema feltiae* "Beijing" strain at the density of 90,000-130,000/m^2 in the field when soil moisture is around 13% for controlling larvae inside the cocoons resulted in efficacy of 54%-77.2%.

References Li Zhenyu et al. 1965; Wang Yonghong, Sun Yihe, and Yin Kun 1997; Xu Zhihua 2006.

(Li Zhenyu, Liu Jin)

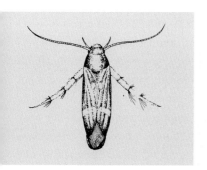

A. *hetaohei* adult (by Zhang Peiyi)

A. hetaohei adult (Li Zhenyu)

A. hetaohei larva and infested nut (edited by Zhang Runzhi)

280 *Stathmopoda masinissa* Meyrick

Classification	Lepidoptera, Heliodinidae
Synonym	*Kakuvoria flavofasciata* Nagano
Common name	Persimmon fruit moth

Distribution Beijing, Hebei, Shanxi, Jiangsu, Anhui, Shandong, Henan, Hubei, Shaanxi, Taiwan; Japan, Sri Lanka.

Hosts *Diospyros kaki* and *D. lotus*.

Damage Larvae feed on fruits as well as young branches. On fruits, they enter from the stems or the base of the stem ends. Larval feeding leads to dropping or wilting ("mummy fruits") of young fruits, and premature yellowing or dropping ("baked persimmon", "red-faced persimmon") of older fruits, which significantly reduce persimmon production.

Morphological characteristics

Adults Females about 7 mm in length, wingspan 15-17 mm. Males slightly smaller. Head and labial palpi yellowish brown, shiny. Compound eyes reddish brown. Antennae filiform, with short hairs on each segment in males. Anterior part of the thorax yellowish in the center. Wings slender, purplish brown, with long marginal hairs. Marginal hairs on hindwings much longer. Each forewing has an oblique yellow stripe near the apical angle. Legs and abdomen yellowish brown. Hind legs long, extending upwards and backwards at rest. Tibiae with dense long hairs similar in color to those found on the wings.

Eggs Oval, about 0.5 mm, milky white first, turning

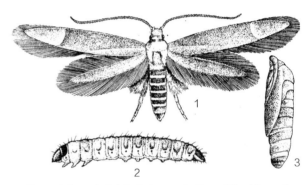

S. masinissa. 1. Adult, 2. Larva, 3. Pupa (Zhu Xingcai)

to pinkish later, with fine longitudinal ridges externally. Ridges covered by short white hairs.

Larvae Newly hatched larvae 0.9 mm in length; head red, thorax brownish. Mature larvae 9-10 mm in length; head brown, pronotum and anal plate dark brown; body purplish dorsally. Meso- and metanotum with an "X"-shaped wrinkled mark, each with a row of transverse verruca in the center. Meso- and metathorax, as well as the first abdominal segment lighter in color with small black dots. There is a transverse wrinkle on each abdominal segment. Each verruca contains a small white hair. Thoracic legs yellowish.

Pupae Length about 7 mm, brown.

Cocoons Oval, about 7.5 mm long, dirty white, attached with tiny wooden pieces and frass.

Biology and behavior

Stathmopoda masinissa occurs two generations a year and overwinters as mature larvae inside cocoons in bark crevices, in 5-10 cm deep soil around base of the trees, or inside infested fruits, with 27.1%, 51.6%, 9.28%, and 2.06% of cocoons found at the base of the trees, main trunks, main branches, inside stem ends, respectively. Larvae start to pupate in mid- to late April the following year. Pupal stage is about 19 days. Adults start to emerge by the end of April or early May with a peak emergence from mid- to late May (7-8 days after initial persimmon flowering to 2-3 days after flowering ends). Larvae of the first generation start to feed on young fruits in late May with most damage found in June. Adults of the first generation appear in

S. masinissa adult (Mu Xifeng)

S. masinissa damage (Mu Xifeng)

S. masinissa infested and non-infested persimmon (Mu Xifeng)

S. masinissa overwintering pupa under bark (Mu Xifeng)

S. masinissa larvae (Mu Xifeng)

S. masinissa damage (Mu Xifeng)

S. masinissa (right) and *Conogethes punctiferalis* (left) on a peach (Mu Xifeng)

early July. The peak emergence period occurs between mid- July and early August. Larvae of the second generation appear in early August. They feed on fruits until harvest. Larvae start to overwinter by the end of August.

Adults are attracted to lights. They stay on the underside of the leaves or dark locations during the day and become active at night. They are most active between 20:00-24:00. Eggs are laid mostly in crevices between stems and stem ends, on the stems, or on the outer margin of stem ends. Eggs are laid singly. Each female can produce about 30 eggs. Egg stage lasts for about 7 days. Newly hatched 1st generation larvae mostly bore into fruits from stems, leaving frass attached by silk outside the entrance holes. Infested fruits turning from green to grayish brown and eventually dry out and remain on the trees. Larvae can migrate to other fruits, especially during high temperature and high humidity. Each larva can damage 3-6 fruits. Mature larvae construct cocoons in crevices on stem ends, inside the fruits, or in the bark crevices for pupation. Larvae of the second generation usually feed on fruit meat under the stem ends, resulting in red and soft fruits and premature dropping.

Prevention and treatment

1. Attract overwintering larvae by covering main trunk and main branches with grass in early August. Remove and burn the grass to destroy the overwintering larvae inside.

2. Till soil to 10 cm deep at the base of the trees before winter to destroy overwintering larvae.

3. Scrape off the old bark before budding, remove and destroy residual stem ends and dry fruits from the trees.

4. Trap adults with black lights during adult period and remove infested fruits between June and August to suppress pest population.

5. Apply 2.5% fenpropathrin at 2,000×, fenvalerate + malathion at 2,500×, or 25% Dimilin at 2,000× as cover spray during peak adult period (larvae hatching period).

6. Protect the pupal parasitoid *Lissonota* sp.

References Wang Pingyuan, Wang Linyao, Fang Chenglai et al. 1983; Xiao Gangrou (ed.) 1992; Zhu Hongfu and Wang Linyao 1997; Ming Guangzeng, Fan Xiumin, Zhao Min et al. 2002; Wang Sheceng, Gao Jiusi, and Xue Minsheng 2008, Zhang Zhixiang 2008.

(Mu Xifeng, Wei Xueqing, Liang Minghan)

281 *Prays alpha* Moriuti
Classification Lepidoptera, Yponomeutidae

Importance *Prays alpha* is an important pest of leaves and tender branches of young ash trees. Larvae bore inside new buds, tender branches, and leaves, resulting in bud and branch tip wilt and leave yellowing from shot holes during light infestations, and abnormal development of young trees such as multiple trunks under heavy infestations. Consecutive infestations may lead to tree mortality.

Distribution Heilongjiang; Japan.

Hosts *Fraxinus mandshurica* in China, and *Juglans mandshurica* in Japan.

Damage Entrance holes on new buds, tender branches, or leaf stalks are covered by silk and brown frass after larval invasion. Infested buds and branches turn black and wilt gradually, leaving multiple heads on each branch commonly known as "five flower heads". Larvae of the first generation slightly roll the leaves with silk and feed on mesophyll and upper epidermis underneath, sparing only the lower epidermis. Larval feeding creates irregular brown spots on the leaves, resulting shot holes that lead to leaf shrinking from moisture loss or leaf wilt.

Morphological characteristics **Adults** Wingspan 12-18 mm, grayish white dorsally and silver ventrally. Legs silver in color. Forewings grayish brown for the anterior half, and white for the posterior half, with a large triangular brown blotch in the center. There is a small brown blotch at the anal corner and scattered brown stripes along the posterior margin. Basal corner of hindwings silver in color, darkening to brown gradually towards the outer margin. Marginal hairs long.

P. *alpha* adult (Li Chengde)

Larvae Length 10-11 mm at maturity. Yellowish green. Dorsal median line brick red and thin. Each body segment contains an irregular brick blotch on either side of the tergum anteriorly.

Biology and behavior *Prays alpha* occurs two generations a year and overwinters as first instar larvae. Overwintering larvae become active in mid-May and bore in tender branches, new buds, and base of leaf stalks from May to June. By mid-June, mature larvae come out of the tunnels and spin thin web-like cocoons between leaf stalks, on small branch joints or on leaf surfaces and pupate inside. Adults of the overwintering generation and larvae of the first generation appear from late June to early July. They feed on leaves by slightly roll the leaves with silk from July to August. Larvae mature in late July and spin thin cocoons on leaf surface for pupation. Adults of the first generation appear from late July to mid-August. Eggs are laid singly on branches and leaves. Larvae of the overwintering generation appear in mid-August and hide inside the scales of top or side buds for overwintering. Infestation is usually heavier in pure forests than in mixed forests. Damage occurs mostly on branch tips.

Prevention and treatment

1. Silviculture. Plant mixed forests. Manchurian ash suffers less from this pest in forests mixed with larch. Increase canopy cover also helps reduce infestations.

2. Mechanic control. Cut off branches and leaves that contain insect nests and burn to destroy them.

3. Biological control. Protect and utilize natural enemies which include spiders and parasitic wasps *Apanteles* spp. in the larval stage and *Apanteles* sp. and species of Pimplinae in the pupal stage. Spread *Beauveria bassiana* during winter. Larvae can also be controlled with 2×10^{10} PIB/mL nuclear polehydrosis virus.

4. Chemical control. Cover spray with 25 g/L lamda-cyhalothrin at 1,500-2,000×, 50% phoxim EC or 50% fenomifos EC at 1000×, 45% profenofos and phoxim mixture at 1,000×, or 12% thiamethoxam and lamda-cyhalothrin mixture at 1,000×.

Reference Xiao Gangrou (ed.) 1992.

(Byun Bong-Kyu, Li Chengde, Qian Fanjun)

282 *Yponomeuta padella* (L.)

Classification Lepidoptera, Yponomeutidae
Synonyms *Phalaena padella* L., *Hyponomeuta variabilis* Zeller
Common name Cherry ermine moth

Importance *Yponomeuta padella* is an important defoliator of fruit trees with frequent outbreaks observed in orchards. Fruit trees completely defoliated during outbreaks appear damaged by forest fire. All fruits on infested trees fall to the ground due to dryness and wilt. Damage by this pest seriously affects fruit production in the coming year, resulting in significantly economic losses.

Distribution Beijing, Tianjin, Hebei, Shanxi, Inner Mongolia, Liaoning, Jilin, Heilongjiang, Jiangsu, Shandong, Shaanxi, Gansu, Qinghai, Ningxia, Xinjiang; Korea Peninsula, Japan, Mongolia, Russia, Europe, and North America.

Hosts *Malus pumila*, *M. spectabilis*, *M. baccata*, *M. asiatica*, *Crataegus pinnatifida*, *Pyrus bretschneideri*, *Prunuspp pseudocerasus*, *P. armeniaca* and other plants in the rose family Rosaceae.

Damage Infested trees contain silk nests that cover the entire crown when population is high. The infested trees appear to be covered with plastic films in foggy mornings.

Morphological characteristics **Adults** Wingspan 19-22 mm. Body white with silk luster. Thorax contains five black spots dorsally. Forewings contain 35-45 small black dots arranged in lines. Hindwings grayish brown.

Larvae Length 13.7 mm at maturity. Body dark grayish brown. Ocelli round and raised. Mandibles five-teethed.

Biology and behavior *Yponomeuta padella* undergoes obligatory diapause and occurs one generation

Y. padella male adult (Xu Gongtian)

Y. padella female adult (Xu Gongtian)

Y. padella young larva (Xu Gongtian)

Y. padella mature larva (Xu Gongtian)

Y. padella damage (Xu Gongtian)

a year around China. It aestivates and overwinters as first instar larvae inside eggshells. Overwintering larvae come out from the eggshells and start to feed between April and May, coincides with flower blooming and separation of inflorescence for apple trees. There are five larval instars. Mature larvae spin thin and transparent silk cocoons inside the nests for pupation between the end of May and early June. Most adults emerge in mid-June. Females lay eggs until early July. Eggs are laid as egg masses on smooth bark of the second year branches. Eggs hatch in about 13 days, with first instar larvae aestivate and overwinter inside eggshells.

Overwintering larvae tie young leaves together with silk and feed on leaf mesophyll. They re-emerge from partially fed leaves and tie new leaves and feed among the tied leaves in the same manner. Larvae consume all leaves in the nest before expanding their nests. Continuous expansion results in gigantic nests over time.

Prevention and treatment

1. Mechanic control. Cut branches and leaves that contain insect nests and burn to destroy them.

2. Biological control. Cover spray with $(36-84) \times 10^8$ spores/mL *B.t.* suspension to control 3rd to 5th instar larvae. Protect and utilize parasitoids such as *Herpestomus brunnicornis*, *Coccygomimus disparis*, and *Iseropus kuwanae*.

3. Chemical control. Cover spray with 25 g/L lamda-cyhalothrin at 1,500×, 2.5% deltamethrin 2,000×, 40% dimethoate 1,500×, or phoxim at 1,000×, 45% profenofos and phoxim mixture at 1,000×, or 12% thiamethoxam and lamda-cyhalothrin mixture at 1,000× to control 1st and 2nd instar larvae.

References   Xiao Gangrou (ed.) 1992; Fang Yan 2000; Zhao Lianji, Zhao Bo, Lu Chengjuan et al. 2000.

(Byun Bong-Kyu, Li Chengde, Bai Jiuwei)

283 *Atteva fabriciella* Swederus

Classification Lepidoptera, Yponomeutidae
Common name Ailanthus webworm

Distribution Guangdong, Guangxi, Sichuan, Yunnan, Macau; Thailand, Indonesia, the Philippines, India, Sri Lanka.

Hosts *Ailanthus excels*, *A. triphysa*, *Brucea javanica*, *B. sumatrana*.

Damage Larvae feed on leaves inside silk nests.

Morphological characteristics **Adults** Length 12.5-13 mm for males and 12.0-12.5 mm for females. Wingspan about 13 mm for males and 14 mm for females. Vertex white but dark brown between antennae. Frons tightly covered by silver scales. Labial palpi white. Antennae scape white; first 2–3 flagellomeres entirely silvery white, the rest of the flagellomeres gray in basal half, white in distal half. Patagium and tegula white, narrowly tinged with orange basally. Mesonotum orange, with a white marking posteriorly. Forewings yellowish brown with white marginal hairs, shiny, and contains 33-46 white spots of various sizes and shapes. Hindwings orange with dirty yellowish brown marginal hairs, lightly colored at the base and yellowish brown on the outer margin. Front legs brownish black, whereas mid- and hind legs lighter colored gradually. Hind legs in males white, with long hairs on tibia and the apical half of the tarsus. Abdomen is the same color as the forewings, with white posterior margin for each segment. The white posterior margins expand to triangle white spots in males.

A. fabriciella adult (edited by Zhang Runzhi)

A. fabriciella larvae (edited by Zhang Runzhi)

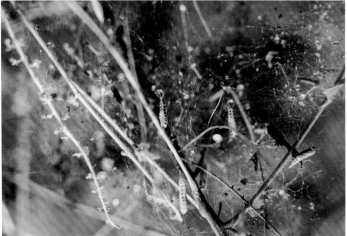
A. fabriciella pupae (edited by Zhang Runzhi)

A. fabriciella eggs (edited by Zhang Runzhi)

Eggs Length 0.73-0.75 mm, width about 0.42 mm, yellowish, irregularly oval and slightly flat.

Larvae Mature larvae 26-28 mm long with a head capsule of 2.0-2.7 mm wide. Yellowish and smooth. Head and thoracic legs brownish black. Antennal base, frons, clypeus, gena and subgena have similar color as the body. The brownish black portion of vertex in females serrate. Thoracic segments with brownish black bands surrounding the body. Each abdominal segment contains four irregularly shaped brownish black blotches that form an incomplete band due to closeness to each other. Thoracic legs well developed. There are five pairs of abdominal prolegs, with multiordinal circle crotches for the first four pairs and multiodinal penellipse crotches for the anal pair.

Pupae Length 15.4-17.1 mm, width 3.05-3.60 mm. Yellowish at first. Head and wing pads slightly colored. Compound eyes reddish brown, with one crescentic black mark on both sides of the head on males. There are four black marks each on abdominal segments 5-8, and two each on segments 9-10. Abdominal black marks linear in males. For females, there is only one pair of small black marks on the tergum of abdominal segment 9. Wing pads of male pupae contain scattered linear black marks.

Biology and behavior *Atteva fabriciella* occurs two generations a year in Guangzhou area with a damage period of May to July. Adults rest inside host plants during the day, which make them difficult to be found. They are active during the night. Females lay eggs 3-4 days after emergence. Eggs are scattered on leaf surface. Egg period lasts for about 13 days. Larval stage lasts for 4-12 days with five instars. Adults live for about 10 days. Newly hatched larvae feed aggregately on leaf mesophyll, leaving only upper epidermis intact. Young larvae start to construct loose nests by spinning silk around leaves and branches and feed on leaves inside. All leaves inside the nests are consumed as larvae grow. Nests are then expanded. Entire branches are converted into loose nests during heavy infestation, with leaves inside consumed completely, leaving leafless branches. Mature larvae spin cocoons and pupate in the silk webs. Pupae hang upside-down in the nests.

Prevention and treatment

1. Mechanic control. Cut branches and leaves that contain insect nests and burn to destroy them.

2. Microbial control. Cover spray with *B.t.* WP (8,000 IU/ul) at 150-200× or *B.t.* var. *galleriae* powder (> 1 × 10^{10} spores/g) at 200-300× resulted in significant control. Better results were achieved when trichlorfon and *B.t.* var. *galleriae* mixtures were used. Larvae can also be controlled with 2×10^{10} PIB/mL nuclear polyhedrosis virus.

3. Chemical control. Cover spray with 80%-90% trichlorfon WP or 90% trichlorfon crystal at 1,000-1,500×, 25 g/L lamda-cyhalothrin at 1,500-2,000×, or 12% thiamethoxam and lamda-cyhalothrin mixture at 1,000× to control 1st and 2nd instar larvae.

References Wu Yousheng and Gao Zezheng 2004.

(Wu Yousheng, Gao Zezheng)

284 Coleophora obducta (Meyrick)

Classification Lepidoptera, Coleophoridae

Synonyms *Protocryptis obducta* Meyrick, *Coleophora dahurica* Falkovitsh, *Coleophora longisignella* Moriuti

Common names Dahurian larch case bearer, Xingan larch case bearer

Importance *Coleophora obducta* is an important larch foliar pest. It is also one of the earliest studied species and was previously recorded as *Coleophora laricella*. Since its first discovery in Liaoning province in 1956, it has been found in several provinces in China, seriously affecting the growth and development of trees, and even resulting in massive tree mortality.

Distribution Hebei, Inner Mongolia, Liaoning, Jilin, Heilongjiang, and Xinjiang; Russia (Far East Region).

Hosts *Larix gmelini*, *L. principis-rupprechtii*, *L. olgensis*, and *L. kaempferi*.

Damage *Coleophora obducta* feeds on larch foliage, resulting in massive grayish white needles at first. Infested needles wilt and become yellow 2-3 days later, appear to be damaged by forest fire.

Morphological characteristics **Adults** Length 2.2-3.3 mm, wingspan 7.9-10.5 mm. Forewings silver with long marginal hairs. Labial palpi slender and dangling downwards with pointed ends. Female ostium bursae and thorn-like signum large, ductus bursae triangular or spear-shaped beyond the spinate belt, tapering towards the end. Middle band inside ductus bursae not well developed and without separation at the end. Male genitalia valves broad and with obvious anal angle, whereas the hump is conspicuous when view from the surface clasper valves. Claspers conspicuous.

Eggs Semisphere, yellow, with 11-13 ridges on surface. Eggshell grayish white after egg hatches.

Larvae Mature larvae yellowish brown. Prothoracic scutum brownish black, shiny, with a mark.

Pupae Fresh red initially, turn to brownish black later. Wing pads for forewings in males obviously reach pass end of abdomen, whereas wing pads for forewings in females generally does not extend beyond the abdomen tip.

Biology and behavior *Coleophora obducta* occurs one generation a year. It overwinters mostly as 3rd instars and sometimes as 2nd instars at the base of short or small branches or rough patches or crevices on branches. Overwintering larvae become active when larch buds start to develop in later April and most overwintering larvae become active by early May. They soon molt into 4th instars which lasts for 12-17 days. Mature larvae pupate around May 10 each year and pupation peaks in mid-May. Pupal stage lasts 16-19 days. Adults emerge in early mornings or late afternoons with a peak emergence in early June. The sex ratio is about 1:1 to 1:1.4 between females and males. Adults live for 3-7 days. Egg period peaks in mid-June. Eggs are scattered singly on

C. obducta adult (Yan Shanchun)

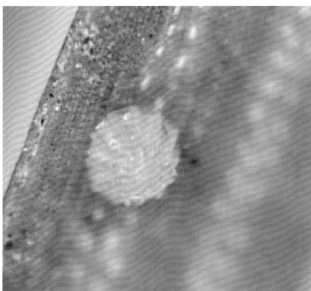

C. obducta egg (Yan Shanchun)

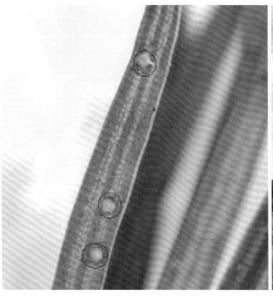
C. obducta eggs (Yan Shanchun)

C. obducta damage (Yan Shanchun)

the back of the needles. Usually only one egg is laid on each needle. However, as many as 9 eggs can be found on a single needle. Each female produces about 30 eggs. Egg stage lasts for about 15 days. Eggs start to hatch in late June with most of them hatch in early July. Newly hatched larvae bore directly into the needles from the bottom of eggs and feed inside until late September to early October to construct cases as 3rd instars (or sometimes 2nd instar). The 4th instar larvae have a big appetite. They are attracted to light. *C. obducta* spends the entire larval stage on larch.

Prevention and treatment

1. Fumigation. Fumigation with lindane during peak adult emergence in early June three times at an interval of 5 days resulted in 78% control. It is only recommended in heavily infested stands to avoid killing associated natural enemies.

2. Killing lamps. Deploy frequency vibration-killing lamps in infested stands to trap and kill adults during adult emergence in early June. However, the relative high cost prevents it from being used over large areas.

3. Plant volatiles. *Coleophora obducta* females are attracted to plant volatiles such as S-a-pinene, S-b-pinene, Phellandrene, 3-carene, Myrcene, and cis-3-Hexen-1-ol at certain concentrations, whereas males are attracted to Ocimene, Myrcene, Camphene, and cis-3-Hexen-1-ol at certain concentrations. Application of plant volatiles attractive to females to healthy larch seedlings showed that there is significant difference in mortality between treatment and control except those treated with 0.004 mol/L Myrcene. The highest mortality of 86.67% was observed when seedlings were treated with S-a-pinene at the concentration of 0.04 mol/L.

4. Biological control. Major natural enemies of *C. obducta* include predatory birds such as *Parus mayor* and *Passer montanus*. Other natural enemies include spiders and wasps.

5. Pheromone trapping. Sex pheromone baited traps significantly reduce pest populations in the field. In one experiment, a total of 5,490 adults were trapped from five study plots between June 4-18 using experimental pheromone traps with a dose of 130 μg per lure.

6. Chemical control. 1) Cover spray with 25% Dimilin at 2,000× resulted in 90% control. 2). Release avermectins fog (mixture of avermectins and No. 0 diesel) in infested stands at the rate of 276 mL/ha resulted in corrected adult mortality of 85%. 3). Use 20% Dimilin + avermectin WP at 600 g/ha and 0.9% avermectins oil suspension at 276 mL/ha to control 1st and 2nd instar larvae resulted in 60% population reduction at 3 days after treatment. Larvae can also be controlled with 2×10^{10} PIB/mL nuclear polyhedrosis virus.

7. Aerial application. Apply 1.8% avermectins at the rate of 120 g/ha provided the best results. The airplane was operated at the speed of 160 km/hour, 5-20 m above the crowns between 3:30-9:30 or 16:30-19:30, with a temperature of 10-25 °C and wind speed of 1-4 m/second.

References Xiao Gangrou (ed.) 1992; Hao Yushan 2003; Li Zhongxiao 2003; Shu Chaoran 2003; Ren Li, 2005; Liu Xiuying 2008; Yan Shanchun 2008, 2009.

(Yan Shanchun, Yang Liming, Yu Enyu)

285 *Coleophora sinensis* Yang

Classification Lepidoptera, Coleophoridae
Common name Chinese larch case bearer

Importance *Coleophora sinensis* is one of the most important pests of *Larix principis-rupprechtii*. It feeds inside the needles which makes early detection difficult. All or most needles on a tree may die from desiccation during outbreaks, resulting in tree mortality under heavy infestation.

Distribution Hebei, Shanxi, Inner Mongolia, and Henan.

Host *Larix principis-rupprechtii*.

Damage Feeding by *C. sinensis* resulting in premature yellowing, desiccation, and leaf fall-off. Infested stands appear red.

Morphological characteristics **Adults** Length 2.5-3 mm, wingspan 6.5-8.5 mm. Body and wings dark gray, silky. Forewings slender and hindwings narrowly lanceolate, both with marginal hairs and no veins. First abdominal segment contains a thin rectangle with longitudinal spines on the notum. Tip of female abdomen triangular with pubescence. Male claspers small, with no protrusion on their ventral sides. Small valves short and wide, with arc-shaped posterior margin.

Eggs Semisphere, with 10-15 uniform ridges on surface. Newly laid eggs beige colored with crystal luster, turn to dark gray eventually.

Larvae Cylindrical, with four instars. Prolegs retrogressed with biordinal biserial bands of crotchets on abdominal segments 3-6 and the anal segment. Mature larva contains a dark brown round mark on both sides of each thoracic segment. Mesonotum contains an elongate oval brown patch.

Pupae Coarctate, dark brown, darker for wing pads.

Biology and behavior *Coleophora sinensis* occurs one generation a year and overwinters as larvae inside cases in bark crevices, buds, or fallen leaves. Overwintering larvae resume feeding on young buds and leaves in early April to early May the following year while carrying their cases. Pupae were found as early as May 4, with most pupae found between late May and early June. Adults start to emerge in early June and reach the peak in mid-June. Adults mate and lay eggs 2-3 days after emergence. Eggs begin to hatch in early July. Newly hatched larvae bore into the needles from the bottom of the eggshells, and construct cases starting in late September.

Adults are attracted to lights and take short flights from one branch to another when disturbed. Most eggs are laid along the mid-vein on the underside of the posterior half of a needle. Eggs are laid singly. Usually one egg was laid on a needle.

Prevention and treatment

1. Natural enemies. Natural enemies such as predatory birds, parasitoids, spiders, ants, and assassin bugs play important roles in regulating *C. sinensis* populations.

2. Fumigation. During windless days in May, fogging with 1:4 1.2% bitter tobacco EC (a biorational insecticide made of extracts from tobacco leaves and sophora and other Chinese herbs) and No. 0 diesel between 4:00-8:00 am or 18:00-22:00 pm at the rate of 600 mL/ha resulted in 95% control.

3. Pheromone trapping. Using sex pheromone (active ingredient Z5-10: OH, release rate of 100 μg/lure) baited traps at the beginning of adult emergence reduced infestation rate by 50.1% and pest population by 81.2%.

4. Chemical control. Cover spray with Dimilin in spring, or mixture of Dimilin II and thiamethoxam or deltamethrin in the fall resulted in good control. Larvae can also be controlled with 2×10^{10} PIB/mL nuclear polyhedrosis virus.

References Shi Guanglu 2002, Hou Deheng 2003, Li Houhun 2003, Li Heming 2006, Ma Yunping 2007, Liang Xiaoming 2008.

(Yan Shanchun, Yang Liming)

Major differences between the two case bearers

Species	*C. sinensis*	*C. obducta*
Forewing color	Dark brown with faint silky luster	Yellowish to grayish brown with strong silky luster
Spiny patches on abdominal nota	Twice as long as wide, with 33 short spines in each patch on the 3rd notum	Four times as long as wide, with 26 short spines in each patch on the 3rd notum
Clasper	Semisphere, slightly depressed on the underside apically, sacculus narrow and small	Near triangular, obtuse apically, sacculus narrow and small
Lateral-back angle on the 8th abdominal segment	> 45°	= 45°
Bursa	With slender and strongly curved hook-shaped central protrusion, length of protrusion equals the width bursa base	Central protrusion robust with slight curve in the middle, about half of the bursa base width
Female genitalia	(Yang Liming)	(Zhang Peiyi)

286 *Casmara patrona* Meyrick

Classification Lepidoptera, Oecophoridae

Importance *Casmara patrona* larvae feed on branches and stems of its host trees, resulting in leaf wilt and branch death from above the galleries. The most severe damage is found on 10-year-old young trees, which usually leads to death of the entire tree.

Distribution Zhejiang, Anhui, Fujian, Jiangxi, Hubei, Hunan, Guangdong, Guangxi, Guizhou, Taiwan; India, Japan.

Hosts *Camellia oleifera* and *C. sinensis.*

Damage Larvae feed on branches, leaving evenly spaced circular defecating holes on the underside of the branch. Yellow frass pellets can be found under the defecating holes. Infested branches usually wilt with empty pith. Heavy infestation occurs mostly in old tea gardens and high-density oil-seed camellia forests.

Morphological characteristics **Adults** Length 12-16 mm, wingspan 32-40 mm. Body covered by grayish brown and grayish white scales. Antennae filiform, grayish white, with an enlarged brown basal segment. Labial palpi sickle-shaped, curved upward and pass the vertex. The second segment robust and covered by dark brown and grayish brown scales; the third segment slender and grayish white with a pointed black tip. Forewings dark brown with six bundles of reddish brown and dark brown erect scales. Three of the six bundles are found at the basal third of the forewing, two bundles are inside the white curved mark at the middle of the wing, and one bundle located outside of the white mark. Hindwings grayish brown. Legs brown. Fore tibia grayish white, with long dark hairs. Hind tibia with alternate brown and gray long hairs. Abdomen brown, shiny with grayish white blotches.

C. patrona adult (Ji Baozhong)

Eggs Length about 1.1 mm, oblong oval, ochraceous, slightly depressed in the middle, with marks on eggshells.

Larvae Length 25-30 mm, yellowish white. Head yellowish brown, pronotum yellowish brown. The last two abdominal nota sclerotized, dark brown.

Pupae Length 16-24 mm, long cylindrical, yellowish brown, with a pair of small protrusions on the ventral surface of the last abdominal segment.

Biology and behavior *Casmara patrona* occurs one generation a year and overwinters as larvae inside infested branches. Larvae continue feeding during the winter when air temperature is above 10℃. Pupation starts between March and April next year and peaks in April and May. Most adults emerge between May and June. Large numbers of larvae appear in mid- to late June. Egg stage lasts for an average of 19.5 days. Larval stage (including overwintering period) spans over nine months, whereas pupal stage is about a month. Adults survive for 4-10 days. Newly hatched larvae crawl to the top of the young shoots and bore inside the shoots from leaf axils. They spin a thin silk layer right above the entrance holes to cover themselves before boring in. Newly hatched larvae feed little, leaving very small galleries. However, infested shoots are emptied by larval feeding due to the small sizes of the shoots, leaving gradually wilting shoots with only the epidermis intact. They then feed inside the sapwood from top down and create a round defecating hole on the underside of the branches at certain interval. The characteristic yellow frass particles are usually found below the extracting holes. Larger oval holes at lower positions on branches sealed with silk indicate pupated mature larvae. Each larva can create a gallery as long as 104 cm inside the branch, with an average of 7-9 and a maximum of 13 defecating holes. Larval gallery increases in size from the top to the bottom, with remains of unfinished extracting holes along the way. The larva is very active inside the gallery. It can easily move forward and backward and change directions. Mature larva chews a slightly larger sub-circular exit hole (about 3.5 mm in diameter) next to the closest defecating hole at the middle or upper portion of the gallery and spins silk to seal it. It then constructs a pupal chamber 3-7 cm below the exit hole for pupation. Pupal chamber is closed on both ends with silk. About three days are needed between silk spinning and pupation. Adults generally emerge near sunset, but can come out in numbers in the afternoon if the weather is hot and humid. Adults mate during the second night after emergence. Mating lasts about 2-3 hours. Most females mate only once in their lifetime. They chose old tea gardens or shady and humid high canopy density oil-seed camellia stands to lay eggs. Eggs are scattered on top of young branches or at the base of top buds. Each female can lay 30-80 eggs. Some females finish oviposition in one day, while others may need 4-5 days. Most eggs hatch between 12:00 and 16:00 pm.

Prevention and treatment

1. Remove infested branches and destroy them. Reduce tree density to 900-1,500 trees/ha through timely pruning and thinning in dense oil-seed camellia stands.

2. Light trapping during peak adult emergence for 2-3 consecutive years can greatly reduce the pest populations.

3. Application of water diluted 20% fenvalerate EC at 30-60 mL/ha at low volume or ultralow volume during egg hatching and larval leaf feeding period.

Reference Xiao Gangrou (ed.) 1992.

(Ji Baozhong, Zhang Kai, Shen Guangpu)

287 Macrobathra flavidus Qian et Liu

Classification Lepidoptera, Cosmopterigidae
Common name Chinese fir cone webworms

Importance *Macrobathra flavidus* often co-occurs with *Dichomeris bimaculatus* in Chinese fir forests in southern China. It affects the production of high-quality Chinese fir seeds.

Distribution Fujian.

Host *Cunninghamia lanceolata*.

Damage Similar to that of *D. bimaculatus*. Larval feeding on cone scales, cone shafts, and seeds, causing the scales to become reddish brown and wilt. Infested young cones become wilted, while larger cones become deformed.

Morphological characteristics **Adults** Wingspan 12-14 mm. Top of head, thorax, and abdomen grayish brown. Compound eyes reddish brown. Labial palpi extremely slender and strongly curved upwards. Antennae dark brown with white blotches. Legs grayish brown, with white blotches on tibia and tarsi. Forewings lanceolate, dark brown, with a brownish yellow band between one sixth and one half of the surface from the basal area. Hindwings grayish brown, slender. Its apex pointed. Marginal hairs long and light colored. **Larvae** Length 8-13 mm, rosy red or purplish red with white bands between segments. Body with alternate red and white rings. Head, pronotum, abdominal notum 9, and anal plate brownish. Anal comb with four serrate teeth.

Biology and behavior *Macrobathra flavidus* occurs one generation a year and overwinters as larvae in Fujian province. Overwintering larvae move to already sporulated currently year male inflorescences for pupation. Adults appear in early May with a peak emergence in mid- to late May. New larvae feed inside wilted current year male inflorescences in June, and migrate into current year cones between August and November. Larvae start to overwinter in mid- to late November inside wilted male inflorescences or galleries in cones.

Prevention and treatment Similar to that of *D. bimaculatus*.

References Qian Fanjun and Liu Youqiao 1997, Qian Fanjun et al. 1998.

(Qian Fanjun)

M. flavidus adult (Qian Fanjun)

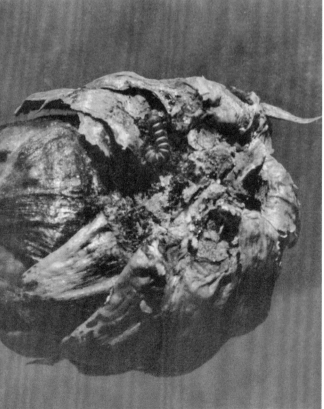

M. flavidus larva and its damage (Qian Fanjun)

288 *Haplonchrois theae* (Kusnezov)

Classification	Lepidoptera, Cosmopterigidae
Synonym	*Parametriotes theae* Kusnezov
Common name	Tea moth

Importance *Haplonchrois theae* is a major pest of oil-seed camellia forests in the rolling country. Heavy infestation leads to damage on 70%-85% of trees in the forests, resulting in 80% reduction of fruiting branches.

Distribution Jiangsu, Zhejiang, Anhui, Fujian, Jiangxi, Henan, Hubei, Hunan, Guangdong, Guangxi, Sichuan, Guizhou, Yunnan, Shaanxi; Japan, Russia, India.

Hosts *Camellia oleifera*, *C. sinensis*, and *C. japonica*.

Damage Larvae bore inside leaf mesophyll, branch tips, and base of leaf stalks, resulting in abnormal growth of young branches, wilting and yellowing leaves, and early death of branch tips. Larval galleries inside branch tips increase in size as the larvae grow, which prevents the formation of normal flower buds on branch tips. Larvae migrate to new branch tips after the branch death, resulting in numerous wilting and yellowing branches on infested trees.

Morphological characteristics **Adults** Females 4-7 mm long with a wingspan of 9-14 mm. Shiny, grayish brown. Antennae filiform with a strong base, as long as the body or slightly shorter than forewings. Labial palpi sickle-shaped and extend laterally. Vertex and gena tightly covered by flat brown scales. Forewings grayish brown, slender and lanceolate with long marginal hairs, shiny and contains scattered small black scales. There is a big black spot in the center close to the posterior margin, and one small black spot at 1/4 distance from the wing tip. Hindwings slender, yellowish basally and grayish black apically. The grayish black marginal hairs longer than the width of the hindwings. Males 4-5 mm long. Lighter in body color and with a narrower abdomen compared to females.

Eggs Oblong oval with slightly truncated ends, milky white and transparent initially, turn to yellowish three days later.

Larvae Length 8-10 mm at maturity. Orange and sparsely covered by thin and short hairs. Head brown, thorax and abdomen yellowish white.

Pupae Length 5-7 mm. Cylindrical, yellowish brown. Head brownish with visible wing pads and antennae. There is a pair of club-shaped protrusion on the ventral surface of the last abdominal segment.

Biology and behavior *Haplonchrois theae* occurs one generation a year and overwinters as larvae inside leaf mesophyll in Anhui, leaving irregular semitransparent yellowish brown blotches on leaf surface. Larvae migrate to young branch tips to feed between March and April the following year. Larvae pupate in early to mid-May and adults emerge 15-24 days later. Adults appear from late May to July. New generation larvae hatch in mid- to late June and mine in leaves or branch tips. Larval stage lasts for as long as 10 months from June to next May. By mid-October, most larvae enter and feed at overwintering sites. Adults usually emerge in the sunny afternoons, with the ambient temperature of 20-25°C and relative humidity of 75%-95%. Adults live for about 10 days. Newly emerged adults are not very active. They begin to fly for suitable locations about 20 minutes after emergence. Adults stay on small branches during the

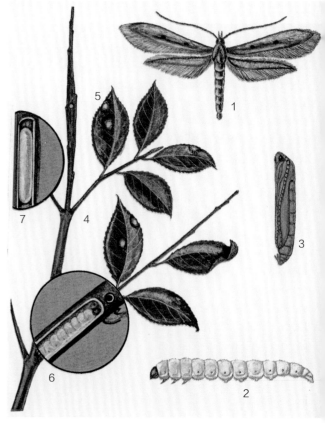

H. theae. 1. Adult, 2. Larva, 3. Pupa, 4. Damage, 5. Damage by young larvae, 6. Larva inside shoot, 7. Cocoon (Tang Shangjie)

H. theae. 1. Adult, 2. Egg, 3. Pupa (Ji Baozhong)

day. They become active and mate at sunset or during the night. Adults are weak fliers and are attracted to lights. Most eggs are laid singly or in small clusters of 1-4 in bark crevices on leaf stalks attaching to small branches. They are generally not very visible. Egg stage lasts for 10-15 days. Each female produces about 50 eggs. Most larvae hatch before 10:00. They migrate to the underside of the leaves in about 30 minutes after hatching. Larvae feed on mesophyll after breaching the upper epidermis. They feed in all directions inside the leaves, creating yellowish brown round blotches. Most blotches are close to the mid vein, slightly convex, and with a diameter of 3-5 mm. Each infested leaf contains 20-50 blotches. Larvae overwinter inside the blotches and migrate to young branch tips the next year. Larvae consume more as they grow when they feed on xylem around the galleries inside branches, leaving only the upper epidermis intact. Infested branches die of desiccation later. Larvae can migrate to new feeding locations during development. Each larva can damage 1-3 spring branch tips.

There are two generations a year and larvae overwinter in buds or leaves in southern Fujian, Guangdong, and Jiangxi provinces except at high elevations. Newly hatched larvae feed on mesophyll inside the leaves, creating yellowish brown blotches. The 2nd and 3rd larvae bore inside young branch tips nearby, leaving short and straight galleries filled with yellowish green frass. Sawdust and frass are visible near infested branch tips. Mature larva chews a round exit hole and pupates below it inside the infested branch.

Prevention and treatment

1. Remove infested branches before adult emergence and place them in cages. Destroy infested branches after emergence of parasitic wasps.

2. Apply 25 g/L lamda-cyhalothrin at 800-1,000×, 90% trichlorfon at 500-1,000×, 2% thiamethoxam at 900-1,000×, or 2×10^{10} PIB/mL nuclear polyhydrosis virus at 1,000-1,500× during larval migration between March and April in heavily infested stands. Apply 45% profenofos and phoxim mixture at 30-60×, or 12% thiamethoxam and lamda-cyhalothrin mixture at 30-60× to tree trunks.

References Xiao Gangrou (ed.) 1992; Chao Jun, Zhan Liming, Lu Jin et al. 2007; Xu Guangyu, Yang Ainong, Huo Tianjun et al. 2007.

(Ji Baozhong, Zhang Kai, Zhu Senhe)

289 Scythropiodes issikii (Takahashi)

Classification Lepidoptera, Lecithoceridae

Synonyms *Depressaria issikii* Takahashi, *Odites issikii* (Takahashi), *Odites plocamopa* Meyrick, *Odites perissopis* Meyrick

Distribution Hebei, Liaoning, Shandong, and Shaanxi in China; Japan.

Hosts *Malus pumila*, *M. asiatica*, *Pyrus* spp., *Vitis vinifera*, *Prunus psudocerasus*, *Prunus salicina*, *Prunus persica*, *Prunus armeniaca*, *Ziziphus jujubal* var. *inermis*, *Poncirus trifoliate*, *Ulmus* spp., *Populus* spp., *Salix* spp.

Damage Newly hatched larvae feed on mesophyll inside galleries in host leaves. Second instar and older larvae feed on leaf margins and spin flat cylindrical nests with leaves. They stay in the nests during the day and feed at night. Heavy infestation leads to complete defoliation of the host trees.

Morphological characteristics **Adults** Females are 8-10 mm long with a wingspan of 17-20 mm, while males are 7-9 mm long with a wingspan of 15.5-19.5 mm. Body yellowish white. Compound eyes black. Labial palpi with a brown blotch externally on the third segment. Forewings brownish, with two round and black blotches near the base. Hindwings grayish white. Dorsal surface of the thorax has a black blotch in the middle. The line on dorsal surface of the abdomen yellowish.

Eggs Oblong, milky yellow at first, turn to yellowish later, surface with fine sculpture.

Larvae Dirty brown when newly hatched. Mature larvae 10-12 mm long, dirty green or dark green. Head and front legs dark brown. Pronotum and abdominal nota dirty brown. Middle and hind legs brownish. Prolegs grayish brown.

Pupae 8-10 mm long, dark reddish brown. Vertex contains a chrysanthemum-flower-shaped protrusion. Abdomen with one reverse-hook-shaped spine on each side.

Biology and behavior *Scythropiodes issikii* occurs two generations a year and overwinters as first instar larvae inside thin cocoons in grass, leaf litter, and bark crevices in Shaanxi and Hebei. Overwintering larvae become active in early April and resume feeding on leaves in mid-April to late May. They pupate inside the cylindrical nests from late May to early June. Pupal stage lasts about 20 days. Adults of the overwintering generation emerge in early to mid-June and lay eggs in mid-June. Egg stages lasts for 10-15 days. Larvae of the first generation start to feed between early to late July. They pupate in mid- to late August. Adults of the first generation emerge, mate, and lay eggs from late August to early

S. issikii adult (Xu Gongtian)

S. issikii adult (with wings spread) (Xu Gongtian)

S. issikii pupa (Xu Gongtian)

S. issikii cocoon (Xu Gongtian)

S. issikii damaged leaf (Xu Gongtian)

September. Most eggs hatch in mid-September. After feeding on leaf mesophyll for a while, the newly hatched second-generation larvae spin thin cocoons to overwinter.

Most adults emerge during the night and are attracted to lights. They stay on branches or the underside of the leaves during the day right. Mating occurs in the next night after emergence. After another 2-4 days, eggs are laid on host leaves along either side of the main veins. Eggs are rarely laid in groups. Each female produces an average of 70 eggs. Adults live for 4-5 days. Egg stage lasts for 10-15 days. There are five instars in the larval stage. Newly hatched larvae crawl to or land on host leaves with the help of the wind and feed on the underside, creating 2-3 mm long linear galleries on leaves. After feeding for 3-5 days, they spin flat nests with leaves. Nests increase in size as larvae grow. When disturbed, larvae exit the nests swiftly and hang in the air with silk threads.

Prevention and treatment

1. Manual. Remove old and loose barks, dead branches, fallen leaves and grass in the winter to reduce overwintering larvae population. Remove larval nests during larval stage. Trap adults with black light at 1 light/100 m.

2. Apply mixture of 5.7% emamectin at 3,000-5,000×, 25% Dimilin at 1,500×, or 5% esfenvalerate or beta cypermethrin EC at 2,000× for larvae, and beta cypermethrin at 1,000×, 25 g/L lamda-cyhalothrin at 800-1,000× for adults. Larvae can be controlled by spraying 2×10^{10} PIB/mL nuclear polyhydrosis virus at 1,000-1,500×.

References   Xiao Gangrou (ed.) 1992; Liu Shixian 2003; Wang Junmin, Zhao Yansheng, Ai Xianqin et al. 2003; Wang Xinxiang 2006, Xu Gongtian et al.2007.

(Wang Hongzhe, Li Menglou, Zhan Youguang)

290 *Anacampsis populella* (Clerck)

Classification Lepidoptera, Gelechiidae
Synonyms *Anacampsis tremulella* Duponchel, *Anacampsis laticinctella* Wood
Common name Sallow leafroller moth

Distribution Hebei, Shanxi, Inner Mongolia, Liaoning, Jilin, Heilongjiang, Shaanxi, Gansu, Qinghai, Ningxia, and Xinjiang; Mongolia, Korea Peninsula, Japan, Russia, Europe, and South Africa.

Hosts *Populus nigra*, *P. tremula*, *Salix caprea*, *S. repens*, *S. alba*, *Betula* spp., and *Acer* spp.

Damage Larvae roll leaves with silk and form cylindrical nests parallel to the main veins and feed inside. The top of the nest is slightly smaller or similar in size as the bottom. Both ends are open to allow for movement of the larvae. Poplars older than 10 years are usually damaged, with single trees, trees in sparse forests or along forest edges more vulnerable to infestation. Heavy infestation results in complete defoliation of the hosts, which seriously affect their growth.

Morphological characteristics **Adults** Wingspan 14.0-19.0 mm. Head brown. Frons grayish white. Labial palpi brown for the 1st and 3rd segments, and grayish white apically for the 2nd segment; the 3rd segment is longer than the 2nd,

A. populella. 1. Adult, 2. Adult head, 3. Eggs, 4. Larva, 5. Pupa (Zhao Ren)

A. populella adult (Li Houhun)

A. populella adult (Xu Gongtian)

grayish white, with brown longitudinal lines ventrally. Thorax, tegulae, and forewings brown with grayish white scales. There is a fuzzy black spot in the middle and the apex of the median cell, as well as at one third and one half of the plica of the forewing. There is a grayish white transverse band at three quarters of the forewings, with several small black spots along the margin anteriorly. Marginal hairs brown. The anterior half of the forewing and the posterior end of thorax sometimes darker in color. Hindwings light in color with gray marginal hairs. Legs grayish brown. Dorsal surface of hind tibia with grayish white scales. Abdomen brown, yellowish brown for nota of segment 2 to 4. Male genitalia uncus wide, with round apex and long marginal hairs. There are two rows of spiky protrusions on the underside of the uncus. Male genitalia gnathos ring like, with slightly phylliform projection in the middle. Claspers short, not reaching uncus. Saccus slender, gradually enlarge and truncate apically. Female genitalia and the 8th abdominal tergum form a triangle or bell-shaped cover. Ductus bursae shorter than ostium bursae. Ostium bursae covered by small spines. Signum bursae saw-like. **Eggs** Oval, 0.8 mm long and 0.4 mm wide, with longitudinal ridges. Yellowish at first, gradually turn to reddish. **Larvae** Yellowish at first, become yellowish green and 12.0-14.0 mm long when mature. Head, pronotum shiny black. Meso- and metanotum each contains two transverse rows of black and round hairy scales. Each of the abdominal segment has two pairs of black and round hairy scales. Anal plate brown. **Pupae** 8.0-10.0 mm long, yellowish brown. Become darker right before emergence. Abdomen covered by fine setae and contains about 20 anal spines.

Biology and behavior *Anacampsis populella* occurs one generation a year and overwinters as eggs inside bark crevices. Eggs hatch in next April when hosts start to blossom. Larvae start to pupate inside rolled leaves in late May. Pupation peaks between mid- to late June. Most adults emerge in late June. Peak egg period falls between late July and early August. Egg stage lasts for about nine months.

Larvae stay in nests and feed on surrounding leaves, with frass deposited inside. They usually feed on several leaves and can feed on more than 10 leaves. Larvae move fast and can jump or move backwards when disturbed. Larval damage can be divided into the following three stages. *The bud stage* – newly hatched larvae bore into the buds, with each bud contains 1-3 larvae. This stage lasts for about one week. *The single-leaf rolling stage* – host leaf starts to expand; larvae roll a single young leaf into a cylinder with silk. This is usually done between 5:00-10:00 am and around 21:00 pm. Larvae feed inside. Leaves become reticulate. Larvae color changes from light to yellowish green. This stage lasts for about 10 days. *The multi-leave rolling stage* – up to four mature leaves are rolled into a nest of multiple layers. Larvae turn to greenish and feed inside the nests. Mature larvae usually spin thin cocoons in the nest and pupate. Rolled leaves turn yellow and gradually fall to the ground together with mature larvae or pupae inside.

Most adults emerge in the morning and stay on coarse barks during the day. They come out at night and are very active between 20:00-23:00. Adults are not attracted to lights. They live for about 40 days. Adults crawl swiftly along the trunks for sugary excreta from hemipterous insects. After mating, they lay eggs in hidden places such as bark crevices on dead branches or branch joints. Eggs are laid in clusters of 10-40 eggs. Each female produces an average of 60 eggs. Mating lasts for about an hour and can occur multiple times. It took about one month from adults to eggs.

Prevention and treatment

1. Remove rolled leaves during light infestation.
2. Apply the mixture of nicotine and matrine at 1,000× or 2×10^{10} PIB/mL nuclear polyhydrosis virus at 1,000-1,500× to control larvae.

References Liu Youqiao and Bai Jiuwei 1979, Li Houhun 2002.

(Li Houhun, Byun Bong-Kyu, Li Chengde, Zhang Zhiyong)

A. populella medium-sized larva (Xu Gongtian)

A. populella mature larva (Xu Gongtian)

A. populella pupa (Xu Gongtian)

A. populella damage to leaf (Xu Gongtian)

A. populella damage to leaf (Xu Gongtian)

291 *Anarsia lineatella* Zeller

Classification	Lepidoptera, Gelechiidae
Synonym	*Anarsia pruniella* Clemens
Common name	Peach twig borer

Distribution Shaanxi and Xinjiang; Japan, Iran, Afghanistan, Turkey, India, Middle East, Europe, North Africa, and North America.

Hosts *Prunus salicina*, *P. spinosa*, *P. domestica*, *P. mume*, *Armeniaca vulgaris*, *Cerasus pseudocerasus*, *Amygdalus communis*, *A. mongolica*, *Malus pumila*, *M. sylvestris*, *M. asiatica*, *Pyrus communis*, *Acer tataricum*, *A. campestre*, and *Diospyros kaki*.

Damage Larvae feed on buds, young branches and fruits, leaving frass and silk at entrance holes. Top of infested branches wilt and dry out gradually. Older larvae feed on young fruits.

Morphological characteristics **Adults** Wingspan 10.0-14.5 mm. Head brown, covered by scattered white scales. Labial palpi dark brown externally and grayish white internally. The second segment of the labial palpus grayish white apically, with rhombus-shaped tufted scales ventrally. The third segment grayish white, brown at one third and one half in females. Thorax and tegulae brown and covered with scattered white scales. Forewings brown with scattered black erect scales. Plica and median cell contain an irregular dark longitudinal mark. A longitudinal long white mark extends backwards near the end of the median cell. Marginal hairs grayish brown. Hindwings and their marginal hairs gray. Legs dark brown, with white rings on tarsi. Hind tibia brownish, with grayish white long scales dorsally. Abdomen gray, brown at the lateral region, grayish white or ochrous at the end. Male genitalia gnathos peach-shaped, slightly concave at the anterior margin and slightly narrowed in the latter half. Tegumen slender and wide in the middle. Claspers are asymmetrical: left clasper wide, near square-shape, with membranous projection and palm-shaped scales, sacculus contains a strong projection at one third of its length. Right clasper narrows basally and parallel laterally, apex enlarged, round, with palm-shaped scales, convex dorsally, ventral side with a slender projection. Female ovipositor large and broad, sclerotized, and inserted at the anterior margin of the 8th abdominal segment. Antephysis short and wide, angle shaped. Ostium bursae oblong oval, funnel-shaped, with round membranous pouch anteriorly. Ductus bursae shorter than ostium bursae. Bursae projection bell-shaped, concave anteriorly and posteriorly, serrulate laterally with strong sclerotized margins. **Eggs** Oval, 0.5 mm long and 0.3 mm wide, with irregular ridges on surface. White at first, gradually turn to yellowish later, and become purple before hatching. **Larvae** Newly hatched larvae 0.7-0.8 mm long,

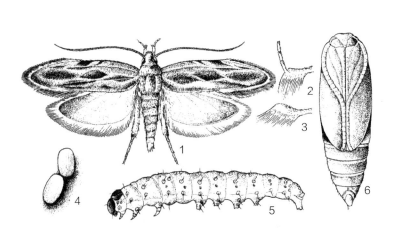

A. lineatella. 1. Adult, 2. Female adult labial palpus, 3. Male adult labial palpus, 4. Eggs, 5. Larva, 6. Pupa (Zhu Xingcai)

A. lineatella. 1. Damage to lateral shoot, 2. Pupation location, 3. Damaged fruit, 4. Damage to main shoot, 5. Larval entrance hole on winter buds, 6. Larva exiting from damaged fruit (Zhu Xingcai)

A. lineatella adult (Li Houhun)

white. Turn to brown after 2-3 hours. Mature larvae 10.0-12.0 mm long. Head, pronotum, and thoracic legs black. Setae and spiracle margins brownish. Caudal plate brownish, proleg crochets in biordinal circle (biordinal biserial bands according to Zhang Xuezu 1980) with three quarters biserial and one quarter uniserial. Anal proleg crochets in biordinal series discontinuous centrally. **Pupae** 5.5-7.0 mm long, brownish and pilose, with many spines near the end of abdomen and ostium bursae. Anal spine hook shaped.

Biology and behavior *Anarsia lineatella* occurs four generations a year and overwinter as young larvae inside winter buds of peach or apricot trees in Xinjiang. Overwinter larvae start to feed in early April when average air temperature reaches 10℃. They mature and pupate after 18-25 days. Pupal stage lasts for 10-12 days. Adults live for 8-12 days. Average duration for the following generations is 6-10 days for eggs, 9-12 days for larvae, 8-13 days for pupae, and 2-9 days for adults.

Larval feeding on leaf buds, flower buds, young fruits, and branches creates boring holes on leaves, resulting in wilting buds and fallen young fruits. Damaged branch tips gradually wilt, droop, and dry out for the portion above the boring holes. Damage period for first generation larvae falls between mid-May to mid-June when they feed mainly on new branch tips, causing wilt, drooping, dry out, or resin flow on the top. Late instar larvae usually attack young fruits directly as branch tips getting old and young fruits becoming bigger. Branch tips on young trees are attacked more frequently than those on old trees. Larvae of the second generation feed exclusively on fruits between late June and late July, resulting in large amount of resin flow on peach fruits. Larvae of the third generation occur in August and are the most damaging generation since they can also feed on fall branch tips in addition to fruits. In general, each fruit contains one larva. Larvae of the fourth generation overwinter in buds in late September. Adults are strongly attracted to sugar and vinegar. Adult longevity is 7-8 days for females and 3-4 days for males. Females start to lay eggs 2-3 days after emergence. Egg period lasts for 4-9 days, while pupal stage is about 10 days. It takes more than one month to finish one generation.

A. lineatella was mistakenly referred as *Anarsia eleagnella* Kuznetzov in 1957. The latter attacks *Hippophae* sp. and *Elaegnus* sp. They partially overlap in distribution.

Prevention and treatment

1. Enhance quarantine measures. Inspection cut branches when transporting them.

2. Maintain orchard sanitation by removing fallen fruits and paint the trunk with white paste.

3. Artificial removal of infested branch tips throughout the year.

4. Trap adults with mixture of sugar and vinegar (1 part sugar, 2 parts vinegar, and 15 parts water). Place lure on branches or on the ground in selected locations.

5. Apply 45% profenofos and phoxim mixture at 30-60×, 12% thiamethoxam and lamda-cyhalothrin mixture at 30-60× to tree trunks. Spray 2×10^{10} PIB/mL nuclear polyhydrosis virus at 1,000-1,500× to control early stage Larvae. In Xinjiang, apply the first spray after peach flowers enlarged, the second spray between late May and early June, and the third spray in early July.

References Bai Jiuwei, Zhao Jianxia, Ma Wenliang 1980; Zhang Xuezu 1980; Li Houhun 2002; Gulibahar Memet et al. 2007.

(Li Houhun, Ma Wenliang)

292 *Anarsia squamerecta* Li et Zheng

Classification Lepidoptera, Gelechiidae

Distribution Shandong and Shaanxi.

Host *Sophora japonica*.

Damage Larvae feed on leaves and cause massive leaf rolling.

Morphological characteristics **Adults** Wingspan 11.5-13.5 mm. Head brown, frons grayish white. Labial palpi dark brown, with scattered white scales, interior surface grayish brown; the tufted scales on ventral surface of the second segment of the labial palpus rectangular; the third segment brown basally and at the center, and grayish brown at one third and the end in the female. Thorax and tegulae brown and covered with scattered white scales. The anterior margin of the forewings slightly convex basally and slightly concave in the middle. Wing tips obtuse, brown and covered by scattered gray scales, with dense black scales erect to form blotches. Anterior margin contains a black and short transverse line, with a semi-circle black blotch in the middle. Forewings contain a black dot at the base. Median cell contains a large black blotch with no clear margins. This blotch extends backwards to the end of the plica. There are two black tufted scales between the plica and the posterior margin of the wing. Wing tip contains lots of black scales. Marginal hairs grayish brown, scattered with brown scales. Males contain long hair bundles ventrally. Hindwings grayish white for the basal half and grayish brown for the anterior half. Marginal hairs grayish brown. Fore- and middle legs dark brown, with white rings on tibia and tarsi. Hind legs brown. Hind femora grayish white apically. Tibia contains dense white long scales dorsally. Spurs dark brown with grayish white tips. Tarsi dark brown with grayish white rings. Male genitalia uncus slender, as wide as the posterior end of tegument, and bends at the end. Uncus concaves in the middle anteriorly and protrudes posteriorly. Tegumen slender and parallel laterally. Tegumen base wide. Claspers asymmetrical. The left clasper oval and contains palm-shaped scales posteriorly and a fine sclerotized projection basally. The projection is longer than the clasper and bent extremely. The right clasper is narrow for the basal half and wide apically, with palm-shaped scales on the apical irregular projection. There is a hairy projection at the end of sacculus that contains a thin projection at two thirds of its length. This projection is about half the size of clasper valve. Aedeagus is strong for the basal two fifth and slender for the apical three fifth. Ovipositor is large and broad, with long setae. Anterior apophysis is very wide, with a length less than half of the posterior apophysis. The 8th abdominal tergite is sclerotized. The 8th abdominal sternite protrudes forward at the center of the anterior margin. The tube-shaped projection is longer than half of the anterior apophysis and narrow apically. Ostium bursae is sclerotized and irregularly shaped, near round, with a serrate pouch at its base.

Eggs Oblong oval, 0.45-0.55 mm long and 0.20-0.25 mm wide. White at first, gradually turn to light brown before hatching.

Larvae Mature larvae 8.0-10.5 mm long. Head, pronotum, and caudal plate black. Abdomen purplish brown or brown, with two rows of dark brown scales on each

Anarsia squamerecta adult
(Li Houhun)

segment. There are four scales in the front row and two in the back row. Each scale contains a seta. Male larvae contain two black spots ventrally on the 4th and 5th abdominal segment. Caudal plate with sparse setae and 5-6 caudal spines; the middle two spines are bigger than the rest. Proleg crochets in biordinal circle discontinuous centrally, whereas anal proleg crochets in two separate groups, with each group contains about 10 crochets.

Pupae 4.5-6.0 mm long, brownish and pilose. The last two segments of the abdomen bends downwards, with dense hooked spines ventrally. Anal spine bifurcated.

Biology and behavior *Anarsia squamerecta* occurs one generation a year and overwinters as the third instar larvae inside branch wounds or bud crevices. Overwintering larvae resume feeding in mid-April and peak in late April. Larvae pupate in early May and pupation peaks in mid-May. Adult emergence begins in mid-May and peaks in late May. Egg stage starts in late May and ends in mid-June. Larvae feed until early to mid-September.

Occurrence of this insect is related to climate and site conditions. Larvae aestivate in mid-April when the average temperature is above 13°C for five consecutive days. Most larvae resume feeding when the average temperature is above 15°C for the same period. Infestation is light on health hosts at fertile sites and severe on weak trees at poor sites. Infestation is heavier in the upper and outer crown compare to the lower and inner crown. It usually co-exists with *Dendrophilia sophora*.

Adults can emerge anytime during the day, with most (69.3%) emerge between 11:00-14:00. One third of the pupal case expose outside the rolled leaves after adult emergence. Adults stay on the underside of the leaves or on branches during the day and become active at night. Mating begins 3-4 days after emergence. Each female produces an average of 46 (38-56) eggs. Adults live for an average of 10.5 days, ranging from 5-16.5 days. Sex ratio is 2.71:1 (females: males). Most eggs are laid among hairs along main veins on the underside of the leaves. Usually one egg is deposited per leaf. Early larval galleries are constructed along the veins, longitudinal, with smooth margins. Later, they branch out over larger areas. Feeding blotch increases to an average area of 12-17 mm^2 before overwintering. There is a respiratory hole at the base of the gallery, where frass is excreted. Larvae can migrate to new leaves. Migration usually occurs between late August and early September by 3rd instars when they reach leaf edges as they feed. Most larvae exit the galleries from respiratory holes and move the neighboring leaves. The 2nd instars start to appear between late June and early July, and the third instars appear from mid-August. Larvae begin to overwinter by mid- to late September. They resume feeding the following spring when new shoots reaching 5-10 cm long. They usually weave 2-7 leaves at branch tips or the ends of the compound leaves. Infested compound leaves become yellow and fall off. Larvae are very active and will retreat quickly back into the leave rolls and suspend themselves with silk threads. Mature larvae exit leaf rolls, crawl to the back of the leaves of neighboring branches, and construct 5 mm diameter rolls, and make thin white cocoons. A few of them pupate inside the existing leaf rolls. Pupal stage lasts for 10.5-16 days, with an average of 13.8 days.

Prevention and treatment Apply 40% monocrotophos or methoate at 1,000×, 50% thiamethoxam 22.5-30 g/ha, or 2×10^{10} PIB/mL nuclear polyhydrosis virus at 1,000-1,500× spray during peak larval period in late April.

References Li Houhun et al. 1998; Li Houhun 2002; Yan Jiahe, Wang Furong, and Li Jipei 2002.

(Li Houhun, Yan Jiahe)

Anarsia squamerecta adults (edited by Zhang Runzhi)

Anarsia squamerecta damage (edited by Zhang Runzhi)

293 *Dendrophilia sophora* Li et Zheng
Classification   Lepidoptera, Gelechiidae

Distribution   Shandong, Shaanxi, and Gansu.

Hosts   *Sophora japonica* and *S. japonica* f. *pendula*.

Damage   Larvae feed on leaves and inflorescence and affect tree growth.

Morphological characteristics   **Adults**   Wingspan 11.0-13.5 mm. Head brown, frons grayish white. First segment and exterior side of the second segment of the labial palpi brown; the 3rd segment and interior side of the second segment grayish white; the 3rd segment black at one third and two thirds of its length. Antennal scape grayish white, dark brown apically; flagellum contains alternate gray and brown rings. Thorax and tegulae brown and covered with scattered white scales. Forewings slender, brown, and covered by grayish white and ocherous scales; ocherous at the base. Anterior margin of the forewings ocherous for the basal one third, with obvious scales at one sixth and one third of its length. Scales at three fifth of its length small. Median cell with a black blotch in the center that extends to anterior margin. A dark brown blotch appears near the anal angle. Tip of forewing scattered with dark scales. Marginal hairs dark gray, mixed with dark brown scales. Hindwings and marginal hairs grayish white. R_4, R_5, and M_1 of forewings share a common basal portion, whereas M_2 and M_3 of the hindwings contain a short branch. Legs black. Tarsi contain white rings. Base of middle and hind tibia ocherous. Abdomen gray on the dorsal surface, dark brown on the lateral surface, and grayish white on the ventral surface. Male genitalia uncus near round, with a middle projection reaching half of the length of the uncus. Gnathos wide at the base, with posterior two thirds slender, and the end sharp. Tegumen broad and concave anteriorly. Claspers slender, the end round. Clasper valve similar to the basal leaf of the aedeagus, its end round and hairy. Saccus slender. Base of penis enlarged, the apical three fifth slender. Female ovipositor large and broad. Anterior apophysis about half of the posterior apophysis in length. The 8th abdominal sternite narrow and sclerotized posteriorly to form a "M" shaped projection. Ostium bursae funnel shaped. Ductus bursae about the same size as ostium bursae, slightly sclerotized. Ostium bursae oblong oval, with slightly sclerotized spiral ridge. Bursae projection near round, with inclining serrulate ridges.

This species is similar to *Dendrophilia neotaphronoma*

D. sophora adult (Li Houhun)

D. sophora medium sized larva (edited by Zhang Runzhi)

D. sophora damage to compound leaf (edited by Zhang Runzhi)

Ponomarenko but can be distinguished by 1) clear marks on forewings; 2) spiral ridges on ostium bursae in female genitalia; 3) near round bursae projection with inclining serrulate ridges.

Eggs Oblong oval, 0.38-0.40 mm × 0.18-0.26 mm. Milky white at first, gradually turn to reddish before hatching.

Larvae Mature larvae 6.42-9.20 mm long. Head and pronotum black. Propleuron contains two black blotches with the larger one on top. End of abdomen reddish to dark red, with light fine hairs. The last few abdominal segments contain black long hairs. There are six caudal spines, the middle two long and robust. Legs generally whitish except brown in front legs and middle and hind tarsi. Proleg crochets in biordinal circle discontinuous centrally. Young larvae similar to mature larvae except yellow and white abdomen. Larvae are greenish in color during leaf rolling or leaf tying. Over 90% of the mature larvae are reddish or deep red in color, the others are yellowish white. Male larvae with a brown blotch on the 5th abdominal tergite.

Pupae 4.82-5.90 mm long, brown and darker on both ends. Covered by dense yellow setae, especially on head, prothorax, and end of the abdomen. Anal opening contains sparse and long hairs. The last 2-3 abdominal segments bend ventrally. Spiracles large and prominent.

Biology and behavior *Dendrophilia sophora* occurs at least two generations a year in Yanglin area in Shaanxi province. Adults of overwintering generation begin to emerge in early April. It occurs three generations and overwinter as 2nd instars in wounds or bud crevices on branches in Shanghe of Shandong province. Overwintering larvae resume feeding by boring into leaf buds in late March. Pupae start to appear in early May and peak between mid- and late May. Adults first appear in late May. Peak egg period falls between early to mid-June. Larvae of the first generation appear in early June with first pupae appear in early July. Peak egg laying period for this generation falls between late July and early August. Larvae of the second generation start to appear in late July, with a peak feeding period of mid- to late July. Pupation occurs in mid-August. Adults of the second generation emerge in late August although they are not as synchronized as those of the first generation. Larvae of the third generation appear in early September. They feed for a short period before constructing thin cocoons inside bud crevices or wounds to overwinter.

Adults are strong fliers and very active. They rest during the day and are active at night. They are attracted to lights. Adults live for 3-5 days normally but can survive for maximum of 14 days when provided with 10% sugar water. The average longevity of the overwintering generation adults is 8.4 days. Adult longevity is 6.0 days (females) and 5.2 days (males) for the first generation, 8.0 days (females) and 7.2 days (males) for the second generation. Sex ratio is 1:1.1 (females: males). Adults can mate and lay eggs within 1-2 days after emergence. Eggs are laid singly (rarely in clusters of 2-3 eggs) inside leaf buds, between hairs at the base of underside of the main leaf veins, underside of young leaves, or flower buds and inflorescence. Egg period lasts for 7-11 days.

Larvae tie two leaves together initially and feed inside. As many as nine leaves are tied together as they grow and feed on the leaves. They quickly exit rolled leaves, spit silk threads, and fall when disturbed. Larvae often break stalks of the top leaflets through feeding, cause leaf wilt within a few days, and result in the presence of one yellow leaflet inside each roll. This is the typical damage symptom of this pest. Larval feeding creates twisted, yellow, and dry leaf residues inside. The affected leaves eventually become yellow and fall. Larvae migrate to other leaflets during development, with 43% of the overwintering larvae move to different leaflets. There are three types of damage from the first generation larvae. 1) Rolling tip leaflets similar to the overwintering generation (60%). 2) Attach two neighboring leaflets and feed along the main veins (40%). This occurs when eggs are laid on the underside of old leaflets. 3) On large trees with inflorescences, most larvae feed on inflorescences due to the lack of tender young leaves on branch tips. Only a small portion of the larvae feed on underside of the mature leaves. Between 40%-90% inflorescences can be damaged, with 1-6 larvae found in each inflorescence. Each larva ties 3-8 flower buds together with silk and feed among them. Damaged flower buds become black and empty. Larval stage lasts for about 28 days. Mature larvae construct thin cocoons inside rolled leaves and pupate between mid- and late May. Pupal stage lasts for 8-10 days. Adults of the first generation appear in late May, with peak emergence in mid-June.

The majority of the second generation larvae tie leaves. A small portion roll leaves on new shoots. Some may also feed on inflorescences due to the long flowering period of the host trees, which could last from mid-June to mid-October. Most young larvae of the third generation mine in the underside of the leaves. The second instar larvae of this generation emerge from the leaves and construct thin cocoons inside bud crevices or wounds on branches to overwinter. The number of larval instars were six (72.1%), five (16.4%), or seven (11.5%). The difference in the number of instars is probably due to hereditary factors. Development time is dependent upon food, temperature, humidity, and photoperiod.

Mature larvae pupate at different locations. Most larvae from the overwintering generation pupate inside the rolled leaves, with a small portion pupate in grass or leaves on the ground. About three quarters of the first generation larvae pupate inside crevices and wounds on branches or grass and leaves on the ground, while the rest remain inside the rolled leaves. For those feed among tied leaves, only 20% pupate among the tied leaves, while 80% pupate inside crevices and wounds on branches or grass and leaves on the ground. For those feed on inflorescences, pupation occurs in crevices and wounds on branches. For larvae of the second generation, about one quarter of those feed among tied leaves pupate among the tied leaves, the rest pupate in crevices and wounds on branches or pith of branches with a diameter > 0.5 cm. *Dendrophilia sophora* are attracted to lights and therefore damage is more severe along forest edges. Up to 100% trees along the roads are damaged during outbreaks. Co-occurrence with *Cyamophila willieti* can cause severe damage.

Prevention and treatment

1. Conserve natural enemies. Major nature enemies of the larval stage include parasitoids such as *Apanteles* spp. (two species), *Itoplectis* sp., *Hyssopus nigritulus*, *Pediobius cassidae*, *Tetrastichus* sp., and *Dimmockia secunda*. Predators include *Hierodula patellifera*, ants, and spiders. Pupal stage natural enemies include parasitoids *Coccygomimus* sp. and *Tetrastichus* sp. The *Apanteles* spp. are important larval parasitoids of the overwintering

D. sophora adult (dorsal view) (edited by Zhang Runzhi)

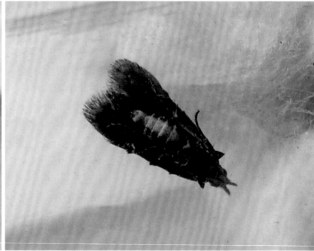

D. sophora adult (ventral view) (edited by Zhang Runzhi)

generation. The highest parasitization rate was 40.6%. The dominate parasitoids to the first generation are *Tetrastichus* sp. and *Hyssopus nigritulus*, with a combined parasitism of 15%-30%. *Apanteles* sp. is the major natural enemy to the larvae of the second generation, with an average parasitism of 40% and maximum of 66.7%. Spiders hunt on *D. sophora* larvae when they come out of rolled leaves for food or migration. Spider predation has significant impact on young larvae of the first generation or second generation before leaf rolling.

2. Application of 20% dinotefuran at 1,000× results in > 95% mortality for larvae. Spray 2×10^{10} PIB/mL nuclear polyhydrosis virus at 1,000-1,500×.

References Li Houhun et al. 1998, Yan Jiahe et al. 2001, Li Houhun, 2002.

(Li Houhun, Yan Jiahe)

D. sophora empty pupa (edited by Zhang Runzhi)

D. sophora mature larva (edited by Zhang Runzhi)

D. sophora pupae (edited by Zhang Runzhi)

294 Dichomeris bimaculatus Liu et Qian

Classification Lepidoptera, Gelechiidae
Common name Chinese fir cone gelechiid

Importance An outbreak of *D. bimaculatus* occurred in Chinese fir forests in southern China with an average damage rate of 11.6% (maximum 43.1%) on cones. It is the most important cone pest of Chinese fir with significant negative impact on seed production in seed orchards.

Distribution Fujian.

Host *Cunninghamia lanceolata*.

Damage Larvae feed on cone scales, cone shafts, and seeds. They cause discoloration of cones, small withered cones, and abnormally shaped cones.

Morphological characteristics **Adults** Length 4-7 mm long with a wingspan of 10-13 mm. Head, antenna yellowish brown. Forewings lanceolate, silver, with two black blotches in the middle. Hindwings trapezoid, silver. Both wings contain gray and long marginal hairs.

Larvae Length 8-12 mm, white. Head, pronotum and anal plate brown. Anterior margin of each thoracic segment and abdominal tergum reddish brown. Body with alternate red and white rings.

D. bimaculatus adult (Qian Fanjun)

D. bimaculatus larva and its damage (Qian Fanjun)

Biology and behavior *Dichomeris bimaculatus* occurs one generation a year and overwinters as larvae in Fujian province. Mature larvae pupate at overwintering sites in March. Adults appear from mid-March to late May, with a peak emergence in mid-May. Larvae of the new generation appear in April. They bore into withered current-year male inflorescences. During June to July, they move into underdeveloped small young cones. They migrate to new cones in August. Each larva can migrate one more time after that. Instars for overwintering larvae are extremely unsynchronized. Larvae overwinter in galleries in infested cones or withered current-year male inflorescences between mid- and late November.

Prevention and treatment

1. Plant resistant asexual Chinese fir varieties such as #431 and #14.
2. Heavy pruning of stands with old trees, high canopy cover, and heavy infestations.
3. Remove all cones (healthy, infested, discolored, and withered) and remaining male inflorescences to destroy overwintering larvae.
4. Apply deltamethrin fog to control adults during peak emergence.

References   Qian Fanjun et al. 1990, 1992, 1994, 1995; Liu Youqiao and Qian Fanjun 1994.

(Qian Fanjun)

D. bimaculatus pupa (Qian Fanjun)

D. bimaculatus damage to cone (Qian Fanjun)

Carposina sasakii Matsumura

Classification Lepidoptera, Carposinidae
Synonym *Carposina niponensis* Walsingham
Common name Peach fruit moth (borer)

Importance *Carposina sasakii* is an important fruit pest of various fruit trees. Infestation reduces fruit quality or renders fruits not suitable for consumption.

Distribution Beijing, Tianjin, Hebei, Shanxi, Inner Mongolia, Liaoning, Jilin, Heilongjiang, Shanghai, Jiangsu, Zhejiang, Anhui, Fujian, Shandong, Henan, Hubei, Hunan, Sichuan, Yunnan, Shaanxi, Gansu, Qinghai, Ningxia, Taiwan; Japan, Korea Peninsula, and Russia.

Hosts *Malus pumila*, *M. asiatica* var. *rinki*, *Zizyphus jujube*, *Crataegus pinnatifida*, *Prunus persica*, *P. armeniaca*, and *Pyrus* spp. in China. *P. amygdalus*, *P. mandshurica*, *C. cuneata*, *M. toringo*, *Sorbus commixta*, *M. micromalusin* in Korea Peninsula. Additional host includes *Phoenix actylifera*.

Damage *Carposina sasakii* infests pome fruits such as apple. Larval feeding inside the fruits usually results in tear-shaped fruit gelatin at the entrance holes. The gelatin dries up and appears as white wax powder later. The entrance holes become black. Larvae usually reach the center of the fruits after entry. They feed in the pulp and excrete frass inside the galleries. Infested young fruits are unable to develop fully, resulting in abnormal fruits referred as "monkey head fruits". Damage on jujube, peach, and hawthorn is usually found in pulp around the core.

Morphological characteristics **Adults** Females: length 7-8 mm, wingspan 16-18 mm. Males: length 5-6 mm, wingspan of 13-15 mm. Body gray in color. Forewings grayish white, with one large dark blue triangular-shaped blotch in the middle near anterior margin, and with seven bundles of bluish brown inclining scales in the middle and the base. Labial palpi short and curve up and forward in males, but long and straight in females. Each antennal segment has lateral villi on the ventral side in males. These kinds of villi are lacking in females. **Eggs** Orange at the beginning, turns to reddish orange or fresh red later. Near global with dense punctures on surface. There are 2-3 rings of "Y" shaped external growth on top. **Larvae** Length 13-16 mm, oblong oval, orange red or peach red. Head yellowish brown. Pronotum and anal plate brown. **Cocoon** Flat oval for cocoons in over-

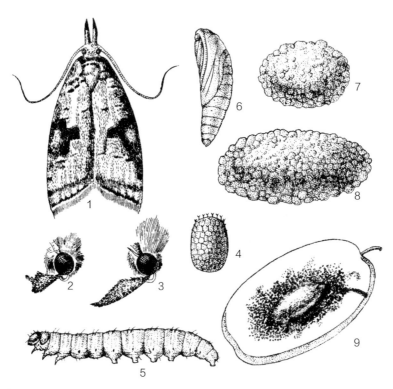

C. sasakii. 1. Adult, 2. Male adult head, 3. Female adult head, 4. Egg, 5. Larva, 6. Pupa, 7. Winter cocoon, 8. Summer cocoon, 9. Damage to fruit (Zhou Defang)

C. sasakii adult (NZM)

wintering generation, 4.5-6.2 mm long, tight and dense. Summer cocoons spindle-shaped, 7.8-9.8 mm long, loose, with a hole on one end.

Biology and behavior *Carposina sasakii* occurs one to two generations a year on apple and jujube; one generation a year on bitter jujube, hawthorn, pear, and apricot; and one to three generations a year on pomegranate along the abandoned Yellow River in Henan, near Nanjing in Jiangsu, western Sichuan, and Huaiyuan County in Anhui. Mature larvae construct flat oval cocoons in 3-10 cm deep soil under the trees. Larvae emerge from the soil the following year and pupate on soil surface. Adults lay eggs at leaf vein joints or depressions on fruit stalks. Larvae feed for 20-30 days and drop out of the fruits. Larvae of one-generation per year population enter soil for overwintering after falling out, whereas those of two-generation construct cocoons on ground surface, pupate, emerge as adults, and start a new generation.

Life cycle of *C. sasakii* differs in different areas. In apple production areas in northeastern China, Shandong, and Shaanxi, overwintering larvae emerge from soil in early May and peak between late May and mid- to late June; whereas for those attack jujube and peach, they emerge from late June to early July; those feed on pear and hawthorns emerge in mid-July. The time from initial infestation to dropping out of the fruits varies from 14-35 days on apple, 9-26 days on jujube, 45 days on hawthorn, and 20 days on pear. Those dropping out of the fruits in late August are usually from single generation populations, whereas those dropping out before late July can develop into the second generation.

Prevention and treatment

1. Silviculture. Remove weeds around base of the trees through deep plowing before overwintering larvae emergence. Cover the tree base with soil to attract overwintering larvae. Expose the overwintering cocoons by spreading the soil during the winter. Turn over 4-cm deep soil under tree crown during summer cocoon construction can reduce damage by the next generation. Put some soil around fruit storage after harvest for median or late maturation varieties can attract larvae exiting from the fruits. Bury and kill the larvae.

2. Mechanic control. Trap adults with black and green lights.

3. Manual control. Bag fruits before adults laying eggs. Remove infested fruits.

4. Biological control. 1) Soil spray of *B.t.* (300 kg/ha) or 25% phoxim micro capsules (2.25 kg/ha) diluted in 150 kg water at tree base when larvae dropping out of fruits, cover the soil with grass. 2) Soil spray of *Steinernema carpocapsae* and *Heterorhabditis bacteriophora* nematodes at the rate of $1-2 \times 10^8$ infective juveniles per 667 m^2. 3) Release parasitic wasp *Chelonos chinensis* in areas not suitable for nematode application. One release is enough if 40%-50% host cocoons become parasitized, and additional release if only 30% cocoons become parasitized. 4) Use attractants made of Z7-20:ket-11 and Z719:ket-11 (19:1) for monitoring and control.

5. Chemical control. 1) Flatten the soil around tree base before larvae emergence, apply chlorpyrifos or triazophos, cover the soil with plastic film, and seal the edge of the film with soil. 2) Foliage spray with 45% chlorpyrifos at 1,000-1,500×, 25 g/L lamda-cyhalothrin at 800-1,000×.

References Beijing Agricultural University 1983; Hua Baozhen 1992, 1996; Hua Lei and Ma Gufang 1993; Hou Wuwei, Ma Youfei et al. 1994; Hua Lei 1995; Bao Jianzhong and Gu Dexiang 1998; Li Dingxu 2002; Lü Peike, Su Huilan et al. 2002; Mu Wei, Zhang Yueliang et al. 2007.

(Liu Li, Tong Dequan)

296 *Cossus cossus chinensis* Rothschild

Classification Lepidoptera, Cossidae
Synonym *Cossus chinensis* Rothschild

Distribution Jiangsu, Zhejiang, Anhui, Fujian, Jiangxi, Shandong, Hubei, Hunan, Guangxi, Sichuan, Yunnan, Shaanxi, Gansu and Ningxia.

Hosts *Robinia pseudoacacia, Populus* spp., *Juglans regia, Carya illinoinensis, Castanea mollissima, Salix babylonica, S. matsudana, S. matsudana, Citrus reticulate, Diospyros kaki.*

Damage *Cossus cossus chinensis* larvae bore in the cambial region and the outer sapwood, creating irregular longitudinal and transverse galleries. They cause cankers and decays on trunks and branches. Infestation weakens growth, resulting branch diebacks and death of crown, and eventually total tree mortality.

Morphological characteristics **Adults** Females 32-39 mm long with a wingspan of 68-87 mm, males 23-32 mm long with a wingspan of 56-68 mm. Antenna pectinate. Tufted hairs on vertex, collar, tegula, and the entire thoracic nota creamy yellow. Forewings without obvious markings except a short black mark along the anterior margin. There is a white cloudy mark below the apex and the anterior half of the median cell, as well as post median cell and before vein 1A. Hindwings dark, with more obvious marks on upper surface. Middle tibia contains a pair of spurs, the hind tibia has two pairs of spur with the median spur located at 1/5 length of the tibia. Hind tarsus normal, not enlarged with reduced pads. **Larvae** Newly hatched larvae pinkish, 3-4 mm long, 0.5-0.8mm wide. Mature larvae 60-92 mm in length and 13-18 mm in width. Body purplish red with luster on the back, alternate yellow and white on the side, and alternate yellowish and pink on the ventral surface. Head purplish black. Pronotum reddish brown, with a large black blotch that is separated by a longitudinal ridge. Abdominal segments with densely distributed small verrucae. Each verruca contains a small brownish or yellowish white hair. Thoracic legs yellowish brown. Prolegs yellowish, with biordinal circular crochets. Anal prolegs well developed.

Biology and behavior *Cossus cossus chinensis* occurs one generation in two years and overwinters as larvae inside the trunk in Kunming region in Yunnan. Mature larvae start to pupae in early March. Adults begin to appear in late March and emergence ends in early June. Newly hatched larvae first seen between mid- and late April. Larvae overwinter by late November.

Adults are attracted to lights. They rest during the day and are active at night. Males are strong fliers that can fly up to a hundred meters each time. Adults can mate right after emergence. Mating occurs mostly between 8:00-9:00 at night and lasts for 20-30 minutes. Eggs are laid singly or in clusters in bark crevices of branches and trunks. Each female produced 385-768 eggs. Egg stage lasts for 15-20 days. Adult longevity is 4-8 days for females and 3-7 days for males. Sex ratio (females: males) is 1:0.88 in wild populations and 1:1.44 for laboratory colonies. Most eggs hatch in the morning. Newly hatched larvae feed aggregately with dozens of individuals per cluster. They feed on the cambium around old larval galleries first, gradually reaching the xylem later. Barks of the damaged area usually protrude and crack and fall off eventually, leading to cankers. Most mature larvae construct cocoons inside the galleries under the bark for pupation. Pupal stage lasts for 19-28 days.

Prevention and treatment

1. Silviculture. Prune infested and withered branches timely, remove infested dying trees. Create mixed plantations.

2. Mechanic control. Trapping with black lights during adult period.

3. Chemical control. Insert aluminum phosphide stripes in larval galleries and seal with mud during larval stage. Spray deltamethrin and fenvalerate on its preferred large branches with rough barks during adult oviposition period or egg-hatch period. Spray 2×10^{10} PIB/mL nuclear polyhydrosis virus at 1,000-1,500×.

References Wu Lin and Huang Zhiyong 1989; Hua Baozhen, Zhou Yao, Fang Deqi et al. 1990; Xu Gongtian 2003; Chen Peng, Liu Hongping, Lu Nan et al. 2006.

C. cossus chinensis adult (Liang Jialin)

(Zong Shixiang)

297 *Cossus orientalis* Gaede

Classification Lepidoptera, Cossidae
Synonyms *Cossus cossus orientalis* Gaede, *Cossus cossus changbaishanensis* Hua et al.

Distribution Beijing, Tianjin, Hebei, Shanxi, Inner Mongolia, Liaoning, Jilin, Heilongjiang, Shandong, Henan, Shaanxi, Gansu, and Ningxia; Russia (Siberia), Korea Peninsula, and Japan.

Hosts *Populus cathayana*, *P. Canadensis*, *Populus × Canadensis* 'Sacrou79', *P. alba* var. *pyramidalis*, *Salix matsudana*, *S. babylonica*, *S. matsudana* f. *toruosa*, *Ulmus pumila*, *Sophora japonica*, *Robinia pseudoacacia*, *Betula platyphylla*, *Malus baccata*, *Fraxinus chinensis*, *Prunus padus*, *Pyrus bretschneideri*, *Prunus persica*, and *Syringa oblate*.

Damage Larvae feed on xylem inside branches, trunks, and roots. They create irregular galleries. Infestations lead to reduced growth, branch dieback or windbreaks, or even total tree mortality.

Morphological characteristics **Adults** Robust and grayish brown. Length 28.1-41.8 mm for females and 22.6-36.7 mm for males. Wingspan 61.1-82.6 mm for females and 50.9-71.9 mm for males. Antennae singly pectinate with wider teeth in the middle, and smaller teeth apically. Tufted hairs on vertex and collar fresh yellow. Tegula and thoracic nota earthy brown. Anterior half of the metanotum dark brown; the posterior half of the metanotum alternate among white, black and yellow. There is a black transverse band on mesonotum. Basal half of forewings silver, with eight short black marks on anterior margin. Median cell contains two short transverse lines. Apical half brown. There is a black stripe starting from the end of Cu_2 extending to and vertical to the anterior margin. Marginal hairs grayish brown. Hindwings light brown. Median cell white. There are obvious brown linear marks on the underside of the hindwings, with a conspicuous dark brown circular blotch in the middle. Middle tibia contains a pair of spurs, whereas there are two pairs of spurs on hind tibia. First tarsal segment enlarged. Central pads of claws vestigial. Adults dimorphic with yellowish brown type and brownish type. **Larvae** Robust and flat cylindrical. Head black. Thorax and abdomen nota purplish red with faint luster. Abdominal sternites peach red. Pronotum contains a large black blotch. There is a white longitudinal line in the blotch that reaches the central point. Metanotum contains a dark brown rectangular blotch, whereas mesonotum contains two brown circular blotches.

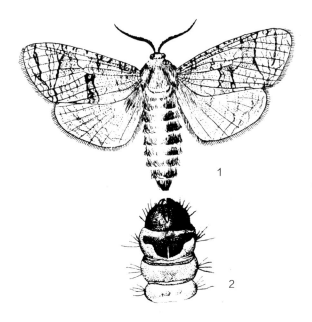

C. orientalis. 1. Adult, 2. Larve (head and pronotum)
(Hu Xingping)

C. orientalis adult (Xu Gongtian)

C. orientalis adult (Xu Gongtian)

C. orientalis larva (Xu Gongtian)

C. orientalis mature larva (Xu Gongtian)

C. orientalis tunnel (Xu Gongtian)

C. orientalis damage (Xu Gongtian)

Anal plate sclerotized. Proleg crochets triserial circular. Anal proleg crochets in biserial bands.

Biology and behavior *Cossus orientalis* occurs one generation in two years in Shandong and Liaoning. It overwinters as larvae. Adult emergence falls between late April and mid-June and peaks from early to mid-May. Adult period varies between years. Adults start to fly, mate, and lay eggs after emergence. Egg stage lasts for 13-21 days. Newly hatched larvae feed in trunks. They overwinter by mid- to late September when after 8-10 molts. Overwintering larvae become active again in late March next year. Peak damage period is from early April to mid- to late September. The 15th-18th instar larvae bore into soil, construct thin cocoons, and overwinter by the end of fall. These overwintering larvae construct the second cocoons in the spring of the third year for pupation. Adults appear in early to mid-April.

Newly hatched larvae prefer feeding on the cambium on trunks and branches aggregately. They enter the xylem later. Infested branches and trunks usually have white or reddish brown frass.

Prevention and treatment

1. Silviculture. 1) Modify forest composition such as replacing vulnerable *Populus cathayana* and *Populus × Canadensis* 'Sacrou79' with *P. alba*. 2) Remove high-pest-density and low value forest belts to reduce source populations. 3) Prune carefully to avoid rough wounds. Avoid pruning during egg-laying period in the spring. 4) Create mixed forest with canopy cover higher than 70%.

2. Pheromone trapping. Trapping with artificial sex pheromone mixture B (5E)-dodeca-5-dienyl acetate.

3. Biological control. Apply nematode *Steinernema bibionis* T335 at the concentration of 1,000 nematodes/mL.

4. Chemical control. Apply 50% phoxim at 1,000-1,500×, or 50% thiamethoxam at 30-40 g/ha for newly hatched larvae before they enter the bark. Inject 25 g/L lamda-cyhalothrin at 20-30×, 45% profenofos and phoxim mixture at 30-60×, or 12% thiamethoxam and lamda-cyhalothrin mixture at 30-60× to tree trunks× into larval galleries.

References Fang Deqi and Chen Shuliang 1982; Fang Deqi et al. 1984a, 1984b, 1986; Hua Baozhen, Zhou Yao, Fang Deqi, et al. 1984; Fang Deqi, Chen Shuliang, Li Xianchen 1992; Xiao Gangrou (ed.) 1992.

(Li Xianchen, Chen Shuliang, Lou Shenxiu, Fang Deqi)

298 *Deserticossus arenicola* (Staudinger)

Classification Lepidoptera, Cossidae
Synonyms *Cossus arenicola* Staudinger, *Holcocerus arenicola* (Staudinger)

Distribution In desert areas in Inner Mongolia, Shaanxi, Gansu, Qinghai, Ningxia, and Xinjiang; Turkey, Russia, Afghanistan, Mongolia, and Iran.

Hosts *Salix psammophila*, *S. microstachya*, *Hippophae rhamnoides*, *Caragana korshinskii*, *Hedysarum fruticosum*.

Damage Larvae of *D. arenicola* feed on roots of *S. psammophila*, especially those exposed main basal roots of trees facing the wind or on top of the sand dunes. They usually bore downwards along the roots for an average of 40-50 cm, reaching up to 110 cm deep in the soil. Longitudinal larval galleries intertwine with each other over large areas. As many as several dozens of larvae are found in a single gallery when population density is high. By the end of the infestation, they empty the entire root system of the host trees and fill the galleries with purplish red wood dusts and frass, resulting in tree mortality.

Morphological characteristics **Adults** Males 20-24 (average 22) mm long with a wingspan of 45-56 (average 50.5) mm, grayish brown. Antennae filiform, robust with no pectinate segments. Labial palpi moderate in size, close to gena surface, reaching almost to the base of the antennae. Collar, tegula, and metanotum brownish gray. Metathorax contains a 1-mm wide black transverse band, which is proceeded by a conspicuous white transverse band of similar width. Abdomen gray, with dense hairs laterally and tufted hairs anally. Shape of wings similar to that of *Yakudza vicarious*. Top angle of the forewings obtuse, slightly concaved centrally along the posterior margin. Forewings 2.2 times as long as wide at the anal angle. The underside of the wings dark gray. Median cell and the anterior two thirds of the wing dark-colored when compared to other parts. There is a small white spot at the end of the median cell and an obvious grayish vein-less area between median cell and 1A. Apical half of the wings covered by extremely fine reticular marks. Subcosta conspicuous but varies among individuals, sometimes it appears as a single black line

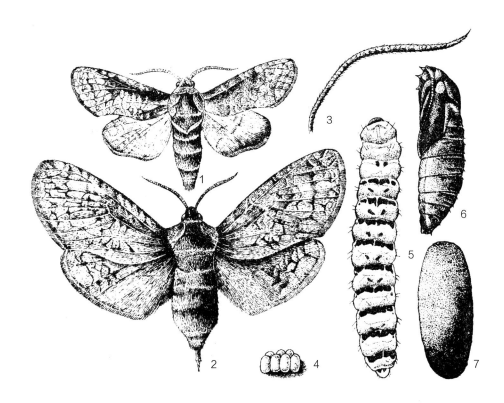

D. arenicola. 1. Male adult, 2. Female adult, 3. Female antenna, 4. Eggs, 5. Larva, 6. Pupa, 7. Cocoon (Zhu Xingcai)

parallel to the costa, or a few circles appears along the subcosta. Marginal hairs with weak cellular marks. The underside of the forewings gray with weak marks. Subcosta white with obvious short black marks, whereas no marks are found on the underside of hindwings. Middle tibia contains a pair of spurs, whereas there are two pairs of spurs on hind tibia, with the inner pair located at one third of the tibial length. Tarsal segments slightly bigger but not enlarged, central pad reduced.

Females 26-33 (average 29.5) mm long with a wingspan of 51-67 (average 59) mm. Antennae relatively short, not reaching center of subcosta of forewings. Abdomen extremely long, with protrude ovipositor at the end. Marks on forewings similar to those of the males. Hindwings contain very weak longitudinal marks.

Eggs 1.4-1.8 mm long and 1.1-1.3 mm wide, oval, grayish white and covered by brown short marks arranged in longitudinal lines initially. Turn to dark gray before they hatch.

Larvae About 3.0 mm long with a head capsule width of 0.42-0.62 mm when hatched. Head and pronotum dark brown, body reddish and covered by soft white hairs. Each abdominal segment contains two peach-colored blotches. Mature larvae 50-60 mm long, head black brown. Coronal suture and both sides of the frons purplish. Body yellowish white. Proscutum hard, with rectangular yellowish red marks that occupy three quarters of the pronotum. Pronotum contains three reddish blotches transversely, with a rectangular middle blotch surrounded by two inverse triangular blotches laterally. Abdomen contains two transverse bands made of reddish blotches on each segment. The front band is wide and darker, while the hind band is narrow and lighter in color, and half the size of the front band. Abdomen yellowish white ventrally, with purplish blotches on each segment. Thoracic legs orange with purplish red tibiae and tarsi. Abdominal prolegs reduced to only the last segment and crochets. Crochets in biserial rings for abdominal prolegs, while biserial bands for anal prolegs.

Pupae 19-37.8 mm long and 5.8-12.7 mm wide, dark brown with lighter abdomen ventrally. Ecdysial suture obvious on thorax and ends at the anterior margin of metanotum centrally. Ratio among the median lines of pronotum, mesonotum, and metanotum is 1:3:0.8. There are many wrinkles on pronotum and mesonotum. Mesonotum truncate posteriorly for the central protrusion. Wing pads for the forewings extend to the posterior margin of the third abdominal segment. There is a serrate tooth on each abdominal segment anteriorly and posteriorly for segments 2-7 in males, with the anterior tooth large and extends beyond the spiracle, while the posterior tooth small and not reaching the spiracle. Segment 8 contains one tooth anteriorly while segment 9 with one tooth centrally. Females are similar to males, except having one large tooth on segment 7 anteriorly.

Cocoons Oblong oval, with a length of 23.1-34.1 mm

D. arenicola adult (Zong Shixiang)

D. arenicola larva (Zong Shixiang)

and maximum width of 10.6-15.1 mm. Cocoons robust with dense texture. They are made of silk and sandy soil, earthy brown externally and white internally. One end is thinner than the other end. The thinner end contains the head of the pupa for easy adult emergence, while thicker end contains the abdomen. This feature can be utilized to locate pupal position inside the cocoon. Male pupae are small with pointed ends, whereas female cocoons are big with large and obtuse head ends that curved slightly in the middle.

Biology and behavior *Deserticossus arenicola* occurs one generation in two years in Qinghai and one generation in four years in Shaanxi. It overwinters as larvae inside the galleries. Mature larvae crawl out of the galleries and enter sandy soil to pupate in May. Adults start to emerge in late May or early June with peak emergence occurs in mid-June. Egg stage lasts for about 25 days. Eggs begin to hatch in late June and early July. Larvae overwinter in late October. Adults first appear in late May and adult emergence ends in mid-July. Peak emergence falls between mid- and late June.

Adults stay in *S. psammophila* bushes during the day and become active at night, especially between 20:30-22:00. Very few adults stay active between 1:00 and 4:00. Mating occurs between 20:30-21:30 and lasts for 6-27 minutes, with an average of 15 minutes. Females mate only once during their lifetime, whereas males can mate multiple times. Females start to search for suitable egg sites from bark crevices on roots or base of the tree near sandy soil after mating. Each female produces 38-665 eggs.

Adult longevity is closely related to air temperature. They live longer under consecutive low temperatures and shorter when high temperature prevails over a period of time, with the shortest adult longevity of 1 day and the longest of 8 days. Females live relatively longer than males. The female to male ratio is 0.68:1. Adults are attracted to lights.

Eggs are arranged tightly as egg masses, with an average of 81 eggs/egg mass, ranging from 15 to 186 eggs each. Egg stage lasts for an average of 24.8 days, ranging from 18 to 40 days. Between 72.6%-100% eggs hatch at the end.

Larvae bore into the bark directly and feed downwards along the bark after hatching, creating 10-20 cm long galleries before overwintering. Most galleries are created in the cambial region, with a few found in the sapwood. Larvae feed inside the heartwood in the second year. Wood dust and frass created during larval boring are extracted out of the tree through the entrance hole at the base of the tree. Larval excreta differ in shape, size, and color among different-year larvae (e.g. yellowish and threadlike for the current-year larvae, brownish and small patches for the second-year larvae, purplish and cylindrical for the third-year larvae, and dark brown and irregular big patches for the fourth-year larvae). Larvae

D. arenicola empty pupa (Zong Shixiang)

D. arenicola larval damage (Zong Shixiang)

migrate to different hosts by the end of the third year or in the fourth year due to high population density and insufficient food resources.

Prevention and treatment

1. Silviculture. Destroy the main trunks and the underground roots with a diameter of 10 cm of 3-5 years old *S. psammophila* through burning before ground thwarts in March.

2. Physical control. Take advantage of the adult phototaxy. Attract and kill large number of adults through trapping with black lights in areas with high population density between June and August.

3. Sex pheromone. Trap *D. arenicola* males using sex pheromone during adult stage. Place triangular traps in *S. psammophila* forests at 1 m above the ground at 100 m intervals. Collect trapped adults every 3-5 days and replace the lures every month. Trapping for four consecutive years resulted in significant reduction of pest damage in *S. psammophila* forests. This is the most effective monitoring control method.

4. Biological control. Important natural enemies include the lizard *Phrynocphalus froutalis* which has a predation rate of 17.78%, and the parasitic wasp *Coccygomimus* sp. with a parasitism of 8.58%. Spray *Beauveria bassiana* suspension in the forest after rain for larval control.

5. Chemical control. Inject chemical insecticides such as 45% profenofos and phoxim mixture, malathion, 25 g/L lamda-cyhalothrin through the trunk or the entrance holes, or 45% profenofos and phoxim mixture at 30-60×, 12% thiamethoxam and lamda-cyhalothrin mixture at 30-60× to tree trunks. Fumigate using avermectin tablets.

References   Hu Zhonglang, Chen Xiaoda, Yang Penghui et al. 1984a, b, 1987; Hua Baozhen, Zhou Yao, Fang Deqi et al. 1990; Guo Zhonghua, Yu Suying, Zhang Jiping et al. 2000; Li Dahai 2004, Xie Wenjuan 2008; Jing Xiaoyuan, Zhang Jintong, Luo Youqing et al. 2010a, b; Yao Donghua, Liu Peihua, Jing Xiaoyuan et al. 2011.

(Zong Shixiang, Luo Youqing, Hu Zhonglang)

299 *Deserticossus artemisiae* (Chou et Hua)

Classification Lepidoptera, Cossidae
Synonym *Holcocerus artemisiae* Chou et Hua
Common name Sagebrush carpenter worm

Importance *Deserticossus artemisiae* outbreaks since 2006 have destroyed sagebrush over wide areas.

Distribution Arid desert areas in western Inner Mongolia, northern Shaanxi, Gansu, and Ningxia.

Hosts *Artemisia ordosica*, *A. sphaerocephala*, and *Peganum harmala*.

Damage Larvae of *D. artemisiae* feed on main stems and roots of sagebrush. Newly hatched larvae bore into the cambium and xylem of roots. Larval feeding empties plant xylem, resulting in death of branches. Heavy infestation leads to mortality of the whole host plants. Infested hosts are easy to be uprooted from the soil due to dead and loose roots.

Morphological characteristics **Adults** Males 19.2-23.9 (average 21) mm long with a wingspan of 36.3-47.0 (average 42.9) mm, grayish brown. Antennae filiform, flat, without pectinate segments; extending beyond two thirds of the anterior margin of the forewings. Labial palpi long, yellowish brown; their tips black and obtuse, pointing towards the compound eye and reach half of its diameter. Tufted hairs on vertex, tegula, and the anterior thoracic portion grayish brown, with two black hair bands at the base of the wings. The posterior portion of the thorax contains two transverse bands with a white one in the front and the black one in the back. Abdomen grayish brown. Forewings with an obtuse apical angle. The underside of the anterior margin yellowish brown, with one row of small black dots. Anterior anal area slightly depressed in the middle. Base of forewings dark brown, forewings yellowish brown to grayish brown. There is an obvious large oval white blotch between median cell and 2A. 1A passes through the blotch. Veins beyond 2A dark brown. There are several dark longitudinal stripes between apical veins. Marginal hairs short, with dark brown marks. Hindwings brownish gray with yellowish brown base. Underside of the forewings dark gray, with a row of conspicuous black dots anteriorly. Marks on the apical half and marginal hairs indistinct. There are no marks on the underside of the hind wings. Middle tibia contains a pair of spurs. There are two pairs of spurs on hind tibia, with the median spur located at the apical two fifth. Hind tarsus slightly enlarged at base, with degenerated central pad.

Females 19.3-27.1 (average 23) mm long with a wingspan of 42.6-57.7 (average 48.4) mm. Abdomen robust, cylindrical, extremely long, and contains a protruded ovipositor. Shape of wings and the mark patterns on wings similar to those of the males. The oval white blotch and black venation in females are not as clear as those in males. Antennae only reach to one third of the anterior margin of the forewings.

Eggs Oval, 1.7-1.8 mm × 1.4-1.5 mm. Egg shells covered by longitudinal ridges and coated by a layer of dark brown sticky material that resembles the color of sagebrush roots.

Larvae Newly hatched larvae 4.8-5.2 mm long, reddish initially, gradually darken later on. Mature larvae fade from red to yellowish white with a shade of pink before pupation, with scattered purplish red blotches. Head dark brown. Proscutum yellow. Dorsal lines yellowish white, with a pair of near square purplish red blotches laterally. Each blotch contains a brown seta. There are irregular purplish red blotches between pleuron and spiracle line. Abdomen lightly colored. Thoracic legs yellow. Abdominal prolegs contain 42 to 69 crochets in uniordinal circle, some with alternate long and short crochets.

Pupae 23.1-34.1 mm long and 10.6-15.1 mm wide, dark brown. Head, thorax, and wing pads dark brown. There are three small projections on the anterior portion

D. artemisiae adult (Zong Shixiang)

D. artemisiae larva and damage (Zong Shixiang)

D. artemisiae eggs (Zong Shixiang)   *D. artemisiae* larva (Zong Shixiang)   *D. artemisiae* pupa (Zong Shixiang)

of the head. Abdomen brown. There are two rows of teeth on each of the 2nd to 5th segments; the teeth in the front row are larger than the teeth in the back row. Only one row of teeth is found on each of the 6th to 8th segments. Tip of the abdomen contains a pair of teeth.

Cocoons Length of 23.1-34.1 mm, width 10.6-15.1 mm. Earthy brown and made of mixture of silk and sandy soil. Oblong oval and slightly curved in the middle. The cocoon wall on one end is thinner than the other end, with the thinner end containing the head of the pupa for easy adult emergence. This feature can be used to identify pupa orientation inside the cocoon.

Biology and behavior *Deserticossus artemisiae* occurs one generation in two years and overwinters as various instars larvae inside the galleries in Ningxia. Mature larvae crawl out of infested sagebrush roots and enter sandy soil and pupate in May. Pupal stage lasts for 16-23 days, with an average of 20 days. Adults start to appear in early June and disappear by the end of August, with three occurrence peaks in early June, early July, and early August. Eggs first appear in mid-June. Larvae start to hatch in late June. Larvae of various instars begin to overwinter in mid-October.

Adults hide inside sagebrush, root collar, or ground cervices during the day. Adult activity starts at 20:00 and peaks between 21:00 to 23:00. Mating and egg laying occur shortly after adult emergence. Females mate only once in their lifetime. Females start to look for suitable locations to lay eggs after mating. They lay eggs over 2-3 hours period at various locations. Each female produces 80-100 eggs. Adults survive for a total of 2-5 days. After mating, the average longevity is 1.5 days for females and 2 days for males. Adults are strongly attracted to lights. Large numbers can be attracted with black lights from 1,000-2,000 m away.

Eggs are scattered in 1-2 cm deep soil around sagebrush, with some eggs attached to host roots. Egg stage lasts for 9-12 days. Newly hatched larvae bore into nearby sagebrush roots and feed initially in the cambial region or between cambium and sapwood. They later migrate to larger root collars to feed on wood and create galleries. Mature larvae enter nearby sandy soil and construct cocoons with soil on root surface for pupation. Adults are not very selective when laying eggs on hosts, with some eggs are laid on recently died plants. Larvae hatched from those eggs can still complete development by feeding on dead host material.

Prevention and treatment

1. Silviculture. Pull and destroy host plants with withered leaves or branches. Those plants most likely contain *D. artemisiae*.

2. Physical control. Attract and kill large number of adults through trapping with 200 W black lights in areas with high population density between June and August.

3. Biological control. Utilize larval parasitoids *Pyemotes* sp. results in moderate control.

4. Sex pheromone. Trap *D. artemisiae* males using sex pheromone during adult stage. Place triangular traps inside sagebrush at 1 m above the ground at 100 m intervals. Collect trapped adults every 3-5 days and replace the lures every month. Trapping for two consecutive years resulted in significant reduction of pest damage in sagebrush.

5. Chemical control. Spray with 45% profenofos and phoxim mixture, 25 g/L lamda-cyhalothrin, or 50% malathion water solution. Apply 45% profenofos and phoxim mixture at 30-60×, 12% thiamethoxam and lamda-cyhalothrin mixture at 30-60× to tree trunks. Spray 2×10^{10} PIB/mL nuclear polyhydrosis virus at 1,000-1,500×.

References Zhen Changsheng 1988a,b; Chen Xiaoda 1989; Hua Baozhen, Zhou Yao, Fang Deqi et al. 1990; Gao Zhaoning 1999; He Dahan and Zhang Kezhi 2000; He Dahan 2004; Xu Zhu 2004; Liu Aiping and Hou Tianjue 2005; Yao Yanfang, Yang Qin, Guo Haiyan et al. 2009; Zhang Jintong Luo Youqing, Zong Shixiang et al. 2009a,b; Zong Shixiang, Luo Youqing, and Cui Yaqin 2009.

(Zong Shixiang, Luo Youqing)

Eogystia hippophaecola (Hua, Chou, Fang et Chen)

Classification Lepidoptera, Cossidae
Synonym *Holcocerus hippophaecolus* Hua, Chou, Fang et Chen
Common names Seabuckthorn carpenter worm, Sandthorn carpenter worm

Importance *Eogystia hippophaecola* is the most important wood borer of seabuckthorns in China. Outbreaks since 2001 have resulted in serious economic loss to the seabuckthorns industry in the country.

Distribution Hebei, Inner Mongolia, Shanxi, Liaoning, Shaanxi, Gansu, Ningxia.

Hosts *Hippophae rhamnoides*, *Ulmus pumila*, *Malus pumila*, *Pyrus* spp., *Amygdalus persica*, *Salix psammophila*, *Prunus sibirica*, *Elaeagnus angustifolia*.

Damage Larvae of *E. hippophaecola* feed on the main stems and roots of seabuckthorns. Infestation on main trunk leaves cottony frass on the surface, resulting in shriveled barks. Infestation to roots produces frass particles around root base, leading to empty roots and tree mortality.

Morphological characteristics **Adults** Body grayish brown; males 21-36 mm long, wingspan of 49-69 mm; females 30-44 mm long, wingspan of 61-87 mm. Antennae of both males and females filiform, extending to the middle of the forewings. Tufted hairs on vertex and collar brownish. Thorax grayish white in the middle, the sides and posterior border and the tegula dark black. Forewings grayish brown, with one row of small black dots in the anterior margin. There are no obvious marks on wing surface. There are only indistinct short stripes between veins near the tip. Hindwings brownish without any marks, similar on both sides. Middle tibia with one pair of spurs. Hind tibia with two pairs of spurs, with the median spur located at three fifth of its length. Tarsus contains many black stylets ventrally. The end of each tarsal segment black. No claw pads on front tarsus.

Larvae Flat cylindrical, 2.02 mm long, head capsule 0.44 mm in width for newly hatched larvae. Mature larvae 60-75 mm long with a head capsule of 6.5-7.5 mm in width. Thoracic nota peach red, pronotum orange red with an orange yellow "W" shaped mark. Abdominal proleg crochets in biordinal circles. Anal proleg crochets in biordinal central bands.

Biology and behavior *Eogystia hippophaecola* occurs one generation in four years and overwinters as larvae inside the galleries in the roots in Jianping of Liaoning province. Mature larvae enter soil to pupate in early to mid-May. Adults start to appear in late May and disappear in early September, with two peak emergences - one in mid-June and one in late July. Eggs begin to hatch in early June. Larvae overwinter in late October.

Most adults emerge between 16:00 and 19:00. They stay on the ground motionlessly right after emergence, and start to move at about 20:00. Mating occurs the same day after emergence and lasts for 15-40 minutes, with a peak mating period observed at 21:30. Females mate only once during their lifetime, but males can mate multiple times. Females lay eggs mostly at night. They usually start to lay eggs between 20:30-22:00 on the second

E. hippophaecola male adult (Zong Shixiang)

E. hippophaecola adult and empty pupa (Zong Shixiang)

E. hippophaecola larvae and root damage (Zong Shixiang)

E. hippophaecola cocoon (Zong Shixiang)

E. hippophaecola pupa (Zong Shixiang)

E. hippophaecola egg (Zong Shixiang)

E. hippophaecola damage (Zong Shixiang)

day after mating. Eggs are laid in masses with a dozen to over one hundred eggs per egg mass. Each female produces 73-617 eggs out of the 134-641 eggs it may produce in the ovary. Egg stage lasts for 9-33 days. Sex ratio between females and males is 0.91:1. Longevity is 2-8 days for males and 3-8 days for females.

Larvae usually feed in groups of a few to over one hundred individuals. Newly hatched larvae feed on eggshells first before moving to cambial region in the trunk. Young larvae migrate from trunk surfaces to trunk bases and roots before overwintering. There are 16 instars in the larval stage. Mature larvae pupate in 10-cm deep soil around trunk base. They construct earthy cocoons and pupate inside after passing a prepupal stage. Pupal stage lasts for 26-37 days.

Prevention and treatment

1. Silviculture. Remove all *Hippophae rhamnoides* and their roots up to 20 cm deep. Kill larvae in roots through burning or chemical treatment.

2. Trapping. Trapping with black lights significantly reduces pest population. Trap *E. hippophaecola* males using sex pheromone. Place triangular traps inside sagebrush at 1 m above the ground at 100 m intervals. Collect trapped adults every 3-5 days and replace the lures every month. Trapping for four consecutive years resulted in significant reduction of pest damage in seabuckthorns.

3. Biological control. Appling nematode *Steinernema carpocapsae* resulted in an infection rate of 13.2%-48%. Application of *Beauveria bassiana* resulted in 11.5%-50% pest mortality. These methods are seldom used due to high cost and low parasitism.

4. Chemical control. Fumigate roots with avermectin, irrigate soil with fenvalerate, apply 45% profenofos and phoxim mixture at 30-60×, 12% thiamethoxam and lamda-cyhalothrin mixture at 30-60× to trunks. Apply 2×10^{10} PIB/mL nuclear polyhydrosis virus at 1,000-1,500×.

References Hua Baozhen, Zhou Yao et al. 1990; Tian Runmin and Tang Mengchang 1997; Luo Youqing, Lu Changkuan, Xu Zhichun et al. 2003a,b; Ma Chaode 2003; Jia Fengyong, Xu Zhichun, Zong Shixiang et al. 2004; Lu Changkuan, Xu Zhichun, Jia Fengyong et al. 2004; Lu Changkuan Zong Shixiang, Luo Youqing et al. 2004; Zong Shixiang, Jia Fengyong, Luo Youqing et al. 2005; Zong Shixiang, Luo Youqing, Xu Zhichun et al. 2006a,b,c.

(Zong Shixiang, Luo Youqing)

301 *Streltzoviella insularis* (Staudinger)

Classification Lepidoptera, Cossidae
Synonyms *Holcocerus arenicola* var. *insularis* Staudinger, *Holcocerus insulans* Staudinger

Importance *Streltzoviella insularis* is an important wood boring pest.

Distribution Beijing, Tianjin, Hebei, Inner Mongolia, Liaoning, Jilin, Heilongjiang, Shanghai, Jiangsu, Anhui, Fujian, Jiangxi, Shandong, Hunan, and Ningxia; Russia.

Hosts *Fraxinus chinensis*, *Broussonetia papyrifera*, *Syringa oblate*, *Ulmus pumila*, *Sophora japonica*, *Ginkgo biloba*, *Salix babylonica*, *Quercus acutissima*, *Malus pumila*, *Magnolia demudata*, *Platanus orientalis*, *Acer truncatum*, *Begonia spectabilis*, *Euonymus japonicas*, *Tamarix chinensis*, *Crataegus pinnatifida* var. *major*, *Toona sinensis*, *Prunus triloba*, and *Spiraea cantoniensis*.

Damage Infestation leads to wind damage, dead branches, or even tree mortality. Frass on branches, trunks, and on the ground reduces aesthetic value of urban trees.

Morphological characteristics **Adults** Grayish brown; females 18-28 mm long, wingspan 36-55 mm; males 14-25 mm long, wingspan 31-46 mm. Antennae filiform, with 58-60 flagella in females and 71-73 in males. Labial palpi grayish brown, extending to the anterior margin of the compound eyes. Tufted hairs on vertex and collar gray. Thorax dark reddish brown. Abdomen

S. insularis adult (Xu Gongtian)

long. Wing length is 2.1 times of the width at the anal angle. Wing surface dark colored with many small and broken longitudinal marks. Marginal hairs gray. There are variations in the marks and venation of the wings. Middle tibia contains a pair of spurs, while hind tibia with two pairs of spurs. Hind tarsus not enlarged, with reduced middle pad.

Larvae Small. Thorax and abdominal terga reddish. Abdominal sternites yellowish white. Head brown. Prothorax dark brown with marks and a white blotch in the middle. Semi-sclerotized marks on meta- and mesothorax brownish.

Biology and behavior *Streltzoviella insularis* occurs one generation in two years and overwinters twice as larvae in Jinan of Shandong province. Overwinter larvae start to pupate in early May. Adult emergence begins in early June and peaks between late June and mid-July. Emergence ends in mid- to late August. Egg hatching starts in early June and ends by mid-September, with most eggs hatch between early to mid-July. Adults mate during the night on the same day of emergence. Egg stage lasts for 9-21 days. Newly hatched larvae feed on eggshells first before entering the cambial region through the bark. Larvae bore in the wood after the third instar and form oval entrance holes. Larvae start to overwinter in October. Overwintering larvae resume feeding inside the hosts next year. They start to pupate inside the galleries in early May. Most larvae pupate between late May and late June. Pupation ends by early August. Adults begin to appear in early June.

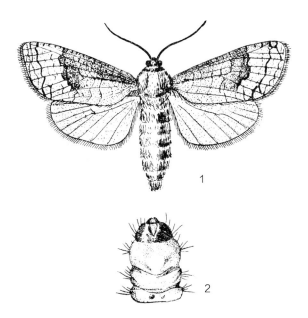

S. insularis.
1. Adult, 2. Larva (head and pronotum) (Hu Xingping)

S. insularis adult (Xu Gongtian)

S. insularis excretion hole from larva (Xu Gongtian)

S. insularis larve expelled material (Xu Gongtian)

S. insularis empty pupae in eclosion holes (Xu Gongtian)

S. insularis damaged tree (Xu Gongtian)

Larvae of *S. insularis* feed gregariously after hatching. As many as a few dozens of larvae can feed aggregately inside one gallery, resulting in severe damage to the trees. Trunks and branches of heavily infested trees are covered by cottony frass around the entrance holes.

Prevention and treatment

1. Silviculture. See methods described for *Cossus cossus orientalis*.

2. Biological control. Application of nematode *Steinernema feltiae* A11at the concentration of 2,000 nematodes/mL resulted in 99% larval mortality.

3. Chemical control. 1) Inject 25 g/L lamda-cyhalothrin or 20% fenvalerate EC at 500× into entrance holes; 2) Fumigate with aluminum phosphide sticks. 45% profenofos and phoxim mixture at 30-60×, 12% thiamethoxam and lamda-cyhalothrin mixture at 30-60× to tree trunks

References Gao Ruitong and Qin Xixiang 1983; Hua Baozhen, Zhou Yao, Fang Deqi et al. 1984; Jiang Sandeng, Fang Deqi, Chen Shuliang et al. 1987; Xiao Gangrou (ed.) 1992; Zhang Jintong and Meng Xianzuo 2001.

(Li Xianchen, Fang Deqi, Jiang Sandeng, Chen Shuliang)

302 *Yakudza vicarius* (Walker)

Classification Lepidoptera, Cossidae
Synonyms *Holcocerus japonicus* Gaede, *Cossus vicarius* Walker
Common name Elm carpenter moth

Importance *Yakudza vicarius* is an important wood boring pest.

Distribution Beijing, Tianjin, Hebei, Inner Mongolia, Liaoning, Jilin, Heilongjiang, Shanghai, Jiangsu, Anhui, Fujian, Jiangxi, Shandong, Henan, Hunan, Sichuan, Yunnan, Shaanxi, Gansu, and Ningxia in China; Russia, Korea Peninsula, Vietnam, Japan.

Hosts *Ulmus pumila*, *Robinia pseudoacacia*, *Quercus acutissima*, *Lonicera japonica*, *Zanthoxylum bungeanum*, *Salix* spp., *Populus* spp., *Juglans regia*, and *Malus pumila*.

Damage Infestation results in honeycomb-shaped damage on host trunk and roots.

Morphological characteristics **Adults** Robust, grayish brown. Females: length 25-40 mm, wingspan 68-87 mm. Males: length 23-34 mm, wingspan 52-68 mm. Antenna filiform, with 71 flagellar segments in males (apical three segments thin and small) and 73-76 segments in females (apical two segments thin and small). Labial palpi extend to the base of the antennae. Collar, tegula, and tufted hairs on vertex dark grayish brown. Tufted hairs on the anterior margin and later half of mesothorax bright white, hairs on scutellum grayish brown, anterior border of scutellum black dark brown. Forewings grayish brown, covered by dense dark brown stripes. Subcostal black and obvious, with a large dark brown blotch between outer cross vein and anterior margin of the median cell. Hindwings grayish without stripes. The underside of the hindwings contains brown stripes and a conspicuous round brown blotch at the center. Female frenulum is made of 11-17 stiff bristles. Middle tibia contains a pair of spurs. Hind tibia contains two pairs of spurs with the middle one located at one quarter of its length apically. Hind tarsus not enlarged, with reduced middle pad.

Larvae Flat cylindrical, fresh red, about 3 mm long when newly hatched, and 63-94 mm long at maturity. Thorax and abdominal terga fresh red. Abdominal sternites lighter in color. Head black. Prothorax brown and sclerotized, with a horizontal "B" shaped ring mark. This mark is dark brown in young instars and becomes lighter past fifth instar. There is a rectangular light blotch in

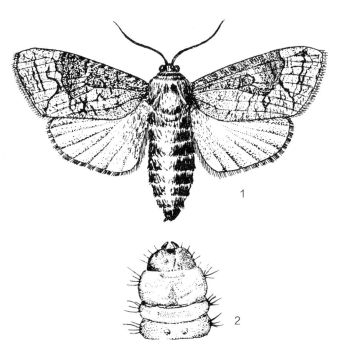

Y. vicarius. 1. Adult, 2. Larva (head and pronotum) (Hu Xingping)

Y. vicarius larva (Xu Gongtian)

front of the ring mark. Metathoracic tergum contains two round blotches. Abdominal prolegs deep orange in triordinal circles with 82-95 crochets, whereas anal prolegs in biordinal bands with 19-23 crochets.

Biology and behavior *Yakudza vicarius* occurs one generation in two years and overwinters twice as larvae in Shandong province. Pupation starts in late April. Adults appear from mid-May to mid-August although some adults are found in mid-September. Adults emerge both during the day and at night, with peak emergence between mid-May and early June. Adults mate on the same day after emergence. Most eggs hatch between mid- and late June. Current year larvae overwinter inside host trunk and roots, whereas second year larvae come out of galleries and enter soft soil by mid- and late October. They construct thin cocoons in the soil and overwinter. By early April, they exit the thin cocoons and construct new cocoons for pupation. A small portion of larvae that are not mature stay inside trunks and roots during the 2nd overwintering period. These larvae resume feeding the next April and construct cocoons for pupation after reaching the last instar.

Young larvae of *Y. vicarius* feed gregariously on eggshells and barks. They enter the cambium and sapwood through wounds and bark crevice when reaching the 2nd and 3rd instars. By the 5th instar, they feed on host roots near the soil surface, creating honeycomb-shaped damage inside the roots the soil surface.

Prevention and treatment

1. Silviculture. See methods described for *S. insularis*.

2. Biological control. 1) Application of 1×10^8 conidia/g suspension of *Beauveria* sp. resulted in larval mortality of 17.85%-100%, or application of 5×10^8 to 5×10^9 conidia/g suspension of *Beauveria* sp. paste to frass excretion holes resulted in 95% larval mortality. 2) Release parasitic wasp *Telenomus holcoceri* and protect predators such as *Erinaceus europaeus*, *Paratenodera sinensis*, and *Phrynocephalus frontalis*.

3. Chemical control. 1) Spray 25 g/L lamda-cyhalothrin at 800-1,000×, 20% fenvalerate at 3,000-5,000× to control newly hatched larvae. 2) Inject 50% malathion emulsifiable suspension or 20% fenvalerate emulsifiable suspension at 100-300× into entrance holes. Apply 45% profenofos and phoxim mixture at 30-60×, or 12% thiamethoxam and lamda-cyhalothrin mixture at 30-60× to tree trunks.

4. Fumigate with aluminum phosphide tablets at tree base or inside galleries for larval control.

References Hua Baozhen, Zhou Yao, Fang Deqi et al. 1984; Fang Deqi et al. 1987; Xiao Gangrou (ed.) 1992.

(Li Xianchen, Fan Deqi, Chen Shuliang)

303 *Polyphagozerra coffeae* (Nietner)

Classification Lepidoptera, Cossidae
Synonym *Zeuzera coffeae* Nietner

Importance Feeding by *P. coffeae* leads to regional wilt on host branches. Most severe damage occurs on fast growing eucalyptus. Infestation leads to headless trees and results in loss of dominate trunks on young trees and negative impact on vigor and volume growth.

Distribution Jiangsu, Zhejiang, Fujian, Jiangxi, Henan, Hunan, Guangdong, Guangxi, Sichuan, and Taiwan; India, Sri Lanka, and Indonesia.

Hosts *Eucalyptus* spp., *Casuarina equisetifolia*, *Castanea mollissima*, *Acacia confuse*, *Sapium sebiferum*, *Sinopimelodendron kwangsiense*, *Toona sinensis*, *Castanopsis hickelii*, *Litchi chinensis*, *Euphoria longan*, *Coffea Arabica*, *Cinchona ledgeriana*, *Metasequoia glyptostroboides*, *Robinia pseudoacacia*, *Psidium guajava*, *Juglans regia*, *Carya cathayensis*, *Pterocarya stenoptera*, *Platanus* spp., *Dalbergia hupeana*, *Citrus reticulate*, *Malus pumila*, and *Pyrus pyrifolia*.

Damage Infestation results in breaking and wilt of the tip of the main trunks, slowing down height growth of the host trees.

Morphological characteristics **Adults** Females 18-20 mm long with a wingspan of 40-50 mm, antennae filiform. Males 11-15 mm long with a wingspan of 33-36 mm, antennae pectinate basally and filiform apically. Body covered by grayish white scales, with six green and blue spots on the back of thorax. Forewings with bluish gray spots on each cell. Hindwings with lighter colored spots and one big bluish black spot in the middle. Each abdominal segment with three longitudinal stripes and one circular spot on each side. There are grayish black tiny spots arranged as a transverse band on each abdominal segment beyond the 3rd abdominal segment. **Eggs** Oblong oval, fresh yellow in color. **Larvae** About 30 mm long at maturity, red with a dark brown head and covered by white pubescence. Pronotum yellowish brown, with a near rectangular dark brown blotch on anterior half, and four lines of black serrate projections at the posterior margin. Anal plate dark brown. **Pupae** Length 18-26 mm. Reddish brown, with several anal spines at the end of the abdomen.

Biology and behavior *Polyphagozerra coffeae* occurs two generations a year and overwinters as larvae inside infested host branches. After overwintering, larvae

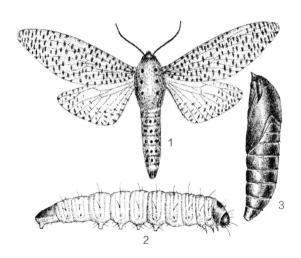

P. coffeae. 1. Adult, 2. Larva, 3. Pupa (Zhang Peiyi)

P. coffeae newly eclosed adult (Wang Jijian)

P. coffeae larva (Wang Jijian)

stay in the same locations or migrate to different branches to feed in late February. Adults of the overwintering generation appear between April and June, whereas first generation occurs between August and October.

Eggs are laid singly on the tips of young branches or axillary buds. Larvae enter the main trunks of young fast-growing eucalyptus and feed in the xylem. They feed around the circumference first then towards the heartwood. Due to the fast growing nature of eucalyptus, the main trunks of infested trees can still grow normally. There are little signs of infestation even when main shoots grow 70-80 cm above the entrance holes. However, these main shoots may break eventually by wind. Larvae finish their development and pupate inside the galleries. A mature larva will create near circular emergence cap under the bark. It chews a small hole on the bark below the cap. Silk mixed with sawdust will fill the galleries between the cap and the small hole. A pupal chamber of about 20-30 mm long is built above the emergence cap. Larva stays in the pupal channel upside down. Pupation occurs within 3-5 days. Pupal stage lasts for 13-37 days. Pupa breaks silk web and the emergence cap and exposes half of its body outside the emergence hole before emergence. Pupal case remains at the emergence hole after adult emergence, which serves as one of the most important characteristics of this pest. Adults can emerge all day long. Most adults mate between 20:00-23:00. Females lay eggs 1-6 hours after mating. Oviposition lasts for 1-4 days. Each female produces about 600 eggs, with most laid inside emergence holes. Egg stage lasts for 9-15 days. Larvae ballooning with silk after hatch and disperse with the help of wind. By late November, larvae stop feeding and overwinter in galleries with both ends sealed with silk, sawdust, and frass.

Prevention and treatment

1. Manual control. Larvae can be easily located in drooping tips of eucalyptus trees. Manually expose the larvae and destroy them with metal wire. Collect broken tips and destroy the larvae by fire.

2. Trapping. Trap adults with black lights during adult stage in heavily infested forests.

3. Chemical control. Spray host trees with 50% fenthion or 40% omethoate EC at 1,500× before larvae entering the xylem. Apply 45% profenofos and phoxim mixture at 30-60×, or 12% thiamethoxam and lamda-cyhalothrin mixture at 30-60× to tree trunks. Apply 50% thiamethoxam at 20-40× to young tree trunks.

References Huang Jinyi and Meng Meiqiong 1986, Xiao Gangrou (ed.) 1992.

(Wang Jijian, Tang Zuting, Qin Danren)

P. coffeae larva and damage (Wang Jijian)

P. coffeae pupa and eggs (Wang Jijian)

P. coffeae damage (Wang Jijian)

304 *Xyleutes persona* (Le Guillou)

Classification Lepidoptera, Cossidae
Synonym *Zeuzera leuconotum* Butler
Common name Oriental leopard moth

Distribution Beijing, Tianjin, Hebei, Shanxi, Shanghai, Jiangsu, Zhejiang, Anhui, Fujian, Jiangxi, Shandong, Henan, Hubei, Hunan, Guangdong, Guangxi, Sichuan, Gansu, Taiwan; Japan, Korea Peninsula, and India.

Hosts Polyphagous species with near 100 host species of broad-leaf trees, ornamentals, and crops, including *Fraxinus chinensis*, *Sophora japonica*, *Robinia pseudoacacia*, *Populus* spp., *Ulmus* spp., *Salix babylonica*, *Platanus* spp., *Koelreuteria paniculata*, *Toona* spp., *Eucommia ulmoides*, *Prunus persica*, *Malus pumila*, *Pyrus* spp., *Diospyros kaki*, *Ziziphus jujube*, *Crataegus pinnatifida*, *Malus seectabilis*, *Lonicera japonica*, and *Chimonanthus praecox*.

Damage Larvae feed in the tips of main branches and leave boring holes on branch surface. Infested branches become empty. Large amount of frass are expelled from the galleries. Infested hosts suffer branch die back, break, and death of all new growth when severe.

Morphological characteristics **Adults** Females 18-30 mm long with a wingspan of 33-46 mm, antennae filiform. Males 18-23 mm long with a wingspan of 29-37 mm, antennae pectinate basally and filiform apically. Body covered by grayish white scales. Wing surface contains many green and bluish black spots. Hindwings lighter in color. Pronotum contains six bright bluish black blotches arranged in two rows. Abdomen with several bluish black blotches of various sizes.

Larvae Newly hatched larvae dark brown, turn to deep red later. Mature larvae about 35 mm long, each segment with a black verruca which bears 1-2 setae. Pronotum with a sclerotized black blotch and a yellow median line. Anal plate usually contains a large bright black blotch.

Biology and behavior *Xyleutes persona* occurs one generation a year and overwinters as larvae inside infested host branches in Shandong, Anhui, and Tianjin. Overwintering larvae start to pupate in early May. Adults emerge from mid-May until early July. Newly hatched larvae appear in late May.

Most adults emerge in the afternoon between 16:00 and 20:00, with half of the pupal cases left inside the emergence holes. Adults mate at night on the same day of emergence. Eggs are usually laid in bark crevices and branch joints, with an average of 400 eggs laid by each female. Egg stage lasts for 20-30 days. Newly hatched larvae bore between the cambium and xylem. Larvae enter the xylem later and create longer galleries with multiple frass extraction holes. Mature larva constructs emergence holes right beneath the cambium. It seals galleries with silk and sawdust to form the pupal chamber. Larva stays in the pupal channel upside down before pupation. Pupal stage lasts for 17-24 days. Adults live 3-7 days in females and 2-5 days in males.

Prevention and treatment

1. Silviculture. Prune and remove Physical control. Trap adults with black lights during adult stage.

X. *persona* adult (Xu Gongtian)

X. *persona* larva (Xu Gongtian)

X. persona pupa (Xu Gongtian)

X. persona empty pupa (Xu Gongtian)

X. persona adult exit hole (Xu Gongtian)

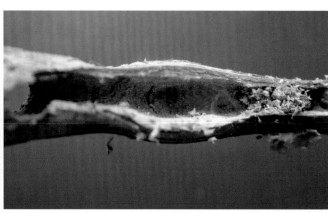
X. persona gallery (Xu Gongtian)

X. persona feces (Xu Gongtian)

X. persona damage (Xu Gongtian)

2. Biological control. Utilization of *Bt*, *Beauveria bassiana*, and *Pyemotes* sp. for larval control.

3. Chemical control. Spray host trees with 5.7% emamectin at 3,000-5,000×, or 50% fenitrothion at 1,000× during egg hatching. Insert cotton balls soaked with 25 g/L lamda-cyhalothrin into entrance holes or insert avermectin tablets into larval galleries and seal holes. Apply 45% profenofos and phoxim mixture at 30-60×, or 12% thiamethoxam and lamda-cyhalothrin mixture at 30-60× to tree trunks.

References Zhu Hongfu et al. 1975, Cao Zigang, Sun Shumei, and Gu Changhe 1978; Wang Yunzun 1989; Hu Qi 2002; Xu Gongtian 2003; Chen Shijun, Li Wenxiu, Wei Tingchun et al. 2006; Lü Yuli, Liu Shiling, Fan Jinrong et al. 2006.

(Zong Shixiang)

305 *Zeuzera multistrigata* (Moore)
Classification Lepidoptera, Cossidae

Importance *Zeuzera multistrigata* is an important pest of beach sheoak and other shelter belt tree species in the coastal areas of China.

Distribution Fujian, Guangdong, and Guangxi; India, Bangladesh, Myanmar, Japan.

Hosts *Casuarina equisetifolia, Acacia mearnsii, Dalbergia balansae, A. confuse, A. auriculaeformis, Grevillea robusta, Evonymus bungeanus, Michelia alba, Euphoria longana, Litchi chinensis, Phyllanthus emblica* in China; and *Cryptomeria fortune, Pyrus bretschneideri, Quercus* sp., *Santalum album, Ilex chinensis* in Japan.

Damage Young larvae bore into young branches and cause branch wilt. Older and mature larvae feed inside main trunks and main roots. Larval galleries destroy xylem and the vascular system of the host. Light infestations lead to stunned new growth and abnormal tree trunks. Severe infestations cause trees susceptible to wind damage or kill the hosts.

Morphological characteristics **Adults** Females 25-44 mm long with a wingspan of 40-70 mm, body grayish white.

Larvae Mature larvae 30-80 mm long. Yellowish or yellowish brown. Pronotum well developed, with a dish-like black blotch at the posterior margin. There are four columns of serrate spines and various small punctures on the blotch.

Biology and behavior *Zeuzera multistrigata* occurs one generation a year and overwinters as late stage larvae inside the galleries at the base of beach sheoak in Fujian. Overwintering larvae resume feeding in late February the following year. Pupation occurs mostly in early June. Pupal stage lasts for 20 days. Peak emergence falls between mid- and late June, whereas most eggs hatch between early and mid-July. Eggs stage lasts for 18 days, and larval stage lasts for 313-321 days.

Adults emerge between 14:00 and early morning of the next day, with most (about 95%) emerge between 16:00 and 20:00. Adults can mate within a few hours post-emergence, normally after 21:00. Females start to lay eggs about half an hour after mating. Eggs are laid in bark crevices at 3-5 different sites. Oviposition lasts for 2-3 days. Each female lays an average of 700 eggs, with 80-130 (average 110) eggs remains inside her body. Egg stage lasts for 16 days. Sex ratio for females and males is 1.46:1. Females live for an average of 6 days while males live for an average of 4 days. Males are more attracted to lights. This makes them the major catch by light trap. Females can only crawl slowly or take short flights due to their large abdomen filled with eggs.

There are 19 instars in the larvae. Newly hatched larvae feed on eggshells. They spin silk which covers the egg mass and stay under the silk film. They disperse two days later to nearby small branches. Larvae migrate multiple times before reaching 10th instar. Entrance holes are constructed at 20-100 cm height on the trunk. Frass are also excreted from the entrance holes. Larvae feed in the pith and form straight galleries, with little frass accumulation inside. During summer, larvae feed towards the roots 10-20 cm below the ground to avoid summer heat. Mature larvae chew a 1 cm-diameter circular emergence cap near bark surface before pupation. The cap is slightly separated from the bark. Most emergence holes are located at the trunk less than 200 cm above ground.

Prevention and treatment

1. Population monitoring and forecasting. Use light traps to attract and kill a portion of the male population, predict peak larval population for control actions.

2. Silviculture. Mixed forests have certain resistance to infestation. *Zeuzera multistrigata* does not feed on needles of slash pine, Japanese black pine, and Masson pine. They cannot bore into the trunk of these trees either. Although it can feed on lemon eucalyptus and chinaberry

Z. multistrigata adult (Xu Gongtian)

Z. multistrigata adults (Huang Jinshui)

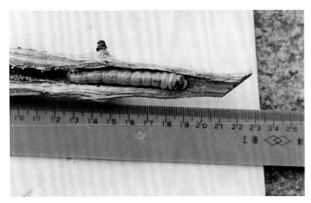

Z. multistrigata larva (Huang Jinshui)

Z. multistrigata eclosion hole (Huang Jinshui)

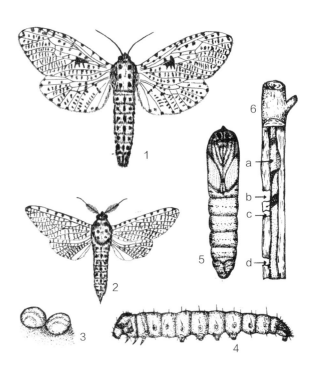

Z. multistrigata. 1. Female adult, 2. Male adult, 3. Eggs, 4. Larva, 5. Pupa, 6. Gallery (a. pupal chamber, b. eclosion hole, c. ventilation hole, d. excretion hole) (Huang Jinshui)

tree trunks at 100 cm above ground to attract males during adult stage resulted in moderate control. Peak trapping period is between 21:30-22:30. Replace females every two days.

5. Biological control. Application of *Beauveria bassiana* paste inside the main trunks is one of the most used methods to control larvae. It resulted in over 95% control. Application of *Steinernema* sp. at 1,000 nematodes/larva into the entrance holes. Ants and *B. bassiana* are the most prevalent natural enemies among the eight recorded species so far. Natural enemies for egg, larval, pupal, and adult stage include *Camponotus* sp., *Technomyrmex* sp., *Hierodula patellifera*, spiders, parasitic flies, *B. bassiana* and a bacterium species based of field surveys.

6. Chemical control. Apply 45% profenofos and phoxim mixture at 30-60×, or 12% thiamethoxam and lamda-cyhalothrin mixture at 30-60× to tree trunks.

References   Huang Jinshui, Huang Yuanhui, He Yiliang 1988; Huang Jinshui, He Yiliang, Lin Qingyuan 1990; Huang Jinshui, Huang Yuanfei, Gao Meiling et al. 1990; Huang Jinshui, Zhen Huicheng, Yang Huaiwen 1992.

(Huang Jinshui, Tang Chensheng)

for a short period, but they do not develop well and cannot finish the life cycle. Use these five trees species for planting mixed forests.

3. Manual control. Insert small sticks (10-15 cm long) into emergence holes to poke death of the pupae or make them unable to emerge through the emergence holes.

4. Pheromone trapping. Place newly emerged but unmated females in rearing cages, and place cages along

306 *Monema flavescens* Walker

Classification Lepidoptera, Limacodidae
Synonym *Cnidocampa flavescens* (Walker)

Importance Outbreaks of *M. flavescens* have been observed in urban forests in Heze city in Shandong province in recent years. Severe damage incurs due to the explosive population growth and rapid spread of this pest.

Distribution All provinces except Guizhou, Tibet, Ningxia, and Xinjiang in China; Japan, Korea Peninsula, and Russia.

Hosts Various tree and fruit species, including *Ziziphus jujube*, *Malus pumila*, *Pyrus pyrifolia*, *Prunus persica*, *P. armeniaca*, *P. salicina*, *Juglans regia*, *Crataegus pinnatifida*, *Diospyros kaki*, *Zanthoxylum bungeanus*, *Citrus reticulate*, *Eriobotrya japonica*, *Pterocarya stenoptera*, *Celtis sinensis*, *Populus* spp., *Salix* spp., *Morus alba*, and *Ulmus* spp.

Damage Feeding by early instar larvae on mesophyll creates transparent blotches on leaves. Late instars consume entire leaves, leaving only leaf stalks intact.

Morphological characteristics Adults Females are 15-17 mm long with a wingspan of 35-39 mm, while males are 13-15 mm long with a wingspan of 30-32 mm. Orange. Forewings yellow for the inner half and yellowish brown for the outer half. There are two dark brown fine marks from the top angle to the posterior margin, with the inner mark separate the inner and outer half of the forewings. Median cell contains a yellowish brown circular spot. Hindwings yellow. Legs brown.

Eggs Flat oval, slightly pointed at one end, 1.5 mm long and 0.9 mm wide, yellowish, with reticulate sculptures on egg membrane.

Larvae Mature larvae 19-25 mm long, robust, near rectangular with slightly wider anterior end. Body fresh yellowish green. There is a large dumb bell-shaped purplish blotch on the back of the late instar larva. Each abdominal segment except the 1st segment contains four branched spines. Upper spiracle line greenish. Lower spiracle line yellowish. Head yellowish brown and concealed under the prothorax.

Pupae Oval, large and robust, 13-15 mm long. Yellowish brown; head and thoracic nota yellow; abdominal nota brown.

Cocoons About 12 mm long, hard. Grayish white, with irregular blackish brown longitudinal marks, resem-

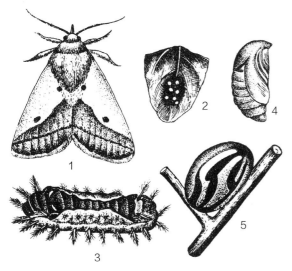

M. flavescens. 1. Adult, 2. Eggs, 3. Larva, 4. Pupa, 5. Cocoon (Zhu Baiting)

M. flavescens adult (Xu Gongtian)

M. flavescens adult (edited by Zhang Runzhi)

bling passerine eggs.

Biology and behavior The insect occurs one generation a year and overwinters as mature larvae inside cocoons at branch joints in northern China. Larvae pupate inside the cocoons in mid-May. Pupal stage lasts for about 15 days. Adults appear in early June and live for 4-7 days. Adults are attracted to lights. They stay on the underside of the leaves during the day and become active at night. Adults mate and lay eggs soon after emergence. Eggs are laid mostly in large masses on the underside of the leaves. Each female produces 50-70 eggs. Egg stage lasts for about 8 days. Most larvae hatch during the day. There are seven instars in the larval

M. flavescens adult (Zhang Runzhi)

M. flavescens cocoon (Li Zhenyu)

M. flavescens larvae (Zhang Runzhi)

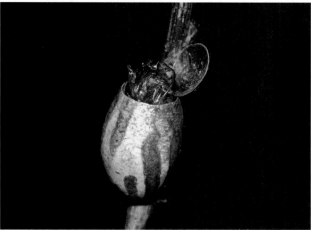

M. flavescens empty pupa (Zhang Runzhi)

M. flavescens young larva (Xu Gongtian)

M. flavescens mature larva (Xu Gongtian)

M. flavescens pupa (ventral view) (Xu Gongtian)

M. flavescens pupa (dorsal view) (Xu Gongtian)

M. flavescens damage (Xu Gongtian)

M. flavescens larva parasitized by braconid wasps (Xu Gongtian)

stage. Newly hatched larvae consume eggshells first before feed on lower epidermis and mesophyll, leaving reticulate leaves with intact upper epidermis and transparent feeding blotches. From 4th instar, larvae create feeding holes. The 5th instars consume entire leaves with only veins intact. Larva contains urticating spines that cause skin rashes and pain. Damage period lasts from mid-July to late August. Mature larvae construct cocoons at branch joints in early September.

Prevention and treatment

1. Biological control. Major natural enemies include *Chrysis shanghaiensis*, *Eurytoma monemas*, praying mantis, NVP. *Chrysis shanghaiensis* is the dominate species with high pupal parasitism. Parasitized pupae are easily recognized by the circular hole on the top end. To protect this parasitoid, collect *C. flavescens* cocoons during the winter and place them inside wooden cages (30 cm × 30 cm × 40 cm) with 3 mm mesh screens in the forest to allow wasp emergence. Destroy *C. flavescens* inside cages to reduce source populations.

2. Trap adults with black light. Deploy black lights in the field between mid-June and mid-July to control adult population of the overwinter generation.

3. Manual control. Destroy larval population to eradicate source population and prevent spread by removing infested leaves during their aggregate feeding period. Collect and destroy mature larvae in mid- to late August when they search for pupation sites on branches.

4. Chemical control. Apply *B.t.* at 600×, 25% Dimilin at 1,000×, 5.7% cyfluthrin EC at 1,000-2,000×, diflubenzuron at 5,000-6,000×, beta cypermethrin at 500-2,000×, or 30% Dimilin + pyridaben WP at 2,000× for larvae control.

References Chinese Academy of Forestry 1983, Zhao Fanggui et al. 1999, Sun Xugen et al. 2001.

(Wang Haiming, Chen Boyao)

Phlossa conjuncta (Walker)

Classification Lepidoptera, Limacodidae
Synonym *Iragoides conjuncta* (Walker)
Common names Slug moth, Cup moth

Importance *Phlossa conjuncta* is a common polyphagous species of many trees.

Distribution Hebei, Liaoning, Jiangsu, Zhejiang, Anhui, Fujian, Jiangxi, Shandong, Hubei, Hunan, Guangdong, Guangxi, Sichuan, Guizhou, Yunnan, and Taiwan in China; Korea Peninsula, Japan, Vietnam, India, Thailand.

Hosts *Ziziphus jujube*, *Pyrus bretschneideri*, *Diospyros kaki*, *Prunus persica*, *Prunus armeniaca*, *Camellia sinensis*, *Juglans regia*, *Malus pumila*, *Ailanthus altissima*, *Ziziphus jujuba* var. *spinosa*, *Paulownia* spp., *Robinia pseudoacacia*, *Prunus Pseudocerasus*, *Mangifera indica*, *Patanus hispanica*, and *Clerodendron trichotomum*.

Damage Larvae feed on leaves, with early instars feed on mesophyll and later instars consume the entire leaves.

Morphological characteristics **Adults** About 14 mm long. Males with a wingspan of 28-31.5 mm while females with a wingspan of 29-33 mm. Antennae short, pectinate in males and filiform in females. Head and collar brownish. Body and forewings reddish brown. Scaly hairs on thoracic nota long, brownish red in the center and brown on both sides. Each abdominal notum contains a patch of upside down "v" shaped reddish brown scaly hairs. Forewings brown at the base and yellowish brown in the middle, containing two near diamond-shaped connected blotches near the outer margin. The anterior blotch brown, and the posterior one reddish brown in color. There is a black dot on the cross vein. Hindwings yellowish brown. **Eggs** Flat oval, 1.2-2.2 mm long, fresh yellow at the beginning, semitransparent. **Larvae** Newly hatched larvae 0.9-1.3 mm long, cylindrical, yellowish with deep-colored backs. Head, the first, and the second abdominal segment each contains a pair of large spines. There are two pairs of spines on anal plate. Mature larvae about 21 mm long, head small and retracted into prothorax. Body yellowish green, with green cloudy marks on the back. Each abdominal segment contains four red and branched spines, with relatively larger ones found on thorax (all four), the middle and the last abdominal segments (central two). There is also a short red hair tuft on each segment laterally. **Cocoons** Oval, 11-14.5 mm long, earthy grayish brown, hard. **Pupae** Oval, 12-13 mm long, yellowish at first, turn to brownish gradually, and brown before adult emergence, with black brown wing pads.

Biology and behavior *Phlossa conjuncta* occurs one generation a year and overwinters as mature larvae inside 7-9 cm deep soil around the base of the host trees in Henan and Hebei. Overwintering larvae pupate in early June the following year. Pupal stage lasts for about 20 days. Adults start to emerge in late June with a peak emergence in July. Most adults emerge between 17:00-23:00. Adults

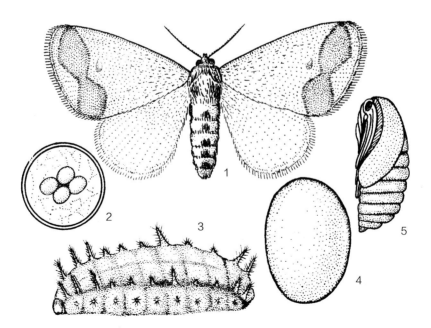

P. conjuncta. 1. Adult, 2. Eggs, 3. Larva, 4. Pupa, 5. Cocoon (Li Guiliang)

P. conjuncta adult (Xu Gongtian) P. conjuncta larva (dorsal view) (Xu Gongtian)

P. conjuncta larva (ventral view) (Xu Gongtian) P. conjuncta damage (Xu Gongtian)

are attracted to lights and live for 1-4 days. They stay on the underside of the leaves during the day. Sometimes they stay upside down by attaching to the leaves, or sticking up with support from both wings. They are not easily disturbed and will stay in position for long times. Adults chase each other and mate at night. Mating time can be more than 15 hours. Females lay eggs the next day after mating. Eggs are laid on the underside of the leaves as cataphracate egg masses. Egg stage lasts for 7-8 days. Newly hatched larvae feed aggregately for a short period of time before dispersing to underside of the leaves. They feed on mesophyll initially, then on whole leaves, leaving only main veins and leaf stalks intact. Most damage occurs between late July and mid-August. Mature larvae gradually crawl down from the trees and enter the soil by late August.

Prevention and treatment

1. Feeding by newly hatched larvae leave white or semitransparent damage blotches, which make them easily discovered. Large numbers of larvae are usually present nearby. Removal of infested leaves and branches will result in significant pest reduction. Trap and kill overwintering larvae by covering the trunks with weed. Overwintering cocoons can also be destroyed manually in the spring and winter through other silviculture activities.

2. High percentage of mature larvae is often parasitized by *Chrysis shanghaiensis*. Parasitized cocoons are easily recognized by the small holes on top of the cocoons. Remove the overwintering cocoons during winter or early spring, save parasitized cocoons, and allow the parasitoid to emerge later.

3. Trap adults with black lights.

4. Chemical control. Apply 90% crystal trichlorfon at 8,000×, 25 g/L lamda-cyhalothrin at 800-1,000×, 50% malathion at 2,000×, or 2.5% deltamethrin at 4,000×. Spray 2×10^{10} PIB/mL nuclear polyhydrosis virus at 1,000-1,500×.

Reference Xiao Gangrou 1992.

(Ji Baozhong, Zhang Kai, Li Guiliang)

308 *Thespea bicolor* (Walker)
Classification Lepidoptera, Limacodidae

Distribution Shanghai, Jiangsu, Zhejiang, Anhui, Fujian, Jiangxi, Hubei, Hunan, Guangdong, Guangxi, Sichuan, Guizhou, Yunnan, and Taiwan; Sri Lanka, India, Myanmar, and Indonesia.

Hosts *Phyllostachys heteroclada* var. *pubescens*, *P. sulphurea* 'Viridis', *P. glauca*, *P. iridescens*, *P. guizhouensis*, *P. edulis* f. *gimmei*, *P. rutile*, *P. makinoi*, *P. nigra*, *P. bambusoides*, *P. glauca*, *P. bissetii*, *P. bambusoides* f. *shouzhu*, *P. vivax*, *P. dulcis*, *P. nidularia* 'Smoothsheath', *P. flexuosa*, *P. fimbriligula*, *P. nuda*, *P. praecox*, *P. praecox* f. *provernalis*, *P. heteroclada*, *P. bambusoides* f. *lacrima-deae*, *P. propinqua*, *Pleioblastus maculosoides*, *Pleioblastus amarus*, *Bambusa pervariabilis*, *B. vulgaris*, *B. textilis*, *B. chungii*, *B. rutile*, *B. multiplex*, *Sinobambusa incana*, *S. tootsik*, *S. farinosa*, and *S. rubroligula*.

Damage Feeding by young larvae on the lower epidermis of bamboo leaves results in wilting and whitening of the leaves. From 4th instar, larvae consume whole leaves. The last instar accounts for 50% of the leaf consumption during the larval stage. Larvae have the ability to devour all bamboo tree leaves and lead to tree mortality under heavy infestation.

Morphological characteristics **Adults** Males are 14-16 mm long with a wingspan of 30-34 mm, while females are 13-19 mm long with a wingspan of 37-44 mm. Vertex and pronotum green, abdomen brownish yellow. Antennae filiform in females, and pectinate with filiform apical two fifths in males. Compound eyes black. Maxillary palpi brownish yellow. Forewings green, with yellowish brown costal margin, outer margin, and marginal hairs. There are two columns of brown spots on subcosta and outer transverse line. Two of the spots on the outer transverse line are relatively large. There are 4-6 small spots on subcosta, sometimes only 2-3 of them are visible. Hindwings brown. Fore- and mid-legs yellow except the outside of fore- and mid-tibia tibia and tarsi, which are brown.

Eggs Flat oval, 1.6 mm for the major axis and 1.3 mm for the minor axis, covered by transparent film. Yellowish at first, gradually turn to milky white. Eggs are arranged in separate masses. Each egg mass contains an average of 10-24 eggs, with a maximum of 150 eggs.

Larvae Newly hatched larvae 1.3 mm long, white. There are eight instars in the larval stage. Head black, usually concealed inside the mesothorax. Body yellowish green with wide greenish gray dorsal lines. There are nine pairs of dark green semicircular blotch between metathorax and abdominal segment 8. There are also eight pairs dark green semicircular blotch between the subdorsal line and upper spiracle line. These blotches correspond to those along the dorsal line. Areas between subdorsal and upper spiracle lines yellowish green. Each area contains one spiny verruca at the center, with a total of 10 verrucae in those areas. Spiracle yellow, circular and located on green spiracle line. There is a spiny verruca between spiracle line and upper spiracle line for each segment, with a green wavy line above the verruca and a yellow line below the spiracle. No spiny verrucae are found on prothorax. Prothorax retracts into mesothorax with the head. It extends out when larva crawls. The spine on the verruca is extremely long on mesothorax, metathorax, and abdominal segment 1, 7, and 8. There is a pair of black pubescent verruca-like hair tufts on abdominal segment 8 laterally and

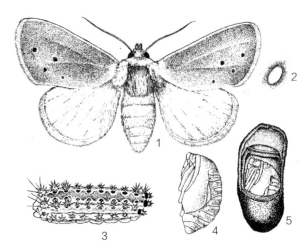

T. bicolor. 1. Adult, 2. Egg, 3. Larva, 4. Pupa, 5. Cocoon (Xu Tiansen)

T. bicolor adult (Xu Tiansen)

T. bicolor young larvae (edited by Zhang Runzhi)

T. bicolor mature larva (edited by Zhang Runzhi)

T. bicolor cocoons (Xu Tiansen)

T. bicolor damage (edited by Zhang Runzhi)

segment 9 posteriorly. Each hair tuft contains a brownish red spiny verruca. Prolegs are reduced to sucking pads.

Pupae 12-16 mm long, milky white at first, turn to brownish yellow gradually. Tip of wing pad reaches the end of abdominal segment 6. Hind tibia surpassed the forewings. There are three pairs of thoracic spiracles. Each abdominal notum contains a broad band of brown spicules anteriorly. Abdomen obtuse apically.

Cocoons 15-21 mm long, double layered. The outer layer grayish brown with a round hole on top. The inner layer gelatinous, hard and bristle, brown with truncated top and a lid for emergence.

Biology and behavior *Thespea bicolor* occurs one generation a year in Jiangsu and Zhejiang, and three generations in Guangdong. It overwinters as mature larvae inside earthy cocoons in the soil. In Zhejiang, overwintering larvae pupate in early to mid-May. Adults start to appear in late May. Adult stage long, with the last adults found in mid- to late August. In Guangdong, adults of each generation appear between mid-April to late May, late June to late July, and early September and early October, respectively; larvae of each generation appear from late April to mid-June, early July to late August, and early September to early November, respectively.

Prevention and treatment

1. Conserve natural enemies. Predators for this pest include *Chrysopa sinica*, *C. Formosa* for eggs; *Sirthenea flavipes*, *Haematiloecha nigrorufa* for larvae; *Chiracantium zhejiangensis*, *Pardosa wuyiensis*, *P. tschekiangensis*, and *Dolomedes sulfureus* for adults. Parasitoids include *Meteorus* sp., *Scenoharops parasae*, *Chlorocryptus* sp. and *Chaetexorista* sp. for larvae.

2. Light trapping. Adults are attracted to lights and can be killed with light traps.

3. Manual control. Remove infested leaves and destroy the larvae.

4. Biological control. Apply *Beauveria* sp. powder to control heavy infestations. Spray 2×10^{10} PIB/mL nuclear polyhydrosis virus at 1,000-1,500×.

References Xu Tiansen and Wang Haojie 2004; Xu Tiansen, Wang Haojie, Yu Caizhu 2008.

(Xu Tiansen)

Parasa consocia Walker

Classification	Lepidoptera, Limacodidae
Synonym	*Latoia consocia* (Walker)
Common name	Green cochild

Importance Larval feeding on leaves leads to complete defoliation of the hosts during heavy infestation.

Distribution Hebei, Shanxi, Inner Mongolia, Liaoning, Jilin, Heilongjiang, Jiangsu, Zhejiang, Anhui, Fujian, Jiangxi, Shandong, Henan, Hubei, Hunan, Guangdong, Guangxi, Shaanxi, and Taiwan; Japan, Korea Peninsula, and Russia.

Hosts *Platanus* sp., *Ulmus laevis*, *Robinia pseudoacacia* var. *pseudoacacia*, *Pyrus bretschneideri*, *Malus pumila*, *Diospyros kaki*, *Ziziphus jujube*, *Quercus wutaishanica*, *Pterocarya stenoptera*, *Quercus acutissima* var. *acutissima*, *Juglans regia*, *Cercis chinensis* f. *chinensis*, *Fraxinus chinensis*, *Populus* spp., *Salix* spp., *Paulownia tomentosa* var. *tomentosa*, *Buxus megistophylla*, *Lagerstroemia indica*, *Cercis chinensis* f. *chinensis*, *Rosa chinensis* var. *chinensis*, *Osmanthus fragrans*, *Cerasus yedoensis*, *Begonia grandis* subsp. *grandis* var. *grandis*, *Camellia japonica*, *Citrus reticulate*, *Paeonia suffruticosa*, *Paeonia lactiflora*, *Prunus persica*, and *Prunus salicina*.

Damage Larvae feed on leaves. Young larvae feed on mesophyll, which results in reticular leaves. Feeding by older larvae causes feeding holes or only veins and stalks are intact.

Morphological characteristics **Adults** Females are 15.5-17 mm long with a wingspan of 36-40 mm. Males are slightly smaller, 12.5-15 mm long with a wingspan of 28-36 mm. Vertex and pronotum greenish. Thorax contains a brown longitudinal line dorsally. Forewings green, with a brown blotch at the base and a yellowish broad band on the outer margin. The band usually contains small brown spots on the inner side, while the rest is green in color. Compound eyes blackish brown. Antennae brown, filiform for females, and pectinate for males for the basal dozens of segments. Legs brown.

Eggs Oval, 1.2-1.3 mm for the major axis and 0.8-0.9 mm for the minor axis, yellowish green.

Larvae Mature larvae 24-27 mm long and 7.0-8.5 mm wide. Head reddish brown, retracted into prothorax. Pronotum black, body fresh green. Prothorax contains a pair of black spiny verrucae. There are two pairs of bluish black blotches on each segment from metathorax to abdominal segment 7. Subdorsal line reddish orange, with 10 pairs of yellowish green spiny verrucae. There is one pair of brown and branched spines on each segment from mesothorax to abdominal segment 9.

Pupae 15-17 mm long, and 7-9 mm wide, oval,

P. consocia adult (edited by Zhang Runzhi)

P. consocia adult with wings spread (Xu Gongtian)

brown.

Cocoons 14.5-16.5 mm long and 7.5-9.5 mm wide, cylindrical, brown.

Biology and behavior *Parasa consocia* occurs 2-3 generations a year and overwinters as larvae inside the cocoons in areas south of the Yangtze River. Overwintering larvae pupate from late April to mid-May the next year. Adults emerge and lay eggs from late May to June. Larvae of the first generation start to appear between June and late July, mature, and construct cocoons in mid-July. Adults of the first generation start to lay eggs in early August. Larvae of the second generation appear from mid-August to September. They construct cocoons in mid-September.

Parasa consocia occurs one generation a year in northeastern China. Overwintering larvae pupate in June. Adults emerge and lay eggs between July and August. Eggs hatch after one week. Mature larvae construct cocoons from late August to late September.

Eggs are laid on the underside of the leaves, with a few dozen eggs arranged cataphracately. Egg stage lasts for 5-7 days. Newly hatched larvae feed on mesophyll. By the third and fourth instars, they feed through the epidermis. The sixth instars feed by encroaching the leaf from the edges towards the center. Larvae aggregate before third instar and disperse afterwards. Larval stage lasts about 30 days. Mature larvae construct cocoons inside loose soil or in the weeds under the crowns. Pupal stage lasts for 5-46 days. Adults live for 3-8 days and are attracted to lights.

Prevention and treatment

1. Silviculture. Destroy overwintering larvae through winter tillage. Combining with winter tillage and fertilization, bury fallen leaves and topsoil. Place 6-9 cm deep layer of soil at 0-30 cm around the root collars to protect roots from lower temperatures. Compact the soil to kill overwintering larvae.

2. Mechanical control. Use light traps to kill adults between 19:00-21:00 during adult period. Manually collect and kill mature larvae, cocoons, and pupae when larvae coming down from the tree for overwintering or during the overwintering period.

3. Biological control. Apply NPV or *B.t.* at 600×.

4. Chemical control. Apply 1.2% mixture of nicotine and matrine EC at 1,000×, or 25% fenoxycarb at 3,000×.

References Chinese Academy of Forestry 1983, Forestry Bureau of Guizhou Province 1987, Xing Tongxuan 1991, Xiao Gangrou (ed.) 1992, Qiu Qiang 2004.

(Li Jilei, Guo Xin, Yan Hengyuan)

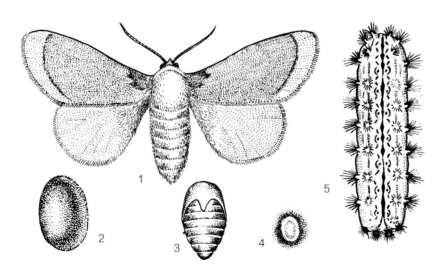

P. consocia. 1. Adult, 2. Cocoon, 3. Pupa, 4. Egg, 5. Larva (Zhang Peiyi)

310 *Thosea sinensis* (Walker)

Classification Lepidoptera, Limacodidae
Common names Flattened eucleid caterpillar, Nettle grub

Importance Larval feeding on leaves leads to complete defoliation of host trees and significant loss in fruit production during heavy infestations.

Distribution Beijing, Hebei, Liaoning, Jilin, Heilongjiang, Jiangsu, Zhejiang, Anhui, Jiangxi, Shandong, Henan, Hubei, Hunan, Guangdong, Guangxi, Sichuan, Yunnan, Shaanxi, and Taiwan; India, Indonesia, Korea Peninsula, Vietnam.

Hosts *Diospyros kaki, Juglans regia, Firmiana platanifolia, Populus* spp., *Morus alba, Zanthoxylum bungeanum, Citrus reticulate, Cinnamomum camphora, Malus pumila, Pyrus bretschneideri, Crataegus pinnatifida* var. *pinnatifida, Prunus armeniaca, Amygdalus persica, Ziziphus jujuba.* var. *jujube, Camellia japonica, Malus spectabilis, Rosa chinensis, Osmanthus fragrans,* and *Euonymus japonica.*

Damage Larval feeding creates many transparent reticular spots under the leaves. Feeding damage results in notched leaves, leaves with feeding holes, or leaves with only stalks intact.

Morphological characteristics **Adults** Females are 16.5-17.5 mm long with a wingspan of 30-38 mm. Males are 14-16 mm long with a wingspan of 26-34 mm. Head grayish brown. Compound eyes black brown. Antennae brown, filiform in females and pectinate in males. Thorax grayish brown. Wings grayish brown, with a brown line from the middle of the anterior margin to the posterior margin on the forewings. Front legs contain a white blotch at each joint. **Eggs** Oblong oval, about 1 mm, with projected back, yellowish green at the beginning, turn to grayish brown later. **Larvae** Newly hatched larvae 1.1-1.2 mm long, light in color. Mature larvae oval and flat, 20-27 mm long, fresh greenish. There is a white longitudinal line runs through the entire body dorsally, which is fanned laterally by bluish green narrow bands containing a column of orange spots. Spinule bundles on either side are extremely small, with depressed dark green diagonal marks in between. Lateral spinule bundles well developed. There is a white diagonal line between dorsal-lateral and lateral-ventral portion of the abdomen for each segment. **Pupae** 10-14 mm long, oval, milky white at the beginning, turn to yellowish brown before emergence. **Cocoons** 13-16 mm long, near sphere-shaped, dark brown.

Biology and behavior *Thosea sinensis* occurs one generation in northern China and two generations a year in the lower Yangtze River area and overwinters as mature larvae inside cocoons in soil. Larvae pupate in mid-May the next year. Adults start to emerge in early June. Damage period by larvae falls between mid-June and late August.

Most adults emerge at twilight, especially between 18:00-20:00. Adults mate and lay eggs right after emergence. Eggs are mostly laid on leaf surface. Newly hatched larvae stay close to eggshells without feeding. They start to feed on eggshells after first molt, then feed on mesophyll, which leaves only the upper epidermis intact. Larvae feed both during the day and at night. They consume entire leaves from 6th instar. At high population density, all leaves except a few apical new ones on host branches are consumed. They usually feed on leaves in the lower crown first and work their way to the upper crown. Larvae go through eight instars. Mature larvae leave trees to construct cocoons in soil between 20:00 and 6:00, with

T. sinensis adult (Xu Gongtian)

T. sinensis adult with wings spread (Xu Gongtian)

T. sinensis mid-sized larva (Xu Gongtian)

T. sinensis mature lava (Xu Gongtian)

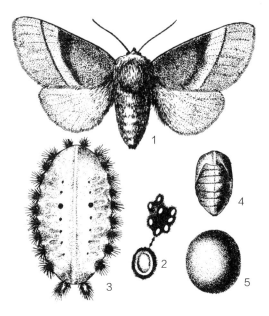

T. sinensis. 1. Adult, 2. Eggs, 3. Larva, 4. Pupa, 5. Cocoon (Zhang Peiyi)

T. sinensis damage (Xu Gongtian)

peak period between 2:00 to 4:00. Depth of cocoons in soil and their distances from tree trunks are related to soil texture. Cocoons are more scattered and buried shallower in clay soil. They are located in deeper positions and closer to trees in soil rich in humus or in sandy soil.

Prevention and treatment

1. Silviculture. Destroy overwintering larvae through winter tillage. Combining with winter tillage and fertilization, bury fallen leaves and topsoil. Place 6-9 cm deep layer of soil at 0-30 cm around the root collars to protect roots from lower temperatures. Compact the soil to kill overwintering larvae.

2. Light trapping. Use light traps to kill adults between 19:00-21:00 during adult period.

3. Manually collect and kill mature larvae, cocoons, and pupae when larvae coming down from the tree for overwintering or during the overwintering period.

4. Biological control. Apply *Bacillus thuringiensis galleriae* at the concentration of 5×10^7 conidia/mL or emulsifiable suspension at 600×. Apply *Beauveria bassiana* (4×10^{11} conidia/g) at 2,000-3,000×. Spray 2×10^{10} PIB/mL nuclear polyhydrosis virus at 1,000-1,500×.

5. Chemical control. Spray 5.7% emamectin at 3,000×, or 25 g/L lamda-cyhalothrin at 1,000×.

References Chinese Academy of Forestry 1983, Xiao Gangrou (ed.) 1992, Xu Gongtian 2003, Forestry Station of Shanghai 2004, Qiu Qiang 2004, Cui Lin et al. 2005, Xu Gongtian and Yang Zhihua 2007.

(Li Jilei, Guo Xin, Yan Hengyuan)

311 *Scopelodes testacea* Butler

Classification Lepidoptera, Limacodidae

Importance Heavy infestations cause complete defoliation of trees and result in severe economic loss to fruit and ornamental industries.

Distribution Zhejiang, Guangdong, Guangxi, Sichuan, and Yunnan; India, Sri Lanka, Malaysia, and Indonesia.

Hosts Fruits such as *Musa nana*, *Musa sapientum*, *Dimocarpus longan*, *Litchi chinensis*, *Mangifera indica*, *Mangifera persiciformis*, *Dracontomelum duperreanum*, *Syzygium samarangense*, *Eugenia malaccensis*, *Cleidiocarpon cavaleriei*; and ornamentals such as *Musa uranoscopos*, *Strelizia reginae*, *Duabanga grandiflora*, *Saraca griffithiana*, *Cephalomappa sinensis*, and *Ardisia* sp.

Damage Larvae of the 2nd to 4th instars nip on the mesophyll and the lower epidermis, leaving semitransparent upper epidermis. Larvae feed on leaf tissue from leaf margins after 5th instar. Infested plants in the family of Musaceae will only have veins intact during heavy infestations.

Morphological characteristics **Adults** Females are 18-23 mm long with a wingspan of 26-30 mm for the forewings. Body yellowish brown. Compound eyes black. Antennae relatively long, filiform for the basal three fifth and pectinate for the rest. Maxillary palpi thin basally and gradually increase in size apically, mostly dark brown except the white portion right below the black brown tip. Forewings gray to yellowish brown with shiny scales. Hindwings light-colored except the dark margins. Legs yellowish brown and covered by brown or grayish white scales. Tarsus contains dark scales. Males are 18-22 mm long with a wingspan of 20-23 mm. Antennae relatively short, double pectinate for the basal one third, and single pectinate for the rest. Forewings grayish brown, shiny, with darker outer margin. Hindwings gray to yellowish brown, with a dirty yellow anal area. Legs covered by grayish black scales except orange underside. Abdomen of both sexes yellow, with an arc-shaped black blotch in the middle of each notum and a pair of black blotches in the middle of sternite for abdominal segment 2-7. Scales between blotches grayish yellow and shiny. End of abdomen contains brush-like black hairs. **Eggs** Oval, 2.30-3.36 mm long and 1.40-2.02 mm wide, yellow with luster. **Larvae** Mature larvae oblong oval, about 40-46 mm long, 20-22 mm wide, and 15 mm high. Body yellowish green to emerald but yellowish ventrally. Head yellowish brown. Head 5.03-5.95 mm wide. Mandibles brown with black brown apical ends. Labral area and ocellar region brownish. Prothorax brownish with well-developed body projections that contain dense spines. Spines brown to dark brown apically. There is one pair of spines on pronotum, propleuron, mesonotum, metanotum, and between mesopleuron and metapleuron, as well as each abdominal notum and pleuron for segment 1-7. The pair of pleural spines on abdominal segment 7 is short and small, with pubescent large black blotch on the back and yellowish brown to dark brown apical end. The dorsal pair of spines on segment 8 with a pubescent large black blotch at the base. Apical ends of all other spines black brown. Body contains indigo spots, with one pair on pro- and mesothorax, three on metathorax, four on each segment from segment 1-5, and three on segment 6. The anterior two of the three spots on segment 6 are surrounded by a near rectangle frame. There is also a backward-inclined near diamond-shaped indigo blotch on the side of each of the ab-

S. testacea male adult (edited by Zhang Runzhi)

S. testacea mature larva (edited by Zhang Runzhi)

S. testacea cocoons (edited by Zhang Runzhi)

S. testacea pupa (edited by Zhang Runzhi)

S. testacea eggs
(edited by Zhang Runzhi)

S. testacea damage
(edited by Zhang Runzhi)

dominal segment 1-6. **Pupae** About 20.2 mm long and 11.6 mm wide, pale yellow, wing buds darker. **Cocoons** About 20-23 mm long and 16-18 mm wide, yellowish to blackish brown.

Biology and behavior *Scopelodes testacea* occurs two generations a year and overwinters as mature larvae inside the cocoons in Guangzhou region. Overwintering larvae become active in early to mid-May the next year. First generation eggs appear in mid-May. Egg period is approximately 6 days. First generation larvae occur between late May and late June. The larval feeding period is about 40 days. Prepupal stage is about 46 days. Larvae begin to construct cocoons by late June and pupate in mid-August. Pupal stage lasts for about 28 days. Adults of the first generation start to appear in mid-August with peak emergence between late August and early September. Eggs of the overwintering generation appear in mid-August. Larvae of this generation appear between mid-August and late November. Larvae feed for about 60 days. Prepupal stage is about 150 days. This generation lasts for about 210 days, with a pupal stage of about 17 days. There are 8-9 larval instars. Few of the larvae have 10 instars. Mature larvae construct cocoons on soil surface or loose soil or leaves near the base of the hosts. Cocoons are also found on wilted but still attached banana leaves. Adult emergence occurs at night. They are attracted to lights. Adults mate on the same day when they emerge. Eggs are laid on the upper surface or underside of the leaves and cataphracately arranged. Newly hatched larvae usually consume most of the eggshells. Larvae feed aggregately under the leaves before the 7th instars and disperse by the 8th instar. Larvae of the 2nd to 4th instars nip on mesophyll and lower epidermis, leaving semitransparent upper epidermis. Larvae feed on host leaves from leaf margins to the inside after the 5th instar. Infested plants in the family of Musaceae with only veins intact during heavy infestations.

Many late stage larvae of the first generation die of NPV infection between June and August each year.

Prevention and treatment

1. Manual control. Remove and destroy infested leaves and branches. Remove cocoons by digging them out from soil around the base of the hosts.

2. Light trapping. Use light traps to kill adults.

3. Protect natural enemies such as *Trichogramma* sp. and species of Ichneumonidae.

4. Biological control. Application of *B.t.* wettable powder (8,000 IU/ul) at 800-1,000× or *Bacillus thuringiensis galleriae* (1×10^{10} conidia/g) at 300-500× results in significant control. Apply *Beauveria bassiana* (4×10^{11} conidia/g) at 2,000-3,000×. Spray 2×10^{10} PIB/mL nuclear polyhydrosis virus at 1,000-1,500×.

5. Chemical control. Apply 5.7% emamectin at 3,000×, or 25 g/L lamda-cyhalothrin at 1,000×.

Reference Wu Yousheng and Gao Zezheng 2004.

(Wu Yousheng, Gao Zezheng)

312 *Fuscartona funeralis* (Butler)

Classification　Lepidoptera, Zygaenidae

Distribution　Shanghai, Jiangsu, Zhejiang, Anhui, Fujian, Jiangxi, Hubei, Hunan, Guangdong, Guangxi, Sichuan, Yunnan and Taiwan in China; Japan, Korea Peninsula, and India.

Hosts　Over 100 bamboo species in the genera of *Phyllostachys*, *Dendrocalamopsis*, *Pleioblastus*, *Pseudosasa*, *Bambusa*, *Sinobambusa*, *Indosasa*, and *Dendrocalamus*.

Damage　Larvae are capable of defoliating bamboo stands at high population density. Light damage affects bamboo growth and shoot production, whereas severe damage leads to death of bamboo trees.

Morphological characteristics　**Adults** Females are 9.5-11.5 mm long with a wingspan of 22.8-25.4 mm, while males are 7.8-9.2 mm long with a wingspan of 17.8-21.5 mm. Body black with greenish blue luster. Antennae filiform and

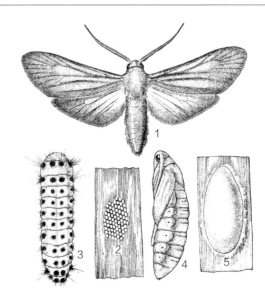

F. funeralis. 1. Adult, 2. Eggs, 3. larva, 4. Pupa, 5. Cocoon (Zhang Xiang)

F. funeralis newly emerged adult (Xu Tiansen)

F. funeralis adult male (Xu Tiansen)

7.5 mm long in females, and pectinate in males. Wings black brown with black base and semitransparent center, and black brown marginal hairs. Forewings slender. Hindwings with acute apical angle. Front tibia with one pair of apical spurs. Hind tibia with two pairs of spurs in the middle and apically. **Eggs** Short cylindrical with slightly obtuse ends. Major axis 0.65-0.78 mm long and short axis 0.46-0.56 mm long. Milky white and shiny at the beginning, turn to bluish before hatch. Eggs are in masses and evenly laid on the underside of the bamboo leaves. Each egg mass contains 25-250 eggs, with some contain near 300 eggs. Females produce an average of 400 eggs per female, with the maximum of 800 eggs. **Larvae** Newly hatched larvae milky white, 0.8-1.0 mm long, and covered by long hairs. Head yellowish, retracted inside the prothorax. Prothorax broad. Pronotum with two visible brown spots by the end of the first instar. There are visible brown marks on metanotum, and abdominal segment 1, 4, 8, and 9. Larvae go through six instars. The 2nd instar larvae contain four verrucae per segment after mesonotum. Verrucae on mesothorax and the first nine abdominal segments as well as those on the spiracle line contain short spines and long hairs. Other verrucae on subdorsal line contain short spines only. Larvae of the other instars are similar in morphology, except with more conspicuous or blighter body color. **Pupae** 8.0-10.0 mm long and 2.0-3.0 mm wide. Male pupae relatively small, flat oval, orange, turn to bluish black before adult emergence. Each abdominal notum contains spinules apically. Those spinules are especially conspicuous on seg-

F. funeralis larva (Xu Tiansen)

F. funeralis cocoons in bamboo (Xu Tiansen)

F. funeralis pupa parasitized by a fly (Xu Tiansen)

F. funeralis eggs (Xu Tiansen)

F. funeralis 1st instar larvae (Xu Tiansen)

F. funeralis pupa (Xu Tiansen)

ment 3-7. There are about 10 anal spines. Antennae and wing pads reach the 4th abdominal segment. **Cocoons** 12-14 mm long, oval or melon seed shaped, brown and leathery, dense and hard externally, and soft and membranous internally. Surface is partially or fully covered by white pubescence.

Biology and behavior *Fuscartona funeralis* occurs three generations a year in Zhejiang and five generations a year in Guangdong. It overwinters as mature larvae inside the cocoons. In Yuhang of Zhejiang, overwintering larvae pupate in mid- to late April. Adults emerge in late April to early May. Adults mate and lay eggs by mid-May. First generation larvae feed between late May and early July, while larvae of the second generation feed between late July and early September and the third generation larvae between mid-September and late October. Larvae construct cocoons to overwinter between late October and early November. Overwintering larvae pupate between January and February in Guangdong, and adults emerge in mid-February. Adult stage for the five generations is from mid-February to early April, late April to early June, late June to late July, early August to early September, and mid-September and mid-November, respectively. The feeding period for larvae in each generation is from early March to late May, mid-May to early July, late June to late August, mid-August to late October, and early October to late December, respectively.

Prevention and treatment

1. Conservation of natural enemies. Predators for this pest includes birds such as *Bambusicola thoracica*, *Garrulax canorus*, and *Cuculus canorus canorus* and spiders such as *Pardosa tschekiangensis*, *Oxyopes* sp., *Pisaura* sp., *Dolomedes pallitarsis*, *Pardosa wuyiensis* for adults; predatory insects such as *Chrysopa kulingensis*, and *Chrysopa formosa* for eggs; and *Coccinella septempunctata*; *Chilocorus kuwanae*, and *Coccinella trifasciata* for eggs and larvae. Parasitoids include *Trichogramma artonae* for eggs; and *Triraphis fuscipennis*, *Meteorus* sp., and *Apanteles* spp. for larvae.

2. Silviculture. Adults stay in sunny locations. Maintain proper canopy cover and species diversity. Do not remove the understory and weeds to protect natural enemies.

3. Manual control. Both eggs and young larvae can be removed and destroyed manually.

4. Biological control. Apply *Beauveria* sp. powder during heavy infestation.

5. Chemical control. Apply 5.7% emamectin at 3,000×, 25 g/L lamda-cyhalothrin at 1000×, or spray 2×10^{10} PIB/mL nuclear polyhydrosis virus at 1,000-1,500× during outbreak.

References Xu Tiansen and Wang Haojie 2004, Xu Tiansen et al. 2008.

(Xu Tiansen)

313 *Histia rhodope* (Cramer)

Classification Lepidoptera, Zygaenidae

Distribution Jiangsu, Zhejiang, Fujian, Hubei, Hunan, Guangdong, Guangxi, Yunnan, and Taiwan; India, Myanmar, and Indonesia.

Host *Bischofia polycarpa*.

Morphological characteristics **Adults** 17-24 mm (average 19 m) long with a wingspan of 47-70 mm (average 61mm). Head small, red with black marks. Antennae black, double pectinate, wider in males compared to females. Pronotum brown, red at the center of the anterior and the posterior margins. Mesonotum blackish brown, red anteriorly, with two red marks that may form a "U" shaped big mark posteriorly. Forewings black with blue underside basally. Hindwings black, bluish green for three fifth of the length from the base to the end of the median cell. There are red basal marks on the underside of both wings. Hind M_2 and M_3 extend to form a tail angle. Abdomen red, with five columns of black marks gradually smaller in size towards the end. Black marks larger in females than in males. The two ventral columns from abdominal segment 1 to 5 or 6 merged into one in the female. End of abdomen truncate and concave in males, whereas female abdomen pointed with exposed black brown ovipositor.

Eggs 0.73-0.82 mm long and 0.45-0.59 mm wide, oval, slightly flat with smooth surface. Milky white at the beginning, turn yellow later, and become grayish right before hatching.

Larvae Slug like, hypertropic and flat, with head concealed inside the pronotum. Crochets on abdominal prolegs in uniserial bands. Body contains scoli and some of them with glands. First instars yellowish with small and branched spines, 1.44 -1.59 mm long. Head dark brown. There are two parallel longitudinal lines laterally from the posterior half of mesothorax to the 8th abdominal segment, which almost converge by the end. Second instars 2.55-2.63 mm long with more obvious scoli. There is only one wider dark brown longitudinal line on each side, which connects posteriorly to form a "U" shape. Third instars 4.40-4.47 mm long, purplish red dorsally with obvious scoli. Head brownish and pronotum yellowish brown. Fourth instars 7.21-7.32 mm long, head brown. The anterior margin on pronotum yellow laterally. Pronotum brown. There are short black transverse bands on the back of each body segment. There are black circular marks between scoli laterally, with corresponding black transverse bands and circular marks between body segments. These bands and marks form three columns of black blotches. Larvae from 5th to 7th instars are similar to the 4th instars with lighter body ranging from pink to dark red. Scoli on the back of the body reddish to peach red in color. Body length is 13.0-13.25 mm for 5th instars, 18.83-20.25 mm for 6th instars, and about 24 mm for 7th

H. rhodope adult (SFDP)

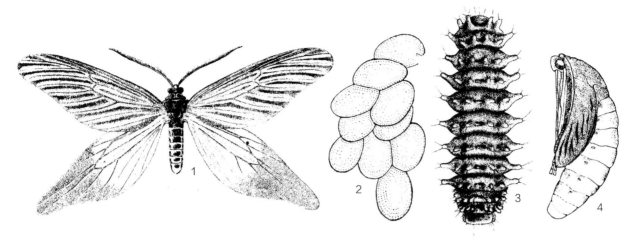

H. rhodope. 1. Adult, 2. Eggs, 3, larva, 4. Pupa, 5. Cocoon (Huang Bangkan)

and 8th instars. There are 10 scoli on meso- and metathorax, and 6 scoli on each abdominal segment from 1 to 8. Abdominal segment 9 contains 4 scoli. Lateral scoli on abdomen brownish yellow and long, whereas dorsal scoli dark purple and short.

Pupae 15.5-20 mm long with an average of 17 mm, entirely yellow at the beginning with a pinkish abdomen. Head turn dark red later. Compound eyes, antennae, thorax, legs, and wing pads peach red. There is a large black blotch on the back and a black blotch on each side on each abdominal segment from 1 to 7. Abdominal segment 6 and 7 exposed under the wing pads and contain two parallel large black blotches.

Cocoons 23-28 mm (average 24.3 mm) long and 7.5-9.0 (average 8.25) mm wide, white or slightly brownish.

Biology and behavior *Histia rhodope* occurs four generations a year in Fuzhou of Fujian and Wuchang of Hubei. It overwinters as various instar larvae under the cork, between cork crevices, depressions on broken branches, or between overlaid leaves on host trees in Fuzhou. A few may also overwinter as mature larvae inside cocoons under the trees. Overwintering larvae can feed in warm winter days. They construct cocoons under the tree in March and April the following year. Larvae pupate in April. Adults start to emerge in mid-April to early May with a peak emergence from mid- to late May.

First generation larvae hatch in late May and exert most damage between early and mid-June. Mature larvae construct cocoons for pupation under the trees between mid-June and early July. Peak adult emergence occurs between late June and early July. Most larvae of the second generation hatch in early to mid-July. They are capable of defoliate an entire tree in 3-4 days by late July. Larvae construct cocoons for pupation in early to mid-August. Most adults emerge between mid- and late August. The peak of the third generation larvae occurs in late August. They often completely defoliate the host trees. Adults of this generation are usually found by mid- to later October. Larvae of the fourth generation appear in mid-November and overwinter in mid-November. Duration for each generation is 61.1 days at 26.2°C for the first, 53.3 days at 30.4°C for the second, 60.2 days at 23.8°C for the third, and 175.4 days at 16.4°C for the fourth, respectively.

In Wuchang of Hubei province, the 1st to 4th generation occurs from late April to late June, mid-June to early August, early August to mid-September, and late September to late April next year, respectively. Among them, larvae of the 2nd and 3rd generation pose the most damage.

Adults emerge during the day especially at noon. Adults take flight between 16:00-18:00. They rest or occasionally crawl in shady areas between branches or leaves during the rest of the day. Adults mate in the same day after emergence or between 14:00-20:00 the next day. Caged females will attract males. Adults fly between trees and flowers for supplemental nutrition. Eggs are laid in masses under the bark. Each female produces about 1,000 eggs. It lays 5-20 eggs (average 10 eggs) each time. Females produced an average of 236-241 eggs per female in the laboratory. All eggs are laid within 5-6 days after emergence. Most eggs hatch between 6:00-10:00. Egg period for the 1st to 4th generation in Fuzhou is 11-16 days (at 22.8°C), 6 days (at 30.4°C), 7 days (at 27.9°C), and 14 days (at 21.7°C), respectively.

Most larvae feed on the underside of the leaves. They can feed on the bark of leaf stalks and young branches when food is scarce. Feeding by the 1st to 3rd instars creates short linear transparent marks with upper epidermis intact. Feeding by the late 4th instars creates shot holes or

H. rhodope adult (Li Zhenyu)

incomplete leaves, turns leaves brown, or defoliates the trees at high population density. Larvae older than the 5th instars usually devote the entire leaves, leaving only branches intact at high population density. Larvae can feed both during the day and at night. Night consumption is 20% more than day consumption. Each larva consumes about 15-20 leaves during the entire larval stage. When disturbed, larvae can secrete a sticky and colorless gland liquid through dorsal scoli. This liquid stays near the gland openings and emits foul smell as a defensive mechanism. Larvae can also feed on their own exuviates.

In Fuzhou, larvae usually begin to feed when air temperature reaches 10°C. Frequent feeding activities occur when temperature is above 13°C. The minimum temperature for larval feeding is 8°C. Larvae go through 6-8 instars, with most having 6 or 7 instars. Larval duration of the 1st to 4th generation in Fuzhou is 30-42 (average 33.3) days (at 25.8°C), 30-38 (average 34.3) days (at 31.1°C), 27-44 (average 33.4) days (at 25.1°C), and 127-151 (average 136.8) days including overwinter (at 15.1°C), respectively. Mature larvae usually drop to the ground to construct cocoons between 6:00-8:00. Some larvae come down to the ground through the trunk. Thin cocoons are usually constructed on surface of large fresh or dried fallen leaves. Some cocoons are found on other thin objects. Very few construct their cocoons on leaves of host trees or nearby bushes.

Pupal stage of the 1st to 4th generation is 9-11 (average 9.8) days (at 28.4°C), 8-12 (average 10.1) days (at 30.1°C), 14-19 (average 15.7) days (at 21.5°C), and 14-24 (average 20.2) days (at 19.6°C), respectively. Few larvae pupate before overwintering. The pupal stage is one and half month to two months. Outbreaks were observed every other year in Fuzhou based on observations from 1951-1958, with scattered populations being observed in non-outbreak years. Various parasitoids attack pest populations during the late stage of outbreaks. Natural control by these parasitoids is probably the major factor for population fluctuation of this pest.

Prevention and treatment

1. Conservation of natural enemies. Egg parasitism in the 2nd generation reached 27.7% in one study. One species of *Apanteles* sp. parasitized 5.8%-16.2% of the *H. rhodope* larvae. Another braconid parasitoid is *Atanycolus* sp. There are two parasitic flies, one of the common species is *Exorista japonica*. Pupal parasitoids include *Goryphus basilaris*. Other larval natural enemies include bacteria and birds.

2. Chemical control. Apply 5.7% emamectin at 3,000×, 25 g/L lamda-cyhalothrin at 1000×, or spray 2×10^{10} PIB/mL nuclear polyhydrosis virus at 1,000-1,500×.

References Huang Qilin 1956, Zheng Hanye 1957, Huang Bangkan 1980.

(Huang Bangkan)

314 *Illiberis ulmivora* (Graeser)

Classification Lepidoptera, Zygaenidae
Synonyms *Inope ulmivora* (Jordan), *Procris pekinensis* Draeseke
Common name Elm leaf worm

Distribution Beijing, Tianjin, Hebei, Shanxi, Shandong, Henan, and Gansu.

Host *Ulmus pumila.*

Damage Larval feeding creates notched leaves. All leaves are devoured at high population density.

Morphological characteristics **Adults** 10-11 mm long with a wingspan of 27-28 mm, black brown. The posterior margin of the abdominal notum, abdominal pluron, end of the abdomen, and legs yellowish brown. Wings semi-transparent.

Eggs Yellow to yellowish brown, oblong oval, 0.5 mm × 0.4 mm in size.

Larvae 1.0-1.1 mm long when newly hatched, and 14-18 mm long at maturity, yellow. Meso- and metathorax, posterior half of the 3rd abdominal segment, and a portion of the 4th and 5th abdominal segments, and the 8th and 9th abdominal segments black. There are five hairy verrucae on the sides of each segment. Hairs on each verruca yellowish. Peritreme yellowish brown. Prolegs with crochets in uniserial band.

Pupae 9-15 mm long, relatively flat, yellowish or yellowish brown. There is a transverse fold at the anterior margin of the first abdominal segment, whereas each of the abdominal segment 2 to 9 contains a line of brown dark brown spines.

Biology and behavior *Illiberis ulmivora* occurs one generation a year and overwinters as mature larvae inside the cocoons in leave layers, various crevices, or trunk

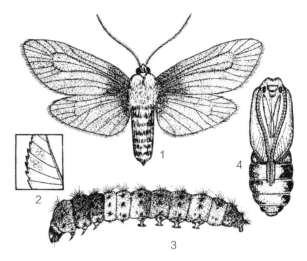

I. ulmivora. 1. Adult, 2. Eggs, 3, larva, 4. Pupa (Liu Qixiang)

hollows in Lanzhou. Adults emerge from late May to late July next year. Eggs stage runs from early June to early August, with a larval stage of mid-June to mid-October. Heavy damage occurs between mid-July and mid-August. Larval stage lasts for about 40 days with eight instars. Pupation begins in early August with a prepupal stage of about 8 days. Most adults emerge around 9:00 with a sex ratio of 1:1.08 (females: males). Eggs are usually laid on the underside of the new leaves and arranged in single, double, or triple layers. Each female produces 18-350 eggs. Egg stage lasts for 7-10 days. Newly hatched larvae arranged neatly at egg sites. The 1st and 2nd instar larvae feed on mesophyll only, creating transparent patches on leaf surface. Larvae disperse after the 3rd instar and can balloon with silk threads. The 7th instar larvae come down from the trees for migration, cocoon construction, pupation, and overwintering during serious food shortages.

Prevention and treatment

1. Remove new shoots in mid-June during outbreaks.

2. Chemical control. Apply 5.7% emamectin at 3,000×, 25 g/L lamda-cyhalothrin at 1000×, or spray 2×10^{10} PIB/mL nuclear polyhydrosis virus at 1,000-1,500× during outbreak.

References Xiao Gangrou (ed.) 1992, Xie Xiaoxi 1994.

I. ulmivora larvae (Xu Gongtian)

(Liu Mantang, Li Menglou, Xie Xiaoxi)

315 *Pryeria sinica* Moore

Classification Lepidoptera, Zygaenidae
Common name Euonymus leaf notcher

Importance Outbreaks of *P. sinica* occurred in various locations in recent years. Heavy infestations cause complete defoliation, affect growth and aesthetic values of the hosts.

Distribution Hubei, Shanghai, Jiangsu, Anhui, Fujian, Shandong, Henan.

Hosts *Euonymus japonicas*, *E. alatus*, *E. bungeanus* and *E. fortunei*.

Damage Newly hatched larvae feed on young branches, creating withered and folded leaves over large areas. Older larvae (> 3rd instars) feed significantly more, and usually consume whole leaves except the stalks. Larvae migrate when food become scarce.

Morphological characteristics **Adults** Females 8-10 mm long with a wingspan of 28-30 mm, slightly flat. Head, compound eyes, antennae, thorax, legs and stigma black. Forewings slightly transparent, yellowish for the basal one third, with sparse black hairs apically. Hindwings lighter in color, with long black hairs basally. There are dark yellow long hairs on coxae and femora. Abdomen orange or orange yellow, with irregular black marks. Thoracic nota, abdominal nota, abdominal pleura, and end of abdomen contain orange long hairs, with the longest hairs found on the last two abdominal segments. Antennae double pectinate with a clubbed apical portion. End of abdomen contains two tufted hairs. Hairs near abdominal sternites black basally and dark yellow apically. Males 7-9 mm long with a wingspan of 25-28 mm. Antennae plumose. Abdomen contains a pair of black brown hair bundles.

Eggs 0.5-0.7 mm long, flat and oval, brownish. Yellowish white at the beginning, turn to pale while gradually. Eggs covered by gelatin and body hairs from the females. Eggs are usually arranged in columns of 3-7 eggs, forming 30-60 mm strips.

Larvae 15-20 mm long at maturity, short and robust, cylindrical. Brownish when newly hatched, turn to yellowish greenish after 2nd instars. Head small, black. Pronotum contains one pair of oval black marks in the center. There is one circular spot on each side of those black marks. Anal plate contains one black mark in the middle, which is surrounded by an oblong circular black mark laterally. There are seven lines among dorsal, subdorsal, upper spiracle, and spiracle lines. There are also body verrucas and short hairs.

Pupae 9-11 mm long, yellowish white initially, turn brown later. There are seven inconspicuous brown longitudinal lines remain on the surface, a line of uniformed spinules on the anterior margin of each segment, and two anal spines.

Cocoons 11-18 mm long and 5-9 mm wide, comprised by grayish white or yellowish brown thin silky

P. sinica adult (SFDP)

P. sinica larvae (SFDP)

P. sinica cocoon (SFDP)

P. sinica larvae (SFDP)

P. sinica damage (SFDP)

material, flat oval, wide anterior and narrow posterior, gourd seed shaped, with semi-transparent silky margins.

Biology and behavior *Pryeria sinica* occurs one generation a year and overwinters as eggs on host branches. Eggs hatch in late February when host terminal buds start to develop. There are four instars in the larvae stage. Newly hatched larvae feed aggregately on the buds, creating reticular damages. The second instar larvae feed on mesophyll and lower epidermis aggregately, leaving the upper epidermis intact. Larvae younger than the 3rd instars can aggregate and balloon with silk. Larvae disperse after the 3rd instars and feed significantly more. Their feeding causes notched leaves, leaves with shot holes, or completely consumed leaves except some veins. Larvae usually migrate and have a clustered distribution. When larvae are mature in early to mid-April, they come down from the trees, construct cocoons with silk among fallen leaves. Adults emerge in mid-October, with most of them emerge around 5:00 pm. Adults are weak fliers and fly between trees during the day. They mate soon after emergence. Mating lasts for 11 hours if uninterrupted. Adults mate only once. Females prefer to lay eggs in masses on 1-2 year-old branches. Eggs are laid within 3-5 minutes interval. Females rub their abdomen after laying eggs and stick some hairs on egg masses. Each female produces 96-196 eggs. Females do not fly even when disturbed during egg-laying and die on the egg masses. Multiple adults usually aggregate on the same branch.

Prevention and treatment

1. Manual control. Hand capture of adults since they are weak fliers. Remove egg masses on small branches, or remove early instar larvae when they are aggregating, or remove and destroy pupae in leaf litter and other materials.

2. Chemical control. Apply 5.7% emamectin at 3,000×, 25 g/L lamda-cyhalothrin at 1000×, or spray 2×1010 PIB/mL nuclear polyhydrosis virus at 1,000-1,500× during outbreak.

References Editorial Committee of Shandong Forest Insects 1993, Wang Haiming et al. 2005.

(Wang Haiming)

316 *Adoxophyes honmai* Yasuda

Classification Lepidoptera, Tortricidae

Synonyms *Archips reticulana* Hübner, *Dichelia privatana* Walker, *Adoxophyes orana orana* (Fischer von Röslerstamm)

Common name Summer fruit tortrix moth

Distribution All provinces except Yunnan and Tibet; Europe, India, Japan, Korea Peninsula, Singapore, Indonesia.

Hosts *Camellia sinensis*, *C. japonica*, *Pyrus bretschneideri*, *Prunus persica*, *P. salicina*, *P. armeniaca*, *P. pseudocerasus*, *Populus* spp., *Salix* spp., *Betula* spp., *Metasequoia glyptostroboides*, *Gossypium* spp., *Triticum* spp., *Vicia faba*, *Medicago* spp., *Glycine max*, *Phaseolus angularis*, *Vigna radiate*, *Arachis hypogaea*, *Sesamum indicum*, Cucurbitaceae, *Crataegus pinnatifida*, *Rosa multiflora*, *Prunus mume*, *Punica granatum*, *Citrus reticulate*, *Litchi chinensis*, *Hibiscus rosa-sinensis*, *Dendrothema morifolium*, *Pittosporum tobira*, *Lagerstroemia indica*, *Malus pumila*, *M. asiatica*, *M. spectabilis*, *Citrus sinensis*, *Lonicera japonica*, *Dimocarpus longan*, *Ginkgo biloba*, *Robinia pseudoacacia*, *Alnus incana*, *Helianthus annuus*, *Hypericum chinense*, *Rubus corchorifolius*, *Prunus triloba*, *Begonia masoniana*.

Damage Larvae infest buds, leaves, flowers and fruits. They secrete silk that prevents leaves from spreading or forces leaves to overlap. Larvae feed among the attached leaves. Leaves are often attached to fruit surfaces by silk. Larvae feed on fruits and form irregular feeding marks. Serious infestations may result in patches of feeding marks on fruit surface and mold growth. A larva uses silk to roll leaves into a nest and feeds inside the nest. Once the leaves are consumed, the larva moves out, builds another nest, and feeds inside.

Morphological characteristics **Adults** Body length 6-8 mm, wingspan 16-20 mm. Body brownish-yellow, forewing dark yellow, hindwing and abdomen light-yellow, the hairs on hindwing margin grey yellow. Male forewing with a fold along the basal 1/3 to 1/2 of the costal margin. Wing base narrow, basal blotch brown, two dark brown bands on forewing, the 2nd band narrower. When wings are closed, the oblique lines in the middle of the two forewings form a "V" shape.

Eggs Length 0.7 mm, width 0.6 mm; elliptical. New eggs light-yellow, semi-transparent. Blackish-brown head capsule is seen on egg surface when mature.

Larvae Mature larvae 13-17 mm long, body slender, head small. Head and pronotum light-yellowish-white, a dark brown mark exists above the ocelli on posterior end of the head. Young larvae light green, mature larvae bright green, prolegs yellow or yellowish-brown, pygidium light-yellow. Anal comb consists of 6-8 bristles. The yellow sex gland of the males begins to be visible on dorsal surface of the abdomen from the 3rd instar.

Pupae Length 9-10 mm, yellowish-brown, slender. The dorsal surface of the 2nd-7th abdominal segments with two

A. honmai adult (Xu Gongtian)

A. honmai adult (Wang Yan)

A. honmai larva (Xu Gongtian)

A. honmai pupa (Xu Gongtian)

A. honmai empty pupa (Xu Gongtian)

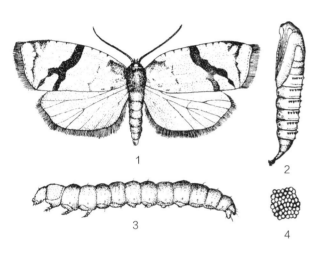
A. honmai. 1.Adult, 2. Pupa, 3. Larva, 4. Eggs (Wang Yongjun, Fan Minsheng)

transverse rows of spines; the first row thick and sparse; the 2nd row small and dense. Anal comb with 8 bristles.

Biology and behavior The insect occurs 3 generations per year in the northeast and central north, 2 generations in Ningxia, 3-4 generations in Shandong, 4 generations in Henan, Anhui, and Jiangsu. The 2nd to 3rd instar move into crevices on tree trunks, branches and overwinter. They are mostly found on tree trunks and main branches of large trees. On small trees, they hide at cut scars or branches with attached old leaves. Overwintering larvae become active when apple trees start to bloom. They crawl to flowers or between leaves and feed. Larger larvae secrete silk and roll up a few leaves and feed inside. Once the leaves are consumed, they move out and continue feeding on young leaves following the same pattern. Those larvae emerged later will damage both leaves and young fruits.

Adult peak emergence is around 17:00. They rest under shade between leaves during the day, and search for mates during night. Adults are attractive to semiochemicals and are slightly phototaxic. Sugar-vinegar and fruit-vinegar mixtures are highly attractive to *A. orata*. There are two synthetic sex pheromones: Z-9-tetradecen-1-ol acetate and Z-11-tetradecen-1-ol acetate. Females lay eggs on smooth fruit surface or upper side of the leaves. Eggs are arranged into fish scale shape masses. Each female produces 1-3 egg masses. The number of eggs per egg mass ranges from dozens to near 200 with an average of 70-80. Egg production is positively correlated to humidity. Therefore, infestation is more serious in years with greater precipitation. Egg development period is around 7 days. Newly hatched larvae scatter around the egg mass and feed on the underside of leaves, between leaves or on fruit surface. The 3rd instar overwinter in September-October.

Prevention and treatment

1. Before buds swell in spring, remove curled barks on trees and burn them.

2. Attract and kill adults using black light traps and sugar-vinegar baited traps.

3. Release *Trichogramma* wasps during emergence of the 1st generation adults, Using sex pheromone baited water bowls to detect adult emergence. *A. orata* starts to lay eggs at 3-5 days after adult emergence. Release *Trichogramma* wasps at this time and repeat every 4-5 days for 4 times. Place one egg card per tree or every other tree. The total release amount is 120,000 eggs per 667 m^2. More egg cards and applications should be considered if experiencing frequent rainfall.

4. Chemical control. Before buds burse in spring, spray calcium polysulfide at 50-100× to kill the overwintering larvae and aphids and mites. Apply 5.7% emamectin at 3,000×, 45% profenofos and lamda-cyhalothrin mixture, 25 g/L lamda-cyhalothrin at 1000×, 50% fenitrothion at 1,000×, or spray 2×10^{10} PIB/mL nuclear polyhydrosis virus at 1,000-1,500×.

References Xiao Gangrou (ed.) 1992, Li Jianhao and Li Dongping 1994, Cao Xiuyun and Liu Yuxiang 2009.

(Ji Baozhong, Zhang Kai, Wang Yongjun, Fan Minsheng)

317 *Cerace stipatana* Walker
Classification Lepidoptera, Tortricidae

Distribution Zhejiang Jiangxi, Fujian, Hunan, Sichuan, Yunnan; India, Japan.

Hosts *Cinnamomum camphora*, *C. glanduliferum*, *Manglietia glauca*, *Persea Americana*, *Euphoria longana*, *Litchi chinensis*, *Liquidambar formosana*, and *Schima superba*.

Morphological characteristics **Adults** Female length 14-17 mm, wingspan 46-54 mm; male length 10-12 mm, wingspan 37-38 mm. Head and humeral plate white; antennae black with white rings. Labial palps black, drooping; the underside of the 1st and 2nd segments and the apex of each segment white. Thorax black, abdomen yellow, end of abdomen black. Forewing purplish black, its front margin with a series of 2-3 mm long, decreasing in size toward the apex, transversely oriented white bands. Under this row of bands are 5 rows of white square spots from base of the wing. Toward the wing tip, the number of rows increase to 8-12. The middle the wing has a reddish brown band that extends to the margin. The band color becomes yellowish brown near the outer margin of the wing and the band becomes much enlarged. Basal part of the hindwing white, its distal margin has a wide black band. Hairs on the wing margin greyish white.

Eggs Round and flat. Egg mass arranged in fish scale pattern. Newly laid eggs white, turning into light-yellow during late stage. A black spot can be seen immediately before hatching.

Larvae Head of the 1st instar black; body light-yellow. Second instar bluish green. Mature larvae pinkish green. Length 29-32 mm, width 3.5-4.5 mm. Each side of the body with a black spot.

Pupae Length 17-21 mm, bluish white to bluish green.

Biology and behavior *Cerace stipatana* occurs 4 generations a year in Zhejiang. It overwinters as pupae inside rolled leaves. Adults emerge during mid- to late March. Egg production occurs during early and mid-April. Egg incubation period is approximately 3 days. Larvae scatter after hatching. They tie leaves with silk and feed inside. The 1st and 2nd instar only feed on a layer of the leaf tissue. Starting from 3rd instar, they can chew holes or chew off pieces of leaves. Once fecal materials filled the web, the larva migrates and makes a new web. It often rolls 3-4 leaves at tip of the shoot. The web has 1-2 escape holes where the larva would escape once disturbed. Larvae go through 5 instars. The development time of larvae, prepupae, and pupae is 13-17, 2-3, and 5-6 days, respectively. Adults mate one day after emergence. They lay eggs on the 2nd day after mating. Eggs are arranged in groups on leaf veins. Each female lays 1-3 egg masses and a total of 128-367 eggs. Female:male ratio is 1:1.7. Females live for 5-7 days. Males live for 3-4 days. The peak emergence of the 1st to 5th generation adults is during early to mi-May, mid- to late June, late July to early August, mid- to late September, and early to mid-November, respectively. Most damage is caused by the 2nd and 3rd generations. Natural enemy population levels significantly increase after the 3rd generation and reduce pest population density. Common egg parasitoid affecting *C. stipatana* is *Trichogramma dendrolimi*. Larval parasitoids include *Elasmus* sp., Braconidae, and *Apanteles* sp. Pupal parasitoids include *Brachymeria lasus*, *Coccygomimus* sp., and *Coccygomimus disparis*. Most abundant parasitoids are *Brachymeria lasus* and *Coccygomimus*. Their parasitization rate is 70%-80%. Predatory birds include *Parus major* and *Oriolus chinensis*.

Prevention and treatment Apply 20 g/L lamda-cyhalothrin at 1,000-1,500×, 50% imidacloprid-monosultap at 600-800×, or 45% profenofos-lamda-cyhalothrin at 1,000-1,500×.

References Tong Xinwang and Lao Guangmin 1984, Jiang Jingfeng and Hu Zhilian 1990, Liu Youqiao et al. 2002.

C. stipatana adult (NZM)

(Liu Yongzheng, Xie Peihua)

318 *Choristoneura lafauryana* (Ragonot)

Classification Lepidoptera, Tortricidae
Synonym *Tortrix lafauryana* Ragonot

Importance *Choristoneura lafauryana* is an important pest in Northeastern China. Larvae feed on leaves and young shoots of *Larix* spp. It also infests *Myrica rubra* and broad-leaved trees and shrubs.

Distribution Liaoning, Jilin, Heilongjiang; Japan, Korea Peninsula, Russia, Europe.

Hosts *Larix* spp., *Myrica rubra*, and broad-leaved trees and shrubs.

Damage Larvae make webs. The following damage was reported: 1) 90% of the shoots broke from bottom; 2) 8% of shoots broke from middle portion; 3) 2% of the shoots broke from the tip. Young stands are more susceptible to infestations. Most damage occurred in 40 cm to 2.5 m tall, 5-10 year-old *Larix* stands. A 17.4% infestation rate was recorded. Occasional infestations were found in greater than 10-year-old stands.

Morphological characteristics **Adults** Wingspan 19-25 mm. Male forewing yellowish brown; with

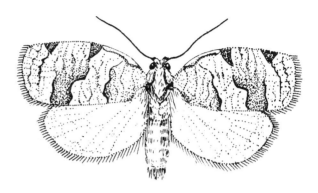

C. lafauryana adult (Hu Xingping)

a fold at the front margin, but the fold is absent near the base of the wing. Basal spot is not obvious. The band at middle of the wing often broken; the stria near the wing margin is conspicuous. Female greyish brown; without colored bands. Hindwing pale greyish yellow.

Eggs Oval, yellowish brown.

Larvae Length 20 mm; body green, with a dark dorsal longitudinal line; head and pronotum greenish brown.

Pupae Length 11 mm, yellowish brown.

Biology and behavior *Choristoneura lafauryana* occurs one generation per year. It overwinters as young larvae. The larvae resume activity in late May. They feed on leaves and young shoots. Side shoots are infested before main shoots. A larva relocates several times during its development. Pupae occur in late June. Adults emerge in early July. Eggs are laid on back of needles. Larvae enter overwintering period in September.

Prevention and treatment Maintain healthy stands and conserve natural enemies. Spray insecticides during larval infestation period.

References Lu Wenmin 1987, Xiao Gangrou (ed.) 1992, Liu Youqiao and Li Guangwu 2002.

(Wu Chunsheng, Lu Wenmin)

C. lafauryana adults (NZM)

319 Cydia pomonella (L.)

Classification Lepidoptera, Tortricidae
Common name Codling moth

Importance *Cydia pomonella* was introduced into China in 1950's. The larvae infest fruits and cause huge economic loss to apple, pear, peach, and walnut production.

Distribution Gansu, Xinjiang; Europe, Asia, Africa, North and South America, Australia, and islands in the Pacific.

Hosts *Malus pumila, M. asiatica, M. spectabilis, Pyrus pyrifolia, P. aromatica, Cydonia oblonga, Crataegus pinnatifida, C. cuneate, Prunus salicina, P. armeniaca, P. amygdalus, P. persica, Juglans regia, Punica granatum, Castanea* spp., *Ficus* spp., *Sorbus* spp.

Damage Larvae bore into fruits, causing decay and loss of fruit.

Morphological characteristics **Adults** Grayish brown and with purple luster; females darker than males; anal area dark brown, with three bronze striae; basal area of wings brown, its outer margin forms a triangular shape, with oblique wavy striae; center of wing pale brown. Underside of wings in males with a large dark area at center; center of hind wing with a dark brown bundle of hairs, with one frenulum. Underside of wings in females without a large dark area; hind wing without a dark brown bundle of hairs, with 4 frenula.

Eggs Flat, its center convex, similar to a drop of wax candle when fresh; old eggs with a pale red ring.

Larvae Pale yellow at beginning, red when mature; dorsal surface darker than ventral surface; mature larvae head yellowish brown, proscutum light-yellow.

Biology and behavior *Cydia pomonella* occurs 1-3 generations a year. It overwinters as mature larvae in cracks on tree trunk and around base of tree in soil. Some larvae may overwinter in boxes, fruit storage rooms. Adults mate within 1-2 days after emergence. Eggs are laid on both sides of leaves. Some eggs are laid on branches and fruits. Leaves and fruits located on top of crown tend to have more eggs. Young larvae will crawl and find a fruit to burrow inside. Mature larvae leave fruits and pupate cracks in soil or on tree trunks or within dead plants. They may also pupate in fruit containers, storage rooms. Adults of the overwintering generation emerge between late April and early May.

Prevention and treatment

1. Quarantine. Prevention of movement of infested fruits to non-infested areas.

2. Control overwintering larvae during January-March and November-December. Scrape off coarse barks and burn them, paint tree trunks with lime sulfur.

3. Attract and kill adults using sex pheromones, place 30 to 60 lures per hectare. Mating disruption by placing 15-30 lures per hectare, place lure about 1.7 m above ground.

4. Remove damaged fruits and fallen fruits, cardboard boxes, wood, bags, bushes, weeds to reduce overwintering sites.

C. pomonella male adult (Zhang Runzhi)

C. pomonella female adult (Zhang Runzhi)

C. pomonella egg (Du Lei)

C. pomonella overwintering larva (Zhang Runzhi)

C. pomonella pupae (Jia Yingchun)

C. pomonella damaged fruit (Zhang Runzhi)

C. pomonella damage to apples (Zhang Runzhi)

5. Conserve natural enemies such as birds, spiders, ground beetles, parasitoids, fungi, bacteria. Release *Trichogramma* sp. or spray *Bacillus thuringiensis*.

6. Wrap tree trunks with jute or *Centranthera cochinchinensis* grass to attract overwintering larvae.

7. Spray 50% phoxim at 1,000-5,000×, 2.5% fenvalerate at 4,000×, 2.5% deltamethrin at 4,000-6,000× or fenoxycarb at 2,000-3,000× twice a year. Each time spray 2-3 times at 7-10 days intervals.

References Huang Yuzhen 2000, Forestry Department of Ministry of Forestry 2005, Qing Zhanyi et al. 2007, Forest Disease and Pest Control Station of the Ministry of Forestry 2008, Zhou Zhaoxu et al. 2008, Yan Yulan 2008.

(Zhao Yuxiang)

320 *Homona issikii* Yasuda

Classification   Lepidoptera, Tortricidae

Distribution   Anhui, Zhejiang, Jiangxi, Hunan, Fujian, Taiwan; Japan.

Host   *Cryptomeria fortune.*

Damage   Larvae fold leaves and form pods and feed on leaves and young shoots. They occasionally feed on cones. Severely infested trees have many pods and all leaves may be consumed. Branches may be killed under heavy infestations.

Morphological characteristics   **Adults** Female length 10-12 mm, male length 8-11 mm; female wingspan 28 mm, male wingspan 24 mm. Antennae filiform, grey. Labial palpi brown, curved upward. The 2nd segment long, the last segment short. Head and thorax brown, abdomen greyish-brown. Forewing greyish yellow, with a purplish-brown spot. Basal spot, median transverse band, and the apical stripe distinct. The median transverse band breaks near the costal margin. Folded area of the male forewing large. Hindwing greyish brown. The posterior end of the male abdomen with yellow hairs.

Eggs Length 1.0-1.2 mm, width 0.6 mm. Flat and oval. New eggs light-yellow, gradually change to darker color, yellowish brown before hatching. Egg shell translucent.

Larvae Mature larvae 21 mm. Head and pronotum of the first instar larvae brown. Mesothorax, metathorax and abdomen light-yellow. The occiput, pronotum, and thoracic legs of the 2nd instar dark reddish-brown. The mesothorax, metathorax, and abdomen of the third to 5th instar pale brown to greenish brown. Anal shield dark brown with multiple setae. Starting from the 3rd instar, dorsal surface of the 5th abdominal segment with a pale red-violet spot. Spiracles oval. Peritreme dark reddish-brown. All segments with multiple brown setae clusters.

Pupae Length 12 mm, width 3 mm; fusiform. Pale yellow at the beginning, turning into dark reddish-brown. The dorsal surface of the 2nd-8th segments with 2 rows of tooth-like structures; those of the front row thick and sparse, the back row slender and dense. Each segment with many hairs. The posterior end obtuse. Anal comb with 8 teeth.

Biology and behavior   *Homona issikii* occurs two generations per year in Zhejiang and overwinters as 1st-2nd instar in hollow brown leaves. They resume activity in mid- to late April. The new shoots development coincides with the infestation peak during early to mid-May. Larvae start to pupate in mid-May. Adults emerge between late May and late July. First generation eggs

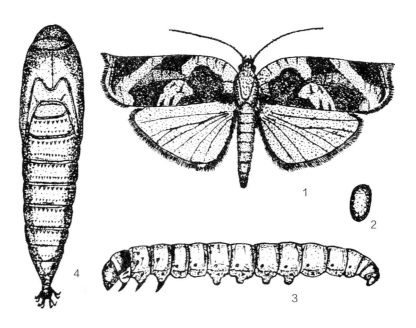

H. issikii. 1. Adult, 2. Egg, 3. Larva, 4. Pupa (Xu Deqin)

H. issikii male adult (NZM)

H. issikii female adult (NZM)

appear from early July to mid-July. Larvae appear from mid-June to early September. Peak infestation period is in August. Pupation starts in mid-August. First generation adults appear during late August and early October. Second generation eggs are laid between early September and early October. Larvae start to appear in late September. They bore into leaves and overwinter.

Adult emergence occurs during day or night, but mostly during 13:00-16:00. They are inactive during the day and rest on branches, leaves or ground cover. They are active during night hours and are attracted to light. Female to male ratio is 1.34:1. They mate 1-2 days after mating, which occurs during day or night but peaks during 01:00-05:00. Before mating, female forewings raise upward and the abdomen bends downward. Males and females only mate once. Mating duration ranges from 3.5-5.5 hours. Female lay eggs the same day after mating. Egg laying period is 2-6 days. Each female lays eggs 2-10 times at 3-18 hours intervals; each female lays 43-254 eggs with an average of 93 eggs. Eggs are laid on young branches and base of leaves. The eggs are laid in imbricate arrangement with 2-13 eggs (average 9 eggs) per egg mass. Average longevity of overwintering females and males are 6.4 and 3.8 days, respectively. Average longevity of the 1st generation females and males are 7.7 and 4.5 days, respectively.

The 1st generation egg period is 8.1 days. A black spot can be seen from the 5-7 days old eggs. Most of the eggs hatch during 6:00-12:00. During hatching, the larva chews open the eggshell and then immediately crawls along branches and needles. It spins a net among a few leaves. The larva feeds inside the leaves. Once a leaf is hollowed by a larva, the larva will move to a neighboring leaf and continues feeding. Therefore, clusters of brown leaves can be seen on branches. These brown leaves can be used to detect overwintering larvae and first instar larvae. The 3rd instar larvae move out of the leaves and spin leaves and young shoots into nests. They feed on whole leaves and epidermis of the branches. One nest may have 1-10 (average 5-6) young shoots. Overwintering larvae first feed on one year old leaves and branches. After 4th instar, they feed on new branches and leaves and may break the branches. First generation larvae usually infest young branches and leaves. Each nest has one larva and each larva may form 3-5 nests. Larvae spin silk and fall after disturbance. Larvae have 5 instars. Mature larvae pupate inside the nests. Pre-pupae period is 2-5 days. Pupal period is 11.2 days for overwintering generation and 5.9 days for the 1st generation.

Infestation is most severe in approximately 10-year-old *Cryptomeria fortune* forests. The activity period of the overwintering larvae is closely correlated with the weather conditions in April. Fewer rainy days and higher temperature will lead to earlier start of the feeding activity. Population levels vary significantly from year to year. In Wencheng of Zhejiang province, heavy infestations occurred in 1985 and 1988 with population density being greater than 100 larvae per tree (maximum was > 500 larvae per tree). However, the population levels in 1986 and 1987 were < 10 larvae per tree.

Many natural enemies of *H. issikii* larvae and pupae were reported. They include *Macrocentrus* sp., *Noplec-*

tis alternans spectabilis, *Campoplex* sp., *Habronyx* sp., *Brachymeria lasus*, *Elasmus* sp, *Pediobius foveolatus*, *Beauveria bassiana*, ants, and spiders. The dominant natural enemies of the overwintering generation are *Macrocentrus* sp. and *B. bassiana*. *Macrocentrus* sp. parasitizes *H. issikii* larvae. When the *H. issikii* larvae mature and spin cocoon, the *Macrocentrus* sp. larvae consume all of the body content of the host and form 10-34 cylindrical yellowish brown cocoons inside the nest. Parasitization rate by *Macrocentrus* sp. reached as high as 26.1%. Parasitization rate by *B. bassiana* reached was as high as 24.5%. The dominant natural enemy of the 1st generation larvae and pupae is *Brachymeria lasus*. Its parasitization rate reached was as high as 34.2%.

Prevention and treatment

1. Biological control. Natural enemies include *Macrocentrus* sp., *Hoplectis alternans spectabilis*, *Campoplex* sp. *Habronyx* sp., *Brachymeria lasus*, *Elasmus* sp. *Pediobius cassidae*, *Beauveria bassiana*, ants, spiders. Overwintering generation is often killed by *Macrocentrus* sp. and *B. bassiana* with reported parasitic rate of 26.1% and 24.5%, respectively. First generation is often parasitized by *Brachymeria lasus* with parasitization rate of 34.2%.

2. Chemical control. Apply 20 g/L lamda-cyhalothrin at 1,000-1,500×, 50% imidacloprid-monosultap at 600-800×, or 45% profenofos-lamda-cyhalothrin at 1,000-1,500×.

References Yang Xiuyuan et al. 1981, Chinese Academy of Sciences - Institute of Zoology 1983, Xu Deqin 1987, Shen Guangpu et al. 1988, Liu Youqiao and Li Guangwu 2002.

(Xu Deqin)

321 *Pandemis heparana* (Denis et Schiffermüller)

Classification Lepidoptera, Tortricidae
Synonym *Tortrix heparana* Denis et Schiffermüller
Common names Dark fruit-tree tortrix, Apple brown tortrix

Importance *Pandemis heparana* is an important pest of forest trees, fruit trees and ornamentals. Its high reproduction potential, multiple generations per year, and overlapping generations make it difficult to control and cause significant economic and ecological loss.

Distribution Beijing, Hebei, Shanxi, Inner Mongolia, Liaoning, Jilin, Heilongjiang, Shanghai, Jiangsu, Zhejiang, Anhui, Shandong, Henan, Hubei, Hunan, Chongqing, Sichuan, Shaanxi, Gansu, Qinghai, Ningxia; Korea Peninsula, Japan, Russia, India, Europe.

Hosts *Malus* spp., *Pyrus* spp., *Prunus armenica*, *P. persica*, *P. pseudocerasus*, *Salix* spp., *Corylus* spp., *Ceryls heterophyllus*, *Fraxinus mandschurica*, *Quercus* spp., *Spiraea* spp., *Alnus hirsute*, *Fagus* spp., *Ulmus* spp., *Tilia* spp., *Sorbus* spp., *Vaccinium* spp., *Lysimachia clethroides*, *Humulus lupulus*, *Morus alba*, *Ligustrum quihoui*, *Pyracantha fortuneana*, *Sophora japonica*, *Ginkgo biloba*, and *Platanus orientalis*.

Damage Young larvae feed on leaf tissue. Damaged leaf surface looks like a sieve. Older larvae scatter and tie 2-3 leaves with silk or roll one leaf a time. They also tie leaves to fruits and feed on fruit surface.

Morphological characteristics **Adults** Wingspan 16-25 mm. Forewing brown; basal spot obvious; middle band starts from middle of the front margin and ends near the posterior angle and becomes wider from top to bottom. Hindwing greyish brown.

Larvae Head and pronotum pale green. Body dark

P. heparana adult (Bai Jiuwei)

P. heparana adult (Li Chengde)

P. heparana adult (lateral view) (Xu Gongtian)

green with whitish color. Most individuals with a pair of black spots on the sides near the posterior margin of the pronotum; color of pinacula paler.

Biology and behavior *Pandemis heparana* occurs two generations a year in Xingcheng (Liaoning province); three generations a year in Changli (Hebei province), Qingdao (Shandong province), and southern Shaanxi; five generations a year in Hefei (Anhui province). Larvae spin white cocoons and overwinter in cracks on tree barks and around the pruning cuts. Larvae resume activity in early May in Xingcheng (Liaoning province). They feed on buds, leaves, and flower buds. In late May, larvae start to roll leaves. Once mature in mid-June, they pupate inside rolled leaves. Adults emerge from late June to mid-July. Peak of oviposition period is during early July. Tens or hundreds of eggs are arranged in fish scale pattern. The 1st generation larvae appear from mid-July to early August. They pupate in mid-August. First generation adults emerge from late August to early September. Second generation (overwintering generation) larvae appear from early September to October. Young larvae start to overwinter in early to mid-October.

Adults are only active during the night. Larvae spin silk and fall once disturbed. When touched, they retreat or jump.

Prevention and treatment

1. Removal. 1) cut off webbed leaves; 2) scrape off coarse and curved bark, dead bark near the pruning cuts; 3) trap adults using light traps during peak of adult emergence; 4) install sugar-vinegar trap in canopy to remove adults. The attractant is a mixture of sugar:liquor:vinegar:water at 1:1:4:16.

2. Biological control. 1) when overwintering larvae resume activity or during the peak of the 1st generation larvae, spray *B.t.* (1×10^{11} spores/mL) at 800×, Apply *Beauveria bassiana* (4×10^{11} conidia/g) at 2,000-3,000×, 2×10^{10} PIB/mL nuclear polyhydrosis virus at 1,000-1,500×; 2) release *Trichogramma dendrolimi* during egg stage. Release parasitoids every other tree or every other row of trees. Release eggs 3-4 times during each generation at 5-day intervals. Release 1,000-2,000 wasps per tree.

3. Chemical control. Spray 1.8% dynamec (a.i. Abamectin) at 3,000-4,000×; 80% DDVP EC, 90% trichlorfon, 48% chlorpyrifos EC, 25% quinalphos, 50% Sumithion, or 50% malathion at 1,000×; 2.5% lamda-cyhalothrin, 2.5% deltamethrin EC, 20% fenvalerate at 3,000-3,500×; 10% bifenthrin EC at 4,000×, 52.25% chlorpyrifos + cypermethrin at 1,500×. Caution should be practiced in orchards.

References Xiao Gangrou (ed.) 1992; Zhao Guorong, Cai Yanping, and Yang Chuncai 1997; Du Liangxiu and Du Chengjin 1999; Liu Youqiao and Li Guangwu 2002.

(Byun Bong-Kyu, Li Chengde, Bai Jiuwei)

322 *Polylopha cassiicola* Liu et Kawabe
Classification Lepidoptera, Tortricidae

Importance *Polylopha cassiicola* is a common pest in Guangxi, Guangdong, and Fujian. It is the major pest of *Cinnamomum cassia*.

Distribution Fujian, Guangdong, Guangxi.

Hosts *Cinnamomum cassia*, *C. camphora*, and *C. parthenoxylum*.

Damage Larvae feed on young shoots of *C. cassia*, causing death of new shoots and development of numerous lateral shoots.

Morphological characteristics **Adults** Wingspan 11-14 mm. Antennae pale brown. Forewing long oval; costa curved; termen oblique. Greyish brown, shiny, mixed with orange reddish brown color, especially in the front margin and the apex. Basal fascia obvious, there are 3-4 rows vertical scales. Hint wing near quadrilateral, without comb setae.

Eggs New eggs milky white, round, diameter 0.1-0.13 mm; turning to blackish brown before hatching.

Larvae Mature larvae 7.3-10.1 mm in length, average length 8.2 mm. Head width 0.53-0.70 mm, average width 0.57 mm. Head and pronotum blackish brown. The pronotum semi-circular and interrupted at the middle. The dorsal surface of the 9th segment with a semi-oval blackish brown spot. Prolegs with a single series of crotchets arranged in a complete circle. The anal proleg with a single series of crotchets arranged in an incomplete circle.

Pupae Length 3.8-8.7 mm, average length 4.4 mm. Yellowish brown, turning black before eclosion. The dorsal surface of the 2nd and 9th abdominal segments each with a transverse row of blackish brown short spines. Each segment between the 3rd and 8th segment has two rows of transverse blackish brown short tubercles; those of the front row conical, the back row small and dense. Anal comb with 4 spines.

Biology and behavior *Polylopha cassiicola* occurs 7 generations in Qinxi (Guangxi) and Huaan (Fujian). The different generations overlap and there is no dormancy or diapause. The 1st generation larvae infest *C. camphora*, young shoots of *C. parthenoxylum*, and occasionally *C. cassia* old shoots. The 2nd generation larvae infest new shoots of *C. cassia* and cause death of new shoots. The 7th generation larvae continue infest *C. cassia*.

Prevention and treatment

1. Monitor population distribution and dynamics. Control the 1st instar larvae during the development of new shoots of *C. cassia*. This is a critical step in preventing the spread of the pest.

2. Plant *C. cassia* in suitable locations where soil layer is thick, loose, with good drainage to ensure the trees' defense capability. Cut down weeds, eliminate or reduce pupae in winter, conserve natural enemies and diversity to facilitate ecological control.

3. Biological control. It was reported that releasing *Trichogramma chilonis* can achieve good control of *P. cassiicola*. Apply *Beauveria bassiana* (4×10^{11} conidia/g) at 2,000-3,000×. Spray 2×10^{10} PIB/mL nuclear polyhydrosis virus at 1,000-1,500×.

4. Chemical control. Apply 20 g/L lamda-cyhalothrin at 1,000-1,500×, 50% imidacloprid-monosultap at 600-800×, or 45% profenofos-lamda-cyhalothrin at 1,000-1,500×.

References Liu Zhicheng and Peng Shibing 1992, Xian Xuxun 1995, Liu Youqiao and Li Guangwu 2002, Zheng Baorong 2007.

(Wu Chunsheng)

P. cassiicola adult (NZM)

323 Ptycholomoides aeriferanus (Herrich-Schäffer)

Classification Lepidoptera, Tortricidae
Synonym *Coccyx aeriferanus* Herrich-Schäffer

Importance *Ptycholomoides aeriferanus* is an important defoliator of *Latrix* spp. Heavy infestations cause large areas of trees look yellow. Young trees may die from consecutive infestations.

Distribution Hebei, Inner Mongolia, Liaoning, Jilin, Heilongjiang; Korea Peninsula, Japan, Russia, Europe.

Hosts *Latrix* spp. *Acer kawakamii*, *Betula* spp.

Damage Lower and middle part of the tree canopy of the infested trees lose almost all leaves when heavily infested with *P. aeriferanus*.

Morphological characteristics **Adults** Wingspan 19-23 mm. Forewing brownish yellow, mixed with whitish grey strigula. Basal fascia, median fascia, and subterminal fascia blackish brown. Hindwing dark brown.

Larvae Head light-yellowish brown, with brown stripes. Pronotum with two pairs of brown spots. Body green, sub-dorsal line dark green or pale green.

Biological and behavior *Ptycholomoides aeriferanus* occurs one generation a year. The newly hatched larvae hide in cracks in the bark, buds, or fallen leaves and overwinter in silk sacks. They bore into buds of mid- and lower part of the tree canopy in mid-April. The larvae feed at base of the leaves with their heads toward the base of the leaves. After the lower part of the tree canopy is consumed, the larvae migrate to middle part of the tree canopy. Larvae mature in mid-May. They pupate among leaves, in cracks of bark, or in fallen leaves. Adults

P. aeriferanus adult (Zhu Xingcai)

emerge in late June.

Adults are attracted to lights. Eggs are laid on front side of the leaves in one row or two rows. They consist of 2-6 eggs and a maximum of 15 eggs. The newly hatched larvae do not feed before overwintering. When disturbed, the larvae waggle and retreat or fall through a silk thread.

Prevention and treatment

1. Silviculture. Plant mixed forests and maintain good sanitation.

2. Physical control. Set up black light traps to kill adults during adult emergence period.

3. Biological control. Aerial spray of 20% *B.t.* at 10×. Apply *Beauveria bassiana* (4×10^{11} conidia/g) at 2,000-3,000×. Spray 2×10^{10} PIB/mL nuclear polyhydrosis virus at 1,000-1,500×. Conserve natural enemies which include *Apanteles* sp., *Ephialtes* sp., and *Dirophanes* spp., and parasitic flies.

4. Chemical control. 1) fogging of matrine at 7.5 kg/ha; 2) aerial spray of 90% trichlorfon EC at 80×, 90% trichlofon EC + 20% thiamethoxam (1:1) at 80×, or 90% trichlofon EC + 25 g/L lamda-cyhalothrin at 100× can achieve greater than 95% control. Apply 20 g/L lamda-cyhalothrin at 1,000-1,500×, 50% imidacloprid-monosultap at 600-800×, or 45% profenofos-lamda-cyhalothrin at 1,000-1,500×.

References Boli County Forest Disease and Pest Control Station of Heilongjiang Province 1974, Xiao Gangrou (ed.) 1992, Liu Youqiao and Li Guangwu 2002.

(Byun Bong-Kyu, Li Chengde, Zhang Runsheng, Liu Youqiao, Bai Jiuwei)

P. aeriferanus adult (Li Chengde)

324 *Ancylis mitterbacheriana* (Denis et Schiffermüller)

Classification	Lepidoptera, Tortricidae
Synonyms	*Phalaena* (*Tortrix*) *mitterbacheriana* Denis & Schiffermüller, *Tortrix retusana* Haworth
Common name	Mitterbach's red roller moth

Distribution Shandong; Russia, Europe.

Hosts *Quercus acutissima, Quercus variabilis*.

Damage Larvae roll up leaves along the middle vein and form pods..

Morphological characteristics **Adults** Male wingspan 18-22 mm, female wingspan 18-27 mm. Frontal area and labial palps grey-white; antennae grey-brown, with white bands in the basal part. Labial palps point upward, the 2nd segment wide; grey-brown; sharp ended. Forewing long and narrow, apical angle distinct, with a sickle shaped color marking toward the tip of the wing. Forewing grey-white, with brown blotches; from base to tip the grey-white and brown alternated parallel stripes form hook-like pattern. From median cell downward, the brown scales form a "W" blotch. Hindwing, legs, and abdomen grey white.

Eggs Flat elliptical. New eggs are red, becomes dark red later and dark grey before hatching.

Larvae Mature larvae 13-15 mm in length. Head brown, thorax and abdomen yellow-white. Pronotum has 6 black marks; the two above the spiracles are the largest, with 4-5 angles; the two median anterior marks are the smallest. Tubercles distinct at each body segment. Those on the methothorax and metathorax are in a single row,

A. mitterbacheriana adult (NZM)

A. mitterbacheriana adult (Wu Chunsheng, Zhang Runzhi)

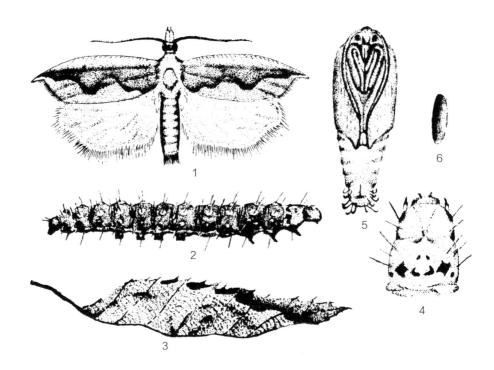

A. mitterbacheriana. 1. Adult, 2. Larva, 3. Damage, 4. Larva and pronotum, 5. Pupa, 6. Egg (Wang Jiashuang)

each tubercle with 2 setae. Those on the abdominal segments arranged in trapezoid pattern, each tubercle with one seta. Proleg crochets are in two uniordinal circles, the outer ring complete, the inner ring incomplete. Anal leg crochets are in two transverse bands.

Pupae Length 7.5-9.5 mm, dark brown. Thorax with a pale dorsal longitudinal line. Each of the 2nd-7th abdominal segments with two rows of spines; the 1st row larger than the 2nd row; each of the 8th-9th abdominal segments with only one row of spines. There are 12 spines at the end of the abdomen; among them, two spines are at each side of the anus.

Biology and behavior *A. mitterbacheriana* occurs one generation per year. Mature larvae overwinter in rolled-up leaves. Pupation occurs in late-January next year. Pupal stage is about 34 days. Adults emerge in early March. Eggs are laid in late March when buds of *Q. acutissima* develop. Larvae mature in October. They fall onto ground along with rolled leaves in November-December and overwinter.

Adults crawl from base of tree branches upward and lay eggs as they climb. Eggs are laid singly. Each female lays 123-236 eggs. Larvae hatch between 10:00-17:00 with peak hatching time at 13:00. Newly hatched larvae are 1.2 mm long. They soon crawl to leaf buds or flower buds on branch tips, spin silk, and move under the scales.

They feed on buds or young branch epidermis. The 1st instar molt into 2nd instar after 11-14 days. They continue to feed on buds or developing leaves that are about 1/3 size of the mature leaves. They fold leaves inward along the edge. The 3rd instar and older larvae fold leaves along the middle vein and form 2.5 cm long and 1 cm wide chambers. One leaf can have 3 chambers. One larva folds 2-4 leaves during its life time. When migrating, they prefer go up and feed on young leaves. A species of aphid live together with *A. mitterbacheriana*. Their relationship is probably mutualistic. The larvae have 6 instars.

Prevention and treatment

1. Conservation of natural enemies. Avoid using insecticides in lightly infested stands.

2. Light trapping during adult emergence period.

3. Clean up leaves and dead branches in January-February.

4. Spray systematic insecticides during larval peak period in heavily infested stands.

References Dong Yancai and Zhu Xinbo 1990, Liu Youqiao and Li Guangwu 2002.

(Wu Chunsheng, Dong Yancai, Zhu Xinbo)

325 *Ancylis sativa* Liu

Classification Lepidoptera, Tortricidae
Synonym *Cerostoma sasahii* Matsumura
Common name Ziziphus leaf roller

Importance *Ancylis sativa* is a serious pest of *Ziziphus jujube* and *Ziziphus jujuba* var. *spinosa* in north and south of China. Heavy infestations cause jujube tree leaves become yellow and reduce fruit production from 40% to as high as 80%-90%.

Distribution Hebei, Shanxi, Shandong, Henan, Hubei, Hunan, Shaanxi.

Hosts *Ziziphus jujube* and *Ziziphus jujuba* var. *spinosa*.

Damage Larvae spin silk and roll leaves into pods. They use silk to attach buds, flowers, leaves and fruits and feed on leaves, flowers, and fruits.

Morphological characteristics **Adults** Length 6-7 mm, wingspan 14 mm, grey brownish-yellow, eyes dark green. Antennae brownish-yellow, 3 mm long. Labial palps directed downward; the last segment small, partially hidden by scales on the 2nd segment. Forewing brownish-yellow. The costal margin has more than 10 black and white alternating stripes; there are three silky white lines originated from the first a few stripes; they point toward the wing tip; the bottom line is the longest and converges with the top one. There are three blackish-brown vertical stripes at the center of the wing. The wing tip curved into sickle shape.

Eggs Elliptical, newly laid eggs white, yellow on the 2nd day, apricot yellow on the 3rd day, and become reddish as they mature. The final color before hatching is orange red.

Larvae Newly hatched larvae have blackish-brown head, yellowish-white abdomen. They become yellow after feeding. Large larvae 15 mm in length. Head pale brown, with blackish-brown spots, thorax and abdomen yellowish-white, pronotum and pygidium brown, body with yellowish-white pubescence. Proleg crochets arranged in double rings, those on anal legs are in two bands.

Pupae Length 7 mm, spindle-shaped, and enclosed in thin cocoons. New pupae green, turning into yellowish-brown later, then blackish brown before adult emergence. There are two rows of tooth-like spines on dorsal surface of each of the abdominal segments. They start and end at the spiracle line. The end has 5 large spines and 12 long hooked setae. Anal comb has 8 spines.

Biology and behavior *Ancylis sativa* occurs 3-4 generations per year. It overwinters as pupae in crevices of tree barks. Adults emerge and lay eggs during mid- and late March next year. Eggs hatch during mid-April to mid-May. The young larvae spin silk along the leaf edge and roll the leaves into pods. They pupate inside the pods. Adults of the 1st and 2nd generations appear from late May to late June and from mid-July to mid-August, respectively. Larvae of the 3rd generation mature in early September and move into cracks on barks, pupate, and overwinter inside the barks.

Adults emerge during the day. After adult emergence, the pupal skin is partially exposed from the cocoon. It was observed that adults emerged successfully from 80-87% of the pupae. Sex ratio is 1:1. Adult rest among leaves during the day. They mate multiple times and lay eggs during the night. Eggs are laid 1-2 days after mating. They are laid singly or as clumps of 4-5 eggs. Adults of the overwintering generation lay eggs on smooth branches. Those of other generations lay eggs on leaves. About 80% eggs are laid on top of the leaves. The first generation females produce most eggs with each female lays an average of > 200 eggs.

Larvae spin silk and attach flowers and leaves and

A. sativa adult (Li Zhenyu)

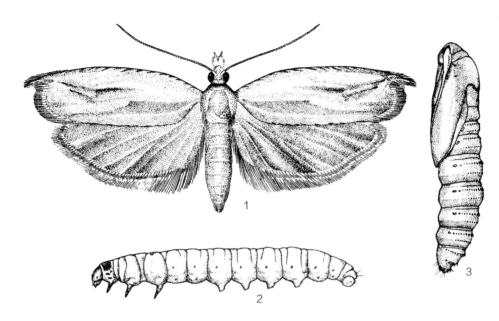

A. sativa. 1. Adult, 2. Larva, 3. Pupa (Zhang Xiang)

feed inside. Once disturbed, they spin silk and fall out. When touched, they often jump a few times and retreat rapidly. The 1st generation feed on undeveloped young buds, causing death of buds and bud regeneration. When infesting fully developed leaves, the larvae roll leaves into pods. The 2nd generation larvae spin silk and attach flowers and leaves. They feed on leaves. The 3rd and 4th generation larvae attach leaves and fruits, feed on leaves and fruits, and cause fruit fall. Larvae molt four times. The 1st-3rd generation larvae pupate in leaves. The 3rd or 4th generation larvae pupate in tree barks, with 70% on tree trunks, 20% on main branches, and some in empty *Cnidocampa flavescens* pupae or pruning wounds. Natural pupal mortality is between 8%-19%.

Outbreaks are closely related to weather conditions. Frequent rain during May-July and humid hot weather promote *A. sativa* outbreaks.

Prevention and treatment

1. Manual control. Wrap the tree trunks near the branch base with 3 cm thick dry grass before September to attract overwintering larvae. Take down the grass in November and scrape off the pupae on the tree trunks and destroy them with fire. Scrape off old barks and cut down broken branches in winter and burn them. Paint tree trunk with calcium oxide (CaO) and fill tree cavities with mud to kill pupae.

2. Biological control. Release *Trichogramma dendrolimi* during egg laying period from mid- and late July. Parasitization rate was 85% after 3,000-5,000 *T. dendrolimi* eggs per 667 m^2 were released.

3. Sex pheromone traps. Use two sex pheromones, E9-12Ac and Z9-12Ac, at 8:2 ratio to attract and kill adults. The number of eggs can be reduced by 68.5-85.2%.

4. Biopesticides. Spray *Bacillus thuringiensis* ($5-10 \times 10^8$ spores/mL) and *Beauveria bassiana* (4×10^{10} spores/g).

5. Chemical control. Apply 5.7% emamectin at 3,000×, 25 g/L lamda-cyhalothrin at 1000×, or spray 2×10^{10} PIB/mL nuclear polyhydrosis virus at 1,000-1,500× during early May, mid-June, and late August.

References Wang Pingyuan et al. 1983; Li Guixiang et al. 1984; Wang Yunzun 1988; Xiao Gangrou (ed.) 1992; Xiao Gangrou et al. 1997; Bao Jianzhong and Gu Dexiang 1998; Liu et al. 2002; Qiu Qiang et al. 2004; Wang Xufen, Zhang Cuiyu, Zhang Xiuhui et al. 2006; Meng Dehui 2007.

(Mi Ying, Jia Yudi, Sun Jinzhong)

326 *Celypha pseudolarixicola* Liu

Classification Lepidoptera, Tortricidae

Importance The host species *Pseudolarix kaempferi* is only found in China. As more *P. kaempferi* trees are planted, *C. pseudolarixicola* infestations become more common. In Longshan forest farm (Lianyuan, Hunan province), 800 km^2 of *P. kaempferi* were infested. All trees in the 27 ha of *P. kaempferi* seed production stand were infested. Most *P. kaempferi* trees in Furongshan forest farm (Anhua, Hunan province) died from infestation and were cut down. *Pseudolarix kaempferi* trees at the tourism sites at Nanyue (Hunan province) and Mount Mogan (Zhejiang province) also suffered infestations in several consecutive years.

Distribution Zhejiang, Jiangxi, Hunan.

Host *Pseudolarix kaempferi*.

Damage Larvae feed on young leaves and buds. They affect tree growth or kill the trees.

Morphological characteristics **Adults** Wing length: male 11-14 mm, female 12-16 mm. Labial palps protrude forward. The basal area of the first two segments yellow. Scales at the tip of the 2nd labial segment enlarged, blackish-brown. Last labial segment short and small, slightly pointed downward. Forewing near rectangular, with dark brown and silver-brown stripes. Its front border with hook-shaped white stripes. Hindwing brown, legs yellow, front tibia brown, tarsal segments with brown rings.

Eggs Elongated oval, 0.5-0.8 mm in length. Newly laid eggs milky or light-yellow, pale red before hatching.

Larvae Young larvae 1-1.4 mm in length, head blackish brown, thorax pale-yellow, abdomen pale-red. Mature larvae 10-12 mm in length, head yellowish-brown, body pale green, prothorax with blackish-brown "V" marking. Crochets form two circles.

Pupae Tanned. Length 6.5-8.5 mm. Each of the abdominal segments with two rows of spikes on the dorsal surface. Those on the front row are large and sparse, those on the back row are small and tight.

Biology and behavior *Celypha pseudolarixicola* occurs one generation per year. Young larvae overwinter under barks. Adults emerge during the next May to June. Female to male ratio is related to population density. In high population density stands, the female: male ratio was 36:64. In low population density stands, female: male ratio was 56:44. Adults are attracted to lights. Their numbers are highest during 8:00-10:00 pm under the light. They rest in the lower to middle part of trees or in the shrubs during cloudy days or before 10:00 am. They occasionally fly during this time. They are active and search for mates in early evening hours. Mating occurs between 7:30 pm - 8:00 am with peak at 8:00 pm - 2:00 am. Copulation duration is 1.3-2.0 hours. Eggs are laid along the main veins on the back of leaves. Eggs are laid individually or several or dozens of eggs are glued together. Each female lays an average of 32 eggs. Young larvae are very active. Once hatched, they search for cracks and wounds on trees and hide inside and enter

C. pseudolarixicola adult (NZM)

C. pseudolarixicola adult (NZM)

C. pseudolarixicola adult (NZM)

diapause. They may feed on barks before overwintering. During the next March when new shoots start to grow, larvae feed on young leaves and hide in young shoots. They spin silk between branches and migrate to new shoots for feeding. Once new leaves are fully spread, larvae use silk to spin lower part of the leaves into cylindrical pods and feed inside the pods. The upper part of the pods is open. Feces are propelled from top of the pods. Damaged shoots are easily spotted. One larva infests 1-3 young shoots a year. One young shoot is only infested by one larva. When population level is high, a shoot may be infested by two larvae.

C. pseudolarixicola occurs mostly in low elevation mountains with 400 m elevation forest stands being most susceptible. At the same elevation, the east slope is more susceptible than the west and south slopes. The north slope is the least susceptible. Pure stands are more susceptible than mixed stands. Edges of a stand is more susceptible than other areas of a stand. Stands of 3-12 years old can be infested. Most serious infestations are found in 2-3 m tall 8-year-old *P. kaempferi* forests.

Prevention and treatment

1. Conservation of natural enemies. There are three parasitic wasps that affect *C. pseudolarixicola* larvae: *Apanteles* sp., *Bracon* sp., and *Scambus* sp. Parasitization rate can reach 66%. Pupal stage has three parasitic wasps: *Itoplectis* sp, *Ephialtes* sp., and *Brachymeria obsurta*.

2. Sex pheromone. Adding synthetic *C. pseudolarixicola* sex pheromone can significantly reduce male moth numbers. At the same trapping period, sex pheromone traps caught significantly more male moths than traps without sex pheromone. The results suggest sex pheromone is a feasible tool for managing *C. pseudolarixicola*.

3. Applying *Beauveria bassiana* 4×10^{10} conidia/g caused 66.1% larval infection and 41% pupal infection.

4. Chemical control. Spray 50% trichlorfon at 500× or 80% DDVP at 1,000× killed 98% of the larvae. DDVP fogging killed 95% of the adults. Apply 20 g/L lamda-cyhalothrin at 1,00-1,500×, 50% imidacloprid-monosultap at 600-800×, 45% profenofos-lamda-cyhalothrin at 1,000-1,500×.

References Liu Youqiao and Li Guangwu 2002; Wang Shufen, Tang Dawu, Ye Cuiceng et al. 2003; Xu Guangyu 2008.

(Wu Chunsheng)

Cryptophlebia ombrodelta (Lower)

Classification Lepidoptera, Tortricidae

Synonyms *Arithrophora ombrodelta* Lower, *Cryptophlebia carophaga* Walsingham, *Argyroploce lasiandra* Meyrick

Common names Litchi fruit moth, Macadamia nut borer

Distribution Hebei, Henan, Guangdong, Guangxi, Hainan, Yunnan, Taiwan; India, Sri Lanka, Nepal, Indonesia, Vietnam, Thailand, Western Malaysia, New Guinea, the Philippines, Japan, Guam, the Caroline Islands, Australia and U. S. A. (Hawaii).

Hosts *Litchi chinensis*, *Citrus* spp., *Acacia franesiana*, *Gieditsia sinensis*, *Averrhoa carambota*, *Saraca dives*, *Sindora tonkiensis*, *Lysidice rhodostegia*, *L. brevicalyx*, *Cassia fistula*, *C. accidentalis*, *Parkinsonia aculenta*, *Bauhinia purpurea*.

Damage Larvae bore into fruits. Frass accumulations inside fruit cause mold development, blackening of the fruits, and early fruit fall.

Morphological characteristics **Adults** Length 6.5-7.5 mm, wingspan 16-23 mm. A pale brown stripe exists near the tip of the forewing. It extends from the costal margin downward. There is sexual dimorphism. Forewing of the females black brown with a black pretornal spot; forewing of males yellowish-brown with a blackish-brown oblique stripe near the posterior margin. Hindwings greyish brown.

Larvae Length 12 mm, body pink in dorsal view and yellowish-white in ventral view.

Biology and behavior *Cryptophlebia ombrodelta* occurs 3 generations per year and overwinters as larvae within legumes and bark crevices. Pupation starts in

C. ombrodelta adult male (NZM)

C. ombrodelta adul femalet (NZM)

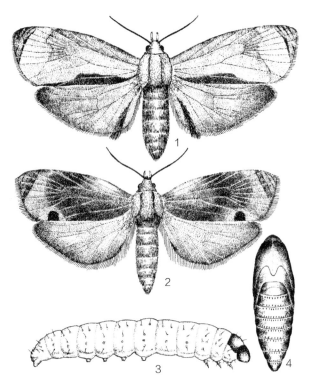

C. ombrodelta. 1. Adult male, 2. Adult female, 3. Larva, 4. Pupa (Zhang Peiyi)

early April and eclosion of adults occurs in early May. Second generation adults appear in mid-June, and third generation adults occur in early July. Larvae are present in May, mid- to late June, and July-September. Adults are active at night and females oviposit on the surface of the legumes, fruits, or nuts. Eggs are laid singly. After hatching, larvae bore into the legume, creating a small pit. Larvae mine on the surface before entering the legumes. Then, they enter the fruits and often result in premature fruit fall. Those legumes remaining on the tree become black and moldy. In early June, the 1st generation mature larvae use silk, frass and debris to make cocoons for pupation in the legumes or between bark crevices. The 2nd generation larvae appear between mid- and late June. The 3rd generation appear in early July and last until September. The larvae exit the legumes and construct cocoons in bark crevices to overwinter. Others overwinter within legumes which may stay on trees, fall to ground or are transported to storage places.

Prevention and treatment

1. Manual control. During late autumn and early spring, remove and destroy infested legumes on the trees. Harvested legumes with pests should also be destroyed.

2. Chemical control. Spray 50% malathion EC or 50% fenitrothion EC at 1,000× when young legumes are infested.

References   Yang Youqian et al. 1982; Xiao Gangrou (ed.) 1992; Liu Youqiao and Li Guangwu 2002; Forest Disease and Pest Quarantine Station of Henan Province 2005; Liu Dongming, He Jun, Zhan Jinsuo et al. 2005.

(Yang Youqian, Li Zhenyu)

328 *Cydia trasias* (Meyrick)

Classification Lepidoptera, Tortricidae
Synonym *Laspeyresia trasias* Meyrick

Distribution Beijing, Tianjin, Hebei, Anhui, Shandong, Henan, Shaanxi, Gansu, Ningxia.

Hosts *Sophora japonica*, *S. japonica* var. *pendula*, *S. japonica* f. *oligophylla*, and *Robinia hispida*.

Damage *Cydia trasias* is common on shade trees along city roads and green belts. Larvae bore into young shoots and feed on petioles of leaves, flowers, and pods. Damaged leaves wilt and fall off the trees, leaving the branches leafless and lower the aesthetic value of the trees.

Morphological characteristics **Adults** Blackish brown. Thorax with bluish purple shinning scales. Forewing greyish brown to greyish black. Costa with a yellowish white line and four black spots. Forewing with cloud-shaped markings. Hindwing blackish brown.

Larvae Mature larvae cylindrical, yellow, somewhat translucent; head dark brown; body with sparse short hairs.

Pupae Yellowish brown. Anal comb with 8 spines.

Biology and behavior *Cydia trasias* occurs three generations a year in Beijing and two generations a year in Hebei. Larvae overwinter in seeds, branches, pods, and cracks of barks. Adult stage occurs between mid-May and mid-June, mid-July and mid-August. Adults emerge mostly in the morning. They are strongly attracted to the sun and lights. Females lay eggs on leaves, petioles, and small branches located on top and edges of the canopy. Larvae appear in late June. Young larvae bore into branches from the base of petioles. They create silk web and feed inside the web before boring into the branches. Blackish brown fecal materials accumulate from the damage sites. Larvae change feeding sites during development. One larva can cause several leaves to fall. When mature, the larva spins a cocoon inside the branch. Larval stage appears between early June and September. Different generations overlap. The 2nd generation emerges over a long period and can cause serious damage. Defoliated branches appear in August. After mid- and late August, most larvae migrate into pods. The seeds may turn black in September. Larvae start overwintering in October. Pure stands often have higher pest populations than mixed stands. Dense and vigorous stands are less likely infested. Low temperature and sudden drop in temperature in the spring can significantly increase the larvae mortality. Hot, humid, and rainy conditions in summer increase the mortality by diseases. Dead larvae inside the tunnels turn into brown, gel like material.

Prevention and treatment

1. Remove the seed pods between October and April in the next year. Trim off infested branches which can be identified by the presence of fecal materials and accumulation of saw dust outside of the tunnels.

2. Black light trapping of adults using sex pheromones, E8E10-12:OH and E8E10-12:AC, at 2:3 ratio. Place one trap every 3-5 trees along walkways. Place the traps on edges of tree canopy at 1.0-1.5 m height. A trap can attract adults from 10 m distance.

3. Attract and conserve *Cyanopica cyana* and other

C. trasias adult (black type) (Xu Gongtian)

C. trasias adult (brown type) (Xu Gongtian)

C. trasias pupa (Xu Gongtian)

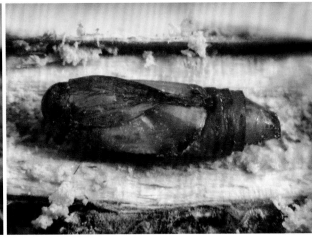
C. trasias pupa (Xu Gongtian)

C. trasias larva (Xu Gongtian)

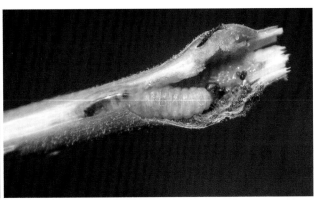
C. trasias damage to stem (Xu Gongtian)

C. trasias excretion (Xu Gongtian)

C. trasias in fruit for overwintering (Xu Gongtian)

C. trasias pheromone trap (Xu Gongtian)

natural enemies.

4. Biological control. Apply *Beauveria bassiana* (4×10^{11} conidia/g) at 2,000-3,000×. Spray 2×10^{10} PIB/mL nuclear polyhydrosis virus at 1,000-1,500×.

5. Chemical control. Spray 20% Dimilin at 1,000×, 1.8% avermectin EC at 1,000-2,000×, 5% imidacloprid EC at 1,000-2,000×, or 20% fenvalerate EC at 1,000-2,000×, 20 g/L lamda-cyhalothrin at 1,000-1,500×, 50% imidacloprid-monosultap at 600-800×, or 45% profenofos-lamda-cyhalothrin at 1,000-1,500×.

References Chen Heming and Qi Runshen 1992; Liu Jinying, Pang Jianjun, Zhai Shanmin et al. 2001; Zhang Guifen 2001; Zhang Guifen, Yan Xiaohua, and Meng Xianzuo 2001; Xu Gongtian and Yang Zhihua 2007; Fang Fang and Liu Jianfeng 2008; Zhao Xiuying, Han Meiqin, Song Shuxia et al. 2008; Zhang Xinfeng, Gao Jiusi, Shi Xianyun et al. 2009.

(Zhang Xudong, Lei Yinshan)

329 *Cymolomia hartigiana* (Saxesen)

Classification   Lepidoptera, Tortricidae
Synonym   *Phalaena hartigiana* Saxesen

Importance   *Cymolomia hartigiana* is a major pest of *Abies* and *Picea* trees. As high as 70% of the trees were infested.

Distribution   Hebei, Jilin, Heilongjiang; Japan, Russia, Europe.

Hosts   *Abies nephrolepis* and *Picea koraiensis*.

Damage   Young larvae feed on leaf tissue. After overwintering, larvae tie leaves with silk and feed inside the web and kill leaves of young shoots.

Morphological characteristics   **Adults** Wingspan 14-15 mm. Body greyish brown. Head with yellowish white long scales. Hairs on the 2nd segment of the labial palps very long. Forewing blackish brown; lower part of the base apricot color with two pairs of greyish white curved stripes; apical part with pale stripes; a white spot appears at the end of the distal cell; costa with 5 pairs of short silver stripes. Hairs on margins and hindwing greyish brown; apical corner of the hindwing darker.

Eggs   Flat oval. Pale yellow when newly laid, then becomes red, changing to pink before hatching.

Larvae   Yellowish green, length 17-18 mm. Head apricot color with two blackish brown spots on the sides; pronotum light-yellow; prolegs brown; pygidium light-yellow; anal comb yellowish brown.

Pupae   Length 9 mm, brownish green, eyes blackish brown.

Biology and behavior   *Cymolomia hartigiana* occurs one generation a year in Dailing region of Heilongjiang province. It overwinters as 3rd instar in leaves. Larvae resume activity in mid- to late-May. They often tie leaves together with silk. They prefer young trees and leaves on new shoots. The shoots may die as a result of heavy infestation. Larvae are active. When disturbed, they quickly hide inside the web. Larvae may relocate during development. Pupation occurs in mid-June. Adults emerge in July.

Females start to lay eggs at 2-5 hours after mating. Eggs are laid singly on back of needles. Each female lays 6-16 eggs with an average of 12 eggs. Egg incubation period is 10-12 days. Hatching rate is 86%. Young larvae bore into leaves and feed inside. The feeding tunnels are

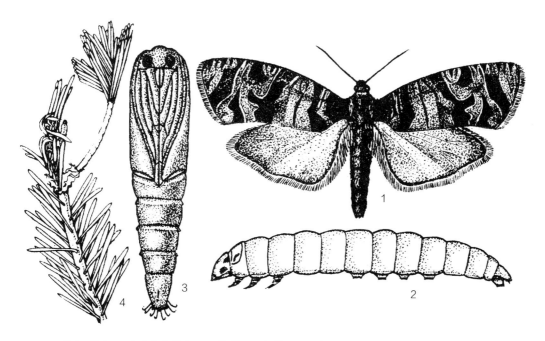

C. hartigiana. 1. Adult, 2. Larva, 3. Pupa, 4. Damage (1-3 by Bai Jiuwei; 4 by Cheng Yicun)

C. hartigiana adult (Wu Chunsheng, Zhang Runzhi)

C. hartigiana adult (NZM)

irregular. Larvae overwinter in mid-October.

Prevention and treatment

1. Plant mixed *Picea koraiensis* forests. Maintain healthy young stands.

2. Propagate and release egg and larval parasitoids.

3. Apply *Beauveria bassiana* (4×10^{11} conidia/g) at 2,000-3,000×. Spray 2×10^{10} PIB/mL nuclear polyhydrosis virus at 1,000-1,500×.

References Ji Yuhe 1992, Xiao Gangrou (ed.) 1992, Liu Youqiao and Li Guangwu 2002.

(Wu Chunsheng, Liu Youqiao, Bai Jiuwei)

330 *Epinotia rubiginosana* (Herrich-Schäffer)

Classification Lepidoptera, Tortricidae

Synonyms *Stegnoptycha rubiginosana* Herrich-Schäffer, *Epinotia rubiginosana koraiensis* Falkovitch

Distribution Beijing, Hebei, Henan, and Shaanxi; Russia, Korea Peninsula, Japan, Europe.

Host *Pinus tabuliformis*.

Damage Early larvae tunnel into old needles near the tip, causing the needles to wither and fall off. Mature larvae tie 6-7 needles together and feed within. The needles turn yellow and fall off. Canopy in late fall and early spring appear yellow.

Morphological characteristics **Adults** Body 5-6 mm long, wingspan 15-20 mm long. Body greyish brown; forewing greyish brown with dark brown basal area, a median transverse band, and striae at terminal of the wing; its anal apex with 6 short, black stripes. Hindwing pale brown.

Larvae Pale brown, 8-10 mm in length.

Biology and behavior *Epinotia rubiginosana* occurs one generation per year. Adults emerge in late March. They lay eggs on needles or young branches. The hatched larvae feed on leaves. The larvae descend to the ground on silken threads and construct overwintering cocoons with pieces of leaves in September.

Adults are most active at dusk. Adult females oviposit on the needles or young shoots, primarily in young or sparse stands, or near the edge of forests. Eggs are laid singly and hatch after 3-7 days. A larva usually enters a needle near the tip and then bores upward and then downward. The larvae usually keep their feeding tunnel free of frass and debris. Infested needles become tubular from larval feeding. After mature, the larva exits the needles, ties 6-7 needles together with silk, and feeds within the needles. The affected needles turn yellow and eventually fall.

Prevention and treatment

1. Silviculture. Plant mixed pine-hardwood forests and

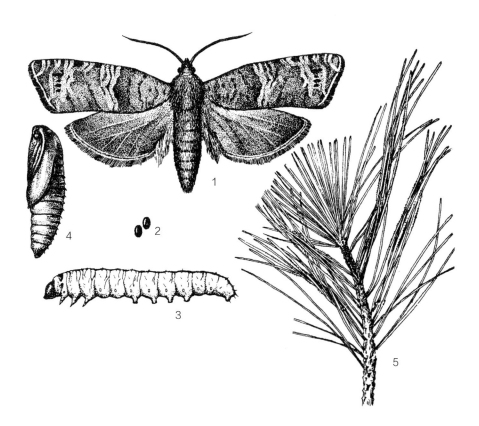

E. rubiginosana. 1. Adult, 2. Eggs, 3. Larva, 4. Pupa, 4. Damage (Zhang Xiang)

E. rubiginosana adult (Li Zhenyu)

E. rubiginosana adult (Xu Gongtian)

promote early crown closure.

2. Chemical control. When early larvae are present apply 50% phoxim EC or 50% malathion EC at 1,000×. Apply 20 g/L lamda-cyhalothrin at 1,000-1,500×, 50% imidacloprid-monosultap at 600-800×, or 45% profenofos-lamda-cyhalothrin at 1,000-1,500×.

References Yang Youqian et al. 1982, Xiao Gangrou (ed.) 1992, Liu Youqiao and Li Guangwu 2002, Forest Disease and Pest Quarantine Station of Henan Province 2005.

(Yang Youqian)

331 *Gatesclakeana idia* Diakonoff

Classification   Lepidoptera, Tortricidae

Importance   *Gatesclakeana idia* larvae feed on leaves, flowers, and fruits of *Averrhoa carambola* and other fruit trees and significantly reduce yield.

Distribution   Zhejiang, Fujian, Jiangxi, Guangxi, Hainan, Taiwan; Southeast Asia.

Hosts   *Averrhoa carambola*, *Dimocarpus longan*, *Litchi chinensis*, and *Sapium sebiferum*.

Damage   First generation larvae tie young leaves from branch tips with silk. Second generation larvae tie flower petals with silk. They infest flowers or young fruit. Third generation larvae primarily feed on fruits.

Morphological characteristics **Adults** Wingspan 14 mm. Head blackish-brown with a cluster of hairs; ocelli red; antennae yellowish-brown. Base of labial palps yellow; tip of the palps blackish brown; the 2nd segment enlarged; the last segment small, drooping. Legs silky-grey; tarsi with blackish brown marking. Middle and hind tibiae with silver grey long hairs; those on hind tibiae longer. Forewing short and wide; blackish brown, mixed with brown, yellow, grey scales. Pink short stripes occur from base to middle section of the front margin. End of the distal cell with a light-yellow spot. Hairs on wing margins blackish-brown. Hindwing greyish black; front margin and the hairs greyish white.

Eggs Flat oval. New eggs light-yellow, gradually changes to darker color. The larva is visible inside the egg before hatching.

Larvae Length 14 mm, yellowish green. Head brownish yellow, its sides with black spots. Legs black.

Pupae Length 8mm.

Biology and behavior   *Gatesclakeana idia* occurs 6 generations per year in Guangfeng county of Jiangxi province. It overwinters as 2nd and 3rd instars in dried fruit bodies or leaves. Overwintering larvae resume activity in mid-April next year. Adults emerge between the end of April and early May, then one generation is completed per month. All developmental stages can be seen any time during the development season.

Most adults mate only once with a small percentage mate multiple times. Eggs are laid singly on surface of

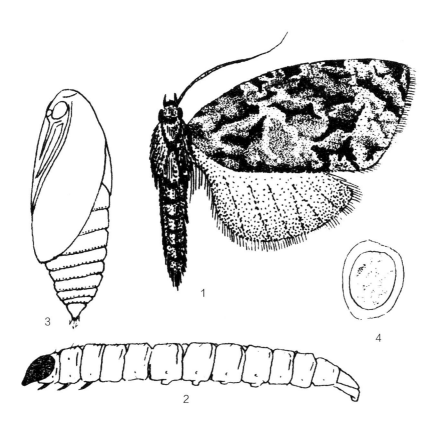

G. idia. 1. Adult, 2. Larva, 3. Pupa, 4. Eggs (Tao Mao)

G. idia adult (Wu Chunsheng, Zhang Runzhi)

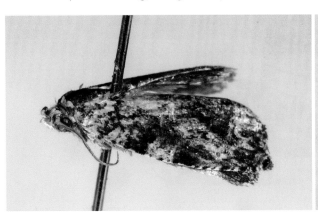

G. idia adult (Wu Chunsheng, Zhang Runzhi)

G. idia adult (NZM)

leaves or fruits. Each female lays eggs 4-6 times. A female lays 50-187 eggs with an average of 130 eggs. Larvae molt 3 times. They do not relocate during development.

Prevention and treatment

1. Clean out the dry fruit bodies and leaves on ground and burn them.

2. Kill adults using light traps.

3. Conservation of natural enemies. Larval parasitoids include *Apanteles changhingensi* and *Mesochorus discitergus*. Pupal parasitoids include a eurytomid wasp (*Bruchophagus gibbus*) and an ichneumonid wasp. The highest parasitization rate among larvae was found in the 1st generation (as high as 31%) and among pupae was in the 4th generation (as high as 27.5%).

4. Chemical control. Apply 20 g/L lamda-cyhalothrin at 1,000-1,500×, 50% imidacloprid-monosultap at 600-800×, or 45% profenofos-lamda-cyhalothrin at 1,000-1,500×.

References Xiao Gangrou (ed.) 1992, Liu Youqiao and Li Guangwu 2002.

(Wu Chunsheng, Shen Guangpu)

Gravitarmata margarotana (Heinemann)

Classification Lepidoptera, Tortricidae
Synonyms *Retinia retiferana* Wocke, *Retinia margarotana* Heinemann
Common name Pine cone moth

Importance *Gravitarmata margarotana* is a widely distributed pest. It not only severely impairs forest regeneration and afforestation, but also affects tree growth and timber value.

Distribution Hebei, Shanxi, Jiangsu, Zhejiang, Anhui, Jiangxi, Shandong, Henan, Hunan, Guangdong, Sichuan, Guizhou, Yunnan, Shaanxi, Gansu, Ningxia; Japan, Korea Peninsula, Turkey, Russia, Germany, Austria, Poland, France, Scotland, and Sweden.

Hosts *Pinus tabulaeformis*, *P. massoniana*, *P. armandii*, *P. bungeana*, *P. koraiensis*, *P. densiflora*, *P. thunbergii*, *P. yunnanensis*, *P. fenzeliana*. According to the literature, the insect can also infest *Pinus strobus*, *P. banksiana*, *P. elliottii*, *Picea* spp., and *Abies* spp.

Damage *Gravitarmata margarotana* damages pine cones and young shoots. Symptoms include:

1. Current year young shoots: During late spring and early summer, the newly hatched larvae bore into the base of leaf buds. Resin and tiny yellow fecal pellets appear at the infested sites. Infested buds gradually wither and fall off. When second-year cones are in short supply, larvae bore into the center of young shoots, resulting in shoot dieback.

2. Current year cones: The newly hatched larvae also infest female flowers and current year cones. They tunnel into the base of the cone stem and cone axis, impairing the tissues and severing the nutrient supply. Infested cone surfaces show resin and yellowish brown feces. Infestation results in the drop of a large number of current year cones and a significant loss in seed yield the following year.

3. Second year cones: Larger larvae bore into second year cones, generally in the middle of the cones, but occasionally at the top or base of the cones. The boring holes are small and irregular, and the tunnel is rough and filled with fecal pellets. The majority of the damaged cones shrink and wither without producing seeds. Cones with lesser damage become malformed due to unbalanced development and produce only a few seeds of poor quality.

Morphological characteristics **Adults** Greyish brown, 6-8 mm in length and 16-20 mm in wingspan. Antennae filiform; each segment densely covered with greyish white short pubescence, annulated. Compound eyes dark brown, subgloboid. Forewings with irregular markings consisted of greyish brown, russet, and blackish brown scales. The apical angle with a curved white stripe. Hindwing greyish brown, outer border dark brown, setae on the border grey.

Eggs Long oval, 0.9 mm in length and 0.7 mm in width. Newly laid eggs milky white, turning dark brown before hatching.

Larvae Newly hatched larvae dirty yellow. Mature larvae 12-20 mm in length; head dark brown; body fleshy red, covered with dense, leathery stripes.

Pupae Reddish brown, 6.5-8.5 mm in length. End of the abdomen fork-like, with four pairs of symmetrical hooks. Cocoons yellowish-brown.

Biology and behavior *Gravitarmata margarotana* occurs one generation per year in Shaanxi, Sichuan, Zhejiang, and Guangdong. It becomes dormant in summer and winter in pupal stage. Eclosion time varies from regions. In Qiaoshan of Shaanxi province, adults emerge in mid-April and peak emergence occurs in late April to

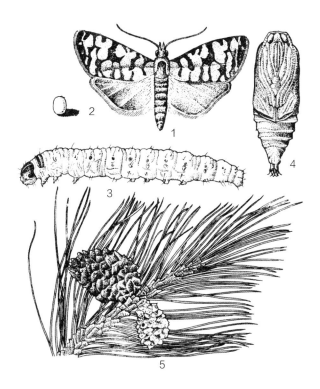

G. margarotana. 1. Adult, 2. Egg, 3. Larva, 4. Pupa, 5. Damage (Zhu Xingcai)

G. margarotana adult (Li Zhenyu)

early May. Larval emergence occurs in early May, peaks in mid-May and ends in late May to early June. Larvae mature in early to mid-June, then leave the cones, spin silk, fall to the ground and construct cocoons for pupation in the leaf litter, weeds, or loose soil. In Sichuan, eclosion of adults begins between late February and late March and peaks in early March. Larval hatch starts in mid- and late March and peaks in early April; mature larvae fall to the ground in mid-May. Pupation occurs in the surface layer of loose soil near the tree trunk or in the lichen layer on the edge of stone blocks or walls. Eclosion of adults occurs in mid-January in Guangdong, and larval infestation occurs in mid-February to late March. Mature larvae begin to make cocoons in mid-March.

Gravitarmata margarotana infestation is greater 1) at lower altitudes than high altitudes, 2) near the bottom of mountains than at the middle and top of the mountains, 3) in pure forests than mixed forests, 4) in open forests than closed, dense forests, 5) in young and half-mature forests than nearly-mature and mature forests, 6) in man-made plantations than natural forests, 7) in forests with abundant cones than forests without cones or with low cone yields, and 8) in *Pinus tabulaeformis* and *P. massoniana* forests than in *P. armandii* and *P. bungeana* forests. The emergence period is earlier on southern slopes than on northern slopes, and at lower altitudes than higher altitudes, and is closely correlated with phytophenology. Larval hatching coincides with blooming of *P. tabulaeformis* and *P. massoniana*.

Prevention and treatment

1. Silviculture. Select resistant tree species such as lacebark pine (*Pinus bungeana*) and *P. fenzeliana*. Maintain mixed forests, increase stands density to reduce infestation.

2. Mechanical control. Remove and burn infested cones and branches before mature larvae drop to soil. Adults can be killed by setting up black light traps.

3. Biological control. At egg stage, releasing egg parasitoids *Trichogramma dendrolimi* 2-3 times can result in 70% control. Conserve the following natural enemies: *Trichogramma dendrolimi*, *Eurytoma appendigaster*, *Phanerotoma kozlovi*, *Macrocentrus resinellae*, *Apanteles laevigatus*, and *A. shemachaensis*. During larval stage, spray *Bacillus thuringiensis* emulsion at 200×.

4. Chemical control. During egg state, spray 50% fenitrothion EC at 100-150× or 20% dimethoate EC at 100×, or 20% fenvalerate EC at 1500×. Apply 20 g/L lamda-cyhalothrin at 1,000-1,500×, 50% imidaclo-prid-monosultap at 600-800×, or 45% profenofos-lamda-cyhalothrin at 1,000-1,500×.

References Li Kuansheng et al. 1974; Liu Youqiao and Bai Jiuwei 1977; Dang Xinde 1979; Li Kuansheng, Tang Guoheng et al. 1981; Dang Xinde 1982; Yates 1986; Forest Research Institute of Shaanxi Province 1990; Xiao Gangrou (ed.) 1992; Li Guixiang et al. 1999; Liu Youqiao and Li Guangwu 2002.

(Li Kuansheng, Li Li)

333 Gypsonoma minutana (Hübner)

Classification Lepidoptera, Tortricidae
Synonym *Tortrix minutana* Hübner
Common name Brindled shoot

Distribution Hebei, Shanxi, Shandong, Henan; Europe, North Africa, Turkey, Russia, Iran, Afghanistan, Mongolia, Japan.

Hosts *Populus* spp., *Salix* spp.

Damage Larvae spin silk, roll leaves together, and feed on leaf surface. Damaged leaves become reticulate and fall.

Morphological characteristics **Adults** Length 5 mm, wingspan 13 mm. Forewing long and narrow with pale or dark brown bands, with a white band between the basal band and median band; costal margin with hooked stigula. Hindwing grey brown.

Larvae Greyish white, 6 mm in length.

Biology and behavior *Gypsonoma minutana* occurs 3-4 generations per year in Henan and overwinters as larvae under barks. The larvae become active next April after budding and leaf expansion. Pupation and adult emergence occur in late April. The peak emergence of the 2nd generation adults is in early June. Overlapping generations are present after June. Larvae of the last generation feed until late October and then overwinter.

Adults are phototaxic, active at night. Females deposit their eggs singly on leaf surfaces. After hatching, larvae

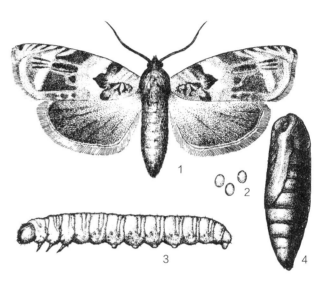

G. minutana. 1. Adult, 2. Eggs, 3. Larva, 4. Pupa (Zhang Xiang)

spin silk which stick leaves together. Damaged leaves become reticulate. Larvae are active and will jump and escape when frightened. Mature larvae spin white, thin cocoons among leaves. Inner branches of 4-5 year-old trees in dense stands are most susceptible to *G. minutana* infestations.

Prevention and treatment

1. Silviculture. Adjust density of the stands and trim to reduce damage.

2. Chemical control. Spray 50% malathion EC or 50% fenitrothion EC at 1,000× or 25% diflubenzuron at 2000× to control early instars.

References Liu Youqiao and Baijiuwei 1977, Yang Youqian et al. 1982, Xiao Gangrou (ed.) 1992, Forest Disease and Pest Quarantine and Control Station of Henan 2005.

(Yang Youqian)

G. minutana adult (Xu Gongtian)

334 *Ophiorrhabda mormopa* (Meyrick)

Classification Lepidoptera, Tortricidae

Synonyms *Lasiognatha mormopa* (Meyrick), *Platypeplus mormopa* Meyrick, *Argyroploce mormopa* Meyrick, *Olethreutes mormopa* Clarke, *Hedya mormopa* Diakonoff

Distribution Fujian, Hainan, Yunnan; Sri Lanka, India, the Philippines, Nepal, Thailand, Vietnam, Malaysia, Indonesia, Brunei.

Hosts *Eugenia* spp., *Nephelium* spp., mangrove, Myrrinaceae, and *Aegiceras corniculatum*, *Syzygium* spp., *Citrus* spp.

Morphological characteristics **Adults** Length 6-8 mm, wingspan 15-17 mm. Head covered with flat scales; ocelli located at posterior end of the head. Proboscis short; labial palps curved and pointed upward; the 2nd segment slightly enlarged; the end of the last segment obtuse. Forewing rather wide, the termen truncated; with distinct basal spot and median band. A pale transverse band lies between the basal spot and the median band. Hindwing pale grey to dark greyish brown. Cu_2 in forewing arises in basal two-thirds of distal cell. Cu_1 in forewing arises at lower angle of the distal cell. M_3 is near Cu_1. M_2 parallels with M_1. The distance between M_2 and M1 is farther than that between M_2 and M_3. R_5 arises from upper angle of the media. Base of R_3, R_4, and R_5 are close. Base of R_2 is between R1 and R_3. R_1 arises from middle point of the top margin of the distal cell. M starts at base and ends below the base of M_2. Hindwing with frenulum. $SC+R_1$ arises from the middle point of the top margin of the distal cell. Rs and M1 arise from same point at top angle of the distal cell. Base of M_2 curved toward the base of M_3. Base of M_3 and Cu_1 are close. Male genitalia uncus long and narrow. Valvae slender. Aedeagus large and thick, with inconspicuous needle at the end. Female ostium bursae is surrounded with dense hairs.

Eggs Elliptical, length 1 mm; short side is 2/3 of the long side. Milky white, lustrous, dark red before hatching.

O. mormopa damage (Huang Jinshui)

O. mormopa larva (Huang Jinshui)

Rolled leaves by O. mormopa larva (Huang Jinshui)

Larvae Length 15-18 mm, width 1.5 mm. Newly hatched larva pale brown; head brownish-red, gradually changes to yellowish green, then dark green or dark grey, light-yellow before pupation. Body with white hairs. Anal prolegs extend backward and clasp-like.

Pupae Length 7-8 mm, width 1.5-2.0 mm. Pale brown, dark brown before eclosion; covered within white cocoon. Dorsal surface of each abdominal segment with two rows of spines. The front row has 4-6 black spines, the back row is also developed. The last segment has 8 spines.

Biology and behavior *Ophiorrhabda mormopa* occurs 7 generations per year in Zhangzhou, Fujian province. It overwinters in pupal stage on *A. corniculatum* leaves.

Adults emerge in May when *A. corniculatum* is in bloom. They are strongly attracted to lights. Eclosion occurs mostly during the night. After emergence, adults become immediately active. They do not need supplementary feeding. Mating process takes about 0.5 hour. Adults are not active during the day. Life span is 4-5 days. Adults select young leaves to lay eggs. They mostly lay one egg per leaf and may lay two eggs per leaf. Egg period is 5 days. A female finishes egg laying in 3-4 days.

Larvae have 5 instars. Larval period is 13-17 days. When disturbed, the larvae may retreat or jump. Young larvae spin silk which holds 2-3 young leaves together and feed within the leaves. Young larvae only feed on leaf tissue and leave the veins intact. From the 3rd instar, they feed on whole leaves. Mature larvae spin silk along leave edges and create pods. Pupation occurs inside the pods. Larvae do not relocate to new pods during development. Infested leaves and branches may die and become dry from feeding. Only one larva exists in each pod. Pupation rate was 75% in one study. Pupal period is 6 days. Once adults emerged, the pupae are exposed and only the end of each pupa is connected to the rolled leaves. When food is scarce, larvae may pupate early.

Prevention and treatment

1. Use light traps to trap and kill adults.

2. Chemical control. Spray Dimilin at 2000×, 20 g/L lamda-cyhalothrin at 1,000-1,500×, 50% imidacloprid-monosultap at 600-800×, or 45% profenofos-lamda-cyhalothrin at 1,000-1,500×.

References Meyrick 1912, 1939; Gates 1958; Diakonoff 1973; Chinese Academy of Sciences - Institute of Zoology 1981; Liu Youqiao and Li Guangwu 2002.

(Ding Bi, Huang Jinshui)

335 *Cydia coniferana* (Saxesen)

Classification Lepidoptera, Tortricidae

Synonyms *Tortrix coniferana* Saxesen, *Grapholitha separatana* Herrich-Schäffer, *Laspeyresia coniferana* (Saxesen)

Importance *Cydia coniferana* is a major pest of pine trees in northern China. Larval feeding causes resin flow, weakening of the trees, and infestation by secondary wood-boring insects.

Distribution Liaoning, Jilin, Heilongjiang; Korea Peninsula, Russia, Europe.

Hosts *Pinus tabulaeformis*, *P. sylvestris* var. *mongolica*, *P. koraiensis*, and *Abies fabri*.

Damage Larvae feed on phloem of the tree trunk and large branches. Larvae bore irregular tunnels and push out brown fecal pellets and shed skins. Damaged sites form resin flow.

Morphological characteristics **Adults** Wingspan 11-14 mm. Females are larger than males. Body greyish black. Head and thorax with long greyish-black scales. Labial palps curved upwards; end of the 2nd segment with white triangular scales. Forewing greyish black and with some white scales. Basal spot not clearly defined. Interior border of the median band with a short oblique white spot both from the costal margin and from the inner margin. Exterior border of the median band with an arched white stripe. A hook-shaped white stripe exists near the apex. Anal veins obvious, with four black and white horizontal stripes. Termen hairs greyish black. Hindwing light greyish black, basal areas lighter in color.

Eggs Flat and oval. New eggs milky white, translucent; gradually changes to creamy yellow; color is pink before hatching.

Larvae Mature larvae 9.8 mm in length. Milky white; head yellowish-brown; pronotum and anal shield greyish-brown.

Pupae Length 6.5-7.0 mm. Anal comb with 8 spines, the 4 in the middle are longer than the rest.

Biology and behavior *Cydia coniferana* occurs one generation per year in Liaoning. It overwinters as 3rd-4th instars in tunnels and spin silk nests. They resume feeding in mid-April. Pupation occurs in late-May. A larva spins a silk chamber in the tunnel or under bark to prepare for pupation. Adults emerge in June. They soon mate and lay eggs. Larvae enter overwintering period in September-October.

When adults emerge from pupae, 2/3 of the empty

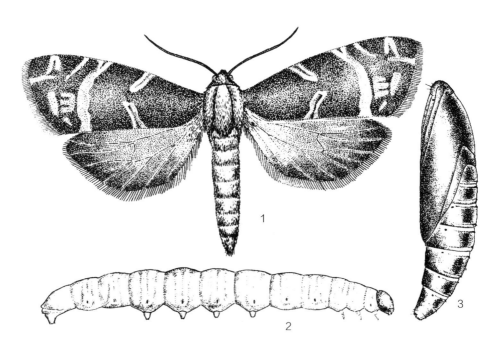

C. coniferana. 1. Adult, 2. Larva, 3. Pupa (Zhang Peiyi)

C. coniferana adult (NZM)

pupae are left on the tree surface. They either tilt up or tilt downward. This is an important characteristics of *C. coniferana*. Adults lay eggs under peeled bark of tree trunks between 0-1 m from the ground. Eggs are laid singly with 1-7 eggs per bark. A female lays an average 28 eggs. Egg incubation period is 10-15 days.

Newly hatched larvae are very active. Soon they will bore into crevices of the tree trunk and hide in silk webs. Their head stick out during feeding. Larvae change feeding sites. The feeding tunnels are smooth and take various shapes. Larvae push fecal pellets and shed skins out of the tunnels during development. Feeding sites form resin flow. This is another identifying characteristics of *C. coniferana*.

Prevention and treatment

1. Trap and kill adults using light traps.

2. To control newly hatched larvae, spray tree trunk with 20% dinotefuran at 1,500-2,000×, 25 g/L lamda-cyhalothrin at 1,000×, or 50% malathion EC at 800-1,000×, 50% imidacloprid-monosultap at 600-800×, or 45% profenofos-lamda-cyhalothrin at 1,000-1,500×. Brush 20% dinotefuran at 5× or 25 g/L lamda-cyhalothrin at 10× to tree trunks.

References Xiao Gangrou (ed.) 1992, Liu Youqiao and Li Guangwu 2002.

(Wu Chunsheng, Song Youwen, Sun Lihua)

Cydia zebeana (Ratzeburg)

336

Classification	Lepidoptera, Tortricidae
Synonyms	*Phalaena* (*Tortrix*) *zebeana* Ratzeburg, *Laspeyresia zebeana* (Ratzeburg)
Common name	Larch bark moth

Importance *Cydia zebeana* is an important boring insect pest of *Larix* forests. The infestation is most serious in seed production plantation and sparse forests.

Distribution Jilin, Heilongjiang, north central region; Russia, Europe.

Hosts *Larix* spp.

Damage Larvae feed on bark and phloem of the new shoots and base of the branches. Wood dusts accumulate at the feeding sites. Resin flow at the damaged sites forms gall-like structure. Young trees damaged by the larvae may have multiple main shoots or form branches from the damage site.

Morphological characteristics **Adults** Wingspan 14 mm. Labial palps slender, slightly drooping. Forewing olive-greenish brown to greyish or greenish brown. Costa stigula is obvious. Anal veins not obvious; with four black markings. Top of the distal cell with a large black spot. Hindwing brown.

Eggs Flat oval; light-yellow, gradually turning to orange yellow.

Larvae Length 7 mm. Dull white, head and pronotum dark brown, shiny.

Pupae Length 7-8 mm. Pale yellow to yellowish brown. Anal comb with 8 spines.

Biology and behavior *Cydia zebeana* occurs one generation every two years in Heilongjiang province. Larvae spin greyish white cocoons and overwinter inside tunnels. They resume activity in next mid-April. Larvae feeding of the bark causes resin flow and abnormal tissue growth and galls. The damage becomes most serious during July and August. Main shoot and lateral branches die from feeding. Larvae enter 2nd overwintering period in October. They mature in next May. Pupation occurs inside galls. Adults appear in June. Eggs are laid at base of young branches on mid- and lower part of the back of the 2nd layer leaves. Young larvae enter base of the 1st year branches and feed inside. They enter overwintering period in October. The pest occurs more often on south side slope, near forest edge, in sparse stands, and on main shoot of young trees (10 m tall), and young branches at mid- and lower part of the trees.

Prevention and treatment

1. Non-chemical methods include quarantine, light trapping of adults, trimming off infested branches. An integrated approach will achieve better control.

2. Fogging using 5% lindane or dusting with 3% lindane in closed canopy mature forests is very effective. Conduct fogging or dusting when wind is not strong. It

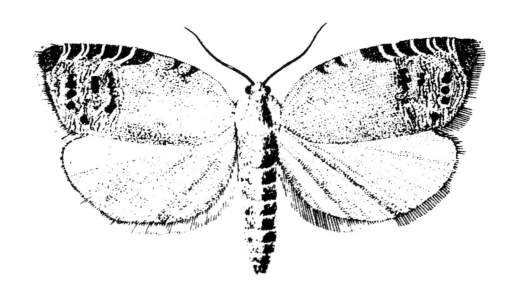

C. zebeana adult (Hu Xingping)

C. zebeana adult (Wu Chunsheng, Zhang Runzhi)

C. zebeana adult (NZM)

is most effective when done during the adult emergence period.

3. In young stands, brush a mixture of 25 g/L lamda-cyhalothrin + 20% thiamethoxam at 1,000× to tree trunks to kill larvae. Apply 20 g/L lamda-cyhalothrin at 1,000-1,500×, 50% imidacloprid-monosultap at 600-800×, or 45% profenofos-lamda-cyhalothrin at 1,000-1,500×.

References Xiao Gangrou (ed.) 1992; Shao Jingwen 1994; Liu Youqiao and Li Guangwu 2002; Shi Tiehao, Sun Zuomin, Wang Wenge et al. 2005.

(Wu Chunsheng, Lu Wenmin)

337 *Pammene ginkgoicola* Liu
Classification Lepidoptera, Tortricidae

Importance *Pammene ginkgoicola* is the most important pest of *Ginkgo biloba*. The infestation rates in Jiangsu, Anhui, and Guangxi can reach 90%-100%, resulting in significant reduction in fruit production.

Distribution Jiangsu, Zhejiang, Anhui, Fujian, Shandong, Henan, Hubei, Hunan, Guangdong, Guangxi.

Host *Ginkgo biloba*.

Damage Larvae mainly feed inside the short branches but can also infest long branches. Leaves and young fruits on the damaged branches wilt. The infested branches cannot regenerate new leaves.

Morphological characteristics **Adults** Wingspan 12 mm. Blackish brown. Head pale greyish brown. Abdomen yellowish brown. Forewing narrow, blackish brown. The middle part with dark stigula. Anal veins obvious, with 4 black stigulae. Front margin of the hindwing pale, distal end brown, termen hair dark brown.

Larvae Length 8-12 mm. Greyish white to light-yellow. Head, pronotum and pygidium blackish brown; sometimes yellowish brown. Each tergum with two pairs of black spots. A color marking also appears above and below the spiracle line of each segment. Anal segment with 5-7 anal spines.

Biology and behavior *Pammene ginkgoicola* occurs one generation a year in Jiangsu province. It overwinters as pupae. Adults emerge between late March and mid-April. They have weak positive phototaxis. Egg laying occurs between mid-April and mid-May. They are laid singly on 1-2 year-old small branches, each branch with 1-5 eggs. Larvae appear from late April to late June. Young larvae feed very little and bore into branches and feed transversely.

When infesting short branches, the larvae enter from base of leaf petiole or between base of leaf petiole and the branch. Each larva may infest two short branches. When infesting long branches (60-150 mm long), larvae enter through the base of a branch and move toward the tip or through middle portion of a branch and move toward the base. Feeding tunnels are 20-50 mm long. Larvae migrate from infested branches to dead leaves during mid-May and mid-June. They roll up the leaves and rest inside the leaves. Occasionally they feed on leaves during night. From mid-May to early July, the mature larvae migrate to cracks on barks and enter diapause. *P. ginkgoicola* overwinters in mid-November in cracks on the bark as pupae. It occurs more often at forest edges than at center of a forest. Weak old trees are more susceptible to *P. ginkgoicola* infestations. Dry conditions during diapause period may reduce the larvae survival.

Prevention and treatment

1. Hand removal of adults every day at 9:00 am when adults are resting on tree trunks. Cut down infested branches during late April and early June.

2. Insecticide bands. Before adult emergence, spray a band of 50% sumithion EC at 250× + 2.5% Deltamethrin EC at 500× to tree trunk or base of branches. Wrap tree trunk with plastic film to kill adults.

3. Spray the canopy. During larvae period from late April to mid-May, spray 25 g/L lamda-cyhalothrin or 50% sumithion EC at 800-1,000×, 2.5% deltamethrin at 3,000-4,000×, 50% imidacloprid-monosultap at 600-800×, or 45% profenofos-lamda-cyhalothrin at 1,000-1,500× to canopy.

References Jiang Dean et al. 1996, 2003; Deng Yinwei et al. 2006; Yang Chunsheng et al. 2006.

(Qian Fanjun)

P. ginkgoicola adult (NZM)

P. ginkgoicola adult (NZM)

338 *Lobesia cunninghamiacola* (Liu et Pai)

Classification Lepidoptera, Tortricidae
Synonym *Polychrosis cunninghamiacola* Liu et Pai

Distribution Jiangsu, Zhejiang, Anhui, Fujian, Jiangxi, Hubei, Hunan, Guangdong, Guangxi, Sichuan.

Host *Cunninghamia lanceolata*.

Damage Larvae infest young branch tips. Both main and lateral branches can be infested. Trees of 3-5 years old, 2.5-4.5 m tall are most likely to be infested. Infestation rate is as high as 80%. Main shoot growth can be reduced by 45 cm or 50% a year. Once main shoot is damaged, lateral shoots form and the tree cannot form a straight trunk.

Morphological characteristics **Adults** Length 4.5-6.5 mm; wingspan 12-15 mm. Antennae filiform. Base of each segment of the antennae apricot yellow; end of each segment of the antennae blackish brown. Labial palps apricot yellow, protrude forward; its 2nd segment enlarged at the end, with brown spot on the lateral side; the last segment drooping. Forewing dark blackish brown, with two parallel basal fascia, one "x" shaped fascia after the basal fascia and a subterminal fascia, which splits to three branches near the apex and front margin of the wing. All fascia apricot yellow with silver stripes. Hindwing pale brownish black, without fascia; its front margin pale grey. Front and middle legs blackish brown. Tibia with three greyish white round markings. Hind legs greyish brown, with four greyish white round markings.

Eggs Flat oval, 0.7-0.8 mm in length; milky white, turning darker color before hatching.

Larvae Length 8-10 mm. Head, pronotum, and pygidium dark red; body purple-reddish brown; each segment with a white ring.

Pupae Length 4.5-6.5 mm. Each abdominal segment with two rows of uneven teeth; the front row larger than

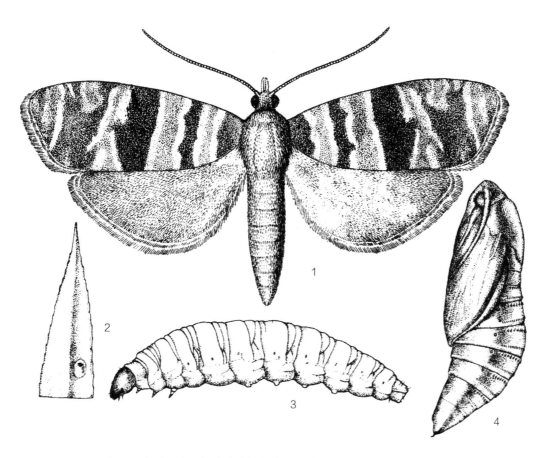

L. cunninghamiacola. 1. Adult, 2. Egg, 3. Larva, 4. Pupa (Zhang Xiang)

the hind row. Anal comb with 8 spines.

Biology and behavior *Lobesia cunninghamiacola* occurs 2-3 generations a year in Jiangsu, Anhui, 2-5 generations in Jiangxi, and 6-7 generations in Hunan. The 1st and 2nd generations cause the most damage. In Hunan, it was reported the 1st and the 2nd generation infestation rates reached 33% and 55%, respectively. A 1978 survey of infestations in Xiashan forest farm (Yugan County of Jiangxi province) showed 1st and 2nd generation infestation rates were 43.5% and 92.8%, respectively. The pupae overwinter in dead shoots. Adults emerge in the end of March to early April next year. The 1st generation larval period is from mid-April to early May; pupation occurs from end of April to mid-May; adults emerge in early and mid-May. The 2nd generation larval period is from end of May to end of June. The 2nd generation starts to show generation differentiation. Most larvae pupate in mid-June. Some larvae postpone pupation to the end of July at the same time as the 3rd generation larvae. Some larvae even postpone to early and mid-September to pupate. Those pupated in mid-June will produce adults in late June. The 3rd generation larvae appear in early to mid-July. If the 2nd generation adults emerge in end July to mid-August, then the 3rd generation will appear in early August. Some pupae will diapause and emerge in next spring. Those 3rd generation adults emerged from the end of July to early August produce 4th generation larvae in early August and infest forests until mid-October. The 2nd to 4th generations overlap from early August and mid-September.

Adults emerge during 10:00-12:00. They are active during night and are attracted to lights. They mate on the 2nd day after emergence. After mating, they start to lay eggs the next day. They prefer lay eggs in sunny, well grown, sparse, 2.5-5.0 m tall, and 4-5 year-old young stands. The next preferred habitat is those below 2.5 m or above 5 m but less than 10 m tall stands. They rarely lay eggs in stands of more than 10 m tall. Eggs are laid singly on back main veins of leaves on young shoots. They usually lay one egg per shoot, sometimes 2-3 eggs per shoot, and rarely 7-10 eggs per shoot. They often lay > 10 eggs continuously on one tree or several surrounding trees. Each female lays approximately 40 eggs. An average of 45 eggs and maximum of 96 eggs were laid

L. cunninghamiacola adult

L. cunninghamiacola adult

L. cunninghamiacola adult

L. cunninghamiacola adult (Wang Jijian)

under laboratory conditions. The mean life span of adults are: overwintering generation female - 8.6 (4-12) days, male - 8.2 (5-12) days; 1st generation - 8 (3-13) days; 2nd generation - 6 days; and other generations - 4 days.

Egg incubation period is approximately one week. They hatch during 5:00-6:00. There are 6 instars. Newly hatched larvae bore into young shoot leaves from edges after about 10 minutes of searching or bore into leaves immediately after hatching. Each larva feeds on 2-3 leaves during the 1st-2nd instar. They only feed on leaf edges. From 3rd instar, larvae bore into shoots. Both feeding and excretion increase significantly from the 3rd instar. The dark reddish brown feces accumulate on infested shoots. The larvae move fast and may relocate 1-2 times. Each larva usually infests two shoots. Each infested shoot only has one larva. It is possible that 2-3 larvae occur in one shoot when population level is high. The 5th-6th instar move slowly. Occasionally they relocate before pupation. Larvae create 2 cm long tunnels. Infested shoots become yellow or red. When migrating, the larvae crawl to a neighboring shoot or are spin silk and use wind to migrate to a new branch. Migration often occurs in the afternoon. Each migration needs 1 hour and another 3 hours to bore into a new shoot. A mature larva bites a hole at about 6 mm from the shoot tip as emergence hole. The larva pupates below the hole in an 8 mm long pupal chamber.

Lobesia cunninghamiacola occurs more often in low hills of under 300 m elevation. Stands of 3-5 years old are more susceptible than those of > 7 years old. Damage is also related to habitat conditions: south side, edge of stands, sparse stands, and pure stands are more susceptible than the north side, center of stands, dense stands, and mixed stands.

Prevention and treatment

1. Conservation of natural enemies. There are many natural enemies of *N. cunninghamiacola*. The egg parasitoids include *Trichogramma confusum*, *T. dendrolimi*. Larval parasitoids include *Brachymeria lasus*, Braconidae, *Elasmus* spp., *Gregopimpla kuwanae*, *Eurytoma* sp., and three parasitic flies. Pupal parasitoids include *Brachymeria* sp., *Apanteles* sp. Pathogens include *Beauveria bassiana* and *Aspergillus* sp. The overwintering generation has the highest disease prevalence.

2. Biological control. Apply *Beauveria bassiana* (4×10^{11} conidia/g) at 2,000-3,000×. Spray 2×10^{10} PIB/mL nuclear polyhydrosis virus at 1,000-1,500×.

3. Chemical control. Apply 20 g/L lamda-cyhalothrin at 1,000-1,500×, 50% imidacloprid-monosultap at 600-800×, or 45% profenofos-lamda-cyhalothrin at 1,000-1,500×.

References Liu Youqiao and Li Guangwu 2002.

(Shen Guangpu, Deng Zhangming)

339 *Cydia strobilella* (L.)

Classification Lepidoptera, Tortricidae
Synonyms *Phalaena strobilella* L., *Pseudotomoides strobilella* L.
Common name Spruce seed moth

Distribution Inner Mongolia, Heilongjiang, Shaanxi, Gansu, Qinghai, Ningxia, Xinjiang; Europe, Russia, North America.

Hosts *Picea asperata*, *Latrix gmellini*.

Damage Larvae infest cones and seeds. Damaged cones become deformed and produce resin. Scales of fallen cones remain closed.

Morphological characteristics **Adults** Length 6 mm, wingspan 10-13 mm. Greyish black; forewing brownish black, basal fascia pale black. The middle of the brownish black median fascia curved forward. Costa has 3-4 groups of strigulae. Hindwing pale brownish black; basal area paler; termen hair yellowish white.

Eggs Round or slightly flat, yellow.

Larvae Length 10-11 mm, yellowish white to yellow; head brown. Occiput shiny. Spiracles small and brown.

Pupae length 4-5 mm, brown. The front raised. The 2nd segment from the end has small bumps which have spines on them. There are four anal spines.

Biology and behavior *Cydia strobilella* occurs one generation a year or one generation every two years. Mature larvae overwinter in cones. Overwintering larvae resume activity in early April. Pupation occurs from early May to mid-June. Adults emerge and lay eggs from early May to mid-June. Eggs hatch from early June to early July. Peak egg hatching period is between mid-June and late June. Larvae overwinter in late July and August.

Adults mate and lay eggs at 1-2 days after emergence. Female to male ratio is 2.36:1. Adult life span is 5 days. Eggs are laid singly on seed scales. Each cone may have 1-29 eggs. Each female lays 34-105 eggs. After hatching, the larva bores into scale and the seed. It feeds on the endosperm of the immature seed and relocates to neighboring seeds as it develops. Mature larvae tunnel into the axis of the cones and overwinter in the tunnel. Few larvae overwinter in scales and seeds. South slope of the hill, south side of the tree canopy, and upper part of the tree canopy suffer more damage. Young stands, sparse and pure stands usually have higher infestation than mid-age and mature, dense, mixed stands. Natural enemies include *Scambus brevicornis*, *Megarhyssa lunator*, *Hyssopus nigritulus*, and one pathogen. Overwintering larvae may suffer 65% mortality from bird predation.

Prevention and treatment

1. Plant mixed forests. During seed harvest, properly dispose the infested cones. Conserve natural enemies. Use sex pheromone to attract adults.

2. During adult emergence period, apply fogging at 15 kg/ha in forests with > 0.6 coverage. During young larvae period, aerial sprays of 2.5% cyhalothrin at 60 g/ha or ULV spray of 2.5% deltamethrin at 100×. Spray 20 g/L lamda-cyhalothrin at 1,000-1,500×, 50% imidacloprid-monosultap at 600-800×, or 45% profenofos-lamda-cyhalothrin at 1,000-1,500×.

References Xiao Gangrou (ed.) 1992; Zhou Jiazi, Qu Bangxuan, Wang Ximeng et al. 1994; Yin Chengrong, Wang Youkui, Lin Hai et al. 2001.

(Xie Shouan, Li Menglou, Wang Jie, Liu Youqiao, Bai Jiuwei)

C. strobilella adult (Zhang Peiyi)

C. strobilella adult (Li Menglou)

340 *Retinia cristata* (Walsingham)

Classification Lepidoptera, Tortricidae
Synonyms *Enarmonia cristata* Walsingham, *Petrova cristata* Obraztsov

Importance *Retinia cristata* is an important boring insect pest of pine trees in China. Larvae infest young shoots and cones, causing dead shoots and reducing seed yield in seed plantations.

Distribution Beijing, Hebei, Shanxi, Liaoning, Jiangsu, Zhejiang, Anhui, Fujian, Jiangxi, Shandong, Henan, Hubei, Hunan, Guangdong, Guangxi, Sichuan, Guizhou, Yunnan, Shaanxi, Gansu; Japan, Korea Peninsula.

Hosts *Pinus massoniana*, *P. thunbergii*, *P. taiwanensis*, *P. tabulaeformis*, *P. densiflora*, *P. elliottii*, *P. kesiya* var. *langbianensis*, *P. serotina*, *P. taeda*, and *P. patustris*.

Damage Infested shoots become wilted above the tunnel entrance. Damaged cones produce resin that mixed with fecal materials. The tunnel entrance funnel-shaped. The tunnels with fecal materials and resin.

Morphological characteristics **Adults** Length 4.6-8.7 mm, wingspan 12.1-19.8 mm. Body yellowish brown. Forewing with a wide silvery median fascia. There are 3-4 silvery cross stripes near the wing base. The silvery tornal marking is elliptical and with three small black spots.

Larvae Length 9.4-15.0 mm, light-yellow, head and pronotum yellowish brown.

Biology and behavior *Retinia cristata* occurs four generations a year. Pupae overwinter in dead shoots and cones. Adult emergence periods of the overwintering generation and the 1st-3rd generations are: early March to late April, late May to mid-July, late July to late August, and September, respectively. The larvae period of the 1st to 4th generations are: late March to mid-June, late June to early August, mid-August to mid-September, and mid-September to mid-November, respectively.

From late March, the 1st generation larvae infest young shoots. From mid-May, they infest the scale and axis of the 2nd year cones. Each larva infests 3-4 scales. Average size of the infested cones is 1.4 cm × 1.2 cm (length × width). Damaged cones are brown. The 1st generation larvae last for 30 days and cause the most damage. The 3rd and 4th generations mainly infest scales. Larvae infest multiple shoots and cones. Mature larvae bore into

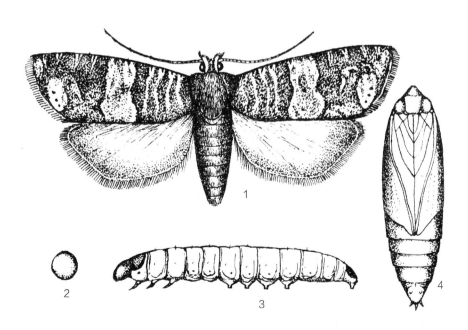

R. cristata. 1. Adult, 2. Egg, 3. Larva, 4. Pupa (Tian Hengde)

R. cristata adult (Li Zhenyu)

the axis of the cones, creating 8.0-11.0 mm long white cocoons and pupate inside the cocoons. The adult emergence peak is during 18:00-20:00. About 1/3-2/3 of each empty pupa is exposed outside of the tunnel entrance. Eggs are laid on needles and base of the scales.

Prevention and treatment

1. Light trapping. The overwintering and 1st generation adults can be trapped and killed with frequency vibration-killing lamps.

2. Conserve natural enemies. *Macrocentrus resinellae* and *Apanteles* sp. are dominant natural enemies during the larval period. Collect infested cones before June and place them in cages. Let the parasitoids emerge between early July and mid-September and disperse into the forest.

3. Chemical control. Spray 2.5% deltamethrin at 1,000× in early April or inject 10% imidacloprid EC at 1:3 into tree trunks. Apply 20 g/L lamda-cyhalothrin at 1,000-1,500×, 50% imidacloprid-monosultap at 600-800×, or 45% profenofos-lamda-cyhalothrin at 1,000-1,500×.

References Zhao Jinnian et al. 1991, 1997, 2004; Liu Youqiao and Li Guangwu 2002.

(Zhao Jinnian, Tian Hengde)

341 *Retinia resinella* (L.)

Classification	Lepidoptera, Tortricidae
Synonyms	*Phalaena* (*Tortirx*) *resinella* L., *Carpocapsa obesana* Laharpe, *Scoparia resinalis* Guenée, and *Pyralis resinana* F.
Common name	Pine resin-gall moth

Importance *Retinia resinella* is an important pest of young *Pinus sylvestris* var. *mongolica* trees. Larvae bore into young shoots, causing death of shoots and even mortality of the infested trees.

Distribution Inner Mongolia, Heilongjiang; Europe and Eastern Russia.

Hosts *Pinus sylvestris* var. *mongolica*, *P. sylvestris*, *P. Montana*, and *P. banksiana*.

Damage Young larvae bore into base of needles. From there they infest the xylem and induce the formation of resin balls. The resin balls are pale and translucent at the beginning, then turning into brown. The empty pupae are partially exposed on surface of the resin balls after adult emergence.

Morphological characteristics **Adults** Wingspan 16-22 mm. Silvery grey; forewing brown. Basal 1/3 of the wing with 3-4 irregular silvery grey stripes. There are two irregular silvery grey "Y" shaped stripes near apex. Hindwing dark grey.

Larvae Mature larvae 8-14 mm. Head brown, pronotum pale brown, thorax and abdomen light-yellowish brown.

Biology and behavior *Retinia resinella* occurs one generation every two years in Inner Mongolia. Larvae overwinter in resin balls on the shoots. They resume activity in late April. Larvae go through 5 instars. Mature larvae pupate inside resin balls in mid-May of the 3rd year. Pupation occurs in mid-May. Adults emerge, mate,

R. resinella adult (Li Chengde)

and lay eggs in June.

Eggs are laid at base of buds of new shoots. Larvae do not migrate during development. Heavier damage occurs at the south facing slope, in sparse stands, and on main shoots of 5-16 year-old trees. Trees of 20 years old suffer less damage and most of the damage occurs on lateral shoots.

Prevention and treatment

1. Silviculture. Young stands should be kept at high density. Cut down infested shoots and destroy them.

2. Biological control. Conserve natural enemies of the larvae. *Scambus* sp. and another un-described parasitic wasp can be found parasitizing *R. resinella*. Apply *Beauveria bassiana* (4×10^{11} conidia/g) at 2,000-3,000×. Spray 2×10^{10} PIB/mL nuclear polyhydrosis virus at 1,000-1,500×.

3. Chemical control. During young larvae period, scrub off the old barks near the bottom of the tree trunk and apply a band of 20% dinotefuran at 10× to control larvae or spray 20% dinotefuran at 1,500× to infested shoots. Apply 20 g/L lamda-cyhalothrin at 1,000-1,500×, 50% imidacloprid-monosultap at 600-800×, or 45% profenofos-lamda-cyhalothrin at 1,000-1,500×.

References Wang Zhiying, Yue Shukui, Dai Huaguo et al. 1990; Xiao Gangrou (ed.) 1992; Liu Youqiao and Li Guangwu 2002.

(Byun Bong-Kyu, Li Chengde, Wang Zhiying, Yue Shukui, Dai Huaguo, Xu Xueen, Ge Jingshan)

R. resinella adult (Li Chengde)

342 *Rhyacionia duplana* (Hübner)

Classification Lepidoptera, Tortricidae
Synonyms *Tortrix duplana* Hübner, *Rhyacionia duplana simulata* Heinrich
Common names Summer shoot moth, Elgin shoot moth

Importance *Rhyacionia duplana* is an important pest of pine trees. It may significantly affect the pine tree development.

Distribution Hebei, Shanxi, Liaoning, Shandong, Henan, Shaanxi; Japan, Europe.

Hosts *Pinus tabulaeformis*, *P. densiflora*, and *P. thunbergii*.

Damage Larvae bore into the center of new shoots, causing wilt of the tips. Trees with multiple-year infestations grow into broom-shape. The infestation site often has thin covers made of silk and resin. When feeding on the bark and outer layer of the wood, the larvae make oval-shaped resin covering which thickens as the larvae develop.

Morphological characteristics **Adults** Wingspan 16-19 mm. Labial palps protruding forward, slightly drooping. The 2nd segment enlarged distally. The terminal segment is partially covered by scales from the 2nd segment; its end obtuse. Head pale brown; thorax, abdomen blackish brown. Forewing greyish brown, the termen rusty brown. Several median fascia are present. The costa with hooked strigulae. Termen very oblique. Hindwing pale greyish brown; apex darker. Termen hair long; greyish white.

Eggs Flat oval; light-yellow, turning red before hatching.

Larvae Young larvae light-yellow, turning to orange yellow as they grow.

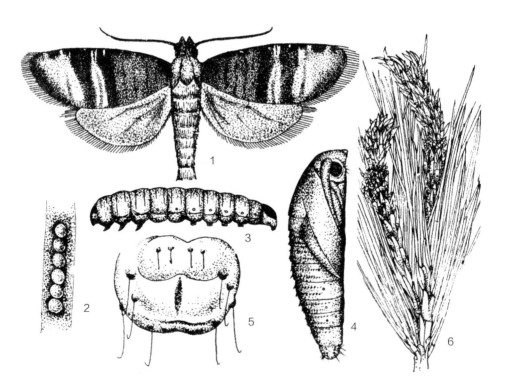

R. duplana. 1. Adult, 2. Eggs, 3. Larva, 4. Pupa, 5. Posterior end of pupa, 6. Damage (Yu Changkui)

R. duplana adult (Wu Chunsheng, Zhang Runzhi)   R. duplana adult (NZM)

Pupae Length 5-7 mm. There are two rows of scoli on the tergite of each of the 2nd-7th segments and one row of scolus on the 8th segment. The end has 8 hooked anal spines.

Biology and behavior *Rhyacionia duplana* occurs one generation a year. Pupae overwinter at base of the trees or at base of the branches inside cocoons. Adults start to emerge in late March. Peak emergence is in mid-April and ends in late April. Adult eclosion peak is between 8:00-10:00. They mate at 2-3 days after eclosion and lay eggs 2-3 days after mating. Eggs are laid at the inner base of new leaves. They are arranged into rows of 15-20 eggs. Each female lays an average (min, max) of 28 (6-50) eggs. Mean (min, max) egg incubation period is 22 (17-27) days. Eggs start to hatch in early May. Young larvae feed on leaf buds first. After about 5 days, they feed on phloem and xylem of the young shoots, causing the shoots to bend and become wilt above the infestation site. Infested shoots are easily broken off by wind. Each larva infests one shoot. When invading a new shoot, the larva uses silk and resin to cover its body. Larvae develop for 25 days inside new shoots before migrating to the base of 3-7 years old pine trees or the base of branches of 7-10 years old trees and feed on bark and phloem. They use silk and resin to cover their body. Larvae start to pupate in early July and overwinter as pupae.

Prevention and treatment

1. Quarantine. Delineate quarantine zones according to infestation levels. Prevent moving or importing un-treated infested trees.

2. Silviculture. Improve sanitation, cutting down infested trees or branches and destroy them. Replant trees in heavily damaged forests using pest-resistant species and mixed tree species.

3. Physical control. Manually remove pests or squeeze pests to death in low, sparse forests. During adult emergence period, attract and kill adults using black light traps. During the fall and winter periods, scrape coarse bark and curved bark to kill the overwintering larvae.

4. Biological control. Conserve, propagate, and release natural enemies. Release parasitic wasps such as *Trichogramma* sp. at 1,000-5,000 per tree for every other tree during peak or late egg stage when pest population is high.

5. Chemical control. Egg hatching period is critical for conducting chemical treatment. Application of 5.7% emamectin at 3,000-5,000× in forests with > 0.6 canopy coverage will achieve good control.

References Xiao Gangrou (ed.) 1992, Liu Youqiao and Li Guangwu 2002.

(Wu Chunsheng, Liu Zhenlu, Zhao Lianguo)

343 *Rhyacionia insulariana* Liu
Classification   Lepidoptera, Tortricidae

Importance   *Rhyacionia insulariana* is a common pest in pine forests in Yunnan province. Old forests or young trees neighboring heavily damaged old trees suffer most damage. The pest lowers the timber quality. Infested trees suffer from slow growth and secondary pests and diseases.

Distribution   Sichuan, Yunnan.

Hosts   *Pinus yunnanensis*, *P. densata*, *P. kesiya* var. *longbianensis*, *P. massoniana*, *armandii*.

Damage   Infested shoots bend toward the infested side and are easily broken by winds or become wilt. The infestation rate of the main shoots is higher than that of the lateral shoots.

Morphological characteristics   **Adults** Small; length 8-11 mm, wingspan 20-32 mm. Head with reddish brown clusters of hairs.

Eggs   Flat oval, size 0.6 mm by 0.4 mm. New eggs orange yellow; turning to light-yellow later.

Larvae   Length 14-15 mm. Pale yellowish brown, head pale reddish brown, pronotum dark castaneous. Setae on each abdominal segment small. Crotchets of prolegs are in two series of complete circles. Anal proleg crotchets are arranged in double series of incomplete circles.

Pupae   Length 9-13 mm, tanned, anal comb with 12 spines.

Biology and behavior   *Rhyacionia insulariana* occurs one generation a year in Hanyuan (Sichuan). Larvae overwinter in buds on branch tips. They resume activity in March. As their feeding rate increases between mid-April and early May, the larvae may relocate to new shoots. From early May to mid-June, mature larvae start

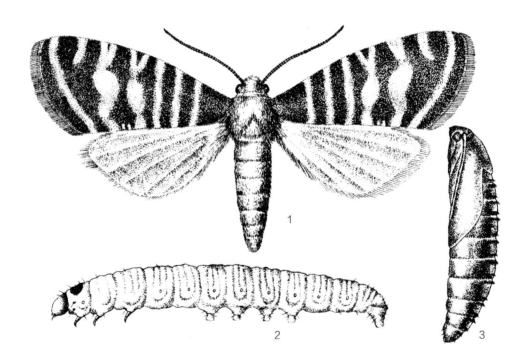

R. insulariana. 1. Adult, 2. Larva, 3. Pupa (Zhang Peiyi)

R. insulariana adult (NZM)

to make cocoons outside of the feeding tunnel, inside male flowers, or in cracks of the coarse bark. Adult emergence and egg laying period is in mid- to late July. Eggs hatch in early August. Young larvae overwinter in late September.

Eggs are laid on leaves, leaf sheath, or young branches. They are laid singly or arranged in rows. Young larvae are very active. They move fast and can spin silk when migrate. Young larvae feed inside leaf sheath and cause leaves wilt. They may also attach several leaves together and feed inside the leave bundle. Adults are attracted to lights.

Prevention and treatment

1. Silviculture. Plant trees at enough density and use a mixture of coniferous and deciduous tree species. Screen for resistant species.

2. Cultural control. Cut off dead branches during winter. Remove infested branches during growing season frequently and place them in cages for collecting parasitic wasps. Install black light traps during adult emergence season to attract and kill them.

3. Biological control. Conserve, mass rearing, and release natural enemies. In stands with high pest populations, release *Trichogramma* sp. or parasitic wasps at 1,000-5,000 per tree every other tree during the egg period. Apply *B.t.* at 0.05-0.2 billion spores per mL to control larvae. Apply *Beauveria bassiana* (4×10^{11} conidia/g) at 2,000-3,000×. Spray 2×10^{10} PIB/mL nuclear polyhydrosis virus at 1,000-1,500×.

4. Chemical control. Apply 25% trichlorfon dust, or spray with 50% DDVP EC at 2,000×, 90% trichlorfon at 1,000×, 50% malathion at 6,000×, 25% phosemet at 800×, or 50% fenamiphos at 400×. Spray 1-2 times to control larvae. In stands with > 0.6 canopy density, ULV application of 2% avermectin is effective. Apply 20 g/L lamda-cyhalothrin at 1,000-1,500×, 50% imidacloprid-monosultap at 600-800×, or 45% profenofos-lamda-cyhalothrin at 1,000-1,500×.

References Xiao Gangrou (ed.) 1992; Liu Youqiao and Li Guangwu 2002; Wang Xiangdong 2005; Yang Xiaofeng, Huwen, Feng Yongxian et al. 2008.

(Li Qiao, Pan Yongzhi, Liu Nengjing)

344 *Rhyacionia pinicolana* (Doubleday)

Classification Lepidoptera, Tortricidae
Synonyms *Retinia pinicolana* Doubleday
Common name Orange spotted shoot moth

Distribution Beijing, Tianjin, Hebei, Shanxi, Liaoning, Jilin, Heilongjiang, Anhui, Fujian, Jiangxi, Hunan, Sichuan, Guizhou, Shaanxi, Gansu; Europe, Japan, Korea Peninsula, Russia.

Hosts *Pinus tabulaeformis*, *P. thunbergii*, *P. sylvestris* var. *mongolica*, and *P. densiflora*.

Damage Larvae feed on new shoots, causing wind damage and height growth.

Morphological characteristics **Adults** Length 6-7 mm, wingspan 19-21 mm. Labial palps protrude forward; its 2nd segment enlarged at middle. Forewing narrow, reddish brown, with more than 10 silvery cross bands; front margin with silvery strigulae. Hindwing dark grey, without marks; the hairs on the wing margin greyish white. This species is similar to *R. puoliana*, but differs from it in the wider silvery bands of the forewing.

Eggs Flat oval, length 1.0-1.2 mm, width 0.8-0.9 mm. Orange yellow.

Larvae Head and pronotum yellowish brown, anal area reddish brown. Proleg crotchets form a single series complete circle of 32-50 crotchets. Mature larvae about 10 mm long.

Pupae Length 6-9 mm, yellowish brown, spindle-shaped, turning to greyish brown before eclosion. Dorsal surface of each of the 2nd to 7th segment with two teeth. The 8th segment with only one tooth. Anal comb with 12 spines.

Biology and behavior *Rhyacionia pinicolana* occurs one generation a year in Xinglongshan (Gansu). Larvae overwinter inside shoots. They resume activity between mid-April and mid-May. Majority of the larvae feed on male flowers. Peak infestation period is in mid- to late

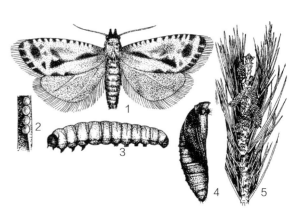

R. pinicolana. 1. Adult, 2. Eggs, 3. Larva, 4. Pupa, 5. Damage (Yu Changkui)

May. From late May to mid-June, larvae migrate into new shoots. The entrance is usually covered with white silk and resin and some feces. Only one larva occurs inside one shoot. Pupae appear inside shoots between early June and early July. Pupal duration is 20-28 days. Adults appear from late June and reach peak emergence during early and mid-June. Mating occurs from two days after emergence. Egg-laying starts from two days after mating. Eggs hatch from late July and early August. Peak egg hatching period is during early and mid-August. Larvae overwinter in early and mid-October.

Prevention and treatment

1. Physical control. Set up black light traps during early and mid-July can effectively reduce the number of eggs and population levels.

2. Chemical control. Spray 2.5% deltamethrin at 5,000× every week will kill > 95% of the larvae. Spray 20% dinotefuran at 1,000× can kill 90% of the larvae. Apply 20 g/L lamda-cyhalothrin at 1,000-1,500×, 5.7% emamectin at 3,000×, 50% imidacloprid-monosultap at 600-800×, or 45% profenofos-lamda-cyhalothrin at 1,000-1,500×.

References Xiao Gangrou (ed.) 1992, Liu Youqiao and Bai Jiuwei 1997, Liu Youqiao and Li Guangwu 2002, Shi Zemei 2006.

R. pinicolana adult (NZM)

(Liu Zhenlu, Li Yingchao, Li Zhenyu)

345 *Strepsicrates coriariae* Oku

Classification Lepidoptera, Tortricidae

Distribution Fujian, Guangdong, Guangxi; Japan.

Hosts *Eucalyptus urophylla*, *E. camaldulensis*, *E. citriodora* and many other species in the genus *Eucalyptus*; *Melaeuca leucadendra*, *M. guingeuneruie s.t.*, *Rhodomytus tomentosa*, *Fristania conferta*.

Damage Larvae attach buds and young leaves into nests and feed inside the nests. Infestation rate can reach more than 80%. Some young trees may have 10-20 nests. The infestation affects growth and shape of the trunk.

Morphological characteristics **Adults** Length 6-7 mm, wingspan 13-14 mm. Antennae filiform, forewing greyish brown, hindwing grey. Wing margin with many long hairs; termen black; front margin with black and trey alternating stripes.

Larvae Length 12-14 mm, cylindrical, pale green. There are three black and two white alternating longitudinal bands on the back.

Pupae Length 5-7 mm, yellowish brown, shiny.

Biology and behavior The insect occurs 8-9 generations a year in Guangdong. There is no overwintering period. Adults mate and lay eggs during the evening hours. Eggs are laid on young shoots, leaves, or petioles. Newly hatched larvae attach young leaves or shoots to form nests and feed inside the nests. The 3rd instar larvae relocate and create new nests. There are five instars. Mature larvae pupate on ground surface or inside nests. The insect infests young trees that are planted before June. Trees of > 1 year-old suffer less infestation as a result of induced resistance from earlier infestations. Drought and other adverse weather conditions may significantly affect the populations.

Prevention and treatment.

1. Silviculture. Plant pest resistant species. A survey of 11 *Eucalyptus* spp. showed the *S. coriariae* population in *E. torelliana* stands was 42 times higher than that in *E. camaldulensi*. *E.* ABL 12 stands suffered little infestation. The peak infestation period is in April and May. Avoid plant trees during this period will minimize damage.

2. Chemical control. Spray 25% Dimilin SC at 1,000-2,000× is effective. It is important the nests become wet from the spray. Spray leaves using 90% trichlorfon at 1,500-2,000×, or 20% fenvalerate EC at 3,000-4,000×, 5.7% emamectin at 3,000×, 20 g/L lamda-cyhalothrin at 1,000-1,500×, 50% imidacloprid-monosultap at 600-800×, or 45% profenofos-lamda-cyhalothrin at 1,000-1,500×.

References Liu Youqiao and Li Guangwu 2002.

(Gu Maobin)

S. coriariae adult (NZM)

346 *Zeiraphera griseana* (Hübner)

Classification Lepidoptera, Tortricidae
Synonyms *Tortrix griseana* Hübner, *Sphaleroptera diniana* Guenée
Common name Larch tortrix

Importance Outbreaks of *Z. griseana* occur periodically in northwestern China. During June and July, heavy infestation may consume all leaves on 100% of the trees in a forest.

Distribution Hebei, Jilin, Gansu, Xinjiang; Japan, Russia, Europe, North America.

Hosts *Larix* spp., *Pinus* spp., *Picea asperata*, *Abies fabri*.

Damage Young larvae burrow into bundles of young leaves and create cylindrical nests. They feed inside the nests until the 3rd instar. Starting from 4th instar, larvae feed freely among branches. They secrete silk as they feed and attach leaf remnants and feces on branches. At later stage of the larval development, majority or all of the needles on a tree are consumed.

Morphological characteristics **Adults** Wingspan 16-24 mm. Second segment of the labial palps enlarged at the end. The last segment of the labial palps small, drooping. Forewing greyish white; basal fascia blackish brown, about 1/3 length of the forewing. The middle part of the outer edge of the basal fascia protruded. The area between basal and median fascia silvery grey, constricted in the middle. The apex silvery grey, with apparent strigulae. Hindwing greyish brown; termen hair yellowish brown.

Eggs Flat oval, light-yellow.

Larvae Length 12-17 mm, dark green; head and pronotum blackish brown; pygidium brown. Setae black.

Pupae Length 9-11 mm, apricot yellow at beginning, turning into brown later. Anal comb with 9-11 spines.

Biology and behavior *Zeiraphera griseana* occurs one generation a year. It overwinters in egg stage. Eggs hatch in late May. Young larvae hide in leaf bundles. They use silk to attach leaves into cylindrical nests. They come out when feeding and retreat into the nest once disturbed. They make a new nest when the size of the nest is too small. When all leaves from a tree are consumed, they spin silk and migrate to neighboring trees with the aid of wind. Mature larvae do not make nests. They feed in the silk web among the branches. Larvae migrate into fallen leaves in late June. They make cocoons and pupate. Adults emerge in late July.

Prevention and treatment

1. Silviculture. Reduce density, prune properly, reduce weeds, and improve tree vigor and natural tolerance to *Z. griseana*.

2. Mechanical control. Burn the fallen leaves and branches in winter will reduce egg numbers. Set up black

Z. griseana adult (Qu Xiaojin)

Z. griseana adult (Wu Chunsheng, Zhang Runzhi)

Z. griseana adult (Li Zhenyu)

light traps in July to attract and kill adults. These methods can maintain pest population under the economic injury level.

3. Biological control. Spray *B.t.* at 800× can reduce 70% of the pest population. Apply *Beauveria bassiana* (4×10^{11} conidia/g) at 2,000-3,000×. Spray 2×10^{10} PIB/mL nuclear polyhydrosis virus at 1,000-1,500×.

4. Chemical control. Spray 2.5% deltamethrin at 2,000× in early June when 95% of the larvae are in 2nd-3rd instar. Apply 20 g/L lamda-cyhalothrin at 1,000-1,500×, 50% imidacloprid-monosultap at 600-800×, or 45% profenofos-lamda-cyhalothrin at 1,000-1,500×.

References Bu Wangui 2000; Liu Youqiao and Li Guangwu 2002; Nie Junqing, Sun Jing, Zhang Yumei et al. 2003.

(Wu Chunsheng, Ma Wenliang, Zang Shouye)

347 *Camptochilus semifasciata* Gaede
Classification Lepidoptera, Thyrididae

Distribution Zhejiang, Fujian, Guangxi.

Hosts *Castanopsis eyrei*, *Castanea mollissima*, *C. henryi*.

Damage Infested leaves are rolled into pods or cylindrical nests. At late infestation stage, the pods turn white.

Morphological characteristics **Adults** Wingspan 33 mm, length 10 mm. Head and labial palps tanned. Antennae filiform, greyish brown. Dorsal surface greyish yellow, ventral surface yellowish brown, lateral sides pink. The inner side of the front tibiae with teeth. Tibia of the middle legs with a pair of spurs. Tibia of hind legs with a pair of spurs at middle and the end. Forewing humeral angle distinct; apex raised. Wing ochrous, reticulate. There is a tanned band extending from apex to middle of hind margin. The band splits below the median cell. There are three yellow markings outside of the oblique band. Hindwing amber color, inner bands double-lined, median band with a single line. The back of the wings shows clear markings. There is an ochrous rectangular marking below the median cell of the forewing.

Eggs Height and width 0.8 mm, yellow. New eggs yellow, turning to darker color later, and becomes blackish brown before hatching. Crown area concave, with 13-15 longitudinal ridges.

Larvae Length 13 mm. Head capsule 3 mm wide. Head black; pronotum blackish brown; middle line of dorsal area yellowish brown; thoracic legs brownish yellow, claws brown. The end of the larvae greenish yellow. Spiracles brownish yellow. Each segment with yellowish white long hairs. Pygidium blackish brown. Crotchets of prolegs brown.

Pupae Length 9-10 mm, width 4 mm, tanned. Anal comb with 4 hooked spines, the inner pair thicker and longer than the outer pair.

Biology and behavior *Camptochilus semifasciata* occurs two generations a year in Songyang (Zhejiang). Pupae overwinter under a shallow layer of soil. Adults emerge between mid-April and early June. The egg incubation period is about 6 days. Eggs start to hatch in late April. The 1st generation larval period is 23-28 days. They can be seen until mid-July. Pupae start to appear in late May. Pupal period is about two months. Adults emerge in late July and lay eggs. The 2nd generation larvae appear from early August. Pupation starts from early September. The pupal period is more than seven months. Adults emerge next spring and live for about 7 days.

C. semifasciata egg (edited by Zhang Runzhi)

C. semifasciata larva (edited by Zhang Runzhi)

C. semifasciata adult (edited by Zhang Runzhi)

Adults usually emerge from pupae during the day. They lay eggs singly on upper surface of leaves. Newly hatched larvae chew edges of leaves and make triangular pods. They feed upon leaf tissue under the leaf surface. Once leaf tissue is consumed, the larva inside each pod abandons the pod and makes a new one. Larger pods are made as the size of the larvae increase. Late instars feed on both leaf tissue and leaf surface. They roll leaves into nests. They do not feed on the outer layer of the cylindrical nests. One larva makes more than 10 nests over its lifetime. As a result, the feeding can affect normal growth of chestnut trees and reduce chestnut yield. Mature larvae fall to the ground along with the nests or crawl to ground. They dig into the soil and spin silk to make smooth cocoons made of soil, sand, and silk. Then they pupate inside the cocoons.

Prevention and treatment

1. Avoid planting *Castanea* spp. near *Castanopsis eyrei*.

2. Collect pods and destroy them.

3. Clean out fallen leaves in winter to reduce overwintering pest population. The collected pods may be placed in cages to allow natural enemies to emerge.

4. For each 10 cm diameter chestnut tree, inject 5 mL of 50% omethoate at 5× to xylem layer below the branches. Increase the number of injection holes for larger trees. Do not use this method within one month before harvesting.

References   Chen Hanlin 1994, Zhu Hongfu and Wang Linyao 1996, Wang Mingyue 2005.

(Chen Hanlin, Ye Jianglin)

348 *Rhodoneura sphoraria* (Swinhoe)
Classification Lepidoptera, Thyrididae

Distribution Hebei, Zhejiang, Sichuan; India.

Hosts *Castanea mollissima, C. henryi, Quercus acuttissima,* and *Q. fabri.*

Damage Infested leaves are rolled into pods. At late infestation stage, the pods turn white.

Morphological characteristics **Adults** Length 8-9 mm, wingspan 22-25 mm. Head tanned, body yellowish brown, ventral surface paler. Front and hindwings castaneous, reticulated. Anterior margin of the forewing with an irregular row of black dotted line; inner band double-lined, wavy; median band double-lined and curvy; becomes narrower from top to bottom; extends from anterior margin to middle of the posterior margin of the wing. The outer stripe protruded outward between M_2 and Cu_1 and curved down near anterior margin. Outer band narrow; connected to the hooked line on the top and reaches to posterior angle; becomes thicken under Cu_1. Hindwing inner band with double lines; inconspicuous. Median band double-lined; narrower from top to bottom. Outer band arched.

Eggs Size 0.6 mm by 0.3 mm. Cylindrical, slightly smaller at the ends. Newly laid eggs white, turning to brownish yellow later, tanned before hatching.

Larvae Mature larvae 12-15 mm in length, yellowish green. Each segment with a ruga at middle; therefore making each segment looks like two sub-segments. Each segment with various sizes of brownish black verrucae; each verruca with one soft thin hair.

Pupae Length 7-10 mm, width 2.5-4 mm, yellowish brown, spindle shaped. Anal comb with a bundle of spines.

Biology and behavior *Rhodoneura sphoraria* occurs 3 generations a year in Zhejiang. Pupae overwinter inside pods that fallen to the ground. Adults emerge from mid-April. First generation larvae emerge from mid-April and are present until mid-June. Pupae appear from early May. Adults emerge from late May. Second generation larvae emerge from early June and pupate from late June. Adults emerge in mid-July. Overwintering larvae are active between August-October. Pupation starts in September.

Adults usually emerge in the early morning. They may visit flowers during the day. They are somewhat attracted to lights. Females lay eggs singly on top of leaves along the veins. They lay eggs along the main vein of small leaves and the lateral veins of large leaves. One egg is usually laid on each leaf. Occasionally two eggs are laid on a leaf. Young larvae spin trumpet-shaped pods and feed inside. One larva resides in each pod. If two larvae are hatched on one leaf, one of them will make a pod using a neighboring leaf. They make larger pods as the larvae body size increases. They only feed on inner layer of the leaves and leave the exterior layer untouched to avoid exposing themselves. When a leaf is consumed, the larva will make a new pod. A mature larva will make a pod from a whole leaf, seal the entrance, and pupate inside the pod. It fixes its body to the silk with the anal comb. The pods fall to the ground in winter. Adult life span is 5-9 days. Egg incubation period is 4-8 days. Larval duration is 19-31 days. The duration for the 1st and 2nd generation pupae is 12-23 days. Duration of the overwintering pupae is 165-221 days.

Prevention and treatment

1. Cultural control. Collect and destroy pods. Remove fallen leaves in winter can reduce overwintering pest population.

2. Biological control. Larval parasitoids include *Microgaster* sp., *Acropimpla* sp., *Xanthopimpla punctate, Eulophus* sp., *Elasmus* sp., and *Sympiesis* sp. Parasitization rate can reach 25%. Collect pods and place them in cages. Allow parasitoids to emerge and disperse into the forests. Apply *Beauveria bassiana* (4×10^{11} conidia/g) at 2,000-3,000×. Spray 2×10^{10} PIB/mL nuclear polyhydro-

R. sphoraria adult (edited by Zhang Runzhi)

R. sphoraria adult (edited by Zhang Runzhi)

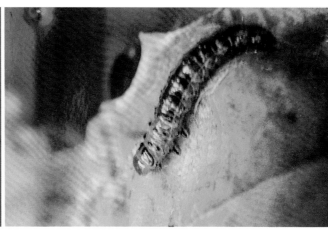
R. sphoraria larva (edited by Zhang Runzhi)

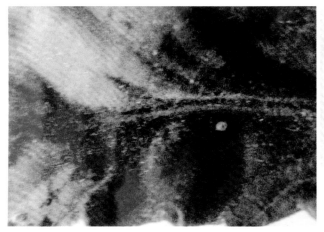
R. sphoraria egg (edited by Zhang Runzhi)

R. sphoraria pupa (edited by Zhang Runzhi)

R. sphoraria damage (edited by Zhang Runzhi)

R. sphoraria damage (edited by Zhang Runzhi)

sis virus at 1,000-1,500×.

3. Chemical control. Inject 5 mL of 50% thiamethoxam diluted at 5× to xylem layer below the branches. Increase the number of injection holes for larger trees. Do not use this method within one month before harvesting. Apply 2 Apply 5.7% eamamectin at 3,000×, 20 g/L lamda-cyhalothrin at 1,000-1,500×, 50% imidacloprid-monosultap at 600-800×, or 45% profenofos-lamda-cyhalothrin at 1,000-1,500×.

References Zhu Hongfu and Wang Linyao 1996; Chen Hanlin, Dong Liyun, Zhou Chuanliang et al. 2002.

(Chen Hanlin, Ye Jianglin, Zhou Jianmin)

349 *Striglina bifida* Chu et Wang
Classification Lepidoptera, Thyrididae

Importance *Striglina bifida* is an important leaf feeder of *Magnolia bilola*. It affects the production of flower, seeds which are used in Chinese medicine.

Distribution Zhejiang.

Host *Magnolia bilola*.

Damage The insect rolls leaves into triangular pods at early stage and cylindrical pods at late developmental stage. Leaves inside the pods are consumed.

Morphological characteristics **Adults** Wingspan 26-41 mm. Head brownish yellow; antennae yellow, with pectinate hairs. Thorax and abdomen reddish brown; the axillary plates and the 1st and 2nd abdominal segments with more red color; the posterior margin of the 4th abdominal segment tanned; the hairs at end of the abdomen yellow; wings yellow, with reticulate brownish red veins; anterior margin of the forewing brownish grey. There is a tanned stripe from middle of posterior margin to apex. The stripe splits near the apex. Median cell with a tanned marking. Hindwing with a tanned stripe extending from posterior margin to anterior margin and connects to the stripe on the forewing. There is another thin brown cross line outside of the stripe.

Eggs Cylindrical, size 1.3 by 0.9 mm. The first generation eggs pale purple. The 2nd generation eggs slightly reddish, with 15-17 longitudinal ridges and many thin rugae between the ridges. The crown region concave, darker. Bottom of the egg flat, with skirt-shaped gelatinous material that fixes the egg to the leaf.

Larvae Newly hatched larvae 3.5 mm in length. Mature larvae 18-24 mm in length. The 1st and 2nd instars red; the 3rd instar and later instars brownish yellow. Head of the mature larvae black. Head densely punctured. Frons and gena sculptured. Ocelli red; pronotum black, with a pale longitudinal groove; abdominal segments tanned; crotchets of prolegs brown, arranged in a complete circle of two series. Thorax and abdominal seg-

Damage to *Magnolia bilola* by *S. bifida* (Jiang Ping)

S. bifida larvae (Ye Yuzhu)

ments with blackish brown pinacula; those at end of the abdomen paler.

Pupae Length 14-16 mm, tanned, without hairs, with fine sculptures. Wing buds color paper, reaching the 4th abdominal segment. Anal comb with 8 hooked spines arranged in two rows. The row near the ventral side has 6 spines. The row near the dorsal surface has 2 spines.

Biology and behavior *Striglina bifida* occurs two generations a year in Songyang (Zhejiang). Pupae overwinter in soil. Adults appear from early April (1st generation) or from late June (2nd generation). They are attracted to lights. Larvae go through 5 instars. They feed inside pods. When one pod is consumed, a new one is made. Larger pods are made by later instars. They feed on leaf tissue, but 4th-5th instar feed on whole leaves and make cylindrical pods. Only the outermost leaf of the pod is not consumed.

Striglina bifida is more common in stands above 700 m elevation.

Prevention and treatment

1. Conservation of natural enemies. Egg parasitoids include *Trichogramma* sp. Larval parasitoids include *Apanteles* sp., *Pediobius cassidae* (hyperparasitoid), *Ceraphron* sp. (hyperparasitoid), and *Eulophus* sp. *Apanteles* sp. is the most common parasitoid.

2. Chemical control. Inject 20% dinotefuran to tree trunk or brush to the tree trunk will achieve good control. Apply 5.7% emamectin at 3,000×, 20 g/L lamda-cyhalothrin at 1,000-1,500×, 50% imidacloprid-monosultap at 600-800×, or 45% profenofos-lamda-cyhalothrin at 1,000-1,500×.

References Chen Hanlin et al. 1993.

(Jiang Ping)

350 *Asclerobia sinensis* (Caradja et Meyrick)
Classification   Lepidoptera, Pyralidae

Distribution   Inner Mongolia, Qinghai, Ningxia.

Host   *Caragana korshinskii*.

Damage   *Asclerobia sinensis* is an important pest of *C. korshinskii* pods and seeds in Inner Mongolia. It lowers the yield and quality of *C. korshinskii* seeds.

Morphological characteristics   **Adults** Length 9-11 mm, wingspan 19-20 mm. Scales on head and labial palps light-yellow; dorsal surface of thorax light-yellow. Forewing with a mixture of greyish black, greyish white, and yellow scales. The end of the scales on outer margin of the forewing white. There is a cross band made of greyish white and greyish black scales before the middle of the forewing. Hindwing pale grey.

Larvae Cylindrical. Yellowish white; head yellowish brown; pronotum of the young larvae with three large sharp black markings; those of the 4th instar and later form a " 人 " pattern. Each side of the prothorax with a black marking. Crotchets of prolegs arranged in a complete circle.

Biology and behavior   *Asclerobia sinensis* occurs one generation a year in western Inner Mongolia. About 2% of them may complete two generations a year. Mature larvae build cocoons in soil and overwinter. Pupation occurs in early April. Adults appear in early May and reach peak emergence around May 20. Eggs appear in mid-May. Larvae appear in late May. They migrate into soil in late June and overwinter. A small percentage of larvae pupate in early July and develop into adults in mid-July. Peak emergence of the 2nd generation adults is in late July. They produce eggs in mid-July. Larvae emerge in late July. They make cocoons in soil in mid-August.

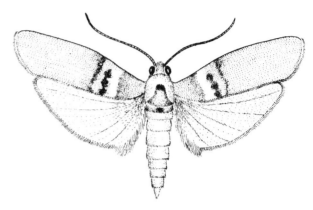

A. sinensis adult (Zhang Peiyi)

After a larva emerged from an egg, it crawls on pods for some periods then bores a hole and enters a pod. It feeds on one side of the seed, then consumes the whole seed or part of the seed and migrates to a different pod. It makes a white cocoon at the entrance holes after migrating into the new pod. Then part of the cocoon is brought into the entrance hole. Some larvae stay in the same pod through its life. Each larva infests about 5 seeds.

Prevention and treatment

1. Set up black light traps or lamps in forests to attract and kill adults.

2. ULV application of 20% dinotefuran: kerosene at 1:4 during adult emergence peak will kill > 90% of the adults. Apply 5.7% emamectin at 3,000×, 20 g/L lamda-cyhalothrin at 1,000-1,500×, 50% imidacloprid-monosultap at 600-800×, or 45% profenofos-lamda-cyhalothrin at 1,000-1,500×.

3. Fogging at 15 kg/ha in densely covered forests during adult emergence period can kill 75% of the adults.

References   Xiao Gangrou (ed.) 1992.

(Chen Hui, Zou Lijie, Liu Naisheng)

A. sinensis adult (NZM)

351 *Cryptoblabes lariciana* Matsumura
Classification   Lepidoptera, Pyralidae

Distribution   Heilongjiang; Japan.

Hosts   *Larix* spp.

Damage   Larvae feed on needles and slow the growth of trees. In 1975, the insect consumed all leaves in more than 6,000 ha forests in Boli county of Heilongjiang province.

Morphological characteristics   Length 6-8 mm, wingspan 18-20 mm; head, the dorsal of thorax and wing base with scattered gray scales; anterior edge of each thoracic segment with a silver gray stripe. Head closely covered with brownish black scales; compound eyes yellowish brown; ocelli black, slightly transparent. Antennae filiform, shorter than forewings, showing black and white alternating rings of hairs. Forewings long and narrow, dark brown. There is a short gray cross stripe near the wing base; the median cross line silver gray; the area between the two lines dark brown and with scattered silver scales. Sub-marginal line with 7-9 irregular black spots, margin hairs light grey. Anal area of hind wing broad, greyish brown, the Sc+R1 veins merge with Rs vein for a short distance beyond the discal cell, M_1 and Rs veins converge share the same stalk. There is a tympanal organ on the side of the 1st abdominal segment.

Eggs   Elongated elliptical, size 1.4-1.7 mm by 0.25-0.45 mm; apricot in color.

C. lariciana adult (Zhang Peiyi)

C. lariciana adult (NZM)

Larvae Mature larvae 12-15 mm in length, head width 1.2-1.4 mm. Head light brown, with a V-shaped black mark. The subdorsal line and supraspiracular line dark brown. Each segment has 4 brown verrucae on the back. There are two setae in front of prothorax spiracle. The verrucae on the side of the 8th abdominal segment located above the spiracle. Base and the end of the thoracic legs brown, the middle with brown rings. Proleg crochets in complete circles, pygidium dark brown.

Pupae Length 7-12 mm width 3-5 mm; reddish brown; the vertex and abdominal end blunt round and smooth. There are eight hook-like cremasters at the end of abdomen, the two in the center are three times longer than the other six cremasters.

Biology and behavior *Cryptoblabes lariciana* occurs one generation a year in Heilongjiang province and overwinters as pupae. Adults emerge from early June next year. Adults rest during the day and are active during early evening. They often fly around canopy. Adults have phototaxis. They oviposit in mid-June. Most eggs are laid near 1/3 position from base of needles and are laid singly. Larvae hatch at the end of June. Young larvae spin leaves with silk and feed inside silk webs. Their feeding increases drastically from August. At the end of August, the larvae become mature and migrate to the ground. They spin cocoons to pupate among dead leaves. Most pupae are located in places with thick litter layer, moist soil, the lower part of sunny slopes, and within 60 cm from the tree base. The population is higher on sunny slopes, forest edge, and forests with canopy coverage below 0.5.

Prevention and treatment

Protect natural enemies. During the pupal stage, parasitic wasps and *Beauveria* sp. caused total mortality up to 54.71%.

Reference   Tian Yu et al. 1990.

(Lu Wenmin)

352 *Dioryctria mongolicella* Wang et Sung
Classification Lepidoptera, Pyralidae

Importance In 1979, *Dioryctria mongolicella* infested 11.8%-12% of the *Pinus sylvestris* var. *mongolica* in Huma county and Aihui county of Heilongjiang province, and 22% the trees in Neijiang county.

Distribution Inner Mongolia, Jilin, Heilongjiang.

Hosts *Pinus sylvestris* var. *mongolica*.

Damage Larvae bore into trunks and large branches, causing resin flow, death of branches, and wind damage.

Morphological characteristics **Adults** Wings 24 mm in length. Head and thorax dark brown, abdomen brownish black. Forewings smooth, the back without vertical scales, base color dark black; basal area and subbasal area line yellowish brown; subbasal line gray; the front and posterior mid-line, and end line greyish white. Hind wings dark brown, slightly black along the outer margin; surface with a few black scales; margin hairs gray.

Larvae Mature larvae 25-30 mm in length. Head yellowish brown. Body greyish green, surface with many brown pinaculums.

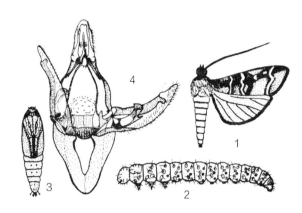

D. mongolicella. 1. Adult, 2. Larva, 3. Pupa, 4. Male genitalia (Qian Fanjun)

Biology and behavior *Dioryctria mongolicella* occurs one generation a year in Heilongjiang. It overwinters as larvae. They resume activity in late April. Large white resin accumulation mixed with brown feces forms at the entrance of galleries. In early and mid-July, mature larvae spin elliptical cocoons under the phloem. Pupal period is 10-15 days. Adults emerge in mid- and late July. Young larvae bore into branches through wounds and feed on phloem. They cause resin flow but the resin mass is small and light white. Larvae may migrate to different sites. After late September, larvae overwinter in galleries.

The damage of *Dioryctria mongolicella* is closely related to the presence of various wounds. Based on survey of 208 damaged trees, the distribution of *D. mongolicella* entry points is pruning wound -38.9%, wound from *Cronartium querccum miyabe* infection -32.2%, mechanical damage -25.5%, wound from *Cronartium flaccidum* infection -5.8%, natural bark laceration -6.3%, and wound from woodpeckers -1.9%. Trees below 9 years old are rarely damaged by the insect. Edge of the forest suffers more damage than the center of the forest.

Prevention and treatment

1. Reinforce quarantine measures to prevent spread of the insect.

2. Avoid mechanical damage of the host during seed harvest, grafting, pruning, and transporting seedlings. Control diseases promptly.

D. mongolicella damage (wind damage due to infestation) (Qian Fanjun)

D. mongolicella infestation induced disease and windbreak of tree trunk (Qian Fanjun)

D. mongolicella damage induced resin accumulation (Qian Fanjun)

D. mongolicella damage (Qian Fanjun)

D. mongolicella larva (Qian Fanjun)

3. Clean up the damaged trees.

4. Manually kill the larvae in galleries during July and August. Trim off the infested branches.

5. Chemical control. In July to August, paint or spray contact insecticide at the trunk wound to kill the larvae.

References Qian Fanjun 1981, 1988; Wang Pingyuan and Song Shimei 1982; Qian Fanjun and Yu He 1984, 1986.

(Qian Fanjun)

353 *Dioryctria pryeri* Ragonot

Classification	Lepidoptera, Pyralidae
Synonyms	*Dioryctria mendacella* Staudinger, *Dioryctria pryeri* Mutuura, *Salebria laruata* Heinrich, *Dioryctria laurata* (Sic), *Salebria laruata* Mutuura, *Salebria laruata* Issiki et Mutuura, *Phycita pryeri* Leech
Common name	Splendid knot-horn moth

Distribution Hebei, Shanxi, Inner Mongolia, Liaoning, Jilin, Heilongjiang, Jiangsu, Zhejiang, Anhui, Shandong, Henan, Sichuan, Shaanxi, Gansu, Qinghai, Xinjiang, Taiwan; Japan, Korea Peninsula.

Hosts *Pinus tabuliformis*, *P. armandii*, *P. taeda*, *P. massoniana*, *P. taiwanensis*, *P. koraiensis*, *P. densiflora*, *P. thunbergii*, *P. sylvestris* var. *mongolica*, *P. bungeana*, *Larix gmelini*, *Picea asperata*, and *Cunninghamia lanceolata*.

Damage *Dioryctria pryeri* larvae bore into the new shoots, staminate flowers, and cones, causing deformation and death of the cones, withering of the shoots and reduced seed production and tree growth.

Morphological characteristics **Adults** 9-13 mm in length, 20-26 mm in wingspan. Body grey to greyish white, with fish scale shaped white spot. Forewings with a reddish brown spot, near base with a grey short cross line; the inner and outer cross lines wavy and greyish white, a dark deep spot rests between the two cross lines; locations where close to leading and trailing edges with light grey cloud spots, discal cell with a crescent white spot, edge hairs greyish-brown.

Eggs Oval, the long diameter is 0.8 mm and the short diameter is 0.5 mm in length. Newly laid eggs milky, become violet black before hatching.

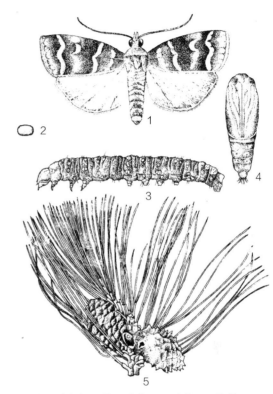

D. pryeri. 1. Adult, 2. Egg, 3. Larva, 4. Pupa, 5. Damage (Zhu Xingcai)

Larvae Mature larvae 15-22 mm in length; body ash black or black blue, glossy. Head, pronotum and the dorsal of 9th and 10th abdominal segments are yellowish brown. Body covered with long setae. Proleg crochets biordinal circle, the posterior proleg crochets biordinal and penellipse.

Pupae Reddish brown or dark reddish brown. Body 11-14 mm in length, head and abdominal end are round and blunt, smooth; the end with 6 hook-like cremasters arrange in an arc.

Biology and behavior *Dioryctria pryeri* occurs one generation a year in Liaoning, Shaanxi and Zhejiang and overwinters as the larvae mainly in the staminate inflorescence, and sometimes in the damaged fruits and

D. pryeri adult (Xu Gongtian)

D. pryeri mature larva (Xu Gongtian)   *D. pryeri* larva in pine shoot (Xu Gongtian)

D. pryeri feces (Xu Gongtian)   *D. pryeri* pupa (Xu Gongtian)

shoots, and in cracks of branches and in barks of tree trunks. The larvae migrate to new staminate inflorescence, shoots, young fruits and two-year-old cones in mid-May in Liaoning and Shaanxi or mid-April in Zhejiang. They pupate in mid-June in Liaoning and Shaanxi and in mid-May in Zhejiang. Adults emerge from late June to early Jul. Larvae emerge in mid-July and feed on the staminate inflorescence, fruits, and shoots. Adults mostly emerge during 8:00-12:00 in Shaanxi and during 16:00-20:00 in Zhejiang. Adults have strong phototaxis. They live for 7-13 days. The insect prefers older trees because older trees produce more staminate inflorescence, which provides food and overwintering habitat for *D. pryeri* young larvae. Thus infestations often start from older forest stands, then spread to mid-aged and young stands. The insect population is also related to fruit production. They mainly damage 2-year-old cones in years with good fruit production. When fruit production is low, they mainly feed on current year shoots.

Prevention and treatment

1. Plant mixed forests. Plant leguminous plants such as *Lespedeza bicolor*, *Amorpha fruticosa* and *Astragalus adsurgens* in open areas of existing stands. At the low-lying areas, plant *Populus* spp. to form mixed forests. In seed plantations, limit the number of male trees to minimize the available food and overwintering sites of *D. pryeri*.

2. In winter and early spring, remove and destroy the damaged fruits, branches and staminate inflorescences.

3. During adult stage, set up black light traps to trap and kill adults.

4. Conservation of natural enemies. Common natural enemies included *Trichogramma dendrolimus* in egg stage. Those in larva stage included *Coccygomimus disparis*, *C. turionella*, *Dusona tenuis*, *Lissonota evetriae*, *Macrocentrus abdominalis*, *M. gibber*, *M. resinellae*, *Phanerotoma kozlovi*, *P. semenovi*, *Protomicroplitis spreta*, *Elasmus ciopkaloi*, *Eurytoma* sp., *Pediobius* sp.,

D. pryeri damaged cone (Xu Gongtian)

D. pryeri early damage to shoots (Xu Gongtian)

D. pryeri damaged pine shoots (Xu Gongtian)

Beauveria bassiana and *Bacillus thuringiensis*. The parasitic rate of *M. resinellae* often reaches about 20% and sometimes reaches to 35%. The natural enemies during pupal stage included *Diadegma* sp., *Dolichomitus* spp., *Hyposter* sp. and *Beauveria bassiana*.

5. Biological control. Spray *Bacillus thuringiensis* at $(1-3) \times 10^8$ spores/mL.

6. Chemical control. Prior to the migration of the overwintering larvae, spray 50% fenamifos EC at 500-1,000×, 50% sumithion EC at 500×, 90% trichlorfon at 300×. Apply 5.7% emamectin at 3,000×, 20 g/L lamda-cyhalothrin at 1,000-1,500×, 50% imidacloprid-monosultap at 600-800×, or 45% profenofos-lamda-cyhalothrin at 1,000-1,500×.

References Li Kuansheng et al. 1964a, b, 1992a,b,c; Dang Xinde 1979, 1982; Zhao Changrun 1981; Zhao Jinnian et al. 1989; Yuan Ronglan et al. 1990.

(Li Kuansheng, Li Li)

354 *Dioryctria rubella* Hampson

Classification Lepidoptera, Pyralidae
Synonyms *Dioryctria splendidella* Herrich-Schäffer, *Phycita rubella* South
Common name Pine shoot moth

Importance The insect affects tree growth and seed quality, resulting in seed yield reduction. It is an important pest in China.

Distribution Beijing, Liaoning, Jilin, Heilongjiang, Jiangsu, Zhejiang, Anhui, Fujian, Jiangxi, Henan, Hubei, Hunan, Guangdong, Guangxi, Chongqing, Guizhou, Yunnan, Shaanxi; Japan, Korea Peninsula, Russia.

Hosts *Pinus tabuliformis*, *P. sylvestris* var. *mongolica*, *P. densiflora*, *P. thunbergii*, *P. massoniana*, *P. taiwanensis*, *P. yunnanensis*, *P. armandii*, *P. elliottii*, *P. taeda*, *P. serotina*, *P. galustris*, *P. griffithii* and *Picea asperata*.

Damage Early and middle instar larvae feed on the winter buds and shoots, respectively. The cylindrical galleries of damaged bud and shoots are filled with white granular dusts. Damaged cones are filled with nearly cylindrical light brown faecal pellets and dry up and crack prematurely.

Morphological characteristics **Adults** Length 10.0-14.0 mm, wingspan 22.0-30.0 mm. Forewings greyish brown and with variant shades of rose reddish brown. Discal cell with a reniform greyish white spot. The inner cross line greyish white, wavy; the side toward the discal cell with a white spot. The outer cross line greyish white. **Larvae** Length 19.0-26.5 mm. Head and pronotum reddish-brown, mesothorax, metathorax and each abdominal segment have 4 pairs of pinacula.

Biology and behavior It occurs two generations a year in Henan, 2-3 generations a year in Zhejiang and three generations a year in Guangxi. In Chun'an county of Zhejiang, it overwinters as early instar and mid- to old instar larvae in winter buds and withered shoots, respectively. In late March, overwintering larvae migrate to shoots. Pupae appear from mid-April to mid-June. Adults appear in the end of June. The 1st generation eggs, larvae, pupae, and adults appear from early May to early July, mid-May to the end of August, early July to early September, and mid-July to mid-September, respectively. The 2nd generation eggs appear from late July to mid-September. Larvae appear from early August and overwinter in shoots. A small percentage of larvae pupate in mid-September, turn to adults in late September, and produce eggs in October. Larvae appear from mid-October.

Adults mostly emerge during daytime, fly during 20:00-22:00, and have phototaxis. Females mostly oviposit on the wound of damaged withered shoots. Eggs are laid singly. The egg stage is about 7 days. The 1st and 2nd generation young larvae crawl into the old galleries, and feed on the wall of galleries about 4 days, then spin silk and migrate to the shoots. A portion of the overwintering larvae bore into the dormant buds of *Pinus massoniana*. Average length of a gallery in the damaged shoots of *P. massoniana* is 18.7cm. Larvae migrate to different buds and shoots during development. The 1st generation larvae bore into seed scales and axis of pine cones. Most seeds in the infested cones become empty. The 1st generation larval stage is about 23 days.

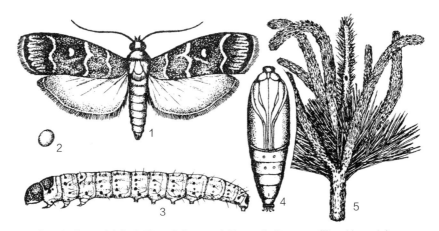

D. rubella. 1. Adult, 2. Egg, 3. Larva, 4. Pupa, 5. Damage (Tian Hengde)

Prevention and treatment

1. Light trapping. During the overwintering generation adult age, set up frequency vibration-killing light traps.

2. Biological control. In late March, apply *Beauveria bassiana* at 6.5×10^9 spores/g and 10 kg per hectare. The pest population reduction was 75%. Spray 2×10^{10} PIB/mL nuclear polyhydrosis virus at 1,000-1,500×.

3. Protect and release natural enemies. Before larvae resume activity in mid-March and during seed harvest in autumn, cut off the damaged shoots and separate out the damaged fruits. Place them in screened cages and let natural enemies emerge. *Macrocentrus watanabei* is the

D. rubella adult (Xu Gongtian)

D. rubella adult (Xu Gongtian)

D. rubella larva (Xu Gongtian)

D. rubella pupa (Xu Gongtian)

D. rubella excretion (Xu Gongtian)

D. rubella tunnel entry (Xu Gongtian)

dominant natural enemy, and its parasitism rate is up to 28.7%. In late April during its adult emergence, avoid spraying insecticides in the woods.

4. Pheromone traps. Use traps baited with (Z)-11-hexadecenyl acetate, cis-11-hexadecenal, and Z, E-9, 11-Tetradecadienyl acetate mixture at 1:7:2 ratio to lure moths.

References Xiao Gangrou (ed.) 1992; Zhao Jinnian et al. 1992, 1995, 1997, 1999; Bao Jianzhong and Gu Dexiang 1998.

(Zhao Jinnian, Tian Hengde, Zhao Changrun)

D. rubella killed shoot (Xu Gongtian)

D. rubella killed young *shoot* (Xu Gongtian)

D. rubella infested *Pinus tabulaeformis* stands (Xu Gongtian)

355 *Etiella zinckenella* (Treitschke)

Classification Lepidoptera, Pyralidae
Common name Pulse pod borer

Distribution Hebei, Shanxi, Liaoning, Shandong, Henan, Hubei, Yunnan, Shaanxi, Taiwan; Japan, Korea Peninsula, Russia.

Host *Rohinia pseudoacacia, Glycine max.*

Damage Larvae feed inside the pods and consume the seeds.

Morphological characteristics **Adults** Length 9 mm, wingspan 22-24 mm; gray yellowish brown; the end of each abdominal segment white. Forewings long and narrow, with mixed dark brown scales and yellow scales; a vertical white stripe present from basal angle to apex angle; an orange yellow cross stripe present along the inner side of the discal cell, a light yellow wide stripe present along the outer side of the discal cell. Hind wings greyish white. **Larvae** Mature larvae 14 mm in length, green, the dorsal red violet.

Biology and behavior *Etiella zinckenella* occurs four generations a year in Henan. It overwinters in larval stage in soil. Larvae pupate in late April next year. Adults emerge in early May. The 1st to the 4th generation adults occur in June, mid-July, early August, and early September, respectively. The last generation larvae crawl into soil to overwinter in October. Adults are active at night. Most eggs are laid singly on the base of calyx. It takes 5-10 days to hatch. After hatching larvae bore into the young pods, subsequently bore into the seeds. They are able to migrate to another pod to feed. After mature larvae exit the pods, they drop to the ground and crawl into soil, spin cocoons, and pupate.

Prevention and treatment Chemical control. In forests with high canopy coverage, apply smoke agent to fumigate during adult appearance period. In addition, spay 50% malathion EC or 50% fenitrothion EC at 1,000×, $2×10^{10}$ PIB/mL nuclear polyhydrosis virus at 1,000-1,500×. during larval period to kill larvae.

References Wang Pingyuan 1980, Yang Youqian et al. 1982, Xiao Gangrou (ed.) 1992, Forest Disease and Pest Prevention and Treatment and Quarantine Station of Henan Province 2005.

(Yang Youqian, Wang Guicheng)

E. zinckenella adult (Wu Chunsheng, Zhang Runzhi)

E. zinckenella adult (NZM)

356 *Euzophera alpherakyella* Ragonot
Classification Lepidoptera, Pyralidae

Distribution Xinjiang; Russia.
Host *Elaeagnus angustifolia*.
Morphological characteristics Adults Body 10-12 mm in length, 18-25 mm in wingspan; scales on the head gray, labipalp gray; antenna filiform and up to 2/3 of the forewing length; the dorsal of thorax dark red, and venter of thorax greyish white; abdomen greyish white; forewings are divided into three parts by two black wavy cross stripes, the location at 2/3 of costal margin greyish white, the base and end dark red, the middle greyish white, margin hairs greyish white; hind wings and margin hairs greyish white. **Eggs** Oval, 0.6-0.8 mm in length, 0.4-0.5 mm in width; bright red, become dark red before hatching. **Larvae** The newly hatched larval head yellowish brown, body white, the dorsal of the abdomen with a red spot; from the 3rd instar, body dark red, head and pronotum yellowish brown; mature larvae 16-18 mm in length, head 1.6-1.8 mm in width; proleg crochets in complete circle and biordinal. **Pupae** 19-11 mm in length, thorax 2.5 mm in width; light green at beginning, later turn into brown orange. Abdominal end with cremasters.

Biology and behavior *Euzophera alpherakyella* occurs two to three generations a year in Urumuqi and three to four generations a year in Akesu Prefecture. It overwinters as mature larvae in cocoons on the trunk and in 5 cm deep soil within 30 cm from trees. Most larvae overwinter on east side of the trees, least number of larvae are found on the northwest side of trees. Larvae pupate in late March to early April next year.

The 1st generation larvae mainly damage the trunk of Elaeagnus angustifolia, and bore between phloem and xylem. The entrance holes are mostly on rough old bark and at mechanical wound. Larvae prefer to feed on stumps. One to more than ten larvae feed on a site. They usually do not migrate when food is adequate. There are yellowish brown or dark brown granular feces outside of the entrance holes. Larvae have 5 instars, their duration is 4, 4-6, 4-5, 6-7, and 7-9 days, respectively. Field observation shows the larval period is 24-57 days. The 2nd to 3rd generation larvae mainly feed on wounds and cracks on branches and shoots; mature larvae spin cocoons under coarse barks; prepupal stage is 2-3 days.

Adults mostly emerge at 18:00-22:00 in Urumqi. A

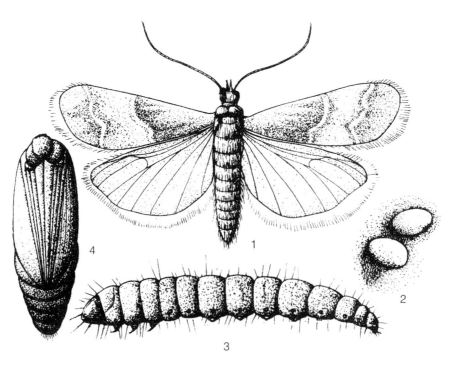

E. alpherakyella. 1. Adult, 2. Egg, 3. Larva, 4. Pupa
(Zhao Jianxia, Wang Yulan)

E. alpherakyella adult
(Wu Chunsheng, Zhang Runzhi)

E. alpherakyella adult
(NZM)

small percentage mate on the same night after emergence, the mating lasts 1 hour and 30 minutes to 3 hours and 50 minutes. They oviposit the next night after mating. A female deposits 41.7% of her eggs on the first night and finishes egg laying in 12 days. Each female lays 14-151 eggs (average is 78.5 eggs). Eggs are laid on the rough barks of trunk, damage site by the larvae, or the scars. Three to ten eggs are oviposited in a small pile. Adults hide in the weeds around the roots or on leaves. They are active at night and are attracted to lights. An unmated female can attract more than 30 males per night.

Prevention and treatment

1. Attract and kill. Collect pupae from the field and let them emerge indoors. Place 2-5 unmated females in each 5 cm diameter and 20 cm long cage; then place a cage over a bowl filled with soapy water to attract and kill males. A cage is able to lure more than 90 males per night.

2. Kill adults using light traps.

3. Conserve natural enemies. The pupal stage parasitoids include Germaria angustata with a parasitization rate of 34.9%-65.8% and Bathyptecles sp. at parasitization rate of 40.0%-55.7%. Larval stage parasitoids include 5 species of Braconidae and 1 species of Chalcididae.

4. Chemical control. Paint 12% thiamethoxam at 1,000× or 50% lamda-cyhalothrin at 100× at the damage site will kill 100% of the larvae.

References Zhao Jianxia et al. 1986, Xiao Gangrou (ed.) 1992.

(Chen Hui, Zhao Jianxia, Wang Yulan)

357 *Euzophera batangensis* Caradja
Classification Lepidoptera, Pyralidae

Distribution Hebei, Fujian, Shanghai, Jiangsu, Zhejiang, Shandong, Hubei, Hunan, Guangdong, Sichuan, Yunnan, Tibet, Shaanxi; Japan.

Hosts *Casuarina equisetifolia*, *Acacia confuse*, *Homalium hainanense*, *Cunninghamia lanceolata*, *Citrus reticulate*, *Eriobotrya japonica*, *Ziziphus jujube*, *Pyrus pyrifolia*, *Pyrus betulaefolia*, *Salix matsudana*, *Salix babylonica*, *Ulmus pumila*, *Robinia pseudoacacia*, *Toona sinensis*, *Populus* spp., *Malus pumila*, *Prunus armeniaca*.

Damage Euzophera batangensis larvae feed between phloem and xylem, causing tissue swelling and bark peeling of Casuarina equisetifolia. Infested trees are susceptible to wind damage. When hundreds of larvae feed around the host phloem, they can cause death of the whole tree.

Morphological characteristics **Adults** Body greyish brown, males and females are similar. Body 5-8 mm in length, 12-15 mm in wingspan. Forewings greyish brown, there is a greyish white wavy cross stripe near the base, in the middle part, and beyond the discal cell; the outer edge of discal cell with black reniform stripe; the outer margin of wings with 5-6 small black spots. Hind wings without stripes and spots, a row of setae present below the discal cell, veins with sparse setae. **Larvae** Body long and thin, about 11 mm in length. Head orange, pronotum reddish brown, thorax and abdomen light brown; body smooth, setae sparse.

Biology and behavior The insect occurs five generations a year in Fujian. It overwinters as larvae in galleries of trunks in mid-December; a small percentage overwinters as pupae. Larvae pupate in early March. Adults appear in early-April. The average longevity of adults is 7.9 days. The average egg stage is 8.54 days; average larval development period is 33 days for the 1st to the 4th generation and 128 days for the overwintering generation; average pupal stage is 8.4 days for the 1st to the 4th generation and 33.8 days for the overwintering generation.

Adults mostly emerge during 18:00-21:00, which represented 82% of total adult emergence in a day. The overwintering generation adults could emerge when temperature is above 13℃. Males mate multiple times. Females start ovipositing next day after mating. Each female lays an average 45.7 eggs. Eggs are laid in bark cracks, on the frass, and at the damage sites from long horned beetles; about 70% of the adults lay eggs at the swollen position of trunks damaged by the larvae.

Larvae have 5 instars. The newly hatched larvae feed on the bark, after the 2nd instar, they bore between phloem and xylem to feed. Two to three larvae often stay together. As they expand their feeding site, hundreds of larvae my feed together. The 1st generation larvae cause greater damage than the other generations.

Prevention and treatment

1. Silviculture. During pruning and other field work, avoid injuring the trunk. The C38, C39, C44 varieties of Casuarina equisetifolia are resistant to Euzophera batangensis. Plant these varieties to minimize Euzophera batangensis occurrence.

2. Manual control. Once discovering swollen tissues and frass on the trunks, scrape off the bark, dig out and

E. batangensis adult (Huang Jinshui)

E. batangensis pupa (Huang Jinshui)

kill the larvae. In early spring, scrape off the warped bark and brush with Bordeaux mixture (1× copper sulphate, 0× calciumoxide and 10× water).

3. Biological control. Apply Beauveria bassiana, Steinernema sp. Beijing and Agriostos strains to control larvae.

References Wang Sizheng, Huang Ju, Song Jizao et al. 1993; Huang Jinshui 1995; Huang Jinshui, Huang Haiqing, Zheng Huicheng et al. 1995; Yang Zhenjiang, Shi Hekui, and Li Yulian 1995; Xu Weidong et al. 2002.

(Huang Jinshui, Tang Chensheng)

E. batangensis damage (Huang Jinshui)

358 *Hypsipyla robusta* (Moore)

Classification Lepidoptera, Pyralidae
Synonym *Hypsipyla pagodella* Ragonot

Distribution Fujian, Guangdong, Guangxi, Hainan, Yunnan; India, Sri Lanka, Australia.

Hosts *Chukrasia tabularis*, *Toona sureni*, *Swietenia macrophylla*, *Swietenia mahogoni*, *Toona microcarpa*, *Spondias lakonensis*, *Acrocarpus fraxinifolius*.

Damage Larvae bore into tender shoots and kill the tender shoots. Infested trees Swietenia trees often fail to grow into tall trees.

Morphological characteristics **Adults** Body 10-13 mm in length, 22-33 mm in wingspan. Whole body covered with greyish brown hairs; ventral surface lighter in color. Forewings greyish brown, with many small spots; the outer and inner cross lines obvious, the outer cross line wave shaped. Hind wings greyish white, the outer margin line gray. **Eggs** 0.67-0.79 mm in length, 0.39-0.47 mm in width. Milky, oval. **Larvae** 15-25 mm in length. Young larvae pink; mature larvae grey or dark red, head bright and dark brown, spiracles red, each segment with blackish brown pinaculum. **Pupae** 10-14 mm in length, 3-4 mm in width; oblong; the dorsal reddish

H. robusta adult (Wu Chunsheng, Zhang Runzhi)

H. robusta adult (NZM)

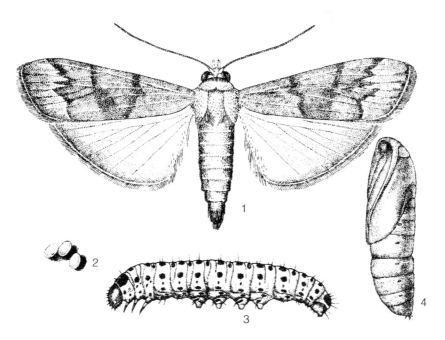

H. robusta. 1. Adult, 2. Eggs, 3. Larva, 4. Pupa
(Zhang Xiang)

brown, segmental venter yellowish brown; spiracles oval, brown; the end with 8 cremasters, the tip of cremasters hook-shaped.

Biology and behavior *Hypsipyla robusta* occurs about 12 generations a year and does not have overwintering phenomenon. When monthly average temperature is 28.5℃, the larval stage is 9-13 days and the pupal stage is 9-11 days; when monthly average temperature is 18.5℃, the larval stage is 18-23 days and the pupal stage is 15-23 days. When the daily average temperature is 29.2℃, the longevity of adults is about 3 days.

Larvae bore into tender shoots. After a larva hollowed one shoot, it migrates to another shoot. The larva chooses a suitable location, spins silk web, then bores into the shoot. The borehole mostly is at base of leaf axils. It takes 25 minutes to 2 hours for a larva to bore into a shoot. The length of the gallery ranges from 2 cm to 28.7 cm. Larvae can migrate to another plant to damage. They are seen migrating both in the daytime and at night. If they don't find a tender shoot, then they gnaw the bark. The damaged tender shoot start withering from the next day. Each larva damages 3-6 shoots. Mature larvae spin cocoons to pupate in the bored shoot.

Adult emergence time is 12:00-21:00. Adults fly fast and do not have phototaxis. They rest during the day in shay places. Eggs are laid singly on tender shoots.

The natural enemies include mantis, spider, ants, Braconidae and parasitic nematodes.

References Guo Bensen 1985, Xiao Gangrou (ed.) 1992.

(Gu Maobin)

359 *Locastra muscosalis* (Walker)
Classification   Lepidoptera, Pyralidae

Importance   *Locastra muscosalis* has wide distribution and strong adaptability. It is able to cause great damage to broad-leaved forests and economic forests such as walnuts and *Toxicodendron verniciflnum*.

Distribution   Beijing, Tianjin, Hebei, Liaoning, Jilin, Zhejiang, Fujian, Jiangxi, Anhui, Jiangsu, Shandong, Henan, Hubei, Hunan, Guangdong, Guangxi, Sichuan, Guizhou, Yunnan, Tibet, Shaanxi, Taiwan; Japan, Vietnam, Laos, Thailand, Myanmar, India, Sri Lanka.

Hosts   Its hosts include 20 species in 9 families: *Choerospondias axillaries*, *Pistacia chinensis*, *Toxicodendron verniciflnum*, *Mangifera indica*, *Rhus chinensis*, *Cotinus coggygria*, *Liquidambar formosana*, *Altingia gracitipes*, *Loropetalum chinense*, *Juglans regia*, *Juglans mandshurica*, *Carya illinoensis*, *Platycarya strobilacea*, *Pterocarya stenoptera*, *Coriaria sinica*, *Acer buergerianum*, *Emmenopterys henryi*, *Alnus cremastogyne*, *Ziziphus jujuba* var. *spinosa*, *Cinnamomum Myanmarnnii*. In Zhejiang it mainly damaged plants of Anacardiaceae, Hamamelidaceae, and Juglandaceae.

Damage   Larvae spin silk and feed on leaves. All leaves may be consumed.

Morphological characteristics   **Adults** Body 13-19 mm in length, 27-43 mm in wingspan. Body reddish brown. Antennae filiform. Forewings chestnut brown, the base dark brown, with a jagged dark brown inner cross line; discal cell with an oval spot consisted of a cluster of dark brown scales; outer cross line brown and wavy; the part between the inner and outer cross lines dark chestnut brown. Hind wings dark brown. The location at 2/3 of the male forewing costal margin with a gland shaped spot. **Eggs** Oval, size 0.8 mm × 0.6 mm; red; egg clutch fish scale shaped. **Larvae** Mature larvae 31-42 mm in length. Head and pronotum dark brown. Dorsal line on abdomen dark brown and wide, subdorsal

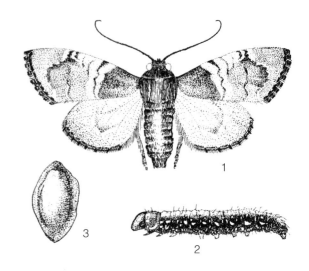

L. muscosalis. 1. Adult, 2. Larva, 3. Cocoon
(Hu Xingping)

L. muscosalis adult (Xu Gongtian)

L. muscosalis adult (Xu Gongtian)

line and spiracular line dark brown mingled with yellowish brown spots; the location below spiracular line brownish yellow to light yellow, pygidium dark brown. **Pupae** 13-19 mm in length, yellowish brown, tail end obtuse. **Cocoons** Brown, flat oval, often with silt attached.

Biology and behavior *Locastra muscosalis* occurs one generation a year in Guizhou, Jiangsu, Shandong, Shaanxi, Hebei and Jilin; two generations a year in Zhejiang, Hunan and Guangxi; three generations a year in Yunnan and Fujian. It overwinters as prepupae in litter or on soil surface under the damaged trees. In Zhejiang, pupation occurs in early April to late May, adults emerge in early May to mid-June. The 1st generation larvae appear in mid-May to early July. They mature in late June to late August and pupate in early July to mid-September pupate in soil. The 1st generation adults emerge in mid-July to mid-October. The 2nd generation larvae appear in late July to mid-October. They leave trees and move to soil from late August. By early November, all overwintering generation larvae crawl into soil or in litter layer, spin cocoons, and turn into prepupae.

Adults emerge at dusk. They are active at night and have phototaxis. Eggs mostly are laid on leaves located on top or outer layer of the crown. Each egg clutch has 100-600 eggs. Adult longevity is 3-5 days. The 1st generation egg stage is 10-11 days; and the 2nd generation egg stage is about 7 days. The newly hatching larvae gregariously crawl around eggshell, and attach leaves with silk. As they grow, they attach more leaves and feed on leaves. After the 5th instar, they leave the aggregation. Mature larvae drop down the ground and bore into 3-8 cm depth soil to spin cocoons. The 1st generation larval stage is 36-43 days, the prepupal stage is 7-17 days; the

L. muscosalis young larva (Xu Gongtian)

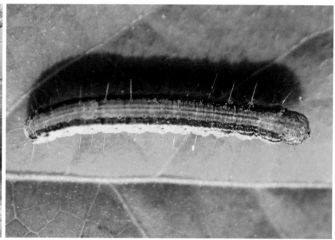

L. muscosalis larva (Xu Gongtian)

L. muscosalis pupa (Xu Gongtian)

L. muscosalis cocoon (edited by Zhang Runzhi)

L. muscosalis mid-aged larva (edited by Zhang Runzhi)

L. muscosalis damage to *Cotinus coggygria* (Xu Gongtian)

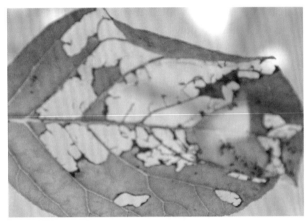
L. muscosalis damage to *Cotinus coggygria* (Xu Gongtian)

L. muscosalis damage to walnut (Xu Gongtian)

L. muscosalis damage to *Juglans mandshurica* (Xu Gongtian)

2nd generation larval stage is 22-23 days, the prepupal stage is up to 7-8 months. The 1st generation pupal stage is 2-26 days and the 2nd generation pupal stage is 11-36 days.

Prevention and treatment

1. Silviculture. Kill the overwintering prepupae by ploughing.

2. Manually remove the silk webs along with the young larvae.

3. Kill adults with light traps.

4. Conserve natural enemies. Parasitoids include Phanerotoma (Phanerotoma) flava, Dolichogenidea sp., Microgaster sp., Amyosoma chinensis, Tropobracan sp., Goniozus japonicus, Villa sp. and Pales pavida. A pathogenic fungus Isaria farinosa may cause 20% mortality of Locastra muscosalis.

5. Spraying 1,000× B.t. at 8,000 IU/uL could limit population density below the economic injury level.

6. Spraying 25% diflubenzuron at 3,000× to control larvae and eggs.

References Chen Hanlin 1988, 1995; Xiao Gangrou (ed.) 1992; Zhang Yuhua, Lai Yongmei, Zang Chuanzhi et al. 2003.

(Chen Hanlin, Ye Jianglin, Ma Guiyan)

360 *Dioryctria yiai* Mutuura et Munroe
Classification Lepidoptera, Pyralidae

Importance *Dioryctria yiai* is a new pest of Masson pine in China. Larval boring inside branch tips and cones results in tip dieback and cone wilt, seriously threatens seed production of Masson pine.

Distribution Hebei, Jiangsu, Zhejiang, Hunan, Guangdong, Sichuan, Shaanxi, Taiwan.

Hosts *Pinus massoniana*, *P. taeda*, and *P. tabulaeformis*.

Damage Larvae feed on male flower tips which results in break off of those flowers. As a result, one to four buds arise from the base of the damaged male buds. They grow to become thin branches. Repeated attack renders the crown to dense and small crown. There is a yellowish white silk leafy cap at the base of damaged cones, with no seeds found inside the empty cavity.

Morphological characteristics **Adults** Body 9.0-12.5 mm long with a wingspan of 20.0-27.0 mm, reddish brown, with a short silver mark at the base of the wing. Median cell contains a sliver crescent at the posterior end, and one wavy silver transverse mark interiorly and exteriorly. There is one cloudy light silver mark near the anterior and posterior margins. **Larvae** Body 13.0-20.0 mm long. Mature larvae black with dark metallic blue. Head reddish brown, pronotum and the first two abdominal nota brown.

Biology and behavior *Dioryctria yiai* occurs one generation a year and overwinters as young larvae inside galleries in the remains of male flowers or cones in Chunan of Zhejiang. Overwintering larvae migrate to and feed in the tips of current year male flowers during the flowering season in early April the following year. A portion of the larval population move into the cones by early May when cones start to increase in sizes. Pupation starts in early May and ends by the end of July. Adults occur between late May and early August. Egg stage of the new generation lasts from early June to mid-August, with first larva appears in mid-June.

Adults emerge mostly during the day, with most emerge between 16:00-20:00. Adults feed on honey for supplemental nutrition and usually attract to lights.

D. yiai adult (Wu Chunsheng, Zhang Runzhi)

D. yiai adult (NZM)

They start to act at twilight with most activities observed around 20:00. Adult longevity is 7-13 days. Most eggs are laid on cones in infested branches. Egg stage lasts for 6-8 days. Newly hatched larvae feed on the dry walls of the old galleries. By 2nd or 3rd instars, larvae gradually abandon those old overwinter galleries and migrate to current-year male flowers on Masson pine or young branch tips of loblolly pine, creating a yellowish white silky cap by constantly pushing feeding sawdust onto the silk web. This cap is used to cover the entrance hole for the larva. Older larvae feed on all parts of the cones except a few scales remains and thin layers of the cone axial, creating gray and stiff dead cones. Larvae crawl swiftly and usually migrate through ballooning with silk threads. Larvae usually pupate inside infested cones, with a few pupae found inside infested branch tips. About 80% larvae pupate, with a pupal stage of 13-25 days.

Dioryctria yiai prefers mature stands than young stands, with heavier infestation usually occur on south-facing slopes compared to north-facing slopes, and plants with more male flowers than those with less male flowers.

Prevention and treatment

1. Trapping. Trapping in seed orchards with vibrating light traps not only provides data for population monitoring, but also destroy a large portion of adult population. Each light trap captures about 2,600 adults every day.

2. Conserving natural enemies. Avoid chemical treatment during adult emergence for dominate natural enemies such as Apanteles sp. and Macrocentrus watanabei between late April and late May, late June and late July, and mid-September and mid-October.

3. Chemical control. Apply 2×10^{10} PIB/mL nuclear polyhydrosis virus at 1,000-1,500×, 2.5% deltamethrin EC or 25 g/L lamda-cyhalothrin at 1,000× during migration for overwinter larvae effectively control larvae population.

References   Zhao Jinnian et al. 1991, 1997, Li Kuansheng 1999, He Junhua et al. 2000.

(Zhao Jinnian)

361 *Propachys nigrivena* Walker
Classification Lepidoptera, Pyralidae

Distribution Zhejiang, Fujian, Jiangxi, Hunan, Guangdong, Sichuan, Yunnan, Taiwan; India.

Host *Cinnamomum camphora*.

Morphological characteristics **Adults** Length 13-16 mm, males 38-44 mm in wingspan, females 48 mm in wingspan. Body bright red; head yellowish brown; compound eyes black. Antennae filiform, yellowish brown; labipalp yellow, closely covered with long black scale hairs, its last segment and the 2nd segment same length. Proboscis long, the base covered with brown scales. The dorsal of thorax dark red. Forewings near square, dark red, veins black, margin hairs red. Hind wings red, the base slight darker, veins not black. Thoracic legs black, the 1st hind tarsus with long hairs. Abdomen black. **Eggs** Diameter 0.5-0.7 mm, oblate, yellowish green. **Larvae** Body 32-37 mm in length, slender. There are two color types. Type 1 is dark brown, the dorsal with a yellowish brown wide stripe, scales on the lateral of body white. Type 2 is yellowish brown with scattered brown spots, scales black, with yellow hairs. Head yellowish brown, with brown spots; thoracic legs well-developed, reddish brown; prolegs short; peritreme dark brown; ventral surface of the body dark brown. **Pupae** 13-16 mm in length, 5-6 mm in width, reddish brown, with many punctures. Head dark brown, compound eyes big, black and convex. Thorax, legs and wings red. Abdomen reddish brown; from the 2nd abdominal segment, the front edge of the dorsal surface with a row of round small punctures; the last abdominal segment black and hardened, two side of the tail end spine-shaped, a long hook present at each spine, with 6 cremasters. **Cocoons** 15-19 mm in length, 9-11 mm in width;

P. nigrivena adult female (NZM)

P. nigrivena adult male (NZM)

oval; silky; very thick; dark brown; soil grains present on cocoons.

Biology and behavior *Propachys nigrivena* occurs three generations a year in Fujian. It overwinters as pupae in soil. Adults emerge in mid- and late April. The 1st generation larvae hatch in late April to early May. They pupate in mid- and late June in soil. Adults emerge in late June to early July. The 2nd generation larvae hatch in mid-July, they pupate from mid-August. Adults appear in late August. The 3rd generation larvae hatch in mid-September. Mature larvae make cocoons and pupate in late October and overwinter.

Prevention and treatment

1. Loosen soil in winter to kill cocoons and reduce the overwintering population size.

2. Kill adults using light traps.

3. In early May, spray 20% deltamethrin at 3,000×, or 25 g/L lamda-cyhalothrin or 15% indoxacarb at 1,000×.

References Wang Pingyuan 1980, Zou Jifu 2000.

(Zou Jifu, Li Yingchao, Li Zhenyu)

362 *Crypsiptya coclesalis* (Walker)

Classification Lepidoptera, Crambidae
Synonym *Algedonia coclesalis* Walker

Distribution Shanghai, Jiangsu, Zhejiang, Anhui, Fujian, Jiangxi, Shandong, Henan, Hubei, Hunan, Guangdong, Guangxi, Sichuan, Shaanxi, Taiwan; Korea Peninsula, Japan, India, Vietnam, Thailand, Myanmar, Indonesia (Java), Borneo.

Hosts *Phyllostachys* spp., *Dendrocalamopsis* spp., *Pleioblastus* spp., *Pseudosasa* spp., *Bambusa* spp., *Dendrocalamus* spp.

Damage Larvae feed on leaves. From the 1st to 4th instar, several larvae often tie current year leaves into pods and feed on leaf surface. Damaged leaves become white and cannot grow. From the 5th instar, each larva ties several new or old bamboo leaves. They feed on whole leaves and cause leaf fall, complete defoliation, and death of bamboo trees which look white color from a distance.

Morphological characteristics **Adults** Males 9-11 mm in length; wingspan 24-28 mm. Females 10-14 mm in length; wingspan 23-32 mm. Body yellow to yellowish brown. Forewing yellow to dark yellow; costa brown; apex and termen form a wide brown band; with three dark brown cross lines with the outer line curved inward and meets the center line. Hindwing paler in color with a wide brown band at the end of the wing and a curved center cross line. Legs silvery white; its exterior surface yellow. Ventral side silvery white.

Eggs Flat oval; size 0.84 mm by 0.75 mm. Newly laid eggs yellow, turning to light-yellow later. The eggs are arranged in fish scale pattern.

Larvae The 1st instar larvae 1.2 mm in length; bluish white. Mature larvae 16-25 mm in length; color varies. Pronotum with 6 black marks; dorsal surface of the meso- and metathorax and each of the abdominal segments with two brown marks. There is also one brown mark above and below the spiracle line on each of the abdominal segment. Larvae have 7-8 instars.

Pupae 12-14 mm in length, orange yellow; raised and forming into two forks at the end. Anal comb with 8 spines and they are located on the two forks; the two spines in the middle are longer than the other spines. Pupae are covered in greyish white elliptical cocoons which are 14-16 mm in length and made of silk and fine soil.

Biology and behavior *Crypsiptya coclesalis* occurs 1-4 generations a year in Zhejiang. Mature larvae pupate inside cocoons. First generation adults lay eggs on young leaves of new bamboo trees. The 1st generation larvae cause most serious damage to bamboo trees. The 3rd and 4th generations cause the least damage. In southern China, all four generations of the insect cause serious damage to *Bambusa textilis* and other similar bamboo species. In Zhejiang, the 1st generation larvae mature in late June to early July. A small percentage of them pupate in pods; the majority make cocoons in soil. A small portion of the larvae pupate and develop into the 2nd generation, the rest of the larvae overwinter or develop into the next generation after a period of diapause. The 2nd

C. coclesalis adult (edited by Zhang Runzhi)

C. coclesalis larva (edited by Zhang Runzhi)

C. coclesalis damage (edited by Zhang Runzhi)

C. coclesalis damage (edited by Zhang Runzhi)

C. coclesalis larva (Xu Gongtian)

C. coclesalis damage (Xu Gongtian)

generation larvae always make cocoons in soil. A small percentage of them pupate. The rest of the 2nd generation larvae overwinter or develop into the next generation after a period of diapause. Similar is true for the 3rd and 4th generations. Adult emergence periods of the 1st to 4th generations are: mid-May to late June, mid-July to mid-August, late August to mid-September, and late September to early October, respectively. The larval occurrence periods of the 1st-4th instars are: late May to late July, late July to early September, early August to mid-October, and early October to early November, respectively.

Prevention and treatment

Conservation of natural enemies. Predatory birds included: *Urocissa erythroryncha*, *Bambusicola thoracica*, *Garrulax canorus*, *Garrulax sannio*, and swallow. Spiders include *Argiope bruennichii*, *Dolomedes pallitarsis*, *D. sulfureus*, *Pardosa tschekiangensis*. Egg predators include *Inocellia crassicornis*, *Chrysopa kulingensis*, and *C. septempunctata*. *Inocellia crassicornis* also predates upon *A. coclesalis* larvae. *Picromerus viridipunctatus*, *Cicindela* sp., *Chlaenius bioculatus*, *Polyrhachis dives*, and *Formica japonica* predate upon *A. coclesalis* larvae and occasionally adults.

Egg parasitoids include *Trichogramma* sp. and *Teienomus* spp. Larvae and pupae parasitoids include *Apanteles* spp., *Macrocentrus* spp., *Aulacocentrum confusum*, *Meteoridea chui*, *Chelonus* spp., *Acropimpla persimilis*, *Goryphus basijaris*, *Xanthopimpla pleuralis pleuralis*, *Xanthopimpla punctate*, *Exorista sorbillans*, and *E. japonica*. Pathogens include *Beauveria bassiana* and *Paecilomyces farinosus*.

(Xu Tiansen)

363 *Circobotys aurealis* (Leech)
Classification Lepidoptera, Crambidae

Distribution Shanghai, Jiangsu, Zhejiang, Anhui, Fujian, Jiangxi, Hubei, Hunan, Guangdong, Guangxi, Taiwan; Korea Peninsula, Japan.

Hosts *Circobotys aurealis* feeds on leaves of bamboo plants in genera *Phyllostachys*, *Pleioblastus*, and *Bambusa*. These include *Phyllostachys heteroclada* var. *pubescens*, *P. sulphurea* 'Viridis', *P. glauca*, *P. iridescens*, *P. makinoi*, *P. bambusoides*, *P. incarnate*, *P. praecox*, *P. vivax*, *P. dulcis*, *P. heteroclada*, *Pleioblastus amarus*, *P. juxianensis*, *Bambusa textilis*, *B. chungii*, *B. vulgaris*, *B. eutuldoides*.

Damage Young larvae feed on the upper surface of the tender bamboo leaves and cause dead spots on bamboo leaves. The older larvae consume whole leaves.

Morphological characteristics **Adults** Adults are dimorphic. Males 11-13 mm long and 29-31 mm in wingspan; females 10-12 mm long and 31-34 mm in wingspan. Head yellow; there are white pubescence between the eyes and frons; antennae filiform, light yellow; body golden yellow; prescutum and scapular covered with long yellow scales; ventral side silvery. Male forewings long and narrow, both wings dark chocolate brown; forewing costal margin and base darker; hind wing base light in color, the margin hairs and outer margin golden yellow. Female forewings slightly long and narrow, covered with thick scales, golden yellow, costal margin and the outer margin darker in color; hind wing light yellow, the outer margin light in color.

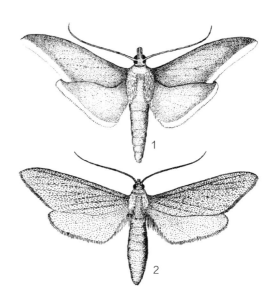

C. aurealis. 1. Male adult, 2. Female adult (Xu Tiansen)

Eggs Oval, 1.2 mm by 1.0 mm in size; milky in color; arranged in single layer egg masses.

Larvae The newly hatched larvae 1.2-1.4 mm in length, milky in color. The mature larvae 25-30 mm in length, light turquoise with slight yellowish color; head light orange yellow, flat; the dorsal line dark green and wide, the subdorsal line ivory; spiracular line thin and milky. Each side of the thoracic segment has a black spot prior to the 4th instar, the pair on the prothorax is the biggest; the black spots on the

C. aurealis male adult (Xu Tiansen)

C. aurealis female adult (Xu Tiansen)

C. aurealis cocoon (Xu Tiansen)   C. aurealis cocoon and pupa (Xu Tiansen)

C. aurealis larva (Xu Tiansen)

mesothorax and metathorax disappear from the 5th instar. The notum of the 8th abdominal segment has three pairs of black spots, the dorsal surface of the 9th abdominal segment has one pair of black spots.

Pupae Length 12-15 mm; yellowish brown; periproct convex and blunt round; 8 cremasters are located on a protuberance, the two cremasters in the middle are two times longer than the other six cremasters.

Cocoons Cylindrical; 26-29 mm in length; leather-like; reddish brown.

Biology and behavior *Circobotys aurealis* occurs one generation a year in Zhejiang province and overwinters as fully developed larvae in cocoons. Pupation occurs during early April and late June next year. Adults start to emerge in late April. Peak adult emergence occurs in mid-May and mid-June. Adults disappear in early July. Eggs are laid during early and mid-May. Larvae appear in mid-May. They become mature either in mid- and late June or early August, then they stop development and overwinter.

Prevention and treatment

Protect natural enemies. The insect has many natural enemies. They include predatory birds such as *Cuculus canorus canorus*, *Garrulax canorus*, *Cyanopica cyana*; spiders such as *Chiracantium zhejiangensis*, *Oxyopex lineatipes*, *Dolomedes pallitarsis*, *Pardosa tschekiangensis*, *Lycosa coelestris*; insects such as *Chrysopa septempunctata*, *Chrysopa Formosa*, and *Picromerus virdiipunctatus*. The parasitoids include *Trichogramma evanescens*, *Apanteles* spp., *Chelonus* spp., *Goryphus basijaris*, and *Xanthopimpla punctate*.

(Xu Tiansen)

364 *Sinibotys evenoralis* (Walker)

Classification Lepidoptera, Crambidae
Synonym *Crocidophora evenoralis* Walker

Distribution Shanghai, Jiangsu, Zhejiang, Anhui, Fujian, Jiangxi, Shandong, Henan, Hubei, Hunan, Guangdong, Guangxi, Sichuan, Shaanxi, Taiwan; Korea Peninsula, Japan, Myanmar.

Hosts *Phyllostachys* plants including *P. heteroclada* var. *pubescens*, *P. bambusoides* f. *shouzhu*, and *P. bissetii*; and *Pleioblastus amarus*.

Damage Larvae weave leaves with silk and feed on leaves. They affect the growth of bamboo shoots and new bamboos.

Morphological characteristics **Adults** Length 9-13 mm and wingspan 26-29 mm; male slight smaller than females; body golden yellow; antennae filiform, light yellow; eyes emerald; wings golden yellow, their margin hairs long and dense, the outer margin with a brown wide stripe, there are 6-7 small black dots between the margin hairs and the outer margin. Forewing costal margin dark in color, three cross lines dark chocolate brown; hind wing with a curved line at the center.

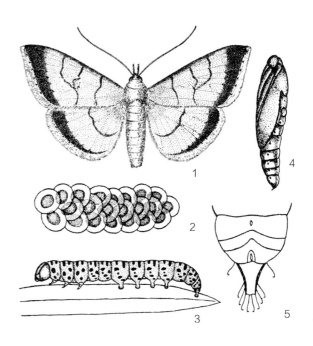

S. evenoralis. 1. Adult, 2. Eggs, 3. Larva, 4. Pupa, 5. Posterior end of pupa (Xu Tiansen)

S. evenoralis adult (Xu Tiansen)

C. evenoralis larva (Xu Tiansen)

C. evenoralis damage (Xu Tiansen)

Eggs Oval, size 1.2 mm by 0.9 mm; light yellow; eggs with an egg mass loosely arranged.

Larvae The newly hatched larvae 1.6 mm in length, milky. The mature larvae 22-30 mm in length, head orange yellow, body light green; the dorsal of each thoracic segment has 3 brown spots, each spot is separated into two parts by the dorsal line, so each segment appears to have 6 spots; the front four spots of prothorax are black; the dorsal of each abdominal segment has 2 brown spots, they are separated into four parts by the dorsal line. There is a big black spot above the front spiracle of prothorax; there are 3 black spots on each side of the mesothorax, metathorax, and abdominal segments. There are 4 black spots on the ventral surface of the 1st, 2nd, 7th, 8th, and 9th abdominal segments.

Pupae Length 13.2-16.5 mm; reddish brown; there are 3 tiny bulges on the cremaster, the middle bulge has two hooks and the other bulges each with one hook. There is a small hook on each side between the cremaster and the last segment; there is one small hook on each side of the cremaster.

Biology and behavior *Crocidophora evenoralis* occurs one generation a year in Zhejiang and Sichuan. It overwinters as the 2nd or 3rd instar larvae among rolled bamboo leaves. Larvae resume activity in February. Each larva attaches three leaves and feeds among them. As it grows, it will make new "leaf pods". Larvae mature and pupate at the end of April. Adults appear during early May and early June. Eggs are laid on new bamboo leaves from mid-May to late May. Young larvae appear in early June. In July the larvae start to estivate in rolled bamboo leaves. This is the only species in the family that overwinters as young larvae.

Prevention and treatment

Protect natural enemies. The predatory birds include *Cissa ergtyrorhyncha*, *Garrulax canorus* and swallow; spiders include *Dolomedes pallitarsis*, *Dolomedes sulfureus*, *Oxyopes sertaus*, *Lycosa pseudoannulata*, and *Lycosa coelestris*; insect predators include *Chrysopa formosa*, *Chrysopa sinica* and *Inocellia crassicornis*, *Chlaenius bioculatus*, *Polyrhachis dives*, and *Formica japonica*. Parasitoids include *Trichogramma* and *Teienomus* spp. on eggs, and *Apanteles* spp., *Meteoridea chui*, *Chelonus* spp., *Xanthopimpla punctate*, *Coccygomimus disparis*, *Acropimpla persimilis* on larvae and pupae.

Reference Xu Tiansen and Wang Haojie 2004.

(Xu Tiansen, Wang Haojie)

365 *Demobotys pervulgalis* (Hampson)
Classification Lepidoptera, Crambidae

Distribution Jiangsu, Zhejiang, Anhui, Fujian, Jiangxi, Hubei, Hunan.

Hosts *Phyllostachys heteroclada* var. *pubescens*, *P. glauca*, *P. iridescens*, *P. praecox*, *P. vivaxthe*.

Damage Larvae feed on leaves and may consume all leaves on a plant.

Morphological characteristics **Adults** Males 8-10 mm long, wingspan 22-26 mm; females 8-11 mm long, wingspan 24-28 mm. Light yellow to yellowish white, venter silvery; head light yellow, compound eyes emerald, antennae filiform and light yellow. Both wings light yellowish white, margin hairs long, 6-7 yellowish brown dots present along the margin hairs. Forewing with three cross lines; the outer line wavy, its lower half not connected with the mid-line; the mid-line light grey. Hind wing mid-line curved, light grey. Legs yellowish white; tibiae with brown rings.

Eggs Elliptical, 1 mm × 0.8 mm in size; milky to light yellow; loosely arranged within each egg mass.

Larvae Newly hatched larvae 1.2-1.4 mm in length, milky. Mature larvae 17-24 mm in length and light yellow to yellowish brown. Head orange yellow; pronotum with 6 black spots; the dorsal of each abdominal segment has two brown spots, the rear spot is separated into three by the subdorsal lines; there is a brown spot above and below each spiracle.

Pupae Length 12-14 mm; orange red in color; cremaster has 8 hooks which are arranged in arc shape.

Biology and behavior *Demobotys pervulgalis* occurs one generation a year in Zhejiang. It overwinters as fully developed larvae (prepupae) in fallen bamboo shoot shells on ground. Pupation occurs during late April and last May next year. Pupal period is 10 days. Adults emerge from late May. Eggs are laid from early June on back of new leaves. Egg incubation period is about ten days. Larvae occur from early June to early August. Larval period is 30-44 days. They spin silk and fall to ground after they mature. They look for ground covers and overwinter within them.

Prevention and treatment

Protect natural enemies. *Demobotys pervulgalis* has many natural enemies. The predatory birds include *Bambusicola thoracica*, *Garrulax canorus* and swallow; spiders include *Oxyopes sertaus*, *Lycosa pseudoannulata*, *Dolomedes pallitarsis*, *Oxtopes macilentus*, and *Clubiona maculate*; insect predators include *Chrysopa kulingensis*, *Chrysopa formosa*, *Chrysopa septempunctata*, *Camponotus japonicus*, *Formica japonica*, carabid beetles, *Cicindela* sp., *Rodolia rufopilosa*, *Adalia bipunctata*, *Propylaea japonica* and *Harmonia axyridis*, *Hierodula patellifera*, *Arma chinensis*; parasitic insects include *Trichogramma dendrolimi*, *T. evanescens*, *Teienomus* spp.

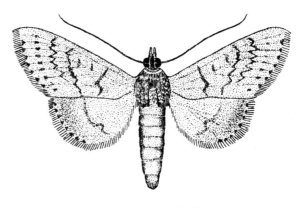

D. pervulgalis adult (Xu Tiansen)

D. pervulgalis 3rd instar larva (Xu Tiansen)

D. pervulgalis pupa (Xu Tiansen)

D. pervulgalis larva (Xu Tiansen)

D. pervulgalis pupa infected by Beauveria bassiana (Xu Tiansen)

D. pervulgalis adult (Xu Tiansen)

D. pervulgalis mature larvae (Xu Tiansen)

D. pervulgalis damage (Xu Tiansen)

which parasitize eggs, and *Apanteles* spp., *Macrocentrus* spp., *Chelonus* spp., *Xanthopimpla stemmator* which parasitize larvae. Their total parasitism rate is about 35%. *Beauveria* sp. caused 5%-27% mortality of the larvae.

Reference Xu Tiansen and Wang Haojie 2004.

(Xu Tiansen, Wang Haojie)

366 *Eumorphobotys obscuralis* (Caradja)

Classification Lepidoptera, Crambidae
Synonym *Calamochrous obscuralis* Caradja

Distribution Jiangsu, Zhejiang, Anhui, Fujian, Jiangxi, Hubei, Hunan, Guangdong, Guangxi, Sichuan, Yunnan; Japan.

Hosts *Phyllostachys bambusoides*, *P. bambusoides* f. *lacrima-deae*, *P. bissetii*, *P. bambusoides* f. *shouzhu*, *P. dulcis*, *P. flexuosa*, *P. heteroclada* var. *pubescens*, *P. sulphurea* 'Viridis', *P. glauca*, *P. iridescens*, *P. praecox*, *P. nuda*, *P. makinoi*, *P. praecox* f. *provernalis*, *P. propinqua*, *P. heteroclada*, *P. amarus*, and *Bambusa textilis*.

Damage Young larvae feed on lower epidermis of bamboo leaves, causing white spots on leaves. From the 4th instar, larvae consume whole leaves.

Morphological characteristics **Adults** Males 12-14 mm in length, 30-35 mm in wingspan; females 13-16 mm in length, 34-37 mm in wingspan. Body greyish yellow, venter silvery; compound eyes emerald when live and black after death; antennae filiform, light yellow. Hairs on scutum of the prothorax long. Male forewings greyish yellow, hind wings ash black, margin hairs yellow; female forewings yellow, margin hairs short, a yellowish brown line present between the margin hairs and the outer margin. Hind wings light yellow and with two ash black spots. Both wings with rosy or yellow luster. Hind tibia with a pair of inner spurs, the outside with one spur which is one third as long as the inner spur. **Eggs** Flat oval, size 1.7 mm × 1.2 mm; milky; arranged in a single layer clutches similar to fish scales. **Larvae** The newly hatched larvae 2 mm in length; milky. Mature larvae 30-35 mm in length; light green; head light orange yellow; the dorsal line dark green, wide, the two sides with light yellowish white lines; spiracle line thin and light yellowish white. A black spot present on the side of each thoracic segment. The dorsal of the 8th abdominal segment with six small black spots arranged in a triangle. **Pupae** 19-21 mm in length, reddish brown; the end with 8 cremasters on a protrusion, the two cre-

E. obscuralis adult male (above) and adult female (below)
(Xu Tiansen)

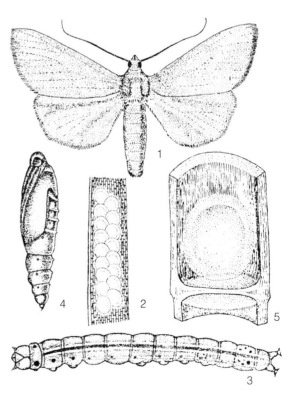

E. obscuralis. 1. Adult, 2. Eggs, 3. Larva, 4. Pupa, 5. Cocoon
(Xu Tiansen)

masters in the middle are longer. **Cocoons** Round, silky, white or ivory; the diameter 32-40 mm; the 1st and 2nd generation cocoons thin, the overwintering generation cocoons thicker.

Biology and behavior *Eumorphobotys obscuralis* occurs two to three generations a year in Zhejiang province. Those occurring two generations a year overwinter as mature larvae in the remnant sheaths of bamboo shoots on the base or the stalk of bamboo plants. They pupate at the end of April. Those insects occurring three generations a year overwinter as larvae on standing bamboo leaves. They resume feeding in April and pupate in mid- and late May. Adults emerge in early May to late May. The 1st generation egg stage is in mid-May to mid-June, larval stage is late May to late July, pupal stage is in early July to mid-August, adult stage is in late July to late August; the 2nd generation egg stage is in late July to late August. Some eggs hatch from early August to the end of August, larvae pupate in early September, adults emerge in early September to the end of September. The

E. obscuralis pupa and cocoon (Xu Tiansen)

E. obscuralis larva killed by parasitoid (Xu Tiansen)

E. obscuralis larva being preyed by *Arma chinensis* (Xu Tiansen)

E. obscuralis young larva under silk cover (Xu Tiansen)

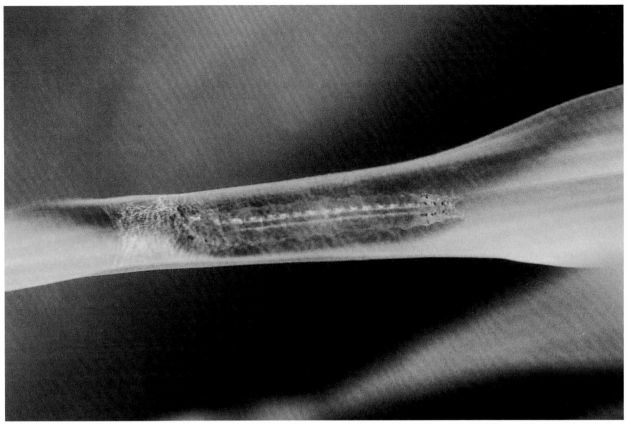

E. obscuralis mature larva (Xu Tiansen)

3rd generation larvae appear from mid-September and overwinter in November. Those late-hatching 2nd generation larvae will not develop into the 3rd generation. They make cocoons in early October and overwinter.

Prevention and treatment Protect natural enemies. The predatory birds include *Cissa ergtyrorhyncha, Garrulax canorus, Pycnonotus sinensis, Eudynamysscolopacea*; spiders include *Chiracantium zhejiangensis, Araneus mitificus, Pardosa tschekiangensis, Dolomedes sulfureus, Lycosa coelestris*; insects include *Chrysopa sinica, C. septempunctata, Polyrhachis dives, Hierodula patellifera*, carabid beetles, *Chlaenius bioculatus, Sirthenea flavipes, Isyndus obscurs, Arma chinensis, Picromerus virdiipunctatus*. The parasitic insects include *Trichogramma dendrolimi, T. evanescens, Apanteles* spp., *Rhogas* sp., *Meteoridea chui, Chelonus* spp., *Acropimpla persimilis* and *Goryphus basijaris*.

References Wang Pingyuan 1980, Xu Tiansen and Wang Haojie 2004.

(Xu Tiansen, Wang Haojie)

367 *Botyodes diniasalis* (Walker)
Classification Lepidoptera, Crambidae

Distribution Hebei, Shanxi, Liaoning, Shandong, Henan, Guangdong, Taiwan; Japan, Korea Peninsula, India, Myanmar.

Hosts *Populus* spp., *Salix* spp.

Damage Botyodes diniasalis tie leaves into pods or fold leaves and consume all leaves on infested branches.

Morphological characteristics **Adults** Length 13 mm, wingspan 30 mm. Forewing and hindwing golden yellow with brown wavy markings. The end of the median cell of the forewing with a brown circular marking; its center is white. **Larvae** Length 15-22 mm, greyish green, each side with a light-yellow longitudinal band along the spiracles.

Biology and behavior *Botyodes diniasalis* occurs 4 generations a year in Henan province. Larvae overwinter in cocoons among fallen leaves, under ground cover, or in bark cracks. They resume activity in early April and pupae in late May and early June. Adults emerge in early June. The peak emergence period of the 2nd to the 4th generation adults are mid-July, mid-August, and mid-September, respectively. Larvae start to overwinter in late October.

Adults are active during the night and are attracted to lights. They lay eggs on back of leaves. Eggs are arranged in a mass or rectangular shape, each egg mass with 50-100 eggs. Young larvae feed on leaf epidermis, then tie leaves into pods or fold leaves and feed inside the rolled leaves. Mature larvae congregate on branch tips and feed on young leaves. They may consume all young leaves on branch tips within 3-5 days. Peak infestation period is during July-August. Larvae jump out the rolled leaves or fall out via a silky line when disturbed. Mature larvae pupate in white cocoons inside the rolled leaves.

Prevention and treatment

1. Light trapping. Black light traps are effective for

B. diniasalis adult (Wu Chunsheng, Zhang Runzhi)

B. diniasalis adult (Xu Gongtian)

B. diniasalis larvae with silk (Xu Gongtian)

B. diniasalis mid-aged larva (Xu Gongtian)

B. diniasalis larva on damaged young shoot (Xu Gongtian)

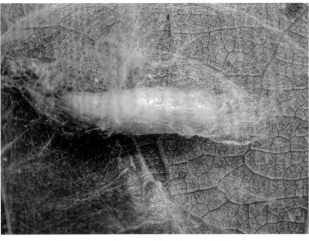
B. diniasalis pupa and cocoon (Xu Gongtian)

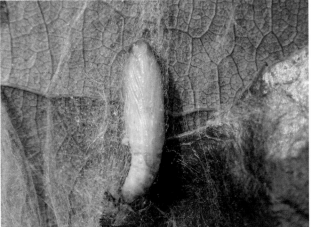
B. diniasalis pupa (Xu Gongtian)

B. diniasalis larva infected with entomopathogenic pathogen (Xu Gongtian)

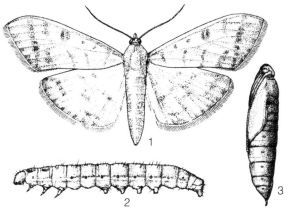
B. diniasalis. 1. Adult, 2. Larva, 3. Pupa (Zhang Xiang)

attracting and killing adults.

2. Biological control. Conserve or propagation of *Trichogramma chilonis*. Apply *Beauveria bassiana* (4×10^{11} conidia/g) at 2,000-3,000×. Spray 2×10^{10} PIB/mL nuclear polyhydrosis virus at 1,000-1,500×.

3. Chemical control. Spray 50% sumithion at 1,000×, or 25% Dimilin at 2,000× during larval periods. Apply 5.7% emamectin at 3,000×, 20 g/L lamda-cyhalothrin at 1,000-1,500×, 50% imidacloprid-monosultap at 600-800×, or 45% profenofos-lamda-cyhalothrin at 1,000-1,500×.

References Yang Youqian et al. 1982, Xiao Gangrou (ed.) 1992, Forest Disease and Pest Quarantine and Control Station of Henan Province 2005.

(Yang Youqian)

368 Cydalima perspectalis (Walker)

Classification Lepidoptera, Crambidae
Synonym *Diaphania perspectalis* (Walker)
Common name Box tree moth

Distribution Beijing, Hebei, Liaoning, Shanghai, Jiangsu, Zhejiang, Fujian, Jiangxi, Shandong, Hubei, Hunan, Guangdong, Sichuan, Guizhou, Tibet, Shaanxi, Qinghai; Japan, Korea Peninsula, India.

Hosts *Buxus sinica*, *Buxus bodinieri*, *Buxus microphylla* var. *koreana*, *Ilex chinensis* and *Euonymus alatus*.

Damage Larvae feed on leaves, young shoots and affect tree growth. They may kill *Buxus trees* when at high population density.

Morphological characteristics **Adults** Length 20-30 mm, wingspan 30-50 mm; head dark brown, scales between antennae white; antennae brown; the 1st segment and the lower half of the 2nd segment of the labial palp white, the upper half of the 2nd segment of the labial palp dark brown; the 3rd segment dark brown. Prothorax, the costal margin, outer margin and posterior margin of forewings and the outer margin of hind wings have a dark brown wide stripe. There are two white spots near costal margin of the forewing, the spot near the outer margin crescent shape, the rest of the wing white, semitransparent. Abdomen white, the end covered with dark brown scales. The end of male abdomen has dark brown tufts of hairs, with one frenulum. Female abdomen slightly larger, without tufts of hairs at the end; with two frenulum. **Eggs** Flat elliptical, size 1.5 mm ×1 mm; bottom flat, the top slightly convex; red violet, dark brown before hatching. **Larvae** Average length of the 1st to 6th instar larvae is: 2.94±0.24, 4.91±0.52, 7.42±1.34, 11.38±1.40, 18.17±1.43 and 34.3±4.20 mm, respectively. The average head width of each instar is: 0.28±0.03, 0.56±0.06, 0.96±0.07, 1.33±0.11, 2.05±0.12 and 2.99±0.11, respectively. Mature larvae head dark brown, thorax and abdomen yellowish green. Dorsal line dark green, subdorsal line and supraspiracular line dark brown, spiracle line light yellowish green; area between the subdorsal line and spiracle line green greyish white. The dorsal of the mesothorax and metathorax each has a pair of dark brown granules. The dorsal of each abdominal segment has two pairs of dark brown granules, the front pair conical, close to each other; the rear pair elliptical and more separated. The side of each segment has a dark brown round granule and setae. **Pupae** 18-20 mm in length; pupae yellowish white, the end of abdomen dark brown, with a row of 8 cremasters.

Biology and behavior *Cydalima perspectalis* occurs three generations a year in Shanghai. It overwinters as the 3rd to 4th instar larvae in weaved leaves. They resume feeding in mid- and late March. Pupation occurs from early May. Adults emerge in late May to early June. The egg stage of the first generation is in late May to late June. The damage period of the 1st generation larvae is in late May to early July. Larvae pupate in late June to

C. perspectalis adult (uniform color) (edited by Zhang Runzhi) *C. perspectalis* adult (bi-color) (edited by Zhang Runzhi)

mid-July, the 1st generation adults emerge in July. The 2nd generation larval damage period is in early July to early August. Pupae and adults appear in late July to early August. The 3rd generation larvae appear in late July and start to overwinter in mid-September. Some individuals occur two generations a year. The 2nd generation larvae undergo diapause in August. Some individuals occur one generation a year. The larvae start diapause in June.

Adults mostly emerge during 19:00-23:00. It only takes about five minutes from the start of pupae contraction to adult emergence. Adults can fly one hour after eclosion. They hide during the day and start activity from 18:00 and reach peak activity during 19:00-23:00. They have weak phototaxis. Adults require supplementary feeding. They start mating about 40 hours after emergence. Each mating lasts 1.5-2 hours. Each female and male only mate once. Females lay eggs from 36 hours after mating.

Eggs mostly are laid on back of the leaves. They are arranged in one layer clutches, sometimes eggs are laid singly. Each egg clutch usually has less than 12 eggs. Each female lays 148-674 eggs (average 330.2 eggs). In *Buxus sinica* and *B. bodinieri* mixed landscape, adults prefer lay eggs on *B. bodinieri*. When the daily mean temperature was 26.9°C, the 1st generation adults lived 3-16 days (average 8.2 days). The egg stage for the 1st and 2nd generation was 5.5 days, the egg stage for the 3rd generation was 3 days.

Young larvae feed on leaves and create holes on leaves. They can consume whole leaves and damage young shoots from the 4th instar. Larvae go through 6 instars, few will have 5 or 7 instars. The total feeding is 8,981±1,874.1 mm2 during larval stage or 44.92±9.74 *Buxus sinica* leaves. The larvae stage of the 1st to the 3rd generation is 40 days, 20 days, and 9 months, respectively. The pupal stage of the 1st generation is 10-12 days and that for the overwintering generation is 12-17 days. Mature larvae mostly spin leaves and make cocoons. From mid-September, the 3rd generation larvae spin silk and attach two or three leaves into a pouch. The larvae build thin cocoons and overwinter. They molt once inside the cocoons.

Prevention and treatment

1. Remove overwintering larvae during winter pruning.

2. Place light traps to attract and kill adults.

3. Protect natural enemies. *Chelonus* sp. and *Dolichogenidea stantoni* often have high parasitism rates.

4. Biological control. Apply Beauveria bassiana

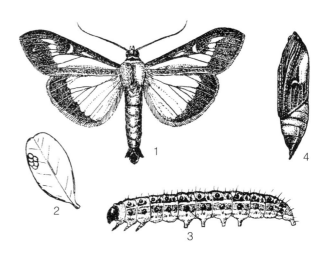

C. perspectalis. 1. Adult, 2. Eggs, 3. Larva, 4. Pupa
(Tang Shangjie)

C. perspectalis adult (Xu Gongtian)

C. perspectalis adult (Xu Gongtian)

C. perspectalis mature larva (Xu Gongtian)

C. perspectalis overwintering larva (Xu Gongtian)

C. perspectalis young larva (Xu Gongtian)

C. perspectalis pupa (Xu Gongtian)

C. perspectalis damage to Buxus bodinieri (Xu Gongtian)

(4×10^{11} conidia/g) at 2,000-3,000×. Spray 2×10^{10} PIB/mL nuclear polyhydrosis virus at 1,000-1,500×.

5. Chemical control. Apply 1,000× diflubenzuron or 40% chlorpyrifos at 1,500×. Apply 5.7% eamamectin at 3,000×, 20 g/L lamda-cyhalothrin at 1,000-1,500×, 50% imidacloprid-monosultap at 600-800×, or 45% profenofos-lamda-cyhalothrin at 1,000-1,500×.

References Xiao Gangrou (ed.) 1992; Chen Hanlin, Gao Zhanggui, Zhou Jianmin et al. 2005; Xu Gongtian and yang Zhihua 2007.

(Tang Shangjie, Chen Hanlin, Zhao Jinnian, Yang Xiuyuan)

Conogethes punctiferalis (Guenée)

Classification	Lepidoptera, Crambidae
Synonym	*Dichocrocis punctiferalis* Guenée
Common names	Peach pyralid, Yellow peach moth

Distribution Beijing, Shanxi, Jiangsu, Zhejiang, Jiangxi, Shandong, Henan, Hubei, Hunan, Sichuan, Shaanxi; Korea Peninsula, Japan, India, Vietnam, Myanmar, Sri Lanka, the Philippines, U.S.A.

Hosts It damages primarily the fruits and seeds of many plants such as *Castanea mollissima*, *Zizyphus jujube*, *Prunus persica*, *Prunus*, *Diospyros kaki*, *Citrus reticulate*, *Eriobotrya japonica*, *Malus pumila*, *Pyrus bretschneideri*, *Punica granatum* and *Crataegus pinnatifida*.

Morphological characteristics Adults 10-14 mm in length, 25-30 mm in wingspan. Body and wings bright yellow. Wings, thorax and abdomen with black spots. Thoracic mid-patagium with a black spot consisted of black scales, the outside of front end and near central of scapula with a black spot, the dorsal surface of thorax with two black spots. Forewings near triangular, with 21-28 black spots; hind wings slightly fan-shaped, with 15-16 black spots. The 1st, 3rd, 4th and 5th abdominal segment each with three black spots, the 6th segment only with one black spot, the 2nd and 7th segment without black spots; the end of the 8th segment black; this characteristics in females sometimes isn't obvious while in males is extremely obvious. Antennae filiform and about half length of the forewing. Compound eyes well developed. Labial palp curled upward, sickle shaped, closely covered with yellow scale hairs; its dorsal surface of the front half with black scale hairs. Proboscis well developed. **Eggs** Oval, 0.6-0.7 mm in length. Newly laid eggs milky, later turn dark red, then reddish-brown before hatching. **Larvae** Mature larvae around 20 mm in length. Body color vary, head dark chocolate brown, pronotum brown. Pygidium greyish brown. Proleg crochets biordinal and penellipse. From the 3rd instar, male larvae have two dark brown sexual glands under the greyish brown spot on the dorsal surface of the 5th abdominal segment. Female larvae do not show sexual glands. **Pupae** 13-15 mm in length, about 4 mm in width. Body yellowish brown. Mandible, mid-leg and antenna longer than 1/2 of the 5th abdominal segment. The end of the abdomen with 6 curly cremasters. **Cocoons** Greyish-white, surface attached with wood dusts.

Biology and behavior *Conogethes punctiferalis* occurs 1-2 generations a year in Liaoning, 3 generations a year in Hebei, Shandong and Shaanxi, 4 generations a year in Henan, Chongqing, Zhejiang and Jiangsu, 4-5 generations a year in Jiangxi and Hubei. In Zhejiang, those feeding on leaves occur 2 generations a year and overwinter as 4th instar larvae in pods made of leaves. They pupate next year during late April and early May. Adults emerge during early and mid-May. Those feeding on wood occur 3 generations a year and overwinter as pupae inside feeding galleries, under the bark or in *Castanea mollissima* nut piles. Adults emerge next year from late April to early. The 1st and 2nd generation larvae appear from June to July and have a habit of feeding on nectar. Eggs are laid singly or in groups of 2-5 eggs per group. On *Pinus massoniana*, eggs are laid on the needles of shoots. Each shoot may have 5-19 eggs. On *Castanea mollissima*, eggs are laid among the spines of the involucre. The average egg period for the 1st and 2nd generation is 12 days and 8 days, respectively. The 1st generation larvae pupate in late July and pupal stage is 14-18 days. Adults emerge during early and mid-August. The 2nd generation larvae are gregarious. They feed inside pods of weaved leaves on pine shoots. They attach frass to the leaf pods as they feed.

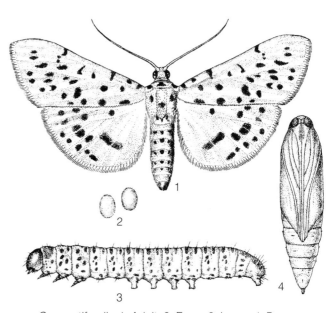

C. punctiferalis. 1. Adult, 2. Eggs, 3. Larva, 4. Pupa

(Zhang Xiang)

C. punctiferalis adult (Xu Gongtian)

C. punctiferalis adult (edited by Zhang Runzhi)

It occurs 4-5 generations a year in Wuhan area of Hubei. Larvae of the overwintering generation pupate in mid-April. Adults emerge from early May to early June. They hide during the day and feed at night, and have some phototaxis. The eggs of the 1st to 5th generations are laid from early May to early June, mid-June to late July, late July to mid-September, late August to late September, and mid-September to late October, respectively. The egg stage ranges from 3 to 10 days. When larvae bore into peach fruits, they secrete yellow transparent sticky material. They bore into pine shoots and result in withering of the shoots. Larvae mature 15-18 days after they bore into the host. They spin silk and make white cocoons inside fruits or under the trees.

Prevention and treatment

1. Timely clean up the fallen fruits and remove the damaged fruits from the trees. Dispose sunflower remains in winter. Burn the involucre of Chinese chestnut and kill the overwintering larvae.

2. During the adult stage, use black light lamp or sugar vinegar mixture to trap adults.

3. Beside the Chinese chestnut plantation, plant *Helianthus annuus* and *Zea mays* to attract adults to oviposit. The eggs on *H. annuus* flowers and *Z. mays* tassels can be killed with insecticides.

4. Attract and kill adults using sex pheromone traps. The mixture of E10-16:Ald, Z10-16:Ald, and 16:Ald at the ratio of 80.4:6.6:13 is attractive to adults.

5. Spray 1,000× 50% phoxim or pyrethroids to kill adults. During the egg hatching period, spray 90% trichlorfon at 300× or 50% sumithrin EC at 500×. Apply 5.7% emamectin at 3,000×, 20 g/L lamda-cyhalothrin at 1,000-1,500×, 50% imidacloprid-monosultap at 600-800×, or 45% profenofos-lamda-cyhalothrin at 1,000-1,500×.

References Xiao Gangrou 1992, Du Guoyu and Gao Limei 2003, Xue Zhicheng 2006.

(Li Zhenyu, Wei Xueqing, Yang Xiuyuan)

370 *Sinomphisa plagialis* (Wileman)

Classification Lepidoptera, Crambidae
Synonym *Omphisa plagialis* Wileman

Distribution Beijing, Hebei, Shanxi, Liaoning, Jiangsu, Zhejiang, Shandong, Henan, Hubei, Hunan, Sichuan, Guizhou, Yunnan, Shaanxi, Gansu; Korea Peninsula, Japan.

Hosts *Catalpa bunger, C. fargesii, C. duclouxii, C. ovate.*

Morphological characteristics **Adults** Body 15 mm in length, 36 mm in wingspan. Body greyish white, the edge of head, thorax, and abdomen slightly brownish; wings white; forewing base with dark brown tooth-shaped double lines; the inner cross line dark brown, the discal cell with a black spot at the inner and outer end; the location below discal cell with a near rectangular big spot; two dark brown wavy stripes present near the outer margin; margin hairs white. Hind wing with a dark brown cross line. **Eggs** Oval, about 1 mm in length, 0.6 mm in width; the newly produced eggs milky, later turn into red and transparent; surface with many dentation. **Larvae** Mature larvae about 22 mm in length; greyish white, pronotum dark brown and divided into two parts; body with black pinaculum. **Pupae** About 15 mm in length; fusiform; yellowish brown.

Biology and behavior *Sinomphisa plagialis* occurs two generations a year in Beijing, Shandong and Henan; two to three generations a year in Jiangxi. It overwinters as mature larvae, or less commonly as large instar larvae. Larvae resume activity in late March, pupate between early April and early June. Adults start to appear between early April. Peak adult emergence period is between end of April and early May. The 1st generation larvae appear in early May. The 2nd generation larvae start to appear from early July to mid-August. Damage from larvae occurs from early May to mid-October. Adults mostly emerge at night, mate at the next night, and feed on nectar for nutrition. Eggs are laid individually at the old leaf scars on limbs, in bark cracks and on the base of new shoots. After hatching, larvae crawl to 1- 3 cm below the tip of shoots and bore into shoots. The period when larvae are exposed is about 2 hours. Larvae feed downward and cause the new shoots to wither. After the new shoots lignify or semi-lignify, spindle-shaped galls form. Some larvae bore into the leaf stalks and pods. A larva can damage a few shoots and within a shoot can have more than one larvae, but individuals are separated by silk membrane. Larvae go through 5 instars. Mature

S. plagialis adult (Xu Gongtian)

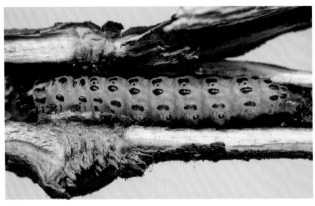
S. plagialis overwintering larva (Xu Gongtian)

Migrating S. plagialis larva (Xu Gongtian)

S. plagialis adult emergence hole (Xu Gongtian)

S. plagialis tunnel (Xu Gongtian)

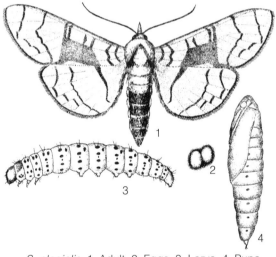
S. plagialis. 1. Adult, 2. Eggs, 3. Larva, 4. Pupa (Zhang Peiyi)

A hole made by S. plagialis larva (Xu Gongtian)

larvae pupate in think cocoons within shoots.

Prevention and treatment

1. Clip off the infested branches during autumn and fall seasons and destroy them.

2. Plug larvae entrance holes with cotton balls soaked with DDVP.

References Guo Congjian, Shao Liangyu, and Yin Wanzhen 1992; Qi Chengjin and Geng Bingtian 1992; Shao Fengshuan and Pang Yongshi 1997; Zhang Zhizhong 1997; Yang Yufa and Liu Zhandong 1999; Zhang Cunli and Li Hongyan 2006.

(Dai Qiusha, Li Zhenyu, Guo Congjian)

371 *Paliga machoeralis* Walker

Classification Lepidoptera, Crambidae
Common name Teak leaf skeletonizer

Importance It is able to damage a variety of fruit trees and is an important pest of coastal forest, *Casuarina equisetifolia,* in southern China.

Distribution Guangdong, Guangxi, Hainan, Yunnan, Taiwan; India, Sri Lanka, Pakistan, Thailand, Myanmar, Malaysia, Indonesia, Australia.

Hosts *Tectona grandis, Callicarpa macrophylla, Callicarpa nudiflora, Casuarina equisetifolia.*

Damage Larvae feed on leaves of hosts.

Morphological characteristics **Adults** Body 10-12 mm in length, 20-25 mm in wingspan. Forewings light yellow, with serval reddish brown wavy stripes, wing margin reddish brown; hind wings light yellow, outer margin reddish brown. Female abdomen short and thick, tail end blunt. Male abdomen long and thin, tail end sharp, abdominal tail end is seen from the dorsal view. **Eggs** Flat oval; milky; the surface with reticulate pattern; 0.63-0.90 mm in length, 0.39-0.61 mm in width. **Larvae** Mature larvae 20-24 mm in length, head about 1.6 mm in width. Upper portion of the spiracular line dark green, lower portion of the spiracular line light green. The dorsal of each abdominal segment with 4 black verrucae arranged in a rectangle; a light yellow spot present on each verruca. **Pupae** 11.5-14.2 mm in length; reddish brown, compound eyes black; the end with 8 cremasters, four of them are longer than the others.

Biology and behavior It occurs 11-12 generations a year in Jianfengling of Hainan. The egg stage is 2 to 3 days, the larval stage is 10-14 days, the prepupal stage is 2 days and the pupal stage is 7-8 days. The period from egg to adult is between 22 and 37 days.

Larvae mostly hatch before 8:00. The 1st and 2nd instar larvae spin silk on back of leaves and feed on the epidermis under the web. The 3rd instar larvae start moving to the upper side of leaves. They pull leaves into concave shape, feed on mesophyll under the web and leave the veins untouched. Each larva creates an escape hole behind its body on the leaf. Once disturbed, it would back

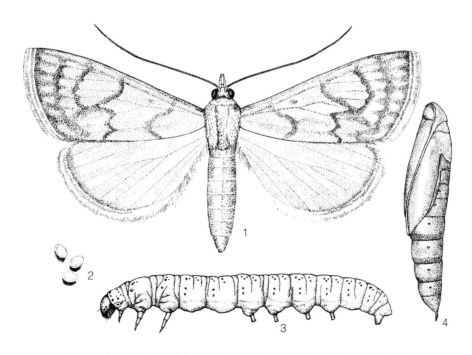

P. machoeralis. 1. Adult, 2. Eggs, 3. Larva, 4. Pupa (Zhang Xiang)

P. machoeralis adult (NZM)

out and escape from the hole or spin silk to drop down through the hole. The total leaf area consumed per larva is 3,044-4,507 mm^2. The amount of feeding on tender leaves is bigger than that on old leaves. Spatial distribution of larvae within the crown is: middle part -59.9%, the bottom -24.8%, and the top -15.3%. Larvae have 5 instars.

Mature larvae create cocoons and pupate within cocoons. Pupation place is mainly on leaves of the host, followed by leaf litter. Adults mostly emerge at night and start mating during the next night. Each adult only mates once. The female lays eggs the next day after mating. Eggs mostly are laid dispersedly on back of leaves. Oviposition period is 3-9 days. Each female lays 107-500 eggs based on laboratory observations. Adults rest on ground cover or on weeds during the day. They have phototaxis. The longevity of females and males is 5-26 days and 7-16 days, respectively. In Hainan, population peaks in July-August when food is most abundant.

Prevention and treatment

1. Biological control. Applying *B.t. galleriae* at 3×10^8 spores/mL and 500 g per hectare reduced population by 80%-90% in 3-4 days.

2. Chemical control. Applying 10% diflubenzuron at 1,500× reduced population by 93.39% after 6 days.

References   Wang Pingyuan 1980, Wu Shixiong, Chen Zhiqing, and Wang Tiehua 1979; Zhang Huaxuan 1980; Zhang Weiyao, Wang Xiaotong, and Xie Daotong 1987; Xiao Gangrou (ed.) 1992.

(Wen Xiujun, Chen Zhiqing, Wu Shixiong)

372 *Cyclidia substigmaria* (Hübner)

Classification Lepidoptera, Drepanidae
Synonym *Enchera substigmaria* Hübner

Importance It is a main defoliator of *Hibiscus cannabinus* and *Alangium chinense*.

Distribution Jiangsu, Zhejiang, Anhui, Fujian, Hubei, Hunan, Guangdong, Guangxi, Sichuan, Guizhou, Yunnan, Shaanxi, Gansu, Taiwan; Japan, Korea Peninsula, Vietnam, Myanmar, India.

Hosts *Firmiana simplex*, *Hibiscus cannabinus*, *Clerodendrum trichotomum*, *Populus tomentosa*, *Alangium chinense*.

Damage Larvae feed on host leaves. Infestations reduce tree vigor. Continuous infestations may result in tree death.

Morphological characteristics **Adults** Females 17-22 mm in length, 50-70 mm in wingspan; males 15-20 mm in length, 46-68 mm in wingspan. Head and antennae black, thorax white, abdomen greyish white, area between segments lighter in color. Forewings white, with light gray spots; there is an oblique line form the apex angle to middle of the hind margin; the outer side of the oblique line is light in color, between the costa and the oblique line is a dark triangle spot near the apex; discal cell with a greyish white reniform spot; hind wing base color is white, the end of the discal cell with a brown round spot; the reverse side of forewings and hind wings white, stripes greyish brown, the end of each discal cell with a dark brown round spot. **Eggs** Elliptical, 0.55 mm in length. Newly produced egg milky, surface smooth; micropyle located on the side of the egg, the center slightly lower, with a white ring and greyish white loops; gray when near hatching. **Larvae** The 1st instar larvae 5-6.5 mm long, head 0.6-0.7 mm in width; body and legs greyish white, whole body with white hairs. Mature larvae (5th instar) 27-35 mm long, head 3.25-3.7 mm in width; dorsal line, subdorsal line and the supraspiracle line dark green; the area between subabdominal line and abdominal line greyish white wide; the area between subdorsal line and the supraspiracle line of each abdominal segment has 4 pairs of yellow and bluish rectangle spots. Thoracic legs greyish white, prolegs greyish yellow, two sides of the anal legs with black spots; peritreme black. **Pupae** 16-18 mm in length, 6-8 mm in width, dark brown. The spiracle location of the 5th to 8th abdominal

C. substigmaria adult (Ji Baozhong, Zhang Kai)

C. substigmaria larva (Ji Baozhong, Zhang Kai)

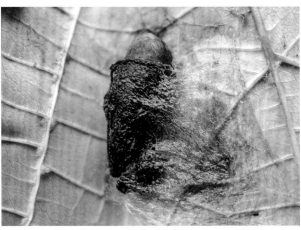

C. substigmaria pupa and thin cocoon (Ji Baozhong, Zhang Kai)

C. substigmaria adult (Hu Xingping)

C. substigmaria pupa (Ji Baozhong, Zhang Kai)

segments convex, segmental venter of the 5th and 6th abdominal segments with proleg scars; the 7th and 8th segments with transverse wrinkles. The tail end blunt, two sides with corniform granules and 9-11 pairs of hooked cremasters; the dorsal of thorax with fine punctures.

Biology and behavior *Cyclidia substigmaria* occurs 2-3 generations a year in Anhui province and overwinters as pupae among fallen leaves. Adults emerge from late April to late May. Larval damage period of the three generations are in May, between early July and late August, and early September, respectively. Adults mostly emerge in early morning before 10:00. Wings fully expand about ten minutes after adult emergence. Adults mate at dusk of the same day and start laying eggs the next day. Eggs are mostly laid on back of leaves near the leaf veins. They are arranged in strips. Each female lays 3 to 5 clutches eggs; each clutch has 10-25 eggs. Adults have phototaxis. They are also seen active during the daytime.

The newly hatched larvae are gregarious. They stay on back of leaves and feed on the mesophyll, leaving the leaf veins untouched. They may drift to other plants through silk thread with the assistance of wind. Larvae after the 2nd instar begin to disperse. Groups of 3-5 larvae attach leaves with silk into nests. They hide and feed in the nests. Mature larvae make cocoons. Prepupal period is one day. Larvae pupate during late September and early November. Adult female longevity is 6-10 days and male longevity is 5-7 days. Egg stage is 7-11 days. The 1st and 2nd generation larval stage is 23-24 days. The 3rd generation larval stage is 170-230 days.

Prevention and treatment

1. Maintain fast growth of trees so the main trunk can be formed rapidly.

2. Clean the ground cover to eliminate the overwintering pupae.

3. Spray 90% trichlorfon at 1,000-2,000× or 50% DDVP at 800-1,000× to kill larvae.

Reference Xiao Gangrou (ed.) 1992.

(Ji Baozhong, Zhang Kai, Zhou Tiying, Wang Linyao)

373 *Apocheima cinerarius* (Erschoff)

Classification Lepidoptera, Geometridae
Common name Spring cankerworm

Importance The spring cankerworm is an early spring leaf-feeding insect pest. Cankerworms develop fast and feed voraciously. Larvae can consume all the young poplar leaves and buds of a tree in a short period of time, resulting in abnormal tree growth, terminal branch death, weakened growth, and triggering to wood-boring insect infestation and large scale tree death. Recently, outbreaks occurred in Heze of Shandong province and became a major threat to forest production.

Distribution Beijing, Hebei, Inner Mongolia, Jiangsu, Anhui, Shandong, Henan, Shaanxi, Gansu, Qinghai, Ningxia, Xinjiang; Russia.

Hosts *Populus* spp., *Salix* spp., *Elaeagnus angustifolia*, *Sophora japonica*, *Morus alba*, *Ulmus pumila*, *Malus pumila*, *Pyrus* spp., *Malus asiatica*, *Populus euphratica*, *Salix cheilophila*.

Damage Newly hatched larvae chew on buds and young leaves; older larvae feed on big leaves, leaving irregular feeding notches. The amount of feeding increases significantly after the third instar and may consume all the tree leaves except the veins.

Morphological characteristics **Adult** Sexually dimorphic. Female moths wingless, 7-9 mm in length; greyish brow, compound eyes black; antenna filiform; dorsal abdomen has lines of black spines, the spine tip blunt, anal plate has lines of black spines. Male moths 10-15 mm in length, wing span 28-37 mm; antenna plumose; forewing light greyish brown to dark brown with 3 brown wavy strips, the middle marking not obvious. **Eggs** 0.8-1 mm in length, oblong, pearl shining with neat striae. Many eggs form an egg mass. Newly laid eggs white grey, become dark purple before hatching. **Larvae** 22-40 mm in length. First instar dark yellow. Subsequent instars' body color varies from brown, green, to grey. Mature larvae greyish brown. One tubercle on each side of A2 segment; ventral line white, spiracle line light yellow. **Pupae** 12-20 mm in length. Greyish yellow, the end has cremasters and splits at the tip. The female pupa has wing remnants and is larger than the male pupa.

Biology

In Heze of Shandong province, spring cankerworm has one generation per year. Pupae overwinter at 10-50 cm, mostly 20-30 cm depth in soil within 3 m radius of a tree trunk. Adults emerge when soil temperature reaches 0°C (or cumulative temperature reaches 3°C) in early to mid-February. Egg laying starts in early February. Eggs hatch between the end of March and early April. Mature larvae pupate in soil from late April to early May.

Moths emerge from soil mainly in the afternoon or evening and hide at base of branches or in tree trunk crevices during day time. Male moths are phototaxic. Mating occurs in the evening. Females lay eggs on tree trunks under 2-2.5 m above ground. Eggs are laid in crevices, on dead branches, or at points of mechanical damage. Each female lays an average of 260 eggs with a range of 170-350 eggs. Each egg mass has 15-220

A. cinerarius adult female (Li Kai)

A. cinerarius adult male (Xu Gongtian)

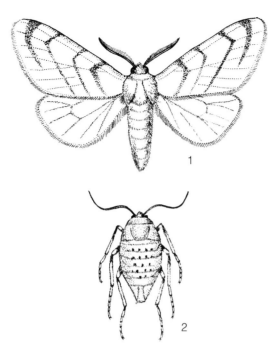

A. cinerarius. 1. Adult male, 2. Adult female
(Zhang Peiyi)

eggs. Egg development is positively correlated with the temperature. Eggs hatch when cumulative temperature reaches 5°C, and reaches the peak at 132°C.

Larvae go through 5 instars. The 1st instar chew on young buds and young leaves. The 2nd and 3rd instars feed on big leaves and form irregular feeding notches. They feed in groups and the amount of feeding is limited. The 4th instar disperse and feed much more than the earlier instars. The 5th instar can eat up all the tree leaves leaving only the veins and can survive without feeding for a period of time. Larvae display thanatosis and disperse through ballooning. The larva developmental stage can be determined by measuring the head capsule width. The measurements are 0.3, 1.0, 1.8, 2.7, and 3.0 mm from the 1st to the 5th instar. Although the head widths are similar between 4th and 5th instars, the capsule is much thicker in the 5th instar than that in the 4th instar.

Pupation peaks in early May. Pupae stand upward. Some may lie horizontally. Adults appear inside pupae by the end of July but do not emerge until the next spring.

Prevention and treatment

1. Pupa destruction. Tillage can kill large number of pupae that inhabit in the soil from summer to next spring. Till soil 30-40 cm deep within 1-1.5 m radius of the trunk in early spring or late fall.

2. Physical barrier. (1) Oleoresin adhesive: since females are wingless and need to climb tree trunks for egg laying, two sticky bands, 5-10 cm wide and 20-40 cm apart, can be applied on tree trunks 1 m above the ground. This is very effective to trap female moths and larvae. Oleoresin adhesive is made of 10 parts rosin, 10 parts castor oil, 1 part butter, and 1 part white wax. To make oleoresin adhesive, boil castor oil, add rosin, then butter and wax, cool till ready to use. Warm to melt before use. Oleoresin adhesive can last for 25-30 days. (2) Plastic ring band. Make a 10-12 cm plastic tape ring/band on tree trunks at 1 m above ground with the upper edge tight and the lower edge loose to form a funnel shape to stop females or larvae from moving up to tree canopy.

3. Trapping. (1) Light traps. Since males are phototaxic, setting up a black light trap in the field during the adult peak time can effectively control the moth population. Sex pheromone lure can also be added to the light trap to enhance the effectiveness. (2) Sugar-vinegar bait-

A. cinerarius adult female (Xu Gongtian)

A. cinerarius egg mass (Xu Gongtian)

ing. In early spring, place premixed sugar and vinegar mixture (sugar: vinegar: wine: water = 3:4:1:3) in the woods with high moth populations to trap moths.

4. Conservation of natural enemies. Minimize insecticide use where pest population or damage is low to protect parasitoid wasps and birds to maintain ecosystem balance.

5. NPV virus. Spray AciNPV at 1.6×10^6 PIB/mL with 0.1% detergent when 80% of the larvae are 1st-2nd instars can achieve satisfactory result.

A. cinerarius larva (brown type) (Xu Gongtian)

A. cinerarius larva (dark color type) (Xu Gongtian)

A. cinerarius larva (light color type) (Xu Gongtian)

A. cinerarius mid-aged larva (Xu Gongtian)

A. cinerarius larva showing bumps on 2nd abdominal segment (Xu Gongtian)

A. cinerarius larvae on poplar leaf (Xu Gongtian)

A. cinerarius damage
(Xu Gongtian)

A. cinerarius damage
(Xu Gongtian)

6. Fogging. When there is minimal air turbulence during 5:00-7:00 or 18:00-20:00, apply 3% high permeation fenoxycarb, 5% beta-cypermethrin, or 1.8% avermectins mixed with diesel at 1: 5-10 ratio depending on the wood thickness.

7. Canopy spray. For small young trees, canopy spray with 5% emamectin at 3,000-5,000×, 25 g/L lamda-cyhalothrin at 800-1,000×.

References Xiao Gangrou 1992, Zhao Fanggui et al. 1999, Sun Xugen 2001.

(Wang Haiming, Wang Ximeng, Feng Wenqi, Jin Buxian)

374 *Ascotis selenaria* (Denis et Schiffermüller)

Classification Lepidoptera, Geometridae
Synonym *Ascotis selenaria dianaria* Hübner
Common name Giant looper

Importance *Ascotis selenaria* is the subspecies of *A. selenaria* found in the Yangtze River region.

Distribution Beijing, Hebei, Jilin, Shanghai, Jiangsu, Zhejiang, Shandong, Henan, Hubei, Hunan, Guangxi, Sichuan, Guizhou, Yunnan, and Tibet; Japan, India, Korea Peninsula, Sri Lanka, Africa, and southern Europe.

Hosts *Metasequoia glyptostroboides*, *Taxodium ascendens*, *Platanus orientalis*, *Pterocarya stenoptera*, *Melia azedarach*, *Paulownia fortuneii*, *Lagerstroemia indica*, *Ligustrum lucidum*, *Morus alba*, *Robinia pseudoacacia*, *Cerasus pseudocerasus*, and *Cerasus serrulata*.

Damage *Ascotis selenaria dianaria* is a serious pest of *M. glyptostroboides*. Larvae feed on young leaves and green twigs from top or bottom of the crown first, and extend to other parts afterwards, resulting damaged or completely defoliated trees.

Morphological characteristics **Adults** Body 16-20 mm long with a wingspan of 18-42 mm, grayish brown with scattered dark brown spots and light-colored scales. Forewings dark gray with inconspicuous white bands. There is an inconspicuous light-colored triangular mark at the top corner, which contains a crescent white mark in the middle. There are also 7-8 inter-connected black crescent marks along the outer margin. Both wings have two inconspicuous black wavy marks from anterior margin to the posterior margin for both wings, and four dark brown stars and two dark brown transverse lines. **Eggs** Oval, green or yellowish brown. Turn to gray or grayish green before hatch, covered by evenly distributed punctures. **Larvae** Body 30-40 mm long at maturity.

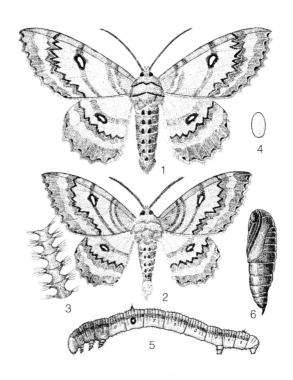

A. selenaria 1. Adult female, 2. Adult male, 3. Male antenna (part), 4. Egg, 5. Larva, 6. Pupa (Hou Boxin)

Body color varies from yellowish green to yellowish brown and dark brown. Cylindrical and smooth, with dense small yellow spots laterally. **Pupae** 17 mm long, dark brown, with projecting verrucae on vertex and a pair of anal spines.

Biology and behavior *Ascotis selenaria* occurs five generations a year and overwinter as pupae in 6-12 cm deep loose soil within 50-85 cm around tree base in Hubei. Most adults emerge between mid- and late April for the overwinter generation, whereas peak emergence for the 1st, 2nd, 3rd, 4th, and 5th generation falls between early and mid-June, early July and mid-July, early August and mid-August, mid-September and late September, and late October and early November, respectively. Larvae stage for the 1st, 2nd, 3rd, 4th, and 5th generation lasts from early to mid-May, mid- to late June, mid- to late July, mid- to late August, and late September to early October, respectively. Larvae can damage *M. glyptostroboides* all year long, but with the heaviest damages caused by the

A. selenaria adult (Xu Gongtian)

A. selenaria adult (Xu Gongtian)

A. selenaria larva (Xu Gongtian)

A. selenaria mature larva (Xu Gongtian)

A. selenaria spinning silk (Xu Gongtian)

A. selenaria pupa (Xu Gongtian)

A. selenaria larva being fed by a bug (Xu Gongtian)

3rd and the 5th generation. Egg stage lasts for 5-10 days, and larval and pupal stage lasts for 9-20 and 8-15 days, respectively. Most adults emerge after dawn. Adults are weak fliers and are strongly attracted to lights. They stay on branches during the day. Adults mate 1-3 days after emergence and mating occurs most past 22:00 at night. Females start to lay eggs 1-2 days after mating. Eggs are laid in clusters of a few dozens and are frequently found on branch joints, trunks, leaf underside and other locations during outbreaks. Each female can produce 400-1,000 eggs. Newly hatched larvae can balloon via silk threads along branches. Larvae are not very active and can mimic the environment, with body color varies greatly. Larvae usually stay at branch joints or between branches, forming triangles that link small branches and trunks. Larvae feed on young leaves before moving to green twigs. Larvae consume large amount of leaves very quickly. A total of 358 larvae were found completely defoliating a small *M. glyptostroboides* tree with a DBH of 6 cm and a height of 5 m in four days based on one observation.

Prevention and treatment

1. Silviculture. Apply selective harvest and strong cultivation measures to improve tree growth for pest resistance. Remove weeds during the winter, deep tillage to reduce overwinter pupal population.

2. Chemical control. Apply 90% trichlorfon at 500-800×, 4.5% beta cypermethrin at 800-1,000×, or 50% Fenitrothion at 1,000× for larvae. Trap adults with vibrating frequency lights and kill them by spray surrounding areas with chemical insecticides.

3. Biological control. Apply B.t. (8,000 IU/ul) at 1,000-1,500× for young larvae results in 90% mortality. Apply *Beauveria bassiana* powder in the winter, or powder or suspension at 1×10^{12} conidia/mL in the spring are effective to larvae, pupae, and adults, with up to 40% mortality. Apply 25% at 1,000× in early May for larval control.

4. Protect natural enemies such as birds *Parus major*, *Dicrurus marocercus*, *Passer montaus*, *Turdus merula*, predatory insects *Calosoma chinense* and *Hierodula patellifera* for population control.

References Chinese Academy of Sciences - Institute of Zoology 1986; Chen Jingyuan and Wu Gaoyun 1988; Ren Shaofu and Wu Gaorong 1990; Xiao Gangrou (ed.) 1992; Ma Qixiang and Jiang Kun 1995; Wang Yongchun, Tang Xiangming, Su Zhihong et al. 1997; Yang Ziqi and Cao Guohua 2002; Zhang Hongsong, He Hongan, Jia Weixi 2002.

(Chen Jingyuan, Li Duxian, Zhou Derong)

375 *Biston marginata* Shiraki

Classification Lepidoptera, Geometridae
Common name Giant tea looper

Importance *Biston marginata* damages mainly *Camelia oleifera*, resulting in premature dropping of fruits and death of the entire trees.

Distribution Anhui, Fujian, Jiangxi, Hubei, Hunan, Guangxi, Taiwan; Japan.

Hosts *Camelia olefera*, *Camelia sinensis*, *Morus alba*, *Castanea mollissoma*, *Citrus reticulate*, and *Acacia richii*.

Damage *Biston marginata* defoliate the entire host trees during outbreaks, resulting in premature drop of fruits. Repeated infestations can kill the trees.

Morphological characteristics **Adults** Body 13-18 mm long with a wingspan of 31-36 mm, short and robust, grayish white with mixed black, grayish yellow, and white scales. Females with lighter body color compared to males. Head small. Compound eyes black with luster. Antennae filiform in females and plumose in males. Forewings slender, with clear outer and inner cross veins, and looming middle cross veins and subcosta that are lighter than the wing in color. Hindwings are the same color as the forewings, relatively smaller with straight outer cross veins. Areas between posterior margin of the outer cross vein and wing base gray for both males and females, whereas areas between the anterior margin of the outer cross vein and the posterior margin of the subcosta dark brown. Forewings contains 6-7 marks along the posterior margin, and grayish white marginal hairs along the posterior and anterior margins. Thorax and underside of the wings grayish white. Female abdomen large with dark brown anal hair clusters, whereas male abdomen relatively slender. **Eggs** Oval, about 0.3 mm long, grassy green when newly laid, turn to green gradually, and become dark brown before hatch. Egg masses covered by dark brown pubescence. Egg masses about 18 mm long. **Larvae** Body 50-55 mm long at maturity, withered yellow and densely covered by dark brown marks. Vertex depressed, with horny projections laterally. Frons contains two black marks. Spiracles purplish red. Thorax and abdomen reddish brown. **Pupae** 11-17 mm long, spindle like, dark reddish brown. Vertex contains two small projections laterally. Tip of abdomen slender and pointed, with one bifurcated anal spine.

Biology and behavior *Biston marginata* occurs one generation a year and overwinters as pupae in loose soil around base of the host trees. Adults start to emerge, mate, and lay eggs from mid- to late February the following year. Peak egg period falls between late February and early March. Eggs hatch in late March. By early to mid-June, mature larvae come down the trees and pupate in soil. The average pupal period is 261.5 days. Females live for an average of 6.25 days, while males live for an average of 4 days. Larval stage lasts for an average of 60 days. Egg stage duration and adult longevity is closely correlated with temperatures. Egg stage lasts for more than one month between mid- to late February, and half a month in mid-March. Adult longevity prolongs when sudden temperature drop occurs before newly emerged females lay their eggs. Eggs hatch between 5:00-13:00, with most of them hatch between 6:00-7:00. Newly hatched eggs feed aggregately and will balloon in the wind with silk threads when disturbed. They disperse

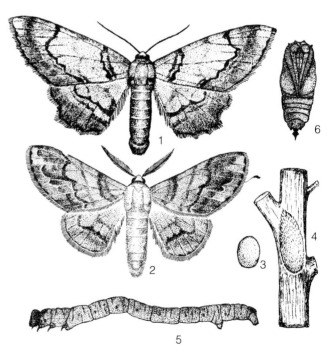

B. marginata. 1. Adult female, 2. Adult male, 3. Egg, 4. Egg mass, 5. Larvae, 6. Pupa
(1-2 by Zhang Xiang; 3-6 by Xu Tiansen)

B. marginata larva (Wang Jijian)

after the 2nd instars. Larvae mimic twigs when at rest. Larvae feed little before the third instars. Leaf consumption increases gradually after the 4th instar. Mature larvae enter soil for pupation one day after they stop feeding. It took them 4 days to pupate after entering soil. Pupation occurs mostly directly under the crown. The preferred sites are loose and moist soil and 15-40 mm in depth. Most adults emerge between 19:00-23:00. Adults can emerge from soil when air temperature reaches 8℃. Mating and egg-laying occurs at night. Most adults mate only once, while very few can mate twice. Mating begins between 2:00-3:00 and ends between 6:00-7:00 and lasts for 3-6 hours. Eggs are laid the next day, with most eggs are laid in one batch.

Prevention and treatment

1. Silviculture. Adopt selective harvest and strong cultivation measures to improve tree growth for pest resistance. Remove weeds during the winter, deep tillage to reduce overwinter pupal population. Clean and burn understory bushes and weeds around host trees to destroy eggs or larvae.

2. Dig pupae out from the ground in early spring or late fall and place them in containers to allow parasitoid emergence. Remove pupae during tillage. Shake tree trunks and collect falling larvae on the ground with plastic films. Wrap trunk base with 5-7 cm wide plastic bands to prevent adult ascending the trees.

3. Apply *B.t.*, *Beauveria bassiana*, NPV for larval control.

4. Trap adults with black lights.

5. Chemical control. Spray 25 g/L lamda-cyhalothrin or trichlorfon at 800-1,000×, 50% fenitrothion at 1,000-1,500×, 2.5% deltamethrin at 2,000-3,000×, 90% trichlorfon crystal at 800-2,000×, 30% fenvalerate at 6,000-8,000×, or 50% phoxim at 2,000×. Use fumigants to control early stage larvae

References Xiao Gangrou (ed.) 1992. Li Dongwen, Chen Zhiyun, Wang Ling et al. 2009.

(Ji Baozhong, Zhang Kai, Peng Jianwen, Zhou Shijuan)

376 *Biston suppressaria* (Guenée)

Classification Lepidoptera, Geometridae
Common name Tung oil tree geometrid

Importance *Biston suppressaria* is a serious pest of *Vernicia fordii*, *Camellia oleifera*, and *Camellia sinensis* in southern China. It can completely defoliate the entire host trees during outbreaks, causing significant economic damage to forest economy and the ecosystem.

Distribution Jiangsu, Zhejiang, Anhui, Jiangxi, Hubei, Hunan, Guangdong, Guangxi, Hainan, Guizhou; India, Myanmar, and Japan.

Hosts *Vernicia fordii*, *Camellia oleifera*, *Camellia sinensis*, *Sapium sebiferum*, *Myrica rubra*, and *Castanea mollissima*.

Damage Feeding by the 1st and 2nd instars creates irregular yellow retinal marks on host leaves. Third instar larvae can punctuate the leaves. By 4th-5th instars, damaged leaves appeared crescent-shaped. Young branches and tree barks can be consumed sometimes as well.

Morphological characteristics **Adults** Females 24-25 mm long with a wingspan of 67-76 mm. Body and wings white with dense bluish black scales. Antennae filiform. Wings contain a black blotch underside. There is a yellow hair cluster at the tip of the abdomen. Males 20-24 mm long, with a wingspan of 55-61 mm. Antennae double plumose. Tip of abdomen without hairs. **Eggs** Oval, about 0.7-0.8 mm, fresh green o brownish in masses, and covered with yellow pubescence. **Larvae** Body 65-72 mm long at maturity, ranging from grayish brown, deep brown, to grayish green. Head contains small granular projections. There is an arc-shaped depression on vertex that forms two acute angles. **Pupae** 19-28 mm, conical, dark brown, with two small dark brown small projections on vertex.

Biology and behavior *Biston suppressaria* occurs 2-4 generations a year and overwinters as pupae in 3-7 cm deep soil around the base of the host trees, with two generations in Anhui (Langxi) and Jiangsu (Shezhu), two to three generations in Hunan (Changsha) and Zhejiang (Lanxi), and three to four generations in Guangdong (Yingde). In Hunan, overwintering pupae begin to emerge in mid-April, with the first generation larvae appear between early May and late June, the second generation larvae between mid-July and early September, and the third generation larvae between mid-September and late October.

Duration for each stage is as the following: 10-15 days for eggs, 30-50 days for larvae, 30-35 days for pupae, and 7-10 days for adults. Duration for the 1st to 6th instars is 3.0-3.8, 2.6-3.6, 3.1-3.7, 3.6-4.1, 6.2-6.4, and 13.3-13.5 days, respectively.

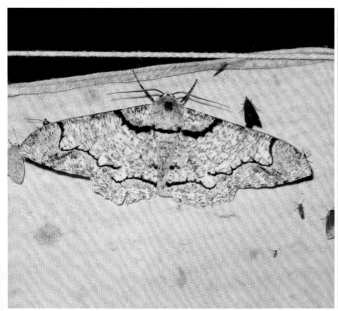

B. *suppressaria* adult (Li Yingchao)

B. *suppressaria* adult (edited by Zhang Runzhi)

B. suppressaria larva (edited by Zhang Runzhi)

B. suppressaria larva (edited by Zhang Runzhi)

B. suppressaria larva on *Bauhinia variegata* (edited by Zhang Runzhi)

B. suppressaria larva (edited by Zhang Runzhi)

B. suppressaria larva on *Castanopsis fissa* (edited by Zhang Runzhi)

B. suppressaria larva on *Acacia confusa* (edited by Zhang Runzhi)

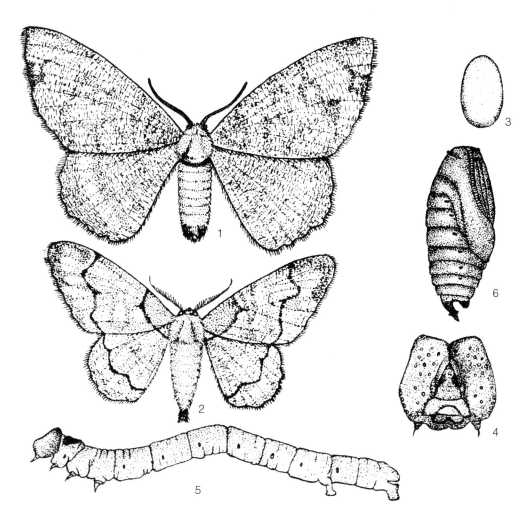

B. suppressaria. 1. Adult female, 2. Adult male, 3. Egg, 4. Head of larva, 5. Larvae, 6. Pupa
(1-2 by Liu Gongpu; 3-6 by Xu Tiansen)

Most adults emerge at around 20:00 and are attracted to lights. Mating occurs right after emergence and eggs are laid the next day. Egg masses are found in bark crevices, wounds, or under branch joints and are covered by yellow pubescence. Each female produces an average of 2,450 eggs, with a maximum of 4,000 eggs.

Natural enemies play an important role in *B. suppressaria* population dynamic, including *Telenomus buzurae*, *Euplectus* sp., *Aleiodes buzurae*, *Cratichneumon* sp., *Therion* sp., and BsNPV.

Prevention and treatment

1. Till to destroy pupae.
2. Adult trapping. Use black light to trap adults during the flight season.
3. Manual control. Manually kill females as they stay on the lower trunk during the day. Remove egg masses from trunk and place them in cages for parasitoids emergence.
4. Biological control. Spray BsNPV at 2×10^7 PIB/mL during the 1st and 2nd instars. Apply *B.t.* at 8,000 IU/μL during 2nd to 5th instars resulted in 83%-100% control.

References Peng Jianwen and Liu Youqiao 1992, Xiao Gangrou (ed.) 1992, Zhang Hanhau and Tan Jicai 2004.

(Tong Xinwang, Wang Wenxue)

Abraxas suspecta (Warren)

Classification	Lepidoptera, Geometridae
Synonym	*Calospilos suspecta* (Warren)
Common name	Euonymus moth

Importance *Abraxas suspecta* outbreaks can lead to complete defoliation of the host in a short period of time, resulting in dead branches that affect host growth. Crawling larvae in the urban areas affect the aesthetics of the urban environment.

Distribution Beijing, Hebei, Shanxi, Inner Mongolia, Liaoning, Jilin, Heilongjiang, Shanghai, jiangsu, Jiangxi, Shandong, Hubei, Hunan, Sichuan, Shaanxi, Taiwan; Japan, Korea Peninsula, and Russia.

Hosts *Euonymus bungeanus*, *Euonymus japonicas*, *Euonymus alatus*, *Ulmus pumila*, *Populus* spp., and *Salix* spp.

Damage Larvae feeding create notches on host leaves. Only leaf veins are intact during heavy infestations.

Morphological characteristics **Adults** Females 12-19 mm long with a wingspan of 34-44 mm. Wings silver white basally, with grayish and yellowish brown marks. There is a contiguous grayish line along the outer margin of the forewings. Outer cross vein appears as a row of grayish marks, forked at the top and with one large reddish brown mark at the bottom. Median cross veins not connected, with one large grayish mark near the median cell that contains a circular mark inside. Base of the wings with a color mark of deep yellow, brown, and gray. Hindwings contain a contiguous grayish mark along the outer margin. Outer cross veins form a wide grayish mark. Median cross veins with disrupted small marks which vary among individuals. Marks on fore- and hindwings become connected when both wings are spread. Marks on the underside of the wings are identical to those on the upper side but without yellowish brown marks. Abdomen golden with nine rows of black marks. Inner side of hind tibia without hair clusters. Males 10-13 mm long with a wingspan of 32-38 mm. Marks on wings identical to those on females. Abdomen golden, with seven rows of black marks. Inner side of hind tibia with a yellow hair cluster. **Eggs** Oval, about 0.6 mm wide and 0.8 mm long, with horizontal and longitudinal marks on the shell. Grayish green initially, turn to grayish black before hatch. **Larvae** Body 28-32 mm long at maturity, black with yellow setae. Head black, with coronal suture and subgenal suture yellowish. Pronotum yellow, contains five black marks with the central three forms a triangle. Dorsal stripe, subdorsal strip, upper spiral stripe, and subventral stripe bluish white. Spiral stripe and ven-

A. suspecta adult (SFDP)

tral stripe yellow and relatively wide. Anal plate black. There is a yellow transverse mark on posterior margin of each segment from thorax to the 6th abdominal segment. Thoracic legs black with yellowish basal areas. Abdominal proleg crotches in biserial central bands. **Pupae** 9-16 mm long and 3.5-5.5 mm wide, spindle shaped. Initially with yellow head and abdomen and greenish thorax, turn to dark red later. Anal spine bifurcated.

Biology and behavior *Abraxas suspecta* occurs two generations in Harbin in Heilongjiang province, two to three generations in Xian in Shaanxi province and Yunkou in Liaoning province, and three generations in Hebei a year. It overwinters as pupae in soil. Adults start to emerge in early to mid-May mostly at night, with a peak emergence in late May. They stay in the crown during the day and take short flights when disturbed. They are active at night. Adults are slightly attracted to lights and do not need supplemental feeding. Mating generally occurs at night and lasts for 6-7 hours. Both males and females mate just once. Mated females begin to lay eggs the same night. Eggs are mostly laid on the back of the leaves in rows along leaf margins. Some eggs are scattered. Each female lays 2-7 egg masses, with each egg mass contains up to 195 eggs. The average fecundity is 258 ± 113 eggs/female, with 15 ± 9 eggs remain in the ovary. Egg stage lasts for 5-6 days. Larvae of the first generation start to hatch in late May. There are five instars during larval stage. Neonates disperse rapidly and seek young leaves as food after hatching. They can balloon or disperse to nearby branches with silk threads when disturbed. Larvae feed on the underside of the leaves. The first and second instar larvae feed on young mesophyll and leave only the upper epidermis intact, or leave feeding holes. They also feed on young buds. Larvae start to feed on leaf margins from the 3rd instar, creating various incised leaves. By the 4th instar, they consume whole leaves except the stalk. The 5th instar larvae consume leaf stalk as well as tender twigs and bark. Larvae feed both during the day and at night. Molting occurs between 3:00-9:00 am. Newly molted larva consumes the exuviate completely except the head capsule. Most mature larvae follow the trunk to the ground to pupate under leaf litter or inside 3 cm deep loose soil around the base. Larvae can also ballooning to the ground with silk. Pupation occurs in mid-June after a 2-3 days long prepupal stage. Larval period of the 1st to 3rd generation is 35, 23, and 25 days, respectively. Larvae from the 2nd and 3rd generations cause the most damage, usually resulting in complete defoliation of the hosts. Most larvae enter 2-3 cm deep soil, pupate, and start overwintering in September.

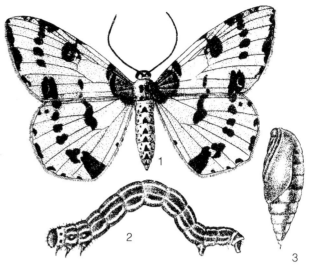

A. suspecta. 1. Adult, 2. Larva, 3. Pupa
(Zhang Peiyi)

Prevention and treatment

1. Manual control. Destroy pupae in the soil by digging them out or loosen the soil. Collect and destroy larvae during their ballooning period.

2. Trapping adults with black lights.

3. Preserving and utilizing natural enemies.

4. Chemical control. Apply 20% Dimilin at 7,000×, *B.t.* EC at 500×, or *B.t. galleriae* (8,000 IU/μL) at 1,000-1,500× to control the 1st generation larvae.

References Yang Ziqi and Cao Huaguo 2002, Xu Gongtian 2007.

(Xu Gongtian, Qiao Xiurong, Yang Xiuyuan, Huang Xiaoyun)

A. suspecta larva (SFDP)

378 *Chihuo zao* Yang

Classification Lepidoptera, Geometridae
Synonym *Sucra jujuba* Chu

Distribution Hebei, Shanxi, Zhejiang, Anhui, Shandong, Henan, and Shaanxi.

Hosts *Ziziphus jujube, Malus pumila,* and *Pyrus* spp.

Damage Larvae feed on young leaves and buds. They can cause complete defoliation under heavy populations.

Morphological characteristics **Adults** Females 12-17 mm long, wingless, spindle shaped, grayish brown. Males 10-15 mm long with a wingspan of 35 mm, light grayish brown. Forewings with clear basal lines and outer cross lines. Hindwings with clear outer cross lines and one black blotch. **Larvae** Body 40 mm long, grayish green with multiple grayish white longitudinal stripes.

Biology and behavior *Chihuo zao* occurs one (occasionally two) generation a year in Henan and overwinters as pupae in the soil. Adult emergence begins in mid-March. Eggs are laid at branch joints or bark crevices. Eggs start to hatch in mid-April. Mature larvae enter the soil in mid-May to pupate for summer aestivation or overwintering. Some pupae stay in soil for two years before emerging as adults.

Females climb up trees in the evening after emergence. Males fly in the forest after emergence and mate with females. Males die two to three days after mating. Mated females lay eggs from the next day after mating. Eggs are laid as masses of up to several hundred eggs per mass at branch joints or bark crevices. Each female produces about 1,200 eggs. Eggs hatch within 15-25 days. Newly hatched larvae disperse quickly and tie tender buds with

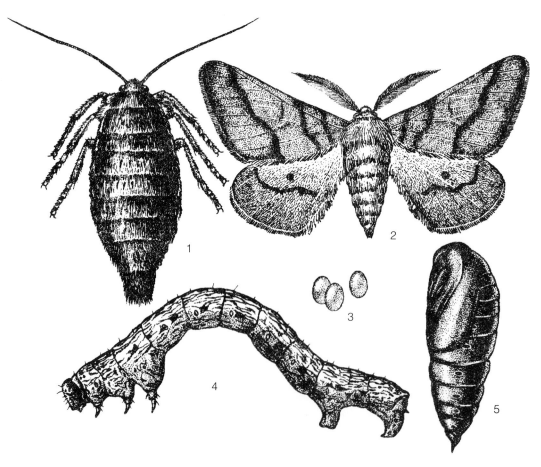

C. zao. 1. Adult female, 2. Adult male, 3. Eggs, 4. Larva, 5. Pupa (Zhang Xiang)

C. zao adult male (Li Zhenyu)

C. zao adult female (Li Zhenyu)

C. zao adult female (Xu Gongtian)

silk to prevent the buds from expansion. Larval feeding on young leaves and buds significantly lower the quantity and quality of the fruits. Larvae can feign death when disturbed by ballooning with silk.

Prevention and treatment

1. Manual control. Destroy pupae in winter or early spring through tillage. Destroy females and prevent them from climbing up trees in early spring by wrapping trunk base with plastic film.

2. Biological control. Protect natural enemies such as *Mesochorus* sp., *Exorista sorbillans*.

3. Chemical control. Apply 25% Dimilin at 2,000×, or 80% trichlorfon at 1,000× during early larval stage.

References Yang Youqian 1982, Xiao Gangrou (ed.) 1992, Henan Forest Pest Management and Quarantine Station 2005.

(Yang Youqian, Li Bangqing)

379 Biston panterinaria (Bremer et Grey)

Classification	Lepidoptera, Geometridae
Synonym	*Culcula panterinaria* (Bremer et Grey)
Common name	Chinese pistacia looper

Importance It is a ferocious polyphagous forest pest. Its characteristic sporadic outbreaks cause dramatic reduction in food and vegetable oil production (50%) and economic forests and fruit production (20%-50%). Trees may die as a result of the infestations.

Distribution Beijing, Hebei, Shanxi, Inner Mongolia, Liaoning, Jilin, Shandong, Henan, Guangxi, Sichuan, Yunnan, Shaanxi, Gansu, and Taiwan; Japan, Korea Peninsula, India, Nepal, Vietnam, and Thailand.

Hosts A total of 170 species within 30 families, including *Pistacia chinensis*, *Juglans regia*, *Castanea mollissima*, *Armeniaca sibirica*, *Malus pumila*, *Amorpha fruticosa*, *Robinia psedoacacia*, *Populus* spp., *Ailanthus altissima*, *Rhus typhina*, *Cotinus coggygria*, *Punica chinensis*, *Crataegus pinnatifida*, *Albizzia julibrissin*, *Paulownia* spp., *Amygdalus triloba*, *Ziziphus jujuba* var. *spinosa*, *Vitex negundo*.

Damage Larvae feed on leaves. At high population density, they cause complete defoliation.

Morphological characteristics **Adults** Body about 30 mm long with a wingspan of 70 mm. Wings white, with gray and orange marks. There is a series of orange and dark brown circular spots on the outer transverse lines on both wings. Forewings with a large orange circular spot at the base. **Eggs** Oval, 0.9 mm, with a layer of yellowish brown hairs. Green in color initially, turn to black before hatching. **Larvae** Body about 70 mm long at maturity, surface rough, usually matches the color of the host such as green, brown, and grayish brown. There are sporadic white spots on the body. Head depresses in the middle and covered with small protrusions. Pronotum with two horn-shaped projections. There

B. panterinari adult (Xu Gongtian)

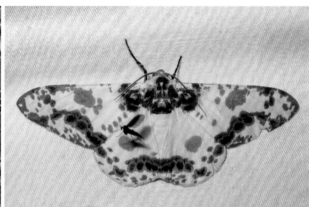

B. panterinari adult (Li Yingchao)

B. panterinari larva (Xu Gongtian)

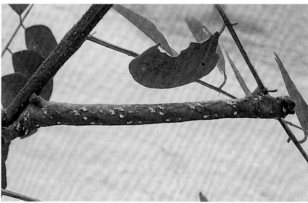

B. panterinari larva (Xu Gongtian)

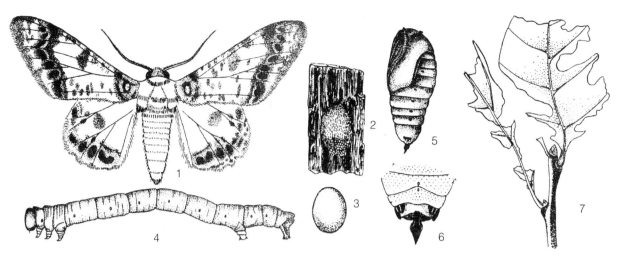

B. panterinari. 1. Adult, 2-3. Egg mass and egg, 4. Larva, 5-6. Pupa and posterior end of pupa, 7. Damage (Xu Tiansen)

are two small white spots on each side of segments from metathorax to the last abdominal segment. **Pupae** About 30 mm long and 8-9 mm wide, spindle shaped. Initially bright green and turn to dark brown later.

Biology and behavior *Biston panterinaria* occurs one generation a year and overwinters as pupae under rocks or in 1-10 cm (mostly 1-5 cm) deep moist soil in Beijing, Hebei, Shaanxi, and Henan. Adult emergence begins in early May, peaks in mid- to late July, and ends by early August. In Beijing, adults start to emerge from June 7 to July 6, peaks between July 16-22, and ends by August 10. During outbreaks, adults usually emerge half a month earlier than usual. Adults are strongly attracted to lights. They are easily discovered during the day as they rest on tree trunks, leaves, or surrounding weeds.

Egg period lasts for about 10 days. More than 90% of the eggs hatch at the ambient temperature of 26.7℃ and RH of 50%-70%. Egg hatching starts in early July, peaks between late July and early August, and ends in late August. Larval stage lasts for about 40 days with six instars. Larvae ballooning with silk when disturbed. Newly hatched larvae usually feed the leaf tips on the outer crown, resulting semi-transparent webbing leaves with only veins intact. By the third instar, larvae are more sedentary with an increased appetite. They consume a whole leaf before moving the next one. When at rest, the body raises into a triangle shape. Complete defoliation usually occurs in July and August. By late September, mature larvae start to fall to the ground in preparation of pupation in soil or under the rocks.

The survival of overwintering pupae is influenced

B. panterinari damage (Xu Gongtian)

mainly by soil moisture, with ideal soil moisture of 12%. A soil moist lower than 10% is detrimental the overwintering pupae. Natural mortality is high among overwintering pupae in years with little snow and dry springs. Pupal mortality is also higher on south facing slopes compared to north facing slopes, and sites with less ground cover. High adult emergence rates and large larval populations were observed in locations with ample precipitation in May.

Most adults emerged between 20:00-23:00, with the ambient temperature of 24.5-25℃. Adults are diurnal and strongly attracted to lights. During the day, they are easily discovered as they rest on trunks, between leaves, in weeds, or on crops. Adult stage lasts for 4-12 days.

B. panterinari mating pair (Xu Gongtian)

Males and females mate right after emergence and females start to lay eggs within 1-2 days after mating. Most eggs are laid as irregular shaped masses in bark cervices, underside of leaves, or on rock surface. Each female can produce an average of 912 eggs, with a maximum of 3,118 eggs.

Prevention and treatment

1. Reduce source population by digging out overwintering pupae in early winter or early spring.

2. Remove egg masses.

3. Trap adults with black lights between June and August.

4. Apply *B.t.* suspension (8,000 IU/ul) at 1,000-5,000×, 20% Dimilin suspension at 7,000×, or NPV to control the 1st to the 3rd instar larvae.

5. Protect natural enemies such as *Trichogramma nitidulus*.

References Wang Yuanmin et al. 1987, Xiao Gangrou (ed.) 1992, Yan Guozeng and Yu Juxiang 2001, Zhang Yuling and Li Jianhong, 2004, Xu Gongtian and Yang Zhihua 2007.

(Wang Jinli, Li Guangwu)

380 *Dilophodes elegans sinica* Wehrli

Classification Lepidoptera, Geometridae

Importance *Dilophodes elegans sinica* is an important defoliator of star anise. Frequent defoliation could reduce its current year production as well as growth in the next year, resulting in economic and ecological damage.

Distribution Guangxi, Yunnan, and Taiwan.

Host *Illicium vercum*.

Damage Larvae feeding creates notches or shot holes on leaves. Larvae consume whole leaves as they grow. Heavy infestation defoliates the entire trees.

Morphological characteristics **Adults** Body 20-25 mm long with a wingspan of 55-60 mm. Antennae filiform. Body and wings greyish white and covered by dense black marks. Females lack hair clusters at the end of the abdomen, and hindwings with one "∧" shaped black spot. Males have black hair clusters at the end of the abdomen, and hind wing with a near circular black spot. **Eggs** Oval, about 1.0 mm × 0.7 mm, shinny. Creamy initially, turn to yellowish later, and with grayish brown before hatching. **Larvae** Body 2.5-8.7 mm long during the 1st and 2nd instars, reddish brown, with dense fine spots, which form brownish body rings alternate with intersegmental membrane. Body 9.4-22 mm long during the 3rd and 4th instars, with apparent spots, green or yellowish green in color. By the 5th and 6 instars, they measure at 24-38 mm, yellowish green, with conspicuous spots and "+" shaped blotches. Mature larvae 35-40 mm long, smooth, yellowish green, hypertrophic, with obvi-

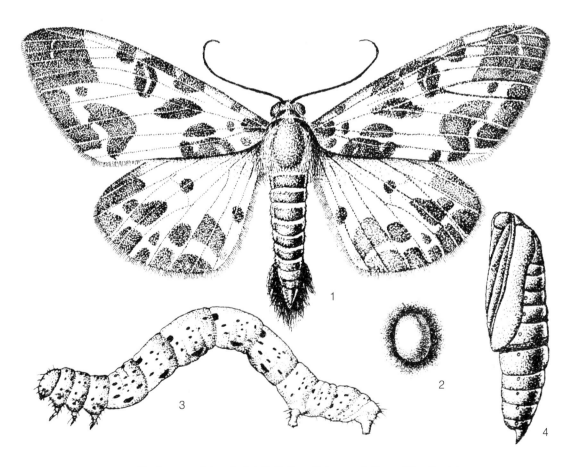

D. elegans sinica. 1. Adult, 2. Egg, 3. Larva, 4. Pupa (Hou Boxin)

ous spots and blotches. There is a large black "+" shaped blotch on each abdominal segment dorsally. Abdomen has two rows of relatively large black blotches laterally and ventrally, each blotch contains a seta. **Pupae** Dark reddish brown, and dark brown before emergence with black blotches. The end of the abdomen with a forked spine.

Biology and behavior *Dilophodes elegans* sinica occurs 4-5 generations a year and overwinters as larvae and pupae in Yulin and Nanning region in Guangxi province. Overwintering larvae resume activity in early February and pupate in mid-March. Adults start to emerge in late March. Larvae of the 1st to 4th generations appear in April-May, June-July, August-September, and October-November, respectively. By early to mid-October, larvae of the 4th generation gradually mature and pupate. Some overwinter as pupae. Others continue to develop into adults and produce the 5th generation and overwinter as larvae by late October and early November. Prepupal stage lasts for 2-6 days, pupal stage lasts for 12-30 days (or 90-120 days if overwinter). Adult longevity ranges from 7 to 20 days. All life stages are found during the year except in January when only larvae or pupae are present.

Most adults emerge between 17:00-21:00 and stay on branches to expand their wings after emerging from the soil. They start to fly the same night after emergence. Adults are attracted to lights and are weak fliers, with each flight lasting for 2-3 minutes and 100 m in distance. Adults stay in bushes, among flowers and weeds and feed on nectars during the day and are active at night. They mate between 1:00-5:00 1-3 days after emergence. Each female can produce 200-300 eggs, with a maximum of 500 eggs. Eggs are laid on the underside of the leaves in the middle or lower crown, usually one egg per leaf. There are six to seven instars during the larval stage. Newly hatched larvae start to feed about 12 hours after hatching. The 1st and 2nd instars feed on mesophyll on the

D. elegans sinica adult (Wang Jijian)

D. elegans sinica adults on a tree (Wang Jijian)

D. elegans sinica eggs and young larvae (Wang Jijian)

D. elegans sinica mature larva (Wang Jijian)

D. elegans sinica pupae (Wang Jijian)

D. elegans sinica prepupae and pupae (Wang Jijian)

D. elegans sinica larva killed by entomopathogen (Wang Jijian)

D. elegans sinica damaged forest (Wang Jijian)

underside, leaving epidermis intact. From the 3rd instar, larvae feed from the edges and consume whole leaves. Each larva can consume 19 leaves during its lifetime. They can also feed on flower buds, young fruits, and tender barks when leaves become unavailable. Young larvae prefer young leaves whereas older larvae are generally found on mature leaves. The tender tips on top of the tree crown are often first consumed. Larvae rest under the leaves or along leaf edges. When disturbed, they become airborne and drop to the ground along silk threads. Larvae return back to the tree through the threads crawl along the trunk a few minutes later. Forest stands with lower canopy coverage suffer more than those with higher canopy coverage. Larvae can resist certain degree of lower temperatures. Mature larvae pupate in 3-4 cm loose soil under the crown.

Prevention and treatment

1. Manual control. Remove pupae from the soil during outbreaks.

2. Microbial control. Apply *Beauveria bassiana* powder between March and April, or *B.t.* or Sengdebao (a mixture of *B.t.* and avermectins) powder or suspension in the summer or fall.

3. Physical control. Attract adults with black light.

4. Chemical control. Apply 2.5% deltamethrin at 3,000-4,000×, or 25 g/L lamda-cyhalothrin at 800-1,000× to control larvae.

References Huang Jinyi and Meng Meiqiong 1986, Xiao Gangrou (ed.) 1992.

(Wang Jijian, Qin Zebo, Zhao Tingkun)

381 *Dysphania militaris* (L.)

Classification Lepidoptera, Geometridae
Common name Common tiger moth

Importance *Dysphania militaris* feeds on corkwood. It can completely defoliate the whole tree under heavy infestation, which affects host growth and its aesthetic value.

Distribution Fujian, Guangdong, Yunnan, Hong Kong, and Macau; Thailand, India, Indonesia, Malaysia, Vietnam, and Cambodia.

Host *Carallia brachiate*.

Damage Young larvae incise leaves whereas larvae older than the 3rd instar feed from the leaf tip or edge and consume whole leaves.

Morphological characteristics **Adults** Body 23-28 mm long with a wingspan of 34-38 mm, apricot with purplish blue marks. Outer half of the forewings purplish blue with two rows of semi-transparent powdery white spots; basal portion apricot with one "E" shaped purplish blue spot; base of the forewings with pits. Hindwings apricot and contain several sporadic purplish blue spots, with a middle large circular spot and one smaller circular spot below. Thorax apricot and contains two purplish blue transverse bands. Abdomen apricot, with a purplish blue transverse band on each segment. **Eggs** Oval, about 1.14-1.20 mm long, 0.91-1.03 mm wide, and 0.72-0.88 mm high. Yellowish when laid, slightly depressed centrally. Base of the egg gradually turn to red with spots appear around. Eggs are surrounded by red bands and turn black before hatching. **Larvae** Newly hatched larva brown, 4.34-6 mm long with a head length of 0.23 mm. It turns yellow with a head width of 0.55-0.64 mm in the 2nd instar. Various spots start to appear on the back of the third instar and increase as it grows. Head width is 0.97-1.44, 1.98-2.08, 2.84-2.91, and 3.98-4.56 mm for the 3rd to 6th instar, respectively. By 6th instar, larva becomes yellow or orange yellow with a lighter colored underside. It measures at 50-60 mm in length; head yellowish brown; thorax contains multiple black spots. Dorsal and spiracular lines bluish green. Dorsal line relatively wide and contains black spots of various sizes. Spiracular line with dense black spots. Spiracles oblong with a grayish black valves. Peritreme black. Anal segment dark and

D. militaris adult female (edited by Zhang Runzhi)

D. militaris larvae (edited by Zhang Runzhi)

D. militaris pupa (edited by Zhang Runzhi)

D. militaris eggs (edited by Zhang Runzhi)

D. militaris larva
(edited by Zhang Runzhi)

D. militaris pupa
(edited by Zhang Runzhi)

without color bands or spots. Proleg crotchets arranged in biordinal uniseries. **Pupae** Body 25-29 mm long and 8-9 mm wide. Yellowish at first, turn grayish brown or dark brown later. Head with one kidney-shaped dark eyespot on each side. There are scattered dark brown spots on the head, thorax, and abdomen. Spiracles dark brown. Anterior border of the last abdominal segment contains protruding serrate dark brown band, which does not meet ventrally. There are eight anal spines, with four in the middle and two on each side. There is also a soft yellowish brown hook in front of the spines.

Biology and behavior *Dysphania militaris* occurs three generations a year and overwinters as pupae. Generations overlap. Adults of the overwintering generation appear in late February to early March. Larvae of the first generation feed on hosts in March, with adults of this generation occurs in mid- to late April. Larvae of the 2nd generation appear in mid-April to May, with adults found in mid- to late May. By mid-May to early June, larvae of the 3rd generation dominate the population. Adults are active during the day. They are slow fliers with an unpleasant odor that deters birds. Eggs are laid singly or in clusters of up to 102 eggs per cluster on leaf surface at night. There are six instars during the larval stage. Front portion of the body of the newly hatched larva curled when at rest. They attach themselves to substrates using their prolegs on the 6th abdominal and the anal segment. The 2nd instar larvae usually hang on twigs or leaves with a silk thread and migrates with the help of wind to other leaves. Feeding by the 2nd instar larvae cause notches on leaves. From the 3rd instar, larvae consume more food and can consume whole leaves and young twigs. When disturbed, larvae usually form an angle with their front portion of the body forming a fist shape. Mature larvae fold leaves with silk and pupate inside folded leaves. Pupal stage lasts for 10-17 days, whereas adults live for 11-17 days.

Prevention and treatment

1. Quarantine. Enforce quarantine to prevent this pest from dispersal.

2. Manual control. Shake branches to dislodge and destroy larvae.

3. Natural enemy protection. Protect egg parasitoids.

4. Chemical control. Apply 45% profenofos and phoxim mixture EC at 1,000-1,500×, 80-90% trichlorfon wettable powder or 25 g/L lamda-cyhalothrin at 1,000-1,500×, or 50% fenitrothion at 1,000-1,500× for larval control.

References Zhu Hongfu et al. 1975, Huang Bangkan et al. 2001, Wu Yousheng and Gao Zezheng 2004.

(Wu Yousheng, Gao Zezheng)

382 *Erannis ankeraria* (Staudinger)

Classification Lepidoptera, Geometridae
Synonym *Erannis defoliaria gigantea* Inoue

Distribution Beijing, Hebei, Inner Mongolia, Heilongjiang, Shaanxi; Hungary.

Hosts *Larix* spp., *Quercus* spp.

Damage An outbreak over 270,000 ha area was recorded in Weichang County in Hebei province in the early 1980s, with the highest density of 20,000 larvae/tree. In addition, outbreaks over 10,000 ha area were also recorded in Saihanba Mechanical Forest as well as Mengluan Forestry Bureau. Occurrence over large areas was also reported in Qinling Mountain in Shaanxi. When the insect occurred in four consecutive years, no significant impact on tree growth was observed in the first year. However, significant impact was observed during the next three years. There was significant impact on diameter growth when population level reached 1,100 larvae/tree and a defoliation rate of 50%. Both diameter and volume growth were significantly impacted at population density of 3,200 larvae/tree and a defoliation rate of 100%.

Morphological characteristics **Adults** Females 12-16 mm long, gray with irregular black spots, spindle shaped. Head dark brown, with a white blotch of setae cluster on the vertex. Antennae filiform, black. Compound eyes black. Thorax with a pair of black blotches on each segment dorsally. The first abdominal segment contains a pair of large black blotch, while other segments contain irregular black spots along dorsal median line. There is a black lateral line running from head to compound eye and the end of the abdomen. Wings greatly reduced to scaly projection. Legs black, long, with 1-2 white rings on each segment. Males 14-17 mm long with a wingspan of 38-42 mm, yellowish brown. Head

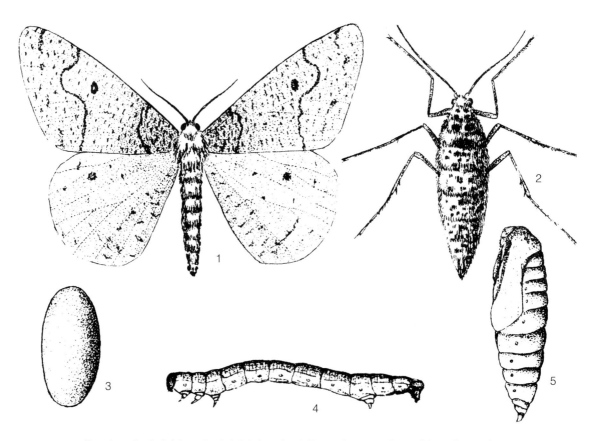

E. ankeraria. 1. Adult male, 2 Adult female, 3 Egg, 4 Larva, 5. Pupa (Yang Cuixian)

E. ankeraria adult male (Liang Jialin)　　　　E. ankeraria adult female (Liang Jialin)

E. ankeraria larva (Liang Jialin)　　　　E. ankeraria pupa (Liang Jialin)

yellowish. Compound eyes black. Antennae pectiniform, short and yellowish with yellowish brown processes. Thorax densely covered by long scales. Wings yellowish. Forewings contain dense irregular brown spots. Median cross vein and kidney-shaped marks brown and very clear. Subcosta light brown. Median cross vein of the hindwings inconspicuous, with one brown circular mark inside. **Larvae** Body 27-33 mm long, yellowish green. Head yellowish brown, with a rough head capsule and reddish brown stripes. Labrum brownish, darker along edges. Antennae yellowish white, with one black spot on the inner side. Body rugous, with 10 interrupted lines dorsally and ventrally. Spiracle line and ventral-median line yellowish green. Spiracle oblong with black edges. **Pupae** Body 14.5 mm long, reddish brown, rugous and resembles orange peel. The 5th abdominal segment contains one dark carina on each side. There is a "Y"-shaped anal spine.

Biology and behavior *Erannis ankeraria* occurs one generation a year and overwinters as eggs in the Greater Khingan area. Eggs hatch in late May. There are five instars during the larval stage, with each lasts for 5-8, 4-6, 4-5, 6-7, and 13-15 days, respectively. Mature larvae enter the soil to pupate after feeding for 35-37days. Pupal stage lasts for 68-79 days. Adults emerge in September with most of them come out in the morning. Females are good crawlers and come up to the tree right after emergence. Males can feign death and drop to the ground when disturbed. Mating occurs at night. Eggs are laid in open cone scales. Eggs overwinter for 230-240 days.

Serve damage usually occurs in closed forest stands, with close to 100% host trees damaged in stands with > 85% canopy cover. Light damage was observed in stands with < 40% canopy cover. No damage was observed in single trees, sunny, and sparse forest. Severe damage was observed in larch plantation at upper slope while no damage in larch-birch mixed forest at low slope on the same site. Heavier damage was also observed at bottom of hills and inside forests compared to forest edges.

References Zhang Lianzhu et al. 1982, Huang Guanhui 1984, Jiang Enyu 1987, Sun Shiying et al. 1987, Fan Meizhen et al. 1988.

(Liu Yuanfu)

383 *Ectropis excellens* (Butler)
Classification Lepidoptera, Geometridae

Distribution Beijing, Hebei, Liaoning, Henan, Guangdong, Sichuan, Taiwan; Japan, Korea Peninsula, and Russia.

Hosts *Robinia pseudoacacia*, *Ulmus* spp., *Populus* spp., *Salix* spp., *Quercus* spp., *Castanea* spp., *Malus pumila*, and *Pyrus* spp.

Damage Young larvae feed on mesophyll of the leaves, leaving only the reticular epidermis intact. Older larvae consume leaves, resulting in incised leaves or leaves with shot holes. Heavy infestation defoliates the entire tree.

Morphological characteristics **Adults** Females 15 mm long with a wingspan of 40 mm, grayish brown, with an obvious black brown spot in the middle of the outer cross vein. Costa with a black color band. Subcosta serrate. Hindwings have wavy transverse bands. Males 13 mm long with a wingspan of 32 mm, with conspicuous marks on the body and the wings. **Larvae** Length 35 mm. Color varies, most larvae are yellowish brown.

Biology and behavior *Ectropis excellens* occurs four generations a year and overwinters as pupae in the soil in Henan. Adults emerge and start to lay eggs in early April. Larvae feed after hatching until early May when they enter the soil to pupate. Adults of the first generation emerge about 10 days later. Adults of the second generation appear in early to mid-July, whereas adults of the third generation occur in mid- to late August. Larvae of

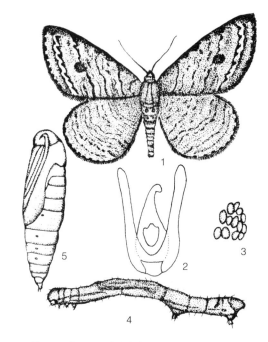

E. excellens. 1. Adult female, 2. Male genitalia, 3. Eggs, 4. Larva, 5. Pupa

the 4th generation mature in mid-September and enter soil for overwintering.

Adults are attracted to lights. Eggs are laid in masses inside bark crevices along the lower trunk. Each female produces about 1,000 eggs. Newly hatched larvae feed

E. excellens adult (Xu Gongtian)

E. excellens adult (showing tufts of hairs on 1st and 2nd abdominal segments) (Xu Gongtian)

E. excellens larva (Xu Gongtian)

E. excellens larva (Xu Gongtian)

E. excellens larva
(Xu Gongtian)

on mesophyll of leaves. Older larvae consume leaves, resulting in incised leaves or leaves with shot holes. Silk webs between branches and among leaves are common during outbreaks. Mature larvae congregate at the base of trees for pupation in soil.

Prevention and treatment

1. Silviculture. Create mixed forest to prevent heavy infestation and damage.

2. Manual control. Destroy pupae from the soil through tillage during winter.

3. Conservation of natural enemies. Parasitoid *Rogas* spp. are commonly found in larvae and should be protected.

4. Chemical control. Apply 25% Dimilin at 2,000×, or aerial application of 25% Dimilin at 600 g/ha for low rate, or 450g/ha for ultra-low rate, or 2×10^{10} PIB/mL nuclear polyhydrosis virus at 150-750 mL/ha.

References Yang Youqian et al. 1982, Xiao Gangrou (ed.) 1992, Henan Forest Insect and Disease Control and Quarantine Station 2005.

(Yang Youqian)

384 *Jankowskia fuscaria* (Leech)
Classification   Lepidoptera, Geometridae

Importance   *Jankowskia fuscaria* is a sporadic pest in the past. The fast expansion of eucalyptus plantations has elevated its status as defoliating pest in recent years. Outbreaks usually bring severe economic damage to the forests.

Distribution   Hunan, Guangdong, and Guangxi.

Hosts   *Eucalyptus* spp., and *Camellia sinensis*.

Damage   Young larvae feed on one side of the leaves. Later they feed on leaf margins and eventually the entire leaf as they grow. Entire forest can be defoliated during heavy infestation.

Morphological characteristics   **Adults** Wingspan is 17-20 mm for males and 21-32 mm for females. Antennae double pectinate with a smooth end quarter for males, and filiform for females. Labial palps short, reaching the frons. Dorsal surface of head and thorax grayish brown, mixed with dark brown marks. The first abdominal segment grayish yellow dorsally. Forewings with an obtuse outer angle and a sharp declining outer margin. Inner line as well as anterior and posterior margin short and black. There is a large grayish brown to yellowish brown mark below M_1 and one row of black spots along the marginal line. Outer margin for hindwings wavy, more so in females, grayish brown to dark brown with lighter base and anterior/posterior margins. Marginal hairs grayish brown mixed with yellow. The underside of the forewings dark brown with cloudy veins. **Larvae** Head capsule 0.25-0.30 mm in width in newly hatched larvae. Dorsal surface blackish brown, with seven white intersegmental bands. Ventral surface with five white lines made of white spots. Mature larvae 48-54 mm long, with a head capsule of 3.2-3.4 mm,

J. fuscaria adults (Wang Jijian)

J. fuscaria larva (Wang Jijian)

J. fuscaria larva (Wang Jijian)

J. fuscaria female pupa and male pupa (Wang Jijian)

J. fuscaria larva parasitized by a braconid wasp (Wang Jijian)

grayish brown. Head, thorax, and anal plate pinkish in some individuals. Head contains scattered black spots.

Biology and behavior *Jankowskia fuscaria* occurs four generations a year and overwinters as larvae on trees in Guangxi. Larvae appear in April, June, August, and October in the forest each year. There are 6-7 instars in the larval stage. Prepupal period lasts for 2 days, while pupal stage lasts for 7-10 days.

Emergence occurs mostly at night. Adults climb to the trunk 2 m above the ground. They spread their wings on eucalyptus barks that are separated from the trunk. Color of wings mimic that of the bark and is not easily separated. Adults mate for several hours the next day. Females then lay about 300 eggs aggregately in one batch on young leaves in the upper crown. Eggs are covered by female body hairs. Up to 100% eggs hatch 3-4 days later.

Newly emerged larvae consume eggshells on top of the egg mass before feeding on one side of the leaves. When disturbed, they drop to the ground with silk threads, or curl into "C" shape to avoid detection. Larvae of the 2nd and 3rd instars start to disperse and feed separately, with one larva occupies one small twig. Mature larva consumes more than one leaf per day. Complete defoliation of host trees occur by late larval stage. Damage to eucalyptus trees in the spring significantly reduces current year growth, leads to tree mortality if coupled with drought and winter damage. Larval frass excreted by middle and late instars is easily recognized. They are greenish brown to dark brown in color, cylindrical (3.5 mm × 2.4 mm), with rough surface and loose texture. Larvae appear bow-like when moving and drop to the ground or nearby branches when disturbed. Mature

J. fuscaria damaged forest (Wang Jijian)

larvae fall or come down from the trunk and pupate in 2 cm deep loose soil or among leaves on the ground within 1 m of the main trunk. Larvae often choose loose soil or crevices on ground to pupate. Pupae are arranged in rows with heads up in soil crevices and separated by at least 3 cm. Horizontal arrangement on hard soil surface is also observed.

Prevention and treatment

1. Silviculture. Plant trees in appropriate density to prevent infestation.

2. Manual control. Beat leaves and branches with a stick and kill the fallen larvae. Destroy pupae in the soil through tillage

3. Protect natural enemies.

4. Microbial control. Apply *Beauveria bassiana* in the spring or *B.t.* in summer.

5. Chemical control. Apply insecticide dust to the crown at high pest density.

Reference   Wang Jijian 2002.

(Wang Jijian)

385 *Meichihuo cihuai* Yang

Classification Lepidoptera, Geometridae

Distribution Beijing, Hebei, Shanxi, Henan, Shaanxi, and Xinjiang.

Hosts *Robinia pseudoacacia*, *Ailanthus altissima*, *Populus* spp., *Salix* spp., *Malus pumila*, *Pyrus* spp., *Amygdalus persica*, *Ziziphus jujube*, *Diospyros kaki*, and *Juglans regia*.

Damage Feeding by young larvae results in notched leaves and shot holes on leaves. Older larvae consume entire leaves that may lead to complete defoliation of forest stand.

Morphological characteristics **Adults** Males brown with a wingspan of 33-42 mm. Antennae plumose. Forewings brownish yellow, with dark brown inner and outer cross veins. Area between those two veins dark brown, whereas area outside the outer vein white. There is a black stripe near the anterior margin. Median cell contains a black dot. Hindwings grayish yellow, with a black dot in the median cell and two brown outer cross veins. Females wingless, 12-14 mm long. **Eggs** Cylindrical, 0.8-0.9 mm × 0.5-0.6 mm, dark brown. **Larvae** Mature larvae about 45mm long. The side of the head capsule with dark spots. Abdomen yellowish, with grayish brown to purplish brown dorsal, subdorsal, upper spiracle, lower spiracle, and subventral lines. Spiracle lines yellowish white. Ventral line yellowish. The 8th abdominal segment contains a pair of yellow projections. **Pupae** Dark reddish brown, 12-18 mm long, with dense round punctures on the anterior half on each segment.

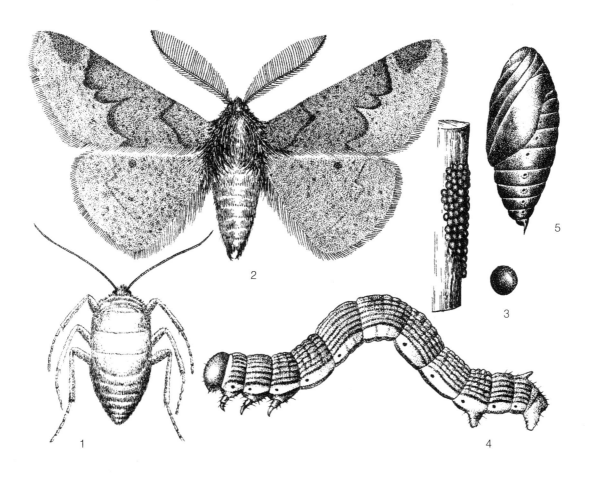

M. cihuai. 1. Adult female, 2. Adult male, 3. Egg mass and egg, 4. Larva, 5. Pupa

(Zhang Peiyi)

Anal plate contains two declining spines. **Cocoons** 15-22 mm long.

Biology and behavior *Meichihuo cihuai* occurs one generation a year and aestivates and overwinters as pupae inside earthy cocoons. Adults emerge between late February and late April. Adult period lasts for more than 50 days. Egg period lasts for 10-31 days. Eggs start to hatch in early April. Larval stage starts in early April and ends in late June. Main damage period is from mid-April to mid-May. Mature larvae come down from the trees between mid-May and mid-June and pupate between late July and mid-August. Prepupal period lasts for about 40 days, whereas pupal stage lasts for about 8 months. Adults are cold hardy and start to emerge when ground thaws in spring. They are attracted to lights and are active at night. Mating occurs right after emergence. Both sexes can mate multiple times. Sex ratio is 2:1 (females: males). Adult longevity is 4-9 days for females and 3-6 days for males. Eggs are laid on the underside of branches. Each female produces an average of 462 eggs, with a maximum of 920 eggs. Newly hatched larvae do not need to feed for a long time and usually balloon with silk threads in the wind. They feed on the tree day and night and drop to the ground when disturbed. Mature larvae spin cocoons in 3-6 cm deep soil within 30 cm around the host base.

Prevention and treatment

1. Use sticky bands around the trunk at 1 m height to control adults during oviposition period.

2. Protect natural enemies. Multiple natural enemies attack various life stages of *M. cihuai*, including parasitic wasps, predatory insects, B. bassiana, and birds. About 18% egg parasitism by *Telenomus* sp. and 10% pupal parasitism by members of Sigalphinae were observed in the field.

3. Apply fumigant at 15-23 kg/ha, or 25% Dimilin II at 1,500×, or 4.5% Beta cypermethrin at 3,000× for larval control.

References   Li Menglou 2002, Han Pinghe 2008.

(Liu Mantang, Li Menglou, Chen Youguang)

M. cihuai adult (Li Zhenyu)

386 *Milionia basalis pryeri* Druce
Classification Lepidoptera, Geometridae

Distribution Guangxi, Hainan.

Host *Nageia nagi*.

Damage Larvae feed on leaves, branches, and bark. They can consume all leaves on a tree.

Morphological characteristics **Adults** Wingspan 49.0-66.5 mm. Antennae double pectinate. Body black, forewing and hind wing black; first antennae segment, posterior border of eyes, later half of the prothorax, posterior border of each abdominal segment, base of Sc vein, and basal half of R and Cu veins with sky blue colored dots. There is a 3 mm wide orange yellow curved band on both side of the forewing. Hind wing also has a large curved orange yellow band similar to that on the forewing; there is a row of seven round or elliptical black marks on the exterior side of the orange yellow band. Ventral surface of hind wing and femora bluish black. **Eggs** Length 1.4-1.6 mm, width 0.8-1.0 mm; elongated oval; white at the beginning, becoming green, red later; greyish brown before hatch. **Larvae** Newly hatched larvae length 3.5 mm, head width 3 mm. Head pale brownish red, dorsal surface and ventral surface dark blue. Front part of the abdomen greyish white, posterior part of the abdomen greyish red. Mature larvae length 36.5-42.3 mm, head width 3.8-4.9 mm. Head, prothorax, the line above the spiracles, ventral side of the thorax, between the 6th prolegs and the last abdominal segment brownish red. Posterior border of the dorsal line on each abdominal segment with a large white mark. Each segment with seven white irregular shaped marks. **Pupae** Reddish brown, spindle shape. Female pupae length 26.0-28.5 mm, width 7.8-8.0 mm; male pupae length 22.5-24.5 mm, width 7.0 mm.

Biology and behavior *Milionia basalis pryeri* occurs three generations per year. They start to overwin-

M. basalis pryeri mating pair (mixed colors) (Wang Jijian)

M. basalis pryeri mating pair (dark brown adults) (Wang Jijian)

M. basalis pryeri silk web made before pupation (Wang Jijian)

M. basalis pryeri larval entrance (Wang Jijian)

ter as pupae between late October and early November. Adults emerge in early April. Peak damage occurs in August and September. Egg period is about one week. Larval period is 25 days. Adult period is 3-7 days. Pupal period (except the overwintering generation) is 7 days.

Adults lay eggs in crevices on barks. They lay a few to dozens of eggs at each place. After hatch, larvae crawl to the tips of shoots, spin silk, and form 30-40 cm diameter round nets that encase the tips of shoots. The nets can be easily seen from a distance. Larvae consume the leaves

M. basalis pryeri pupa (Wang Jijian)

M. basalis pryeri eggs (near hatching) (Wang Jijian)

M. basalis pryeri mating pair (yellowish brown color) (Wang Jijian)

M. basalis pryeri empty pupa (Wang Jijian)

M. basalis pryeri pupae (From left to right: male, female, pupae close to eclose) (Wang Jijian)

M. basalis pryeri galls and damaged trunks (Wang Jijian)

inside the nets. From the 3rd instar, larvae scatter and feed on leaves below the shoots. The larvae can also feed on the bark when numbers are high. Mature larvae spin silk, move to ground through a silk line or through the trunk. They pupae in dead leaves or in soil about 2 cm depth. The pupation site is mostly within 1 m from the tree trunk. Adults rest on trunk after eclosion. After one day, they can fly and mate. Mating duration can be more than two hours. They lay eggs the same day or the next day after mating.

Prevention and treatment

1. Conservation of natural enemies. Birds are important predators and should be protected.

2. Manual control. Use a bamboo stick to disturb the leaves and let the larvae fall to ground, then destroy the larvae.

3. Apply *Beauveria bassiana* during larval stage.

4. Chemical control. Spray 2.5% deltamethrin at 3,000× to foliage can reduce population by more than 95%.

References Wang Jijian, Yang Xiuhao, Liang Chen et al. 2014.

(Wang Jijian, Liang Chen, Feng Guangmiao)

387 *Percnia giraffata* (Guenée)

Classification Lepidoptera, Geometridae
Common name Large black-spotted geometrid

Importance Larvae prefer leaves on fruits trees, especially persimmon trees. It's one of the most important pests of the fruit trees. Heavy infestation weakens tree growth, which usually reduces fruit production and results in economic damage.

Distribution Beijing, Hebei, Shanxi, Anhui, Henan, Hubei, Sichuan, and Taiwan; Japan, Korea Peninsula, Vietnam, Myanmar, India, Indonesia, and Russia.

Hosts *Diospyros kaki*, *D. lotus*, *Malus pumila*, *Prunus salicina*, *P. armeniaca*, *Crataegus pinnatifida*, *Ziziphus jujube* var. *spinosa*, *Ulmus pumila*, *Morus alba*, *Sophora japonica*, *Pistacia chinensis*, and a variety of tree species, oil and agriculture crops during outbreaks.

Damage Newly hatched larvae feed only mesophyll. Older larvae disperse and usually feed on the entire leaves and flowers.

Morphological characteristics **Adults** Body about 25 mm long with a wingspan of 75 mm. Prothorax yellow with a near square black mark. Wings white and densely covered by dark brown spots. Abdomen yellow with grayish brown and black spots. **Eggs** Oval, about 0.8-1.0 mm long, bright green initially, turn to dark brown later. **Larvae** Body about 55 mm long at

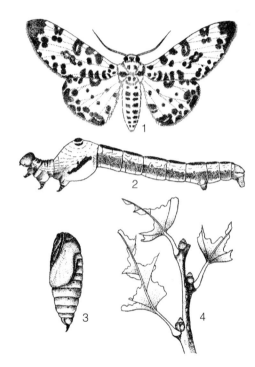

P. giraffata. 1. Adult, 2. Larva, 3. Pupa, 4. Damage (Xu Tiansen)

P. giraffata adult (Li Zhenyu)

P. giraffata adult (Li Zhenyu)

maturity. Head yellowish brown and covered by white particles. Dorsal line dark brown, with one wide yellow band on each side that contains a black line. The 4th and 5th abdominal segments are especially enlarged, and each has a pair of black oval eyespots at the enlarged portion. **Pupae** Dark brown, about 23 mm long. The enation on anterior thorax contains a "+" formed by the cross of the transverse carina and central thoracic carina. Anal plate has one spine.

Biology and behavior *Percnia giraffata* occurs two generations a year and overwinters as pupae in soil. Pupae start to emerge in late May and adult emergence peaks from late June to early July. Females begin to lay eggs in early June. Egg-hatching starts in early June and peaks in mid- to late July. Mature larvae begin to pupate in mid-July with most pupate in early August. Adults of the first generation appear in late July, with most adults emerge between early to mid-August. Larvae of the second generation occur in early August and peak in mid- to late August. Larvae mature in early September and pupation ends in early October.

Newly hatched larvae feed on mesophyll. Older larvae disperse and feed all day long separately. When disturbed they balloon with silk thread, although they may come back to the same sites later. Mature larvae prefer loose and moist soil for pupation, resulting in more pupae in northern sites.

Adults are attracted to lights or water (slightly). They are active at night and rest during the day on twigs, rock surface. Mating occurs right after emergence. Eggs are laid in masses on the underside of the leaves or young branch tips.

Prevention and treatment

1. Silviculture. Proper fertilizer and water management to improve tree vigor, proper pruning for better sunlight penetration, timely removal of infested branches and weeds for reduced source population, and proper temperature and humidity control.

2. Manual removal of larvae during outbreaks. Pupae removal through tillage during the winter and spring.

3. Biological control. Apply *B.t.* or NPV suspension for larvae control.

4. Chemical control. Apply 20% Dimilin at 7,000× before larvae reaching the 3rd instar.

References Zhu Hongfu, Wang Linyao, Fang Chenglai 1979, Xu Yiling et al. 1984, Xu Gongtian 2003, Qiu Qiang 2004, Xu Gongtian and Yang Zhihua 2007.

(Zhang Xue, Li Guangwu)

388 *Semiothisa cinerearia* (Bremer et Grey)

Classification Lepidoptera, Geometridae
Synonym *Chiasmia cinerearia* (Bremer et Grey)

Distribution Beijing, Hebei, Liaoning, Jiangsu, Zhejiang, Anhui, Jiangxi, Shandong, Henan, Tibet, Shaanxi, Gansu, and Taiwan; Japan.

Hosts *Sophora japonica*, *S. japonica* var. *pendula*, and *Robinia pseudoacacia*.

Damage Larval feeding often completely defoliates the affected trees.

Morphological characteristics **Adults** Body yellowish brown with dark brown spots, 12-17 mm long with a wingspan of 30-45 mm. Forewings dark-colored outside, with a black triangle mark near the top corner. Hindwings with darker and protruded outside, which contains a small black spot. **Eggs** Oblong oval, 0.58-0.67 mm long and 0.42-0.48 mm wide, truncated at one end, with small honey comb shaped depressions on the surface. **Larvae** Body 20-40 mm long at maturity. Yellowish brown initially, turn to green later. Diphormic, with one form green from the 2nd to the 5th instars, and the other form contains lateral dark brown stripes or blotches from the 2nd to the 5th instars and purplish red in the 5th instars. **Pupae** Purplish brown, 16 mm long and 5.6 mm wide, with one anal spine which contains two setae. Those two setae slightly about parallel in males but those in females direct towards the lateral sides.

Biology and behavior *Semiothisa cinerearia* occurs three to four generations a year and overwinters as pupae in Beijing. Each female produces an average of 420 eggs. Eggs start to hatch in early May. Most damage (93% leaf consumption) is caused by the 5th and 6th instar larvae. Larvae can drop and spit silk threads when dis-

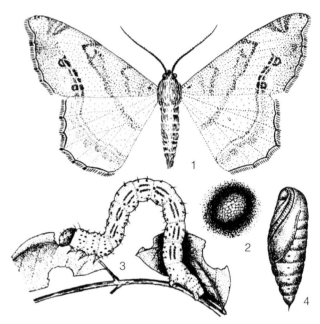

S. cinerearia. 1. Adult, 2. Egg, 3. Larva, 4. Pupa (Zhang Peiyi)

S. cinerearia adult (Xu Gongtian)

S. cinerearia adult (Xu Gongtian)

S. cinerearia eggs (Xu Gongtian)

S. cinerearia 1st generation young larva (Xu Gongtian)

S. cinerearia mid-aged larva (Xu Gongtian)

S. cinerearia 1st generation young larva (Xu Gongtian)

S. cinerearia larva (Xu Gongtian)

S. cinerearia damage (Xu Gongtian)

turbed. Mature larvae enter soil for pupation. Males are attracted to lights. Pupae enter diapause during the winter.

Prevention and treatment

1. Manually removal of the overwintering pupae from shallow soil.

2. Protect natural enemies such as *Casinaria* sp.

3. Apply 50% Dimilin I at 10,000× or 2% imidacloprid at 2,000× during egg hatching for controlling the 1st generation.

References Xiao Gangrou (ed.) 1992, Capital Reforestation Commission 2000, Yu Guoyue 2015.

(Huang Jingfang, Li Zhenyu)

389 *Thalassodes quadraria* Guenée
Classification   Lepidoptera, Geometridae

Importance   Heavy infestation leads to complete defoliation of the host trees, which seriously impacts host growth and reduces their ornamental and ecological functions. Outbreaks were observed in areas in Guangdong in the 1980's.

Distribution   Shanghai, Jiangsu, Zhejiang, Fujian, Jiangxi, Hunan, Guangdong, Guangxi, Chongqing, Yunnan, and Taiwan; India, Japan, Thailand, Malaysia, and Indonesian.

Hosts   *Cinnamomum camphora*, *Litchi chinensis*, *Dimocarpus longan*, *Mangifera indica*, and *Camellia sinensis*.

Damage   Larvae feed on host leaves. The 1st and 2nd instars feed on mesophyll on leaf surface, leaving leaf veins and lower epidermis intact. Feeding by the 3rd instars creates notches or shot holes on host leaves. Leaf consumption increases significantly by the 4th instar, by which time they start feeding along the edges. The 5th instars feed on the entire leaf, leaving only the stalk and main vein intact.

Morphological characteristics   **Adults** Body 12-14 mm long with a wingspan of 33-36 mm. Head gray. Compound eyes black. Antennae grayish yellow. Thorax and abdomen bright green dorsally, and gray ventrally and laterally. Forewings bright green and covered by white sparkling broken marks. Forewings grayish white mixed with yellow at the margins, and lighter for the underside. Abdomen grayish white mixed with yellow laterally. There are two straight fine cross lines on both wings. **Eggs** Oval, smooth, 0.6 mm in diameter. Yellowish white initially, turn to yellowish later, and purple before hatching. **Larvae** There are six instars. Newly hatched larvae similar to leaf buds, yellowish. They resemble tender twigs from the 2nd instar, purplish red with a hint of green. Mature larvae 38-40 mm long, purplish green. They hang on twigs with their anal prolegs with thorax and abdomen projected and mimic twigs. Larva with a large head and slightly tapping abdomen. Head yellowish green, two sides of the vertex protruded, posterior border of vertex with grooves. Frontal area depressed. Abdomen yellowish green. Spiracle line yellowish, obvious. Other body lines blurring. Tip of the abdomen pointed. Thoracic legs and prolegs yellowish green. **Pupae** Spindle-shaped, 17-21 mm long, purplish brown,

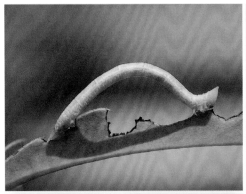

T. quadraria larvae (edited by Zhang Runzhi)   *T. quadraria* adult (Liu Xingping)

turn to purplish green later. Wing buds bright green. Anal spine bifurcated.

Biology and behavior *Thalassodes quadraria* occurs four to six generations a year in Nanchang in Jiangxi, province and four generations a year in Shanghai city and Nanping in Fujian province with overlapping generations. It overwinters as late instar larvae on host branches. Overwintering larvae resume feeding between late February and late March the next year. For those with four generations a year, adult emergence occurs between mid-April and mid-May for the overwintering generation. The adult period for the 1st to the 3rd generations are mid- to late June, early to mid-August, and early to mid-October, respectively. Larvae of the 4th generation overwinter by mid- to late October.

Most adults emerge at night with the peak emergence occurs between 17:00-23:00. They stay between leaves in the crown during the day and start to take flights at dusk. They are attracted to lights and new growth. Mating occurs the same night after emergence and lasts for about an hour. Most females mate only once while some can mate twice. They start to lay eggs the next day or even the same day. Eggs are laid individually in bark crevices, branch joints or the underside of the leaves. Each female produces an average of 168 eggs, ranging from 87-513 eggs. Egg stage lasts for about 7 days. Larvae can migrate for food source.

Prevention and treatment

1. Silviculture. Destroy overwintering pupae through tillage or prune off infested branches.

2. Light trapping. Trap adults with lights, especially for the overwintering generation.

3. Biological control. Protect natural enemies such as

T. quadraria adult (Gui Aili)

Apanteles sp., *Cantheconidea furcellate*, Vespidae, Scoliidae, and sparrows.

4. Chemical control. Apply 90% crystal trichlorfon at 800×, 25 g/L lamda-cyhalothrin suspension or 40% isocarbophos at 1,000×, 18% bisultap water suspension at 500×, 90% Monosultap or cartap at 1,500-2,000×, 2.5% deltamethrin or 10% cypermethrin, 25% Dimilin at 2,000×, 0.36% Matrine water suspension at 1,000×, or 8,000 IU/uL *B.t.* suspension at 1,000× for controlling young larvae. Also apply *B.t.* galleriae (3×10^8 spores/g) at 1,000× 2-3 times every 7 days.

References Zhang Shimei and Hu Meicao 1984; Chen Shunli, Li Yougong, Li Qinzhou et al. 1989; Huang Zhongliang 2000; Pei Feng, Sun Xingquan, Ye Lihong et al. 2008.

(Liu Xingping, Zhang Shimei, Hu Meicao)

390 *Zamacra excavate* (Dyar)

Classification Lepidoptera, Geometridae
Synonym *Apochima excavate* Dyar

Importance This polyphagous pest causes reduced growth during light infestations, complete defoliation or host mortality during heavy infestations, and serious damage to fruit production.

Distribution Beijing, Hebei, Liaoning, Jilin, Inner Mongolia, Shandong, Henan, Hubei, Shaanxi, Ningxia, and Xinjiang; Korea Peninsula, and Japan.

Hosts *Robinia pseudoacacia*, *Populus tomentosa*, *Sophora japonica*, *Ulmus pumila*, *Juglans regia*, *Koelreuteria paniculata*, *Morus alba*, *Salix* spp., *Acer truncatum*, *Fraxinus chinensis*, *Lonicera maackii*, *Philadelphus pekinensis*, *Malus prunifolia*, *Pyrus bretschneideri*, *Syringe* spp., *Chaenomeles speciose*, *Malus pumila*, *Amygdalus persiea*, *Citrus reticulate*, *Vitex negundo* var. *heterophylla*, *Ailathus altissima*, *Castanea mollissima*, *Ziziphus jujube*, *Zizyphus jujube* var. *spinosa*, *Prunus sibirica*, and *Ligustrum obtusifolium*.

Damage Larvae feed on flower buds, leaves, and young fruits. Heavy infestation can result in complete host defoliation, leaving only stalks for leaves. They cause fruits fall prematurely.

Morphological characteristics **Adults** Females grayish brown, 14-15 mm long with a wingspan of 40-50 mm, antenna filiform, fore wing silver with three grayish brown transverse lines. Wings roof-like at rest. Males dark brown, 12-14 mm long with a wingspan of 38 mm, abdomen contains one hair cluster apically, antennae pectinate. **Eggs** Oval, grayish brown and depressed in the middle. **Larvae** Yellowish blue, 30-35 mm long.

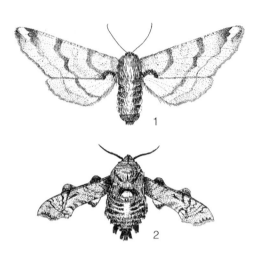

Z. excavate. 1. Adult female, 2. Adult male (Hu Xingping)

Head brown. Prothorax yellow laterally. Dorsal surface of the abdominal segments 1-4 with ochreous spines. Those on segment 2-4 relatively long. There is a pair of brownish green spines on the dorsal surface of the 8th abdominal segment, and one greenish spine on each side of the abdominal segments 2-5. Intersegmental membrane yellow. Subdorsal lines on abdominal segments 4-8 greenish. Spiracle line dark green. **Pupae** Reddish brown, spindle shape, 14-17 mm long, with two hard anal spines. **Cocoons** Oval, grayish.

Biology and behavior *Zamacra excavate* occurs one generation a year and overwinters inside cocoons under soil surface at trunk base or on bark in Beijing,

Z. excavate adult (Xu Gongtian)

Z. excavate adult (Xu Gongtian)

Z. excavate eggs (Xu Gongtian)

Z. excavate eggs (Xu Gongtian)

Z. excavate young larva (Xu Gongtian)

Z. excavate mid-aged larva (Xu Gongtian)

Z. excavate mature larva (Xu Gongtian)

Hebei, Shanxi, and Hubei. Adults emerge in mid-March. Mating occurs right after emergence. Adults can feign death. They fall to the ground after being disturbed. Adults are weak fliers and live for about 7 days. Eggs are rectangular in shape and laid on host branches. Eggs hatch 15-20 days later. Larvae of the first generation appear in early to mid-April. There are four instars in the larval stage. The 1st and 2nd instars are generally active at night and rest during the day. By the 3rd and 4th instars, they feed both during the day and at night. They gradually move from the top of the crown to the lower crown as they grow. When disturbed, larvae will hide their heads under the body. Larvae of all instars can drop with silk threads after being disturbed, or during migration for more food when population density becomes too high.

Z. excavate damage to jujube tree (Xu Gongtian)

Prevention and treatment

1. Quarantine. Inspect root system of seedlings to prevent disperse of the overwintering pupae.

2. Physical control. Trap adults with black lights during adult flight season.

3. Manual control. Dig out overwintering pupae from the base of the tree, or allow poultry to feed on them. Remove egg masses or remove infested branches by hand.

4. Chemical control. Apply 20% Dimilin suspension at 7,000×, or *B.t.* suspension (8,000 IU/μL) at 1,000-5,000× for larval control.

References Xiao Gangrou (ed.) 1992, Niu Jianzhong et al. 1999, Niu Jianzhong 2000, Xu Gongtian 2007.

(Liu Huan, Li Suo, Zhao Huaiqian, Zhang Shiquan)

391 *Xerodes rufescentaria* (Motschulsky)

Classification Lepidoptera, Geometridae

Synonyms *Zethenia rufescentaria* Motschulsky, *Endropia consociaria* Christoph, *Zethenia rufescentaria chosenaria* Bryk

Importance *Xerodes rufescentaria* is a sporadic pest of larch forest in northern China with outbreaks recorded in some areas in Heilongjiang in the 1990s.

Distribution Beijing, Shanxi, Liaoning, Jilin, Heilongjiang, Jiangxi, and Taiwan; Russia, Korea Peninsula, and Japan.

Hosts *Larix* spp.

Damage Larval feeding on leaves defoliates the hosts under heavy infestation. Consecutive defoliations will kill young trees.

Morphological characteristics **Adults** Wingspan 29-41 mm. Forewings with an acutely projected top corner. Outer margin depressed between top corner and M_2. M_2 extends inward with an angle. Hindwings with strong wavy outer margin, with Cu_1 extends directly to the anal corner. **Eggs** Elongated elliptical; 0.5 by 0.3 mm in size; surface with rows of sculptures; green at beginning, puplish brown after 2-3 days. **Larvae** The middle and late stage larvae have an inverted Y line on the head and a diamond-shaped transverse mark on the tergum of each segment. The diamond-shaped mark is also marked by two small white spots anteriorly and one black spot in the middle. There is also an oval yellowish green spot on each side of each body segment. **Pupae** 12-15 mm long; reddish brown; head with fine pubescence.

Biology and behavior Xerodes rufescentaria occurs one generation a year and overwinters as cocoons under leaf litters in Heilongjiang. Emergence starts in mid-May and peaks from late May to early June, with most adults emerge between 11:00-17:00. Adults rest on weeds in forests and fly 2-3 m above the ground when disturbed. They are attracted to lights. Males live for 1.5-6.5 days while females live for 3-9 days. Mating occurs at night 1-2 days after emergence. Eggs are laid singly a few days later on host needles, in bark crevices, or on branches. Each female produces 2-65 eggs, with an average of 20 eggs. Egg period lasts for 8-14 days, with an average of 12 days. Eggs begin to hatch in late June. Newly hatched larvae feed on chorion and can balloon with silk threads when disturbed. Larvae are very active. Larvae go through six instars. Mature larvae leave trees, enter leaf litter and pupate. Larvae pupate in 1-2 cm deep

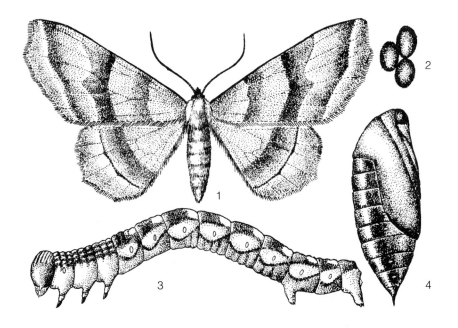

X. *rufescentaria*. 1. Adult, 2. Eggs, 3. Larva, 4. Pupa (Zhang Peiyi)

X. rufescentaria adult (Li Chengde)

soil if the leaf litter layer is too thin.

Prevention and treatment

1. Silviculture. Maintain proper canopy coverage and growth of the trees.

2. Physical control. Trap adults with black lights during adult fly season and trap overwintering pupae by covering tree base with weeds for later destruction.

3. Conservation of natural enemies. Protect nature enemies that include 23 species of ichneumonids, two species of tachnids, two species of predatory birds, one species of spiders, one species of ants, three species of microorganisms.

4. Biological control. Apply 2.0×10^9 PIB/mL ZrNPV to control the 2nd and 3rd instar larvae.

5. Chemical control. Application of 25% Dimilin suspension at 2,500× (a.i. 150 g/ha) during the 3rd instar period resulted in 89.85% efficiency.

References Xiao Gangrou (ed.) 1992; Yue Shukui, Yue Hua, Fang Hong et al. 1994; Liu Jiazhi, Zhang Guocai, Yue Shukui et al. 1996; Xi Jinghui, Pan Hongyu, Chen Yujiang et al. 2002; Liu Linjun 2005; Fu Haibin, Li Junhuan, Jiangli et al. 2007; Wang Jianguo, Lin Yujian, Hu Xueyan et al. 2008; Tao Wanqiang, Pan Yanping, Liu Huan et al. 2009.

(Han Huilin, Li Chengde, Liu Xuanji)

392 *Epicopeia mencia* Moore

Classification Lepidoptera, Epicopeiidae
Common name Elm caterpillar

Distribution Beijing, Hebei, Liaoning, Jilin, Heilongjiang, Jiangsu, Zhejiang, Jiangxi, Shandong, Henan, Hubei and Guizhou.

Host *Ulmus pumila*.

Morphological characteristics **Adults** Body 19-22 mm long with a wingspan of 60-85 mm. Maxillary palp red. Antennae pectinate. Wings brown. Anal corner of the hind wing tail-like, with two crescent or round red spots on its outer margin. Dorsal surface of the abdomen black, the lateral side of the abdomen red except the area near the spiracles black. Each of the two humeral plates with one red spot. Posterior border of the last few segments red. **Eggs** Round, yellow with luster. **Larvae** Body 44-58 mm long at maturity, Head black. Body greenish. The entire body is covered by thick wax which makes the insect white in color. Dorsal line yellow. There is a black circular spot on each segment posteriorly. Subdorsal line and upper spiracle line are made of brown marks. Spiracle line yellow. Spiracle valves black. Intersegment membrane yellow. There is a near triangular brown mark laterally on the prolegs. Proleg crotchets black. Body setae yellowish.

Pupae Dark brown covered by oval earthy cocoons.

Biology and behavior *Epicopeia mencia* occurs one generation a year and overwinter as pupae in soil in Shandong. Adults begin to emerge in June. They look like swallowtails and are active during the day. Eggs are

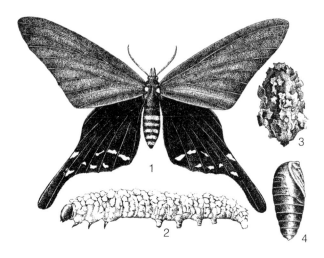

E. mencia. 1. Adult, 2. Larva, 3. Cocoon, 4. Pupa
(Liu Shiru, Zhu Xingcai)

E. mencia adult (Xu Gongtian)

E. mencia larva (Xu Gongtian)

E. mencia larva (Xu Gongtian)

E. mencia larvae (Xu Gongtian)

laid singly on leaf surface. Newly hatched larvae feed on mesophyll. Old larvae consume entire leaves and prefer young leaves on branch tips. Most damage occurs in August. After larvae become mature in September, they enter soil, pupate, and overwinter.

Prevention and treatment

1. Manual control. Kill larvae after dislodging them from the tree or kill mature larvae on the ground before they entering soil.

2. Apply 25 g/L lamda-cyhalothrin at 1,000× or 90% trichlorfon at 800× to control newly hatched larvae.

References Guizhou Forestry Bureau 1987, Xiao Gangrou (ed.) 1992, Wang Lizhong et al. 1998.

(Xie Yingping, Liu Shiru)

393 *Epiplema moza* Butler

Classification Lepidoptera, Uraniidae

Importance *Epiplema moza* is an important defoliator of paulownia. Larval feeding on host results in notched leaves and leaves with only veins intact when populations are high. Infestation reduces tree growth and even leads to tree mortality with significant damage to local economy. An infestation of 5% of the trees will result economic damage. It has become an important pest of paulownia forests since the 1980s.

Distribution Anhui, Fujian, Jiangxi, and Hunan; India, Japan.

Host *Paulownia fortunei*.

Damage Larvae feed on host leaves. The 1st and 2nd instar larvae feed on mesophyll. Feeding by the 3rd instars creates notched leaves or shot holes on the leaves. The 4th instar larvae consume entire leaves, with only stalk and main veins intact.

Morphological characteristics **Adults** Body 6.2-7.4 mm long with a wingspan of 20-27.5 mm. Forewings gradually increase in width from base to the tip, with an acute top corner and projected anal corner. There is an angular projection in the middle of the outer margin. Medial cross vein and outer line dark brown and arc-shaped, with one dark brown round mark near the posterior margin of the outer line. Outer marginal line with an oval black mark near the top corner, which contains two black spots in females and one black spot in males. Hindwings with two black brown arc-shaped dark brown marks, and two small tail-like projections on the outer margin. **Eggs** Bun-shaped and slightly depressed in the middle, about 0.7 mm in diameter. Creamy white initially and turn to yellowish brown later. They turn to reddish before hatching. There are various radiating carinae on the surface. **Larvae** Creamy white initially, turn to greenish two days later. Each thoracic segment contains six black papillae, with the pair on prothorax more conspicuous than others. There are also six black papillae arranged in two rows (4 in the front and 2 in the back) on each abdominal segment. The 3rd instar larvae green, prothorax with 12 black papillae in two rows (6 in each row), and 8 black papillae in one row on meso- and

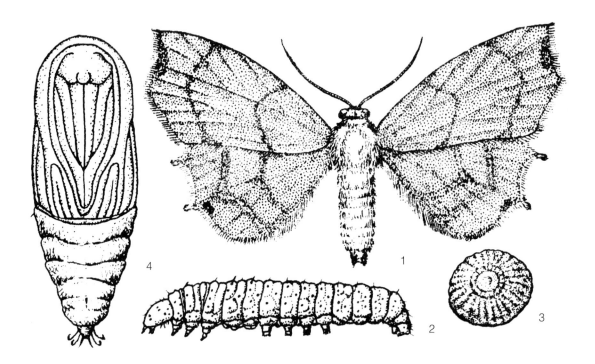

E. mencia. 1. Adult, 2. Larva, 3. Egg, 4 Pupa (Gui Aili)

metathorax. There are 10 black papillae in two rows (6 in the front and 4 in the back) on each abdominal segment except the first segment, which contains only 8 papillae. Each papilla contains one black or yellowish brown seta. Mature larvae 13.6-17.8 mm long with a head capsule of 1.4 mm in width. They turn to yellowish green and purplish green from green before pupation with reduced black papillae on thorax and abdomen.

Pupae Body 8.5-9.6 mm long and 3.4-4.2 mm wide. There is a large dark brown bifurcated anal spine. There are two twisted papillae on each side of the spine and one papilla on the dorsal side of the spine. Wing buds reach the middle of the 4th abdominal segment.

Biology and behavior *Epiplema moza* occurs three to five generations (mostly five generations) a year and overwinters as pupae in soil in Nanchang, Jiangxi. Adults of the overwintering generation start to emerge in early April and later May, lay eggs between mid-April and early June. Eggs hatch between late April and mid-June. Various life stages of different generations co-exist in the field until early November. Larvae of the 3rd and 4th generation that pupate before mid-September can emerge as adults and lay eggs within the same year. Larvae from those eggs continue development into the 4th and 5th generation, respectively. Part of those larvae pupate by late September and overwinter as diapause pupae. Some individuals can overwinter for up to eight months. Egg stage generally lasts for 3-7 days. Larval, pupal, and adult stage lasts for 10-16, 10-14, and 4-8 days respectively.

Emergence occurs mostly between 15:00-22:00 and peaks between 16:30-18:00. Adults stay on the underside of leaves or on shady branches during the day and become active at night or in the afternoon in cloudy days. They are attracted to lights. Mating occurs mostly between 19:00-23:00 one day after emergence. Females lay their eggs singly on the back of the leaves the next evening. Each female produces 30-45 eggs. Newly hatched larvae feed on mesophyll. Larvae older than the 3rd instars consume significant more leaves. Larval population clumps in paulownia forests, which usually defoliate the hosts. Damaged leaves with only the main veins intact.

E. mencia adult (Liu Xingping)

Mature larvae fall to the ground through silk threads and construct cocoons for pupation. The cocoons are covered with soil particles. Larvae mainly infest trees of 2-7 m tall. They prefer leaves in the middle and lower part of the crown. Trees in forest suffer more than those on the edges.

Prevention and treatment

1. Destroy pupae through tillage during pupal stage to prevent adult emergence.

2. Trap adults with black lights.

3. Conserve natural enemies. Major natural enemies include *Chrysopa pallens*, *Temelwcha stangli*, *Ommatius chinensis*, ants, frogs, birds, and *Beauveria bassiana*.

4. Application of trichlorfon, phosmet, deltamethrin, omethoate, or chlordimeform suspension at label rate during peak larval populations of the 2nd and 4th generation resulted in 95% control.

References Zhang Shimei and Hu Meicao 1983, Deng Xiuming et al. 2002.

(Liu Xingping, Zhang Shimei, Hu Meicao)

394 *Mirina christophi* (Staudinger)

Classification Lepidoptera, Endromidae

Distribution Liaoning, Jilin, and Heilongjiang; Europe.

Hosts *Lonicera maackii* and *L. ruprechtiana*.

Morphological characteristics **Adults** Females are 15-20 mm long with a wingspan of 39-48 mm. Males are 16-17 mm long with a wingspan of 36-43 mm. Body yellowish white, robust, and covered by thick scales. Head small, hidden inside prothorax. Antennae double pectinate, gray, with yellowish white stem. Compound eyes black. Ocelli absent. Proboscis not well developed. Maxillary palp curved upwards. Forewings yellowish white. Median cell contains a black kidney-shaped mark, and a black circular spot above it. Hindwings darker with three arc-shaped brown bands. Median cell contains one black mark. Femur and tibia covered by long hairs. There is a tibial epiphisis on the front leg internally. Abdomen yellowish white, lighter than thorax. **Eggs** Oblong oval, 1.9-2.0 mm long and 1.5-1.6 mm wide, slightly depressed centrally. Greenish initially, turn to purplish red gradually, and purplish black with yellow marks before hatching. **Larvae** Body color varies with instars. First instars with black body spines. Body of the 2nd instars black dorsally and dark red laterally. Body and spines below the spiracle line greenish in the third instars. Fourth instars have a tri-colored (black, white and red) band on the back and black tipped bright green spines. The lateral side of each of the segment in the 5th instar larvae contains a declivous greenish mark, abdomen brownish green ventrally. There is one body spine above the front legs laterally, and four spines laterally on prothorax and mesothorax. The two spines closest to the median thoracic line are the longest among all. There are five spines on each side of the abdominal segments 1 and 2, four spines on each side of the segments 3-7, two spines and on each side and one dorsal spine on segment 8, and three spines on each side of segment 9. Length of the body spines reduced from thorax towards end of the abdomen. There is an eversible gland at the base of the proleg laterally on abdominal segments 3 and 4, which can extend when disturbed. Proleg crochets in biordinal semi-circular. Body length is 6.2-8.1, 10.3-11.4, 16.5-18.1, 27.0-30.5, and 39.0-53.0 mm in the 1st, to the 5th instars, respectively. Head capsule width of the 1st to the 5th instars is 0.8-0.9, 1.2-1.3, 1.6-1.9, 2.3-2.5, and 3.1-3.4 mm, respective-

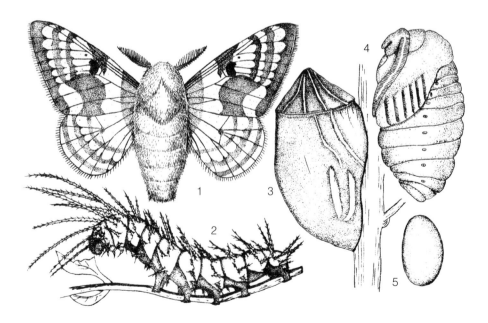

M. christophi. 1. Adult, 2. Larva, 3. Cocoon, 4. Pupa, 5. Egg (Sun Lihua)

M. christophi adult
(Xu Gongtian)

ly. **Pupae** Females 18.0-20.5 mm long and 8.0-9.0 mm wide. Smoky black, smooth without hairs but winkled. Antennae shorter than forewings. Males 15-18 mm long and 7.0-7.5 mm wide, antennae equal to or longer than forewings. Tip of abdomen smooth without anal spines. **Cocoons** Body 20-24 mm long and 9-11 mm wide, urniform, yellowish brown, grayish brown or dark brown with white carinae. There are carinae on the vertex.

Biology and behavior *Mirina christophi* occurs one generation a year and overwinters as pupae in Liaoning. Adults start to emerge and lay eggs in late April. Egg stage lasts for 11-13 days. Eggs hatch in early May. Larval stage lasts for 39-41 days with five instars, with each instar lasts for 8, 6, 7, 7, and 11 days, respectively. Mature larvae begin to construct cocoons for pupation in late June and overwinter.

Most adults emerge between 12:00-15:00. After emergence, adults wave their wings as they crawl until wings rest on the back roof-like. They stay like this with antennae hiding beneath the wings before mating. Peak emergence falls at early May. Adults mate about 6 hours after emergence. Mating flights generally take place at 19:00. Males are most active at 20:00 when they fly between branches and leaves looking for females. Mating starts at 21:00 and lasts for 6-8 hours. Eggs are laid singly on the back of the leaves of small branches at the same night. Each female can lay eggs eight times, with the most eggs laid in the first time, and gradually reduce in numbers toward the last time. Each female can produce an average of 143 eggs, ranging from 122-163 eggs. Sex ratio is 1:1 (female:male). Adults are attracted to lights and can mimic death. Males live for 8-10 days and females for 10-13 days.

Most eggs hatch between 6:00-12:00. Eggs become transparent before hatching. Hatching larva chews a hole near the insemination hole for about an hour before coming out. Newly hatched larvae crawl on leaf surface for 1-2 minutes until reaching the resting area, where they stay for 2-3 hours before feeding. They feed more than 20 times in a 24-hour cycle. When disturbed, they'll rapidly extend their body spines and wave their heads. Feeding times increase as the larvae grow.

Mature larvae stop feeding and find branch tips for pupation two days before pupation. Ideal pupation locations are branches of 2 cm in diameter and 10 cm above the ground. Cocoons are constructed by the mature larvae. Most cocoons are constructed before 6:00. It takes a larva more than 2 hours to finish the cocoon. Prepupal stage is about 8 days. Outbreaks occur most along the edges of larch forest.

Prevention and treatment

1. Manually remove branches containing cocoons.
2. Protect and utilize natural enemies including *Araniella displicata*, *Pholcus* sp., and *Xysticus ephippiatus*.
3. Trap adults with lights.

Reference Sun Lihua et al. 1987.

(Sun Lihua)

395 *Pyrosis eximia* Oberthür

Classification Lepidoptera, Lasiocampidae
Synonym *Bhima eximia* Oberthür

Importance This polyphagous pest is able to cause serious damage during outbreaks.

Distribution Shanxi, Jiangsu, Hunan, and Shaanxi; Korea Peninsula, and Russia.

Hosts *Quercus* spp., *Castanea mollissima*, *Juglans regia*, and *Malus pumila*.

Damage Larvae feed on leaves. They can defoliate the whole trees during outbreaks.

Morphological characteristics **Adults** Wingspan is 67-74 mm in females and 45-56 mm in males. Wings dark red. Antennae black. Forewings with yellowish white inner and outer cross lines and sub-marginal line. Area between inner and outer cross lines dark brown. There is a grayish white mark at the end of the median cell. Hindwings contains two wide yellow bands in the center, and three protruding dark brown marks on the outer half. **Eggs** Cylindrical, grayish brown and cov-

P. eximia. 1. Adult female, 2. Adult male
(Hou Boxin)

ered by thick scales. **Larvae** Mature larvae 5-6.5 mm long, grayish brown, pink, or greenish, with dark brown marks on the head. Anterior of prothorax with a row of setae one each side. The two rows of verrucae along the spiracle line and below spiracle line with short setae. The rest of the body is sparsely covered by long setae. **Pupae** Length 3.5-3.8 mm in females and 2.2-3.1 mm in males. Head, thorax, and back of the abdomen dark brown. Abdomen slightly reddish brown ventrally. Head and back of the thorax covered by dense villi.

Biology and behavior *Pyrosis eximia* occurs one generation a year and overwinters as eggs. Eggs hatch in early April. Mature larvae construct cocoons by early September. Adults start to emerge and mate during the day in late October. One male can mate with two females. Females lay eggs on branches in the lower crown after mating. Each female produces 345-599 eggs in 3-4

P. eximia adult female (above) and adult male (below)
(Li Menglou)

egg masses. Adults live for 3-4 days. Egg stage lasts for 148-154 days. There are 7-8 instars in the larval stage which lasts for 146-163 days. Duration for each of the seven instars is 9-10, 5-6, 5-6, 5-6, 9-11, 17-20, 27-89, and 68-79 days, respectively. Some mature larvae start to pupate at 7th instars. Larvae younger than the 3rd instars feed aggregately on leaf surface, whereas older instars feed separately. Mature larvae construct cocoons with leaves on branches.

Prevention and control

1. Plant mixed forests with proper thinning practice.
2. Manually remove eggs, larvae, and cocoons.
3. Trap adults with light traps.
4. Biological control. Apply mixture of *Beauveria bassiana* and *B.t.* ($2-3\times10^8$ spores/mL) at 10× for larval control.
5. Chemical control. Apply 20% Dimilin suspension at 240-300 mL/ha, 20% cypermethrin at 30-45 mL/ha, 24.5% emamectin benzoate and buprofezin at 73.5-88 g/ha.

References Gao Desan and Zhang Yiyong 1998, Liu Youqiao and Wu Chunsheng 2006.

(Xie Shouan, Li Menglou, Tong Xinwang)

396 *Pyrosis idiota* Graeser

Classification Lepidoptera, Lasiocampidae
Synonym *Bhima idiota* (Graeser)

Distribution Beijing, Hebei, Shanxi, Inner Mongolia, Liaoning, Jilin, Heilongjiang, Anhui, Henan, Hubei, Guangdong, Shaanxi; Korea Peninsula, Russia, and Japan.

Hosts *Malus pumila*, *Pyrus* spp., *Prunus armeniaca*, *Ponulus* spp., *Salix* spp., *Xanthoceras sorbifolium* and *Acer saccharum*.

Damage Larval feeding creates notched leaves or leaves with shot holes.

Morphological characteristics Adults Males 22-26 mm long with a wingspan of 45-53 mm, dark brown. There is a circular or triangular white mark on the tip of the median cell on the forewings, and two yellowish bands connected by a light yellow mark on the outer portion of the hindwings. Females 32-37 mm long with a wingspan of 65-71 mm. **Eggs** Oval, 1.6 mm long, yellowish. **Larvae** Body 70-78 mm long, yellowish brown, with yellow marks on the head. Subdorsal line dark brown. Meso- and metathorax each contains a black rectangular mark made of black hair bundles. There is also a black hair bundle on the dorsal surface of the 8th abdominal segment. Prothorax contains two verrucae anteriorly. Metathorax, mesothorax, and abdominal segment 1-8 each contains a verruca laterally. Each verruca

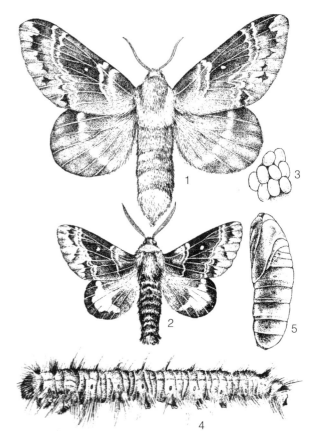

P. idiota. 1. Adult female, 2. Adult male, 3. Eggs, 4. Larva, 5. Pupa (Zhu Xingcai)

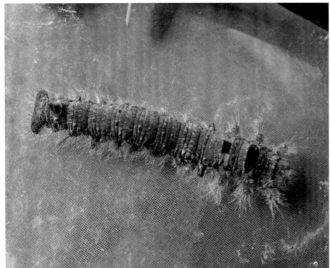

P. idiota larva (Qi Runshen)

P. idiota larva (Qi Runshen)

P. idiota adult female (Xu Gongtian)

P. idiota adult male (Xu Gongtian)

contains yellowish white long hairs and black hairs. Newly hatched larvae black and covered by white hairs. **Pupae** Brown to black. **Cocoons** oblong oval, 39-53 mm long, dark gray or earthy yellow and covered by hairs.

Biology and behavior *Pyrosis idiota* occurs one generation a year and overwinters as prepupae aggregately after mature larvae constructing cocoons inside tree hollows, bark crevices, or ground cover. Pupation begins in mid-April to mid-May. Pupal stage lasts for 15-30 days. Adults emerge between mid-May and late June. Mating occurs right after emergence. Eggs are laid on 1-2 year-old branches the next day. Each female produces 250-370 eggs. Adults rest during the day and are active at night. They are attracted to lights and can live for 4-6 days. Egg stage lasts for 10-15 days. Eggs hatch between early June and early July. There are eight instars in larval stage. Newly hatched larvae feed aggregately near egg sites, and disperse afterwards. They further disperse when they are in 5-6th instars. They feed during the night. Larval stage lasts for more than 100 days. By mid- to later September until early October, mature larvae come down to construct cocoons for overwintering.

Prevention and treatment See methods for *Pyrosis eximia*.

References Xiao Gangrou (ed.) 1992, Zhou Jiaxi et al. 1994, Liu Youqiao and Wu Chunsheng 2006.

(Xie Shouan, Li Menglou, Li Rongbo)

397 Cosmotriche inexperta (Leech)

Classification Lepidoptera, Lasiocampidae
Synonym *Crinocraspeda inexperta* Leech

Distribution Zhejiang, Anhui, Jiangxi, Fujian.

Hosts *Pinus taiwanensis*, *Pseudolarix amabilis*, *Pinus massoniana*, and *Pinus thunbergii*.

Damage Larval feeding on leaves defoliates the entire tree during outbreaks.

Morphological characteristics **Adults** Females 16-24 mm long with a wingspan of 34-48 mm, while males 16-22 mm long with a wingspan of 34-38 mm. Head dark reddish brown, abdomen gray. Antennae plumose in males and pectinate in females. Thorax contains a group of silver scales laterally. Hairs on tibia three times the length of those on other leg segments. There are brown and white rings on tarsi. Forewings brown with white and a wavy inner cross vein. Medial cross vein not obvious but white wavy beyond Cu. Median cell with a white crescent spot. Outer cross vein appears as indistinct grayish band. Marginal hairs alternate between dark brown and grayish white. Color of males lighter than females. There is a grayish white basal area on the forewings. **Eggs** Oval, 1.5-2.0 mm for the long axis and 1.1-1.4 mm for the short axis, bright green initially, turn to yellow 3-5 days later, and black before hatching. **Larvae** Mature larvae 39-59 mm long, head brownish yellow. Clypeus grayish white. Frons contains a black longitudinal mark in the middle. Ecdysial suture yellowish. Body dark reddish brown with purplish red marks. Prothorax with one bluish black and forward long hair bundle laterally. Meso- and metathorax with one blackish gray and upright hair bundle dorsally. There is a large diamond-shaped bluish black mark on each abdominal segment 1-7 dorsally, with each mark covered randomly by long scales. Subdorsal line orange yellow. Upper spiracle line grayish brown, with brown, red, or yellow declivent marks below. Abdomen purplish red. **Pupae** Body 15.5-21.0 mm long, dark reddish brown. Cocoons oblong oval, 18.0-23.5 mm long, grayish brown and covered by black setae.

Biology and behavior *Cosmotriche inexperta* occurs two generations a year and overwinters as pupae under trees in Zhejiang. Adults start to emerge between mid-April and late June with the peak emergence in early May. Eggs of the first generation appear from late April to mid-June and peak in mid- to late May. Larval stage of the first generation falls between early May and early August. Mature larvae begin to pupate from mid-July and end in late August. Adults of the first generation

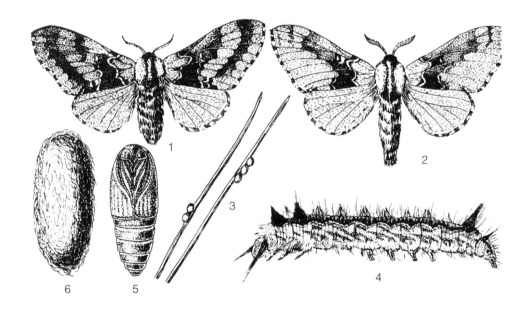

C. inexperta. 1. Adult female, 2. Adult male, 3. Eggs, 4. Larva, 5. Pupa, 6. Cocoon

(Yuan Ronglan)

C. inexperta adult female
(NZM)

C. inexperta adult male
(NZM)

emerge between early to late August. Eggs of the second generation occur between early August to early September, and larvae of the second generation feed between mid-August and early November.

Most eggs are laid singly on host needles. Occasionally two or three eggs are laid together. The average egg incubation period is 12.8 days at 18°C and 7.4 days at 25°C. Most larvae pass through five instars, with some pass through six instars. Hatching occurs mostly in early mornings. Newly hatched larvae feed on chorion first before moving to old needle tips. Larvae of the 1st and 2nd instars feed top down towards the edges, which results in dry and dead needles. By third instars they consume entire needles. Young instars (before the 4th instars) do not feed much, accounts for about 5.4% consumption of the entire larval stage. Larval feeding increases after the 4th instars. Feeding by the last instar larvae accounts for 60.6% of total needle consumption. Mature larvae construct cocoons under the rocks or in leaf litters. Pupal stage last for about 20 days for the first generation.

Most adults emerge during the day with peak emergence observed between 13:00-16:00. Adults stay in bushes, weeds or leaf surface after emergence during the day. Females mate once, rarely twice; while most males copulate twice, and some can mate up to seven times. Mating lasts for an average of 308 minutes, ranging from 180 to 445 minutes. Females start to lay eggs right after mating. Each female produces an average of 113.6 eggs, ranging from 55 to 215 eggs. Eggs are mostly laid on needles in the lower crown (> 60%), with some in the middle crown, and very few on the top crown.

Prevention and treatment See methods for *Pyrosis diota* Graeser.

References Hou Taoqian 1987, Xiao Gangrou (ed.) 1992, Liu Youqiao and Wu Chunsheng 2006.

(Zhang Yongan, Yuan Ronglan)

398 *Kunugia xichangensis* (Tsai et Liu)

Classification Lepidoptera, Lasiocampidae
Synonyms *Dendrolimus xichangensis* Tsai et Liu, *Cyclophragma xichangensis* (Tsai et Liu)
Common name Xichang pine caterpillar

Distribution Hunan, Sichuan, Guizhou, Yunnan, and Shaanxi.

Hosts *Pinus yunnanensis*, *Quercus varibilis*, *Quercus dentate*, *Quercus aliena* var. *acutiserrata*, *Adinandra bockiana*, *Roegneria japonensis* var. *haekliana*, and *Potentilla* sp.

Morphological characteristics **Adults** Females 37-45 mm long with a wingspan of 78-95 mm, brownish with pectinate antennae. Forewings yellowish brown, curved anteriorly with a slightly declivent outer margin. There is a 6-8 mm brown transverse band in the middle of the wing. Sub-marginal spots blackish and arranged in a wavy line. Median cell contains a white spot at the end. Hindwings brownish. Males 34-40 mm long with a wingspan of 60-90 mm, brown with plumose antennae. Transverse band on wing 1-6 mm, orcherous. **Eggs** Oblong oval, 1.7 mm on the long axis and 1.49 mm on the short axis, smooth, with mixed red and white marks on surface. **Larvae** Bluish black for newly hatched, 5-10 mm long. By the 2nd instars, each thoracic segment contains a verruca with a dark brown hair laterally. Larvae older than the 5th instars reddish brown with a brownish yellow head. Meso- and metathorax each contains black hair bundles dorsally. Abdomen with inconspicuous marks dorsally and black hair bundles laterally and ventrally. Mature larvae 80-100 mm long. **Pupae** Green initially, turn to brown one day later. Vertex and abdominal segments densely covered by golden short hairs. **Cocoons** oval, dark brown with black setae on surface.

Biology and behavior *Kunugia xichangensis* occurs one generation a year and overwinters as 4th or 5th instar larvae in Sichuan. Larvae resume feeding from mid- to late February and pupate between early June and early August. Peak pupation occurs between late June and early July. Adults emerge and lay eggs between mid-July and late August with the peak appears between early July and early August. Eggs hatch between early August and mid-September and peaks between mid- and

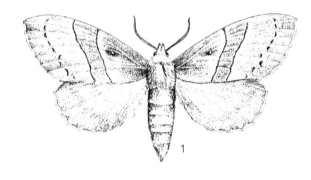

K. xichangensis. 1. Adult, 2. Larva
(Zhu Xingcai)

K. xichangensis adult male (NZM)

K. xichangensis adult female (NZM)

K. xichangensis adult (Li Zhenyu)

late August. Larvae overwinter by early to mid-December. The development time for eggs, larvae, prepupae, and pupae are 11-20 (average 14.9), 298-312, 3-6 (average 4.2), and 21-32 (average 26.6) days, respectively. Adults live for 5-12 days. Unmated females live for an average of 11 days. Each egg mass is made of several dozens to hundreds of eggs. Eggs hatch between 10:00-16:00, with the most come out around 14:00. Larvae start to feed 4-6 hours after hatching and can ballooning with silk threads during migration. The 1st and 2nd instar larvae feed only on mesophyll, whereas the 3rd instars consume the entire needles. Feeding occurs mostly in early mornings or at night. They usually come down the trees by 9:00 and rest in weeds on the north side of the trees and go up the trees to feed around 17:00. They curl up and drop to the weeds when disturbed. Each instar lasts for 22-28 days with an average of 25 days before overwintering. Each instar lasts for 33-38 days with an average of 35 days after overwintering. Larvae overwinter in early to mid-December when temperatures drop below 12°C. Some larvae can feed on leaves during overwinter when air temperature is above 12°C. Overwintering larvae resume feeding in mid- to later February. Mature larvae construct cocoons in weeds or leaf litters for pupation.

Most (56.2%) adults emerge between 21:00-24:00 with an average daily temperature of 22.1-22.6°C and a RH of 80%. Mating occurs 1-2 days after emergence between 20:00 and day break, with most occurs between 20:00-24:00, and lasts for 18-30 (average 20) hours. Females can mate 1-5 times. They lay eggs right after mating mostly on weeds, with some on the underside of *Quercus varibilis* and *Adinandra bockiana*. No eggs were found on *Pinus yunnanensis*. About 50% of the eggs are laid after the first mating, with an average of 425 eggs/female, ranging from 261-513 eggs. Oviposition lasts for about 8 days (6-11 days). Adults rest in weeds or underside of the leaves during the day and become active at dusk, especially around 22:00. Adults are slightly attracted to lights.

Prevention and control Protect natural enemies such as egg parasitoids Telenomus dendrolimusi and Trichogramma dendrolim; larval *parasitoids Apanteles lacteicolor*, *Casinaria nigripes*, and *Xanthopimpla predator*; and predatory birds *Cuculus fugas hyperythrus*, *Parus major artatus*, and *Garrulax canorus canorus*.

References Hou Taoqian 1987, Xiao Gangyou 1992, Liu Youqiao and Wu Chunsheng 2006.

(Zhang Yongan, Chen Sufen)

399 *Dendrolimus grisea* (Moore)

Classification Lepidoptera, Lasiocampidae
Synonym *Dendrolimus houi* Lajonquiére
Common name Yunnan pine caterpillar

Distribution Zhejiang, Fujian, Jiangxi, Hubei, Hunan, Guangxi, Sichuan, Guizhou, Yunnan, and Shaanxi; India, Thailand, Vietnam.

Hosts *Pinus yunnanensis*, *Cryptomeria fortune*, *Sabina chinensis*, *Biota orientalis*, *Keteleeria fortune*, and *Pinus langhianensis*.

Morphological characteristics **Adults** Females 36-50 mm long with a wingspan of 110-120 mm, body densely covered by grayish scales. Antennae pectinate, yellow in the center. There are four brown transverse bands on the forewings that extend from the front to the posterior margin. Inner cross vein and middle line inconspicuous. Two outer cross veins curved at the top and wavy at the bottom. There are nine grayish-brown crescent sub-marginal marks between the outer cross vein and the margin, which arranged in an arc for marks no.1-5 and a straight line for marks no. 6-9. Median cell without a conspicuous mark. Hindwings without marks or stripes. Abdomen robust. Males smaller than females, 34-42 mm long with a wingspan of 70-87 mm, darker and densely covered by reddish brown scales. Antennae plumose. Marks on the wings similar to those of the females except for a relative conspicuous mark in median cell. Abdomen small. **Eggs** Globose with a diameter of 1.5-1.7 mm, grayish brown, with three yellow and white ring-like belts. There is a grayish brown spot on both side of the middle belt. **Larvae** The first instars 7-8 mm long, grayish brown. Head brown. Thoracic segments contain dark brown longitudinal marks dorsally, with dense black brown hair bundles laterally. There is a pair of dark brown spots dorsally on each abdominal segment. Each spot contains setal fascicles. The second instars 8.0-13.5 mm long. Orange yellow, head dark brown, meso- and metanotum each with one dark brown mark. White hairs are present between the marks. Zonate marks also appear on each abdominal segment dorsally. Dorsal surface of

D. grisea adult male (Zhang Peiyi)

D. grisea adult female (Zhang Peiyi)

the 4-5 abdominal segments has two conspicuous grayish white bell-shaped marks. The 3rd instars 11.5-19.0 mm long, brighter in color. The 4th instars 18.0-30.0 mm long, with four white spots arranged in a square on each abdominal segment dorsally. The anterior two spots are brighter in color than the posterior two. The 5th instar 21.5-37.0 mm long, with darker body and more dark-brown surface marks. Each abdominal segment contains two black setal fascicles dorsally. There are also dense white long hairs laterally. Mature larvae (6-7th instars) robust, 90-116 mm long, black, with less clear bell-shaped abdominal mark dorsally. **Pupae** Spindle-shaped, about 35.5-50.5 mm long. Brownish initially, darken gradually to dark brown later. Each segment contains sparse reddish short hairs. There is a cremaster at the end of the abdomen. **Cocoons** cylindrical, 60-80 mm long, grayish white initially, turn to withered yellow later, and covered by black setae from larvae.

Biology and behavior *Dendrolimus grisea* occurs two generations a year and overwinters as eggs and larvae in southern Yunnan. Overwintering larvae start to pupate in mid-May. Adults appear in late June. Larvae of the first generation hatch in mid-July. They begin to pupate in mid-September. Adults of the first generation appear in mid-October. Larvae of the second generation occur in mid-December. Overwintering larvae in the Simao area pupate in late April, with adults emerge in late May. Larvae of the first generation pupate in early September. Adult emergence and egg-laying occur in early October. Some eggs hatch into larvae before overwintering, while others overwinter.

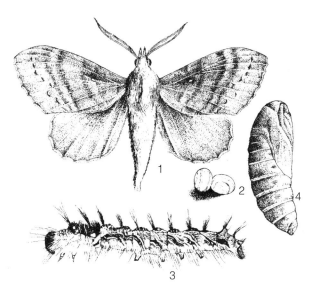

D. grisea. 1. Adult, 2. Eggs, 3. Larva, 4. Pupa
(Zhu Xingcai)

Most adults emerge in the afternoon or at night (18:00-22:00). Mating occurs right after emergence. Each female produces an average of 500 eggs, ranging from 400-600 eggs. A few females start to lay eggs 3-7 days after emergence. Adults stay between needles, in weeds, or on the north side of the trees during the day. They are weakly attracted to light and are active at night. Adult longevity differs based on generation and sex, with an average of 7-9 days. In general, adults from the overwintering generation live longer than those of the first generation, and females live longer than males. Most eggs are laid on host needles, with more found on the new growth. A few eggs are found on branches or weeds, especially in defoliated pine forests. Each egg mass contains 3-300 eggs. Females prefer to lay eggs on trees younger than 20 years old, especially those on the forest edges in heavily infested areas.

Newly hatched larvae feed aggregately and can ballooning with silken threads. Leaf consumption increases dramatically by the 3rd and 4th instar, with most feeding occurs between 9:00-11:00. Larvae hide at the base of needle bundles to avoid high temperatures at noon. The 5th and 6th instars feed all day, with most occurs between

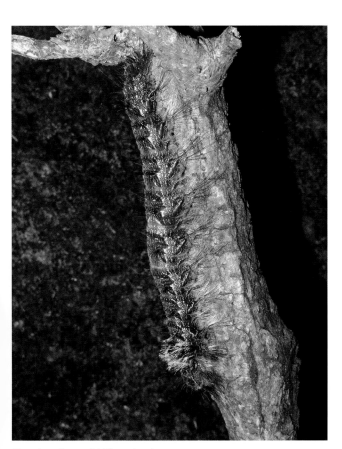

D. grisea larva (Li Yingchao)

D. grisea damage (Zhang Yongan)

15:00-17:00. Mature larvae generally construct cocoons between needles or on branches for pupation. Cocoons are found all over the place on branches, in bark crevices, broad leaved trees, or bushes during heavy infestation.

Prevention and control

1. Utilize pupal parasitoid, *Eucepsis* sp. or egg parasitoid, *Anastatus* sp.

2. Apply *Dendrolimus grisea* NPV powder at 15-30 kg/ha during the 3-5th instars for larval control. Apply *Beauveria bassiana* at 1×10^8 conidia/mL suspension or powder at 24°C with a RH of 95%.

3. Apply the mixture of 25% Dimilin I and *Beauveria bassiana*.

References   Hou Taoqian 1987; Jiang Yeqin, Zhang Yakun, Chen Wenjie et al. 1989; Liu Deli 1990; Xiao Gangrou (ed.) 1992; Long Furong, Tang Yongjun, Huang Huiping et al. 2004; Duan Xiaohong 2006; Liu Youqiao and Wu Chunsheng 2006; Wu Xiaomin and Zhang Wenfeng 2008.

(Zhang Yongan, Zhao Congli, Xu Weiliang)

400 *Dendrolimus kikuchii kikuchii* Matsumura

Classification Lepidoptera, Lasiocampidae
Synonym *Dendrolimus kikuchii* Matsumura
Common name Simao pine caterpillar

Distribution Zhejiang, Anhui, Fujian, Jiangxi, Hubei, Hunan, Guangdong, Guangxi, Hainan, Sichuan, Yunnan, Gansu and Taiwan; Vietnam.

Hosts *Pinus langhianensis*, *P. yunnanensis*, *P. armandi*, *P. massoniana*, *P. ikedai*, *P. taiwanensis*, *Keteleeria evelyniana*, and *Pseudolarix kaempferi*.

Morphological characteristics Adults Males 22-27 mm long with a wingspan of 53-65 mm, brown to dark brown. There are four dark brown wavy lines on the forewings between wing base and the outer margin. There are eight near circular yellow marks forming a line under the outer margin. Top corner contains three white marks. Median cell contains one white mark that connects to the anal corner by a large and conspicuous kidney-shaped yellow mark. Male genitalia knife-shaped, curved backwards on the tip, and contains long teeth on the posterior half, with large and densely arranged teeth on the back side. Clasper highly sclerotized and pointed, curved downwards like a sickle. Claspette twice as long as clasper, strongly curved downward from the middle, and protrudes upwards at the end. Clasper flat at the end, with teeth on both sides. Females 25-31 mm long with a wingspan of 68-75 mm, lighter than males and without kidney-shaped yellow mark. White mark in median cell and the four dark wavy lines conspicuous. **Eggs** Near circular, 1.89 mm long and 1.64 mm wide. There are three yellow rings on chorion, with one brown spot surrounded by a white ring on both side of the middle ring. **Larvae** Newly hatched larvae 5-6 mm long. Head and back of the thorax orange. Metanotum and mesonotum black with yellowish central area. Dorsal line yellowish white. Subdorsal line made of yellow and black marks. Spiracle line and upper spiracle line yellowish. Prothorax contains two hair bundles laterally, with hairs longer than half of the body length. Marks on larvae of the 2nd to 5th instars more conspicuous. Body of the the 5th instars 35 mm long, head orange. Hair bundles on prothorax still exist, black with white tip. Meta- and mesothorax contain hair bundle as well, with one orange bundle in the middle. Dorsal line made of upside down orange triangular marks, whereas subdorsal line made of dark brown and yellowish diamond-shaped marks. Upper spiracle line orange. Spiracle line yellow and interrupted by orange marks on prolegs. Marks remain unchanged after the 6th instars except that yellowish hairs start to appear laterally. By the 7th instar, lateral setae on the back grow, with long black setae on black marks. Abdomen hairs longer in the front than those in the back. Dorsal line made of black and deep orange upside-down triangular marks, with more black marks than orange ones. Setae on meta- and mesothorax long. Mature larvae deep red in color. **Pupae** Oblong oval, brown. Greenish initially. Female pupae about 36 mm long on average. Male pupae about 32 mm long. **Cocoons** Diamond-shaped, gray, turn to dirty brown before adult emergence, covered by setae. Female cocoons about 67 mm × 20 mm in size, and male cocoons about 58 mm × 20 mm in size.

Biology and behavior *Dendrolimus kikuchii kikuchii* life cycle varies based on location. In Kunming of Yunnan, most occurs one generation a year but some

D. kikuchii kikuchii adult male (Li Zhenyu)   *D. kikuchii kikuchii* adult female (Li Zhenyu)   *D. kikuchii kikuchii* larva (Zhang Yongan)

have two generations. Body length and head capsule width is 34.6 and 3.5 mm, respectively in the 5th instar larvae of the 2nd generation. Body length and head capsule measured 21.8 and 2.38 mm, respectively in the 5th instar from population undergoes one year life cycle. Eggs hatched after July 20 can finish only one generation per year. Eggs from the same female may also diverge in the number of generations for unknown reasons. Two generations were observed in Jingdong of Yunnan and Zhejiang, and three generations in Guangdong and Hainan. All generations overwinter as larvae. In some areas in Yunnan where temperatures are high in winter, larvae can still feed around noon.

Overwintering larvae pupate in late April in Jingdong, Yunnan but they can be seen until late July. Adults start to appear in late May and lay eggs in mid-June. Larvae of the first generation appear in late June and can be seen as late as mid-September. First generation larvae start to construct cocoons by late August. Adults of the first generation start to appear in mid-September and lay eggs in late September. Larvae of the second generation appear in mid-October and overwinter. Most mature larvae construct cocoons among host needles, with a few in bark crevices, on among weeds and bushes, or under rocks. Larvae feeding ends one day before cocoon construction.

Most adults emerge between 18:00-22:00, with 80% emerge between 18:00-20:00. They mate and lay eggs right after emergence. Unfertilized eggs will not hatch. Adults hide during the day and are active at night with most activities observed at dusk. They are somewhat attracted to lights.

Most eggs are laid in clusters on host needles. Female fecundity is positively correlated with pupae weight. Each female can produce a maximum of 303 eggs and a minimum of 52 eggs. In Jingdong of Yunnan, most first-generation eggs hatch around 9:00 (24.6%), with 65.5% eggs hatch between 8:00-11:00. Hatching rate for the second-generation eggs is 64.4%-92.1%. Hatching seems to be correlated with high humidity during the day. Peak hatching occurs one hour after the peak humidity.

Newly hatched larvae consume chorion. They feed on host needles aggregately at first, leaving notches along needle edges. Larvae are very active and can jump or balloon with silk threads when disturbed. Mature larvae raise their heads and thoracic setae when disturbed.

Leaf consumption increases as the larvae grow. By the 6th instar, each larva consumes 878.5 cm needles in

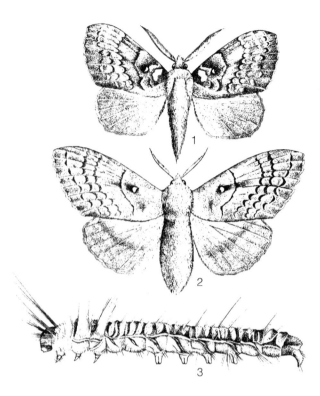

D. kikuchii kikuchii. 1. Male, 2. Female, 3. Larva
(Zhu Xingcai)

a 24-hour cycle. Larva of the 7th instar consumes 1095.5 cm needles in one 24-hour cycle, with a four-day leaf consumption of 4039.5 cm in total. Leaf consumption at night accounts for 68.8%, or 2,780 cm of needles.

Prevention and treatment

1. Forest closure and modification of pure pine forests.
2. Microbial control with $5-10\times10^7$ spores/mL B.t. or $1-2 \times 10^8$ conidia/mL *Beauveria bassiana* water solution, or 5×10^8 conidia/mL *Beauveria bassiana* oil suspension, emulsifiable suspension, or water solution. Apply DCPA(A) together with *B.t.* and *Beauveria bassiana*.
3. Chemical control. Fogging with trichlorfon-malathion at 1500-2250 g/ha, or pyrethroids ultralow volume cover spray at 15-30 mL/ha, or Dimilin cover spray.

References Hou Taoqian 1987; Xiao Gangrou (ed.) 1992; Zhang Chaoju 2002; Lu Bin, Bu Lianggao, and Shu Weiqi 2003; Liu Youqiao and Wu Chunsheng 2006; Zhang Hong 2006.

(Zhang Yongan, Chen Changjie)

401 *Dendrolimus punctata punctata* (Walker)

Classification Lepidoptera, Lasiocampidae
Synonyms *Dendrolimus punctata* Kirby, *Dendrolimus punctatus* Walker
Common name Masson pine caterpillar

Distribution Jiangsu, Zhejiang, Anhui, Fujian, Jiangxi, Henan, Hubei, Hunan, Guangdong, Guangxi, Hainan, Chongqing, Sichuan, Guizhou, Yunnan, Shaanxi, Taiwan; Vietnam.

Hosts *Pinus massoniana, P. elliottii, P. taeda, P. yunnanensis*, and *P. latteri*.

Morphological characteristics **Adults** Body color varies from gray to grayish brown, yellow brown, and brown. Females lighter than males, 20-32 mm long with a wingspan of 42.8-80.7 mm. Males 21-32 mm with a wingspan of 36.1-62.5 mm. Head small, labial palps conspicuous. Compound eyes yellowish green. Antennae short pectinate in females, and plumose in males. Forewings wide, curved along the outer margin, with 3-4 inconspicuous bow-shaped transverse bands. The inner side of the dark brown marks along the outer margin brownish. Hindwings triangular without marks. Male genitalia dagger-shaped, with dense teeth on the front portion. Claspette 1/4 to 1/3 of the clasper in length. End of the clasper highly sclerotized and curved upwards. **Eggs** Oval, about 1.5 mm long and 1.1 mm wide. Reddish or yellowish red initially, turn to purplish brown before hatching. Surface smooth. **Larvae** Body color varies greatly before the 3rd instars. First instar larvae yellowish green to yellowish gray, gray laterally, with four conspicuous dark brown spots laterally on abdominal segments 2-5. Second instar larvae dark reddish brown mixed with white spots. Meta- and mesothorax

D. punctata punctata adult (edited by Zhang Runzhi)

D. punctata punctata larva (edited by Zhang Runzhi)

D. punctata punctata cocoon (edited by Zhang Runzhi)

D. punctata punctata pupae (edited by Zhang Runzhi)

D. punctata punctata eggs (edited by Zhang Runzhi)

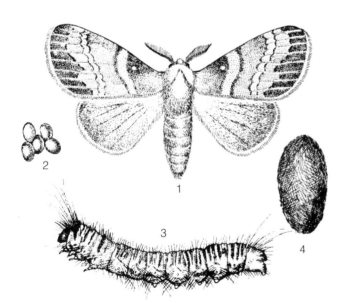
D. punctata punctata. 1. Adult male, 2. Eggs, 3. Larva, 4. Cocoon (Zhang Peiyi)

contain two black setal bands dorsally. Body and setae color change little in 4th and 5th instar larvae. Mature larvae brownish red or black, 38-88 mm long with a head capsule of 3.4-3.7 mm wide. There are two color types: brownish red or black. Body covered with spindle-shaped silver white or sliver yellow scales close to the body. Setae on abdominal hair clusters flat, with teeth at the end. There are a lot of grayish white hairs along the body laterally. Those on the head especially long. There is a longitudinal band along the entire body laterally, with one white spot above the line from metathorax to the 8th abdominal segment. There is a transparent light circular mark near 2/3 of the 9th abdominal tergum in larvae older than 3rd instar. This circular mark is surrounded by a brown area. Males can be separated from females by having an orange or yellowish center in the circular marks. **Pupae** Spindle-shaped, 22-37 mm long, brown to reddish brown and covered by yellow villi. Anal spine slender, yellowish brown, forms a hook at the end. **Cocoons** 30-46 mm long, oblong oval, grayish white. Turn to dirty brown before adult emergence and covered by sparse setae.

Biology and behavior *Dendrolimus punctata punctata* occurs 2-4 generations a year depending on locations. It has two generations in Xinyang of Henan, two to three generations a year in Yangtze River basin, three to four generations a year in Guangdong, Guangxi, and Fujian, and four to five generations a year in Hainan.

Overwintering larvae resume feeding in early February. In the Yangtze River basin, overwintering larvae construct cocoons in mid- to late April, Adult emerge and lay eggs in early May, with first generation larvae appear in mid- to late May. They pupate in early July and adults emerge and lay eggs in mid-July. Larvae of the 2nd generation appear in late July, with some pupate in early September and then emerge as adults, lay eggs to start the third generation, and overwinter as larvae in mid-November. A portion of the population diapauses in mid- to late August until November before overwintering. Duration of each stage varies as well. Overwintering period lasts for < 60 days in Guangdong, about 90 days in Hunan, and as long as 120 days in Henan. Egg stage lasts for 11, 7, and 7 days for the 1st, 2nd, and 3rd generation in Hunan, respectively; and 8, 6, and 6 days in Guangxi, respectively. Larval stage lasts for 49, 63 days for the 1st and 2nd generation, respectively in Hunan; 54, 46, and 54 days for the three generations, respectively, in Guangxi. Pupal stage lasts an average of 21, 16, and 13 days for the overwintering, 1st, and 2nd generation in Hunan; and 16, 12, 13, and 17 for the overwintering, 1st, 2nd, and 3rd generation, respectively in Guangxi. Adult stage lasts for 7.5, 7, and 8 days for overwintering, 1st, and 2nd generation, respectively, in Hunan; and 8, 7, 7, and 7 days for overwintering, 1st, 2nd, and 3rd generation, respectively, in Guangxi.

Most larvae overwinter inside pine needles in the upper crown of young and middle age trees. Some in bark crevices of old trees. No obvious overwintering in areas where winter is warm. Pupation occurs at locations vary from crowns, to bark crevices, to bushes and ground covers. Eggs are laid in masses of a few dozens to a few hundred eggs on needles or twigs, usually with 300-400 eggs per mass. Newly hatched larvae consume chorion first before feed aggregately on nearby needles. The 1st and 2nd generation can balloon with wind when disturbed. By the 3rd and 4th instars, they disperse to feed and can drop to the ground when disturbed. The 5th and 6th instars migrate. During outbreaks, mature larvae are usually seen on forest floor. They'll raise their head when disturbed.

Larvae of the 1st and 2nd generation molt 5-6 times. The overwintering generation molt 7-8 times, or even 9 times.

Prevention and treatment

1. Biological control. Apply *Beauveria bassiana* 7.5 kg/ha powder at 1×10^{10} spores/g, 1.5 l/ha oil suspension at 1×10^{10} spores/mL, or 2.25 l/ha at 6×10^{9} spores/mL suspension; apply *B.t. galleriae* at 1,500 g/ha; apply *Trichogramma dendrolimi* at 75,000-150,000 wasps/ha;

2. Adult trapping with black light.

3. Apply 25% Dimilin at 240-300 mL/ha, 20% lamda-cyhalothrin at 30-45 mL/ha, or 20% fenvalerate at 30-60 mL/ha.

References   Xiao Gangrou (ed.) 1992, Hou Taoqian 1987, He Jinyi and Xu Guangyu 2009, Chen Zongping and Xu Guangyu, Liu Youqiao and Wu Chunsheng 2006.

(Zhang Yongan, Xiao Gangrou, Peng Jianwen, He Jietian, Yan Jingjun, Hou Taoqian)

402 *Dendrolimus punctata tehchangensis* Tsai et Liu

Classification Lepidoptera, Lasiocampidae
Common name Dechang pine caterpillar

Distribution Sichuan, Yunnan, and Gansu.

Hosts *Pinus yunnanensis*, *P. armandii*, *P. massoniana*, and *P. yunnanensis* var. *pygmaea*.

Morphological characteristics **Adults** Females 23-31 mm long with a wingspan of 53-71 mm. Grayish brown or yellowish brown. Antennae pectinate, short, yellowish with brown teeth. Forewings slender, curved anteriorly. Outer margin dark brown and arc-shaped and wavy sometimes. Cross veins inconspicuous, brown. The Sub-marginal mark no.8 and 9 reach to wing margin. There is an incomplete brown cross stripe between the outer cross line and sub-marginal marks. Median cell contains a white spot at the end. Males 17-30 mm with a wingspan of 41-51 mm, reddish brown or dark brown. Antennae with yellow or brown flagella. Surface marks on wings similar to those of the females but darker. **Eggs** Oblong, about 1.5 mm long and 1.1 mm wide, color varies from pinkish red to red, yellowish red, yellowish green, and partially light-yellow and partially grayish white. **Larvae** Newly enclosed larvae yellow or green with a black head, gray laterally, with four obvious dark brown spots laterally on abdominal segment 2-5. Mature larvae 45-90 mm long, head yellowish brown, with two bundles of shinny bluish black or purplish black setae on meta- and mesothorax. The rest of the setae silver or yellowish white. There are two bundles of setae on each abdominal segment. Lateral villi silver white, especially long near the head. Ventral surface brownish yellow and reddish brown at the central area. Dorsal surface of the abdomen grayish white, mixed with golden, black, and purple, or alternating white and gray colors. **Pupae** Spindle-shaped, 20-40 mm long, brown and covered

D. punctata tehchangensis adult female (Li Zhenyu)

by dense villi, tip with hooked yellow hairs. Cocoons oblong, yellowish brown, and covered by bluish black setae.

Biology and behavior *Dendrolimus punctata tehchangensis* occurs one or two generations a year. It overwinters as 4th or 5th instar larvae at needle base or under ground covers in young forests, and inside bark crevices or needle base or ground covers in mature forests. Larvae resume feeding in early to mid-February. They pupate in early March after 3-4 molts with peak pupation between mid- to late May. Adults start to emerge in late April with a peak emergence between early to late June. Eggs of the first generation appear in early May and hatch in mid-May with the peak in late June. About 70% of the first generation larvae pupate by mid-July. First generation pupation peaks between mid- to late August. Adults emerge and lay eggs in early August with a peak between mid- and late September. Larvae of the second generation appear in late August with a peak between late September and early October. Larvae overwinter by early December. About 30% of the first generation larvae overwinter in late October.

Pupal stage lasts for 25.7 (21-29) and 23.7 (20-29) days for the overwintering and first generation, respectively. Adults females of the overwintering generation live for 8.1 (6-13) and 9.3 (4-13) days, respectively. Adult males live for 7.5 (4-12) and 8.1 (6-9) days, respectively. Egg stage lasts for 18.6 (14-24) and 13.1 (10-16) days for the overwintering and the first generation, respectively.

Neonates consume chorion first before feeding on nearby needles, resulting in notches on needles. Damaged needles become yellow. First instar larvae can balloon with silk threads. The 2nd instars migrate through crawling. By third instars, larvae feed separately on entire needles and drop when disturbed. Fourth instar larvae can jump if disturbed. The 5th and 6th instars are less active. They often stay on pine branches. When disturbed, they raise their head and expose the thoracic setae. Mature larvae construct cocoons for pupation among needles or at needle base. Occasionally cocoons are found in depression of trunk, in bark crevices, under rocks or on ground covers.

Pupal stage is 20-29 days. Adults emerge between 17:00-6:00. Majority of adults emerge between 18:00-24:00. Adults can mate several hours after emergence. Most mate the 2nd day after emergence. Mating occurs between 24:00-6:00, with most (65%) between 2:00-4:00.

In general females lay most of their eggs after the initial mating and a few more later on. Some females lay a few eggs before mating. Some even lay eggs without mating at all. Most eggs (70%) are laid before 21:00, with the rest laid before day break. Very few eggs are laid during the day. Fecundity is 485 (34-808) and 242 (96-388) eggs/female for the overwintering and the first generation, respectively. Males are strongly attracted to lights while females are less so. Adults die off gradually after mating and egg-laying.

Eggs are laid in masses on pine needles, or sometimes at needle bases or the crown or branch joints. Each egg mass contains an average of 50 (5-85) eggs. Females prefer edges of fully closed forests for egg-laying. Eggs are mostly found 1-5 m (especially 2-4 m) above the ground. Egg hatching rate is 81.2% and 82.3% for the first and second generation, respectively.

Prevention and treatment

1. Apply DPtCPV.

2. Apply Hulinshen powder to control 3rd and 4th instar larvae.

References   Hou Taoqian 1987; Wu Jialin and Chen Youfen 1992; Xiao Gangrou (ed.) 1992; Zhang Yufa, Yang Dongming, and Zhang Kun 2003; Liu Youqiao and Wu Chunsheng 2006.

(Zhang Yongan, Chen Suwen)

403 *Dendrolimus punctata wenshanensis* Tsai et Liu

Classification Lepidoptera, Lasiocampidae
Common name Wenshan pine caterpillar

Distribution Guangxi, Guizhou, and Yunnan.

Hosts *Pinus yunnanensis*, *P. langhianensis*, *P. massoniana*, *P. armandii*, and *P. elliottii*.

Morphological characteristics **Adults** Males 24-29 mm with a wingspan of 45-56 mm. Antennae plumose. There are five dark brown wavy cross lines between the base and sub-marginal line on the forewing, with weak costal antemedian line and transverse line. The sub-marginal line consists of nine dark brown marks. Of which the last two marks are connected. Innerside of the marks brown. Inner and middle cross lines wide and dark in color. Median cell with a relatively clear white spot at the end. Hindwings similar in color to forewings but without marks. Genitalia claspers large and long, cylindrical with a longitudinal groove and obtuse tip. Surface covered by setae of various length. Claspettes short, small, and pointed, one third as long as the claspers and covered by setae at the base and the lower half. Aedeagus long, blade like with small teeth anteriorly. Clasper with a highly sclerotized and curved end which is decorated by a row of tooth-like projections. Females mostly brown but vary greatly in color among individuals. Head, thorax, and forewings mostly brown or brownish. Some are grayish brown or dark gray. Body 26-31 mm long with a wingspan of 58-73 mm. Antennae pectinate. The five transverse lines on forewings are clear except the sub-marginal line and the out cross line. Inner and middle transverse lines wide and form two transverse bands. The white spot in the median cell not always distinct. Hindwings slightly lighter than forewings in color. Abdomen large and covered by brown villi. Lamella antevaginalis tongue-like, enlarged for the basal two thirds, semi-circular anteriorly with a depressed center. Lateral appendages small and wrinkled, with a pouch and setae. **Eggs** Oval, 1.5 mm long and 1.0 mm wide. Pinkish, reddish, or purplish red. **Larvae** Mostly black or dark brown, some dark reddish brown. First instars yellowish green, 4-6 mm long with a head capsule of 0.9 mm in width. Urticating hair bands start to appear between thoracic segment 2 and 3. Second instars reddish brown, yellowish brown, or yellowish white, 6-12 mm long with a 1.1 mm wide head capsule. There is a grayish white mark between abdominal segment 4 and 5. Third instars 13-15 mm long with a head capsule of 1.3 mm in width. Hair bands on meso- and metathorax widened. Hairs on the sides of abdomen long. Setae on the dorsal surface of each segment strong. The fifth instars 25-34 mm long with a head capsule of 2.8-3.0 mm in width, head dark brown. Thoracic and abdominal hairs bluish dark brown and shiny, especially those on meta- and mesothorax. All body segments are covered by yellowish white long hairs laterally. The 6th and 7th instars 40-55 and 58-65 mm long with a head capsule of 3.5-3.8 and 4-5 mm in width, respectively. **Pupae** Spindle-shaped, brown. Cocoons oblong oval, grayish brown or white with one tapering end, covered by shiny black hairs. Female cocoons 38 mm long, and male cocoons 31 mm long.

Biology and behavior *Dendrolimus punctata wenshanensis* occurs two generations a year and overwinters as 4th or 5th instar larvae. Larvae resume feeding in mid- to late February and pupate in mid-April. Adults emerge in mid-May. Eggs of the first generation appear

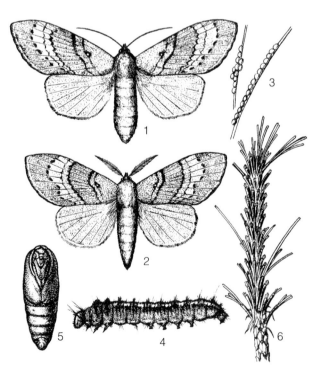

D. punctata wenshanensis. 1. Adult female, 2. Adult male, 3. Eggs, 4. Larva, 5. Pupa, 6. Damage (Zhang Xiang)

D. punctata wenshanensis adult (Li Zhenyu)

in mid-May and hatch in early June. Larvae of the first generation pupate in late August, and adults emerge in mid-September. Second generation eggs appear in mid-September and hatch in early October. Larvae of this generation overwinter in needles or under the bark by mid- to late December. The can still feed during sunny and warm days. Overwintering period lasts for more than three months. The occurrence periods of the above stages are 10-15 days earlier in forests below 1,400 m elevation.

Most eggs hatch between 7:00-9:00. Newly hatched larvae consume chorion. There are generally seven instars in the larval stage. Some larvae go through six or nine instars. First instars feed aggregately, resulting in notched needles. Most second instars feed aggregately too. First and second instars can balloon with silk threads when disturbed. Needle consumption increases in the 3rd and 4th instars. They'll jump if being touched. By 5th instars, they have well-developed setae which can cause red and swelling skins to people. Larvae migrate when food become scant. Mature larvae spit green body fluid when disturbed and show off their urticating hairs. Mature larvae construct cocoons in weeds at the base of the trees. Very few construct cocoons on branches. Those cocoons on branches usually are thin and small, atrophic or parasitized by natural enemies. Larvae pupate inside the cocoons 3-5 days later. Pupal stage lasts for about 20 days.

Most adults emerge between 19:00-20:00 under high humidity. Dry weather delays emergence. Adults stay put during the day and are active at night. Females can lay eggs same night after mating. Each female produces 338 (230-406) eggs on average in 3-5 times, ranging from several to 100-200 eggs each time. There are residual eggs inside dead females. Most eggs are laid in masses on health needles on forest edges. Egg stage lasts for half a month. Adults are good fliers with a longevity of 7-8 days for females and 6 days for males.

Prevention and treatment   Apply mixture of NPV and CPV, *B.t.* or insecticide fog.

References   Hou Taoqian 1987; Xiao Gangrou (ed.) 1992; Feng Yuyuan, Ren Jinlong, Yang Lin 2002; Xu Guolian, Chai Shouquan, Xie Kaili et al. 2002; Zha Guanglin 2003; Liu Youqiao and Wu Chunsheng 2006.

(Zhang Yongan, Luo Congfu)

404 *Dendrolimus spectabilis* Butler

Classification Lepidoptera, Lasiocampidae
Synonym *Dendrolimus punctatus spectabilis* Zhao et Wu
Common name Japanese red pine caterpillar

Distribution Hebei, Liaoning, Jiangsu, Shandong, and Henan; Korea Peninsula, and Japan.

Hosts *Pinus densiflora, P. thunbergii, P. tabulaeformis*, and *P. sylvestris* var. *mongolica*.

Morphological characteristics **Adults** Wingspan 48-69 mm for males and 70-89 mm for females. Body color varies from grayish white to grayish brown and reddish brown. Middle and outer cross veins on the forewings white, with white marks on the outer side of the sub-marginal marks. There is a wide dark brown band between middle and outer cross veins on male forewings. **Eggs** Oval, 1.8 mm long and 1.3 mm wide. Greenish initially, turn to pinkish later, and purplish brown before hatching. **Larvae** Newly enclosed larvae 4 mm long with yellow back and black head. Body hairs inconspicuous. Dorsal marks become obvious in the 2nd instar, and become yellowish brown, dark brown, and black by the 3rd instar. Mature larvae 80-90 mm long, with obvious black setae on the 2nd and 3rd abdominal segment. **Pupae** Spindle-shaped, 35-45 mm long, dark reddish brown. Cocoons grayish white and covered by setae.

Biology and behavior Dendrolimus spectabilis occurs one generation a year and overwinters as larvae. Its occurrence period varies according to climates at various locations. Larvae resume activity by climbing up the host trees in early March in Shandong Peninsula. Pupation occurs in mid-July. Adults start to emerge in late July and peaks between early and mid-August. Females start to lay eggs after emergence. Eggs hatch in mid-August and peak between late August and early September. Larvae overwinter in late October. Occurrence period in central Shandong is 20 days earlier compared to that in the peninsula, 10 days later in Liaoning than that in Shandong or Hebei. The cocoon period in Hebei is 20 days earlier than that in Shandong and 10 earlier than that in Liaoning. The overwintering period in Liaoning is 10 days later than in Shandong and Hebei.

Most adult emerge between 17:00-23:00 and peak between 19:00-20:00 when 57.8% adults emerge. Generally, males emerge an hour earlier than females. They mate right after emergence with heads hanging downwards. Mating lasts from 10 to 24 hours. Males live for an average of 8 days, and females for 7 days.

Most eggs are laid between 18:00-23:00 with a peak at 20:00. Mot females only oviposit once, with a few for 2-3 times. Unmated females lay a few unfertilized eggs. Each female produces an average of 622 eggs, ranging from 241 to 916 eggs.

D. spectabilis adult male (Li Zhenyu)

D. spectabilis adult female (Xu Gongtian)

D. spectabilis adult male (Xu Gongtian)

D. spectabilis mature larva (Xu Gongtian)

D. spectabilis mature larva (Xu Gongtian)

D. spectabilis cocoon (Xu Gongtian)

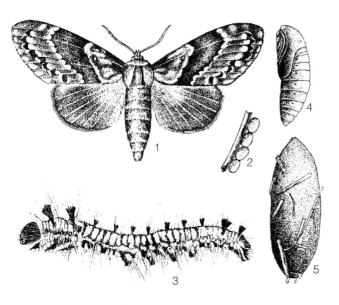

D. spectabilis. 1. Adult, 2. Eggs, 3. Larva, 4. Pupa, 5. Cocoon (Zhu Xingcai)

Eggs hatch between 4:00-8:00 within 10 days after being deposited, with most hatch between 4:00-5:00. Hatching rate is 63.1% in Shandong and 83%-100% in Liaoning. Neonates consume chorion first before feeding on nearby needles, resulting in notches on needles. Damaged needles become yellow. By the third instar, larvae consume the entire needles. The 1st and 2nd instar larvae can balloon with silk threads. The 2nd instar feed separately and do not produce silk threads anymore. They can jump when disturbed. Mature larvae are less active and stay on pine branches when not feeding. They expose their thoracic setae when disturbed. Larvae feed until early November and then climb down along the trunk to find bark crevices, weeds, or rocks for overwintering. Most larvae overwinter at sunny sites.

Larvae molts 7-8 times with females molt one more time than males. Mature larvae construct cocoons among needles and pupate in two days. Pupal stage lasts for an

D. spectabilis adult damage (Zhang Runzhi)

average of 17 days, ranging from 13 to 21 days.

Prevention and treatment

1. Apply mixture of CPV and *B.t.*

2. Apply mixture of abamectin and Dimilin at 6,000 L/ha, 450 g/ha, or 300g/ha for the control of 4th and 5th instars.

References Hou Taoqian 1987; Xiao Gangrou (ed.) 1992; Chen Suwei et al. 1999; Liu Youqiao and Wu Chunsheng 2006; An Baiguo, Wang Xinan, Duan Chunhua et al 2007.

(Zhang Yongan, Sun Yujia)

D. spectabilis damage (Zhang Yongan)

405 *Dendrolimus suffuscus suffuscus* Lajonquiere

Classification Lepidoptera, Lasiocampidae
Synonyms *Dendrolimus superans* Tsai et Liu, *Dendrolimus suffuscus* Lajonquiere

Importance *Dendrolimus suffuscus suffuscus* is an important defoliator that can cause tree mortality during outbreaks.

Distribution Hebei, Jiangsu, Jiangxi, Shandong, and Henan.

Hosts *Platycladus orientalis*, *P. densiflora*, *P. tabuliformis*, and *P. bungeana*.

Damage The insect can defoliate entire trees and even consume bark of young branches during outbreaks.

Morphological characteristics **Adults** Body color varies from grayish brown to dark brown based on location. Females 31-39 mm long with a wingspan of 77-90 mm. Males 25-35 mm long with a wingspan of 60-69 mm. Antennae grayish brown, short bipectinate in females and long bipectinate for males. Humeral plate well developed, almost reaches posterior margin of the mesothorax. The median, outer, and sub-marginal lines dark brown, with one brownish serrate mark outside the outer transverse line as well as inside the sub-marginal line. The outside one is particularly conspicuous. The white spot on median cell conspicuous or inconspicuous. There is an obscure black transverse mark along M_1 between the tip and the outer cross line. Hindwings grayish brown, covered by dense hairs basally. There is a grayish brown arc-shaped transverse mark in the center on both wings. Legs grayish brown, with one pair of spurs in the middle and hind legs. **Eggs** Oval, 1.92-2.36 mm long

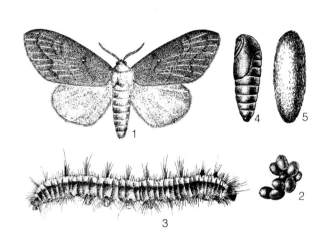

D. suffuscus suffuscus
1. Adult, 2. Eggs, 3. Larva, 4. Pupa, 5. Cocoon (Zhang Peiyi)

and 1.46-1.88 mm wide. Greenish initially, turn to grayish brown later, and dark before hatching. **Larvae** Newly hatched larvae yellowish green, 5.18-9.00 mm long, with setal bands on meta- and mesothorax and dark brown transverse bands between each abdominal segment. Mature larvae dark gray, 56.8-104.6 mm long with a head capsule of 4.02-6.38 mm in width. Newly molted larvae fresh in color, head round and covered by yellowish brown villi. Coronal suture white, with one reddish brown spots on each side in the middle. There is a dark

D. suffuscus suffuscus adult female (Li Xianchen)

D. suffuscus suffuscus adult male (Li Xianchen)

D. suffuscus suffuscus adult female (Li Xianchen)

D. suffuscus suffuscus adult male (Li Xianchen)

brown mark on the frons. There are two black longitudinal marks on the back of prothorax and one greenish blue scale bundle and two pink and white transverse bands on meta- and mesothorax. Lateral verruca well developed with brownish or dark brownish hairs, especially on the thorax. Abdomen reddish brown ventrally, lighter for the last four segments, with dark brown or brown marks on each segment. Thoracic legs dark brown. Prolegs reddish brown with crochets in biserial incomplete circle. **Pupae** Spindle-shaped, 28-42 mm long, green initially, turn to brown or dark brown later. Abdomen slightly curved with a pointed end. Abdominal segment 5 movable. Anal spine hooked. Gonopore oval and depressed and located on the 9th abdominal segment in males, and double sectioned and located on abdominal segment 8 in females.

Biology and behavior *Dendrolimus suffuscus suffuscus* occurs one to two generations a year in Shandong with overlapping generations. Larvae of both generations can be seen in the same habitat or even the same host tree. It overwinters as 3rd to 8th instar larvae under the rocks, in leaf litters, or in weeds inside the forest. Larvae resume feeding in early to mid-March with all of them on host trees by early to mid-April. Larvae from one year generation in shaded locations may have delayed life cycle. Pupation occurs in mid-April and peaks in late April. Adults of the overwintering generation emerge in mid-May. First generation larvae appear in early June and pupate in early July, with adults occur between mid-July and early September. Larvae of the second generation appear between mid-August and mid-September, and start to overwinter in early October. All larvae overwinter by late October.

Newly hatched larvae consume chorion first. They then feed on needles aggregately. They balloon with silk threads. Larvae feed heads down mostly at night on branches. Mature larvae stop feeding 1-2 days before constructing cocoons for pupation. Most adults emerge at night or in the early morning. They mate and lay eggs right after emergence.

Prevention and treatment

1. Silviculture. Improve stand conditions to promote natural control by natural enemies.

2. Manual control. Adult trapping with black lights between June and July, cocoon removal between July and August, larval removal through trunk cover with weeds in October, and larval hunt under the rock or bark during overwintering.

3. Biological control. Egg parasitoids such as *Trichogramma dendrolimi*, *Anastatus japonicus*, *Anastatus albitarsis*, *Pachyneroa nawi* exert 40%-60% natural parasitism; predatory insects such as *Tenodera aridifolia sinensis*, birds such as *Parus major* and *Cyanopica cyana interposita*. Application of *Beauveria bassiana* resulted in 71%-96% larval control.

4. Chemical control. 1) Apply 25% Dimilin at 1,500-2,000×; 2) Apply toxic bands containing 2.5% cyhalothrin or 5% esfenvalerate around tree trunk can cause 95% control over a month.

References Fang Deqi 1980; Xiao Gangrou (ed.) 1992; Chen Shuliang, Li Xianchen, Xu Yanqiang et al. 1993; Liu Youqiao and Wu Chunsheng 2006.

(Li Xianchen, Fang Deqi)

406 *Dendrolimus superans* (Butler)

Classification	Lepidoptera, Lasiocampidae
Synonyms	*Odonestis superans* Butler, *Eutricha fentont* Butler, *Dendrolimus stbirtcus* Tschetvcrikov, *Dendrolimus jezoensis* Matsumura, *Dendrolimus albolineatus* Matsumura
Common name	Larch caterpillar

Distribution Beijing, Hebei, Inner Mongolia, Liaoning, Jilin, Heilongjiang, and Xinjiang; Russia, Korea Peninsula, Japan, and Mongolia.

Hosts *Larix gmelini*, *Larix olgensis* var. *koreana*, *Pinus koraiensis*, *Abies nephrolepis*, *Picea jezoensis* var. *komarovii*, *Picea jezoensis* var. *microsperma*, *Pinus tabulaeformis*, *Pinus thunbergii*, *Pinus sylvestris* var. *mongolica*, *Picea obovate*, *Picea koraiensis*, *Abies fabri*, and *Abies nephrolepis*.

Morphological characteristics **Adults** Females 28-45 mm long with a wingspan of 70-110 mm, antennae pectinate. Males 24-37 mm long with a wingspan of 55-86 mm, antennae plumose. Body color varies from grayish white to grayish brown, reddish brown, and dark brown. Forewings wide with straight outer margin; inner, middle, and out cross veins serrate. There are eight black markings arranged in "3" shape on the submarginal line, with the last two marks parallel to the outer margin if connected. White mark on median cell large and obvious. Hindwings ocherous. **Eggs** Oval, 2.5 mm long and 1.8 mm wide, greenish initially, turn to yellowish, red, or dark red later. **Larvae** Mature larvae 55-90 mm long, grayish brown with yellow markings and covered by silver or golden hairs. Meta- and mesothorax contain two bluish black setal bands. The eighth abdominal segment contains dark blue long hairs dorsally. **Pupae** Body 30-45 mm long, dark brown to black and covered by dense yellow villi. Cocoons white or grayish brown, covered by bluish black larval setae on surface.

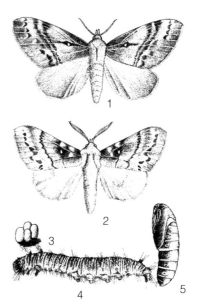

D. superans. 1. Adult female, 2. Adult male, 3. Eggs, 4. Larva, 5. Pupa (Zhu Xingcai)

Biology and behavior *Dendrolimus superans* occurs one generation a year or one generation in two years and overwinters as larvae under leaf litter. In Altai, Xinjiang, it mostly undergoes two-year life cycle, with about 15% finish the generation in one year. Larvae of the two-year generation become adults in June, whereas those from the one-year generation emerge as adults in August. Offspring of the one-year generation switch to two-year generation afterwards, whereas the offspring of the two-year generation will undergo one-year life cycle.

D. superans adult male (Xu Gongtian)

D. superans adult female (Xu Gongtian)

D. superans 3rd instar overwintering larva (Xu Gongtian)

D. superans eggs (Xu Gongtian)

D. superans damage (Zhang Yongan)

Therefore, it essentially occurs two generations in three years. In areas with higher accumulative temperatures, larval development accelerates and results in higher proportion of one-year generation. Both one-year and two-year generation exist in Changbai region in northeastern China, whereas one-year generation is most prevalent in southern Liaoning.

Overwintering larvae resume feeding in the spring when average air temperature is above 8-10℃. It feeds on leaf buds first before moving to expanded needles. Feeding starts at the tip of the needle. It will drop to the ground and curve up when disturbed. Some of the two-year generation larvae resume feeding in the third spring and pupate in late May or early June after half a month's feeding, whereas others need to feed longer before pupation. They aggregate in the crown before pupation. Pre-pupal stage lasts for 4-8 days, with a pupal stage of 18-32 days. Pupal stage for one-year generation is shorter than those of the two-year generation. Adult emergence starts in late June and peaks between early to late July.

Some emerge in August. Adult emergence for the one-year generation is more synchronized compared to the two-year generation, which can last up to two months. Adults are strongly attracted to lights. Mating usually occurs at sunset or clear night. Mating occurs one day after emergence. Females fly to dense needles in the middle or low crown to lay eggs. Eggs are laid in masses and arranged irregularly on needles or twigs. Each female produces 128-515 eggs. Adults live for 4-15 days. Eggs hatch in 12-15 days. Newly hatched larvae concentrate on branch tips and can balloon to other branches with silk threads if disturbed. Second instar larvae disperse and will fall to the ground when disturbed. There are 7-9 instars in the larval stage. *Dendrolimus superans* overwinters as 3rd to 4th instar larvae for those undergoes one-year life cycle. Those with two-year life cycle overwinters as 2nd to 3rd instar in the first year, and 6th to 7th instars in the 2nd year. Most (95%) leaf damage is caused by the last two instars.

Prevention and treatment

1. Monitoring population using a lure containing E5,Z7-12:Alol and E5,Z7-12:OH.

2. Install a plastic band on the trunk to prevent overwintering larvae from returning to crown. Apply Dimilin III powder at 450-600 g/ha.

3. Release *Trichogramma* sp. at 300,000/ha.

References Hou Taoqian 1987, Xiao Gangrou (ed.) 1992, Li Xueling 2006, Liu Youqiao and Wu Chunsheng 2006, Pei Xinchun 2009.

(Zhang Yongan, Zhang Runsheng, Ma Wenliang)

407 *Dendrolimus tabulaeformis* Tsai et Liu

Classification Lepidoptera, Lasiocampidae
Synonym *Dendrolimus punctatus tabulaeformis* (Tsai et Liu)
Common name Chinese pine caterpillar

Distribution Beijing, Hebei, Shanxi, Liaoning, Shandong, Henan, Chongqing, Sichuan and Shaanxi.

Hosts *Pinus tabulaeformis*, *P. sylvestris* var. *mongolica*, *P. armandii*, *P. densiflora*, *P. massoniana*, and *P. bungeana*.

Morphological characteristics **Adults** Females 23-30 mm long with a wingspan of 57-75 mm. Body grayish brown to dark brown with clear markings. The inconspicuous white spot at the end of median cell on the forewings in females located above the inner transverse line or slightly outside. Costa slightly curved. Out margin curved and wavy sometimes. Transverse lines brown, with inconspicuous inner line, slightly curved middle line, and strongly curved wavy outer line. Submarginal markings black, crescent with brownish inner marks. The first six marks form an arc, and the 7-9 marks form an oblique line. The 8th mark located in the second cell. The 9th mark consists of two smaller marks and located in the first cell. Antennae pectinate with yellowish flagella and brown teeth. Lamella antevaginalis large, diamond shaped. Lamella vaginalis diamond-shaped with carinal depression in the middle. Lamella lateralvaginalis oblong round with a pouch at the end. Males 20-28 mm long with a wingspan of 45-61 mm. Males darker in color with conspicuous white spot at the median cell compared to females. Cross lines clear. The inner side of the submarginal marks brown. Antennae pectinae with yellowish or brown flagella and brown teeth. Clasper cone-shaped with an obtuse tip that depressed laterally. Claspette half the size of the clasper, pointed posteriorly. Genitalia blade-shaped and slightly curved upwards, depressed in the middle, enlarged at the base, and tapered towards the end. The small hook on the tip of the genitalia contains sclerotized teeth anteriorly, which is denser near the blade. The sclerotized clasper end curved upwards, with one row of teeth on the outside, and one row curved teeth inside. **Eggs** Oval, 1.75 mm long and 1.36 mm wide, greenish on the end with insemination opening and pinkish on the other end. There are 7-12 (9 on average) unguiform patches around the fertilization hole. **Larvae** Grayish black with long lateral hairs and clear markings. Head brownish yellow with a dark brown mark at the center of the frons. Urticating hairs on dorsal surface of thorax obvious. Spindle-shaped lodging scales absent on the back of the abdomen, while there are spindle-shaped flat setal bundles on each abdominal segment anteriorly. Short setae are also found at the base of each bundle. Body contains an interrupted longitudinal band on each side and inconspicuous white markings on the bands. There is also a declivent mark extends from the anterior portion to the ventral surface of the abdomen on each segment. Mature larvae 55-72 mm long. **Pupae** Brown to dark reddish brown with a short anal spine, which is slightly curved or curved into a circle; 24-33 mm long for females and 20-26 mm for males. Cocoons white or brownish and covered by black setae.

Biology and behavior *Dendrolimus tabulaeformis* generally occurs one to three generations a year and overwinters as larvae. Larvae resume feeding from mid-March to early April in Beijing. Pupation starts

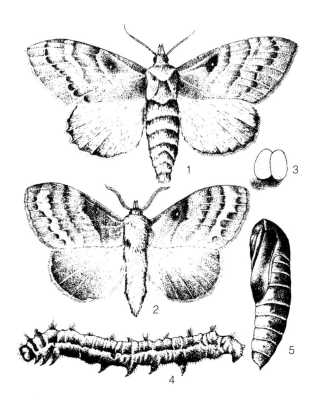

D. tabulaeformis. 1. Adult female, 2. Adult male, 3. Eggs, 4. Larva, 5. Pupa (Zhu Xingcai)

D. tabulaeformis larvae (Zhang Yongan)

from mid-May to early June, peaks between mid- to late June, and ends in late July. Adult emergence begins in early June, peaks between early to mid-July, and ends in mid-August. First generation eggs appear in early June, peaks in early to mid-July, and ends in mid-August. First generation larvae appear in mid-June, with a portion enters diapause and overwinter in early to mid-October to complete a one-year life cycle. The other portion continues development and pupates in late July and become adults in mid-August. Females lay eggs and eggs hatch between late August and mid-September. Larvae overwinter between mid- and late October to complete two generations in a year.

Egg stage for the first generation lasts for an average of 8 days, whereas the larval stage lasts for about 45 days with 6-7 (mostly 6) instars. The average duration for each instar is 6-8 days for the 1-5th instar, and 10 days for the 6th instar. Pupal stage lasts for an average of 19 days, ranging from 15-23 days. Females live for an average of 7 days while males live for 6 days. Larval stage for the overwintering generation is 11 months over two years with 8-9 (mostly 8) instars for those with one generation a year, and about 10 months with 7-8 (mostly 7) instars for those with two generations a year. Pupal stage for the overwintering generation is 19 days on average, ranging from 15-23 days. Females of this generation live for 6-7 days and males live for 4-5 days. Egg stage lasts for an average of 9 days.

Eggs are laid on pine needles as eggs masses. They are reddish initially, turn to purplish red before hatching. Most eggs hatch in the early morning.

Newly hatched larvae aggregate on needles near the egg masses. They start to feed on needle edges a few hours later, resulting in notched and dried needles. Larvae disperse after the 2nd instars. First and second instars can balloon in the wind aided by silk threads. By the third instars, larvae consume entire needles. Needle consumption increases dramatically by the 4th instars. Mature larvae stay on twigs most of the time. They display their bluish black thoracic setae when disturbed. Migration increases when food becomes scarce.

Larvae come down from the tree by mid- to late October to overwinter in bark crevices, soil fissures, under the leaves or rocks around the base, with most found at sunny locations within 30 cm or between 30-65 cm around the base. Most larvae overwinter as 4th or 5th instars. Overwintering mortality is generally about 10%. Larvae resume feeding between mid-March and early April

D. *tabulaeformis* adult male (Li Zhenyu)   D. *tabulaeformis* adult female (Li Zhenyu)

D. *tabulaeformis* cocoon (Xu Gongtian)   D. *tabulaeformis* ldamaged pine needles (Xu Gongtian)

when air temperature reaches 5℃ and above. Mature larvae pupate on needles, twigs, weeds and bushes 2-3 days after feeding stops, with most found in the lower crown in young stands.

Adults come out from the dorsal suture between the head and thorax during emergence. Flight does not take place until body dries up and wings harden sometime later. Most adults emerge between 20:00-22:00 and mate right after emergence. Adults usually mate once although some can mate 2-3 times. Mating lasts for about 24 hours. Females begin to lay eggs the same night after mating. Each female produces dozens to 500 eggs. The majority of the eggs are laid on the first day. The rest of the eggs are laid over a 3-day period. There are few residual eggs in females which mated normally. Sizable residual eggs remain in unmated females or those mated several days after emergence. Sex ratio ranges from 1:0.8 to 1:1.6 (females: males).

Prevention and treatment

1. Use high voltage electrocution trap during adult emergence. Traps are set at 2 m above the ground at forest edge.

2. Use plastic band (3.5 cm wide and 0.1 mm thick) as a physical barrier before larvae resume feeding in mid- to late March.

3. Monitoring population using a lure containing E5,Z7-12:OH, E5,Z7-12:OAc, and E5,Z7-12:OPr at 100:47:29 raio. Apply *B.t.* powder between early May to early June in locations with no access to water.

4. Apply 50% acetamiprid 3,000×.

References Hou Taoqian 1987, Xiao Gangrou (ed.) 1992, Ma Yuting 2002, Liu Youqiao and Wu Chun-sheng 2006, An Wenyi 2007.

(Zhang Yongan, Yan Jingjun)

Gastropacha populifolia (Esper)

408

Classification	Lepidoptera, Lasiocampidae
Synonyms	*Bombyx populifolia* Esper, *Gastropacha tsingtauica* Grunberg, *Gastropacha angustipennis* Walker, *Gastropacha populifolia* f. *fumosa* Lajonquiere, *Gastropacha populifolia* f. *rubatrata* Lajonquiere, *Gastropacha populifolia* (Esper) Lajonquiere
Common name	Poplar lasiocampid

Importance *Gastropacha populifolia* is a serious defoliator of many plants, such as *Populus* L., *Salix* spp., *Malus* spp., *Pyrus* spp., and *Prunus persica*. Larvae cause the most damage. They defoliate leaves and affect tree development, impact integrity of landscape and cause economic loss.

Distribution Beijing, Hebei, Shanxi, Inner Mongolia, Liaoning, Heilongjiang, Jiangsu, Zhejiang, Anhui, Jiangxi, Shandong, Henan, Hubei, Hunan, Guangxi, Sichuan, Yunnan, Shaanxi, Gansu, Qinghai; Europe, Russia, Japan, Korea Peninsula.

Hosts *Populus* spp., *Salix* spp, *Juglans regia*, *Pyrus* spp., *Prunus persica*, *Malus* spp., *M. asiatica*, *Prunus salicina*, *P. armeniaca*, *P. mume*.

Damage The 1st and 2nd instar larvae feed gregariously, causing holes on leaves.

Morphological characteristics **Adults** Males 38-61 mm in length, females 54-96 mm in length. Body and wings yellowish-brown, forewing narrow and long, inner margin short, outer margin arc wave-shaped, forewing with five black intermittent wavy stripes, the end of the discal cell with a black spot. Hind wing with three obvious stripes, costal margin orange-yellow, outer margin light yellow. Forewings and hind wings covered with few black scale hairs. Body color and forewing spots vary greatly, dark yellowish-brown to yellow, the spots and stripes on wings sometimes vague or disappear. **Eggs** 2 mm in length, greyish-white, with black pattern, egg clutch covered with greyish-yellow fluffs. **Larvae** Mature larvae 80-85 mm in length, head brown and slight flat. Body greyish-brown, dorsal surface of mesothorax and metathorax with a bule-black spot, a reddish yellow cross stripe is located behind the spot. The 8th abdominal segment with a big verruca, its periphery black

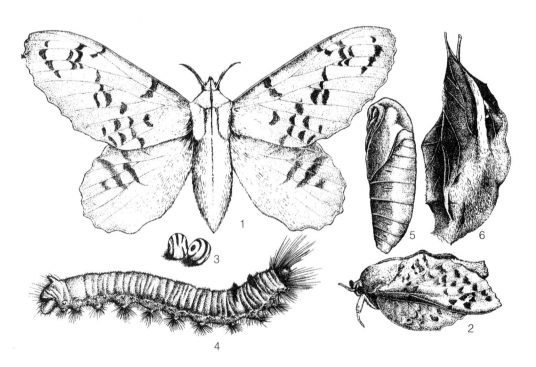

G. populifolia. 1-2. Adults, 3. Eggs, 4. Larva, 5. Pupa, 6. Cocoon

(4 by Zhang Peiyi; 1-3, 5-6 by Zhu Xingcai)

G. populifolia adult (Xu Gongtian)

G. populifolia adult (Xu Gongtian)

G. populifolia overwintering larva (Xu Gongtian)

G. populifolia mature larva (Xu Gongtian)

and the top greyish-white. The subdorsal line on the 11th segment with a round granule. Dorsal line brown, lateral line with upside down "v" shaped dark chocolate-brown stripe. Each segment with a pair of brown verrucae of different sizes, its edge black and the top with earthy yellow tufts of hairs. Above each verruca with a black "V" shaped spot. Spiracle black, peritreme yellowish-brown. Thoracic legs and prolegs greyish-brown, area between prolegs with brown cross stripes. **Pupae** Brown. **Cocoons** Greyish-brown, with larval chaetae on it.

Biology and behavior *Gastropacha populifolia* occurs two generations and rarely three generations a year in Henan. It overwinters as larvae on the trunk. Lar-

G. populifolia mature larva (Xu Gongtian)

G. populifolia larva (8th abdominal segment with a verruca) (Xu Gongtian)

G. populifolia larva (mesothorax and metathorax with a black spot) (Xu Gongtian)

G. populifolia cocoon (Xu Gongtian)

vae start feeding in mid- and late March, pupate in mid-April to mid-March, and adults emerge in mid-March to early June. The 1st generation egg hatch in mid-May, larvae pupate in mid-June to mid-July, and adults enclose in late June to late July. The 2nd generation larvae hatch in early and mid- July, pupate in late August, and adults emerge in early September. The 3rd generation larvae hatch in mid-September. Larvae stop feeding in mid- and late October. They overwinter as the 4th and 5th instars on the trunk. Some of the 2nd generation larvae hatch in early May, feed until mid- or late September, overwinter as the 6th or 7th instars. The 1st and 2nd instars feed gregariously and cause holes on leaves. Larvae older than the 3rd instar disperse. Mature larvae weave leaves with silk to make cocoons and pupate on the trunk. Eggs are laid on leaves. Each female lays 200-300 eggs.

Gastropacha populifolia occurs one generation a year in Beijing, it overwinters as larvae on branches, trunks or among dead leaves. Larvae become active in April, they spin cocoons to pupate on the trunks or branches in June. Adults start emergence in early July. Adults have phototaxis. They lay eggs on leaves. Each female lays 200-300 eggs. Eggs hatch in July and the egg stage is about 12 days.

Prevention and treatment

1. Mechanical control. Artificial catch and kill larvae on the branches and trunks.

2. Physical control. Trap adults with black light.

3. Biological control. During severe damage period, apply *B.t.* at 1×10^{10} spores/mL

4. Chemical control. Apply 3% acetamiprid at 1,000× to kill larvae.

References He Shiyuan 1984, Li Gefang et al. 1984, Zhu Hongfu et al. 1984, Xiao Gangrou (ed.) 1992, Liu Youqiao and Wu Chunsheng 2006, Xu Gongtian 2007.

(Guan Ling, Gao Liniu)

409 *Lebeda nobilis sinina* Lajonquiere

Classification Lepidoptera, Lasiocampidae
Synonym *Lebeda nobilis* Walker

Distribution Jiangsu, Zhejiang, Anhui, Fujian, Jiangxi, Henan, Hubei, Hunan, Guangxi, Shaanxi, Taiwan.

Hosts *Camellia oleifera, Pinus massoniana, P. elliottii, Quercus fabri, Q. acutissima, Castanea mollissima, C. sclerophylla, C. henryi, Myrica rubra, Platycarya strobilacea, P. orientalis, Fagus longipetiolata, Pterocarya stenoptera*.

Damage In Daoxian, Jiangyong and Jiang Hua of Hunan Province, local outbreaks occur every year. When *Camellia oleifera* was damaged by L. *nobilis sinina*, its twigs withered to death, trees stopped blooming and fruiting. Outbreaks caused huge economic loss to *Camellia oleifera* fruits growers. In recent years, it posed serious damage to *Pinus massoniana* in Liuyang County. This insect often eats up old leaves of *P. massoniana*, causing serious impact on tree growth.

Morphological characteristics **Adults** Female length 75-95 mm, wingspan 50-80 mm. Body color varies greatly, including yellowish-brown, reddish-brown, dark brown and greyish-brown. Males are darker than females in general. Forewing with two light brown diagonal cross stripes, the end of discal cell with a silvery spot, metagonia with two dark chocolate-brown spots; hind wings reddish-brown, the middle with a light brown cross stripe. **Eggs** Greyish-brown, spherical, 2.5 mm in diameter, both upper and lower spherical surface with a brownish black round spot surrounded by a greyish-white loop. **Larvae** The 1st instar dark chocolate-brown, head dark black, glossy, covered with sparse white seta. Dorsal thorax yellowish-brown, dorsal abdomen bluish violet, each dorsal segment with two tufts of hairs, tufts on the 8th segment are longer than others, lateral abdomen greyish-yellow, body 7-13 mm in length. The 2nd instar black blue, mixed with greyish-white stripe; dorsal thorax start showing tufts of hairs in two colors of black and yellow. The 3rd instar greyish-brown, tufts of hairs on dorsal thorax wider than those of the 2nd instar. From the 1st to the 8th dorsal abdomen of the 4th instar, each segment with two tufts of hairs in alternating light yellow and dark black colors. When at rest, the front tuft often overlaps the rear tuft. The yellow and black tufts on dorsal thorax of the 5th instar turn into bluish green. The 6th instar greyish-brown, ventral abdomen light gray, densely covered with reddish-brown spots. The 7th instar 113-134 mm in length. **Pupae** Oblong, the end of abdomen slight thin, dark reddish-brown. Vertex and abdominal internodes densely covered with yellowish-brown fluffs. Female pupae 43-57 mm in length, 24-27 mm in width; male pupae 37-48 mm in length, 20-24 mm in width.

Biology and behavior *Lebeda nobilis sinina* occurs one generation a year in Hunan and overwinters as larvae in eggs. It hatches in early and mid-March. Larvae have 7 instars and the development duration is 123-160 days.

During hatching, larva bites through a hole from one end of the egg and eats off 1/3 to 1/2 of the eggshell, then crawls slowly out of the egg. Hatching occurs mostly during 6:00-8:00 and 16:00-17:00. Newly hatched larvae feed in groups, larvae after the 3rd instar disperse and feed day and night. When temperature is high in the summer, they stop feeding during the day and often rest at the shady side of tree trunk base. They crawl out and feed at dusk or dawn. Larvae molt 6 times.

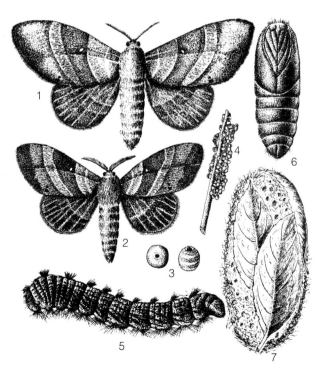

L. nobilis sinina. 1. Adult female, 2. Adult male, 3. Eggs, 4. Egg mass, 5. Larva, 6. Pupa, 7. Cocoon (Hou Boxin)

L. nobilis sinina adult (Li Zhenyu)

Most mature larvae spin cocoons and pupate in *Camellia oleifera* leaves and pine needles, some spin cocoons and pupate in bushes. Cocoon is yellowish-brown, and attached with thick poison hairs. Surface of the cocoon has irregular reticular holes. The prepupal stage is about 7 days and the pupal stage is 20-25 days. The newly emerged adults rest for 4-5 minutes, wings slightly vibrate, then unfold, and wings stay closely to the back. Adults mate 6-8 hours after emergence. Mating mostly occurs during 4:00-5:00. Oviposition mostly takes place at night. On average each female lays 170 eggs in two or three times. Eggs are laid on the twig of *Camellia oleifera* and needles of *Pinus massoniana*. Adults have strong phototaxis. They hide during the daytime and are active at night.

Lebeda nobilis sinina often occurs in the low hilly land. They are rare in mountains above 500 m in elevation. According to the survey at a mountain with elevation of over 300 m in Dao County in Hunan Province, the insect population is high at the foot of the mountain. On average, the density is 2.6 cocoon/tree and 1.23 egg cluster/tree. While at the middle of the mountain, the density is 0.59 cocoon/tree and 0.19 egg cluster/tree. The insect density at the top of the mountain is 0.045 cocoon/tree and 0.14 egg cluster/tree. With the increase of elevation, the population density significantly decreased.

Also, *Lebeda nobilis sinina* population density is closely related to forest stand structure. More severe damage occurred in *Camellia oleifera* and *Pinus massoniana* mixed forests, while the pest population density is smaller in pure *Camellia oleifera* forests.

Natural enemies of eggs include *Trichogramma dendrolimi* Matsumura, *Telenomus lebedae* Chen et Tong, *Anastatus japonicas*, *Tetrastichus* and *Pteromalus*; Larval natural enemy is *Lebeda nobilis* sinina Nuclear Polyhedrosis Virus; Pupal natural enemies include *Xanthopimpla pedator* Fabricius, *Theronea zebra diluta* Gupta, *Itoplectis naranyae* (Ashmead) and *Lioprocata beesoni*.

References Peng Jianwen 1959, Liu Youqiao and Wu Chunsheng 2006.

(Peng Jianwen, Ma Wanyan, Li Zhenyu)

410 *Malacosoma dentata* Mell

Classification Lepidoptera, Lasiocampidae
Synonym *Malacosoma neustria dentata* Mell

Importance *Malacosoma dentata* causes serious damage mainly to *Liquidambar formosana* due to its feeding behavior. During outbreaks, *M. dentata* cause complete defoliation of *L. formosana* and death of branches and upper shoots.

Distribution Zhejiang, Anhui, Fujian, Jiangxi, Hunan, Guangdong, Guangxi, Sichuan.

Hosts *Liquidambar formosana*, *Celtis sinensis*, *Castannea seguinii*, *Quercus* sp.

Morphological characteristics **Adults** Males 24-30 mm, females 32-38 mm in wingspan. Body and wings brown or yellowish-brown; females are lighter in color than males, light brown. **Larvae** Mature larvae 43-50 mm in length, orange and blue, head dark brown.

Biology and behavior *Malacosoma dentate* occurs one generation a year in regions south of the Yangtze River, it overwinters as eggs on twigs. In Fujian, eggs hatch in early March. The peak damage season is between late April and early May. Adults emerge and lay eggs in late May. Egg duration is 9 months. In Anhui, newly hatched larvae appear in late April. The peak damage season is between mid- and late May. The last instar larvae exuviate in the webbed nests and crawl down trees, and then spin cocoons to pupate. Adults start to emerge, mate, and oviposit in early and mid- June.

Adult eclosion occurs all day, but reaches peak at

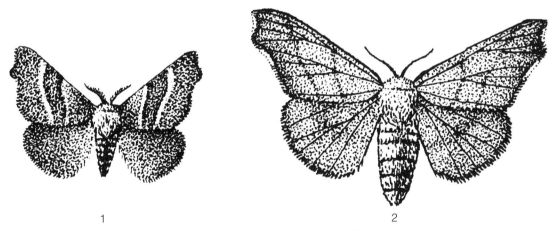

M. dentata. 1. Adult male, 2. Adult female (Chen Ruijin)

M. dentata adults (left: female, right: male) (Huang Jinshui)

dusk. Each female lays 150-160 eggs, mostly on tender shoots of less than 2.5 cm in diameter. Eggs are arranged tightly as an egg clutch in a circle around the branchlet. The surface of eggs is greyish-white. The average adult longevity is 6.5 days. The peak emergence period is from ate May to early June. Adults have strong phototaxis.

Larvae have 6 instars (few have 5 instars). Newly hatched larvae cluster in groups and spin white silk to form webs. Young larvae eat little, only feed on tender twigs and leaves around egg clusters. Larvae after the 3rd instar move to tree crotches or lower part of trunks to spin silk and make webs. For trees younger than eight years old, larvae gather at the base of the trunk, while for older trees above 15 years, larvae always gather at the forks of branches. Larvae feed at night and gregariously hide in the web during daytime. During its outbreaks in Anhui, each web had 4,000-8,000 larvae and the maximum number was 12,000 larvae per web. Larvae exuviate in the web. Mature larvae eat more. They crawl towards tree canopy in a line. During outbreak, they could defoliate the whole *Liquidambar formosana*, only leave the main veins. They also feed on *Castannea seguinii*, *Quercus* sp. and *Celtis sinensis*. They go down the tree and spin cocoons to pupate two to six days later. Cocoons can be found among dry leaves and bark cracks.

Prevention and treatment

1. Mechanical control. Remove egg clusters and kill mature larvae in the web on small *Liquidambar formosana* trees. Pick and kill pupae in cocoons at the base of host trees manually.

2. Biological control. Protect and utilize tachinids and nuclear polyhedrosis virus. Spray 2×10^9 PIB/mL nuclear polyhydrosis virus at 1,000×.

3. Physical control. Use black light to trap adults.

4. Chemical control. When population is high, spray 2.5 % deltamethrin at 1,000-2,000× or 25% Dimilin at 3,500× or 20% dinotefuran at 800×. Block and kill larvae that going down trees by installing pyrethroid poison rope on the trunk. Spray webs with kerosene to kill larvae inside the webs.

References Huang Jinshui 1989, Wang Mingfeng and Chen Bolin 1997, Tao Weichang and Wang Mingfeng 2004, Liu Youqiao and Wu Chunsheng 2006.

(Huang Jinshui, Tang Chensheng, Hou Taoqian)

M. dentata larvae (Huang Jinshui)

411 *Malacosoma neustria testacea* (Motschulsky)

Classification Lepidoptera, Lasiocampidae
Common names Tent caterpillar

Distribution Beijing, Hebei, Shanxi, Inner Mongolia, Liaoning, Jilin, Heilongjiang, Jiangsu, Anhui, Jiangxi, Shandong, Henan, Hubei, Hunan, Guangdong, Sichuan, Shaanxi, Gansu; Japan, Korea Peninsula.

Hosts *Crataegus pinnatifida*, *Malus pumila*, *Pyrus* spp., *Prunus armeniaca*, *P. salicina*, *P. persica*, *M. spectabilis*, *P. Pseudocerasus*, *M. asiatica*, *Populus* spp., *Ulmus* spp., *Quercus* spp., *Larix* spp., *Phellodendron amurensa*, *Juglans regia*, *Salix* spp., *Batula Platyhylla*, *Corylus* spp., *Sorbus pohuashanensis*.

Damage Larvae feed on tender buds and leaves, they can spin silk and make large webbed nests between branches. Younger larvae feed gregariously inside webs. During severe damage period, it can completely defoliate the whole tree, cause tree mortality, and reduce yield of fruits and tree growth.

Morphological characteristics **Adults** Males 24-32 mm in wingspan, yellowish-brown in color. Females 29-39 mm in wingspan, brown in color, the abdomen is darker. Forewings dark brown, the middle with two light yellowish-brown cross line shaped stripes; hind wings light brown, stripes not obvious. **Eggs** Greyish-white, oval, the mid-top concave. Eggs are laid on twigs, forming a "thimble" ring shape. **Larvae** Mature larvae 55 mm in length. Lateral side with bright bluish grey, yellow or black stripe. Dorsal side with obvious white stripe, along stripe with orange-yellow cross lines. Each dorsal segment with black long hairs, lateral side

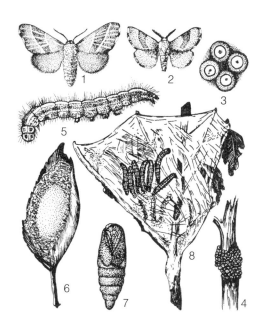

M. neustria testacea. 1. Adult female, 2. Adult male, 3. Eggs, 4. Egg mass, 5. Larva, 6. Cocoon, 7. Pupa. 8. Damage (Chen Ruijin)

with light brown long hairs, the hairs on abdomen short. Head bluish grey, with dark spots. **Pupae** 13-24 mm in length, dark chocolate-brown, with golden yellow hairs. **Cocoons** Greyish-white, silky and bilayer.

Biology and behavior *Malacosoma neustria testacea* occurs one generation a year and overwinters as

M. neustria testacea adult male (Xu Gongtian)

M. neustria testacea adult female (Xu Gongtian)

M. neustria testacea adult male (Xu Gongtian)

M. neustria testacea larva (Zhang Peiyi)

M. neustria testacea egg mass (Xu Gongtian)

M. neustria testacea 1st instar larvae (Xu Gongtian)

M. neustria testacea 2nd instar larvae (Xu Gongtian)

M. neustria testacea larvae (Xu Gongtian)

M. neustria testacea larvae (Xu Gongtian)

M. neustria testacea larva (Xu Gongtian)

M. neustria testacea cocoon (Xu Gongtian)

M. neustria testacea cocoon parasitized by wasps (Xu Gongtian)

eggs. Eggs hatch in spring, young larvae gather on tender buds near the egg cluster and feed on leaves, they weave webbed nests. The 1st instar last for ten days, the 2nd instar larvae move toward branches and trunks, they weave webs at the fork of branches, they rest in the webbed nests during daytime and feed at night. The webbed nest looks like a tent, so the larvae are called "tent caterpillars". Larvae have six instars. Mature larvae disperse and cause severe damage. Larvae spin cocoons and pupate among leaves, weeds and shrubs, and in tree holes. The pupal stage is 12-15 days. Adults mate the same day when emerge. Eggs are laid at the end of twigs, 150-400 eggs per egg cluster. The egg clutch resembles a thimble, so the insect is also called "thimble caterpillar". Adults have strong phototaxis.

In Wugang area, Henan Province, eggs hatch in early March, larvae mature in mid-April and pupate in late April, adults emerge in early and mid- May. Damage occurs in late May to early June.

Prevention and treatment

1. Manual control. Remove egg clusters and kill young larvae in webs.

2. Protect and utilize natural enemies, such as *Tachina* sp., *Xanthopimpla punctate*, NPV and birds.

3. Biological control. Spray 2×10^9 PIB/mL nuclear polyhydrosis virus at 1,000× to kill the 1st and 2nd instar larvae in webs.

4. Chemical control. During the 2nd and 3rd instars, spray 25% Dimilin at 1,500-2,000× or 3% fenoxycarb at 3,000-4,000×.

References Xiao Gangrou 1992, Zhang Zhizhong 1992, Ran Yali 2001, Liu Youqiao and Wu Chunsheng 2006.

(Wan Shaoxia, Hou Taoqian)

412 *Malacosoma rectifascia* Lajonquière

Classification Lepidoptera, Lasiocampidae

Distribution Shanxi.

Hosts *Betula* spp., *Populus davidiana*, *Rosa xanthina*, *Hippophae rhamnoides*, *Quercus liaotungensis*.

Damage: Larvae feed on leaves causing notches and holes. They completely defoliate trees when population density is high.

Morphological characteristics **Adults** Females 33-38 mm, males 26-30 mm in wingspan. Females yellowish-brown, the middle of the forewing with two parallel brown cross lines, margin hairs brown except those at the indent part greyish-white, the middle of the hind wing with a dark spot. The middle of male forewings with dark brown wide stripes, inner and outer sides of the stripe with light brown striae. **Eggs** Greyish-white, 1.1-1.4 mm in length, 0.6-0.7 mm in diameter, the top flat, the middle depressed. **Larvae** The dorsal side of young larvae greyish-black, abdomen yellowish-brown, head black. Mature larvae 30-47 mm in length, the upper spiracle line bright yellow, dorsal seta yellowish-brown. **Pupae** Yellowish-brown, 13-17 mm in length, covered with yellowish-brown short hairs. Cocoons ivory in color.

Biology and behavior *Malacosoma rectifascia*

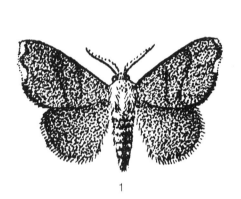 

M. rectifascia. 1. Adult male, 2. Adult female (Chen Ruijin)

M. rectifascia adult female (Li Zhenyu)   *M. rectifascia* adult male (Li Zhenyu)

M. rectifascia egg mass (Li Zhenyu)

M. rectifascia larvae

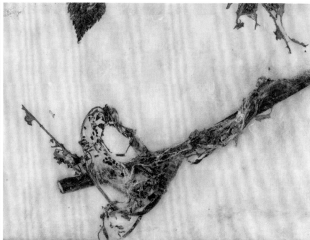

M. rectifascia damage

occurs one generation a year in Shanxi and overwinters as eggs. Larvae hatch and feed in April. They spin cocoons and pupate in early July. Adults emerge in late July. They mate and oviposit soon after emergence. Males live for 4 days, females live for 6.8 days. Eggs are mostly laid on new twigs, egg cluster ring-shaped and covered with spongy secretion. Newly hatched larvae are gregarious and weave webs. They start to feed two days later. The last instar larvae disperse. A larva consumes 0.3-0.5 g leaves per day. Mature larvae spin cocoons and pupate in leaf litter, among tree roots and rock cracks. During pupal stage, 20% of them are killed by *Beauveria* sp. and ants.

Prevention and treatment Refer to Pyrosis eximia and *Trabala vishnou gigantina*.

References Xiao Gangrou (ed.) 1992, Liu Youqiao and Wu Chunsheng 2006.

(Xie Shouan, Li Menglou, Hou Taoqian)

Paralebeda plagifera (Walker)

Classification Lepidoptera, Lasiocampidae

Synonyms *Lebeda plagifera* Walker, *Paralebeda urda backi* Lajonquière, *Odonestis plagifera* Walker, *Odonestis urda* Swinhoe

Distribution Zhejiang, Anhui, Fujian, Jiangxi, Guangdong, Guangxi, Sichuan, Tibet; India, Nepal, Vietnam, Thailand.

Hosts *Metasequoia glyptostroboides*, *Ginkgo biloba*, *Phoebe nanmu*, *Cupressus funebris*, *Quercus* spp.

Morphological characteristics **Adults** Females 45-52 mm in length, 115-130 mm in wingspan, antennae filiform, labipalp dark reddish-brown. Males 40-45 mm in length, 83-100 mm in wingspan, labipalp dark chocolate brown. Adults body brown or russet in color. The middle of the forewing with a brown diagonal cross stripe, its front end wide and the rear end narrow, it turns lighter in color from the costal margin to the inner margin; edge of the cross stripe greyish-white. Spots on the sub-margin merge into a thick wave-shaped stripe, inner side of the anal angle on the tail end with an obvious oval spot, inner cross line not obvious. The middle of the hind wing with two inconspicuous stripes. **Eggs** Ivory, round, 2.0-2.1 mm in diameter, with fine sculptures, the top with depressed yellowish-brown spots. **Larvae** Mature larvae 110-125 mm in length, head yellowish-brown, body greyish brown in color, flat and wide. The dorsal side of mesothorax and metathorax with yellowish-brown poisonous hair bands; each of the 3rd to 6th abdominal segment with a white spot on the dorsal surface, the 8th segment with a brownish black brush shaped hair tuft. **Pupae** Yellowish-brown, 60-80 mm in length, the end of abdomen with a pair of cremasters. **Cocoon** yellowish-brown in color, 70-90 mm in length.

Biology and behavior *Paralebeda plagifera* occurs one generation a year in Sichuan and overwinters as larvae. Larvae become active and feed in late March and early April. They spin cocoons and pupate in early and mid-July. Adults emerge in early and mid-August and oviposit in late August. Larvae hatch in mid-September and overwinter in early November.

Adult peak emergence period is in late August, mostly occur at 20:00 to 22:00. They mate and oviposit ten minutes after emergence. Adults have strong flight ability and phototaxis. They rest on trunks or on the back of branches or leaves, with their wings stay closely to the back. Body color is similar to the color of barks or withered leaves. Females lay eggs singly or in groups on trunks or on the back of leaves of *Metasequoia glyptostroboides* and *Ginkgo biloba*. Unmated females lay sterile eggs. Newly hatched larvae usually feed gregariously and then disperse after the 3rd instar. Larvae mostly feed at night, and rest on trunks or branches during the daytime. Their body color mimics the bark color. The 3rd and 4th instar overwinter in bark cracks in late November. Their mor-

P. plagifer adult (Zhang Peiyi)

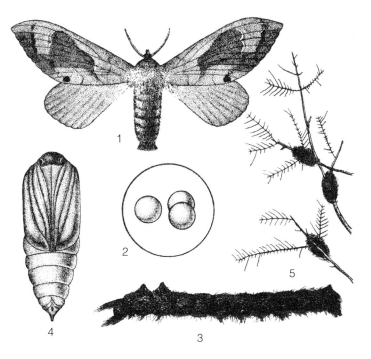

P. plagifer. 1. Adult, 2. Eggs, 3. Larva, 4. Pupa. 5. Cocoons and damage
(Wang Jianchuan)

tality is 30%-40% due to low temperature. Overwintering larvae become active and feed in late March. They often cause huge damage. Larvae mature in middle and late July. They look for thick canopy, weave leaves to make cocoon and pupate. The duration of each stage is: eggs 8-15 days, larvae 300 days, and pupae 10-15 days. The longevity of adults is 15-20 days. Each female lays 213 eggs on average, the most is 304 eggs.

The mortality of larvae infected with NPV under natural condition is 10%-20%. There is a tachinid can parasitize larvae and pupae, and the parasitic rate reaches 5%-10%.

References Zhu Hongfu 1979, Chen Zhiqing et al. 1982, Liang Dongrui et al. 1986, Liu Youqiao and Wu Chunsheng 2006.

(Jing Heming, Huang Dingfang, Li Zhenyu)

414 *Suana concolor* (Walker)

Classification Lepidoptera, Lasiocampidae

Synonyms *Cosmotriche davisa* Moore, *Lebeda bimaculata* Walker, *Suana ampla* Walker, *Suana cervina* Moore

Distribution Fujian, Jiangxi, Hunan, Guangdong, Guangxi, Hainan, Sichuan, Yunnan; India, Sri Lanka, Vietnam, Myanmar, Thailand, the Philippines, Malaysia, Indonesia.

Hosts Eucalyptus trees such as *Eucalyptus exserta*. *Casuarina equisetifolia* Forst., pomegrade, mango, jackfruit.

Damage Heavy infestations completely defoliate the host trees, resulting in stunned tree growth.

Morphological characteristics **Adults** Females 38-45 mm long with a wingspread of 84-116 mm, antennae grayish white, body and wings brown. Forewings contain a large white oval mark at the end of the median cell. Hindwings brownish and without marks. Males smaller, antennae black brown, body and wings reddish brown. Forewings contain a rectangle white mark at the end of the median cell. **Eggs** Oval, 1.8-2.2 mm long,

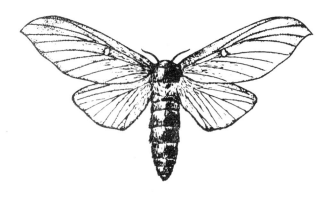

S. concolor adult (Qu Xiaojin)

S. concolor larvae (Gu Maobin)

S. *concolor* larvae (Gu Maobin)

grayish white, smooth and shiny. **Larvae** Mature larvae 45-136 mm long, grayish white, yellowish brown, or black brown and covered by black brown setae. Setae laid flat usually but erect for defense when disturbed. There is a black rectangle black hair mark at the center of the dorsal surface of meta- and mesothorax. There is a hair bundle on the verruca below the spiracle on each thoracic and abdominal segment. **Pupae** Females 27-55 mm long while males 27-34 mm long, black brown or dark red with a short, hooked anal spine. Cocoons spindle-shaped, grayish white and covered with dense black setae.

Biology and behavior *Suana concolor* occurs two generations a year in southern China. Females are weak fliers. Males are strong fliers. Mating occurs at night and lasts for 12-24 hours. Each female produces an average of 500 eggs. Eggs are laid in egg masses. The 4th-5th instar larvae feed in the crown at night and rest on tree trunk or the underside of the branches during the day. Mature larvae construct cocoons on the trunk, in the weeds, or rock fissures and pupate 3-6 days later.

Prevention and treatment

1. Physical removal of egg masses or larvae.

2. Trap adults with black lights.

3. Apply Dimilin suspension at 1,000-2,000×, 90% trichlorfon at 1,000-1,500×, or 20% sumicidin at 2,000-3,000× for larvae control.

4. Conserve natural enemy, *Blepharipa schineri* Walker (Diptera: Tachinidae). Its natural parasitism rate could reach 33%-48%.

References Liu Lianren 1990, Xiao Gangrou (ed.) 1992, Liu Youqiao and Wu Chunsheng 2006.

(Gu Maobin, Liu Lianren, Li zhenyu)

415 Trabala vishnou vishnou (Lefebure)

Classification Lepidoptera, Lasiocamlidae
Synonyms *Trabala vishnou* (Lefebure), *Gastropacha vishnou* Lefebure
Common name Tent caterpillar

Importance It is a sporadic, polyphagous defoliator. The damage is serious during outbreak.

Distribution Zhejiang, Fujian, Jiangxi, Guangdong, Hainan, Sichuan, Yunnan, Shaanxi, Taiwan; India, Myanmar, Sri Lanka, Pakistan, Japan.

Hosts *Quercus* spp., *Begonia aptera*, *Liquidambar formosana*, *Eucalyptus citriodora*, *Syzygium* spp.

Damage Larvae feed on leaves. When the population density is high, trees can be completely defoliated.

Morphological characteristics **Adults** Females 25-38 mm in length, 60-95 mm in wingspan, light yellowish green to orange yellow in color. Wings margin hairs dark chocolate-brown, inner line dark chocolate-brown, outer line wave-shaped and dark chocolate-brown, sub-marginal line consists of 8-9 dark brown spots, the posterior discal cell with a big yellowish-brown spot, the end of the abdomen with a clump of ivory setae. Males yellowish green to green in color, the location between outer marginal line and marginal line yellowish white, the posterior discal cell with a black dark chocolate-brown spot. **Eggs** Oval, 0.3 mm in length, greyish white in color, with reticulated sculpture. **Larvae** 65-84 mm in length, females covered with long dark yellow hairs, males covered with greyish-white hairs. Head with dark brown stripes, both sides of cranial median sulcus with a dark chocolate-brown longitudinal stripe. Mid-proscutum with a dark chocolate-brown "×" shaped stripe, both sides of prothorax with a black verruca bearing black long hairs, both sides of the body with black verrucae bearing setae, others of body with ivory hairs. **Pupae** Reddish-brown, 28-32 mm in length. Cocoons 40-75 mm in length, greyish-yellow in color, saddle shaped.

Biology and behavior *Trabala vishnou vishnou* occurs 3-4 generations a year in Guangzhou, the last generation occurs in late November. It occurs 4 generations a year in Taiwan and 5 generations a year in Hainan, and has no dormancy phenomenon. It occurs one generation a year in Shanxi, Shaanxi and Henan and overwinters as eggs. In southern China, male larvae have 5 instars and the period is 30-41 days, while female larvae have 6 instars and the period is 41-49 days. In northern China, larval stage is 80-90 days, larvae mature in July and spin cocoons and pupate on branches and trunks, pupal stage is 9-20 days, adults emerge in late July and August. Newly hatched larvae feed gregariously on mesophyll, they spin silk and fall off when disturbed, larvae disperse after the 2nd instar. Adults hide during the daytime and

T. vishnou vishnou adult (edited by Zhang Runzhi)

T. vishnou vishnou adult and cocoon (Li Menglou)   *T. vishnou vishnou* larva (edited by Zhang Runzhi)

T. vishnou vishnou cocoon (edited by Zhang Runzhi)

become active at night; they have strong flight ability and phototaxis; most eggs are laid on branches or trunks; each female lays an average of 327 eggs.

Prevention and treatment

1. Physical control. When population size is small, pick egg clusters in winter and spring. During larvae occurrence period, catch and kill gregarious larvae. Trap and kill adults using black light traps.

2. Chemical control. When the damage is serious, spray 8×10^9 spores/g *Beauveria bassiana* powder at 7.5 kg per 1 ha, or spray 25% Dimilin at 1,000× or 25 g/L thiamethoxam at 1,000×.

References Yang Zhirong and Liu Shigui 1991, Xiao Gangrou (ed.) 1992, Liu Youqiao and Wu Chunsheng 2006.

(Xie Shouan, Li Menglou, Chen Zhiqing, Wu Shixiong)

416 *Trabala vishnou gigantina* Yang
Classification Lepidoptera, Lasiocampidae

Importance It is a sporadic, polyphagous defoliator. The damage is serious when it outbreaks.

Distribution Beijing, Shanxi, Inner Mongolia, Henan, Shaanxi, Gansu.

Hosts *Quercus* spp., *Castanea mollissima*, *Juglans regia*.

Damage Larvae feed on the leaves. When the damage was serious, they generally ate up the leaves.

Morphological characteristics **Adults** 22-38 mm in length, 54-95 mm in wingspan, yellowish green. Thorax and the middle part of forewings yellow, discal cell with a close triangle dark chocolate-brown small spot, wings posterior area with a near quadrilateral dark chocolate-brown big spot, subouter margin line wave-shaped and consists of 8-9 small dark chocolate-brown spots. **Eggs** Oval, 0.3-0.35 mm in length, 0.22-0.28 mm in width, greyish-white. **Larvae** Mature larvae 65-84 mm in length, females closely covered with dark yellow long hairs, males covered with greyish-white hairs. Head yellowish-brown, pronotum with a dark chocolate-brown spot, each side of the front edge of the pronotum with a black verruca bearing black long hairs. Two sides of body with black verrucae attached with clusters of setae. **Pupae** Fusiform, reddish-brown, 28-32 mm in length, with sparse black short hairs.

Biology and behavior *Trabala vishnou gigantina* occurs one generation a year and overwinters as eggs on the trunks and small branches. Eggs hatch in late April to late May, the newly hatched larvae feed on eggshells, later feed on the mesophyll. The 1st to 3rd instar larvae feed gregariously, they spin silk and drop off when disturbed. Larvae after the 4th instar disperse and feed a lot, they raise and swing their heads when disturbed. Mature larvae spin cocoons to pupate on the branches, among shrubs, weeds and rocks in late August. Pupal stage is

T. vishnou gigantina adult male (Li Zhenyu)

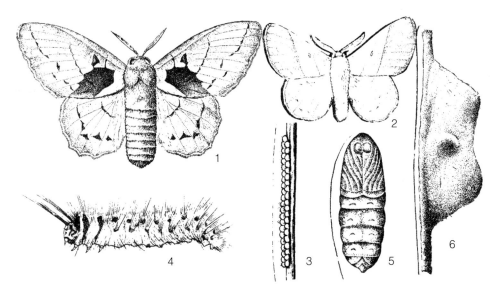

T. vishnou gigantina. 1. Adult female, 2. Adult male, 3. Eggs, 4. Larva, 5. Pupa, 6. Cocoon
(Zhu Xingcai)

T. vishnou gigantina adult female (Li Zhenyu)

9-20 days. Adults emerge and mate in mid-August to mid-September, they oviposit the same night or the next day. Eggs are laid in two rows. Each female lays 290-380 eggs. Adults have phototaxis and the average longevity is 4.9 days. The parasitic rate of parasitic wasps is 24% during the pupal stage. The infection rate by *Beauveria* spp. and nuclear polyhedrosis virus is 18% during the larval stage.

Prevention and treatment

1. Cultural control and Physical control. Plant mixed forests. Maintain proper canopy coverage to reduce damage. Manually pick eggs and cocoons, trap and kill larvae, trap and kill adults using black light traps.

2. Biological control. Spay *B.t.* at 1,000× or nuclear polyhedrosis virus solution.

3. Chemical control. During larval stage, foliar spray of 25% chlorbenzuron at 1,000×, or 25 g/L lamda-cyhalothrin at 1,000-1,500×, 2.5% deltamethrin EC at 5,000-8,000×, or 50% fenitrothion at 1,000×.

References Xiao Gangrou (ed.) 1992, Liu Youqiao and Wu Chunsheng 2006.

(Xie Shouan, Li Menglou)

417 *Rondotia menciana* Moore
Classification Lepidoptera, Bombycidae

Distribution Beijing, Hebei, Shanxi, Liaoning, Jiangsu, Zhejiang, Jiangxi, Shandong, Henan, Guangdong, Hubei, Hunan, Sichuan, Shaanxi, Gansu; Korea Peninsula.

Hosts *Morus* spp.

Morphological characteristics **Adults**cFemales 8.0-10.8 mm in length, 39.0-47.1 mm in wingspan. Males 8.6-9.6 mm in length, 29.4-30.7 mm in wingspan. Body and wings yellow, antennae plumose, dark chocolate brown. There is an arc-shaped indent below the apex angle of the forewing outer margin. Wing surface with two black wave-shaped cross stripes; the location between two cross stripes with a black short stripe. Hind wings also with two black cross stripes. Males darker in color, the abdomen slender and directed upward. Female's abdomen fat and big, directed downward. Ventral abdomen of the females that lay overwintering eggs is covered with dark brown hairs. **Eggs** Flat oval, the middle slightly concave. Major axis 0.50-0.85 mm, average 0.72 mm, minor axis 0.50-0.65 mm, average 0.64 mm. Eggshell surface with dense polygon-shaped granules. Non-overwintering eggs white, turn into pink before hatch. Newly laid overwintering eggs yellowish white, later turn brown. Non-overwintering eggs are laid in 3-10 rows, with 6-14 eggs in each row and multiple layers. Overwintering eggs are arranged in egg clutches, oval, 5-12 mm in diameter. The middle convex. The top covered with brownish black hairs. Larvae 24 mm in length. Head black. Thorax and abdomen milky in color. Each segment with transverse wrinkles, between wrinkles with black spots, the spots disappear when larvae mature. Young larvae are covered with a layer of white powder. Then body turn into cauliflower-yellow after three molts. The dorsal surface of the 8th abdominal segment with a black caudal horn. **Pupae** Long cylindrical. Females 10.0-15.4 mm, males 7.6-9.6 mm in length. Milky in color. Vertex, compound eyes and spiracles dark brown. Body turn yellow and black stripes appear on wings two days prior to adult emergence. Ventral abdomen dark brown in some pupae. Cocoon light yellow, oblong, 12-16 mm in length, cocoon shell loose.

Biology and behavior *Rondotia menciana* has univoltinism, bivoltinism and trivoltinism. They all overwinter as egg clusters with lids on branches and stems of *Morus* sp. Eggs hatch in early June, and peak at late

R. menciana adult (Li Zhenyu)

R. menciana adult (Li Zhenyu)

June. This generation larvae are called "First Huang". They pupate in mid-July, adults emerge and oviposit in late July. Then the univoltine moths lay egg clusters covered with lids. While the bivoltine and trivoltine moths lay eggs without lids that covering the egg clusters; eggs hatch in early August; these larvae are called "Second Huang". They pupate in late August, and adults emerge and oviposit in early September. The bivoltine moths lay clutches of eggs with lids. The trivoltine moths lay egg clutches without lids; eggs hatch in mid-September, the larvae are called "Third Huang". They pupate in early October; adults emerge in late October and lay eggs with lids in clutches. In Jiangsu and Zhejiang area, the majority is bivoltine, so the "Second Huang" causes the most damage. The univoltine moths are less common and the trivoltine moths are the least common. In Shangdong province, the majority is univoltine. Most damage occurs in early and mid-August. Bivoltine population only accounts for 1%-2%.

Adults emerge during the day. They are strong fliers and are active at night, and they have phototaxis. Generally, adults mate 3 hours after emergence, but some won't mate until the 6th day after emergence. They oviposit two hours after mating, and oviposition peak is during 10:00-14:00. The no-lid egg clusters are mostly laid on the back of leaves, few are laid on branches. While almost all eggs with lids are laid on the main and side branches and one year old twigs of *Morus* spp. Generally for the medium sized Morus spp., eggs are mostly laid on the branches, some are laid on one year old twigs, and rarely on the trunks; whereas for taller *Morus* spp., eggs are mainly laid on the trunks and branches, and rarely on the one year old twigs. Most eggs are laid on the downside of oblique branches or the exterior side of upright branches. On average, each egg clutch with lid has 120-140 eggs, each egg clutch without lid has 280-300 eggs. Adult longevity is: the overwintering generation 3-4 days; the 1st generation 3-4 days; the 2nd generation 5-6 days. Maximum longevity is 10 days. Male's longevity is longer than female's; females producing eggs with lids

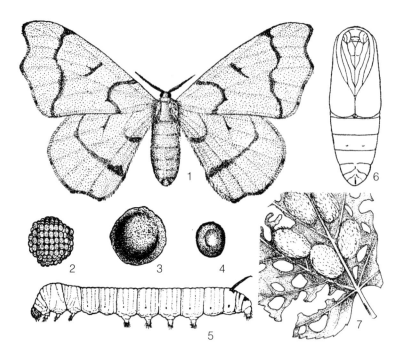

R. menciana. 1. Adult female, 2. Egg mass without cover, 3. Egg mass, 4 Egg, 5. Larva, 6. Pupa, 7. Cocoon and damage
(Zhejiang Agricultural University)

live longer than females producing eggs without lids.

Overwintering egg period is: univoltinism 338 days, bivoltinism 286 days, and trivoltinism 246 days. Egg period of the 1st generation is 9 days, and that for the 2nd generation is 12 days.

Larvae mostly hatch during 6:00-9:00. The hatching rate of eggs without lids is 95%-100%, the highest hatching rate of eggs with lids is 81% with the average being 60.57%.

The newly hatched larvae feed on the lower epidermis and mesophyll tissue of leaves and leave the transparent upper epidermis. After the first molt, larvae are able to eat through leaves and create holes on leaves. Larvae have five instars. The larval period is 18-33 days. The 1st and 2nd generation mature larvae spin cocoons on the downside of leaves, the 3rd generation ones spin cocoons on branches. The pupal duration is 6-17 days.

Prevention and treatment

1. Physical control. Pick and kill larvae and cocoons, scrape off egg clusters. Change the rearing arrangement of the silkworm. First, shake branches often to collect and kill disturbed larvae during the 2nd and 3rd instars. Second, pick cocoons and kill pupae on August 2 of each year. Third, scape off egg clusters from early October to late March. Fourth, in serious infested areas, try to rear early fall silkworms or rear fall silkworms ahead of time and pick leaves earlier to reduce R. menciana population.

2. Biological control. Protect *Telenomus (Aholcus) rondotiae*: eggs of the *R. menciana* are usually parasitized by *T. (Aholcus) rondotiae*. Save scraped eggs under low temperature, and move them to "protectors" before mulberry sprout in spring. Place the devices in seriously infested stands, so that hatched *T. rondotiae* would fly out and seek new hosts. The hatched *R. menciana* larvae would drown in the protectors. Protect *Coccygomimus luctuosus*. *Coccygomimus luctuosus* parasitizes *R. menciana* cocoons. Put the collected cocoons in protectors, so that the emerged parasitic wasps can fly out and seek for new hosts, but emerged *R. menciana* cannot fly out of the protectors.

3. Chemical control. During the peak egg hatching period, apply 80% DDVP or 50% phoxim at 1,000-1,500× on leaves and trunks of *Morus* spp. Leaves are safe for silkworms 5 days after spraying.

4. Utilize sex pheromone of R. menciana (*E10, Z12*-hexadeca-10, 12-dien-1-yl acetate). At mulberry fields with low *R. menciana* population, apply one trap with 100 µg dose per 667 m^2.

References Zhang Xiaozhong, Zhang Jilong and Wang Fuying 1998; Guo Yuliang and Xiao Gangrou 2006.

(Peng Guandi, Li Zhenyu, Gao Zuxun)

418 *Eupterote chinensis* Leech

Classification Lepidoptera, Eupterotidae
Common name Chinese processionary moth

Importance *Eupterote chinensis* feed on a large variety of broad-leaved trees. It can defoliate the whole tree and cause huge damage and loss to landscape and forestry.

Distribution Hunan, Sichuan, Yunnan.

Hosts *Paulownia fortune, Liriodendron chinense, Kalopanax septemlobus, Amygdalus* spp., *Toona sinensis, Ziziphus jujuba* var. *spinosa, Platanus orientalis.*

Damage Larvae feed on leaves, only the midribs and petioles are left intact.

Morphological characteristics **Adults** Females 65-86 mm in wingspan, males 55-72 mm in wingspan. Whole body light yellow. Compound eyes dark chocolate brown. Antennae yellowish-brown, female antennae unipectinate and thin, filiform at the first glance, male antennae plumose. Female labipalp reddish-brown or its tip slightly yellowish; legs brown, legs and thorax with light yellow hair tufts, females have especially more hair tufts. Scales on wings thin, wing surface with 5-6 intermittent red wave-shaped stripes, stripes at the forewing costal margin area thicker, the forewing apex angle and hind wing inner margin with big and irregular reddish-brown spots, males spots obvious, the reddish-brown spots vary among individuals. **Eggs** Round, 1.3 mm in diameter, yellow at the beginning and paler later. **Larvae** Seven instars in total. A common characteristic shared by various instars is the body has dense dull white long hairs. The newly hatched larvae light yellow, about 3.5 mm

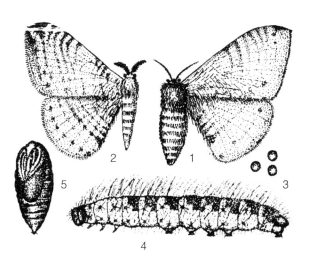

E. chinensis. 1. Adult female, 2. Adult male, 3. Eggs, 4. Larva, 5. Pupa (Zhou Lijun)

in length, head 0.6 mm in width, dark chocolate brown, the middle of dorsal body with two rows of dark chocolate-brown granules, primary setae yellowish-white. Body color turn into yellowish green after feeding, the 3rd instar turn into yellow. The head of the 4th instar black thread shape, body dull yellow, granules on mid-dorsal side light red. The 5th instar head shows Y shaped pattern, granules with scolus and long hairs in the same colour, internode dusty blue. The 6th instar lighter in colour than mature larvae. Mature larvae body 45-70 mm in length, head 5.2 mm in width; head in alternating black and dark yellow colours, body yellow with grey in colour; body hairs yellowish-white and 15 mm in length, each dorsal segment of the abdomen with a pot of black pubescence mixed with yellowish-white short hairs. **Pupae** 20-24 mm in length, 7-10 mm in diameter, dark brown, internodes light brown. **Cocoons** fusiform, thin, consist of black and grey silk; often glued to fallen leaves and grass.

Biology and behavior *Eupterote chinensis* occurs one generation a year in Huaihua, Hunan. They spin cocoons to overwinter in leaf litter, tree holes, cracks and loose surface soil after mid-September. Adults emerge in early June and reach emergence peak in mid- June.

E. chinensis adult (Zhang Peiyi)

E. chinensis larva (Zhang Peiyi)   E. chinensis larva (Zhang Peiyi)

Some emerge in late June. Adults become active during the daytime. They take mating flights around hosts especially before 10:00 and after 16:00. They are strong flyers, some can fly in moderate rain. Their longevity is 3-5 days. Adults are not attracted to lights at night. Eggs are laid at the edge of the back of leaves, they are always laid in hundreds.

Newly laid eggs are light yellow, then turn darker after 10 days, they turn into greyish-yellow around 15 days and a dark dot appears on the top, and then hatch in 3 more days. The egg stage is around 18 days.

Larvae are gregarious, they feed at night. The 1st and 2nd instars cluster together on the back of leaves during day and night. After the 3rd instar, larvae cluster together on the mid-trunks or big branches in the daytime. With the increase of the instars, the height of their rest places in the daytime gradually decrease, and some of the 4th instar move downward to the lower part of the trunk. After the 5th instar, larvae mostly rest on the trunks below 2 m and most of them are on the shade side. Larvae on a tree always cluster into one or two groups. A large larvae cluster can reach 80 cm in length and 20 cm in width, the number of larvae within a cluster is huge. At dusk when the sun goes down at around 19:30, larvae start move in several parallel lines towards leaves. Once they arrive at twigs, larvae move forward in single lines, end to end. Once they reach leaves, they arrange neatly at the leaf margin, head being outward, and eat up whole leaves but the petioles. They stop feeding at 4:30 the next day and crawl down trees. By 5:00 in the morning, most larvae rest at the shelter place. It is common to find completely defoliated trees. The larval stage last 70 days and larvae enter prepupal stage. After another 7 days, they pupate and overwinter.

Prevention and treatment

1. Silviculture. Avoid planting several hosts of *Eupterote chinensis* together. Plant some nectar plants to help parasitic wasps survive and reproduce.

2. Physical control and chemical control. Hit larvae clusters with a hammer or board. Spray 80% DDVP or trichlorfon at 800× using a hand-held sprayer.

3. Biological control. Protect predators such as mantis, and parasitic wasps such as *Exorista japonica* and *Exorista sorbillans*.

References   Chinese Academy of Sciences - Institute of Zoology 1983, Xiao Gangrou (ed.) 1992.

(Zhang Lijun, Zhou Lijun)

419 *Eupterote sapivora* Yang
Classification Lepidoptera, Eupterotidae

Importance *Eupterote sapivora* is one of the main pests of *Sapium sebiferum*. Since the mid-1970s, periodic outbreaks of *E. sapivora* occurred. Larvae are covered with urticating hairs, which often cause dermatitis and inflamed skin.

Distribution Guizhou.

Hosts *Sapium sebiferum*, *Paulownia* spp., *Toona sinensis*, *Populus* spp., *Cinnamomum camphora*, *Vernicia fordii*, *Prunus persica*, *Prunus* spp.

Damage It feeds on leaves, tender twigs and sepals.

Morphological characteristics **Adults** Females 24-29 mm in length, 75-88 mm in wingspan. Body yellow, head vertex and the dorsal labipalp reddish-brown; cteniis on each side of antennae flagellum short and small, their length barely longer than flagellum width. Thorax yellow. Most part of legs yellow, dorsal tarsus brown; dorsal forelegs reddish-brown from the femur, only the joints at the end of mid- and hind legs reddish-brown. Wings yellow; with less reddish-brown stripes than the male's; there is a row of spots along outer margin line. Forewings and hind wings with 6-7 spots of various sizes, some spots almost fade; lines inconspicuous; subouter margin line appears as a discrete, thin, wave-shaped stripe; lines present near the costal margin of forewings and inner margin of hind wings; wing margin hairs yellow. Brown spots on underside of wings are similar to those on top of the wings but more inconspicuous, some almost all yellow. Males 20-29 mm in length, 70-80 mm in wingspan. Head reddish-brown; the first two antennal segments yellow and the back densely covered with reddish-brown hairs, the flagellum brown, cteniis on both sides greyish-brown. Thorax yellow, only the base of prothorax reddish-brown. Leg color same as females. Wings yellow, with reddish-brown stripes. Forewing stripes only obvious at costal margin and outer margin, the ones near the apex angle the most obvious; stripes on the costal margin with three obvious ripples, each with three crests and extend downward but don't reach M_2 vein; the location from inner side of ripples to wing base with four small spots arranged in equidistance, the outer two often combine into one big spot; the location from outer side of ripples to wing apex with a big triangle brown spot; outer margin with brown broad stipes, gradually narrower downward and intermittently separate from the subouter margin line, inner side of the line with three spots, the mid-spot is the biggest; inner margin without obvious stripes, only few brown scales represent indistinctly stipes. Hind wing with a row of brown spots along outer margin line, two spots at the front and three at the rear; other stripes inconspicuous; sometimes the subouter margin line and mid-line indistinct. Wing margin hairs yellow, but the hairs outside of brown strip on outer margin of forewings brown. Stripes on the underside of wings are similar to ones on the top side, the base of forewing costal margin with purple wide stripe, costal margin of hind wing with four obvious stripes. Abdomen yellow. **Larvae** There are seven instars. Their morphology varies slightly with different environments. Body covered with yellowish white to light brown long hairs; pleopod crochets biordinal mesoseries;

E. sapivora adult (Yu Jinyong)

E. sapivora adult (Yu Jinyong)

antennae brown, 3 segmented; with 5 ocelli. The first instar larvae light yellow to yellowish green, the dorsal side with 13 black granules arranged in a line, body 2-2.5 mm in length, head 0.8-0.9 mm in width. After the 4th instar, larvae turn into yellow or dark yellow, 22.2-35.7 mm in length, head 4.1-5.1 mm in width, forehead "ʌ" shaped, the granules on the back black, there is a line of orange round spots below the spiracular line, the internodes greyish-white. After the 5th instar, the internodes turn darker, the "ʌ"shaped mark with scattered short hairs. Color of the 6th instar larvae lighter than the mature larvae. Mature larvae yellowish-brown to brown-black, 63-72.5 mm in length, below the subdorsal line with granules arranged in a line, the granules densely covered with thick dark grey and brownish black urticating hairs. Head 5.8-6.8 mm in width, brownish black, frontal suture depressed and reddish-brown.

Biology and behavior *Eupterote sapivora* occurs one generation a year in Guizhou and overwinters as pupae on the base of trunk and among the weeds, soil surface and rock cracks around the tree. Few overwinter in the cracks of bark. Adults emerge and oviposit in early May. Larvae hatch in early June. Larvae pupate from late September to mid-October.

Eggs are generally laid on the back of leaves, each egg clutch has 120-760 eggs (average 424 eggs). Egg stage is 16-22 days (average 19.2 days) at 23.9℃. The natural egg hatching rate was 88.5%-97.6% in woods. The 1st and 2nd instars gather on the back of leaves during daytime, but after the 3rd and 4th instars, larvae gather on the trunks below 2m. They crawl up the tree to feed at around 22:00, and crawl down the tree and gather at the same spot where they rest during the daytime. Larvae crawl up and down the tree in a neat line. The larval periods from the 1st to the 7th instars is 14, 15, 13.5, 15.5, 16.5, 14 and 16 days, respectively. They can survive after starving for a maximum of 3, 3-4, 3-4, 3-6, 4-7, 5-8, and 6-9 days, respectively. The 1st instar appear from late June to early July, and feed on the mesophyll and epidermis. The 2nd and 3rd larvae appear in early July and August, and feed on the epidermis and leaf margin. The 4th and 5th instars appear in late July to early September, and feed on the front half of leaves and remain the midribs. The 6th and 7th instars appear in late July to early September, and feed on the whole leaves and only leave the petioles. The 5th to 7th instars cause serious damage. The peak damage is from the end of August to September.

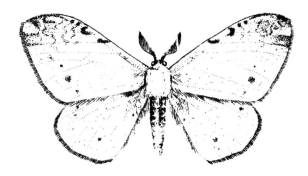

E. sapivora adult male (Hu Xingping)

Adults emerge in early to mid- June, averaged eclosion rate was 73.56%. The sex ratio of females to males in woods was 1:1.09. They start mating 24 hours after emergence, and most occur at dusk. Females oviposit four hours after mating. Peak oviposition occurs on the second and third days. Peak flight time is during 8:00-13:00. They fly around hosts in the open places in the woods. Females are not attracted to lights. Females and males live for 8.3 and 5.9 days, respectively.

The insect occurs more often in the low altitude area (500-1,000m) than high altitude area (above 1,700m), at the lower slope than the upper slope, at the old aged trees than young trees, and it prefers *Sapium sebiferum*.

Prevention and treatment

1. Physical control. Scrape and bury gathered larvae at the tree base into a pit.

2. Silviculture. Select pest resistant cultivars such as *Paulownia fortune* with sticky glands.

3. Conserve natural enemies. In larval stage, its natural enemies include *Hierodula patellifera*, *Eysarcoris ventralis*, *Parena rufotestacea*, *Chrysopa pallens*, *Formica rufibarbis sinae*, *Apantelis*, *Camponotus japonicas*, *Anastatus japonicas*, *Brachymeria* westwood, *Copidosoma floridanum*, Ichneumqnidae, *Beauveria bassiana*, *Eupterote sapivora* nuclear polyhedrosis virus; in egg stage, its natural enemies include *Telenomus euproctidis* and ants; in pupal stage, its natural enemy includes *Beauveria bassiana*.

4. Chemical control. Treat tree trunks with a mixture of 12% thiamethoxam and lamda-cyhalothrin at 1,000-1,500×.

Reference Xiao Gangrou (ed.) 1992.

(Yu Jinyong, Zhou Xianming, Zhang Jinguo)

420 *Actias ningpoana* Felder

Classification Lepidoptera, Saturniidae
Common name Chinese moon moth

Distribution Hebei, Jiangsu, Zhejiang, Jiangxi, Henan, Hubei, Hunan, Guangdong, Sichuan, Guizhou, Yunnan, Taiwan; Malaysia, India, Sri Lanka, Myanmar.

Hosts *Populus* spp., *Salix* spp., *Jugians regia*, *Cerasus pseudocerasus*, *Malus pumila*, *Prunus armeniaca*, *Elaeagnus angustifolia*, *Pterocarya stenoptera*, *Ulmus pumila*, *Alnus cremastogyne*, *Cinnamomum camphora*, *Liquidambar formosana*, *Hibiscus syriacus*, *Camptotheca acuminate*.

Morphological characteristics **Adults** 32-38 mm in length, 100-130 mm in wingspan, body strong, white, between antennae with a purple cross stripe. Antennae yellowish-brown plumose; compound eyes big, spherical and black. The front edge of scapulet base on the dorsal thorax with a dark purple cross stripe, wings nattierblue, the base with white flocculent scale hairs, wing veins greyish-yellow and obvious, margin hairs light yellow; forewing costal margin with a longitudinal stripe in white, purple and brownish black colors, connecting with the dorsal purple cross stripe. Hind wing anal angle long-tail shaped, 40 mm in length, the edge of hind wing caudal horn with light yellow scale hairs, some hairs slightly with purple. Discal cell of mid- forewings and hind wings with an oval eye shaped spot. The mid- spot with a transparent cross stripe. Segmental venter light in color, almost brown. Legs purple red. **Larvae** Mature larvae 62-80 mm in length, head small and greenish brown, body yellowish green. The verrucae on the dorsal mesothorax, metathorax and the 8th abdominal segment large, their top yellow, the base black. Other verrucae small, the top reddish orange, the base brownish black. Above the 1st to 8th abdominal segment spiracle line reddish-brown, below yellow. Mid- pygidium and the rear edge of caudal legs brownish red, larvae turn tan in color before pupation. **Pupae** Dark brown, fusiform. **Cocoons** Oval, silk texture coarse and greyish brown to yellowish-brown, the outside covered with leaves.

Biology and behavior *Actias ningpoana* occurs two generations a year. It overwinters as pupae in cocoons in late October, adults emerge in mid- and late March and early and mid- May. The peak larval damage of each generation is: mid-March to mid-June for the 1st generation; mid- and late July for the 2nd generation, late

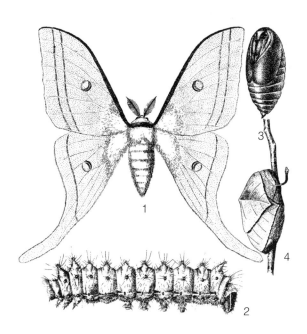

A. ningpoana. 1. Adult; 2. Larva; 3. Pupa; 4. Cocoon
(Zhang Peiyi)

A. ningpoana adult (Li Zhenyu)

September to early October for the 3rd generation.

Adults are active at night, two activity peaks occur during 20:00-21:00 and 0:00-1:00. Adults are attracted to black light. Adults mate the same night when they emerge and oviposit the next day. On average, each female lays 165 eggs.

Generally, larvae have 5 instars. Larvae feed on eggshells to supplement nutrition after hatch, and feed on

A. ningpoana adult (edited by Zhang Runzhi)

A. ningpoana adult (edited by Zhang Runzhi)

A. ningpoana larva (edited by Zhang Runzhi)

A. ningpoana 3rd instar larva (Xu Gongtian)

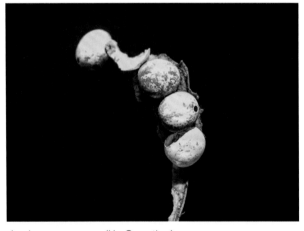
A. ningpoana eggs (Xu Gongtian)

leaf margins hours later. Larval aggregate before the 2nd instar, often tens of larvae gather together and feed on the back of leaves. Larvae disperse after the 3rd instar, but they are slow. Their feeding increases greatly after the 5th instar, and they eat more at night than during the daytime, the damaged leaves only with a part of the petioles. According to an investigation, each larva consumes up to 1,760 cm^2 leaves. Larvae feed less before the 3rd instar, and they feed a lot after the 5th instar. Mature larvae weave leaves with silk to make cocoons on original hosts, they pupate among the nearby brushes or weeds. Adults have strong flight ability, they lay eggs on undamaged hosts.

Prevention and treatment

1. Physical control. During winter time, combining with landscape management such as pruning and cutting, remove cocoons hang on the branches and trunks; timely clear cocoons on the mixed shrub or weeds. Manually catch and kill larvae when the population is low. Trap and kill adults by using black lights.

2. Biological control. In egg stage, release *Trichogramma dendrolimi*, the infection rate reaches 60% to 70%. In old instar larval period, spray 2×10^{10} PIB/mL nuclear polyhydrosis virus at 1,000-1,500×.

3. Chemical control. Treat tree trunks with a mixture of 12% thiamethoxam and lamda-cyhalothrin at 1,000-1,500× during the 1st and 2nd generation.

References Bi Xianzhang 2002; Xu Gongtian 2003; Yuan Haibin et al. 2004; Yuan Bo, Mo Yiqin 2006; Chen Bilian et al. 2006; Yuanfeng, Zhang Yalin, Feng Jinian, 2006.

(Li Jing, Li Zhenyu, Lian Yueyan, Fang Huilan)

421 *Antheraea pernyi* (Guérin-Méneville)

Classification Lepidoptera, Saturniidae
Common names Chinese (oak) tussar moth, Chinese tasar moth, Temperate tussar moth

Distribution Hebei, Liaoning, Jilin, Heilongjiang, Jiangsu, Zhejiang, Shandong, Henan, Hubei, Hunan, Sichuan, Guizhou.

Hosts *Quercus dentate*, *Quercus liaotungensis*, *Quercus mongolica*, *Quercus acutissima*, *Juglans regia*, *Cinnamomum camphora*, *Salix viminalis*.

Morphological characteristics **Adults** Females 35-45 mm in length, 150-180 mm in wingspan, body orange-yellow, body densely covered with yellowish-brown scale hairs. Forewings and hind wings near triangle, mid- wing with a beautiful eye shaped stripe. Antennae long and narrow, 13 mm in length, 2.5 mm in width; abdomen thick, 7 segments are visible. Males 30-35 mm in length, 130-160 mm in wingspan; lighter in color than female, antennae developed and wide, 14 mm in length, 7 mm in width; abdomen small, 8 segments are visible. **Eggs** 2.2-3.2 mm in length, 1.8-2.6 mm in width, oval, slightly flat, dark brown or brown or light brown. After egg showering or disinfection treatment, eggs turn greyish-white. **Larvae** Five instars in total. The 1st instar black, head capsule purplish red. The 2nd instar yellow or green or bluish yellow due to different varieties, setae sparser than the 1st instar. After the 4th instar, side of body above the spiracles on the 1st to 7th abdominal segments with a light brown or puce with white upper spiracle line; mid- segmental venter with a purple red medioventral line; subdorsal line, upper spiracle line and lower spiracle line on each segment with three pairs of verrucae, generally bilateral symmetry, verrucae on the dorsal side bigger than others and with seven long setae, six of them arranged in hexagon shape, the other one in the middle. **Pupae** Females 45 mm in length, 22 mm in width, oval, dark brown or tan. Mid- segmental venter of the 8th abdominal segment with a longitudinal groove, "X" shaped. Males 38 mm in length, 19 mm in width, mid- segmental venter of the 9th abdominal segment with an umbilicate dot. Antenna wide and protuberant. **Cocoons** Oblong, 12-16 mm in length, cocoon shell loose. The end of the handle slightly sharp, another end slightly blunt. Spring cocoons white, easy to be loosened up. Fall cocoons mostly yellowish-brown, not as easy as the spring ones to be loosened up. Cocoon color is closely related to diet type, leaf quality and the climate conditions.

Biology and behavior *Antheraea pernyi* occurs one or two generations a year and overwinters as pupae. It only occurs one generation a year in Guangxi, Guizhou, Sichuan and Henan; two generations a year in Shandong, Liaoning and Heilongjiang. Photoperiod is a major external factor determine the number of generations. The 5th instar is very sensitive to light. For *Antheraea pernyi* occurring one generation a year, adults emerge and oviposit in March to early April, larvae hatch in early and mid- April, and spin cocoons to pupate in mid- and late May, then enter diapause, adults emerge in

A. pernyi adult male (Li Zhenyu)

A. pernyi adult female (Li Zhenyu)

April the next year. Antheraea pernyi occurring two generations a year have spring silkworms and autumn ones. For spring silkworms, adults emerge and oviposit in early April, larvae hatch in late July and early August, which are called autumn silkworms. Autumn silkworms spin cocoons, pupate and overwinter in mid- and late September. Larvae have direct drinking habits, both adults and larvae have strong phototaxis.

Not only *Antheraea pernyi* can help making silk, its pupae have been used to rear *Chouioia cunea*, a biological control agent for controlling *Hyphantria cunea* and other lepidopterous species in recent years. Each pupa can support 5,000 wasps.

Prevention and treatment

1. Egg surface sterilization. This is a key measure for prevention of silkworm diseases. At present, there are two commonly used methods, one is formaldehyde sterilization: wash eggs in 20℃ water, then sterilize them for 30 minutes with 3% formaldehyde solution under 23-25℃; the other is sterilization in hydrochloric acid and formaldehyde mixture solution: soak eggs in 0.5%-1% sodium hydroxide solution for a minute, then take them out and clean them with water immediately, later sterilize them with 3% formaldehyde and 10% hydrochloric acid mixture solution under 20-22℃ for 10 minutes, then mix the mixture with 20℃ water.

2. Control pebrine. 1) Try to use diseases free eggs. 2) Take microscopic examination, supplementing with visual inspection, strictly keep out eggs with diseases; 3) Sterilization. Sterilize the rearing room and all utensils with 3% formaldehyde or hydrochloric acid and formaldehyde mixture solution; also, strictly dispose pathogens to prevent cross infection in silk farms.

3. Control nuclear polyhedrosis virus disease of *Antheraea pernyi*. 1) select and breed disease-resistant varieties, extend the 1st generation hybrids; 2) strictly sterilize the rearing room, utensils and egg surface; 3) keep the optimal density at rearing process, provide silkworms with good quality and enough diet leaves to keep them healthy and strong.

4. Control the empty-gut disease of *Antheraea pernyi*. Sterilizing the egg surface with 1% caustic soda or hydrochloric acid (sulfuric acid). In addition, adjust foods in order to prevent silkworm farms from reinfection.

5. Control the "spit white water" softening disease.

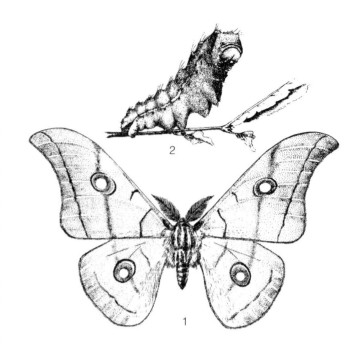

A. pernyi. 1. Adult, 2. Larva
(1 by Qu Xiaojin; 2 by Hong Jiaqi)

After sterilizing the eggs surface, soak eggs in 8-32 × "Candele" diluent for 10 minutes, and place them under the shade to allow them to dry; during the 3rd instar peak feeding period, feed silkworms with leaves sprayed with 8-32 × "Candele" for 24 hours, then feed them with new leaves.

6. Control parasitization by tachinid flies. Soak silkworms with 800× "Destroying Exorista sorbillans" EC solution on sunny days to kill the endoparasites.

7. Control nematodosis of *Antheraea pernyi*. Spray 0.03%-0.05% Phenamiphos No.1 or 0.01% Phenamiphos No.2 seven days after raining to kill parasitic nematodes in silkworm bodies.

8. Control grasshoppers in silkworm farms. Place toxic bait in the silkworm farm can kill 96% of the grasshoppers.

9. Control ground beetles in silkworm farms. Spread a 15 cm wide ring around *Xylosma racemosum* with insecticide powder.

References   Xiao Gangrou 1992, Ren Xianwei 1995, Wang Lianzhen 2001.

(Dai qiusha, Li Zhenyu, Hong Jiaqi, Tian Rongle)

422 *Caligula japonica* (Moore)

Classification	Lepidoptera, Saturniidae
Synonyms	*Dictyoploca kurimushi* Voolshow, *Dictyoploca regina* Staudinger, *Dictyoploca japonica* Moore
Common name	Japanese giant silkworm

Distribution Hebei, Liaoning, Jilin, Heilongjiang, Zhejiang, Fujian, Jiangxi, Shandong, Hubei, Hunan, Guangdong, Guangxi, Hainan, Chongqing, Sichuan, Guizhou, Yunnan, Shaanxi, Taiwan; Japan, Korea Peninsula, Russia (Siberia and far east coast area). Endemic species of East Asia.

Hosts *Juglans regia*, *J. mandshurica*, *Rhus verniciflua*, *Toxicodendron succedaneum*, *Ginkgo biloba*, *Cinnamonum camphora*, *Pterocarya stenoptera*, *Populus* spp., *Salix* spp., *Betula* spp., *Castanea* spp., *Liquidambar formosana*, *Ulmus* spp., *Quercus* spp., *Malus pumila*.

Damage Larvae feed on leaves. From 4th instar, they often eat up the whole leaves of a tree.

Morphological characteristics **Adults** Male body 28 mm in length, about 56 mm in wingspan; female body 34 mm in length, 63 mm in wingspan. Antennae yellowish brown, male's plumose, female's pectinate; wings greyish brown or orange yellow, location between scapulet and prothorax with greyish-brown cross stripe; forewing apex angle extrude outward, tip blunt and round, the front edge of apex angle with a fusiform black spot, inner line puce and arc-shaped. Two outer margin lines wavy shaped, distal half of the wing and wing base is darker than that at other areas. Hindwing has a wide red area in the middle, the end of discal cell with a big eye-shaped black spot, two subouter margin lines wavy. Base of forewing and hindwing with long puce fluffs. **Eggs** Major axis 2.0-2.5 mm, minor axis 1.2-1.5mm; oblong, often upright, the top with round black dot. Newly produced egg ivory or green, later greyish-brown or greyish-white, ash black before hatching. **Larvae** The newly hatched larva black, covered with long black fluffs, later gradually become dense, body color turns into ultramarine or greenish-yellow after the 3rd instar (larvae feed on *Rhus verniciflua* and *Juglans regia* turn black); each segment densely covered with long white fluffs; verrucae on the back and two sides of the body each with 1-2 black long setae. Spiracles ultramarine. Legs light yellow, the tip black. **Pupae** Female 45-50 mm in length, male 35-42 mm in length, fusiform, reddish-brown to reddish orange, female usually darker than male. **Cocoons** 50-70 mm in length, 25-30 mm in width, oblong, yellowish brown to dark brown. Cocoon silk is thick, and the texture is hard, the interspaces among cocoon silk are big, cage shape. Pupa in the cocoon is visible, the cocoon end with less silk is the emergence hole.

Biology and behavior The insect occurs one generation a year in Jilin, Liaoning, Henan, Hubei, Jiangxi and Sichuan. It overwinters as eggs. Most eggs are deposited at the base of tree trunks and on low, young trees or on the ground near the host. Eggs are laid in clutches or in piles, each clutch has dozens of eggs, they are easy

C. japonica adult female (Xu Gongtian)

C. japonica adult male (Xu Gongtian)

detected. The overwintering eggs hatch in early May in Liaoning, larvae occur from May to June, pupae occur from mid-June to early July, adults emerge between mid- and late August. In Hubei, eggs begin to hatch when the temperature is 18-22 °C in April, larvae occur from April to July, pupae occur from early June to early October, and adults occur from mid-August to late October; the emergence peak is in mid-September. In Jiangxi and north Guangxi, eggs hatch in late March to early April, larvae occur from April to June, pupate occur from late May to mid-June, adults occur from late August to early November. Adult longevity is 5-12 days, egg stage is 160-220 days, larva stage is 40-75 days, and pupa stage is 112-148 days.

Adults mostly eclose at night and mate at the same night or the next night, mating lasts for 12-24 hours, A female deposits all its eggs in 3-5 times, each female lays 100-600 eggs (average about 300 eggs). Adults hide in the shade close to the pupae and cocoons in the daytime and start being active at dusk. Eggs are mostly laid in co-coons, empty pupae, soil surface under grass, and cracks of trunks. Larvae crawl upward along the trunk after hatching, and often feed in groups on leaves close to the ground. Several or tens of the 1st and 2nd instar larvae usually congregate in a group on the back of a leaf, with

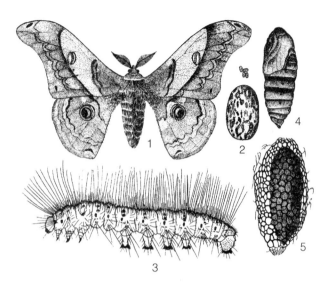

C. japonica. 1. Adult, 2. Eggs, 3. Larva, 4. Pupa, 5. Cocoon (Xu Tiansen)

their heads oriented toward the leaf margin. Sometimes one single leaf has as many as 40-60 larvae. Larvae disperse after the 3rd instar and feed singly, their damage occurs all over the canopy. They often migrate in groups to other trees when the food is insufficient. Their appetite is very big after the 4th instar, larvae often eat up whole

C. japonica adult (Li Zhenyu)

C. japonica adult cocoon (Xu Gongtian)

C. japonica pupa parasitized by wasps (Xu Gongtian)

plant leaves, and they stop feeding activity after they enter the 6th instar. Most of them pupate from late May to mid-June. After larvae mature, they choose a thin 2nd year branch, weave leaves and spin cocoons mixed with a few leaves. When they spin cocoons, they weave very sparse outer rings around their bodies, then add silk and sew them together. Most larvae choose to spin cocoons at places 1.0-1.5m above the ground on short plants that are close to their host plants. Few larvae make cocoons in the grass or in cracks of tree crotches 2-3 m high. They enter aestivation after pupation.

Prevention and treatment

1. Mechanical control. Combining with forest management or pest survey in early spring, scrape off egg clutches and remove cocoons. Place eggs and pupae into a basket and cover with screen to collect natural enemies, then release natural enemies in the woods. This is a convenient, easy, and important control method.

2. Light traps. Trap and kill male adults using black light. Killing large number of males can reduce the amount of fertilized eggs and then control them before their damaging.

3. Protect natural enemies in the forest. These include: *Trichogramma* sp., *Telenomus* sp., *Anastatus* sp. and *Pseudanstatus albiguttata* in egg stage; *Exorista sorbillans* in larval stage; and *Xanthopimpla predator* in pupal stage. Also, varuiou birds such as *Parus major*, *Pycnonotus sinensis*, *Garrulax canorus*, and *Pica pica* can predate upon larval stage.

4. Chemical control. (1) Drill holes and inject pesticide at the base of the trunk to kill larvae before the 3rd instar. For trees with the diameter at breast height (DBH) over 20 cm, drill four to five 10 mm diameter, 100-200 mm depth holes at different directions at the base of the trunk with 45° tilting degree. Inject 20% imidacloprid agent into the holes at 0.8 mL/cm of DBH, then seal holes with wet yellow mud. (2) Spray *B.t.* or 1.2% Matrine EC during the larvae gregarious period. Focus on the trunks from ground to middle height and tender leaves at the lower part of the canopy.

References Liu Ruiming 1984, Wei Ping 1989, Sun Qionghua 1991, Zhu Hongfu and Wang Linyao 1996, Li Guoyuan 2001, Jiang Dean 2003, Ding Dongsun et al. 2006.

(Ding Dongsun, Qin Zebo, Sun Qionghua, Zhao Tingkun)

423 *Eriogyna pyretorum* (Westwood)

Classification Lepidoptera, Saturniidae
Synonym *Eriogyna tegusomushi* Sasaki

Importance There are 3 subspecies in China, including *Eriogyna pyretorum pyretorum* Westwood distributed in Northeast and North China and overwintering as pupae, *Eriogyna pyretorum cognata* Jordan distributed in East China and overwintering as pupae or eggs, and *Eriogyna pyretorum lucifea* Jordan distributed in Sichuan and overwintering as egg. It is a defoliator pest, but it is also a resource insect that spins silk and makes cocoons. The quality of silk produced by worms feeding on *Cinnamomum camphora* leaves is the best. Their silk is smooth, transparent, sturdy and water tolerant. It is pure transparent in water, so it is a superior fishing material. It is also made for superior surgical sutures.

Distribution Hebei, Inner Mongolia, Liaoning, Jilin, Heilongjiang, Jiangsu, Zhejiang, Anhui, Fujian, Jiangxi, Shandong, Henan, Hubei, Hunan, Guangdong, Guangxi, Hainan, Sichuan, Guizhou, Shaanxi, Gansu; Russia, India, Myanmar, Vietnam.

Hosts *Pyrus bretschneideri*, *Ulmus pumila*, *Cinnamomum camphora*, *Acer* sp., *Quercus acutissima*, *Castanea mollissima*, *Pterocarya stenoptera*, maple, *Liquidambar formosana*, *Elaeagnus angustifolia*, *Pyrus pyrifolia*, *Camellia olefera*, *Ginkgo biloba*, *Paulownia* spp., *Juglans regia*, *Betula* spp., *Sassafras tsumu*, *Punica granatum*, *Camptotheca acuminate*, *Acer truncatum* Bunge, *Ilex chiensis*, *Sapium sebiferum*, *Toxicodendron vernicifluum*, *Citrus reticulate*, *Eriobotrya japonica*, *Lithocarpus glaber*, *Rosa* sp., *Psidium guajara*.

Damage The 1st to the 3rd instar larvae fed aggregatively, they feed leaves into incisions. They feed dispersedly after the 4th instar. Larvae feed fiercely, and they often defoliate the whole tree and impact trees growth.

Morphological characteristics **Adults** Female body about 35 mm in length, 118 mm in wingspan. Wings greyish brown, the middle of forewings and hind wings with an oval eye shaped spot, its end

E. pyretorum adult male (Ji Baozhong, Zhang Kai)

near the wing base slight bigger; the outside stripe of the eye spot blueish black, the exterior side of the inner layer with a light blue semi-circle stripe; the innermost layer is an earthy yellow circle, inner side of it dark reddish-brown; the middle is a crescent-shaped speculum. Forewing base with a triangle dark spot, the apex with 2 red-violet stripes, inside the apex with 2 black short stripes; the inner cross line brownish black, the outer cross line brown and double saw tooth-shaped; subouter margin shown as intermittent black spots, outer margin line greyish-brown, between them is a white cross band. Hind wings are similar to forewings, but color slightly lighter, the eye spots slightly smaller. Ventral and dorsal surface densely covered with greyish-white fluffs, tail densely covered with blueish brown scale hairs, abdominal internodes with white fluff rings. Male body about 25 mm in length, 88 mm in wingspan. Body color slightly darker than the female, stripes pattern basically the same as the female, eye spot smaller than the female, the crescent-shaped speculum in the eye spot of the hindwing unclear. **Eggs** 2 mm in length, cylindric, milky with tiny blue, several or a dozen eggs closely arranged in a clutch, covered with a layer of thick greyish-brown female tail hairs. Egg stage is about 20 days under 20℃ eggs are light ash black when they are about to hatch. **Larvae** 8 instars in total, larval stage is about 80 days. The 1st instar larva body 5-7 mm, body black, head and mandible black, glossy, head with clumps of long and thin white hairs. The dorsal side and two sides of each segment with several cylindric granules, each segment of thorax with 8 granules, each abdominal segment from the 1st to the 8th with 6 granules, the 9th abdominal segment with 4 granules, the last segment with 2 granules, each granule with several thin hairs. The 2nd instar larva body 10-13 mm, body bluish green, head still black and glossy; dorsal line, subdorsal line, upper spiracle line and lower spiracle dark blue, granules with hard spines. The 3rd instar larva body 16-24 mm, body color lighter than the 2nd instar, with sparse small black dots. The 4th-6th instars are 31-35mm, 36-39mm, and 43-46mm in length, respectively; the 7th instar body 52-58 mm in length, the dorsal side yellow, venter bluish green. The 8th instar body 62-65 mm, hard spines on granules stand up in groups, soft and glossy. Mature larvae slightly transparent, head green, body yellowish green. Each segment of the body has sarcomas, the 1st thoracic segment with 6 sarcomas, the other segments each with 8 sarcomas, each sarcoma with 4-8 brown hard spines. Epiproct with 3 black spots,

E. pyretorum adult female (edited by Zhang Runzhi)

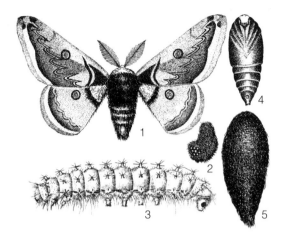

E. pyretorum. 1. Adult, 2. Eggs, 3. Larva, 4. Pupa, 5. Cocoon
(Zhang Peiyi)

E. pyretorum egg mass (Wang Yan)

E. pyretorum cocoon (Wang Yan) E. pyretorum larvae (edited by Zhang Runzhi)

E. pyretorum larva (edited by Zhang Runzhi)

E. pyretorum cocoons (edited by Zhang Runzhi)

paraproct with a big black spot. **Pupae** Body 27-35 mm in length, tan, fusiform, whole body hard, frontal area with an unconspicuous almost square shaped light color spot, with 16 cremasters. **Cocoons** 35-40 mm in length, greyish-brown, oblong.

Biology and behavior The *Eriogyna pyretorum* occurs one generation a year and overwinters as pupae in cocoons in cracks of limbs and barks, few overwinter as eggs. Adult eclosion peak is mid-February in Guangdong, late February to early March in Fujian and late March to early April in Zhejiang. Adults mostly emerge at dusk or early in the morning, they hide in the daytime, and have phototaxis. They mate soon after emergence. Mating mostly take place at night and last for 5-6 hours,

E. pyretorum larvae (edited by Zhang Runzhi)

E. pyretorum pupa (edited by Zhang Runzhi)

E. pyretorum damage to Camphor tree
(edited by Zhang Runzhi)

E. pyretorum damage to *Liquidambar formosana*
(edited by Zhang Runzhi)

most last until dawn. They oviposit 1-2 days after mating, mostly early in the morning, rarely in the daytime. Eggs are mostly laid in piles on the trunk or branches, few are laid dispersedly, and each pile has more than 50 eggs. Each female lays 250-420 eggs. Egg pile is densely covered with black fluffs of the female. The 1st to the 3rd instars larvae feed gregariously and the 4th instar larvae disperse. As the larvae grow, leaves cannot support their weight. They always crawl to petioles or branches. The body closely huddles up when frightened. Larvae often crawl on the trunk and move to other places to feed around noon. Mature larvae spin cocoons on the trunk or at the crotch from the dusk or the afternoon, it takes 1-2 days to finish. After 8-12 days prepupae stage, it starts to pupate.

Prevention and treatment

1. Mechanical control. Hand remove cocoons in the wintertime. Destroy larvae based on their habits of gregariousness and crawling up and down on the trunk during the occurrence period.

2. Physical control. Trap adults with black light in the adult eclosion peak.

3. Biological control. Apply *Beauveria bassiana* in early raining season, or spray with fluid out of dead insect bodies that were killed by *B.t.* or nuclear polyhedrosis virus.

4. Chemical control. To control the 1st to the 4th instar larvae, spray a mixture of 12% thiamethoxam and lamda-cyhalothrin at 1,000-1,500×.

References Xiao Gangrou (ed.) 1992; Yin Anliang, Zhang Jiasheng, Zhao Junlin et al. 2008.

(Ji Baozhong, Zhang Kai, Fang Huilan, Lian Yueyan)

424 Samia cynthia (Drury)

Classification Lepidoptera, Saturniidae
Synonym *Philosamia cynthia* Walker et Felder
Common name Ailanthus silkmoth

Distribution Beijing, Hebei, Liaoning, Jilin, Shanghai, Jiangsu, Zhejiang, Anhui, Fujian, Jiangxi, Shandong, Henan, Hubei, Hunan, Guangdong, Guangxi, Sichuan, Guizhou, Yunnan, Tibet, Gansu, Taiwan; Korea Peninsula, Japan, U.S.A., France.

Hosts *Sapium sebiferum*, *Ailanthus altissima*, *Ilex chinensis*, *Platanus hispanica*, *Rhus chinensis*, *Cinnamomum camphora*, *Citrus reticulate*, *Michelia figo*, *Firmiana platanifolia*, *Juglans regia*, *Pterocarya stenoptera*, *Robinia pseudoacacia*, *Zanthoxylum bungeanum*, *Paulownia* spp., *Ricinus communis*.

Damage Larvae often cause complete defoliation of *Sapium sebiferum* and *Zanthoxylum bungeanum* and reduce seed yields of *Sapium sebiferum* and *Zanthoxylum bungeanum*.

Morphological characteristics **Adults** Female body 25-30 mm in length, male body 20-25 mm

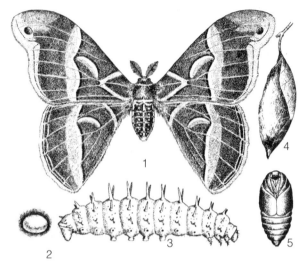

S. cynthia. 1. Adult, 2. Egg, 3. Larva, 4. Cocoon, 5. Pupa (Zhang Peiyi)

S. cynthia adult (Xu Gongtian)

S. cynthia eggs (Xu Gongtian)

S. cynthia larva (Xu Gongtian)

S. cynthia late stage larva (Xu Gongtian)

S. cynthia mature larva (Xu Gongtian)

S. cynthia pupae (Xu Gongtian)

S. cynthia pupae in cocoon (Xu Gongtian)

S. cynthia cocoons (Xu Gongtian)

S. cynthia pupa parasitized by natural enemy (Xu Gongtian)

in length, 115-125 mm in wingspan. Body greenish brown; head, the front end of cervix, the posterior edge of prothorax, the abdominal dorsal line, lateral line and tail end white. Forewings brown, apex angle round and convex, pinkish purple, with a black eye shaped spot, upside of the spot white arc-shaped; the middle of forewings and hindwings each with a crescent-shaped spot, the upper edge of the crescent-shaped spot dark brown, the middle translucent, the lower edge earthy yellow; outside of the crescent spot is a wide band passing through the whole wing, the middle of the wide band pink, the outside white, the inner side dark brown; basal angle brown, the edge of it with a white bent stripe. **Eggs** About 1.5 mm in length, greyish-white, with brown spot, flat oval. **Larvae** Mature larva body 55-60 mm in length. Turquoise, covered with white powder. The subdorsal line, upper spiracle line and lower spiracle line of each segment respectively with a row of conspicuous scoluses, those on the subdorsal line bigger than those on the other two rows. Between subdorsal line and upper spiracle line, the rear of spiracles, below the spiracle line, the bases of thoracic legs and prolegs with black spots. Sieve plate of spiracle light yellow, peritreme black, thoracic legs yellow. Prolegs turquoise, the tip end yellow. **Pupae** Body 26-30 mm in length, 14 mm in width, dark reddish-brown. **Cocoons** greyish-white, olive shape, about 50 mm in length. Upper end with a hole, the cocoon handle 40-130 mm in length, half of the cocoon is often warped with a leaf.

Biology and behavior *Samia cynthia* occurs 2-3 generations a year and overwinters as pupae. For those occurring 2 generations a year, overwintering generation adults emerge and oviposit in early and mid-May. Egg stage is about 12 days. The 1st generation larvae hatch in mid- and late May. Larva stage is around 30 days. Larvae spin cocoons and pupate in late June. The 1st generation adults emerge and oviposit in August to early and mid-September. Adult longevity is 5-10 days. The 2nd generation larval damage period is from September to November; later larvae successively pupate to overwinter.

Adults have phototaxis and strong flight ability. Each female lays 300 eggs on average. Males are strongly attracted to females. An unmated female in a cage can continuously attract males. However, indoor reared adults cannot mate easily. Cutting female wings will promote their mating. Eggs are laid in piles on the back of the host leaves. The first instar larvae are gregarious. After molting, larvae often eat up the shed skins. Mature larvae weave leaves and spin cocoons on the trunks, overwintering generation larvae often spin cocoons on shrubs.

Natural enemies of larvae include *Apanteles* sp., *Gregopimpla himalayensis*, *Coccygomimus parnarae* and *Xanthopimpla konowi*.

Prevention and treatment

1. Mechanical control. Destroy larvae and hand remove cocoons.

2. Physical control. Trap adults with black light.

3. Biological control. Apply $1-2\times10^8$ spore/mL *B.t.* or 1×10^8 NPV/mL to kill larvae of *Philosamia Cynthia*.

4. Chemical control. Apply 12% thiamethoxam and lamda-cyhalothrin at 1,000-1,500× to kill 1st to 3rd instar larvae.

Reference Wang Guang 1957, Xiao Ganrou (ed.) 1992, Yu Guoyue 2015.

(Fang Huilan, Lian Yueyan)

425 *Attacus atlas* (L.)

Classification	Lepidoptera, Saturniidae
Synonyms	*Phalaena atlas* L., *Saturnia atlas* Walker, *Attacus indicus maximus* Velentini, *Platysamia atlas* Oliver
Common name	Atlas moth

Distribution Fujian, Jiangxi, Hunan, Guangdong, Guangxi, Hainan, Guizhou, Yunnan, Taiwan; India, Myanmar, Indonesia.

Hosts *Sapium sebiferum*, *Cinnamomum camphora*, *Salix babylonica*, *Albizia lebbeck*, *Berberis silvatarou-cana*, *Dioscorea esculenta*, *Setaria viridis*, *Malus pumila*, *Ilex chinensis*, *Betula* spp., Paulownia, *Pittosporum tobira*, *Ficus microcarpa* var. *pusillifolia*, *Schima superba*, *Neolamarskia cadamba*, *Camellia oleifera*, *Phyllanthus emblica*, *Cinnamomum cassia*, Acer, *Punica granatum*, Carpinus, *Bischofia polycarpa*, *Camellia sinensis*, *Quercus variabilis*.

Damage Larvae feed on the leaves and tender shoots. They feed leaves into incisions or hole when the population is small, and can also defoliate the whole tree when the population is big.

Morphological characteristics **Adults** Reddish-brown in color, the largest moth, 30-40 mm in length, 250-300 mm in wingspan. **Larvae** Mature larvae 75 mm in length, head and thoracic legs yellow, thorax light green and covered with white powder, each abdominal segment with 6 symmetrical spines, between spines with dark chocolate-brown spots.

Biology and behavior *Attacus atlas* occurs two

A. atlas adult (edited by Zhang Runzhi)

A. atlas adult wing (Zhang Peiyi)

A. atlas adult (Zhang Peiyi)

A. atlas adult (Zhang Peiyi)

A. atlas damage (edited by Zhang Runzhi)

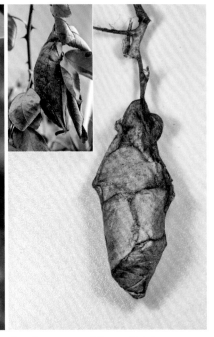
A. atlas cocoon (Zhang Peiyi)

generations a year in northern Fujian, three generations a year in Jinghong City, Yunnan. The newly hatched larvae aggregate along the midrids on the back of leaves. Larvae slight shape like a "C", they nibble up the leaves, only the midribs and petioles are left intact. The 3rd and 4th instars feed the whole leaves but the petioles, and the 5th instar larvae eat up whole leaves and turn the branch bold, then move to another branch.

In Jinghong, Yunnan, overwintering pupae emerge in early May. The first generation eggs first occur in mid-May, the first generation larvae first seen in early June. The peak damage of the first generation larvae is during mid- to late June. The 1st generation adults emerge in early July, adult stage last until late July. The 2nd generation larvae occur in late August.

Prevention and treatment

1. Artificial control. Remove eggs from trees. Cut off cocoons with pupae on branches before new shoots sprout. Look for and catch larvae on the trees based on the frass location.

2. Physical control. Trap and kill adults with black light.

3. Biological control. Protect and utilize *Theronia zebradiluta* and *Xanthopimpla pedator*.

References   Wang Wenxue 1986, Zhu Hongfu, Wang Linyao 1996.

(Liu Jianhong, Pan Yongzhi)

426 *Notonagemia analis* (Felder)

Classification Lepidoptera, Sphingidae
Synonym *Meganoton analis* (Felder)
Common name Grey double-bristled hawkmoth

Distribution Shanghai, Jiangsu, Zhejiang, Anhui, Fujian, Jiangxi, Hubei, Hunan, Guangdong, Guangxi, Hainan, Sichuan, Yunnan, Tibet, Shaanxi, Taiwan; India, Nepal.

Hosts *Magnolia officinalis* spp. *biloba*, *M. officinalis*, *M. cylindrical*, *M. liliflora*, *M. denudata*, *Liriodendron chinense*, *Sassafras tsumu*.

Damage During outbreak, it can consume all leaves on host plants in large areas and cause the forests look like being burned by wild fire.

Morphological characteristics **Adults** 83-140 mm in wingspan. Head greyish brown, thoracic back strong, outer edge of scapulae with thick black longitudinal line, rear edge with a pair of black spots; abdominal dorsal line brown, two sides of it with wide reddish brown longitudinal band and intermittent white stripe; thoracic and abdominal venter white. Forewings reddish brown, densely covered with white dots; inner line inconspicuous, midline reddish black and obvious, outer line discontinuous; outer margin white; anterior of the oblique line of apex angle with a near triangular reddish black spot, near the apex on M_1 vein with oval spots; discal cell with a white dot, and with a wide reddish black oblique line leading to the area between R_3 and M_3. Hindwings ochre yellow, near anal angle with a separate reddish black spot, also with an inconspicuous cross band reaching middle of hindwing. **Eggs** Near round, 1.7 mm in diameter. Light green when it was newly produced, later turn into yellowish brown. **Larvae** Mature larvae 70-85 mm in length. Head green, with yellow granules. Pronotum sclerotized strongly. Prothorax and mesothorax each with 5 annelets, metathorax with 6 annelets, the 1st to 7th abdominal segments each with 8 annelets. Caudal horn 11-13 mm in length, coniform, tilted up and backward. Body color varies, the part above spiracle line green in most individuals, the part below spiracle line and ventral surface greyish green; each side of the 7th abdominal segment with a yellow oblique stripe extending from the front border to the base of caudal horn; dorsal surface of some individuals yellow, the front end of the 2nd to 4th (or the 2nd to 7th) abdominal dorsal surface with purple "V" shaped pattern. **Pupae** 40-65mm in length, long cylindrical. Thorax is the widest, abdomen gradually thinner towards the tail end. Yellowish green at early pupation, and turns tan later. Proboscis protrudes at the thoracic venter, the end of rostrum connects with mid-thoracic venter, handle shaped. Cremaster hard, coniform.

Biology and behavior *Notonagemia analis* generally occurs 3 generations (few 2 generations) a year in Zhejiang. They overwinter as pupae. Pupae eclosion starts in late April and ends in late May. The first generation larvae start pupating in late May, some larvae enter pupal stage in late June. The first generation adult emergence starts in late June, some adults emerge in mid-July. The first generation pupal stage lasts 16-20 days, some could reach over 60 days, they stay dormant in summer as pupae, and diapause until mid-August, then emerge together with the second generation adults. The second generation larvae pupate during late July to mid-August, they eclose in mid-August to early September. Most overwintering generation larvae pupate in mid- and late September, adults emerge in April and May next year after 201-241 days pupal stage.

Adults mostly emerge between 19:00-21:00. They hide in shelters during the daytime, and rarely fly even when they are forced. Adults always stay on damaged trunks, mimic branch scars on trunk, and are not easily recognized. They are active at night and have phototaxis. Eggs are laid dispersedly on the lower surface of leaves. Gen-

N. analis adult (edited by Zhang Runzhi)

N. analis adult (Zhang Peiyi)

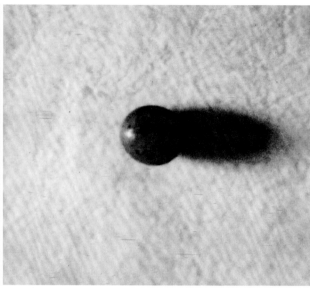
N. analis egg (edited by Zhang Runzhi)

N. analis larva (edited by Zhang Runzhi)

N. analis pupa (edited by Zhang Runzhi)

erally, each leaf receives one egg. When the population density is high, each leaf could have more than ten eggs. Adult's longevity is 3-7 days. Egg stage is 5-6 days. Neonates feed on eggshells, and they start to feed on foliage several hours later. The 1st instar larvae gnaw leaves into small holes only near the eggshell. From the 2nd instar, larvae start eating from leaves margin. Consumption increases as they grow. The 5th instar larvae consume significantly much more and is the gluttony period. This period is the longest, the infested forests show clear damages only during this stage. They often defoliate all the leaves of large areas of trees when the population density is high. Larvae mostly feed on the lower surface of leaves, and crawl to the veins to rest after they eat. When mature, they crawl to the ground and dig into the shallow soil or stay under dead leaves to make loose cocoons and pupate.

Prevention and treatment

1. Protect and utilize natural enemies is very important to control *Notonagemia analis*. In egg stage, natural enemies include *Trichogramma* sp., *Anastatus japonicus*, *A. albitarsis*, *Anastatus* sp., *Mesopolobus tabatae*, and *Telenomus* sp.; in larval and pupal stages, natural enemies include *Blepharipa zebina*, *Drino lugens* and *Drino argenticeps*. The parasitic rate of eggs could reach over 90% during pest outbreaks.

2. Injecting systemic pesticide to the trunk could achieve good control results during the larval stage.

References Chinese Academy of Sciences - Institute of Zoology 1983; Chen Hanlin 1990, 1993.

(Chen Hanlin, Pan Zhixin)

427 *Callambulyx tatarinovii* (Bremer et Grey)

Classification Lepidoptera, Sphingidae
Synonym *Smerinthus tatarinovi* Bremer et Grey

Importance It primarily damages host leaves. When the damage is serious, all leaves can be eaten up and tree vigor is impaired.

Distribution Beijing, Hebei, Shanxi, Inner Mongolia, Liaoning, Jilin, Heilongjiang, Shanghai, Zhejiang, Fujian, Shandong, Henan, Hubei, Hunan, Shaanxi, Gansu, Ningxia, Xinjiang; Japan, Korea Peninsula, Russia.

Hosts *Ulmus pumila*, *Hemiptelea davidii*, Ulmaceae, *Salix matsudan*, Salicaceae, *Zelkova schneideriana*, *Euonymus alatus*, *Populus* sp. (occasionally damaged).

Damage Early instar larvae feed on the epidermis of plant leaves. They chew leaves into holes or incisions. Old instar larvae eat up the whole leaves and only leave parts of leaf veins and petioles. When the damage was serious, branches turned bald and tree vigor was impaired severely. Large feces are often seen under trees, so they are easy to be detected.

Morphological characteristics **Adults** Body 30-33 mm in length, 75-79 mm in wingspan. Antennae upper part white, lower part brown. All femurs light green. Wing surface pinkish green, dorsal surface of thorax dark green. Forewing costal margin at apex angle with a big triangle dark green spot; between the median line and the outer line is a dark green patch; the outer line appears as two wavy stripes. Hindwing red, posterior margin white, outer margin light green, anal angle with dark cross stripes. Dorsal surface of thorax

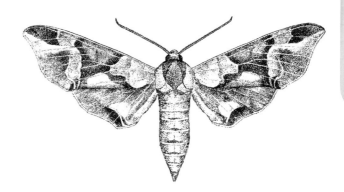

C. tatarinovii adult (Zhang Xiang)

dark green, abdominal dorsal surface pinkish green, the posterior edge of each segment with a yellowish brown stripe. **Eggs** Oval. The newly laid eggs yellowish green, brownish red before hatching. **Larvae** Mature larvae 60-70 mm in length, head length longer than width, almost triangle, dark green, two sides with yellow edges. Body green, densely covered with light yellow granules. Prothorax not divided into sub-segments, each side from the 1st to the 8th abdominal segments with 7 light green oblique lines composed of big granules; the top and bottom margin of oblique lines puce (the 2nd, the 4th, and the 6th lines thin, slightly yellow), the 7th line connects with the caudal horn. Caudal horn upright, extends backward into cone-shape, about more than 10 mm in length, pur-

C. tatarinovii adult (Xu Gongtian)

C. tatarinovii adult (Li Zhenyu)

C. tatarinovii larva (Xu Gongtian)

C. tatarinovii larva (Xu Gongtian)

plish green. Thoracic legs ochre-brown, prolegs color similar to body color, the end puce, with a white narrow ring on it. Body color varies in young stage, it has two types: green type-whole body green, the upper and lower edge of oblique lines puce; red spot type-body yellowish green, oblique lines reddish orange. Regardless of the color type, larvae before the 3rd instar especially the 1st instar are pinkish green and densely covered with white granules. **Larval** feces: fresh feces dark green, oval. **Pupae** 36.2-39.0 mm in length. Dark chocolate-brown, ventral surface of the abdomen slightly lighter in color.

Biology and behavior *Callambulyx tatarinovii* occurs two generations a year in Beijing, and the larval occurrence peak is in June and August. The 2nd generation mature larvae crawl down the trees and build chambers in the soil near hosts to pupate. Generally, they dig about 5 cm in soil. The overwintering pupae eclose in mid-June the next year. Adults have strong phototaxis. Females start oviposition one day after mating. Eggs are laid singly on the shoots of new branches. Egg stage is generally 7-10 days, but the 2nd generation egg stage is only 5-7 days. Adult longevity is 18-21 days. Larvae have 5 instars. The 1st-5th instar period is: 5-6, 6-7, 6-7, 8-9, and 7-8 days, respectively. The prepupa stage is 2-3 days and the whole larval stage is 33.5-40 days.

Prevention and treatment

1. Mechanical control. 1) During pupal stage, rake the soil, hoe weeds or turn up the soil around the trees, will kill overwintering pupae. 2) Larvae fall off trees when disturbed. Knock down larvae from hosts and destroy them when they occur. 3). Hand-remove larvae on trees.

2. Physical control. Based on adult phototaxis, trap adults with black light or frequency vibrancy-killing

C. tatarinovii larva parasitized by braconid wasps (Xu Gongtian)

lamp during adult occurrence period.

3. Conservation of natural enemies such as mantis, wasps and Braconidae.

4. Biological control. Before the 3rd instar, apply 16000IU/ul *B.t.* wettable powder at 1,000-1,200×. Larvae will die and decay gradually on trees instead of directly drop onto ground, which can protect natural enemies and prevent environmental pollution.

5. Chemical control. For the 3rd-4th instars larvae, spray 20% diflubenzuron SC at 3,000-3,500×, 25% chlorbenzuron SC at 2,000-2,500×, 20% tebufenozide SC at 1,500-2,000×, or 12% thiamethoxam and lamda-cyhalothrin at 1,000-1,500×.

References Wang Xujie 1985, Xiao Gangrou (ed.) 1992, Fan Di 1993, Zhu Hongfu and Wang Linyao 1997.

(Yu Juxiang, Wang Shanshan, Li Yajie, Lin Jihui)

428 *Hyles hippophaes* (Esper)

Classification Lepidoptera, Sphingidae
Synonym *Celerio hippophaes* (Esper)

Distribution Beijing, Inner Mongolia, Tibet, Shaanxi, Ningxia, Xinjiang; Spain, France, Germany, Russia.

Hosts *Elaeagnus angustifolia*, *Populus* spp., *Salix* spp., *Vitis vinifera*.

Damage Larvae feed on host leaves and may cause complete defoliation.

Morphological characteristics **Adults** Body 31-39 mm in length, 66-70 mm in wingspan. The dorsal side of antennae white, area between vertex and frons and the two sides of scapulae with white scales, two sides of the 1st to 3rd abdominal segments with black and white spots. Forewing with a dark brown triangle patch towards the outer margin; wing base black, area from apex angle to mid- inner margin appears as a greyish white oblique stripe. Hind wing base black, red in the middle, anal angle with a big white spot. **Eggs** Short elliptical, green. **Larvae** Body 70 mm in length, the dorsal side green, densely covered with white dots. Both sides of the thorax and abdomen with a white stripe. Ventral surface light green. Caudal horn thin, the dorsal side black, with spinulae, the ventral side light yellow. **Pupae** 43 mm in length, light brown. Head and thorax slightly green, rear end of abdomen gradually darker in color, tail end sharp.

H. hippophaes adult (Li Menglou)

H. hippophaes adult (Li Menglou)

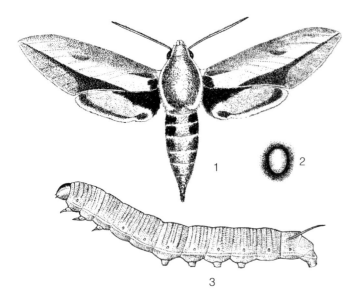

H. hippophaes. 1. Adult, 2. Egg, 3. Larva
(Zhang Peiyi)

H. hippophaes pupa (Li Menglou)

Biology and behavior *Hyles hippophae*s occurs 1-2 generations a year in Ningxia and overwinters as pupae in the soil. The overwintering pupae eclose in May, adults lay eggs on leaves of *Elaeagnus angustifolia*, each female lays about 500 eggs. Eggs hatch within 10-20 days. Larvae peak occurs in late June, they enter soil to pupae in mid-July. The 2nd generation larvae occur in mid- and late August, and they enter soil and pupate in September. Overwintering sites mostly are at high places around trees. Adults have phototaxis.

Prevention and treatment

1. Trap and kill adults using black light. Since larvae fall off trees when disturbed, shake trees and catch larvae when the population is small. During pupal stage, hoe the soil under the crown to exterminate pupae.

2. During the 1st and 3rd instar stage when the infestation is serious, spray a mixture of 12% thiamethoxam and lamda-cyhalothrin at 1,000-1,500×. Apply 25 g/L lamda-cyhalothrin smoke at 15-23 kg smoke agent per 1 hm^2 in dense forests.

References Chinese Academy of Forestry Institute of Zoology 1987, Xiao Gangrou (ed.) 1992, Guan Ling and Tao Wanqiang (ed.) 2010.

(Li Menglou, Xu Zhaoji, Wang Ximeng)

429 *Clanis bilineata bilineata* (Walker)

Classification   Lepidoptera, Sphingidae
Synonyms   *Clanis bilineata* (Walker), *Basiana bilineata* Walker
Common name   Two-lined velvet hawkmoth

Distribution   Hebei, Zhejiang, Hunan, Hainan; India.

Hosts   Leguminosae (*Pueraria* spp., *Mucuna* spp.), *Robinia pseudoacacia*.

Damage   Larvae feed on leaves. During outbreak, it can cause complete defoliation.

Morphological characteristics   Adults 60-65 mm in wingspan. Body and wings brown; the dorsal side of antennae pink, the ventral side brown; the dorsal lines of head and thorax puce; the dorsal side of abdomen greyish-brown, two sides dry yellow; the posterior border of the 5th to 7th abdominal segments brown; exterior surface of the middle tibia and hind tibia silver-white. Forewings brown, the mid- costal margin with a greyish white near triangle spot; inner, mid- and outer cross lines brown, the location near costal margin of apex angle with brown oblique stripes, the part below the stripes lighter in color, the longitudinal band at R_3 vein brownish black; hindwings brownish black, costal margin and near anal angle yellow, middle part with a thin greyish black cross band. Underside side of forewing and hindwing yellow, each cross line obvious, greyish black, center of the forewing base with a black long streak, the apex near costal margin with a dull white triangle spot. **Eggs** 2-2.5 mm in diameter, ivory, spherical. **Larvae** Mature larvae 80-90 mm in length, head dark green, mouthparts orange brown; body light green, prothorax with yellow granules, mesothorax with 4 wrinkles, metathorax with 6 wrinkles. Two sides from the 1st to the 8th abdominal segments with yellow oblique stripes, with small winkles and white spine-shaped granules on the back; caudal horn yellowish green, pointing back and downward; sieve plates of spiracles light yellow, peritreme yellowish-brown, thoracic legs orange brown. Head coronal suture raised, forming one peak, near triangle from anterior view. **Pupae** About 50 mm in length, 18 mm in width. Reddish brown. The front of spiracles from the 5th to the 7th abdominal segments each with a cross groove, cremaster triangle, the tip has no furcation.

Biology and behavior   *Clanis bilineata bilineata* occurs 1-2 generations a year and overwinters as mature larvae in 10 cm deep soil. The 1st generation larvae start

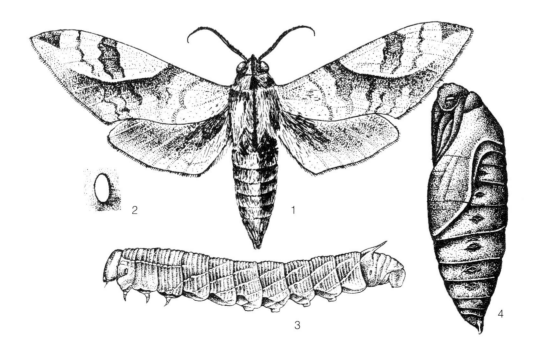

C. bilineata bilineata. 1. Adult, 2. Egg, 3. Larva, 4. Pupa
(1 by Hu Xingping; 2-4 by Zhu Xingcai)

C. bilineata bilineata adult (Li Zhenyu)

C. bilineata bilineata adult (edited by Zhang Runzhi)

pupation in early and mid- June the next year, and their eclosion peak is in mid- and late July, the larval occurrence peak is in early and mid- August, mature larvae generally burrow into soil to overwinter in early September, their body appears in horse shoe shape in soil.

Adults are motionless during the daytime and they don't fly far even after disturbed. They start activity at night. They have phototaxis. Females generally start oviposition 3 hours after mating. They prefer laying eggs on the leaf's lower surface, very few lay eggs on the upper surface. The average oviposition period is 3 days. Each female lays 200-450 eggs, the average is 350 eggs. Adult longevity is 7-10 days. Egg stage is 6-8 days. The newly hatched larvae spin silk and hang themselves on silk. Their natural mortality rate is high. Larvae avoid lights and may migrate to other plants. Before the 4th instar, larvae mostly hide on the lower surface of leaves during the daytime; after 5th instar, their weight cannot be supported by leaves, so they move to limbs. They feed mostly at night and can feed all day when the weather is cloudy.

The 1st - 3rd instar period is 14.5 days. The 4th instar larval period is 10 days, the 5th instar period is the same as the first 3 instar period. The average larval period is around 39 days.

Prevention and treatment

1. Set up neon lamp traps to trap adults.

2. Spray detergent at 1,000× to control the 1st instar larvae.

References   Wang Pingyuan et al. 1983; Ai Yuxiu 1987; Wang Xujie et al. 1985; Xiao Gangrou et al. 1992, 1997; Zhu Hongfu et al. 1997.

(Mi Ying, Zhao Huaiqian, Zhang Shiquan)

430 *Dolbina tancrei* Staudinger

Classification Lepidoptera, Sphingidae
Synonyms *Dolbina leteralis* Matsumura, *Dolbina curvata* Matsumura

Distribution Beijing, Hebei, Shanxi, Heilongjiang, Jiangsu, Zhejiang, Shandong, Henan, Sichuan, Tibet, Shaanxi, Gansu; Korea Peninsula, Japan, India.

Hosts *Ligustrum obtusifolium*, *L. lucidum*, *Corylus heterophylla*, *Osmanthus fragrans*, *Fraxinus chinensis*, *Syringa oblate*, *Forsythia suspense*.

Damage Larvae feed on leaves. They could eat up all the leaves at high density.

Morphological characteristics **Adults** Body 26-34 mm in length, 50-82 mm in wingspan. Body greyish white, with yellowish white stripes. The end of forewing discal cell with a white spot. Inner and outer lines are composed of 3 saw tooth shaped stripes, wing base with a group of brown stripes, subouter margin line light white, outer margin with a row of brown spots. Hind wings brown. Abdominal dorsal middle line black. **Eggs** Oval, 2.3 mm×1.9 mm, emerald green or light green. **Larvae** Mature larva 64-70 mm in length, emerald green or dark green. Both sides of the head with white border, the dorsal side of each thoracic segment with two transverse rows of white small spines; thoracic legs reddish brown, their exterior side with small red spots; each abdominal segment with white stripes tilted toward the caudal horn. **Pupae** Dark chocolate brown, body 41-44 mm in length.

Biology and behavior *Dolbina tancrei* occurs 2 generations a year in Shanxi and 4 generations a year in Sichuan. It overwinters as pupae in the soil. Adults emerge in early and mid-April the next year. Adults emergence time for the 1st, 2nd, and 3rd generation are early to mid-June, late July to early August, and early to mid- September, respectively. The generations overlap.

D. tancrei adult (Xu Gongtian)

D. tancrei adult (Hu Xingping)

D. tancrei adult (Li Menglou)

Pupa stage is up to over 160 days, larva stage is 20-30 days, pupa stage is about 12-15 days, prepupa stage is 2-3 days, and egg stage is 6 days. Adults emerge, mate and oviposit at night, they have phototaxis. Females lay eggs on the lower surface of leaves from the next night right after emergence to the 8th night, each leaf mostly receives one egg and each female lays 100-300 eggs. Larvae have 5 instars. A single larva could eat up a whole leaf when it reaches the 3rd instar. Larvae hide on the lower surface of leaves when they are not feeding, and they tend to fall off when disturbed. Egg stage is 10 days when the average temperature is 19.1℃ and 5 days when the average temperature is 27.3℃, larva stage is 20-22 days when the average temperature is 25℃ and 30 days when the average temperature is 19.1℃.

Prevention and treatment

1. Trap and kill adults using black light during adult occurrence period. Release *Trichogramma* sp. during egg stage. Shake hosts and collect larvae during larva stage. Turn up the soil to kill pupae during pupa stage. Spray 1×10^{10} spores per mL *B.t.* at 600× during larva stage.

2. For larvae before the 3rd to 4th instars, spray 12% thiamethoxam and lamda-cyhalothrin at 1,000-1,500×, or apply smoke agent at 15-23 kg per hectare to control larvae before the 3rd instar.

Reference Xiao Gangrou (ed.) 1992.

(Liu Mantang, Li Menglou, Wu Cibin)

431 *Psilogramma menephron* (Cramer)

Classification Lepidoptera, Sphingidae
Common names Privet hawk moth, Large brown hawkmoth

Distribution Beijing, Hebei, Jiangsu, Henan, Hunan, Guangdong, Sichuan, Shaanxi, Taiwan; Myanmar, Japan, India, Indonesia, Korea Peninsula, Sri Lanka, the Philippines, Oceania.

Hosts *Paulownia* spp., *Platanus* spp., *Catalpa bungei*, *Melia azedarach*, *Syringa* spp., *Ligustrum obtusifolium*.

Morphological characteristics **Adults** Body 15-50 mm long with a wingspread of 105-130 mm, dark gray. Median line on forewings double wavy, brownish black. There are two black markings below the median cell and one black line on the top corner. Hindwings brown. **Larvae** Length 92-110 mm. Green, color dimorphic. One type contains one lateral white declivent mark running from prothorax to the 8th abdominal segment on both sides and a green anal horn; the other contains brown body marks and brown anal horn.

Biology and behavior *Psilogramma menephron* occurs two generations a year and overwinters as pupae in soil in Henan. Adults begin to emerge and lay eggs in early April. Second generation adults start to emerge in September and October, with larvae of this generation feed until the end of October. Then they overwinter in soil as pupae.

Adults are strongly attracted to lights and are active mostly at night. Most eggs are laid on the underside of the leaves on large trees. Newly hatched larvae feed on the epidermis of the leaves, while older larvae encroach the leaves by creating notches or feeding holes. Most damages are found between June and July, evident by large amount of leaf portions and frass under the tree.

Prevention and treatment

1. Physical control. Destroy pupae in soil by tillage in the winter. Using a stick to beat leaves and destroy the fallen larvae.

2. Biological control. Preserve praying mantis *Hierod-*

P. menephron adult (Xu Gongtian)

P. menephron adult (edited by Zhang Runzhi)

P. menephron adult (Li Yingchao)

ula patellifera.

3. Chemical control. Apply 12% thiamethoxam and lamda-cyhalothrin at 1,000-1,500× to control larvae.

References Yang Youqian 1982, Chinese Academy of Sciences Institute of Zoology 1987, Xiao Gangrou (ed.) 1992, Henan Forest Pest Control and Quarantine Station 2005.

(Yang Youqian)

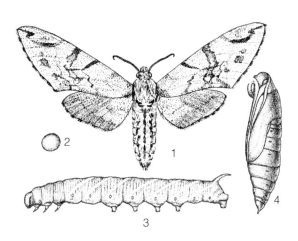

P. menephron. 1. Adult; 2. Egg; 3. Larva; 4. Pupa (Zhang Xiang)

P. menephron larva on Chinese ash tree (edited by Zhang Runzhi)

P. menephron larva on paulownia tree (edited by Zhang Runzhi)

P. menephron larva on willow tree (edited by Zhang Runzhi)

P. menephron mature larva (Xu Gongtian)

P. menephron mature larva (Xu Gongtian)

432 *Smerinthus planus* Walker

Classification Lepidoptera, Sphingidae
Synonym *Smerithus planus planus* Walker
Common name Oriental eyed hawkmoth

Importance *Smerinthus planus* mainly damage *Populus* spp., *Salix* spp. and some fruit trees. It is a common defoliator of forests. When it occurs seriously, it could eat up all leaves of damaged trees, and significantly affect the tree growth.

Distribution Beijing, Hebei, Shanxi, Inner Mongolia, Liaoning, Jilin, Heilongjiang, Jiangsu, Zhejiang, Anhui, Fujian, Jiangxi, Shandong, Henan, Hubei, Shaanxi, Gansu, Qinghai, Ningxia; Korea Peninsula, Mongolia, Japan, Russia.

Hosts *Populus* spp., *Salix* spp., *Prunus persica*, *P. pseudocerasus*, *P. mume*, *P. salicina*, *Malus spectabilis*, *M. pumila*, *Elaeagnus angustifolia*.

Damage *Smerinthus planus* damages trees mainly by larval feeding on leaves. The young larvae dispersedly feed on tender leaves and bite the leaves into incisions. Older larvae consume whole leaves and cause complete defoliation at high density.

Morphological characteristics **Adults** Body 32-36 mm in length, 85-92 mm in wingspan. Wings greyish brown. Antennae light yellow. Compound eyes large, dark green. Center of thoracic notum brown. Forewing base greyish yellow, outer margin shallow saw tooth shape, margin hairs very short. Two dark brown spots present in the front and rear part between the mid- and outer lines. The front part of discal cell with a "J" light color stripe, outer cross lines appear as 2 dark wavy stripes, outer margin below apex angle dark in color. Hindwing light yellowish brown, with a big blue eye spot in the middle, surrounding the spot is a greyish white circle, the outer most circle is bluish black, the upper part of blue eye spot pink. The blue eye spot on the reverse side of hindwing inconspicuous. **Eggs** Major axis about 1.8 mm, oval. The newly laid eggs emerald green, glossy, later turn into yellowish green. **Larvae** Mature larva body 70-90 mm in length. Head small, 4.5-5.0 mm in width, green, near triangle shape, two sides light yellow; thorax turquoise, each segment with thin tucks; prothorax with 6 cross rows of granules; mesothorax with 4 annelets, both right and left of each annulet with a big granule. Abdomen greenish, two sides of the 1st to 8th abdominal segment with light yellow oblique stripes, the last oblique stripe reaches caudal horn, caudal horn pointed upward and rearward. Sieve plates of spiracles light yellow, peritreme black, the front often with a purple spot, abdominal venter slightly dark in color, thoracic legs brown, prolegs green, the end brown. **Pupae** 28-35mm in length, dark red at beginning, later turn into dark brown. Wing buds short, the tip only reaches 2/3 of the 3rd abdominal segment, anal angle obviously protruding to inner margin.

Biology and behavior *Smerinthus planus* occurs two generations a year in Beijing, Hebei, Qinghai (Xining) and Gansu (Lanzhou), 3 generations a year in Henan and Shaanxi (Xi'an), 4 generations a year in Jiangsu. Adult appearance periods for two generations area are: late April to May, and mid- and late June to July; for three generations area are: mid- and late April, July, and August; for four generations area are: mid-April, late June, early August, and mid-September. Adults mostly

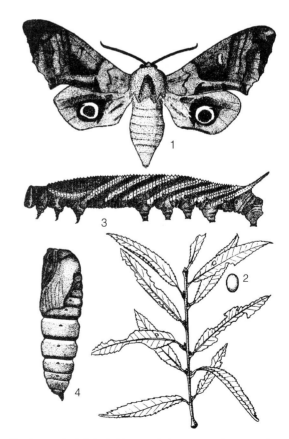

S. planus. 1. Adult, 2. Egg and damage, 3, larva. 4. Pupa
(Xu Tiansen)

S. planus adult (edited by Zhang Runzhi)

S. planus larva (Xu Gongtian)

S. planus larva (Xu Gongtian)

S. planus larva parasitized by braconid wasps (Xu Gongtian)

emerge at night. It takes 50 minutes from breaking the shell to fully spreading the wings. Adults hide in the daytime and come out at night, and have very strong flight abilities and phototaxis. Adults mate the next day after emergence and mostly at night, mating duration could reach 5 hours. Adults start oviposition the next day after mating, eggs are mostly laid on the lower surface of leaves, and egg stage is about 15 days. Eggs are laid singly, occasionally laid into a string, each female lays 200-400 eggs. The 1st and 2nd instars larvae dispersedly feed on tender leaves and bite them into incisions; the 4th to 5th instars consume much more. The 5th instars larvae consume the most. Mature larvae crawl down trees and pupate in the soil near the hosts.

Prevention and treatment

1. Trap adults with black light during adult occurrence period.

2. Physical control. Larvae fall to the ground when disturbed. Shake the host trees, collect and destroy the larvae. Dig out the overwintering pupae from soil during winter.

3. Protect and utilize natural enemies such as *Hierodula patellifera*.

4. Chemical control. During larval stage, spray 12% thiamethoxam and lamda-cyhalothrin at 1,000-1,500×.

References   Wang Pingyuan, Wang Linyao, Fang Chenglai et al. 1983; Wang Xujie, Xu Zhihua, Dong Xuhui et al. 1985; Yuan Decan, Huang Pinlong, Wang Kejian et al. 1988; Xiao Gangrou 1992; Zhu Hongfu and Wang Linyao 1997; Ma Qi, Qidefu, and Liu Yongzhong 2006; Xu Gongtian and Yang Zhihua 2007; Zhang Zhixiang 2008.

(Zhao Jiali, Dai Qiusha)

433 *Amplypterus panopus* (Cramer)

Classification Lepidoptera, Sphingidae
Synonyms *Sphinx panopus* Cramer, *Calymnia panopus* Rothschild et Jordan, *Amplypterus panopus* Moore
Common name Mango hawk moth

Distribution Hunan, Hainan, Yunnan; India, Malaysia, the Philippines, Sri Lanka.

Hosts Mangifera, Anacardiaceae, Guttiforae, *Calophyllum inophyllum*.

Damage Larvae feed on leaves and cause defoliation or notched leaves.

Morphological characteristics **Adults** Body yellow, thick and big, 75 mm in wingspan. **Larvae** There are five instars. The 1st to the 4th instar larvae head sharp, vertex dichotomous, densely covered with granulated bumps; the 5th instar larvae head sharp, vertex without furcation. Larvae hatched among bronze-colored young leaves are mostly yellow; while larvae hatched among light green leaves are mostly green.

Biology and behavior *Amplypterus panopus* occurs 3 generations a year in Xishuangbanna of Yunnan province. The overwintering generation adults emerge from mid- and late February to March. The 1st generation adults appear in mid-May. The 2nd generation adults occur in late September. They last until late October.

Females prefer laying eggs on covert tender leaves. Eggs laid on the upper surface of leaves are readily washed off by rain, while eggs laid on the lower surface of leaves could survive and hatch.

Prevention and treatment

1. Silviculture. Combining with loosening soil and fertilizing management in winter, exterminate overwintering pupae in the soil around the trunk base.

2. Mechanical control. Look for and remove the leaf damaging larvae based on the feces location on the ground.

3. Trap with black light. Trap and kill adults using black light traps.

4. Protect and utilize natural enemies. *Isyndus reticulatus* and *Sycanus croceovittatus* prey on the young larvae.

5. Chemical control. During young larvae period, apply DDVP EC or spray trichlorfon crystal diluent. The pest population reduction is up to 90%.

References Situ Yingxian 1983, Huang Fusheng 1987, Zhu Hongfu and Wang Linyao 1997.

(Liu Jianhong, Pan Yongzhi)

A. panopus adult (edited by Zhang Runzhi)

A. panopus adult (edited by Zhang Runzhi)

434 *Daphnis nerii* (L.)

Classification Lepidoptera, Sphingidae
Common names Oleander hawk moth, Army green moth

Importance *Daphnis nerii* is a serious defoliator of Nerium indicum and Rauvolfia romitoria in the recent years. Outbreaks occurred in some areas, causing significant ecological and economic loss to medicinal plants and landscaping plants.

Distribution Shanghai, Fujian, Guangdong, Guangxi, Sichuan, Yunnan, Macao, Taiwan.

Hosts *Nerium indicum, Alstonia seholaris, Rauvolfia romitoria, Catharanthus roseus, Allamanda neriifolia, Theretia peruviana.*

Damage It damages shoots and leaves. It causes damage to both seedlings and adult trees.

Morphological characteristics **Adults** 47 mm in wingspan, body 55 mm in length. Vertex dark green, area between the base of the two antennae with a dull white "∞" shaped pattern; labipalp dark green, inner side of palpiger with white stripes; compound eyes big, convex, reddish brown, with black sparse spots; the ventral side of antennae yellow, the dorsal side green; the dorsal side of thorax dark green; the front and rear border of prothorax with yellow cross stripes, outer edges of pronotum with white cilium; thorax and abdomen brownish green. Legs green, the part below tibia slightly light, greyish brown. The front and rear border of the tergum of the 1st abdominal segment dark green, its center with dull white cross band; the other segments greyish-brown, rear border of other segments with green and white alternating thin cross stripes, forming internode rings. Ventral side of abdomen slightly lighter in color, its center with a yellow longitudinal line; two sides and mid- of abdominal end with green spots. Forewings dark green, wing base with a yellowish brown spot, inside of it with a black spot; inner line dull white in color and protruded outward as an arc, the location between inner line and wing base with an dark green shield-shaped spot; tortuosity of the mid-line large, the location between inner and mid-line with a pink wide oblique stripe, discal cell with a triangle spot, outer line dull white with double lines, but only reach the upper of M_3 vein, the location between M_3 and Cu_2 veins with longitudinal brownish green area; apex angle sharp and protrudes outward, inside of it with a green half-moon shaped spot, and with a white oblique line extends inward along the M vein, the area from the upper of the oblique line to costal margin yellowish green. Hindwings green, costal margin and inner margin light in color, middle area with a dull yellow arch pattern; the reverse side of forewing and hindwing brownish green; apex angle, two sides of anal angle and outside of discal cell with rusty yellow spots; from apex angles to inner side of anal angle of forewing and hind wing with a straight white line, discal cell with indistinct white reniform stripe; hind wing base rusty red, obscure, mid-line gray, tortuosity large, outer line white. Male is similar to female in shape, key distinction is that male has no "∞" shaped pattern between the two antennae, the 7th abdominal segment is visible from the dorsal view, the sides of this segment with dark green long stripes. **Eggs** Near spherical, light yellow, 0.8 mm in diameter. **Larvae** Newly hatched larvae light yellowish green, head 0.76-0.82 mm in width, caudal horn pointed upward, black, long and thin, about 3/5 of the body length. Mature larvae 85-86 mm in length, head 4.80-5.09 mm in width, 5.51-5.80 mm in height, head and thorax yellowish green, thoracic legs reddish brown, body and prolegs greyish-green. The upper spiracle line white and thick, upper and lower side of it with various sized white and yellow small dots with indigo blue thin edges. Both sides of the front edge of

D. nerii larva (edited by Zhang Runzhi)

the metathoracic dorsal side with an eye-shaped spot, the middle of the spot is white and sapphire, the edge of the spot with black thick frame. Spiracles narrow and long, sieve plates of spiracle black. Caudal horn yellow, short and thick, the tail end blunt, pointing backward, slightly bend downward. **Pupae** 70-85 mm in length, 15-17 mm in width. Head, ventral surface light yellowish brown, dorsal surface of abdomen light brown, abdomen scattered with dark chocolate brown spots, sides from the 2nd to 8th abdominal segment each with a near round black spot. The middle of wing bud area on the ventral side with a black longitudinal stripe. The dorsal side has a black blue longitudinal stripe, which is continuous on thorax segments and intermittent on abdominal segments.

Biology and behavior *Daphnis nerii* occurs three generations a year in Guangzhou of Guangdong Province. The overwintering generation adults appear in April. The 1st generation larvae appear in May. They pupate in early June. The 1st generation adults appear in mid-June. The 2nd generation eggs appear between late

D. nerii adult male (edited by Zhang Runzhi)

D. nerii adult female (edited by Zhang Runzhi)

D. nerii 1st instar larva (edited by Zhang Runzhi)

D. nerii egg (edited by Zhang Runzhi)

D. nerii mature larva ready to pupate (edited by Zhang Runzhi)

D. nerii pupa (edited by Zhang Runzhi)

June and July. Larvae appear from early July to early August. Pupal stage is in early and mid-August. The 2nd generation adults appear in late August. The overwintering egg, larva, and pupa stage is from mid- to late August, late August to late September, and from mid- to late September, respectively.

Egg stage is about 2-3 days. Larvae have 5 instars. In Hengyang of Hunan province, the 1st generation larval period from 1st to 5th instar is: 4.5-6, 3-3.5, 3.5-5, 4-4.5, and 8-12 days, respectively. Pupal stage lasts for 11-14 days.

Adults are active at night. Eggs are laid singly on lower surface of leaves, petioles, or shoots. Larvae consume eggshell after hatching, then feed on the epidermis of leaves, lower surface, and a little of mesophyll. From the 2nd instar, larvae start to feed on leaf margin, causing small incisions. As the larvae increase in size, they start to feed from leaf tip to the base until they eat up the whole leaf. Larvae crawl down to the ground to look for appropriate sites before pupation and pupate in shallow surface soil layer. Pupa twists its body when touched.

Prevention and treatment

1. Trap and kill adults with black light.

2. Larvae fall to ground when disturbed. Based on this habit, shake the hosts and collect larvae.

3. Biological control. (1) Protect and utilize natural enemies such as *Bracon* sp., *Apanycolus* sp., *Apanteles* sp., *Telenomus* sp., *Vespa* sp. and birds. (2) spray 1.2% matrine at 2,000×, 2,000 IU/μL *B.t.* at 200×, or $1-3\times10^{10}$ *Bacillus thuringiensis galleriae* powder at 300-500×. The last one has synergism with trichlorfon.

4. Chemical control. Spray 80%-90% trichlorfon at 800-1,000×, 50% malathion EC at 1,000-1,500×, or 12% thiamethoxam and lamda-cyhalothrin at 1,000-1,500×.

References   Zhu Hongfu and Wang Linyao 1997; Wang Yuqin, Wu Yousheng, Gao Zezheng et al. 2004; Zhang Lixia 2007; Lei Yulan, Lin Zhonggui 2010.

(Wu Yousheng, Gao Zezheng)

435 *Besaia anaemica* (Leech)

Classification Lepidoptera, Notodontidae
Synonyms *Pydna anaemica* Kiriakoff, *Mimopydna insignis* (Leech)

Distribution Shanghai, Jiangsu, Zhejiang, Fujian, Jiangxi, Hubei, Hunan, Sichuan, Yunnan.

Hosts *Phyllostachys aureosulcata* f. *aureocaulis*, *P. aurita*, *P. bambusoides* f. *Lacrima-deae*, *Phyllostachys heteroclada* var. *pubescens*, *P. bambusoides*, *P. sulphurea* 'Viridis', *P. glauca*, *P. praecox*, *P. iridescens*, *P. nidularia* 'Smoothsheath', *P. nidularia* f. *galbro-vagina*, *Bambusa multiplex*, *B. textilis*.

Damage Larvae feed on leaves. When the population is high, they could eat up all the bamboo leaves.

Morphological characteristics **Adults** Male body 28.5-31.4 mm in length, 53.5-60.5 mm in wingspan; female body 23.5-28.8 mm in length, 64-70 mm in wingspan. Body color and stripes on forewings vary considerably, yellowish white or yellowish brown. Head greyish white; compound eyes big, dark chocolate brown; male antennae short pectinate, greyish brown; female antennae filiform, yellowish white. Jugular sclerite of prothorax and proscutum are densely covered with thick and long fluffs, yellowish white in male, darker in female. Forewing scales thick, greyish yellow in males and white in females; margin hairs short and thick, between margin hairs and outer margin with a row of black dots, or if the dots are obscure, 2-4 black dots show at the outer margin near apex angle. Outer margin near apex angle, the middle of costal margin and the edge of inner margin light greyish brown, discal cell with two rows of black dots paralleling with the outer margin, the outer row with more than ten dots, the inner row with 8-9 dots, the black dots on females are bigger than those on males. Posterior margin with scattered big brown spots. Male hindwing brown, female hindwing yellowish white. Center of the reverse side of male wings brown, that of females yellowish white. Legs are covered with long and dense yellowish white fluffs, the end of fore tibia, the middle and end of mid- and hind tibia with a pair of spurs, one spur is long and the other one is short. Abdomen dark yellowish white; male lanky, abdominal end surpasses the wings when wings are folded; female dumpy, abdominal end flat. **Eggs** Spherosome, major axis 1.74 mm, minor axis 1.52 mm. Milky, egg-shell smooth, glossy, no stripes, 2-5 eggs are laid into a strip on the lower surface of bamboo leaves. **Larvae** The newly hatched larvae body 3.2-4.0 mm in length, light yellowish green. Mature larvae body 58-70 mm in length, emerald green and light yellow; head lemon yellow; lines and stripes complex in color, ecdysial line light greyish green, the stripe from labrum to subdorsal line dark grey; the stripe from labrum to the upper spiracle line red; the stripe from labrum to the spiracle line dark black, the part below it yellow. Dorsal line bluish green, the location from subdorsal line to the upper spiracle line with 3 longitudinal light turquoise lines; the spiracle line yellow, with a green line closely adjoining above it; the lower spiracle line yellowish green or yellow. Thoracic legs red, prolegs green, last segment and crochets red, spiracle yellowish-brown. **Pupae** 25-31 mm in length, fusiform, the shape is special in the Notodontidae pupae. Head small, very sharp, metathoracic segment and the 1st abdominal segment are the widest, the width is about 1/3 of the length, the abdomen toward the end sharpens like a straight line, wing buds reach near the end of the 4th abdominal segment. Cremasters arch cutter-shaped, flat, with very short tooth-shaped protuberance, no hookes; viridis when it newly pupates, later gradually turns into red and dark red from the dorsal side, then the whole body becomes dark black. Mature larvae spin cocoons in 2 cm deep soil, cocoons 35-40 mm in length, composed

B. anaemica adult female (Xu Tiansen)

B. anaemica adult male (Xu Tiansen)

B. anaemica mature larva (Xu Tiansen)

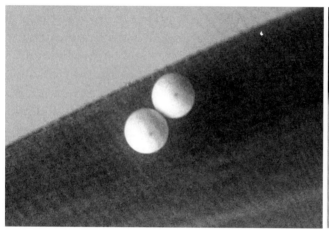
B. anaemica eggs (Xu Tiansen)

B. anaemica egg parasitized by a wasp (Xu Tiansen)

of silk and fine soil particles.

Biology and behavior *Besaia anaemica* occurs 4 generations a year in Zhejiang and overwinters as pupae. Adults emerge in early May, the 1st generation larvae occur from mid-May to mid-July, and pupate from early July to late July; the 2nd generation adults emerge in mid-July, larvae period is from late July to mid-September, pupal period from mid-August to mid-September; the 3rd generation adults emerge from late August, larvae period is from late August to mid-October, pupae period is from mid-September to late October; the 4th generation adults emerge from late September, larvae period is from early October to mid-May the next year, larvae pupate from early May and end in late May.

Prevention and treatment Protect natural enemies. Predatory animals include birds (such as *Cissa ergtyrorhyncha*, *Garrulax canorus*, and *Cuculus canorus canorus*), spiders (such as *Argiope bruennichii*, *Lycosa coelestris*, and *Lycosa pseudoannulata*) prey on adults and larvae; insect predators include *Hierodula patellifera*, *Isaria cicadae*, *Picromerus virdiipunctatus* and *Sirthenea flavipes*. Parasitoids include *Teienomus euproctidis* in egg stage and *Rhogas* sp., *Coccygomimus luctuosus*, *Campoplex* sp., *Colpotrochia (Colpotrochia) pilosa pilosa*, and *Exorista civilis* in larval stage; parasitoids that parasitize during the host larval stage and emerge during host pupal stage include *Tetrastichus* sp. and *Enicospilus* sp. Virus could infect larvae and the parasitic rate is high.

References Wu Chunsheng and Fang Chenglai (eds.) 2003, Xu Tiansen and Wang Haojie 2004.

(Xu Tiansen, Wang Haojie)

436 *Besaia goddrica* (Schaus)

Classification Lepidoptera, Notodontidae
Synonyms *Pydna goddrica* Schaus, *Besaia rubiginea simplicior* Gaede

Distribution Jiangsu, Zhejiang, Anhui, Fujian, Jiangxi, Henan (South), Hubei, Hunan, Guangdong, Guangxi, Sichuan, Shaanxi.

Hosts *Phyllostachys heteroclada* var. *pubescens*, *P. aurita*, *P. bambusoides*, *P. rutile*, *Phyllostachys sulphurea* 'Viridis', *P. heteroclada*, *P. glauca*, *P. iridescens*, *P. nidularia* 'Smoothsheath', *P. meyeri*, *P. praecox*, *P. dulcis*, *P. nuda*, *Bambusa multiplex*, *Bambusa textilis*, *Bambusa pervariabilis*, *Bambusa multiplex* var. *riviereorum*.

Damage Larvae feed on bamboo leaves. They caused death of *Phyllostachys heteroclada* var. *pubescens* and reduction of new bamboos the next year.

Morphological characteristics **Adults** Male body 19-25 mm in length, 43-51 mm in wingspan; female body 20-25 mm in length, 50-58 mm in wingspan. Body greyish yellow to greyish brown, the front hair tufts and basal hair tufts on head are very long. Female forewings yellowish white to greyish yellow, there is a gray oblique stripe running from apex angle to outer line, anal angle area below the oblique stripe greyish brown; male forewings greyish yellow, costal margin yellowish white, there is a dark greyish brown longitudinal stripe in the middle, below it with light yellowish white margin. There are 5-6 black dots among veins of outer margin line, hindwings dark greyish brown. **Eggs** Globular, major axis 1.4 mm, minor axis 1.2 mm. Milky, eggshell smooth, glossy, without stripes. **Larvae** The newly hatched larvae 3 mm in length, light yellowish green. Mature larvae 48-62 mm in length, viridis, body is covered with white powder, dorsal line, subdorsal line and the upper spiracle line pinkish cyan, embedded on viridis body. Spiracle line dark yellowish-brown, thinner behind metathorax in some individuals; spiracle yellowish white, the location near prothoracic spiracle brownish red, there is a yellow dot above the spiracle. There is a yellow dot behind mesothoracic, metathoracic, and abdominal spiracles. The number of instars vary from 5-7 with 7 instars being the least common. **Pupae** 20-26 mm in length, reddish-brown to dark chocolate brown, with 8 cremasters arranged in two rows of 6 and 2 cremasters. **Cocoons** 35 mm in length, consisted of silk and soil in 2 cm deep soil. Thin, the inside greyish white, smooth, the outside attached with soil grains.

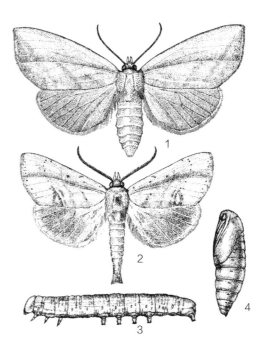

B. goddrica. 1-2. Adult, 3. Larva, 4. Pupa (Zhang Xiang)

B. goddrica adult female (Xu Tiansen)

Biology and behavior *Besaia goddrica* occurs 4 generations a year in Zhejiang Province and overwinters as larvae on bamboo. Larvae generally don't eat anything when daily average temperature is below 3°C. They feed

B. goddrica adult female (Xu Tiansen)

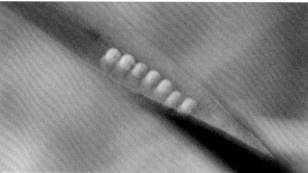
B. goddrica eggs (Xu Tiansen)

B. goddrica 4th instar larva (Xu Tiansen)

B. goddrica mature larva (Xu Tiansen)

at noon when daily average temperature is above 8℃. Their feeding increases in March. Larvae pupate in early April. The 1st generation adults emerge from early April, peak from late April to early May, and disappear in early June. Egg stage is in mid-April to early June. Larval period is from late April to early July, and pupa stage is from late May to mid-July. The 2nd generation adult stage is from early June to late July, egg stage is from mid-June to late July, larval period is from late June to late August, and pupal stage is from late July to early September. The 3rd generation adult, egg, larva, and pupa stage is from early August to mid-September, early and mid-August to late September, mid-August to mid- and late October, early September to late October, respectively. The 4th generation adult stage is in mid-September to early November, egg stage is in early October to early and mid-November, larvae hatch in early October and forage until the 2nd and 3rd instars, then begin to overwinter.

Prevention and treatment Protect natural enemies. Predatory animals include birds such as *Garrulax canorus* and *Cuculus canorus canorus*, and spiders such as *Argiope bruennichii* (Scopoli), *Oxtopes macilentus*, *Oxyopes sertaus*, and *Chiracantium zhejiangensis* which prey on adults and larvae. Insect predators include *Hi-*

B. goddrica pupa and cocoon (Xu Tiansen)

erodula patellifera, *Arma chinensis*, and *Picromerus virdiipunctatus*. Parasitoids include *Trichogramma dendrolimi*, *T. evanescens*, and *Telenomus lebedae* in egg stage and *Rhogas* sp., *Coccygomimus luctuosus*, *Campoplex* sp., *Colpotrochia* (*Colpotrochia*) *pilosa pilosa*, *Exorista japonica*, and *Exorista civilis* in larval stage; parasitoids that parasitize in larval stage and emerge in pupal stage include *Tetrastichus* sp. and *Enicospilus* sp.

References Wu Chunsheng and Fang Chenglai 2003, Xu Tiansen and Wang Haojie 2004.

(Xu Tiansen, Wang Haojie)

437 *Armiana retrofusca* (de Joannis)

Classification	Lepidoptera, Notodontidae
Synonym	*Norraca retrofusca* de Joannis

Distribution Shanghai, Jiangsu, Zhejiang, Anhui, Fujian, Jiangxi, Hubei, Hunan, Guangdong, Sichuan; Vietnam.

Hosts *Phyllostachys heteroclada* var. *pubescens*, *P. sulphurea* 'Viridis', *P. glauca*, *P. iridescens*, *P. vivax*, *P. praecox*, *P. nuda*, *P. heteroclada*, *P. bambusoides*, *Bambusa multiplex*.

Damage Larvae feed on bamboo leaves.

Morphological characteristics **Adults** Male 21.2-25.5 mm in length, 48.2-63.5 mm in wingspan; female 19.5-24.5 mm in length, 56.0-68.5 mm in wingspan. Body light yellow. Head greyish-brown; compound eyes greyish green; male antennae short pectinate, yellowish white, one side of ctenii greyish black; female antennae filiform, light yellow. Hairs of prothoracic squamula dense and strong, Center of notum with a greyish brown longitudinal line extending to the vertex. Forewing costal margin normal, base of outer margin parallel with costal margin, the posterior arc-shaped and connects with costal margin, inner margin very short, anal angle near right angled and making forewing narrow and long. Male forewing yellowish white, the costal margin cell with two rows of discontinuous greyish brown spots, the position of sub-middle fold with a greyish brown spot, the inside of hindwing's costal and inner margins with a light brown spot; female forewing light yellow, hindwing base greyish to reddish brown, outer margin with a row of indistinct small dots. **Eggs** Globous, 1.25-1.43 mm in diameter, 1.15-1.31 mm in height, pure white, bright, opaque, enamel-like textured, like small ping pong balls. Top of egg shows a black dot before hatching. Egg is laid singly on the tip of upper surface of bamboo leaves. **Larvae** The newly hatched larvae 3 mm in length, cinerous, pinaculum obvious; head dull yellow, top of cheek with a black circle, legs black. Mature larvae 62-70 mm in length, body color yellow or gray. Head flesh yellow or flesh white. Mandible flesh yellow, level with ecdysial line, with a black or dark grey long streak. A strip of black spots run from mouthparts along prothoracic spiracle to metathorax, separated at the end of metathoracic segmen; one upward to the dorsal line and the other downward to the lower spiracle line. There is a yellow

A. retrofusca adult male (Xu Tiansen)

A. retrofusca newly molted 5th instar larva (Xu Tiansen)

A. retrofusca cocoon and pupa (Xu Tiansen)

A. retrofusca mature larva (Xu Tiansen)

spot between the dorsal line and spiracle line on each of the 1st to 3rd thoracic segments. The black subdosal line and the upper spiracle line are visible inside of the spot. The area above the subdorsal line dark in color, mixed with light flesh red or cinerous or purplish grey. The upper spiracle line thin, cinerous or light gray. Spiracle line black, base line bright yellow. Thoracic legs with black circles, ventral surface and prolegs black. Caudal horn short, black, the rear flesh yellow. There are some black granular dots below the dorsal line from metathorax to the 3rd abdominal segment. **Pupae** 18-24 mm in length, bright reddish brown in early pupal stage, later turn into dark brown. Tip of antennae reaches under middle legs; the male ctenii is visible; female smooth. 8 cremasters, the middle 6 form one long bundle; the other 2 lean to the dorsal side and separated, directed toward right and left, and are 1/2 of the length of the other cremasters; hook of cremaster obvious, bright red, glossy. **Cocoons** 32 mm in length, soil cocoon is composed of silk and soil. Very thin, the inside smooth, the outside attached with soil grains.

Biology and behavior *Armiana retrofusca* occurs 3-4 generations a year in Zhejiang and overwinters as pupae under soil surface. Adults emerge from the end of

A. retrofusca egg (Xu Tiansen)

A. retrofusca mature larva (Xu Tiansen)

A. retrofusca mating pair (Xu Tiansen)

March to early April for insects that have 3 generations a year; for those with 4 generations a year, adults emerge in middle April. The adult occurrence periods of the 1st - 4th generations are from: early April to early May, early June to early July, late July to late August, and late September to mid-October, respectively. The egg stage of the 1st-4th generations are from: mid-April to mid-May, mid-June to early July, early August to early September, early and mid- October, respectively. The egg stage of the 1st-4th generations are from: mid-April to mid-June, mid-June to early August, mid-August to late October, and mid-October to late November, respectively. The 1st generation pupa stage is from late May to mid-June, the 2nd generation is from mid-July to late August, the 3rd generation for insects with 4 generations a year is from early September to mid-October, the overwintering generation pupa of insects with 3 generations a year is from mid-October to mid-April next year, and for insect with 4 generations a year is from late November to early May next year.

Prevention and treatment Protect natural enemies. Predatory animals include birds such as *Cyanopica cyana* and *Garrulax canorus*, and spiders such as *Clubiona maculate*, *Pardosa tschekiangensis*, *Oxyopes sertaus*, and *Chiracantium zhejiangensis* which prey on adults and larvae. Insect predators include *Isaria cicadae*, *Cosmolestes* sp., *Sirthenea flavipes* and *Picromerus virdiipunctatus*. Parasites include *Teichogramma closterae*, *Trichogramma dendrolimi* and *Telenomus dendrolimusi* in egg stage and *Rhogas* sp., *Xanthopimpla predator*, *Xanthopimpla stemmator*, *Campoplex* sp., *Exorista sorbillans* and *Exorista civilis* in larva stage; parasites which parasitize it in larva stage and emerge in pupa stage include *Tetrastichus* sp.

References Wu Chunsheng and Fang Chenglai 2003, Xu Tiansen and Wang Haojie 2004.

(Xu Tiansen, Wang Haojie)

438 *Cerura menciana* Moore

Classification Lepidoptera, Notodontidae
Common name Poplar prominent

Distribution Beijing, Hebei, Inner Mongolia, Liaoning, Jilin, Heilongjiang, Jiangsu, Zhejiang, Fujian, Jiangxi, Shandong, Henan, Hubei, Hunan, Hainan, Sichuan, Yunnan, Tibet, Shaanxi, Gansu, Ningxia, Taiwan; Europe, the former Soviet Union (South), Japan, Korea Peninsula, Vietnam.

Hosts Plants of Salicaceae.

Morphological characteristics **Adults** Body 28-30 mm in length, 75-80 mm in wingspan. Whole body greyish-white, head and thorax with a little purplish-brown. Thorax with 8-10 black dots arranged in two rows. The base of forewing with two black dots. The outside of the discal cell with several rows of black wavy stripes, outer margin with eight black dots. Hind wing black and white slightly with purple color, veins dark chocolate-brown, cross veins black. Pronotum with a red-violet triangle spot, the lateral sides of the 4th abdominal segment with white stripes. Caudal legs extend into a pair of long caudal horns with reddish-brown microtrichiae. **Eggs** Steamed bread shaped, 3 mm in diameter, reddish-brown, the middle with a black dot. **Larvae** Mature larvae 50 mm in length, 6 mm in width. Head brown, two cheeks with black spots, body leaf green in color, dorsal thorax with triangle upright granules, stripes red-violet, the location near the rear edge of the 4th thoracic segment with a white stripe and the front of it has a brown edge, a pair of caudal legs degenerated into tail shape, with microtrichiae. **Pupae** 25 mm in length, 12 mm in width, reddish-brown. Body stout, the upper end with a colloid sealed emergence hole. **Cocoons** 37 mm in length, 22 mm in width.

Biology and behavior *Cerura menciana* occurs two generations a year and overwinters as pupae in most areas. It occurs three generations a year in Xian, Shaanxi. In two generations occurrence areas, the overwintering adults successively emerge in April, larvae hatch in late

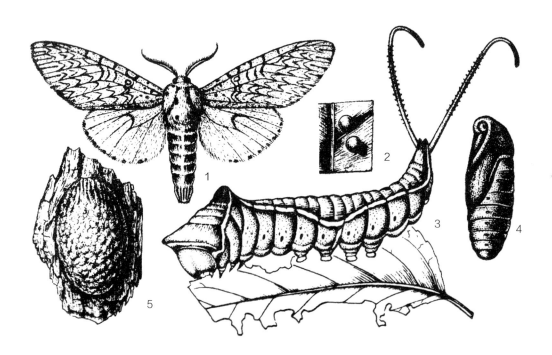

H. hippophaes. 1. Adult, 2. Eggs, 3. Larva, 4. Pupa. 5. Cocoon

(Shao Yuhua)

C. menciana larva (Xu Gongtian)

C. menciana young larva (Xu Gongtian)

C. menciana adult (Xu Gongtian)

C. menciana cocoon
(edited by Zhang Runzhi)

C. menciana egg (Xu Gongtian)

C. menciana larva (Xu Gongtian)

C. menciana cocoon (Xu Gongtian)

C. menciana scar left from cocoon (Xu Gongtian)

C. menciana larva parasitized by wasps (Xu Gongtian)

May, and the 1st generation larvae damage period is from the end of May to early July. The 2nd generation larval period is from August to early September. Larvae mature and spin cocoons and overwinter after mid-September. Adult longevity is 7-10 days. Adults rest during the daytime and become active at night, they have phototaxis. Adults start emergence at around 16:00, the emergence peak is at 18:00. They mate in the current night and most mating occurs during 2:00-3:00. Eggs are laid individually on the leaves the current night. Most eggs are laid on the back of leaves, some are laid on the leaves top and small twigs. Egg stage is around 15 days and egg hatching rate is around 95%. Larvae spins silk and drift away with the wind after it grows older, and eat leaves into incisions. They feed a lot during the 4th and 5th instars. It shakes its red eversible and shrinkable gland when disturbed. Larvae have five instars. Mature larvae gnaw the bark on tree branches or trunks. They weave wooden pieces with silk to make cocoons. Cocoons are tough in texture, and attached tightly to the trunks, their color blends well with the bark.

Prevention and treatment

1. Physical control. Manually kill pupae to reduce the 1st generation population. Manually catch and kill larvae. Trap and kill adults using black light traps.

2. Biological control. *Trichogramma* wasps carry virus. Using the wasps can cause epidemic of virus disease.

3. Chemical control. During larval stage, spray 12% thiamethoxam and lamda-cyhalothrin at 1,000-1,500×.

References Forestry Pest and Disease Control Station of Henan 1991, Li Yajie and Lin Jihui 1992, Wu Chunsheng and Fang Chenglai 2003, Shanghai Forestry Station 2004, Yu Jun 2001, He Chengjiang et al. 2006.

(Guo Xin, Li Zhenyu, Li Yajie, Lin Jihui)

Clostera anachoreta (Denis & Schiffermüller)

439

Classification Lepidoptera, Notodontidae
Synonyms *Pygaera anachoreta pallida* Staudinger, *Pygaera mahatma* Bryk
Common name Scarce chocolate-tip

Distribution Beijing, Hebei, Shanxi, Liaoning, Jiangsu, Shandong, Henan, Sichuan, Shaanxi; Europe, India, Indonesia, Japan, Korea Peninsula, Sri Lanka, Russia, Vietnam.

Hosts *Populus* spp., and *Salix* spp.

Damage Young larvae feed aggregately on the underside of the leaves. Larger larvae tie leaves together to form feeding pouches and consume entire leaves.

Morphological characteristics **Adults** Females 15-20 mm long with a wingspread of 38-42 mm, males 13-17 mm long with a wingspread of 23-37 mm. Body grayish brown. There are four grayish white wavy cross stripes on the wings, a fanlike mark at the top corner of the wings, and a black spot below the fanlike mark. Hind wing grayish brown. **Larvae** Body 32-40 mm long, abdomen grayish white, with a ring of eight orange verruca

C. anachoreta adult (Xu Gongtian)

C. anachoreta adult (Li Zhenyu)

C. anachoreta adult (edited by Zhang Runzhi)

C. anachoreta mating pair (Xu Gongtian)

C. anachoreta mature larva (Xu Gongtian)

C. anachoreta mature larva (edited by Zhang Runzhi)

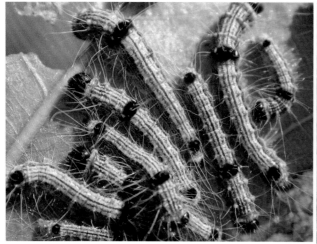
C. anachoreta larvae (Xu Gongtian)

C. anachoreta damage (Xu Gongtian)

C. anachoreta eggs (Xu Gongtian)

C. anachoreta eggs and young larvae (Xu Gongtian)

C. anachoreta pupa in a leaf roll (Xu Gongtian)

C. anachoreta pupa (Xu Gongtian)

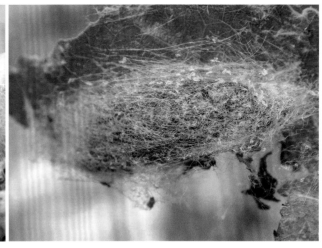
C. anachoreta cocoon (Xu Gongtian)

C. anachoreta damage (Xu Gongtian)

C. anachoreta damage (Xu Gongtian)

on each segment.

Biology and behavior *Clostera anachoreta* occurs 3-4 generations a year and overwinters as pupae in Henan. Adults begin to emerge and lay eggs in mid-March. Adults for the 1st-4th generation appear in early to mid-June, early to mid-July, early to mid-August, and mid-September, respectively. Larvae of the 4th generation feed until the end of October before overwintering as pupae inside the soil.

Adults are active at night and are attracted to lights. Adults of the overwintering generation appear before full leaf flush and lay eggs mostly on branches. Eggs of the other generations are laid mostly on the underside of the leaves, usually with several hundred eggs in a single layer at one spot. Newly hatched larvae feed aggregately on epidermis with heads line up in the same direction. Larger larvae tie leaves separately to form leaf pouches and feed inside until the entire leaf is consumed. Mature larvae construct cocoons inside the pouches for pupation. Mature larvae of the last generation construct cocoons on the ground, on dead leaves, on the bark, on ground cover, or in top soil to overwinter as pupae.

Prevention and treatment

1. Manual control. Remove feeding pouches in nursery or young forest stands to kill larvae.

2. Biological control. Preserve parasitoids such as *Trichogramma closterae*, *Iseropus* sp., and *Exorista civilis*.

3. Chemical control. During larval period, spray 12% thiamethoxam and lamda-cyhalothrin at 1,000-1,500×.

References Yang Youqian 1982, Xiao Gangrou (ed.) 1992, Wu Chunsheng and Fang Chenglai 2003, Henan Forest Pest Control and Quarantine Station 2005.

(Yang Youqian, Chen Zhiqing)

440 *Clostera anastomosis* (L.)

Classification Lepidoptera, Notodontidae
Synonym *Neoclostera insignior* Kiriakoff

Importance It is a main defoliator of Populus spp. In the central and western regions of China. It often outbreaks in artificial *Populus* spp. Forest in large areas. It can significantly affect tree growth.

Distribution Beijing, Hebei, Inner Mongolia, Liaoning, Jilin, Heilongjiang, Shanghai, Jiangsu, Zhejiang, Anhui, Fujian, Jiangxi, Hubei, Hunan, Sichuan, Yunnan, Shaanxi, Gansu, Qinghai, Xinjiang; Japan, Korea Peninsula, Russia, Mongolia, India, Indonesia, Sri Lanka, Europe and North America.

Hosts *Populus* spp., *Salix* spp., *Betula* spp.

Damage Larvae feed on leaves. It can defoliate whole trees in large areas in a short period of time.

Morphological characteristics **Adults** 13-17 mm in length, 30-50 mm in wingspan. Body grey-

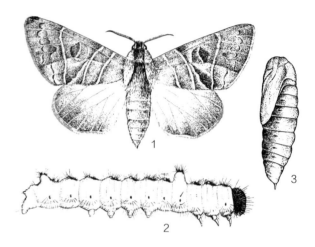

C. anastomosis. 1. Adult, 2. Larva, 3. Pupa
(1 and 3 by Zhu Xingcai; 2 by Zhang Peiyi)

C. anastomosis adult (Xu Gongtian)

C. anastomosis adult (Yan Shanchun)

C. anastomosis larvae (Yan Shanchun)

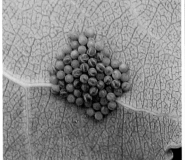
C. anastomosis eggs (newly laid) (Yan Shanchun)

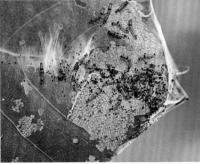
C. anastomosis newly hatched larvae (Yan Shanchun)

C. anastomosis eggs (near hatching) (Yan Shanchun)

ish-brown, the dorsal side of the head and thorax with a dark reddish-brown wide longitudinal band. Forewing greyish-brown, with four light cross stripes. The end of the discal cell with a dark rounded spot, the middle with a white short line dividing the rounded spot into two halves. **Eggs** Near steamed bread shaped, light green, 0.6 mm in diameter. **Larvae** Mature larvae 35-40 mm in length. Body slightly clavate, slightly bent, the dorsal dark chocolate-brown, with red granules and white small dots, the dorsal of 1st and 8th abdominal segment respectively with a big granule inserted 4 black small seta verrucae, body covered with white fine hairs. **Pupae** 15-18 mm in length. Reddish-brown, glossy, near conical.

Biology and behavior *Clostera anastomosis* occurs one generation a year in Great Khingan of the Northeast. It overwinters as the 3rd instar larvae in the litter under the host trees. The overwintering larvae resume activity in late May next year. They are gregarious and spin cocoons to pupate in mid- and late June. Adults emerge, mate and oviposit in early July. Eggs hatch in mid-July. Larvae leave hosts in early August and spin white oval cocoons to overwinter. It occurs 6-7 generations a year in Shanghai area and overwinters as eggs on the limb of *Populus* spp.; few individuals overwinter as the 3rd-4th instars larvae and pupae. The overwintering eggs hatch from early April next year. Larvae gnaw bud scales and bark of tender branches and feed on leaves with unfolding of leaves. Larvae pupate in mid- and late May. Adults emerge, mate and oviposit in mid-May to early June. Some larvae develop slowly in November and overwinter as the 3rd-4th instar larvae in litter layer. While most larvae will develop into adults which oviposit during late November and early December and overwinter as eggs.

In Great Khingan, the development time for the 1st to 7th instar larvae are: 4, 8-11, 294-302, 5-7, 5-9, and 6-10 days. The total larval stage is 326-338 days. Pupal stage is 17-19 days. Adult longevity is 4-19 days. Egg stage is 10-11 days.

Eggs are laid on the underside of *Populus* spp. leaves. They hatch during 4:00-6:00. Eggs of the same egg clutch generally are hatched on the same day, hatch rate was observed as 93%-98%. The newly hatched larvae are gregarious on the blade and start feeding on the mesophyll after some time. Leaves become net-like and the blade become withered and yellow. Larvae feed on leaf margin after the 2nd instar. Young larvae spin silk when fall from leaves and spread with the aid of wind. Their feeding increases significantly after the 4th instar. They start to eat whole blades and easily fall to the ground when frightened. Larvae rest on twigs when not feeding. The insect is abundant in sparse woods. When leaves of whole branches are eaten up, they migrate to feed on nearby trees of *Salix raddeana* and *Betula platyphylla*.

About 3/4 of the adult emergence takes place during the daytime. Adults are inactive during the day and rest on branches and leaves of *Populus* spp. or shrubs. They are active at night. They have phototaxis. Adults mate and oviposit several hours after emergence. Eggs are laid in egg masses. A female lays up to 1,682 eggs with an average of more than 500 eggs.

Prevention and treatment

1. Silviculture. Planting mixed forests and maintain good canopy coverage to create conducive environment for natural enemies. Cultivate pest resistant varieties.

2. Physical control. Trap adults with black light lamp during the overwintering generation and 1st generation adult emergence periods. In areas without power supply, set up campfires to kill adults. Manually collect eggs and pupae.

3. Biological control. The granulosis virus of *Clostera anastomosis* is one of the key factors that shorten the damage period. When the *Clostera anastomosis* population is high, the virus infection rate often reaches around 30%. Applying 3×10^3 or 5×10^3 virus dilution resulted in good control of the 3rd and 4th instar larvae.

Applying *B.t.* at 100-200× is able to keep the pest population low for a long-term and has no adverse effect on natural enemies such as the robber flies, spiders and chalcidoid. Protection and attraction of beneficial birds are also effective biological control measures. Crow and magpie feed on larvae and pupae of *Clostera anastomosis* in woods.

4. Chemical control. Spray 20% dinotefuran at 800×, 12% thiamethoxam and lamda-cyhalothrin at 1,000-1,500×, Apply an insecticide band the above insecticides. Spread the mixture to tree trunk to form a 3-5cm width closed ring.

References Liu Bo, Yang Jun, and Zhang Xitang et al. 1992; Xiao Gangrou (ed.) 1992; Tian Feng, Zhang Yanmin, Wang Wenzhi et al. 1993; Wang Fuwei, Niu Yanzhang, Hou Liwei et al. 1993; Wang Zhiming, Pi Zhongqing, and Ning Changlin 1993; Li Yanmei 1996; Yu Jingru 1998; Li Li, Sun Xu, and Meng Huanwen

C. anastomosis larva (Yan Shanchun)

C. anastomosis pupa (Yan Shanchun)

C. anastomosis damage
(Yan Shanchun)

2000; Yang Jun and Lin Sen 2000; Li Chengde 2003; Gao Rong, Di Xudong, Yang Zhende et al. 2004; Li Yana 2006; Fang Jie, Cui Yongsan, Zhao Boguang et al. 2007; Zhang Lanying, Han Dan, Li Xue et al. 2008.

(Yan Shanchun, Xu Chonghua, Tang Shangjie)

441 *Euhampsonia cristata* (Butler)

Classification Lepidoptera, Notodontidae
Synonym *Trabala cristata* Butler

Distribution Beijing, Hebei, Shanxi, Inner Mongolia, Liaoning, Jilin, Heilongjiang, Jiangsu, Zhejiang, Anhui, Shandong, Henan, Hubei, Hunan, Hainan, Sichuan, Yunnan, Shaanxi, Gansu, Taiwan; Japan, Korea Peninsula, Russia, Myanmar, Laos, Thailand.

Hosts *Castanea mollissima*, and *Quercus* spp.

Damage Young larvae feed on epidermis that leads to reticular damage on leaves. Larger larvae create notches and feeding holes or consume entire leaves.

Morphological characteristics **Adults** Body yellowish brown, 35-38 mm long with a wingspread of 75-85 mm. There are two dark brown transverse bands on forewings. A pair of equal-sized white circular spots appear between the two transverse bands. **Larvae** Body 60-70 mm long, green and smooth. Head large. Mature larvae purplish red.

Biology and behavior *Euhampsonia cristata* occurs two generations a year and overwinters as pupae in the soil. Adults are strong fliers and are attracted to lights. They start to emerge in early June and lay eggs in clusters of 3-4 eggs/cluster on the underside of the leaf. Each female produces 370-610 eggs. Egg stage lasts for 4-5 days. Eggs start to hatch in mid-June. Newly hatched larvae disperse through ballooning in the wind with silk threads and feed on leaf epidermis, creating reticular leaf surface. Older larvae can consume entire leaves quickly, leaving only veins intact. Heavy damage mostly coincides with humid and high temperature period between June and July. Mature larvae drop to the ground in

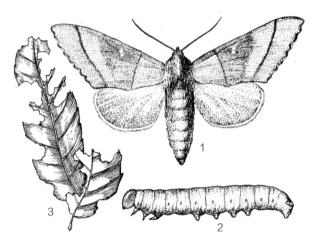

E. cristata. 1. Adult, 2. Larva, 3. Damage
(Zhang Xiang)

E. cristata adult female (Ji Baozhong)

E. cristata adult male (Ji Baozhong)

E. cristata adult male (NZM)

E. cristata eggs, larva, pupa, female, male (From left to right) (Ji Baozhong)

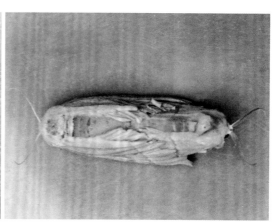
E. cristata mating pair (Ji Baozhong)

E. cristata prepupae (Ji Baozhong)

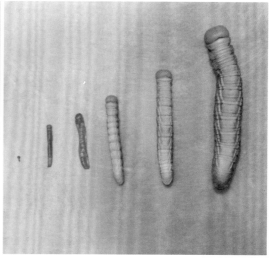
E. cristata 1st to 6th instar larvae (Ji Baozhong)

mid-July to construct pupal chambers in the soil, with a portion stay inside for overwintering, and the rest continue development to adults in early August. These adults mate and lay eggs for another generation. Mature larvae of this generation pupate in soil between September and October for overwintering.

Prevention and treatment Chemical control. Spray 12% thiamethoxam and lamda-cyhalothrin at 1,000-1,500× for larvae control.

References Yang Youqian 1982, Xiao Gangrou (ed.) 1992, Wu Chun-sheng and Fang Chenglai 2003, Henan Forest Pest Control and Quarantine Station 2005.

(Yang Youqian)

442 *Periergos dispar* (Kiriakoff)

Classification Lepidoptera, Notodontidae
Synonym *Loudonta dispar* (Kiriakoff)

Distribution Jiangsu, Zhejiang, Anhui, Fujian, Jiangxi, Hubei, Hunan, Guangxi, Sichuan, Guizhou, Yunnan.

Hosts 24 species and varieties in *Phyllostachys*; *Pleioblastus amarus*, and *P. juxianensis*.

Damage Larvae feed on bamboo leaves, kill bamboo plants, and reduce bamboo shoot yield.

Morphological characteristics **Adults** Female body 16-23 mm in length, 46-54 mm in wingspan. Males 12-18 mm in length, 35-42 mm in wingspan. Head and body yellowish-white, compound eyes black, antennae filiform. Forewing long and narrow, apex of the wing protruded, color and stripes vary considerably, most are yellow and orange; the color from the costal margin to the outer margin dark. Wing surface scattered with obscure brown cloud point spots, the middle with a dark reddish-brown spot. Male body 12-18 mm in length, 35-42 mm in wingspan. Antennae bipectinate, body yellowish-brown. Forewing rusty yellow, with greyish-brown cloud points, the middle with a black dot. A small percentage of both male and female individuals have the forewing with an oblique line of light brown spots from wing base to apex. **Eggs** Oblate, 1.69-1.88 mm in diameter, orange red; the top flat, the center sinks into persimmon shape. **Larvae** The newly hatched larvae 3 mm in length, milky, whole body covered with primordial setae. Larvae have 5-7 instars. Color of different instars vary considerably from dull yellow to viridis. Mature larvae 45-60 mm in length, the lower spiracle line yellowish-white, front edge of prothorax with a black velvet colored spot. **Pupae** 18.5-26.5 mm in length. Dark viridis at beginning, dark chocolate-brown before emergence. Wing buds reach the end of the 4th abdominal segment; the end of the abdomen with a semicircle margin on the back, fine setae, and 8 cremasters. **Cocoons** Soil cocoons 30 mm in length; the interior surface smooth,

P. dispar. 1. Adult female, 2. Adult male, 3. Egg, 4. Larva, 5. Pupa (Xu Tiansen)

P. dispar adult (above: males, below: females) (Xu Tiansen)

P. dispar eggs parasitized by wasps (Xu Tiansen)

P. dispar 2nd instar (Xu Tiansen)

P. dispar 5th instar (Xu Tiansen)

P. dispar larva being predated by a bug (Xu Tiansen)

P. dispar cocoon and pupa (Xu Tiansen)

the exterior surface with soil grains.

Biology and behavior *Periergos dispar* occurs 3-4 generations a year in Zhejiang and 4 generations a year in Hunan. It overwinters as pupae when 3 generations occur a year or overwinters as mature larvae when 4 generations occur a year. The overwintering larvae pupate in late March and end until early May. Adults emerge in mid-April to mid-May and oviposit in late April to the end of May. Egg masses in a single layer are laid on the upper side of bamboo leaves, few eggs are laid on the underside of bamboo leaves.

Larval feeding period is in early May to late June. Larvae pupate in the end of May to the end of June. The 2nd generation adults emerge in mid-June to early July and oviposit in late June to mid-July. Larval feeding period is in late June to mid-August, and larvae pupate in late July to late August. The 3rd generation adults emerge in early August to early September and oviposit in mid-August to mid-September. Larval feeding period is in mid-August to early October. A portion of larvae develop into the 4th generation in mid-September to mid-October. The 4th generation adults emerge in late September to late October and oviposit in early October to late October, and eggs hatch in mid-October. Larvae become mature in February next year. In Hunan, adult occurrence periods of the 1st to the 4th generation are: early April to mid-

P. dispar larva (Xu Tiansen)

P. dispar larva (Xu Tiansen)

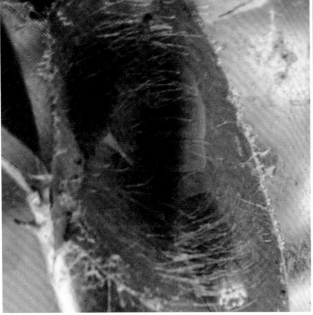
P. dispar larva in cocoon (Xu Tiansen)

P. dispar eggs parasitized by wasps (Xu Tiansen)

May, mid-June to early August, late July to early September, and early September to early November, respectively. The larval periods of the 1st to the 4th generation are: late April to early June, early July to mid-August, early August to early October, and mid-September to early November, respectively.

Prevention and treatment Protect natural enemies. The reported natural enemies of the *Periergos dispar* include predatory birds such as *Cuculus canorus*, *Cyanopica cyana*, and *Garrulax canorus*; spiders such as *Clubiona maculate*, *Oxyopex lineatipes*, *Lycosa pseudoannulata*, *Lycosa coelestris*, *Dolomedes sulfureus*, and *Dolomedes pallitarsis*; insect predators such as *Isaria cicadae*, *Hierodula patellifera*, *Polyrhachis dives*, *Formica japonica*, *Isyndus obscurus*, *Haematoloecha nigrorufa*, *Cosmolestes* sp., *Sirthenea flavipes*, *Picromerus viridipunctatus*, and *Arma chinensis*. Parasitoids include *Teichogramma closterae*, *Trichogramma dendrolimi*, *Teichogramma ivelae*, *Telenomus dasychili*, *Telenomus dendrolimusi* and *Telenomus lebedae* in egg stage; *Rhogas* sp., *Enicospilus lineolatus*, *Coccygomimus parnarae*, *Xanthopimpla predator*, *Coccygomimus aethiops*, *Campoplex* sp., *Colpotrochia (Colpotrochia) pilosa pilosa*, *Coccygomimus luctuosus*, *Exorista japonica*, and *Exorista civilis* Rondaniin in larval stage; and *Tetrastichus* sp. and *Enicospilu*s sp. in pupal stage.

References Wu Chunsheng and Fang Chenglai 2003, Xu Tiansen and Wang Haojie 2004.

(Xu Tiansen, Wang Haojie)

443 *Micromelalopha sieversi* (Staudinger)

Classification Lepidoptera, Notodontidae

Synonyms *Micromelalopha troglodyte* (Graeser), *Micromelalopha populivona* Yang et Lee, *Pygaera sieversi* Staudinger

Distribution Beijing, Hebei, Shanxi, Liaoning, Jilin, Heilongjiang, Jiangsu, Zhejiang, Anhui, Jiangxi, Shandong, Henan, Hubei, Hunan, Sichuan, Yunnan, Tibet; Japan, Korea Peninsula, Russia.

Hosts *Populus* spp., *Salix* spp.

Damage Larvae feed on leaves. All leaves on a tree can be devoured.

Morphological characteristics **Adults** Body 11-14 mm in length, 24-26 mm in wingspan. Body greyish-brown. Forewing with 3 yellowish-white cross lines, the inner cross line bifurcate, the outer cross line wavy. Hind wing yellowish-brown.

Larvae Body 21-23 mm in length. Body greyish-brown. Two sides of body each with a yellow longitudinal band, the dorsal of the 1st and 8th abdominal segments each with a large granule.

Biology and behavior *Micromelalopha sieversi* occurs 4 generations a year in Henan province and overwinters as pupae in bark cracks. Adults begin to emerge in mid-April next year. The adult period of the 1st to the 4th generations are: mid- and late May, late June, mid- and late July, and mid- and late August, respectively. Larvae from this generation occur in September and October.

Adults mate and oviposit during night. They are attracted to lights. Most eggs are laid on leaves, arranged in egg clutches, each egg clutch with 300-400 eggs. Young larvae are gregarious on leaves and feed on the epidermis, the damaged leaves net looking. Later, they scatter and cause incisions or holes on leaves. They consume all leaves when the population density is high. Larvae move slowly. They hide in rough bark cracks of trunks and between tree crotches during the daytime. Mature larvae spin silk and weave leaves to create thin cocoons. Larvae

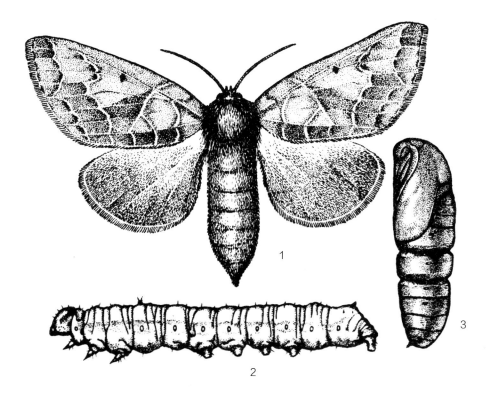

M. sieversi. 1. Adult, 2. Larva, 3. Pupa

(Zhang Xiang)

M. sieversi adult (Xu Gongtian)

of the last generation spin cocoons in bark cracks or in soil and overwinter.

Prevention and treatment

1. Biological control. *Trichogramma closterae*, *Exorista civilis* and *Iseropus* sp. have the potential to cause high mortality of *M. sieversi* eggs and larvae. They can be mass reared and released for pest suppression.

2. Chemical control. In early larval occurrence period, spray 12% thiamethoxam and lamda-cyhalothrin at 1,000-1,500×.

References Yang Youqian et al. 1982, Xiao Gangrou (ed.) 1992, Wu Chunsheng and Fang Chenglai 2003, Forest Disease and Pest Prevention and Treatment and Quarantine Station of Henan Province 2005.

M. sieversi adult (Xu Gongtian)

(Yang Youqian, Li Zhenyu)

444 *Cerura tattakana* Matsumura

Classification Lepidoptera, Notodontidae
Synonyms *Neocerura wisei* (Swinhoe), *Neocerura tattakana* (Matsumura)

Importance *Cerura tattakana* is a major pest of *Homalium hainanense*, host tree of *Laccifer lacca*. Larval feeding on leaves leads to complete defoliation of the host trees under heavy infestation. Mature larvae construct pupal chambers between bark and the sapwood, creating physical wounds that make trees vulnerable to wind damage.

Distribution Jiangsu, Zhejiang, Fujian, Hubei, Guangdong, Guangxi, Hainan, Sichuan, Yunnan, Shaanxi, Taiwan; Japan, Vietnam, India, Sri Lanka, and Indonesia.

Hosts *Homalium hainanense*, *Populus* spp., and *Salix* spp.

Morphological characteristics **Adults** Females 28-32 mm long with a wingspread of 65-87 mm, while males 24-30 mm long with a wingspread of 55-67 mm. Body near grayish white. In females, the 7th and 8th abdominal segments white, with black margins; the 7th segment with a black ring at center and a black spot in the ring. Males have a small ring on the 7th segment; its 8th segment white and with a semi-circular black ring; posterior border black. **Larvae** Mature larvae about 50 mm long, head reddish brown, body fresh green. Dorsal surface of metathorax forms a single peak. Prothorax large and hard, purplish red with one broad pinkish central band with white margins. This band extends to abdominal segments 4-5 where it enlarges to form a diamond. Abdomen ends with one pair of purplish red serrate cau-

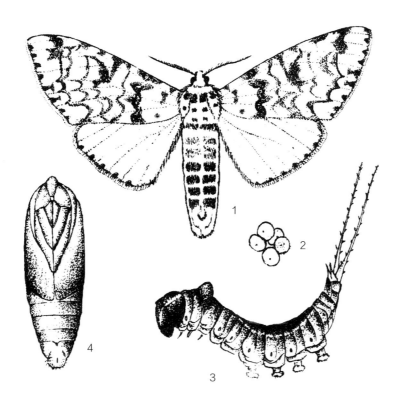

C. tattakana. 1. Adult, 2. Eggs, 3. Larva, 4. Pupa (Yang Cuixian)

C. tattakana adult (Zhang Peiyi)

C. tattakana adult (Huang Jinshui)

C. tattakana adult (Huang Jinshui)

dal horns.

Biology and behavior *Cerura tattakana* occurs three generations a year and overwinters as pupae inside cocoons at the base of host trees in Fujian. Larvae period of the 1st, 2nd, and 3rd generation is between early April to mid-May, mid-June to late August, mid-September to late November, respectively. Mature larvae construct cocoons by mid-November to overwinter as pupae. Adults emerge between mid- and late February. It has 5-6 generations a year in Jianfengling of Hainan where it feed year around without overwintering.

Most adults emerge at dusk after leaving pupal cases inside the hard cocoons. Females start to lay eggs a few hours after emergence. Each female produces an average of 198 eggs. Adults live for 6-11 days and are attracted to lights. Eggs are laid singly on leaf surface. Egg stage lasts for 7-15 days with a hatching rate of 95%. The 3rd instar and older larvae feed on leaves from the tip to the base. Older larvae consume 8-10 leaves each per night. There are five instars in the larval stage. Mature larvae chew pupal chambers between the bark and the sapwood and construct hard cocoons with silk and sawdust inside. Larvae pupate in about 7 days. Pupal stage lasts for 15-30 days with an adult emergence rate of 87.5%.

Prevention and treatment

1. Manual control. Destroy pupae at the base of the trees.

2. Biological control. Utilize natural enemies such as predator *Creobroten femmatus* or *Beauveria bassiana* for larvae control.

References Chen Zhiqing et al. 1978, Huang Jinshui et al. 1982, Wu Chunsheng and Fang Chenglai 2003.

(Huang Jinshui, Tang Chensheng, Li Zhenyu)

Phalera bucephala (L.)

Classification Lepidoptera, Notodontidae
Synonyms *Phalera* (*Noctua*) *bucephala* L., *Phalera bucephala infulgens* Graeser
Common name Buff-tip moth

Distribution Inner Mongolia, Jilin, Heilongjiang, Xinjiang; Korea Peninsula, Russia (Siberia), Europe, Northeast Africa and East Asia.

Hosts *Salix* spp., *Populus* spp., *Ulmus* spp., *Betula* spp., *Quercus* spp., *Corylus* spp., *Acer* spp., *Tilia* spp., *Sorbus* spp., *Juglans regia*, *Fagus* spp., *Citrus aurantium*, *Malus* spp., *Pyrus* spp., *Prunus* sp.

Damage Larvae feed on leaves and can consume all leaves on some branches of a tree.

Morphological characteristics **Adults** 52-64 mm in wingspan; males smaller than females; robust. Forewing greyish-brown, apex area with a large light yellowish-white spot; hind wing yellowish-white. **Larvae** Head black and glossy. The dorsal surface orange yellow, with light grayish yellow hairs. The dorsal line, the subdorsal line, the upper spiracle line and the lower spiracle line black. Spiracles black. Ventral surface black, abdominal line wide and yellow. The exterior side of prolegs black, the interior side yellow.

Biology and behavior *Phalera bucephala* occurs one generation a year in Xinjiang. It overwinters as pupae in weeds or soil. Adults emerge in mid-May to mid-July. Adults have strong phototaxis. They lay eggs on leaves of isolated trees. Eggs are arranged in single layer clutches. Each female produces an average of 238 eggs based on indoor observation. The egg stage is approximately 9 days. The newly hatched larvae are gregarious on the lower surface of leaves. They feed on the lower epidermis and mesophyll. The leaves look become semitransparent. The damaged leaves become dry.

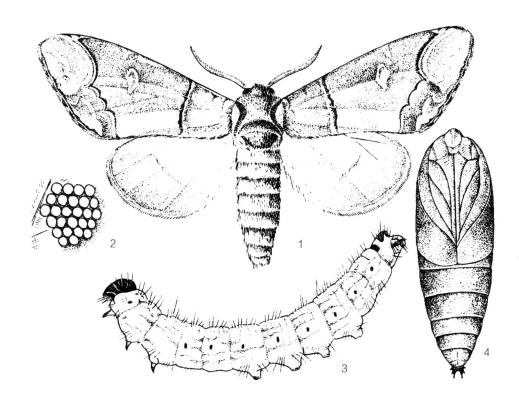

P. bucephala. 1. Adult, 2. Eggs, 3. Larva, 4. Pupa (Zhu Xingcai)

P. bucephala adult (Li Chengde)

Larvae often weave leaves with silk. When at rest, their abdomen tilts up. Their heads raise and repeatedly move from side to side when they are frightened. Older larvae scatter and feed individually. They can eat up the leaves of some branches. Larvae play death and they will fall to the ground when they encounter vibration. Larval development speed is affected by the host. Development time is shortest when feeding on *Salix* spp. Larval damage period is in June to the end of September.

Prevention and treatment

1. Silviculture. Cut down isolated host trees.

2. Manual control. Remove leaves with gregarious young larvae and destroy them. For old larvae, kill them after knocking them off from hosts. Trap adults with light traps. Dig soil around the root collars of trees to kill pupae.

3. Chemical control. During the 1st instar period, spray 5.7% emamectin at 3,000×, 90% trichlorfon at 1,000-1,200×, 25 g/L lamda-cyhalothrin at 1,000-1,200×, or 50% malathion EC at 1,000×, or 50% fenitrothion EC at 1,000×.

References Xiao Gangrou 1992, Wu Chunsheng and Fang Chenglai 2003, Fang Honglian and Cao Yuping 2006.

(Han Huilin, Zhang Xuezu)

446 *Phalerodonta bombycina* (Oberthür)

Classification Lepidoptera, Notodontidae

Synonyms *Phalerodonta albibasis* (Chiang), *Notodonta albibasis* Oberthür, *Ochrostigma albibasis* Chiang

Importance Outbreaks of *Phalerodonta bombycina* in *Quercus* woodlands can seriously affect tree growth, acorn yield, and tussah production.

Distribution Jilin, Jiangsu, Zhejiang, Anhui, Jiangxi, Shandong, Hubei, Hunan, Sichuan, Shaanxi.

Hosts *Quercus fabri*, *Q. acutissima*, *Q. chenii*, *Q. variublis*.

Morphological characteristics **Adults** 18-22 mm in length, male 38-48 mm in wingspan, female 48-52 mm in wingspan. Body yellow or greyish-brown, antennae yellowish brown, pectinate. Forewing greyish brown, costal margin and base dark chocolate brown, sub-base line saw-tooth shaped. Hind wing greyish brown, with an inconspicuous outer cross line. Female abdominal end with yellowish brown and dark chocolate brown cluster of hairs. Male body color darker than females, their abdominal end without dark chocolate brown hairs. **Eggs** Oblate, greyish white or slightly yellow, egg clutch covered with brown hairs. **Larvae** Body 30-40 mm long. Head reddish orange, thorax and abdomen light green, the dorsal and two sides of body with purplish-brown spots, crochets uniordinal mesoseries. **Pupae** 15-20 mm in length, dark brown, abdominal end smooth and blunt.

Biology and behavior *Phalerodonta bombycina* occurs one generation a year and overwinters as eggs. The eggs hatch in early and mid-April next year. Newly hatched larvae cluster on twigs and feed on the mesophyll of tender leaves. After the 3rd instar, they feed on whole leaves. The ground may be covered with frass.

P. bombycin adult (Yang Chuncai, Tang Yanping)

After the 4th instar, larvae scatter and feed individually. They migrate to another plant after they eat up leaves of a tree. When they are frightened, they raise their heads and their abdomen, and spit black liquid from their mouthparts. Larvae have 5 instars and larval stage is 42-52 days. Larvae mature in late May to early June; they crawl down the tree, spin cocoons, and pupate in 3-10 cm deep loose soil around trunk base. Pupal stage is more than four months, adults emerge from late October to early November. Adult emerge between 13:00-18:00. Adults rest on shrubs, weeds and trunk bases during the daytime and are active dusk. They have phototaxis. Eggs are laid on twigs located in the mid- and lower crown. Most of the eggs are arranged into egg clutches which have 4-6 rows. Egg clutches are covered with dark chocolate brown fluffs.

Prevention and treatment

1. Silviculture. Plant mixed forests to suppress the insect populations.

2. Manual control. Remove egg clutches and gregarious larvae. Trap adults with light traps in late October to early November.

3. Biological control. Apply 5×10^7 spores per 1 mL *Beauveria bassiana* or 5×10^7-1×10^8 spores per 1 mL *B.t.* liquid to spray in late April to early May.

4. Chemical control. Before the 4th instar, fogging with 10:1 mixture of 10% cypermethrin and diesel oil at 40-50 mL per 667 m² in woods.

References Chinese Academy of Sciences – Institute of Zoology 1982, Editorial Board of Anhui Forest Diseases and Pests 1988, Xiao Gangrou (ed.) 1992, Wu Chensheng and Fang Chenglai. 2003.

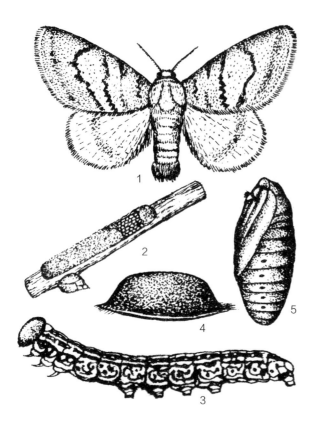

P. bombycin. 1. Adult, 2. Eggs, 3. Larva, 4. Cocoon, 5. Pupa (Tian Hengde)

(Yang Chuncai, Tang Yanping, Tian Hengde)

447 *Pterostoma sinica* Moore

Classification Lepidoptera, Notodontidae
Synonym *Pterostoma grisea* Graeser

Importance Larvae often occur together with *Semiothisa cinerearia*. They often consume all leaves of hosts at high population density.

Distribution Beijing, Hebei, Shanxi, Liaoning, Shanghai, Jiangsu, Zhejiang, Anhui, Fujian, Jiangxi, Shandong, Hubei, Hunan, Guangxi, Sichuan, Yunnan, Tibet, Shaanxi, Gansu; Japan, Korea Peninsula, Russia.

Hosts *Sophora japonica*, *Robinia pseudoacacia*, *Maackia amurensis*, *Wisteria floribunda*.

Damage Larvae feed on host leaves.

Morphological characteristics **Adults** Male 21-27 mm in length, 56-64 mm in wingspan; female 27-32 mm in length, 68-80 mm in wingspan. Head and thorax straw yellow, the front and rear edges of cervix brown. Dorsal surface of abdomen dark greyish brown, the end yellowish brown; abdominal venter light greyish yellow, the middle with 4 dark brown longitudinal lines. Forewing straw yellowish brown to greyish and yellowish white, comb-shaped hair tuft of hind margin dark brown to dark chocolate-brown, double strands and sawtooth shape. Hind wing light brown to dark brown, veins dark brown, inner margin and base straw yellow; outer line is a vague straw yellow band; terminal line dark brown; margin hairs at the end of the veins straw yellow. **Eggs** Light yellowish green, round, the bottom slightly flat, looks like steamed bread. **Larvae** Mature larvae 56-58 mm long, 9 mm wide. Body smooth and slightly flat, head and thorax smaller than abdomen; the dorsal surface pinkish green, venter dark green, inter-segment yellowish green; spiracle line yellowish white, about 1 mm in width, with a black margin, extends forward to the head; spiracles white. **Pupae** About 30 mm in length, about 9 mm in width, dark brown, glossy, oval, with 4 cremasters. **Cocoons** About 45 mm in length, about 25 mm in width, oblong, dull gray, rough.

Biology and behavior *Pterostoma sinica* occurs

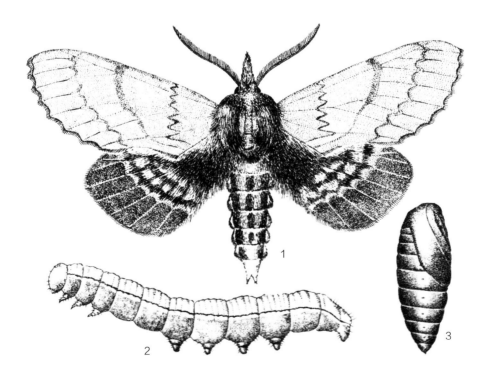

P. sinica. 1. Adult, 2. Larva, 3. Pupa (Zhang Peiyi)

P. sinica adult (Xu Gongtian)

P. sinica adult (Xu Gongtian)

P. sinica larva showing black spots on head (Xu Gongtian)

P. sinica pupa (Xu Gongtian)

P. sinica arva showing yellow band (Xu Gongtian)

P. sinica larva showing black rings on abdominal legs (Xu Gongtian)

P. sinica larva (Li zhenyu)   P. sinica larva (Xu Gongtian)

two generations a year in Beijing. They overwinter as pupae in soil. The overwintering generation adults emerge in late May next year. Eggs are dispersedly laid on leaf surface. Egg stage is 6-8 days. The larval stage of the two generations occur in May to July and in August to September, respectively. The first generation adults appear in July to September.

Adults are attracted to light. When at rest, larvae do not raise their heads and abdomen. The first generation larvae mostly spin cocoons at the base of walls, under bricks and stones, and beside tree roots, while the overwintering generation larvae mostly bore into the soil and pupate.

Prevention and treatment

1. Silviculture. Turn up the soil in winter. Dig the soil near the wall and beside tree roots to kill the overwintering pupae.

2. Manual control. Hand remove young larvae and eggs.

3. Trap adults with light traps.

4. Protect and utilize natural enemies such as mantis, *Trichogramma closterae* and *Exorista civilis*.

5. Chemical control. Spray 5.7% emamectin at 3,000-5,000×.

References   Chinese Academy of Sciences - Institute of Zoology 1978, Editorial Board of the Hebei Forest Insects 1985, Sun Yujia 1989, Xiao Gangrou (ed.) 1992, Zheng Leyi and Gui Hong 1999, Wu Chunsheng and Fang Chenglai 2003, Wu Shiying 2005, Xu Gongtian and Yang Zhihua 2007.

(Wang Jinli, Li Jing, Zhao Huaiqian)

448 Syntypistis cyanea (Leech)

Classification Lepidoptera, Notodontidae

Synonyms *Quadrialcarifera cyanea* (Leech), *Quadrialcarifera fransciscana* Kiriakoff, *Somera cyanea* Leech

Importance It is a main defoliator of *Carya cathayensis*. Outbreaks occurred periodically (once per 10 years). They cause fruits of *C. cathayensis* drop prematurely and limb die back.

Distribution Zhejiang, Fujian, Jiangxi, Hubei, Guangdong, Yunnan, Taiwan; Japan, Vietnam, Korea Peninsula.

Host *Carya cathayensis*.

Damage Larvae are voracious feeders of leaves. They can devour leaves of the hosts in large areas.

Morphological characteristics **Adults** Body 20-25 mm in length, male slight smaller than female. Male 39-46 mm in wingspan, female about 50 mm in wingspan. Forewing slightly light reddish brown, mixed with yellowish white and yellowish green scales, greyish white along costal margin, inner and outer margins light brown and indistinct. Hind wing greyish brown, costal margin dark, with an obscure outer band. Antennae plumose, the end filiform. **Eggs** Round, size similar to rape seed, yellow. **Larvae** Body 25-40 mm long, young larvae turquoise before the 3rd instar, yellowish green after the 4th instar. Mature larvae with a red or violet dorsal line, spiracles red, epiproct red. Head pinkish green, with a white small granule; a yellow ring present between the head and the thorax. **Pupae** 20-30 mm in length, yellowish-brown or dark brown.

Biology and behavior *Syntypistis cyanea* occurs four generations a year. In late September to early October, mature larvae bore into loose and moist soil, pupate, and overwinter. Adults emerge in mid-April next year. Eggs are laid on the underside of leaves, few are laid on barks. Eggs are arranged in egg clutches, each egg clutch has 10-150 eggs. Each female lays 50-500 eggs. It takes 5-7 days for eggs to hatch. Newly hatched larvae are gregarious and feed near the egg clutches. They feed on leaf margin and cause incisions. After the 3rd instar, larvae on

S. cyanea adult

S. cyanea eggs

whole leave. It takes approximately 25 days for larvae to become mature. Larvae can be seen crawling up and down on tree trunks during 8:00-10:00. Mature larvae crawl into the soil and pupate. Larval periods of the four generations are: early May to late June, mid-July to late July, early August to late August, and early September to early October, respectively. Adults have strong phototaxis, and they hide on trunks and mate on the same night or next night after emergence. Adult appearance periods of the four generations are early and mid-April, late June, late July, and late August, respectively.

Prevention and treatment

1. Biological control. Release *Trichogramma* wasps to control eggs. During larval stage, apply *Beauveria bassiana*.

2. Trap adults with black light traps.

3. Apply a 10 cm wide insecticide band on tree trunks. The insecticide can be dimethoate and grease at 1:4 ratio.

4. Spray 2.5% fenvalerate at 3,0000× or 80% DDVP at 1,000-1,500× to control larvae. For large area infestations, apply trichlorphon and malathion fogging at dawn or at dusk on cloudy and moist days.

References   Wu Chunsheng and Fang Chenglai 2003, Hu Guoliang and Yu Caizhu 2005.

(Wang Yiping, Li Zhenyu)

449 *Stauropus alternus* Walker

Classification	Lepidoptera, Notodontidae
Synonym	*Neostauropus alternus* Kiriakoff
Common name	Lobster caterpillar

Distribution Guangdong, Guangxi, Hainan, Yunnan, Hong Kong, Taiwan; India, Sri Lanka, Myanmar, Indonesia, the Philippines, Vietnam, Malaysia.

Hosts *Casuarina equisetifolia*, *Citrus reticulate*, *Mangifera indica*, *Euphoria longana*, *Camellia sinensis*, *Coffea Arabica*, *Cassia javanica*, *Cassia glauca*, *Cassia fistula*, *Acacia confuse*, *Phoenix hanceana*, *Anacardium occidentale*.

Damage *Stauropus alternus* feed on branches and can cause significant damage to hosts. Outbreaks of *Stauropus alternus* occurred in Wenchang County of Hainan province during 1975-1977. They caused severe damage to 667 ha areas of *Casuarina equisetifolia*.

Morphological characteristics **Adults** Females 24-32 mm in length, 55-67 mm in wingspan, antennae filiform; males 20-22 mm in length, 38-46 mm in wingspan, antennae plumose. Head and thorax brownish gray, abdominal dorsal greyish brown, the last four segments light in color and close greyish white. Male forewing greyish brown, dark in color between the inner and outer cross lines, outer margin pale brown. Female forewing greyish and reddish-brown, female and male forewing bases brownish gray. Male hind wing costal margin area and inner margin area dark brown, the other areas greyish white; female hind wing greyish and reddish brown. **Eggs** Oblate, 1.2-1.5 mm wide, light yellow. Top surface slightly flat, the center with a round pit, surface covered with reticulate punctures. **Larvae** The 1st and 2nd instars 5-10 mm long, reddish brown, looks like an ant. The 3rd instar larvae 10-15 mm long, dark brown, the dorsal of the 5th and 6th abdominal segments with a distinct white spot. The 4th instar larvae 15-23 mm long, thorax with a distinct white line, head capsule with 2 longitudinal bands. The 5th instar larvae 22-33 mm long, the dorsal line white. The 6th instar larvae 28-45 mm long. The 7th instar larvae 40-45 mm long, dull yellow, orange, blackish green or greyish-white; head capsule black, mid-leg long and thin and is 5 times as long as the foreleg; the dorsal of the 1st to the 5th abdominal segments with a pair of granules. Pygidium enlarged, caudal legs appear as caudal horns. When at rest, the body looks like a boat. **Pupae** Male 19-26 mm in length, female 22-29 mm in length. Bright yellow at beginning, later turn into reddish brown, dark brown before eclosion.

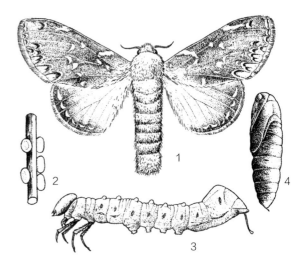

S. alternus. 1. Adult, 2. Eggs, 3. Larva, 4. Pupa
(Zhu Xingcai)

Biology and behavior *Stauropus alternus* occurs 6-7 generations a year in Wenchang County of Hainan province and has no overwintering phenomenon. The development time of the 1st generation eggs, larvae, prepupae and pupae are: 8-9, 23-30, 2-3, and 12-14 days, respectively. Those for the 3rd to 5th generations are: 5, 22-26, 2, and 8-9 days, respectively. Larvae have 7 instars in total. The 1st to the 7th instar development time for the 1st generation are: 1-3, 2-4, 3-4, 3-4, 3-4, 4-5, and 5-8 days, respectively.

The newly hatched larvae are gregarious. They bite small branches and cause incision. Larvae bite off small branches after the 3rd instar. Larval feeding increases significantly after the 4th instar. They mostly feed from the middle or close 1/3 from the base of branchlets and bite off the branches. This special feeding method greatly increases the severity of its damage.

The spatial distribution of the larvae in the crown was found to be: the lower layer -50%, the middle layer -40%, and the upper layer -10%. Mature larvae spin yellowish brown oval cocoons on branches. Adults mostly emerge during the first half of the night. Generally, males

S. alternus adult (Zhang Peiyi)

S. alternus cocoon in leaf pod (Zhang Peiyi)

S. alternus larva (Zhang Peiyi)

S. alternus larva (Zhang Peiyi)

emerge earlier than females. Adults mate the same night after emergence. Mating mostly takes place during 4:00-5:00. Each mating lasts around 40 minutes. Adults only mate once. Adults begin to oviposit at dusk after mating and eggs are mostly laid on small branches and arranged in moniliform. Each female lays 122-297 (average 176) eggs. Adults hide on trunks during the day. About 90% of them on trunks below 2 m. They are strong fliers. Male to female ratios of one overwintering generation was 1.7:1, and one of the 2nd generation was 4:1. Male longevity is 5-15 days, while female longevity is 5-13 days.

When it occurred in Wenchang County of Hainan province, it spread from north to south. It infested vigorous woods of *Casuarina equisetifolia* along seaside in 3 years. The population crashed and reoccurrence never happened. The pruned and thinned forest stands suffered more damage than the unpruned forest stands, and population density on forest edge was higher than in the woods. The insect preferred the windward and sunny sites. This is one reason why it mostly occurred in the seaside.

Prevention and treatment

1. Conservation of natural enemies. The known parasitoids include two species of egg parasites, two species of braconid, Xanthopimpla predator, one species of Brachymeria, and one species of Tachinidae. Predators include *Evarcha* spp., praying mantis, ants, and birds. In Huwei forest farm of Wenchang County in 1975, the parasitic rate was above 90%. As a result, the pest population fell sharply next year. Pest outbreak occurred over thousands of acres of forests in Daodong forest farm in 1977, the pest became almost extinct the next year as a result of disease epidemics.

2. During larval stage, apply 50% fenitrothion EC at 1,500-2,000×.

References Chen Zhiqing 1977, Chen Zhiqing et al. 1982.

(Chen Zhiqing Wu Shixiong)

450 *Phalera flavescens* (Bremer et Grey)

Classification Lepidoptera, Notodontidae

Synonyms *Pygaera flavescens* Bremer et Grey, *Trisula andreas* Oberthür, *Phalera flavescens kuangtungensis* Mell

Importance *Phalera flavescens* is omnivorous. It often causes significant damage to horticulture, forestry and landscaping, and results in ecological and economic losses.

Distribution All provinces in China except Xinjiang, Guangxi, Guizhou, Tibet, Gansu, Qinghai, and Ningxia; Korea Peninsula, Russia, Japan, Myanmar.

Hosts *Malus* spp., *Pyrus* spp., *Eriobotrya* spp., *Photinia* spp., *Prunus* spp., *Crataegus* spp., *Pyracantha fortuneana*, *Castanea mollissima*, *Ulmus* spp.

Damage Larvae feed on leaves. They often consume all leaves of the hosts, affecting the fruit yield and growth of the host.

Morphological characteristics **Adults** Males 18-23 mm in length, 35-50 mm in wingspan. Females 18-26 mm in length, 45-55 mm in wingspan. Head and thoracic dorsal light yellowish white, abdominal dorsal yellowish-brown. **Larvae** Mature larvae about 50 mm long. Head black and glossy. Thoracic and abdominal dorsal purple black, ventral surface red-violet. Body densely covered with grayish and yellowish white long hairs. Head and abdomen tilted upward when at rest.

Biology and behavior *Phalera flavescens* occurs one generation a year and overwinters as pupae. In north China, adults begin to emerge in mid- and late June next year. In south China, they emerge in early and mid-August. The egg stage is 7-10 days, and larva stage is around 40 days. Larvae mature in mid- and late August in north China. They mature in late September to late October in South China.

Adults hide during the daytime and come out at night. They exhibit strong phototaxy. They lay eggs on the underside of leaves. The eggs are arranged in monolayer egg clutches. The 1st and 2nd instar larvae are gregarious

P. flavescen adult (Xu Gongtian)

P. flavescen adult
(Li Zhenyu)

and feed on the mesophyll. After the 3rd instar, larvae scatter. Larvae feed in the early morning and at night. They spin silk and fall from the tree when they are frightened. After maturity, they pupate in 5-7 cm deep soil near the tree base.

Prevention and treatment

1. Manual control. Cut off branches and leaves along with the gregarious young larvae. Shake larvae off from trees and destroy them. Turn up the soil in the fall or in spring to kill the overwintering pupae.

2. Light trapping. During adult emergence period, set up light traps to trap adults.

3. Biological control. During larval damage period, spray *B.t.* When mature larvae crawl to soil to pupate, apply *Beauveria bassiana* to the ground under tree crowns, and then loosen the soil layer to facilitate infection of the pathogen.

References   Cai Rongquan 1979; Zhang Hongxi, Cao Zhongchen, Zhao Xianglan et al. 1996; Wu Chunsheng and Fang Chenglai 2003.

(Tang Yanping, Yang Chuncai)

451 *Phalera takasagoensis* Matsumura

Classification	Lepidoptera, Notodontidae
Synonyms	*Phalera takasagoensis ulmovora* Yang et Lee, *Phalera takasagoensis matsumurae* Okano
Common name	Narrow yellow-tipped prominent

Distribution Beijing, Hebei, Liaoning, Jiangsu, Zhejiang, Fujian, Jiangxi, Henan, Hunan, Shaanxi, Gansu, Taiwan; Japan, Korea Peninsula.

Hosts *Ulmus* spp., *Pyrus bretschneideri*, *Castanea mollissima*, *Prunus pseudocerasus*.

Damage Larvae feed on leaves and may consume all leaves of the host.

Morphological characteristics Adults Females: 20-24 mm in length, 53-60 mm in wingspan. Males: 18-23 mm in length, 42-53 mm in wingspan. Yellowish brown. Forewing greyish-brown, apex angle with an obvious light yellow spot, margin black. **Larvae** Body 50 mm long. The back with a longitudinal cyan black stripe. The subdorsal lines, the upper spiracle line and the lower spiracle line white; the ventral line yellowish white; the middle of each segment with a red loop, which is covered with dense light yellow long hairs.

Biology and behavior *Phalera takasagoensis* occurs one generation a year in Henan Province and overwinters as pupae in soil. Adults emerge in May and June

P. takasagoensis adult

P. takasagoensis adult (Ji Baozhong)

P. takasagoensis adult female (NZM)

P. takasagoensis adult male (NZM)

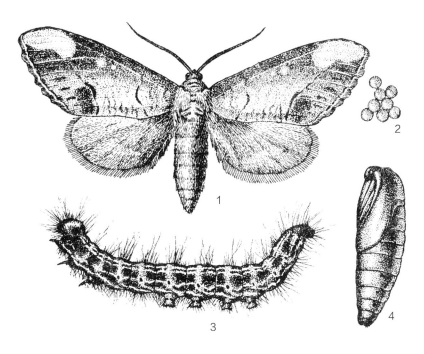

P. takasagoensis. 1. Adult, 2. Eggs, 3. Larva, 4. Pupa
(Zhang Xiang)

P. takasagoensis larva (Xu Gongtian)

next year. They are attracted to light. Eggs are laid on the lower surface of leaves. They are arranged in monolayer egg clutches. When larvae are resting, their heads face one direction and their abdomen raised. When they are frightened, they spin silk and fall. They cause most serious damage in August and September. Larvae migrate to the ground and pupate in soil.

Prevention and treatment Chemical control. Spray 12% thiamethoxam and lamda-cyhalothrin at 1,000-1,500×, to control the larvae.

References Yang Youqian et al. 1982, Xiao Gangrou (ed.) 1992, Forest Disease and Pest Prevention and Treatment and Quarantine Station of Henan Province 2005, Wu Chunsheng and Fang Chenglai 2010.

(Yang Youqian, Fan Zhongmin)

452 Agrotis tokionis Butler

Classification	Lepidoptera, Noctuidae
Synonym	*Trachea tokionis* (Butler)
Common names	Giant cutworm

Importance It is one of the most important underground pest for agriculture and forestry in China.

Distribution All provinces in China, mainly occurs along the mid- and lower part of the Yangtze River. It often occurs together with *Agrotis ipsilon*. It also distributes in Russia, Korea Peninsula and Japan.

Hosts *Gossypium* spp., *Zea mays*, *Sorghum vulgare*, *Ipomoea batatas*, Fabaceae, Brassicaceae, Cucurbitaceae, *Cannabis* spp., *Nicotiana tabacum*, *Camellia sinensis*, *Populus* spp.

Damage Larvae cause damage to leaves and stems of plants through feeding.

Morphological characteristics **Adults** Body 20-25 mm in length, 43-58 mm in wingspan; head and thorax brown, the 2nd segment of labipalp with a black spot. Abdomen and forewing greyish brown. Costal area and discal cell inside the outer cross line dark brown; base lines are double, brown and reach the sub-median fold; inner cross lines are double, wavy and black; sword mark with narrow black margin, the round mark and the reniform mark brown and with black margin; mid-cross line brown; outer cross lines are double, zigzag, and brown; submarginal line zigzag, light brown; marginal line with a row of black dots. Hind wing light yellowish brown.

Eggs Hemispheric, 1.8 mm in diameter, 1.5 mm in height, light yellow in early stage, later gradually turn into yellowish brown, greyish brown before hatch.

Larvae Body about 40-62 mm in length, yellowish brown, with many wrinkles, granules not obvious. Head brown, the middle with a pair of dark brown longitudinal stripes. clypeus triangle, the base longer than other sides. The front pair of the two pairs of pinacula on dorsal side of the 1st to 8th abdominal segment are similar to or slightly smaller than the rear pair in size. Spiracle oblong, black. Pygidium dark brown except that the yellowish brown area beside the two setae near the end.

Pupae 22-29 mm in length, light yellow in early stage, later turn into yellowish brown. The front margin of 4th to 5th segments densely covered with punctures. Cremaster of abdominal end triangle, with a pair of short spines, black.

Biology and behavior *Agrotis tokionis* occurs one generation a year and overwinters as larvae in weeds and on soil surface of green fields. It emerges from underground and causes damages in early March in Yangtze River basin. It often occurs with *Agrotis ipsilon* in April to May. When air temperature is higher than 20°C, it will enter diapause. Pupation occurs in mid-September.

A. tokionis adult (Xu Gongtian)

A. tokionis adult (Xu Gongtian)

Adults emerge in early and mid-October. Adult occurrence period is more than one month. Some adults can still be found in mid-November. The peak oviposition period is at the end of mid-October. Each female is able to lay 1,000 eggs. The peak of egg hatch period is in early November. Egg stage is 11-24 days, and larval stage is more than 30 days. Larvae go through 6-7 instars. The 1st to 3rd instar larvae hide below soil surface under leaves of hosts or between leaf layers during the daytime, and feed on leaves and form many small transparent windows at night. Later instars will bite off portions of leaves or the stems. The newly hatched larvae can live 6.5-10.5 days (average 8.5 days) without feeding. The 3rd instar larvae overwinter in soil. When soil temperature rises above 6°C, overwintering larvae resume feeding. April is the peak damage period in a year. At the end of mid-June, mature larvae cease feeding, and about 15 days later, they build soil chambers at 3-5 cm below soil surface and over summer for about three months.

Pupae are more abundant in corn and soybean fields than in other fields. Pupal stage is 25-32 days. Adult phototaxis is strong. They are attracted to sugar, vinegar, wine, and water solution at the ratio of 6:3:1:10. Adults mate and oviposit at night, each female lays 34-1,084 eggs (average 454 eggs). About 54% of the eggs are laid during the first two days. Female longevity is 2-14 days (average 7.8 days), while male longevity is 3.5-19 days (average 11.5 days). Eggs are laid on soil surface (44.9%), leaf litter (20.4%), or plants (34.7%).

Prevention and treatment

1. Agricultural control. During the peak oviposition period, remove weeds from the field and dispose them; ploughing to reduce the number of larvae and eggs.

2. Manual control. In fields of low *A. tokionis* population density, manually catch the larvae in the early morning.

3. Chemical control. Prepare 15% chlorpyrifos and phoxim mixture treated bait using crushed rapeseed. Spread the bait to the base of maize and cotton seedlings at 60-75 kg bait per hectare at dusk. The control effect was above 90%. After maize and cotton seeds germinated, spray the ground with 12% thiamethoxam and lamda-cyhalothrin at 1,000-1,500×,

References Zhu Hongfu et al. 1964; Chen Wenkui 1985; Wei Hongjun, Zhang Zhiliang, and Wang Yinchang 1989; Xue Fangsen, Shen Rongwu, and Zhu Xingfen 1990; Chen Chang 1999a,b.; Xu Gongtian and Yang Zhihua 2007.

(Shi Juan)

453 *Agrotis ipsilon* (Hufnagel)

Classification Lepidoptera, Noctuidae
Synonyms *Agrotis ypsilon* (Rott.), *Noctua suffusa* (Denis et Schiffermüller)
Common names Black cutworm

Importance It is a common pest in China. Infestation is serious in flood land and irrigated land along rivers and lakes in the south. It is a major pest of crops, fruit trees and tree seedlings.

Distribution All provinces in China; U.S.A., Europe, Canada, Japan, New Zealand, South Africa, South America, and the Pacific.

Hosts *Camellia sinensis*, *Populus* spp., *Larix* spp., *Cryptomeria fortune*, *Pinus koraiensis*, *Platycladus orientalis*, *Thujopsis dolabrata*, *Pinus sylvestnis* var. *mongolica*, *Fraxinus mandschurica*, *Juglans mandshurica*, *Pinus massoniana*, *Pinus elliottii*, *Pinus taeda*. Most crops are also attacked.

Damage Larvae attack seeds and seedlings.

Morphological characteristics **Adults** Body 17-23 mm in length, 40-54 mm in wingspan. Head, dorsal side of thorax dark brown, legs brown, foreleg tibia and outer edge of tarsus greyish brown, the end of each segment of mid- and hind legs with greyish brown stripe. Forewing brown; costal margin area dark brown; the area inside the outer margin mostly dark brown; base lines are double, slight brown, not wavy; inner cross lines are double, black, wavy; the circular mark, reniform spot, and sword pattern black, obvious. Hind wing greyish white.

Eggs Steamed bun shape; 0.5 mm in diameter, 0.3 mm in height; the surface with longitudinal and cross carinae; milky at the beginning, later turn into yellow.

Larvae Cylindrical; body 37-50 mm in length, 5-6

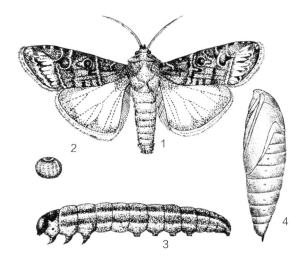

A. ipsilon. 1. Adult, 2. Egg, 3. Larva, 4. Pupa
(Yu Changkui)

mm in width. Head brown, with dark chocolate-brown irregular reticular pattern; body greyish brown to dark brown; body surface coarse, closely covered with granules which are uneven size and separated from each other. Dorsal line, subdorsal line and spiracle line dark brown, but not obvious. Pronotum dark brown, pygidium yellowish brown and with two obvious dark brown longitudinal bands. Thoracic legs and prolegs yellowish

A. ipsilon adult (Xu Gongtian)

A. ipsilon adult (Xu Gongtian)

brown.

Pupae Body 18-24 mm in length, 6-7.5 mm in width; reddish brown, glossy, with two cremasters. End of mouthparts and wing buds reach the rear margin of 4th abdominal segment. The first five abdominal segments cylindrical, their width almost same as that of the thorax.

Biology and behavior *Agrotis ipsilon* occurs 3-4 generations a year. The overwintering generation adults begin to appear in late March to early April, and reach peak in late April. They may also appear around October. Adults hide during the daytime and come out at night, they hide in soil cracks or under withered grass. Peak activity is during 19:00-22:00. Adults begin to oviposit 4-6 days after emergence. Eggs are mostly laid on weeds that are short and have dense leaves. Few eggs are laid on dead leaves or in soil cracks. Eggs are generally laid on leaves close to the ground. The highest oviposition site is 13 cm high. Adults are strongly attracted to black lights and have strong chemotaxis. They are attracted to sour, sweet, and liquor odor. Preferred habitat is moist and sticky soil. High temperature is disadvantageous for the growth of *Agrotis ipsilon*. When relative humidity is less than 45%, egg hatch rate and larvae survival rate were very low. Larvae overwinter mainly in un-ploughed fields with abundant weeds. Larvae have 6 instars, sometimes they go through 7-8 instars. The 1st to 2nd instar larvae are gregarious on weeds or tender leaves of seedlings. They feed both during the day and night. They disperse after the 3rd instar. The damage is severe when heavy dew is present before dawn. Larvae drag the cut-off stems of seedlings into soil for food. After seedlings lignify, larvae feed on tender leaves. After the 4th instar, larvae hide between wet and dry layers of soil surface in the daytime, come out at night, and bite off seedlings on the ground, then drag them into soil holes, or feed on non-germinated seeds.

The occurrence time vary with climates of various regions, but all serious damages are done by the 1st generation larvae. The 1st generation larval stage is generally 30-40 days. Mature larvae mostly hide in 5 cm deep soil to build soil chamber to pupate, and pupal stage is about 9-19 days.

Prevention and treatment

1. Cultural control. During winter, sweep fallen leaves on the ground into the ditch, cut off branches with diseases and pests, scrape off diseased barks and bury them.

2. Sex pheromone trap. Field experiments show three components Z7-12:Ac, Z9-14:Ac, and Z11-16:Ac are attractive to *A. ipsilon*. Installation of traps baited with this mixture is effective in reducing *A. ipsilon* damage.

3. Chemical control. The 1st- 3rd instars are susceptible to insecticide treatment as they are exposed on ground surface. Spray 40% chlorpyrifos and phoxim mixture at 800-1,000. Apply 500-750× B.*t.* strain SC at 2,000 IU/μL to vegetable roots is also effective.

References Wu Xinmin 1959; Li Yongxi 1964; Zhai Yongjian 1966; Yang Xiuyuan and Wu Jian 1981;Wei Hongjun, Zhang Zhiliang, and Wang Yinchang 1989; Xiao Gangrou (ed.) 1992; Picmibon et al. 1997; Chen Yixin (ed.) 1999; Gemeno et al. 2000; Li Fang, Chen Jiahua, and He Rongbin 2001; Xu Gongtian and Yang Zhihua 2007; Guo Xiuzhi, Deng Zhigang, and Mao Hongjie 2009; Xiang Yuyong, Yang Maofa, and Li Zizhong 2009; Qiu Shumei, Luo Runquan, and Yaoyong 2010; Zheng Xu 2010.

(Shi Juan, Wu Cibin, Fan Zhongmin)

454 *Agrotis segetum* (Denis and Schiffermüller)

Classification Lepidoptera, Noctuidae
Synonyms *Euxoa segetum* Schiffermüller, *Agrotis segetum* Schiffermüller
Common names Cutworm, Turnip moth

Distribution All provinces in China except in Hainan, Guangxi, and Guangdong; Africa, Europe, India, Japan, Korea Peninsula.

Hosts *Triticum aestivum*, *Brassica campestris*, *Raphanus sativus*, *Spinacia oleracea* and many other crops and landscape plants.

Damage Larvae feed on leaves and stems of plants and kill plants by breaking off the stems.

Morphological characteristics **Adults** Body 15-18 mm in length, 32-43 mm in wingspan; yellowish brown. Forewing with inconspicuous base line, inner and outer cross lines and mid-cross line; the reniform pattern and rod-like pattern obvious, without a wedge pattern; dark chocolate-brown side. Hind wing white, slightly yellowish brown.

Eggs Length 0.44-0.49 mm, width 0.69-0.73 mm. Oblate, yellowish brown, the surface with cross and longitudinal ridges.

Larvae Length 35-45 mm, width 5-6 mm. Yellow, glossy, body surface with many wrinkles; back of the abdomen with 4 pinacula, those on the front larger; a yellowish brown spot present on each side of the pygidium.

Pupae The dorsal side of the 4th abdominal segment with small and sparse punctures, punctures on the 5th- to 7th abdominal segments small and dense.

Biology and behavior The *Agrotis segetum* occurs two generations a year in northern Xinjiang; three generations in Hebei, Inner Mongolia, Shaanxi, Gansu and southern Xinjiang and Huang Huai area; and 3-4 generations in Shandong. It overwinters as pupae or mature larvae in soil about 10 cm depth. In northern Xinjiang, 89.2% of the population overwintered as mature larvae and a small percentage overwintered as the 4th - 5th instar larvae. In Inner Mongolia and Shandong, the peak damage period is in May to June. There two damage periods (spring and autumn) in Xinjiang. Adults lay eggs on plants above ground or in soil where vegetation is sparse. In general, 3-4 eggs are laid on the lower surface of each

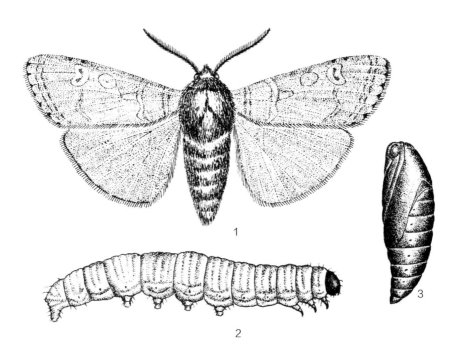

A. segetum. 1. Adult, 2. Larva, 3. Pupa (Zhang Peiyi)

A. segetum adult (Xu Gongtian)

A. segetum adult (Xu Gongtian)

leaf. Occasionally, eggs are laid on the upper surface of leaves or leaf tips or young stems. Adults have phototaxis and are attracted to sugar and vinegar solution.

The 1st instar larvae feed on the mesophyll, leaving the epidermis intact. They cluster on the tender shoots. The 2nd instar larvae are able to break off the young tips. The 3rd instar larvae often break off young stems. The 4th instar and older larvae will bite off young stems near the ground. The 6th instar larvae can 1-3 seedlings per larvae in one night. Larvae can still bore into the hardened stems around the circumference and cause withering and death of the plants.

Larvae stop feeding and overwinter when daily mean temperature is 2.8-2.9 ℃ in late November to early December. Larval density varies with the landscape characteristics and crop species. Preferred crops include *Triticum aestivum*, *Brassica campestris*, *Raphanus sativus*, and *Spinacia oleracea*. Higher density occurs at ridges and edges of fields where weeds are abundant. Larvae overwinter in soil at 3-7 cm depth and up to 14 cm depth. Pupae appear in March and April next year. Peak of adult emergence occurs between late April and early May.

Prevention and treatment

1. Attract and kill adults. Using black light traps or sugar-vinegar-liquor mixture to attract adults during the peak of adult stage is an effective and convenient control method.

2. Eliminate weeds. Areas with weeds are the main oviposition sites of *Agrotis segetum*. Weeds provide food to young larvae of *A. segetum*. Clear weeds before spring seedlings emerge or when larvae are in the 1st to 2nd instars. The weeds should be promptly burned or transported to compost site.

3. Manually catch and kill larvae. In early morning when broken seedlings are found, dig the soil and kill the larvae.

4. Biological control. The *Agrotis segetum* granulosis virus (AsGV) can be used as a biological insecticide.

5. Chemical control. Spray crops with 300× 20% thiamethoxam, or 1,000× 45% profenofos and phoxim mixture. Use a mixture of 25-40 kg fresh grass and 40% chlorpyrifos and phoxim mixture at 800-1,000× to attract and kill larvae.

References Li Zuolong 1964; Dong Jiantang 1983; Wei Hongjun, Zhang Zhiliang, Wang Yinchang 1989; Xia Zhixian and Ding Fulan 1989; Xiao Gangrou, 1992; Ai Xiulian, Ji Qing, Long Tao et al. 1995; Chen Yixin 1999; He Mei and Liu Zhuangjun 2000; Xu Gongtian and Yang Zhihua 2007; Yin Feifei, Wang Manli, Tan Ying et al. 2008; Nakanishi, Goto, Kobayashi et al. 2010.

(Shi Juan, Wu Cibin, Fan Zhongmin)

455 *Apamea apameoides* (Draudt)

Classification Lepidoptera, Noctuidae
Synonyms *Oligia apameoides* Draudt, *Oligia vulgaris* Butler

Distribution Zhejiang, Jiangsu, Anhui, Fujian, Jiangxi, Henan, Hubei, Hunan, Guangdong, Guangxi, Sichuan, Guizhou, Yunnan, Shaanxi, Taiwan; Japan.

Hosts *Phyllostachys iridescens*, *P. sulphurea* 'Viridis', *P. glauca*, *P. praecox*, *P. vivax*, *P. meyeri*, *P. praecox* f. *provernalis*, *P.e dulis* f. *gimmei*, *P. dulcis*, *P. decora*, *P. elegans*, *P. bambusoides* f. *lacrima-deae*, *P.b ambusoides*, *P. heteroclada* var. *pubescens*, *P. bissetii*, *P. nuda*, *P. pubescens* f. *gracillis*, *P. aureosulcata* f. *pekinensis*, *P. aureosulcata* f. *pekinensis*, *P. glabrata sinobambusa*, *P. heteroclada*, *P. nidularia* f. *galbro-vagina*, *P. sulphurea*, *Pleioblastus amarus*. Larvae also feed on intermediate host such as grass family including *Roegneria ciliaris*, *Roegneria japonensis*, *Pennisetum alopecuroides*, *Alopecurus aegualis*, *Avena tatua*, *Cleistogenes caespitosa*, *Imperata cylindrical* var. *mojor*, *Eragrosis pilosa*, var. *imberbis*. They also feed on plants of Cyperaceae including *Carex lanceolata*, *Carex brownie*, *C. breviculmis* var. *fibrillose*, *C. glaucaeformis*, and plants of Equisetaceae including *Equisetum hiemale*, *E. arvense*, *E. ramosissimum*, *E. palustre*, *E. telmateia*, and *Funaria sinensis*.

Damage Most damage occurs on small bamboo shoots. Larvae feed within shoots. Most damaged bamboo shoots can't grow into bamboo. The light damaged bamboo shoots are able to grow into bamboo, but the bamboo texture will be hard and brittle.

Morphological characteristics **Adults** Females 14-20 mm in length, 36-48 mm in wingspan; males 11-16 mm in length, 30-39 mm in wingspan. Body brown, female compound eyes dark brown, male eyes light brown. Antennae filiform, greyish-yellow. Scales on prothoracic tegula thick and long. Forewing purplish brown, margin hairs neatly arranged, with slight notches; reniform mark yellowish white, those of the males' are more obvious than those of the females; outer and inner cross lines black, double wavy; terminal line, circular lines light yellow, but not obvious. Hind wing ash black, wing base light in color. **Eggs** Oblate, major axis 0.66-0.74 mm, 0.46-0.54 mm in height, the top slightly concave, with reticulate sculptures. Milky at the beginning, later turn into light yellow. Eggs are neatly arranged on

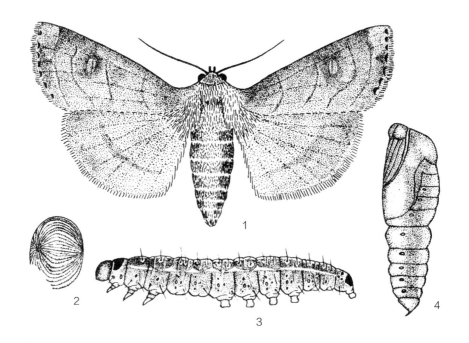

A. apameoides. 1. Adult, 2. Egg, 3. Larva, 4. Pupa (Xu Tiansen)

leaf edges. Leaves with eggs curl up and cover the eggs. **Larvae** The newly hatched larvae 1.5 mm long, light yellowish brown, sparsely covered with white setae. Mature larvae 26-40 mm long, light purplish brown or purplish grey. Head orange red. Dorsal line very thin, subdorsal line thick; all dirty white; subdorsal line has many broken points. Prescutum and pygidium pitch black. There are 6 black spots in front of the pygidium, the two spots in the middle bigger. **Pupae** Length 11-21 mm female pupae bigger than male pupae; reddish brown; with 4 cremasters, the two middle cremasters slightly thicker and longer.

Biology and behavior *Apamea apameoides* occurs one generation a year and overwinters as eggs. Eggs hatch from early February to the end of March in Henan province. Larvae bore into bamboo shoots in early April to mid-May. They become mature in mid-May to mid-June. Adults emerge from early June to early July. Eggs hatch from early February to early April in Zhejiang. Larvae crawl into weeds to feed after hatch. They migrate to bamboo shoots in mid-April to early May. Larvae mature and spin cocoons in mid- and late May. Adults appear from early June to early July.

Prevention and treatment

Protect natural enemies. Adult predators include various birds, a spider (*Lycosa pseudoannulata*), and a ground beetle (*Calathus* sp.). *Trichogramma* spp. species are found as egg parasitoids. Toads and *Calathus* sp. may kill young larvae when they migrate from weeds to bamboo shoots. Centipedes may follow young larvae into bamboo shoot sheathes and kill the young larvae. *Apanteles* sp. and *Exorista* sp. parasitize larvae and pupae, but the parasitization rates are low.

References Chinese Academy of Sciences Institute of Zoology 1987, Xu Tiansen and Wang Haojie 2004.

(Xu Tiansen, Wang Haojie)

A. apameoides cocoon and pupa (Xu Tiansen)

A. apameoides male (above) and female (below) (Xu Tiansen)

A. apameoides larva (Xu Tiansen)

456 *Kumasia kumaso* (Sugi)

Classification Lepidoptera, Noctuidae
Synonym *Apamea kumaso* Sugi

Distribution Shanghai, Jiangsu, Zhejiang, Anhui, Fujian, Jiangxi, Hubei, Hunan, Guangdong, Guangxi, Sichuan, Guizhou, Yunnan; Japan.

Hosts *Phyllostachys dulcis, P. glauca, P. glabrata sinobambusa, P. praecox, P. aureosulcata* f. *pekinensis, P. iridescens, P. heteroclada* var. *pubescens, P. vivax, P. rubromarginata, P. aureosulcata* f. *pekinensis, P. aureosulcata, P. nigra* var. *henonis, P. meyeri, P. nidularia* f. *galbro-vagina, P. bambusoides. Pseudosasa amabilis, Pseudosasa guanxianensis.*

Damage Larvae feed in bamboo shoots. The feeding galleries are filled with frass. Damaged bamboo shoots may become rotten and die.

Morphological characteristics **Adults** Females 17.5-20.5 mm in length, 40-45 mm in wingspan; males 15.5-18.5 mm in length, 38-41 mm in wingspan. Body light yellowish brown, antennae filiform, compound eyes dark brown. Hairs of frontal tufts, basal tufts and squamula long and thick. Forewing light brown, margin hairs wavy; terminal line light greyish white, inside of it is a row of triangle black small spots; subterminal line wavy; sword stripe dark brown; reniform stripe light brown, its inner edge dark brown and outer edge greyish white; circular stripe oval, with an obvious black margin; wedge stripe obvious, located below the circular stripe. Hind wing without spots, dark gray. Legs greyish brown, tarsi with light brown rings.

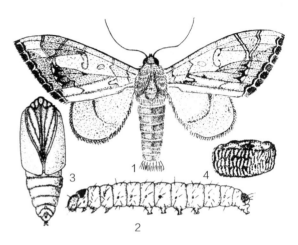

K. kumaso. 1. Adult, 2. Larva, 3. Pupa, 4. Egg
(Xi Rong)

Eggs Oblate, major axis 0.91-0.97 mm, minor axis 0.49-0.52 mm. Milky at beginning, later turn to light yellow, light brown before hatch.

Larvae The newly hatched larvae 1.5 mm long, light purplish brown. Mature larvae 34-48 mm long, body light grey purple, head orange, pronotum shield black. Body smooth, with an indistinct light colored dorsal line, pronotum and pygidium yellowish brown, spiracles black, crochets uniordinal and single row.

Pupae Length 18-21 mm, reddish brown, with 4 cremasters, the two middle cremasters slightly longer.

Biology and behavior *Kumasia kumaso* occurs one generation a year and overwinters as eggs. In Jiangsu, the overwintering eggs hatch in early and mid-April next year. Larvae bore into bamboo shoots in early and mid-May. They feed 15-25 days in bamboo shoots and then pupate in late May. Adults emerge in early and mid-June. In Zhejiang, eggs hatch in late March to early April. Larval feeding period is in early April to late May, and adult occurrence period is in early June to late June. In Guangdong, eggs hatch in late February to March. Larvae bore into bamboo leaf buds to feed. They are able to move to another leaf bud when they eat up a leaf bud. At late stage of the 2nd instar, larvae leave leaf buds to

K. kumaso mature larva (Xu Tiansen)

K. kumaso adult (Xu Tiansen)

K. kumaso cocoon and pupa (Xu Tiansen)

look for shelters such as leaf sheaths that are about to fall off. They spin several silk threads and attach themselves to the leaf sheaths or drop down to the ground to look for appropriate places to rest for 2-10 days. Larvae on bamboo drop to ground in late March. They enter bamboo shoots at the end of March or in early April. Larvae feed about 13 days in bamboo shoots, and then in mid-April larvae become mature and come out of bamboo shoots to pupate in soil. Pupal stage is about 18 days. Adults emerge from the end of April to mid-May. Adult longevity is 5-10 days.

Prevention and treatment

1. Strengthen bamboo forest management. Minimize weed growth, especially the Poaceae and Cyperaceae weeds, is the key to reduce pest populations. Remove small or damaged bamboo shoots to reduce the pest population.

2. Light traps. Adults of *K. kumaso* have phototaxis. Set up black light lamps to trap adults.

3. Protect natural enemies. *Trichogramma* sp. and tachinid flies are found parasitizing *K. kumaso* eggs or larvae.

4. Chemical control. In bamboo forests with high pest density, spaying 1,000× 80% DDVP EC diluent, 3,000× 2.5% deltamethrin, or 200× 8% cypermethrin liquid when bamboo shoots emerge.

References Xiao Gangrou (ed.) 1992, Xu Tiansen and Wang Haojie 2004.

(Xu Tiansen, Wang Haojie)

457 Sapporia repetita (Butler)

Classification Lepidoptera, Noctuidae
Synonyms *Apamea repetita* Butler, *Agrotis conjuncta* Leech

Distribution Jiangsu, Zhejiang, Fujian, Jiangxi, Henan, Hunan, Guangdong, Sichuan; Japan.

Hosts *Phyllostachys nidularia* 'Smoothsheath', *P. nigra* var. *henonis*, *P. iridescens*, *P. platyglossa*, *P. vivax* f. *huanvenzhu*, *P. sulphurea* 'Viridis', *P. glauca*, *P. vivax*, *P. makinoi*, *P. flexuosa*, *P. bambusoides* f. *lacrima-deae*, *P. heteroclada* f. *solida*, *P. heteroclada*, *P. nigella*, *Pleioblastus amarus*, *Pseudosasa amabilis*.

Damage Larvae bore into bamboo shoots of thin diameters and feed in them. Damaged bamboo shoots appear withered or die.

Morphological characteristics **Adults** Females 14-19 mm in length, 35-46 mm in wingspan; males 12-16 mm in length, 28-37 mm in wingspan. Body yellowish brown to dark brown. Female compound eyes dark brown, male compound eyes light brown. Antennae filiform, greyish yellow. Most part of the forewing dark brown, the rest area yellowish brown. Terminal lines are double, black; the inner line is made of black dots. There are two light spots from terminal line to the outer cross line, the spot near the costal margin lighter than another one, the two spots are separated by one black line. Reniform spot distinct, white in female and partially white in male. Margin hairs black, zigzag.

Larvae Mature larvae 26-38 mm long, light purplish brown or purplish grey. Head orange red. Pronotum black, divided into two parts by the dorsal line; dorsal line rust yellow or orange red, subdorsal line slight thick, white or dirty white, discontinued before the front end of 2nd abdominal segment. Pygidium black, also divided into two parts by the dorsal line.

Pupae Length 12-20 mm, female pupae larger than male pupae, reddish-brown, with 4 cremasters, the two middle cremasters slightly thicker and longer.

Biology and behavior *Sapporia repetita conjuncta* occurs one generation a year and overwinters as eggs. In Zhejiang, the overwintering eggs hatch in February to March. Larvae bore into bamboo shoots in early and mid- April, and feed in bamboo shoots until mid- and late May for about 25 days. Mature larvae leave bamboo shoots, construct thin cocoons in soil, and pupate in mid- and late May. Adult occurrence period is in early to late June. They are not active after emergence. They don't need to replenish nutrition before mating and oviposition.

Prevention and treatment

Protect natural enemies. *Apanteles* sp. and *Exorista japonica* are known parasitoids of the *A. repetita conjuncta* larvae.

Reference Xu Tiansen and Wang Haojie 2004.

(Xu Tiansen, Wang Haojie)

S. repetita 4th instar larva (Xu Tiansen)

S. repetita adult male (Xu Tiansen)

458 *Bambusiphila vulgaris* (Butler)

Classification Lepidoptera, Noctuidae
Synonym *Polydesma vulgaris* Butler

Distribution Shanghai, Jiangsu, Zhejiang, Anhui, Fujian, Jiangxi, Shandong, Henan, Hubei, Hunan, Guangdong, Guangxi, Sichuan, Yunnan, Shaanxi; Japan.

Hosts Most species in *Phyllostachys, Neosinocalamus affinis, Neosinocalamus affinis* 'Chrysotrichus', *Pleioblastus juxianensis, Pleioblastus simony, Pseudosasa guanxianensis*. Intermediate hosts include *Roegneria kamoji, Poa faberi* Rendle, *Poa acroleuca* Steud, *Avena fatua, Alopecurus aequalis, Eragrostis pilosa* var. *imberbis, Carex lanceolata, Riccaridia latifrons*, and *Carex brownie*.

Damage Larvae feed in bamboo shoots. Feeding causes shortened nodes, broken heads, broken shoots, heart rot, and brittle timber.

Morphological characteristics **Adults** Female body 17-25 mm in length, 35-54 mm in wingspan; male body 15-22 mm in length, 32-50 mm in wingspan. Body greyish brown, female body color lighter. Compound eyes dark brown. Antennae filiform, greyish yellow. Wings brown, margin hairs zigzag; terminal line black, with a row of 7-8 small black dots; reniform mark light yellow; an inverted dark brown triangular mark appears near the apex angle; apex angle yellowish white. Wing base dark brown, loop line obvious. Hind wing color light. Male wings greyish white, terminal line composed of 7-8 black dots. The inverted triangle spot near the

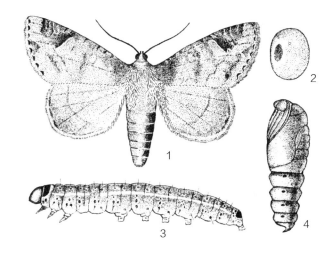

B. vulgaris. 1. Adult, 2. Egg, 3. Larva, 4. Pupa (Xu Tiansen)

B. vulgaris male (above) and female (below) (Xu Tiansen)

B. vulgaris larva (Xu Tiansen)

B. vulgaris pupa (Xu Tiansen) B. vulgaris larva and pupae of a parasitic fly (Xu Tiansen)

B. vulgaris damage (Xu Tiansen) B. vulgaris damage (Xu Tiansen)

apex angle light brown, hind wing greyish brown, wing base color light. Legs dark grey, the rear end of each tarsus with a light yellow spot.

Eggs Nearly round, major axis 0.72-0.88 mm, minor axis 0.65-0.81 mm, milky.

Larvae The newly hatched larvae about 1.6 mm in length, light purplish brown. Mature larvae 36-50 mm in length. Head orange red. Dorsal line and subdorsal line white, dorsal line very thin and distinct; subdorsal line wider. Pronotum and pygidium pitch black, divided into two parts by an orange red dorsal line from the middle. There are 6 small black spots on the dorsal surface of the 9th abdominal segment and the front of the pygidium. They are arranged in a triangle on two sides of dorsal line, the two spots near the dorsal line are bigger than the others.

B. vulgaris young larva (Xu Tiansen)   *B. vulgaris* damage (Xu Tiansen)

Pupae Length 14-24 mm in length, viridis in early pupal stage, later become reddish brown; with 4 cremasters, the two middle cremasters thick and long.

Biology and behavior *Bambusiphila vulgaris* occurs one generation a year and overwinters as eggs. In Zhejiang Province, adult emergence period is in mid-June to early July. Adults are able to mate on the same day after emergence. They lay eggs the day after mating. Most eggs are laid in rows on herbaceous plants. The leaves with eggs roll up and enclose the eggs. Some eggs are laid on withered bamboo plants, on the ground, or at base of bamboo leaves. Larvae hatch out in January to February of next year. They crawl onto stems of weeds to feed. When bamboo shoots emerge from soil in early and mid- April, larvae immediately migrate from weeds to bamboo shoots. Larval damage period is in early April to late May. Larvae mature and pupate in mid-May.

Prevention and treatment

Protect natural enemies. Natural enemies of adults include predators such as various birds, a spider (*Oxyopes sertaus*), and *Chlaenius bioculatus*. There is an egg parasitoid, *Trichogramma evanescens*, Young larvae are often preyed by a toad and *Calathus* sp. when they migrate from weeds to bamboo shoots. Larval parasitoids include *Apanteles* spp., *Winthemia diversa* and *Exorista japonica*. The parasitic rates are not high.

References   Chen Yixin 1999, Xu Tiansen and Wang Haojie 2004.

(Xu Tiansen, Wang Haojie)

459 *Hyblaea puera* Cramer

Classification Lepidoptera, Hyblaeidae
Common name Teak moth

Distribution Jiangsu, Hubei, Guangdong, Guangxi, Hainan, Yunnan, Taiwan; India, Myanmar, Sri Lanka, Malaysia, Indonesia, Japan, South America, Southern Africa.

Hosts Verbenaceae, Araliaceae, Bignoniaceae, Juglandaceae and Oleaceae. Preferred hosts include *Tectona grandis*, *Gmelina hainanensis*, *Vitex microphylla*, *Vitex agnus-castus*, *Callicarpa* sp., *Callicarpa nudiflora*, *Cassia fistula* and *Oroxylum indicum*.

Damage An estimated 200 ha *Tectona grandis* forest was damaged seriously in Dehong of Yunnan in 1983.

Morphological characteristics **Adults** Body 13-16 mm in length, 28-42 mm in wingspan. Head and thorax light grey to reddish brown, abdomen dark brown with orange yellow clitellum. Forewing dark brown, with an arched grey or reddish brown band; the underside of the forewing with a large brown spot. Hind wing dark brown, with a yellow band in the middle; its posterior margin with a small orange red spot near the anal angle; its venter orange red, with a black spot near the apex and two black spots near the anal angle. **Eggs** 1.0 mm in length, 0.4 mm in width, oblong; bright yellow, but milky on leaves of *Tectona grandis*. **Larvae** Mature larvae 35-40 mm in length. The newly hatched larvae milky, head black, body green after they feed. The 2nd to the 3rd instars ash black. After the 4th instar, those reared in groups in the laboratory ash black to black, verrucae black; subdorsal line white, the rear border of the 8th abdominal notum with a rectangular white spot. Those larvae reared singly and those in the wild have black head, bright color; pronotum trapezoid, yellowish brown, the margin dark green. Dorsal side from prothorax to anal segment yellowish green, with two yellowish brown longitudinal stripes, the stripes are most obvious on thorax and after the 5th abdominal segment. Two sides of the dorsal surface of each segment with a black spot and a yellow spot. Dorsal line and subdorsal line greyish white, subspiracular line light yellow; pygidium with small black spots. **Pupae** Males 15-19 mm in length, female 16-20 mm in length. Light green in early pupation stage, dark brown near emergence.

Biology and behavior *Hyblaea puera* occurs 12 generations a year in Jianfengling of Hainan. Population size peaks in mid-June to mid-July. Later the numbers drop as the leaves of *Tectona grandis* grow old and hard. Damage only appears in nurseries and young branches. The duration from eggs to adults is 18 days in May, 37 days in January to February. Larvae have 6 instars, and larval stage is 7-15 days. Pupal stage is 6-8 days.

Most larvae hatch in the morning. Larvae only feed on tender leaves. Larvae chew semi-circular notches on margin of leaves, roll up the leaves, and stay inside them. When larvae are disturbed, they retreat to their original hiding places or spin silk to drop down. Only a few veins of a leaf may be left after feeding. Mature larvae weave leaves and pupate inside. During outbreaks, they also can pupate on shrubs and weeds.

Adults emerge at night. They mate the next night after emergence. Mating requires 3-4 hours. They begin to oviposit the night after mating. Eggs are laid singly on leaves. Most of them are on the lower surfaces of leaves. Each female lays 204-843 eggs (average 477 eggs). Adults hide during the day and are active during the night. They are attracted to lights. The observed longest longevity of males and females are 17 and 13 days, respectively.

Natural enemies include one species in Brachymerinae and Braconidae. Four Ichneumonid wasps were reported: *Camptotypus rugosus*, *Echthromorpha agrestoria notulatoria*, *Sychnostigma hyblaenum* and *Friona okinawana*.

References Chen Zhiqing et al. 1978, 1984.

(Chen Zhiqing)

H. puera adult (NZM)

460 *Gadirtha inexacta* (Walker)

Classification Lepidoptera, Nolidae
Synonym *Iscadia inexacta* Walker

Distribution Jiangsu, Zhejiang, Fujian, Jiangxi, Hubei, Hunan, Guangdong, Guangxi, Hainan, Guizhou, Taiwan; India, Indonesia, Myanmar, Singapore, South Pacific islands.

Host *Sapium sebiferum*.

Morphological characteristics **Adults** Body 20-24 mm in length, about 45-51 mm in wingspan. Head and thorax mostly grey, mingled with brown color. Labipalp about 2 mm in length. Forewing inner side with a black spot, middle section with many black and brown dots; subouter margin greyish white; outer margin made of a row of black dots. Hind wing light yellowish brown to brown, margin hairs yellowish white.

Larvae Mature larvae 28-31 mm in length. Head yellowish green, vertex with raised granules. Body light yellow, with fine and long hairs. Dorsal line black or consisted of discontinuous black spots; subdorsal line appears as a yellow band; the supraspiracular line yellow, spiracle line yellow and not obvious. Setae very long, the longest seta is about 1/3 of the body length, setae black or white. Spiracles oval, brown.

Biology and behavior *Gadirtha inexacta* occurs four generations a year in Fujian province, and overwinters as eggs on limbs or the lower surface of leaves. Larvae begin to hatch in early May next year. The larval damage periods of the four generations are in early May to mid-June, early July to early August, early August to early September, and mid-September to mid-October, respectively. The adult emergence periods are in late June, late July, early September, and mid-October, respectively.

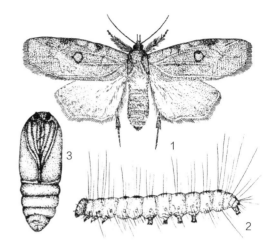

G. inexacta. 1. Adult, 2. Larva, 3. Pupa
(Huang Jinshui)

Most adults emerge at night. They mate and oviposit 2-3 days after emergence. Adults hide during the day and are active at night. They have phototaxis. Adult longevity is 8 days. Eggs are laid singly on the lower surfaces of leaves or on limbs. Egg stage is 9-13 days. After the 3rd instar, larval feeding increases significantly. They prefer feeding on young leaves. When disturbed, they often jump away. Larvae have 6 instars.

Prevention and treatment

1. Manual control. Manually remove pupae. Shake off larvae and destroy them.

2. Biological control. Conserve ants which can prey upon pupae.

References Huang Jinshui et al. 1987, Chen Yixin 1999.

(Huang Jinshui, Tang Chensheng)

G. inexacta adult (NZM)

461 *Selepa celtis* Moore

Classification Lepidoptera, Nolidae
Synonym *Subrita curviferella* Walker

Importance *Selepa celtis* becomes a frequent pest in Guangdong in recent years.

Distribution Jiangsu, Zhejiang, Fujian, Jiangxi, Hubei, Guangdong, Guangxi, Sichuan; India, Sri Lanka, Indonesia, the Philippines.

Hosts *Duabanga grandiflora*, *Camellia sinensis*, *Garcinia tinctoria*, *Lagerstroemia speciosa*, *Castanea mollissima*, *Pyrus bretschneideri*, *Psidium guajara*, *Mangifera indica*, *Eriobotrya japonica*, *Bischofia javanica*.

Morphological characteristics **Adults** Females 9-11 mm in length, 24-26 mm in wingspan; males 8-9 mm in length, 20-22 mm in wingspan. Forewing greyish brown, the center with a spiral ring and 3 upward clusters of scales in the ring; the one near the center greyish white in most part, the other two brown. There are 3 obvious brown scale clusters near the anal angle. Hind wing greyish white. **Eggs** Steamed bun shape, 0.25 mm in diameter and height; light yellow and slightly reddish; the top with a round pit, its margin with 18 vertical carinae that are connected by small cross carinae. **Larvae** Mature larvae head 1.8-2.0 mm in width, 14-22 mm in length. The dorsal surface of the 2nd and 7th abdominal segments each with a black spot. Sides of the 2nd and 6th abdominal segments each with a black dot; each of these dots becomes two as the larvae become mature. Sides of the abdomen with 2 gray longitudinal stripes when mature. Base of setae are white. **Pupae** Oval, beige. Female pupae 10.0-11.0 mm in length, male pupae 8.5-9.0 mm in length. The venter of the 6th abdominal segment, the dorsal surface of the 9th abdominal segment each with a longitudinal ridges. Cocoons flat oval, the surface covered with many soil particles, 15.0-20.0 mm in length, 6.0-7.0 mm in width.

Biology and behavior *Selepa celtis* occurs about seven generations a year in Guangzhou. Peak numbers appear in April to October. When reared at 22.60 °C to 28.5 °C daily average temperature, the egg stage was 6-9 days, larval stage was 13-19 days, pupal stage was 8-13 days, pre-oviposition period was 2-3 days, and life cycle was 33-43 days. When reared at 27 °C daily average temperature, all the 1st to the 4th instar larval periods were 3 days, and the 5th instar larval period was 3.5-4.5 days. Adults emerge at night. They mate on the second or third

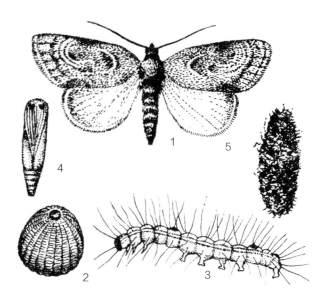

S. celtis. 1. Adult, 2. Egg, 3. Larva, 4. Pupa. 5. Cocoon (Wu Jianfen)

night after emergence. Most of them oviposit on the third night. Each female generally lays one egg clutch, each egg clutch has 30-100 eggs. Most eggs are laid on leaf surfaces covered by adjacent leaves. Hatching rate generally was around 88%. Larvae generally have 5 instars, few have 4 instars. Larvae are gregarious. The 1st to the 4th instars only feed on the epidermis and mesophyll on the underside of the leaves, while the 5th instar larvae eat leaves into holes or incisions, or consume whole leaves with only veins left. After larvae become mature, they crawl down to the ground, and then spin cocoons on soil surface or bases of trunks.

Adult longevity generally is over 10 days. Larval feeding amount is 4,747 mm^2 of leaves throughout its lifetime. The percent consumption by the 1st to the 5th instar is: 0.5, 2.1, 7.6, 31, and 58.8%, respectively. Natural enemies mainly include *Cantheconidea concinna*, praying mantis, and ants.

Prevention and treatment

1. Silviculture. Shake trees and manually remove the larvae fallen to the ground. Remove ground cover to reduce overwintering sites.

2. Trap and kill adults with light traps.

3. Biological control. Conserve natural enemies such

S. celtis pupa (edited by Zhang Runzhi)

S. celtis larva (edited by Zhang Runzhi)

S. celtis adult (NZM)

as ants, lady beetles, wasps, Tachinidae, Chrysopidae, birds, and bats.

4. Chemical control. For young instar larvae, spray 50% fenitrothion EC at 1,000×, or 25 g/L lamda-cyhalothrin at 1,000×, 20% thiamethoxam or malathion at 800-1,000×, 45% profenofos and phoxim mixture or 50% imidacloprid + monosultap at 600-800×, 40% isocarbophos or 25% carbaryl wettable powder at 1,500×, 2.5% deltamethrin or 20% Sumicidin at 1,000-1,500×. Apply 150-300 mL 10% cypermethrin EC per hectare, or 20% diflubenzuron at 2,000-3,000×. For larvae within shoots, inject 10× 40% omethoate EC into trunks, or brush infested area with 20-40× 50% sumithion EC or 50% 2-dichloroethyl dimethyl phosphate EC.

References Wu Jianfen et al. 1984, Lu Chuanchuan et al. 1985, Chen Yixin 1999.

(Wu Jianfen)

462 *Carea angulata* (F.)

Classification Lepidoptera, Nolidae

Synonyms *Carea subtilis* Walker, *Carea innocens* Swinhoe, *Carea subilis* Hampson, *Carea intermedia* Swinhoe

Importance *Carea angulata* is an important pest of *Eucalyptus citriodora*. During outbreaks, it can consume all leaves of large areas of eucalyptus forest.

Distribution Guangdong, Guangxi, Hainan, Yunnan; India, Indonesia, Sri Lanka.

Hosts *Eucalyptus citriodora*, *Syzygium cumini*, *Syzygium jambos*.

Damage Larvae feed on leaves and can consume all leaves of the host trees.

Morphological characteristics **Adults** Body 15-20 mm in length, 31-41 mm in wingspan. Head and thorax reddish brown, labipalp brown, legs white. Abdomen pink, dorsal side and the rear end brown. Forewing reddish brown, frons and inner cross line dark brown. There is a dark brown oblique stripe from the front end of outer cross line to the rear end of inner cross line. Hind wing white and slightly transparent.

Eggs Round, diameter 1.4 mm, the bottom flat, surface with many longitudinal ridges. Light yellow at beginning, later turn into greyish white.

Larvae Length 24-37 mm, thorax raised upward into a 10 mm diameter sphere, bluish green and glossy. Head dark brown. Dorsal side of each abdominal segment greyish yellow. Pygidium dark brown, cerci and anal legs light yellow.

Pupae Length 14-18 mm, width 6-7 mm. Dorsal side dark brown, abdomen light yellow at early time, turn into dark brown prior to emergence. End of the abdomen smooth, without cremaster.

Biology and behavior *Carea angulata* occurs six generations a year in Guangdong. Adults mate and oviposit at night, and eggs are laid singly on the lower or upper surfaces of leaves. Each female lays 324-488 eggs. Adults are strong flyers and exhibit strong phototaxis. They hide at places of dense branches and leaves. Their wings spread and abdominal ends raised when at rest. The newly hatched larvae feed on the mesophyll. From the 3rd instar, larvae feed on whole leaves. They spit brown liquid when disturbed. Mature larvae weave leaves and spin cocoons. When leaves are eaten up, they crawl to the ground and spin cocoons among weeds. Drought in spring and summer often cause outbreaks of this insect. Natural enemies include birds, spiders, assassin bugs, preying mantis and wasps.

Prevention and treatment

1. Silviculture. Replace *E. citriodora* trees with other eucalyptus species.

2. Trap adults. Trap adults with black light traps during 19:00-21:00.

3. Biological control. Apply 1×10^{10} spores/g *B.t.* powder at 1,000-1,500× on cloudy days or before dusk on a sunny day.

4. Chemical control. Apply 25 g/L lamda-cyhalothrin at 800-1,000×, Apply 5.7% eamamectin at 3,000-5,000×.

References Chen Yixin 1999, Gong Mingqin, Gu Maobin, Chen Peizhen et al. 2007.

C. *angulata* adult (NZM)

(Gu Maobin)

463 *Earias pudicana* Staudinger

Classification Lepidoptera, Nolidae
Synonym *Earias pudicana pupillana* Staudinger

Distribution Beijing, Tianjin, Hebei, Shanxi, Liaoning, Jilin, Heilongjiang, Jiangsu, Zhejiang, Jiangxi, Shandong, Henan, Hubei, Sichuan, Shaanxi, Ningxia; Japan, India (North), Korea Peninsula, Russia.

Hosts *Salix babylonica*, *Salix integra*.

Morphological characteristics Adults Body 8-10 mm in length, 20-23 mm in wingspan. Head and thorax pink green, antennae dark brown, labipalp greyish brown. Forewing yellowish green, base of costal margin yellow and with red halo, the end of discal cell with an obvious purplish brown orbicular spot, outer margin and margin hairs dark brown. Hind wing greyish white, slightly transparent, margin hairs white. Abdomen and legs white, tarsi purplish brown.

Eggs Diameter 0.4 mm, steamed bun shaped, greyish blue.

Larvae Mature larvae 15-18 mm in length; the 1st instar larvae light greyish yellow to dark brown; head and pronotum dark brown; thoracic and abdominal dorsal surface greyish yellow; dorsal surface of thorax and abdomen with 2 and 3 elongated spots, respectively. Subdorsal line dark purple. The sides of the dorsal surface of the 2nd, 3rd, 5th, and 8th, abdominal segments each with purple black verrucae and one seta per verruca. The lower spiracle line white. Thoracic and abdominal surfaces covered with scattered granules and short hairs on them, ventral surface greyish white.

Pupae Body 8-11 mm in length, 2 mm in width; dorsal side dark brown or slightly with greenish, two sides and segmental venter light brown, rear end round and blunt. Cocoons white or greyish brown.

Biology and behavior *Earias pudicana* occurs two generations a year in Heilongjiang, Jilin, Liaoning, Ningxia and Beijing and 3-4 generations a year in Shaanxi and Hubei. It overwinters as pupae. Adults emerge in late March next year in Hubei and in mid- and late April in Beijing. Larvae can be seen from May to September. They start to overwinter in early and mid-September to early October. After emergence, adults rest on the lower surface of leaves in the daytime, fly at

E. pudicana adult (Xu Gongtian)

E. pudicana adult
(Xu Gongtian)

E. pudicana larva (Xu Gongtian)

E. pudicana larva (Xu Gongtian)

E. pudicana cocoon (Xu Gongtian)

E. pudicana damage to salix (Xu Gongtian)

E. pudicana (Hebei Forestry School)
1. Adult, 2. Egg, 3. Larva, 4. Pupa, 5. Cocoon, 6. Damage

night, and they have strong phototaxis. Adults lay eggs on tips of tender leaves or tender shoots. The newly hatched larvae spin silk and weave tender leaves into nests, in which they live and feed. Feeding by older instar larvae creates notches. Each larva can damage 3-4 tender shoots throughout its life. Non-overwintering mature larvae spin cocoons to pupate in leaf rolls. While overwintering mature larvae spin cocoons to pupate in fallen leaves or within ground cover or in bark cracks or hidden areas of limbs.

Prevention and treatment

1. Artificially remove leaves that are rolled up by larvae.

2. Trap and kill adults with black light traps.

3. Chemical control. Spray 7,000× 20% diflubenzuron SC during larval stage.

References Xu Lianfeng, Shi Shaolin, Zhao Lingquan et al. 2004; Zhang Bairui, Zheng Jun, and Chen Changjie 2004; Xu Gongtian and Yang Zhihua 2007.

(Li Zhenyu, Wei Xueqing, Zhang Shiquan)

464 *Eligma narcissus* (Cramer)

Classification Lepidoptera, Nolidae
Synonym *Phalaena* (*Bombyx*) *narcissus* Cramer

Distribution Beijing, Hebei, Shanghai, Zhejiang, Fujian, Jiangxi, Shandong, Henan, Hubei, Hunan, Sichuan, Guizhou, Yunnan, Shaanxi, Gansu; Japan, India, Indonesia, Malaysia, the Philippines.

Hosts *Ailanthus altissima*, *Tooan sinensis*, *Koelreuteria paniculata*, Adults damage mature fruits of *Pyrus* spp., *Malus pumila* and *Citrus reticulata*.

Damage *Eligma narcissus* feed on leaves or mature fruits and can cause severe damage to the host.

Morphological characteristics **Adults** Body 26-38 mm in length, about 67-80 mm in wingspan. Head and thorax light greyish brown, abdomen apricot yellow. Forewing long and narrow, ash black, with a white longitudinal band from wing base to wing apex. Greater part of hind wing apricot yellow, terminal part purplish cyan.

Eggs Close round, milky.

Larvae Body 38-47 mm in length. Head black, body orange yellow, dorsal surface with dark brown spots and covered with white long and thin setae.

Pupae Length 26-27 mm. Body flat, fusiform, reddish brown. Cocoons 50-64 mm in length.

Biology and behavior *Eligma narcissus* occurs two generations a year in Henan and overwinters as pupae. Adults emerge and lay eggs in early and mid-March. The 2nd generation adults appear in early and mid-July. This generation larvae cause severe damage to hosts.

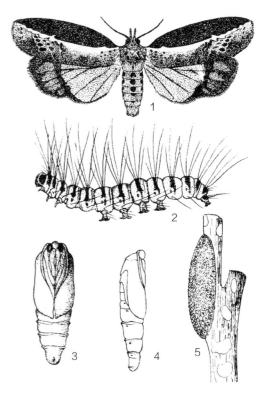

E. narcissus (Xu Tiansen)
1. Adult, 2. Larva, 3、4. Pupa, 5. Cocoon

They are able to consume all leaves on a tree and only veins and petioles are left during heavy infestations. Larvae mature in August to September and spin cocoons on barks of branches and trunks.

Larvae stay and feed on back of leaves. They spring to the ground or to other leaves when frightened. Mature larvae crawl to trunks and spin gray thin cocoons and pupate. Adults have phototaxis. They hide during the day in shady and dark areas such as trunk or underside of leaves.

Prevention and treatment

1. Silviculture. In winter, scrape off overwintering cocoons on trunks. In late September to February of the following year, manually remove the larvae and cocoons.

2. Trap adults with frequency vibrancy-killing lamps.

3. Protect natural enemies. During pupal stage, *Pimpla*

E. narcissus adult (Xu Gongtian)

E. narcissus adult (Xu Gongtian)   E. narcissus larva (Xu Gongtian)

E. narcissus pupa (above) (Xu Gongtian)

E. narcissus pupa (below) (Xu Gongtian)

E. narcissus empty cocoon (Xu Gongtian)

E. narcissus empty cocoon (Xu Gongtian)

E. narcissus larvae infected wich Beauveria bassiana (Xu Gongtian)

inctuosa, *Sympiesis* sp., and *Tachina* sp. are common parasitoids. Collect and place pupae in insect cages and let parasitoids fly out in spring of the next year.

4. Biological control. In larval stage, spray 1×10^{10} spores/mL *B.t.* emulsion at 500-1,000×.

5. Chemical control. Applying 25 g/L lamda-cyhalothrin at 800-1,000 × to control the 5th to 6th instar larvae. The control effect can reach 100%. The control effect of 1,000× 90% trichlorfon spray can reach 94%.

References Xiao Gangrou 1992; Zhang Zhizhong 1992; Chen Yixin 1999; Lin Xiaoan, Pei Haichao, Huang Weizheng et al. 2005.

(Wan Shaoxia, Zhou Jiaxi)

465 *Camptoloma interiorata* (Walker)

Classification Lepidoptera, Nolidae
Synonyms *Numenes interiorata* Walker, *Camptoloma erythropygum* Felder
Common name False tiger moth

Importance It is an important defoliator of *Quercus* plants and can eat up all leaves. They affect silkworm breeding.

Distribution Beijing, Hebei, Liaoning, Jilin, Heilongjiang, Jiangsu, Zhejiang, Anhui, Fujian, Shandong, Henan, Hubei, Hunan, Guangdong, Guangxi, Sichuan, Yunnan; Korea Peninsula, Japan, Russian Far East.

Hosts *Morus alba*, *Salix* spp, *Castanea mollissima*, *Castanopsis sclerophylla*, *Quercus acutissima*, *Quercus aliena*, *Q. glandulifera*, *Q. variabilis*, *Q. liaotungensis*, *Q. mongolica*, *Sapium sebiferum*.

Damage Larvae feed on buds and leaves of various Quercus plants. Damaged plants are unable to bloom and grow leaves.

Morphological characteristics **Adults** Body 10 mm in length, 26-36 mm in wingspan, body orange yellow. Forewing yellow, with 6 black lines radially extending from anal angle area to costal margin, two lines near wing base "V" shaped, the 3rd line located at the discal cell and short. The rear half of outer margin with two groups of vermilion spots, each group has 2 branches extending to the wing base. There are 3 black spots on the margin near anal angle. Hind wing orange yellow. Female abdominal end covered with thick pink fluffs.

Eggs Oblate, light yellow.

Larvae Mature larvae about 30-35 mm in length, head dark brown, pronotum dark brown, divided into 4 pieces by yellowish white fine lines. Thorax and abdomen greyish yellow, with 13 brown longitudinal lines, each segment covered with several white long hairs. Proleg base and pygidium are dark brown.

Pupae About 10-12 mm in length, slightly fusiform, dark brown, abdominal end with a circle of toothed protuberances.

Cocoons Dark yellow.

Biology and behavior *Camptoloma interiorata* occurs one generation a year in Gansu, Shandong and Jiangsu and overwinters as the 3rd instar larvae in galls. It occurs one generation a year in Shenyang of Liaoning and overwinters as the 10th instar larvae in soil. In March to April next year, the overwintering larvae become active and feed on buds and tender leaves. They often bite 1-2 round holes on buds and bore into buds and feed. The feeding kills buds and has serious impact on blooming

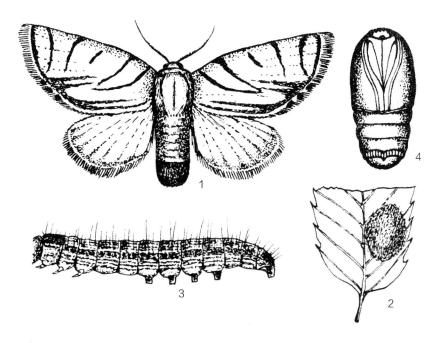

C. interiorata. 1. Adult, 2. Egg mass, 3. Larva, 4. Pupa (Tian Hengde)

C. interiorata adult (Xu Gongtian)

and leaf growth. Larvae mature in May. They crawl to leaf litter layer or under stones, spin cocoons, and pupate. Pupal stage is about one month. Adults emerge, mate and oviposit from June. Eggs hatch in July to August. Larvae begin to overwinter in October and November.

Adults mostly emerge during the daytime. They mate at dusk. Males often fly over females and generate noise. After about five minutes, females also start flying. After females and males dance together for about 8-10 minutes, they start mating. Mating lasts about 30 minutes. A male only mates once. Most females oviposit during the night, few females oviposit in the next morning. Most eggs are laid on back of leaves located in the middle and lower part of the crown. Eggs are arranged in monolayer round or oval egg clutches which are covered with pink hairs. Each egg clutch has about 231 eggs. Female longevity is about 5-8 days and male longevity is about 4-6 days. Males exhibit stronger phototaxis than females.

Egg stage is 2-3 weeks. Newly hatched larvae spin silk under the egg clutch and compose a greyish white gall and attach the petiole to the branch in order to prevent the leaf from falling off. The 1st instar larvae feed on leaves and eggshell in the galls. Larvae come out of the galls to feed after the 2nd instar. Larvae discard feces, exuviae, and head capsules out of the galls before they feed. Larvae come out to feed together. When disturbed, they shake their heads sideways a few times and produce "rustling" sound.

Larvae spin silk on their foraging trails. They follow the silk trails when going out for food or returning to the galls. If silk trails are destroyed, they often cannot locate their galls and form clusters with their heads facing each other. As the instar increases, the area of the galls enlarges. When a gall is damaged, they are able to repair it soon. Each gall has an average of 848 larvae and a maximum of 3,000 larvae. The galls may move toward the feeding site. New galls may form when larvae create new galls.

Prevention and treatment

1. Manual control. Between November and March of the next year, scrape off the overwintering galls on trunks and branches. Apply kerosene to the two ends of galls. In autumn, search for the "white leaves" damaged by the insect, destroy the larvae on leaves. In the 5th instar period,

C. interiorata larvae (Xu Gongtian)

wrap tree trunks with straws or hays at 1 m height. After mature larvae pupate in them, then destroy the straws or hays.

2. Trap adults with black light traps.

3. The following insecticides are effective for Camptoloma interiorata management: 90% trichlorfon at 2,000×, 80% DDVP EC at 1,000-2,000×, 50% sumithrin EC at 800×, 75% phoxim EC at 2,500-3,000×, 2.5% deltamethrin at 8,000×, 50 mg/kg Dimilin solution, and *B.t.* The best treatment time is in late April.

References Xiao Gangrou 1992; Geng Yilong, Xu Heguang, Li Dongjun et al. 1998; Xi Zhongcheng 1999; Zheng Wenyun and Gao Desan 2001.

(Ji Baozhong, Zhang Kai, Tian Hengde)

466 *Hypocala subsatura* Guenée

Classification Lepidoptera, Erebidae
Synonyms *Hypocala aspersa* Butler, *Hypocala subsatura* var. *limbata* Butler

Distribution Beijing, Hebei, Inner Mongolia, Liaoning, Jiangsu, Zhejiang, Fujian, Shandong, Henan, Guangdong, Hainan, Yunnan, Guizhou, Tibet, Shaanxi, Gansu, Taiwan; Bangladesh, India, Japan.

Hosts *Diospyros kaki*, *Malus pumila*, *Pyrus* spp., *Prunus salicina*, *Quercus* spp.

Damage Larvae spin silk and attach leaves of seedlings and saplings. They feed on leaves and buds, causing death of branches. They can reduce the quality of seedlings and growth of young trees significantly.

Morphological characteristics **Adults** Body 18-21 mm in length, 38-42 mm in wingspan, whole body brown, antennae filiform and covered with pubescence. Forewing purplish brown, covered with dense dark brown fine spots; outer cross line and inner cross line brown, wavy; with a black margined reiniform mark. Hind wing brownish black, the basal half part of the posterior margin with a large yellow spot; there is a smaller yellow spot near the anal angle, the middle of outer margin, and the central part of the hind wing, respectively.

Eggs Diameter 0.5 mm, semi-globous, the surface with ridges in a radial pattern.

Larvae Mature larvae 30-35 mm in length. Head yellowish brown, body black. Spiracular line from prothorax to the 8th abdominal segment consisted of discontinuous and irregular round yellow spots. Proleg crochets in uniordinal mesoseries.

Pupae Length 14-19 mm, width 4.5-6.2 mm; reddish-brown to dark brown; with 4 cremasters, those two in the middle longer than the others.

H. subsatura. 1. Adult, 2. Larva (Hu Xingping)

H. subsatura adult (Tang Yanping, Yang Chuncai)

Biology and behavior *Hypocala subsatura* occurs two generations a year in Yangtze River basin and overwinters as pupae in soil chambers. Adults appear in early May next year. The 1st generation larval period is in mid-May to early July. Mature larvae crawl down the trees, dig into soil, and construct pupal chambers. The 2nd generation larvae appear during July and August. They crawl down the trees and overwinter in mid- and late August. Most adults emerge during 20:00-21:00. They have weak flight ability and phototaxis. They hide in weeds or on leaves during the daytime. Adults feed before laying eggs. Most eggs are laid singly on shoots and leaves of the upper crown. The newly hatched larvae crawl to the top of shoots, bore into the terminal buds, and feed on buds. The damaged buds are filled with fine granular frass. They wither rapidly. The larvae then crawl down to feed on tender leaves or bite off the terminal buds. Young larvae roll leaves up and hide themselves inside the leaves. The 1st instar larvae often spin silk and drift with the aid of wind. Mature larvae bore into the soil, construct chambers and pupate.

Prevention and treatment

1. Attract and kill. During adult occurrence period, attract adults with sugar and vinegar liquid.

2. Chemical control. In early larval damage period, spray 1,500× 20% sumicidin EC or 800× 10% imidacloprid EC.

References Zhu Hongfu, Wang Linyao, and Fang Chenglai 1979; Chinese Academy of Sciences – Institute of Zoology 1982; Xiao Gangrou (ed.) 1992; Zhao Jinnian and Chen Sheng 1993; Chen Yixin (ed.) 1999.

(Tang Yanping, Yang Chuncai, Wei Qiyuan)

467 *Ericeia inangulata* (Guenée)

Classification Lepidoptera, Erebidae
Synonyms *Hulodes inangulata* Guenée, *Girpa inangulata* Moore

Distribution Fujian, Hunan, Guangdong, Guangxi, Hainan, Yunnan, Tibet; Australia, Bangladesh, India, Myanmar, Sri Lanka.

Hosts *Acacia mangium*, *Acacia confuse*, *Dalbergia* spp., *Mimosa* spp., *Acacia mearnsii*.

Morphological characteristics **Adults** Length 20 mm, width 50 mm. Whole body greyish brown, wing base with small black dots, forewing with scattered black fine dots; inner line slightly black and wavy, median line obscure and brown, circular pattern appears as a black mark, reniform mark dark brown, outer line black and wavy, double lined. Female submedian fold with a round black spot, the position near the apex angle with a grey wavy stripe, apex angle with a black oblique stripe, outer margin is a row of black dots.

Larvae Mature larvae 50-55 mm in length, the front end more slender, the 1st to the 3rd abdominal segments often bend into bridge shape; prolegs 4 pairs, the 1st pair of prolegs the smallest. Head greyish yellow, two sides of vertex each with a white spot, the side with black irregular reticular pattern. Body reddish yellow or ash black, covered with yellowish brown or black spots, subdorsal line and spiracular line black, ventral line black, pygidium yellowish brown. Thoracic legs yellowish brown, prolegs color same as body color.

Biology and behavior *Ericeia inangulata* overwinters as pupae. In Shenzhen, adults can be seen from April to September. Larvae feed on leaves. They tend to cluster on tender leaves of tree tips. They rest on branches of the lower parts of plants during the day and they are not easily found. After October, mature larvae crawl down to the ground, weave leaves into balls, and pupate in shallow soil.

Prevention and treatment

1. Manual control. Catch and kill larvae when they crawl down trees to pupate. Use a rake to remove leaves and leaf balls containing pupae in soil. Wrap straws around the base of *Acacia mangium* trees and reduce the number of larvae pupating in soil.

2. Chemical control. Spray 150-200× 8,000 IU/μL *B.t.* wettable powder or 600-1,000× 16,000 IU/μL *B.t.* wettable powder, or 3,000-4,000× 25% diflubenzuron suspension, 25 g/L lamda-cyhalothrin at 800-1,000×, or pyrethroid insecticides, or botanical insecticide such as 600-800× 0.3% azadirachtin.

References Chen Yixin 1999, Feng Huiling and Xie Haibiao 2003.

(Feng Huiling, Li Yingchao, Li Zhenyu)

E. inangulata larva
(edited by Zhang Runzhi)

468 *Episparis tortuosalis* Moore

Classification Lepidoptera, Erebidae

Distribution Guangdong, Guangxi, Hainan; Bangladesh, India.

Hosts *Chukrasia tabularis*, *Michelia champaca*.

Morphological characteristics **Adults** Female length 21 mm, wingspan 49-55 mm; male length 18 mm, wingspan 49-50 mm. Brown, thorax and abdomen with thick setae; males darker than females; thorax and venter of the anterior of the abdomen white. Female antennae filiform, male antennae feathery for the basal 2/3 part. Forewing costal margin straight, male forewing marginal line, sub-marginal line, inner cross line white; the middle cross line and basal line also white, but not obvious; with a white new moon shaped mark; these stripes not obvious in females. Ventral surface of forewing has a coffee color triangular mark near apex; hind wing outer margin with a reddish yellow mark, the ventral surface with a white wide stripe, a coffee color spot and two brown spots present inside the stripe; females do not have brown spots.

Eggs Transparent, with radiating lines; contents of the eggs green; eggs are covered with gel material; steamed bread shape, with honeycomb marks.

Larvae Mature larvae 45 mm long; yellow, with black setae and tubercles, and some brown marks.

Pupae Length 22 mm. Reddish brown, the end with ridges arranged in a circle.

Biology and behavior *Episparis tortuosalis* occurs 4-6 generations a year in Hainan province. It overwinters as pupae or larvae. It finishes one generation in 37 days in summer and 78 days in winter. The development period of eggs, larvae, and pupae in winter is 8, 50, and 20 days, respectively. The development period of eggs, larvae, and pupae in summer is 5, 25, and 11 days, respectively. Larvae go through 8 instars. Adults emerge during 16:00-24:00. They mate and lay eggs during the night. Each female lays an average of 581 eggs. They are slightly attracted to lights. Larvae feed on old leaves. They make cocoons before pupation.

References Xiao Gangrou 1992, Chen Yixin 1999.

(Gu Maobin)

E. tortuosalis adult (NZM)

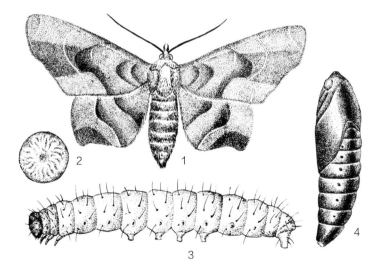

E. tortuosalis. 1. Adult, 2. Egg, 3. Larva, 4. Pupa (Zhang Peiyi)

469 *Aloa lactinea* (Cramer)

Classification Lepidoptera, Erebidae
Synonyms *Amsacta lactinea* Cramer, *Phalaena lactinea* Cramer
Common name Red costate tiger moth

Importance *Aloa lactinea* is a polyphagous pest. It can significantly reduce crop yield. Up to 30% yield reduction was reported.

Distribution Beijing, Tianjin, Hebei, Shanxi, Inner Mongolia, Liaoning, Heilongjiang, Jiangsu, Zhejiang, Anhui, Fujian, Jiangxi, Shandong, Henan, Hubei, Hunan, Guangdong, Guangxi, Hainan, Sichuan, Yunnan, Tibet, Shaanxi, Taiwan; Nepal, Myanmar, India, Vietnam, Japan, Korea Peninsula, Sri Lanka, Indonesia.

Hosts A total of 109 species in 26 families were reported to be the host of *Aloa lactinea*. They included 26 species of crops, 16 species of trees, and 67 herbaceous plants. Examples are: *Zea mays, Glycine max, Phaseolus vulgaris, Setaria italic, Gossypium* spp., *Sesamum indicum, Sorghum vulgare, Helianthus annuus, Vigna radiate, Brassica rapa pekinensis, Raphanus sativus, Brassica oleracea, Solanum melongena, Allium fistulosum, Morus alba, Musa* spp., *Dendranthema morifolium, Zinnia elegans, Gomphrena globosa, Celosia cristata, Prunus mume, Yucca gloriosa, Hibiscus syriacus, Kerria japonica, Ailanthus altissima, Quercus* spp., *Melia azedarach, Tagetes erecta, Citrus sinensis, Amorpha fruticosa*.

Damage Larvae feed on leaves, flowers and fruits. They consume all leaves of hosts when at high population density.

Morphological characteristics **Adults** Body about 18-20 mm in length, 46-64 mm in wingspan. Head and cervix red, abdominal dorsal orange, venter white. Forewing white, costal margin with distinct red margin line, the upper angle of discal cell with a black dot. Cross stripe of hind wing black crescent-shaped, outer margin with 1-2 black spots in males and 4-5 black spots in females. **Eggs** Sphere shape, light yellow. **Larvae** Mature larvae about 40 mm in length, reddish-brown to black, with black verrucae, yellowish brown long hairs present on verrucae. **Pupae** Oblong, dark brown.

Biology and behavior *Aloa lactinea* occurs one generation a year in Hebei and overwinters as pupae. Adults emerge in May to June. They hide during the daytime and come out at night, and are attracted to lights. Eggs are laid in egg clutches on lower surface of leaves. Larvae have 7 instars, the first instar larvae are gregarious. After the 3rd instar, they no longer aggregate. They spin silk and drop when they encounter disturbance. Feeding on leaves by the 3rd instar cause notches on leaves. In autumn, mature larvae spin cocoons, pupate, and overwinter in soil, among dead leaves, and in cracks.

Prevention and treatment

1. Manual control. Timely remove the eggs and young larvae and destroy them. In pupal stage, dig pupae out from soil to reduce the number of overwintering pupae.

2. Physical control. Trap adults with black light traps.

3. Biological control. During peak of egg stage or young larvae stage, spray *B.t.* solution.

4. Chemical control. Aerial spray or ground spray of 25% chlorbenzuron or 20% diflubenzuron, or 20% triflumuron to control larvae.

Reference Fang Chenglai 2000.

(Liu Huan)

A. lactinea adult (Li Zhenyu)

A. lactinea adult (Li Zhenyu)

470 *Lemyra phasma* (Leech)

Classification Lepidoptera, Erebidae
Synonyms *Thyrgorina phasma* Leech, *Alphaea phasma* Leech, *Diacrisia phasma* Leech

Importance The *Lemyra phasma* damages many kinds of economic value trees, food crops, medicinal plants, fruits, vegetables and ornamental plants. Its hosts are up to 55 families, 94 genera and 111 species. When the insect damage seriously occurs, leaves are eaten up, and it seriously impacts on the growth and development of hosts.

Distribution Hubei, Hunan, Sichuan, Guizhou, Yunnan, Tibet.

Hosts *Prunus persica*, *Pyrus bretschneideri*, *Catalpa ovate*, *C. duclouxii*, *Ligustrum lucidum*, *Zea mays*, *Glycine max*.

Damage Feeding by *Lemyra phasma* cause incisions on leaves. Damaged leaves become curled and withered, and then turned into dark reddish brown.

Morphological characteristics **Adults** Males 32-42 mm in wingspan, 16 mm in length; females 38-42 mm in wingspan, 20 mm in length. Wings milky, dull yellow to dark brown. Forewing with a sub-terminal band, hind wing cross veins with black dots, base color dark.

Eggs Egg clutches oblong or irregular shape, light red or dark yellow at beginning, later turns into reddish brown.

Larvae Head light rosy, body dark grey and with metallic luster, and with yellow spots and a dorsal line, verrucae light brown and covered with black and white long hairs.

Pupae Reddish brown, head and thoracic dorsal dark color, cremasters with reddish brown spines.

Cocoons Oblong, white, light yellow or light red.

Biology and behavior *Lemyra phasma* occurs one generation a year in Kunming of Yunnan and overwinters as pupae. Adults emerge and oviposit from early and mid-May next year, larvae hatch in early and mid-June, larvae have 7 instars in total. They spin cocoons in mid- and late October.

Larvae feed on leaves. After the 3rd instar, larvae crawl up and down on the trunk or branches, or migrate to other plants by silk. Mature larvae spin cocoons under fallen leaves, in holes and cracks on walls, in indoor stacked books or folded clothes, on windows and door frames, and other shelters. Adults are active during the night, and mate on host leaves. Females mostly lay eggs on the lower surface of leaves. Based on indoor observation, a female oviposits 5 times in total. Each female lays eggs over one week period and lays about 500 eggs. It takes 10-23 days for eggs to hatch.

Prevention and treatment

1. Manual control. In winter and spring, use a shovel scrape off old barks with pupae, sweep fallen leaves, then burn them. Trap adults with light traps. Cut off leaves with egg clutches and destroy them. In larval stage, cut off leaves with gregarious young larvae and burn or bury them, or immerse them in pesticide solution. After the 5th instar, wrap tree trunks with straws or hays at 1 m height. After mature larvae pupate in them, then remove the straws or hays and destroy them.

2. Biological control. Protect and utilize parasitoids of Lemyra phasma larvae. They include *Coccygomimus disparis*, *Eriborus* sp., *Rhogas* sp., *Monodontomerus minor*,

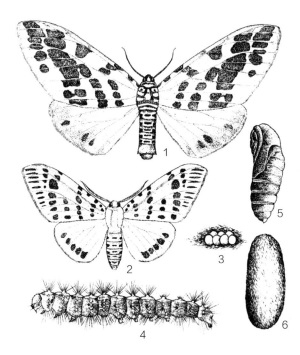

L. phasma. 1. Adult female, 2. Adult male, 3. Eggs, 4. Larva, 5. Pupa, 6. Cocoon (Zhu Xingcai)

L. phasma adult male (NZM)

L. phasma adult female (NZM)

L. phasma adult male (NZM)

and *Maxexoristops bicolor*.

3. Chemical control. For younger than the 4th instar larvae, spray 90% trichlorfon at 2,000×, 80% DDVP EC at 1,000-2,000×, or 50% sumithrin EC at 800×, or 75% phoxim EC at 2,500-3,000×, or 2.5% deltamethrin at 8,000×, or 50 ppm Dimilin.

References Cheng Liang 1976, Xiao Gangrou 1992, Chen Erhou 1999, Fang Chenglai 2000.

(Li Qiao, Pan Yongzhi)

Hyphantria cunea (Drury)

471

Classification	Lepidoptera, Erebidae
Synonyms	*Cycnia cunea* Drury et Hübner, *Hyphantria cunea* Drury et Fitch, *Hyphantria cunea* Drury et Hampson, *Hyphantria textor* Harris, *Phalaena cunea* Drury, *Spilosoma cunea* Drury
Common name	Fall webworm

Importance *Hyphantria cunea* is a worldwide quarantine pest. It was listed as a quarantine pest by Ministry of Agriculture of the People's Republic of China, General Administration of Quality Supervision, Inspection and Quarantine of the People's Republic of China and State Forestry Administration of the People's Republic of China. It has potential to cause significant loss in wood production, ecosystem, and aesthetic values of the trees.

Distribution Beijing, Tianjin, Hebei, Liaoning, Jiangsu, Shandong, Henan; Japan, Korea Peninsula, Turkey, Hungary, Czech, Slovakia, Yugoslavia, Romania, Austria, the former Soviet Union, Poland, France, Germany, Bulgaria, Italy, Greece, U.S.A., Canada, Mexico.

Hosts American white moth infests hundreds of plants including: *Acer mono*, *A. negundo*, *A. truncatum*, *Aesculus chinensis*, *Ailanthus altissima*, *Albizia julibrissin*, *Amorpha fruticosa*, *Betula platyphylla*, *Brassica pekinensis*, *Broussonetia papyrifera*, *Buxus sinica*, *Campsis grandiflora*, *Catalpa bungei*, *Cercis chinensis*, *Clerodendron trichotomum*, *Cornus alba*, *Cornus walteri*, *Cotinus coggygria* var. *cinerea*, *Crataegus pinnatifida*, *Cucurbita moschata*, *Dendranthema morifolium*, *Dioscorea opposita*, *Diospyros kaki*, *Diospyros lotus*, *Eucommia ulmoides*,

H. cunea adult female and egg mass (Zhang Runzhi)

H. cunea adult male (Xu Gongtian)

H. cunea overwintering generation male (Zhang Runzhi)

H. cunea mating pair (overwintering generation) (Xu Gongtian)

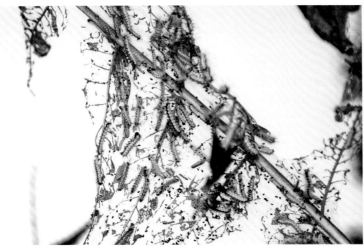

H. cunea damage (Zhang Runzhi)

H. cunea larvae (Zhang Runzhi)

H. cunea larva (Xu Gongtian)

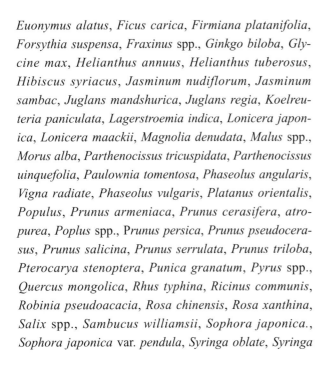

H. cunea adult female (Xu Gongtian)

H. cunea mating pair (1st generation) (Xu Gongtian)

Euonymus alatus, Ficus carica, Firmiana platanifolia, Forsythia suspensa, Fraxinus spp., *Ginkgo biloba, Glycine max, Helianthus annuus, Helianthus tuberosus, Hibiscus syriacus, Jasminum nudiflorum, Jasminum sambac, Juglans mandshurica, Juglans regia, Koelreuteria paniculata, Lagerstroemia indica, Lonicera japonica, Lonicera maackii, Magnolia denudata, Malus* spp., *Morus alba, Parthenocissus tricuspidata, Parthenocissus uinquefolia, Paulownia tomentosa, Phaseolus angularis, Vigna radiate, Phaseolus vulgaris, Platanus orientalis, Populus, Prunus armeniaca, Prunus cerasifera, atropurea, Poplus* spp., *Prunus persica, Prunus pseudocerasus, Prunus salicina, Prunus serrulata, Prunus triloba, Pterocarya stenoptera, Punica granatum, Pyrus* spp., *Quercus mongolica, Rhus typhina, Ricinus communis, Robinia pseudoacacia, Rosa chinensis, Rosa xanthina, Salix* spp., *Sambucus williamsii, Sophora japonica., Sophora japonica* var. *pendula, Syringa oblate, Syringa oblate* var. *affini, Tilia mongolica, Toona sinensis, Ulmus* spp., *Vitis vinifera, Zanthoxylum bungeanum, Zea mays, Ziziphus jujuba.*

Damage The 1st and the 2nd instar larvae cluster on the lower surface of leaves and feed on the mesophyll. Larvae create webbed nests as they feed. The 3rd to the 5th instars bite through leaves and feed on the leaf margin. The 6th and the 7th instars often consume whole leaves.

Morphological characteristics Adults Females 9.5-15.0 mm in length, 30.0-42.0 mm in wingspan; males 9.0-13.5 mm in length, 25.0-36.5 mm in wingspan. Male antennae venter dark brown, bipectinate; female antennae pectinate. Body white. Proboscis short and thin; labipalp small, its side and the end dark brown. Wings white, male forewing with dense brown spots, female forewing often without spots. Coxae and femur of front legs orange, tibia and most part of the tarsus black.

Eggs Round, 0.50-0.53 mm in diameter. The surface

H. cunea 1st larvae (Xu Gongtian)

H. cunea overwintering pupae (Xu Gongtian)

H. cunea female pupa (Xu Gongtian)

H. cunea larvae and web (Xu Gongtian)

H. cunea male pupa (Xu Gongtian)

H. cunea web made by larvae (Xu Gongtian)

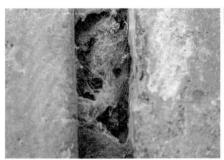

H. cunea damage to sycamore tree (Xu Gongtian)

H. cunea shed skin (Xu Gongtian)

H. cunea cocoon and pupa (Xu Gongtian)

H. cunea larva parasitized by wasps (Xu Gongtian)

A *Chouioia cunea* wasp laying eggs on *H. cunea* larva (Xu Gongtian)

with many regular small punctures. The newly laid eggs light green or yellowish green, glossy; the eggs turn into greyish green or greyish brown before hatching; the top dark brown. Egg clutch 2-3 cm^2 in size, the surface covered with hairs and scales from female abdomen.

Larvae Head of the mature larvae black, glossy, 2.4-2.7 mm in width; 22.0-37.0 mm in length. Width of the head longer than the height. Body slender, cylindrical, the dorsal with a black longitudinal band, verrucae on each segment well developed, verrucae with white or greyish white hair tufts mingled with black and brown setae. Verrucae on the side of the body orange. Spiracles white, oblong, their margin dark brown. Segmental venter yellowish brown or light gray. Head of the 1st instar larvae 0.3 mm in width, body length 1.8-2.8 mm. Head black and glossy, body yellowish green. Head of the 2nd instar larvae 0.5-0.6 mm in width, body length 2.8-4.2 mm. Body color similar to the 1st instar larvae. The dorsal verrucae black, each verruca with a thick and long black seta and short and thin white hair tufts, abdominal crochets begin to appear. Head of the 3rd instar larvae 0.8-0.9 mm in width, body length 4.0-8.5 mm. Head black and glossy. Thorax light yellow, its dorsal surface with two rows of verrucae; dorsal surface of the abdomen with two rows of black verrucae. The D1 and D2 black, D2 with a black long seta and 3-4 short black and white hairs. Proleg crochets uniordinal, homoideous mesoseries. The 4th instar larvae are same as mature larvae.

Cocoons Greyish white, thin, loose, mingled with larval body hair, oval.

Pupae Body about 9.0-12.0 mm in length, 3.3-4.5 mm in width. Light yellow at the beginning, turn into orange, brown and dark reddish-brown later, the dorsal surface of the mesothorax slightly concave. Cremasters are composed of 8-15 spinelets; the end of each spinelet is expanded.

Biology and behavior *Hyphantria cunea* occurs one generation a year in northern United States and Canada, and 2-4 generations a year in the central and South of United States. It was first detected in Dandong of Liaoning of China in 1979. It occurs two generations a year in most areas of China, and it is able to finish three generations a year in Beijing area and overwinters as pupae. The overwintering generation adults occur in May to July, while the summer generation adults occur in late July to early August. In Beijing, the peak of overwintering generation adult stage appears in early May, the peak of 1st generation adult stage appears in early and mid-July, and the peak of 2nd generation adult stage appears in late August and early September. Male emerges 2-3 days earlier than females. Adults are active at dusk and dawn. They rest on the underside of the host leaves and in grass during the day. Females lay eggs on the underside of leaves 1-2 hours after mating. Eggs are arranged into egg clutches that are covered with white scales. Each female lays 2-3 batches of eggs over 2-4 days period. The first batch has the most eggs. The overwintering generation adults mostly lay eggs at middle and lower parts of the crown, while the summer generation adults mostly lay eggs at the middle and upper parts of the crown. Adults are weak flyers; they are attracted to lights.

The newly hatched larvae feed on eggshell, and they spin a silky web around the eggshell. The 1st to 3rd instars larvae feed gregariously on the mesophyll tissue of leaves, with veins and the upper epidermis intact. The 1st to 4th instar larvae construct webbed nests made of silk and leaves, which can reach 1-2 m long. After the 5th instar, larvae begin to abandon the webbed nest and disperse. Feeding by the last instar accounts for 50% of a larva's total consumption. Larvae of 5th instar or older can live up to 8-12 days without feeding. Mature larvae

H. cunea damage to sycamore tree (Xu Gongtian)

H. cunea damage to black locust tree (Xu Gongtian)

Weed band to attract H. cunea larvae (Xu Gongtian)

Weed band to attract H. cunea larvae (Xu Gongtian)

crawl down the trees and pupate under stones and tiles, in leaf litter, in bark cracks, in tree holes and in cracks of roof under the crown.

Prevention and treatment

1. Quarantine. Inspect timber, seedlings, fruits, vegetables, agriculture and forestry crops, packaging materials, and transportation tools coming from H. cunea infested areas.

2. Trap adults with black light traps.

3. Manually remove webbed nests before the 4th instar stage.

4. For tall trees, bundle the trunk with straws at 1.0-1.5 m above the ground to attract larvae during pupation period. After the pupation period is over, destroy the straws along with the pupae. Alternatively, create an insecticide ring at the base of the tree trunk to kill larvae passing the trunk.

5. Release *Chouioia cunea* during late larval period and early pupation period.

6. Microbial control. Ground spray or aerial spray of *B.t.* (4×10^6 IU/mL) and HcNPV (1×10^9 PIB/mL) mixture.

7. Sex pheromone traps. Set trap pheromone traps at 2-2.5 m height, one trap per 100 m distance to attract and kill males.

8. Chemical control. The following treatment is effective against *H. cunea*: Ground spray of 25% chlorbenzuron suspending at 1,500-3,000g/ha or aerial spray at 600-900 g/ha; ground spray of 20% diflubenzuron suspension at 640-900 g/ha or aerial spray at 300-450g/ha; ground spray of 20% triflumuron suspension at 105-120 g/ha.

References Xiao Gangrou 1991, Ministry of Forestry of People's Republic of China 1996, Zeng Dapeng 1998, Fang Chenglai 2000, Tao Wanqiang 2008.

(Tao Wanqiang, Guan Ling, Li Yufan, Ai Dehong, Wang Jingwen, Yu Huyong)

Calliteara axutha (Collenette)

Classification Lepidoptera, Erebidae
Synonym *Dasychira axutha* Collenette

Importance *Calliteara axutha* is the second most important defoliator of pine trees after pine caterpillars. It can cause severe damage to pine forests.

Distribution Beijing, Liaoning, Heilongjiang Zhejiang, Anhui, Jiangxi, Hubei, Hunan, Guangdong, Guangxi, Shaanxi; Japan.

Hosts *Pinus massoniana, P. elliottii, P. taeda, P. caribaea, P. yunnanensis, P. kesiya* var. *langbianensis, P. tropicalis, P. ikeda, P. rigida, P. oocarpa, P. kesiya, P. taebulaeformis, Keteleeria fortunei.*

Damage Young larvae feed on the side of leaves and cause leaves to curl. The later stages of larvae consume whole leaves and can consume 100% of the leaves on a tree.

Morphological characteristics **Adults** Females 18-25 mm in length, 40-60 mm in wingspan. Males 13-18 mm in length, 30-40 mm in wingspan. Greyish black. Forewing whitish grey with a tint of brown color; sub-base line brown black, zigzag; inner and outer cross lines brownish black; sub-outer margin line brown, wavy; outer margin line dark brown; margin hairs brownish gray alternated with dark brown color. Hind wing greyish white in females, ash black in males. Cross veins and outer cross line dark brown.

Eggs Greyish brown, hemispheric, diameter about 1 mm; the middle concave, the center with a black dot.

Larvae Newly hatched larvae 4-5 mm in length; mature larvae 35-45 mm in length, head width 3.5-5.0 mm. Head reddish brown, body yellowish brown, mixed with irregular red and dark brown pattern, and covered with black hairs. Thorax and abdomen with verrucae, each verruca with brownish black long hairs. Two sides of pronotum and the center of the dorsal surface of the 8th abdominal segment with bundles of black long hairs, they are pointed toward the front and the rear end, respectively. The dorsal surface of the 1st to the 4th abdominal segment are covered with brush-shaped yellowish brown tufts of hairs. Eversible gland is located at the center of the dorsal surface of the 7th abdominal segment.

Pupae Length 14-28 mm; dark red, body surface covered with yellow hairs, the dorsal surface closely covered with yellowish brown tufts of hairs, abdominal end with stiff cremasters.

Cocoons Length 20-35 mm, oval, greyish brown, cocoons thin and loose, attached with poisonous hairs.

Biology and behavior *Calliteara axutha* occurs four generations a year in the south of Guangxi, three generations a year in the north of Guangxi, Hunan, Zheji-

C. axutha adult female (Wang Jijian)

ang and Anhui, and overwinters as pupae. In areas where it occurs four generations a year, it overwinters as pupae or mature larvae. The overwintering pupae develop into adults during mid-March and early April next year. Subsequent adult emergence periods are in mid-June, mid-August, and mid-October, respectively. The 1st to the 4th generation larvae appear in April to May, July to August, September to October, and November to December, respectively. In mid-December, a portion of the larvae pupate while others (5th or 6th instars) hibernate among needles. These larvae eventually pupae before early February. In areas where it occurs three generations a year, it overwinters as pupae. Adults emerge in mid- and late April of next year, the 1st generation larvae and adults appear in May to June and early July, respectively. The 2nd generation larvae and adults appear in June to August and mid-September, respectively. The overwintering generation larvae appear in mid- and late September. They are active until October, then spin cocoons and overwinter in early to mid-November. The egg stage is 4-10 days. Larvae have 8 instars; few of them have 9 instars. The 1st to the 9th instar larval period is: 4-12, 3-7, 4-8, 4-9, 4-13, 5-16, 6-11, 4-12 and 7-10 days, respectively. The larval stage is 40-65 days (excluding the overwintering

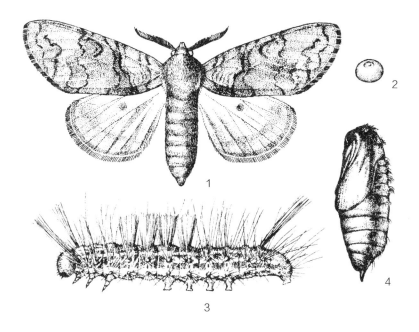

C. axutha. 1. Adult, 2. Egg, 3. Larva, 4. Pupa (Zhang Xiang)

C. axutha adult female (Wang Jijian)

C. axutha larva (Wang Jijian)

973

C. axutha cocoon (Wang Jijian)   *C. axutha* male pupa (Wang Jijian)

C. axutha damage (Wang Jijian)   *C. axutha* damage (Wang Jijian)

larval stage which is 100-120 days). Pupal stage is 13-18 days (excluding the overwintering pupal stage which is 80-120 days). Adult longevity is 3-8 days.

Adults mostly emerge at dusk. They are able to fly about 1 hour after emergence. Adults are active at night and they have strong phototaxis. Adults mate on the same night or the second night after emergence. They lay eggs on needles. Eggs are laid over 2-3 days period, each females lays 250-500 eggs. The eggs are in egg masses. During outbreaks, eggs may be laid on all tree species, telephone poles, or shrubs. It takes 4-6 days for eggs to hatch. Young larvae feed on one side of needs and the damaged needles curl and dry. The 1st and 2nd instars may drift to other trees with the aid of wind as they spin silk and drop off from trees. After the 3rd instar, larvae consume whole leaves. They often cut off needles and leave a small portion at the base, thus cause severe damage to hosts. Mature larvae crawl to the ground and spin cocoons in the litter layer, on weeds and shrubs, around trunks, among needles, or in bark cracks. Several or dozens of cocoons often attach to each other.

Prevention and treatment

1. Manual control. Manually remove egg clutches, groups of young larvae, and pupae.

2. Protect natural enemies. Natural enemies include *Telenomus* sp., *Trichogramma* sp. and *Anastatus* sp. in egg stage; *Casinaria nigripes* and *Rogas* sp. in larval stage; *Xanthopimpla predator*, *Brachymeria* sp., *Blepharipa zebina* and *Carcelia rasella* in pupal stage.

4. Biological control. *Beauveria bassiana* and *B.t.* are effective biological pesticides to control *Calliteara axutha*.

5. Chemical control. During outbreaks and when larvae are in the early stages, apply insecticide sprays, dusts, or fogging.

References   Huang Jinyi and Meng Meiqiong 1986, Xiao Gangrou 1992, Zhao Zhongling 2003.

(Wang Jijian, Yang Youqian, Wei Lin, Qin Zebo)

473 *Arna bipunctapex* (Hampson)

Classification   Lepidoptera, Erebidae
Synonyms   *Euproctis bipunctapex* (Hampson), *Nygmia bipunctapex* Hampson
Common name   Tallow tree brown tail moth

Importance   The insect is an important pest of *Sapium sebiferum*. During an outbreak in Wushan County of Chongqing City in 1989, about 900,000 *S. sebiferum* trees were damaged. After the larvae consumed all *S. sebiferum* leaves, they migrated to crop fields. About 343 ha of *Brassica campestris* field were damaged, and 8.7 ha area had to be replanted. About 580 people visited doctors due to contact with the urticating hairs from the larvae.

Distribution   Shanghai, Jiangsu, Zhejiang, Anhui, Fujian, Jiangxi, Henan, Hubei, Hunan, Guangdong, Guangxi, Hainan, Chongqing, Sichuan, Guizhou, Yunnan, Tibet, Taiwan; Myanmar, Singapore, India, Kashmir area.

Hosts   More than 42 species in 21 families, 19 orders, 3 classes, and 2 phyla are recorded as hosts. The most preferred host is *Sapium sebiferum* (Euphorbiaceae). The second preferred host is *Vernicia fordii*. Sporadic occurrence was recorded from other hosts. The reported hosts are: *Cryptomeria fornunei*, *Liquidambar formosana*, *Ficus microcarpa*, *Morus alba*, *Myrica rubra*, *Castanea mollissima*, *Caastanopsis calathiformis*, *Castanopsis eyrei*, *Quercus* spp., *Alnus cremastogyne*, *Cinnamomum camphora*, *Camellia oleifera*, *Camellia sinensis*, *Schima superba*, *Bischofia javanica*, *Sapium sebiferum*, *Vernicia fordii*, *Vernicia montana*, *Cucurbita moschata*, *Brassica campestris*, *Brassica chinensis*, *Brassica napus*, *Raphanus sativus*, *Populus* spp., *Salix* spp., *Amygdalus persica*, *Eriobotrya japonica*, *Malus pumila*, *Prunus salicina*, *Pyrus pyrifolia*, *Pyrus* spp., *Acacia confusa*, *Glycine max*, *Robinia pseudoacacia*, *Citrus reticulata*, *Diospyros kaki*, *Ligustrum lucidum*, *Osmanthus fragrans*, *Solanum tuberosum*, *Paulownia fortunei*, *Catalpa ovata*, and *Dioscorea esculenta*.

Damage   Larvae feed on leaves, buds, and young branches of *Sapium sebiferum*. They affect the growth of host trees and reduce seed reduction, or tree death. Their urticating hairs case swelling, pain, and itchiness of the skin.

Morphological characteristics   **Adults** Males

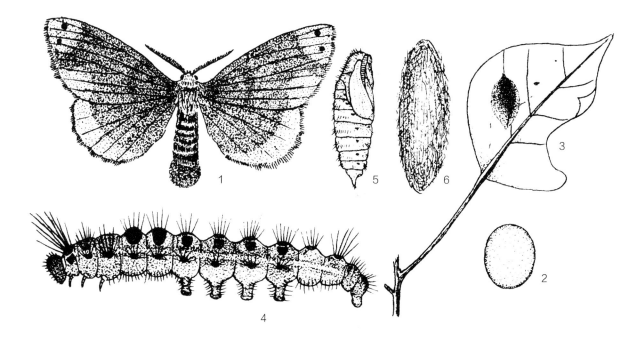

A. bipunctapex. 1. Adult, 2-3. Egg, 4. Larva, 5. Pupa, 6. Cocoon (Xu Tiansen)

A. bipunctapex adult (NZM)

9-11 mm in length, 26-38 mm in wingspan. Females 13-15 mm in length, 36-42 mm in wingspan. Body yellow and with reddish brown stripes. Wings brown, forewing apex angle with a large yellow triangular spot and two blackish brown spots within the triangular spot. Triangle spot at the anal angle of forewing and the border of the hind wing yellow.

Eggs Light green or yellowish green, oval, diameter about 0.8 mm. Eggs are arranged in 3-5 layers in the size of 25 mm × 6 mm × 4 mm, the egg masses are covered with dark yellow hairs.

Larvae Mature larvae 24-30 mm in length. Head dark brown. Thorax, abdomen yellowish brown, covered with greyish white long hairs. Both the dorsal surface and the sides of thorax and abdominal segments with black verrucae; the dorsal surface of mesothorax and metathorax each with two verrucae; each abdominal segment with 4 verrucae; those on the 1st, 2nd and 8th abdominal segments especially obvious and joined in the middle. Verrucae covered with yellowish brown and white long hairs. Verrucae on metathorax and the eversible gland reddish orange. Body color and verrucae color vary with different instars and generations.

Pupae Body 10-15 mm in length, yellowish brown to brown, closely covered with short hairs, cremasters with a clump of hook spines.

Cocoons 15-20 mm in length, yellowish brown, covered with white poisonous hairs.

Biology and behavior *Arna bipunctapex* occurs two generations a year and overwinters as the 3rd to 5th instar larvae. Larvae resume activity in late March to early April next year. They pupate in mid- and late May. Adults emerge and oviposit in early and mid-June. The 1st generation larvae hatch in late June to early July, pupate in mid- and late August. The 1st generation adults emerge and oviposit in early and mid-September. The 2nd generation larvae hatch in mid- and late September. They start overwintering in November.

Mature larvae spin cocoons and pupate in loose surface soil, in seams of stones, weeds, or in bark cracks. The pupal stage of the overwintering generation and the 1st generation is about 13 and 10 days, respectively. Adults emerge mostly during 17:00-20:00. They hide on trees or on grass during the day and are most active during 18:00-22:00. Adults have strong phototaxis and are strong fliers. They lay eggs shortly after mating. Oviposition occurs mostly between 22:00 and 6:00 the next day. Eggs are usually laid on back of leaves. Each female usually lays 200-300 eggs. The maximum number of eggs laid by a female is 600 eggs. Eggs are grouped in egg masses. Adult longevity is 2-4 days, the longest is 7 days. Egg stage for the 1st generation and the 2nd generation is 10-16 days and 13-16 days, respectively.

Larvae have 10 instars in total. The larval stage for the 1st and 2nd generation is about 56 and 107 days, respectively. Most eggs hatch around 8:00. The newly hatched larvae cluster around the egg clutch, and feed on eggshells, later they feed on host leaves. Larvae are gregarious. The size of the aggregation becomes smaller as the larvae grow. Thousands of larvae may be found in an aggregation. The 1st to 3rd instars larvae mostly rest on the back of leaves and feed from the center of leaves to leaf margin. Before the 3rd instar, larvae only feed on mesophyll, causing leaves change color and fall. After the 3rd instar, they feed on whole leaves. The 4th instar larvae often weave leaves and feed inside the leaves. The 5th instar larvae cluster on two sides of the leaves, with head to the head, and feed from the leaf tip to the petiole. They aggregate on the shady are of the tree trunk during the hot days and crawl up the tree to feed again in 16:00-17:00. During heavy rain, some larvae cluster on main branches or on tree trunk for shelter.

Hundreds to thousands of overwintering larvae usually cluster in 3-10 layers at base of branches, cracks of lower trunks, or base of tree trunk. They are covered with 0.2-2.0 mm thick silk web. The height of the overwintering larvae is related to tree age. For 35-40 years old *Sapium sebiferum*, most larvae are at 3-7 m high branches. For 20-30 years old *S. sebiferum*, most larvae are on less than 3 m high branches; for 4-15 years old *S. sebiferum*, most larvae are on trunk bases.

This insect occurs mostly in mountainous areas. Young and old *S. sebiferum* trees are less susceptible to *A. bipunctapex* damage. Trees around farm land are less likely attacked than those on hills. High temperature and high humidity climate favors the outbreak of *A. bipunctapex*.

The recorded natural enemies include *Telenomus* (*Aholcus*) *euproctidis* in egg stage; *Charops bicolor* (*Szepligeti*), *Hyposoter* sp., *Eriborus* sp., *Casinaria* sp., and *Aleiodes euproctis*, in larval stage. Three parasitoids are recorded from India: *Erycia bezzii*, *Pales aurescens*, and *Isosturmia picta*. The last one is widely distributed in China. *Brachymeria lasus* is found in pupal stage.

Prevention and treatment

1. Manual control. During winter when harvesting seeds of *S. sebiferum*, remove the overwintering larvae on trees.

2. Attract and kill. Wrap the tree trunks with hay to attract larvae, then destroy them.

3. Kill adults using light traps.

4. Biological control. Apply *B.t.* (8,000 IU/μL) at 1,500-6,000 mL/ha or *Beauveria bassiana*.

5. Chemical control. Effective insecticides include 50% imidacloprid + monosultap at 600-800×, 25% diflubenzuron at 4,000×, or 0.3% azadirachtin at 2,000×.

References   Lin Zhong 1957, People's Committee of Dongyang County 1958, Ying Yanlong et al. 1958, Tonglu County Forest Bureau 1959, Zhejiang Forest Bureau 1959, Zhao Zhongling 1978, Yang Xiuyuan and Wu Jian 1981, Office of the Jiangsu Forest Disease and Pest Survey 1982, Chinese Academy of Sciences - Institute of Zoology 1987, Forest Protection Division of the Ministry of Forestry 1988, He Junhua and Chen Xuexin 1990, Shaanxi Forest Research Institute 1990, Shaanxi Forest Research Institute and Hunan Forest Research Institute 1990, Wang Xiangjiang 1990, Fujian Forest Research Institute 1991, Xiao Youxing 1992, Yu Siqin and Sun Yuanfeng 1993, Xue Wanqi and Zhao Jianming 1996, Fang Zhigang and Wu Hong 2001, Zhao Zhongling 2003, Hua Lizhong 2005.

(Wu Juwen)

474 *Euproctis flava* (Bremer)

Classification Lepidoptera, Erebidae

Synonyms *Aroa flava* Bremer, *Aroa subflava* Bremer, *Artata subflava* Bremer, *Nygmia subflava* Bremer

Importance *Euproctis flava* is an important pest of ornamental plants and economically important trees. Cherries, apple trees, and *Amygdalus triloba* are the most susceptible. During outbreaks, it seriously affects the aesthetic value of the trees and reduces the fruit yield.

Distribution Beijing, Hebei, Shanxi, Inner Mongolia, Liaoning, Jilin, Heilongjiang, Jiangsu, Zhejiang, Anhui, Fujian, Jiangxi, Shandong, Henan, Hubei, Hunan, Guangdong, Guangxi, Sichuan, Guizhou, Yunnan, Shaanxi, Gansu; Japan; Korea Peninsula; Russia.

Hosts *Cerasus pseudocerasus*, *Pyrus* spp., *Malus pumila*, *Prunus persica*, *Rosa chinensis*, *Rosa rugosa*, *P. mume*, *Prunus* spp., *Begonia evansiana*, *Diospyros kaki*, *Camellia japonica*, *Hypericum* spp., *Cornus alba*, *Quercus mongolica*, *Quercus aliena*, *Quercus wutaishanica*, *Quercus acutissima*, *Fagus longipetiola*, *Castanea mollissima*, *Larix gmalini*, *Malus pumila*, *Eriobotrya japonica*, *Punica granatum*, *Robinia pseudoacacia*, *Alnus japonica*, *Wisteria sinensis*, *Boehmeria silvestrii*, *Radix Notoginseng*, *Camellia sinensis*, *Acer* spp., *Platycladus* spp., *Pinus* spp.

Damage Larvae aggregate on leaves and build silk nets. They chew holes and notches on leaves and may consume all leaves.

Morphological characteristics **Adults** Males 25-33 mm in wingspan, females 35-42 mm in wingspan. Antennae pectinate, yellow; the main shaft yellowish white. Body yellow or light orange yellow. Forewing yellow, middle part with a brown wide cross band, which lean outward from costal margin to discal cell and lean inward at break angle of inner margin. Wing apex area with two brown round dots, margin hairs light yellow. Hind wing without spots, the base light in color, outer margin dark in color, margin hairs light yellow.

Eggs Diameter 0.5-0.6 mm, oval, light-yellow.

Larvae Mature larvae 30-40 mm in length. Head dark brown. Body yellow or orange yellow. Dorsal line orange yellow, thin, but wider on mesothoracic and metathoracic segments, break off at the black spot of the dorsal surface.

E. flava adult (Li Yingchao)

E. flava adult (Li Yingchao)

The dorsal surface of the thorax and each of the 5th to 10th abdominal segments with two black longitudinal bands; the bands widest at the front of the thorax and toward the end of the abdomen. Pygidium black, the dorsal surface from the 8th abdominal segment black. All segments with dark yellowish brown verrucae; those verrucae on the dorsal surface of the 1st, 2nd and 8th abdominal segments are large and black; verrucae with yellowish brown or light dark brown long hairs. Thoracic legs brown, shiny. Prolegs light dark brown, with light brown long hairs.

Pupae Length about 15 mm; yellowish brown; the dorsal surface covered with short hairs; the end of the cremaster with hooks.

Cocoons Length 25-30 mm, oval, greyish-brown.

Biology and behavior *Euproctis flava* occurs one generation a year in Liaoning, two generations a year in north China, three generations a year in Hunan. It overwinters as the 8th to 9th instar larvae in tree holes or cracks of trunk bases or weeds or under fallen leaves. Larvae start activity when air temperature reaches above 6℃ around the end of April and early May. They feed on buds and young leaves. Adults appear in late June to early July. The 1st generation larvae hatch in late July. The, the 2nd generation adults emerge at the end of August. They start to overwinter in late September. Adult exhibit strong phototaxis. They are most active during 19:00-23:00.

Eggs are laid on back of leaves. They are arranged in 3-4 layers. Each egg clutch is oblong, the outside is covered with yellow hairs, and has 400 eggs in average. Egg stage is 13-15 days. The newly hatched larvae cluster on leaves near the ground, and they begin to disperse after the 1st instar. The 1st instar larvae only feed on the mesophyll, with veins intact. After the 2nd instar, larvae consume part of whole leaves.

Prevention and treatment

1. Manual control. In winter, clear fallen leaves and weeds; scrape rough barks and kill the overwintering larvae. In summer, timely remove egg clutches, manually destroy the gregarious larvae.

2. Kill adults with light traps.

3. Spray *Euproctis flava* NPV to control larvae can reduce 92% of the population.

4. Protect and utilize natural enemies. Larval natural enemies include *Carcelia ammphion*, *Exorista japonica*, *Protichneumon* sp., *Apanteles* sp., *Hexamermis microamphidis* and *Hexamermis arsenoidea*.

References Zhao Zhongling 1978, 2003; Wu Peiyu 1982, 1984; Chinese Academy of Science Institute of Zoology 1987; Chinese Forest Protection Division of the Ministry of Forestry 1988; Yi Boren, Kang Zhixian, Wei Juxiang et al 1992; Yin Boren and Kang Zhixian 1992.

(Pan Yanping, Shui Shengying)

E. flava adult male (Xu Gongtian)

E. flava larva (Xu Gongtian)

E. flava larva (Xu Gongtian)

475 *Arna pseudoconspersa* (Strand)

Classification	Lepidoptera, Erebidae
Synonyms	*Artaxa conspersa* Butler, *Nygmia pseudoconspersa* Strand, *Euproctis pseudoconspersa* (Strand)
Common name	Tea tussock moth

Importance *Arna pseudoconspersa* is a common pest throughout tea production areas in China. Affected tea trees fail to bloom or the leaves are consumed. The urticating hairs of the larvae cause skin itchiness and swelling.

Distribution Jiangsu, Zhejiang, Anhui, Fujian, Jiangxi, Hubei, Hunan, Guangdong, Guangxi, Sichuan, Guizhou, Yunnan, Tibet, Shaanxi, Gansu, Taiwan; India, Japan, Korea Peninsula, Vietnam.

Hosts *Camellia oleifera*, *Camellia sinensis*, *Vernicia fordii*, *Sapium sebiferum*, *Citrus reticulate*, *Eriobotrya japonica*.

Damage The 1st instar larvae cluster together and feed on the mesophyll. After the 3rd instar, they feed along the leaf margin and consume whole leaves. When food is scarce, they also feed on tender branches and the pericarp.

Morphological characteristics **Adults** Females 10-12 mm in length, 30-35 mm in wingspan, yellowish brown, wing surface completely covered with dark

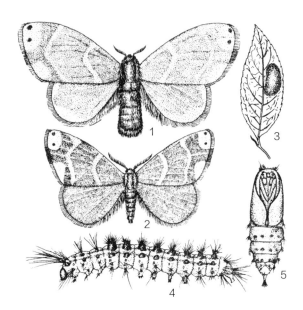

A. pseudoconspersa. 1. Adult female, 2. Adult male, 3. Egg mass, 4. Larva, 5. Pupa (Liu Gongfu)

A. pseudoconspersa adult (Xu Gongtian)

A. pseudoconspersa adult (Li Yingchao)

A. pseudoconspersa adult (Xu Gongtian)

brown scales. Males 10 mm in length, 20-26 mm in wingspan. Wing color varies with generations, wings of the 1st generation dark brown, wings of the 2nd to the 3rd generation mostly yellowish brown; two black spots at the apex angle.

Eggs Near spherical, light yellow. The middle of each egg clutch with 2-3 layers of eggs, the egg clutch is covered with yellow hairs.

Larvae Mature larvae 20-26 mm in length, yellow to yellowish brown, each segment except the last one with 4 pairs verrucae.

Biology and behavior *Arna pseudoconspersa* occurs 2 generations a year in north central of Zhejiang, north Jiangxi, and north of the Yangtze river, 3 generations a year in south Zhejiang, south central Jiangxi, Hunan and Guangdong, 3 generations a year at high mountains of Fujian and 4 generations a year at low mountains, and 5 generations a year in Taiwan. It overwinters as egg clutches on the lower surface of old leaves located at middle and lower canopy. In Hunan, overwintering eggs hatch in early April of next year, larvae pupate in mid-May, and adults emerge in early June. The 2nd generation larvae hatch in late June, the 3rd generation larvae hatch in late August, and adults emerge and oviposit in mid- and late October.

Larvae feed on leaves and spin silk. After they consume all leaves on a host, they line up and migrate in groups to another plant. Mature larvae crawl into gaps of soil or among leaves and pupate.

Most adults emerge from dusk to 22:00. They can mate on the same night. Each female usually lays 100-200 eggs. The preferred oviposition site is *Camellia oleifera* forest.

Recorded natural enemies include *Telenomus* (*Aholcus*) *euproctidis* in egg stage; *Dolichogenidea lacteicolor*, *Enicospilus flavocephalus*, *Enicospilus pseudoconspersae*, *Exorista japonica*, *Drino facialis*, *Townsend* and *Isosturmia picta* in larval stage. Predators include Carabidae, *Arma custos*, praying mantis, and spiders.

Prevention and treatment

1. Silviculture. Maintain vigorous growth of the *Camellia oleifera* forest. Till soil and bury the overwintering pupae.

2. Manual control. Remove eggs and congregating young larvae and destroy them or place eggs in cages and allow parasitic wasps emerge.

3. Trap adults with light traps between 19:00-23:00.

4. Biological control. Spray 0.5×10^6-1.0×10^7 PIB/mL EpNPV or 2×10^8 spores/mL *B.t.* or 1×10^8 spores/mL *Beauveria bassiana* can achieve satisfactory result.

5. Chemical control. Before larvae reaching the 3rd instar, spray 300-500× 2.5% rotenone, 1,000-15,000× 2% matrine, 2.5% deltamethrin or 25 g/L lamda-cyhalothrin at 800-1,000×.

References Peng Jianwen and Liu Youqiao 1992, Xiao Gangrou 1992, Zhao Zhongling 2003, He Junhua et al. 2004, Zhang Hanhao and Tan Jicai 2004.

(Tong Xinwang, Wang Wenxue)

476 *Ivela ochropoda* (Eversmann)

Classification Lepidoptera, Erebidae
Synonym *Stilpnotia ochropoda* Eversmann
Common name Yellow-legged tussock moth

Distribution Beijing, Hebei, Shanxi, Inner Mongolia, Liaoning, Jilin, Heilongjiang, Shandong, Henan, Shaanxi, and Ningxia; Russia, Korea Peninsula, Japan.

Hosts *Ulmus* spp.

Damage Newly hatched larvae feed on mesophyll, leaving epidermis intact. Feeding by older larvae results in notched leaves, leaves with feeding holes, or complete host defoliation.

Morphological characteristics **Adults** Females 12-15 mm long with a wingspread of 25-40 mm, pure white. Front femur from anterior portion to tarsi orange yellow, anterior portion of tibia and tarsi on middle and hind legs orange yellow.

Larvae Body grayish yellow, 30 mm long. White verruca present on each body segment. There is a black brown eversible gland on the back of abdominal segment 7 and 8.

Biology and behavior *Ivela ochropoda* occurs two generations a year and overwinters as larvae inside bark crevices in Henan. Larvae resume feeding on new leaves between April and May. Mature larvae construct cocoons on the underside of the leaves, in bushes, or on weeds by mid-June. Adults emerge about 10 days later. Adults are strongly attracted to lights and lay eggs in rows on twigs or the underside of the leaves. Eggs hatch in about 10 days. Newly-hatched larvae feed on mesophyll and creates feeding holes. Older larvae consume entire leaves. First generation adults appear in early Au-

I. ochropoda adult (Xu Gongtian)

I. ochropoda larva (Xu Gongtian)

I. ochropoda adult (Li Zhenyu)

I. ochropoda adult (Xu Gongtian)

I. ochropoda larva (Xu Gongtian)

I. ochropoda mature larva (Xu Gongtian)

I. ochropoda pupa (Xu Gongtian)

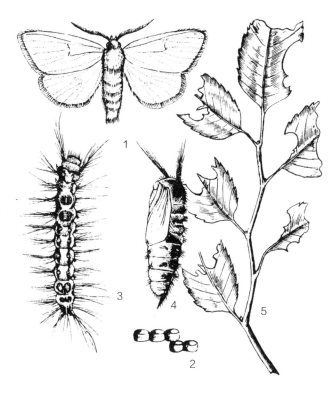
I. ochropoda. 1. Adult, 2. Eggs, 3. Larva, 4. Pupa, 5. Damage (Zhu Xingcai)

gust, while adults of the second generation occur in early September. Larvae from the second generation feed until the end of October before overwintering in bark crevices.

Prevention and treatment

1. Manual control. Trap adults with black lights.

2. Conserve natural enemy, *Schinera tergesina* Róndani (Diptera: Tachinidae).

3. Chemical control. Apply 50% malathion or 50% fenitrothion at 1,000×, or 2.5% deltamethrin at 2,000× for larvae control.

References Yang Youqian 1982, Xiao Gangrou 1992, Zhao Zhongling 2003, Henan Forest Pests Control and Quarantine Station 2005.

(Yang Youqian, Liu Zhenlu, Xue Xianqing)

477 *Lymantria dispar* (L.)

Classification Lepidoptera, Erebidae
Synonyms *Bombyx dispar* L., *Phalaena dispar* L., *Hypogymna dispar* L., *Larie dispar* L.
Common name Gypsy moth

Importance *Lymantria dispar* is a common pest in China.

Distribution All parts of China; Japan, Korea Peninsula, Europe and America. It is mainly distributed between the latitude of 20°-58°N.

Hosts It damages about more than 500 species of plants. Commonly damaged hosts include: *Diospyros kaki*, *Prunus armeniaca*, *Larix principis-rupprechtii*, *Populus* spp., *Salix* spp., *Juglans regia*, *Fraxinus chinensis*, *Lonicera japonica*, *Euonymus alatus*, *Syringa oblata*, *Ziziphus jujuba*, *Elaeagnus angustifolia*, *Elaeagnus pungens*, *Betula* spp. and *Tilia* spp.

Damage Larvae feed on leaves, cause notches or consume whole leaves.

Morphological characteristics **Adults** Dimorphic. Males 16-21 mm in length, 37-54 mm in wingspan. Head, compound eyes black, hind tibia with two pairs of spurs. Forewing greyish brown or brown, with a dark zigzag cross line, the center of the discal cell with a dark brown dot, cross vein with a crooked dark brown stripe. The lower surface of the forewing and hind wing yellowish brown. Females 22-30 mm in length, 58-80 mm in wingspan. Forewing yellowish white, the cross vein of the discal cell with an obvious dark brow pattern; the outer margin of both wings with dark brown spots. Female abdomen larger than male, tail end with yellowish brown hair tuft.

Eggs Drum shape, one side slightly depressed; diameter about 1.3 mm; greyish white at the beginning, later becomes purplish brown, glossy. Hundreds of and even thousands of eggs are glued together. The size of each egg clutch is about 2-4 cm long. Eggs are covered with very thick yellowish brown hairs.

Larvae Mature larvae 52-68 mm in length, head

L. dispar adult female (Xu Gongtian)

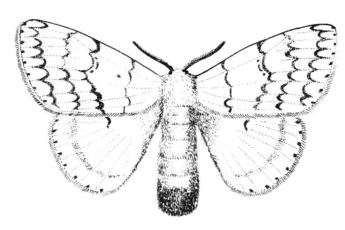

L. dispar adult (Zhang Peiyi)

L. dispar adult male (Xu Gongtian)

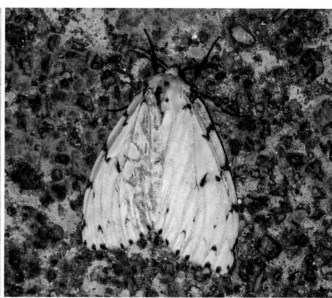

L. dispar adult female (Li Yingchao)

yellowish brown. Abdomen brown, dorsal line greyish yellow, each segment with 6 verrucae that are located on the subdorsal line, supraspiracular line and subspiracular line. The front 5 pairs of verrucae on the subdorsal line are blue, the rear 7 pairs of verrucae are red; setae on them are greyish brown and long.

Pupae Length 19-34 mm. Female pupae larger than male pupae. Body reddish brown or dark brown, covered with rusty yellow hair tufts.

Biology and behavior Gypsy moth occurs one generation a year and overwinters as eggs at depressed or cracked areas of the tree trunk, between rocks or on structures. Larvae hatch in late April of next year to early May. Newly hatched larvae stay on the egg clutch for a while, and then they cluster on a leaf. They feed and move at night. When frightened, they would spin silk and drop. After the 2nd instar, larvae disperse. They hide in bark cracks or crawl into the soil or gaps of stones under trees during the daytime. They crawl up trees in groups at dusk. After the 4th instar, their feeding increases significantly. Damage from their feeding occurs mostly from May to June. Larval stage is around 45 days. They go through 6 instars and mature in early and mid-June, and then they crawl into bark cracks or weeds, the soil and gaps of stones under trees and spin cocoons. Pupal stage is 11-16 days. Adults appear in mid-June. Adult males fly during the day. Females do not fly. Females attract males through sex pheromone. Females lay eggs one day after

L. dispar adult male (Xu Gongtian)

L. dispar adult female with egg mass (Xu Gongtian)

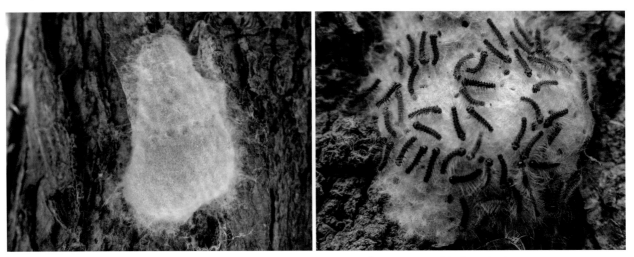

L. dispar egg mass (Xu Gongtian)

L. dispar 1st instar larvae (Xu Gongtian)

L. dispar mature larva (Xu Gongtian)

L. dispar mature larva (front view) (Xu Gongtian)

L. dispar 2nd instar larva (Xu Gongtian)

L. dispar mid-aged larva (Xu Gongtian)

L. dispar pupa (Xu Gongtian)

L. dispar pupa parasitized by natural enemy (Xu Gongtian)

they mate. Most eggs are laid on 8-25 cm thick branches or in shady areas in soil and rock cracks. Adults have strong phototaxis.

Occurrence of Gypsy moth is closely related to canopy density, degree of soil fertility, and size of the forest. Poor soil areas often suffer more serious damage. Areas with high canopy coverage and larger forests suffer less damage.

Prevention and treatment

1. Manual control. Collect and destroy the egg masses from August to April of next year.

2. Protect natural enemies. In China, there are 91 known natural enemies of *Lymantria dispar*. The parasitoids include Ichneumonidae, Braconidae, Chalcididae, Eulophidae, Torymidae, Pteromalidae, Encyrtidae, Eupelmidae and Tachinidae.

3. Light traps. Set up light traps at 500 m intervals to kill adults.

4. Utilize (+)-(7R, 8S)-disparlure sex pheromone for population monitoring, killing male moths, and mate disruption.

5. Biological control. Apply 3.75×10^{11}-7.5×10^{11} PB/ha. LDNPV to kill larvae.

6. Chemical control. Apply 7,000× 20% diflubenzuron to kill larvae.

References Xiao Gangrou 1992; Yan Jingjun, Yao Defu, and Liu Houping et al. 1994; Li Zhenyu, Yao Defu, Chen Yongmei et al. 2001; Zhao Zhongling 2003; Shi Juan, Li Zhenyu, Yan Guozeng et al. 2004; Zhang Yongan 2004; Xu Gongtian and Yang Zhihua 2007; Hou Yaqin, Nan Nan, and Li Zhenyu 2009.

(Li Zhenyu, Qiao Xiurong, Xu Wenru, Li Yajie)

478 *Lymantria mathura* Moore

Classification Lepidoptera, Erebidae
Synonyms *Liparis mathura* Swinhoe, *Lymantria aurora* Swinhoe

Distribution Hebei, Shanxi, Liaoning, Jilin, Heilongjiang, Jiangsu, Shandong, Henan, Hubei, Hunan, Sichuan, Yunnan, Shaanxi, Taiwan; Korea Peninsula, Japan, India.

Hosts *Quercus* spp., *Malus pumila*, *Pyrus bretschneideri*, *Castanea* spp., *Rhus sylvestris*, *Cyclobalanopsis glauca*, *Zeikova schneideriana*.

Morphological characteristics **Adults** Males 50 mm in wingspan, shaft of antennae white brown, pectina brown. Labipalp light orange yellow. Thorax and legs light orange yellow, with dark brown spots. Abdomen dark orange yellow, the sides slightly reddish. There are black spots between abdominal segments. Anal setae yellowish white. Forewing greyish white, with dark brown spots, veins white. Basal line dark brown. Inner cross line bends outward at the middle part. The center of the discal cell with a round spot. Cross veins dark brown. Median cross line zigzag. Outer cross line consisted of a row of crescent spots, slightly lean outward from costal margin to Cu_2 vein, and then bends inward to the inner margin. Submarginal line is a row of crescent spots, ending at A_1 vein. Terminal line consisted of a row of small dots embedded among veins. Margin hairs greyish white. Hind wing dark orange yellow, cross veins brown, submarginal line brown; terminal line consisted of a row of dark brown dots; margin hairs yellowish white.

Females 80 mm in wingspan, greyish white. Labipalp pink; center of the thoracic notum with a black dot and two pink dots; the front half of the abdomen pink, the rear half white, two sides with black spots; legs pink, with black spots. Forewing subbasal line black, with pink and black spots. Inner cross line brown, sawtooth shaped. Median cross line brown, wavy, forming a brown semicircle near the costal margin. Cross veins brown. Outer

L. Mathura adult male (Li Zhenyu)

L. Mathura adult female (Li Zhenyu)

cross line brown, sawtooth shaped. Submarginal line brown, sawtooth shaped, ending at A_1 vein. Marginal line consisted of a row of brown dots. Margin hairs pink, area between veins brown. Costal margin and outer margin pink. Hind wing light pink; cross veins greyish brown; submarginal line consisted of a row of greyish brown spots; marginal line consisted of a row of brown dots, margin hairs pink.

Eggs Spherical, brown or greyish yellow.

Larvae Mature larvae 50-55 mm in length. Head yellowish brown with dark blackish brown round dots. Body dark brown, with yellowish white spots. Dorsal line white on prothorax, black on other segments. Spiracle line black; subspiracular line greyish white. Dorsal surface of prothorax with two large verrucae and dark brown or greyish brown setae. Center of mesothorax and metathorax with yellowish brown longitudinal stripes. Granules of other segments yellowish brown, with dark brown and greyish brown hair tufts. Body venter yellowish brown. Thoracic legs reddish brown, glossy. Prolegs reddish brown, outer side with black spots. Eversible gland red.

Pupae About 28 mm in length, greyish brown, head with a pair of black short hair tufts, abdominal dorsal side with short hair tufts.

Biology and behavior *Lymantria mathura* occurs one generation a year in Heilongjiang, and overwinters as eggs. Larvae hatch in May of next year, the newly hatched larvae cluster nearby eggshells. Mature larvae spin cocoons to pupate among weeds, branches, or leaves in late July. Adults emerge in early August. Female moths are not active during the day. Male moths fly under the shade during the daytime. Females lay eggs on tree trunks, each egg clutch has about 200 eggs. Egg clutches are covered with greyish white body hairs of females.

Prevention and treatment

1. Scrape off egg clutches and destroy them.
2. Trap adults with light traps.
3. Protect natural enemies such as Ichneumonidae, Braconidae and Tachinidae.
4. Spray 1,000× 25% diflubenzuron to control young instar larvae.

References Zhao Zhongling 1978, 2003; Xu Gongtian and Yang Zhihua 2007.

(Li Zhenyu)

Lymantria monacha (L.)

479

Classification	Lepidoptera, Erebidae
Synonyms	*Ocneria monacha* L., *Bombyx monacha* L., *Noctua heteroclita* Müller
Common names	Nun moth, Black arched tussock moth

Distribution Beijing, Hebei, Shanxi, Liaoning, Jilin, Heilongjiang, Zhejiang, Shandong, Sichuan, Yunnan, Guizhou, Shaanxi, Gansu, Taiwan; Japan, Russia, Austria, Germany, Czech, Slovakia, Poland.

Hosts *Keteleeria fortune*, *Pseudotsuga sinensis*, *Picea asperata*, *Abies fobri*, *Tsuga chinensis*, *Pinus denniflora*, *Pinus armandii*, *Pinus yunnanensis*, *Larix gmelini*, *Quercus acutissima*, *Carpinus cordata*, *Fagus longipeliolata*, *Tilia* spp., *Betula* spp., *Salix* spp., *Populus davidiana*.

Morphological characteristics **Adults** Females 25-28 in length, 50-60 mm in wingspan. Males 15-17 mm in length, 30-45 mm in wingspan. Female forewing greyish white, with 4 black zigzag cross bands, the apex of the discal cell with a black stripe. Hind wing greyish white, without stripes. Wing stripes of males are similar to those of females, but are more distinct. Ventral surface of thorax and abdomen covered with dense pink hairs.

Eggs Size 1 mm × 1.2 mm, yellowish white at beginning, later turn to brown.

Larvae Mature larvae 43-45 mm in length. Head yel-

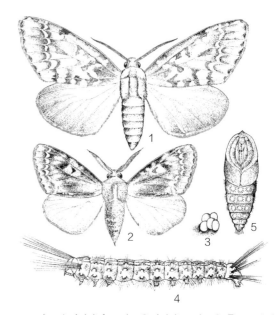

L. monacha. 1. Adult female, 2. Adult male, 3. Eggs, 4. Larva, 5. Pupa (Zhu Xingcai)

L. monacha adult female (Xu Gongtian)

L. monacha adult male (Li Zhenyu)

lowish brown, legs yellow. Mature larvae color vary considerably from light purple, ivory yellow to dark gray. Prothorax with two large granules on the back, each with black hair tufts pointing forward. The 6th and 7th abdominal segments each with a small yellowish red eversible gland.

Pupae Body 18-25 mm in length. Brown, glossy, end of cremaster with hooks.

Biology and behavior *Lymantria monacha* occurs one generation a year and overwinters as eggs. The overwintering eggs hatch in mid- and late March in Yunnan, April to May in Beijing and northeastern China. The newly hatched larvae feed on egg shell, and are able to migrate with the aid of wind. Mature larvae construct cocoons and pupae in tree holes, bark cracks and weeds during May and June. Pupal stage lasts 15-20 days. Adults mainly oviposit on the lower part of tree trunks with a diameter above 20 cm. Each female lays an average of 200 eggs, 15-20 eggs generally are glued together in an egg clutch, and the egg clutches are covered with yellowish white material.

Prevention and treatment

1. Kill adults with light traps.

2. Spray 20% imidacloprid at 3,000× during larval stage from April to June.

References   Zhao Zhongling 1978, Xiao Gangrou 1992, Zhang Zhizhong 1997, Zhao Zhongling 2003, Xu Gongtian and Yang Zhihua 2007.

(Li Zhenyu, Xu Gongtian)

Lymantria xylina Swinhoe

Classification Lepidoptera, Erebidae
Common name Casuarina moth

Importance *Lymantria xylina* is a major defoliator of Casuarina in the shelter forest in coastal southeastern China. Larval feeding on the bark of twigs or branches results in stunned growth or tree mortality over large area under heavy infestation. It was first discovered in Pingtan County, Fujian province in the 1970s. Since then areas infested by this pest have increased each year. It becomes one of the major forest pests in Fujian today. Damage by this pest over large areas significantly reduced the function of coastal shelter forest, which negatively influences the safety and livelihood of local population.

Distribution Zhenjiang, Fujian, Hunan, Guangdong, Guangxi, Taiwan; Japan, India.

Hosts *Casuarina equisetifolia*, *Acacia confuse*, *Psidium guajava*, *Litchi chinensis*, *Dimocarpus longgana*, *Myrica rubra*, *Eriobotrya japonica*, *Diospyros kaki*, *Punica granatum*, *Pyrus* spp., *Ficus carica*, *Castanea mollissima*, *Mangifera indica*, *Ricinus communis*, *Salix babylonica*, and *Camellia sinensis*.

Morphological characteristics **Adults** Females 22-33 mm long with a wingspan of 30-40 mm, yellowish long white with red and white hairs on vertex. Subcostal line exists on the forewings. Inner cross line obvious only on anterior margin of the forewings. Outer cross line grayish brown, wide. Males 16-25 mm long with a wingspan of 24-30 mm, grayish white. Antennae plumose, black.

Larvae Body blackish gray or yellowish brown, 38-62 mm long with a head capsule of 5.2-6.5 mm wide. Coronal suture surrounded with upside down Y-shaped black marks. Ocellar triangle with C-shaped black markings. Eversible gland reddish brown, cone-shaped and depressed on top.

Biology and behavior *Lymantria xylina* occurs one generation a year and overwinters as fully-developed larvae in eggs in Fujian. Eggs hatch between March and April the following year. Newly hatched larvae concentrate on the surface of the egg masses or hide in the back under direct sunshine or strong winds. They balloon in the wind with silk threads to other branches a few days later. Early feeding by young larvae creates notches on twigs. By the 3rd instar, they feed along the twig towards the tip on one side, then downwards to the base on the

L. xylina adult and egg mass (Huang Jinshui)

L. xylina larva (Huang Jinshui)

L. xylina pupa (Huang Jinshui)

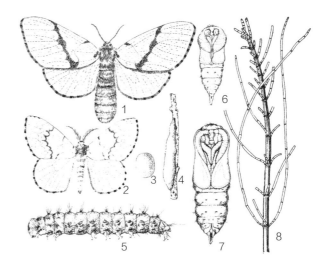

L. xylina. 1. Adult female, 2. Adult male, 3-4. Egg and egg mass, 5. Larva, 6-7. Pupa, 8. Damage (Li Yougong)

other side. They usually cut the twigs off in the middle. There are seven instars in larval stage. Few larvae go through 6 or 8 instars. Larval stage lasts for 45-64 days. Mature larvae attach themselves to host twigs, branch joints, or trunks in mid- to late May with a few silk threads and the anal spine. Pupation occurs 1-3 days later. Pupal stage lasts for 5-14 days.

Adult emergence starts in late May and ends in early June. Most females emerge between 12:00-18:00. Females usually are less active and stay on branches or craw slowly. They are capable of flying short distance. Most males emerge between 18:00-24:00 and are very active. They are strongly attracted to lights and are capable of long flights especially during mate-seeking period. Each female produces a single egg mass, which contains an average of 1,019 eggs. Most eggs are laid on twigs, with a few on tree trunks. About 70% of the eggs are found 2-4 m above the ground on trees. Some eggs are laid as high as 9 m above the ground.

Prevention and treatment

1. Silviculture. Select resistant varieties for forest regeneration and create mixed stands to prevent pest outbreaks.

2. Manual control. Destroy eggs masses or aggregated young larvae through pruning, traps adults with black lights.

3. Biological control. Apply *Beauveria bassiana*, NPV, *B.t.* for larvae before the 4th instars. Preserve natural enemies such as *Ooencyrtus* spp. and *Xanthopimpla pedator*.

4. Chemical control. Apply organic phosphates, carbamates, pyrethroids such as 10% bifenthrin, 2.5% deltamethrin, 2.5% Dimilin, 25 g/L lamda-cyhalothrin.

References   Li Yougong and Chen Shunli 1981; Huang Jinshui and He Yiliang 1988; Huang Furong 2000; Zhao Zhongling 2003; Wei Chujiang, Xie Dayang, Zhuang Chenhui et al. 2004; Xu Yaochang 2005; Zhu Junhong and Zhang Fangping 2004.

(Huang Jinshui, Cai Shouping)

481 *Orgyia antiguoides* (Hübner)

Classification Lepidoptera, Erebidae
Synonyms *Notolophus ericae* Germar, *Orgyia caliacrae* Caradja, *Bomby antiguoides* Hübner

Distribution Beijing, Hebei, Liaoning, Jilin, Heilongjiang, Shandong, Shaanxi, Gansu, Ningxia, and Qinghai; Russia and Europe.

Hosts *Populus* spp., *Salix* spp., *Pinus* spp., *Ulmus pumila*, *Betula* spp., *Quercus* spp., *Hedysarum scoparium*, *Hedysarum leave*, *Hdysarum leave*, *Fagus longipetiolata*, *Elaeagnus angustifolia*, *Malus pumila*, *Pyrus* sp., *Prunus salicina*, *Crateagus pinnatifida*, *Caragana korshinskii*, *Calligonum mongolicum*, *Agriophyllum squarrosum*, *Haloxylon ammodendron*, *Myrica rubra*, *Rhamnus davurica*, *Tamarix chinensis*, *Rosa rugose*, *Malus spectabilis*, *Rosa hybrida*, *Rhododendron simsii*, *Ziziphus jujube* var. *spinosa*, *Glycine max*, *Ammopiptanthus mongolicus*, and *Hippophae rhamnoides*.

Damage As a common forest defoliator, young larvae of *Orgyia ericae* feed on young branches and flowers, resulting in damage to young seeds and seed pods. Entire trees can be defoliated during heavy infestation.

Morphological characteristics **Adults** Females 14.0-16.3 mm long with reduced wings, yellowish brown and covered by white circular villi. Males 8-10 mm long with a wingspan of 21-28 mm, black brown. Antennae bipectinate. Forewings ochraceous brown with a triangular purplish gray mark anteriorly. Cross veins ochraceous brown, crescent, and with a clear white mark on the out margin. Marginal hairs yellowish. Hindwings dark ochraceous brown, with dense long hair at the base. Marginal hairs yellowish. **Eggs** Flat oval, about 1 mm in diameter, white and with one small brown spot in the middle. **Larvae** Mature larvae 24.4mm long, yellowish green with black dorsal lines. Prothorax contains a long and forward inclining black feathery hair bundle laterally. There is a whitish yellow hair-brush dorsally on abdominal segments 1-4. Abdominal segment 8 contains a long black feathery hair bundle dorsally. Head and legs black. Verruca withered yellow and contains grayish long hairs. Eversible gland withered yellow. **Pupae** Body 13.9 mm long and yellowish brown for females, and 11.0 mm long and dark brown for males, with three groups of short white villi dorsally. **Cocoons** Body 9-15 mm long, grayish white or grayish yellow, with some setae and broken leaf portions tied by silken threads.

Biology and behavior *Orgyia antiguoides* occurs two generations a year and overwinters as eggs inside cocoons in Inner Mongolia. Overwintering eggs hatch between mid-May and early June. Newly hatched larvae stay inside cocoons for 5-7 days before emerging from the ostium and feed on young leaves and fruits or migrate to

O. antiguoides adult male (Li Zhenyu)

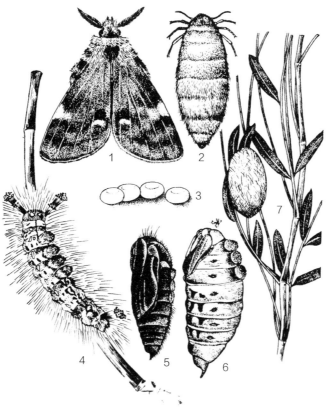

O. antiguoides. 1. Adult male, 2. Adult female, 3. Eggs, 4. Larva, 5-6. Pupa, 7. Cocoon (Zhu Xingcai)

other branches aided by wind with silk threads. Mature larvae construct cocoons for pupation on branches between early June and late July. Adults emerge, mate, and lay eggs between late June and early July. Males usually emerge about a week earlier than females. First generation larvae appear between mid- to late July, pupate in mid- to late August, emerge, mate, and lay eggs from late August to late September. Eggs overwinter inside cocoons.

Egg stage lasts for 8-14 days (about 8 months for overwintering generation). Larval stage lasts for 18-42 days with an average of 30 days. There are four to five instars in larval stage. Pupal stage lasts for 11-14 days. Adults live for 1-3 days for males and 4-11 days for females. Males actively seek females during the day and mate with females inside cocoons through the ostium. Females have reduced wings and cannot fly. They stay inside cocoons for their entire life. Females produce powerful sex pheromones to attract males for copulation. Females lay eggs inside the cocoons after mating and die shortly after. Each female produces 104-415 eggs, with an average of 250 eggs. Egg hatching is well synchronized for the first generation, but not so for the second generation.

Larvae stay away from direct sunlight and high temperature and feed mostly in early mornings and late afternoons during cool weather. They choose to stay on the back of the leaves or fruits when air temperature is higher than 34°C. Larvae of the 1st and 2nd instars can balloon in the wind with silk threads during migration. The 5th and 6th instar larvae will curve their body and fall to the ground when disturbed.

Prevention and treatment

1. Manual control. Remove cocoons and eggs in early spring, June to August, and winter to reduce source populations.

2. Trapping adults with black lights or semiochemicals.

3. Preserve and utilize natural enemies such as parasitoids *Telenomus dalmani*, *Tetrastichis* spp., *Aprostocetus* sp., *Monodontomerus monor*, *Pimpla disparis*, *Itoplectis viduata*, *Pales pavida*, *Exorista larvarum*, Pathogen OeNPV, predators *Pica pica sericea*, *Cyanopica cyana*, *Harmonia axyridis*, *Coccinella septempunctata*, *Arma chinensis*, *Argiooe bruennichii*, and *Araneus ventricosus*.

4. Chemical control. Apply 20% Dimilin I at 8,000× or 20% mixture of fenvalerate and fenitrothion at 2,000×, Apply 5.7% eamamectin at 3,000-5,000× to control 1st and 2nd instars.

References Zhao Zhongling 1978; Liu Fabang, Zhao Jixing, Yu Chengren et al. 1991; Wang Xinming and Xu Zhaoji 1992; Zu Aimin and Dai Meixue 1997; Wang Xiong and Liu Qiang 2003; Su Mei, Yang Fenyong, Liu Xiufeng et al. 2007; Li Haiyan et al. 2009.

(Mu Xifeng, Li Suo, Wang Xinming, Xu Zhaoji)

482 *Orgyia antiqua* (L.)

Classification Lepidoptera, Erebidae
Synonyms *Phulaena gonostigma* L., *Gynaephora recens* Hübner, *Orgyia gonostigma* (L.)
Common names Vapourer moth, Top spotted tussock moth

Distribution Beijing, Tianjin, Hebei, Shanxi, Inner Mongolia, Liaoning, Jilin, Heilongjiang, Jiangsu, Zhejiang, Shandong, Henan, Hubei, Hunan, Chongqing, Sichuan, Guizhou, Shaanxi, Gansu, and Ningxia; Korea Peninsula, Japan, Russia and Europe.

Hosts *Salix* spp., *Populus* spp., *Betula* spp., *Alnus cremastogyne*, *Corylus heterophylla*, *Fagus longipetiolata*, *Quercus* spp., *Pyrus* sp., *Malus pumila*, *Sorbus pohuashanensis*, *Crateagus pinnatifida*, *Larix* sp., *Rubus corchorifolius*,*Ulmus pumila*, *Platanus* sp., *Paulownia* sp., *Carpinus turczaninowii*, *Prunus salicina*, *Prunus mume*, *Prunus pseudocerasus*, *Zanthoxylum bungeanum*, *Amelanchier* sp., *Rosa chinensis*, *Malus spectabilis*, *Camellia japonica*, *Magnolia denudate*, *Robinia hispids*, *Castanea mollissima*, *Juglans regia*, *Picea asperata*, *Pinus* sp., *Cannabis sativa*, *Arachis hypogaea*, *Glycine max*, *Lespedeza bicolor*, *Prunus persica*, and *Prunus armeniaca*.

Damage Larvae feed on young buds, leaves and flowers.

Morphological characteristics **Adults** Sexually dimorphic. Females 12-25 mm long, oblong oval with vestigial wings. Body covered by gray and yellowish white villi. Males 11-15 mm long with a wingspan of 25-36 mm, body grayish brown. Forewings reddish brown with a yellow mark on the top corner. Hindwings contain a white crescent mark at the anal corner.

Eggs Near round and depressed at the micropyle, 0.7-0.9 mm in diameter and 0.6 mm in height, greenish initially, turn to yellowish later, and grayish brown before hatching.

Larvae Mature larvae 33-40 mm, dark gray, both sides of the thorax with a forward-pointing black hair bundle. Abdominal segment 8 contains a backward-pointing black hair bundle. Back of abdominal segments 1-4 contain a 1.8-2.8 mm long brownish yellow hair-brush in the center.

Pupae Females spindle-shaped, 9-11 mm long, yellowish brown to dark brown with three groups of white shirt villi. Males cylindrical, 13.0-13.9 mm long with sparse yellow villi dorsally.

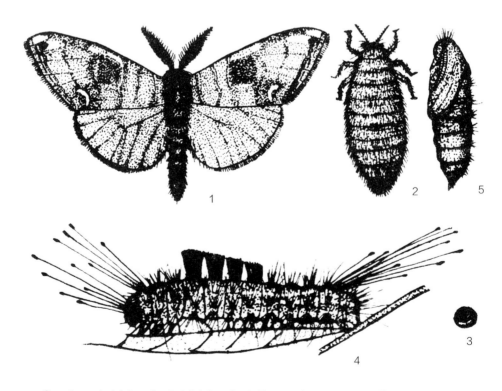

O. antiqua. 1. Adult male, 2. Adult female, 3. Eggs, 4. Larva, 5. Pupa (Sun Qiaoyun)

O. antiqua adult male (Yu Juxiang)

O. antiqua larva (Yu Juxiang)

Cocoons Grayish yellow with a loose out layer and dense inner layer. Female cocoons 12-20 mm long, cylindrical, while male cocoons 11 mm long and spindle-shaped.

Biology and behavior *Orgyia antiqua* occurs three generations a year and overwinters as 2nd or 3rd instar larvae inside tied leaves on trees, among leaves on the ground, or other materials. Larvae resume feeding during leaf expansion in the spring, with most damage observed between late March and early April. Pupation starts in mid-April when mature larvae tie up two to three leaves. Pupal stages lasts for 10-13 days. Adults appear between late April and early May. Males live for about two days. Females lay eggs in masses near cocoons, with 300-500 eggs/mass. Egg stage lasts for 11-13 days. First generation larvae appear in mid-May, with most damage occur on new branches and leaves in early June. First generation pupae appear in mid-June, with a pupal period of 7-8 days. Adults emerge in late June. Egg stage for this generation lasts for 10-11 days. Larvae of the second generation appear in early July, with most damage occurs in mid-July. Second generation pupae appear in late July, and adults emerge in early August. After hatching in late August, larvae feed on host until late September or early October before overwintering. Larvae can balloon with silk threads in the wind. They usually stay on the underside of the branches or at the base of the buds and become active after dawn.

Prevention and treatment

1. Physical control. Remove loose barks and leaves, weeds in the winter to reduce larvae overwintering sites, remove infested branches during the growing season.

2. Population monitoring with semiochemical such as

O. antiqua eggs, cocoon, and adult female (Yu Juxiang)

(Z,Z)-6,9-heneicosadien-11-one.

3. Biological control. Protect natural enemies such as *Exorista larvarum*, *Bracon* sp., *Pimpla disparis*, *Itoplectis riduata*, *Aprostocetus* sp., *Tetrastichus* sp., *Dibrachys cavus*, *Monodontomenus minor*, and *Casinaria nigripes*. Apply OgNPV can result in 83.3% larval mortality.

4. Chemical control. Apply 5 °Bé lime sulphur during budding, apply 20% Dimilin at 5,000-8,000×, 25 g/L lamda-cyhalothrin at 1,000-1,500×, or apply 5.7% eamamectin at 3,000-5,000× during outbreaks.

References Zhao Zhongling 1978; Sun Qiaoyun, Tai Xiaoling, and Xu Longdi 1985; Sun Qiaoyun 1992; Wu Qingbiao and Bai Yinzhen 2005; Li Haiyan, Zong Shixiang, Sheng Maoling et al. 2009; Chen Guofa, Li Tao, Sheng Maoling et al. 2011;

(Mu Xifeng, Li Suo, Li Zhenyu)

Orgyia postica (Walker)

483

Classification Lepidoptera, Erebidae
Synonyms *Lacida postica* Walker, *Orgyia ceylanica* Nietner
Common names Cocoa tussock moth, Hevea tussock moth

Distribution Fujian, Guangdong, Guangxi, Yunnan, and Taiwan; Sri Lanka, India, Myanmar, the Philippines, Indonesia, Malaysia, Australia, and Japan.

Hosts *Camellia sinensis, Hevea brasiliensis, Mangifera indica, Eucalyptus* spp., *Acacia mearnsii, Albizzia julibrissin, Citrus reticulate, Morus alba, Malus pumila, Prunus persica, Pyrus bretschnerderi, Sapium sebiferum, Casuarina equisetifolia, Canarium album, Dalbergia balansae, Vitis vinifera, Ricinus communis, Gossypium* spp., *Fagopyrum esculentum, Glycine max, Arachis hypogaea, Ipomoea batatas, Solanum tuberosum, Brassica oleracea, Solanum melongena,* and *Allium fistulosum.*

Morphological characteristics **Adults** Males with a wingspan of 22-25 mm. Body and legs dark red, antennal shaft brownish, antennal teeth brownish black. Forewings dark reddish brown with black declivent basal line and wavy inner cross line. Cross veins brown with black and white margins. Outer cross line wavy, black. Area between posterior margin of the median cell and the inner line gray. Sub-marginal line doubled, wavy and black. Sub-marginal area gray with longitudinal black marks. Marginal line consists of a series of broken dark brown lines. Marginal hairs dark brown with dark brown marks. Hindwings dark brown with brown marginal hairs. Females about 15 mm long with reduced wings, yellowish white with a darker end. Head and thorax small. Eggs visible inside the abdomen.

Eggs Sphere, truncate on top, 0.7 mm in diameter, with brownish punctuations on surface.

Larvae Mature larvae about 36 mm long, yellowish and covered by sparse villi. Dorsal and sub-dorsal line brown. There is a long brown hair bundle on prothorax laterally and the 8th abdominal segment dorsally, and one yellow hair-brush dorsally from abdominal segment 1 to 4, and one gray long hair bundle laterally on abdominal segments 1-2. Head orange. Eversible gland reddish brown.

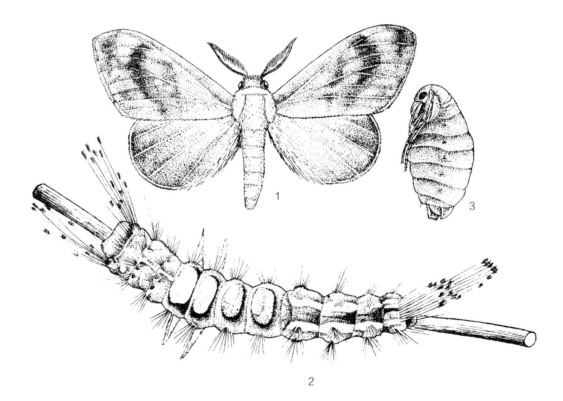

O. postica. 1. Adult male, 2. larva, 3. Pupae (Zhu Xingcai)

O. postica larva (edited by Zhang Runzhi)

O. postica larva (edited by Zhang Runzhi)

Pupae Yellowish brown and about 18 mm long.

Cocoons Yellow, oval, rough, and with black setae on the surface.

Biology and behavior *Orgyia postica* occurs six generations a year in Sanshui County in Guangdong province, with the 1st to 6th generation appear between late March to early May, early May and mid-June, mid-June to late July, mid-July to late September, late September to mid-November, and late December to late March of the following year, respectively. Generations overlap with each other. All life stages are seen between June and August. Overwintering larvae can resume feeding if temperature is high. They start to pupate in early March. Females lay eggs on cocoons or other nearby materials. Each female produces an average of 383 eggs. Egg stage lasts for 6-9 days in the summer and 17-27 days in the winter. Larval stage lasts for 8-22 days in the summer and 24-61 days in the winter. Pupal stage lasts for 4-10 days in the summer and 15-25 days in the winter. Each generation lasts for 40-50 days.

Newly hatched larvae feed on host aggregately first then disperse. They can completely defoliate the hosts during outbreaks. Natural enemies such as *Apanteles liparids*, *Brachymeria obscurata*, *Exorista sorbillans*, *Exorista larvarum*, and *Zenilla modicella* are abundant after outbreaks.

References Li Fengsun et al. 1933, Zhao Zhongling 1978.

(Chen Qingxiong)

484 *Pantana phyllostachysae* Chao
Classification   Lepidoptera, Erebidae

Importance   *Pantana phyllostachysae* is one of the most important bamboo pests in southern China. Defoliation during outbreaks results in reduction for new shoots. Continuous damage leads to tree mortality.

Distribution   Jiangsu, Zhejiang, Anhui, Fujian, Jiangxi, Hubei, Hunan, Guangdong, Guangxi, Chongqing, Sichuan, Guizhou, Yunnan, and Tibet.

Hosts   *Phyllostachys pubescens, P. viridis, P. glauca, P. iridenscens, P. praecox, P. nuda, P. propinque, P. bissetii, P. sulphurea,* and *Pleioblastus amarus*.

Damage   Larvae cause host defoliation similar to fire damage during outbreaks.

Morphological characteristics   **Adults** Wingspan 32-35 mm for females and 26-30 mm for males. Body yellow in color, lighter for females. Antennae pectinate, shaft yellowish whit, teeth grayish black. Antennal teeth short and sparse in females. Forewings yellowish for females and yellowish to brownish yellow for males. There is an orange red mark at the middle of the posterior margin. Hindwings lighter in color.

Eggs Drum-shaped. 0.9 mm in height, yellowish white, with a brownish ring on top.

Larvae Mature larvae 20-25 mm long, grayish brown and covered by black and yellowish white long hairs. There is a verruca on both sides of the prothorax. Each verruca contains a forward-pointing black hair bundle consists of feathery scales. There are reddish brown hairbrushes in the middle of abdominal tergum for segment 1-4 and 8. Hairbrush on abdominal segment 8 with feathery scales that forms a backward-pointing long hair bundle.

Pupae Body 9-17 mm long, yellowish brown to reddish brown. Each segment is covered by yellowish white hairs. There are about 30 anal spines with hooked ends. Cocoons oval, thin, and covered by setae from larvae.

Biology and behavior   *Pantana phyllostachysae* occurs three generations a year in Lishui in Zhejiang, Nanping in Fujian, Qimen in Anhui, and Jinping in Yunnan; and four generations a year in Shangrao in Jiangxi and in Sichuan. It overwinters as eggs and $1^{st}/2^{nd}$ instar larvae on host trees. Overwintering larvae resume feeding in late February in Zhejiang. Overwintering eggs begin to hatch at the same period. Adults of the overwintering generation start to emerge in late May and peak in early June. Adults of the first generation start to emerge in late July and peak in mid-August. Adults of the second generation start to emerge in early October and peak in late October. Larvae damage period is from late February

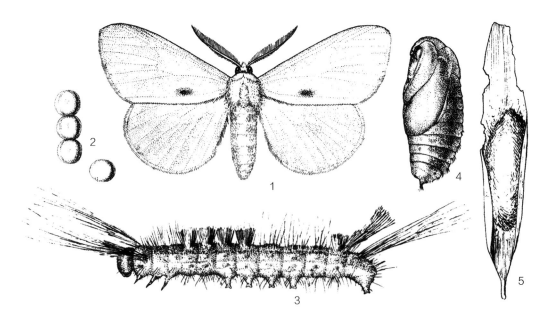

P. phyllostachysae. 1. Adult male, 2. Eggs, 3. Larva, 4. Pupa, 5. Cocoon (Zhang Xiang)

to mid-June for the overwintering generation, mid-June to early August for the first generation, mid-August to early October for the second generation, early October to mid-November for the third generation. By November, eggs and the newly hatched larvae overwinter. In Jiangxi where there are four generations a year, damage period is from mid-March to late April, late May to late June, July, and mid-August to early October, respectively.

P. phyllostachysae adult female (Xu Tiansen)

P. phyllostachysae adult male (Xu Tiansen)

P. phyllostachysae adult female (edited by Zhang Runzhi)

P. phyllostachysae adult male (edited by Zhang Runzhi)

P. phyllostachysae 4th instar larvae (Xu Tiansen)

P. phyllostachysae cocoons and pupae (Xu Tiansen)

P. phyllostachysae mature larva (Xu Tiansen)

Parasitic fly larva emerged from *P. phyllostachysae* larva (Xu Tiansen)

P. phyllostachysae damage (edited by Zhang Runzhi)

P. phyllostachysae eggs (edited by Zhang Runzhi)

Adults are strongly attracted to lights. They stay in the crown or in weeds during the day and seek mates, mate, and lay eggs at night. Females lay eggs about five hours after mating. Egg-laying period lasts for about two days. Adults live for 8-13 days. Egg period lasts for 8-15 days for the first and second generation, but for as long as 4-5 months for the overwintering generation. Eggs of the overwintering generation can hatch before winter, contain well developed embryos but do not hatch until the following spring, or finish development in the spring before hatching. Newly hatched larvae can balloon with silk threads when disturbed. Larvae older than the 3rd instars will curve and fall to the ground after being dis-

P. phyllostachysae larva (edited by Zhang Runzhi)

Braconid cocoons on *P. phyllostachysae* larva (edited by Zhang Runzhi)

P. phyllostachysae cocoon (edited by Zhang Runzhi)

turbed. Mature larvae usually construct cocoons between bamboo leaves, with some on the trunk or in the bushes. Larvae in cocoons lose their long body hairs with only brown hair brushes on abdominal segment 1-4 and 8 as well as villi intact and enter prepupal period. Larval stage lasts for 40-46 days for the 1st and 2nd generation. Larval stage for the overwintering generation is about three months for those hatched in the spring and six months those overwinter as larvae. Prepupal stage lasts for 2-3 days. Pupal stage lasts for 6-14 days.

Prevention and treatment

1. Stop moving infested materials.

2. Trap adults with kerosene lights at 15-30 lights/ha, or black lights.

3. Conserve natural enemies. Natural enemies are important factors regulating pest population. Egg parasitoids such as *Telenomus* sp. and *Anastatus* sp. can parasitize 50%-70% eggs after outbreaks, whereas larval parasitoid *Apanteles* sp. can parasitize 40% of the larvae. Other natural enemies include birds, amphibians are also important in pest population dynamics.

4. Apply *Beauveria bassiana* at 5×10^9 conidia/mL or *B.t.* at 8,000 IU/μL resulted in 60%-70% population reduction.

5. Apply 2.5% deltamethrin or fenvalerate through fogging for larvae and adults.

References Zhao Zhongling 1977; Jiangxi Forest Pest Control Station Entomology Section 1979; Chen Hanlin 1980, 1983; Xiao Gangrou 1992; Jiang Jiadan 2001; Xu Tiansen and Wang Haojie 2004; Jiang Ping and Xu Zhihong 2005.

(Xu Tiansen, Chen Hanlin, Xu Zhenwang, Liang Xiangmei, Xing Longping)

485 *Pantana sinica* Moore
Classification Lepidoptera, Erebidae

Distribution Shanghai, Jiangsu, Zhejiang, Anhui, Fujian, Jiangxi, Hubei, Hunan, Guangdong, Guangxi, Chongqing, Sichuan, Guizhou, Yunnan.

Hosts *Phyllostachys heteroclada* var. *pubescens*, *P. aureosulcata*, *P. aureosulcata* f. *aureocaulis*, *P. bissetii*, *P. dulcis*, *P. flexuosa*, *P. guizhouensis*, *P. heteroclada*, *P. heterocyla* 'Pubescens', *P. nidularia* 'Smoothsheath', *P. iridescens*, *P. glauca*, *P. nigra*, *P. rutile*, *P. sulphurea* 'Viridis', *P. ivax*, *P. heterolada* var. *funhuaensis*, and *P. praecox*.

Damage Larval feeding on leaves could result in tree mortality and reduced shoot production under heavy infestation. Larval setae are irritating to some people.

Morphological characteristics **Adults** There are three adult forms: females, winter males, and summer males. Females 12-16 mm long with a wingspan of 35-39 mm, Antennae grayish white with short and black tooth. Compound eyes black. Labial palps orange yellow. Head, forewings, and abdomen grayish white or slightly brownish. Forewings white with brown scales at the base, and on the anterior and out margins. There is a black mark between M_2 and M_3, M_3 and Cu_1, Cu_1 and Cu_2, and Cu_2 and Cu, respectively. Hindwings milky white. Abdomen and legs grayish white with some brownish yellow. Winter males 9-13 mm long with a wingspan 29-35 mm. Antennae plumose, black or grayish black. Compound eyes black. Labial palps rusty yellow. Head and prothorax grayish white or grayish yellow. Abdomen black. Villi on the sides and posterior end of metathorax gray. Front margin, area between center cross line and outer line of the forewings partially or completely black. Light colored along the veins. Marks on forewings similar to those on females. Hind wings white, dark gray at the base and apical corner in a few individuals. Summer males similar in size to winter males, but completely black, Cu and 2A veins brown. Abdomen and legs grayish white with trace of brown.

Eggs Oval, slightly flat, 0.9 mm wide, and 0.8 mm in height. Flat on top, depressed at the center and surrounded by a brownish ring.

Larvae Newly hatched larvae 2.5 mm long, yellowish white with black scales. The verruca on the lateral side of prothoracic contains a black long hair bundle. Third instars yellowish white or gray, with hairbrush on the back of the abdominal segment 1-4. Head of the 4th instars dark, with longer hairbrushes. Dorsal line, subdorsal line grayish black and interrupted by grayish white hairs. Mature larvae dark yellowish brown, 20-31 mm long. Verruca on the lateral side of prothoracic with a conspicuous black and long hair bundle. Dorsal line black and wide. Subdorsal line and upper spiracle line grayish white. Hair brushes on abdominal segment 1-4 brownish red. Abdominal segment 8 contains a backwards back and long hair bundle which accompanied by two black hairy verrucae at the base. Verrucae on each segment and the subdorsal line contain short hair bundles. There are 5-7 (mostly 6) instars for the 1st and 3rd generation, and 6-8 (mostly 7) instars for the second generation.

Pupae Length 11-15 mm for males and 16-19 mm for females. Orange yellow. There is a seta on both sides of the frons. Body covered by dense yellowish white short hairs dorsally. Hairs on the back of thorax long. Antennae of males large and cover 2/5 of the forewing pads, and reach to the middle of the legs. Pads of the forewings reach to the end of the 4th abdominal segment. Anal spines with a lot of hooks; antennae on females stout and wide, as long as maxilla. Pads for forewings reach only the middle of the 4th abdominal segment.

Cocoons Spindle-shaped, 18-26 mm long, grayish yellow or yellowish brown, silky; thin for the summer generation and double-layered for the winter generation. The outer layer for the winter generation very fine, grayish black and covered with

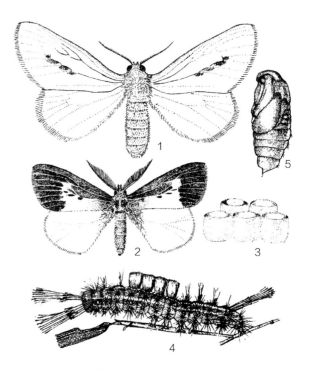

P. sinica. 1. Adult female, 2. Adult male, 3. Eggs, 4. Larva, 5. Pupa (Xu Tiansen)

P. sinica adults (From top to bottom: summer type male, winter type male, female) (Xu Tiansen)

P. sinica male pupa (Xu Tiansen)

P. sinica mature larva (Xu Tiansen)

P. sinica larva killed by ichneumonid wasp (Xu Tiansen)

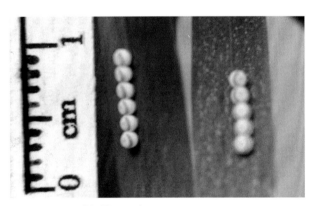

P. sinica eggs (Xu Tiansen)

a few hairs, while the inner layer loose and yellowish brown.

Biology and behavior *Pantana sinica* occurs three generations a year and overwinters as pupae in Zhejiang. Adults start to emerge in late April and emergence ends in late May. Egg stage of the 1st to the 3rd generations appears from late April to late May, early June to early August, and late August to late September, respectively. Larval stage of the 1st to the 3rd generations appears between early May and mid-July, early July and early August, early September and early December, respectively. Pupal stage of the 1st to the 3rd generations appears from early June to mid-July, early August to late September, and early October to early May of the following year, respectively. Adult stage of the 1st to the 3rd generations appears from mid-June to early August and mid-August to late September, respectively. Mating and egg-laying occur right after emergence. Egg stage lasts for 11-13, 7-8, and 7-9 days for the 1st, 2nd, and 3rd generation, respectively. Larval stage lasts for 34-49, 28-42, and 30-74 days for the 1st, 2nd, and 3rd generation, respectively. Pupal stage lasts for 10-21, 10-13, and 168-192 days for the 1st, 2nd, and 3rd generation, respectively. Adults live for 7-8, 5-14, and 4-15 days for the 1st, 2nd, and 3rd generation, respectively. Average generation time is 31, 64, and 248 days for the 1st, 2nd, and 3rd generation, respectively.

Prevention and treatment Conserve natural enemies. Egg parasitoids include *Trichogramma dendrolimi*, *Trichogramma ivelae*, *Telenomus dasychili*, *Telenomus euproctidis*. Larval parasitoids include *Apanteles liparidis*, *Enicospilus* sp., *Rogas* sp. and *Coccygomimus luctuosus*. There is a hyperparasitoid, *Mesochorus* sp. Parasitism by *Trichogramma* spp. after release in the field is between 28%-45%. Natural parasitism by *Telenomus* spp. is between 8.8%-28.5%. Larval parasitoids include two species. The parasitism rate is 4%-6% by *A. liparidis* and 1%-3% by *Rogas* sp.

References Zhao Zhongling 2003, Xu Tiansen and Wang Haojie 2004.

(Xu Tiansen, Wang Haojie)

486 *Parocneria furva* (Leech)

Classification Lepidoptera, Erebidae
Synonym *Ocneria furva* Leech
Common name Juniper tussock moth

Importance Damaged hosts suffer stunted growth, reduced vigor and wilt gradually. The infested hosts are susceptible to secondary pests such as long-horned beetles and bark beetles. Severe infestation can lead to tree mortality.

Distribution Beijing, Hebei, Shanxi, Inner Mongolia, Liaoning, Jilin, Heilongjiang, Jiangsu, Zhejiang, Anhui, Fujian, Shandong, Henan, Hubei, Hunan, Sichuan, Guizhou, Shaanxi, and Gansu; Japan.

Hosts *Platycladus orientalis*, *Juniperus chinensis*, *Sabina vulgaris*, and *Fokienia hodginsii*.

Damage Larvae feed on needle tips to cause damage. Larvae of the overwintering generation feeds on young leaf buds.

Morphological characteristics **Adults** Wingspan is 19-27 mm for males and 20-34 mm for females. Male antennae plumose and black. Body and wings grayish brown. Forewings with thin, inconspicuous black marks, inner line bends outward between median cell and Cu_2. Outer line and sub apical line inconspicuous, serrate, distinct at the posterior of M_1 and Cu_2, and area around Cu_2 grayish white. Marginal hairs black mixed with gray hairs. Hindwings lighter with gray marginal hairs. Females with lighter body color and clearer wing marks.

Eggs Oblong oval, 0.9-1.0 mm in diameter. Fresh green and shiny initially, gradually turn to yellowish brown, and dark brown before hatching, with punctures on surface.

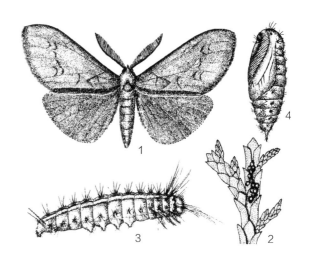

P. furva. 1. Adult, 2. Eggs, 3. Larva, 4. Pupa (Zhang Peiyi)

Larvae Mature larvae 20-30 mm long. Head grayish black or yellowish brown with dark brown marks. Body greenish gray or grayish brown. Dorsal line dark green. Each body segment contains brownish white verruca with yellowish brown and black setae. Dorsal surface of the abdominal segment 3, 7, 8, and 11 white. Subdorsal line dark green between abdominal segment 4 and 11. Area between subdorsal line and spiracle line contains

P. furva adult (Xu Gongtian)

P. furva adult (Xu Gongtian)

white marks. Abdomen yellowish brown ventrally. There is one reddish eversible gland on the back of abdominal segment 6 and 7.

Pupae Length 10-13 mm, green and turn to brown before emergence. Each body segment contains 8 white marks with a few white hairs. Spiracle black. Anal spine brown.

Biology and behavior *Parocneria furva* occurs two generations a year and overwinters as larvae in scales, twigs, or bark crevices in Beijing. Larvae resume feeding between late March and early April the following year. Adults of the first generation emerge in late June, while adults of the second generation emerge between late August and early September.

P. furva adult (edited by Zhang Runzhi)

P. furva adult female (Xu Gongtian)

P. furva eggs (Xu Gongtian)

P. furva eggshells (Xu Gongtian)

P. furva larva (Xu Gongtian)

P. furva mature larva (Xu Gongtian)

P. furva pupa (Xu Gongtian)

P. furva pupa (Xu Gongtian)

P. furva damage (Xu Gongtian)

P. furva damage (Xu Gongtian)

Most adults emerge in the morning and are strongly attracted to lights. They stay inside the crown during the day and mate and lay eggs at night. Most eggs are laid on scale surface in groups of 2-40 eggs on the sunny side of the crown or on trees of the forest edges. Larvae stay in bark crevices or between leaves during the day and feed at night. Young larvae feed on the tip or the edge of the needle and create notches. Larvae after the third instar consume entire needles. Most damage occurs between July and August. Larvae stop feeding and enter aestivation when temperature is too high in the summer. They resume feeding when temperature drops in the fall. Mature larvae secrete silk between leaves before pupation.

Occurrence of *Parocneria furva* in the field correlates with environmental factors. Heavy infestation usually occurs in stands with fewer species, high canopy cover. Sparse forests, young stands, Chinese thuja and black locust mixed forests rarely suffer damage from this pest.

Prevention and treatment

1. Silviculture. Avoid monoculture, timely cultivation to improve tree vigor can reduce pest occurrence.

2. Trap adults with light traps.

3. Conservation of natural enemies. Protect parasitoids such as *Telenomus* spp. in egg stage, *Exorista sorbillans* and *Carcelia* spp. in larval stage, *Brachymeria lasus* and *Apanteles* spp. in pupal stage; and predators such as *Arma chinensis* and *Vespa* spp.

4. Biological control. Apply *B.t.* at 8,000 IU/μL at 1,000-1,500× for control of older larvae.

5. Chemical control. Apply 20% Dimilin at 6,000-8,000× to control young instars,

References Wang Yongjun 1992, Zhang Zhongling 2003, Xu Gongtian 2007.

(Pan Yanping, Huang Pan, Wang Yongjun)

487 *Parocneria orienta* Chao
Classification Lepidoptera, Erebidae

Distribution Zhejiang, Fujian, Hubei, Hunan, Sichuan.

Hosts *Cupressus funebris*, *Platycladus orientalis*.

Damage Larvae feed on needle tips or new needles could lead to complete defoliation during outbreaks, resulting in branch dieback.

Morphological characteristics **Adults** Males 12-15 mm long with a wingspan of 29-35 mm. Antennae brownish with dark brown flagella. Head and thorax grayish brown. Legs grayish brown with white marks. Forewings whiter or brownish white with brown to dark brown marks. Marginal hairs white mixed with brown to dark brown hairs. Females 18-20 mm long with a wingspan of 33-45 mm, lighter than males with clearer marks. Hindwings grayish white with brown to dark brown margins.

Eggs Oblong oval, 0.6-0.8 mm in diameter and depressed at the center. Dark green initially, turn to yellow, grayish brown, or dark brown.

Larvae Body 22-42 mm long, head brownish black. Body green with grayish white and grayish brown marks dorsally and laterally. Verruca red with white and black setae.

Pupae Body 12-20 mm, green or grayish green with yellowish white marks on the abdomen.

Biology and behavior *Parocneria orienta* occurs two generations a year and overwinters as eggs or larvae in Sichuan. Larvae become active in the early February and cause most damage between late April and mid-May. Adult emergence perks between early and mid-June. Larvae of the first generation appear in late May and cause most of the damage between mid-August and mid-September. Peak adult emergence falls between early and later October. Eggs of the second generation hatch between October and December and overwinter as larvae afterwards.

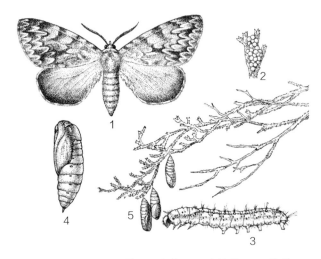

P. orienta. 1. Adult, 2. Eggs, 3. Larva. 4-5. Pupae, 6. Damage (Zhang Xiang)

P. orienta adult (FPC)

P. orienta larva (FPC)

P. orienta eggs (FPC)   *P. orienta* eggs (Wang Yan)

P. orienta pupa (Wang Yan)   *P. orienta* damage (FPC)

Adults are strongly attracted to lights and are active at night. They live for 8.1 days for the overwintering generation and 10.9 days for the first generation. Most eggs are laid on the back of the scales, twigs, or branch joints in the middle or lower crown. Each female produces 240-613 and 180-260 eggs for the overwintering and the first generation, respectively. Egg stage is 14.3 days for the first generation. There are six instars in the larval stage. Mature larvae pupate between leaves or on branches upside down with some silk.

Prevention and control

1. Spray POGV for larvae control. Release *Telenomus* sp. Natural parasitization rate by *Telenomus* sp. on eggs was 74% for the overwintering generation and 80% for the first generation; *Apanteles liparidis* is a common larval parasitoid with a parasitism rate of 20%-30%. Other natural enemies include parasitic flies, POGV.

2. Biological control. Spray *B.t.* (8,000 IU/μL) at 1,000-1,500× to control larvae.

3. Chemical control. Spray 5.7% emamectin at 3,000-5,000×, or 10% imidacloprid at 1,000× for larvae control.

References   Xiao Gangrou 1992.

(Li Menglou, Zeng Chuihui, Li Yuanxiang)

488 *Somena scintillans* (Walker)

Classification	Lepidoptera, Erebidae
Synonyms	*Artaxa scintillans* Walker, *Euproctis scintillans* Walker, *Porthesia scintillans* Walker, *Nygmia scintillans* Walker
Common name	Yellow tail tussock moth

Distribution Zhejiang, Fujian, Henan, Hunan, Guangdong, Guangxi, Sichuan, Yunnan, Shaanxi, Taiwan; India, Indonesia, Myanmar, Malaysia, Pakistan, Singapore, Sri Lanka.

Hosts *Acacia mearnsii*, *Robinia pseudoacacia*, *Camellia sinensis*, *Citrus reticulate*, *Pyrus bretschneideri*, *Dalbergia hupeana*, *Euphoria longana*, *Eriobotrya japonica*, *Claoxylon polot*, *Michelia alba*, *Ricinus communis*, *Zea mays*, *Gossypium* sp., *Litchi chinensis*, and some members of the Cruciferae family.

Damage Larvae feed on leaves, leaf stalk, and inflorescences of *Acacia mearnsii*, which negatively affect its growth.

Morphological characteristics **Adults** Females 9.0-12.3 mm long with a wingspan of 25-37.3 mm, while males 8.0-11.0 mm long with a wingspan of 20.1-27.2 mm. Antennae with yellowish white to yellowish pedicel and yellowish brown flagella. Labial palps orange yellow. Compound eyes large, black. Head and collar plate orange yellow. Thorax and abdomen yellowish brown. Anal hair cluster orange yellow. Legs yellowish with long yellow hairs. Forewings reddish brown, shiny purplish. Inner and outer cross lines yellow and curved outwards although not clear on some individuals. Anterior margin, out margin, and marginal hairs orange yellow, with the yellow out margin and marginal hairs subdivided into three sections by brown portions. Hindwings yellow.

Eggs Flat oval, 0.64-0.72 mm in diameter and 0.46-0.60 mm in height, depressed at the center. Surface smooth and shiny. Yellow initially, turn to reddish brown gradually.

Larvae Newly hatched larvae 3.19-4.40 mm in length with a head capsule of 0.30-0.48 mm in width. Head brown, body yellowish with yellow dorsal lines. Mature larvae 13.4-23.5 mm in length with a head capsule of 1.63-2.40 mm in width. Head brownish, body dark

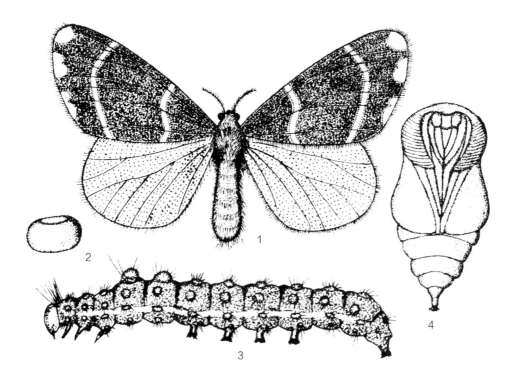

S. scintillans. 1. Adult, 2. Egg, 3. Larvae, 4. Pupa (Chen Shunli)

S. scintillans adult (NZM)

brown. Prothorax contains three yellow longitudinal marks dorsally with orange red, forward protruding lateral verruca. Mesothorax with two yellow longitudinal marks and three yellow transverse marks dorsally. The 2nd verruca on mesothorax orange red. The 2nd and 3rd verruca on abdominal segments 1-8 black and contain dark long brown hairs and short white hairbrushes. Abdominal segment 9 contains a "Y" shaped yellow mark dorsally. Yellow dorsal line between abdominal segments 3-7 wide with dark red fine line in the middle. Sub-spiracle line orange yellow. Eversible glands yellow. Spiracle oval, brownish. Thoracic legs dark brown. Prolegs orange yellow with grayish black external margins. Crotchets in uniserial semi-circles.

Pupae Oval, dark brown, 10.9-13.8 mm long for females and 8.7-10.8 mm for males. Prothorax hairy dorsally but hairs not in clusters. Mesothorax protruding in the middle with one central carina and surrounded by two long setal clusters. Anal spine cone-shaped, with 26 spinets apically.

Cocoons Oblong oval, reddish brown, silky, tight, with sparse setae on surface. Males 11.8-18.2 mm and females 16.0-23.6 mm in length.

Biology and behavior *Somena scintillans* occurs seven generations a year and overwinters as third or older instar larvae on *Acacia mearnsii* in Nanping of Fujian. Larvae can feed during the winter if air temperature is high. Larvae pupate in late March the following year. Adults emerge in mid-April. Larval damage period is between early to mid-May, early to mid-June, mid- to late July, mid- to late August, late September to early October, early to mid-November, and early to mid-March for the 1st to the 7th (overwintering) generation, respectively.

Most adults emerge between 15:00-19:00, with males generally emerge 1-2 days before the females. Emergence rate was 86.4, 82.8, 78.5, 90.1, 84.6, 88.8, and 80.2% for the 1st to the 7th generation, respectively. Adults mate between 0:00-4:00. They can mate the same day after emergence. Mating lasts for 9-20 hours. Females mate only once and start to lay eggs the next day. Most eggs are laid in groups and covered by villi. Females lay eggs at night on twigs and leaflets of *Acacia mearnsii*. Egg-laying period lasts for 3-9 days. The average fecundity is 214 eggs/female (range: 96-368). Adults live for 3-10 days.

Egg stage for the 1st to the 7th generation is 5.8 (5-8), 4.1 (3-5), 3.8 (3-5), 3.1 (3-4), 5.6 (4-7), 6.2 (5-7), and 7.1 (6-11), respectively. Most eggs hatch between 17:00-19:00, with an average hatching rate of 98.1%.

There are five (a few four) instars in the larval stage. Larvae development time of the 1st to the 7th generation is 21.9 (20-24), 17.4 (17-18), 20.6 (18-25), 22.9 (18-34), 23.1 (18-31), 29.1 (26-33), and 90.4 (85-98) days, respectively. Newly hatched larvae consume chorion first before feeding on young leaves. Feeding by young larvae creates small notches on compound leaves. By the third instars, they feed separately and consume entire leaflets, leaf stalks, and inflorescences. They may also migrate to other trees. Larvae feed all day long especially when it is cloudy. Each larva consumes an average of 1.076 g of host leaves based on laboratory observations. Feeding by the 5th instar accounts for 73.7% of the total consumption. Mature larvae come down along the trunk and construct cocoons among weeds or in leaf litters. Some

S. scintillans adult (NZM)

S. scintillans adult male (NZM)

cocoons are located in crevices at tree base. Feeding by mature larvae during the last couple of days increased dramatically. Prepupal stage lasts for 1-4 days. Pupation rate is between 81.6-95.2%. Pupal stage lasts for 9.5 (6-12), 6.8 (6-7), 6.1 (4-7), 7.1 (4-12), 12.4 (7-30), 29.1 (21-51), and 12.5 (10-15) days for the 1st to the 7th generation, respectively.

Pest occurrence correlates with winter temperature, site conditions, and canopy cover. Low winter temperature negatively impacts survival of the overwintering larvae. Low temperature also prevents pupation for the 6th instar larvae and reduced female fecundity. In Youxi of Fujian, open forests, isolated or street trees, and tree on the sunny side suffered more than closed forests, trees in pure stands, and tree on the shady side.

Prevention and treatment

Conserve natural enemies. Major natural enemies include *Enicospilus* sp., *Apanteles* sp., *B.t.*, *Beauveria bassiana*, NPV. Survey results in 1986 showed that 33.4, and 4.3% of the 6th and 7th generation larvae were infected by *B.t.*, respectively. NPV infection was usually found in the 2nd and 5th generation and cause local epidemics.

References Zhao Zhongling 1978, 2003; Shi Mubiao et al. 1984; Chen Shunli et al. 1989.

(Chen Shunli, Lu Wenmin)

489 Leucoma candida (Staudinger)

Classification Lepidoptera, Erebidae
Synonyms *Stilpnotia candida* Staudinger, *Liparis salicis* Bremer
Common name Willow moth

Distribution Beijing, Hebei, Shanxi, Inner Mongolia, Liaoning, Jilin, Heilongjiang, Fujian, Jiangxi, Shandong, Henan, Hubei, Hunan, Sichuan, Yunnan, Tibet, Gansu, Qinghai, Xinjiang; Korea Peninsula, Japan, Mongolia, Russia.

Hosts *Populus davidiana, P. nigra, P. berolinensis, P. simonii, P. pseudo-simonii, Alnus japonica, Salix* spp., *Betula platyphylla,* and *Corylus heterophylla.*

Morphological characteristics **Adults** Body 14-23 mm long with a wingspan of 35-52 mm, covered by white villi and slightly shiny. Pedicel white with brown marks. Wings covered by dense and wide scales. Male genitalia with lots of fine teeth on the external margin. Uncus enlarged at the base. Cornutus on vesica made of various spines. Plate over ostium bursae in females "M" shaped.

Eggs Round, dark brown.

Larvae Mature larvae 30-50 mm long, brownish black. Dorsal line blackish and surrounded by yellowish brown area and a grayish black strip laterally. There are black transverse bands on abdominal segment 1, 2, 6, and 7 dorsally. Spiracle line grayish brown. Spiracles brown. Peritreme black. Each body segment contains 8 red or black verrucae that arranged in a row, with dense yellowish brown long hairs and a few black or brown short hairs on each of them. Ventral side of the abdomen greenish brown. Thoracic legs brown. Eversible glands reddish brown.

Pupae Body 16-26 mm long, brownish black with brownish yellow setae. Surface rough with dense pits and punctures.

Biology and behavior *Leucoma candida* occurs one generation a year in Heilongjiang and Gansu and two generations a year in Linzhou of Henan. It overwinters as young larvae inside cocoons in bark crevices. Overwintering larvae resume feeding in mid-April the following year and begin to pupate in bark crevices or under soil chunks in late May. Peak pupation falls in early June. Adult emergence ends in mid-July. Females start to lay eggs in early June and peaks in mid-June. Most larvae of the first generation appear in early July, with a peak pupation period in mid-August. Pupation ends in late August. Peak adult emergence falls in late August. Larvae of the second generation appear in early September. Young larvae of this generation overwinter.

Adult emergence occurs mostly at night. Adults are attracted to lights and stay during the day. Most eggs are laid in random groups on bark, underside of the leaves, and twigs. Egg stage lasts for 14-18 days. Newly hatched larvae do not feed immediately but rather stay hiding. They feed on young twigs or leaves before ballooning in the wind with silky threads. Larvae strongly avoid direct lights. Mature larvae feed at night and stay hiding during the day. All larvae stop feeding 2-3 days before molting and one day after molting. Mature larvae usually roll leaves in the lower crown before molting. They pupate after three days of prepupal stage.

Prevention and control

1. Physical control. Trap adults with black light traps.

2. Trap adults with semiochemical 3Z-cis-6, 7-cis-9, 10-D-i-epoxy-heneicosene.

3. Biological control. Conserve natural enemies (see natural enemies of *Stilprotia salicis*). Apply SsNPV (see prevention and treatment for *S. salicis*). Spray *B.t.* 8,000 IU/μL at 1,000-1,500×, or *B.t. galleriae* at 400-600×.

4. Chemical control. Apply 25 g/L lamda-cyhalothrin at 1,000-1,500×, Apply 5.7% eamamectin at 3,000-5,000×.

References Zhao Zhongling 1978, 2003; Chinese Academy of Sciences - Institute of Zoology 1987, Wang Jianquan 1992; Xiao Gangrou (ed.) 1992, Shaefer et al. 2000; Li Guishan et al. 2001; Li Shuji, Zhang Yuxiao et al. 2007.

L. candida adult (Xu Gongtian)

(Shui Shengying, Li Zhenyu)

Leucoma salicis (L.)

Classification Lepidoptera, Erebidae
Synonyms *Stilprotia salicis* (L.), *Bombyx salicis* L., *Liparis salicis* L., *Phalaena salicis* L.
Common names White satin moth, Satin moth

Distribution Hebei, Inner Mongolia, Liaoning, Jilin, Heilongjiang, Jiangsu, Shandong, Henan, Tibet, Shaanxi, Gansu, Qinghai, Ningxia, and Xinjiang; Mongolia, Korea Peninsula, Russia, Japan, Europe, North America.

Hosts *Populus* spp., *Salix* spp., *Ulmus* spp., *Betula platyphylla*, *Corylus heterophylla*, *Fraxinus chinensis*, and *Toxicodendron vernicifluum*.

Morphological characteristics **Adults** Body 11-20 mm long with a wingspan of 33-55 mm, covered by white silky villi. Compound eyes round, black. Pedicel white. Scales on wing surface broad and sparsely arranged. Outer margin of the claspers of the male genitalia smooth, uncus enlarged to triangle shape at the base. Cornutus on vesica made of hooked hard plate. Plate over ostium bursae in females trapezoid. Tibia and areas between tarsal segments with black and white rings.

Eggs Flat oval, greenish, in masses and covered by white foam.

Larvae Yellow with dark brown subdorsal line, orange or brownish yellow dorsal verruca, and black thoracic legs.

Pupae Smooth, brownish black with blackish yellow marks.

Biology and behavior *Leucoma salicis* occurs one generation a year in Heilongjiang and Inner Mongolia, two generations a year in Shandong, and 2-3 generations a year in Urumqi of Xinjiang. It overwinters as 2nd-3rd instar larvae in bark crevices, tree hollows, leaf litters or weeds under the tree in early September in Inner Mongolia. Larvae resume feeding in mid-May the following year. They start to pupate in late June. Adults emerge and lay eggs in early July. Eggs hatch in late July. By late September and early October, larvae of the second generation begin to overwinter.

Eggs are laid on tree trunks and host leaves, with each

L. salicis adult (Li Zhenyu)

L. salicis adult (Xu Gongtian)

L. salicis larva (Xu Gongtian)

L. salicis larva (Xu Gongtian)

L. salicis damage (Xu Gongtian)

L. salicis pupa (Xu Gongtian)

L. salicis larva killed by virus (Xu Gongtian)

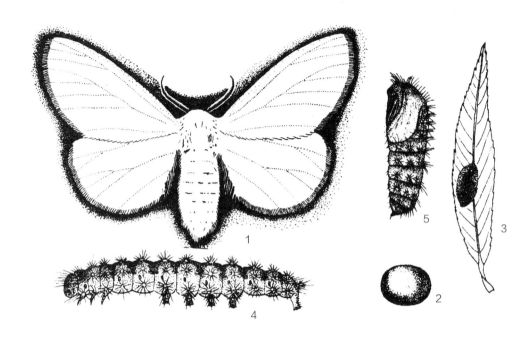

L. salicis. 1. Adult, 2-3. Egg and egg mass, 4. Larva, 5. Pupa (Xu Tiansen)

female produces an average of 300 eggs. Eggs hatch within half a month. Larvae feed mostly during the day but could also feed at night. Larvae from the 1st to the 3rd instars feed aggregately and can balloon in the wind with silk threads. Larvae disperse after the 4th instar. Mature larvae roll leaves and pupate inside. Pupal stage lasts for 9-19 days. Peak adult emergence falls between 20:00-22:00. Adults are attracted to lights. They start to fly and mate six hours after emergence.

Prevention and control 1. Physical control. Trap adults with black lights.

2. Biological control. Conserve natural enemies such as *Exlorista anoena*, *Telenomus* spp., *Charops* sp., *Trichogramma closterae*, *Beauveria bassiana*, *Metarhizium flavoviride*. Apply SsNPV $6.0-7.5\times10^{11}$ PB/ha mixed with activated carbon, or *B.t. galleriae* at 400-600× to control 2nd to 4th instar larvae.

3. Chemical control. Apply 25 g/L lamda-cyhalothrin at 1,00-1,500× to control larvae.

References Zhao Zhongling 1978, 2003; Chinese Academy of Sciences - Institute of Zoology 1987; Xu Longjiang 1992; Liu Zhenqing et al. 1995; Su Mei and Yang Fenyong 2000; Liang Wenxiu et al. 2001.

(Shui Shengying, Li Zhenyu, Xu Longjiang)

L. salicis adult (edited by Zhang Runzhi)

491 *Cifuna locuples* Walker
Classification Lepidoptera, Erebidae

Distribution Beijing, Hebei, Shanxi, Inner Mongolia, Liaoning, Jilin, Heilongjiang, Shanghai, Jiangsu, Zhejiang, Anhui, Fujian, Jiangxi, Shandong, Hubei, Guangdong, Guangxi, Sichuan, Guizhou, Yunnan, Tibet, Shaanxi, Gansu, Qinghai, Ningxia; Russia, Korea Peninsula, Japan, India, Vietnam.

Hosts *Liquidambar formosana*, *Aleurites fordii*, *Platanus orientalis*, *Populus* spp., *Salix* spp., *Zelkova scheneideriana*, *Ulmus pumila*, *Cerasus*, *Diospyros kaki*, *Paulowina fortune*, *Camellia sinensis*, *Chaenomeles sinensis*, *Pinus massoniana*, *Sophora japonica*, *Populus tomentosa*, *Camellia oleifera*, *Wisteria sinensis*, *Robinia pseudoacacia*, *Acacia mearnsii*, *Rosa chinensis*, *Pyrus pyrifolia*, *Eriobotrya japonica*.

Damage Larvae feed on leaves. Their urticating hairs cause dermatitis, macules, pain, and itchiness.

Morphological characteristics **Adults** Males 12-17 mm in length, 24-40 mm in wingspan, antennae plumose. Females 18-22 mm in length, 45-53 mm in wingspan, antennae short pectinate. Head and thorax dark yellowish brown, abdomen brownish yellow; front and hind wings yellowish brown, hind wings slightly lighter in color. Metathorax and the dorsal surface of the 2nd and the 3rd abdominal segment each with a black short hair tuft. Forewing inner line appears as a brown band, inner side of the band with a fine white line; cross veins reniform, yellowish brown, with dark border.

Larvae Mature larvae 30-43 mm in length, head black, body dark brown. Subdorsal line and lower spiracle line brownish orange, discontinuous. Pronotum with

C. locuples adult male (Xu Gongtian)

C. locuples larva and pupa (Huang Jinshui)

C. locuples larva (Huang Jinshui)

C. locuples adult and eggs (Huang Jinshui)

a large black verruca on each side. Verrucae on other segments brown, bearing tan white hairs. The dorsal surface of the 1st to the 4th abdominal segment each with a dark yellowish brown short hairbrush; the 8th abdominal dorsal surface with dark brown hair bundles; anal segment with light brown and long hair clumps.

Biology and behavior *Cifuna locuples* occurs three generations a year in Fujian province. It overwinters as mature larvae in fallen leaves. Larvae become active in early March to early April next year, and pupate in late April to mid-May. Adults emerge in early May to late May. The third generation larvae appear from mid-September to early May next year. Larvae begin to overwinter in early or late November. The pupal stage is 6-15 days. The peak adult emergence period of the three generations is in mid-May, mid-July, and mid-September, respectively.

Most adults emerge during the night. They can mate and oviposit the next day after emergence. Adult longevity is 5-11 days. Eggs are laid on leaves in egg masses. A female produces 2-5 egg masses and a total of 450-650 eggs. Larvae have 6-8 instars. Mature larvae spin silk cocoons on leaves or branches. The overwintering generation mature larvae crawl into fallen leaves at tree bases and overwinter at the end of October.

Prevention and treatment

1. Manual control. In autumn and early spring, collect and dispose dry branches and fallen leaves. At the end of peak oviposition period, manually remove leaves with egg clutches.

2. Trap and kill adults using light traps.

3. Chemical control. Spray insecticides such as 2,000-3,000× diflubenzuron, 3,000× 1% avermectin EC, 3,000× 30% fenvalerate EC, or 5.7% emamectin at 3,000-5,000×.

References Fujian Forest Research Institute 1991; Sun Xingquan, Zhou Lina, Chen Xiaolin et al. 2002.

(Huang Jinshui, Cai Shouping)

Perina nuda (F.)

Classification Lepidoptera, Erebidae

Synonyms *Bombyx nuda* F., *Acanthopsyche bipars* Matsumura, *Perina basalis* Walker, *Euproctis combinate* Walker

Distribution Zhejiang, Fujian, Jiangxi, Hubei, Hunan, Guangdong, Guangxi, Chongqing, Sichuan, Tibet, Hongkong, Taiwan; Japan, India, Nepal, Sri Lanka.

Hosts Members of the *Ficus* genus such as *Ficus microcarpa* and *F. virens*.

Morphological characteristics **Adults** Males with a wingspan of 30-38 mm. Antennae with brown pedicel and dark brown flagellum. Labial palps, head, front tibia, underside of the thorax, and anal hair clusters orange. Thorax and base of the abdomen grayish brown. Prothorax grayish brown. Abdomen dark brown with grayish brown intersegment. Legs grayish brown. Forewings transparent with dark brown veins. Base of the wings and posterior margin except anal corner dark brown. Hindwings dark brown with transparent apical corner. Posterior margin light grayish brown. Females with a wingspan of 41-50 mm. Antennae with yellowish pedicel and grayish yellow flagellum. Head, legs, and anal hair cluster yellow. Both wings yellowish. There are scattered brown scales along the posterior edge of the median cell on the forewings. **Larvae** Length 21-36 mm, grayish green with green, white, and red patchy marks. There are brown hair clusters on abdominal segment 1 and 2 dorsally. Each abdominal segment contains three pairs of red, hairy body protrusions, with the lateral pair relatively big. Dorsal line yellow and wide. Mature larvae green with dark dorsal line. **Pupae** Body about 21 mm long, spindle-shaped, with a large head and tapered end, and reddish brown and dark brown marks on the surface.

Biology and behavior *Perina nuda* occurs five to six generations a year and overwinters as 5th-6th instar larvae on host leaves in Beipei in Chongqing. Overwintering larvae without obvious diapause and can feed if the temperature is high. Active larvae and pupae can be found between later December and early January. Overwintering larvae increase feeding by early March and start to pupate in early April. Pupation peaks in mid-April. Adults start to emerge in mid-April and peaks in late April. Adults are not attracted to lights. Most adults are found between July and August. Females lay eggs one day after emergence, with 260-420 eggs laid by each female on leaf surface, leaf stalk, or young branches.

Prevention and treatment Chemical control. Apply 5.7% emamectin at 3,000×.

References Zhao Zhongling 1978, 2003; Wu Weiwen, Wei Lüyan, and Li Xueru 2002; Yang Bin 2010.

(Wu Weiwen, Li Yingchao, Li Zhengyu)

P. nuda adult (Zhang Peiyi)　　　P. nuda adult (Zhang Peiyi)

493 Polyura athamas (Drury)

Classification Lepidoptera, Nymphalidae
Synonym *Polyura athamas athamas* (Drury)
Common name The common nawab

Distribution Fujian, Guangxi, Yunnan; Myanmar, Malaysia, Thailand.

Hosts *Acacia mearnsii*, *A. dealbate*, *Leucaena leucocephala*.

Morphological characteristics **Adults** Length 17-20 mm, wingspan 53-69 mm. Antennae black; head, thorax, abdomen and wing blackish brown. The center of the forewing and hindwing has a large kidney shaped pale green spot, 23-34 mm long. The forewing has a small oval pale green spot near the apex. The outer margin of the hindwing has 8 pale green spots and a pair of tails at the posterior end. **Eggs** Oval, milky color at beginning, pale green later, black before hatching. **Larvae** Mature larvae 25-39 mm, dark green; vertex with two long and two short green spines; pronotum and mesonotum each with one yellow circle; metathorax to 8th abdominal segment each with a curved yellow line. There is an intermittent yellowish green line under the spiracles. Body covered with green tubercles. **Pupae** Pale green, smooth, elongated grape shape, with black markings; the end sharp.

Biology and behavior *Polyura athamas* occurs 4 generations a year in Fuzhou (Fujian). It usually overwinters as 2nd to 3rd instar larvae on leaves. The larvae spit silk on leaf surface and rest on the silk in winter. When temperature is high during noon, they crawl to other leaves to feed and come back and rest on the silk. Mortality in winter is high. Those overwintering as pupae can develop into adults in mid-April. Overwintering larvae resume normal feeding in late March and early April. They pupate in mid- to late May. Adults emerge from mid-May to early June. First generation eggs,

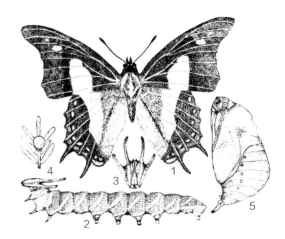

P. athamas. 1. Adult, 2. Larvae, 3. Front view of larval head, 4. Egg, 5. Pupa (He Xueyou)

larvae, pupae, and adults appear from late May to mid-June, mid-June to late July, early July to late July, and mid-July to late July, respectively. Second generation larvae, pupae, and adults appear from late July to late August, late August to late September, and early September to early October, respectively. Third generation eggs, larvae, pupae, and adults appear from early September to late September, early September to late October, early October to early November, and early October to early November, respectively. Fourth generation eggs and larvae appear in early October. Eggs can be seen until mid-November. Larvae overwinter in early December.

Adults lay eggs on leaves that are at least 2 m above the ground. Eggs are laid singly on both sides of leaves. Young larvae consume partial or all eggshells, then feed on leaves. From the 2nd instar, the head shows two long and two short green protuberances. The body shows light-yellowish green stripes. Larvae do not feed 1-2 days before molting and 1-2 days after molting. The 5th instars molt into pupae on leaf petioles or back of leaves. They hang themselves from leaf petioles using the end of the abdomen. Pupal period of the overwintering generation is about 2 weeks, the period for 2nd and 3rd generations is about 1 week. An egg parasitoid, *Trichogramma* sp., was reported.

References Shang Dieren 1981, Zhao Xiufu 1981, Zhou Yao et al. 1994.

P. athamas larva (Wang Jijian)

(Li Yunwei, Yang Jiahuan, He Xueyou)

494 *Polyura narcaea* (Hewitson)

Classification Lepidoptera, Nymphalidae
Synonyms *Polyura narcaea narcaea* (Hewitson), *Polyura narcaea meghaduta* (Fruhstorfer)

Distribution Hebei, Shanxi, Jiangsu, Zhejiang, Anhui, Fujian, Jiangxi, Shandong, Henan, Hubei, Hunan, Guangdong, Guangxi, Sichuan, Guizhou, Yunnan, Shaanxi, Gansu, Taiwan; India, Myanmar, Thailand, Vietnam.

Hosts *Albizia kalkora*, *Dalbergia hupeana* et al.

Morphological characteristics **Adults** There are two biotypes. Spring type is larger, 25 mm long, wingspan 70 mm. Dorsal surface with black pubescence. Vertex with 4 golden yellow round spots, they are arranged in a square. Wing green, the front margin of the forewing black; there are two black bands (marginal and submarginal bands) located at the outer margin and below the outer margin; the area between the two bands green. The cross vein of the median cell black; there is a black club-shaped band below the median cell, it extends near the submarginal band. Hindwing outer margin black; its posterior margin has two tails. Submarginal band black, extends to the posterior angle, which is yellow.

Summer type smaller, 20 mm long, wingspan 60 mm. Color and the bands are similar to the spring type. The differences are: there is a row of green round spots between the marginal and submarginal bands; the club-shaped black band below the median cell connects to the submarginal band; area from base of the wing to cross vein of the median cell greyish black; the hindwing has a black wide band from the base to the posterior angle.

Eggs Near round, 1.5 mm wide, pale green; the top flat, with dark brown round stripe.

Larvae Mature larvae 35-48 mm, green; each segment with fine creases, light-yellow spots exist within the creases. Head green, its sides light-yellow; vertex with 3 pairs of spines, the middle pair very short, brown; the other two pairs green; 10 mm and 7 mm long, respectively; there are two rows of small spines on the large spines. Spiracle line light-yellow, reaches to the anal angle; one pair of anal angles, triangular, light-yellow.

Pupae Length 18-22 mm, width 11-15 mm. Vertex round and smooth, body curls ventrally, dorsal surface with 8 longitudinal yellow lines.

Biology and behavior *Polyura narcaea* occurs 2 generations a year. It overwinters as pupae. Adults emerge in late April next year. They lay eggs in early May. The 1st

P. narcaea adult (dorsal view) (Ji Baozhong)

P. narcaea adult (ventral view) (Ji Baozhong)

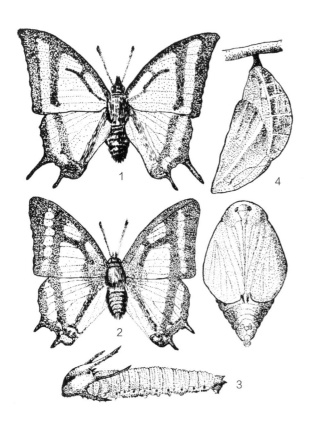

P. narcaea. 1. Spring type adult, 2. Summer type adult, 3. Larva, 4. Pupa (Zhou Tiying)

generation egg, larva, prepupa, and pupa development period is 9-11, 43-56, 2-3, and 11-13 days, respectively. The 2nd generation egg, larva, and pupa development period is 6-8, 44-61, and 180-220 days, respectively. Adults live for 1 month. The egg laying period is long, therefore, various instars may appear on hosts.

Adult emergence time is mostly during 7:00-9:00. They fly 10-15 minutes after emergence. They are active from noon to 14:00. Adults often stay on rotten stumps located on sunny side and warm animal dung. When disturbed, they fly suddenly and sometimes return to the original spots later. Adults lay eggs individually on leaf surfaces. Young larvae consume most of the egg shells after hatching. They start feeding on leaves from 2nd day. They feed from leaf edges, leave the veins and lower epidermis. The 2nd instars can consume one leaf a time, but most of them only feed partial of the leaves. The 3rd and 4th instars can feed 4-5 leaves a time. Mature larvae stop feeding one day before pupation. They hang from small twigs with head pointed downward and the abdomen fixed to the twig.

References Zhou Tiying et al. 1983, 1986; Zhou Yao et al. 1994; Xu Zhihua 2013.

(Zhou Tiying, Zhong Guoqing)

495 *Aporia crataegi* (L.)

Classification Lepidoptera, Pieridae
Synonyms *Ascia crataegi* L., *Pieris crataegi* L., *Pontia crataegi* L.
Common names Pear white, Black-veined white

Distribution Beijing, Hebei, Shanxi, Inner Mongolia, Liaoning, Jilin, Heilongjiang, Zhejiang, Anhui, Shandong, Henan, Hubei, Sichuan, Tibet, Shaanxi, Gansu, Qinghai, Ningxia, Xinjiang; Korea Peninsula, Japan, Russia, western Europe, Mediterranean, north Africa.

Hosts *Crateagus pinnatifida*, *Malus pumila*, *M. asiatica*, *Pyrus pyrifolia*, *P. armeniaca*, *P. salicina*, *Cerasus pseudocerasus*, *Pyrus betulifolia*, *Malus baccata*, *M. prunifolia*, *Cotoneaster* spp., *Spiraea salicifolia*, *Sorbaria sorbifolia*, *Malus micromalus*, *Syringa oblata*, *Euphorbia milii* var. *splendens*, *Ulmus pumila*.

Damage Larvae feed on buds, leaves, and young flowers. They reduce fruit production. Severe infestations will cause complete defoliation.

Morphological characteristics **Adults** Length 22-25 mm, wingspan 64-76 mm. Body black; head, legs with light-yellowish white to greyish white scales; antennae black, its tip light-yellowish white; wings white, veins black, outer margin with a series of triangular smoky spots except the end of the anal vein. Forewing scales unevenly distributed, some areas semi-translucent. Hindwing veins clearly black, scales thicker than those in forewing; greyish white. Female abdomen thick; thorax yellowish white, with sparse thin hairs. Male abdomen slender; thorax with abundant thin hairs.

Eggs Cylindrical, height 1.0-1.6 mm, width 0.5 mm; the tip similar to bullet head. Egg masses consisted of dozens to more than 100 eggs. There are 12-14 longitudinal ridges on each egg. There are no transverse ridges. Top of the egg with protruded edges. New eggs golden yellow, the top becomes black, translucent before hatching.

Larvae Mature larvae 40-45 mm, with 5 instars. First instar greyish brown; head, thorax, and anal region black. Dorsal surface with 3 black longitudinal bands, between them are 2 yellowish brown bands. The sides the body grey; ventral surface purple grey; head, anterior area of the thoracic legs, pronotum, and spiracle plates black. Body scattered with small black spots, which have yellowish white thin hairs on them.

Pupae Length 25 mm, width 7 mm. Two types of pupae exist: black type and yellow type. Black type has yellowish white color; with many large black spots; head, mouthparts, legs, antennae, ridges on dorsal surface, wing margin, and abdomen black; tubercles on vertex yellow; upper margin of the eyes with a yellow spot. Yellow type with yellow color; less black spots; smaller size. The black:yellow type ratio is 32:68.

Morphological characteristics *Aporia crataegi* occurs one generation a year. It overwinters as 2nd-3rd instars in nests located on tree crown. They resume activity when temperatures reach 10-12 ℃ next spring. They first feed in groups on buds, then on young flowers, leaves, and petals. The feeding can significantly reduce fruit yield. At low temperature, during cloudy or rainy days and nights, they hide in nests. They disperse after 5th instar and will not return to nests during the nights or cloudy or rainy days. Each 5th instar larva consumes 3-4 *Crateagus pinnatifida* leaves. The 4th and 5th instars are not active. They play death when disturbed and fall to ground, curl into balls. Feeding usually occurs during the day at 16:00-18:00. Mature larvae search for a suitable place to pupate. The pupation sites can be on the trees, shrubs, weeds, or crops. They make cocoons, using anal legs to fix to the substrate. They use a strand of silk to wrap the 1st abdominal segment to the branch, then pupate. The development period from the start of activity in spring to pupation is 40 days. The pupation dates in

A. crataegi adult (Xu Gongtian)

Shanxi is between May 2-May 18, and May 10-May 20 in Liaoning. Pupal period is 14-19 days.

Adults usually emerge during the day. They fly at forest edge, open field, flowers, weeds, and feed on nectars. They gather at puddles, ponds, or irrigation channels to obtain water. Mating occurs the same day after emergence. Females lay eggs 3 days after mating. Eggs are mostly laid during noon. They are laid neatly on back of leaves, each egg mass has 25-30 eggs. Each female lays 200-500 eggs during its lifetime. Females live for 6-7 days. Males live for 3-4 days.

Prevention and treatment

1. Cultural control. Remove nests in winter in winter and before budding in Spring.

2. Nuclear polyhedrosis virus (NPV). Natural NPV infestation rate is 37%-42%. Crush 12-13 NPV killed 3rd to 4th larvae, add 15 litre water, then filter out the debris. The liquid can be sprayed to one tree. Add 5% glycerol to the liquid and store in refrigerator for future use.

3. Conservation and use of natural enemies. Collect *Apanteles glomeratus* pupae from *Aporia crataegi* infested forests and release in infested areas. Place 10-20 cocoons on each tree before 3rd instar. *Brachymeria lasus* is a common egg parasitoid of *A. crataegi*. Avoid using insecticides when they are emerging. Plant poplar, elm, and pine trees at the edge of orchards to provide alternative hosts for this parasitoid. Place crop stalks on sunny, warm place in orchards to provide overwintering habitats for the parasitoid. Other natural enemies include egg parasitoid *Pteromalus puparum*, pupal parasitoids *Coccygomimus disparis* and *Pteromalus puparum*.

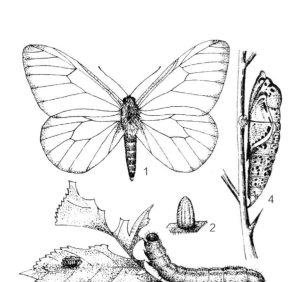

A. crataegi. 1. Adult, 2. Egg, 3. Larva, damage, and egg mass, 4. Pupa (Yu Changkui)

A. crataegi mid-aged larvae (Xu Gongtian)

A. crataegi mature larva (Xu Gongtian)

A. crataegi pupa (Xu Gongtian)

A. crataegi damage (Xu Gongtian)

A. crataegi larva killed by pathogen (Xu Gongtian)

A. crataegi lava parasitized by braconid wasp (Xu Gongtian)

A. crataegi larvae and pupae (Yan Shanchun)

4. Chemical control. Spray insecticides to crown when overwintering larvae start to feed and when young larvae emerge. Using 20% fenvalerate or 2.5% deltamethrin at 2,000× can achieve > 90% control.

References Xiao Gangrou 1992, Jiang Hanlin 2001, Li Chengde 2004, Wen Xuefei and Zou Jimei 2007.

(Yan Shanchun, Liu Zhenlu)

496 *Papilio elwesi* Leech

Classification	Lepidoptera, Papilionidae
Synonym	*Agehana elwesi* (Leech)
Common name	Wide-tailed swallowtail

Distribution Fujian, Jiangxi, Hubei, Hunan, Guangdong, Guangxi, Sichuan, Guizhou, Shaanxi.

Hosts *Sassafras tzumu*, *Liriodendron chinense*, *Magnolia officinalis*.

Damage Larvae feed on leaves.

Morphological characteristics **Adults** Wingspan 132-150 mm; black, thorax and base of wings with abundant hairs. Forewing narrow, scales thin, blackish brown, veins clear, scales along the veins thicker, scales at the outer margin thicker. The black scales in median area of the hindwing thin, with or without white patch. If a white patch is present, the size and color vary. The scales from end of the median cell to the end of the tail thick and felt-like; outer margin wavy; tails wide and long, shoe-shaped, 20 mm long, 14 mm wide; M_3 and Cu_1 are in the tail; anal angle round, red, its center black; at the end of each cell with 1-2 red new moon shaped spots; area before the tail black.

Eggs Round, 1.8 mm in diameter; smooth; green at beginning, shiny, turning to yellow and darker later.

Larvae There are 5 instars. The 1st - 4th instars are similar to bird droppings. Mature larvae green, smooth and hairless. Length 40-70 mm, head width 6 mm. Head blackish brown. Pronotum with a half circle pale black spot. Metanotum with a pair of blackish brown eye spots. The dorsal surface of the 1st abdominal segment with two black tri-lobed spots. The sides of the 3rd to 7th abdominal segments with symmetrical, connected, irregular, blackish brown spots; within spots are 3 pale blue stripes. Ventral surface and the legs pale brown, similar to the markings on the sides of the body. Crochets reddish brown; spiracles elongated round, bluish grey; scent gland "Y" shape, greyish, normally not exposed.

Pupae Length 43 mm, blackish grey, similar to broken dead branch; uneven on the dorsal surface, center of the dorsal surface with a convex spot. The tubercles

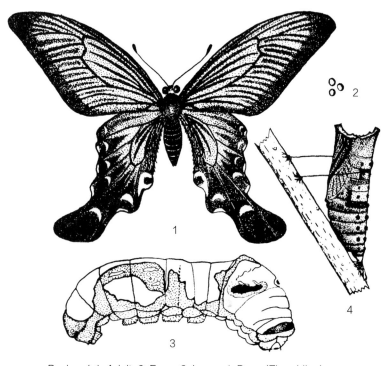

P. elwesi. 1. Adult, 2. Eggs, 3. Larva, 4. Pupa (Zhou Lijun)

P. elwesi adult (NZM)

on the thoracic and abdominal segments are arranged in rows; center of abdomen with tint of bronze green, tapered at the end; the end attaches to the silk pad; the waist of the pupa with a black silk strand.

Biology and behavior *Papilio elwesi* occurs two generations a year in Hunan. It overwinters as pupae on branches, tree trunk from ground to 1 mm height, and bushes. Sometimes they overwinter on building surfaces. Pupae form a 45° angle with the substrate. Adults emerge in mid- and late April. First generation eggs and larvae appear in early May. Pupae emerge in mid-June. Adults emerge in early July and produce eggs in late July. In mid-September, most larvae pupate and overwinter. The developmental period of each stage is: egg – 7 days, larva – 32-35 days, 1st generation pupa – 17-20 days, overwintering pupa – 190-220 days, prepupae – 2 days, and adult – half a month.

Adults emerge during the day. They can fly about 2 hours after eclosion. Adults feed from one day after emergence. Eggs are laid singly along the main veins of leaves on new shoots. Adults lay eggs around noon time. They are strong fliers. They glide gracefully during flight.

Larvae consume the eggshells after hatching. After 10 hours, they make notches on leaves. Mature larvae consume whole leaves; a larva consumes 150-305 mg food within 24 hours period. They rest on leaf surface when not feeding. The leaf surface is laced with silk; its margins roll to the center slightly; the larva contrasts to arrowhead shape. When a larva enters into pre-pupa period, it empties its content, the body becomes reddish brown. They search for a place to pupate; then the head and the end of the abdomen is fixed to the substrate. After one day, it changes to pupa.

Prevention and treatment *Papilio elwesi* only occurs in China. Although widely distributed, but it never reaches high population densities and does not require control. Natural enemies include egg parasitoid (*Trichogramma confusum*), predators (*Vespa magnifica*, *V. mandarinia mandarinia*), birds (*Pycnonotus sinensis*, *Parus major*), and entomopathogenic bacteria.

References Northwest Agricultural College- Department of Plant Protection 1978, Li Chuanlong 1984, Xiao Gangrou 1992, Zhou Yao et al. 1994.

(Zhang Lijun, Zhou Lijun)

497 Papilio epycides epycides Hewitson

Classification Lepidoptera, Papilionidae
Synonym *Chilasa epycides epycides* (Hewitson)
Common name Camphor swallowtail

Importance *Papilio epycides epycides* is a defoliator. Although widely distributed, it never caused significant damage.

Distribution Zhejiang, Fujian, Jiangxi, Hunan, Sichuan, Yunnan, Taiwan; Bhutan, Myanmar, India.

Hosts *Cinnamomum camphora*, *C. parthenoxylon*, *C. bodinieri*.

Morphological characteristics **Adults** Mid-sized, wingspan 80-90 mm; black, with evenly distributed white spots; head, thorax, anterior part of the abdomen, femora, and interior margin of the hindwing with dense long hairs; black scales on wings thin, blackish brown, those along the veins thicker; areas between the veins light-yellow. The forewing has 9 light-yellow elliptical spots arranged in an arc near the outer margin. Each cell along the hindwing outer margin has 1-2 near round spots that arranged in two arcs; anal angle with an orange red round spot. Forewing has same colors on the two sides. The underside of the hindwing does not have the clear patterns as the front side or only the anal angle shows a spot; color brown.

Eggs Diameter 1 mm, round. Pale yellowish green, darker later.

Larvae Greyish black at beginning, yellowish brown after started feeding. The head width and body length of the instars are – 1st: 0.55 mm wide, 2.5-6 mm long; 2nd: 1 mm wide, 6-10 mm long; 3rd: 1.55 mm wide, 10-16 mm long; 4th: 2.75 mm wide, 16-28 mm long; 5th: 3.7 mm, 28-44 mm long. The 3rd instar greyish brown, surface with indistinct yellowish decorations and distinct black tubercles. The 4th instar darker, blackish brown, shiny, with distinct yellow markings. Pronotum and anal plate of the mature larvae orange yellow; color varies; with green, yellow, and light-yellow stripes. Body with irregular stripes bordered with black color. The stripes on the center of the dorsal surface form chains; each segment with 6 pale blue round spots. Spiracles small, greyish white; spiracle line black, the line above it yellowish

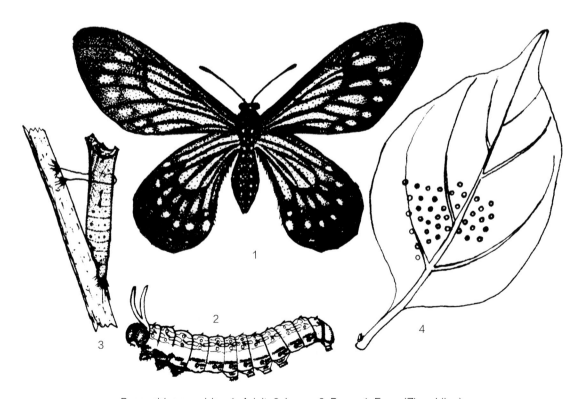

P. epycides epycides. 1. Adult, 2. Larva, 3. Pupa, 4. Eggs (Zhou Lijun)

P. epycides epycides adult (NZM)

white. Thoracic legs and ventral surface of the thorax black; prolegs and ventral surface of the abdomen yellowish white, semi-transparent; crochets reddish brown. Spines 1.5 mm long. Head black, scent gland "Y"-shaped, orange yellow, easily exposed.

Pupae Diameter 7 mm, length 25-28 mm; hard, near cylindrical; greyish black or pale brown, similar to broken dead branch; dorsal surface with "3" shaped symmetrical depressions. Surface with tubercles, those on the center of the dorsal surface more distinct.

Biology and behavior *Papilio epycides epycides* occurs one generation a year in Hunan. It overwinters as pupae around the host on shrubs or buildings. Adults emerge in early April, eggs appear in mid-April, larvae appear in late April, and pupae appear in late May. The development period of each stage is: eggs – 9 days, larvae – 26 days, pupae – 315 days, adults – 15 days.

Adults emerge during the day. They mate after supplementary feeding. Mating lasts for 1 hour. Each female contains 180 eggs. They prefer depositing eggs on dorsal surface of young leaves. Eggs are arranged sparsely and neatly as a group of dozens or > 100. Larvae consume the eggshells after emergence. They make notches on leaves. After 2nd instars, they only leave main veins and petioles left. They are gregarious before 5th instars. Mature larvae cease feeding, change to reddish brown, fix themselves using silk to branches, and enter into pre-pupae. After 1-2 days, they change to pupae. The pupae and branches form a 45° angle. There is a black silk at the waist. Pupae become soft and black the day before eclosion.

Prevention and treatment Predators include birds, praying mantis, vespid wasps. The 1st-4th instars can be hand-removed since they are gregarious. Those on tall trees can be sprayed with 80% DDVP at 800×.

References Li Chuanlong 1992, Xiao Gangrou 1992, Zhou Yao et al. 1994.

(Zhang Lijun, Zhou Lijun)

498 Papilio slateri Hewitson

Classification Lepidoptera, Papilionidae
Synonym *Chilasa slateri* Hewitson
Common name Blue striped mime

Importance *Papilio slateri* larvae damage host leaves and reduce fruit yield.

Distribution Fujian, Hunan, Hainan; Myanmar, Thailand, Malaysia.

Hosts *Litsea cubeba*, *Cinnamomum camphora*, *Connamomum longepaniculatum*.

Damage It infests leaves and only leave the main veins and petioles.

Morphological characteristics Adults Mid-size, length 27 mm, wingspan 67-120 mm; body and wings black, with scattered white spots; eyes brown; median cell of the forewing with 1-4 blue spots, each cell with one blue spot, arranged into an arc; hindwing color lighter than the forewing, brownish black, with a black bordered orange yellow spot near the anal angle; underside brown, without spots or vague white spots except for the anal spot, some may have 1-2 dull white spots in each cell.

Eggs Round, 1 mm in diameter, yellowish green, darker before hatching.

Larvae Head black, scent gland orange yellow, body with neatly arranged spines, similar to sea cucumber, with 5 instars. First instar 2.5-5.5 mm long, head width 0.6 mm, black at beginning, orange yellow before molting; second instar 5.5-9.0 mm long, head width 1 mm, yellowish brown; third instar 9-16 mm long, head width 1.8 mm, with pale markings; fourth instar 16-26 mm long, head width 2.6 mm, dark orange yellow, body with evenly distributed light-yellow round spots, dorsal median line is composed of "V" shaped decorations; fifth instar 26-40 mm long, head width 4 mm, black, with light-yellow markings, spines 1.3 mm long.

Pupae Near cylindrical, similar to broken limb, yellowish to greyish brown, the dorsal surface has "3" shaped depressions.

Biology and behavior *Papilio slateri* occurs one generation a year in Hunan. It overwinters as pupae in shrubs, edge of forests, or on buildings. Adults emerge in early May. Eggs appear in mid-May. Eggs are laid on underside of young leaves. They are sparely spaced and neatly arranged. One leaf may have dozens to more than a hundred eggs. Egg period is 8 days. Larvae are gregarious, dozens of them are on one leaf. Each instar lasts for 5 days. The last instar period is 6 days. Prepupae appear in early June. Their color is reddish yellow. Head and tail are fixed by silk to substrates. Waist area has a black silk. Prepupae period is 2 days. Pupae and substrates form a 45° angle. Adults emerge from pupae after about 335 days.

Prevention and treatment In yards, larvae can be collected and destroyed. In large forest, spray 25 g/L lamda-cyhalothrin at 1,000-1,500× or other insecticides can achieve good results. They usually do not cause severe damage.

References Li Chuanlong 1992, Xiao Gangrou 1992, Zhou Yao et al. 1994.

(Zhang Lijun, Zhou Lijun)

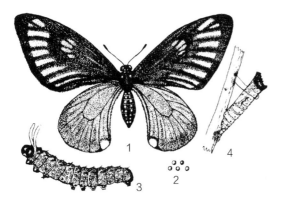

P. slateri. 1. Adult, 2. Eggs, 3. Larva, 4. Pupa (Zhou Lijun)

P. slateri adult (NZM)

499 *Papilio machaon* L.

Classification Lepidoptera, Papilionidae
Common name Yellow swallowtail butterfly

Importance *Papilio machaon* is a widely distributed species but does not occur in large numbers. The larvae are used as Chinese medicine.

Distribution Beijing, Hebei, Shanxi, Liaoning, Jilin, Heilongjiang, Shanghai, Zhejiang, Fujian, Jiangxi, Shandong, Henan, Guangdong, Guangxi, Sichuan, Yunnan, Tibet, Shaanxi, Gansu, Qinghai, Xinjiang, Taiwan; Asia, Europe, North Africa, North America.

Host *Foeniculum vulgare*, *Daucus carota* var. *sativa*, *Apium graveolens*, *Saposhnikovia divaricate*, *Heracleum hemsleyanum*, *Torilis scabra*.

Damage It infests leaves and flowers, heavy infestation may consume the whole leaves except for the petioles.

Morphological characteristics **Adults** Wingspan 90-120 mm, body yellow, dorsal median lin with a black wide longitudinal stripe. Outer margin of wings with a black wide band, forewing with 8 yellow elliptical spots inside the black band, the end of the median cell with two black transverse spots; hindwing with 6 yellow crescent spots inside the black band. The inner side of each yellow spot with a yellow spot. Anal angle with one red round spot.

Eggs Height 1 mm, diameter 1.2 mm, round, light-yellow at beginning, purplish black before hatching.

Larva Young larvae black with white spots, similar to bird dropping; mature larvae 50 mm long, elongated oval, smooth and hairless, light-yellowish green, each segment with a wide black band separated by yellow spots.

Pupae Yellowish green at beginning, yellowish brown before eclosion, with stripes, head with two horns, dorsal surface and the sides of thorax also have projections.

Biology and behavior *Papilio machaon* occurs 2 generations a year at high elevations and 3-4 generations a year in temperate region. Adults feed on flower nectar during the day. They lay eggs on tip of leaves, flowers, or buds. A female only lays one egg a time and flies away. Larvae go through 5 instars. They feed during the night. After disturbed, they expose the scent glands, emit fluid and odor to protect themselves. Larvae rest on main veins of leaves. Adults rest on thick stems. Mature larvae pupate on hidden branches or back of leaves.

Prevention and treatment

1. Silviculture. After harvesting the crops, remove the weeds and host plants to reduce overwintering populations.

2. Mechanical control. Hand remove larvae.

3. Conservation of parasitic wasps.

4. Chemical control. Spray *B.t.* 8,000 IU/μL at 1,000-1,500×, 10% imidacloprid at 3,000×, or 25 g/L lamda-cyhalothrin at 1,000-1,500× for 2-3 times at one week intervals.

References Zhou Yao et al. 1994, Wang Zhicheng 1999, Wu Chunsheng 2001, Zhang Weiwei and Li Yuansheng 2011.

(Huang Pan)

P. machaon adult (Li Zhenyu)

Papilio xuthus L.

Classification Lepidoptera, Papilionidae
Common names Asian swallowtail, Chinese yellow swallowtail, the Xuthus swallowtail

Distribution Hebei, Shanxi, Liaoning, Jilin, Heilongjiang, Jiangsu, Zhejiang, Fujian, Shandong, Hubei, Hunan, Guangdong, Guangxi, Hainan, Sichuan, Guizhou, Yunnan, Shaanxi, Gansu, Taiwan; Japan, Korea Peninsula, former Soviet Union, Philippines, Myanmar, India, Vietnam, Sri Lanka, Malaysia, Australia.

Hosts *Citrus reticulate*, *Zanthoxylum bungeanum*, *Phellodendron amurense*.

Damage Larvae damage make incisions on leaves or consume whole leaves.

Morphological characteristics **Adults** Wingspan 61-95 mm, wing color light-yellowish green, vein margins black; both wings with black wide band along the outer margins, within the band are some crescent spots; anal angle with an orange spot, within the spot is a black spot.

Eggs Globular, diameter 1.2-1.5 mm, yellow, to purplish grey to black.

Larvae First instar black, spinulae abundant; the 2nd to 4th instar blackish brown, with many scoli and white oblique stripes, similar to bird droppings. The 5th instar 45 mm long, yellowish green; the dorsal surface of the metathorax with an eye spot on each side. Metathorax, 1st, 4th -6th abdominal segments with bluish black oblique bands. Subspiracular line with a white spot at each segment. Scent gland orange yellow.

Pupae Length 29-32 mm, bright green, with brown spots, color varies with the environment. The mesonotum raised, high and sharp. Vertex with two projections and emarginated in the middle. Yellowish green, dorsal surface of the metathorax with eye spots on the two sides,

Biology and behavior *Papilio xuthus* occurs 3 generations a year along the Yangtze River and the north, 4 generations a year in Jiangxi, 5-6 generations a year in Fujian and Taiwan. They overwinter as pupae on limbs, back of leaves, etc. Adult periods in Zhejiang are: overwintering generation – May-June, 1st generation – July-August, 2nd generation September-October. The adult periods of the overwintering to 5th generations in Guangdong are: March-April, April-May, late May to June, late June to July, August-September, October-November, respectively. Adults are active during the day, feed on nec-

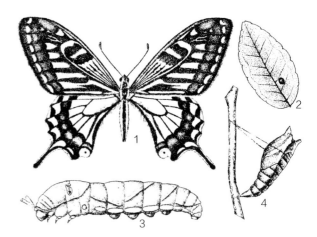

P. xuthus. 1. Adult, 2. Eggs, 3. Larva, 4. Pupa (Zhang Peiyi)

P. xuthus adult (dorsal view) (Li Zhenyu)

P. xuthus adult (ventral view) (Li Zhenyu)

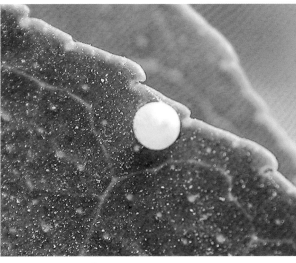
P. xuthus egg (Xu Gongtian)

P. xuthus young larva (Xu Gongtian)

P. xuthus mature larva exposing glands (Xu Gongtian)

P. xuthus pupa (Xu Gongtian)

tar, lay eggs singly on buds and back of leaves. Egg stage is 7 days, larvae with 5 instars. Young larvae feed on egg shells first, then buds and leaves. Mature larvae secrete silk at hidden places and pupate. Natural enemies include *Pteromalus puparum* and *Brachymeria lasus*.

Prevention and treatment

1. At low density, hand remove larvae and pupae. They can be placed in screened cages to let parasitic wasps emerge.

2. Biological control. Place mature larvae and pupae in cages and let parasitoids such as *Brachymeria lasus* and *Pteromalus puparum* to emerge. At high population density, spray *Bacillus thuringiensis galleriae* 8,000 IU/μL at 1,000-1,500×.

3. Chemical control. Spray 5.7% emamectin at 3,000-5,000×, 20% cypermethrin at 30-45 mL/ha, 25 g/L lamda-cyhalothrin at 1,000-1,500× to control 1st to 4th instar larvae.

References Li Chuanlong 1966; Shaanxi Academy of Agriculture and Forestry- Forest Research Institute 1977, Northwest Agricultural College- Department of Plant Protection 1978; Editorial Board of Diseases and Pests of Crops in China 1981; Chinese Academy of Agriculture- Citrus Research Institute 1985; Huang Jinyi et al.1986; Chinese Academy of Sciences - Institute of Zoology 1987; Xiao Gangrou (ed.) 1992; Yuan Yu, Lü Longshi, Jin Dayong 2001; Hua Chun, Yu Weiyan, Chen Quanzhan et al. 2007.

(Li Menglou, Li Zhenyu, Liu Lianren)

501 *Matapa aria* (Moore)

Classification Lepidoptera, Hesperidae
Common name Common redeye

Distribution Zhejiang, Fujian, Jiangxi, Guangdong, Guangxi, Hainan, Sichuan, Hong Kong; India, Sri Lanka, the Philippines, Myanmar, Laos, Malaysia, Indonesia.

Hosts *Bambusa blumeana, B. flexuosa, B. ventricosa, B. multiplex, B. multiplex* var. *riviereorum, B. multiplex* 'Fernleaf', *B. textilis* var. *purpurascens, B. vulgaris* 'Wamin', *Dendrocalamopsis giganteus, D. oldhami, Phyllostachys heteroclada, Phyllostachys sulphurea* 'Viridis'.

Damage Larvae roll leaves, feed inside rolled leaves, affect bamboo growth and their aesthetic value.

Morphological characteristics **Adults** Females 13.4-16.8 mm long, males 14.8-20.7 mm long, wingspan 40.5-45.8 mm. Dark brown, head brownish yellow, maxillary palpi thick; eyes red, pale brown after dead. Antennae brownish grey, greyish white between segments. Wings dark brown to black, without markings; the costal margin of forewing slightly expanded at 1/3 of the wing length and depressed at 2/3 of the wing length. Front margin and basal area brownish yellow, margin hairs yellowish white, with brownish yellow iridescence. Underside of wings dark brownish yellow, ventral surface of abdomen densely covered with yellow hairs. End of the abdomen truncated, slightly arched, yellowish white.

Eggs Dome shape, diameter 1.6-1.9 mm, height 1.0-1.2 mm. Greyish white at beginning, greyish brown or greyish green later, with radiant ridges from tip to the side. The top shows a red spot before hatching. Their surface often with body scales from the female and is often mistakenly seen as spines.

Larvae Young larvae 2 mm long. Dark red, head black, shinny; 3rd instar larvae yellowish green. There are 5 instars, head width of the 1st-5th instar are: 0.86, 1.08, 1.41, 2.13, and 2.94 mm, respectively. Mature larvae 26.5-37.5 mm long, light-yellowish green, each segment is divided into 3-4 segments, covered with thick white powder. Anterior of the prothorathic spiracle with a narrow black band, extending upward and to the anterior part of the other spiracle. Spiracle black, the 1st and 9th spiracles are larger than others.

Pupae Length 19.5-25.2 mm. Pale yellow at beginning, milky later; pronotum greyish black after 3 days, eyes red; head and thorax greyish black later, then all parts greyish black. Black prior to eclosion, eyes red. Proboscis reaches the end of the 6th abdominal segment, base of wing buds with a pair cremasters, dorsal surface with one large projection on each side, the projection has 3 clusters of hairs.

Biology and behavior *Matapa aria* occurs 5 generations a year in Guangdong. It overwinters at 2nd and 3rd instars in leaf pots. They resume feeding in early and mid-March. Pupae appear in mid-April. Adults appear from late April to early May. The larval periods of the 1st-5th generations are: early May to early June, mid-June to late July, late July to late August, late August to late September, late October to late April next year, respectively. The adult periods of the 1st-5th generations are: late April

M. aria adult (ventral view) (Xu Tiansen)

M. aria adult (dorsal view) (Xu Tiansen)

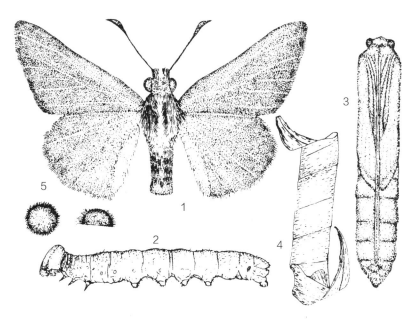

M. aria. 1. Adult, 2. Larva, 3. Pupa, 4. Damage, 5. Egg (Huang Jinshui, Lin Qingyuan)

M. aria pupa (Xu Tiansen)

M. aria larva (Xu Tiansen)

to late May, mid-June to early July, mid-July to early August, mid-August to mid-September, late September to late October, respectively. Generations overlap.

Prevention and Treatment Conserve natural enemies. Predatory enemies include *Pisaura* sp., *Oxyopes* sp., *Harpactor* sp., and *Hierodula patellifera*. They prey upon larvae when they migrate from one pod to another pod. Parasitoids include Trichogramma, *Coccygomimus parnarae* and *Exorista japonica* which parasitize *Matapa aria* larvae.

References Xu Tiansen and Wang Haojie 2004.

(Xu Tiansen, Wang Haojie, Huang Jinshui, Lin Qingyuan)

502 *Acantholyda erythrocephala* (L.)

Classification	Hymenoptera, Pamphiliidae
Synonym	*Tenthredo erythrocephala* L.
Common name	Pine false webworm

Importance Pine false webworm is a serious defoliator in northern coniferous forests in the world. It started to occur in large numbers in Shenyang of Liaoning province since 2000 and caused significant economic and ecological loss.

Distribution Liaoning, Heilongjiang; United Kingdom, U.S.A., Russia, central Europe, northern Europe, Korea peninsula.

Hosts *Pinus tabuliformis*, *P. armandii*, *P. bungeana*, *P. densiflora*, *P. sylvestris* var. *mongolica*, *P. koraiensis*.

Damage Larvae hide in nets. Outbreaks occur rapidly and may consume all leaves in a short period. The leaves in the next year small, often diseased. Tree growth is slow or the trees even die after infestation.

Morphological characteristics **Adults** Females 12-15 mm long. Head reddish brown. Males 11-13 mm, head black.

Larvae Young larvae head yellowish white, body purplish brown, later become olive color, head light-yellow, then darker. Mature larvae 22-26 mm long.

Biology and behavior *Acantholyda erythrocephala* occurs one generation a year in Liaoning. Pupation starts in early April and peaks in late April. Adults emerge from late April, peaks in early May. Eggs are laid from late April. Larvae start to emerge in early May and peak emergence is in mid-May. Mature larvae fall to soil and overwinter in mid- and late June.

There are more males than females during the early adult emergence period. Adults prefer flying under the sun. They do not fly during the rain. After disturbed, they

A. erythrocephala adult male (Wei Meicai)

A. erythrocephala adult female (Wei Meicai)

fall to ground or lower part and escape. Adults live for 3-10 days.

Eggs are laid on sunny side of the leaves, singly or in rows. Egg period is about 10 days. After emergence, larvae enclose themselves in a net on limbs. First instars chew off the needles and feed the basal part of the leaves. They also attach surrounding leaves to the net and feed on the leaves. The leftover leaves and feces gradually accumulate on the net and enlarge the net. The net expands as the larvae feed and eventually may connect with other nets. Larvae have 5 instars, each instar is 3-5 days long, larval period is 16-25 days.

Mature larvae enter into soil in mid- and late June and make soil chambers. They overwinter as pre-pupae. Each soil chamber has one insect. Most of them are located in soil directly under the canopy. Depth is 0-10 cm. Pupation occurs in next April. Pupal period is 15-18 days.

Prevention and treatment

1. Silviculture. Plant mixed forests. Keep forest canopy coverage at least 0.7. Turn soil in early spring and winter to expose the overwintering insects.

2. Mechanical control. Collect nets and destroy the insects within the nets.

3. Biological control. Natural enemies during egg stage include *Trichogramma* sp., *Camponotus* sp., and *Harmonia axyridis*; larvae stage include *Coccygomimus disparis*, *Sphedanolestes annulipes*, *Harpactor* sp., *Vespula* sp., and *Camponotus* sp.; pupal stage include *Beauveria bassiana*; adult stage include *Camponotus* sp., *Cyanopica cyana*, and *Lanius* sp.

4. Chemical control. Larval stage can be controlled by any insecticides labelled for defoliators. Examples are Apply 45% profenofos and phoxim mixture at 1,000-1,500×, 25 g/L lamda-cyhalothrin at 1,000-1,500×, and 10% pyrethrum at 8,000×.

References Xiao Gangrou 1992; Wang Guiqing 2000; Xiao Keren, Chen Tianlin et al. 2005.

(Wang Hongbin, Goa Xingrong, Zhang Zhen)

503 *Acantholyda flavomarginata* Maa

Classification Hymenoptera, Pamphiliidae
Synonym *Acantholyda guizhouica* Xiao

Importance *Acantholyda flavomarginata* is a serious defoliator in southern China.

Distribution Fujian, Jiangxi, Hunan, Guangxi, Sichuan, Guizhou, Taiwan.

Hosts *Pinus massoniana*, *P. yunnanensis*, *P. armandii*, *P. morrisonicola*.

Damage Larvae feed on leaves and make nets. Infested trees look like burned by fire. Trees may die after 2-3 years consecutive infestations.

Morphological characteristics **Adults** Wingspan 8-15 mm, females 12-16 mm long. The side of the dorsal surface of abdomen, posterior end of the ventral surface yellow to reddish yellow. Male 10-12 mm. Antennae 26-32 mm.

Larvae Body pale green before 5th instar, then yellow. Mature larvae reddish yellow. Dorsal line indistinct. Mature larvae 23-25 mm.

Biology and behavior *Acantholyda flavomarginata* occurs one generation a year in Sichuan and infests *P. yunnanensis*. It overwinters as pre-pupae in soil under the tree canopy. Pupation occurs in April. Adults emerge from late April and peak emergence occurs in early May. Eggs are mostly laid between early and mid-May and hatch mostly in mid-May. Larvae start to go down the trees in early July and all enter soil in late July.

Acantholyda flavomarginata occurs one generation a

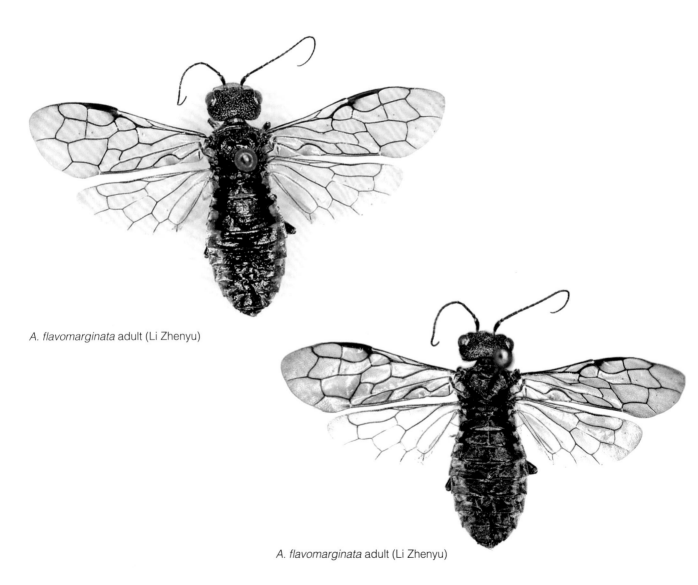

A. flavomarginata adult (Li Zhenyu)

A. flavomarginata adult (Li Zhenyu)

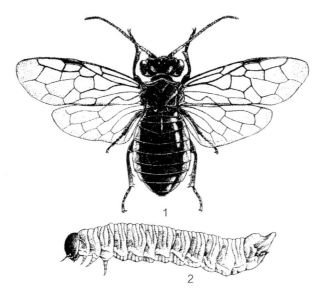

A. flavomarginata. 1. Adult, 2. Larva (Zhu Xingcai)

year in Guizhou and infests *P. armandii*. It overwinters as mature larvae in soil. They pupate in mid-March next year and reach peak in late March. Adults emerge from early April to late April.

Adults are active in mid- and upper part of the tree canopy. They mate and lay eggs from the peak of adult emergence. Females can produce young parthenogenetically. Eggs are glued to the inner side of leaves with their long axis same direction as the leaves. Eggs are in a line. Adults prefer lay eggs on low trees. Egg period is 10-15 days. First and 2nd instars feed at base of young leaves. Larval stage is 6-7 days. Larvae chew off leaves and drag them into the nests and feed. They also tend to hang on silk and feed. Larval period is 40 days. In late July, mature larvae abandon the nests and fall to ground, enter soil and overwinter.

Prevention and treatment

1. Silviculture. Remove infested limbs and trees in early April. Plant mixed forests.

2. Mechanical control. Turn oil in late fall and early winter. Dig soil under tree canopy to remove pupae.

3. Biological control. Spray *B.t.* (8,000 IU/μL) at 1,000-5,000×, to kill young larvae. Conserve natural enemies such as *Polyrhachis dives*, *Carcelia rasella*, *Apanteles ordinarius*.

4. Chemical control. In early May, 25 g/L lamda-cyhalothrin at 1,000-1,500×. In late May, apply 5.7% emamectin at 3,000-5,000×.

References Xiao Gangrou 1992, 2002; Du Juan, Zhang Chaoju, Zhao Wende et al. 2003; Xu Jiandong, Zheng Hongjun, Yuan Chaoxian et al. 2007.

(Wang Hongbin, Gao Xingrong, Zhang Zhen, Xiao Gangrou)

504 *Acantholyda peiyingaopaoa* Hsiao
Classification Hymenoptera, Pamphiliidae

Importance *Acantholyda peiyingaopaoa* is a serious pest of *Picea koraiensis* in Baiyin Aobo (Chefeng city, Inner Mongolia).

Distribution Inner Mongolia.

Host *Picea koraiensis*.

Damage Larvae infest needles. They make nets and hide inside nets.

Morphological characteristics **Adults** Body black, shiny. Females 13-14 mm long, antennae 28-34 segmented; males 10-12 mm long, antennae 30-32 segmented.

Larvae Mature larvae 18 mm long, olive color, with purple iridescence, dorsal line dark green, ventral line pale purple.

Pupae Brightly green at beginning, ventral surface of abdomen and legs slightly with purplish red, become greyish brown before eclosion.

Biology and behavior *Acantholyda peiyingaopaoa* occurs one generation a year and overwinters as prepupae. They pupate in early May to late May. Adult emergence period is from early June to mid-July. Eggs are laid from early June to late June. Larvae appear from mid-June. They fall to ground, make soil chambers, and turn to prepupae from mid-July and reach peak in mid-August. Eggs are laid singly, one on each leaf, usually on inner side of old leaves. Each nest contains one larva. The nest is tubular, consists of silk and secretion. Lara is visible from outside of the nest. Feces and leaf debris are not present in nests. Larvae expel feces to the ground.

Prevention and treatment

1. Silviculture. Manage the forests properly and maintain canopy coverage.

2. Chemical control. It is best to spray when larvae start to hatch. Spray125 ppm Dimilin at 750 kg/ha can reduce population by 97%. For older larvae, apply 2.5% lamda-cyhalothrin at 5,000×, or 45% profenofos and phoxim mixture at 1,000-1,500×.

References Xiao Gangrou 1992, 2002; Lu Changkuan, Ai Qinggang, Fan Yunzhong et al. 2001.

(Wang Hongbin, Gao Xingrong, Zhang Zhen, Xiao Gangrou)

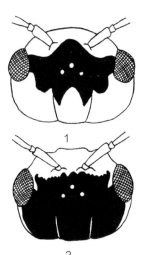

A. peiyingaopaoa. 1. Adult female head, 2. Adult male head (Zhang Peiyi)

A. peiyingaopaoa adult female (Zhang Peiyi)

A. peiyingaopaoa adult male (Zhang Peiyi)

505 *Acantholyda posticalis* Matsumura

Classification Hymenoptera, Pamphiliidae

Synonyms *Acantholyda pinivora* Enslin, *Tenthredo stellata* Christ, *Lyda nemoralis sensu* Thomson

Importance *Acantholyda posticalis* outbreaks occur often in its distribution areas.

Distribution Hebei, Shanxi, Heilongjiang, Shandong, Henan, Shaanxi; Europe, Mongolia, Japan, Korea Peninsula, Siberia.

Hosts *Pinus tabuliformis*, *P. densiflora*, *P. sylvestris* var. *mongolica*, *P. koraiensis*, *P. nigra*, *P. sylvestris*.

Damage Larvae spread fast. They make silk nests, chew off leaves and drag partial leaves into nests. They may consume all leaves on new shoots within 20 days. Infestations weaken *Pinus tabulaeformis* growth or even kill them.

Morphological characteristics **Adults** Females 13-15 mm long, body black, with yellow, light-yellow or yellowish white marks.

Eggs Length 3.5-4.0 mm, half-moon or boat shape, milky at beginning, flesh red and the sharp end becomes black before hatching.

Larvae Length 15-23 mm. Yellowish green at beginning, mature larvae light-yellow to brownish yellow.

Pupae Females yellow, 15-19 mm; males light-yellow, 10-11 mm long; black before hatching.

Biology and behavior *Acantholyda posticalis* occurs one generation a year and overwinters as prepupae in soil. Pupation occurs in late April to mid- May in

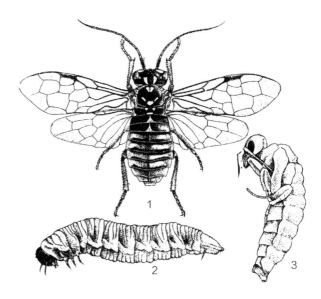

A. posticalis. 1. Adult, 2. Larva, 3. Pupa (Zhu Xingcai)

Shanxi. Adults emerge mostly from early June to mid-June. Larvae start to leave trees in early July and finish in late July. The occurrence period in Heilongjiang is 10 days later than that in Shanxi, and the occurrence period in Henan is 15 days earlier than in Shanxi. Adults emerge in early and mid-July in Pingshan (Hebei) and some areas in Shanxi; the species name in these areas still needs to be investigated. Adults can fly about half a day after emergence. Adults fly in mid- and upper layer of canopy, or crawl on limbs. After disturbed, they fall and run on the ground. Mating usually occurs at 10:00-16:00. Each adult mates multiple times. Eggs are deposited on dorsal side of leaves. They lay eggs first on last year's leaves, then on current year leaves. Eggs are laid singly. It takes a female 2-5 minutes to lay one egg and stops laying eggs after 3-4 days. Each female lays about 20 eggs. Young larva makes a nest at the base of 3-5 needles. It breaks leaves and brings them into the nest to feed. After the 3rd instar, it migrates to base of a young shoot and makes a dense tubular nest with openings on both ends. After the 4th instar, feeding increases significantly. One larva consumes 160 cm long needles. After disturbed, it

A. posticalis adult (FPC)

A. posticalis larva (FPC)

A. posticalis nest (FPC)

A. posticalis pupae (FPC)

A. posticalis overwintering larva (FPC)

retreats to nest and may drop via a silk and hangs itself on silk.

Prevention and treatment

1. Monitoring. Conduct regular sampling on the ground and predict occurrence based on weather and other biological indictors. Use pheromone traps to monitor populations.

2. Silviculture. The insect often occurs in poor grown, sparse forests. Proper management of the forest will help reduce pest occurrence.

3. Mechanical control. Turn soil to destroy pupal chambers. Hand remove nests.

4. Conservation of natural enemies. The parasitoid *Trichogramma* sp. and an ant species *Formica fukaii* are important natural enemies.

5. Chemical control. Apply 25 g/L lamda-cyhalothrin at 1,000-1,500×, Apply 5.7% emamectin at 3,000×, 20 g/L lamda-cyhalothrin at 1,000-1,500×, 50% imidaclo-prid-monosultap at 600-800×, Dimilin spray, or matrine fogging.

References Liu Jing, Sun Qiwen, Pang Xianwei 1990; Lou Wei, Hou Aiju, Dai Yuwei 1990; Xiao Gangrou 1992; Xiao Gangrou, Huang Xiaoyun, Zhou Shuzhi 1992; Liu Qinyun 2002; Xiao Gangrou 2002; Wang Wenlin 1998; Cheng Lansheng and Chu Yuejin 2002; vLiang Zhonggui, Li Jianjun, Liu Jianqiang 2007; Yang Dahong, Wang Xiaoji, Zhou Mingqing 2007.

(Zhang Zhen, Liu Xiaolian, Kong Xiangbo, Zhao Ruiliang, Li Jinpeng, Hou Aiju, Lou Wei)

506 *Cephalcia tienmua* Maa

Classification Hymenoptera, Pamphiliidae
Synonym *Cephalcia tianmua* Maa, Xiao et Chen

Importance *Cephalcia tienmua* occasionally becomes a pest and can affect tree growth.

Distribution Zhejiang.

Host *Pinus massoniana*.

Damage Severely infested trees become completely defoliated and many nests exist on trees.

Morphological characteristics **Adults** Male wing span 22-24 mm, female wingspan 26-34 mm. Head black, with yellowish white markings; antennae white except that the base and the tip black. Thorax black, with yellowish white markings. Wings transparent, with yellow appearance; stigma black, its base to posterior margin with a dark transverse band. Legs yellowish brown except for coxa and trochanter which are black. Abdomen black, sides of the dorsal surface, posterior border of the ventral surface yellowish white.

Eggs Elongated elliptical, 2.5 mm long, 0.7 mm wide; slightly curved; milky, covered with black gelatinous materials.

Larvae Mature larvae 23-31 mm. Green, dorsal line pale green. Head reddish brown, pronotum brown, lamina supraanalis light-yellow.

Pupae Length 14-22 mm, new pupae bright green, then yellow, and black before eclosion.

Biology and behavior *Cephalcia tienmua* occurs one generation a year. It overwinters as prepupae in soil. Pupation occurs during mid-February and late March. Adults emerge during mid-March and mid-April. Eggs are laid from late March to late April. Larvae hatch from mid-April to mid-May. They enter soil from early June and all larvae are in soil in early July.

Adults stay in soil chambers for 4-12 days after eclosion. They emerge from soil in clear weather. They can immediately mate after emergence. Eggs are laid on old needles. Females prefer laying eggs on well-grown trees with thick leaves. A female saws a slit on a needle and lays eggs on it. Adults live for 7-14 days. Half their life is spent in soil chambers.

Eggs can only survive on live needles. When needs die, eggs cannot hath. Egg period is 15-22 days. Larvae migrate to base of needles after hatching, secrete silk, make nests, and feed on leaves. They feed on old leaves rather than current year leaves. Young larvae only consume margins of the leaves. As they grow, they consume whole leaves. Sometimes they break off needles and drag the leaves to the nest and feed there. They attach the feces to the silk nest and form nest like balls. A ball can have multiple larvae. When the surrounding leaves are consumed, they build passages to neighboring branches and feed. The nest size increases as the larvae grow. Af-

C. tienmua adult
(edited by Zhang Runzhi)

C. tienmua larva (edited by Zhang Runzhi)   C. tienmua pupa (edited by Zhang Runzhi)

C. tienmua damage
(edited by Zhang Runzhi)

C. tienmua eggs (edited by Zhang Runzhi)

ter 35-52 days, larvae mature and enter into 5 cm deep soil and build chambers. Most larvae are concentrated to the base of the trees. Prepupal period is 9 months. Pupal period is 21-31 days.

Prevention and treatment

1. Silviculture. Plant mixed forests and reduce pest populations.

2. Conservation of natural enemies. Predatory enemies include *Polyrhachis divers*, *Pheidole nodus*, *Chrysoperla sinica*, *C. savioi*, *Chrysoper septempunctata*, *Hormonia yedoensis*, *H. axyridis*, *Propylea japonica*. The parasitic rate of prepupae by *Myxexoristops blondeli*, *Anisotacrus* sp. can reach 47.45%.

3. Release *Steinernema carpocapsae* "Beijing" strain at 2,000/mL into soil under tree canopy can reduce pest population by 70%.

4. Spray 25% Dimilin at 400× or 25 g/L lamda-cyhalothrin at 1,000-1,500×.

References Xiao Gangrou and Chen Hanlin 1993; Chen Hanlin 1996; Chen Hanlin, Zhao Renyou, Jin Genming et al. 1997; Xiao Gangrou 2002.

(Chen Hanlin)

507 *Cephalcia yanqingensis* Xiao

Classification Hymenoptera, Pamphiliidae

Importance *Cephalcia yanqingensis* was found in the early 1980's in Yanqing county of Beijing.

Distribution Beijing.

Host *Pinus tabuliformis*.

Damage Larvae feed on leaves and may kill the host at high population density.

Morphological characteristics **Adults** Females 13-19 mm, wingspan 22-34 mm. head orange red, shinny. Antennae reddish yellow, the last 2-3 segments with black color. There is a suture behind the eyes. Eyes, ocelli area, tip of mandibles and mandibular teeth black; palpi dark reddish brown. Clypeus raised in center, anterior border truncated. Cervix, pronotum reddish yellow, basalare yellowish brown. Forewing from stigma to area exterior of 1r-m, 1m-Cu, Cu_1 pale black, with dark blue iridescence; area inner this line dark grey except for the costal cell. C and S_C veins dark yellow; middle of R vein blackish brown, the two ends dark yellow. Base of M, 1A dark yellow, their margins black. End of veins black. 1r-m, 1m-Cu, Cu_1, stigma black, end dark yellow. Hindwing costal cell yellowish white, other areas dark grey; C and S_C veins dark yellow, other veins black (base of M and 1A veins dark yellow). Legs reddish brown; abdomen reddish brown. Antennae 27-29-segmented.

Males 10-16 mm, wingspan 21-25 mm. Head black; antennae dark yellow, 27-29 segmented, the last 2-3 segments with black color; palpi dark yellow. Thorax black, somewhat shiny; pronotum dark yellow except for the black spot in the center, cervix dark yellow except the base; most part of the epimeron of mesonotum yellowish white. Wings dark grey; stigma and other veins black. Legs dark orange red, abdomen black, sides of the dorsal surface black, genitalia dark greyish yellow.

Eggs Length 2.0-2.5 mm, width 1.0-1.1 mm; near elliptical, center slightly curved, one end tapered. Yellow at beginning, yellowish brown near hatching.

Larvae Length 16-26 mm. Reddish yellow; ocelli area brown; mandibles blackish brown; frons with 3 black spots. Pronotum with a dark brown spot at the posterior border and four blackish brown smaller spots on the sides. Mesothorax and metathorax with two black brown spots on each side. There are two black brown spots between the two thoracic legs on each segment.

Pupae Females 13.0-21.6 mm long, males 12.5-15.5 mm long. Yellowish brown. Eyes blackish brown. Antennae yellowish white, palpi milky, frons with three black

C. yanqingensis adult female (left) and male (right) (Yan Guozeng)

spots arranges in a triangular shape.

Biology and behavior *Cephalcia yanqingensis* occurs one generation a year but can also take two years to finish one generation. They overwinter as mature larvae in soil. Pupation occurs in May. Adults emerge from mid-May to early June. Adults lay eggs from early June and end in early July. Pupal period is about 15 days. Males appear earlier than females. Eclosion period is about 10 days. Adults are active during clear, warm days at 10:00-15:00. They rest among needles and weeds. They often fly around canopy or among trees, searching for mates, lay eggs on needles. A female lays 5-9 eggs on each needle and lays 24-40 eggs. Adults live for 20-30 days. Female:male ratio is 1:1.25. Egg period is 26-30 days. Hatch rate is 96%.

Young larvae secrete silk and make tubular nests. They break off leaves and drag them to the nest. After 3rd instar, larvae migrate to new locations and feed individually with one larva per nest. Each larva consumes 20 sheaths of needles. After disturbed, larvae retreat inside the nests. Larvae have 6 instars, larval period is 28-47 days. Mature larvae fall to ground in early August and burrow into 1-10 cm soil and overwinter.

Prevention and treatment

1. Dig soil under tree canopy to destroy overwintering larvae.

2. Attract and kill adults using red glue boards.

3. Conservation of natural enemies. An unknown parasitic fly is found parasitizing mature larvae.

4. Apply 5% cypermethrin smoke to kill adults before they start laying eggs.

References   Xiao Gangrou 2002.

(Yan Guozeng)

C. yanqingensis male (Li Zhenyu)

C. yanqingensis female (Li Zhenyu)

508 *Chinolyda flagellicornis* (Smith)

Classification Hymenoptera, Pamphiliidae
Synonyms *Lyda flagellicornis* Smith, *Cephalcia flagellicornis* (Smith)

Importance *Chinolyda flagellicornis* often cause high defoliation rate to hosts and even kill the hosts.

Distribution Zhejiang, Fujian, Hubei, Chongqing, Sichuan.

Hosts *Cupressus funebris*, *Sabina chinensis*, *S. chinensis* 'Pyramidalis', *Juniperus formosana*, *Platycladus orientalis*, *Chamaecyparis obtuse*, *Cryptomeria fortune*.

Damage Larvae feed on leaves or bark of young limbs and cause death of limbs. Heavily infested forests appear red and similar to fire burn. Some trees may even die from infestations.

Morphological characteristics **Adults** Females 11-14 mm, wingspan 24-29 mm. Body reddish brown, mandible tips, base and end of the antennae flagellum, occelli area, epimeron of mesothorax black.

Eggs Pale yellow, elongated elliptical, 1.5 mm long, slightly curved.

Larvae Length 18-23 mm. Head reddish brown, thorax and abdomen with several white longitudinal stripes.

Pupae Brightly green at beginning, dark brown or yellowish brown near eclosion.

Biology and behavior *Chinolyda flagellicornis* occurs one generation a year and overwinters as prepupae in 2-20 cm soil. Pupation occurs in March. Adults

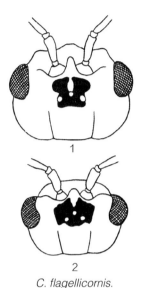

C. flagellicornis.
1. Adult female, 2. Adult male (Zhang Peiyi)

emerge in mid-March to late April. They lay eggs from late March to May. Larval period is from early April to mid-June. They enter soil from early May to early June. After eclosion, adults stay for 1-3 days before emerging to soil surface. They crawl for short periods on ground, then move to shrubs or hosts and mate. They lay eggs 3-5 hours after mating. Eggs are laid on dorsal side of

C. flagellicornis adults (FPC)

C. flagellicornis mating pair (FPC)

C. flagellicornis larvae (FPC)

C. flagellicornis eggs (FPC)

leaves. Young larvae secrete silk and make a nest. After 6-7 hours they feed on epidermis of leaves. When leaves near the nest are consumed, they migrate to neighboring branches and make new nests. Larvae have 6-7 instars. Mature larvae feed singly.

Prevention and control

1. Silviculture. Properly manage forests where trees are sparse and are not grown well. Plant mixed forests. The broad-leaved trees such as *Quercus* spp., *Alnus cremastogyne*, *Cinnamomum camphora* are suitable.

2. Mechanical control. Turn soil to destroy prepupae. Remove nests on canopy.

3. Biological control. Spray 6×10^9 conidia/g *Metarhizium anisopliae* at 60 kg/ha to kill the 3rd instar or younger larvae.

4. Chemical control. Spray 25 g/L lamda-cyhalothrin at 800-1,000×.

References Yang Dasheng et al. 1987; Xiao Gangrou, Huang Xiaoyun, Zhou Shuzhi 1991; Xiao Gangrou 1992, 2002; Xiao Yugui and Guo Hengxiao 2003; Zheng Yongxiang, Peng Jialong, Wang Mingsheng et al. 2009.

(Zhang Zhen, Liu Xiaozhen, Kong Xiangbo, Yang Dasheng, Zeng Lin)

C. flagellicornis damage (FPC)

509 *Cephalcia kunyushanica* Xiao

Classification Hymenoptera, Pamphiliidae

Importance *Cephalcia kunyushanica* was found in 1983 in Shandong. It can cause significant damage to hosts.

Distribution Shandong.

Hosts *Pinus koraiensis, P. densiflora, P. thunbergii.*

Damage Infested hosts have brown canopy, reduced growth. Severely infested hosts may nearly die.

Morphological characteristics **Adults** Females 15 mm long, body reddish brown, antennae 30 segmented, yellowish brown, its last 1-2 segments blackish. The following parts are black: eye; base of ocelli; anterior part of the mandibles; the prescutum, scutum inner and outer margin, scutellum, side of scutellum, and postnotum of the mesonotum; metanotum and scutellum of metathorax. Forewing transparent. Hindwing apex and outer margin pale smoky brown, other parts of the wing light-yellow, transparent; stigma black. Head, thorax sculpture small and sparse, medium depth. Posterior border of abdominal segments with small and sparse sculpture. Slender hairs yellow, short, sparse.

Males 13 mm long. Head black, antennae light-yellow, the end blackish. Side of the antennae, inner margin of the eye sockets, clypeus, frons, antennae sockets, cheek, and palpi yellow. Thorax black; sides and posterior border of pronotum, the posterior of the episternum, epipleurite yellow. Legs yellow; coxa, trochanter, and femur with black color on ventral surface. Wings transparent, with pale brown color, costa yellow. Abdomen notum black, sides yellow, ventral surface yellow, anterior border of each segment with a black transverse band. The 9th abdominal segment and the copulatory organ dark yellow. Clypeus raised in center, anterior border truncate. Head and thorax hairs yellow and thin, medium length, sparse. Posterior of ocelli, above the eyes with medium size and density sculpture. Middle of dorsal surface with very fine sculpture.

Eggs Length 2.5 mm, boat shape, one end thicker, two sides with a depressed line. Tea color at beginning, shiny, turns to brown later.

Larvae Mature larvae length 21-24 mm, light-yellow-

C. kunyushanica adult male (Zhang Peiyi)

ish brown, head reddish brown. Pronotum with a raised brown spot on the back. Sides of thorax with brown spots. Each segment is further divided into 4 small segments. Spiracles brown.

Pupae Pale yellow.

Biology and behavior *Cephalcia kunyushanica* occurs one generation a year in Shandong and overwinters as mature larvae in soil. They pupate during late April and early June next year. Adults start to emerge in late May and end in late June. They soon lay eggs after emergence. Larvae emergence period is from early June to early July. Damage period is from late June to August. Larvae start go down trees in mid-August and peak in late August. All larvae go down trees in early September.

Adults usually emerge from soil during 9:00-15:00. After short period of crawling, they can fly to trees, mate, and lay eggs without feeding. A female lays eggs over 3 days period, life span is 7 days. Males are stronger fliers than females. A male lives for 5-8 days. Females emerge 3-5 days later than males.

Eggs are laid singly on upper surface of needles. Each leaf has 4-10 eggs. A female lays 4-20 eggs, with an average of 12 eggs. The egg hatch rate is 85%.

Larvae go through 4-5 instars. Young larvae are slow, feed in aggregations on scales of needles, then feed on needles on young shoots. The 2nd instar feed more, secrete silk at base of needles, feed inside the silk net, and form nests. Once the leaves around the nest are consumed, the larvae expand the nest and may connect 2-3 nests. They break needles from the base, drag the needles to the nest entrance and feed from there. They feed from the breaking point to the tip of the leaves. After 3rd instar, they feed much more and can consume all leaves on 1-2 year-old shoots. The canopy may be full of nests.

Prevention and treatment

Spray 45% profenofos and phoxim mixture at 1,00-1,500×.

References   Xiao Gangrou 1987, 2002; Xiao Gangrou et al. 1991; Wang Chuanzhen, Wang Jinggang, Yang Jun et al. 2000; Yang Jun, Ling Song, Liu Deling et al. 2001.

(Wang Hongbin, Gao Xingrong, Zhang Zhen)

C. kunyushanica adult female (Zhang Peiyi)

510 *Sirex rufiabdominis* Xiao et Wu
Classification Hymenoptera, Siricidae

Importance *Sirex rufiabdominis* is a serious wood boring insect pest to forestry and urban landscape.

Distribution Jiangsu, Zhejiang, Anhui, Shandong.

Hosts *Pinus massoniana, P. tabulaeformis.*

Damage Larvae feed inside tree trunks. More than 3,000 larvae may infest a host. Infested trees are distributed sporadically, rarely in patches. Infested trees may die from death of cambium, fungal infections.

Morphological characteristics **Adults** Females 18-34 mm long; antennae black; head, thorax, dorsal surface of abdominal 1st-9th segments, coxa, trochanter, femur, and 5th tibia segments bluish black; with metallic iridescence. Males 14-22 mm long; similar to females in color. Antennae black, 19 segmented. The dorsal plates of the 1st-2nd abdominal segments bluish black, the anterior border of the 3rd dorsal plate black. Other segments reddish brown.

Eggs Length 1.7 mm, width 0.3 mm, spindle shape, white.

Larvae Tubular, light-yellow, head milky, mouthparts brown, abdomen with a semi-transparent dorsal median line. There are many brown short setae on the fan-shaped anal plate; the end with tooth-like projections. Thoracic legs degenerated. Mature larvae 14-30 mm long.

Pupae Length 12-34 mm. Pale yellow, brown before hatching.

Biology and behavior *Sirex rufiabdominis* occurs one generation a year. It usually occurs in elevations below 400 m and in > 20 years old forests. Adults lay eggs from mid-September to early September. Eggs hatch after 21-24 days. After a few days feeding activity inside tree trunks, larvae will overwinter. They resume activity in next March. They feed upward or downward along the tree trunks. Larval period is about 10 months. Mature larvae usually pupate at 5-20 mm near the edge of the wood with head pointing outward. Pupal period is 17-20 days. After adult emergence, it feeds for 3-7 days, chews a 5-8 mm diameter emergence hole. Adult emergence starts in early September, ends in late November. They mate soon after emergence from the holes. A female mates 1-4 times, males only mate one time. Adults are not phototaxic. They prefer flying on clear days. They are found

S. rufiabdominis adult male (Zhang Peiyi)

S. rufiabdominis adult female (Zhang Peiyi)

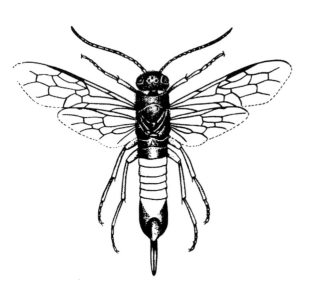

S. rufiabdominis adult female (Zhang Peiyi)

more often in pure pine forests than in mixed forests, forests with weak trees, damaged trees from other insects or diseases, sparse forests, and sunny slopes.

Prevention and treatment

1. Silviculture. Cleaning the forest before the end of the August.

2. Baiting. Place bait logs in the forest. After larvae hatched, debark the bait logs and destroy larvae.

3. Conservation of natural enemies. The following are important natural enemies: *Megarhyssa praecellens*, praying mantis, spiders, and birds (*Lanius* sp. and *Cyanopica cyana*).

Reverences Wu Xiazhong 1985, Xiao Gangrou et al. 1991, Xiao Gangrou 1992.

(Kong Xiangbo, Zhang Zhen, Chen Dafeng, Wu Xiazhong)

511 *Tremex fuscicornis* (F.)

Classification Hymenoptera, Siricidae
Common name Tremex wasp

Importance *Tremex fuscicornis* is a wood boring insect pest to many broad-leaved trees. They mainly infest *Salix* spp. and poplar trees and can cause tremendous economic and ecological damage.

Distribution Beijing, Tianjin, Liaoning, Jilin, Heilongjiang, Hebei, Shanxi, Inner Mongolia, Shanghai, Jiangsu, Zhejiang, Fujian, Jiangxi, Shandong, Hunan, Tibet, Shaanxi; Japan, Korea Peninsula, Australia, Western Europe.

Hosts *Populus tomentosa*, *P. davidiana*, *P. lasiocarpa*, *P. simonii*, *Salix babylonica*, *Ulmus* spp., *Zelkova schneideriana*, *Fagus longipeliolata*, *Quercus* spp., *Pterocarya stenoptera*, *Prunus armeniaca*.

Damage Larvae feed on phloem and xylem. Adults chew 1-6 mm diameter exit holes. The barks split and peel longitudinally along the exit holes and may detach from the trunks. Xylem may also split from loss of water and the tree subsequently dies.

Morphological characteristics **Adults** Females 16-43 mm long, wingspan 18-46 mm, blackish brown. The middle of the antennae dark to black, especially on the ventral side. Base of the front tibiae yellowish brown; mid-tibiae and hind-tibiae, basal half of the hind tarsi, the 2nd, 3rd and 8th abdominal segments, and anterior border of the 4th-6th segment yellow. Other parts black. Pronotum, mesonotum, ovipositor sheath reddish brown.

Males 11-17 mm long, metallic; the basal three antennae segments, front and mid-tibiae and tarsi, the 5th tarsal segment of the hind tarsi reddish brown. Thorax and abdomen black. Wing light-yellowish brown, transparent.

Eggs Length 1-1.5 mm, elliptical, slightly curved, the front end slender, milky.

Larvae Length 12-46 mm, tubular, milky. Head yellowish brown, thoracic legs small, the end of the abdomen brown.

Pupae Female length 16-42 mm, milky, head light-yellow. Males 11-17 mm long.

Morphological characteristics *Tremex fuscicornis* occurs one generation a year in Shaanxi as larvae in galleries. They resume activity in mid- to late March. Pupation occurs from mid-April. Adults emerge from late May. Females live for 8 days, males live for 7 days. Each female lays 13-28 eggs; egg period is 28-36 days. Larvae

T. fuscicornis adult female (Wang Yan)

T. fuscicornis adult (Wang Yan)

T. fuscicornis damage (Wang Yan)

T. fuscicornis eclosion holes (Wang Yan)

T. fuscicornis larva (Wang Yan)

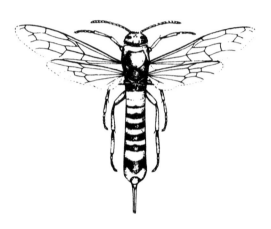

T. fuscicornis adult female (Zhang Peiyi)

emerge from mid-June. They overwinter in December. Larvae have 4-6 instars; larval period is 9-10 months. Adults are active during the day and are not phototaxic. They can fly up to 15 m high. They mate one day after emergence. Males mate one time, females lay eggs 1-3 days after mating. Each egg depression on trees usually yields 9 larvae and forms many galleries. Mature larvae make pupal chambers near 10-20 mm from edge of the wood. Weak trees have more numbers of the pest than healthy trees.

Prevention and treatment

1. Silviculture. Plant tolerant tree species and mixed forests. Promptly remove weak and infested trees. Take out cut trees or debark soon after cutting.

2. Baiting. Plant bait wood in forests. Debark during egg hatching period.

3. Conservation of natural enemies. Parasitic wasp (*Meganhyssa parccelleus*), praying mantis, spiders and birds (*Lanius* sp., and *Cyanopica cyana*) are important natural enemies.

4. Chemical control. Spray 45% profenofos and phoxim mixture at 1,000×, 2.5% lamda-cyhalothrin at 1,000-1,500× to kill adults.

References Xiao Gangrou, Huang Xiaoyun, Zhou Shuzhi 1991; Xiao Gangrou 1992, Zhao Zhenzhong, Zhang Baikui, Xing Xiuqing 2003.

(Kong Xiangbo, Zhang Zhen, Chen Dafeng, Han Chongxuan, Chen Xiaoda, Tang Guoheng)

512 *Urocerus gigas gigas* (L.)

Classification Hymenoptera, Siricidae
Synonyms *Urocerus gigas taiganus* Benson, *Urocerus taiganus* Benson
Common name Giant woodwasp

Importance *Urocerus gigas gigas* is an important wood boring insect pest to *Larix* spp., *Picea* spp, and *Abies* spp. It infests weak, dead, and newly cut trees and stumps.

Distribution Hebei, Shanxi, Liaoning, Heilongjiang, Shandong, Gansu, Qinghai, Xinjiang; Russia, Poland, Finland, Norway.

Hosts *Larix gmellini, Picea asperata, Abies fabri.*

Damage Leaves of infested trees become yellow to reddish brown and eventually fall. Larvae bore into tree trunks and make galleries. Their length is about 20 cm and diameter is about 5.8 mm. The end of the galleries are pupal chambers. Wood dusts fill the galleries. Adults chew open round exit holes which are 5-7 mm in diameter.

Morphological characteristics **Adults** Females 23-37 mm long. Black, shiny. Head and thorax densely punctured. Antennae filiform, dark yellow or yellowish brown. Thorax black. Wings membraneous and transparent, light-yellowish brown. Legs yellow; coxa, trochanter, femur black. Sheath of ovipositor brown. Males 12.5-33.0 mm long. Similar color as females, but scape of antennae black, other segments reddish brown. Abdomen color varies significantly, each side of the dorsal plate of the 9th abdominal segment with a large yellow round spot.

Larvae Mature larvae 20-32 mm, tubular, milky. Head light-yellow, without ocelli. Antennae 3 segmented, very short. Thorax with 3 degenerated legs, short, and not segmented. End of abdomen with a brown short spine, base of the spine with small spinulae.

Eggs Length 1.5 mm, width 0.20-0.25 mm. Near conical, pale milky, front end obtuse, rear end sharp.

Pupae Length 12.5-30.0 mm, milky, head light-yellow, eyes and mouthparts pale brown.

Biology and behavior *Urocerus gigas gigas* occurs one generation a year in Liaoning and overwinters as larvae in galleries. They resume activity in late April. Pupae appear from early May to early June. Pupal period is 90 days. Adults emerge from late May. Their peak occurs from mid-June to mid-July. Adult period is 90 days. Eggs are mostly laid during mid- and late June. Egg period is 100 days. Larvae start to appear in mid-June. Peak hatching period is during early and mid-July. Larvae start overwintering in mid-October.

Prevention and treatment

1. Quarantine. Prevent movement of infested logs.
2. Sanitation. Bury infested stumps.
3. Use bait wood to attract egg-laying adults. Debark the bait wood or fumigate the bait wood when eggs are hatching.
4. Chemical control. Infested logs can be fumigated

U. gigas gigas adult female (Xu Gongtian)

U. gigas gigas adult male (Xu Gongtian)

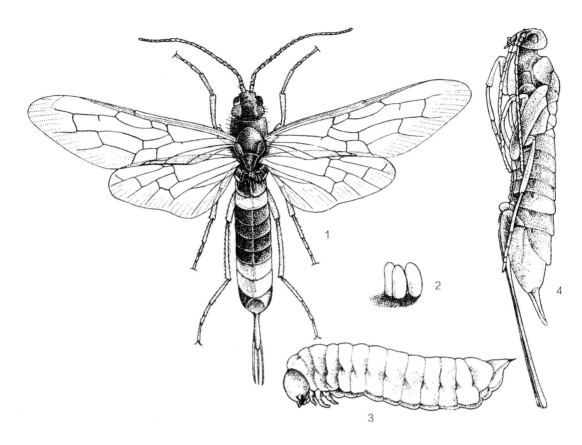

U. gigas gigas. 1. Adult, 2. Eggs, 3. Larva, 4. Pupa (Zhu Xingcai)

with methyl bromide at 15 g/m^3 or aluminum phosphide at 9 g/m^3 for 24 hours. Treat infested stumps with 45% profenofos and phoxim mixture or 15% indoxacarb.

References Ren Zuofo 1964; Xiao Huang Xiaoyun, Zhou Shuzhi 1991; Xiao Gangrou 1992; Wang Yingmin, Zhao Shidong, Ye Shuqin 2001; Ye Shuqin, Sun Jianwen, Xu Shuiwei 2002.

(Kong Xiangbo, Zhang Zhen, Chen Dafeng, Ren Zuofo)

513 *Arge captiva* Smith

Classification Hymenoptera, Argidae
Common name Japanese elm sawfly

Distribution Beijing, Hebei, Liaoning, Jilin, Jiangxi, Shandong, Henan; Japan.

Hosts *Ulmus* spp.

Damage *Arge captiva* often occurs on seedlings in plantations. Larvae feed on young leaves and may consume all leaves.

Morphological characteristics **Adults** Length 9-12 mm, wingspan 17-25 mm; metallic, head bluish black, wings smoky brown, legs bluish black.

Larvae Length 21-26 mm, yellowish green; each segment with three transverse rows of brown sarcoma.

Biology and behavior *Arge captiva* occurs 2 generations a year in Henan. It overwinters as larvae in soil. Pupae and adults appear in April and May. Each female lays 35-60 eggs. Larvae appear in June. They gather in groups during early morning and evening hours and feed on young leaves of seedlings. All leaves on a seedling can be consumed in a short period of time. Mature larvae enter soil and pupate in June and July. The 2nd generation adults emerge in July and August. The larvae mature in August and September. They crawl to hidden places in soil and overwinter in cocoons.

Adults are weak fliers. They chase each other on ground and mate. Eggs are laid on leaves of middle or lower part of the seedlings. When laying eggs, a female grasps its legs on a leaf, creates a slit between the upper and lower epidermis at the leaf edge, and deposits one egg. After finish laying one egg, it moves forward and lays another

A. *captiva* larvae (Xu Gongtian)

A. *captiva* adult (Xu Gongtian)

A. captiva larvae (Xu Gongtian)

A. captiva mature larva (Xu Gongtian)

A. captiva adult laying eggs on leaf edge (Xu Gongtian)

A. captiva cocoons (Xu Gongtian)

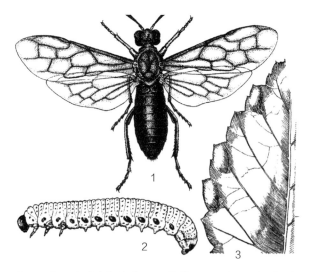
A. captiva. 1. Adult, 2. Larva, 3. Egg marks on leaf edge (Zhang Xiang)

egg. Once most of the leaf edge is laid with eggs, the female will choose another leaf and continues laying eggs. The leaf part where eggs are laid swell. After 7-9 days, eggs hatch. Newly hatched eggs feed on young leaves only. Larvae feed during day and night. When disturbed, they curl and fall to ground. Mature larvae enter soil, make cocoons and pupate.

Prevention and treatment

1. Mechanical control. Use a bamboo stick to knock off the larvae from seedlings during the morning or evening. Destroy the larvae that fall off.

2. Chemical control. Spray 50% sumithion at 1,000×, 50% malathion at 1,000×, Apply 45% profenofos and phoxim mixture at 1,000-1,500× when larvae occur.

References Yang Youqian et al. 1982; Xiao Gangrou, Huang Xiaoyun, Zhou Shuzhi et al. 1991; Xiao Gangrou 1992; Forest Disease and Pest Control Station of Henan Province 2005.

(Yang Youqian)

514 *Agenocimbex ulmusvora* Yang
Classification Hymenoptera, Cimbicidae

Importance It is an important pest of *Ulmus* spp. in recent years. The pest can cause complete defoliation, affect tree vigor and their aesthetic value.

Distribution Zhejiang, Anhui, Gansu.

Hosts *Ulmus pumila, U. davidiana, U. parifolia.*

Damage Larvae feed on leaves, cause notches or consume whole leaves.

Morphological characteristics **Adults** Length 18-22 mm, wingspan 41-46 mm. Body yellowish brown. Antennae 8 segmented, 6th-8th segments larger, segments not obvious. Head with long brownish yellow and black hairs; thorax with medium length hairs, and abdomen with short yellow hairs. Similar to vespid wasps.

Larvae Mature larvae 35-40 mm long, head width 4 mm. Body freshly green, ocelli black, with many folds.

Cocoons Length 20-26 mm, width 8-10 mm, elongated elliptical, constricted at 2/5 of the length. Reddish brown, attached with fallen leaves and soil particles.

Biology and behavior *Agenocimbex ulmusvora* occurs one generation a year and overwinters as pre-pupae inside cocoons in shallow soil, grass roots, or fallen leaves. They pupate in late March. Adults emerge in early and mid-April. Larvae mature in late May. About 20% of them enter into diapause.

Adults stay in cocoons for 1-3 days after eclosion. They emerge from cocoons during 14:00-17:00. After 1-3 days, they mate and deposit eggs under epidermis on edge of leaves. They prefer lay eggs on medium aged leaves on small limbs. Eggs are laid singly. Some leaves may have two eggs. Each female lays 60-70 eggs. Larvae have 5 instars and 30-35 days development period. Larval feeding starts from leaf edge, the 1st-2nd instar feed up to the mid-vein, 3rd instar feed on whole leaves and only leave the petiole. One larva consumes 350-400 cm^2 leaf area or 32-42 *Ulmus pumila* leaves during its lifetime. When feeding, larvae keep their body curled; when at rest, they curled into a circle on back of leaves. Single trees, edge of forest, and lower canopy suffer damage first.

Prevention and treatment

1. Silviculture. Plant local tree species and mix them with *Ulmus* spp.

2. Mechanical control. Plough soil during June and December to kill cocoons in soil.

3. Biological control. Apply *Beauveria bassiana* powder or water solution to roots and under tree canopy to kill cocoons will kill 80% of the cocoons.

4. Chemical control. Spray 10% imidacloprid at 800× in mid-May when larvae are at 3rd instar can reduce population by 95%.

References Xiao Gangrou, Huang Xiaoyun, Zhou Shuzhi 1991; Yang Chuncai and Li Min 1996; Yang Chuncai, Wang Wenge, Yin Xiangyu et al. 1996; Wu Xingyu, Gaosong, Chen Ruisheng et al. 2001; Li Menglou and Wu Xingyu 2003.

(Yang Chuncai, Liu Jianzhong, Tang Yanping)

A. ulmusvora adult (Wei Meicai)

515 *Cimbex connatus taukushi* Marlatt
Classification　Hymenoptera, Cimbicidae

Importance　It is an important pest to poplar trees. At high density, they can defoliate the whole tree and kill the trees.

Distribution　Beijing, Inner Mongolia, Liaoning, Jilin, Heilongjiang, Shaanxi; Russia, Japan, Korea Peninsula.

Hosts　*Populus euramericana*, *P. simonii*, *P. berolinensis*, *Salix matsudana*, *S. gordejevii*, *S. fragilis*.

Damage　Larvae feed on leaves, cause notches or consume whole leaves with only veins and petioles left.

Morphological characteristics　**Adults** Females 14-26 mm long, yellowish brown or reddish brown. Males 22-30 mm, reddish brown.

Larvae Mature larvae 38 mm long, light-yellowish green, thorax and abdomen with a thick bluish black dorsal band, body with many creases.

Biology and behavior　*Cimbex connatus taukushi* occurs one generation a year in Heilongjiang. It overwinters as mature larvae inside cocoons in soil or fallen leaves. Pupation starts in mid-May. Pupal period is 11-14 days. Adults emerge from late May to early July. Eggs hatch in mid-June. Mature larvae start go down trees in mid-August.

Adults emerge during the days. They can mate the same day after emergence. They sometimes reproduce parthenogenetically. Females live for 6.9 days, males live for 4.2 days. Eggs are laid in leaves, singly or one row along each side of the main vein. They lay 5-6 eggs on each leaf. Larvae feed in the afternoon. When not feeding, they curl into circles on leaf surface. They are gregarious. Each 4th instar larva consumes 40 mm^2 a day. After disturbed, the anus will release a defensive fluid. Mature larvae make cocoons in 6 cm deep soil. Surface of the cocoons has leaves and soil particles. Overwintering larvae may enter into diapause for up to 3 years.

Prevention and treatment

1. Mechanical control. Shake small trees and collect fallen larvae. Dig soil to collect cocoons. Hand remove leaves with eggs.

2. Conserve natural enemies. An undescribed parasitic wasp species is an important natural enemy.

3. Chemical control. Spray 45% profenofos and phoxim mixture at 1,000×, 20% fenvalerate or 2.5% deltamethrin at 3,000-5,000× to kill larval populations.

References　Xiao Gangrou, Huang Xiaoyun, Zhou Shuzhi 1991; Xiao Gangrou 1992; Zhang Junsheng and Bai Zhongwen 2000; Guan Ling, Tao Wanqian (eds.) 2010.

(Li Chengde, Qi Mujie, Hu Yinyue)

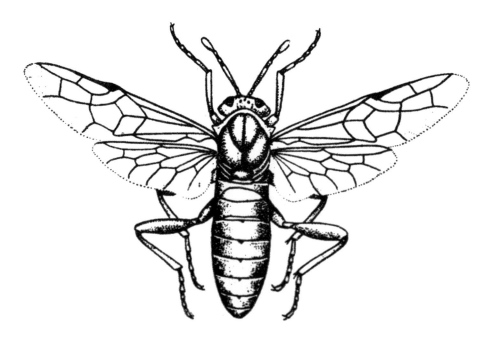

C. connatus taukushi adult male (Zhang Peiyi)

C. connatus taukushi adult male (Li Chengde)

C. connatus taukushi adult female (Li Chengde)

C. connatus taukushi adult (Li Zhenyu)

516 *Augomonoctenus smithi* Xiao et Wu
Classification   Hymenoptera, Diprionidae

Importance   It is an important pest to *Cupressus funebris*. It reduces seed yield significantly.

Distribution   Chongqing, Sichuan.

Host   *Cupressus funebris*.

Damage   Larvae feed on cones, cause cones dry and fall off.

Morphological characteristics   **Adults** Females 6-7.5 mm long, males 5-7 mm long. Body bluish black, metallic.

Eggs Purplish red, length 9-11.7 mm, width 3.3-5.1 mm.

Larvae Mature larvae head width 1.41 mm.

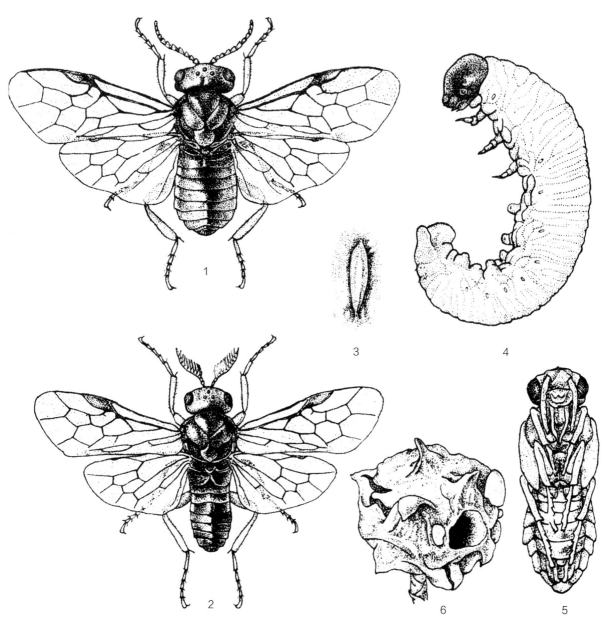

A. smithi. 1. Adult female, 2. Adult male, 3. Egg, 4. Larva, 5. Pupa, 6. Damaged fruit (Zhang Xiang)

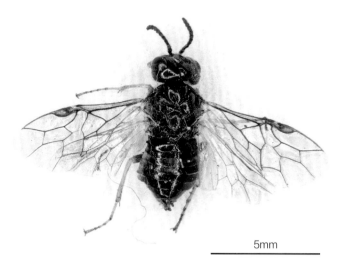

A. smithi adult female (Li Zhenyu)

A. smithi adult male (Li Zhenyu)

Pupae Young larvae light-yellow, later become bluish black.

Biology and behavior *Augomonoctenus smithi* spends 1-2 years to complete one generation. The 7th instar larvae make cocoons in loose soil, lichen, fallen leaves during mid- and late June. They overwinter as pre-pupae inside cocoons. Pupae occur in late March and late April. Adults occur from early April to early May. Pupae occur from mid-April to late June; 50% of the prepupae will pupate the same year; those occur one generation in two years will pupate in March next year, and adults appear in April. Adults eclose during 11:00-16:00. Males are better flyers than females. Mating occurs mostly at 15:00-19:00. Adults mate multiple times. Females lay eggs 1-8 (average 3) days after eclosion. Each female lays 6-117 (average 66) eggs. Eggs are laid singly on cones. Each cone usually has one egg. Eggs most hatch before 10:00 during a day. About 90% of the eggs hatch successfully. Young larvae enter into cones about 10-12 hours after hatching. A larva transfers to a new cone after an old cone is consumed. A larva infests 5-12 (average 8) cones during its lifetime. Mature larvae enter soil and make cocoons. It usually occurs at 350-1,200 m elevation with 600-900 m elevation being the most preferred.

Prevention and treatment

1. Silviculture. Plant mixed forests, remove dead trees, prevent from human destruction, and improve trees' natural resistance.

2. Cultural control. Collect infested cones and cocoons and place them in screened cages. Place the cages in forests to allow parasitic wasps emerging from the pests to return to forests.

3. Chemical control. Spray Dimilin or pyrethroid insecticides or release matrine smoke to control the pest.

References Xiao Gangrou and Wu Jian 1983; Xiao Gangrou 1992.

(Zhang Zhen, Wang Hongbin, Zeng Chuihui)

517 *Diprion jingyuanensis* Xiao et Zhang
Classification Hymenoptera, Diprionidae

Importance Outbreaks of *Diprion jingyuanensis* was first reported in 1989 in Jingyuan county of Gansu province. Later its outbreaks occurred in Qinyuan county of Shanxi province.

Distribution Shanxi, Gansu.

Host *Pinus tabulaeformis.*

Damage Larvae feed on leaves and can consume all leaves on a tree. Their feeding cause stunned growth or tree mortality.

Morphological characteristics **Adults** Females 12 mm long, wingspan 25 mm. Head 21-23 segmented. Scape, pedicel yellowish brown, flagellum black. Thorax black, pronotum yellow.

Larvae First instar pale greyish black or greyish green, head light-yellowish white, shiny. Eye area and mandibles black. The 2nd instar light-yellow and become yellow as they develop. The 4th instar dorsal line distinct, become thicker and darker in color as they grow. The 7th instar average 27.92 mm long and maximum 30 mm long. There are black dorsal lines and lateral lines across the body; black short spines thick. The 8th instar size shortened, fresh yellow, smooth, without spines, the dorsal and lateral lines are broken into series of black spots. Legs and head black.

Biology and behavior *Diprion jingyuanensis* occurs one generation a year in Shanxi. It overwinters as prepupae in cocoons. Pupae appear in early May and reach peak in late May. Adults emerge from late May to mid-June. Each female lays 120-229 eggs. They are laid inside leaves that located on top of shoots. Eggs hatch between mid-June and early July. Larvae feed in groups. When disturbed, their heads and tails raise and secret yellow fluid. Between late August and late October, mature larvae fall to ground and make cocoons in fallen leaves. Cocoons are cylindrical, white at beginning, then yellowish brown to mahogany brown. Some individuals may diapause and finish one generation in two years. Females may reproduce parthenogenetically. Sparse forest, pure forest, and northern slopes have higher populations. Warm and moist weather in October favors the development of cocoons. High temperature and low humidity in July help the outbreak of *D. jingyuanensis*. Too much precipitation and too low precipitation will help the outbreak of *D. jingyuanensis*.

Prevention and treatment

1. Silviculture. Plant mixed forests, remove dead trees, prevent from human destruction, improve trees' natural resistance.

2. Cultural control. Collect larvae and cocoons and place them in screened cages. Place the cages in forests to allow parasitic wasps emerging from the pests to return to forests.

D. jingyuanensis adult male (Zhang Peiyi)   *D. jingyuanensis* adult female (Zhang Peiyi)

D. jingyuanensis adult females (above) and males (below) (Zhang Peiyi)

D. jingyuanensis damage (FPC)

D. jingyuanensis damage (FPC)

3. Biological control. Spray 1×10^9/mL *B. bassiana* at 22.5 kg/ha or *B.t.* at 2.5 kg/ha. A virus from this pest is also a promising biological control agent.

4. Chemical control. Spray imidacloprid + monosultap at 800× or release matrine smoke to control the pest.

References Xiao Gangrou and Zhang You 1994; Li Fengyao, Wang Lizhong, Zhao Jin et al. 1995; Ma Riqian, Gao Shushan, Guo Yuyong et al. 1998; Li Fengyao, Liu Suicun, Huo Luyuan 2000; Huo Lüyuan, Wang Lizhong, and Liu Suicun et al. 2001; Chen Guofa, Zhou Shuzhi, and Zhang Zhen 2003; Tong Ying 2003; Fan Lihua, Xie Yingping, Yuan Guisheng et al. 2005; Xue Wangsheng 2005; Zhang, Wang, Chen et al. 2005; Zhang Yun, Zhang Zhen, Wang Hongbin 2006.

D. jingyuanensis cocoons (FPC)

(Zhang Zhen, Wang Hongbin)

518 *Diprion liuwanensis* Huang et Xiao

Classification　Hymenoptera, Diprionidae

Importance　It is an important pest to *Pinus massoniana* and affects tree growth.

Distribution　Anhui, Jiangxi, Guangdong, Guangxi.

Host　*Pinus massoniana*, *P. taiwanensis*.

Damage　Larvae feed on leaves in groups and consume all leaves and only leave the leaf sheath.

Morphological characteristics　**Adults** Females 9-11 mm. Antennae black, the 1st and 2nd segment yellowish brown. Head black; palpi yellowish brown; clypeus, ocelli, rear of occiput reddish brown. Thorax black, pronotum yellowish brown. Wings transparent, area before stigma smoky brown in the forewing, veins blackish brown, base of stigma black, end of stigma smoky brown. Hindwing tip slightly smoky brown in color. Legs black, ventral surface of front tibiae and tarsi paler. The anterior border of the 1-3 abdominal dorsal plates yellow; the middle of the anterior border of 4-7 abdominal dorsal plates and their sides, the sides of the 8th abdominal segment yellowish white; the ventral surface of 2-6 abdominal segments yellow.

Males 7-9.5 mm long; 1st and 2nd segment of antennae, palpi, clypeus, tip of femur, tibiae, tarsi reddish brown to dark reddish brown. Veins light-yellow. Stigma blackish brown, its tip yellowish brown. Ventral surface of the abdomen with dense sculpture. Antennae 23 segmented, nearly as wide as head, each segment except the 2nd and 3rd segment has two long pectinates. Others similar to females.

Eggs　Boat shaped, 1.0-1.2 mm long, 0.4-0.5 mm in diameter. Pale yellow at beginning, pale green and with a black spot near hatching.

Larvae　Mature larvae 26-28 mm long. Head dark brown and shiny. Thorax pale brown, dorsal surface with a few dark longitudinal creases, each crease with a band of black spots and setae, the spots at the 1st and the last segment are more abundant. Thoracic legs black, prolegs light-yellow.

Cocoons and Pupae　Pupae 7-10 mm, spindle shape,

D. liuwanensis adult male (Zhang Peiyi)

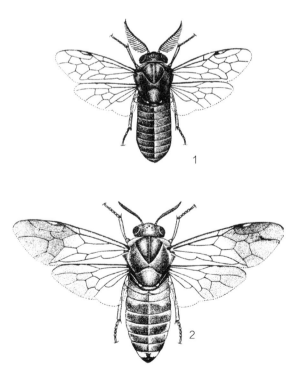

D. liuwanensis. 1. Adult male, 2. Adult female (Zhang Peiyi)

D. liuwanensis adult female (Zhang Peiyi)

light-yellow, shiny, subdorsal line obvious. Cocoons cylindrical, white at the beginning, dark brown later, 8-10 mm long, 4-6 mm wide.

Biology and behavior *Diprion liuwanensis* occurs two generations a year in Jiangxi. It overwinters as prepupae in fallen leaves and among grasses. Adults emerge from early June. Larvae appear from late June. Cocoons appear in early August. First generation adults lay eggs in mid-August. The 2nd generation larvae appear in mid-September. They make cocoons in late September. Adults emerge during day and night, mostly during 12:00-13:00. Adults spread their wings 2-3 minutes after emerging from pupae. They excrete green fecal material around 10 minutes after emergence, then fly and crawl. After about half hour, they can mate. Mating occurs mostly on branches, each mating lasts 15-20 minutes. They lay eggs on the 3rd day in leaves of young shoots. Each leaf may have 10-14 eggs. A female of the 1st generation lays 60-80 eggs, a female of the 2nd generation lays 80-120 eggs. Eggs hatch mostly during 5:00-8:00. The 1st-4th instars are gregarious. They feed during day and night. They migrate to new spots in groups when leaves are consumed.

Prevention and treatment

Cut off branches along with congregating larvae. Spray 20% dinotefuran at 800×, 50% acetamiprid at 2,000×, or 50% imidacloprid + monosultap at 800×.

References Xiao Gangrou, Huang Xiaoyun, Zhou Shuzhi 1991; Xiao Gangrou 1992.

(Zhang Zhen, Zhang Peiyi, Sun Yonglin, Liang Xinqiang)

519 *Diprion nanhuaensis* Xiao

Classification Hymenoptera, Diprionidae

Importance It is an important pest to *Pinus* spp. At high density, it can consume all leaves and affect tree growth.

Distribution Henan, Guizhou, Yunnan.

Hosts *Pinus yunnanensis*, *P. yunnanensis*, *P. massoniana*.

Damage Larvae feed on leaves gregariously and only the leaf sheath is left un-attacked.

Morphological characteristics **Adults** Female length 9-11 mm, black. The following parts are yellow: pronotum, upper of scapularia, pre-scutum edge, scutum front edge, scutellum posterior border, and metanotum except the scutellum. Ventral side of the front and median coxae, trochanter, tip of femur, front tibiae, tarsi, and median tibia yellow or light-yellow; claws reddish brown. Males 7-9 mm long, black. Antennae 22 segmented, the 1st and 2nd segments slightly reddish brown. The tip of the coxae, trochanter with yellowish brown color; tip of femur, tibiae, tarsi, and claws yellowish brown. Wings transparent, veins dark brown. Basal half of stigma blackish brown, the rest yellowish brown. The

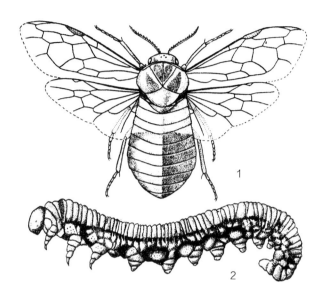

D. nanhuaensis. 1. Adult, 2. Larva (Zhang Peiyi)

D. nanhuaensis adult (Li Zhenyu)

center of the dorsal surface of the 2nd abdominal segment yellowish brown.

Eggs Length 1.5 mm, width 0.5 mm; kidney shape, slightly flattened, egg shell soft and thick; greyish white at beginning, become milky yellow later, and greyish black before hatching.

Larvae Mature larvae length 22-36 mm, width 3.6-4.0 mm. Head black, shiny, thorax pale green, dorsal line and supraspiracular line black. Each segment with a wide black transverse band and 3-5 narrow black transverse bands.

Pupae Length 8-12 mm, yellowish green.

Biology and behavior *Diprion nanhuaensis* occurs two generations a year in Nanhua of Yunnan. It overwinters as pupae within cocoons in fallen leaves. Adults emerge in from late March to early April. The first generation eggs appear in early and mid-April, larvae appear from mid-April to mid-July. Cocoons appear from early to mid-July, adults appear from mid- to late July. The 2nd generation eggs, larvae, cocoons appear from late July to early August, early August to early November,

D. nanhuaensis larvae
(Yang Zaihua)

D. nanhuaensis larvae
(Yang Zaihua)

and early November to late March, respectively. It occurs one generation a year in Yuxi of Yunnan.

Adults mostly emerge from cocoons during 12:00-13:00. After a few minutes, they can fly. Adults lay eggs in a line inside each leaf. Each needle has 20-30 eggs. A female can lay 70-150 eggs. Larvae feed gregariously.

Prevention and treatment

1. Silviculture. Keep canopy coverage above 0.7 will help reduce pest populations.

2. Cultural control. Hand-remove larvae or cocoons.

3. Biological control. Spray a virus or *Beauveria bassiana*.

4. Chemical control. Spray 20% dinotefuran at 800×, 50% acetamiprid at 2,000×, or Apply 5.7% emamectin at 3,000×, 20 g/L lamda-cyhalothrin at 1,000-1,500×, or 50% imidacloprid-monosultap at 800×.

References Liu Guangyao 1986; Fang Wencheng 1988; Xiao Gangrou, Huang Xiaoyun, Zhou Shuzhi 1991; Xiao Gangrou 1992; Yang Miaomiao, Li Menglou, Qu Liangjian et al. 2007; Chen Haishuo, Ma Xiaoqi, Nie Hongshan et al. 2009.

(Zhang Zhen, Wang Hongbin, Fang Wencheng)

520 *Gilpinia massoniana* Xiao
Classification Hymenoptera, Diprionidae

Importance Outbreaks start to occur in the 1980's in *Pinus massoniana* forests. Mean population density during outbreak was 1,000 per tree. Maximum density was 3,000 per tree. Tree growth was affected.

Distribution Anhui.

Hosts *Pinus massoniana*, *P. thunbergii*, *P. taeda*, *P. elliottii*.

Damage Larvae feed on leaves and cause leaves yellow. At high density, 90% of the leaves can be consumed.

Morphological characteristics **Adults** Females 9.0-12.5 mm long, wingspan 18.0-22.6 mm. Antennae single pectinate, 25 segmented, basal area dark yellow, tip black. Males 8.0-12.2 mm long, wingspan 16.2-20.5 mm. antennae double-pectinate, 30 segmented.

Larvae Mature larvae 16-35 mm long, head width 2.2-2.3 mm. Head blackish brown, exterior surface of thoracic legs black, interior surface of thoracic legs light-yellow. Body pale green, with black subdorsal line, supraspiracular line, and subabdominal line. Supraspiracular line wide.

Cocoons Elliptical, silky, shiny, yellowish brown, length 8.8-13.9 mm, width 4.2-4.6 mm.

Biology and behavior *Gilpinia massoniana* occurs two generations a year in Anhui. It overwinters as prepupae within cocoons in fallen leaves, among grass roots, in loose soil, etc. Pupae occur in early April. Adults emerge in mid-April. The 1st generation adults emerge in early September. Larvae make cocoons in November and overwinter.

There are about 10% of the individuals in each generation that will enter into diapause. Adults lay eggs under the epidermis in the groove of needles. They feed during day and night. The 1st and 2nd instar feed on edges and form notches. The 3rd instar feed on whole leaves. A larva consumes 600 cm^2 leaf area or 20 bundles of leaves. The first generation larvae only feed on current year leaves.

Prevention and control

1. Silviculture. Plant mixed forests.

2. Biological control. Apply *Beauveria bassiana* powder or spray before April or after September.

3. Chemical control. Apply thermal fog of 10% fenvalerate and diesel mixture at 100:200 will reduce population by 95%.

References Liu Ping, Shi Jin, Zhu Guangyu 1992; Xiao Gangrou 1992, Wu Xiazhong, Song Minhao, Zheng Dahu 1994.

(Tang Yanping, Yang Chuncai)

G. massoniana adult male (Zhang Peiyi)

G. massoniana adult female (Zhang Peiyi)

521 *Neodiprion dailingensis* Xiao et Zhou
Classification Hymenoptera, Diprionidae

Importance *Neodiprion dailingensis* is an important pest to pine and spruce trees. Large areas of trees may be damaged.

Distribution Liaoning, Heilongjiang.

Hosts *Pinus koraiensis*, *P. tabuliformis*, *P. densiflora*, *Picea* spp.

Damage Larvae feed on leaves. Feeding by young larvae causes leaves curl. After the 3rd instar, larvae consume whole leaves.

Morphological characteristics **Adults** Females 6.5-9.0 mm long. Basal two segments of the antennae reddish brown, other segments blackish brown. Head and thorax reddish brown, abdomen yellowish brown. Males 6.0-6.5 mm long. Head blackish brown, thorax black, dorsal surface of the abdomen blackish brown, ventral surface brown.

Eggs Elongated elliptical, greyish yellow.

Larvae Mature larvae 20-28 mm long. Head black, thorax, and abdomen greyish green; dorsal surface with two grey lines.

Cocoons Golden yellow, elongated oval.

Biology and behavior *Neodiprion dailingensis* occurs one generation a year. A small portion requires two years to finish one generation. It overwinters as eggs. Eggs hatch in mid-May. Larvae feed gregariously. One small branch may have 30-50 larvae. One needle may have 2-4 larvae. They move to a new location after leaves are consumed. When disturbed, their head and thorax tilt. Mature larvae make cocoons from late June to early July under fallen leaves or occasionally on branches. Adults emerge and lay eggs from late August to mid-September. Some mature larvae enter diapause, overwinter within cocoons, and eclose in early October next year. These individuals spend two years to finish the cycle.

Adult emergence occurs mostly during 13:00-15:00. They mate the next day. Males can mate multiple times,

N. dailingensis adult male (Zhang Peiyi)

N. dailingensis adult female (Zhang Peiyi)

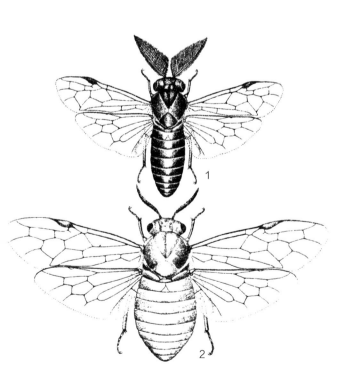

N. dailingensis 1. Adult male, 2. Adult female (Zhang Peiyi)

each mating lasts 40-60 minutes. Adults are good flyers. They lay 2-8 eggs per needle at ridges of leaves. Each female lays 40-87 eggs (average 59). Females live for 8-12 days; males live for 6-8 days.

Prevention and treatment

1. Silviculture. Prevent human disturbance. Remove infested limbs and trees. Plant mixed forests.

2. Mechanical control. Cut off larvae infested branches or knock off larvae and destroy them. Collect cocoons and place them in cages to allow ichneumonid wasps to emerge and return to forests.

3. Biological control. A virus of the pest can be used.

4. Chemical control. Spray Dimilin or 45% profenofos and phoxim mixture at 30-60×, or 12% thiamethoxam and lamda-cyhalothrin mixture at 1,000×, or fogging with matrine.

References   Xiao Gangrou, Huang Xiaoyun, Zhou Shuzhi 1991; Wang Yongzhe, Zhang Xiaolong, Wang Dong 1992; Xiao Gangrou 1992; Wang Yongzhe, Chen Tianlin, and Ning Jiwen 1995; Wang Yingmin, Zhang Yingwei, and Yang Weiyu 1999.

(Zhang Zhen, Zhang Peiyi, Zhang Shimin)

522 *Neodiprion huizeensis* Xiao et Zhou
Classification Hymenoptera, Diprionidae

Importance Outbreaks occurred in Huize county of Yunnan province. All tree leaves were consumed and tree growth was affected.

Distribution Yunnan.

Hosts *Pinus armandii.*

Damage Young larvae feed on leaves and cause notches. After 3rd instar, they can consume all leaves in a bundle.

Morphological characteristics **Adults** Females length 7.9-8.2 mm. Reddish brown, mandibles blackish brown. The following are black: antennae except the 1st, 2nd, and base of the 3rd segment; areas around ocelli, the posterior end of the suture behind ocelli, anterior border and inner margin of the scutum of mesothorax, the spot on dorsal surface of the 1st abdominal segment. Wings transparent. Males 6.5-6.6 mm ling, black, shiny, with reddish black palpi at anterior half of mandibles; femora, tibiae, tarsi, ventral surface of abdomen reddish brown; veins darker than females, center of stigma transparent; punctures on head and thorax dense, punctures on ventral surface of abdomen and sculpture similar to female. Antennae with 22 segments.

Eggs 1.6-2.4 mm long. Banana shape, light-yellow.

Larvae Pale yellow at beginning, dark brown before hatching.

Cocoons and Pupae Paly yellow, eyes black, antennae and legs yellowish white. Cocoons elongated elliptical. Pale yellow.

Biology and behavior *Neodiprion huizeensis* occurs two generations a year. It overwinters as prepupae in cocoons. Adult emergence starts in late March. Eggs and larvae appear in April. Larvae go through 6 instars. They leave trees and make cocoons in mid- and late June. Adults start to emerge in late July. Eggs hatch in late August. The 2nd generation larvae leave trees in mid-November and overwinter under ground cover.

When an adult emerges, it chews open a cap on the cocoon. After a few hours of waiting, it starts to fly. Females live for 2-4 days; males live for 3-5 days. A female attracts a few males, but only mates with one male. It

N. huizeensis larvae (Yang Zaihua)

N. huizeensis larva (Yang Zaihua)

N. huizeensis larvae (Yang Zaihua)

lays eggs immediately after mating. Females usually lay eggs on old leaves and sunny side of branches. They lay one egg per leaf or occasionally two eggs per leaf. All leaves in a bundle may have eggs. Eggs are covered with gelatinous material. Young larvae feed on leaf tip and make notches. They can migrate between trees to find food. Fifth instar larvae crawl to fallen leaves to make cocoons.

Prevention and treatment

1. Silviculture. Maintain canopy coverage above 0.7 will help keep pest population low

2. Mechanical control. Hand remove cocoons in July.

3. Chemical control. Spray 20% dinotefuran at 800×. Apply 5.7% emamectin at 3,000×, 20 g/L lamda-cyhalothrin at 1,000-1,500×, or 50% imidacloprid-monosultap at 600×, or 50% acetamiprid at 2,000×.

References Xu Zhenghui, Xu Zhiqiang, and Wu Yi 1990; Xiao Gangrou, Huang Xiaoyun, Zhou Shuzhi 1991; Ying Hongtao and Luo Zhengfang 2002.

(Zhang Zhen, Zhang Peiyi)

523 *Neodiprion xiangyunicus* Xiao et Zhou
Classification Hymenoptera, Diprionidae

Importance Outbreaks occurred in Huize county of Yunnan province. All tree leaves were consumed and tree growth was affected.

Distribution Sichuan, Guizhou, Yunnan.

Hosts *Pinus yunnanensis*, *P. armandii*.

Damage *Neodiprion xiangyunicus* is a major pest of *Pinus yunnanensis*. Heavy infestation can cause complete defoliation and affect tree growth.

Morphological characteristics **Adults** Mid-size, length 9-10 mm. Females yellowish brown; wings translucent, with unevenly distributed dark yellow color; veins and stigma dark yellow; antennae 27 segmented; head and thorax with medium length, long, and yellow pubescence. Males vein, stigma blackish brown, shiny; antennae 32 segmented; dorsal surface of the 1st to 8th abdominal segments with sculpture.

Larvae Head of young larvae black or brown, head of mature larvae dark or brown; supraspiracular line distinct; each body segment (except prothorax) is divided into three sub-segments, each sub-segment with rows of setae.

Biology and behavior *Neodiprion xiangyunicus* occurs one generation a year in Xichang of Sichuan. They overwinter as pupae in soil under the tree canopy. Adults emerge in late May. Eggs are most laid in early July. Young larvae appear in mid- and late August. Larvae go through 8 instars and 160 days. Larvae are gregarious. Their head secretes white sticky fluid when disturbed. Adults prefer lay eggs in sparse forest, single trees, edges of forests, and upper part of canopy. Outbreaks often occur at elevation above 1,700 m.

N. xiangyunicus adult female (Zhang Peiyi)

N. xiangyunicus mid-aged larva (Xu Zhenghui)

N. xiangyunicus mature larva (Xu Zhenghui)

N. xiangyunicus mature larva (Xu Zhenghui)

N. xiangyunicus mid-aged larva (Xu Zhenghui)

Prevention and treatment

1. Silviculture. Prevent forests from human disturbance. Plant mixed forests.

2. Mechanical control. Hand remove infested branches along with the larvae. Dig soil to remove cocoons.

3. Chemical control. Spray 2.5% deltamethrin at 2,000× during mid- and late October to control young larvae. Apply fogger in forests with greater than 0.6 canopy coverage.

References Xiao Gangrou 1992, Su Zhiyuan 2004.

(Guo Hengxiao)

524 *Nesodiprion huanglongshanicus* Xiao et Huang
Classification Hymenoptera, Diprionidae

Importance It is an important pest to *Pinus tabuliformis* and *P. armandii*. Young forests suffer more than older forests.

Distribution Shaanxi.

Hosts *Pinus tabuliformis, P. armandii.*

Damage Young larvae of *Nesodiprion huanglongshanicus* feed on leaf edge and cause notches. After 2nd instar, they consume whole leaves.

Morphological characteristics **Adults** Females 8-9 mm long, wingspan 14-16 mm. Body black, with weak blue color; antennae black; maxillae, labipalp dark yellow; posterior angle of pronotum, tip of coxae, trochanter, and the spots on side of the 7th and 8th abdominal dorsal surface light-yellowish white; wings transparent; veins, stigma blackish brown; costa yellow. Head, pronotum, and scutellum of mesothorax with sparse coarse sculpture; pre-scutum and scutum of mesothorax with fine evenly distributed sculpture; dorsal surface of the 1st abdominal segment very sparsely punctured. Antennae 22 segmented; starting from the 3rd segment double pectinate, the last two segments single pectinae; the 1st segment is 2x as long as the 2nd segment; the 3rd segment shorter than the other segments. Lancet long, its ventral surface straight, with 11 rings.

Males 6.0-7.5 mm long, wingspan 10-12 mm. Color similar to females except that the 7th and 8th abdominal dorsal surface without yellowish white spots. Sculpture similar to males. The end of penis valve nearly truncated, its ventral projection small, tip round, with 12 spines.

Eggs Length 1.5 mm. Pale yellow at beginning, purple red on one end and milky on the other end before hatching.

Larvae Mature larvae 18-23 mm long. Head golden yellow, shiny; ocelli area black; thorax and abdomen yellowish green. The dorsal surface with black two longitudinal lines; each side has one thick black longitudinal line. Exterior of thoracic legs black; coxae and tip of trochanter yellowish green.

Cocoons Cylindrical, brownish yellow, smooth or with grass or soil particles. Length 7 mm, width 2.5 mm. Maximum size is 11 mm × 5 mm, smallest size is 5.5 mm × 1.5 mm.

Pupae Females 7.5-9.5 mm long. Eyes greyish black at beginning, yellowish green, antennae and legs light-yel-

N. huanglongshanicus adult (Wei Meicai)

N. huanglongshanicus adult female (Zhang Peiyi)

lowish white. Before eclosion, head, antennae, thorax black and shiny; abdomen greyish black, yellowish green between segments; the sides of the 7th-8th abdominal segments with light-yellowish white spots; ventral surface brownish yellow. Male pupae 6.5-8.0 mm long. Color similar to that of the females. The sides of the 7th-8th abdominal segments without light-yellowish white spots.

Biology and behavior *Nesodiprion huanglongshanicus* occurs one or two generations a year in Shaanxi. The latter is more common. They overwinter as pre-pupae within cocoons. Pupation occurs from early May to late May. Adults emerge from early June to late July. Eggs appear from early June to late June. Eggs hatch from mid-June. Peak damage period is in early July. The 2nd generation eggs appear in early August. They hatch in mid-August. Mature larvae make cocoons in soil around tree trunks, under fallen leaves, under rocks, or on exposed soil from late September to early October. The east and south side of the canopy is often more abundant than the west and north side of the canopy.

Male pupae occur earlier than female pupae. During peak time, male to female ratio is almost 1:1. Under natural conditions, pupation rate is 82%. Those located on south side pupate earlier than those located on north side. On the same slope, those located at upper part of the slope pupate earlier than those located at lower part of the slope. Those larvae under sparse ground cover pupate earlier than those under dense ground cover.

Adult emergence period is mostly during 10:00-17:00. Eclosion rate is 84.4%. Males emerge 2-3 days earlier than females; those on south side eclose 1-2 days earlier than those on the north side. They start to fly a few hours after eclosion. Males are stronger flyers than females. They fly on top of canopy. They rest on leaves or twigs in early evening and go down trees at sunset. They can mate same day after eclosion. A female lays one egg a time. They deposit eggs in the tissue of current year leaves. The 1st generation female lays 0-31 (average 14) eggs. The 2nd generation female lays 0-45 (average 18) eggs. The leaf tissue swells 1-2 days after an egg is laid. Females may reproduce parthenogenetically. The female:male ratio is correlated with population density. The female:male pupae ratio at light, medium, and heavy population density is: 1:1.1, 1:1.7, and 1:2.3, respectively. Adult female:male ratio at light, medium, and heavy population density is: 1:1.2, 1:1.6, and 1:2.2, respectively.

Larvae are gregarious. Young larvae feed on edges of current year leaves and form notches. From the 2nd instar, they feed on one year old leaves and consume whole leaves. They migrate to another tree once the leaves are consumed. The 1st generation larvae consume 949.9 cm long. The 2nd generation larvae consume 701 cm long. Larvae feed mostly during 10:00-14:00.

Larvae molt 4-5 times and molt mostly during the early morning and afternoon. They are tolerant to starvation. The 5th instar can crawl for 15.6 cm after 10 days of starvation. The 6th instar larvae can crawl 90.1 cm after 9 days starvation. They are also tolerant to low temperature. When temperature is -3 to -2 °C during October, they can still feed.

Prevention and treatment.

1. Silviculture. Promote the growth of young forests and their natural resistance.

2. Mechanical control. Hand collect larvae and cocoons and destroy them.

3. Conservation of natural enemies. Important natural enemies include assassin bugs, parasitic flies, *Beauveria bassiana*, black ants, and birds (*Parus major*, *Phoenicurus auroreus*).

4. Chemical control. Spray 90% trichlorfon or 25 g/L lamda-cyhalothrin at 2,000-2,500×, or Apply 5.7% eamamectin at 3,000×, 20 g/L lamda-cyhalothrin at 1,000-1,500×, 50% imidacloprid-monosultap at 600-800×, 20% dinotefuran at 1,000-1,500× to control 1st to 3rd instar larvae. Apply fog of chitin synthesis inhibitor before 3rd instar in dense young forest is also effective.

References Xiao Gangrou, Huang Xiaoyun, Zhou Shuzhi 1981.

(Zhu Jian)

525 *Nesodiprion zhejiangensis* Zhou et Xiao
Classification   Hymenoptera, Diprionidae

Importance   *Nesodiprion zhejiangensis* mainly infests young trees. During outbreaks, 100% of the trees are infested.

Distribution   Zhejiang, Anhui, Fujian, Jiangxi, Hubei, Hunan, Guangdong, Guangxi, Sichuan, Guizhou, Yunnan.

Hosts   *Pinus massoniana, P. teada, P. elliottii, P. thunbergii.*

Damage   Larvae feed on needles and may cause complete defoliation, affect tree growth, or even tree mortality.

Morphological characteristics   **Adults** Females 6.5-7.8 mm. Black; the 1st and 2nd antennae segments, maxillae, labial palpi dark yellow; abdomen shiny, with bluish color.

Eggs Boat shape, 1.2-1.5 mm × 0.3 mm, light-yellow.

Larvae Mature larvae 20-25 mm long, head yellow, antennae black. Thorax and abdomen yellowish green, dorsal line nearly white, sub-dorsal line brown, supraspiracular line black.

Pupae Female pupae length 7 mm, widh 2.5 mm, frontal area with three projections arranged near triangular shape. Male pupae length 5-6 mm, frontal area with four projections. Pupae yellowish white, antennae and legs white.

Cocoons Silky, cylindrical, milky at beginning, yellowish brown later, slightly shiny.

Biology and behavior   *Nesodiprion zhejiangensis* occurs 3-4 generations a year in Changsha of Hunan province, 2 generations a year in Guizhou, and 3 generations a year in Nanping of Fujian province. It overwinters as prepupae within cocoons on needles. Larvae appear from late April to early May. The 2nd to 4th generation larvae appear in July, August, and September,

N. zhejiangensis larvae (SFDP)

respectively. The 4th generation larvae make cocoons and overwinter in mid-October. Larvae go through 4-5 instars and 24-40 days. Larvae feed on 2-3 year-old trees. *Pinus teada* is preferred over *P. elliottii*. From 3rd instar, larvae consume whole leaves. The 1st to 3rd generation pupal period is 9-12 days. Adults lay eggs under epidermis on leaves, each leaf with 2-3 eggs. A female lays 14-18 eggs. Leaves swell 5-6 days after eggs are deposited. After another 3-4 days, the egg site splits. Adults live for 3-5 days. They sometimes are parthenogenetic. The parasitic rate of the 2nd generation pupae in 1986 was 27.8%-54.7%. The 1st and 2nd generation populations are usually high, the numbers go down during 3rd and 4th generations.

Prevention and treatment

1. Silviculture. Proper management of young trees will help reduce damage.

2. Mechanical control. Knock down larvae or remove larvae infested branches. Collect cocoons and place them in cages. Place cages in forests to allow parasitic wasps emerging from the cocoons return to forests.

3. Biological control. Spray *B. bassiana* at 1×10^{11}/mL at 22.5 kg/ha or *B.t.* to control larvae can reduce population by 80% and 90%, respectively. A virus is also promising control agent. *Diprion jingyuanensis* sex pheromone is useful for monitoring pest populations.

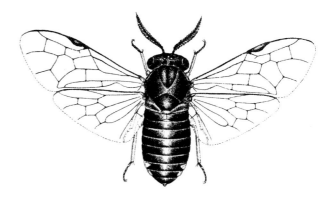

N. zhejiangensis adult female (Zhang Peiyi)

4. Chemical control. Spray Dimilin, pyrethroids or use matrine fogging.

References   Xiao Gangrou, Huang Xiaoyun, Zhou Shuzhi 1991; Xiao Gangrou 1992; Yu Peiwang 1998; Song Shengyin, Wang Yong, Ou Zhengquan et al. 2008; Li Shilan, Zheng Shan, Wang Junhuai et al. 2009.

(Zhang Zhen, Zhang Peiyi, Wu Youzhen)

Anafenusa acericola (Xiao)

Classification Hymenoptera, Tenthredinidae
Synonym *Messa acericola* Xiao

Importance *Anafenusa acericola* was first recorded in Taian of Shandong province in 1988. It is a potential leaf pest. During an outbreak, 48.4% of the leaves were damaged in 30-year-old *Acer truncatum* stands in a scenic site in Taian. The infestations significantly affected the aesthetic values of the forests.

Distribution Shandong.

Host *Acer truncatum*.

Damage Larvae feed within leaf tissue. Damaged leaves show semi-translucent yellowish brown scars. Each scar contains one larva. Several scars may merge into a large scar, causing leaves dry, black, curled, and fall prematurely.

Morphological characteristics **Adults** Females 3.7 mm in length, wingspan 7.2 mm. Head reddish yellow, with luster; anterior part of the mandibles black, mandibular palps yellowish brown; the 1st and 2nd segments of the antennae reddish yellow at base and black at the remainder area; eyes black. Thorax black, with luster; posterior angles of pronotum and pteralia yellowish brown; Wings translucent, veins black, stigma greyish yellow; legs yellowish brown. Abdomen black, with luster; its 1st notum with a yellowish white area. Scupture on scutum of the mesothorax and abdominal notum sparse and fine.

Eggs Oval, 0.4-0.7 mm long, milky.

Larvae Mature larvae 6.5 mm, yellowish green. Head light-yellow; anterior part of the mandibles dark. Prolegs degenerated. Each of the dorsal surface of the 1st to 8th abdominal segments with two light brown transverse stripes; the 1st stripe is shorter. The 2nd to 8th abdominal segment with one pair of small tubercles on each segment.

Pupa Lengh 4-5 mm, yellowish brown.

Cocoons Length 4.4 mm, width 2.5 mm, elongated elliptical.

Biology and behavior *Anafenusa acericola* occurs one generation a year. It overwinters as prepupae in soil. Pupation occur in late April. Adults emerge in early May. Eggs are laid in mid-May and larvae emerge in late May. Mature larvae appear in late June and migrate down the host trees. All larvae overwinter by late July. The development time is related to elevation. Parthnogenesis exists. Adults are most active during 9:30-16:30. They are active on leaf surface during the day and rest on weeds or ventral side of the leaves. Adults lay egg singly on upper surface of leaves under the epidermis. They prefer laying eggs along veins and angles of old leaves. One egg is usually laid per leaf. More than 10 eggs could be laid on one leaf at high population densities. A female lays 15-32 eggs. Adults live for 3-4 days. They play death when disturbed. Egg incubation period is 7.3 days. Newly hatched larvae feed on leaf tissue under epidermis, rendering the leaf surface into thin film form. As the larvae feed, feces accumulate within the feeding scars and form black spots. The underside of the leaves remains intact and green. Larvae go through 6 instars. Feeding during 4th-5th instars constitutes 2/3 of its entire life consumption. One larva consumes about 1/4 of a whole leaf. Higher relative humidity and precipitation favors the population growth.

Prevention and treatment

During the 2nd-3rd instar stage, create insecticide bands or inject insecticides into tree trunks. During 4th to 5th instar stage, spray appropriate insecticide to tree canopy.

References Liu Jing 1992, Xiao Gangrou 1992.

(Liu Jing, Lu Xiuxin, Liu Shiru)

A. acericola adult (Zhang Peiyi)

527 *Fenusella taianensis* (Xiao et Zhou)
Classification Hymenoptera, Tenthredinidae

Distribution Beijing, Liaoning, Jiangsu, Shandong.

Hosts *Populus simonii, P. pseudo-simonii, P. beijingensis, P. simonii × P. nigra* var. *italic*.

Damage Larvae damage laves by feeding on leaf tissue.

Morphological characteristics **Adults** Female length 4.0 cm, wingspan 9.5 cm. Head black. Antennae 9 segmented, yellowish brown except base of the 1st and 2nd segment, as well as the dorsal surface of the 3rd and 4th segments are darker. Ventral surface light-yellow; posterior border of pronotum and pteralia yellowish white. Abdomen black, posterior end of the notum and ventral surface with narrow yellowish white bands. Coxa brown to pale brown. Ventral surface of tibia and tarsi yellowish white, dorsal surface of tarsi brown. Male length 3.8 mm, wingspan 8.0 m; more slender than female.

Eggs Elliptical, milky, semi-transparent.

Larvae Mature larvae 5.6-6.0 mm, milky. Thoracic legs with one brown claw at the end of each leg. Prono-

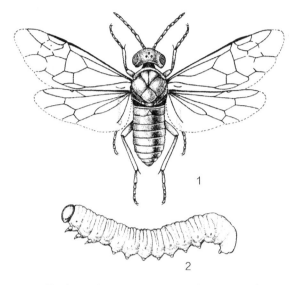

F. taianensis, 1. Adult, 2. Larva (Zhang Peiyi)

F. taianensis adult (Li Zhenyu)

F. taianensis adult (Xu Gongtian)

F. taianensis larva from a cocoon (Xu Gongtian)

F. taianensis cocoons (Xu Gongtian)

tum with two large pale purple marks, Mesonotum with two small purple marks.

Cocoons Elliptical, 6 mm long, 3.5 mm wide; made of mud, interior surface smooth.

Biology and behavior *Fenusella taianensis* occurs one generation a year. Pupae appear in early March in Shandong. Adults emerge from mid-March. Female:-male ratio is 3:1. Females lay eggs 2-3 days after mating. They lay eggs singly. One to three eggs are laid on each leaf. A female lays 38-40 eggs. Larvae feed inside leaves and leave the upper and lower epidermis untouched. Damaged leaves become yellowish brown. Larvae go through 3 instars. Mature larvae burrow 5 cm deep into soil in late April. They make cocoons and become dormant and overwinter.

Prevention and treatment

1. Conserve *Celata populus* parasitoid. Natural parasitization rate can reach 90%.

2. Spray 5% cypermethrin at 100× to control adults.

References Xiao Gangrou, Huang Xiaoyun, Zhou Shuzhi 1991; Xiao Gangrou 1992; Zhang Qiusheng, Chen Shirong, and Lan Shumei 1996; Yan Jiahe, Dong Changming, Wang Furong et al. 1998; Tao Wanqiang, Guan Ling, Yu Juxiang et al. 2003.

(Li Zhenyu, Li Bingsheng)

528 *Caliroa annulipes* (Klug)

Classification Hymenoptera, Tenthredinidae

Distribution Zhejiang, Fujian; Britain, Ireland, Northern Europe, Russia.

Hosts *Quercus acutissima, Q. fabri, Q. glandulifera.*

Damage Larvae feed on leaf tissue. Damaged leaves become dry and white, with net appearance.

Morphological characteristics **Adults** Length 4.5-5.5 mm, wingspan 10 mm. Front femur tip, front and middle tibiae and tarsi, basal half of hind tibia, base of the 1st hind tarsal segment, and a triangular area near the posterior border of the 1st abdominal segment white, other areas black, with luster; front wing smoky brown with its tip pale grey. Hind wing pale grey. Stigma, veins blackish brown. Antennae filiform, 9 segmented, the 3rd segment the longest. Hindwing with two closed discal cells.

Eggs Oval, 0.6 mm in diameter, grass green. Egg shell membranous and soft.

Larvae Mature larvae 12 mm in length. Thorax wide and narrows toward abdomen. The maximum width is 2.3 mm. Body is covered with sticky material that is similar to that on slugs, white jade color, translucent, the food and digestive duct visible from the back. Head small, yellow, globous, hidden under the prothorax. Black ocelli spots obvious.

Pupae Lengh 4-5 mm. Brown, eyes black.

Biology and behavior *Caliroa annulipes* occurs 3-4 generations a year in Zhejiang. It overwinters as prepupae. The overwintering generation pupate from late March to mid-April. Adults emerge from early April to late April. The 1st generation developmental periods are: eggs – early April to early May, larvae – mid-April to mid-May, pupae – May, adults – mid-May to early June. The 2nd generation development periods are: eggs – mid-May to early June, larvae – late May to late June, pupae – mid-June to early July, adults – mid-June to mid-July. The 3rd generation development periods are: eggs – mid-June to mid-July, larvae – late June to mid-July. All larvae turn to prepupae in late July and become dormant. Most of them will overwinter. A small portion of the

C. annulipes adult (edited by Zhang Runzhi)

C. annulipes cocoon (edited by Zhang Runzhi)

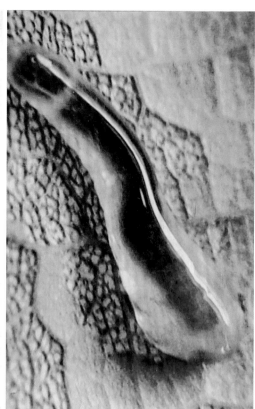

C. annulipes larva (edited by Zhang Runzhi)

prepupae molt into pupae in late September to early October. Adults emerge in October. The 4th instar larvae enter soil and turn to prepupae from mid-October to early November and overwinter.

Adults rest on host leaves close to the ground. They lay eggs same day after emergence. Eggs are laid singly on upper surface of leaves under epidermis. The leaf surface where eggs are laid become slightly raised. One to 18 eggs are laid on each leaf. Adults prefer lay eggs on leaves close to ground. Therefore small trees are more susceptible to infestations. Adults live for 1-4 days. Larvae feed on leaf tissue and epidermis from the back of leaves. They prefer mature leaves. At high density, they may consume all leaf tissue of a leaf.

Prevention and treatment

1. Silviculture. Thinning properly to improve light condition and air circulation.

2. Conservation of natural enemies. *Olesicampe* sp. is an important parasitoid of larvae.

3. Spray pyrethroids during larval infestation period.

References   Chen Hanlin, Liu Jinli, Wang Jiashun et al. 1999.

(Chen Hanlin)

529 *Dasmithius camellia* (Zhou et Huang)

Classification Hymenoptera, Tenthredinidae

Importance *Dasmithius camellia* is an important pest of *Camellia oleifera*. It can cause significant economic loss to seed production.

Distribution Fujian, Jiangxi, Hunan.

Host *Camellia oleifera.*

Damage Larvae feed on leaves and may cause complete defoliation. Infested trees suffer reduced seed yield.

Morphological characteristics **Adults** Female length 6.5-8.5 mm. Black, shiny. Antennae black, labial palps dark brown; the apical three mandibular palps brown; posterior border of the pronotum and pteralia whitish yellow. Coxa, femur black, trochanter white, tibia and tarsi brown. Wings translucent, with yellow luster; stigma black, its base with yellow color; veins blackish brown. Clypeus with coarse punctures. The frons and the triangular area around the ocelli with indistinct punctures, and with fine striae. Clypeus, frons, eye sockets, and prothorax with whitish yellow long fine hairs. Wings with many black hairs. Sheath long and slender from top view, round from side view; lancet highly sclerotized with 16-17 rings.

Male length 5.8-7.5 mm. Color similar to females except the two sides of the posterior end of the pronotum yellowish white, pteralia blackish brown, the light color area of the legs had darker color than the female or are blackish brown. Body and legs with dense fine hairs.

Eggs Length 1.0-1.2 mm. Elongated oval, milky. One end is slightly slender than the other end. Color becomes light-yellow before hatching.

Larvae Pale yellow; color turns to green after 3rd instar. Mature larvae 20-22 mm long, bluish gray, frons to triangular area near ocelli bluish black, spiracular line and legs white.

Pupae Length 6-8 mm. Pale green at beginning, turns to light-yellow to yellowish brown later.

Biology and behavior *Dasmithius camellia* occurs one generation a year in Hunan. It overwinters as prepupae within cocoons in soil. Pupation occurs between December and January in the next year. Adults emerge from late February to early March. Eggs hatch from early March and reach peak in mid-March. Adults rest on host leaves and branches. Adult sex ratio is about 1:1. They mate during the day. Eggs are laid on swelled buds. One egg is laid on each bud. Under heavy infestation, one bud may have five eggs. Egg period is 7-11 days. Young larvae feed on leaves in buds, the 3rd instar start feeding on young leaves, the 4th instar feed on both

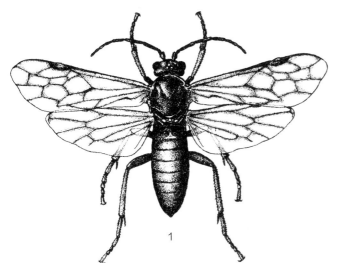
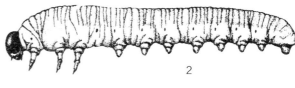

D. camellia 1. Adult, 2. Larva (Zhang Xiang)

D. camellia larva (Wang Jijian)

young and old leaves. More than 400 larvae were found per *C. oleifera* when population was high. Larval period is 22-26 days. Feeding peak lasts for 9-11 days. Larvae enter dormancy in late April. A small portion may undergo diapause.

Population of *Dasmithius camellia* is closely associated with ecological diversity of the forests. Areas further away from mixed forests have higher pest populations.

Prevention and treatment

1. Larvae play death when disturbed. During late March and early April when larvae are in 3rd instar, place plastic sheets under trees, shake trees, and destroy fallen larvae.

2. Spray 45% profenofos and phoxim mixture or pyrethroid insecticide to kill larvae.

References Xiao Gangrou, Huang Xiaoyun, Zhou Shuzhi 1991; Xiao Gangrou 1992; Huang Dunyuan, Yu Jiangfan, Hao Jiasheng et al. 2010.

(Zhang Peiyi, Zhang Zhen, Weng Yuexia, Xu Tiansen)

530 *Eutomostethus longidentus* Wei

Classification Hymenoptera, Tenthredinidae

Importance Outbreaks of *Eutomostethus longidentus* occurred in Zhejiang. It caused complete defoliation of bamboo forests. Consecutive infestations can cause mortality of bamboo trees in large areas.

Distribution Hebei, Zhejiang, Fujian, Jiangxi, Guangxi, Chongqing.

Host *Phyllostachys pubercens.*

Damage Larvae feed on leaves of the host plant.

Morphological characteristics **Adults** Length 6-8 mm. Body and legs black, with luster. Wings smoky brown, the tip paler, stigma and veins black. Hairs blackish brown. Body smooth, the posterior end of the scutellum with fine sculpture. Eyes large, inner border strongly constricted downward. Antennae filiform, shorter than the C vein of the front wing, 9 segmented; the 2nd segment larger than wide, the 3rd segment is 1.5 times as long as the 4th segment. Front tibiae inner spur with distinct membranous lobe. Claws small, inner tooth small, the end tooth long and oblique. Hind wing with closed M cell. Anal cell 1.5 × long as the cu-a vein.

Eggs Length 1.1 mm, width 0.8 mm, elliptical, pale blue.

Larvae Mature larvae 18-21 mm. Head orange yellow, ocelli mark black. A round black spot exists near the end of the vertex. Antennae conical, dark brown, 4 segmented. Mouthparts dark brown. Body brownish yellow. Larvae greenish after feeding, thorax paler, abdomen darker. Each abdominal segment with 6 small rings, the 2nd ring of each segment with 6 setae.

Pupae Length 6.5 mm, width 2 mm. Milky yellow at

E. longidentus adult (edited by Zhang Runzhi)

E. longidentus eggs (edited by Zhang Runzhi)

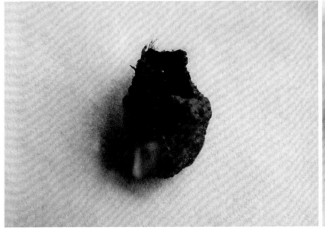

E. longidentus cocoon (edited by Zhang Runzhi)

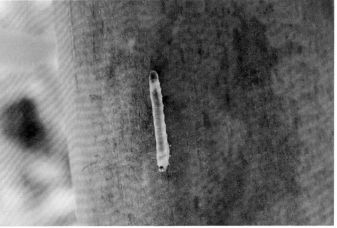

E. longidentus larva (edited by Zhang Runzhi)

E. longidentus pupa (edited by Zhang Runzhi)

E. longidentus damage (edited by Zhang Runzhi)

beginning, turning to brownish yellow later. Cocoons 9 by 5 mm in size, they are made of soil.

Biology and behavior *Eutomostethus longidentus* occurs one generation a year or one generation every two years. It overwinters or stays dormant in summer as prepupae in soil. Pupation occurs from early May. Adult emergence spans from early May to late June. Larval feeding occurs from late May to mid-July. Larvae pupate from early June.

Adults emerge from cocoons around noon. They fly on canopy or within forests and rest on host leaves or other plants. They rarely fly in cloudy and rainy days. Females lay eggs the same day after emergence. They lay eggs on new leaves. Eggs are laid inside leaf tissue and are visible under the raised leaf epidermis. Each leaf usually has one egg mass which arranged in one row of several or more than 10 eggs. Occasionally more than one egg masses are laid on one leaf. Each female lays a few egg masses within 3-10 days period. Adults live for 6-14 days.

Egg stage is 4-9 days. Larvae feed on leaf margin. Late stage larvae often migrate downward along the bamboo trunks. Some may rest on weeds. Mature larvae molt, defecate brown feces, shrink to 10 mm in length. They fall to the ground or migrate to the ground and pupate in soil. Larval period is 17-22 days. The prepupae period is 11 months for those occur one generation a year and 23 months for those occur one generation every two years. Larvae of the same developmental period will produce similar numbers of adults in the 2nd year and 3rd year. Pupal period is one week.

Prevention and treatment

During outbreak, injection of imidacloprid insecticide inside bamboo trunks can achieve good results. Fogging can reduce 80% of the populations. Applying microencapsulated insecticides to bamboo trunks are also effective.

Reference Hua Baosong, Xu Zhenwang, Liao Lihong et al. 2004.

(Chen Hanlin, Xu Zhenwang, Hua Baosong)

531 *Eutomostethus nigritus* Xiao

Classification Hymenoptera, Tenthredinidae
Synonym *Amonophadnus nigritus* (Xiao)

Importance Outbreaks of *E. nigritus* occurred in 26.7 ha in 1985 in Zhejiang. About 1,300 bamboo trees died from the outbreak.

Distribution Zhejiang, Anhui, Fujian.

Hosts *Phyllostachys pubercens, P. bambusoides, P. glauca.*

Damage Larvae feed on leaves of the host plant.

Morphological characteristics **Adults** Length 7-9 mm. Black, with sky blue luster; antennae black, 9 segmented, with dense pubescence. Front wing pale smoky brown, stigma black, with yellowish color at center; veins black. Front and middle femur, tibia, 1st-2nd tarsi or only 1st and 3rd tarsi, hind femur tip and tibia yellowish white. Antennae sulcus deep. Head and thorax with fine and sparse sculpture. Scapularia without sculpture. Head and thorax with black hairs. Sheath black.

Male length 5-7 mm. Antennae 9 segmented; the ratio between the length of the 3rd and the total length of the 4th plus 5th segment ratio is 1:1.1. Scutum of mesothorax with dense sculpture. Other characters similar to females.

Eggs Length 2 mm, width 0.8 mm. Elongated oval, newly laid eggs pink, gray before hatching.

Larvae Newly hatched larvae light-yellow, head black. There are two black spots below spiracular line on each of the abdominal segment during the 5th instar. The spots change to tubercles during the 6th instar. Mature larvae yellow, shinny; spiracles blackish brown, abdomen with two rows of transverse spines. Epiproct with > 30 spines.

Pupae Length 10 mm. Pale yellow, brownish black before elosion.

Cocoons Length 11 mm, width 8 mm. Elliptical.

Biology and behavior *Eutomostethus nigritus* occurs one or two generations a year. It overwinters as prepupae in soil. Pupation occurs in mid-May. Adults emerge from late May to late June. Eggs are laid from

E. nigritus adult (Wei Meicai)

early June and reach peak production in mid-June. Eggs hatch in mid-June. The 7th instar larvae appear in mid-July and migrate down the host and enter into soil. Some prepupae will pupate in late June. Adults emerge in early September. Eggs are laid from early September and hatch in late September. Mature larvae leave bamboo trees in late October. The pupal period is 8-9 days, adult period is 3-9 days, egg stage is 9-12 days.

Larvae go through 7 instars. For those with two generations a year, the 1st generation needs 50 days, the 2nd generation needs 200 days. Adults are male biased with males representing for 78% of all adults at the beginning. During peak emergence period, females represent for 70% of the adults. Eggs are laid on upper surface of leaves along the main vein. A row of 3-56 eggs are laid per leaf inside epidermis. The leaf surface where eggs are deposited is elevated. Egg hatch rates were 50%-60% for the 1st generation and 20%-100% for the 2nd generation. Larvae arrange in a row along the leaf edge when feeding. Their heads direct to the leaf base. The end of the body hangs over the head of the larva behind it. They feed from leaf tip to leaf base and only leave the main vein untouched. Then they migrate to another leaf. From the 4th instar, they feed separately. One larva consumes about 43 cm^2 leaf area.

Prevention and treatment

1. Silviculture. Remove *Polygonum cuspidatum* from the forest. Its flowers are the food source for *A. nigritus* adults.

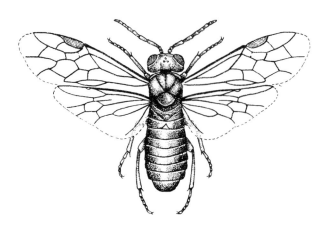

E. nigritus adult (Zhang Peiyi)

2. Spray *Bacillus thuringiensis* at 1×10^9 spores/mL, or *Beauveria bassiana* at 1×10^{10} conidia/mL.

3. Spray 2.5% deltamethrin at 5,000×, or 20% fenvalerate at 3,000-6,000×.

References   Xiao Gangrou, Huang Xiaoyun, Zhou Shuzhi 1991; Xiao Gangrou 1992; Liu Qiaoyun 1999; Chen Jiuchun 2006.

(Wang Maozhi, Zhu Jian, Tu Yonghai, Xu Sishan, Li Zhenyu)

532 *Eutomostethus deqingensis* Xiao
Classification Hymenoptera, Tenthredinidae

Importance *Eutomostethus deqingensis* is an important pest of bamboo trees in China.

Distribution Zhejiang, Jiangxi.

Hosts *Phyllostachys heterocycla*, *P. muda*.

Damage Larvae feed on leaves. Damaged leaves often only have their main veins left after infestation.

Morphological characteristics **Adults** Female length 7.0-8.0 mm. Antennae 9 segmented, with the 3rd segment the longest. Head black, eyes grayish black; antennae and their setae black; thorax black; wings smoky brown, anterior area paler, stigma and veins black. Hindwing with a closed M cell. Legs black; abdomen black, with purple luster. The median of the anterior border of clypeus raised. Male length 5.0-6.0 mm.

Eggs Length 1.8 mm, width 0.6 mm. Elongated elliptical.

Larvae There are 6 instars. Young larvae white. Length 2.7-3.3 mm. Mature larvae paly green, head yellowish brown, epiproct black. Dorsal surface with many black setae.

Pupae Length 6.5-7.7 mm. Thorax yellowish green, abdomen bright green, legs white; black before eclosion.

Cocoons Elliptical, 9.0 mm in length.

Biology and behavior *Eutomostethus deqingensis* occurs one generation a year. It overwinters as mature larvae in cocoons. They pupate from late April to mid-May. Adults emerge from early May to early June. Larvae appear from late May to early July. Mature larvae enter soil in mid-July and make cocoons. Adults emerge during the day. More adults emerge during morning hours. Adults rest on bushes and weeds. They mate multiple times and lay eggs in bamboo leaves. They insert the epidermis and lay eggs inside the leaves, causing leaf surface to rise slightly. A row of 1-12 eggs (most have 2 to 6 eggs) are arranged neatly arranged on a leaf. After hatching, larvae feed on leaves. Their feeding increases significantly from the 4th instar. The 5th instar larvae consume the most. The 6th instar larvae do not feed. After molting, they enter soil and make cocoons. They rest at 1- 5 cm in soil. Their occurrence is related to slope direction and position on the slope. Lower part of a slope suffers greater damage than upper part of a slope, northern slopes suffers greater damage than southern slopes. Forest edge suffers higher infestation than the center of forests.

Prevention and treatment

1. Silviculture. During the fall and winter months, dig soil to expose larvae and cocoons. Many of them will die from natural enemy predation or mechanical destruction.

2. Chemical control. When more than 1/3 of the plants were infested, control measures can be implemented. Spray 50% methamidophos at 2,000× to control adults.

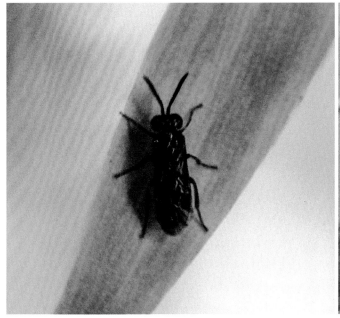

E. deqingensis adult (Xu Tiansen)

E. deqingensis eggs (Xu Tiansen)

E. deqingensis 3rd instar larva (Xu Tiansen)

E. deqingensis mature larva (Xu Tiansen)

Spray 2.5% deltamethrin at 400×, 25% fenvalerate-dimethoate mixture at 4,000×, inject 50% methamidophos (1.5 mL/plant) into bamboo trunks to kill 1st-4th instar larvae. Population reduction can reach 95%.

References Xiao Gangrou 1993; Zhu Zhijian, Tu Yonghai, Xu Sishan et al. 1994, 1995, 1996; Wang Tianlong, Luo Dahua, Yao Qingui 1997, 2000.

(Zhang Peiyi, Zhang Zhen)

533 *Hemichroa crocea* (Geoffroy)

Classification Hymenoptera, Tenthredinidae
Common names Alder sawfly, Striped alder sawfly

Importance *Hemichroa crocea* is a pest of broad-leaved trees. Larvae and adults have small activity areas. Their distribution is narrow and they are easily controlled.

Distribution Sichuan; Europe, North America.

Hosts *Alnus cremastogyne*, *Betula*, *Corylus*, *Carpinus*, *Salix*.

Damage Larvae feed gregariously on leaves. They migrate to other leaves after one leaf is consumed.

Morphological characteristics **Adults** Female length 6-8 mm. Body reddish yellow; antennae, mouthparts, prothorax, ventral side of mesothorax, metathorax, posterior border of the 1st abdominal notum, coxa, femur, anterior part of tibia, and tarsi black; wings pale smoky brown; veins black, with luster; stigma, costa and some veins brownish.

Male length 5-6 mm. Antennae, head, thorax black or pitch black; pteralia black; anterior part of femur, tibia reddish brown or yellowish brown; abdomen pitch black; pteralia smoky brown. Area outside of the stigma translucent. Posterior part of stigma and its center with some brownish yellow color. Costa and veins near the front with yellow color. Veins near wing base pitch black.

Eggs Length 1 mm, width 0.4 mm. Kidney shape. Milky, turning to white and transparent before hatching.

Larvae Mature larvae 13-20 mm long. Head black, with luster. Orange yellow; upper spiracular line and lower spiracular line black; thoracic leg claws red.

H. crocea adult (Wei Meicai)

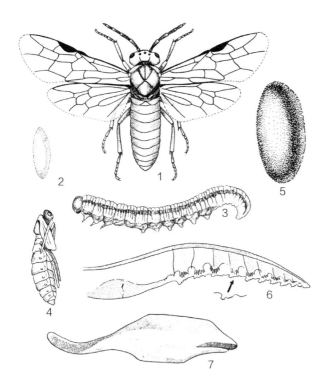

H. crocea. 1. Female adult, 2. Egg, 3. Larva, 4. Pupa, 5. Cocoon, 6. Lancet, 7. Valviceps (Zhang Peiyi)

Pupae Length 10 mm, width 3-5 mm. Pale yellow.

Cocoons Elliptical, constricted in the middle. Length 13 mm, width 6 mm. Dark brown, covered with sturdy mud and sands.

Biology and behavior *Hemichroa crocea* occurs two generations a year in Sichuan. It overwinters as prepupae in soil. Pupation starts in mid-February. Adults emerge when temperatures rise above 12°C in early March. Larvae appear in mid-March. The 3rd-4th instars appear in mid-April. Larvae migrate down the trees in late April and become dormant. They pupate in late September. The 2nd generation adults emerge in late October when temperatures are around 15°C. The 1st instar larvae appear in early November. Mature larvae migrate down the trees in late November. Adults can fly soon after emergence. Peak emergence period is 12:00-18:00. They can lay eggs same day after emergence. Females are primarily parthenogenetic. They lay eggs under epidermis along leaf veins. Each leaf may have 5-30 eggs. A female lays 33-100 eggs (average 74 eggs). Larvae go through 5 instars. Larval development period is 27-39 days for the 1st generation and 24-30 days for the 2nd generation. Pupal period is 20-25 days for the 1st generation and 20 days for the 2nd generation. Adult period is 5-8 days for the 1st generation and 4-6 days for the 2nd generation. Larvae prefer feeding at 15-25°C. The end of the abdomen curls downward when feeding. From 3rd instar, the abdomen raises every 3-5 seconds during feeding. Mature larvae dig into 3-5 cm deep soil and make cocoons. Individual cocoons or groups of 3-5 cocoons are found in soil.

Prevention and treatment

1. Cultural control. Turn soil when larvae are dormant and destroy larvae, prepupae, and pupae.

2. Biological control. Protect birds. Conserve natural enemies such as *Camponotus japonicas* and *Brachymeria* sp.

3. Chemical control. Spray 25% bisultap at 800×, 40% fenvalerate at 12,000× to control 3rd to 4th instar larvae. Spray contact insecticide on the ground directly under the canopy to kill mature larvae.

References Xiao Gangrou 1990, 1992; Xiao Gangrou, Huang Xiaoyun, Zhou Shuzhi 1991.

(Zhang Peiyi, Zhang Zhen, Xiao Kanglin)

534 *Megabeleses liriodendrovorax* Xiao

Classification Hymenoptera, Tenthredinidae

Distribution Zhejiang, Jiangxi.

Host *Liriodendron chinense.*

Damage Larvae feed on leaves. At high density, they can consume all leaves on a tree.

Morphological characteristics Adults Length 9-11 cm; front wing 9-11 cm long. Head black, shiny. Antennae black, 9 segmented. Anterior border of clypeus slightly emarginated. Thorax black, shiny. Wings transparent; front with smoky brown color; stigma and veins black. Legs black, front and middle tibia and tarsi brown. Abdomen black, with sky blue luster. Sheath black, extended beyond the end of the abdomen. Body hair light-yellow. Wings with many black short setae.

Eggs Length 1.7 mm, width 0.5 mm. Translucent, soft.

Larvae Mature larvae 23-26 mm. Yellowish green, head black.

Pupae Length 12-14 mm, width 4-5 mm. Milky, turning to black before eclosion.

Biology and behavior *Megabeleses liriodendrovorax* occurs one generation a year. It overwinters as prepupae in soil cocoons. In Zhejiang, pupation occurs from early April to mid-May. Adults emerge between late April and mid-June. Eggs are laid from early May to mid-June. They hatch between mid-May and late June. Larvae enter into soil and turn to prepupae from early June to mid-July. In Jiangxi, adults emerge from mid-April and reach peak emergence in late April. Larvae mature in late May and turn to prepupae.

After emergence from cocoons, adults appear on ground in warm clear days. They fly to canopy and rest on back of leaves. They do not fly during night, early morning and rainy days. They are active during day time on clear days. Adults lay eggs on back of leaves. Eggs are laid in leaf tissue along secondary veins in groups of 2-43 eggs. Each leaf usually contains one egg mass. Occasionally a leaf may have 2-3 egg masses. Adults live for 7-10 days after emergence from soil.

Egg period is about 7 days. Young larvae first feed around the egg mass, creating small holes. Then they move to branches and congregate along margins of young leaves. After young leaves are consumed, they migrate to old leaves. Larvae are active on back of leaves.

M. liriodendrovorax adult (edited by Zhang Runzhi)

M. liriodendrovorax cocoon (edited by Zhang Runzhi)

M. liriodendrovorax larva (edited by Zhang Runzhi)

M. liriodendrovorax damage (edited by Zhang Runzhi)

Young larvae raise their abdomen, when disturbed, they swing their abdomen. From 3rd instar, the body curl into C shape. They grasp the leaf margin with thoracic legs. Larval period is 21-37 days. They go through 5 instars. Mature larvae enter into soil, make 16 by 10 mm sized elliptical soil chambers. The body shrinks to 10-14 mm long and turn to prepupae. Prepual period is about 10 months. Pupal period is about 20 days.

Prevention and treatment

1. Silviculture. Loosen soil during fall and winter to destroy the prepupae.

2. Chemical control. Inject systematic insecticide in tree trunks to kill larvae.

References   Ouyang Guiming and Huang Weihe 1993, Xiao Gangrou 1993, 1997, Chen Hanlin 1995.

(Chen Hanlin, Lin Xiuming, Ye Changlong)

535 *Moricella rufonota* Rohwer
Classification Hymenoptera, Tenthredinidae

Importance *Moricella rufonota* is an important pest of *Cinnamomum camphora* in southern China.

Distribution Shanghai, Jiangsu, Zhejiang, Anhui, Fujian, Jiangxi, Hubei, Hunan, Guangdong, Guangxi, Sichuan, Guizhou, Taiwan.

Host *Cinnamomum camphora.*

Damage The 1st instar larvae feed on leaf tissue and leave the epidermis intact. The 2nd instar feed on young leaves and shoots. From 2nd instar, they feed on whole leaves. At high density, they can consume all leaves on a tree.

Morphological characteristics **Adults** Female length 8-10 mm. Head black with luster. Antennae filiform. Sides of pronotum, prescutum, scutum, scutellum, and scapularia of mesothorax brownish yellow with luster; postnotum, metathorax, ventral surface of mesothorax and abdomen black. Front coxa, middle of femur of the front and middle legs, the base of femur, tip of tibia, and tarsi of hind legs black. The other area of the legs yellowish white. Wings transparent, with some pale brown color. Stigma, veins black. Male length 6-8 mm; other characteristics similar to female.

Eggs Length 0.9-1.4 mm, width 0.4-0.5 mm. Elliptical, curled at one end, milky.

Larvae Mature larvae 15-18 mm. Pale green, with many creases. Head black. From 4th instar, thorax and the notum of the 1st and 2nd abdominal segments with many small dots and pale green stripes.

Pupae Length 7.5-10.0 mm. Pale yellow, turning to dark yellow later. Eyes black. Enclosed in silky cocoons which have attached soil particles. Blackish brown, elongated elliptical.

Biology and behavior *Moricella rufonota* occurs three generations a year in Hubei, 2-3 generations a year in Shanghai and Jiangsu, 1-3 generations south of Jiangxi, 1-2 generations in Guangdong, Zhejiang, Anhui, and Sichuan. It overwinters in soil as prepupae. After the 1st generation larvae turn to prepupae, some will enter into diapause. Others continue to develop. In Guangzhou, adults emerge from overwintering prepupae in late February. The 1st generation larvae hatch in early March. Mature larvae enter soil in mid-March. Some of them

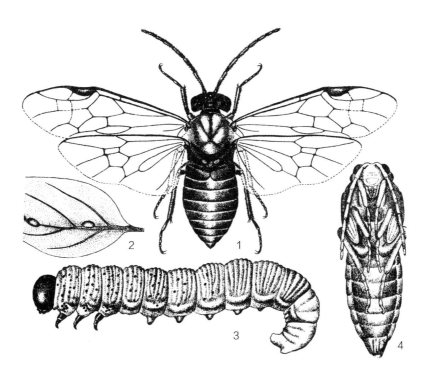

M. rufonota. 1. Adult, 2. Eggs, 3. Larva, 4. Pupa (Zhang Peiyi)

M. rufonota adult (Li Zhenyu)

M. rufonota adult (Li Zhenyu)

M. rufonota larva and damage (Wang Jijian)

will undergo diapause until the next year. The others will pupate. Adults emerge in late March and lay eggs. The 2nd generation larvae hatch in early April. Mature larvae enter soil in mid-April. Adults emerge in mid- and late May and lay eggs. The 3rd generation larvae appear from mid-May to early June. Mature larvae enter soil from early June to mid-July. Larvae period is 11-16 days and go through 4 instars. Few larvae are present in mid-July.

In years of warm winter and spring, adults of the overwintering generation emerge between mid- and late February. Larvae enter soil and make cocoons starting from late February. The 1st generation pupal period is 12 days. The 2nd generation pupal period is about 20 days.

Adults emerge during the day and mate the same day as emergence. Eggs are laid under epidermis on young leaves and flower buds. They lay up to 16 eggs on a leaf. A female lays 75-158 eggs during its lifetime. Adults are parthenogenetic. They live for 4 days and lay eggs over 3-6 days period.

Prevention and treatment

1. Biological control. Spray *Bacillus thuringiensis* at 0.5×10^8-1.5×10^8 spores/mL.

2. Chemical control. Spray 2.5% deltamethrin at 5,000×, or 10% cypermethrin at 1,000×. Other low toxicity chemicals such as azadirachtin, bisultap, chlorfluazuron are also effective against 1st and 2nd instars.

3. Integrated pest management. Clean up weeds and dead branches and turn soil in winter will reduce large numbers of overwintering population. Spray to control 1st generation larvae. Hand remove larvae in nursery or young stands.

References Xiao Gangrou, Huang Xiaoyun, Zhou Shuzhi 1991; Xiao Gangrou 1992; Li Qingyun 2000; Liu Yongsheng 2001; Fan Min, Xu Weiyu, Guan Liqin et al. 2006; Sun Xingquan, Jiang Gendi, Gu Yanfei et al. 2006; Jiang Zhengmin, Ma Junfei, Xu Guangyu et al. 2008; Wang Xiaojuan 2008; Yuan Shenghua 2008.

(Zhang Peiyi, Zhang Zhen, Chen Zefan)

536 *Nematus prunivorous* Xiao

Classification Hymenoptera, Tenthredinidae

Importance Larval feeding causes tree death, slow tree growth, and loss of yield.

Distribution Zhejiang.

Host *Prunus armeniaca*.

Damage Larvae feed on leaves. Only veins remain after larval feeding.

Morphological characteristics **Adults** Length 8.0-13.0 mm. Black, with purple blue luster. Antennae black. Wings yellow, transparent; stigma black, with an indistinct yellow spot at base.

Larvae Length 20.0-25.0 mm. Body blackish brown, from the 5th abdominal segment to the end of the body light-yellow; area along the dorsal glands on the prothorax and mesothorax, metathorax milky.

Biology and behavior *Nematus prunivorous* occurs two generations a year. It overwinters as mature larvae in soil. Pupae appear from mid-March to early April. Adults emerge from early and mid-April. The 1st generation eggs and larvae appear in April, pupae appear from late April to mid-May, adults emerge in mid-May. The 2nd generation eggs appear in May, larvae appear from early May.

Adults fly about 10 minutes after emergence from soil. They are active during 9:00-11:00 in clear non-windy days. After mating, females rest on back of leaves. They lay eggs along veins from tip of the leaves. The ratio of eggs laid along the main vein and secondary veins is 1.5:1.0.

Adult lifespan is 2-4 days. Eggs hatch in 6-8 days. Hatch rate under natural conditions is about 94.6%. The 1st and 2nd instar larvae congregate along leaf margin with their heads facing the same direction. Their abdominal ends curl inward. When disturbed, their heads raise up. From 3rd instar, they scatter and feed on different leaves. Larvae go through 5-6 instars. The 1st to 6th instar period is 3.3, 2.8, 2.1, 3.1, 3.3, and 3.2 days, respectively. Larvae mature within 14-18 days and then fall to the ground. They dig into soil to about 4 cm deep. The 1st generation larvae will turn to pupae the 2nd after they enter into soil. Cocoons often have attached soil particles. Pupal period is 4-7 days.

Prevention and treatment

1. Manual control. Loosen the soil around host trees in winter to kill overwintering larvae. Remove leaves along with larvae.

2. Chemical control. At high population density, spray 2.5% lamda-cyhalothrin at 4,000×, 20% fenvalerate at 2,000×, and 2.5% deltamethrin at 3,000-4,000× can reduce 95% of the pest population.

References Xiao Gangrou 1995, Zhao Jinnian 1998.

N. prunivorous adult (Wei Meicai)

(Zhao Jinnian)

537 *Pachynematus itoi* Okutani

Classification　Hymenoptera, Tenthredinidae
Common name　Larch sawfly

Importance　*Pachynematus itoi* is an important pest of coniferous trees in northern China.

Distribution　Hebei, Liaoning, Jilin, Heilongjiang; Japan, Korea Peninsula, Austria.

Hosts　*Larix olgensis*, *L. kaempferi*, *L. gmelinii*, *L. sibirica*, *L. olgertsis*.

Damage　Larvae feed on leaves. Only veins remain after larval feeding.

Morphological characteristics　**Adults** Female length 7-8 mm. Yellowish brown, with light-yellow pubescence. The following are black: the antennae flagellum (or 4th to 9th antennae segments), border along the eyes, posterior of the ocelli, a triangular mark on prescuttum of pronotum, the "L" mark on scutum of mesothorax, scutum of metathorax, side of metathorax. Front leg, coxa and femur of middle leg, and hind leg are black; side of coxa, tip of hind femur yellowish brown. Wings light-yellow, transparent. Center of notum of the 1st abdominal segment, anterior border of notum of the 2nd abdominal segment black; other areas yellowish brown. The abdominal segments 2-7 had indistinct pale brown marks on the sides.

Male length 5.5-6.0 mm. Head yellowish brown, mandible tip brown, other parts of the mouthparts yellowish brown. Dorsal area of the head, areas between the eyes (ocelli and post-ocelli areas) with black spots. Thorax black; sides of pronotum, pteralia yellowish brown.

P. itoi adult female (Zhang Peiyi)

Eggs　Elliptical, 1.46 mm × 0.58 mm. Yellow, becoming yellowish brown near hatching.

Larvae　Head bright brown. Body pale green or pale greenish gray. Semi-transparent to transparent. Body with black or brown spots. Each segment has square spots on the dorsal line and lateral line. There are two almost connected irregular oval spots at coxa under each thoracic spot. Thoracic legs black.

Pupae　Length 8-9 mm. Milky, eyes red.

Cocoons　Elliptical, dark red.

Biology and behavior　*Pachynematus itoi* occurs 1-3 generations a year. In Jilin province, it occurs three generations a year and overwinters as prepupae in fallen leaves. Pupae appear in early May. Adults emerge in mid-May. Eggs hatch in early June. Adults emerge from early July. The peak of the 2nd and 3rd generation larvae feeding period is in late July and late August, respectively. Larvae migrate to fallen leaves from mid-September to overwinter.

Young larvae feed gregariously. They scatter from the 3rd instar. Larvae prefer feeding on old leaves. Their distribution pattern is in clusters. An outbreak lasts for about two years. Each outbreak cycle is about 8-10 years. When 20%-30% of leaves are consumed, hosts can over compensate the loss. When almost all leaves are consumed, the 2nd year timber growth loss will be 65.6%.

P. itoi adult (FPC)

P. itoi larvae (FPC)

P. itoi damage (FPC)

Prevention and treatment

1. Cultural control. Manually destroy clusters of young larvae.

2. Conserve natural enemies: *Argrotheruteutes abbreviates*, *A. macroincubitor*, *Endasys liaoningensis*, *Aptesis chosiensis*, *Arenetra genangusta*, *Bathythrix cilifacialis*, *Pleolophus setiferae*, *Tritneptis klugii*, *Cleptes semiauratus*, *Drino inconspicua*, *Myxexoristops blondeli*.

3. Chemical control. Spray 40% phoxim, fogging with 5% esfenvalerate, spray with 20% triflumuron, 10% cypermethrin at 800-1,00×, 50% malathion at 800-1,000, 48% dusban at 1,000×, 20% fenvalerate at 2,000×, 3% fenoxycarb at 3,000-4,000×, or 25% abamectin + Dimilin at 1,500-2,000 is effective.

References Xiao Gangrou, Huang Xiaoyun, Zhou Shuzhi 1991; Xiao Gangrou (ed.) 1992; Zhu Chuanfu, Yang Wenxue, Wang Jinguo et al. 1994; Sheng Maoling, Gao Lixin, Sun Shuping et al. 2002; Jin Meilan 2004; Wang Zhiming, Liu Guorong, Cheng Bin et al. 2006; Yu Qing and Zhang Weiyao 2008.

(Zhang Peiyi, Zhang Zhen, Niu Yanzhang, Wang Fuwei)

538 *Pristiphora beijingensis* Zhou et Zhang

Classification Hymenoptera, Tenthredinidae
Common name Poplar sawfly

Importance *Pristiphora beijingensis* is an important pest of poplar nurseries. Their feeding affects growth of seedlings.

Distribution Beijing, Hebei, Liaoning.

Hosts *Populus* spp.

Damage Larvae feed on leaves. Only veins remain after larval feeding.

Morphological characteristics **Adults** Female length 5.8-7.6 mm. Head, dorsal surface of the body black, ventral surface pale brown; wings transparent, with dense pale brown pubescence. Body with dense pubescence. Male length 4.7-5.9 mm, similar to female.

Eggs White, transparent, oval, 1 mm long.

Larvae Young larvae transparent, head gray. From 2nd instar, occiput and thorax black, body yellowish green. As the larvae grow, the body color become greener. After 3rd instar, each segment has some black spots. Mature larvae 11-13 mm long, head width is 1.3-1.5 mm.

Cocoons Yellowish brown, elongated elliptical, silky, 7-10 mm long.

Pupae Pale green. Legs, antennae, and wing buds white.

Biology and behavior *Pristiphora beijingensis* occurs 8-9 generations a year in Beijing. Eggs appear from early April. Larvae overwinter by mid-October. Adults are active during 6:00-18:00. Mating time is 3-226 seconds (mean 81.1 seconds). They can reproduce parthenogenetically. Unfertilized eggs develop into males.

P. beijingensis adult (Zhang Peiyi)

P. beijingensis adult (Zhang Peiyi)

P. beijingensis adult (Zhang Peiyi)

P. beijingensis egg-laying adults (Zhang Zhen)

P. beijingensis eggs on leaf edge (Zhang Peiyi)

P. beijingensis mature larvae (Zhang Zhen)

P. beijingensis young larvae (Zhang Zhen)

P. beijingensis cocoons and pupa (Zhang Zhen)

P. beijingensis larvae (Zhang Peiyi)   *P. beijingensis* cocoons (Zhang Peiyi)

P. beijingensis larva being fed by a predatory bug (Zhang Zhen)

P. beijingensis larvae (Zhang Peiyi)

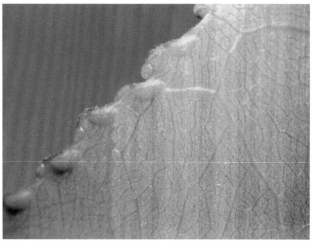
P. beijingensis eggs (Zhang Peiyi)

P. beijingensis adults (Zhang Peiyi)

Females lay eggs around noon and in the afternoons. It takes 46.3 seconds to lay one egg. Eggs are laid inside the leaf tissue around young leaf edge. Young larvae feed on leaf edge and make the leaves net looking and yellow. They gradually feed toward center of the leaves. From 3rd instar, they feed on whole leaves. They migrate to other leaves from 4th instar and can consume all leaves on a tree. Males have 4 instars, females have 5 instars. Prepupae period is 1-6 days (mean 4.1 days). Pupae period is 3-7 days (mean 4.6 days). Forest edge suffers lighter infestations.

Prevention and treatment

1. Cultural control. Remove leaves along with larvae.

2. Silviculture. Plant resistant varies.

3. Biological control. The following products are effective: *Bacillus thuringiensis* subsp. *Dendrolimus* and *B.t* subsp. *kurstaki*, and nematode *Steinernema feltiae*.

4. Chemical control. Spray 5.7% eamamectin at 3,000×, 20 g/L lamda-cyhalothrin at 1,000-1,500×, or 50% imidacloprid-monosultap at 600-800×.

References Zhou Shuzhi and Zhang Zhen 1993, Zhou Shuzhi, Huang Xiaoyun, and Zhang Zhen 1995, Xu Gongtian and Yang Zhihua 2007.

(Zhang Zhen, Wang Hongbin)

539 *Pristiphora conjugata* (Dahlbom)

Classification Hymenoptera, Tenthredinidae
Common name Poplar sawfly

Distribution Inner Mongolia, Northeastern China, Xinjiang; Russia, Britain, Italy, U.S.A.

Hosts Most preferred hosts are: *Populus × beijingensis, P. nigra, P. tremula, Populus ×berolinensis*. Preferred hos is *Salix matsudana*.

Damage Larvae feed on leaves. Only leaf veins are left.

Morphological characteristics **Adults** Female length 7-8 mm; male length 5-6 mm. Wingspan 12 mm. Body yellowish brown, with luster and white short pubescence. Front border of clypeus truncated. Eyes and antennae blackish brown. Head black except for the labrum, labial palpi, base of mandibles yellowish brown. Wings translucent, covered with dense fine hairs; stigma blackish brown, center dark yellow; base of C, R, and A veins light-yellow, other veins blackish brown. Abdomen yellow. There are black spots on notum of the 1st-8th abdominal segments.

Eggs Elliptical, 1.3-1.5 mm long, 0.3 mm wide. Milky, smooth.

Larvae Length of the 1st instar 2.5-2.8 mm, 2nd instar 3.5-4.0 mm, 3rd instar 5.5-7.0 mm, 4th instar 9.0-11.0 mm, mature larvae 15-17 mm. Head blackish brown, clypeus brown, truncated, labrum yellow. Body yellowish green, the 3rd thoracic segment with 7 rows transverse black spots on the back, base of thoracic legs brown, each of the lateral side of the 1st-7th abdominal segment with 5 black spots, the base of thoracic legs and prolegs on the 1st-7th abdominal segments each with two black spots. The dorsal surface of the 7th and 8th abdominal segment each with two transverse rows of small black spots.

Pupae Length 6.0-7.5 mm, green. Head orange yellow, eyes reddish brown. Antennae and thoracic legs milky. Abdomen yellow.

Cocoons Female 8.0-8.5 mm long, male 4-6 mm long. Oval. Grayish white, turning to brown later.

Biology and behavior *Pristiphora conjugata* occurs 4-5 generations a year in Xinjiang. It overwinters as prepupae in cocoons in 2-3 cm deep soil. Pupation occurs in early April. Adults emerge in mid- and late April. Eggs hatch in late April. Pupae appear in mid-May. Adults appear in late May. The adult occurrence periods of various generations are: mid-April to mid-May, late May to early June, late June to early July, early to late August, and early to late September. The development period of each generation is 25-30 days. The larval infestation period of each generation is late April to mid-May, late May to mid-June, early July to mid-July, early August to late August, early September to late September. The various generations overlap. The most damage occurs during July and September. Larvae enter into soil at the end of September and early October.

Adults of the overwintering generation emerge in mid-April when mean daily temperature reaches 15°C. Males emerge about 2 hours earlier than females. Most adults appear during 14:00-18:00. Males are more active than females. They are weak flyers and are swift. They can fly about 2.5 m. When disturbed, they fall to the ground with ventral side facing upward, they occasionally make some noises. After turning over, they fly away or walk away. Adults do not play death and are not phototaxic. They feed on nectars for nutrition. Sex ratio is 1:1. They

P. conjugate adult (Xu Gongtian)

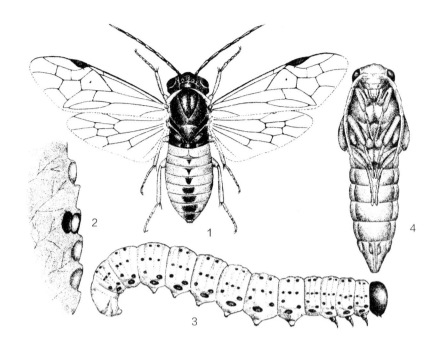

P. conjugate. 1. Adult, 2. Eggs, 3. Larva, 4. Pupa (Zhang Peiyi)

mate same day as emergence. Mating occurs most during 12:00-17:00. Males often compete for mates and attack each other. They only mate once. After about 12 hours, females can lay eggs. They lay eggs in the morning. Females lay eggs into leaf tissue. A female spends about 1-3 minutes to lay one egg. Each female lays eggs 1-3 times, each time about 59 eggs (range: 43-75). Females live for 2-4 days. Females can reproduce parthenogenetically. Eggs are laid along leaf edges. Each leaf may have 6-28 eggs. Egg incubation period is 4-5 days, 90% of them can hatch.

Young larvae feed gregariously on young leaves located at upper portion of the seedlings. The feeding causes small notches. Their abdomen bends downward or the end rises up. When disturbed, their abdomen rises. The 3rd-5th instar scatter at mid- and lower part of the canopy. They often consume whole leaf tissue except the veins. Larvae play death and may attack each other. When defecating, they raise the abdomen. Larvae go through 5 instars, 1-2 days apart between instars. After 7-10 days feeding, larvae mature and migrate down into soil and become prepupae. After another 5-6 days, they turn into pupae. Pupal period is 10-12 days.

Prevention and treatment

1. Silviculture. Between late October and mid-April, turn soil to destroy overwintering prepupae.

2. Cultural control. Manually remove congregating young larvae and eggs. Place a plastic film under tree canopy, shake trees and collect the fallen larvae.

3. Biological control. Protect and utilize *Arma chinensis*.

4. Biopesticides. Spray 25% Dimilin or 1.8% abamectin at 1,500-2,000 to control 1st-3rd instar larvae.

5. Chemical control. Spray 2.5% deltamethrin at 2,000-4,000×, 40% fenvalerate-malathion at 2,000×, or 50% sumithion at 800-1,000.

References   Yang Xiuyuan 1981, Wang Aijing et al. 1983, Xiao Gangrou 1992.

(Wang Aijing)

540 *Pristiphora erichsonii* (Hartig)

Classification Hymenoptera, Tenthredinidae
Common name Larch sawfly

Importance *Pristiphora erichsonii* is an important pest of *Larix* spp. and may occur in large areas.

Distribution Beijing, Hebei, Shanxi, Inner Mongolia, Liaoning, Jilin, Heilongjiang, Shaanxi, Gansu, Ningxia; Russia, Northern Europe, Britain, U.S.A., Canada.

Hosts *Larix principis-rupprechtii, L. gmemlinii.*

Damage Larvae feed on host leaves, cause curled shoots and death of branches.

Morphological characteristics **Adults** Female length 8.5-10.0 mm. Black with luster, the 1st-5th abdominal notum, anterior border of the 6th abdominal notum orange yellow. Male length 8-9 mm.

Eggs Length 1.3 mm, width 0.4 mm. Elongated elliptical. Yellowish white, yellowish green before hatching.

Larvae Mature larvae 15-20 mm, head width 2-3 mm. Blackish brown, thorax, dorsal surface of abdomen grayish green, ventral surface of abdomen pale gray, prolegs blackish brown.

Cocoons Brown, silky. Length 9-15 mm, width 5 mm.

Pupae Length 9-10 mm. White, brownish black before eclosion.

Biology and behavior *Pristiphora erichsonii* occurs one generation a year in Beijing. It overwinters as prepupae in soil under the tree canopy. Pupation occurs in early May. Adults emerge in mid-May. Females are predominantly parthenogenetic. They lay eggs from 4 hours after emergence. Eggs are laid on tip of new shoots in two alternating rows per shoot. Each shoot may have

P. erichsonii larvae (FPC)

P. erichsonii adult (FPC)

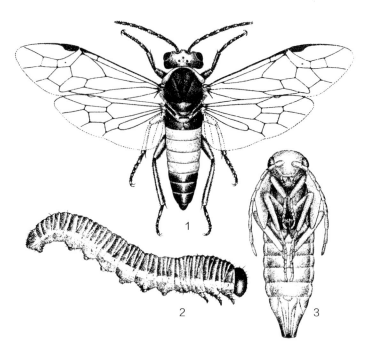

P. erichsonii. 1. Adult, 2. Larva, 3. Pupa (Zhang Peiyi)

5-80 eggs (mean 50 eggs). A female lays eggs in 1-5 new shoots. Each female produces 20-120 eggs. As the eggs develop, the affected shoot curls to a hook shape. The cambium and phloem where eggs are deposited dry out and split. Egg period is 3-10 days. Young larvae are gregarious. They feed on young leaf tissue and cause clusters of dry leaves. From the 2nd instar, they feed on whole leaves. From the 3rd instar, they migrate to old leaves, feed from top to base of leaves. They feed on lower canopy leaves first, then move to the top canopy. Larvae make cocoons in soil during late June to mid-July. The pupation and adult emergence period in northern Shaanxi is about 10 days later than Beijing. All larvae will overwinter from early August. The optimum larval development temperature is 16-20 ℃. The sparse, weak forest stands usually have higher pest population density.

Prevention and treatment

Monitor pest populations. When pest feeding causes more than 40% leaf loss, then a combination of tools and methods should be considered.

1. Silviculture. Plant resistant tree species. Keep canopy coverage at 0.7-0.8.

2. Manual control. Collect and destroy adults and larvae. Dig soil to remove cocoons. Clean up weeds and leaves in the fall and spring.

3. Biological control. Spray *Steinernema feltiae* to leaves. Apply abamecctin or *B.t.* dust. When mature larvae go down the trees, apply *Beauveria bassiana* or *Paecilomyces* sp. to kill larvae.

4. Chemical control. Spray or fogging with pyrethroids or Dimilin are effective.

References Xie Qingke 1990, Xiao Gangrou, Huang Xiaoyun, and Zhou Shuzhi 1991, Li Huicheng, Wang Jianzhong, and Guofei 1992; Li Menglou, Liu Chaobin, and Wu Dingkun 1992; Li Menglou, Wu Dingkun, and Liu Chaobin 1992; Wu Dingkun, Liu Chaobin, and Jiang Xinong 1992; Xiao Gangrou 1992; Zhou Shuzhi, Huang Xiaoyun, Zhang Zhen et al. 1995; Wang Gong 2009; Xiao Gucheng, Yu Zhijia, Fan Yapeng et al. 2009.

(Zhang Zhen, Wang Hongbin, Li Huicheng, Wang Jianzhong)

541 *Stauronematus compressicornis* (F.)

Classification Hymenoptera, Tenthredinidae
Synonym *Lygaeonematus compressicornis* (F.)

Distribution Inner Mongolia, Liaoning, Jilin, Heilongjiang, Xinjiang; Russia, Japan, Korea Peninsula, Britain.

Hosts Most preferred hosts are: *Populus* × *beijingensis*, *P. nigra*, *P. tremula*, *Populus* ×*berolinensis*. Preferred hos is *Salix matsudana*.

Damage Larvae feed on leaves. Only leaf veins are left.

Morphological characteristics **Adults** Female length 7-8 mm; male length 5-6 mm. Black, with luster and white short pubescence. Antennae brown, laterally compressed. Pronotum, pteralia yellow. Wings transparent, stigma blackish brown, veins pale brown. Legs yellow, hind tibia and end of tarsi black. The inner and tooth of the claws parallel with each other, their base enlarged in to a lobe. Sheath reaches the end of cerci, similar width as the basal tarsi, round, the end pointed, the side with curled setae.

Eggs Elliptical, 1.3-1.5 mm long, 0.3 mm wide. Milky, smooth.

Larvae Newly hatched larvae fresh green. Length of the 1st to 5th instar length are: 1.8-2.0 mm, 2.5-3.5 mm, 4.0-4.5 mm, 9.0-11.0 mm, and 12.0-14.0 mm, respectively. Head blackish brown, vertex green, clypeus truncated. There are 4 black spots on each side of each thoracic segment. Thoracic legs yellowish brown. Body with many brown spots of various sizes.

Pupae Length 6.0-7.5 mm. Grayish green, head orange yellow. Mouthparts, antennae, wings, legs milky. The posterior border of the 1st to 8th abdominal notum green.

Cocoons Female 7-8 mm long, male 4-6 mm long. Milky, turning to brown later.

Biology and behavior *Stauronematus compressicornis* is very similar to *Pristiphora conjugata*. They often occur simultaneously. The following are the differences: each female lays eggs in 5-6 leaves, each leaf with 3-15 eggs; a female lays a total of 30-60 eggs; larval pe-

S. compressicornis adult (Xu Gongtian)

riod is 8-20 days, pupal period is 6-18 days.

It was observed that the larvae secrete foaming material which turns to wax threads before feeding. The wax threads are 3 mm long. They are arranged into 1-3 rows around the feeding marks.

Prevention and treatment

Same as that for *Pristiphora conjugata*.

References   Yang Xiuyuan 1981, Wang Aijing et al. 1983, Xiao Gangrou 1992.

(Wang Aijing)

S. compressicornis cocoons (Xu Gongtian)

S. compressicornis larva (Xu Gongtian)

S. compressicornis damage (Xu Gongtian)

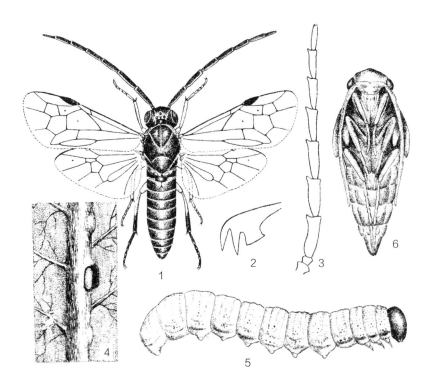

S. compressicornis. 1. Adult, 2. Claw, 3. Antenna, 4, egg, 5. Larva, 6. Pupa (Zhang Peiyi)

542 *Dryocosmus kuriphilus* Yasumatsu

Classification Hymenoptera, Cynipidae
Common name Chestnut gall wasp

Distribution Beijing, Tianjin, Hebei, Liaoning, Jiangsu, Zhejiang, Anhui, Fujian, Jiangxi, Hubei, Hunan, Shandong, Henan, Guangdong, Guangxi, Shaanxi; Japan, Korea Peninsula.

Hosts *Castanea mollissima*, *C. sequinii*, *C. henryi*.

Damage Infested buds form galls and cannot develop into new shoots or flowers. The heavily infested branches may die. Fruit yield may decrease for several years after infestation.

Morphological characteristics **Adults** Length 2.5-3.0 mm. Brown with luster. Antennae 14 segmented; the scape, pecidel thick. Scutellum near round, raised. Ovipositor brown.

Eggs Elliptical, milky. Length 0.15-0.17 mm. Its end with a slender pedicel of 0.5-0.7 mm long. The end of the pedicel enlarged.

Larvae Mature larvae 2.5-3.0 mm long. Milky, yellowish white when mature. Mouthparts brown. Smooth, with distinct separation between thorax and abdomen.

Pupae Length 2.5-3.0 mm. Milky, blackish brown near eclosion, eyes brownish red.

Biology and behavior *Dryocosmus kuriphilus* occurs one generation a year. It overwinters as newly hatched larvae in buds. In south Jiangsu, larvae feed when buds start to grow in early April. Bright green to reddish brown galls are formed during this time. Their shape is somewhat round. The size depends on the number of larvae inside. Average size is 1.0-2.5 mm by 0.9-2.0 mm. The chamber size at the late development stage of the galls is 1.0-3.1 mm by 1-2 mm. The inner walls of the galls are lignified and hard. Each gall has 1-16 larvae, with 2-5 larvae being the most common. Larvae live for 30-70 days inside galls. Pupae start to appear in early May and reach peak in late May. Adults emerge from early June and reach peak in mid-June. Adults stay inside galls for 10-15 days after eclosion. They chew 1 mm diameter holes when exit from inside the galls. Adults emerge from galls most during late June in Jiangsu. They are not strong flyers. They rest on back of leaves during night and do not feed and are not phototaxic. Adult life

D. kuriphilus adult (Li Zhenyu)

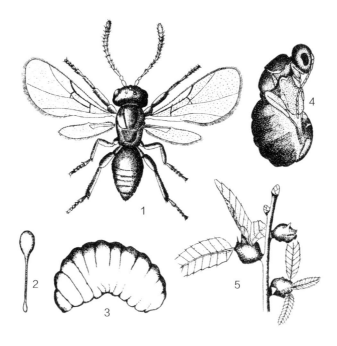

D. kuriphilus. 1. Adult, 2. Egg, 3. Larva, 4. Pupa, 5. Damage

span is 5.5 days (average 3.1 days) after emergence from galls. There are no males. Adults lay eggs during 6:00-17:00. A female lays 2-4 eggs a time. Eggs are laid inside buds. Dissection of females reveals each female contains about 200 eggs. Larvae hatch most during late August. The newly hatched larvae feed for a short period of time inside bud tissue. The tissue surrounding the larvae swells a little. From late October, larvae overwinter.

Natural enemies and precipitation are important factors limiting the population growth. There are 24 species of reported parasitoids, including *Torymus sinensis*, *T. geranii*, *Eurytoma setigera*, etc. *Torymus sinensis* is the most widely distributed. It overwinters in galls as mature larvae. Adults lay eggs on larvae. Parasitization rate of 7.03% was reported in 1978 when branch infestation rate was 56.83%. During 1979-1982, parasitization rate of 63.23%, 81.15%, 68.85%, 67.7% were recorded, respectively. The branch infestation rate was 24.24%, 9.9%, 5.0%, and 1.2% over the four years, respectively. Other notable parasitoids include *Eupelmus urozonus*, *Megastigmus maculipennis*, and *M. nipponicus*.

The species, variety, and type of rootstock of the host is associated with pest populations.

Prevention and treatment

1. Quarantine. Avoid bringing in infested seedlings and grafts.

2. Silviculture. When infestation is heavy, cut off one year old branches above the dormant buds will reduce pest populations. *Dryocosmus kuriphilus* does not lay eggs in dormant buds.

3. Cultural control. For infested young trees, remove galls before May.

4. Chemical control. Spray 50% sumithion at 500× twice in early and mid-June can reduce 90% of the population. Spray 25 g/L lamda-cyhalothrin at 2,000×, 2.5% deltamethrin at 3,000×, or 10% imidacloprid WP at 0.5g/L during adult period.

References Sun Yongchun et al. 1965; Zhi Cunxun 1979; Ao Xianbin et al. 1980; Zhang Changhui 1983; Luo Youqing 1985, Huang Jingfang and Luo Youqing 1988, Li Chengwei 1998; Wei Libang 1990; Yi Yehua 2004; Tong Xinwang et al. 2005; Ren Shuang 2009.

(Sun Yongchun, Xu Fuyuan, Jiang Ping)

543 *Megastigmus cryptomeriae* Yano

Classification Hymenoptera, Megastigmidae

Importance *Megastigmus cryptomeriae* is an important pest of cedar fruits.

Distribution Zhejiang, Fujian, Jiangxi, Hubei, and Taiwan; Japan.

Hosts *Cryptomeria fortunei, C. japonica*.

Damage The larvae bore into the cedar seeds, and consume the seed endosperm, and result in hollow seeds.

Morphological characteristics **Adults** Female length 2.4 to 2.8 mm, forewing length 2.1 to 2.5 mm. There is a black tumor-like protrusion (pterostigma) on the front margin. The head is protruded upward with fine transverse striae. Thorax length is twice of its width. Pronotum length is 1.1 to 1.3 times of its width, its front border slightly concave, covered with fine transverse striae and scattered black setae. The front edge of scutellum of the mesonotum has imbricated texture and transverse striae, its rear edge with rugose. The basal cell of the forewings has hairs at the distal end, its lower part is almost covered by a row of hairs on the cubital vein. The adaxial surface of the distal portion of the costal chamber has a row of hairs, whereas the abaxial surface of the basal half portion has a row of hairs. Male length is 2.1 to 2.6 mm, forewing length 2.1 to 2.2 mm, with pterostigma. Thorax is thin, length is 2.3 to 2.6 times of its width. The pronotum significantly narrows at the front portion. Ocelli area has dark brown spots. The abdominal dorsum is black, or with a wide dark strip only on the second to fifth segments of the abdomen.

Larvae Milky white, length is 2.0-2.8 mm, C-shaped. The body is thick in the middle and thin in both ends. The dorsum is smooth without cysts. Mandible elongated triangle in shape with 4-5 teeth.

Pupae Exarate pupae. Length is 2-2.8mm, milky white, the compound eyes are red, yellowish brown before adult emergence.

Biology and behavior *Megastigmus cryptomeriae* occurs one generation per year in Zhejiang province. The insect overwinters as mature larvae in residual seeds in the field, or in the seeds in storage facilities. The overwintering larvae become active in early March of the following year. Peak pupation time is in mid- and late March. Adult emergence occurs during mid-April and mid-May. Larvae begin to hatch in mid-May. They bore into the young fresh fruits and become mature after consuming the seed. Then they begin to diapause. A larva only consumes one seed during its larval stage.

Prevention and treatment

1. Reinforce quarantine procedure. Fumigation of infested seeds in a confined environment using aluminum phosphide as fumigant can effectively kill the larvae.

2. Elimination of the residual insects in the forest. In combination with annual seed harvest, complete removal of all of the fruits (including all of the fruits found on the ground and on the trees) can effectively reduce the overwintering larval population.

3. Warm water treatment. Soaking the seeds in 50-60℃ warm water for 15 minutes will kill the insect larvae. At the same time, the germination rate of the seeds is not affected.

4. Chemical control. The conifer stand with heavy infestation can be treated with the emamectin at the wasp emergence time to manage the adult population.

References He Junhua 1984, Zhang Jinfang 1989, Yang Shaoli et al. 1997.

(Jiang Ping, He Junhua, Xu Deqin)

M. cryptomeriae adult female (Li Zhenyu)

544 *Leptocybe invasa* Fisher et La Salle

Classification Hymenoptera, Eulophidae
Common name Blue gum chalcid

Distribution Fujian, Jiangxi, Guangdong, Guangxi, Hainan; Iran, Iraq, Lebanon, Cambodia, India, Vietnam, Laos, Israel, Jordan, Syria, Thailand, Turkey, Uganda, Kenya, Tanzania, Algeria, Morocco, South Africa, France, Italy, Greece, Spain, Portugal, New Zealand, Australia, U.S.A.

Hosts *Eucalyptus botryoides*, *E. bridgesiana*, *E. camaldulensis*, *E. globulus*, *E. gunii*, *E. Grandis*, *E. robusta*, *E. saligna*, *E. tereticornis*, *E. viminalis*, *E. urophylla*, *E. cinerea*, *E. pulverulenta*, *E. grundis* × *E. tereticornis* DH201, *E. exserta*, *E. grandis* × *E. urophylla*, *E. maidie*.

Damage There are three types of wasp damage on gum trees: 1) no damage or gall formation; 2) visible oviposition wound, abnormal shoots and foliage; and 3) the formation of galls, which causes serious damage to the host plants. The somatic line DH201-2 is the most susceptible. The most damage occurs on 1 to 3-year-old gum saplings. Larvae develop in leaf veins, petioles and young shoots. The infested eucalyptus tissue form distinct insect galls. High insect population may cause the leaves to curl, and the surface of the leaves and young shoots are covered with tubular galls. Infestations slow tree growth and development and may cause death of new growth, i.e., tip of young shoots and leaves.

Morphological characteristics **Female Adults** Length 1.1 to 1.4 mm. Head and body brown, mixed with metallic blue to green. Antennal scape yellow, scape length is approximately half of the pedicel segment; flagellum brown to light brown, with four ringed segments, three rectangular segments and another three rectangular segments that form a club. There is deep groove at the triangular region surrounding the ocelli. Edge of the mouthparts pale brown to yellow, coxae of the front legs yellow, mid- and hind legs have the same color as the body, femur and tarsus yellow, the last segment of the tarsus brown. Abdomen short, and oval in shape. Lower anal plate extends to half of the abdomen; ovipositor sheath is short, which does not reach the end of the abdomen. **Male Adults** The body length is 0.8 to 1.2 mm; head and body metallic brown and blue to green, abdomen brown with light metallic color dorsally, femur light-yellow, coxae of the mid- and hind legs have a metallic hue; antennal scape yellow, the length is 3× of its width; pedicel yellow, length is 1.5× of its width. The base of pedicel is relatively dark dorsally; the last six segments yellow, the length of the segment forming the club is 2.5× of its width. The three segments prior to the club with a few bristles.

Biology and behavior *Leptocybe invasa* occurs 5 to 6 generations per year in Guangxi province with overlapping generations. Larvae overwinter in galls. Adults emerge from the galls in late February. The main emergence period

L. invasa adults (Wang Jijian)

L. invasa damage to eucalyptus tree (Wang Jijian)

is from 8:00 to 14:00, which is 87% of the total daily adult emergence. A female lays an average of 139 eggs. This wasp reproduces mainly parthenogenically; they can also reproduce sexually. The sex ratio is 150-200:1. Males represented more than 1% in Bobai, Guangxi province.

Prevention and treatment

1. Quarantine. Prevent the transport of any of wasp-infested nursery materials into a new region that has no record of the wasp.

2. Planting wasp-resistant eucalyptus, such as *Eucalyptus henryi*, *E. citriodora*, *E. tessllaris*, *E. cloeziana*, *E. pellita*, *E. microcorys*, *E. pilularis*, *E. robusta*, *E. coolabah*, *E. globulus*, *E. smithii*, *E. moluccana*, and *Eucalyptus polycarpa*. These species suffer less wasp damage than the others.

3. Utilizing yellow sticky plastic cards. The yellow sticky cards can trap a large number of wasps. The control efficacy is up to 81%.

4. Protection and preservation of natural enemies. The parasitoid, *Aproslocelus* sp., and a variety of spiders should be protected.

5. Chemical control. Spraying 0.02% imidacloprid solution is effective. Repeat application after 15 to 20 days.

References Chen Shangwen, Liang Yiping, and Yang Xiuhao 2009; Chang Runlei and Zhou Xudong 2010; Wang Jijian et al. 2010, 2011; Wu Yaojun, Li Dewei, Chang Mingshan et al. 2010; Luo Jitong, Jiang Jinpei, Wang Jijian et al. 2011.

(Wang Jijian, Li Zhenyu)

L. invasa damage to eucalyptus trees (Wang Jijian)

L. invasa elosion holes (Wang Jijian)

L. invasa galls (Wang Jijian)

545 *Quadrastichus erythrinae* Kim

Classification Hymenoptera, Eulophidae
Common name Erythrina gall wasp

Distribution Fujian, Guangdong, Guangxi, Hainan, Taiwan; Mauritius, Reunion island, U.S.A., Singapore, India, Thailand, the Philippines, American Samoa, Guam and Japan.

Hosts *Erythrina indica*, *Erythrina variegate*, *E. variegate* var. *orientalis*, *E. coralloidendron*, *E. crista galli*, *E. indica* var. *picta*, *E. abyssinica*, *E. berteroana*, *E. fusca*, *E. sandwicensis*.

Morphological characteristics **Female Adults** Dark brown with yellow spots, head yellow, postgena area is brown. There are three ocelli, which are red, and arranged in a triangular shape. Pronotum is dark brown, with 3 to 5 short setae, and a concaved light yellow spot in the middle. Scutellum is brownish yellow, with two pairs of setae, a few with 3 pairs, and there are two light-yellow longitudinal stripes in the middle. Wings are colorless, and transparent. The cilia on the wing surface is dark brown, and veins are brown. The coxae of the front and hind legs is yellow, while coxae of the middle legs are white. The dorsum of the first segment of the abdomen is light-yellow. The dorsum of the second segment of the abdomen has two light-yellow spots. They extend towards the midline from both sides, and stop at the fourth segment. Anal plate is relatively long, up to 0.8 to 0.9 times of the abdomen length, which can reach the inner edge of the sixth abdominal segment.

Male Adults The body is white to light yellow, with brown spots. Head and antennae are light-yellow; there are three red ocelli. Pronotum is dark brown, there is a yellowish white spot in the middle. The scutellum is light-yellow, there are two light yellow white longitudinal lines in the middle. The legs are yellowish white. The dorsal part of the abdomen is light-yellow, while the ventral part of the abdomen is dark brown.

Biology and behavior *Quadrastichus erythrinae* occurs 9 to 10 generations per year in Shenzhen, Guangdong province with overlapping generations. Adults can mate shortly after their eclosion. Females deposit eggs into the epidermal tissue of the new leaves, leaf petioles, and young tender shoots. The larvae feed on the epidermal tissue after hatch, and form wasp galls. The majority of the wasp galls have only one larva per gall, a few galls have two larvae per gall. Larvae complete development and pupate inside the galls. Adults emerge from emergence holes. The life cycle is short with a generation per month. The insect has high reproductive capacity, and once the trees are infested, the damage can soon be spread to the whole tree.

Prevention and treatment

1. Quarantine. Prohibition of transporting the infested nursery materials to the non-infested regions.

2. Manual removal of the overwintering wasp galls can be effective. The overwintering galls on tree branches can be removed and burned to eliminate the sources of *Q. erythrinae*.

3. Utilization of adult phototropism. Light traps can be placed at the scale of 1 hm^2 per lamp to trap and reduce adult population.

4. Chemical control. The spraying of 10% imidacloprid at 3,000×, thiamethoxam at 4,000×, and 5% emamectin benzoate at 4,000× during larval stage can be effective.

References Huang Fengying et al. 2005; Yang Weidong et al. 2005; Du Yuzhou et al. 2006; Chen Xiaojun et al. 2010; Forest Disease and Pest Control Station of the Ministry of Forestry 2010.

(Xiong Huilong, Li Zhenyu)

Q. erythrinae damage (Li Zhenyu)

Q. erythrinae adult male (Li Zhenyu)

Q. erythrinae adult female (Li Zhenyu)

546 *Aiolomorphus rhopaloides* Walker

Classification Hymenoptera, Eurytomidae

Distribution Jiangsu, Zhejiang, Anhui, Fujian, Jiangxi, Hubei, and Hunan; Japan.

Hosts The bamboo *Phyllostachys heteroclada* var. *pubescens*, denudata *Sinarundinaria nitida*, the Pleioblastus *Pleioblastus amarus*, the yellow bamboo *Phyllostachys sulphurea*, and water bamboo *Phyllostachys conesta*.

Damage The larvae feed on base of leaf buds, causing abnormal growth of petioles. At high population density, they cause bent branch tips, falling leaves, death of bamboo plants, and reduced bamboo quality. The emergence of the bamboo shoots in the following year is also reduced.

Morphological characteristics **Adult** The body length 7.50 to 8.52 mm, black and shiny, scattered with grey-yellowish white long hairs. Head is slightly wider than the thorax. Mandible and labial palpi are reddish-brown. The compound eyes are black. Ocelli are arranged in an obtuse triangle and dark brown. Antennae are long, with 11 segments, flagellum shaped, attached to the middle of the frons; the scape, pedicel, and the club are reddish brown. The thorax is robust, the notum is covered with dense sculpture. Prothorax is large, the width is 1.5 times of its length. The longitudinal groove on the mesonotum is obvious; propodeum is flat, with a longitudinal groove. Wings are transparent, light yellowish brown, and pteralia and veins are reddish brown. The stigma vein on the forewings is about half of the length of costa vein, the subcostal vein is shorter than the costa vein, which is 1.6 to 1.7× of the stigma vein. Ventral side of the adult body is orange in color.

Eggs Elongaged oval, 0.5-0.6 mm in length, 0.13-0.18 mm in width; one end sharper than the other end. Milky white at beginning, yellowish before hatching.

Larvae The body length of newly hatched larvae is 0.8 to 1.0 mm, and milky white. Larvae have 5 instars. The head capsule width of each instar is 0.04, 0.17, 0.29, 0.42, and 0.54 mm, respectively. Mature larva is 7 to 9 mm in length, milky white in color, and covered with short dark brown hair, and mouthparts are dark brown.

Pupae Length is 7.5 to 9.5 mm. New pupa is milky white. The head, thorax and dorsal surface abdomen are black before emergence.

Biology and behavior *Aiolomorphus rhopaloides* has one generation per year in Zhejiang province.

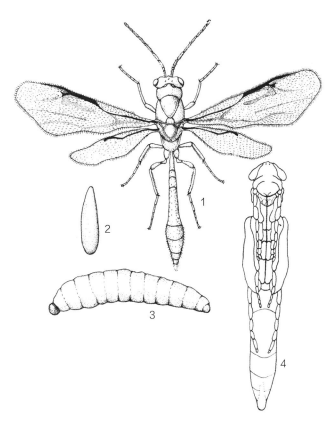

A. rhopaloides.
1. Adult, 2. Egg, 3, larva, 4. Pupa (Li Yougong)

A. rhopaloides adults (Xu Tiansen)

A. rhopaloides damaged bamboo shoot (Xu Tiansen)

A. rhopaloides damaged bamboo stem (Xu Tiansen)

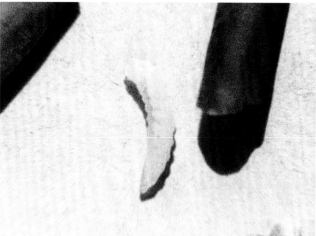
A. rhopaloides larva (Xu Tiansen)

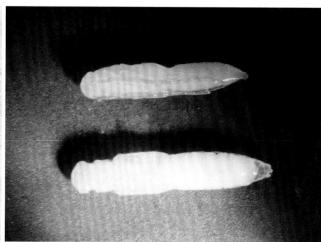
A. rhopaloides pupae (Xu Tiansen)

It overwinters in pupal stage. Adult emergence starts in mid-February of the following year, reaches peak in mid-March, and finishes in mid-April. After eclosion, adults rest inside the galls. In mid- and late March, when the average daily temperature is above 10℃, adults begin to emerge from the galls. At the end of March and early April, most of the wasps emerge from the galls. Adult emergence from the galls ends in early May. The egg stage occurs between the end of March and early May. Larvae occur initially in early April, and peaks in late April to early May. Larvae mature and pupate in early and mid-September, then they overwinter.

Adults oviposit at the base of the leaf buds when the old leaves fall off and new buds sprout and enlarge. Female adults deposit 1 to 3 eggs per bud, and ultimately one larva survives in each petiole. The oviposition sites are concentrated, which can result in almost all of the buds being infested on a newly grown small shoot.

Prevention and treatment

Protection and preservation of natural enemies can be an effective management strategy for this wasp. The most common parasitoid wasp is *Diomorus aiolomorphi*. There are also other parasitoid wasps have been found from the bamboo galls, e.g., *Ormyrus punctiger*, *Norbanus aiolomorphi*, *Megastigmus sinensis*, *Homoporus japonicus*, and *Eupelmus urozonus*.

References Fujian College of Forestry Forest Protection Research Group 1977, Liu Yongzheng 1979, Xu Tiansen and Wang Haojie 2004.

(Xu Tiansen, Wang Haojie, Li Yougong, Liu Yongzheng)

547 *Eurytoma laricis* Yano

Classification Hymenoptera, Eurytomidae

Distribution Shanxi, Inner Mongolia, Liaoning, Jilin, Heilongjiang; Russia, Mongolia, and Japan.

Hosts *Larix gmelini*, *L. kaempferi*, *L. olgensis*, and *L. princips-rupprechtii*.

Morphological characteristics **Adults** Female length approximately 3 mm, and black without metallic hue. Compound eyes are reddish brown. Mouthparts and the end of femur, as well as tibia and tarsus of the legs are yellowish brown. Head, thorax, distal end of the abdomen, legs and antennae are covered with dense fine white hairs. Head is sphere shaped, and slightly wider than thorax. The three ocelli are arranged in an obtuse triangle. The antennae are located above the bottom of the compound eyes, with 11 segments; scape long, pedicel short, the last three segments almost fused, forming a club. Thorax length is greater than its width. Prothorax is slightly narrower than the mesothorax. Scutellum is protruded upward, and in oval shape. Forewings are longer than the abdomen, and the wing length is about 2.5 times of its width; the costa vein length is 2× of the stigma vein, the posterior costa vein is longer than the stigma vein, and the tip of the stigma vein is enlarged, and shaped like a bird head. Forewing is covered with fine hairs. The distal end of the costa vein of the hind wing has a set of three hamuli. The dorsal surface of tibia, as well as the first and the second segments of tarsus on the hind legs are covered with thick silver gray bristles. The bristles are arranged in a row on the tibia. The abdomen is apparently flattened laterally, and the abdomen length is greater than the total length of head and thorax. The fourth abdominal segment is the longest. The ovipositor sheath is protruded and with the last abdominal sections collectively form a slightly upturned plow-like protrusion. The males are slightly smaller than the females, body length is 2 mm. The shape of the antennae and abdomen of the males differs from the females. The antennae have 10 segments, club is made of two segments which are almost fused together. The first segment of the abdomen is elongated, and rod-shaped, and the rest part is near spherical.

Eggs Length about 0.1 mm, milky white, oblong in shape. There is an egg stalk that is slightly longer than

E. laricis adult
(Li Zhenyu)

1mm

E. laricis. 1. Adult, 2. Larva (Zhang Peiyi)

the egg body.

Larvae Body length is 2-3 mm; white, maggot-like, in "C" shape, without legs; the head is extremely small; mandibles are well developed, their tips are reddish brown.

Pupae The pupa is 2-3 mm long, milky white; the compound eyes are red. The pupae become black before adult emergence.

Biology and behavior *Eurytoma laricis* finishes a generation in 1-3 years. It overwinters in seeds as mature larvae. In early May of the following year, the larvae pupate. Those larvae with continuous diapause do not pupate and maintain their larval stage in seeds and overwinter in the second and the third winter. Adult emergence begins in early June, and peaks in mid- and late June. They oviposit on young and tender pine cones, and the larvae hatch in early July, bore into the fruits to feed on seeds.

Each seed has only one larva, and each larva only damages one seed. The damaged seeds do not show any visual signs until appearance of emergence holes. Emergence holes are round with a neat edge. Adults can emerge at any time of a day, but mostly occur in the mornings. Most of the adults can chew off the seed coat and emerge, but a small number of adults are unable to chew off the seed coat, and die in the seeds. More males are found at the early emergence, while more females are found later. Under the laboratory conditions and provided with sugar solution, the life span of the females is 17 to 34 days, while the males is 14 to 18 days.

Field survey showed that in Daxinganling forest region, the trees at the middle of the hills showed more damage than the trees on either top or base of the hills, sunny side suffers more damage than the shady side, and mature trees have more damage than the young trees.

References Gao Buqu et al. 1983, Liao Dingxi et al. 1987.

(Gao Buqu)

548 *Eurytoma maslovskii* Nikolskaya

Classification Hymenoptera, Eurytomidae

Importance *Eurytoma maslovskii* is an important pest on peach and apricot fruits. It causes fruit drop and damaged seeds. The average damage rate on hickory was 92.8%, the highest damage rate was 100%. On apricot, average infestation rate was 49.3%, highest damage rate was 98.7% in a single plant. The larvae can be transported over long distances by seed transportation. The peach seed wasp has been an added as the new forest quarantine insect pest for Hebei and Shandong Provinces.

Distribution Beijing, Tianjin, Hebei, Shanxi, Inner Mongolia, Liaoning, Henan, Shandong; Russia, Korea Peninsula, Japan, and India.

Hosts *Prunus armeniaca*, *P. armeniaca* var. *ansu*, *P. persica*, *P. davidiana*, *P. davidiana* var. *rubra*, *P. davidiana* var. *alba*, *P. mume*.

Damage The adult ovipositor can penetrate through the fruit exocarp, the nut, and seed coat. Gum flow occurs at the oviposition site of the peach fruit, but not on apricot fruits. The oviposition site leaves brown dots on the milky white kernel husk. 60.9% of the oviposited fruits fall off shortly after the oviposition (before larvae hatch). Even if the infested fruits do not fall, the seeds would be consumed by the larvae, and the fruits shrink and become greyish black.

Morphological characteristics **Adults** Females and males have dimorphism. Females 4.83-8.01 mm (average 7.1 mm) long. Black; both basal and distal ends of femur and tibia of front and middle legs, and the end of femur, and both ends of tibia, and tarsus of the hind legs are yellow to brown. Head and thorax are covered with coarse sculpture. Wings transparent. There is a thick vein near the anterior border. End of femoa, tibiae and tarsi yellowish brown. Tarsi 5-segmented, with two claws. End of abdomen with long hairs. There are 8 visible abdominal segments.

Males 4.13-7.26 mm long, similar to females except the antennae an abdomen. Antennae geniculate, the 7th funicular segment raised dorsally, each segment bearing longer hairs on top and bottom than that in females; hairs at the last segment shorter.Thorax larger than abdomen, abdomen smaller than female; first abdominal segment pedicle shape, other segments semi-globulous.

Eggs Oblong, slightly curved, and 0.15 mm × 0.35 mm in size, milky white in color, and nearly transparent. The front end has a backward curved short stalk, whereas the

E. maslovskii adult (Li Zhenyu)

E. maslovskii adult (Li Zhenyu)

rear end has an elongated egg stalk with multiple curves. The length of the egg stalk is 4 to 5× of the egg length.

Larvae Length are 6 to 7 mm, milky white, fusiform-shaped, slightly flat, both ends are bent ventrally. Legless, small head is light yellow, and mostly retracted into thorax, mandibles are hard and brown in color with a pair of molars. The larval body has 13 segments, the distal segment is small, often retracted under the previous segment. Each segment has brown cone-shaped, slightly curved setae. Of these segments, thorax has 10 setae, and the remaining segments have 8 setae per segment. In addition, there are two setae between the dorsal midline to the lateral lines, and one on the lateral line of the larvae, and one between the lateral and the ventral midline. There are more setae on the last segment. Spiracles are round, yellowish brown in color, with a total of nine pairs on the second to the tenth segments of the larval body.

Pupae The body is long and similar to the adults, and slightly fusiform in shape. New pupae are milky white, and gradually become yellowish brown, turn into black before adult emergence.

Biology and behavior Majority of *Eurytoma maslovskii* occurs one generation per year in Chengde, Hebei province. About 9.9% larvae go into diapause in the second year, and need two years to complete a life cycle. The mature larvae can estivate and overwinter inside seeds. Overwintering larvae in the shady areas start to pupate in late March and early April, the pupation peaks in early April. Adult emerge between the end of April and the end of May. Females start laying eggs in early May, and reach peak oviposition in mid-May. Larvae begin to hatch and damage seeds in mid-May. Larvae gradually mature and begin estivation and overwintering in late June and early July. The insect activities in the sunny area are 5 to 7 days earlier than in the shady area.

The pupa stage lasts for an average of 28 days, emergence rate is 98.7%. After its eclosion, the adult chews a hole on the top of the endocarp of 1.5-2 mm in diameter, and crawls out of the endocarp of the fruits. The process needs an average of 3.8 days in apricot, and 7.6 days in *Prunus davidiana*. The sex ratio of adults (female:male) in peach fruits is 2:1, and in apricot is 2:3. When *Spiraea trilobata* begins flowering, the adult wasps begin to emerge, the peak begins 6 days later, the peak time is nine days after it was first seen. Life span of the adults is about 30 days, the wasps are inactive on rainy days, and are most active at noon time in the sunny and warm days. Adults are often active one hour after sunrise. They fly on the sunny side over the tree canopy, and frequently land on leaves. They search for glands on petioles and feed on gland secretion.

The majority of adults can mate and then oviposit after one to three days of excreting the milky white fluid inside its abdomen. During oviposition period, adults fly inside the canopy to search for suitable oviposition sites on young fruits. Once a site is located, the female pierces into the young fruit with its ovipositor and oviposit inside fruit tissue. It withdraws its ovipositor immediately after oviposition and crawl transversely with several circles, and then leave. When adults oviposit, the ovipositor pierces through exocarp, mesocarp, endocarp, endosperm, and deposits eggs into the seed endosperm that has not been absorbed by the cotyledons, the wasp can deposit one to several eggs per fruit. It lays an average 1.3 eggs (maximum 5 eggs) in each apricot fruit and 4.3 eggs (maximum 14 eggs) in each *Prunus davidiana* fruit. The egg distribution pattern inside the forest is aggregated. The egg distribution within a tree is also aggregated at the beginning. Later they are evenly distributed. Most of the fruits with early oviposition fall off before egg hatch, and 60.9% of the prematurely dropped fruits are infested. Eggs hatch in 10 days. Newly hatched larvae move toward the tip of the seeds in the endosperm that has not been absorbed by the cotyledons. The larval path is clearly marked by the small air bubbles.

When encountering the growing cotyledons, the larva feeds and moves toward to the center of the cotyledons. Then, when the larva reaches the middle of the cotyledons, it (bores) consumes cotyledon tissue around to make living space for itself. One larva only consumes a fruit (nut). Cannibalism has also been reported for larvae within the same fruit. Only one larva completes its larval development in a fruit. The larval stage lasts approximately 40 days. At the end of June to early July, the larva gradually consumes the fruit, and only leaves remnants of the seed coat. The damaged fruits gradually wilt and shrink, and become black and hard. At the ripening time, most of the damaged fruits gradually drop off, but a small number of the damaged shrunk fruits still hang on the branches until the harvest time or until the following spring before the new growth occurs.

Prevention and treatment

1. Performing thorough monitoring. Conducting a thorough survey between July and March to monitor the larvae inside the damaged and shriveled fruits on the trees or underneath the trees at the estivation and overwintering stages. Alternatively, the survey of the old fruit nuts on the ground for the adult emergence holes to determine the range of the wasp distribution. Also, based on the dissection of the young peach and apricot fruits, and by identifying the brown dots (oviposition sites) on the milky white exocarp of the new fruits, and existence of larva inside the nut, and the damage on the cotyledons, the infestation rate of the new fruits can be determined. Subsequently, the trees with 3.5% damaged fruits should be treated at the adult emergence time in the following year.

2. Promotion of utilizing insect resistant fruit tree cultivars. For example, a wild apricot cultivar, 'wasp resistant number 3' can reduce the wasp damage by 88.6%. The economic value is 2.65 times of the average cultivar and 20.7 times of the susceptible wild apricot cultivar.

3. A predatory mite, *Pediculoides ventricosus* can prey on the overwintering larvae, pupae and adults inside the nut (endocarp). The parasitization rate is 6.1% and can be up to 9.5%. Some of the mites can parasitize the mature larvae of the same year. A specie of *Beauveria bassiana* can infect the overwintering larvae, and the parasitization rate is up to 22.1%. There is also a reddish colored *B. bassiana* can infect the overwintering larvae, but the infection rate is not as good as the species of *B. bassiana* that is white in color.

4. Silviculture. The removal and destroy of the infested fruit nuts on the trees and on the ground can be effective in reducing the insect sources for the next growing season.

5. Reinforcing strict quarantine measures to stop its dispersal. Utilizing water immersion method for quarantine inspection and elimination of the infested fruit nuts. The ratio for the fruit nuts and liquid volume is 1:5. During the quarantine inspection, only the floating fruit nuts need to be dissected. For insect elimination, destroying of the floating nuts is sufficient. Nuts with different maturity periods should be examined in solutions with different density for accuracy. For the newly hulled wet fruit nuts, a sodium chloride (NaCl) solution with the density of 1.01 to 1.05 should be used; for elimination of the infestation, the density of 1.01 to 1.03 should be used. The dried nuts can be examined using ethanol solution with the density 0.86 to 0.82 for quarantine purpose. Using water immersion for 23 to 25 hours can eliminate the infested nuts.

6. Adult control. From the beginning to the peak of the adult emergence period, the spraying of the following insecticides can be effective: 2.5% deltamethrin at 150 to 600× with the addition of a slow-releasing material.

References Tang Guanzhong, Lu Rongyan, Guo Zhenping et al. 1999; Tang Guanzhong, Niu Jingsheng, Liu Yufen et al. 1999; Tang Guanzhong 2002; Tang Guanzhong, Peng Jinyou, Guo Zhenping et al. 2005, 2006; Guan Wenchen, Tang Guanzhong, Pen Jinyou et al. 2006; Tang Guanzhong, Yan Hailiang, Guo Zhenping et al. 2007.

(Tang Guanzhong)

549 *Eurytoma plotnikovi* Nikolskaya
Classification Hymenoptera, Eurytomidae

Distribution Hebei, Shanxi, Henan, Shaanxi; Russia, Kazakhstan, Iran, and south-western European countries.

Host *Pistacia chinensis*.

Damage *Eurytoma plotnikovi* larvae feed on seeds of the host plants and reduce seed production.

Morphological characteristics **Adults** Female length 3.0 to 4.5 mm, head, propodeum and the first segment of abdomen are black, and posterior end of the abdomen has black spots on both sides, and the rest of the abdomen is reddish brown. Legs and scape and pedicel of antennae are dark yellow, but the club is in light color. Veins are yellow. The coxae, and the end of tibia and tarsus are yellow, but the end of tarsus, claw and base of the arolium are brown, whereas distal end of the arolium is yellow. Head is oblong, slightly wider than thorax. The length of the antennae is 1.2-1.4 mm; the length of pedicel is greater than its width, but is shorter than the first funicle. The length of each funicle is greater than its width. The length of its first sub-segment is more than twice of its width. The length of the fifth flagellum sub-segment is 1.5 times of its width. Sculptures on head and thorax are not deep, which are covered with white hairs. Prothorax is transversely rectangular in shape, and the mesothorax has an apparent longitudinal groove. The scutellum is narrow in front and wide in the back, length and width are approximately equal. Abdomen is shorter than thorax, smooth, slightly flat laterally, oval shaped; promodeum is small and expanede transversely, with a thorn-like protrusions on both sides. The dorsum of the fourth abdominal segment is the longest, slightly longer than the third segment. The end of the abdomen is only slightly in pear-shaped. Ovipositor is just visible. Male body length is 2.6 to 3.3 mm, body is black. Funicles are stalk like and connected on the side way. Propodeum is long, the length is almost three times of its width. Legs are yellow. Femur of the hind legs is relatively dark. Antennae are 0.9 to 1.2 mm long.

Eggs Size 0.3 mm×0.1 mm, and milky white in color, and oblong in shape. The egg has a thread like white egg stalk. The length of the egg stalk and egg body are approximately in the same length.

Larvae The body length is 4.3-5.0 mm. Mature larvae are pointed at both ends, but wide in the middle. Head and thorax are bent ventrally. Newly hatched larvae

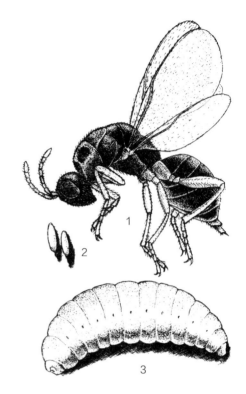

E. plotnikovi. 1. Adult 2. Eggs, 3. Larva (Zhu Xingcai)

milky white, and become yellowish white when they are mature. Head is extremely small, and sclerotized. Mandibles are well developed, sickle-shaped, and yellowish brown in color.

Pupae The size is 3.2-4.0 mm×1.2-1.6 mm. The pupae are initially white to beige. The eyes become orange to red orange in color at pre-emergence. The body is yellowish brown.

Biology and behavior *Eurytoma plotnikovi* is univoltine in Hebei and Henan Provinces. A small portion of the insects occurs one generation every two years. The insects overwinter as mature larvae inside fruits. The larvae pupate in mid-April of the following year. Pupal stage lasts 15 to 20 days. The adult eclosion starts at the end of April to the beginning of May. Adult eclosion peaks in mid- and late May. Most of the adult eclosion is found between 07:00 to 12:00, which represents 91.6% of the total number of the eclosion of the day. After eclosion, the adults chew out of the fruits. The newly emerged adults can fly away within a few seconds, and rarely crawl on fruit surface. Adults fly around the cano-

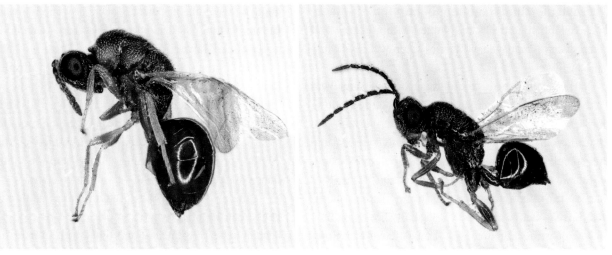

E. plotnikovi adult female (Li Zhenyu) *E. plotnikovi* adult male (Li Zhenyu)

py, mate and oviposit at day time, and rest on the abaxial side of the leaves at night. When the temperature arises to 18℃, the adults start to crawl, and at 20℃ they start to fly. The adults are inactive on windy or rainy days with low temperature. Adult mating occurs mostly in the afternoon. Before mating, the male uses its antennae to touch the female's antennae, and start mating after contacting several times. The mating lasts about 12 seconds. Adult life span is 7 to 12 days, and the longest is 18 days.

The beginning of the oviposition period is generally in early May, and ovipositon period peaks in mid- and late May, and ended in mid-June. The oviposition period usually lasts for approximately 60 days. The beginning of the oviposition time varies. If the adults emerge early and ambient temperature is low, the pre-oviposition period is then longer than usual. In contrast, when the adults emerge late and the ambient temperature is high, the pre-oviposition period is short. The shortest is over one day. Before ovipositing, the females crawl on the fruit surface, inspect the fruit surface with antennae, and select oviposition sites; about 88% of the eggs are deposited around the young fruit suture. The female uses its ovipositor to pierce through the fruit surface, and deposits its eggs inside fruit tissue. In general, one egg is deposited in each fruit. In special circumstances, up to 14 eggs are deposited per fruit. Heavy infestations cause 100% fruit drop off. The fecundity of the females is 10-31 eggs with an average of 18 eggs. The egg stage usually lasts 3 to 5 days.

After hatch, if there are multiple eggs in a fruit, the larva hatched first feed on either eggs or cannibalize on the larvae hatched later. Only one larva survives in a fruit. There are five instars in the larval stage, and the larval development can be clearly divided into three stages: slow growth phase, rapid developing phase, and resting phase. Before the enlargement of the seed embryo in the pistachio fruit, the larvae are active between the exocarp (fruit surface) and the embryo, irrespective of hatching time. Larvae feed on the inner exocarp and the spongy tissue outside of the embryo. They develop slowly and stay in the first instar, thus, it is called the slow growth phase. The period from the end of oviposition to larval boring (feeding) into the embryo lasts for 20 to 30 days. There is little damage during this phase. When the embryo starts to expand, and cotyledons start to grow in mid-July, the larvae bore through the seed surface, and feed on embryos and the developing cotyledons. They quickly molt into the 2nd instar larvae. After approximately half a month, the larvae will consume cotyledons, and develop into the fifth instar larvae. This is called the rapid development phase. The loss of cotyledons leads to yield reduction or a complete loss of the crop. After completely consuming the cotyledons in the seeds, the larvae mature and enter into their dormant stage. After September, most of the infested fruits fall to the ground, the larvae begin to overwinter.

References Liao Dingxi et al. 1987, Jin Xingrui et al. 1988.

(Jin Xingrui, Tian Shibo, Zhao Shuer)

Prevention and treatment of eurytomid wasps

1. Establishment of seed nursery. Reinforcing and strengthening management tactics to protect seed yield and quality.

2. Harvest all seeds to eliminate the overwintering insects. In the area with severe infestation, deep plowing after harvest can eliminate overwintering larvae in the infested fruits. In the years with low fruit bearing, the complete removal of the blooms can be effective to eliminate the oviposition sites of the adults. At the fruit harvest time, the removal and destroy (crushing) of infested fruits can be effective. At the early flowering time of the locust tree, the removal of locust blooms can prevent the seed wasp damage on locust seeds. In the year with low fruit production year, keep 15 to 30 new shoots per hectare and use them to attract wasp oviposition, then remove them next year (high fruit production year). Delayed planting of *Caragana* spp. (June in Inner Mongolia) can stop the dispersal of the overwintering wasps. Before planting, the scalding of locust seeds with boiling water can kill larvae.

3. Strengthening seed quarantine, and prohibiting import and export of the infested seeds. When the seed moisture content is below 10% to 20%, the seeds should be fumigated prior to be exported. When the room temperature is above 15°C, fumigate with chloropicrin at $30 g/m^3$ for 80 hours; or using sulfuryl fluoride fumigation at 40 to 50 g/m^3 for 3 to 4 days can be effective.

4. The use of dichlorvos fogging during adult emergence can be effective. The dose is 7.5 to 15 kg per hectare, repeat treatment 2 or 3 times with an interval of 5 to 7 days is effective. The ultra-low volume spraying of 5-fold diluted dichlorvos at 1.5 to 2.3 kg per hectare, or conventional spray of 25 g/L lamda-cyhalothrin at 1,500×, or 20% fenpropathrin EC at 4,000 to 5,000× is also effective.

550 *Tetramesa phyllotachitis* (Gahan)
Classification Hymenoptera, Eurytomidae

Distribution Jiangsu, Zhejiang, Anhui, Fujian, Jiangxi; Japan, U.S.A.

Hosts Straight bamboo, *Phyllostachy sulphurea* 'Viridis', *P. lofushanensis*, *P. verrucosa*, *P. aureosulcata* f. *pekinensis*, *P. glauca*, *P. meyeri*, *P. praecox*, *P. propinqua*, *P. aurita*, *P. bambusoides* f. *lacrima-deae*.

Damage *Tetramesa phyllotachitis* larvae feed inside petioles of host leaves, stimulate petiole abnormal growth, and cause swelling, thickening, and bamboo leaves fall off gradually. When larvae are mature, the wall of bamboo leaf petiole thin and brittle even though the petioles look thick and long. After eclosion, the adults chew off holes and fly out from the petioles. The leaves then wither and die. Damaged bamboo plants grow poorly, bamboo fiber is dry and brittle with reduced quality.

Morphological characteristics **Adults** Male length 7.0 mm, female length 7.6 to 8.2 mm; black, shiny, with scattered yellowish white long hairs. Head width is greater than length. Antennae are located in the middle of the frons, filliform, long, which is the same as the total length of head and thorax. The scape and pedicel segments are mainly yellow to reddish brown. Compound eyes are protruded prominently, with silverfish grey to yellowish black in color, and smooth and hairless. The ocelli are arranged in an obtuse triangle. Prothorax is large, its width is 1.5 times of its length. The front of the longitudinal groove on the mesonotum is apparent, and disappears posteriorly. Wings are transparent, veins are light-yellowish brown, covered with brown hairs. The costal vein is thick, the length is 2.3 times of the pterastigma.

T. phyllotachitis larva (Xu Tiansen)

T. phyllotachitis eclosion holes (Xu Tiansen)

T. phyllotachitis adults (Xu Tiansen)

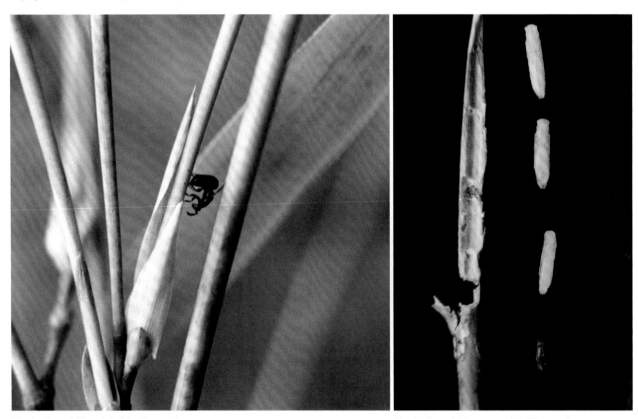

T. phyllotachitis adult (Xu Tiansen)

T. phyllotachitis pupae (Xu Tiansen)

Larvae The newly hatched larval body length is 1.5 mm, milky white, body segments are diffused. The mature larva body length is 7.5 mm, and creamy yellow in color.

Pupae Length is 5.2-6.2 mm, milky white. Head transversely wide. The compound eyes are prominent, and protruded. The middle groove of mesonotum is deep. Abdomen is 1.6× longer than head and thorax. Tarsi of the hind legs located at the end of the fourth abdominal segment of a pupa.

Biology and behavior In Hangzhou of Zhejiang Province, there is one generation per year. It overwinters in pupa stage. Adults emerge at the end of March to early April. They oviposit in petioles of the new tender leaves. Larvae appear in early May. They feed on the inner wall of bamboo petioles, which stimulates the growth and thickening of the petioles. A leaf petiole can contain three to five larvae, each in a separate chamber. The mature larvae pupate and overwinter in late October.

Prevention and treatment

Protection of natural enemies. Several parasitoid wasps have been reported, including *Ormyrus punctiger*, *Norbanus aiolomorphi*, and *Homoporus japonicus*.

Reference Xu Tiansen and Wang Haojie 2004.

(Xu Tiansen Wang Haojie)

551 *Solenopsis invicta* Buren

Classification Hymenoptera, Formicidae
Common name Red imported fire ant

Importance *Solenopsis invicta* is an introduced pest originated from South America. It is a dangerous pest of human health and environment safety. Their stings cause pain and blister. They may cause allergic reactions and even death in rare cases. In Euclytus forests and crop fields, it disrupts the native ecosystem and causes significant difficulties in management.

Distribution Fujian, Hunan, Guangdong, Guangxi, Hong Kong, Macau, Taiwan; U.S.A., Brazil, Paraguay, Argentina, Australia, New Zealand, Malaysia.

Hosts *Solenopsis invicta* feeds on many plants seeds, fruits, roots, young buds, stems. The hosts include *Prunus persica*, *Lichi chinensis*, *Arachis hypogaea*, *Euclyptus* spp., *Helianthus annuus*, *Citrus reticulata*, *Zea mays*, *Glycine max*, *Vitis vinifera*. *S. invicta* is a predator of pests such as ticks, chigger mites, pests Pyralid pests on *Saccharum officinarum*, corn earworm. *S. invicta* may also predate upon natural enemies, compete with other predatory ants, and change the ecosystem composistion.

Damage *Solenopsis invicta* is polyphagous. They feed on arthropods, vertebrates, invertebrates, and many plants.

Morphological characteristics **Workers** The size varies significantly. Body length 2.0-7.2 mm, head length 0.9-1.5 mm, head width 0.7-1.32 mm. Head, thorax, legs, and antennae brownish red. Abdomen black or near black. Head somewhat rectangular. Eyes black. Antennae 10-segmented, the last two segments forming a club. Anterior border of clypeus with a median tooth. Anterior part of the pronotum raised. The suture between mesonotum and propodeum is distinct. There are two nodes; the 1st one is triangular on side view, sharp, higher than the 2nd nodes; the 2nd one is round conical, shiny on top. Abdomen oval, with a sting.

Alates Length 8.1-9.0 mm, larger than workers; females larger than males. A pterostigma present at the middle of the front margin of the front wing; a pterostigma also present at middle of hind wing and 2/3 of the front margin.

Eggs Length 0.24 -0.3 mm (average 0.26 mm), width 0.16-0.24 mm (average 0.20 mm), elliptical, creamy white.

Larvae Asexuals 1.7-2.8 mm long (average 2.16 mm), head width 1.0-1.6 mm (average 1.29 mm). Sexuals 4.0-5.0 mm long (average 4.48 mm), head width 1.5-3.5 mm (average 2.85 mm). White, grub-like.

Pupae Asexuals 2.7-3.5 mm long (average 3.22 mm), sexuals 4.9-6.1 mm long (average 5.19 mm). Creamy white, later turns to yellowish brown. Males turn to black before adults emerge. Females and worker pupae become brown or reddish brown.

Biology and behavior The development period is 7-14 days (average 9.6 days) for eggs, 6-15 days for larvae (average 10.6 days), 8-13 days (average 11.4 days) for worker pupae, 12-16 days (average 13.6 days) for female pupae, 11-15 days (average 12.5 days) for male pupae. Males die soon after swarming and mating. Females shed their wings after mating and search for a suitable place to dig a nest and lay eggs.

Solenopsis invicta builds mounds of up to 60 cm in diameter and height. They often appear in sunny places.

S. invicta workers (Zhang Runzhi)

S. invicta nest (Zhang Runzhi)

S. invicta workers on nest (Zhang Runzhi)

There are no openings on the top of the mounds. Before swarming, workers open a hole on the mound. Each nest contains one or more queens. Mature colonies may contain hundreds of alates and 100,000 to 500,000 workers. They swarm at the end of spring and early summer. Mating occurs at 90-300 m height. Males die soon after mating. Females fly up to 3-5 km away and dig nests. Females lay eggs within 24 hours after mating. They first lay 10-15 eggs. A queen in a mature colony produces 1,500-5,000 eggs per day. Queens live for 6-7 years; workers live for 1-6 months.

Prevention and treatment

1. Physical control. Pour 4-8 L of boiling water to each nest. Treat 3-4 times at 5-10 days intervals.

2. Chemical control. 1) Apply baits containing hydramethylnon, fipronil, or methoprene. Mix active ingredients with vegetable oil, biscuits, and bread crumbs. Broadcast baits in areas with high *S. invicta* density. Individual mounds may be treated by applying 5-20 g bait around each mound. After 7-10 days, spray a suitable insecticide to each mound. Do not apply bait indoors to avoid attracting ants into rooms. 2) Trenching. Pour 5-25 L bifenthrin, cypermethrin, fipronil, or chlorpyrifos insecticides from top of mound or 30 cm away from perimeter of mound.

3. Medical treatment after being stung by *S. invicta*. Rinse the stung sites with water or soap water. Raise the stung site. Apply anti-itch or anti-inflammatory cream to relieve symptoms. Place ice pack at the injured site. Avoid scratching the skin. If the victim is allergic to fire ant stings, the victim needs to be transported to a hospital immediately. Educate children not to touch *S. invicta* mounds or play around *S. invicta* mounds. Wear rubber gloves and long rubber boots; apply Vaseline to boots when working in infested areas.

References Liu Dong 2006; Lü Lihua, He Yurong, Liu Jie et al. 2006, Zhang Runzhi 2007; Chen Jing 2008.

(Gu Maobin, Li Zhenyu)

References

Abdulla Akbar, Abduwahap Aziz, Adil Sattar, Wang Yan, 2017. The tropism of *Carpomya vesuviana* Costa to different yellow sticky boards [J]. Journal of Xinjiang Agricultural University, 40(4): 283-287.

Adili Shateer, He Shanyong, Tian Chengming et al, 2008. Occurrence and spatial distribution of *Carpomya vesuviana* Costa in Turufan region of Xinjiang [J]. Plant Quarantine, 22(5): 295-297.

Ai Xiulian, Ji Qing, Longtao et al, 1995. Function of the initiation gene of *Agrotis segetum* granulosis virus [J]. Xinjiang Agricultural Science, (4): 161-164.

Ai Yuxiu, 1987. Study on the bionomics of *Clanis bilineata* (Walker) [J]. Journal of Guizhou Agricultural College, (1): 93-96.

Akanbl, M. O, 1971. The biology, ecology and control of Phalanta phalantha Drury (Lepidoptera: Nymphalidae), a defoliator of Populus spp. in Nigeria [J]. Bulletin of Entomological Society of Nigeria, 3(1): 19-26.

An Baiguo, Wang Xinan, Duan Chunhua et al, 2007. Field evaluation of 10% avermectin and tebufenozide mixture against *Dendrolimus spectabilis* [J]. Shandong Forest Science and Technology, 2: 69-70.

An Ruijun, Li Xiuhui, and Zhang Dongmei, 2005. Bionomics of *Ambrostoma quadriimpressum* Motschulsky [J]. Forest Science and Technology, 30(5): 18-20.

An Wenyi, 2007. Field evaluation of 1.2% "Kushen"fogging against *Dendrolimus tabulaeformis* [J]. Shaanxi Forest Science and Technology, (3): 92-93.

Ao Xianbin et al, 1980. Biology and natural enemies of *Dryocosmus kuriphilus* Yasumatsu [J]. Fruit Tree Science and Technology, (4): 17-29.

Appleby, J. E, 1999. The pine shoot beetle and the Asian longhorned beetle, two new exotic pests New and re-emerging pests [J]. Symposium, Saint-Jean-sur-Richelieu, Quebec, Canade, Phytoprotection 80: 97-101.

Atwals, A. S, 1976. Agricultural pests of Indian and South-East Asia [J]. Kalyani publishers, 231-232.

Bai Jintao, 1986. Preliminary study on *Argyresthia sabinae* Moriuli [J]. Shandong Forestry Science and Technology, (4): 35-36.

Bai Jintao, 1990. Preliminary study of the bionomics of *Ophrida xanthospilota* (Baly) [J]. Forest Pest and Disease, (2): 5-6.

Bai Jiuwei, Zhao Jianxia, and Ma Wenliang, 1980. Bionomics of *Anarsia lineatella* Zeller [J]. Scientia Silvae Sinicae, S1: 127-129, plates IV-V.

Bai Shuilong, 1960. Colored pictorial guide to Taiwan butterflies [M]. Baoyushe.

Balock, J. W. and T. T. Koyuma. Notes on the biology and ecnmonic importance of the mango weevil Sterno chetus mangiferae (Fabricius) (Coleoptera: Curculionidae), in Hawaii [J]. Proc Hawaiian Entomol Soc., 18: 353-364.

Bao Jianzhong and Gu Dexiang, 1998. China biological control [M]. Taiyuan: Shanxi Science and Technology Publishing House.

Bao Qimin, Bao Wenbin, and Jin Jueyuan, 2000. Bionomics of *Micadina zhejiangensis* Chen et He [J]. Journal of Zhejiang Forest College, 17(2): 166-169.

Bao Qimin, Jin Jueyuan, and Bao Wenbin, 2000. Bionoics and control of *Sinophasma crassum* Chen et He [J]. Forest Pest and Disease, (3): 15-17.

Beijing Agricultural University, 1983. Fruit tree entomology [M]. Beijing: Agriculture Publishing House.

Beijing Forest College, 1979. Forest entomology [M]. Beijing: China Forestry Publishing House.

Beijing Forest College, 1980. Forest entomology [M]. Beijing: China Forestry Publishing House.

Beijing Forest College, 1985. Forest entomology [M]. Beijing: China Forestry Publishing House.

Bi Xianghong, 1989. Control threshold for *Ambrostoma quadriimpressum* [J]. Forest Science and Technology, (3): 24.

Bi Xianzhang, 2002. Bionomics and control of *Actias selene ningpoana* Felder on Cornus officinalis Sieb. et Zucc [J]. Anhui Forestry, (1): 19.

Bose, P. K., Y. S. Sankranarayanan, and S. C. Gupya, 1963. Chemistry of lac [J]. Indian Lac Research Institute, Ranchi. pp.1-68.

Bu Wanggui, Yin Chenglong, and Zhang Yaorong, 1996. Biological characteristics and control of *Melolontha hippocastani mongolica* [J]. Journal of Gansu Forestry Science and

Technology, (2): 28-33.

Bu Wangui, 2000. Bionomics of *Zeiraphera griseana* (Hübner) [J]. Gansu Forest Science and Technology, 25(1): 36, 51.

Bureau of Animal and Plant Inspection and Quarantine of the People's Republic of China, 1997. Quarantine pests of imported plants of China [M]. Beijing: China Agriculture Publishing House.

Cai Banghua and Chen Ningsheng, 1963. New termites from south China [J]. Acta Entomologica Sinica, 12(2): 167-198.

Cai Banghua and Li Zhaolin, 1959. Preliminary study on the fauna of bark beetles in northern China. pp. 73-117. In Collections of entomological papers [M]. Beijing: Science Publishing House.

Cai Banghua et al, 1946. Study on Chinese gall (2)- observations on mutualistic galls [J]. Guangxi Agriculture, (6): 45-47.

Cai Banghua, 1963. Study on *Serica orientalis* Motschulsky [J]. Acta Entomologica Sinica, (4): 490-505.

Cai Rongquan, 1979. Economic insect fauna of China Fasc. 16. Lepidoptera: Notodontidae [M]. Beijing: Science Publishing House.

Cao Chengyi, 1981. *Sinoxylon japonicum* Lesne - a new record in China [J]. Acta Entomologica Sinica, 3(2): 118.

Cao Kecheng, 1966. *Lampra limbata* Gebler [J]. Shanxi Agriculture, (7): 32.

Cao Xiuyun and Liu Yuxiang, 2009. Study on the *Adoxophyes orana orana* control techniques [J]. Anhui Agricultural Science, 37(8): 3500-3501.

Cao Zigang, Sun Shumei, and Gu Changhe, 1978. Preliminary observation on *Zeuzera leuconotum* Butler [J]. Entomological Knowledge, (4): 120-121.

Capital Reforestation Commission, 2000. Diseases, pests, rodents of urban trees [M]. Beijing: China Forestry Publishing House.

Chai Ximin and Jiang Ping, 2003. Occurrence and control of pine wilt disease [M]. Beijing: China Agriculture Publishing House.

Chai Ximin, Bao Qimin, and Cheng Jingfu, 1997. Moths that infest pine cones and shoots in Zhejiang [J]. Zhejiang Forestry Science and Technology, 17(2): 1-11.

Chai Xiushan and Liang Shangxing, 1990. Bionomics and control of *Pissodes* sp [J]. Entomological Knowledge, 27: 352-354.

Chai Xiushan and Liang Shangxing, 1992. Bionomics of *Pissodes punctatus* Langor et Zhang [J]. Western Forestry Science, (2): 44-45.

Chan Hanlin and Huang Shuisheng, 1993. Biological characteristics of *Meganoton analis* (Felder) [J]. Forest Pest and Disease, (4): 20-21.

Chang Baoshan, Liu Suicun, Zhao Xiaomei et al, 2001. A study on the occurrence of red burpentine beetle [J]. Shanxi Forestry Science and Technology, (4): 1-4.

Chang Runlei and Zhou Xudong, 2010. Research status on *Leptocybe invasa* Fisher & La Salle in foreign countries [J]. China Forest Pest and Disease, 29(1): 22-25.

Chao Jun, Zhan Liming, Lu Jin et al, 2007. Bionomics and control of *Parametriotes theae* Kuznetzov [J]. Jiangxi Plant Protection, 30(3): 119-120.

Chen Bilian, Sun Xingquan, Li Huiping et al, 2006. Biological characteristics and control of *Actias selene ningpoana* Felder [J]. Journal of Shanghai Jiaotong University (Agricultural Science), 24(4): 389-391.

Chen Chang, 1999a. Biological characterisitcs of the great black cutworm (*Agrotis tokionis* Butler) [J]. Jilin Animal Husbandry and Veterinary Medicine, (8): 37.

Chen Chang, 1999b. Biological characteristics and control of the great black cutworm (*Agrotis tokionis* Butler) [J]. Agricultural Science and Technology, (4): 31.

Chen Erhou, 1999. Checklist of parasitoids of pine caterpillar and *Lemyra phasma* (Leech) and description of some species [J]. Yunnan forest Science and Technology, 88(3): 48-51.

Chen Fangjie, 1937. Four new species of scale insects [J]. Insects and Plant Diseases, (5): 18-19.

Chen Guofa, Li Tao, Sheng Maoling et al, 2011. Studies on the sex pheromone of *Orgia ericae* [J]. Forestry Science and Technology Development, 25(1): 73-76.

Chen Guofa, Zhang Qinghe, Wang Yanjun et al, 2009. Aggregation pheromone of bark beetle *Ips duplicatus* [J]. Journal of Northeast Forestry University, 37(7): 96-98.

Chen Guofa, Zhou Shuzhi, Zhang Zhen et al, 2003. Application technique of *Diprion jingyuanensis* Xiao et Zhang sex pheromone traps [J]. Journal of Northeast Forestry University, 4(7): 16-17.

Chen Haishuo, Ma Xiaoqi, Nie Hongshan et al, 2009. Occurrence and control of *Diprion nanhuaensis* Xiao [J]. Henan Forestry Science and Technology, 29(2): 17-19.

Chen Hanlin and Huang Shuisheng, 1993. Studies on *Striglfna bifida* Chu et Wang [J]. Entomological Knowledge, (6): 339-341.

Chen Hanlin, Dong Liyun, Zhou Chuanliang et al, 2002. Bionomics of *Rhodoneura sphoraria* (Swinhoe) [J]. Forest Pest and Disease, 21(4): 9-11.

Chen Hanlin, Gao Zhanggui, Zhou Jianmin et al, 2005. Study on the biology of *Diaphania perspectalis* (Walker) [J]. Jiangxi Plant Protection, 28(1): 1-4.

Chen Hanlin, Liu Jinli, Wang Jiashun et al, 1999. Bionomics of *Caliroa annulipes* (Klug) [J]. Forest Pest and Disease, (1): 6-8.

Chen Hanlin, Zhao Renyou, Jin Genming et al, 1997. Study on the bionomics of *Cephalcia tianmua* Maa (Hymenoptera, Symphyta, Pamphiliidae) [J]. Scientia Silvae Sinicae, 33(3): 279-282.

Chen Hanlin, 1980. A preliminary report on the control of *Capnodium theae* Hara using omethoate band on tree trunks [J]. Zhejiang Forestry Science and Technology, (4): 2-4.

Chen Hanlin, 1980. Biological characteristics of *Pantana phyllostachysae* Chao [J]. Zhejiang Forestry Science and Technology, (3): 17-20.

Chen Hanlin, 1982. Bionomics and control of *Atysa cinnamomi* [J]. Lishui Region Science and Scientia Silvae SinicaertTechnology, (2): 59-65.

Chen Hanlin, 1983. Control of *Pantana phyllostachysae* Chao [J]. Lishui Forestry Science and Technology, (2): 54-58.

Chen Hanlin, 1988. Study on *Locastra muscosalis* Walker [J]. Zhejiang Forest Diseases and Pests, (1): 5-9.

Chen Hanlin, 1990. A new pest of magnolia - *Meganoton analis* (Felder) [J]. Forest Pest and Disease, (2): 21-22.

Chen Hanlin, 1990. Identification of *Chilocorus rufitarsus* Motschulsky [J]. Zhejiang Forestry Science and Technology, 10(6): 48-49.

Chen Hanlin, 1994. Brief notes on Thyrididae (Lepidoptera) from Zhejiang province [J]. Entomological Journal of East China, 3(1): 21-24.

Chen Hanlin, 1995. A new pest of *Magnolia officinalis* Rehd. et Wils. - *Asterococcus muratae* Kuwana [J]. Forest Research 8(special issue), 136-138.

Chen Hanlin, 1995. Occurrence and control of *Locastra muscosalis* Walker [J]. Plant Protection, 21(4): 24-26.

Chen Hanlin, 1995. Study on *Megabeleses liriodendrovorax* Xiao [J]. Journal of Zhejiang Forestry Science and Technology, 15(2): 49-52.

Chen Hanlin, 1996. Infectivity of *Steinernema carpocapsae* Beijing strain against 3 pest insects of pine [J]. Chinese Journal of Biological Control, 12(2): 96-97.

Chen Heming and Qi Runshen, 1992. A study on *Cydia trasias* (Meyrick) [J]. Plant Protection, (3): 8-10.

Chen Jiaoda, Hu Zhonglang, Yang Pengju, Ji Xiangzhou, 1983. Preliminary observations on *Agrilus ratundicollis* Saunders [J]. Forest Pest and Disease, (2): 8-10, 20.

Chen Jingyuan and Wu Gaoyun, 1988. A new pest of *Metasequoia glyptostroboides - Ascotis selenaria dianaria* Hübner [J]. Forest Pest and Disease, (1): 28.

Chen Jinxiu and Fang Minggang, 2001. Effect of natural enemies on *Kermes nawae* Kuwana and their utilization [J]. Anhui Agricultural Science, 29(1): 68-71.

Chen Jiuchun, 2006. Preliminary study on *Eutomostethus nigritus* Xiao [J]. Anhui Agricultural Science Bulletin, (3): 12-14.

Chen Jun and Cheng Huizhen, 1997. Occurrence and control of *Sinoxylon japonicum* Lesne (Coleoptera: Bostrichidae) [J]. Entomological Knowledge, 34(1): 20-21.

Chen Minghong, Chen Youqing, Zhang Hong et al, 2009. Culture of lac insects and lac production [M]. Beijing: China Forestry Publishing House.

Chen Peichang and Chen Shuchun, 1997. Checklist and keys to genus *Sinophasma* Gunter [J]. Forest Pest and Disease, (2): 12-14.

Chen Peichang and Chen Shuchun, 1997b. Identification, biology, and control of major phasmid insects in China [J]. Jurnal of Beijing Forestry University, 19(9): 70-75.

Chen Peichang, Chen Shuchun, Wang Jijian et al, 1998. Food consumption and control of *Sinophasma largum* Chen et Chen [J]. Forest Pest and Disease, (4): 10-12.

Chen Peng, Liu Hongping, Lu Nan et al, 2006. Risk analysis of *Cossus chinensis* Rothschild in Yunnan province [J]. Western Forest Science, 35(2): 121-123.

Chen Qinhua, Tian Liming, Lin Shucai et al, 2008. Bionomics of larch cone flies in Qiqihar area [J]. Protection Forestry Science and Technology, (1): 20-22.

Chen Shijun, Li Wenxiu, Wei Yanchun et al, 2006. Occurrence of *Zeuzera leuconotum* Butler on *Lonicera japonica* and its control techniques [J]. China Plant Protection, 26(5): 30-31.

Chen Shixiang et al, 1959. Economic insect fauna of China Fasc. 1. Coleoptera: Cerambycidae [M]. Beijing: Science Publishing House.

Chen Shixiang, 1978. *Atysa cinnamomi* - a new Galerucine beetle injurous to camphor trees in Fujkien [J]. Acta Entomologica Sinica, 21(1): 55-56.

Chen Shixiang, 1986. Fauna of China Fasc. 2: Insecta, Coleoptera, Hispidae [M]. Beijing: Science Publishing House.

Chen Shuchun and He Yunheng, 1985. Species and distribution of the genus *Sinoplasma* (Phasmatodea: Heteronemiidae) in China [J]. Journal of Beijing Forest College, (3): 33-38.

Chen Shuchun and He Yunheng, 2008. Phasmida of China [M]. Beijing: China Forestry Publishing House.

Chen Shuchun and Wang Jijian, 1993. A new species of Macellina Uvarov from Guangxi (Phasmida: Heteronemiidae) [J]. Acta Entomologica Sinica, 36(4): 328-330.

Chen Shuchun et al, 1994. Bionomics of *Baculum minutidentatum* Chen et He [J]. Jilin Forestry Science and Technology, (2): 21-23.

Chen Shuchun, 1986. An important pest, *Sinophasma brevipenne* Günther [J]. Forest Pest and Disease, (2): 37.

Chen Shuchun, 1994. Current research on stick insects in China and future recommendations [J]. Forest Pest and Disease, (3): 38-40.

Chen Shuliang, Li Xianchen, Xu Yanqiang et al, 1993. Studies on *Dendrolimus suffuscus* Lajonquiere [J]. Shandong Forestry Science and Technology, (2): 38-42.

Chen Shunli, Li Yougong, and Huang Changxiao, 1989. A study on castor tussock moth *Porthesia scintillans* (Walker) [J]. Journal of Fujian Forest College, (9): 1-9.

Chen Shunli, Li Yougong, and Li Qinzhou, 1989. Preliminary study on *Thalassodes quadraria* Guenée [J]. Forest Pest and Disease, 3: 14-15.

Chen Shunli, Wu Fuhua, and Hou Qinwen, 2004. Bionomics of *Hemiberlesia pitysophila* Takagi [J]. Fujian Forest Science and Technology, 31(2): 1-4.

Chen Suwei, Chi Renping, Xu Heguang et al, 1999. Control of pine caterpillars with mixture of cytoplasmic polyhedrosis virus and Bt [J]. Shandong Forestry Science and Technology, (6): 31-34.

Chen Weimin, 2004. Occurrence and control of a new pest of *Myrica rubra* - *Aleurotrachelus camelliae* Kuwana [J]. China Fruit News, 20(8): 46.

Chen Weimin, 2006. A pest of *Myrica rubra* - *Aleurotrachelus camelliae* Kuwana [J]. Zhong Zhi Yuan Di, (7): 13.

Chen Weimin, 2008. Occurrence and control of a new pest of *Myrica rubra* - *Aleurotrachelus camelliae* Kuwana [J]. China Fruit News, 25(4): 25-26.

Chen Wenkui, 1985. Dynamic changes of some internal organs and physiological conditions during the summer-diapause-period of the great black cutworm (*Agrotis tokionis* Butler) [J]. Journal of Nanking Agricultural University, (2): 47-58.

Chen Xian, Chen Jiwen, Wang Churong et al, 2004. A preliminary study on the effect of 16% ChongXianQing against *Brontispa longissima* [J]. Guangdong Agricultural Science, 2: 37-39.

Chen Xiaoda, Dang Xinde, Li Feng et al, 1990. Bionomics and control of *Curculio* sp. [J]. Forest Pest and Disease, (4): 1-2.

Chen Xiaoda, 1989. A study of the distribution of cossids from Shaanxi province and the biology of *Holcocerus artemisiae* Chou et Hua [J]. Shaanxi Forest Science and Technology, (4): 71-73.

Chen Xiaojun, Xu Hanhong, Yang Yizhong et al, 2010. Infestation, spread, and control of erythrina gall wasp [J]. Forest Pest and Disease, (4): 28-32.

Chen Xiaojun, Yu Daojian, Jiao Yi et al, 2005. Occurrence, damage and quarantine of the erythrina gall wasp [J]. Plant Protection, 31(6): 36-38.

Chen Xiaoming and Feng Ying, 1991. Discussion on the natural mortality and its mechanism of lac insect (*Kerria lacca*) [J]. Forest Research, 4(5): 582-584.

Chen Xiaoming, Chen Yong, Ye Shoude et al, 1997. Distribution patterns of *Ericerus pela* on host plants [J]. Forest Research, 10(4): 415-419.

Chen Xiaoming, Wang zili, Chen Yong et al, 2007a. The impact of environmental factors on the wax secretion by Chinese white wax scale (*Ericeus pela* Chavannes) [J]. Acta Ecologica Sinica, 27(1): 103-112.

Chen Xiaoming, Wang zili, Chen Yong et al, 2007b. The main climate factors affecting was excretion of *Ericerus pela* Chavannes (Homoptera: Coccidae) and an analysis of its ecological adaptability [J]. Acta Entomologica Sinica, 50(2): 136-143.

Chen Xiaoming, Wang Zili, Chen Yong, 2008. Age-specific life table of Chinese wax scale (*Ericerus pela*) natural population and analysis of death key factors [J]. Scientia Silvae Sinicae, 44(9): 87-94.

Chen Xiaoming, 1998. The resources conservation of lac insects *Kerria* and its utilization [J]. Chinese Biodiversity, 6(4): 287-290.

Chen Xiaoming, 2005. Diversity of lac insects [M]. Kunming: Yunnan Science and Technology Publishing House.

Chen Yiqun, Huang Honghui, Lin Mingguang et al. 2004. Occurrence and control of *Brontispa longissima* in Hainan province [J]. Plant Quarantine, 18(5): 280-281.

Chen Yongge and Gu Dexiang, 1995. Study on the spatial distribution pattern of *Hemiberlesia pitysophila* Takagi [J]. Acta Scientiarum Naturalium Universitatis Sunyaseni (Natural Science edition), 34(4): 122-124.

Chen Yongge and Gu Dexiang, 1997. Population dynamics of Bionomics of *Hemiberlesia pitysophila* Takagi [J]. Ecological Science, 16(2): 18-20.

Chen Yongling, 1978. Controlling *Polychrosis cunninghamiacola* Liu et Pai with smoking agent [J]. Forest Science and Technology, (8): 18-19.

Chen Yu, 1958. Observation of *Chrysomela populi* Linnaeus in Xinjiang [J]. Entomological Knowledge, 4(5): 218-219.

Chen Yuanqing, 1984. Changes in the names of *Cryptorhynchus magniferae* and *C. gravis* [J]. Plant Quarantine, (6): 3-4

Chen Yuanqing, 1990. Keys to common genera and species of Attelabidae in China [J]. Forest Pest and Disease, (2): 12-14.

Chen Zhiqing, 1978. Studies of *Clostera anachoreta* (F.) and *Neocerura wisei* (Swinhoe) than infest *Homalium hainanense* [J].

Tropical Forestry Science and Technology, (1): 12.

Chen Zhiqing and Chen Peizhen, 1989. Study on the control of *Hemiberlesia pitysophila* Takagi by pruning and thinning [J]. Forest Research, (4): 388-394.

Chen Zhiqing and Wu Shixiong, 1978. Preliminary study on *Hyblaea puera* Cramer [J]. Tropical Forest Science and Technology, (3): 19-22.

Chen Zhiqing and Wu Shixiong, 1982. Bionomics of *Stauropus alternus* Walker [J]. Acta Entomologica Sinica, 25(3): 342-344.

Chen Zhiqing and Wu Shixiong, 1984. Preliminary observation on *Hyblaea puera* Cramer [J]. Entomological Knowledge, 21(4): 161-163.

Chen Zhiqing, 1973. Biology and control of a psylid infesting *Homalium hainanense* Gagnep [J]. Tropical Forestry Science and Technology, (3): 18-25.

Chen Zhiqing, 1977. A new pest of *Casuarina equisetifolia* - *Stauropus alternus* Walker [J]. Entomological Knowledge, 14(3): 91-92.

Chen Zhong and Xusu, 1987. A study on the control of *Curculio davidi* Fairmaire [J]. Plant Protection, 5: 13-15.

Chen Zongping and Xu Guangyu, 2008. Effectiveness of deltamethrin against *Dendrolimus punctata punctata* and other pine shoot pests [J]. Agricultural Technology Service, 25(10): 68-69.

Cheng Guifang and Yang Jikun, 1997a. A new quarantine pest discovered in Beijing - *Opogona sacchari* (Bojer) [J]. Plant Quarantine, 11(2): 95-101.

Cheng Guifang and Yang Jikun, 1997b. A new pest of *Draceana fragrans* - *Opogona sacchari* (Bojer) [J]. Plant Protection, (1): 33-34.

Cheng Guifang, Lu Qi, and Yang Jikun, 1998. A study on the factors associated with the *Opogona sacchari* (Bojer) outbreaks and management strategies [J]. Plant Quarantine, 2: 95-97.

Cheng Guifang, 1997. Occurrence of *Opogona sacchari* in China [J]. Plant Protection, 23(6): 46.

Cheng Hong, Yan Shanchun, Sui Xiang et al, 2006. Relationship between tannin content of main poplar strains and damage of *Xylotrechus rusticus* (L.) in Heilongjiang province [J]. Journal of Northeast Forestry University, 34(2): 31-33.

Cheng Lansheng and Chu Yuejin, 2002. Damage of *Acantholyda posticalis* and its control [J]. Anhui Forestry, (3): 17.

Cheng Liang, 1976. Preliminary study on *Lemyra phasma* (Leech) (Lepidoptera: Arctiidae) [J]. Acta Entomologica Sinica, 19(4): 410-416.

Cheng Lichao, 2007. Effect of poplar bark volatiles on *Xylotrechus rusticus* (L.) adults [M]. Northeast Forestry University Master thesis.

Chi Defu, Sun Fan, Qiao Runxi et al, 1999. Chemical control of *Strobilomyia* spp [J]. Journal of Northeast Forestry University, 27(1): 28-31.

Chi Xingzhen, Wang Jijian, Chen Peichang et al, 1998. Biology of *Sinophasma pseudomirabile* Chen et Chen [J]. Forest Pest and Disease, 2: 27-29.

China Forest Seed Company, 1988. Handbook of forest seed and fruit pest control [M]. Beijing: China Forestry Publishing House.

China Ministry of Forestry, 1996. Quarantine forest plants of China [M]. Beijing: China Forestry Publishing House.

China Real Estate Management Association - Termite Control Committee, 2008. Integrated termite management training manual [M]. Nanjing: Nanjing University Publishing House.

Chinese Academy of Agriculture - Citrus Research Institute, 1985. Pictorial guide to citrus tree diseases and pests [M]. Chengdu: Sichuan Science and Technology Publishing House.

Chinese Academy of Agriculture - Fruit Tree Research Institute, 1994. Guide to fruit tree diseases and pests (2nd edition) [M]. Beijing: China Agriculture Publishing House.

Chinese Academy of Agriculture and Forestry - Institute of Science and Technology Information, 1974. Review of forestry industry in foreign countries [M]. Beijing: Science Publishing House.

Chinese Academy of Forestry, 1983. China forest insects [M]. Beijing: China Forestry Publishing House.

Chinese Academy of Sciences - Institute of Zoology, 1987. Yunnan forest insects [M]. Kunming: Yunnan Science and Technology Publishing House.

Chinese Academy of Sciences - Institute of Zoology, 1977. Preliminary study on sex attractant of *Polychrosis cunninghamiacola* Liu et Pa [J]. Forest Science and Technology, (3): 16-17.

Chinese Academy of Sciences - Institute of Zoology, 1978. Pictorial guide to insect natural enemies [M]. Beijing: Science Publishing House.

Chinese Academy of Sciences - Institute of Zoology, 1981. Illustrated guide to Chinese moths (I) [M]. Beijing: Science Publishing House.

Chinese Academy of Sciences - Institute of Zoology, 1981. Illustrated guide to Chinese moths (II) [M]. Beijing: Science Publishing House.

Chinese Academy of Sciences - Institute of Zoology, 1981. Illustrated guide to Chinese moths (III) [M]. Beijing: Science Publishing House.

Chinese Academy of Sciences - Institute of Zoology, 1981. Illustrated guide to Chinese moths (IV) [M]. Beijing: Science Publishing House.

Chinese Academy of Sciences - Institute of Zoology, 1986. Agricultural insects of China (1) [M]. Beijing: Agriculture Publishing House.

Chinese Academy of Sciences - Institute of Zoology, 1987. Agricultural insects of China (1) [M]. Beijing: Agriculture Publishing House.

Chong Xiulin, Fan Li, 2001. Infestation and control of *Gastrolina depressa* [J]. Journal of Jiangsu Forestry Science and Technology, 28(2): 39.

College of Central South Forest College, 1987. Economic forest entomology [M]. Beijing: China Forestry Publishing House.

Communist Labor University of Jiangxi, 1975. A preliminary study on *Polychrosis cunninghamiacola* Liu et Pai [J]. Entomological Knowledge, 12(3): 41-44.

Cui Guangcheng, 1987. Mole crickets found in Tibet [J]. Entomological Knowledge, 24(1): 46.

Cui Jingyue, Li Guangwu, and Li Zhongxiu, 1996. Control of subterranean pests [M]. Beijing: Jindun Publishing House.

Cui Lin and Liu Yuesheng, 2005. The occurrence of *Thosea sinensis* (Walker) in tea plantation and its control [J]. China Tea, 27(2): 21.

Cui Wei and Gao Baojia, 1995. Major scale insects on trees of economic importance in north China and their control [M]. Beijing: China Forestry Publishing House.

Dai Xiangguang, 1987. Studies on the bionomics of *Macrotermes bayneyi* Light [J]. Scientia Silvae Sinicae, 23(4): 498-502.

Dang Guojun, 2007. Occurrence of *Mallambyx raddei* Blessig and control strategies [J]. Jilin Forestry Science and Technology, 36(6): 28-35.

Dang Xinde, 1979. Description of parasitoids of Shaanxi forest pests [J]. Shaanxi Forest Science and Technology, (4): 57-68.

Dang Xinde, 1982. Description of chalcidoid parasitoids of Shaanxi forest pests (2) [J]. Shaanxi Forest Science and Technology, (3): 75-77, 74.

Deng Guofan, Liu Youqiao, and Sui Jingzhi, 1983. Beijing: China agricultural insects [M]. Agriculture Publishing House.

Deng Xiuming, Zhu Quanmin, Su Shichun et al, 2002. Spatial distribution of larvae of Epiplema moza Butler [J]. Fujian Forestry Science and Technology, 29(2): 17-20.

Deng Xun, Ma Xiaoqian, Wei Xia et al, 2009. Evaluation of three plant derived insecticides against *Pissodes validirostris* [J]. Forest Science and Technology, 34(1): 26-27.

Deng Yinwei, Li Xiaotie, Zhou Haiping et al, 2006. Application of integrated disease and pest management for *Ginko biloba* trees [J]. Forest Science and Technology Development, (1): 63-66.

Diakonoff, A, 1973. The south asiatic Olethreutini (Tortricidae) [J]. Zoologische monographieen van het Rijksmuseum van Natuurlijke Historie, no. 1.

Ding Bi, Huang Jinshui, Huang Haiqing et al, 1996. Study on asexual propagation of *Casuarina* spp. [J]. Protection Forest Science and Technology, (1): 4-7.

Ding Dongsun and Shi Mingqing, 1997. Differences between *Semanotus bifasciatus* (Motschulsky) and *Semanotus sinoauster* (Gresit) [J]. Plant Quarantine, (5): 293-296.

Ding Dongsun, Zhong Longyun, Chen Hua et al, 2006. Biology of *Dictyoploca japonica* and least toxic control methods [J]. Jiangxi Forestry Science and Technology, (4): 21-22.

Ding Fuzhang, Zhang Zehua, Zhang Lisheng et al, 2006. Control effect of *Metarhizium anisopliae* on red palm weevil [J]. Journal of Southwest Agricultural University (Natural Science edition), 28(3): 454-456.

Ding Jitong, Adil Sattar, Cheng Xiao-tian, Huang Xue-yuan, Amina, 2014. Supercooling points and freezing points of *Carpomya vesuviana* Costa [J]. Acta Agriculturae Boreali-occidentalis Sinica, 23(11): 163-167.

Ding Qianping, 2004. Integrated pest management strategies for managing *Diorhabda elongata deserticola* Chen [J]. Journal of Gansu Forestry Science and Technology, (3): 57-58.

Ding Shaojiang, Li Chaoxu, Liang Minguo et al, 2007. Biological control of *Brontispa longissima* by releasing Tetrastichus brontispae in urban zones of Shenzhen [J]. China Journal of Biological Control, 23(4): 306-309.

Dong Cunyu, 1997. Bionomics of *Phloeosinus aubei* [J]. Jiangxu Forestry Science and Technology, 24(4): 34-35.

Dong Jiantang, 1983. Bionomics of *Agrotis segetum* [J]. Entomological Knowledge, (1): 14-17.

Dong Yancai and Zhu Xinbo, 1990. Bionomics and occurrence of *Ancylis mitterbacheriana* (Denis et Schiffermüller) [J]. Entomological Knowledge, 27(3): 153-154.

Du Guoyu and Gao Limei, 2003. Damage by peach fruit borer and its integrated control [J]. Disease and Pest Control, (6): 31.

Du Juan, Zhang Chaoju, Zhao Wende et al, 2003. Bionomics of *Acantholyda flavomarginata* Maa and control experiments [J]. Sichuan Forestry Science and Technology, 24(4): 42-45.

Du Liangxiu and Du Chengjin, 1999. Study on *Pandemis heparana* (Denis et Schiffermüller) [J]. Forest Science and Technology Development, (5): 35-36.

Du Pin, Ren Fang, and Mei Liru, 1999. Bionomics and control of *Podagricomela shirahatai* (Chǔjǒ) [J]. Entomological

Knowledge, 36(6): 335-337.

Du Xiangge and Zhang Youyan, 2003. Occurrence and control of *Proagopertha lucidula* in cherry orchard [J]. China Fruit Trees, (3): 26-28.

Du Yuzhou, Guo Jianbo, Zheng Fushan et al, 2006. Predicting likely areas of the erythrina gall wasp, *Quadrastichus erythrinae* in China [J]. Plant Protection, 32(1): 63-66.

Duan Xiaohong, 2006. Occurrence and integrated management of *Dendrolimus yunnanensis* in Lincang City [J]. Lincang Science and Technology, (2): 40-42.

Duan Zhaoyao and Lei Guilin, 1998. A preliminary study on the infestation characteristics of *Pissodes punctatus* Langor et Zhang [J]. Yunnan Forestry Science and Technology, (3): 81-85.

Eaton, C. B. and R. R. Lara, 1967. Red turpentine beetle Dendroctonus valens LeConte, in A. G. Davidson and R. M. Prentice (eds.). Important Forest Insects of Mutual Concern to Canada, the United States and Mexico [M]. Canada Department of Forestry and Rural Development Pub. 1180. Ottawa, 248.

Editorial Board of Hebei Forest Insects, 1985. Hebei Forest Insects [M]. Shijiazhuang: Hebei Science and Technology Publishing House.

Editorial Board of the Chinese Agricultural Diseases and Pests, 1981. Chinese agricultural diseases and pests (2) [M]. Beijing: Agriculture Publishing House.

Editorial board of the handbook of forest protection, 1971. Handbook of forest protection [M]. Beijing: Agriculture Publishing House.

Editorial committee of the Atlas of Shandong Forest Pests. Forest Insects of Shandong province [M]. Beijing: China Forestry Publishing House.

Fan Di, 1985. Preliminary studies on *Phloeosinus aubei* Perris [J]. Forest Pest and Disease, (2): 14-15.

Fan Di, 1995. Research advances on bag worm, *Eumeta preyeri* (Leech) [J]. Shandong Forestry Science and Technology, (5): 50-53.

Fan Jun, Huangxia, Zhang Yawen et al, 2007. Infestations of *Acantholyda posticalis* Matsumura in Luonan County and its control strategies [J]. Shaanxi Forest Science and Technology, (1): 34-35, 37.

Fan Lihua, Xie Yingping, Yuan Guisheng et al, 2005. Study on the control of *Diprion jingyuanensis* Xiao et Zhang with Beauveria bassiana [J]. Forest Pest and Disease, 24(4): 29-32.

Fan Meizhen, Guo Chao, Feng Wei et al, 1988. A study on the *Erannis ankeraria* nuclear polyhedrosis visus [J]. Shaanxi Forest Science and Technology, (1): 42-45.

Fan Min, Xu Weiyu, Guan Liqin et al, 2006. Study on occurrence and control of *Orthaga achatina* Bulter and *Moricella rufonota* Rohwer [J]. Acta Agriculture Shanghai, 22(3): 51-54.

Fan Renjun, Dong Jinming, Cao Man et al, 1994. Occurrence and control of *Asias halodentri* Pallas in *Hippophae rhammoides* [J]. Hippophae, 7(1): 24-27.

Fan Zhongmin, 1962. Preliminary study on *Cinara pinea* Mordusiko [J]. Scientia Silvae Sinicae, 7(4): 312-313.

Fan Zide, 1988. Economic insect fauna of China Fasc 37. Diptera: Anthomyiidae [M]. Beijing: Science Publishing House.

Fan Zide, 1992. Keys to the common flies in China (2nd edition) [M]. Beijing: Beijing Science and Technology Publishing House.

Fang Chenglai, 2000. Fauna of China Fasc [M]. 19: Insecta, Lepidoptera, Arctiidae. Beijing: Science Publishing House.

Fang Deqi and Chen Shuliang, 1982. Identification of four species of Cossidae (1) [J]. Entomological Knowledge, 19(2): 30-32.

Fang Deqi and Chen Shuliang, 1986. A survey of the species and distribution of carpenter moth [J]. Shandong Forestry Science and Technology, (4): 44-46.

Fang Deqi and Chen Shuliang, 1986. Integrated pest management of carpenter moth [J]. Forest Pest and Disease, (4): 41-44, 48.

Fang Deqi and Chen Shuliang, 1987. Bionomics of ealm carpenter moth [J]. Scientia Silvae Sinicae, (Entomology edition): 72-77.

Fang Deqi, Chen Shuliang, and Li Xianchen, 1992. Advances in Cossidae research in China [J]. Shaanxi Forestry Science and Technology, (2): 29-35.

Fang Deqi, Chen Shuliang, Jiang Sandeng et al, 1984. A study on the biological characteristics of *Cossus cossus orientalis* (Gaede) [J]. Entomological Knowledge, 21(6): 257-259.

Fang Deqi, 1980. Biological characteristics of *Dendrolimus suffuscus* Lajonquiere [J]. Entomological Knowledge, 17(5): 211-212.

Fang Fang and Liu Jianfeng, 2008. Preliminary report on the bionomics and control of *Cydia trasias* (Meyrick) (Lepidoptera: Tortricidae) [J]. Modern Agricultural Science and Technology, (22): 111.

Fang Honglian and Cao Yuping, 2006. Bionomics of *Phalera bucephala* (Linnaeus) [J]. Xinjiang Agricultural Science and Technology, (2): 27.

Fang Jie, Cui Yongsan, Zhao Boguang et al, 2007. Research advances of *Clostera anastomosis* in China [J]. Forest Pest and Disease, 26(2): 28-31.

Fang Junren, 1999. Chrysomelidae in Shanxi province [M]. Beijing: China Forestry Publishing House.

Fang Sanyang, 1982. A study on *Pineus cembrae pinikoreanus* Zhang et Fang [J]. Journal of Northeast Forest University, (supplement): 133-138.

Fang Sanyang, 1982. Study on the relationship between light intensity and population density of Adelgids in Shaoshinganling [J]. Acta Ecologica Sinica, (2): 147-150.

Fang Wencheng, 1988. Biology and control of *Diprion nanhuaensis* Xiao [J]. Journal of Southwest Forest College, 8(2): 196-202.

Fang Yan, 2000. On parasitic enemy of *Yponomeuta padella* in Shenyang region I: Ichneumodidae [J]. Journal of Liaoning Forestry Science and Technology, (6): 5, 19.

Fang Zhigang and Wu Hong, 2001. A checklist of insects from Zhejiang [M]. Beijing: China Forestry Publishing House.

FAO, 2009. Global review of forest pests and Diseases [J]. FAO Forestry Paper 156: 1-222.

Feng Huiling and Xie Haibiao, 2003. Damage of *Ericeia inangulata* (Guenee) and its control [J]. Plant Protection, 29(4): 55.

Feng Shiming and Situ Yingxian, 2004. Occurrence of *Pissodes punctatus* Langor et Zhang in Yunnan and its quarantine techniques [J]. Forest Pest and Disease, 23(1): 29-31.

Feng Shiming, Zeng Shusheng, Yang Lingxuan et al, 1999. A primary study on the *Sphecia siningensis* (Lepidoptera: Aegeriidae) [J]. Journal of Southwest Forestry College, 19(4): 231-234.

Feng Ying, Chen Xoaoming, Chen Yong et al, 2001. Studies on the nutritive and health care value of white wax scale (*Ericerus pela* Chavannes) and evaluation of its functional factors [J]. Forest Research, 154 (3): 322-327.

Feng Yuyuan, Ren Jinlong, and Yang Lin, 2002. Bionomics of *Dendrolimus punctata wenshanensis* Tsai et Liu and its control using fogging technique [J]. Yunnan forest Science and Technology, 2: 56-58.

Flecher, B, 1914. Some south indian insects and other aninals of inportance [J]. Madras: Printed by superintendent government press, 341.

Forest Bureau of Taojiang County, 1979. Bionomics and control of *Polychrosis cunninghamiacola* Liu et Pai [J]. Hunan Forest Science and Technology, (1): 19-21.

Forest Disease and Pest Control Station of Boli County of Heilongjiang, 1974. Occurrence of *Ptycholomoides aeriferanus* (Herrich-Schäffer) and its control [J]. Entomological Knowledge, 11(1): 38-40.

Forest Disease and Pest Control Station of the Ministry of Forestry, 2008. Chinese forestry pests overview [M]. Beijing: China Forestry Publishing House.

Forest Disease and Pest Control Station of the Ministry of Forestry, 2010. Forest pest control (1). pp. 222-223 [M]. Beijing: China Forestry Publishing House.

Forest Disease and Pest Control Station of the Ministry of Forestry, 2005. Forest quarantine pests and operation procedures [M]. Beijing: China Forestry Publishing House.

Forest Disease and Pest Control Station of the Ministry of Forestry, 2008. General introduction of forest pests in China [M]. Beijing: China Forestry Publishing House.

Forest Disease and Pest Prevention and Treatment and Quarantine Station of Henan Province, 2005. Control techniques for pests in Henan [M]. Zhengzhou: Henan Huanghe Shuili Publishing House.

Forest Industry College of Nanjing, 1975. A preliminary study on *Polychrosis cunninghamiacola* Liu et Pai [J]. Southern Forest Science and Technology, (3): 21-26.

Forest Industry Research Institute of the Chinese Academy of Forestry and Agriculture, 1978. Influence of height growth of Chinese fir after tip damage by *Polychrosis cunninghamiacola* Liu et Pai [J]. Forest Science and Technology, (3): 16-18.

Forest Protection Division of the Ministry of Forestry, 1988. Checklist of forest diseases and pests (1).

Forest Protection Station of Jiangxi, 1979. Preliminary study on *Pantana phyllostachysae* Chao [J]. Forest Pest and Disease, (3): 5-8.

Forest Research Academy of Hebei, 1959. Control of *Chrysomela populi* Linnaeus [J]. Forest Science and Technology Reports, (6): 5.

Forest Research Institue of Shaanxi Province, 1977. Pictorial guide to Shaanxi forest diseases and pests (vol. 1) [C]. Xian: Shaanxi Science and Technology Publishing House.

Forest Research Institue of Shaanxi Province, 1984. Pictorial guide to Shaanxi forest diseases and pests (vol. 2) [M]. Xian: Shaanxi Science and Technology Publishing House.

Forest Research Institute of the Chinese Academy of Forestry, 1978. Control of *Polychrosis cuninghamiocola* Liu et Pai (Lepidoptera: Tortricidae) using *Trichogramma dendrolimi* Matsumura [J]. Scientia Silvae Sinicae, (4): 42-45.

Forestry Bureau of Guizhou Province, 1987. Forest diseases and pests in Guizhou [M]. Guizhou: Guizhou People's Publishing House.

Forestry Bureau of Henan Province, 1988. Henan forest insects [M]. Beijing: China Agricultural Science Publishing House.

Fu Haibin, Li Junhuan, Jiang Li et al, 2007. An investigation of pests found in world horticulture exhibit in Shenyang [M].

Journal of Liaoning Forestry Science and Technology.

Fujian College of Forestry Forest Protection Research Group, 1977. Wasps infesting bamboo trees [J]. Scientia Silvae Sinicae, (4): 56-62.

Fujian Forest Research Institute, 1991. Fujian Forest Insects [M]. Beijing: China Agriculture Science and Technology Publishing House.

Fuping, Yang Lixin, Gong Xiaoling et al, 2001. Bionomics of *Lampra limbata* Gebler [J]. Forest Pest and Disease, s1: 22.

Gangolly, S. R, 1957. The mango. New Delhi: S. N. Guna Ray Press Ltd., 495.

Gan Jiasheng, 1982. Preliminary study on *Poecilocoris latus* Dallas and a parasitoid in Scelionidae [J]. Western Forest Science, (1): 61-63.

Gao Buqu, Xu Shanbin, Zhang Rong et al, 1983. Biology and ecology of *Eurytoma laricis* Yano (Hymenoptera: Eurytomidae) [J]. Journal of Northeast Forest College, 11(3): 51-56.

Gao Changqi, Ren Xiaoguang, Wang Dongsheng et al, 1998. Population dynamics of *Ips subelongatus* Motschulsky and its prediction [J]. Journal of Northeast Forestry University, 26(1): 24-28.

Gao Changqi, Wang Zhiming, and Yu Enyu, 1993. Studies on the artificial rearing of *Arma chinensis* (Fallou) [J]. Jilin Forestry Science and Technology, (2): 16.

Gao Changqi, 1987. Recommendations on integrated pest management of *Ambrostoma quadriimpressum* Motschulsky [J]. Jilin Forest Science and Technology, (4): 29-30.

Gao Desan and Zhang Yiyong, 1998. Observations on the bionomics of *Pyrosis eximia* Oberthür [J]. Journal of Shenyang Agricultural University, (2): 123-126.

Gao Huanting and Zhang Guolong, 2007. Bionomics and integrated management of *Agrilus zanthoxylumi* Li [J]. Shaanxi Agricultural Science, (2): 183-184.

Gao Jingshun, 1995. Damage of *Icerya purchasi* to *Acacia mearnsii* and its control [J]. Economic Forest Researches, 14(supplement): 46, 50.

Gao Rong, Di Xudong, Yang Zhende et al, 2004. Bioactivity of *Podophyllotoxin* and its analogues against the 3rd instar larvae of *Clostera anastomosis* [J]. Chinese Journal of Pesticides, 43(9): 424-426.

Gao Ruitong and Qin Xixiang, 1983. Preliminary study on *Streltzoviella insularis* (Staudinger) [J]. Forest Pest and Disease, (1): 3-6.

Gao Ruitong, Zhao Tonghai, Liu Houping et al, 2004. Investigation of the distribution and infestation of *Agrilus marcopoli* Obenberger in China [J]. Transactions of China Pulp and Paper, (supplement): 363-365.

Gao Zhaoning, 1999. Illustrated guide of Ningxia Agricultural Insects (vol. 3) [M]. Beijing: China Agriculture Publishing House.

Gates, J. F. G, 1958. Catalogue of the type specimens of microlepidoptera in the British museum (Natural History) described by Edward Meyrick [J]. Vol. III. 495, pl. 246, figs 1-16, 2-26.

Ge Siqin, Yang Xingke, Wang Shuyong et al, 2003. Revision on the three subspecies of *Gastrolina depressa* (Coleoptera: Chrysomelidae: Chrysomelinae) [J]. Acta Entomologica Sinica, 46(4): 512-518.

Gemeno, C., A. F. Lutfallah, and K. Haynes, 2000. Pheromone blend variation and cross-attraction among populations of the black cutworm moth (Lepidoptera: Noctuidae) [J]. Annals of the Entomological Society of America 93(6): 1322-1328.

Geng Yilong, Xu Heguang, Li Dongjun et al, 1998. Bionomics and control techniques of *Camptoloma interiorata* (Walker) [J]. Shandong Forest Science and Technology, (5): 31-33.

Gong Mingqin, Bu Maobin, Chen Peizhen et al, 2007. In Diseases and pests of euclyptus in China [M]. Guangzhou: Guangdong Science and Technology Publishing House.

Gu Maobin and Chen Peizhen, 1987. A preliminary study of *Phalanta phalanta phalanta* Drury [J]. Scientia Silvae Sinicae, 23(1): 105-108.

Gu Maobin, 1983. Biology and control of *Catopsilia pomona* Fabricius [J]. Acta Entomologica Sinica, 26(2): 172-178.

Gu Maobin, 1987. Population dynamics *Catopsilia pomona* Fabricius under various environments [J]. Tropical Forest Science and Technology, (4): 68-70.

Gu Ping, Zhou Lingqin, Xu Zhong, 2004. Biological characteristics and control of *Periphyllus koelreuteria* (Takahashi) in Shanghai district [J]. Journal of Shanghai Jiaotong University (Agricultural Science), 24(4): 389-391.

Guan Wenchen, Tang Guanzhong, Peng Jinyou et al, 2006. Eurytoma maslovskii *Nikolskaya quarantine* and control techniques [J]. Hebei Forest Science and Technology, (6): 1-3.

Guan Yongqiang and Zhang Yan, 2001. Control of *Cinara* sp [J]. Inner Mongolia Forestry, (7): 29.

Guan Yongqiang, 1992. Infestation and control of *Diaspidiotus slavonicus* (Green) [J]. Inner Mongolia Forestry, (7): 26-27.

Gui Bingzhong and Dai Mingguo, 2007. Bionomics and control of *Hartigia viator* (Smith) [J]. Proceedings of the 8th Annual Meeting of the Entomological Society of China. pp. 551-552.

Guizhou Forest Disease and Pest Survey Office, 1982. Checklist of the diseases and pests of major tree species in Guizhou.

Gulibahar Memet et al, 2007. Control methods for *Anarsia lineatella* Zeller [J]. Plant Protection, (2): 35.

Guo Congjian, Liang Zhenhua, Yang Yuzhen et al, 1965. A study on the control of *Curculio davidi* Faust [J]. Scientia Silvae Sinica, 10(3): 239-246.

Guo Congjian, Shao Liangyu, and Yin Wanzhen, 1992. Bionomics and control of *Omphisa plagialis* Wileman [J]. Scientia Silvae Sinicae, (3): 213-219.

Guo Xiuzhi, Deng Zhigang, and Mao Hongjie, 2009. Behavior and control of *Agrotis ypsilon* (Rott.) [J]. Jilin Forest Science and Technology, 38(4): 54.

Guo Yuliang, 2005. Bionomics of *Rondotia menciana* Mooreand and field application of its sex pheromone [D]. Shandong Agricultural University Master thesis.

Guo Zaibin, Zhao Aiguo, Li Xifu et al, 2000. Bionomics and control of *Cinara tujafilina* (Del Guercio) [J]. Henan Forestry Science and Technology, (3): 16-17.

Guo Zhonghua, Yu Suying, and Zhang Jiping, 2000. Cutting to restore growth of shrubs on sand and prevent infestation from borers [J]. Shaanxi Forest Science and Technology, (4): 37-39.

Guy, S. and A. K. Marshall, 1935. New Indian Curculionidae (Col.) [J]. Indian Forest Record 1: 263-281.

Haack, R. A., Jendek, E. Liu, H. et al, 2002. The emerald ash borer: a new exotic pest in North America [J]. Newsletter of the Michigan Entomological Society 47: 1-5.

Haack, R. A., R. L. Kenneth, and V. C. Mastro, 1997. New York's battle with the Asian long-horned beetle [J]. Journal of Forestry 95: 11-15.

Han Pinghe, 2008. Occurrence, infestation, and quarantine of *Meichuo cihuai* Yang [J]. Hebei Forestry Science and Technology, (3): 67.

Han Xuejian, 2005. Damage of *Aromia bungii* Faldermann and its control [J]. Plant Doctor, 18(5): 18.

Hansen, J. D., J. W. Armstrong, and S. A. Brown, 1989. The distribution and biologic observations of mango weevil Crytorrhynchus mangiferae (F.) (Cole.: Gurulionidae), in Hawaii [J]. Proceedings of the Hawaii Entomological Society 29:21-39.

Hao Yushan, 2003. The effect of ultra-low volume aerial spray with 1.8% AV bacteriophage on *Coleophora dahurica* [J]. Journal of Northeast Forestry University, 31(2): 25-27.

He Dahan and Zhang Kezhi, 2000. Moving sand management and insect diversity [M]. Xian: Shaanxi Normal University Publishing House.

He Dahan, 2004. Management of moving sand and pests [M]. Beijing: Science Publishing House.

He Decheng, Jiangheng, Tang Xianchun et al, 2006. Preliminary study of using biological agent to control *Cerura menciana* Moore [J]. Xinjiang Agricultural Science, 43(3): 224-227.

He Jinyi and Xu Guangyu, 2009. Effectiveness of dimilin III against *Dendrolimus punctata* and its impact on parasitoids [J]. Agricultural Technology Service, 26(6): 64-65.

He Junhua and Chen Xuexin, 1990. Studies on the species of *Aleiodes* Wesmael parasitic on the forest insect pests in China [J]. Acta Zoologica Sinica, 15(2): 201-208.

He Junhua and Wang Aijing, 1998. *Aspidocolpus erythrogaster* Tobias, a new recorded genus and species of Braconidae (Hymenoptera) from China [J]. Journal of Xinjiang Agricultural University, 21(2): 153-154.

He Junhua et al, 2000. Fauna of China Fasc. 18: Insecta, Hymenoptera, Braconidae (1) [M]. Beijing: Science Publishing House.

He Junhua et al, 2004. Parasitic wasps in Zhejiang [M]. Beijing: Science Publishing House.

He Junhua, 1984. A new pest of *Cryptomeria fortunei* - *Megastigmus eryptomeriae* Yano [J]. Forest Pest and Disease, (4): 11-13.

He Mei and Liu Zhuangjun, 2000. *Agrotis segetum* granulosis virus genome [J]. Acta Biochimica et Biophysica Sinica, (1): 93.

He Shanyong, Wen Junbao, Adil Satar, Tian Chengming, 2010. Research progress of quarantine pest *Carpomya vesuviana* [J]. Scientia Silvae Sinicae, 46(7): 147-154.

He Shiyuan, 1984. Flora of Beijing (1) [M]. Beijing: Beijing Publishing House.

He Wei, Ren Lanhua, Chao Guode et al, 2003. Dominant cricket species and their occurrence in southwestern Shandong farm lands [J]. Plant Protection Technology and Extension, (6): 13-15.

He Zhikun and Wang Aijing, 1999. Preliminary study on the pest resistance of white elm and its insect resistant materials [J]. Xinjiang Agricultural Science, (6): 283-284.

Hill, D, 1975. Agricultural insect pest of tropics and their control [J]. Cambridge university press, 389-390.

Hou Deheng, 2003. Study on biological properties and prevention techniques of Northern China's larch casebearer [J]. Sci-Tech Information Development & Economy, 13(6): 98-99.

Hou Qichang and Yang Youqian, 1998. Study on the biological characteristics and control measures of *Urostylis yangi* Maal [J]. Acta Agriculturae Universitatis Henanensis, 32(3): 268-270, 292.

Hou Taoqian, 1987. China pine caterpillars [M]. Beijing: Science Publishing House.

Hou Wuwei, Ma Youfei, Gao Weiceng et al, 1994. Phototaxis

of *Carposina niponensis* Walsingham [J]. Acta Entomologica Sinica, 37(2): 165-170.

Hou Yaqin, Nan Nan, and Li Zhenyu, 2009. Advances on *Lymantria dispar* research [J]. Hebei Fruit Tree Research, 24(4): 439-444.

Hou Yaqin, Wang Xiaojun, Li Jinyu et al, 2009. Bionomics of *Rhynchaenus empopulifolis* Chen [J]. Forest Pest and Disease, 28(2): 32-34.

Hu Changxiao, 2003. Occurrence and advances in control of *Semanotus sinoauster* Gressitt (Coleoptera: Cerambycidae) [J]. Plant Protection Technology and Extension, 23(1): 39-41.

Hu Chaoxiao, Ding Yonghui, and Sun Ke, 2007. Adcances on the study of *Aromia bungii* (Faldermann) in China [J]. Agriculture and Technology, 27(1): 63-66.

Hu Guoliang and Yu Caizhu, 2005. Pictorial guide to pecan diseases and pests and their control [M]. Beijing: China Agriculture Publishing House.

Hu Heling and Wang Liangyan, 1976. A study on *Matsucoccus massonianae* (1) [J]. Acta Entomologica Sinica, (4): 383-392.

Hu Keming, 1988. Biological and ecological characteristics of *Poecilocoris latus* Dall [J]. Tea Science, 8(2): 43-46.

Hu Longsheng, Qi Changjiang, Zhu Yinfei, Tian Cheng-ming, Ren Ling, 2012. Determination on larva instars of Ber fruit fly [J]. Journal of Xinjiang Agricultural University, 35(2): 137-139.

Hu Qi, 2002. Importance of quarantine of flowers and trees purchased from other regions [J]. Tianjin Agricultural and Forest Science and Technology, (5): 34-36.

Hu Shengchang, 1989. Bionomics of *Odontopera urania* (Wehrli) [J]. Entomological Knowledge, 26(6): 348-350.

Hu Weiping, Liang Yankang, 2008. Prediction and control of *Scythropus yasumatsui* [J]. Shanxi Agriculture, 14(1): 41.

Hu Xuwen, 2005. Study on *Periphyllus acerihabitans* Zhang [J]. Anhui Agriculture Bulletin, 11(4): 110-111.

Hu Yimin, Ye Anding, and Hu Huaying, 1996. Occurrence and control of *Kurisakia sinocaryae* Zhang [J]. Economic Forest Researches, (3): 43-44.

Hu Yinyue, Dai Huaguo, Hu Chunxiang et al, 1982. Prelinary study on *Quadraspidotus gigas* (Thiem et Gerneck) [J]. Scientia Silvae Sinicae, 18(2): 160-169.

Hu Zhonglang, Chen Xiaoda, and Yang Penghui, 1984a. Control studies on *Holcocerus arenicola* Staudinger [J]. Shanxi Forestry Science and Technology, (4): 9-20.

Hu Zhonglang, Chen Xiaoda, and Yang Penghui, 1984b. Identification of five species of Cossidae in Yulin region [J]. Shanxi Forestry Science and Technology, (3): 33-36.

Hu Zhonglang, Chen Xiaoda, and Yang Penghui, 1987. Studies on *Holcocerus arenicola* (Staudinger) [J]. Acta Entomologica Sinica, 30(3): 259-265.

Hu Zhonglang, Zhao Zongling, and Yang Penghui, 1988. A preliminary study of *Diorhabda rybakowi* Weise [J]. Shaanxi Forest Science and Technology, (2): 51-56.

Hu Zuodong, Zhang Fuhe, Wang Jianyou et al, 1998. Biological studies on *Eriosoma lanuginosum dilanuginosum* (Zhang) [J]. Journal of Northwest Agricultural University, 26(4): 25-28.

Hua Baosong, Xu Zhenwang, Liao Lihong, Hu Jingeng, Wu Jun, 2004. The morphology and life cycle of *Eutomostethus longidentus* [J]. Entomological Knowledge, 41(6): 589-592.

Hua Baozhen, Zhou Yao, Fang Deqi et al, 1990. Cossidae of China [M]. Tianze Press. Yangling.

Hua Baozheng, 1992. On the scientific name of the peach fruit borer [J]. Acta Entomologica Sinica, 14(4): 312-313.

Hua Baozheng, 1996. Peach fruit borer host biotype formation and species differentiation mechanism [M]. Northwest Agricultural University Doctoral dissertation.

Hua Chun, Yu Weiyan, Chen Quanzhan et al, 2007. Bionomics of *Papilio xuthus* Linnaeus in Nanjing [J]. Anhui Agricultural Science Bulletin, 13(24): 103-104.

Hua Fengming, Ni Liangcai, and Jin Minxin, 1999. Efficacy of several insecticides against *Icerya purchasi* Maskell on *Liquidambar formosana* [J]. Zhejiang Forestry Science and Technology, 19(6): 46-47, 50.

Hua Lei and Ma Gufang, 1993. Emergence of overwintering *Carposina niponensis* Walsingham larvae on different hosts [J]. Entomological Knowledge, 30(1): 22-25.

Hua Lei, 1995. Current research on *Conogethes punctiferalis* [J]. Shaanxi Forest Science and Technology, (2): 57-62.

Hua Lizhong, 2000. List of Chinese insects [M]. Guangzhou: Zhongshan (Su Yat-sen) University Press.

Hua Lizhong, 2005. List of Chinese Insects, Vol.III [M]. Guangzhou: Sun Yat-sen University Press.

Huang Bangkan, 1980. Biology and control of *Histia rhodope* Cramer [J]. Journal of Fujian Agricultural College, 9(1): 61-79.

Huang Bankan, 2001. Insects of Fujian (vol. 5) [M]. Fuzhou: Fujian Forestry Science and Technology Publishing House.

Huang Dazhuang, Li Liangming, Tang Xiaoqin et al, 2004. Forest diseases, pests, and rodents in southern Tibet [J]. Forest Pest and Disease, 23(5): 31-33.

Huang Dunyuan, Yu Jiangfan, Hao Jiasheng et al, 2010. Comparison of occurrence and damage degrees of *Dasmithius camellia* in camellia forest of different habitats [J]. Journal of Central South University of Forestry & Technology (Natural Science edition), 30(1): 59-64.

Huang Fayu, Liang Guangqin, Liang Qiong Chao et al, 2000. Quarantine and elimination of *Brontispa longissima* [J]. Plant Quarantine, 14(3): 158-160.

Huang Fengying, Fang Yuanwei, Huang Jian et al, 2005. A new introduced pest in China - *Quadrastichus erythrinae* Kim [J]. Entomological Knowledge, 42(6): 731-733.

Huang Furong, 2000. Studies on the control of *Lymantria xylina* Swinhoe with mixtures of *Beauveria bassiana*, *Bacillus thuringiensis*, and deltamethrin [J]. Fujian Forestry Science and Technology, 27(3): 55-58.

Huang Fusheng, Li Guixiang, and Zhu Shimo, 1989. Taxonomy and biology of *Chinese termites* [M]. Yangling: Tianze Press.

Huang Fusheng, 2000. Fauna of China Fasc. 17: Insecta, Isoptera [M]. Beijing: Science Publishing House.

Huang Fusheng, 2002. Hainan Forest Insects [M]. Beijing: Science Publishing House.

Huang Guanhui, 1984. Study on *Erannis ankeraria* Staudinger nuclear polyhedrosis virus [J]. Forest Science and Technology, (1): 29-30.

Huang Jiade, 1993. Biology and control of *Chalcocelis albiguttata* (Snellen) [J]. Guangxi Plant Protection, (4): 8-10.

Huang Jingfang and Luo Youqing, 1988. Study on the natural enemies of chestnut gall wasp in China [J]. Scientia Silvae Sinicae, 24(2): 162-169.

Huang Jinshui and Chen Yunpeng, 1987. Bionomics of *Gadirtha inexacta* Walker (Lepidoptera: Noctuidae) and its control [J]. Fujian Forestry Science and Technology, (1): 37-40.

Huang Jinshui and He Yiliang, 1988. Studies on the resistance of three species of Casuarina to *Lymantria xylina* [J]. Forest Science and Technology, (8): 15-17.

Huang Jinshui and Huang Shuishi, 1982. Preliminary observation on *Cerura tattakana* Matsumura [J]. Fujian Forestry Science and Technology, (2): 23-26.

Huang Jinshui, He Xueyou, Cai Tiangui et al, 1989. Bionomics and control of *Malacosoma dentata* Mell [J]. Fujian Forestry Science and Technology, 16(1): 34-37.

Huang Jinshui, He Xueyou, Ding Bi et al, 1999. Study on the planting of pest resistant strains of *Casuarina* spp. through asexual propagation [J]. Journal of Fujian Forestry Science and Technology, 26(supplement): 129-132.

Huang Jinshui, He Xueyou, Ye Jianxiong et al, 2000. Review on research on control techniques of the major trunk borers of *Casuarina* spp. in protection forest [J]. Protection Forest Science and Technology (special issue), 1-6.

Huang Jinshui, He Yiliang, and Lin Qingyuan, 1990. Beauveria bassiana paste and its application in forests [J]. China Journal of Biological Control, 6(3): 121-123.

Huang Jinshui, Huang Haiqing, Zheng Huicheng et al, 1995. Bionomics of *Euzophera batangensis* Caradja [J]. Fujian Forestry Science and Technology, 22(1): 1-8.

Huang Jinshui, Huang Yuanfei, Gao Meiling et al, 1990. Determination of larval instars of *Zeuzera multistrigata* Moore (Lepidoptera: Cossidae) [J]. Entomological Knowledge, 27(1): 25-27.

Huang Jinshui, Huang Yuanhui, and He Yiliang, 1988. Study on *Zeuzera multistrigata* Moore (Lepidoptera: Cossidae) [J]. Forest Research, 14(2): 1-15.

Huang Jinshui, Xing Longping, and Lan Siwen, 1991. *Euproctis bipunctapex* (Hampson) [C]. In Fujian Forest Insects. Beijing: China Agricultural Science and Technology Publishing House.

Huang Jinshui, Zheng Huicheng, and Yang Huaiwen, 1992. A study on the application of *Steinernema* spp. for control of *Zeuzera multistrigata* Moore [J]. Scientia Silvae Sinicae, 28(1): 39-46.

Huang Jinshui, Zheng Huicheng, Yang Huaiwen et al, 1992. Pathogeneity of eight strains of *Steinernema feltiae* against *Arbela baibarana* Mats [J]. Forest Pest and Disease, (1): 11-13.

Huang Jinshui, 1995. A study on the occurrence and integrated management of *Euzophera batangensis* [J]. Scientia Silvae Sinicae, 31(5): 421-427.

Huang Jinyi and Meng Meiqiong, 1986. Illustrated guide to forest disease and pest control [M]. Nanning: Guangxi People's Publishing House.

Huang Ketuan, Li Yanbo, and Lu Zhenming, 1984. Occurrence of *Chaetodacus ferrugineus* Fabricius in Western Guangxi [J]. Plant Quarantine, (2): 46-48.

Huang Li, Xi Yong, Ma Jun et al, 1994. Bionomics and control of *Eulecanium gigantea* (Shinji) [J]. Xinjiang Agricultural Science, (2): 82-84.

Huang Qilin, 1956. On some morphological features of the Zygaenid larvae [J]. Acta Entomologica Sinica, 6(2): 193-201, plates I-IX.

Huang Qinghai and Chen Wanfang, 1989. Preliminary study on *Lampra limbata* Gebler [J]. Journal of Hebei Forest College, 4(3): 39-46.

Huang Wanbin, 2007. Biological characteristics of *Castanes mollissima's Crecoccus candidus* Wang and corresponding control techniques [J]. Forest Inventory and Planning, 32(3): 146-149.

Huang Xiaoqing and Zhang Xianmin, 2007. Propagation and release of *Tetrastichus brontispa* and an Eulophidae wasp for controlling *Brontispa longissima* [J]. Modern Agricultural Science and Technology, (17): 87-88.

Huang Yanqing, 2000. Survey of diseases and pests of *Casuarina equisetifolia* in Quanzhou city and their control methods [J]. Protection Forest Science and Technology, S2: 6.

Huang Yazhi and Pei Rukang, 1986. A preliminary study on the biology and control of *Sternochetus frigidus* and *Sternochetus mangiferae* [J]. Yunnan Tropical Plant Science and Technology, 33(1): 20-24.

Huang Yonghuai, 2004. Bionomics of *Xylotrechus rusticus* L. and its control techniques [D]. Northeast Forestry University Master thesis.

Huang Yuzhen, 2000. Occurrence and control of *Cydia pomonella* [J]. Plant Protection Technology and Extension, 20 (5): 20.

Huang Zhenyu, 2006. Chemical control of *Hemiberlesia pitysophila* Takagi [J]. Journal of Nanjing Forestry University 30(5): 119-122.

Huang Zhongliang, 2000. The interactions of population dynamics of *Thalassodes quadraria* and the plant community structure and climate factors in Dinghushan [J]. Chinese Journal of Ecology, 19(3): 24-27, 31.

Hunan Forest Research Institute, 1979. Survey of *Polychrosis cunninghamiacola* Liu et Pai in Henshan county [J]. Hunan Forestry Science and Technology, (2): 25-29.

Hunan Forestry Department, 1992. Iconography of Hunan forest insects [M]. Changsha: Hunan Science and Technology Publishing House.

Huo Lüyuan, Wang Lizhong, Liu Suicun et al, 2001. *Diprion jingyuanensis* Xiao et Zhang control techniques [J]. Shanxi Forestry Science and Technology, (2): 27-29.

Huo Wenquan, Zhao Guojun, and Han Jingbo, 1989. Studies on *Lamellcossus terebra* (Schiffermüller) [J]. Shandong Forest Science and Technology, (4): 36-40.

Integrated Pest Management Team of *Hemiberlesia pitysophila*, 1989. Bionomics and occurrence of *Hemiberlesia pitysophila* Takagi [J]. Forest Science and Technology, (5): 3-7.

Ji Baozhong, Wang Yinchang, Qian Fanjun et al, 1996. Effect of diflubenzuron on reproductivity and nucleic acid content of *Batocera horsfieldi* Hope [J]. Journal of Nanjing Forest University, 20(2): 5-8.

Ji Rui, Xiao Yutao, Luo Fang et al, 2010. Efficacy tests of nine kinds of pesticides for controlling *Corythucha ciliae* [J]. Entomological Knowledge, 47(3): 543-546.

Ji Weirong, Zhang Liyan, and Zhang Yawen. 1995, Observations on the bionomics and control of *Podagricomela shirahatai* (Chǔjǒ) [J]. Journal of Shanxi Agricultural University, 15(4): 383-385.

Ji Yuhe, 1992. A preliminary study on *Cymolomia hartigiana* (Saxesen) [J]. Jilin Forestry Science and Technology, (5): 31-33, 39.

Jia Fengyong, Xu Zhichun, Zong Shixiang et al, 2004. Chemical control of *Holcocerus hippophaecolus* larvae [J]. Forest Pest and Disease, 23(6): 16-19.

Jia Yanmei, Ye Yongcheng, Guo Zhonghua et al, 1998. The preliminary research on spatial distribution pattern of *Rhagoletis batava obseceriosa* and its use [J]. Shaanxi Forest Science and Technology, (3): 78-81.

Jiang Baoping, Liu Shuping, Fang Jianwen et al, 2007. Occurrence regularity and control techniques of *Acantholyda posticalis* in south of Shaanxi [J]. Shaanxi Forest Science and Technology, (1): 31-33.

Jiang Dean, Liu Yongsheng, Li Guoyuan et al, 2006. Occurrence and control of *Pammene ginkgoicola* Liu [J]. Forest Science and Technology, (11): 26-28.

Jiang Dean, 2003. A preliminary report on the efficacy of B.t. against *Pammene ginkgoicola* Liu [J]. Forest Science and Technology Development, (6): 46-47.

Jiang Dean, 2003. Occurrence and control of *Dictyoploca japonica* Butler [J]. Forest Science and Technology, 28(1): 25-27.

Jiang Hanlin, 2001. Biology and control of *Aporia crataegi* [J]. Entomological Knowledge, 38(2): 198-199.

Jiang Jiadan, 2001. Biological characteristics and control of *Pantana phyllostachysae* Chao [J]. Forest Research and Technology, (2): 12-14.

Jiang Jie, Lin Yuyin, and Wu Zhiyuan, 1990. Study on *Paranthrene actinidiae* Yang et Wang affecting *Actinidia* spp. [J]. Scientia Silvae Sinicae, 26(2): 117-125.

Jiang Jingfeng and Hu Zhilian, 1990. Preliminary study on the biology of *Cerace stipatana* Walker [J]. Plant Protection (supplement), 35-36.

Jiang Jinwei and Ding Shibo, 2008. Occurrence and damage by introduced pest, sycamore lace bug [J]. Plant Quarantine, 22(6): 374-376.

Jiang Li, Xu Hui, Zhang Qiyu et al, 2009. Bionomics of *Hapsifera barbata* Christoph [J]. Shandong Agricultural Science, (10): 54-57.

Jiang Ping and Xu Zhihong, 2005. Colored pictorial guide to bamboo diseases and pests [M]. Beijing: China Agriculture Science and Technology Publishing House.

Jiang Sandeng, Fang Deqi, and Chen Shuliang, 1987. Bionomics and control of *Streltzoviella insularis* (Staudinger) [J]. Shandong Forest Science and Technology, (3): 32-35.

Jiang Sandeng, 1990. Bionomics and control of *Meliboeus cerskyi* Obenberger [J]. Shandong Forestry Science and Technology, (4): 51-55.

Jiang Shunan, Pu Fuji, and Hua Lizhong, 1985. Economic insect fauna of China Fasc. 35. Coleoptera: Cerambycidae (3) [M]. Beijing: Science Publishing House.

Jiang Shunan, 1989. Larvae of Cerambycidae in China [M]. Chongqing: Chongqing Publishing House.

Jiang Yeqin, Zhang Yakun, Chen Wenjie et al, 1989. Study on *Dendrolimus houi* [J]. Journal of Fujian Forest College, 9(1): 22-27.

Jiang Yu, Liu Weiquan, Jin Aijun et al, 2002. Laboratory rearing of *Baculum minutidentatum* Chen et He and observation on its ability to regrow appendages [J]. Journal of Jilin Agricultural University, 24(4): 61-63.

Jiang Yucai and Wang Shusen, 1989. Survey of *Pityogenes chalcographus* Linnaeus [J]. Liaoning Forestry Science and Technology, (2): 39-41.

Jiang Yuen, 1987. Impact of *Erannis ankeraria* Staudinger outbreaks on forest tree growth [J]. Hebei Forestry Science and Technology, (2): 34-36.

Jiang Zhengming, Ma Junfei, Xu Guangyu et al, 2008. Bionomics and integrated management of *Mesoneura rufonota* Rohwer [J]. Hebei Agricultural Science, 12(8): 43-44.

Jiao Qiyuan et al, 1940. The Chinese gallnut in Sichuan (2) [J]. Journal of Jinling, 9(1-2): 85-93.

Jiao Qiyuan, 1938. Chinese galls in Sichuan [J]. New Agriculture and Forestry, (514): 10-18.

Jin Gesi and Wuer Lan Han, 2006. Bionomics of *Xylotrechus rusticus* and its integrated management [J]. Practical Forestry Techniques, (7): 28-29.

Jin Gesi and Wuer Lanhan, 2006. Bionomics and control of *Xylotrechus rusticus* [J]. Xinjiang Forestry, 39-40.

Jin Meilan, 2004. Bionomics of *Pachynematus itoi* Okutani and its control methods [J]. Jilin Forest Science and Technology, 33(1): 39-40.

Jin Ruozhong, Ruan Qingshu, and Yun Lili, 2005. A study on the biology of *Agrilus marcopoli* Obenberger [J]. Liaoning Forestry Science and Technology, (5): 22-24.

Jin Xiaoyuan, Zhang Jintong, Luo Youqing et al, 2010. Circadian rhythms of sexual behavior and pheromone titers of Holcocerus arenicola (Lepidoptera: Cossidae) [J]. Acta Entanologica Sinica 53: 307-313.

Jin Xingrui, Tian Shibo, Zhao Shuer et al, 1988. Bionomics and control of *Eurytoma plotnirovi* Nikolskaya [J]. North China Agriculture, 3(3): 71-76.

Jing Xiaoyuan, Zhang Jintong, Luo Youqing et al, 2010. Synthesis and biological activity evaluation of sex attractant for *Holcocerus arenicola* (Lepidoptera: Cossidae) [J]. Scientia Silvae Sinicae, 46(4): 87-92.

Ju Ruiting and Li Bo, 2010. Sycamore lace bug, *Corythucha ciliata*, an invasive alien pest rapidly spreading in urban China [J]. Biodiversity Science, 18(6): 638-646.

Jun Xuan, 1984. Occurrence and control of *Poecilocoris latus* Dallas [J]. China Tea, (6): 13-14.

Kang Le, 1993. African mole cricket should be named Oriental mole cricket in China [J]. Entomological Knowledge, 30(2): 124-127.

Kang Wentong, 1998. Bionomics and control of *Lepidarbela bailbarana* (Mats.) [J]. Entomological Journal of East China, 7(2): 41-44.

Kumata, T. and H. Kuroko, 1988. Japanese species of the Acroeercops-group (Lepidoptera: Gracillariidae), Part II [J]. Insecta Matsuma (New Series). 40: 1-133.

Langor, D. W., Y.-X. SiTu, and R.-Z. Zhang, 1999. Two new species of Pissodes (Coleoptera: Curculionidae) from China [J]. The Canadian Entomologist 131: 593-603.

Lan Jie, Qin He, Kang Xiaolong, Li Huaqing et al, 2001. Control methods for *Anomala corpulenta* Motschulsky [J]. Sichuan Forest Science and Technology, 22(3): 78-79.

Lan Siwen, Weng Xizhao, Huang Jincong et al, 1993. A study on magnolia leaf miner, *Acrocercops urbana* Meyrick [J]. Journal of Fujian Forest College, 13(4): 371-380.

Lan Yuguang, 2008. A major pest of eucalyptus seedlings - *Brachytrupes portentosus* Lichtenstein and its control [J]. Agricultural Science, (6): 133.

Lei Chaoliang and Zhou Zhibo, 1998. Checklist of insects in Hubei province [M]. Wuhan: Hubei Science and Technology Publishing House.

Lei Guilin, Duan Zhaoyao, Feng Zhiwei et al, 2003. Pest risk analysis of *Pissodes punctatus* Langor et Zhang [J]. Journal of Northeast Forestry University, 31(3): 62-63.

Li Changyuan and Wang Huiping, 2006. Study on the biological control of *Diorhabda elongata deserticola* Chen [J]. Practical Forestry Technology, (4): 27-28.

Li Chaoxu, Tan Weiquan, Huang Shanchun et al, 2008. Effectiveness of releasing parasitoids for controlling *Brontispa longissima* (Gestro) [J]. Forest Science and Technology Development, 22(1): 41-44.

Li Chengde, 2003. Forest entomology [M]. Beijing: China Forestry Publishing House.

Li Chengwei, 1998. Bionomics of *Dryocomus kuriphilus* (Yasumatsu) [J]. Forest Science and Technology Development, 4: 36-38.

Li Chuanlong, 1958. Butterflies [M]. Beijing: Science Publishing House.

Li Chuanlong, 1966. Swallowtails [J]. Bulletin of Biology, (2): 25-29.

Li Chuanlong, 1984. A revision of the Chinese rare papilionid butterfly *Agehana elwesi* (Leech) [J]. Acta Zoologica Sinica, 9(3): 335.

Li Chuanlong, 1992. Pictorial guide to Chinese butterflies [M]. Shanghai: Far East Publishing House.

Li Chuanren, Xia Wensheng, and Wang Fulian, 2007. First record of *Corythucha ciliata* (Say) in China (Hemiptera: Tingidae) [J]. Acta Zoologica Sinica, 32(4): 944-946.

Li Chunfeng and Miao Xiaoxing, 2009. Occurrence and control of *Megastigmus sabinae* Xu et He in Makehe forest area [J]. Journal of Qinghai University (Natural Science edition), 27(2): 66-68.

Li Cuifang, Zhang Yufeng, and Zhou Zhifang. Bionomics and control of *Gastropacha populifolia* (Esper) [J]. Hebei Fruit Tree, (3): 21-22.

Li Dahai, 2004. Investigation of diseases and pests of three plants in sandy areas and their control techniques [J]. Shaanxi Forest Science and Technology, (3): 42-45.

Li Dingxu, 2002. Control techniques for *Carposina niponensis* Walsingham [J]. Plant Protection, 28(3): 18-20.

Li Dongwen, Chen Zhiyun, Wang Ling et al, 2009. Bionomics of *Biston marginata* Shiraki [J]. Guangdong Forest Science and Technology, 25(2): 55-59.

Li Fang, Chen Jiahua, and He Rongbin, 2001. Review of utilization of natural enemies of *Agrotis ypsilon* (Rott.) [J]. Natural Enemies of Insects, 23(1): 43-38.

Li Fasheng, 2011. Psyllidae of China (2) [M]. Beijing: Science Publishing House.

Li Fengyao, Liu Suicun, Hu Lüyuan et al. Study on the biology and occurrence of *Diprion jingyuanensis* Xiao et Zhang [J]. Shanxi Forestry Science and Technology, (2): 10-16.

Li Fengyao, Wang Lizhong, Zhao Jin et al, 1995. Occurrence and control of *Diprion jingyuanensis* Xiao et Zhang [J]. Forest Research 18(special issue): 184.

Li Guangming, Chen Ming, Han Chongxuan et al, 2007. Current research on *Mallambyx raddei* Blessig and prospects [J]. Shaanxi Forest Science and Technology, (2): 48-51.

Li Guishan, Yang Zhanqiang, Wang Wenming et al, 2001. Delayed effect of Dimilin III on *Leuoma candida* Staudinger [J]. Forest Pest and Disease, (2): 15-16.

Li Guixiang, 2002. Chinese termites and their control [M]. Beijing: Science Publishing House.

Li Guoyuan, 2001. Occurrence and control of *Dictyoploca japonica* Butler [J]. Hubei Agricultural Science, (5): 58-59.

Li Haiyan, Zong Shixiang, Sheng Maoling et al, 2009. Investigation of the natural enemies of *Orgia ericae* [J]. Scientia Silvae Sinicae, 45(2): 167-170.

Li Heming, Zhou Jingqing, and Cai Shengguo, 2006. Fogging to control *Coleophora sinensis* Yang [J]. Hebei Forestry, (4): 40.

Li Hongchang and Xia Kailing, 2006. Fauna of China Fasc. 43: Insecta, Orthoptera, Acridoidea, Acrididae [M]. Beijing: Science Publishing House.

Li Hongtao, Zhaobo, and Liu Xiaoming, 2004. Bionomics of *Phassus excrescens* Butler and its control techniques [J]. Jilin Forestry Science and Technology, 33(4): 27, 47.

Li Houhun, 2002. Gelechiidae of China (1) [M]. Tianjin: Nankai University Press.

Li Houhun, 2003. Morphological study and redescription of *Coleophora sinensis* Yang and *Coleophora obducta* (Meyrick) (Lepidoptera: Coleophoridae) [J]. Entomotaxonomia, 25(4): 295-302.

Li Huicheng, Wang Jianzhong, and Guofei, 1992. Bionomics of larch sawfly and its control methods [J]. Scientia Silvae Sinicae, 24(4): 317-322.

Li Jianglin, 1985. Occurrence and control of *Melolontha tarimensis* Semenov [J]. Entomological Knowledge, 22(4): 156-158.

Li Jianhao and Li Dongping, 1994. Bionomics of *Adoxophyes orana orana* [J]. Sichuan Forestry Science and Technology, 15(3): 56-57.

Li Jiayuan, 1980. A preliminary report on the bionomics and control of the *Galerucid (Atysa) cinnamomi* Chen [J]. Acta Entomologica Sinica, 23(3): 338-340.

Li Jishun, Chang Guobin, Song Yushuang et al, 2001. Causes of red burpentine beetle outbreaks and management strategies [J]. Forest Pest and Disease, (4): 41-44.

Li Keming, Tan Weiyuan, Li Chaoxu et al, 2008. Effect of cold front on *Brontispa longissima* (Gestro) and its parasitoids [J]. Anhui Agricultural Science Bulletin, 14(18): 88-90.

Li Kuansheng et al, 1964. Bionomics of *Dioryctria mendacella* Staudinger [J]. Collection of technical papers on Shaanxi Forest Science and Technology, 4: 102-111.

Li Kuansheng et al, 1966. Brief description of *Shirahoshizo coniferae* Chao [J]. In "Collection of Shaanxi forest science and technology papers, 6: 164-166.

Li Kuansheng et al, 1992. A preliminary study on *Dioryctria mendacella* Staudinger [C]. In "Control techniques of cone and seed pests of Pinus tabulaeformis". Xian: Shaanxi Science and Technology Publishing House.

Li Kuansheng et al, 1992. Investigation on the overwintering sites of *Dioryctria mendacella* Staudinger [C]. In "Control techniques of cone and seed pests of Pinus tabulaeformis". Xian: Shaanxi Science and Technology Publishing House.

Li Kuansheng, Tang Guoheng, Jin Buxian et al, 1981. A study on the control of *Gravitarmata margarotana* Hein using *Tri-*

chogramma spp [J]. Shaanxi Forest Science and Technology, (4):33-46.

Li Kuansheng, Zhang Yudai, and Li Yangzhi, 1974. A preliminary study on *Gravitamata margarotana* Hein (Lepidoptera: Olethreutidae) in Shensi province [J]. Acta Entomologica Sinica, 17(1): 16-28.

Li Kuansheng, Zhang Yudai, and Li Yangzhi. Identification of three cone pests of *Pinus tablaeformis* [J]. Entomological Knowledge, (5): 211-213.

Li Kuansheng, 1959. Bark beetles of *Pinus armandii* and observations on the life history of *Dendroctonus* sp [J]. Shaanxi Agricultural Science, (7): 277-286.

Li Kuansheng, 1989. *Dendroctonus armandi* Tsai et Li [C]. In "Encyclopedia of Chinese agriculture (Forestry 1)". Beijing: Agriculture Publishing House.

Li Kuansheng, 1992. Seed and cone pests in China [M]. Beijing: China Forestry Publishing House.

Li Kuansheng, 1999. Seed and cone pests of coniferous trees [M]. Beijing: China Forestry Publishing House.

Li Li, Sun Xu, and Meng Huanwen, 2000. Study on the biological characteristics and control of *Costera anastomosis* L. (Lepidoptera: Notodontidae) [J]. Journal of Inner Mongolia Agricultural University 21(3): 18-21.

Li Lianchang, Dong Haifu, Lin Guoqiang et al, 1984. Identification, synthesis, and field application of *Ancylis sativa* Liu sex pheromone [J]. Journal of Shanxi Agricultural University 4(2): 149-161.

Li Lianchang, 1965. A study on *Aporia crataegi* (L.) in Shanxi [J]. Acta Entomological Sinica 14(6): 545-550.

Li Menglou and Wu Xingyu, 2003. A new species of *Agenocimbex* (Hymenoptera: Cimbicidae) from Gansu, China [J]. Scientia Silvae Sinicae, 39(1): 103-104.

Li Menglou, Cao Zhimin, and Wang Peixin, 1989. Cultivation of *Zanthoxylum bungeanum* Maxim and its disease and pest control [M]. Xian: Shaanxi Science and Technology Publishing House.

Li Menglou, Liu Chaobin, and Wu Dingkun, 1992. Control threshold of larch sawfly [J]. Journal of Northwest Forest College, 7(4): 107-113.

Li Menglou, Wu Dingkun, and Liu Chaobin, 1992. Bionomic strategy and control tactics of larch sawfly [J]. Journal of Northwest Forest College, 7(4): 114-120.

Li Menglou, 2002. General forest entomology [M]. Beijing: China Forestry Publishing House.

Li Qing, 2004. Control of *Cinara* sp. using environmentally compatible insecticides [J]. Practical Forestry Technology, (8): 27-28.

Li Qingxi, Jiang Xinmin, and Liu Fangzheng, 1987. A preliminary study on *Scolytus schevyrewi* Semenov [J]. Journal of Bayi Agricultural College, (3): 12-18.

Li Qingyun, 2000. Occurrence, damage, and control of *Mesoneura rufonata* (Rohwer) in Wuxi area [J]. Jiangsu Afforestation, (6): 34.

Li Shilan, Zheng Shan, Wang Junhuai et al, 2009. Survey on life history of *Nesodipron zhejiangensis* and techniques for control [J]. Forest Inventory and Planning, 34(1): 80-82.

Li Shuangcheng, Li Yonghe, Ma Jin et al, 2001. Study on control threshold of *Pissodes punctatus* Langor et Zhang [J]. Yunnan Forestry Science and Technology, (1): 51-53.

Li Shuangcheng, Ma Jin, Gui Junwen et al, 2000. Spatial distribution pattern of the eggs and pupae of *Pissodes punctatus* Langor et Zhang [J]. Yunnan Forestry Science and Technology, (4): 62-65.

Li Shuji, Zhang Yuxiao, and Zhang Li, 2007. Biology and control of *Stilpnotia candida* Staudinger [J]. Henan Forestry Science and Technology, 27(2): 46-48.

Li Suping, Chen Xiulong, Han Guozhu et al, 2006. Bionomics and control of *Ricania shantungensis* [J]. Forest Pest and Disease, 25(3): 36-38.

Li Tiesheng, 1985. Economic insect fauna of China Fasc. 30. Hymenoptera: Vespoidea [M]. Beijing: Science Publishing House.

Li Xiaohua and Gao Pei, 2005. Bionomics and control techniques of *Cinara* sp [J]. Shaanxi Forest Science and Technology, (4): 35-36.

Li Xiushan, Bing Jicai, Xu Xiao et al, 2002. Study on the bionomics and control of *Baculum chongxinense* [J]. Scientia Silvae Sinicae, 38(6): 159-163.

Li Xueling, 2006. Insecticide fogging to control *Dendrolimus superans* [J]. Xinjiang Forestry, (1): 44.

Li Yabai, Li Shuyuan, and He Weile, 1981. Bionomics and integrated control of *Pissodes ualidirostris* Gyll [J]. Inner Mongolia Forestry Science and Technology, (2): 3-18.

Li Yajie, 1959. Preliminary study on the chemical control of three leaf miners affecting poplar trees [J]. Scientia Silvae Sinicae, (4): 321-324.

Li Yajie, 1978. Poplar leaf beetle and its control [J]. New Agriculture, (10): 12.

Li Yajie, 1983. China poplar tree pests [M]. Shenyang: Liaoning Science and Technology Publishing House.

Li Yan and Chen Shuchun, 1998. Present research outline on biology and ecology of the Phasmatodea [J]. Journal of Forestry, 20(1): 59-66.

Li Yana, 2006. Infestation of *Costera anastomosis* L. (Lepidoptera: Notodontidae) and its control techniques [J].

Hebei Forestry, 2(2): 42.

Li Yanmei, 1996. Infestation of *Costera anastomosis* L. (Lepidoptera: Notodontidae) and its control techniques [J]. Protection Forest Science and Technology, (2): 46-48.

Li Yanzheng, Zhao Ren, and Jin Buxian, 1964. Bionomics of three geometrid insects and their control in Louguantai forest [J]. Entomological Knowledge, 8(4): 167-170.

Li Yonghe, Xie Kaili, and Cao Kuiguang, 2002. Integrated pest management of major pests of *Pinus armandii* [J]. Forest Pest and Disease, 21(3): 13-16.

Li Yongxi, 1964. Behavior and control of *Agrotis ypsilon* (Rott.) [J]. Entomological Knowledge, (1): 2-3.

Li Yougong and Chen Shunli, 1981. Studies on the lymantriid moth *Lymantria xylina* Swinhoe [J]. Acta Entomologica Sinica, 24(2): 174-183.

Li Yougong, Chen Shunli, Li Liu et al, 1990. Bionomics and occurrence of *Stiphanitis laudata* Drake et Poor [J]. Entomological Knowledge, 27(2): 104-107.

Li Zhanwen, Zhang Aiping, Sun Yaowu et al, 2007. Damage by *Quadraspidiotus perniciosus* (Comstock) and its control techniques [J]. Plant Quarantine, 21(3): 163-164.

Li Zhenyu, Wu Peiheng, and Guo Guangzhong, 1989. Study on the sex pheromone of *Sesia siningensis* (Hsu) [J]. Journal of Beijing Forestry University, 13(1): 24-29.

Li Zhenyu, Yao Defu, and Chen Yongmei et al, 2001. Parasitoids and alternate hosts of Gypsy moth in Beijing area [J]. Journal of Beijing Forestry University, 23(5): 39-42.

Li Zhenyu, Zhang Zhizhong, and Huang Jingfang. Occurrence and control of *Atrijuglans hetauhei* [J]. Bulletin of Biology, (4): 30-33.

Li Zhiyong, 1983. Occurrence and control of *Sesia siningensis* (Hsu), a pest of poplar trees [J]. Journal of Shanxi Agricultural University, 3(1): 74-80.

Li Zhonghuan, Hu Weihong, and Wang Aijing, 1995. Prediction of occurrence of green leaf hopper [J]. Xinjiang Agricultural Research, (6): 260-262.

Li Zhongxiao, 2003. Biological control techniques of *Coleophora obducta* [D]. Northeast Forestry University Master thesis.

Li Zhongxin and Liu Yusheng, 2004. Effect of Temperature on development of egg parasitoid Trissolcus halyomorphae and the eggs of its host, *Halyomorpha halys* [J]. China Journal of Biological Control, 20(1): 64-66.

Li Zuolong, 1964. Review of foreign literature on the study of *Agrotis segetum* [J]. Xinjiang Agricultural Science, (5): 179-182.

Li, H. and Z. Zheng, 1998. A taxonomic study on the genus Anarsia Zeller from the mainland of China (Lepidoptera: Gelechiidae) [J]. Acta Zoologicae Academiae Scientiarum Hungaricae 43: 121-132.

Liang Chengfeng, 2003. Prediction and control of major forest diseases and pests in southern China [M]. Beijing: China Forestry Publishing House.

Liang Chengjie and Zhaoling, 1987. Biology and control of *Hegesidemus habrus* Darke [J]. Scientia Silvae Sinicae, 23(3): 376-382.

Liang Wenxiu, Deng Biao, Wang Shiqi et al, 2001. Research and application prospect of *Stilpnotia salicis* NPV [J]. Forest Pest and Disease, (5): 25-27.

Liang Xiaoming, 2008. Morphological characteristics and bionomics of *Coleophora sinensis* Yang [J]. Forestry of Shanxi, (6): 29-30.

Liang Yiping, Yang Xiuhao et al, 2009. A study on using sticky cards to control Leptocybe invasa Fisher et LaSalle [J]. Modern Education and Teaching Exploration,12.

Liang Zhonggui, Li Jianjun, Liu Jianqiang et al, 2007. Research progress on *Acantholyda posticalis* [J]. China Plant Protection, (5): 14-17.

Liao Dingxi et al, 1987. Economic insect fauna of China Fasc. 34. Hymenoptera: Chalcidoidea (1) [M]. Beijing: Science Publishing House.

Lin Shunde, 2003. Infestation of *Brachytrupes portentosus* Lichtenstein in eucalytus forests and control [J]. Practical Forestry Technology, (8): 32.

Lin Yanmou, Fu Yueguan, and Zhang Junhong, 1991. Bionomics and control of *Chalcocelis albiguttata* [J]. Tropical Agricultural Science, (2): 33-34.

Lin Yougiao and Qian Fanjun, 1994. A new species of the genus Dichomeris to china fir (Lepidoptera: Opelechiidae) [J]. Entomologia sinica (4): 297-300.

Lin Zhong, 1957. Importance of controllng caterpillars to *Sapium sebiferum* production [J]. Zhejiang Forestry Bulletin, (5): 38-39.

Liu Aiping and Hou Tianjue, 2005. Diseases and Pests of grass lands and their control [M]. Beijing: China Agricultural Science Publishing House.

Liu Bo, Yang Jun, Zhang Xitang et al, 1992. Insecticide trials against *Costera anastomosis* L. (Lepidoptera: Notodontidae) [J]. Jilin Forestry Science and Technology, 21(6): 35-36.

Liu Delin, 1990. Description of two parasitoids of Yunnan pine caterpillar in Shaanxi [J]. Journal of Northwest Forest College, 5(4): 52-53.

Liu Dongming, Gao Zezheng, and Xing Fuwu, 2003. Bionomics and control of *Batocera rubus* (L .) [J]. Forest Pest and Disease, 22(6): 10-12.

Liu Er and Li Decheng, 2009. Risk analysis of *Saperda populnea* (Linnaeus) [J]. Forestry Science and Technology Development, 23(2): 70-72.

Liu Fabang, Zhao Jixing, Yu Chengren et al, 1991. A new leaf pest of *Hippophae rhamnoides* Linn. - *Orgyia ericae* Germar [J]. Forest Pest and Disease, (3): 34-35.

Liu Guangyao, 1986. A brief report on the bionomics and control of *Diprion nanhuaensis* Xiao [J]. Forest Science and Technology, (9): 27-28.

Liu Hongguang, Wang Aijing, Zhao Yingying et al, 2000. Predition of population of *Xylotrechus namanganensis* Heydel [J]. Xinjiang Agricultural Science, (3): 143-146.

Liu Huiying and Zhou Qingjiu, 1994. Studies on *Synanthedon castanevora* Yang et Wang [J]. Journal of Hebei Forest College (supplement): 76-84.

Liu Huiying, Zhou Qingjiu, and Wu Dianyi, 1989. Preliminary studies on *Synanthedon castanevora* Yang et Wang [J]. Forest Research, 2(4): 381-387.

Liu Jiazhi, Zhang Guocai, Yue Shukui et al, 1996. Studies on the chemical control of *Xerodes rufescentaria* (Motschulsky) [J]. Journal of Northeast Forestry University, 24(3): 26-31.

Liu Jiazhu, 2003. Evaluation of the efficacy of several insecticides for the control of *Lachnus tropicalis* (Van der Goot) eggs [J]. Shandong Forestry Science and Technology, (5): 35.

Liu Jing, Sun Qiwen, and Pang Xianwei, 1990. Bionomics of *Acantholyda posticalis* Matsumura and economic analysis of pest management [J]. Shandong Forest Science and Technology, (2): 38-41.

Liu Jing, 1992. Bionomics and control of *Anafenusa acericola* (Xiao) [J]. Entomological Journal of East China, 1(1): 41-45.

Liu Jinying, Pang Jianjun, and Zhai Shanmin, 2001. Sustainable control techniques of major pests of *Sophora japonica* L [J]. Tianjin Construction Science and Technology, (Horticulture edition) 122-123.

Liu Juhua, Xiao Yinsong, Luo Fangzheng et al, 2005. Relationship between occurrence of *Pissodes punctatus* pest and the growth status of host trees [J]. Journal of Southwest Forestry College, 25(2): 53-55.

Liu Kui, Peng Zhengqiang, and Fu Yueguan, 2002. Advances in red palm weevil research [J]. Tropical Agricultural Science, 22(2): 70-77.

Liu Liling, Zhong Junhong, and Dai Zirong, 1991. Apterous neotenic formation in laboratory colonies of termite *Reticulitermes flaviceps* (Oshima) and scanning electron microscopic observation of the external morphology [J]. Natural Enemies of Insects, 13(1): 35-42.

Liu Linjun, 2005. General characteristics of pests and their dynamics in Guandi forest region [J]. Shanxi Forestry Science and Technology, (4): 30-31, 43.

Liu Mingtang, 1989. Infestation and biology of *Shansiaspis sinensis* Tang [J]. Entomological Knowledge, 26(1): 24-26.

Liu Ping, Shi Jin, and Zhu Guangyu, 1992. A study on *Gilpinia massoniana* [J]. Forest Research, 5(2): 196-202.

Liu Qiaoyun, 1999. Control experiments on *Eutomostethus nigritus* Xiao [J]. Journal of Fujian Forestry College, 19(2): 129-132.

Liu Ruiming, 1984. A brief report on the life history of *Dictyoploca japonica* Butler [J]. Jiangxi Plant Protection, (4): 18.

Liu Shiqi, 1988. Illustrated monograph of Anhui forest pest and disease [M]. Hefei: Anhui Science and Technology Publishing House.

Liu Shixian, 2003. Study on *Odites issikii* (Takahashi) and its control [J]. Northwest Horticulture, (1): 43-44.

Liu Shouli, Yang Shuiqiong, Yang Yuekui et al, 2005. Bionomics and integrated management of *Pissodes punctatus* Langor et Zhang [J]. Practical Techniques of Forestry, (8): 29-30.

Liu Suyun, 2002. Occurrence and control of *Acantholyda posticalis* Matsumura [J]. Hebei Forest Science and Technology, (3): 33.

Liu Xiuying, 2008. Monitoring and control of *Coleophora obducta* with sex attractant [J]. Forest Pest and Disease, 27(5): 9-10.

Liu Yongsheng, 2001. Bionomics and control techniques of *Mesoneura rufonota* Rohwer [J]. Forest Science and Technology, 26(5): 20-22.

Liu Yongsheng, 2002. Occurrence and chemical control of *Asterolecanium castaneae* Russell [J]. China Southern Fruit Trees, 31(3): 70-71.

Liu Yongzheng, 1979. Preliminary study on *Aiolomorphus rhopaloides* [J]. Forest Pest and Disease, (3): 3-4.

Liu Youqiao and Bai Jiuwe, 1977. A new Trotricid moth on *Cunninghamia lanceolata* (Lamb.) Hook (Lepidoptera: Tortricidae) [J]. Acta Entomologica Sinica, 20(2): 217-218, 220, Plate I.

Liu Youqiao and Bai Jiuwei, 1977. Economic insect fauna of China Fasc. 11. Lepidoptera: Tortricidae (1) [M]. Beijing: Science Publishing House.

Liu Youqiao and Bai Jiuwei, 1979. A study on the Gelechiid months in Tai-ling forest areas of Heilongjiang province [J]. Scientia Silvae Sinicae, (4): 276-280, IV, V.

Liu Youqiao and Li Guangwu, 2002. Fauna of China, Volume 27: Insecta, Lepidoptera, Tortricidae [M]. Beijing: Science

Publishing House.

Liu Youqiao and Wu Chunsheng, 2006. Fauna of China Fasc. 47: Insecta, Lepidoptera, Lasiocampidae [M]. Beijing: Science Publishing House.

Liu Youqiao, 1992. A study of Microlepidoptera (Oecophoridae and Xyloryctidae) damaging *Bombax* and *Cinnamomum* [J]. Forest Research, 5(2): 203-206.

Liu Yuanfu and Fengcai, 1974. A preliminary observation on *Ticera castanea* Swinhoe [J]. Tropical Forestry Science and Technology, (2): 20-25.

Liu Yuanfu and Gu Maobin, 1977. A preliminary study on *Aristobia hispida* (Saunders) [J]. Tropical Forestry Science and Technology, (3): 18-30.

Liu Yuping, Liu Guifeng, Suhui et al, 2005. The study on the natural enemies in scale insects [J]. Journal of Inner Mongolia University for Nationalities (Natural Science edition) 20(3): 285-288.

Liu Zheng, Zhang Guilan, Li Yan et al, 1993. Biological control of peach red necked longihorn *Aromia bungii* with entomopathogenic nematodes [J]. Bulletin of Biological Control, (4): 186-187.

Liu Zhenqing, Wang Xiaoqing, Wang Shiqi et al, 1995. A study on the bionomics of *Stilpnotia salicis* [J]. Forest Pest and Disease, (3): 18-19.

Liu Zhicheng and Peng Shibing, 1992. Control of the Chinese cinnamon tip moth, *Cophoprora* sp. (Lepidoptera: Tortricidae) with Trichogramma chilonis [J]. China Journal of Biological Control, (2): 61-63.

Liu Zili, Huang Lei, Yi Junji et al, 2005. Field evaluation of hexaflumuron treated paper for controlling *Coptotermes formosanus* Shiraki [J]. Forest Pest and Disease, 24(4): 44.

Long Furong, Tang Yongjun, Huang Huiping et al, 2004. Control effect of *Dendrolimus houi* NPV dust in the forest [J]. Forest Pest and Disease, 23(4): 36-38.

Lou Shenxiu, Li Zhongxi, and Wu Yanru, 1991. Studies on *Cerceris rufipes* Evecta (Hym.: Sphecidae), a parasitoid of adult weevils [J]. Chinese Journal of Biological Control, (2): 94.

Lou Wei, Hou Aiju, and Dai Yuwei, 1990. Study on the management of *Acantholyda posticalis* [J]. Protection Forestry Science and Technology, (2): 49-58.

Lu Bin, Bo Lianggao, and Shu Weiqi, 2003. Occurrence and control of *Dendrolimus kikuchii kikuchii* Matsumura [J]. Anhui Forest Science and Technology, (1): 29.

Lu Changkuan, Ai Qinggang, Fan Yunzhong et al, 2001. Chemical control of mature larvae of *Acantholyda peiyingaopaoa* Hsiao [J]. Inner Mongolia Forestry Science and Technology, (supplement): 41, 44.

Lu Changkuan, Xu Zhichun, Jia Fengyong et al, 2004. Laboratory virulence and infectivity of *Steinernema carpocapsae* to *Holcocerus hippophaecolus* [J]. Chinese Journal of Biological Control, 20(4): 280-282.

Lu Changkuan, Zong Shixiang, Luo Youqing et al, 2004. Sex attraction of *Holcocerus hippophaecolus* (Lepidoptera: Cossidae) [J]. Journal of Beijing Forestry University, 26(2): 79-83.

Lü Changren and Zhan Zhongcai, 1989. Studies on the biology of *Matsucoccus shennongjiaensis* Young et Liu (Homoptera, Margarodidae) [J]. Scientia Silvae Sinica, 25(6): 577-581.

Lu Chuanchuan and Wen Ruizheng, 1985. Biology and control of *Selepa celtis* Moore [J]. Entomological Knowledge, 22(2): 78-80.

Lu Chuanchuan and Wu Huixiong, 1991. Bionomics and control of *Parasaissetia nigra* (Nietner) [J]. Natural Enemies of Insects, 13(3): 101-106.

Lu Chuanchuan, 1985. Preliminary study on *Chaleocelides albigutata* Snelle [J]. Entomological Knowledge, 22(1): 25-26.

Lü Peike, Su Huilan et al, 2002. Colored pictorial guide to Chinese fruit tree diseases and pests [M]. Beijing: Huaxia Publishing House.

Lu Rongchun, 2008. Ecology of *Tomicus yunnanensis* and *Tomicus minor* and their effective attractant [M]. Beijing Forestry University Master thesis.

Lü Ruoqing, 1988. A preliminary observation on *Glyphipterix semiflavana* Issiki [J]. Forest Pest and Disease, 4: 13-14.

Lu Shuitian et al, 1993. Monograph of longhorned beetles in Xinjiang [M]. Urumuqi: Xinjiang Science and Technology Publishing House.

Lu Wenmin, 1987. Investigation of six tortricid pests in Mudanjiang forest area [J]. Forest Science and Technology, (3): 16-19.

LÜ Xiaohong, Wu Huanggui and Yang Zhou, 2000. Biology and control of *Phloeosinus aubei* [J]. Shanxi Forestry Science and Technology, 3(1): 28-30.

Lu Yingyi and Fang Minggang, 1992. Infestation of *Cyllorhynchites ursulus* Roelofs and its control [J]. Anhui Forestry Science and Technology, (4): 28-30.

Lü Yuli, Liu Shiling, Fan Jinlong et al, 2006. A new forest pest, *Zeuzera leuconotum* Butler [J]. Tianjin Agricultural and Forest Science, (4): 44.

Lu Zhiwei, Li Zhonghuan, and Wang Aijing, 1994. Occurrence of green leaf hopper and integrated pest management [J]. Xinjiang Agricultural Science, (4): 164-166.

Luo Jitong, Jiang Jinpei, Wang Jijian et al, 2011. Bionomics of *Leptocybe invasa* in Bobai of Guangxi Autonomous region [J].

Forest Pest and Disease, 30(4): 10-12.

Luo Yizhen, 1995. Soil insects [M]. Beijing: China Agriculture Publishing House.

Luo Youqing, Lu Changkuan, and Xu Zhichun, 2003. Control strategies on *Holcocerus hippophaecolus* [J]. The Global Seabuckthorn Research and Development, 1(1): 31-33.

Luo Youqing, Lu Changkuan, and Xu Zhichun, 2003. Control strategies on a new forest pest - seabuckthorn carpenter worm [J]. Forest Pest and Disease, 22(5): 25-28.

Luo Youqing, 1985. Review of natural enemies of chestnut gall wasp [J]. Journal of Beijing Forestry University 2: 82-92.

Ma Chaode, 2003. Control of *Holcocerus hippophaecolus* Hua et chou in seabuckthorn forest in China [J]. Hippophae, 16(2): 15-17.

Ma Qi, Qi Defu, and Liu Yongzhong, 2006. Bionomics of *Smerinthus planus* (Walker) (Lepidoptera: Sphingidae) [J]. Journal of Qinghai University (Natural Science Edition), 24(2): 69-72.

Ma Qixiang and Jiang Kun, 1995. Iconography of diseases, pests, weeds of cotton [M]. Zhengzhou: Henan Science and Technology Publishing House.

Ma Riqian, Gao Shushan, Guo Yuyong et al, 1998. Bionomics of *Diprion jingyuanensis* Xiao et Zhang and control effectiveness of aerial insecticide spray [J]. Shanxi Forestry Science and Technology, (2): 4-5, 8.

Ma Shirui, 1983. Preliminary observations on *Shansiaspis sinensis* Tang and discussions on its control [J]. Inner Mongolia Forestry Science and Technology, (4): 21-24.

Ma Xiaoqian, Wei Xia, Zhou Qi et al, 2008. Repellent effect of two repellents on *Pissodes validirostris* [J]. Forestry Science & Technology, 33(5): 28-29.

Ma Xingqiong, 2007. Wood boring pests of mulberry and their control [J]. Sichuan Sericulture, (3): 27.

Ma Yanfang, Xie Zongmu, Zhang Yongqiang et al, 2011. Chemical control study on *Melolontha hippocastani mongolica* Ménétriés [J]. Forest Pest and Disease, 30(6): 41-42.

Ma Yunping, 2007. Bionomics and control techniques of *Coleophora sinensis* Yang [J]. Forestry of Shanxi, (2): 32-33.

Ma Yuting, 2002. Occurrence and management of *Dendrolimus tabulaeformis* [J]. Sci-Tech Information Development & Economy, 12(3): 144-145.

McManus, M. L., B. Forster, M. Knizek, and W. Grodzki, 1999. The Asian longhorned beetle, a newly introduced pest in the United States. Methodology of forest insect and disease survey in Central Europe. Proceedings of the Second Workshop of the IUFRO Working Party 7.03.10. Sion-Chateauneuf, Switzerland, 20-23 April, pp. 94-97.

Meng Dehui, 2007. How to control *Ancylis sativa* Liu [J]. Forest and Fruit Trees, (20): 15.

Meng Gen, Han Bao, and Han Tiejuan, 1990. *Population surveillance* and control of *Monochamus urussovi* Fisher [J]. Journal of Northeast Forestry University, 18(3): 26-30.

Meng Qingying, 2007. Biology and host selection by *Gastrolina depressa* Baly [D]. Shandong Agricultural University master thesis.

Meng Xiangzhi, Ji Yuhe, and Sun Xiufeng, 2000. Study on biological characteristics of *Pissodes* sp [J]. Jilin Forestry Science and Technology, 29(6): 6-11.

Meng Xiangzhi, 2002. Chemical control techniques for *Xylotrechus rusticus* L [J]. Journal of Changchun University, 12(6): 12-14.

Meng Yongcheng and Wang Zhiping, 1984. Bionomics and control of *Pimeus sichunanus* Zhang [J]. Sichuan Forest Science and Technology, (1): 15-17.

Meyrick Edward, 1912. Descriptions of indian microlepidoptera XV [J]. Journ Bombay Natural History Society 21: 852-877.

Meyrick, E, 1939. New microlepidoptera, with notes on others [J]. Transantions of the Royal Entomological Society of London 89: 47-62.

Miao Chunsheng, Cui Jingyue, and Zhang Hui, 1966. Observations on the annual activity and damage of mole crickets [J]. Entomological Knowledge, 10(3): 134-137.

Miao Jiancai, Chi Defu, Chang Guoshan et al, 2006. Demonstration of *Brontispa longissima* control using abamectin and diflubenzuron [J]. Practical Forestry Technology, (12): 26-27.

Miao Jiancai, Chi Defu, Chang Guoshan et al, 2009. Green food production through control of *Brontispa longissima* using abamectin and diflubenzuron [J]. Practical Forestry Technology, (4): 36-37.

Miao Zhenwang, Zhou Weimin, Huo Lüyuan et al, 2000. Study on bionomics of *Dendrocnotus valens* Leconte [J]. Shanxi Forestry Science and Technology, (1): 34-40.

Ming Guangzeng, Yuan Xiumei, Zhao Min et al, 2002. Occurrence and control of *Stathmopoda masinissa* Meyrick [J]. Plant Protection, 28(2): 56-57.

Mu Wei, Zhang Yueliang, Chen Zhaoliang et al, 2007. The efficiency and degradation dynamicas of insecticides for controlling the overwintering larvae of *Carposina niponensis* Walsingham [J]. Acta Phytophylacica Sinica, 34(1): 91-95.

Mu Xifeng, Sun Jingshuang, Lu Wenfeng et al, 2010. Bionomics and control of *Obolodiplosis robiniae* in Beijing [J]. Forest Pest and Dieseasae, 29(5): 15-18.

Nakanishi, T., C. Goto, M. Kobayashi et al, 2010. Comparative studies of Lepidopteran Baculovirus-Specific Protein FP25K: Development of a vovel Bombyx mori nucleopolyhedrovirus-based vector with a modified fp25K gene [J]. Journal of Virology 84: 5191-5200.

Nengnai Zhibu and Bai Wenhui, 1980. A serious pest of *Caragana korshinskii*, *Kytorrhinnus immixtus* Moschulsky [J]. Entomological Knowledge, 17(5): 212-213.

Nie Junqing, Sun Jing, Zhang Yumei et al, 2003. Bionomics and control of *Zeiraphera griseana* (Hübner) [J]. Qinghai Agriculture and Forestry Science and Technology, (supplement): 68, 41.

Niu Jianzhong, Zhang Jibin, and Liang Jialin, 1999. Observation on biological features of Zamacra excarata [J]. Hebei Journal of Forestry and Orchard Research, 14(4): 355-357.

Niu Jianzhong, 2000. Occurrence of *Zamacra excarata* Dyar on apple trees and its control techniques [J]. China Fruit News, 16(5): 44.

Northwest Agriculture and Forestry University - Department of Plant Protection, 1978. Economic Insect Fauna of Shaanxi Province (Lepidoptera, Butterflies) [M]. Xi'an: Shaanxi People's Publishing House.

Office of the Jiangsu Forest Disease and Pest Survey, 1982. Checklist of forest diseases, pests, and natural enemies in Jiangsu.

Oszi, B., M. Landanyi, and L. Hufnagel, 2005. Population dynamics of the sycamore lace bug, Corythucha ciliata (Say) (Heteroptera: Tingidae) in Hungary [J]. Applied Ecology and Environmental Research, 4: 135-150.

Ou Bingrong and Hong Guangji, 1984. Bionomics of lac insects [J]. Acta Entomologia Sinica, 27(1): 70-78.

Ou Bingrong and Hong Guangji, 1990. Description of a new species of *Kerria* (Homoptera: Lacciferidae) in Yunnan province [J]. Entomotaxonomia, 7(1): 16-17.

Ouyan Guiming and Huang Weihe, 1993. Study on the biological characteristics of *Megabeleses liriodendrovorax* Xiao [J].Forest Pest and Disease, (2): 12.

Ouyan Qunqing and Zheng Zhiying, 1991. A study on the control of *Tonica niviferana* Walker [J]. Guangdong Landscape and Architecture, (2): 40-42.

Pan Jianguo, Tian Zejun, and Bai Jinshi, 1985. Relationship between harvest time and quality of *Malaphis chinensis* Bell [J]. Forest Science and Technology, (9): 8-10.

Pan Wuyao, Tang Ziying, and Chen Zepan et al, 1989. Bionomics of *Hemiberlesia pitysophila* Takagi [J]. Forest Pest and Disease, (1): 1-6.

Pan Wuyao, Tang Ziying, Lian Junhe et al, 1987. Study on the control of *Hemiberlesia pitysophila* using resin and diesel suspension [J]. Forest Pest and Disease, (1): 14-17.

Pan Wuyao, Tang Ziying, Xie Guolin et al, 1987. Study on a new pest in southern China - *Hemiberlesia pitysophila* Takagi (Hemiptera: Diaspididae) [J]. Collections of entomological reserch papers, (7): 177-189.

Pan Wuyao, Tang Ziying, Xie Guolin et al, 1993. Introduction and application of *Coccobius azumai* Tachikawa (Hymenoptera: Aphelinidae) [J]. Forest Pest and Disease, (1): 15-18.

Pang Zhenghong, 2006. Economic forest disease and pest control techniques [M]. Nanning: Guangxi Science and Technology Publishing House.

Pei Feng, Sun Xingquan, Ye Lihong et al, 2008. Occurrence and control of *Thalassodes quadraria* Guenée [J]. Anhui Agricultural Science Bulletin, 14(22): 98-99.

Pei Xinchun, 2009. Control of *Dendrolimus superans* (Butler) in *Pinus silvestris* var. *mongolica* forests using plastic bands [J]. Inner Mongolia Forestry Investigation and Design, 32(1): 92-93.

Peng Jianwen, 1959. Preliminary observations on the life history and habits of *Lebeda nobilis* Walker - a pest on *Camella oleose* in Hunan [J]. Acta Entomologica Sinica, 9(4): 336-341.

Peng Longhui, Xu Yongqing, Tang Yanlong et al, 2007. Stand factors and risk assessment of harm extent of *Hylobitelus xiaoi* Zhang [J]. Acta Agricultruae Universitatis Jiangxiensis, 29(5): 745-749.

People's Committee of Dongyang County, 1958. Experiences of controlling *Euproctis bipunctapex* (Hampson) (Lepidoptera: Lymantriidae) [J]. Zhejiang Forestry, (4): 38-40.

Picimbon, J. F., C. Gadenne, J. M. Bécard et al, 1997. Sex pheromone of the French black cutworm moth, Agrotis ipsilon (Lepidoptera: Noctuidae): Identification and regulation of a multicomponent blend [J]. Journal of Chemical Ecology 23(1): 211-230.

Pingxiang Science and Technology Bureau, 1976. Preliminary report on the studies of *Oides leucomelauena* Weise [J]. Guangxi Forest Science and Technology Papers, (2).

Pu Fuji, 1980. Economic insect fauna of China Fasc. 19. Coleoptera: Cerambycidae (2) [M]. Beijing: Science Publishing House.

Pu Yonglan, Yang Shizhang, Lin Lin et al, 2003. Bionomics and control of *Alcidodes juglans* Chao [J]. Entomological Knowledge, 40(3): 262-264.

Qi Chengjin and Geng Bingtian, 1992. Study on the larval instars of *Omphisa plagialis* Wileman [J]. Entomological Knowledge, (4): 241-242.

Qi Chengjin, 1999. Cerambycidae of Shandong province [M]. Jinan: Shandong Science and Technology Publishing House.

Qi Jie, Luo Youqing, Huang Jingfang et al, 1999. Construction of the life table of *Anoplophora glabripennis* (Motschulsky) [C]. In "Advances of major forest diseases and pests in China".

Qi Jingliang and Wang Yuying, 1981. Studies on *Matsucoccus yunnanensis* Ferris (1) [J]. Scientia Silvae Sinicae, 17(1): 20-25.

Qi Qinglan, 1987. Life history and control of *Phloeosinus aubei* [J]. Entomological Knowledge, 24(1): 34-36.

Qi Xinhua, 2000. Bionomics and control of *Podagricomela shirahatai* (Chŭjŏ) [J]. Forest Pest and Disease, (6): 26-27.

Qian Fanjun and Ji Baozhong, 1998. Bioactivity of insect growth regulator on *Batocera horsefieldi* (Hope) adults [J]. Journal of Nanjing Forestry University, 22(1): 91-94.

Qian Fanjun and Yu He, 1986. Study on the bioecology of the borers of two species of *Pinus sylvestris* var. *mongolica* [J]. Journal of Northeast Forestry University, 14(4): 60-66.

Qian Fanjun and Yuhe, 1984. Field keys to borers of *Pinus silvestris* var. *mongolica* [J]. Forest Pest and Disease, (4): 36-38.

Qian Fanjun, Weng Yuzhen, and Yu Rongzhuo, 1994. Major pests in Chinese fir seed orchard and their integrated control strategies [J]. Fujian Forestry Science and Technology, 21(3): 76-80.

Qian Fanjun, Weng Yuzhen, Yu Rongzhuo et al, 1992. The study of effect of cone pests and discoloration on output and quality of seed in Chinese fir seed orchard [J]. Journal of Nanjing Forest University, 16(1): 31-34.

Qian Fanjun, Weng Yuzhen, Yu Rongzhuo et al, 1992. The study of major pests and their occurrence in Chinese fir seed orchard [J]. Journal of Central South Forestry University, 12(2): 152-156.

Qian Fanjun, Yu Rongzhuo, Zhang Fushou et al, 1995. A study on the bioecology of *Dichomeris bimaculatus* Liu et Qian [J]. Journal of Nanjing Forest University, 19(4): 27-32.

Qian Fanjun, Yun Junjie, and Du Xisheng, 1997. Distribution of *Batocera horsfieldi* (Hope) eggs on poplar tree trunk [J]. Journal of Central China Forest College, 17(3): 82-85.

Qian Fanjun, Yun Junjie, Ye Zhongya et al, 1996. Food sources for adult *Batocera horsfieldi* (Hope) and their effect on spread of the pest [J]. Journal of Central China Forest College, 16(2): 62-64.

Qian Fanjun, 1981. A preliminary report on *Dioryctria* sp. Forest Science and Technology, (9): 27-30.

Qian Fanjun, 1988. Survey of borers of *Pinus silvestris* var. *mongolica* [J]. Forest Pest and Disease, (3): 35-37.

Qian Tingyu, 1964. Study on two borers on lychee trees [J]. Acta Entomologica Sinica, 13(2): 159-167.

Qiang Zhonglan and Ma Changlin, 2007. Occurrence of *Cyamophila willieti* (Wu) (Hemiptera: Psyllidae) and control experiments [J]. Journal of Gansu Forestry Science and Technology, 32(1): 60-63.

Qiao Shichun, Peng Aijia, and Li Yanfeng, 2008. Study on the bionomics of *Diorhabda elongata deserticola* Chen [J]. Practical Forestry Technology, (12): 27-28.

Qin Changsheng, Xu Jinzhu, Xie Penghui et al, 2008. Screening of *Metarhizium anisopliae* compatible insecticides and control of red palm weevil with mixed insecticide mixtures [J]. Journal of South China Agricultural University, 29(2): 44-46.

Qin Xiuyun, Li Fenghua, Zhang Liqing et al, 2000. Application of omethoate-sugar-vinegar solution to control larch cone flies [J]. Forest Science and Technology, (4): 28.

Qin Zhanyi, Liu Shenghu, Yue Caixia, 2007. Bionomics and integrated pest management of *Cydia pomonella* at Dunhuang region in Gansu province [J]. Plant Quarantine, 21(3): 170-171.

Qinghai Forest Bureau, 1982. Collection of forest disease and pest survey results in Qinghai (1980-1982).

Qiu Qiang et al, 2004. Colored pictorial guide to Chinese fruit tree diseases and pests [M]. Zhengzhou: Henan Science and Technology Publishing House.

Qiu Shumei, Luo Runquan, and Yao Yong, 2010. Occurrence and control techniques for *Agrotis ipsilon* in Nanchang county [J]. Modern Horticulture, (5): 54-55.

Qu Bangxuan, Chen Hui, Liu Fudai et al, 1991. Study on *Melanophila decastigma* Fabricius II - occurring regularity and its control [J]. Journal of Northwestern Forestry College, 7(1): 28-33.

Qu Hong, 1981. Preliminary study on the physical and chemical properties of white wax [J]. Chemistry and Industry of Forest Products, 1(4): 44-48.

Rao Wencong, Luo Yongsen, Ding Wenhua et al, 2006. Efficacy of "Lesao" insecticide on *Paurocephara psylloptera* [J]. Guangdong Silkworm Industry, (2): 28-30.

Ren Li, 2005. Bionomics and control techniques of *Coleophora obducta* [J]. Henan Forest Science and Technology, 25(4): 24-25.

Ren Shaofu and Wu Gaorong, 1990. A pest of forest trees, *Ascotis selenaria dianaria* Hübner [J]. Entomological Knowledge, (5): 303-304.

Ren Shuang, 2009. Study on the parasitic wasp species of *Dryocosmus kuriphilus* and their control in Chongqing region [J]. Anhui Agricultural Science, 37(7): 3055-3056, 3103.

Ren Xianwei, 1995. Dendrology [M]. Beijing: China Forestry Publishing House.

Ren Xianwei, 2008. Dendrology (northern edition) [M]. Beijing: China Forestry Publishing House.

Ren Yuegang, Cui Jianye, and Li Yaohui, 2001. Life history and integrated control of *Cinara* sp. [J]. Inner Mongolia Forest Science and Technology, (4): 35-36.

Ren Zuofo, 1959. A preliminary survey of *Dendroctonus armandi* Tsai et Li at Qinling [C]. In "Forest pest research report". Beijing: Science Publishing House.

Ren Zuofo, 1959. A preliminary suvey of *Dendroctonus armandi* Tsai et Li at Qinling [C]. In "A preliminary study report on forest pests". Beijing: Science Publishing House.

Ren Zuofo, 1964. Brief description of *Sirex gigas* L. [J]. Entomological Knowledge, 8(6): 260-261.

Rong Xuanxiong, Chen Murong, Deng Xianghui et al, 2003. Efficacy studies of five insecticides against *Brontispa longissima* (Gestro) [J]. Guangdong Forestry Science and Technology, 19(4): 49-50.

Run Guixin, Song Fuping, and Zhang Jie, 2007. Biological control of grubs [C]. Proceedings of the annual meeting of the Chinese society of plant protection. pp. 582-588.

Run Haike, Li Haiqiang, Zhang Yaozeng et al, 2007. Occurrence and control of *Byctiscus omissus* Voss [J]. Shaanxi Forest Science and Technology, (3): 81-82.

Schaffer, P. W., Yao Defu, and You Dekang, 2000. Capture of Stilpnotia candida males in traps baited with "Leucomalure": the synthetic sex pheromone of Leucoma salicis [J]. Forest Pest and Disease (5): 39-41.

Seo, S. T., D. L. Chambers, M. Komura, and C. Y. L. Lee, 1970. Mortality of mango weevil treated by dieletric heating [J]. Journal of Economic Entomology, 63: 1977-1978.

Shaanxi Forest Research Institute and Hunan Forest Research Institute, 1990. Illustrated guide to parasitoids of forest pests [M]. Yangling: Tianze Press.

Shang Dieren, 1981. Butterflies of Taiwan [M]. Taiwan Natural Science and Culture Enterprise Co.

Shang Fengshuang and Pang Yongshi, 1997. Bionomics and control of *Omphisa plagialis* Wileman [J]. Jilin Forestry Science and Technology, (4): 13-14.

Shanghai Forest Station, 2004. Forest disease and pest control [M]. Shanghai: Shanghai Science and Technology Publishing House.

Shangluo Forest Station, 1975. Preliminary study on *Curculio davidi* [J]. Shaanxi Forest Science and Technology, (2): 71.

Shao Jingwen, Meng Fanrong, Zhou Haiying et al, 1994. Bionomics of *Cydia zebeana* (Ratzeburg) [J]. Journal of Northeast Forestry University, (5): 100-104.

Shao Qianghua, 1990. Studies on the integrated control tests of *Shansiaspis sinensis* [J]. Journal of Beijing Forestry University, 12(3): 74-82.

Shao Yanzhen, Yan Xuebin, Chen Zhenhua et al, 2001. Control of *Monochamus urussovi* Fisher using phoxim [J]. Forest Science and Technology, 26(1): 26-27.

Shen Fuyong, Zhu Yuxing, and Zhang Yujun, 2001. Quarantine techniques for *Sonsaucoccus sinensis* (Chen) [J]. Plant Quarantine, 15(2): 87-88.

Shen Guangpu and Liu Youqiao, 1988. Small moths from Jiangxi [J]. Acta Agriculturae Universitatis Jiangxiensis, (special issue): 68-69.

Shen Guangpu, Xiao Xiaoling, and Zhu Yisu, 1985. Ecology and the spatial distribution of the *Trioza camphorae* Sasaki eggs and larvae [J]. Forest Science and Technology, (2): 24-29.

Shen Jie, Zhang Qingrong, and Xu Zhihong, 2002. Quarantine and control of a dangerous pest of flowers and trees - *Opogona sacchari* (Bojer) [J]. Zhejiang Forestry Science and Technology, 22(3): 238-242.

Shen Ping, Chang Chengxiu, Zhang Yongqiang et al, 2008. Morphological characteristics and population dynamics of *Cyamophila willieti* (Wu) [J]. Gansu Forestry Science and Technology, 33(1): 30-33.

Shen Qiang, Wang Juying, Liu Jianding et al, 2007. Bionomics and control of *Ricania shantungensis* [J]. Entomological Knowledge, 44(1): 116-118.

Shen Qiang, Xu Qiyao, Zhang Yifeng et al, 1993. Bionomics of *Chionaspis saitamensis* Kuwana (Hemiptera: Diaspididae) [J]. Forest Pest and Disease, (4): 22-23.

Shen Qiang, Xu Qiyao, Zhang Yifeng et al, 1994. Study on *Chionaspis saitamensis* Kuwana (Hemiptera: Diaspididae) [J]. Entomological Knowledge, 31(6): 353-356.

Sheng Maoling and Sun Shupiing, 2010. Natural enemies of wood borers in China - Ichneumonidae [M]. Beijing: Science and Technology Publishing House.

Sheng Maoling, Gao Lixin, Sun Shuping et al, 2002. Parasitoids of *Pachynematus itoi* Okutani and their control effect [J]. Liaoning Forest Science and Technology, (2): 1-3.

Sheng Maoling, 2005. Systematic studies on Ichneumonidae parasitizing woodboring insects in Palearctic region in China [D]. Doctoral dissertation. Beijing Forestry University.

Sheng Maoling, 2005. Taxonomic studies on the natural enemies of wood borers - Ichneumonidae (Hymenoptera) in Palearctic Region [D]. Beijing Forestry University Doctoral dissertation.

Shi Guanglu, Wang Zhihong, Zhao Lilin et al, 2002. Study on the applying technology of ecological regulation and management of chemistry of *Coleophora sinensis* Yang [J]. Journal of Shanxi Agricultural University, 22(4): 307-310.

Shi Juan, Li Zhenyu, Yan Guozeng et al, 2004. Stand description factors and risk analysis of harm extent of the Gypsy moth [J]. Scientia Silvae Sinicae, 40(1): 106-109.

Shi Mubiao, Xian Bingcai, Dai Guanqun et al. A preliminary study on the nuclear polyhedrosis virus of three pests of ornamental plants [J]. Forest Science and Technology, (6): 27-28.

Shi Shuqing, 2012. Systematic studies on Chinese Aseminae and Disteniidae (Coleoptera: Cerambycoidea) [D]. Doctoral dissertation. Southwestern University.

Shi Tiesong, Sun Zuomin, and Wang Wenge, 2005. Chemical control of *Cydia zebeana* (Ratzeburg) [J]. Protecion Forest Science and Technology, (2): 79-80.

Shi Weixiang, 2007. Control of subterranean pests in flower beds [J]. Modern Agricultural Science and Technology, (13): 91.

Shi Zemei, 2006. Study on biological characteristics and control of *Rhyacionia pinicolata* (Doubleday) in Xinglong Natural Reserve of Gansu province [J]. Journal of Gansu Forestry Science and Technology, 31(2): 54-55.

Shu Chaoran, 2003. Current research on *Coleophora obducta* (Meyrick) and prospects of future research [J]. Liaoning Forest Science and Technology, (1): 177-189.

Shukla, R. P, 1985. Bioecology and management of mango weevil Sternochetus mangiferae (F.) [J]. International Journal of Tropical Agriculture 3: 292-303.

Sichuan Disease and Forest Pest Survey Office, 1982. Checklist of forest insects in Sichuan.

Sichuan Province Wu Bei Zi Research Team, 1985. Study on the quality of Chinese galls [J]. Forest Science and Technology, 52(4): 1-9.

Singh, B. L, 1960. The mango [J]. London: Leonard Hill (Book) Ltd., 391-321.

Situ Yingxian and Yang Bing, 1989. Research on *Stermochetus olivieri* Faust I: morphology and biology [J]. Journal of Southwest Forest College, 9(3): 36-46.

Situ Yingxian, Aiyonghua, and Feng Shiming, 2000. Description of the morphology of larvae and pupae of five weevils in Yunnan province [J]. Journal of Southwest Forest College, 20(2): 100-106.

Situ Yingxian, 1983. A study on *Compsogene panopus* (Cramer) [J]. Tropical Agricultural Science and Technology, 4(2): 53-54.

Situ Yingxian, 1991. Research on Stermochetus olivieri Faust I: integrated pest management [J]. Journal of Southwest Forest College, 11(1): 72-78.

Situ Yingxian, 1992. Integrated pest management in mango orchard [J]. Journal of Southwest Forestry College, 12(2): 180-184.

Situ Yingxian, 1993. The spread of mango weeviles (Curculionidae) and their identification [J]. Journal of Southwest Forestry College, 13(3): 177-181.

Smith, R. H, 1961. Red turpentine beetle. Forest Pest Leaflet 55 (revised). Washington DC: U. S. Department of Agriculture. Forest Service, 8.

Sohn, J. C., and Wu, C. S, 2013. A taxonomic review of Attevidae (Lepidoptera: Yponomeutoidea) from China with descriptions of two new species and a revised identity of the Ailanthus webworm moth, Atteva fabriciella, from the Asian tropics [J]. Journal of Insect Science, 13: 1-16.

Song Jinfeng, 2004. Control of *Scythropus yasumatsui* on ziziphus seedlings [J]. Practical Forest Technology, (6): 29.

Song Liwen, Ren Bingzhong, Sun Shouhui et al, 2005. Field trapping test on semiochemicals of pine shoot beetle *Tomicus piniperda* L. [J]. Journal of Northeast Forestry University, 33(1): 38-40.

Song Liwen, 2005. Studies on aggregation pheromone of *Ips subelongatus* Motschulsky - bioactivity and application of synthetic lure [D]. Northeast Normal University master thesis.

Song Shengyin, Wang Yong, Ou Zhengquan et al, 2008. Bionomics of *Nesodiprion zhejiangensis*, a new record in Guizhou province [J]. Forest Pest and Disease, 27(3): 22-23, 41.

South China Agricultural University and Guangzhou Landscape Bureau. 1985. Control of diseases and pests of ornamental plants [M]. Guangzhou: Guangdong Science and Technology Publishing House.

South China Tropical Crop College, 1980. Control of diseases and pests of tropical crops [M]. Beijing: Agriculture Publishing House.

Su Mei and Yang Fenyong, 2000. Techniques of controlling *Stilpnotia salicis* with ultra low volume spray of diflubenzuron [J]. Journal of Inner Mongolia Forestry Science and Technology, (Supplement): 59-62.

Su Mei, Yang Fenyong, Liu Xiufeng et al, 2007. Survey techniques for parasitoids of *Orgia ericae* [J]. China Suburban Science and Technology, (2): 19-20.

Su Shiyou and Zhoubo, 1988. Bionomics and control of *Phloeosinus sinensis* Schedle [J]. Scientia Silvae Sinica, 24(2): 239-241.

Su Xing and Wu Jianfen, 1977. Preliminary study on three wood borers infesting *Casuarina equisetifolia* [J]. Entomological

Knowledge, 14(5): 152-154.

Su Xiufeng and Wang Hongmei, 2005. Integrated control of *Saperda carcharias* (L.) [J]. Xinjiang Forestry, (3): 26.

Su Zhiyuan, 2004. Pictorial guide of forest pests in Liangshan [M]. Chengdu: Sichuan Forest Science Publishing House.

Sun Dexiang et al, 1989. Spatial distribution pattern of *Parthenolecaniun corni* (Bouche) [J]. Journal of Inner Mongolia Forestry College, (1): 102-107.

Sun Dexiang, Wang Jianyi, Wu Guangrong et al, 1993. A preliminary study on *Prodiaspis tamaricicola* Young [J]. Journal of Ningxia Agricultural College (Agricultural Science edition), 14(4): 28-31.

Sun Fenghai, Fang Aicheng, Sun Xianhua et al, 1994. Observation on the biology of *Asias halodendri* (Pallas) [J]. Shandong Forestry Science and Technology, (3): 40-41.

Sun Jianghua, Roques, and Fang Sanyang, 1996. Some biological and ecological characteristics of *Strobilomyia melaniola* (Fan) (Diptera: Anthomyiidae) [J]. Scientia Silvae Sinicae, 32(3): 238-242.

Sun Jimei, Ding Shan, Xiao Hua et al, 1997. A study on control of *Monochamus alternatus* using *Beauveria bassiana* [J]. Forest Pest and Disease, (3): 16-19.

Sun Lihua, Song Youwen, and Shan Lihua, 1987. A preliminary study of *Mirina chystophi* Staudiager [J]. Acta Entomologica Sinica, 30(4): 455-457.

Sun Qiaoyun, Tai Xiaoling, and Xu Longdi, 1985. Preliminary observation on the biology of *Orgia gonostigma* L. [J]. Jiangsu Forestry Science and Technology, (1): 30-31.

Sun Qionghua, 1991. Biology and control of *Dictyoploca japonica* [J]. Forest Research, 4(3): 273-279.

Sun Shaofang, 2003. A study on bio-control of Paecilomyces farinosus on *Cyllorhynehites ursulus* and *Curculio bimaculatus* [J]. Yunnan Forestry Science and Technology, (1): 19-23.

Sun Shiying, Lü Zexun, Huang Guanhui et al, 1987. Epidemics of *Erannis ankeraria* Staudinger NPV [J]. Forest Pest and Disease, (1): 23-24.

Sun Xiaoling, Chengbin, Gao Changqi et al, 2006. Current research on the occurrence and control of *Mallambyx raddei* Blessig [J]. Journal of Jilin Normal University (Natural Science edition), (1): 54-56.

Sun Xiaoling, Gao Changqi, Cheng Bin et al, 2006. Adult emergence rule of *Ips typographus* Linnaeus monitored by pheromone [J]. Journal of Northeast Forestry University, 34(3): 7-8.

Sun Xiaoling, 2006. Study on Semio-chemicals of *Ips typographus* L. [D]. Northeast Forestry University Doctoral dissertation.

Sun Xingquan, Jiang Gendi, Gu Yanfei et al, 2006. Bionomics of *Mesoneura rufonota* and the effect of temperature on larval development period [J]. Forest Pest and Disease, 25(2): 7-9.

Sun Xuehai, 2003. Infestation characteristics of *Opogona sacchari* on *Broussonetia papyrifera* and its control [J]. Plant Protection Technology and Extension, (11): 15-16.

Sun Xugen and Li Zhanpeng, 2001. Forest and fruit pest control [M]. Beijing: China Forestry Publishing House.

Sun Yongchun and Fan Minsheng, 1965. Preliminary observations on *Dryocosmus kuriphilus* (Yasumatsu) [J]. Entomological Knowledge, 9(5): 286-289.

Sun Yongchun, 1963. Preliminary study on *Curculio davidi* [C]. In "Proceedings of the Chinese Forest Association". Beijing: Agriculture Publishing House.

Sun Yujia and Zhang Zhaoyi, 1989. Study on *Hapsifera barbara* (Christoph) [J]. Acta Entomologica Sinica, 32(3): 350-353.

Sun Yujia, 1989. A study on Pterostoma sinicum Moore [J]. Shandong Forestry Science and Technology, (1): 47.

Sun Zhangding, 1942. Production and sales of Chinese galls in western Hunan and eastern Guizhou [J]. Xi Nan Shi Ye Tong Xun, 5(4): 95-104.

Tan Jicai, 2002. Control of diseases and pests on tea plants [M]. Beijing: China Agriculture Publishing House.

Tan Juanjie et al, 1980. Economic insect fauna of China Fasc. 18. Coleoptera: Chrysomeloidea (1) [M]. Beijing: Science Publishing House.

Tan Weiquan, Chen Siting, Huang Shanchun et al, 2006. Occurrence and control of *Brontispa longissima* in Hainan province [J]. China Southern Fruit Trees, 35(1): 46-47.

Tan Weiquan, Zhao Hui, and Han Chaowen, 2002. Occurrence of red palm weevil in Hainan and its control [J]. Yunnan Tropical Crop Science and Technology, 25(4): 29-30.

Tang Fangde and Hao Jingchu, 1995. The Margarodidae in China [M]. Beijing: China Science and Technology Publishing House.

Tang Fangde, Zhou Jing, Sun Zhanxian et al, 1980. Infestation of Diaspididae scales on norther forest belts [J]. Shanxi Agriculture Science, (4): 8-11.

Tang Fangde, 1977. Chinese scale insects in horticulture (vol. 1). Shenyang City Horticulture Research Institute.

Tang Fangde, 1984. Chinese scale insects in horticulture (vol. 2). Shangxi Agricultural University.

Tang Fangde, 1986. Chinese scale insects in horticulture (vol. 3). Shangxi Agricultural University.

Tang Fangde, 1991. Chinese Coccidae [M]. Taiyuan: Shanxi Gaoxiao Lianhe Publishing House.

Tang Guanghui, Jiang Zhili, Zhang Wenfeng et al, 2006.

Efficacy of trunk injection with insecticides to control red palm weevil [J]. Forest Pest and Disease, 25(4): 39-41.

Tang Guangzhong, 2002. Survey of *Eurytoma maslovskii* Nikolskaya resistant wild apricot [J]. Hebei Forestry Science and Technology, 2: 13-14.

Tang Guanzhong, Lü Rongyan, Guo Zhenping et al, 1999. Spatial distribution of *Eurytoma maslovskii* Nikolskaya eggs [J]. Hebei Forest Science and Technology, (4): 10-12.

Tang Guanzhong, Niu Jingsheng, Liu Yufen et al, 1999. Bionomics of *Eurytoma maslovskii* Nikolskaya [J]. Forest Pest and Disease, (3): 5-7.

Tang Guanzhong, Pen Jinyou, Guo Zhenping et al, 2006. Eurytoma maslovskii *Nikolskaya resistant* wild apricot and mechanism of resistance [J]. Hebei Forestry Science and Technology, (1): 7-8.

Tang Guanzhong, Peng Jinyou, Guo Zhenping et al, 2005. Morphology of the Eurytoma maslovskii Nikolskaya and Eurytoma samsonowi Wassiliew larvae and differences in their infestation characteristics [J]. Hebei Forest Science and Technology, (3): 24.

Tang Guanzhong, Yan Hailiang, Guo Zhenping et al, 2007. Control of *Eurytoma maslovskii* Nikolskaya adults [J]. Hebei Forest Science and Technology, (1): 1-3.

Tang Guanzhong, 2005. Integrated management of *Cryptorrhynchus lapathi* Linnaeus [J]. Hebei Forest Science and Technology, (2): 1-3.

Tang Jue et al, 1987. Study on the biology of Chinese horned gall aphid - *Schlechtendalia chinensis* (Hemiptera: Pemphigidae), overwintering generation can develop into apterous parthnogenetic adults [J]. Resource Insects, 2(1-2): 14-16.

Tang Jue, 1956. Chinese galls [J]. Entomological Knowledge, 2(3): 113-116.

Tang Jue, 1976. Studies on Chinese galls (2) - observations on mutualistic aphids [J]. Acta Entomologica Sinica, 19(3): 282-296.

Tang Li, Zhao Yumei, Tang Cai, 2005. Study on functional response of *Nephus Rhuguus* (Kamia) to Oracella acuta (Lobdell) [J]. Natural Enemies of Insects, 27(1): 27-31.

Tang Weiqiang, 2000. Bionomics and control of *Xyleborus mutilatus* Blandford [J]. Journal of Zhejiang Forest College, 17(4): 417-420.

Tang Xinpu, 1996. Practical control technologies of fruit diseases and pests [M]. Beijing: China Agriculture Science and Technology Publishing House.

Tang Yanlong, Wen Xiaosui, Xu Yongqing et al, 2007. Application of grey relational grade analysis to studying the influence level of environmental factors in slash pines infested by *Hylobitelus xiaoi* Zhang [J]. Acta Agriculturae Universitatus Jiangxiensis, 29(3): 356-359.

Tangjue and Cai Banghua, 1957. Study on the Chinese gallnut in Meitan of Guizhou [J]. Acta Entomologica Sinica, 7(1): 113-140.

Tao Jiaju, 1943. Identification of Chinese gall aphids in Sichuan [J]. New Agriculture and Forestry, 2(3): 17-21.

Tao Mei, Chen Guohua, and Yang Benli, 2003. Parasitoids of *Crescoccus candidus* Wang [J]. Journal of Yunnan Agricultural University, 18(4): 413-415.

Tao Wanqiang, Guan Ling, Yu Juxiang et al, 2003. Pest risk analysis and risk management of *Messa taianensis* [J]. Forest Pest and Disease, 22(4): 8-9.

Tao Wanqiang, Pan Yanping, Liu Huan et al, 2009. Insect fauna of Lepidoptera in Songshan Natural Reserve of Beijing [J]. Anhui Agricultural Research, 37(6): 2592-2595.

Tao Wanqiang, 2008. Bionomics of Hyphantria cunea in Beijing [J]. Forest Pest and Disease, (2): 9.

Tao Weichang and Wang Mingfeng, 2004. *Malacosoma dentata* Mell infestation and its control [J]. Anhui Forestry Science and Technology, (1): 18-19.

Tavella, L. and A. Arzone, 1987. Investigations on the natural enemies of Corythucha ciliate (Say) (Rhynchota Heteroptera) [J]. Redia 70: 443-457.

Tea Research Institute of Chinese Academy of Agricultural Sciences, 1974. Tea tree disease and pest control [M]. Beijing: China Agriculture Publishing House

Technical Group for Aerial Application of Insecticides for Forest Pests, 1978. Summary of the techniques using aerial application to control forest pests [J]. Ningxia Agricultural Science and Technology, (1): 38-46.

Tian Feng, Zhang Yanmin, Wang Wenzhi et al, 1993. Natural enemies of *Costera anastomosis* L. (Lepidoptera: Notodontidae) [J]. Natural Enemies of Insects, 15(4): 155-156.

Tian Guanghe, 1991. Insecticide treated rope to control *Scythropus yasumatsui* Kono et Morinoto [J]. Henan Forest Science and Technology, (2): 10.

Tian Runmin and Tang Mengchang, 1997. A preliminary study on *Holcocerus hippophaecolus* (Lepidoptera: Cossidae) [J]. Journal of Inner Mongolia Forestry Science and Technology, (1): 36-38.

Tian Yu and Liu Mengying, 1990. The location and ultrastructure of pheromone gland of the female yellow peach moth (*Dichocrocis punctiferalis* Guenee) [J]. Acta Entomologica Sinica, 33(2): 254-255.

Tian Zejun et al, 1985. Study on six hosts of overwintering

Schlechtendalia chinensis (Bell) [J]. Animal World, 2(1): 55-58.

Tian Zejun et al, 1986. Investigation of the resources, distribution, production of Chinese galls [J]. Resource Insects, 1(1): 20-28.

Tian Zejun, Pan Guangquan, Pan Yanzheng et al, 1988. Bionomics and population prediction of four species of Chinese gall on *Rhus chinensis* [J]. Animal Research, 9(4): 401-408.

Tian Zejun, 1996. Cultivation techniques for Chinese galls [M]. Beijing: Jindun Publishing House.

Tong Chaoran, Li Xinzheng, Liu Anmin et al, 1985. Biological characteristics and control of *Parthenolecaniun corni* (Bouche) [J]. Forest Science and Technology, (5): 29-32.

Tong Guojian, Tang Ziying, Pan Wuyao et al, 1988. A preliminary study on the population dynamics of *Hemiberlesia pitysophila* [J]. Forest Science and Technology, (2): 6-11.

Tong Xinwang and Lao Guangmin, 1984. A pest of *Cinnamomum camphora* - *Creace stipatana* Walker [J]. Hunan Forest Science and Technology, (4): 44-45.

Tong Xinwang, Fu Youbin and Ni Lexiang, 2005. The species and protection and utilization of parasitic wasps of *Dryocosmus kuriphilus* (Yasumatsu) in Hunan [J]. Hunan Forestry Science and Technology, 32(2): 35-37.

Tong Xinwang, 2004. Study and control of termites in Hunan province [M]. Changsha: Hunan Science and Technology Publishing House.

Tong Ying, 2003. Integrated control of *Diprion jingyuanensis* Xiao et Zhang [J]. Sci-Tech Information Development and Economy, 13(3): 162-163.

Tonglu County Forest Bureau, 1959. Control of tallow tree brown tail moth [J]. Zhejiang Forestry, (5): 32.

Tremblay, E. and C. Petriello, 1984. Possibilities of rational chemical control of Corythucha ciliate (Say) (Rhynchota, Tingidae), on the basis of phenological data [J]. Difesa delle Piante 7: 237-244.

Tsai, P. H. et al, 1946. The classifications of three genera and six new species from Meitan, Kweichow [J]. Transactions of the Royal Entomological Society of London 97: 405-418.

Varshney, R. K, 1976. Taxonomic studies on lac insects of India [J]. Oriental Insect Supplement (5): 1-97.

Varshney, R. K, 1984. A review of family (Kerridae) in the Orient (Homoptena: Coccoidea) [J]. Oriental Insect, 18: 361-385.

Wan Shaoxia and Zhang Lifeng, 2004. Occurrence and control of *Halyomorpha picus* on fruit trees [J]. Hebei Agricultural Science and Technology, (8): 18.

Wan Shaoxia and Zhang Lifeng, 2005. Control wood borers with Dapnne genkwa [J]. Forest Pest and Disease, 24(2): 29.

Wan Wenlin, 1998. Bionomics and control of *Acantholyda posticalis* [J]. Journal of Henan Institute of Education, (1): 56-58.

Wang Aijing and Bai Jiuwei, 1984. Study on a new pest of elm - *Bacculatrix* sp. [J]. Journal of Northeast Forest College, 9(4): 376-380.

Wang Aijing et al, 2000. A new species of parasitoid, *Leluthia* sp., on Turanoclytus namangensis [J]. Xinjiang Forest Science and Technology, 2: 28-29.

Wang Aijing, Liu Hongguang, and Deng Kerong, 1999. Study on the losses caused by *Xyloterchus namanganensis* Heydel [J]. Scientia Silvae Sinicae, 35(2): 72-76.

Wang Aijing, Liu Hongguang, Deng Kerong et al, 1999. Study on spatial distribution model and its application on controlling *Xylotrechus namanganensis* [J]. Forest Research, 13(6): 684-687.

Wang Aijing, Lixia, Liu Hongguang, et al, 1999. Integrated pest management of *Xylotrechus namanganensis* Heydel [J]. Xinjiang Agricultural Science, (5): 223-225.

Wang Aijing, 1983. Study on *Pristiphra congugata* Delht and *Stauronematus compressicomis* Fabricius [J]. Xinjiang Agricultural Science, (6): 20-22.

Wang Aijing, 1984. Biological studies on *Agelastica alni orientalis* Baly [J]. Xinjiang Forest Science and Technology, 14(3): 24-27.

Wang Aijing, 1995. Control threshold of the green leaf hopper, *Cicadella viridis* (L.) [J]. Scientia Silvae Sinicae, 31(1): 81-85.

Wang Aijing, 1995. Efficacy of BA-1 EC against *Agelastica alni orientalis* Baly [J]. Xinjiang Forestry Science and Technology, 43(1): 28-30.

Wang Aijing, 1995. Spatial distribution of the green leaf hopper eggs (*Cicadella viridis* (L.)) [J]. Journal of Northeast Forestry University, 23(1): 40-45.

Wang Aijing, 1996. Biological characteristics of the green leaf hopper eggs (*Cicadella viridis* (L.)) [J]. Xinjiang Agricultural Science, (4): 186-188.

Wang Aijing, 1998. A preliminary study on *Zombrus siostedti* (Fahringer) [J]. Xinjiang Agricultural Science, (5): 220-221.

Wang Aijing, 2001. A study on bionomics of *Xylotrechus namanganensis* [J]. Forest Research, 14(5): 560-565.

Wang Anjing, Wang Chenxiang, and Li Zhonghuan, 1996. Description of the ornamental plant hosts of green leaf hopper [J]. Xinjiang Forestry Science and Technology, (2): 21-23.

Wang Chuanzhen, Wang Jinggang, Yan Juan et al, 2000. Bionomics of *Cephalcia kunyushanica* Xiao [J]. Forest Pest and Disease, (4): 20-22.

Wang Fengyin, Zhang Chuangling, and Li Xuxuan, 2007. Preliminary study on the bionomics and control of *Eulecanium kuwanai* (Kanda) [J]. Liaoning Forestry Science and Technology, (4): 56-57.

Wang Fulian, Li Chuanren, Liu Wanxue et al, 2008. Advance in biological characteristics and control techniques of the new invasive sycamore lace bug (Corythucha ciliata) [J]. Scientia Silvae Sinicae, 44(6): 137-142.

Wang Fuwei, Niu Yanzhang, Hou Liwei et al, 1993. Study on the biological characteristics of *Costera anastomosis* L. (Lepidoptera: Notodontidae) [J]. Jilin Forestry Science and Technology, 22(6): 30-42.

Wang Ge, 2006. Monitoring and integrated management of Pissodes punctatus Langor et Zhang in Tianbao project area of Yunnan province [J]. Forest Inventory and Planning,31(supplement): 167-168, 169.

Wang Gong, 2009. Occurrence and control of *Pristiphora erichisonii* (Hartig) in Xinglong Natural Reserve of Gansu province [J]. Journal of Gansu Forestry Science and Technology, 34(3): 54-56.

Wang Guicheng, Liu Aitu, and Xie Zuying, 1965. Biology and control of *Trioza magnisetosa* Log [J]. Scientia Silvae Sinica, 10(3): 247-256.

Wang Guiqing and Zhou Changhong, 2003. Occurrence and management of *Baculum minutidentatum* Chen et He [J]. Forest Pest and Disease, 22(6): 31-33.

Wang Guiqing, 2000. A preliminary report of the sawfly, *Acantholyda erythrocephala* (Hymenoptera: Pamphiliidae) in Shenyang [J]. Scientia Silvae Sinicae, 36(4): 110-111.

Wang Hailin, Cao Kuiguang, and Chen Gang, 1998. Preliminary study on *Crescoccus candidus* Wang [J]. Journal of Yunnan Forest College, 8(2): 238-241.

Wang Haiming, Niu Yingfu, Liu Baodong et al, 2003. Experiment on the control of *Eriococcus lagerostroemiae* and *Apriona swainsoni* with three kinds of pesticides [J]. Forest Pest and Disease, (2): 27-30.

Wang Haiming, 2005. Biological characteristics and control of *Pryeria sinica* Moore [J]. Shandong Forestry Science and Technology, (2): 46-47.

Wang Jianguo, Lin Yujian, Hu Xueyan et al, 2008. A list of Ennominae from Jiangxi province (Lepidoptera: Geometridae) [J]. Jiangxi Plant Protection, 31(1): 43-48.

Wang Jiangzhu, 1997. Diseases, pests, and weeds of apple and pear trees [M]. Beijing: China Agricultural Publishing House.

Wang Jianquan, 1992. Bionomics and biological control of Stilpnotia candida Staudinger [J]. Gansu Forest Science and Technology, (1): 18-21.

Wang Jianyi, Wu Sanan, Tang Hua et al, 2009. Scale insects in Ningxia and their control [M]. Beijing: Science Publishing House.

Wang Jianyi, Wu Sanan, Tang Hua et al, 2009. Scale insects of Ningxia and their natural enemies [M].Beijing: Science Publishing House.

Wang Jijian, 2002. Bionomics and control of *Jankowskia fuscara* Leech [J]. Guangxi Forest Science, (1): 32-40.

Wang Jijian and Chen Shuchun, 1998. Bionomics and control of *Aruanoidea flavescens* Chen et Wang [J]. Forest Pest and Disease, 27(2): 77-79.

Wang Jijian and Kang Fujuan, 1990. A preliminary study on *Entoria bobaiensis* Chen [J]. Forest Pest and Disease, (2): 3-4.

Wang Jijian and Tan Fujuan, 1993. Bionomics and control of *Euzophera batangensis* Caradja [J]. Forest Pest and Disease, (3): 1-2.

Wang Jijian and Tang Fujuan, 1992. Bionomics of *Macellina digitata* Chen et He [J]. Forest Pest and Disease, 2: 15-17.

Wang Jijian, Chen Jiang, Luo Jitong et al, 2010. Report on the infestation of *Leptocybe invasa* Fisher & La Salle [J]. Guangxi Forestry Science and Technology, 39(4): 208-210.

Wang Jijian, Huang Duankun, Chen Jinning et al, 1997. A study on the scab of *Illicium verum* and its control [J]. Forest Pest and Disease, (4): 6-9.

Wang Jijian, Huang Duankun, Chen Jinning et al, 1998. Two thrips that infesting *Illicium verum* [J]. Guangxi Agricultural Science, (4): 191-193.

Wang Jijian, Jiang Jinpei, Luo Jitong et al, 2011. Research studies on *Leptocybe invasa* Fisher & La Salle in foreign countries [J]. Forest Pest and Disease, 30(1): 12-14.

Wang Jijian, Yang Xiuhao, Liang Chen et al, 2014. *Milionia basalis* Walker, an important pest of *Nageia nagi* [J]. Guangxi Plant Proetection, (2): 22-23.

Wang Jijian, Yang Xiuhao, Luo Jitong et al, 2015. Biological studies of *Endoclyta signifer* Waller [J]. Forest Pest and Disease, (1): 22-24.

Wang Jijian, 1988. Preliminary observation on the bionomics of *Sinophasma mirabile* Günther [J]. Forest Pest and Disease, (1): 11-12.

Wang Jijian, 1989. A new pest of Pinus elliottii, *Philus antennatus* (Gyllenhal) [J]. Guangxi Forest Science and Technology, (3): 18-20.

Wang Jijian, 1998. *Cinara* sp. and its natural enemies [J]. Guangxi Forestry, (2): 28.

Wang Jinyou and Li Zhixing, 1995. Color plates of deciduous fruit tree diseases [M]. Beijing: Jindun Publishing House.

Wang Junmin, Zhao Yansheng, Ai Xianqin et al, 2003. Preliminary report on the chemical control of *Odites issikii* (Takahashi) [J]. Plant Protection Technology and Extension, 23(12): 31-32.

Wang Lianzhen, 2001. Techniques for mass rearing of *Antheraea pernyi* Guerin-Meneville in the 1990s in China [J]. Liaoning Silk, (2): 34-36.

Wang Lichun, Zhang Guocai, Xu Xueen et al, 1990. Chemical control of *Petrova resinella* (L.) [C]. In Yu Shukui. Seed and cone pests of *Pinus silvestris* var. *mongolica* (1). Harbin: Northeast Forestry University Publishing House.

Wang Lizhong, Han Yuguang, Zhou Weimin et al, 1998. Bionomics and control of *Epicopeia mencia* Moore [J]. Forest Pest and Disease, (2): 38-39.

Wang Mingfeng and Chen Bolin, 1997. *Malacosoma dentata* Mell infestation and its control [J]. Forest Science and Technology Development, (4): 51-52.

Wang Mingyue, 2005. Preliminary study on the bionomics of *Camptochilus semifasciata* Gaede [J]. Forest Pest and Disease, 24(2): 16-17.

Wang Na, 2003. Studies on the control of *Melanophila decastigma* (Coleoptera: Buprestidae) [J]. Xinjiang Forestry, (2): 41.

Wang Na, 2003. Study on the control methods of *Melanophila picta* Pallas [J]. Xinjiang Forestry, (2): 41.

Wang Nianci, Li Zhaohui, Liu Guilin et al, 1990. Studies on the bionomics of *Periphyllus koelreutoriae* (Takahashii) and its control [J]. Journal of Shandong Agricultural University, 21(1): 47-50.

Wang Nianci, Li Zhaohui, Liu Guilin et al, 1991. Morphological characters and population dynamics of *Periphyllus koelreutoriae* (Takahashii) [J]. Journal of Shandong Agricultural University, 21(1): 47-50.

Wang Pingyuan and Song Shimei, 1982. Description of a new species of *Dioryctria zellr* on *Pinus sylvestris* var. *mongolica* from northeast China, with establishment of a new species group [J]. Acta Entomologica Sinica, 25(3): 323-326.

Wang Pingyuan, 1980. Economic insect fauna of China Fasc. 21. Lepidoptera: Pyralidae [M]. Beijing: Science Publishing House.

Wang Runxi, 1978. Control of *Lepyrus japonicus* Roelofs using Beauveria bassiana [J]. New Agriculture, (17): 17.

Wang Sheceng, Gao Jiusi, and Xue Minsheng, 2008. Occurrence of *Kakivoria favofasciata* Nagano (Lepidoptera: Heliodinidae) and its integrated management [J]. Modern Agriculture Science and Technology, (9): 81, 83.

Wang Shufen, Tang Dawu, Ye Cuiceng et al, 2003. Application of sex pheromone of *Celypha pseudolarixicola* Liu [J]. Journal of Central South Forestry University, 23(4): 85-87.

Wang Tianlong, Luo Dahua, and Yao Qingui, 1997. A preliminary report on the biological characteristics of *Dutomostethus deqingensis* Xiao [J]. Forest Pest and Disease, (2): 35-36.

Wang Tianlong, Luo Dahua, and Yao Qingui, 2000. Effect of *Dutomostethus deqingensis* Xiao infestation on the growth of *Phylostachys pubescens* [J]. Jiangxi Forestry Science and Technology, (4): 17-18.

Wang Weiyi, Wang Weizhong, 1988. Occurrence and control of *Gastrolina depressa* thoracica Baly [J]. Journal of Liaoning Forestry Science and Technology, (6): 55.

Wang Wenkai, 2000. Latin names of the two Batocera species in China [J]. Entomological Knowledge, 37(3): 191-192.

Wang Xiangdong, 2005. Occurrence and control of *Rhyacionia insulariana* Liu in Liangshan forest area of Sichuan province [J]. Agricultural Science and Technology, (7): 43.

Wang Xiangjiang, 1990. Outbreaks of *Euproctis bipunctapex* (Hampson) in Wushan County of Sichuan province [J]. Plant Protection, 16(5): 51.

Wang Xiaojuan, 2008. Preliminary study on *Mesoneura rufonota* Rohwer [J]. Anhui Agricultural Science Bulletin, 14(23): 178, 203.

Wang Xiaojun, Yang Zhongqi, and Wang Xiaoyi, 2006. Biology and control of *Rhynchaenus empopulifolis* in Beijing [J]. Entomological Knowledge, 43(6): 858-863.

Wang Xingde and Bi Qiaoling, 1988. Preliminary observation of *Cinara tujafilina* (Del Guercio) [J]. Entomological Knowledge, (3): 161.

Wang Xingjian, 1990. Discussion on the identification of Tephritidae of quarantine importance [J]. Plant Quarantine, 4(6): 440-446.

Wang Xinxiang, 2006. Occurrence and control of *Odites issikii* (Takahashi) [J]. Hebei Fruit Trees (6): 36-37.

Wang Xiong and Liu Qiang, 2002. Study on *Orgyia ericae* Germar, a pest of an endangered plant Ammopiptanthus mongolicus [J]. Journal of the Normal University of Inner Mongolia (Natural Science edition), 31(4): 374-377.

Wang Xiumei, Zang Liansheng, Zou Yunwei et al, 2012. Predatory effect of *Harmonia axyridis* Pallas on *Ambrostoma quadriimpressum* Motschulsky [J]. Journal of Northeast Forestry University, 40(1): 70-72.

Wang Xixin, Zhao Minyang, Zhu Zongqi et al, 2001. Bionomics and control techniques of *Ptilinus fuscus* Geoffroy [J]. Gansu Forest Science and Technology, (1): 15-17.

Wang Xixin, 2000. Control of *Ptilinus fuscus* Geoffroy [J]. Forest Research 13(2): 209-212.

Wang Xueshan, Ning Bo, Pan Shuqin et al, 1996. Bionomics and control of *Proagopertha lucidula* Faldermann [J]. Entomological Knowledge, 33(2): 111-112.

Wang Xufeng, Zhang Cuiyu, Zhang Xiuhui et al, 2006. Occurrence and control of *Ancylis sativa* Liu in Binzhou of Shandong province [J]. Plant Protection, 32(2): 109.

Wang Yan, 2007. Forest diseases and pests [M]. Shanghai: Shanghai Science and Technology Publishing House.

Wang Yanping, Wu Sanan, Zhang Runzhi, 2009. Pest risk analysis of a new invasive pest, *Phenacoccus solenopsis*, to China [J]. Chinese Bulletin of Entomology, 46(1): 101-106.

Wang Yian, 1985. A field test report on controlling fungus-growing termites by using poisonous bait [J]. Journal of Nanjing Forestry University, (1): 52-59.

Wang Yin and Yang Jikun, 1994. A new species of genus *Bembecia* (Lepidptera: Sesiidae) [J]. Journal of Northwest Forestry College, 9(3): 31-33.

Wang Yingmin, Zhao Shidong, Ye Shuqin et al, 2001. Distribution of *Urocerus gigas taiganus* Benson emergence holes on larch trunks [J]. Forest Pest and Disease, 20(3): 15-16.

Wang Yingming, Zhang Yingwei, and Yang Weiyu, 1999. Spatial distribution of *Neodiprion dailingensis* Xiao et Zhou eggs on canopy [J]. Liaoning Forestry Science and Technology, (5): 45-46.

Wang Yong, Zeng Juping, 2013. Pests of camphor trees in Jiangxi and their integrated pest management [J]. Biological Disaster Science, 36(3): 304-315.

Wang Yongchun, Tang Xiangming, Su Zhihong et al, 1997. Occurrence and control of *Ascotis selenaria dianaria* Hübner on cherry trees [J]. Deciduous Fruit Trees, (4): 52.

Wang Yonghong, Sun Yizhi, and Yinkun, 1997. Study on the control of *Atrijuglans hetauhei* Yang with *Steinernema feltiae* "Beijing" strain [J]. Shaanxi Agriculture Science and Technology, (1): 5-7.

Wang Yongzhe, Chen Tianlin, and Ning Jiwen, 1995. Spatial distribution pattern and sampling techniques of *Neodiprion dailingensis* Xiao et Zhou [J]. Liaoning Forestry Science and Technology, (1): 31-33, 59.

Wang Yongzhe, Zhang Xiaolong, Wang Dong et al, 1992. Biological characteristics and control of *Neodiprion dailingensis* Xiao et Zhou [J]. Entomological Knowledge, 29(5): 279-281.

Wang Yuanmin and Xujun, 1987. Larval morphology of the geometrid *Culcula panterinaria* Bremer et Grey, with a discussion on an unnamed seta of prothoracic shield (Lepidoptera: Geometridae) [J]. Acta Entomologica Sinica, 30(3): 323-326.

Wang Yulin, Yang Xueshe, Tan Lijiang et al, 2002. Chemical control techniques for *Xylotrechus rusticus* L. [J]. Forest Science and Technology, 27(3): 1.

Wang Yunzun, 1988. Biology and control of *Ancylis sativa* Liu [J]. Shandong Forest Science and Technology, 3: 34-38.

Wang Yunzun, 1989. Biology and control of *Zeuzera leuconotum* Butler [J]. Deciduous Fruit Trees S1: 175-177.

Wang Yuqing, Wu Yousheng, and Gao Zezheng et al, 2004. Occurrence and control of *Deilephila nerii* Linnaeus [J]. Journal of Guangdong Horticulture, (supplement): 155-156.

Wang Zhicheng, 1999. Monograph of butterflies in Northeast [M]. Changchun: Jilin Science and Technology Publishing House.

Wang Zhicheng, 2003. Monograph of long-horned beetles in Northeast [M]. Changchun: Jilin Science and Technology Publishing House.

Wang Zhigang, Yan Junjie, Liu Yujun et al, 2003. Investigation of *Anoplophora glabripennis* in southern Tibet [J]. Journal of Northeast Forestry University, (4): 70-71.

Wang Zhigang, 2004. Population dynamics and control strategies for *Anoplophora glabripennis* (Moltschulsky) in China [D]. Northeast Forestry University Doctoral dissertation.

Wang Zhiming and Ni Hongjin, 1991. Observation on *Megopis sinica* (White) and its control [J]. China Fruit Trees, (1): 34-35.

Wang Zhiming, Liu Guorong, Cheng Bin, 2006. Biology of *Pachynematus itoi* Okutani and its influence on larch growth [J]. Journal of Northeast Forestry University, 34(5): 13-15.

Wang Zhiming, Pi Zhongqing, Ning changlin et al, 1993. Control of *Clostera anastomosis* (Linnaeus) using Bacillus thuringiensis [J]. Jilin Forest Science and Technology, 22(5): 31-32.

Wang Zhiying, Yue Shukui, Dai Guohua et al, 1990. Biology and ecology of *Petrova resinella* (L.) [C]. In Yu Shukui 1990. Seed and cone pests of *Pinus silvestris* var. *mongolica* (1). pp. 96-101. Harbin: Northeast Forestry University Publishing House.

Wang Zhuhong, Huang Jian, Liang Zhisheng et al, 2004. Introduction and application of *Coccobius azumai* Tachikawa (Hymenoptera: Aphelinidae) [J]. Journal of Fujian Agriculture and Forestry University (Natural Science Edition), 33(3): 313-317.

Wang Zili, Chen Xiaoming, Chen Yong et al, 2003. Parthenogenesis of *Ericerus pela* [J]. Forest Research, 16(4): 386-390.

Wang Ziqing, Yao Defu, Cui Shiying et al, 1982. A new species of *Laccifer*, with preliminary studies on the biological characteristic (Homoptera: Lacciferidae) [J]. Scientia Silvae Sinicae, 18(1): 53-57.

Wang Ziqing, 1982. Description of a new genus and a new species of Lecanodiaspididae (Homoptera: Coccoidea) [J]. Acta Entomologia Sinica, 25(1): 85-88.

Wang Ziqing, 2001. Fauna of China Fasc. 22: Insecta, Isoptera, Coccoidea [M]. Beijing: Science Publishing House.

Wang Zongkai, 1964. Preliminary study on *Poecilocoris latus* Dallas [J]. Entomological Knowledge, 8(2): 69-70.

Wang, Jijian, 1994. Observation on the Biology and behavior of *Longivalva hainanensis* [J]. Agricultural Sciences of Guangxi Province, (6): 281-282.

Wei Chujiang, Xie Daxyang, Zhuang Chenhui et al, 2004. Plague division of *Lymantria xylina* and its application in Fujian province [J]. Acta Agriculturae Universitatis Jiangxiensis, 26(5): 774-777.

Wei Hongjun, Zhang Zhiliang, and Wang Yinchang, 1989. Subterranean pests of China [M]. Shanghai: Shanghai Science and Technology Publishing House.

Wei Hongjun, 1979. Integrated management of subterranean pests [C]. In "Integrated management of major pests in China". Beijing: Science Publishing House.

Wei Hongjun, 1990. *Agriotes subrittatus* Motschulsky [C]. In Wu Fuzheng and Guang Zhihe.Encyclopedia of Agriculture - Insects. Beijing: Agriculture Publishing House.

Wei Jiqian, Mo Jianchu, Xu Wen et al, 2010. Advances in research on *Reticulitermes chinensis* (Isoptera: Rhinotermitidae) in China [J]. China Journal of Vector Biology and Control, 21(6): 635-637.

Wei Libang, 1990. Study on the control of chestnut gall wasp [J]. Journal of Southwest Forestry College, 10(1): 86-93.

Wei Meicai, 1997. Further studies on the tribe fenusini (Hymenoptera: Tenthredinidae) [J]. Acta Zootaxonomica Sinica 22(3): 286-300.

Wei Ping, 1989. Behavior and control of *Dictyoploca japonica* Butler [J]. Entomological Knowledge, (6): 347.

Wei Qiyuan, 1985. Preliminary study on *Poecilocoris latus* Dallas [J]. Entomological Knowledge, 22(1): 21-23.

Wei Xiangdong, Cheng Xuhui, Wang Youfeng, 1991. Biological observations on *Basiprionota bisignata* Boheman [J]. Journal of Gansu Forestry Science and Technology, (3): 46-47.

Wen Bing and Wang Aijing, 1998. Entomopathogens of *Xylotrechus namanganensis* Heydl [J]. Xinjiang Agricultural Science, (2): 77-78.

Wen Shouyi and Xu Longjiang, 1959. Preliminary observation on life habits of Buprestidae insects [J]. Xinjiang Agricultural Science, (2): 59-63.

Wen Shouyi and Xu Longjiang, 1965. A preliminary study on Bupresidae insects [C]. In Proceedings of the Poplar Conference. Beijing: China Agriculture Publishing House.

Wen Shouyi and Xu Longjiang, 1987. Prevention and control of forest insect pests [J]. Urumuqi: Xinjiang People's Publishing House.

Wen Shouyi et al, 1959. Preliminary report on forest pests [M]. Beijing: Science Publishing House.

Wen Shouyi et al, 1959. Survey of bark beetles in spruce forest at Nanshan of Urumuqi and biology of *Ips duplicatus* Sahalberg [C]. In "Preliminary Reports on Forest Pests. Beijing: Science Publishing House.

Wen Shouyi et al, 1984. Prevention and control of forest insect pests [M]. Urumuqi: Xinjiang People's Publishing House.

Wen Xiaosui, Kuang Yuanyu, Shi Mingqing et al, 2004a. *Hylobitelus xiaoi* (Coleoptera: Curculionidae) adult feeding, oviposition, and behavior [J]. Acta Entomologica Sinica, 47(5): 624-629.

Wen Xiaosui, Shi Mingqing, and Kuang Yuanyu, 2004. Occurrence of a new pine borer, *Hylobitelus xiaoi* Zhang and its ecological control [J]. Acta Agriculturae Universitatis Jiangxiensis, 26(4): 495-498.

Wen Xiaosui, Shi Mingqing, and Kuang Yuanyu, 2005. Influence of humidity on feeding fecundity and survival of *Hylobitelus xiaoi* Zhang [J]. Acta Agricultruae Universitatis Jiangxiensis, 27(1): 89-92.

Wen Xiaosui, Wang Hui, Shi Mingqing et al, 2007. Experiment on the efficacy of "Hulinshen II DP" against *Hylobitelus xiaoi* [J]. Forest Pest and Disease, 26(1): 37-39.

Wen X. S., Kuang, Y. Y., Shi, M. Q. et al, 2004b. Biology of *Hylobitelus xiaoi* Zhang (Coleoptera: Curculionidae), a new pest of slash pine, Pinus elliottii Engelm [J]. Journal of Economic Entomology 97: 1958-1964.

Wen, X. S., Kuang, Y. Y., Shi, M. Q, et al, 2006. Effect of pruning and ground treatment on the populations of *Hylobitelus xiaoi*, a new debarking weevil in slash pine plantations [J]. Agricutural and Foresty Entomology 8: 263-265.

Wen, X. S., Shi, M. Q., Haack, R. A., et al, 2007. *Hylobitelus xiaoi* (Coleoptera: Curculionidae) adult feeding, oviposition, and egg and pupal development at constant temperatures [J]. Journal of Entomological Science 42: 28-34.

Wen Xuefei and Zou Jimei, 2007. The occurrence and control of *Aporia crataegi* [J]. Northern Horticulture, (9): 218-219.

Wen Zhenhong, 2003. Chemical control and environmentally friendly control techniques for *Dendrolimus superans* (Butler) and Phassus excrescens Butler [D]. Northeast Forestry University Master thesis.

Wu Chunshen and Fang Chenglai, 2003. Fauna of China Fasc. 31: Insecta, Lepidoptera, Notodontidae [M]. Beijing: Science Publishing House.

Wu Chunsheng and Fang Chenlai, 2010. Insect Fauna of Henan: Lepidoptera [M]. Beijing: Science Publishing House.

Wu Chunsheng, Meng Xianlin, Wang Heng et al, 2007. Handbook on Chinese butterflies [M]. Beijing: Science Publishing House.

Wu Chunsheng, 1988. Bionomics of *Shirahoshizo coniferae* Chao [J]. Journal of Southwest Forestry College, 8(1): 83-86.

Wu Chunsheng, 2001. Fauna of China Fasc. 25: Insecta, Lepidoptera, Papilionidae [M]. Beijing: Science Publishing House.

Wu Cibin, 1989. *Ericerus pela* and white wax production [M]. Beijing: China Forestry Publishing House.

Wu Da Zhang, 1951. Study on *Holotrichia titanis* Reitter [J]. Acta Entomologica Sinica, (4): 379-401.

Wu Dingkun, Liu Chaobin, and Jiang Xinong, 1992. Bionomics of larch sawfly in Qinling mountain [J]. Journal of Northwest Forestry College, 7(4): 77-82.

Wu Faji, 1982. Artificial rearing of Chinese gall aphid [J]. Entomological Knowledge, 19(3): 33-34.

Wu Fuzhen and Guang Zhihe, 1990. Encyclopedia of Agriculture in China (Insects) [M]. Beijing: Agriculture Publishing House.

Wu Fuzheng and Gao Zhaoning, 1978. Illustrated guide of Ningxia agricultural insects (revision) [M]. Beijing: China Agriculture Publishing House.

Wu Hai, 2006. Bionomics and control of *Agrilus zanthoxylumi* Li [J]. Entomological Knowledge, 43(2): 236-239.

Wu Haiwei, Luo Youqing, Tang Wandi et al, 2006. Infestation of *Asemum amurense* Kraatz -an important forest pest [J]. China Forest Pest and Disease, 25(4): 15-18.

Wu Hong, 1995. Insects at Baishanzu of eastern China [M]. Beijing: China Forestry Publishing House.

Wu Hongguang, Ren Yulin, and Luo Daimo, 1984. Studies on *Augomonoctenus smithi* [J]. Sichuan Forestry Science and Technology, (3): 40-45.

Wu Jiajiao, Chen Naizhong, 2008. Quarantine pests in genus *Carpomya* (Diptera: Tephritidae) [J]. Plant Quarantine, 22(1): 32-34.

Wu Jialin and Chen Youfen, 1992. A preliminary study on *Dendrolimus punctatus dechangensis* cytoplasmic polyhedrosis virus [J]. Journal of Southwest Forestry College, 12(1): 58-62.

Wu Jian, Wang Changlu, 1995. The ants of China [M]. Beijing: China Forestry Publishing House.

Wu Jianfen and Huang Zenghe, 1984. A preliminary report on the study of *Selepa celtis* Moore [J]. Forest Science and Technology, (4): 28.

Wu Jianfen, 1990. Morphology of Bionomics of *Hemiberlesia pitysophila* Takagi [J]. Guangdong Forest Science and Technology, (6): 3-5.

Wu Junxiang, 1999. Agricultural entomology [M]. World Book Publishing House.

Wu Juwen, 1972. African mole cricket [C]. In "Handbook of forest diseases and pests". Beijing: Agriculture Publishing House.

Wu Juwen, 1972. *Euproctis bipunctapex* (Hampson) [C]. In Handbook of Forest Diseases and Pests Control. Beijing: China Forestry Publishing House.

Wu Kunhong and Yu Fasheng, 2001. A preliminary investigation on red palm weevil [J]. Tropical Forestry, 29(3): 141-144.

Wu Lin and Huang Zhiyong, 1989. A preliminary study on the biology and control of the yellow brisket wood moth (Cossus chinensis Rothsch.) [J]. Journal of the Southwest Forest College, 9(1): 47-54.

Wu Peiheng, Li Zhenyu, and Chen Zhouxian, 1988. Studies on *Synanthedon castanevora* Yang et Wang [J]. Forest Pest and Disease, 2: 4-5.

Wu Peiyu, 1982. study on the control and natural enemies of *Euproctis flava* (Bremer) [J]. Entomological Knowledge, (2): 29.

Wu Peiyu, 1984. Study on *Euprocits flava* Bremer, a pest of *Quercus palustris* [J]. Sericulture, 10(4): 225-229.

Wu Qing, Zeng Ling, Kong Jingchen et al, 2006. Effectiveness of field application of *Metarhizium anisopliae* for controlling red palm weevil [J]. Journal of Shandong Agricultural University (Natural Science Edition), 37(4): 568-572.

Wu Qingbiao and Bai Yinzhen, 2005. Occurrence of *Orgia gonostigma* L. and its control techniques [J]. China Plant Protection, 25(5): 27-28.

Wu Sanan, Zhang Runzhi, 2009. A new invasive pest to cotton production - *Phenacoccus solenopsis* Tinsley [J]. Chinese Bulletin of Entomology, 46(1): 156-162.

Wu Shijun, 1983. Biological characteristics of *Bambusaspis hemiphaerica* Kuwana and its control [J]. Entomological Knowledge, 20(2): 77.

Wu Shixiong, Chen Zhiqing, and Wang Tiehua, 1979. Preliminary study on *Pyrausta machoeralis* Walker [J]. Acta Entomologica Sinica, 22(2): 156-162.

Wu Shiying, 2005. Pictorial guide to urban tree diseases and pests [M]. Shanghai: Shanghai Science and Technology Publishing House.

Wu Weiwen, Wei Lüyan, and Li Xueru, 2002. Biological studies on *Perina nuda* (F.) [J]. Proceedings of the Annual

Meeting of the Entomological Society of China. pp. 514-517.

Wu Xian Xiang, 2001. Integrated control methods for *Oides leucomelauena* Weise [J]. Guangxi Forestry, (1): 32.

Wu Xiaomin and Zhang Wenfeng, 2008. Effect compared with different control methods for *Dendrolimus houi* Lajonquiere [J]. Hubei Forestry Science and Technology, (2): 28-32.

Wu Xiaoying, Zhong Yihai, Li Hong et al, 2004. Biology of *Brontispa longissima* and efficacy of insecticides [J]. Plant Qurantine, 18(3): 133-140.

Wu Xiazhong, Song Minhao, and Zheng Dahu, 1994. A study on *Gilpinia massoniana* [J]. Scientia Silvae Sinicae, 30(3): 383-392.

Wu Xiazhong, 1985. A preliminary study on *Sirex rufiabdomins* [J]. Scientia Silvae Sinicae, 21(3): 315-318.

Wu Xingyu, Gao Song, Chen Duansheng et al, 2001. Preliminary research on *Agenocimbex elmina* Li et Wu [J]. Journal of Northest Forestry University, 16(4): 50-51.

Wu Xinmin, 1959. Integrated management of *Agrotis ypsilon* (Rott.) [J]. Cotton, (3): 22-23.

Wu Yaojun, Li Dewei, Chang Mingshan et al, 2010. Bionomics of *Leptocybe invasa* Fisher & La Salle [J]. China Forest Pest and Disease, 29(5): 1-4, 10.

Wu Yiyou, Liang Jian, Liu Xianbao et al, 2001. Bionomics of Curculio davidi adults and least toxic control methods [J]. Hubei Forestry Science and Technology, 116: 27-31.

Wu Yiyou, Yang Jian, Xie Puqing et al, 1999. Field evaluation of various *Bacillus thuringiensis* insecticides to control *Curculio davidi* Fairmaire [J]. Economic Forest Researchs, 17(3): 40-42.

Wu Yousheng and Gao Zezheng, 2004a. Preliminary study on biology of *Dysphania militaris* Linnaeus [J]. Forest Pest and Disease, 23(5): 25-26.

Wu Yousheng and Gao Zezheng, 2004b. Occurrence of three pests of leafblotch miners on garden plants in Guangzhou [J]. Entomological Knowledge, 41(4): 328-330.

Wu Yousheng and Gao Zezheng, 2004c. A new pest of tropical fruit trees - *Scopelodes testacea* Butler [J]. China Souther Fruit Trees, 33(5): 47-48.

Wu Yousheng and Gao Zezheng, 2004d. A report on the morphology, biology and control of *Atteva fabriciella* (Swederus) [J]. Practical Forestry Technology, (5): 30.

Wu Yousheng, Dong Danlin, Liu Dongming et al, 1998. Occurrence of red palm weevil on palm trees [J]. Guangdong Horticulture, (1): 38.

Wu Zongxing, Liu Zhifu, Yu Mingzhong et al, 2003. Studies on the major pests and diseases of *Zanthoxylum bungeanum* and their control [J]. Sichuan Forestry Science and Technology, 24(4): 58-61.

Xi Fusheng, Luo Jitong, Li Guiyu et al, 2007. Pests and diseases of eucalyptus trees and the natural enemies of pests [M]. Nanning: Guangxi Science and Technology Publishing House.

Xi Jinghui, Pan Hongyu, Chen Yujiang et al, 2002. Checklist of the Geometrid insects [J]. Journal of Jilin Agricultural University, 24(5): 53-57.

Xi Yong, Bai Yulong, Song Yinghua et al. Control threshold for *Eulecanium gigantea* (Shinji) [J]. Forest Pest and Disease, (6): 15-17.

Xi Yong, Ren Ling, and Liu Jibao, 1996. Occurrence and integrated pest control of *Trioza magnisetosa* Log [J]. Xinjiang Agricultural Science, (5): 228-229.

Xi Yong, 1996. Occurrence of Eulecanium gigantean Shinji on *Amygdalus communis* L [J]. Entomological Knowledge, 33(5): 273.

Xi Zhongcheng, 1999. Bionomics and integrated control techniques of *Camptoloma interiorata* (Walker) [J]. Gansu Forestry Science and Technology, 24(2): 35-37.

Xia Gucheng, Yu Zhijia, Fan Yapeng et al, 2009. Experiment on the spray of insecticides against *Pristiphora erichsonii* with dispenser and pesticide bomb [J]. Gansu Forest Science and Technology, 34(3): 50-51, 54.

Xia Junwen, 1987. Study on *Phyllonorycter populifoliella* (Treitschke) [J]. Entomological Knowledge, 24(4): 221.

Xia Meiyan and Li Ji, 2001. Characteristics of climatic change in recent years and its impact on agricultural diseases and pests [J]. Green book on disease and pest reduction in agriculture and forestry, (3): 19-24.

Xia Wensheng, Liu Chao, Dong Likun et al, 2007. Occurrence and bionomics of sycamore lace bug [J]. Plant Protection, 33(6): 142-145.

Xia Xina, 2004. Environmentally friendly control of diseases and pests of ornamental plants [M]. Beijing: China Agriculture Publishing House.

Xia Zhixian and Ding Fulan, 1989. Characteristics of *Agrotis segetum* and integrated control [J]. China Cotton, (2): 45-46.

Xia Zhonghui, 1990. Observations on the infestation of *Plagioder versicolora* on poplar trees [J]. Hunan Science and Technology, (1): 40-41.

Xian Shenghua, Kang Shangfu, Yuyong et al, 1998. Bionomics and control of *Apion collare* Schilsky [J]. Eucalyptus Science and Technology, (2): 47-49

Xian Shenghua, Yu Yong, Liang Xueming et al, 2002. A preliminary study on the control of *Anomala cupripes* Hope in eucalyptus forests [J]. Eucalyptus Science and Technology, (2): 53.

Xian Xuxun, 1995. Biology and control of *Polylopha cassiicola* Liu et Kawabe [J]. Entomological Knowledge, 32(4): 220-223.

Xiang He, 1980. Studies of Chinese gall-nut aphids on *Rhus potaninii* Maxim [J]. Acta Entomologica Sinica, 2(4): 303-313.

Xiang Yuyong, Yang Maofa, and Li Zizhong, 2009. Sex pheromone components of the female black cutworm moth in China: Identification and field trials [J]. Zoological Research, 30(1): 59-64.

Xiao Caiyu et al, 1981. Handbook on the identification of Hemiptera in China (vol. 2) [M]. Beijing: Science Publishing House.

Xiao Caiyu, 1981. Handbook on the identification of Hemiptera in China [M]. Beijing: Science Publishing House.

Xiao Gangrou and Chen Hanlin, 1993. Description of males of *Cephalcia tianmua* Maa (Hymenoptera: Pamphiliidae) [J]. Forest Research, (6): 65-67.

Xiao Gangrou and Wu Jian, 1983. A new species of the genus *Augomonctenus* (Hymenoptera: Diprionidae) [J]. Scientia Silvae Sinicae, 19(2): 141-143.

Xiao Gangrou and Zhang You, 1994. A new species of the genus *Diprion* (Hymenoptera: Diprionidae) from China [J]. Forest Research, 7(6): 663-665.

Xiao Gangrou, Huang Xiaoyun, and Zhou Shuzhi, 1981. Three new species of *Nesodiprion* from China (Hymenoptera, Symphyta, Diprionidae) [J]. Scientia Silvae Sinicae, 17(3): 247-249.

Xiao Gangrou, 1987. Four new species of Cephalciidae (Hymenoptera: Pamphiliidae) from China [J]. Scientia Silvae Sinicae, (Insects edition): 1-4.

Xiao Gangrou, 1992. Forest insects of China (Second Edition) [M]. Beijing: China Forestry Publishing House.

Xiao Gangrou, 1992. Two new sawflies of the genus *Gilpinia* in China (Hymenoptera, Symphyta, Diprionidae) [J]. Forest Research, 5(2): 193-195.

Xiao Gangrou, 1993. A new species of the genus *Megabeleses* from China (Hymenoptera, Tenthredinidae) [J]. Forest Research, 6(2): 148-150.

Xiao Gangrou, 1995. A New Species of the Genus *Nematus* (Hymenoptera: Tenthredinidae) from China [J]. Forest Research, 8(5): 497-499.

Xiao Gangrou, 2002. Pamphiliidae of China. pp. 37-40 [M]. Beijing: China Forestry Publishing House.

Xiao Ganrou et al, 1997. A Latin-Chinese-English dictionary of insects, ticks, spiders and nematodes [M]. Beijing: China Forestry Publishing House.

Xiao Guangrou, Huang Xiaoyun, Zhou Shuzhi et al, 1991. Economically important sawflies of China (1) (Hymenoptera, Symphyta) [M]. Beijing: Science Publishing House.

Xiao Keren, Chen Tianlin, Wang Qi et al, 2005. Bionomics and control of *Acantholyda erythrocephala* [J]. Forest Pest and Disease, 25(1): 30-32.

Xiao Weiliang and Li Guixiang, 1995. Species and morphological characters of termites from Xishuangbanna [J]. Termite Science and Technology, 12(3): 6-10.

Xiao Youxing, 1992. *Euproctis bipunctapex* (Hampson) [C]. In Iconography of Hunan Forest Insects. Changsha: Hunan Science Publishing House.

Xiao Yugui and Guo Hengxiao, 2003. Infection path and pathogenicity of *Metarhizium anisopliae* on *Chinolyda flagellicornis* larva [J]. Forest Pest and Disease, 22(1): 12-14.

Xiao Yuyu, Wang Feng, Ju Ruiting et al, 2010. Life history and occurrence of *Corythucha ciliata* (Say) in Shanghai district [J]. Entomological Knowledge, 47(2): 404-408.

Xie Guolin and Yan Aojin, 1983. Study on *Nesticoccus sinensis* Tang [J]. Acta Entomologica Sinica, 26(3): 268-277.

Xie Guolin, Hu Jinlin, Li Quhuo et al, 1984. Preliminary report on the survey of *Hemiberlesia pitysophila* Takagi in Guangdong province [J]. Forest Pest and Disease, (1): 39-41.

Xie Guolin, Pan Wuyao, Tang Ziying et al, 1997. Effectiveness of *Coccobinus azumai* Tachikawa on the population control of *Hemiberlesia pitysophila* Takagi and its sustainable effect [J]. Acta Entomologica Sinica, 40(2): 135-144.

Xie Qingke, 1990. Occurrence period and population density of *Pristiphora erichsonii* (Hartig) [J]. Entomological Knowledge, 27(6): 354-355.

Xie Wenjuan, 2008. Bionomics and control of *Holcocerus arenicola* Staudinger [J]. Qinghai Agricutural and Forest Science and Technology, (4): 52-53.

Xie Wentian, Xu Qingliang, and Song Youwen, 2001. Species and control of bark beetles attacting Juglans regia [J]. Forest Pest and Disease, (4): 33-35.

Xie Xiaoxi, 1994. Study on *Illiberis ulmivora* Graeser [J]. Gansu Forest Science and Technology, (2): 26-30.

Xie Yingping, Xue Jiaoliang, and Zheng Leyi, 2006. Ultra fine structure and chemical components of wax excretion by scale insects (Hemiptera: Coccidae) [M]. Beijing: China Forestry Publishing House.

Xie Yingping, 1998. Scale insects of forests and fruit trees in Shanxi [M]. Beijing: China Forestry Publishing House.

Xie Zhenlun, 1996. Isolation and identification of virus disease of *Chaleocelides albigutata* Snellen [J]. Guangdong Tea, (4): 26-28.

Xing Tongxuan, 1991. Persisting effect of the granulosis virus in controlling *Parasa consocia* (Lepidoptera: Limacodidae) [J].

Chinese Journal of Biological Control ,7(4): 186-187.

Xiong Shansong et al, 2005. Integrated pest management for long-horned beetles on poplar trees [M]. Yinchuan: Ningxia Publishing House.

Xu Deqin, 1987. Bionomics and control of *Homona isskii* Yaouda [J]. Journal of Zhejing Forestry Science and Technology, 4(1): 50-55.

Xu Fuyuan, 1994. *Monochamus alternatus* Hope adult supplementary feeding and its control in Nanjing [J]. Forest Research, 7(2): 215-218.

Xu Gongtian and Yang Zhihua, 2007. The horticultural pests of China [M]. Beijing: China Forestry Publishing House.

Xu Gongtian, 2003. Colored pictorial guide to the control of diseases and pests of ornamental plants [M]. Beijing: China Agriculture Publishing House.

Xu Guangyu, Li Duoxiang, Li Tubiao et al, 2008. Life history and control of *Podontia lutea* Olivier in western Anhui forests [J]. Agricultural Technology Service, 25(8): 163.

Xu Guangyu, Yang Ainong, Qu Tianjun et al, 2007. Study on bio-characteristics of tea shoot borer and its control [J]. Journal of Hebei Agricultural Science, 11(6): 28-29.

Xu Guangyu, 2008. Biological characteristics of *Celypha pseudolarixicola* Liu and its control techniques [J]. Agricultural Technology Service, 25(7): 142.

Xu Guolian, Chai Shouquan, Xie Kaili et al, 2002. Bionomics of *Dendrolimus punctata wenshanensis* Tsai et Liu and efficacy of two bio-pesticides against *Dendrolimus punctata* [J]. Forest Pest and Disease, 21(5): 15-18.

Xu Ji, Li Yue, Gao Jinna et al, 1983. A preliminary study on scale insects on bamboo trees [J]. Shaanxi Forest Science and Technology, (4): 47-56.

Xu Jiandong, Zheng Hongjun, Yuan Chaoxian et al, 2007. Bionomics of *Acantholyda flavomarginata* Maa and control experiment [J]. Guizhou Forestry Science and Technology, 35(3): 14-16.

Xu Lianfeng, Shi Shaolin, Zhao Lingquan et al, 2004. Control of diseases and pests of *Populus davidiana* × *P. bolleana* Loucne [J]. Protection Forest Science and Technology, (3): 69-70.

Xu Minghui, 1993. Control of diseases and pests of ornamental plants [M]. Beijing: China Forestry Publishing House.

Xu Shouzhen, Wu Jiangong, Meng Changxiao et al, 1996. Biology and control of *Sesia siningensis* (Hsu) [J]. Entomological Knowledge, 33(6): 338-340.

Xu Shuiwei, Zhu Jianzhou, Wang Liming et al, 2004. Studies on the bionomics and control of *Gastrolina depressa* Baly [J]. Forest Science and Technology, (3): 9.

Xu Tiansen and Wang Haojie, 2004. Major pests of bamboo in China [M]. Beijing: China Forestry Publishing House.

Xu Tiansen et al, 1975. Checklist of bamboo insect pests [J]. Subtropical Forestry Science and Technology, (1-2): 18-26.

Xu Tiansen, Wang Haojie, and Yu Caizhu, 2008. Illustrated guide to the identification and control of bamboo diseases and pests [M]. Hangzhou: Zhejiang Science and Technology Publishing House.

Xu Tiansen, 1987. Handbook of forest disease and pest control [M]. Beijing: China Forestry Publishing House.

Xu Weidong, 2002. Investigation on a new shoot pest *Euzophera batangensis* Caradja of loquat [J]. Entomological Journal of East China, 11(1): 107-108.

Xu Xiaoping, Liu Bisheng, Wang Shengsen et al, 1998. Occurrence and control of *Paranthrene actinidiae* Yang et Wang [J]. Agricultural Technology Service, (5): 25.

Xu Yaochang, 2005. Integrated control of *Lymantria xylina* Swinhoe in Zhangzhou [J]. Fujian Forest Science and Technology, 32(3): 15-19.

Xu Yiling et al, 1984. Compilation of information on forest disease and pest survey in Hubei province [M]. Hubei Forestry Department.

Xu Ying, Song Jingyun, and Yao Dianjing, 2001. Fogging to control *Strobilomyia laricicola* (Kårl) and cost benefit analysis [J]. Heihe Science and Technology, (2): 10-12.

Xu Zhaoji, 1963. A preliminary study on *Trioza magisetosa* Log [C]. Proceedings of the first meeting of the Ningxia Agricultural Society. Ningxia: Ningxia People's Publishing House.

Xu Zhenghui, Xu Zhiqiang, and Wu Yi, 1990. Bionomics of *Neodiprion huizeensis* Xiao et Zhou [J]. Journal of Southwest Forestry College, 10(2): 203-208.

Xu Zhenguo, 1981. Description of a new species of the genus *Specosesia* Hampton (Lepidoptera: Sesiidae) [J]. Scientia Silvae Sinicae, 20(2): 165-170.

Xu Zhihong and Jiangping, 2001. Pictural guide to diseases and pests of Chinese chestnut [M]. Hangzhou: Zhejiang Science and Technology Publishing House.

Xu Zhihua, 2006. Illustrated guide to diseases and pests of ornamental flowers [M]. Beijing: China Agriculture Publishing House.

Xu Zhizhong, 2009. Biological characteristics of *Semanotus sinoauster* Gressitt (Coleoptera: Cerambycidae) [J]. Anhui Agricultural Science Bulletin, 15(3): 170-199.

Xu Zhu, 2004. Handbook of forage grass [M]. Beijing: Chemical Industry Publishing House.

Xue Fangsen and Shen Rongwu, 1990. Preliminary study on

Phauda triadum Walker [J]. Plant Protection, 16(5): 7-8.

Xue Fangsen, Shen Rongwu, and Zhu Xingfen, 1990. Biology and summer diapause of the great black cutworm (*Agrotis tokionis* Butler) [J]. Jiangxi Plant Protection, (2): 6-8.

Xue Guishou, Mao Jianping, Pu Guanqin et al, 2007. Checklist of mulberry pests from China (II) [J]. Sericulture, 33(4): 629-633.

Xue Wangqi and Zhao Jianming, 1996. Flies of China (2) [J]. Liaoning Science and Technology Publishing House, 23(1): 5-7.

Xue Wansheng, 2005. The experiment of controlling Diprion jingyuanensis Xiao et Zhang with the matridine smoking agent [J]. Sci-Tech Information Development & Economy, 15(24): 245, 249.

Xue Xianqing et al, 1978. Occurrence and control of *Melanophila picta* (Pallas) [J]. China Forest Research, (2): 27-41.

Xue Yonggui, 2008. Study on the biological characteristics and control of *Ips nitdus* Eggers [J]. Anhui Agricultural Science Bulletin, (13): 162-163.

Xue Zhicheng, 2006. Methods for controlling *Dichocrocis punctiferalis* Guenée [J]. Hunan Forestry, (4): 18.

Xun Xingquan, Zhou Lina, Chen Xiaolin et al, 2002. Preliminary study on the biology and control of *Cifuna locuples* Walker damaging rose in Shanghai [J]. Journal of Shanghai Jiaotong University (Agricultural Science edition), 20(supplement): 84-88.

Yan Aojin, Ji Baozhong, Qian Fanjun et al, 1997. A study on *Batocera horsfieldi* (Hope) [J]. Journal of Nanjing Forest University, 21(1): 1-6.

Yan Guozeng and Yu Juxiang, 2001. A major pest of fruit trees in surburban Beijing - *Culcula panterinaria* (Bremer et Grey) [J]. Afforestation and Living, (2): 25-26.

Yan Jiahe, Bai Lulin, Li Jipei et al, 2001. A new pest, *Dendrophilia sophora*, on pagoda tree [J]. Entomological Knowledge, 38(6): 444-449.

Yan Jiahe, Dong Changming, Wang Furong et al, 1998. Bionomics of *Messa taianensis* Xiao et Zhou [J]. Forest Pest and Disease, (3): 20-22.

Yan Jiahe, Wang Furong, and Li Jipei, 2002. Preliminary study on a new pest of *Sophora japonica* - *Anarsia squamerecta* Li et Zheng [J]. Entomological Knowledge, 39(5): 363-366.

Yan Jingjun, Yao Defu, Liu Houping et al, 1994. Checklist of natural enemies of *Lymantria dispar* (L.) in China [J]. Forest Science and Technology, (5): 25-27.

Yan Junjie and Yan Huahui, 1999. Ecological control model of *Anoplophora glabripennis* (Motschulsky) [J]. Journal of Hebei Agricultural University, 22(4): 102-106.

Yan Junjie, Yu Xiulin, and Ren Chaozuo, 1989. Numerical analysis of insect population and spatial distribution pattern and sampling methods of glabrous spotted willow borer [J]. Journal of Biomathmatics, 4(2): 102-106.

Yan Shanchun, Chi Defu, and Sun Jianghua, 2004. Current status and prospects of environmentally friendly control of *Strobilomyia* spp. [J]. Forest Science and Technology Management, (1): 34-35.

Yan Shanchun, Hu Yinyue, Sun Jianghua et al, 1999. Relationship between larch cone volatile profile and the damage by cone fly (*Strobilomyia* spp.) in northeastern China [J]. Scientia Silvae Sinicae, 35(3): 58-62.

Yan Shanchun, Hu Yinyue, Zhen Dingfan et al, 1997. Research advances of *Strobilomyia* spp. in northeast China [J]. Journal of Northeast Forestry University, 25(2): 53-58.

Yan Shanchun, Jiang Haiyan, Li Liqun et al, 1998. The development regular of larch cone flies and its chemical control [J]. Journal of Northeast Forestry University, 26(2): 73-76.

Yan Shanchun, Jiang Xinglin, Xu Fangling et al, 2002. The trapping effects of two different color cup traps on larch cone flies [J]. Journal of Northeast Forestry University, 30(1): 30-32.

Yan Shanchun, Li Jinguo, Wen Aiting et al, 2006. Association between the damage of *Xylotrechus rusticus* (Coleoptera: Cerambycidae) and the compositions and contents of amino acids in different poplar strains [J]. Acta Entomologica Sinica, 49(1): 93-99.

Yan Shanchun, Sun Jianghua, Chi Defu et al, 2003. The repellency effects of plant volatiles to *Strobilomyia* spp. damaging larch cones [J]. Acta Ecologica Sinica, 23(2): 314-329.

Yan Shanchun, Sun Jianghua, Roques et al, 2000. Blue cup trapping of larch cone flies (*Strobilomyia* spp.) [J]. Entomological Knowledge, 37(4): 197-199.

Yan Shanchun, Yang Hui, Gao Lulu et al, 2009. Response of *Coleophora obducta* to larch volatile compositions [J]. Scientia Silvae Sinicae, 45(5): 94-101.

Yan Shanchun, Zhang Xudong, Hu Yinyue et al, 1997. Visual trapping experiment of the flies, *Strobilomyia* spp. (Diptera: Anthomyiidae) damaging larch cones [J]. Journal of Northeast Forestry University, 27(1): 28-31.

Yan Yulan, 2008. Behavior and control of *Cydia pomonella* (L.) [J]. China Agricultural Technology Extension, 24(3): 49-50.

Yang Bin, 2010. Effectiveness of two formulations of emamectin benzoate on the control of *Perina nuda* (F.) [J]. Nong Jia Zhi You, (3): 50-52.

Yang Chuncai and Li Min, 1996. A new species of the genus

Agenocimbex (Hymenoptera: Cimbicidae) from China [J]. Journal of Anhui Agricultural University, 23(1): 5-7.

Yang Chuncai, Wang Wenge, Yin Xiangyu et al, 1996. Study on *Agenocimbex ulmusvora* Yang [J]. Forest Research, 9(4): 376-380.

Yang Chuncai, Yu Wansu, and Wang Guixiang, 1995. Study on *Periphyllus acerihabitans* Zhang [J]. Journal of Anhui Agricultural University, 22(3): 233-238.

Yang Chunsheng, Zhu Shufang, Huang Hongyun et al, 2006. Bionomics of *Pammene ginkgoicola* Liu and its control techniques [J]. Guangxi Forest Science, 35(1): 14-17.

Yang Dahong, Wang Xiaoji, Zhou Mingqing et al, 2007. Control experiments of *Acantholyda posticalis* Matsumura [J]. Shaanxi Forest Science and Technology, (1): 36-37.

Yang Dasheng et al, 1987. Preliminary study on *Chinolyda flagellicornis* (F. Smith) [J]. Forest Pest and Disease, (11): 22-24.

Yang Haixiu, Kan Hongkun, Zheng Changcheng et al, 2001. Use of "Sutieling" (a Bt insecticide) against *Malacosoma neustria* [J]. Forest Pest and Disease, (2): 27-29.

Yang Jikun and Li Fasheng, 1986. Five new species of the genus Paurocephalinae and a new genus and a new species (Homoptera: Psyllidae: Paurocephalinae) [J]. Wuyi Science, (6): 45-58.

Yang Jikun and Wang Yin, 1989. A new genus and six species of clear-wings damaging forest and fruit tree [J]. Forest Research, 2(3): 229-238.

Yang Juan, Ling Song, Liu Deling et al, 2001. Bionomics and control techniques of *Cephalcia kunyushanica* Xiao [J]. Shandong Forestry Science and Technology, 3(3): 41-43.

Yang Jun and Lin Sen, 2000. Observation of the ultra fine structure of *Costera anastomosis* L. nuclear polyhedrosis virus (Lepidoptera: Notodontidae) [J]. Jilin Forestry Science and Technology, 29(1): 17-18.

Yang Leifang and Liu Guanghua, 2009. Occurrence and control of *Clytus validus* Fairmaire [J]. Sichuan Forestry Science and Technology, 30(4): 92-95.

Yang Miaomiao, Li Menglou, Qu Liangjian et al, 2007. Discovery of *Diprion nanhuaensis* nuclear polyhedrosis virus and its virulence determination [J]. Scientia Silvae Sinicae, 43(7): 138-141.

Yang Minsheng, 1984. Hosts of *Lawana imitata* Melichar and its natural enemies [J]. Tropical Forestry Science and Technology, (1): 26-30.

Yang Penghui, Hu Zhonglang, Zhao Zonglin et al, 1996. Biology of *Bembecia hedysari* Wang et Yang [J]. Entomological Knowledge, 33(3): 162-163.

Yang Pinglan and Ren Zunyi, 1974. Bionomics and control of *Matsucoccns matsumurae* Kuwana [J]. Forest Science and Technology, (8): 9-12.

Yang Pinglan, Hu Jinlin, and Ren Zunyi, 1976. *Matsucoccus sinensis* Chen [J]. Acta Entomological Sinica, 19(2): 199-204.

Yang Pinglan, Hu Jinlin, and Ren Zunyi, 1980. On pine needle scales [J]. Acta Entomologica Sinica, 23(1): 42-46.

Yang Pinglan, Lü Changren and Zhan Zhongcai, 1987. A new species *Matsucoccus shennongjiaensis* sp. n. (Homoptera, Margarodidae) [J]. Collections of Entomological Papers, (6): 195-198.

Yang Pinglan, 1982. General introduction to the taxonomy of Chinese scale insects [M]. Shanghai: Shanghai Science and Technology Publishing House.

Yang Qian, Xie Yingping, Fan Jinhua et al, 2013. Genetic diversification of three geographical populations of *Matsucoccus matsumurae* (Kuwana) [J]. Scientia Silvae Sinicae, 49(12): 88-96.

Yang Qian, Xie Yingping, Fan Jinhua, et al, 2013. Genetic differentiation of *Matsucoccus matsumurae* from three geographic populations in China [J]. Scientia Silvae Sinicae, 49(12): 88-96.

Yang Shanchun, Yang Hui, Gao Lulu et al, 2008. Influences of larch volatiles and seven kinds of environmentally safe insecticides on olfactory and oviposit responses of Coleophora obducta [J]. Scientia Silvae Sinicae, 44(12): 83-87.

Yang Shaoli, Chen Huizhen, Chen Fei et al, 1997. Observations on *Megastigmus cryptomeri* Yano and its control [J]. Zhejiang Forestry Science and Technology, 14(4): 30-33.

Yang Weiyi, 1964. Economic insect fauna of China Fasc. 2. Hemiptera: Pentatomidae [M]. Beijing: Science Publishing House.

Yang Xiaofeng, Huwen, Feng Yongxian et al, 2008. Bionomics and infestation of *Rhyacionia insulariana* Liu [J]. Sichuan Forest Science and Technology, 29(5): 43-44.

Yang Xinyuan, Yang Shirong, and Guo Chunhua, 2000. Occurrence and control of *Melolontha hippocastani mongolica* [J]. Plant Doctor, (5): 30.

Yang Xiuyuan and Wu Jian, 1981. Checklist of Chinese forest insects [M]. Beijing: China Forestry Publishing House.

Yang Yanyan, Li Zhaohui, Wang Rugang et al, 2004. A study on the predation of *Phloeosinus aubei* Perris by *Tillus notatus* Klug [J]. Shandong Agricultural Science, (6): 40-42.

Yang Yongzhong and Chen Xingfu. 1999, Bionomics of *Asterolecanium castaneae* Russell and its control [J]. Zhejiang Forestry Science and Technology, 19(5): 31-34.

Yang Youlan, Wang Hongwu, and Lü Xiaohu, 2002. Bionomics

and control of *Cyamophila willieti* (Wu) [J]. Entomological Knowledge, 39(6): 433-436.

Yang Youqian and Li Xiusheng, 1982. Forest disease and pest control [M]. Zhengzhou: Henan Science and Technology Publishing House.

Yang Youqian and Zhang Aifen, 1986. Study on *Fiorinia japonica* Kuwana [J]. Forest Pest and Disease, (1): 18-20.

Yang Youqian, Si Shengli, Wang Gaoping et al, 1995. Control technology of major pests of chestnut in Xinyang area [J]. Henan Forestry Science and Technology, 1: 8-11.

Yang Youqian, 1999. Preliminary study on the bionomics of *Sinoxylon japonicum* Lesne (Coleoptera: Bostrichidae) [J]. Forest Pest and Disease, (2): 2-3.

Yang Youqian, 2000. Bionomics and control of pests in *Fraxinus chinensis* [J]. Forest Pests and Disease, (5): 17-18.

Yang Youqian, 2000. *Cryphalus tabulaeformis* Tsai et Li - a new pest on *Pinus tabulaeformis* [J]. Forest Pest and Disease, (6): 22-23.

Yang Yufa and Liu Zhandong, 1999. Bionomics and control of *Omphisa plagialis* Wileman [J]. Jilin Forestry Science and Technology, (5): 26-31.

Yang Yunhan, 1986. *Podagricomela shirahatai* (Chŭjŏ) [J]. Plant Protection, 12(5): 20-21.

Yang Zhende, Tian Xiaoqing, and Zhao Boguang, 2006. Threshold and effective accumulative temperature for the development of *Plagiodera versicolora* [J]. Journal of Beijing Forestry University, 28(2): 139-141.

Yang Zhenjiang, Shi Hekui, and Li Yulian, 1995. Bionomics and control of *Euzophera batangensis* Caradja [J]. Entomological Knowledge, 32(6): 340-342.

Yang Zhirong and Liu Shigui, 1991. Isolation and identification of *Trabala vishnou* Lefebure nuclear polyhedrosis virus [J]. Virologica Sinica, 6(4): 376-378.

Yang Zhong Qi, Wang Xiaoyi, Cao Liangming, Yao Yanxia, Tang Yanlong, 2014. Re-description of *Sclerodermus guani* and revision of the genus (Hymenoptera; Bethylidae) in China [J]. Chinese Journal of Biological Control, 30(1): 1-12.

Yang Zhongqi, Qiao Xiurong, Bu Wenjun et al, 2006. An important invasive pest - *Obolodiplosis robiniae* (Haldemann) (Diptera: Cecidomyidae) [J]. Acta Entomologica Sinica, 49(6): 1050-1053.

Yang Zhongqi, Wang Xiaoyi, Cao Liangming et al, 2014. Re-description of *Sclerodermus guani* and revision of the genus (Hymenoptera: Bethylidae) in China [J]. Chinese Journal of Biological Control, 30(1): 1-12.

Yang Zhongqi, 1996. Parasitic wasps of bark beetles in China [M]. Beijing: Science Publishing House.

Yang Ziqi and Cao Guohua, 2002. Pictorial guide to the control of diseases and pests of ornamental plants [M]. Beijing: China Forestry Publishing House.

Yao Donghua, Liu Peihua, Jing Xiaoyuan et al, 2011. Integrated management of *Holocerus arenicola* Staudinger [J]. Modern Agricultural Science and Technology, (2): 194-195.

Yao Yanfang, Yang Qin, Guo Haiyan et al, 2009. Investigation of two borers infesting *Artemisia* spp. [J]. Inner Mongolia Forestry Investigation and Design, 32(4): 103-104.

Yao Yanxia and Yang Zhongqi, 2008. Three species of genus *Pteromalus* (Hymenoptera: Pteromalidae) parasitizing *Rhynchaenus empopulifolis* (Coleoptera: Curculionidae), with description of a new species from China [J]. Scientia Silvae Sinicae, 44(4): 90-94.

Yates, H. O, 1986. Checklist of insect and mite species attacking cones and seeds of world conifers [J]. Journal of Entomological Science, 21(2): 142-168.

Ye Hui, Lü Jun, and Lieutier, 2004. On the bionomics of *Tomicus minor* (Hartig) (Coleoptera: Scolytidae) in Yunnan province [J]. Acta Entomologica Sinica, 47(2): 223-228.

Ye Mengxian, 1983. Study on *Lampra limbata* Gebler [J]. Northern Fruit Trees, (3): 27-28.

Ye Shuqin, Sun Jianwen, and Xu Shuiwei, 2002. Burying stumps to control *Urocerus gigas taiganus* Benson [J]. Liaoning Forest Science and Technology, (1): 43-44.

Yellow: corrected by Wang. Red: need to check for accuracy.

Yi Boren, Kang Zhixian, Wei Juxiang et al, 1992. A preliminary study of *Euproctis flava* (Bremer) [J]. Nothern Horticulture, 6: 37-39.

Yi Yehua, Li Yizhen, and Zheng Zhulong, 2004. Control techniques of *Dryocomus kuriphilus* (Yasumatsu) [J]. Guangdong Forestry Science and Technology, 20(2): 47-50.

Yin Anliang, Zhang Jiasheng, Zhao Junlin et al, 2008. Bionomics and control of *Eriogyna pyretorum* (Westwood) [J]. Forest Pest and Disease, 27(1): 18-20.

Yin Chenglong, Wang Youkui, Lin Hai et al, 2001. Natural enemies of pests in *Picea crassifolia* forests and their protecion [J]. Journal of Beihua University (Natural Science edition), 2(1): 44-46.

Yin Chunchu, 2003. Bionomics and control of *Acalolepta sublusca* (Thomson) [J]. Hunan Agricultural Science, (1): 54-56.

Yin Feifei, Wang Manli, Tan Ying et al, 2008. A functional F analogue of Autographa californica nucleopolyhedrovirus GP64 from the Agrotis segetum Granulovirus [J]. Journal of Virology, 82(17): 8922-8926.

Yin Haisheng and Liu Xianwei, 1995. Synopsis on the classification of Grylloidea and Gryllotalpoidea from

China [M]. Shanghai: Shanghai Science and Technology Literature Publishing House.

Yin Huifen, Huang Fusheng, and Li Zhaolin, 1984. Economic insect fauna of China Fasc. 29. Coleoptera: Scolytidae [M]. Beijing: Science Publishing House.

Yin Huifen, 2000. The synopsis on morphological and biological characters of *Dendrocnotus valens* Leconte [J]. Acta Zoologica Sinica, 25(1): 120, 43.

Yin Huifeng, Huang Fusheng, and Li Zhaolin, 1984. Economic Insect Fauna of China Fasc. 29. Coleoptera: Scolytidae [M]. Beijing: Science Publishing Housing.

Yin Shicai, 1982. A preliminary study on Grylloidea Holmgren [J]. Scientia Silvae Sinicae, 18(1): 58-63.

Yin Xiangyu, 2004. Preliminary study report on *Curculio davidi* [J]. Anhui Agricultural Science Bulletin, 12(5):205.

Yin Xinming, 1994. Description of *Philus antennatus* (Gyllenhal) larvae and the taxonomic status within Cerambycidae [D]. Southwest Agricultural University Doctoral dissertation.

Yin Yanbao, Zhao Qikai, and Jiang Xinglin, 2005. Prediction of *Strobilomyia laricicola* (Kårl) infestations in Jiagedaqi area [J]. Journal of Northeast Forestry University, 33(4): 14-16.

Ying Hongtao and Luo Zhengfang, 2002. Bionomics and control of *Neodiprion huizeensis* Xiao et Zhou [J]. Yunnan Forestry Science and Technology, 98(1): 65-67.

Ying Yanlong and Huang Bingzhao, 1958. Field studies on the control of black dotted tussock moth in Tonglu county [J]. Rapid Reports of Forest Science and Technology, (5): 3.

You Shijuan, Liu Jianfeng, Huang Dechao et al, 2013. A review of the mealybug Oracella acuta: Invasion and management in China and potential incursions into other countries [J]. Forest Ecology and Management 305: 96-102.

Yu Chengming, 1959. Preliminary survey of bark beetles in *Pinus silvestris* var. *mongolica* at Hulun Buir [C]. In Beijing Forestry College "Preliminary investigation of forest pest". Science Publishing House.

Yu Enyu et al, 1984. Control of *Ambrostoma quadriimpressum* Motschulsky using JLY-8401 poison band [M]. Database on research in China forestry science and Technology.

Yu Enyu, Gao Changqi, Wang Zhiming, 1987. Biology and field release of *Asynacta ambrostomae* Liao (Hymenoptera: Trichogrammatidae) [J]. Forest Science and Technology, (12): 16-18.

Yu Enyui, Gao Changyi, and Wang Zhiming, 1987. Bionomics of *Trichogramma* sp. on *Ambrostoma quadriimpressum* Motschulsky and release experiment [J]. Forest Science and Technology, (12): 18-20.

Yu Fangbei, 1988. Preliminary studies on *Gastrolina depressa* [J]. Forest Pest and Disease, (3): 12-13.

Yu Feng, Kong Wenjun, Ma Jianhong, Mahmuti Niyazi, Albrecht Abrac, 2011. Occurrence and control tactics of *Carpomya vesuviana* Costa in Turpan [J]. Plant Protection, 37(1): 166-167.

Yu Guiping and Gao Bangnian, 2005. Bionomics of *Aromia bungii* Faldermann [J]. Forest Pest and Disease, 24(5): 15-16.

Yu Jingru, 1998. Control of *Clostera anastomosis* (L.) with cypermethrin treated band [J]. Hebei Forestry Science and Technology, (1): 39-40.

Yu Jun, 2001. Bionomics and control of *Cerura menciana* Moore [J]. Anhui Forestry, (2): 24.

Yu Junde, 1943. Plant derived tannin resources in China [J]. Scientific World, 10(6): 343-352.

Yu Lichen, Liang Lairong, Ao Xianbin et al, 1997. Studies on the morphology and biology of *Pyemotis scolyti* (Oudemans) [J]. Acta Arachnologia Sinica, 6(1): 46-52.

Yu Peiwang, 1998. Bionomics of *Nesodipron zhejiangensis* [J]. Journal of Fujian Forestry Science and Technology, 25(2): 15-19.

Yu Peiyu, Wang Shuyong et al, 1996. Economic insect fauna of China Fasc. 54. Coleoptera: Chrysomeloidea (2) [M]. Beijing: Science Publishing House.

Yu Peiyu, Wang Shuyong, Yang Xingke, 1996. Chinese Economic Insects Book 54 - Coleoptera: Chrysomelidae (Part II) [M]. Beijing: Science Publishing House.

Yu Qing and Zhang Weiyao, 2008. Study on the biological characteristics of *Pachynematus itoi* Okutani [J]. Inner Mongolia Forestry Investigation and Design, 31(5): 84-85.

Yu Siqin and Sun Yuanfeng, 1993. Fauna of agricultural insects in Henan [M]. Beijing: China Agricultural Science and Technology Publishing House.

Yuan Bo and Mo Yiqin, 2006. Artificial rearing of *Actias selene ningpoana* Felder [J]. Anhui Agricultural Science, 34(6): 1092.

Yuan Feng, Zhang Yalin, Feng Jinian, 2006. Insect taxonomy (2nd edition) [M]. Beijing: China Agricultural Publishing House.

Yuan Guisheng, Xie Yingping, Niu Yu et al, 2006. Pathogenecity of *Beauveria bassiana* to *Dendrolimus tabulaeformis* in Shanxi forest [J]. China Journal of Biological Control, 22(2): 118-122.

Yuan Haibin, Liu Ying, Shen Dishan et al, 2004. Morphology and bionomics of *Actias selene ningpoana* Felder [J]. Journal of Jilin Agricultural University, 26(4): 431.

Yuan Ronglan, Lai Zhenliang, Wu Ying et al, 1990. Bionomics of *Dioryctria mendacella* Staudinger [J]. Journal of Zhejiang Forest College, 7(2): 147-152.

Yuan Shenghua, 2008. Chemical control of *Mesoneura rufonota* Rohwer [J]. Science and Technology Innovation Herald, (33): 235.

Yuan Yu, Lü Longshi, and Jin Dayong, 2001. Biological and ecological characteristics of *Papilio xuthus* Linnaeus [J]. Agriculture and Technology, 21(3): 19-22.

Yuang Jiahuan, 1986. Bionomics of *Athyma ranga serica* Leech [J]. Forest Science and Technology, (1): 18-19.

Yue Shukui, Yue Hua, Fang Hong et al, 1994. Bionomics of *Xerodes rufescentaria* (Motschulsky) [J]. Journal of Northeast Forestry University, 22(5): 38-43.

Ze Sangzi, Yan Zhengliang, Zhang Zhen et al, 2010. The diversity of volatile monoterpenes released from *Pissodes punctatus* at different part of *Pinus armandii* and their influences on the weevil's behavior [J]. Journal of Environmental Entomology, 32(1): 36-40.

Zeng Aiguo, 1981. Control of *Anomoneura mori* Schwarz [J]. Northern Sericulture, (2): 50-51.

Zeng Dapeng, 1998. Quarantine diseases and pests of imported forest plants of China [M]. Beijing: China Forestry Publishing House.

Zeng Ling, Zhou Rong, Cui Zhixin et al, 2003. Effect of host plants on development and survival of *Brontispa longissima* (Gestro) [J]. Journal of South China Agricultural University (Natural Science Edition), 24(4): 37-39.

Zeng Ling, Zhou Rong, Cui Zhixin et al, 2003. Effect of hosts on red palm weevil development [J]. Journal of South China Agricultural University (Natural Science edition), 24(4): 37-39.

Zeng Saifei, 2008. A preliminary report on the control of *Anomala cupripes* Hope in eucalyptus forests [J]. Fujian Forestry Science and Technology, 35(4): 175-177.

Zenglin, Ma Xiying, Zhang Bing et al, 2004. Screening of least toxic insecticides for control of *Curculio davidi* [J]. Liaoning Forestry Science and Technology, 3: 16-17.

Zha Guanglin, 2003. Using mixture of *Dendrolimus punctatus wenshanensis* NPV, CPV to control pests [J]. Forest Inventary and Planning, 28(2): 105-108.

Zhai Yongjian, 1966. Survey of overwintering *Agrotis ypsilon* (Rott.) [J]. Entomological Knowledge, (3): 170.

Zhang Bairui, Zheng Jun, and Chen Changjie, 2004. Damage to willow seedlings by *Earias pudicana pupillana* Staudinger and its control [J]. Jilin Forestry Science and Technology, (11): 39-43.

Zhang Bingyan, 2006. Pictorial guide to the identification and control of diseases and pests of *Zanthoxylum bungeanum* Maxim [M]. Beijing: Jindun Publishing House.

Zhang Changhui, 1983. Preliminary analysis of the causes of outbreaks of chestnut gall wasp [J]. Bulletin of Fruit Tree Science and Technology, (1): 30-34.

Zhang Chaoju, 2002. Bionomics and control of *Dendrolimus kikuchii kikuchii* Matsumura [J]. Entomological Journal of East China, 11(2): 74-78.

Zhang Cunli and Li Hongyan, 2006. Control experiments for *Omphisa plagialis* Wileman [J]. Anhui Agricultural Science, (13): 3115-3172.

Zhang Dandan, Chi Defu, Jiang Haiyan et al, 2001. The control effects of forest management measures on the damage of larch cone flies [J]. Journal of Northeast Forestry University, 29(6): 18-19.

Zhang Dehai and Wang Enguang, 1998. Efficacy of insecticides against *Adelgis laricis* Vallot [J]. Journal of Qinghai Agriculture and Forestry Science and Technology, (3): 26-28.

Zhang Ensheng, Li Yanhong, and Yuan Deshui, 2000. Preliminary study on the control of *Strobilomyia laricicola* (Kårl) by injecting insecticide to base of tree trunk [J]. Hebei Forestry Science and Technology, (2): 15-16.

Zhang Fangping, Fu Yueguan, Peng Zhengqiang et al, 2006. Bionomics and control of *Parasaissetia nigra* (Nietner) [J]. Tropical Agricultural Science, 26(1): 36-41.

Zhang Guangxue and Fang Sanyang, 1981. A description of *Pineus cembrae pinikoreanus* Zhang et Fang subsp [J]. nov. (Homoptera: Adelgidae). Journal of Northeast Forest University, (4): 15-18.

Zhang Guangxue and Xu Tiesen, 1983. Economic insect fauna of China Fasc. 25 Homoptera: Aphids (1) [M]. Beijing: Science Press.

Zhang Guangxue, Qiao Gexia, Zhong Tiesen et al, 1999. Economic insect fauna of China Fasc. 14. Homoptera: Mindaridae and Pemphigidae [M]. Beijing: Science Publishing House.

Zhang Guangxue, 1999. Aphid pests on agricultural crops and forestry in northwestern China [M]. Beijing: The Chinese Environmental Science Press.

Zhang Guifen, Yan Xiaohua, and Meng Xianzuo, 2001. Effects of the sex pheromone traps on capture of *Cydia trasias* (Meyrick) (Lepidoptera: Olethreutidae) male moth [J]. Scientia Silvae Sinicae, (4): 388-394.

Zhang Haijun, Mao Liren, Qin Dezhi et al, 2005. Infestations of *Curculio davidi* Faust and its control strategies [J]. Northern Fruit Free Science, (5): 37.

Zhang Hanhao and Tan Jicai, 2004. Tea pests in China and their non-toxic management [M]. Hefei: Anhui Science and Technology Publishing House.

Zhang Hong, 2006. Infestation of *Dendrolimus kikuchii kikuchii* Matsumura and its control methods [J]. Forest Inventory and Planning, 31(supplement): 178-180.

Zhang Hongsong, He Hongan, and Jia Weixi, 2002. Occurrence and control of *Ascotis selenaria dianaria* Hübner [J]. Henan Forestry, 3: 27.

Zhang Hongxi, Cao Zhongchen, Zhao Xianglan et al, 1996. Study on biological characteristics of *Phalera flavescense* in east Hebei [J]. Journal of Hebei Agricultural Technology Normal University, 10(4): 75-77.

Zhang Huan, Xu Songyue, Gao Zairun et al, 1995. Life table analysis of *Baculum minutidentatum* Chen et He [J]. Forest Pest and Disease, (4): 10-11.

Zhang Huan, Xu Songyue, Gao Zairun, et al. 1996. Food selection and food consumption of *Baculum minutidentatum* Chen et He [J]. Forest Pest and Disease, (2): 23-25.

Zhang Huaxuan, 1980. Preliminary study on the control of *Pyrosta machaeralis* Walker using *Bacillus thuringiensis* galleriae [J]. Tropical Forestry, (1): 19-23.

Zhang, H., Ye, H., Haack, R. A. and Langor, D. W, 2004. Biology of Pissodes yunnanensis (Coleoptera: Curculionidae), a pest of Yunnan pine in southwestern China [J]. The Canadian Entomologist, 136(5): 719-726.

Zhang Jinfang, 1989. A new pest of *Cryptomeria fortunei* seed - *Megastigmus cryptomeriae* Yano [J]. Forest Science and Technology, (8): 30-31, 13.

Zhang Jintong and Meng Xianzuo, 2001. Sexual behavior of *Holcocerus insularis* and circadian rhythm of its sex pheromone production and release [J]. Acta Entomologica Sinica, 44(4): 428-432.

Zhang Jintong, Jin Xiaoyuan, Luo Youqing, et al, 2009. The sex pheromone of the sand sagebrush Carpenter worm, Holcocerus artemisiae (Lepidoptera, Cossidae) [J]. Z. Naturforsch, 64c, 590-596.

Zhang Jintong, Luo Youqing, Zong Shixiang et al, 2009. Analysis, synthesis, and biological activities of sex attractant for *Holcocerus artemisiae* (Lepidoptera: Cossidae) [J]. Scientia Silvae Sinicae, 45(9): 106-110.

Zhang Junsheng and Bai Zhongwen, 2000. Study on *Cimbex taukushi* Marlatt [J]. Forest Science and Technology, 25(6): 25-27.

Zhang Li Xin and Yang Lixin, 2006. Bionomics of *Lampra limbata* Gebler [J]. Inner Mongolia Agriculture Science and Technology, (1): 77.

Zhang Li, 1983. Occurrence and control of *Lampra limbata* Gebler [J]. Northwest Horticulture (Fruit tree issue), (4): 27.

Zhang Lianqin and Liang Xiongfei, 1992. Dispersal distance of *Monochamus alternatus* Hope [J]. Forest Science and Technology, (12): 26-27.

Zhang Lianzhu and Zhang Guangsheng, 1982. Control of *Erannis ankeraria* Staudinger [J]. Hebei Forestry Science and Technology, (3): 37.

Zhang Lixia, Guan Zhibin, Fu Xianhui et al, 2002. Occurrence and control of *Hypomeces squamosus* Fabricius [J]. Plant Protection, 28(1): 59-60.

Zhang Lixia, 2007. A preliminary report on the infestation of *Rauvolfia vomitoria* by Deilephila nerii L. [J]. Plant Protection, 33(1): 138.

Zhang Mengqi, 1977. Two species of long-horned beetles infesting *Robinia pseudoacacia* [J]. Entomological Knowledge, 14(2): 56-57.

Zhang Peikun, 1989. Population dynamics and control methods of *Oides leucomelauena* Weise [J]. Guangxi Plant Protection, (2): 15-17.

Zhang Qinghe, Liu Huifang, Sun Yujian et al, 1990. Vertical distribution of *Ips subelongatus* Motschulsky on burned larch tree [J]. Journal of Northeast Forestry University, 18(4): 14-17.

Zhang Qiusheng, Chen Shirong, and Lan Shumei, 1996. A study on *Mesa taianensis* Xiao et Zhou in Jiangsu [J]. Jiangsu Forestry Science and Technology, 23(2): 43-44.

Zhang Runzhi, Chen Xiaoda, and Dang Xinde, 1992. A new weevil infesting *Hippophae rhamnoides* seeds, Curculio hippophes Zhang [J]. Scientia Silvae Sinicae, 28(5): 412-414.

Zhang Runzhi, Ren Li, Wang Chunlin et al, 2001. Research progress on mango weevils (Coleopterza: Curculionidae) [J]. Entomological Knowledge, 38(5): 342-345.

Zhang Runzhi, Renli, Sun Jianghua et al, 2003. Morphological differences of the cocconut pest insect, *Rhynchophorus ferrugineus* (Oliver) [J]. China Forest Pest and Disease, 22(2): 3-6.

Zhang Runzhi, Wang Fuxiang (Ed.), 2010. *Phenacoccus solenopsis* Tinsley [M]. Beijing: China Agricultural Publishing House.

Zhang Runzhi, Wang Xingjian, Adili Shataer, 2007. Identification and risk analysis of quarantine pest, *Carpomya vesuviana* Costa [J]. Chinese Bulletin of Entomology, 44(6): 928-930.

Zhang Runzhi, 1997. A new species *Hylobitelus xiaoi* sp. nov. (Coleoptera: Curculionidae) [J]. Scientia Silvae Sinicae, 33(6): 541-545.

Zhang Shengfang and Liu Yongping, 1991. Identification of *Acanthoscelides pallidipennis* Motschulsky [J]. Forest Pest and Disease, (1): 42-43.

Zhang Shengfang, Liu Yongping, Wu Zengqiang, 1998. Beetles associated with stored products in China [M]. Beijing: China

Agriculture Science Technology Publishing House.

Zhang Shimei (Ed.), 1985. Chinese Economic Insects Book 31-Hemiptera (Part I) [M]. Beijing: Science Publishing House.

Zhang Shimei and Hu Meicao, 1983. Preliminary report on the study of *Epiplema moza* Butler [J]. Jiangxi Forestry Science and Technology, (5): 13-18.

Zhang Shimei and Hu Meicao, 1984. Biological observations of two geometrid infesting camphor tree - *Thalassodes quadraria* Guenee and *Trigonoptila latimarginaia* Leech in Nanchang [J]. Forest Pest and Disease, (4): 18-20.

Zhang Shimei and Zhao Yongxiang, 1996. Geographical distribution of agricultural insects in China [M]. Beijing: China Agriculture Publishing House.

Zhang Shimei et al, 1985. Economic insect fauna of China Fasc. 21. Hemiptera (1) [M]. Beijing: Science Publishing House.

Zhang Shimei et al, 1988. Agricultural diseases, pests, and weeds in Tibet [M]. Lasa: Tibet People's Publishing House.

Zhang Shimei et al, 1995. Economic insect fauna of China Fasc. 50. Hemiptera (2) [M]. Beijing: Science Publishing House.

Zhang Shiquan, Zhang Yong, and Cuiwei, 1976. A study on *Tischeria decidua* Wocke [J]. Acta Entomologica Sinica, 19(1): 61-71.

Zhang Weiyao, Wang Xiaotong, and Xie Daotong, 1987. Preliminary report on the control of *Pyrausta machoeralis* Walker using dimilin III [J]. Tropical Forestry Science and Technology, (3): 38-43.

Zhang Xiankai and Zuo Yuxiang, 1986. Preliminary study on *Thylactus simulans* Gahan [J]. Entomological Knowledge, (5): 208-210.

Zhang Xiaozhong, Zhang Jilong, and Wang Fuying, 1998. Occurrence and integrated management of mulberry white caterpillar in Yangcheng County of Shanxi province [J]. Northern Sericulture, 19(2): 27-28.

Zhang Xinfeng, Gao Jiusi, Shi Xianyuan et al, 2009. Occurrence and integrated management of *Cydia trasias* (Meyrick) [J]. Modern Agriculture Science and Technology, (18): 168, 172.

Zhang Xingyue and Luo Youqing. Major biological disasters of Chinese forests [M]. Beijing: China Forestry Publishing House.

Zhang Xuewu and Wang Jinrui, 1996. The outbreak law and comprehensive harnessing of *Atysa marginata* [J]. Journal of Fujian Forestry Science and Technology, 23(2): 55-58.

Zhang Xuezu and Qu Bangxuan, 1983. A preliminary study on *Quadraspidiotus slavonfeus* (Green) [J]. Journal of Xinjiang Agricultural University, (4): 1-7.

Zhang Xuezu, 1980. Identification of several fruit borers in Xinjiang [J]. Xinjiang Agricultural Science, (1): 25.

Zhang Yanqiu and Liu Wei, 2002. Occurrence and control of *Apriona germari* (Hope) [J]. Plant Doctor, 15(4): 6-7.

Zhang Yifeng, Wang Juying, and Shen Qiang, 2000. Integrated control of *Kermes castaneae* [J]. Forest Pest and Disease, 19(6): 32-33.

Zhang Yingjuan and Yang Chi, 2000. Comparative analysis of genetic diversity in the endangered shrub *Teraena mongolica* and its related congener *Zygophyllum xanthozylon* [J]. Acta Phytoecologica Sinica, (4): 425-429.

Zhang Yinglan, Han Dan, Li Xue et al, 2008. Control techniques of *Costera anastomosis* L. (Lepidoptera: Notodontidae) [J]. Journal of Mudanjiang Normal University (Natural Science edition), 63(2): 16-17.

Zhang Yining, 1999. A study of bionomics and control of *Pissodes* sp. (Coleoptera: Curculionidae) [J]. Journal of Southwest Forestry College, 19(2): 118-121.

Zhang Yongan, 2004. Industrial production of insect virus products for forest pests [J]. Practical Forestry Technology, (11): 1.

Zhang Yubao, Li Jinguo, An Kun et al, 2006. Relationship between reducing sugar content in different poplar strains and damage of *Xylotrechus rusticus* (Coleoptera: Cerambycidae) [J]. Journal of Northeast Forest University, 34(2): 35-37.

Zhang Yufa, Yang Dongming, Zhang Kun et al, 2003. Control of *Dendrolimus punctata tehchangensis* Tsai et Liu with "Hulinshen" dusts [J]. Practical Forestry Technology, (3): 30.

Zhang Yuhua, Lai Yongmei, Zang Chuanzhi et al, 2003. Occurrence and control of *Locastra muscosalis* Walker on Pistacia chinensis [J]. Shaanxi Forest Science and Technology, (1): 48-50.

Zhang Yujun, Zheng Junshan, Pang Xuhong, 2013. Studies on least toxic methods for control of *Ambrostoma quadriimpressum* Motschulsky [J]. Jilin Forestry Science and Technology, 42(1): 37-38, 46.

Zhang Yuling and Li Jianhong, 2004. Occurrence and control of *Culcula panterinaria* (Bremer et Grey) [J]. Sci-Tech Information Development and Economy, 14(7): 183-184.

Zhang Yun, Zhang Zhen, and Wang Hongbin, 2006. Influence of climate factors on outbreaks of *Diprion jingyuanensis* [J]. China Journal of Applied Environmental Biology, 12(5): 660-664.

Zhang Zhen and Zhang Xudong, 2009. Red turpentine beetle [C]. In Wan Fanghao, Guo Jianying, Zhang Feng et al. Studies on the invasive pests in China. Beijing: Science Publishing House.

Zhang Zhen, Wang Hongbin, and Kong Xiangbo, 2009. Red burpentine beetle [C]. In Wan Fanghao, Zheng Xiaobo, and Guo Jianying. Biology and control of major invasive pests in Agriculture and Forestry in China. Beijing: Science Publishing House.

Zhang Zhen, Wang Hongbin, Chen Guofa, et al, 2005. Sex pheromone for monitoring flight periods and population densities of the pine sawfly, Diprion jingyuanensis Xiao et Zhang (Hym., Diprionidae) [J]. Journal of Applied Entomology, 129(7): 368-374.

Zhang Zhiguang and Shi Yuliang, 1958. A study on *Eulecanium kunoensis* (Kuw.) [J]. Journal of Shandong Agricultural College, 3: 1-12.

Zhang Zhili, 1984. Economic Insect Fauna of China Fasc. 28. Coleoptera: Scarabaeoidea [M]. Beijing: Science Publishing House.

Zhang Zhiti, Zhang Lijun, Zhao Changbin et al, 1981. Observations on the life history of *Gryllotalpa unispina* Saussure [J]. Plant Protection, 7(4): 10-11.

Zhang Zhixiang, Xu Hanhong, and Jiang Dingxin, 2008. Study and application of Yejiaqing eluviation powder in a hung bag used for controlling *Brontispa longissima* [J]. Guangdong Agricultural Science, (2): 65-68.

Zhang Zhizhong, Chen Xueying, Li Zhenyu et al, 1982. Preliminary report on a study of the population dynamics of *Cinara* sp [J]. Journal of Beijing Forestry University, (3): 68-80.

Zhang Zhizhong, 1959. Observations on the biology and control of *Sympiezomias lewisi* Roelofs [C]. In "Preliminary Report on Forest Pests". Beijing: Science Publishing House.

Zhang Zhizhong, 1997. Forest entomology [M]. Beijing: China Forestry Publishing House.

Zhang Zuoshuang, Xiong Deping, and Cheng Wei, 2008. Control of stem borers by a parasitoid, *Pyemotes trittci*, Largereze-Fossot & Montane [J]. Chinese Journal of Biological Control, 24(1): 1-6.

Zhao Changrun, 1981. Preliminary study on *Dioryctria mendacella* Staudinger [J]. Entomological Knowledge, 18(1):20-22.

Zhao Fanggui, Li Jipei and Sun Xugen, 1999. Identification and control of common forest and fruit pests [M]. Jinan: Jinan Publishing Housing.

Zhao Fanggui, 1965. Observation of *Aphrophora flavipes* Uhler [J]. Entomological Knowledge, 9(1): 44-46.

Zhao Guorong, Cai Yanping, and Yang Chuncai, 1997. Study on the damage of ornamental trees by *Pandemis heparana* (Denis et Schiffermüller) [J]. Forest Science and Technology, (10): 30-32.

Zhao Jianxia and Wang Yulan, 1986. A preliminary study on the bionomics and control of *Euzophera alpherakyella* Ragonot [J]. Entomological Knowledge, 23(6): 273-275.

Zhao Jinnian and Cao Bin, 1987. Bionomics of *Phloeosinus perlatus* Chapuis and its control [J]. Entomological Knowledge, 24(4): 227-230.

Zhao Jinnian and Chen Sheng, 1993. Bionomics and control of *Hypocala subsatura* (Guenee) [J]. Forest Research, 6(3): 341-344.

Zhao Jinnian and Huang Hui, 1989. Influence of the development of cones and staminate branches of *Pinus massoniana* by *Dioryctria pryeri* Ragonot [J]. Forest Research, 2(3): 300-303.

Zhao Jinnian and Huang Hui, 1997. Damage by larvae of *Dioryctria yiai* Mutuura et Munroe and its population estimation [J]. Scientia Silvae Sinicae, 33(3): 247-251.

Zhao Jinnian and Huang Hui, 1998. Bionomics of *Nematus prunivorous* Xiao [J]. Entomological Knowledge, 35(2): 83-84.

Zhao Jinnian and Yingjie, 1988. Population dynamics of *Shirahoshizo patruelis* (Voss) [J]. Forest Pest and Disease, (4): 4-6.

Zhao Jinnian et al, 1999. Survey of pests and natural enemies in massonia pine forests [J]. Scientia Silvae Sinicae, 35(special issue): 62-65.

Zhao Jinnian, Chen Sheng, and Huang Hui, 1991. A study on *Dioryctria yiai* Mutuura et Munroe [J]. Forest Research, 4(3): 291-296.

Zhao Jinnian, Chen Sheng, and Huang Hui, 1992. Studies on occurrence and methods of control of *Dioryctria rubella* Hampson [J]. 28(2): 131-137.

Zhao Jinnian, Chen Sheng, and Tang Zhilin et al, 1991. Aggregation, dispersal and control of *Blastophagus piniperda* L. [J]. Forest Science and Technology, (7): 31-33.

Zhao Jinnian, Chen Sheng, Huanghui et al, 1991. A study on *Petrova cristata* in Masson pine seed orchard [J]. Forest Research, 4(6): 662-668.

Zhao Jinnian, Huanghui, and Zhoushishui, 1997. A study on the insect and rodent pests in Masson pine seed orchard [J]. Forest Research, 10(2): 173-181.

Zhao Jinnian, Jiang Jingmin and Shen Keqin, 1995. Effect of *Diocyctria rubella* infestation on the height growth of *Pinus taeda* [J]. Forest Pest and Disease, (3): 25-27.

Zhao Jinnian, Lin Changchun, Jiang Liyuan et al, 2001. Studies on the effect of M99-1 attractant on *Monochamus alternatus* Hope [J]. Forest Research, 14(5): 523-529.

Zhao Jinnian, Liu Ruoping, and Zhou Mingqin, 1988. A preliminary study on *Phassus nodus* Chu et Wang [J]. Scientia Silvae Sinicae, 24(1): 101-105.

Zhao Jinnian, Ying jie, and Caobin, 1988. A preliminary study on *Phloeosinus sinensis* Schedle [J]. Forest Research, 1(2): 186-190.

Zhao Jinnian, Yingjie, and Tang Weiqiang, 1987. Preliminary

study on *Niphades verrcosus* (Voss) [J]. Practical Forest Technology, (10): 12-14.

Zhao Jinnian, 1990. Hepialidae insects of China and research advances [J]. Plant Protection (supplement), 53-54.

Zhao Jinnin et al, 2004. Wood boring insects of massonian pine and their integrated management [J]. Forest Research, 17(special issue): 62-65.

Zhao Junfang, 2006. Infestation and control of *Hegesidemus harbrus* on poplar trees in northern Henan [J]. Practical Forestry Technology, (3): 26-27.

Zhao Lianjie, Zhao Bo, Lu Chengjuan et al, 2000. Bionomics and control of *Yponomeuta padella* (Linnaeus) [J]. Jilin Forestry Science and Technology, 29(13): 12-13, 56.

Zhao Ling and Liang Chenjie, 1989. The mite *Allotrombium* sp. - a new predator of *Hegesidemus habrus* Darke [J]. Forest Research, 2(1): 42-46.

Zhao Lingai and Zhuming, 1999. Occurrence of *Dasineura datifolia* Jiang in central Shaanxi [J]. Shaanxi Forest Science and Technology, (3): 57-58.

Zhao Shifeng, Zhao Ruiliang, Zhang Zhiyong et al, 1984. Forest pests of Shanxi province (vol. 1) [M]. Beijing: China Forestry Publishing House.

Zhao Suihua, 1985. Preliminary study on the biology of *Ophrida xanthospilota* (Baly) [J]. Plant Protection, 3(11): 15-16.

Zhao Tieliang, Sun Jianghua, Yan Shanchun et al, 2002. Bionomics of *Strobilomyia* spp. and infestation characteristics [J]. Forest Pest and Disease, 21(3): 6-8.

Zhao Wenjie, Mao Haolong, Yuan Shiyun et al, 1994. A study on the bionomics of *Adelgis laricis* Vallot and its control [J]. Journal of Gansu Forestry Science and Technology, (2): 32-34.

Zhao Xiufu, 1981. Insects in Fujian province [M]. Fuzhou: Fujian Science and Technology Publishing House.

Zhao Xiuying, Han Meiqin, Song Shuxia et al, 2008. Preliminary report on the occurrence of *Cydia trasias* (Meyrick) [J]. Hebei Forestry Science and Technology, (3): 25.

Zhao Yangchang and Chen Yuanqing, 1980. Economic insect fauna of China Fasc. 20. Coleoptera: Curculionidae (1) [M]. Beijing: Science Publishing House.

Zhao Yangchang, 1963. Economic insect fauna of China Fasc. 4. Coleoptera: Tenebrionidae [M]. Beijing: Science Publishing House.

Zhao Yanpeng, 1985. Studies on *Aphrophora intermedia* Uhler (Hemiptera: Aphrophoridae) [J]. Liaoning Forestry Science and Technology, 14(2): 28-32.

Zhao Yumei, Tang Cai, and Lan Cuiyu, 2008. Study on the control effect of the *Verticillium lecanii* on *Oracella atuta* [J]. Journal of Shandong Agricultural University (natural science edition), 39(2): 183-187.

Zhao Zhenzhong, Zhang Baikui, and Xing Xiuqing, 2003. Occurrence and control of *Tremex fuscicornis* (Fabricius) [J]. Hebei Agricultural Technology, (8): 19.

Zhao Zhongling, 1978. Economic insect fauna of China Fasc. 12. Lepidoptera: Lymantriidae [M]. Beijing: Science Publishing House.

Zhao Zhongling, 1993. Study on the genus Eumeta in China and description of new species [C]. In Collections of papers on systematic evolution. Beijing: China Science and Technology Publishing House.

Zhao Zhongling, 2003. Fauna of China Fasc. 30: Insecta, Lepidoptera, Lymantriidae [M]. Beijing: Science Publishing House.

Zhashui County Yingpan Forest Farm, 1973. Preliminary study on *Sonsaucoccus sinensis* (Chen) [J]. Shaanxi Forest Science and Technology, (3): 10-11.

Zhejiang Agricultural University, 1987. Plant quarantine [M]. Shanghai: Shanghai Science and Technology Publishing House.

Zhejiang Forest Bureau, 1959. *Euproctis bipunctapex* (Hampson) [J]. Zhejiang Forestry, (1): 40-41.

Zhejiang Songganjie Control Research Group, 1976. Preliminary studies on *Matsucoccus massonianae* Y.H.R [J]. Forestry Science and Technology, (1): 18-23.

Zhen Changsheng, 1988a. Study on sagebrush carpenter worm [J]. Grassland of China, (1): 40-42.

Zhen Changsheng, 1988b. A preliminary study on Artemisia borer [J]. Journal of Inner Mongolia College of Agriculture and Animal Husbandry, 9(2): 74-81.

Zhen Zhixian, Chi Defu, Zhang Xiaoyan et al, 2001. Advances in research on *Phassus excrescens* Butler [J]. Journal of Hebei Forestry and Orchard Research, 16(2): 178-182.

Zheng Baorong, 2007. Population dynamics of *Polylopha cassiicola* Liu et Kawabe and its integrated management [J]. Fujian Forestry Science and Technology, 34(2): 10-13, 31.

Zheng Chuanjiang and Shen Qiang, 2000. Integrated *Chionaspis saitamensis* Kuwana control techniques [J]. China Fruit News, 16(12): 45.

Zheng Hanye, 1957. Study on Histia rhodope Cramer [J]. Acta Entomologica Sinica, 7(1): 355-359.

Zheng Leyi and Gui Hong, 1999. Insect Taxonomy [M]. Nanjing: Nanjing Normal University Publishing House.

Zheng Leyi, Gui Hong, 1999. Insect taxonomy (II) [M]. Nanjing: Nanjing Normal University Publishing House.

Zheng Wenyun and Gao Desan, 2001. Bionomics of *Camptolo-*

ma interiorrata Walker, a pest of Quercus palustris [J]. Forest Science and Technology, 26(4): 22-25.

Zheng Xu, 2010. Discussions on integrated pest management for pear diseases and pests [J]. Modern Horticulture, (5): 55.

Zheng Yongxiang, Peng Jialong, Wang Mingsheng et al, 2009. Study on the damage threshold and sampling technique of *Chinolyda flagellicornis* [J]. Journal of Nanjing Forestry University (Natural Science edition), 33(5): 142-146.

Zheng Zhemin, Xia Kailing et al, 1998. Fauna of China Fasc. 10: Insecta, Acridoidea,Oedipodidae and Arcypteridae [M]. Beijing: Science Publishing House.

Zhi Cunxun, 1979. Breeding of *Dryocosmus kuriphilus* Yasumatsu resistant chestnut [J]. Fruit Tree Science and Technology, (4): 55-59.

Zho Yajun, 1965. Chemical control of *Parthenolecaniun corni* (Bouche) [J]. Acta Entomologica Sinica, 9(5): 280.

Zhong Hongfu and Chen Yixin, 1964. Economic insect fauna of China Fasc. 3. Lepidoptera: Noctuidae (1) [M]. Beijing: Science Publishing House.

Zhong Yihai, Liu Kui, Peng Zhengqiang et al, 2003. A new high risk pest - *Brontispa longissima* [J]. Tropical Agricultural Science, 23(4): 67-71.

Zhou Jiaxi, Li Houhun, Sun Qinhang et al, 1997. A study on *Cryhalus tabulaeformis chienzhuangensis* Tsai et Li [J]. Journal of Northwest Forestry University,12(supplement): 85-88.

Zhou Jiaxi, Qu Bangxuan, Wang Ximeng et al, 1994. Forest Pests in northwest and their control [M]. Xian: Shaanxi Science Publishing House.

Zhou Mingkuan, Luo Xiuwen, and Zhu Yan, 1993. Bionomics of *Alcidodes sauteri* (Heller) and its control [J]. Entomological Knowledge, 30(6): 344-345.

Zhou Rong, Zeng Ling, Cui Zhixin et al, 2004. Morphological observations on *Brontispa longissima* [J]. Plant Quarantine, 18(2): 84-85.

Zhou Rong, Zeng Ling, Liang Guangwen et al, 2004. Life history of the laboratory population of the palm leaf beetle *Brontispa longissima* [J]. Entomological Knowledge, 41(4): 336-339.

Zhou Shuzhi and Zhang Zhen, 1993. A new species and a new Chinese record in Tehthredinidae [J]. Forest Research, 6(special issue): 57-59.

Zhou Shuzhi, Huang Xiaoyun, Zhang Zhen et al, 1995. Bionomics of larch sawfly and its control methods [J]. Forest Research, 8(2): 145-151.

Zhou Shuzhi, Huang Xiaoyun, Zhang Zhen et el, 1995. Studies on *Pristiphora beijingensis* Zhou et Zhang [J]. Forest Research, 8(5): 556-563.

Zhou Tiying, Xu Weijin, and Zhong Guoqing, 1983. A preliminary study of *Polyura narcaea* [J]. Forest Pest and Disease, (4): 26.

Zhou Tiying, Xu Weijin, and Zhong Guoqing, 1986. A preliminary study of *Polyura narcaea* [J]. Entomological Knowledge, 23(1): 24-25.

Zhou Yao et al, 1994. Fauna of Chinese butterflies (2) [M]. Zhengzhou: Henan Science and Technology Publishing House.

Zhou Yao, Lu Jinsheng, Huang Ju et al, 1985. Economic insect fauna of China Fasc. 36. Homoptera: Fulgoroidea [M]. Beijing: Science Publishing House.

Zhou Yao, 1985a. *Hemiberlesia pitysophila* Takagi [C]. In Diaspididae of China. pp. 402-403. Xian: Shaanxi Science Publishing House.

Zhou Yao, 1985b.Chinese Diaspididae (vol. 2) [M]. Xian: Shaanxi Science and Technology Publishing House.

Zhou Yao, 2000. Fauna of Chinese butterflies [M]. Zhengzhou: Henan Science and Technology Publishing House.

Zhou Yousheng, Shen Farong, Zhao Huanping et al, 1995. A preliminary study on the biology and control of *Sternochetus frigidus* Fabricius [J]. Journal of Southwestern Agricultural University, 17(5): 456-459.

Zhou Yuemei, 1979. Preliminary study on *Trioza camphorae* Sasaki [J]. Forest Diseases and Pests, (4): 7.

Zhou Yushi, Wan Chengzhu, Zeng Weiguo, 1990. A preliminary study on using non-toxic insect glue to control *Ambrostoma quadriimpressum* Motschulsky [J]. Liaoning Forestry Science and Technology, (1): 38-39.

Zhou Zhaoxu, Luo Jincang, Chen Ming, 2008. Biological characteristics and dynamics of the codling moth, Cydia pomonella, in the Zhangye area [J]. Plant Protection, 34(4): 111-114.

Zhou Zhijuan, 1981. Biology and control of the camellia weevil *Curculio chinensis* Chevrolat [J]. Acta Entomologica Sinica, 24(1): 48-52.

Zhou Zhongming and Huang Jingfang, 1983. Disease and pests of seedlings [M]. Beijing: China Forestry Publishing House.

Zhu Chuanfu, Yang Wenxue, Wang Jinguo et al, 1994. A preliminary study on *Pachynematus itoi* Okutani [J]. Forest Science and Technology, 19(1): 25-26.

Zhu Hongfu and Wang Linyao, 1996. Fauna of China Fasc. 5: Insecta, Lepidoptera, Bombycidae, Saturniidae, and Thyrididae [M]. Beijing: Science Publishing House.

Zhu Hongfu and Wang Linyao, 1997. Fauna of China Fasc. 11: Insecta, Lepidoptera, Sphingidae [M]. Beijing: Science Publishing House.

Zhu Hongfu et al, 1975. Pictorial guide to moths [M]. Beijing: Science Publishing House.

Zhu Hongfu et al, 1984. Pictorial guide to moths (2) [M]. Beijing: Science Publishing House.

Zhu Hongfu et al, 2004. Fauna of China Fasc. 38: Insecta, Lepidoptera, Hepialidae and Epoplernidae [M]. Beijing: Science Publishing House.

Zhu Hongfu, Wang Linyao, and Fang Yonglai, 1979. Pictorial guide to larvae of moths (1) [M]. Beijing: Science Publishing House.

Zhu Jian, 1966. A preliminary study on *Nesodiprion* sp, [J]. Scientia Silvae Sinicae, 11(1): 52-62.

Zhu Junhong and Zhang Fangping, 2004. Lymantriid pests of tropical fruit trees and their control techniques [J]. China Southern Fruit Trees, 33(3): 37-40.

Zhu Shude and Lu Ziqiang, 1996. Horticulture entomology [M]. Beijing: China Agricultural Science and Technology Publishing House.

Zhu Tianhui, 2002. Control of diseases and Pests of seedlings [M]. Beijing: China Forestry Publishing House.

Zhu Yi and Wang Xinyan, 2006. Infestation of *Salix koriyanagi* Kimara by *Plagiodera versicolora* (Laicharting) and its control [J]. Journal of Anhui Agricultural Research, 34(12): 2780, 2852.

Zhu Zhijian, Tu Yonghai, Xu Sishan et al, 1995. A study on the biological characteristics and control of *Dutomostethus deqingeneis* Xiao [J]. Journal of Zhejiang Forest College, 11(3): 291-296.

Zhu Zhijian, Tu Yonghai, Xu Sishan et al, 1995. A study on the control threshold of *Dutomostethus deqingeneis* Xiao [J]. Journal of Zhejiang Forest College, 12(3): 271-275.

Zhu Zhijian, Tu Yonghai, Xu Sishan et al, 1996. Parameters of spatial distribution of *Dutomostethus deqingeneis* Xiao and their application [J]. Collection of papers on bamboo research, 15(1): 39-44.

Zong Shixiang, Jia Fengyong, Luo Youqing et al, 2005. Harm characteristics and population dynamics of *Holcocerus hippophaecolus* (Lepidoptera: Cossidae) [J]. Journal of Beijing Forestry University, 27(1): 70-74.

Zong Shixiang, Luo Youqing, and Cui Yaqin, 2009. Damage characteristics of three boring pests in Artemisia ordosica [J]. Forestry Studies in China, 11 (1): 24-27.

Zong Shixiang, Luo Youqing, and Lu Changkuan et al, 2006. Bionomics of *Holcocerus hippophaecolus* (Lepidoptera: Cossidae) [J]. Scientia Silvae Sinicae, 43(1): 102-107.

Zong Shixiang, Luo Youqing, Xu Zhichun et al, 2006. Field trapping trials of sex pheromone for *Holcocerus hippophaecolus* (Lepidoptera: Cossidae) [J]. Journal of Beijing Forestry University, 28(6): 109-112.

Zong Shixiang, Luo Youqing, Xu Zhichun et al, 2006. Preliminary study on different larval instars of seabuckthron carpenter moth, *Holcocerus hippophaecolus* [J]. Entomological Knowledge, 43(5): 626-631.

Zong Shixiang, Luo Youqing, Xu Zhichun et al, 2006. Spatial distribution regularities of *Holcocerus hippophaecolus* pupae [J]. Acta Ecologica Sinica, 26(10): 3232-3237.

Zong Shixiang, Yao Guolong, Luo Youqing et al, 2005. Niche of main boring pests in *Hippophae rhamnoidea* [J]. Acta Ecologica Sinica, 25(12): 3264-3270.

Zou Gaoshun, Li Jiayuan, and Chen Siming, 1985. Bionomics of *Tischeria decidua* Wocke (Lepidoptera: Tischeriidae) and its control [J]. Tropical Forest Research, (3): 31-34.

Zou Jifu, 2000. Bionomics of *Propachys nigrivena* Walker [J]. Journal of Zhejiang Forest College, 17(4): 414-416.

Zou Lijie, Liu Naisheng, He Feiyue et al, 1989. A study on *Kytorrhinnus immixtus* Motschulsky [J]. Forest Pest and Disease, (4): 1-3.

Zu Aiming and Dai Meixue, 1997. The bioassay and yield evaluation of the nuclear polyhedrosis virus from the *Orgia ericae* [J]. China Journal of Biological Control, 13(2): 57-60.

Zuo Tongtong, Chi Defu, and Wang Muyuan, 2008. Repellent effects of phenolic acids in different strain poplars to *Xylotrechus rusticus* [J]. Acta Phytophylacica Sinica, 35(2): 160-164.

Host Plant Names

A

Abies fabri　冷杉
Abies firma　日本冷杉
Abies holophylla　辽东冷杉（杉松）
Abies koreana　朝鲜冷杉
Abies nepholepis　臭冷杉
Abies　冷杉属
Abutilon avicennae　苘麻
Acacia auriculaeformis　大叶相思
Acacia cleaibata　银荆树
Acacia confusa　台湾相思,相思树
Acacia franesiana　金合欢
Acacia mangium　马占相思
Acacia mearnsii　黑荆树
Acacia　相思树属
Acer buergerianum　三角枫
Acer campestre　栓皮槭
Acer fabri　红翅槭
Acer kawakamii　尖叶槭
Acer mono　色木槭,五角枫
Acer negundo　复叶槭
Acer palmatum　红枫
Acer pseudoplatanus　桐叶槭
Acer saccharum　糖槭
Acer spp.　槭
Acer tataricum　鞑靼槭
Acer truncatum　平基槭,元宝槭
Acer　枫属
Achras zapota　人心果
Acrocarpus fraxinifolius　顶果木
Actinidia arguta　软枣猕猴桃
Actinidia chinensis　猕猴桃,中华猕猴桃
Actinidia eriantha　毛花猕猴桃
Actinidia fulvicoma　棕毛猕猴桃
Adenanthera pauonina　孔雀豆
Adinandra bockiana　四川杨桐
Aegiceras corniculatum　桐花树
Aesculus chinensis　七叶树
Agave sisalana　剑麻
Aglaia odorata　米兰
Agriophyllum squarrosum　沙米

Ailanthus altissima 'Qiantouchum'　千头椿
Ailanthus altissima　臭椿
Alangium chinense　八角枫
Albizia lebbeck　大叶合欢
Albizzia falcata　南洋楹
Albizzia julibrissin　合欢
Albizzia kalkora　山槐
Albizzia procera　白格
Aleurites fordii　三年桐
Allamanda neriifolia　软枝黄蝉
Allium fistulosum　葱
Allium tuberosum　韭菜
Alnus cremastogyne　桤木
Alnus hirsuta　毛赤杨
Alnus japonica　赤杨,日本桤木
Alnus nepalensis　旱冬瓜
Alnus sibirica　辽东桤木
Alnus　桤木属
Alopecurus aequalis　看麦娘
Alpinia　山姜属
Alstonia scholaris　盆架子,糖胶树
Alstonia yunnanensis　鸡骨常山
Altingia gracitipes　细柄蕈树
Ambrosia artemisiifolia　豚草
Amelanchier sp.　唐棣
Ammopiptanthus mongolicus　沙冬青
Amorpha fruticosa　紫穗槐
Amygdalus communis　巴旦杏
Amygdalus mongolica　蒙古扁桃
Anacardiaceae　漆树科
Anacardium occidentale　腰果
Angelica sinensis　当归
Annona muricata　刺果番荔枝
Annona squamosa　番荔枝
Anthocephalus chinensis　团花（黄梁木）
Apiaceae　伞形科
Apium graveolens　旱芹（芹菜）
Araceae　天南星科
Arachis hypogaea　花生
Araliaceae　五加科

Archontophoenix alexandrae 假槟榔
Ardisia sp. 密鳞紫金牛
Areca catechu 槟榔
Arecastrum romanzoffianum 山葵
Arenga pinnala 桄榔
Armeniaca sibirica 山杏
Artemis iascoparia 黄蒿
Artemisia annua 黄花蒿
Artemisia capillaris 茵陈蒿
Artemisia integrifolia 柳叶蒿
Artemisia ordosica 黑沙蒿
Artemisia princeps 野艾蒿
Artemisia sphaerocephala 圆头沙蒿
Artemisis halodendron 盐蒿
Artocarpus heterophyllus 波罗蜜, 木菠萝
Artocarpus lingnanensis 桂木
Arundinaria hsienchuensis 仙居苦竹
Aspidium spp. 三叉蕨
Astragalus membranaceus 地黄芪
Avena nuda 青稞
Avena tatua 野燕麦
Averrhoa carambola 杨桃

B

Baechea frutescens 岗松
Balaka 巴拉卡棕属
Bambusa albolineata 花竹
Bambusa blumeana 簕竹
Bambusa cerosissima 单竹
Bambusa chungii 粉箪竹
Bambusa dissemulator var. *hispida* 毛簕竹
Bambusa dissimulator 坭簕竹
Bambusa eutuldoides 大眼竹
Bambusa flexuosa 小簕竹
Bambusa gibboides 鱼肚腩竹
Bambusa indigena 乡土竹
Bambusa intermedia 绵竹
Bambusa lenta 藤枝竹
Bambusa multiplex 'Fernleaf' 凤尾竹
Bambusa multiplex var. *riviereorum* 观音竹
Bambusa multiplex 孝顺竹
Bambusa pachinensis 米筛竹
Bambusa pervariabilis 撑篙竹
Bambusa ramispinosa 坭黄竹
Bambusa remotiflora 甲竹
Bambusa rutila 木竹
Bambusa surrecta 油竹
Bambusa textilis var. *fasca* 椽竹
Bambusa textilis var. *glabra* 光秆青皮竹

Bambusa textilis var. *gracilis* 崖州竹
Bambusa textilis var. *purpurascens* 紫秆竹
Bambusa textilis 青皮竹
Bambusa tulda 马甲竹
Bambusa ventricosa 小佛肚竹
Bambusa vulgaris 'Vittata' 黄金间碧玉
Bambusa vulgaris 'Wamin' 大佛肚竹
Bambusa vulgaris 龙头竹
Bambusa wenchouensis 大木竹
Bambusa 簕竹属
Bambusoideae 竹
Bauhinia uariegata 羊蹄甲
Begonia masoniana 十字海棠
Bentinckiopsis 肖斑棕属
Berberis silvataroucana 小檗
Beta vulgaris 甜菜
Betula davurica 黑桦
Betula luminifera 光皮桦
Betula platyphylla 白桦
Betula pumila 矮桦
Betula spp. 桦
Bignoniaceae 紫葳科
Biota orientalis 扁柏
Bischofia javanica 秋枫
Bischofia polycapa 重阳木
Boehmeria nivea 苎麻
Boehmeria silvestrii 赤麻
Bombax malabaricum 木棉
Borassus flabellifer 糖棕
Brassica campestris 油菜
Brassica chiensis 白菜, 青菜
Brassica napus 欧洲油菜
Brassica oleracea 甘蓝
Brassicaceae 十字花科
Bridelia tomentosa 土蜜树
Broussonetia papyrifera 构树
Brucea javanica 鸦胆子
Burretiodendron hsienmu 蚬木
Buxus bodinieri 雀舌黄杨
Buxus microphylla var. *koreana* 朝鲜黄杨
Buxus sinica 黄杨
Buxus sinica 小叶黄杨

C

Caesalpinia sappan 苏木
Caesalpinia sepiaria 云实
Cajanus cajan 木豆
Calamus 省藤属

Callicarpa macrophylla 大叶紫珠	*Casuarina eguisetifolia* 木麻黄
Callicarpa nudiflora 裸花紫珠	*Catalpa bungei* 楸树
Callicarpa sp. 白花紫珠	*Catalpa duclouxii* 滇楸
Calligonum mongolicum 沙拐枣	*Catalpa fargesii* 灰楸
Callistemon rigidus 红千层	*Catalpa ovata* 梓树
Calophyllum inophyllum 红厚壳	*Catharanthus roseus* 长春花
Camellia 山茶属	*Cedrus deodara* 雪松
Camellia japonica 茶花	*Celosia cristata* 鸡冠花
Camellia japonica 山茶	*Celtis bungeana* 黑弹朴（小叶朴）
Camellia olefera 油茶	*Celtis sinensis* 朴树
Camellia sinensis 茶树	*Cephalomappa sinensis* 肥牛木
Camellia vietnamensis 越南油茶	*Cerasus pseudocerasus* 樱桃
Campsis grandiflora 凌霄	*Cerasus* spp. 樱花
Camptotheca acuminata 喜树	*Cercis chinensis* 紫荆
Canarium album 橄榄	*Chaenomeles sinensis* 木瓜
Canna indica 美人蕉	*Chaenomeles speciosa* 贴梗海棠
Cannabis sativa 大麻	*Chamaecyparis obtusa* 日本扁柏
Cannabis spp. 麻	*Chamaecyparis* 扁柏属
Caragana korshinskii 柠条	*Chamaedaphne* 甸杜属
Caragana microphylla 小叶柠条	Chenopodiaceae 藜科
Caragana sinica 锦鸡儿	*Chimonanthus praecox* 蜡梅
Caragana 锦鸡儿属	*Choerospondias axillaries* 南酸枣
Carallia brachiata 竹节树	*Chrysalidocarpus lutescens* 散尾葵
Carica papaya 番木瓜	*Chukrasia tabularis* 麻楝
Carpinus cordata 千金榆	*Cinchona ledgeriana* 金鸡纳
Carpinus turczaninowii 鹅耳枥	*Cinnamomum bodinieri* 猴樟
Carpinus 鹅耳枥属，千斤榆属	*Cinnamomum burmannii* 阴香
Carya cathayensis 山核桃	*Cinnamomum camphora* 香樟，樟树
Carya illinoinensis 美国山核桃，薄壳山核桃	*Cinnamomum cassia* 肉桂，桂皮
Carya ovata 粗皮山核桃	*Cinnamomum glanduliferum* 云南樟
Caryota ochlandra 鱼尾葵	*Cinnamomum longepaniculatum* 油樟
Caryota urens 孔雀椰子	*Cinnamomum ovatum* 香梓楠
Caryota 鱼尾葵属	*Cinnamomum parthenoxylon* 黄樟
Cassia accidentalis 野扁豆	*Cinnamomum* 樟属
Cassia fistula 腊肠树	*Cirsium souliei* 大蓟
Cassia glauca 粉绿决明	*Citrullus lanatus* 西瓜
Cassia javanica 爪哇决明	*Citrus aurantium* var. *amara* 玳玳
Cassia siamea 铁刀木	*Citrus aurantium* 酸橙
Cassia surattensis 黄槐	*Citrus grandis* 沙田柚，柚树
Castanea henryi 锥栗	*Citrus medica* var. *sarcodactylis* 佛手
Castanea mollissima 板栗	*Citrus reticulata* 柑橘
Castanea seguinii 毛栗，茅栗	*Citrus sinensis* 橙，脐橙
Castanopsis calathiformis 丝锥（丝栗）	*Citrus* spp. 柑橘类，橘类
Castanopsis carlesii 米锥，小红栲	*Claoxylon indicum* 桐树
Castanopsis eyrei 甜槠	*Claoxylon polot* 白桐
Castanopsis fargesii 丝栗栲	*Clausena lansium* 黄皮，黄皮果
Castanopsis hickelii 红锥	*Cleidiocarpon caraleriei* 蝴蝶果
Castanopsis sclerophylla 苦槠	*Cleistogenes caespitosa* 丛生隐子草
Castanopsis spp. 栗，槠	*Clerodendron trichotomum* 海洲常山

Clerodendrum serratum　大青叶
Cocos nucifera　椰子
Coffea arabica　咖啡
Colutea　膀胱豆属
Convolvulaceae　旋花科
Corchorus spp.　黄麻
Coriaria sinica　马桑
Cornus alba　红瑞木
Cornus walteri　毛梾
Cornus wilsoniana　光皮树
Coronilla　小冠花属
Corylus heterophylla　平榛, 榛
Corylus mandshurica　毛榛
Corylus　榛属
Cotinus coggygria　黄栌
Cotoneaster spp.　栒子
Crataegus cuneata　野山楂
Crataegus pinnatifida var. *major*　山里红
Crataegus pinnatifida　山楂
Croton tiglium　巴豆
Cryptomeria fortunei　柳杉
Cryptomeria japonica　日本柳杉
Cucumis sativus　黄瓜
Cucurbita moschata　南瓜, 窝瓜
Cucurbitaceae　葫芦科
Cunninghamia lanceolata　杉木
Cupressaceae　柏科
Cupressus funebris　柏木, 垂柏
Cycas revoluta　苏铁
Cyclobalanopsis glauca　青冈栎
Cydonia oblonga　榅桲

D

Dalbergia balansae　南岭黄檀
Dalbergia hainanensis　海南黄檀
Dalbergia hupeana　黄檀
Dalbergia obtusifolia　牛肋巴
Dalbergia odorifera　降香黄檀（花檀）
Dalbergia siemaoensis　思茅黄檀
Dalbergia　黄檀属
Daucus carota var. *sativa*　胡萝卜
Delonix regia　凤凰木
Dendrocalamopsis asper　马来甜龙竹
Dendrocalamopsis beecheyana var. *pubescens*　大头典竹
Dendrocalamopsis beecheyana　吊丝球竹
Dendrocalamopsis daii　大绿竹
Dendrocalamopsis giganteus　龙竹
Dendrocalamopsis minor　吊丝竹

Dendrocalamopsis oldhami f. *revolute*　花头黄竹
Dendrocalamopsis oldhami　绿竹
Dendrocalamopsis stenoaurita　黄麻竹
Dendrocalamopsis tomentosus　毛龙竹
Dendrocalamopsis yunnanicus　云南龙竹
Dendrocalamopsis　绿竹属
Dendrocalamus latiflorus　麻竹
Dendrocalamus strictus　牡竹
Dendrocalamus　牡竹属
Dendronthema morifolium　菊花
Dianthus caryophyllus　香石竹（康乃馨）
Dictammnus dasycarpus　白藓
Dimocarpus longan　龙眼
Dioscorea esculenta　甘薯
Dioscorea opposita　山药
Diospyros kaki　柿
Diospyros lotus　黑枣（君迁子）
Dolicho lablab　扁豆
Draceana fragrans　巴西木
Draceana marginata　金边香龙血树
Dracontomelum duperreanum　人面子
Duabanga grandiflora　八宝树

E

Elaeagnus angustifolia　沙枣
Elaeagnus angustifolia　银柳
Elaeagnus pungens　胡颓子
Elaeagnus umbellata　牛奶子
Elaeis guineensis　油棕
Elaeocarpus sylvestris　杜英, 山杜英
Emmenopterys henryi　香果树
Equisetum arvense　问荆
Equisetum hiemale　木贼
Equisetum ramosissimum　节节草
Eragrosis pilosa var. *imberbis*　无毛画眉草
Erioaena spectabilis　泡火绳
Eriobotrya japonica　枇杷
Ervatamia divaricata　狗牙花
Erythrina abyssinica　毛刺桐
Erythrina berteroana　马提罗亚刺桐
Erythrina coralloidendron　珊瑚刺桐（龙牙花）
Erythrina cristagalli　鸡冠刺桐（美丽刺桐）
Erythrina indica var. *picta*　黄脉刺桐
Erythrina variegata　刺桐, 杂色刺桐
Erythrina variegata var. *orientalis*　金脉刺桐
Erythrophloeum fordii　格木
Eucalyptus botryoides　葡萄桉
Eucalyptus bridgesiana　苹果桉
Eucalyptus camaldulensis　赤桉

Eucalyptus cinerea 灰桉
Eucalyptus citridora 柠檬桉
Eucalyptus exserta 窿缘桉
Eucalyptus globulus 蓝桉
Eucalyptus grandis 巨桉
Eucalyptus grandis × *Eucalyptus urophylla* 巨尾桉
Eucalyptus grundis × *Eucalyptus tereticornis* 巨圆桉
Eucalyptus gunii 西达桉
Eucalyptus maidie 直杆桉
Eucalyptus pulverulenta 银叶山桉
Eucalyptus robusta 大叶桉
Eucalyptus saligna 柳桉
Eucalyptus spp. 桉树
Eucalyptus tereticornis 细叶桉
Eucalyptus urophylla 尾叶桉
Eucalyptus viminalis 多枝桉
Eucalyptus 桉属
Eucommia ulmoides 杜仲
Eugenia malaccensis 玫瑰苹果
Eugenia 番樱桃属
Euonymus alatus 卫矛
Euonymus bungeanus 丝棉木
Euonymus fortunei 扶芳藤
Euonymus japonica var. *albo-marginatus* 银边黄杨卫矛
Euonymus japonica var. *aureo-marginatus* 金边黄杨卫矛
Euonymus japonica var. *aureo-variegatus* 金心黄杨卫矛
Euonymus japonicus 冬青卫矛
Euonymus kiautshovicus 胶东卫矛
Euphorbia heterophylla 猩猩草
Euphorbia milii var. *splendens* 刺梅
Euphorbia pulcherrima 一品红
Eurya japonica 柃木

F

Fabaceae 蝶形花科
Fagus longipetiolata 山毛榉, 水青冈
Fagus spp. 山毛榉
Fagus 水青冈属（山毛榉属）
Ficus carica 无花果
Ficus cunia 偏叶榕
Ficus elastica 橡皮树
Ficus erecta 天仙果
Ficus hispida 对叶榕
Ficus microcarpa var. *pusillifolia* 小叶榕
Ficus microcarpa 榕树
Ficus microcarpa 细叶榕
Ficus pumila 薜荔
Ficus racemosa 聚果榕
Ficus religiosa 菩提树
Ficus virens 笔管榕, 大叶榕, 黄葛树
Ficus 榕属
Firmiana platanifolia 青桐, 梧桐
Flacourtia indica 刺篱子
Foeniculum vulgare 茴香
Fokienia hodginsii 福建柏, 建柏
Forsythia suspensa 连翘
Fortunella margarita 金橘
Fragaria ananassa 草莓
Fraxinus bungeana 小叶白蜡
Fraxinus chinensis var. *rhychophylla* 大叶白蜡
Fraxinus chinensis 白蜡
Fraxinus mandschurica 水曲柳
Fraxinus rhynchophylla 花曲柳
Fraxinus 白蜡属
Fristania conferta 红胶木

G

Garcinia tinctoria 人面果
Gardenia jasminoides 栀子
Ginkgo biloba 银杏
Gleditsia japonica 山皂荚
Gleditsia sinensis 皂角
Glochidion sp. 算盘子
Glycine max 大豆
Glycyrrhiza pallidiflora 刺果甘草
Glycyrrhiza uralensis 甘草
Gmelina arborea 云南石梓
Gmelina hainanensis 海南石梓
Gomphrena globosa 千日红
Gossypium hirsutum 棉花
Grevillea robusta 银桦
Grewia biloba 扁担杆子
Guttiferae 藤黄科

H

Haloxylon ammodendron 梭梭
Hdysarum leave 杨柴
Hedara nepalensis var. *sinensis* 常春藤
Hedysarum fruticosum 踏郎
Hedysarum scoparium 花棒
Helianthus annuus 向日葵
Helianthus tuberosus 洋姜
Hemiptelea davidii 刺榆
Heracleum barbatum 老山芹
Heracleum hemsleyanum 独活

Hevea brasiliensis 橡胶
Hibiscus rosasinensis 扶桑
Hibiscus syriacus 木槿
Hibiscus 木槿属
Hippophae rhamnoides 沙棘
Homalium hainanense 母生（红花天料木）
Homalium laoticum 老挝天料木
Hordeum spp. 大麦
Humulus lupulus 啤酒花, 蛇麻
Hupericum chinense 金丝桃
Hyophorbe lagenicaulis 酒瓶椰子

I

Ilex chiensis 冬青
Ilex crenata 龟甲冬青
Illicium verum 八角
Illicium 八角属
Imperata cylindrica var. *mojor* 白茅
Indocalamus migoi 天目箬竹
Indocalamus tessellatus 箬竹
Indosasa 大节竹属
Ipomoea batatas 红薯
Itoa orientalis 伊桐

J

Jasminum nudiflorum 迎春
Jasminum sambac 茉莉花
Juglandaceae 胡桃科
Juglans mandshurica 核桃楸
Juglans regia 核桃
Juglans sigillata 泡核桃
Juniperus formasana 刺柏

K

Kalopanax septemlobus 刺楸
Kerria japonica 棣棠
Keteleeria davidiana 铁坚油杉
Keteleeria evelyniana 云南油杉
Keteleeria fortune 油杉
Koelreuteria bipinnata var. *integrifoliola* 黄山栾树
Koelreuteria bipinnata 复羽叶栾树
Koelreuteria paniculata 栾树

L

Lagerstroemia indica 紫薇
Lagerstroemia speciosa 大叶紫薇
Lamiaceae 唇形科
Larix gmelini 落叶松

Larix kaempferi 日本落叶松
Larix koreana 朝鲜落叶松
Larix olgensis var. *koreana* 黄花落叶松
Larix olgensis 长白落叶松
Larix potaninii 红杉
Larix principis-rupprechtii 华北落叶松
Larix sibirica 西伯利亚落叶松
Larix sibirica 新疆落叶松
Larix 落叶松属
Latania lontaroidea 红棕榈
Lauraceae 樟科
Laurus nobilis 月桂
Leguminosae 豆科
Leonurus artemisia 益母蒿
Lespedeza bicolor 胡枝子
Leucaena leucocephala 'Salvador' 新银合欢
Lichi chinensis 荔枝
Ligustrum lucidum 女贞
Ligustrum obtusifolium 水蜡
Ligustrum quihoui 小叶女贞
Ligustrum sinense var. *stauntonii* 卵叶小蜡
Liliaceae 百合科
Linum usitatissimum 亚麻
Liquidamba formosana 枫香
Liquidambar sytraciflua 胶皮枫香树
Liriodendron chinense 鹅掌楸, 马褂木
Lithocarpus glaber 石栎
Litsea cubeba 山苍子
Litsea ichangensis 宜昌木姜子
Livistona chinensis 蒲葵
Lonicera japonica 金银花, 忍冬
Lonicera maackii 金银忍冬, 金银木
Lonicera ruprechtiana 长白忍冬
Loropetalum chinense 檵木
Lycium chinense 枸杞
Lycopersicon esculentum 番茄
Lysidice brericalyx 短萼仪花
Lysidice rhodostegia 仪花
Lysimachia clethroides 珍珠菜

M

Maackia amurensis var. *buergeri* 怀槐
Maackia amurensis 朝鲜槐（山槐）
Magnolia cylindrical 黄山木兰
Magnolia denudata 玉兰, 木兰, 白玉兰
Magnolia grandiflora 广玉兰, 荷花玉兰
Magnolia liliflora 紫玉兰
Magnolia officinalis subsp. *biloba* 凹叶厚朴
Magnolia officinalis 厚朴

Magnolia sieboldii　天女花
Magnolia soulangeana　二乔玉兰
Mahonia bealei　十大功劳
Mallotus apelta　白背野桐
Mallotus barbatus　红帽顶，毛桐
Malus asiatica var. *rinki*　海红
Malus asiatica　花红，林檎，沙果
Malus baccata　山荆子
Malus micromalus　西府海棠
Malus prunifolia　海棠，楸子
Malus pumila　苹果
Malus spp.　海棠
Malus sylvestris　小苹果
Malus toringo　三叶海棠
Malus　苹果属
Malvaceae　锦葵科
Mangifera indica　杧果，印度芒
Mangifera siamensis　暹逻芒（本地小芒）
Mangifera sylvatica　野生芒
Manglietia fordiana　木莲
Manglietia glauca　灰木莲
Manihat esculenta　木薯
Medicago sativa　紫花苜蓿
Medicago sp.　苜蓿
Melaluca leucadendron　白千层
Melaluca quingueneruia　白树油树
Melia azedarach　苦楝，楝树
Melilotus suaveolens　草木犀
Metasequoia glyptostroboides　水杉
Metroxylon sagus　西谷椰子
Michelia alba　白兰
Michelia champaca　黄兰
Michelia figo　含笑
Michelia macclurei　火力楠
Michelia platypetala　乐昌含笑
Mimosa sepiaria　筋仔树
Mimosa　含羞草属
Miscanthus sinensis　芒
Moghania macropylla　大叶千斤拔
Morus alba　桑
Morus bombycis　山桑
Morus spp.　桑树
Mucuna　黎豆属
Musa nana　香蕉
Musa sapientum　大蕉
Musa uranoscopos　红花蕉
Musa　香蕉属
Myrica rubra　杨梅
Myrinaceae　紫金牛科

N

Nandina domestica　南天竹
Neodypsis decaryi　三角椰子
Neolamarskia cadamba　黄梁木
Neosinocalamus affinis 'Chrysotrichus'　黄毛竹
Neosinocalamus affinis　慈竹
Nephelium　红毛丹属
Nerium indicum　夹竹桃
Nicotiana spp.　烟草
Nicotiana tabacum　烟草
Nitraria sibirica　西伯利亚白刺
Nitraria sphaerocarpa　大果白刺
Nitraria tangutorum　唐古特白刺
Nyssa sinensis　蓝果树

O

Olea europaea　油橄榄
Olea sp.　木犀榄
Oleaceae　木犀科
Ormosia hosiei　红豆树
Oroxylum indicum　木蝴蝶
Oryza sativa　水稻
Osmanthus fragrans　桂花
Osmanthus sp.　木犀

P

Pachira macrocarpa　发财树（马拉巴栗）
Paeonia lactiflora　芍药
Paeonia suffruticosa　牡丹
Paramichelia baillonii　合果木
Parashorea chinesis　擎天树
Parkinsonia aculenta　扁轴木
Parthenocissus tricuspidata　地锦
Parthenocissus uinquefolia　五叶地锦
Paulownia fortunei　泡桐，白花泡桐
Paulownia　泡桐属
Peganum harmala　骆驼蓬
Pelargonium hortorum　天竺葵
Pennisetum alopecuroides　狼尾草
Persea americana　油梨
Phaseolus angularis　赤豆
Phaseolus vulgaris　菜豆，豆角
Phellodendron amurensa　黄波罗，黄檗
Philadelphus pekinensis　太平花
Phoebe zhennan　楠木
Phoenix daclylifera　椰枣，海枣
Phoenix hanceana var. *formosana*　台湾海枣
Phoenix hanceana　刺葵
Phoenix sylvestris　银海枣

Photinia davidisoniae 椤木石楠	*Phyllostachys nuda* 石竹
Photinia spp. 石楠	*Phyllostachys parvifolia* 安吉金竹
Phyllanthus emblica 余甘子	*Phyllostachys platyglossa* 灰水竹
Phyllostachys acuta 尖头青竹	*Phyllostachys praecox* f. *provernalis* 雷竹
Phyllostachys angusta 黄古竹	*Phyllostachys praecox* f. *viridisulcata* 花秆早竹
Phyllostachys arcana 石绿竹	*Phyllostachys praecox* 早竹
Phyllostachys arcane f. *luteosulcata* 黄槽石绿竹	*Phyllostachys prominens* 高节竹
Phyllostachys atrovaginata 乌芽竹	*Phyllostachys propinqua* 沙竹, 早园竹
Phyllostachys aurea 罗汉竹	*Phyllostachys pubescens* f. *gracillis* 金毛竹
Phyllostachys aureosulcata f. *aureocaulis* 黄秆京竹	*Phyllostachys pubescens* f. *gracillis* 金丝毛竹
Phyllostachys aureosulcata f. *pekinensis* 金镶玉竹	*Phyllostachys pubescens* 楠竹
Phyllostachys aureosulcata f. *pekinensis* 京竹	*Phyllostachys rigida* 硬头青竹
Phyllostachys aureosulcata 黄槽竹	*Phyllostachys robustiramea* 芽竹
Phyllostachys aurita 毛环水竹	*Phyllostachys rubromarginata* 红边竹
Phyllostachys bambusoides f. *lacrimadeae* 斑竹	*Phyllostachys rutila* 衢县红壳竹
Phyllostachys bambusoides f. *mixta* 黄槽斑竹	*Phyllostachys sulphurea* 'Houzeau' 黄皮绿筋竹
Phyllostachys bambusoides f. *shouzhu* 寿竹	*Phyllostachys sulphurea* 'Viridis' 刚竹
Phyllostachys bambusoides 五月季竹	*Phyllostachys sulphurea* 黄皮竹
Phyllostachys bissetii 白夹竹	*Phyllostachys sulphurea* 金竹
Phyllostachys circumpilis 毛壳花哺鸡竹	*Phyllostachys tianmuensis* 天目早竹
Phyllostachys decora 白皮淡竹	*Phyllostachys verrucosa* 长沙刚竹
Phyllostachys dulcis 白哺鸡竹	*Phyllostachys viridi-glaucescens* 绿粉竹
Phyllostachys edulis f. *gimmei* 黄槽毛竹	*Phyllostachys viridis* var. *youngii* 黄皮刚竹
Phyllostachys fimbriligula 角竹	*Phyllostachys vivax* f. *aureocaulis* 黄秆乌哺鸡竹
Phyllostachys flexuosa 甜竹	*Phyllostachys vivax* f. *huanvenzhu* 黄纹竹
Phyllostachys glabrata sinobambusa 花哺鸡竹	*Phyllostachys vivax* 乌哺鸡竹
Phyllostachys glauca f. *yuozhu* 筠竹	*Phyllostachys yunhoensis* 云和哺鸡竹
Phyllostachys glauca 淡竹, 花皮淡竹	*Phyllostachys* 刚竹属
Phyllostachys guizhouensis 贵州刚竹	*Phylslotachys elegans* 甜笋竹
Phyllostachys heteroclada f. *solida* 实心竹	*Picea abies* 挪威云杉
Phyllostachys heteroclada var. *pubescens* 毛竹	*Picea asperata* 云杉
Phyllostachys heteroclada 水竹	*Picea balfouriana* 川西云杉
Phyllostachys heterocycal 'Tao' 花毛竹	*Picea brachytyla* 麦吊云杉
Phyllostachys heterocycla 'Obliquinoda' 强竹	*Picea crassifolia* 青海云杉
Phyllostachys heterocycla 'Pubescens' 方秆毛竹	*Picea jezoensis* var. *komarovii* 长白鱼鳞松
Phyllostachys heterolada var. *funhuaensis* 奉化水竹	*Picea jezoensis* var. *microsperma* 鱼鳞云杉
Phyllostachys incarnata 红壳雷竹	*Picea koraiensis* 红皮云杉
Phyllostachys iridensis 红壳竹	*Picea likiangensis* 丽江云杉
Phyllostachys iridescens 红竹	*Picea meyeri* 白杆
Phyllostachys kwangsiensis 假毛竹	*Picea mongolia* 沙地云杉
Phyllostachys lofushanensis 大节刚竹	*Picea obovata* 新疆云杉, 西伯利亚云杉
Phyllostachys makinoi 台湾桂竹	*Picea purpurea* 紫果云杉
Phyllostachys meyeri 浙江淡竹	*Picea schrenkiana* 雪岭云杉
Phyllostachys nidularia 'Smoothsheath' 篦竹	*Picea wilsonii* 青杆
Phyllostachys nidularia f. *galbro-vagina* 光箨篦竹	*Picea* 云杉属
Phyllostachys nigella 富阳乌哺鸡竹	Pinaceae 松科
Phyllostachys nigra var. *henonis* 毛金竹	*Pinus armandii* 华山松
Phyllostachys nigra 紫竹	*Pinus banksiana* 北美短叶松
Phyllostachys nuda f. *localis* 紫蒲头石竹	*Pinus banksiana* 短叶松

Pinus bungeana　白皮松
Pinus caribaea var. *caribaea*　本种加勒比松
Pinus caribaea var. *hondurensis*　洪都拉斯加勒比松
Pinus caribaea　加勒比松
Pinus clausa var. *immuginata*　裂果沙松
Pinus denniflora　赤松
Pinus densata　高山松
Pinus densiflora　日本赤松
Pinus echinata　萌芽松
Pinus elliottii　湿地松
Pinus faiwarensis　黄山松
Pinus fenzeliana　葵花松（海南五针松）
Pinus galustris　长叶松
Pinus glabra　光松
Pinus griffithii　乔松
Pinus ikedai　海南松
Pinus kesiya var. *langbianensis*　思茅松
Pinus kesiya　卡西亚松（卡锡松）
Pinus koraiensis　红松
Pinus kwangtungenesis　华南五针松（广东松）
Pinus latteri　南亚松
Pinus luchuensis　琉球松
Pinus massoniana　马尾松
Pinus montana　中欧山松
Pinus morrisonicola　台湾五针松
Pinus nigra　欧洲黑松
Pinus oocarpa　卵果松
Pinus parviflora　日本五针松
Pinus patula　展叶松
Pinus pinea　意大利五针松
Pinus pumila　矮松，偃松
Pinus rigida var. *serotine*　晚松
Pinus rigida　刚松
Pinus strobes　北美乔松（北美五针松）
Pinus sylvestris var. *mongolica*　樟子松
Pinus sylvestris　欧洲赤松
Pinus tabuliformis　油松
Pinus taeda　火炬松
Pinus thunbergii　黑松，日本黑松
Pinus tropicalis　热带松
Pinus yunnanensis var. *pygmaea*　地盘松
Pinus yunnanensis　云南松
Pinus　松属
Pistacia chinensis　黄连木
Pisum sativum　豌豆
Pittosporum tobira　海桐
Plagiomnium acutum　湿地匐灯藓
Plagiomnium maximoviczii　侧枝匐灯藓
Plagiomnium rhyhcnophorum　钝叶匐灯藓

Plagiomnium succlenlum　大叶匐灯藓
Plagiomnium vesicatum　圆叶匐灯藓
Platanus hispanica　二球悬铃木（英桐）
Platanus occidentalis　一球悬铃木（美桐）
Platanus orientalis　法桐（三球悬铃木）
Platanus spp.　悬铃木
Platycarya strobilacea　化香
Platycladus orientalis 'Sieboldii'　千头柏
Platycladus orientalis　侧柏
Platycladus　侧柏属
Pleioblastus amarus　苦竹
Pleioblastus gozadakensis　秋竹
Pleioblastus hsienchuensis　云和苦竹
Pleioblastus juxianensis　衢县苦竹
Pleioblastus maculosoides　丽水苦竹
Pleioblastus simony　川竹
Pleioblastus　苦竹（大明竹）属
Poa sphondylodes　硬质早熟禾
Poaceae　禾本科
Podocarpus macrophyllus　罗汉松
Polygonaceae　蓼科
Polygonum orientale　红蓼
Poncirus trifoliata　栀子
Populus × *beijingensis*　北京杨
Populus × *canadensis* 'Sacrau 79'　沙兰杨
Populus × *canadensis* 'Robusta'　健杨
Populus × *canadensis* 'Serotina'　晚花杨
Populus alba var. *pyramidalis*　新疆杨，银白杨
Populus alba　白杨
Populus berolinensis　中东杨
Populus canadensis 'I-214'　I-214 杨
Populus canadensis × *Populus cathayana*　加青杨
Populus candicans　欧洲大叶杨
Populus canescens　银灰杨
Populus cathayana　青杨
Populus dakauensis　大官杨
Populus davidiana　山杨
Populus diversifolia　胡杨
Populus guariento　107 杨
Populus hopeiensis　河北杨
Populus lasiocarpa　大叶杨
Populus laurifolia　苦杨
Populus nigra var. *italica*　钻天杨
Populus nigra var. *thevestine*　箭杆杨
Populus nigra　黑杨，欧洲黑杨
Populus pseudo-simonii　小青杨
Populus simonii　小叶杨，小叶青杨
Populus spp.　杨树
Populus suaveolens　甜杨

Populus szechuanica var. *tibetica*　藏川杨
Populus tomentosa　毛白杨
Populus tremula　欧洲山杨
Populus ussuriensis　大青杨
Populus yunnanensis　滇杨
Populus × *canadensis*　加杨
Populus × *euramericana*　欧美杨
Populus × *xiaohei*　小黑杨
Populus × *xiaozhuanica* 'Beicheng'　白城杨
Populus × *xiaozhuanica* 'Balizhuangyang'　泰青杨
Populus × *xiaozhuanica* 'Opera'　合作杨
Portulaca oleracea　马齿苋
Potentilla sp.　委陵菜
Prunus armeniaca　杏
Prunus cerasifera f. *atropurpurea*　紫叶李
Prunus cerasifera var. *atropurpurea*　红叶李
Prunus davidiana var. *alba*　白花山桃
Prunus davidiana var. *rubra*　红花山桃
Prunus davidiana　山桃
Prunus domestica　欧洲李
Prunus dulcis　巴旦木，扁桃
Prunus japonica　郁李
Prunus mume　梅，酸梅
Prunus padus　稠李
Prunus persica f. *rubra-plena*　碧桃
Prunus persica　桃树
Prunus salicina　李
Prunus spinosa　黑刺李
Prunus spp.　李树
Prunus triloba　榆叶梅
Prunus yedoensis　日本樱花
Pseudolarix kaempferi　金钱松
Pseudosasa amabilis　茶秆竹
Pseudosasa guanxianensis　笔秆竹
Pseudosasa　矢竹属
Pseudotsuga menziesii　北美黄杉
Pseudotsuga sinensis　黄杉
Pseudotsuga　黄杉属
Psidium guajara　番石榴
Ptarocarya stenoptera　枫杨
Pueraria lobata　葛根，葛条
Pueraria tonkinensis　铁藤
Pueraria wallichii　马鹿花
Pueraria　葛属
Punica granatum　石榴
Pyracantha fortuneana　火棘
Pyrus berulaefolla　杜梨
Pyrus bretschneideri　梨，白梨
Pyrus calleryana　豆梨

Pyrus communis var. *satiea*　西洋梨
Pyrus prifolia var. *culte*　酥梨
Pyrus pyrifolia　沙梨
Pyrus spp.　梨树
Pyrus ussuriensis　秋子梨

Q

Quercus acutissima　麻栎
Quercus aliena　槲栎
Quercus chenii　小叶栎
Quercus dentata　槲树（波罗栎）
Quercus fabri　白栎
Quercus glandulifera　枹树
Quercus laurifolia　桂叶栎
Quercus liaotungensis　辽东栎
Quercus mongolica　柞树（蒙古栎）
Quercus spp.　栎类
Quercus variabilis　栓皮栎，粗皮青冈

R

Radix notoginseng　山漆
Raphanus sativus　萝卜
Rauvolfia vomitoria　催吐萝芙木
Ravenea rivularis　国王椰子
Rhamnus davurica　鼠李
Rhapis excelsa　棕竹
Rhododendron simsii　杜鹃
Rhododendron spp.　杜鹃类
Rhodomytus tomentosa　桃金娘
Rhus chinensis var. *roxburghii*　滨盐肤木
Rhus chinensis　盐肤木
Rhus typhina　火炬树
Ricinus communis　蓖麻
Robinia hispids　江南槐
Robinia pseudoacacia 'Idaho'　香花槐
Robinia pseudoacacia　刺槐
Roegneria ciliaris　纤毛鹅观草
Roegneria japonensis var. *haekliana*　细叶鹅观草
Roegneria kamojiohwi　鹅观草
Rosa chinenses　月季
Rosa hybrida　丰花月季
Rosa laevigata　金缨子
Rosa multiflora　蔷薇
Rosa rugosa　玫瑰
Rosa sp.　野蔷薇
Rosa xanthina　黄刺玫
Rosa　蔷薇属
Rosaceae　蔷薇科
Roystonea regia　大王椰子

Rubus corchorifolius 树莓, 悬钩子
Rutaceae 芸香科

S

Sabina chinensis 'Pyramidalis' 塔柏
Sabina chinensis 桧柏, 圆柏
Sabina covallium 密枝圆柏
Sabina kovarovii 塔枝圆柏
Sabina przewalskii 祁连圆柏
Sabina saltuaria 方枝圆柏
Sabina tibetica 大果圆柏
Sabina virginiana 铅笔柏
Sabina vulgaris 沙地柏
Saccharum sinense 甘蔗
Salicaceae 杨柳科
Salix alba var. *tristis* 银中杨
Salix alba 白柳
Salix babylonica 垂柳, 绢柳
Salix capitate 白皮柳（圆头柳）
Salix caprea 黄花柳
Salix fragilis 爆竹柳
Salix gordejevii 黄柳
Salix integra 杞柳
Salix matsudana f. *toruosa* 龙爪柳
Salix matsudana 旱柳
Salix microstachya 小红柳
Salix paraplesia var. *subintegra* 左旋柳
Salix psammophila 沙柳
Salix repens 匍匐柳
Salix spp. 柳树
Salix viminalis 蒿柳, 伪蒿柳
Salix wallichiana 皂柳
Sambucus williamsii 接骨木
Santalum album 檀香
Sapindus mukorossi 无患子
Sapium sebiferum 乌桕
Saposhnikovia divaricata 防风
Saraca dives 无忧树
Saraca griffithiana 无忧花
Sassafras randaiensis 台湾檫木
Sassafras tzumu 檫树
Sataria faberii 大狗尾草
Sataria glauca 金色狗尾草
Sataria itelica 小米
Sataria viridis 狗尾草
Scheffera octophylla 鹅掌柴
Schima superba 木荷
Schisandra 五味子属
Schisandra chinensis 五味子

Securinega suffruticosa 叶底珠
Semiarundinaria lubrica 浙东四季竹
Sesamum indicum 胡麻, 芝麻
Setaria italica 谷子, 粟
Sinarundinaria nitida 箭竹
Sindora tonkinensis 东京油楠
Sinobambusa edulis 黄间竹
Sinobambusa incana 毛环唐竹
Sinobambusa rubroligula 红舌唐竹
Sinobambusa tootsik 唐竹
Sinobambusa 唐竹属
Solanaceae 茄科
Solanum melongena 茄子
Solanum tuberosum 马铃薯
Sophora japonica var. *pendula* 龙爪槐
Sophora japonica 槐树
Sorbaria sorbifolia 珍珠梅
Sorbus pohuashanensis 花楸
Sorbus spp. 花楸
Sorghum vulgare 高粱
Spinacia oleracea 菠菜
Spiraea cantoniensis 麻叶绣球
Spiraea salicifolia 绣线菊
Spondias lakonensis 岭南酸枣
Strelizia reginae 鹤望兰
Swietenia macrophylla 大叶桃花心木
Swietenia mahogoni 桃花心木, 小叶桃花心木
Syagrus schizophylla 克利椰子
Symplocos spp. 山矾
Syringa oblata var. *affinis* 白丁香
Syringa oblata 丁香, 紫丁香
Syringa 丁香属
Syzygium cumini 乌墨（海南蒲桃）
Syzygium jambos 蒲桃
Syzygium samarangense 莲雾
Syzygium 蒲桃属

T

Tagetes erecta 万寿菊
Tamarix chinensis 柽柳
Tamarix ramosissima 多枝柽柳
Taxodiaceae 杉科
Taxodium ascendens 池杉
Taxus cuspidate var. *nana* 矮紫杉
Tectona grandis 柚木
Terminalia catappa 榄仁树
Tetraena monglica 四合木
Theobroma cacao 可可
Theretia peruviana 黄花夹竹桃

Thujopsis dolabrata 罗汉柏
Tilia mandschurica 糠椴
Tilia mongolica 蒙椴
Tilia spp. 椴
Toona ciliata 红椿
Toona microcarpa 小果香椿
Toona sinensis 香椿
Toona 香椿属
Toona sureni 红楝子
Torilis scabra 窃衣
Toxicodendron succedaneum 野漆
Toxicodendron verniciflum 漆树
Trachycarpus fortunei 棕榈
Trema tomentosa 山黄麻
Triticum aestivum 麦，冬麦，小麦
Tsoongiodendron odor 观光木
Tsuga chinensis 铁杉

U

Ulmaceae 榆科
Ulmus davidiana var. *japonica* 春榆
Ulmus densa 圆冠榆
Ulmus fulva 粗枝榆
Ulmus laevis 大叶榆，新疆大叶榆
Ulmus macrocarpa 大果榆，黄榆
Ulmus parvifolia 榔榆
Ulmus pumila 'Pendula' 垂枝榆
Ulmus pumila 'Pyramidalis' 钻天榆
Ulmus pumila 白榆，榆树
Ulmus spp. 榆

V

Vaccinium mandarinorum 米饭花
Vaccinium spp. 越橘
Verbenaceae 马鞭草科
Vernicia fordii 油桐
Vernicia montana 木油桐（千年桐）
Viburnum odoratissimum 法国冬青，珊瑚树
Vicia faba 蚕豆
Vigna radiata 绿豆
Vigna unguiculata 豇豆
Vitaceae 葡萄科
Vitex agnus-castus 淡紫花牡荆
Vitex microphylla 小叶牡荆
Vitex negundo var. *heterophylla* 荆条
Vitis vinifera 葡萄

W

Washingtonia filifera 华盛顿椰子
Washingtonia robusta 光叶加州蒲葵
Weigela florida 锦带花
Wisteria sinensis 紫藤
Wistevia floribunda 多花紫藤

X

Xanthoceras sorbifolia 文冠果
Xylosma japonicum 柞木

Y

Yucca elephantipes 巨丝兰
Yucca gloriosa 凤尾兰
Yushania niitakayamensis 玉山竹
Yushania polytricha 滑竹
Yushania qiaojiaensis 海竹

Z

Zanthaxylum bungeanum 花椒
Zea mays 玉米
Zeikova schneideriana 榉
Zelkova serrata 光叶榉
Zentoxylum bungeanus 北方花椒
Zinnia elegans 百日草
Ziziphus jujuba var. *spinosa* 酸枣
Ziziphus jujuba 枣，红枣

Natural Enemy Names

Insects

A

Acerophagus coccois　火炬松短索跳小蜂
Acrocormus ulmi　榆痣斑金小蜂
Acrocormus wuyingensis　五营痣斑金小蜂
Acropimpla persimilis　螟蛾顶姬蜂
Acropimpla　顶姬蜂属
Adalia bipunctata　二星瓢虫
Adialytus salicaphis　杨腺溶蚜茧蜂
Adonia variegata　多异瓢虫
Agrothereutes abbreviatus　短翅田猎姬蜂
Agrothereutes macroincubitor　大田猎姬蜂
Aiolocaria hexaspilota　六斑异瓢虫
Aiolocaria mirabilis　奇变瓢虫
Aleiodes euproctis　乌桕毛虫脊茧蜂
Amyosoma chinensis　中华茧蜂
Anacallocleonymus gracilis　柏蠹长体刺角金小蜂
Anaggrus nesticoccus　巢粉蚧长索跳小蜂
Anagyrus dactylopii　粉蚧长索跳小蜂
Anastatus albitarsis　白跗平腹小蜂
Anastatus disparis　舞毒蛾卵平腹小蜂
Anastatus japonicus　舞毒蛾平腹小蜂
Anastatus sp.　平腹小蜂
Anastatus sp.　无斑平腹小蜂
Anicetus annulatus　软蚧扁角跳小蜂
Anicetus benificus　红蜡蚧扁角跳小蜂
Anicetus ceroplastis　蜡蚧扁角跳小蜂
Anisolemnia dilatata　十斑大瓢虫
Anisotacrus sp.　短唇姬蜂
Anthemus aspidioti　盾蚧长缨跳小蜂
Anthocoris pilosus　蒙新原花蝽
Anysis saissetiae　黑盔蚧长盾金小蜂
Apanteles laevigatus　小卷蛾绒茧蜂
Apanteles liparidis　毒蛾绒茧蜂
Apanteles ordinaries　松毛虫绒茧蜂
Apanteles shemachaensis　尺蛾绒茧蜂
Apanteles sp.　纵卷叶螟绒茧蜂
Apanteles　绒茧蜂属

Aphaenogaster smythiesi　史氏盘腹蚁
Aphaenogaster　盘腹蚁属
Aphelinus　蚜小蜂属
Aphycus sp.　柿绒蚧跳小蜂
Aphytis fisheri　红圆蚧金黄蚜小蜂
Aphytis proclia　桑盾蚧黄金蚜小蜂
Aphytis　黄蚜小蜂属
Aprostocetus ceroplastae　蜡蚧啮小蜂
Aprostocetus dendroctoni　木蠹长尾啮小蜂
Aprostocetus fukutai　天牛卵长尾啮小蜂
Aprostocetus sp.　毒蛾长尾啮小蜂
Aptus mirmicoides　阿姬蝽
Archenomus longicornis　条棒短索蚜小蜂
Arenetra genangusta　狭颊刻姬蜂
Arma chinensis　蠋蝽
Aspidocolpus erythrogaster　红腹盾长茧蜂
Asymactu ambrostomae　榆紫叶甲异赤眼蜂
Aulacocentrum confusum　混腔室茧蜂

B

Bathyptectes　象甲姬蜂属
Bathythrix cilifacialis　毛面泥甲姬蜂
Bledius sp.　隐翅虫
Blepharipa zebina　蚕饰腹寄蝇
Bothrocalvia albolineata　细纹裸瓢虫
Brachymeria excarinata　无脊大腿小蜂
Brachymeria lasus　广大腿小蜂
Brachymeria secundaria　次生大腿小蜂
Brachymeria　大腿小蜂属
Bracon　茧蜂属，小茧蜂属
Braconidae　茧蜂科

C

Callocleonymus ianthinus　紫色小蠹刺金小蜂
Calosoma chinense　中华金星步甲
Calosota conifera　针叶树丽旋小蜂
Calosota koraiensis　红松丽旋小蜂
Calosota longigasteris　长腹丽旋小蜂
Calosota pumilae　榆小蠹丽旋小蜂

Calosota qilianshanensis　祁连山丽旋小蜂
Calvia quinquedecimguttata　十五星裸瓢虫
Camponotus japonicus　日本弓背蚁
Camponotus　弓背蚁属
Campoplex sp.　高缝姬蜂
Camptotypus rugosus　皱弯姬蜂
Cantheconidea concinna　厉蝽
Carabus　步甲属
Carcelia ammphion　灰腹狭颊寄蝇
Carcelia rasella　松毛虫狭颊寄蝇
Carcelia　狭颊寄蝇属
Casinaria nigripes　黑足凹眼姬蜂
Casinaria　凹眼姬蜂属
Celata populus　杨潜姬蜂
Ceraphron　分盾细蜂属
Cerceris rufipes evecta　叉突节腹泥蜂
Chaetexorista klapperichi　苹绿刺蛾寄蝇
Chaetexorista sp.　刺蛾寄蝇
Charops bicolor　螟蛉悬茧姬蜂
Charops　悬茧姬蜂属
Cheiloneurus claviger　长缘刷盾跳小蜂
Cheiloneurus claviger　刷盾长缘跳小蜂
Cheiropachus cavicapitis　小蠹凹面四斑金小蜂
Cheiropachus juglandis　核桃小蠹四斑金小蜂
Cheiropachus quadrum　果树小蠹四斑金小蜂
Chelonus spp.　甲腹茧蜂
Chilocorus geminus　李斑唇瓢虫
Chilocorus hupehanus　湖北红点唇瓢虫
Chilocorus kuwanae　红点唇瓢虫
Chilocorus rufitarsus　宽缘唇瓢虫
Chilomenes sexmaculata　六斑月唇瓢虫
Chlaenius bioculatus　双斑青步甲
Chlorocryptus　绿姬蜂属
Chouioia cunea　白蛾周氏啮小峰
Chrysis shanghaiensis　上海青蜂
Chrysopa formosa　丽草蛉
Chrysopa kulingensis　牯岭草蛉
Chrysopa perla　欧洲草蛉
Chrysopa phyllochroma　叶色草蛉
Chrysopa septempunctata　大草蛉
Chrysopa sinica　中华草蛉
Chrysopa　草蛉属
Chrysoperla savioi　松氏通草蛉
Chrysoperla sinica　中华通草蛉
Cicindela　虎甲属
Cleonymus pini　松蠹短颊金小峰
Cleonymus ulmi　榆蠹短颊金小峰
Cleptes semiauratus　暗尖胸青蜂
Cleridae　郭公甲科

Coccinella longifaseciata　纵条瓢虫
Coccinella septempunctata　七星瓢虫
Coccinella transversoguttata　横斑瓢虫
Coccinella trifasciata　横带瓢虫
Coccobius azumai　松突圆蚧花角蚜小蜂
Coccophagus chengtuensis　成都软蚧蚜小蜂
Coccophagus hawaiiensis　夏威夷软蚧蚜小蜂
Coccophagus ishii　赛黄盾食蚧蚜小蜂
Coccophagus japonicus　日本软蚧蚜小蜂
Coccophagus silvestrii　闽粤食蚧蚜小蜂
Coccphagus pulchellus　软蚧食蚜小蜂
Coccygomimus disparis　舞毒蛾黑瘤姬蜂
Coccygomimus luctuosus　野蚕黑瘤姬蜂
Coccygomimus parnarae　稻苞虫黑瘤姬蜂
Coccygomimus　黑瘤姬蜂属
Coceophagus yoshidae　黑色食蚧蚜小蜂
Coeloides qinlingensis　秦岭刻鞭茧峰
Colpotrochia pilosa pilosa　毛圆胸姬蜂指名亚种
Compieriella bifasciata　双条巨角跳小蜂
Cosmolestes sp.　勺猎蝽
Creobroter femmatus　花螳螂
Crossocosmia schineri　梳胫饰腹寄生蝇
Crossocosmia tibialis　柞蚕饰腹寄蝇
Cryptolaemus montrouzieri　孟氏隐唇瓢虫
Cybocephalus niponicus　日本方头甲
Cydnocoris russatus　淡红（艳红）猎蝽
Dastarcus helophoroides　花绒寄甲
Dastarcus lingulus　花绒坚甲
Diaeretiella　蚜茧蜂属

D

Dibrachys cavus　黑青金小蜂
Dibrachys yunnaneniis　云南黑青金小蜂
Dimmokia pomaceus　稻苞虫兔唇姬小蜂
Dinotiscus aponius　大痣小蠹狄金小蜂
Dinotiscus armandi　松蠹狄金小蜂
Dinotiscus eupterus　方痣小蠹狄金小蜂
Diomorus aiolomorphi　竹瘿冯长尾小蜂
Dolichogenidea lacteicolor　茶毛虫长绒茧蜂
Dolichogenidea sp.　长绒茧蜂
Dolichogenidea stantoni　绢野螟长绒茧蜂
Dolichomitus populneus　兜姬蜂
Dolichomitus tuberculatus　天牛兜姬蜂
Dolichomitus　兜姬蜂属
Dolichus haleensis　赤胸步甲
Drino argenticeps　银颜赘寄蝇
Drino facialis　狭颜赘寄蝇
Drino inconspicua　平庸赘寄蝇
Drino lugens　忧郁赘寄蝇

Dufouriellus ater　黑沟胸花蝽
Dusona tenuis　细都姬蜂

E

Echthromorpha agrestoria notulatoria　松毛虫恶姬蜂
Elasmus ciopkaloi　螟虫扁股小蜂
Elasmus　扁股小蜂属
Elatophilus nipponensis　松干蚧花蝽
Encarsia berlesei　桑盾蚧恩蚜小蜂
Encarsia sp.　恩蚜小蜂
Endasys liaoningensis　辽宁思姬蜂
Enicospilus flavocephalus　黄头细颚姬蜂
Enicospilus lineolatus　细线细颚姬蜂
Enicospilus pseudoconspersae　茶毛虫（茶毒蛾）细颚姬蜂
Enicospilus　细颚姬蜂属
Entedon ulmi　榆小蠹灿姬小蜂
Ephialtes　长尾姬蜂属
Epicauta ruficeps　红头芫菁
Epirhyssa hyblaeana　蛾上皱姬蜂
Eriborus sp.　钝唇姬蜂
Erycia bezzii　贝氏埃里寄蝇
Eulophus　姬小蜂属
Eupelmidae　旋小蜂科
Eupelmus sp.　竹蝉旋小蜂
Eupelmus urozonus　栗瘿旋小蜂
Eupelmus urozonus　小蠹尾带旋小蜂
Eurytoma appendigaster　悬腹广肩小蜂
Eurytoma esuriensi　榆平背广肩小蜂
Eurytoma juglansi　小蠹长柄广肩小蜂
Eurytoma longicauda　小蠹长尾广肩小蜂
Eurytoma monemae　刺蛾广肩小蜂
Eurytoma morio　普通小蠹广肩小蜂
Eurytoma regiae　核桃小蠹广肩小蜂
Eurytoma ruficornis　小蠹红角广肩小蜂
Eurytoma scolyti　小蠹圆角广肩小蜂
Eurytoma yunnanensis　小蠹长体广肩小蜂
Eurytoma　广肩小蜂属
Exochomus mongol　蒙古光瓢虫
Exorista amoena　毛虫追寄蝇
Exorista civilis　伞裙追寄蝇
Exorista fallax　红尾追寄蝇
Exorista japonica　日本追寄蝇
Exorista larvarum　古毒蛾追寄蝇
Exorista sorbillans　家蚕追寄蝇
Exorista　追寄蝇属

F

Forficula tomis　托球螋

Formica .　蚁属
Formica fukaii　深井凹头蚁
Formica gagatoides　亮腹黑褐蚁
Formica japonica　日本黑褐蚁
Formicidae　蚂蚁科
Friona okinawana　柚木夜蛾弗姬蜂

G

Germaria angustata　阿古蕾寄蝇
Goniozus japonicus　日本棱角肿腿蜂
Goniozus sinicus　中华棱角肿腿蜂
Goryphus basilaris　横带驼姬蜂
Gregopimpla himalayensis　喜马拉雅聚瘤姬蜂
Gregopimpla kuwanae　桑蟥聚瘤姬蜂

H

Habronyx sp.　软姬蜂
Haematoloecha nigrorufa　黑红猎蝽
Harmonia axyridis　异色瓢虫
Harmonia dimidiata　红肩瓢虫
Harmonia obscurosignata　隐斑瓢虫
Harpactor　真猎蝽属
Hemerobius humuli　全北褐蛉
Herpestomus brunnicornis　棕角巢蛾姬蜂
Heydenia angularicoxa　基角长胸肿腿金小蜂
Heydenia scolyti　小蠹长胸肿腿金小蜂
Hierodula patellifera　广腹螳螂
Himacerus mirmicoides　希姬蝽
Hippodamia tredecimpunctata　十三星瓢虫
Histeridae　阎魔甲科
Homoporus japonicus　纹黄枝瘿金小蜂
Hoplectis alternans spectabilis　松毛虫埃姬蜂
Hyperaspis sinensis　中华显盾瓢虫
Hyposoter　镶颚姬蜂属
Hyssopus nigritulus　球果平胸姬小蜂

I

Inocellia crassicornis　粗角盲蛇蛉
Ipideurytoma acuminati　六齿小蠹广肩小蜂
Iseropus sp.　舟蛾群瘤姬蜂
Isosturmia picta　多径毛异丛毛寄蝇
Isyndus obscurs　茶褐猎蝽
Isyndus reticulatus　锥盾菱猎蝽
Itoplectis viduata　寡埃姬蜂
Itoplectis　埃姬蜂属

K

Kimminsia sufuensis　疏附齐褐蛉

L

Lasius niger　黑毛蚁
Leluthia (Euhecabalodes) sp.　脊虎天牛莱洛茧蜂
Lemnia saucia　黄斑盘瓢虫
Leskia aurea　金黄莱寄蝇
Lestodiplosis pentagona　桑盾蚧盗瘿蚊
Lestodiposis sp.　松蚧瘿蚊
Leucopis atratula　黑蚜斑腹蝇
Lissonota sapinea　枞缺沟姬蜂
Lissonota sp.　柿蒂虫缺沟姬蜂
Litomastix maculatus　多胚跳小蜂

M

Macrocentrus abdominalis　黄长距茧蜂
Macrocentrus gibber　卷蛾圆瘤长体茧蜂
Macrocentrus resinellae　球果卷蛾长体茧蜂, 松小卷蛾长体茧蜂
Macrocentrus watanabei　渡边长体茧蜂
Macrocentrus　长距茧蜂属
Macromesus persichus tuberculatus　桃小蠹长足金小蜂
Maequartia tenebricosa　阴叶甲寄蝇
Marietla picta　豹纹花翅蚜小蜂
Marietta carnesi　瘦柄花翅蚜小蜂
Marietta sp.　斑腿花翅蚜小蜂
Meganhyssa praecellens　翅斑马尾姬蜂
Megastigmus maculipennis　斑翅大痣长尾小蜂
Megastigmus niponicus　日本大痣长尾小蜂
Megastigmus sinensis　中华大痣小蜂
Mesochorus　菱室姬蜂属
Mesocomys orientalis　松毛虫短角平腹小蜂
Mesopolobus subfumatus　松毛虫白角金小蜂
Metacolus sinicus　华肿脉金小蜂
Metacolus unifasciatus　双斑肿脉金小蜂
Metapelma zhangi　张氏扁胫旋小蜂
Metaphycus pulvinariae　绵蚧阔柄跳小蜂
Meteoridea chui　祝氏鳞跨茧蜂
Meteorus narangae　螟蛉黄茧蜂
Meteorus sp.　黄茧蜂属
Microterys clauseni　球蚧花翅跳小蜂
Microterys ericeri　白蜡虫花翅跳小蜂
Microterys postmarginis　后缘花翅跳小蜂
Microterys speciosus　蜡蚧花翅跳小蜂
Microterys yunnanensis　云南花翅跳小蜂
Microterys zhaoi　赵氏花翅跳小蜂
Microterys　花翅跳小蜂属
Monodontomerus monor　齿腿长尾小蜂
Myxexoristops blondeli　扁尾撵寄蝇

N

Nabis pseudoferus　拟原姬蝽
Nabis sinoferus　华姬蝽, 华姬猎蝽
Nealsomyia quadrimaculata　四斑尼尔寄蝇
Neophryxe psychidis　简须新怯寄蝇
Nephus ryuguus　圆斑弯叶毛瓢虫
Nephus　弯叶毛瓢虫属
Norbanus aiolomorphi　竹瘿长角金小蜂

O

Oenopia conglobata　多星瓢虫, 菱斑巧瓢虫
Oodera pumilae　榆蝶胸肿腿金小蜂
Ooencyrtus longivenosus　长脉卵跳小蜂
Opius　蝇茧蜂属
Orius vicinus　邻小花蝽
Orius　小花蝽属
Ormyrus punctiger　点腹刻腹小蜂
Oxysychus sphaerotrypesi　球小蠹奥金小蜂

P

Pales aurescens　金黄栉寄蝇
Pales pavida　蓝黑栉寄蝇
Paratrechina　立毛蚁属
Pediobiua　叶甲姬小蜂属
Pediobius cassidae　柄腹姬小蜂
Phanerotoma flava　黄色白茧蜂
Phanerotoma kozlovi　考氏白茧蜂
Phanerotoma semenovi　球果螟白茧蜂
Pheidole nodus　宽结大头蚁
Phleudecatoma cunnighamiae　杉蠹黄色广小蜂
Phleudecatoma platycladi　柏蠹黄色广肩小蜂
Picromerus lewisi　益蝽
Picromerus virdipunctatus　绿点益蝽
Pimpla aethiops　满点瘤姬蜂
Pimpla disparis　古毒蛾黑瘤姬蜂
Pimpla luctuosa　野蚕瘤寄蜂
Pimpla sp.　龙眼裳卷蛾黑瘤姬蜂
Pimpla turionella　卷蛾瘤姬蜂
Plagiolepis rothneyi　罗思尼氏斜结蚁
Platygerrhus piceae　云杉小蠹璞金小蜂
Platygster sp.　瘿蚊腹细蜂
Pleolophus setiferae　毛瘤角姬蜂
Polistes rothneyi grahami　陆马蜂
Polyrhachis dives　双齿多刺蚁
Prococcophagus　原食蚧蚜小蜂属
Propylaea japonica　龟纹瓢虫
Propylaea quatuordecimpunctata　方斑瓢虫
Prospaltella berlesei　桑蚧寡节小蜂
Protomicroplitis spreta　斑螟大距侧沟茧蜂

Pseudanstatus albiguttata　白跗平腹小蜂
Pseudophycus malinus　粉蚧短角跳小蜂
Pteromalus miyunensis　密云金小蜂
Pteromalus procetus　皮金小蜂
Pteromalus semotus　瑟茅金小蜂
Pycnetron curculionidis　松扁腹长尾金小蜂
Pygolampis bidentara　双刺胸猎蝽

R

Respa germanica　德国黄胡蜂
Rhaphitelus maculatus　小蠹棍角金小蜂
Rhogas　内茧蜂属
Rhopalicus guttatus　隆胸罗葩金小蜂
Rhopalicus tutela　长痣罗葩金小蜂
Rodolia cardinalis　澳洲瓢虫
Rodolia limbata　红环瓢虫
Rodolia pumila　小红瓢虫
Rodolia rufopilosa　大红瓢虫
Roptrocerus ipius　西北小蠹长尾金小蜂
Roptrocerus mirus　奇异小蠹长尾金小蜂
Roptrocerus qinlingensis　松蠹长尾金小蜂
Roptrocerus xylophagorum　木小蠹长尾金小蜂
Roptrocerus yunnanensis　云南小蠹长尾金小蜂

S

Sandalus sp.　竹蝉履甲
Scambus　曲姬蜂属
Scambus brericornis　短角曲姬蜂
Scambus eurygenys　宽颊曲姬蜂
Scambus punctatus　密点曲姬蜂
Scambus sagax　沙曲姬蜂
Scambus sudeticus　球果象曲姬蜂
Scambus　曲姬蜂属
Scenoharops parasae　竹刺蛾小室姬蜂
Schreineria ceresia　蜡天牛蛀姬蜂
Schreineria populnea　青杨天牛蛀姬蜂
Sclerodermus guani　管氏肿腿蜂
Scymnus (*Neopullus*) *babai*　黑背毛瓢虫
Sirthenea flavipes　黄足猎蝽
Sospita chinensis　华鹿瓢虫
Spathius agrili　白蜡吉丁柄腹茧蜂
Sphedanolestes annulipes　双环猛猎蝽
Sphedanolestes gularis　红缘猛猎蝽
Staphylinidae　隐翅甲科
Sycanus croceovittatus　中黄猎蝽
Sympherobius matsucocciphagus　松蚧益蛉
Sympherobius weisong　卫松益蛉
Sympiesis　羽角姬小蜂属
Synharmonia conglobata　菱斑和瓢虫

T

Tachina　寄蝇属
Technomyrmex　狡臭蚁属
Telenomus (*Aholcus*) *euproctidis*　茶毛虫（茶毒蛾）黑卵蜂
Telenomus dalmani　达氏黑卵蜂
Telenomus dasychiri　松茸毒蛾黑卵蜂
Telenomus dendrolimusi　松毛虫黑卵蜂
Telenomus holcoceri　蠹蛾黑卵蜂
Telenomus lebedae　油茶枯叶蛾黑卵蜂
Telenomus nitidulus　杨毒蛾黑卵蜂
Telenomus sp.　浙江黑卵蜂
Telenomus　黑卵蜂属
Tenodera sinensis　中华大刀螂
Tetrastichus armandii　松蠹啮小蜂
Tetrastichus brontispa　椰心叶甲啮小蜂
Tetrastichus clavicornis　刺角卵腹啮小蜂
Tetrastichus cupressi　柏小蠹啮小蜂
Tetrastichus juglansi　核桃小蠹啮小蜂
Tetrastichus planipennisi　白蜡吉丁啮小蜂
Tetrastichus sp.　毒蛾卵啮小蜂
Tetrastichus sp.　粉蚧啮小蜂
Tetrastichus sp.　蓝绿啮小蜂
Tetrastichus sp.　桑木虱啮小蜂
Tetrastichus sp.　舟蛾啮小蜂
Tetrastichus telon　长腹木蠹啮小蜂
Tetrastichus thoracicus　隆胸小蠹啮小蜂
Tetrastichus　啮小蜂属
Theocolax phlaeosini　小蠹蚁形金小蜂
Theronia zebra diluta　松毛虫匙鬃瘤姬蜂
Tillus notatus　二带赤颈郭公虫
Tiphia　钩土蜂属
Tomicobia liaoi　廖氏截尾金小蜂
Tomicobia seitneri　暗绿截尾金小蜂
Torymidae　长尾小蜂科
Triaspis sp.　杨跳象三盾茧蜂
Trichodes sinae　红斑郭公虫
Trichogramma artonae　斑蛾赤眼蜂
Trichogramma chilonis　螟黄赤眼蜂
Trichogramma chilonis　玉米螟卵赤眼蜂
Trichogramma closterae　舟蛾赤眼蜂
Trichogramma dendrolimus　松毛虫赤眼蜂
Trichogramma evanescens　广赤眼蜂
Trichogramma ivelae　毒蛾赤眼蜂
Trichogramma polychrosis　杉卷蛾赤眼蜂
Trichogramma　赤眼蜂属
Trigonoderus fraxini　水曲柳长体金小蜂
Triraphis fuscipennis　暗翅三缝茧蜂
Triraphis sichuanensis　四川三缝茧蜂
Trissolcus halyomorphae　茶翅蝽沟卵蜂

Tritneptis klugii　克禄格翠金小蜂
Tropobracon sp.　丛螟茧蜂

V

Vespa　胡蜂属
Vespa bicolor bicolor　黑盾胡蜂
Vespa magnifica　大胡蜂
Vespa mandarinia mandarinia　金环胡蜂
Vespa tropica leefmansi　大金箍胡蜂
Vespa velutina nigrihorax　墨胸胡蜂
Vespula　黄胡蜂属
Villa sp.　绒蜂虻

W

Winthemia diversa　变异温寄蝇

X

Xanthopimpla konowi　樗蚕黑点瘤姬蜂
Xanthopimpla pedator　松毛虫黑点瘤姬蜂
Xanthopimpla pleuralis pleuralis　侧黑点瘤姬蜂指名亚种
Xanthopimpla punctata　广黑点瘤姬蜂
Xanthopimpla stemmator　螟黑点瘤姬蜂

Y

Yolinus albopustulatus　淡裙猎蝽

Z

Zarhopalus debarri　迪氏跳小蜂
Zenilla modicella　中室彩寄蝇
Zolotarewskya robusta　核桃消颊齿腿金小蜂
Zombrus bicolor　两色刺足茧蜂

Birds

Acrocephalus orientalis　东方大苇莺
Anthus hodgsoni　树鹨
Apus apus　楼燕
Asio flammeus　短耳鸮
Bambusicola thoracica　灰胸竹鸡
Caprimulgus indicus calonyx　普通夜鹰
Cissa ergtyrorhyncha　长尾蓝雀
Corvus spp.　乌鸦
Cuculus canorus canorus　大杜鹃指名亚种
Cuculus canorus　大杜鹃
Cuculus micropterus　布谷鸟（四声杜鹃）
Cuculus micropterus　四声杜鹃（布谷鸟）
Cuculus spp.　杜鹃
Cyanopica cyana　灰喜鹊
Dendrocopos canicapillus　星头啄木鸟
Dendrocopos leucotos　白背啄木鸟
Dendrocopos major　大斑啄木鸟
Dendrocopos minor kamtschakensis　小斑啄木鸟（新疆亚种）
Dendrocopos minor tianshanicus　斑啄木鸟（新疆亚种）
Dendrocopos spp.　啄木鸟
Dendrocops hyperythrus subrufinus　棕腹啄木鸟
Emberiza cioides　三道眉草鹀
Eurystomus orientalis　三宝鸟（佛法僧）
Falco tinnunculus　红隼
Falco vespertinus amurensis　红脚隼
Garrulax canorus　画眉
Lanius cristatus　红尾伯劳
Lanius spp.　伯劳
Motacilla alba　白鹡鸰
Oriolus chinensis diffuses　黑枕黄鹂
Otus scops　红角鸮
Parus major artatus　白脸山雀
Parus major　大山雀
Passer montanus　麻雀（树麻雀）
Pericrocotus divaricatus　灰山椒鸟
Phoenicurus auroreus　北红尾鸲
Pica pica sericea　喜鹊（普通亚种）
Pycnonotus sinensis　白头鹎
Sturnus cineraceus　灰椋鸟
Turdus merula　乌鸫
Upupa epops　戴胜

Spiders

Araneus mitificus　丽园蛛
Araneus ventricosus　大腹圆蛛
Argiope bruennichii　横纹金蛛
Chiracanthium spp.　红螯蛛
Chiracanthium zhejiangensis　浙江红螯蛛
Clubiona maculata　斑管巢蛛
Dolomedes pallitarsis　宽条狡蛛
Dolomedes sulfureus　黄褐狡蛛
Lycosa coelestris　黑腹狼蛛
Misumenops tricuspidatus　三突花蛛
Oxtopes macilentus　细纹猫蛛
Oxtopes spp.　猫蛛
Oxyopes sertaus　斜纹猫蛛
Oxyopex lineatipes　线纹猫蛛
Oxytate parallela　平行绿蟹蛛
Pardosa pseudoannulata　拟环纹豹蛛
Pardosa tschekiangensis　浙江豹蛛
Pardosa wuyiensis　武夷豹蛛
Peucetia spp.　松猫蛛
Pisaura ancora　锚盗蛛
Pisaura spp.　盗蛛

Plexippus paykulli　黑色蝇虎蛛

Allothrombium pulvinum　小枕异绒螨
Aschersonia duplex　双生座壳孢
Aspergillus candidus　亮白曲霉菌
Bacillus popilliae　乳状菌（日本金龟芽孢杆菌）
Bacillus thuringiensis galleriae　青虫菌
Bacillus thuringiensis　苏云金杆菌
Beauveria bassiana　球孢白僵菌
Cladosporium cladosporioides　芽枝状枝孢霉
Cordyceps sinensis　竹蝉虫草（冬虫夏草）
Erinaceus europaeus　刺猬
Heterorhabditis spp.　异小杆线虫

Hexamermis arsenoidea　强壮六索线虫
Hexamermis microamphidis　细小六索线虫
Isaria cicadae　竹蝉蝉花菌
Metarhizium anisopliae　绿僵菌
Pediculoides ventricosus　虱形螨
Phrynocephalus frontalis　榆林沙蜥
Pyemotes spp.　蒲螨属
Pyemotes tritici　麦蒲螨
Steinernema bibionis　毛纹斯氏线虫
Steinernema feltiae　夜蛾斯氏线虫
Steinernema glaseri　格氏线虫
Steinernema spp.　斯氏线虫
Verticillium lecanum　蜡蚧轮枝菌

Scientific Names Index

A

Abraxas suspecta **(Warren) 762**
Acalolepta sublusca **(Thomson) 326**
Acanthococcus lagerstroemiae **Kuwana 201**
Acanthoecia larminati **(Heylaerts) 542**
Acantholyda erythrocephala **(L.) 1038**
Acantholyda flavomarginata **Maa 1040**
Acantholyda guizhouica Xiao 1040
Acantholyda peiyingaopaoa **Hsiao 1042**
Acantholyda pinivora Enslin 1043
Acantholyda posticalis **Matsumura 1043**
Acanthopsyche bipars Matsumura 1021
Acanthoscelides pallidipennis **(Motschulsky) 382**
Acanthoscelides plagiatus Reiche et Saulcy 382
Acrocercops ficuvorella Yazaki 556
Acrocercops urbana Meyrick 554
Actias ningpoana **Felder 859**
Adelges laricis laricis **Vallot 170**
Adoxophyes honmai **Yasuda 642**
Adoxophyes orana orana (Fischer von Röslerstamm) 642
Aegeria apiformis Clerck 562
Aegosoma sinicum **White 358**
Aenaria pinchii **Yang 271**
Agehana elwesi (Leech) 1028
Agelastica alni glabra **(Fischer von Waldheim) 386**
Agelastica orientalis Baly 386
Agenocimbex ulmusvora **Yang 1061**
Agonoscena xanthoceratis **Li 125**
Agrilus boreoccidentalis Obenberger 310
Agrilus cersnii Obenberger 310
Agrilus charbinensis Thery 310
Agrilus klapperichi Obenberger 310
Agrilus marcopoli Obenberger 304
Agrilus moerens **Saunders 306**
Agrilus nipponigena Obenberger 310
Agrilus pekinensis Obenberger 310
Agrilus pekinensis pekinensis **Obenberger 310**
Agrilus planipennis **Fairmaire 304**
Agrilus ratundicollis Saunders 306
Agrilus subrobustus **Saunders 308**
Agrilus viduus Kerremans 308

Agrilus zanthoxylumi **Zhang et Wang 309**
Agriotes fuscicollis Miwa 318
Agriotes subvittatus **Motschulsky 318**
Agrotis conjuncta Leech 944
Agrotis ipsilon **(Hufnagel) 936**
Agrotis segetum **(Denis and Schiffermüller) 938**
Agrotis segetum Schiffermüller 938
Agrotis tokionis **Butler 934**
Agrotis ypsilon (Rott.) 936
Aiolomorphus rhopaloides **Walker 1121**
Alcidodes juglans Chao 428
Aleurotrachelus camelliae **(Kuwana) 142**
Algedonia coclesalis Walker 727
Aloa lactinea **(Cramer) 964**
Alphaea phasma Leech 965
Alucita sacchari Bojer 535
Amatissa snelleni **(Heylaerts) 538**
Ambrostoma quadriimpressum **(Motschulsky) 388**
Amonophadnus nigritus (Xiao) 1093
Amplypterus panopus **(Cramer) 889**
Amplypterus panopus Moore 889
Amsacta lactinea Cramer 964
Anacampsis laticinctella Wood 586
Anacampsis populella **(Clerck) 586**
Anacampsis tremulella Duponchel 586
Anafenusa acericola **(Xiao) 1084**
Anarsia lineatella **Zeller 588**
Anarsia pruniella Clemens 588
Anarsia squamerecta **Li et Zheng 590**
Ancylis mitterbacheriana **(Denis et Schiffermüller) 655**
Ancylis sativa **Liu 657**
Anomala corpulenta **Motschulsky 288**
Anomala cupripes **(Hope) 291**
Anomoneura mori **Schwarz 127**
Anoplistes halodendri **(Pallas) 344**
Anoplophora chinensis **(Forster) 328**
Anoplophora glabripennis **(Motschulsky) 330**
Antheraea pernyi **(Guérin-Méneville) 861**
Aonidiella perciciosus (Comstock) 242
Apamea apameoides **(Draudt) 940**
Apamea kumaso Sugi 942

Apamea repetita Butler 944
Aphis craccivora Koch 144
Aphis odinae (van der Goot) 147
Aphis populialbae Boyer de Fonscolombe 148
Aphis robiniae Macchiati 144
Aphrophora costalis Matsumura 109
Aphrophora flavipes Uhler 110
Aphrophora horizontalis Kato 107
Apion collare Schilsky 427
Apocheima cinerarius (Erschoff) 751
Apochima excavate Dyar 792
Aporia crataegi (L.) 1025
Apriona germari (Hope) 332
Apriona rugicollis Chevrolat 332
Apriona swainsoni (Hope) 335
Archips reticulana Hübner 642
Arge captiva Smith 1059
Argyroploce lasiandra Meyrick 661
Argyroploce mormopa Meyrick 674
Aristobia approximator (Thomson) 337
Aristobia hispida (Saunders) 338
Arithrophora ombrodelta Lower 661
Armiana retrofusca (de Joannis) 897
Arna bipunctapex (Hampson) 975
Arna pseudoconspersa (Strand) 980
Aroa flava Bremer 978
Aroa subflava Bremer 978
Aromia bungii (Faldermann) 340
Artata subflava Bremer 978
Artaxa conspersa Butler 980
Artaxa scintillans Walker 1012
Aruanoidea flavescens Chen et Wang 79
Ascia crataegi L. 1025
Asclerobia sinensis (Caradja et Meyrick) 702
Ascotis selenaria (Denis et Schiffermüller) 755
Ascotis selenaria dianaria Hübner 755
Asemum amurense Kraatz 342
Asemum striatum (L.) 342
Asiacornococcus kaki (Kuwana) 199
Asias halodendri (Pallas) 344
Aspidiotus gigas Thiem et Gerneck 240
Aspidiotus multiglandulatus Borchsenius 240
Aspidiotus perniciosus Comstock 242
Asterococcus muratae (Kuwana) 223
Asterolecamium hemisphaerica Kuwana 225
Atrijuglans hetaohei Yang 567
Attacus atlas (L.) 873
Attacus indicus maximus Velentini 873
Atteva fabriciella Swederus 573
Augomonoctenus smithi Xiao et Wu 1064

Aulacaspis pentagona Cockerell 238
Aulacaspis rosarum (Borchsenius) 227
Aulacaspis sassafris Chen, Wu & Su 229
Aularches miliaris (L.) 58
Aularches miliaris scabiosus (F.) 58

B

Baculum chongxinense Chen et He 67
Baculum intermedium Chen et Wang 71
Baculum minutidentatus Chen et He 69
Bambusaspis hemisphaerica (Kuwana) 225
Bambusiphila vulgaris (Butler) 945
Basiana bilineata Walker 881
Basiprionota bisignata (Boheman) 390
Batocera davidis Deyrolle 346
Batocera horsfieldi (Hope) 348
Batocera lineolata Chevrolat 348
Batocera rubus (L.) 351
Besaia anaemica (Leech) 893
Besaia goddrica (Schaus) 895
Besaia rubiginea simplicior Gaede 895
Bhima eximia Oberthür 802
Bhima idiota (Graeser) 804
Biston marginata Shiraki 757
Biston panterinaria (Bremer et Grey) 766
Biston suppressaria (Guenée) 759
Blastophagus minor Hartig 505
Blastophagus piniperda L. 506
Bomby antiguoides Hübner 994
Bombyx dispar L. 984
Bombyx monacha L. 990
Bombyx nuda F. 1021
Bombyx populifolia Esper 832
Bombyx salicis L. 1016
Bostrichus pinastri Bechstein 490
Botyodes diniasalis (Walker) 738
Brachytrupes portentosus Lichtenstein 52
Brontispa castanea Lea 391
Brontispa javana Weise 391
Brontispa longissima (Gestro) 391
Brontispa reicherti Uhmann 391
Brontispa selebensis Gestro 391
Brontispa simmondsi Maulik 391
Bruchus collusus Fall. 382
Bruchus perplexus Fall. 382
Buprestis conspersa Gyllenhal 317
Buprestis picta Pallas 315
Buprestis plebeia (Herbst) 317
Buprestis rustica Herbst 317
Buprestis tenebrionis (Panzer) 317

Buprestis tenebrionis Schaeffer 317
Buprestis variolosa Paykull 317
Byctiscus omissus Voss 414
Byctiscus rugosus (Gelber) 414

C

Calamochrous obscuralis Caradja 735
Caligula japonica (Moore) 863
Caliroa annulipes (Klug) 1087
Callambulyx tatarinovii (Bremer et Grey) 877
Callichroma bungii Kolbe 340
Callidiellum villosulum (Fairmaire) 354
Calliteara axutha (Collenette) 972
Caloptilia chrysolampra (Meyrick) 553
Caloptilia dentata Liu et Yuan 551
Calospilos suspecta (Warren) 762
Calymnia panopus Rothschild et Jordan 889
Camptochilus semifasciata Gaede 696
Camptoloma erythropygum Felder 957
Camptoloma interiorata (Walker) 957
Carea angulata (F.) 952
Carea innocens Swinhoe 952
Carea intermedia Swinhoe 952
Carea subilis Hampson 952
Carea subtilis Walker 952
Carpocapsa obesana Laharpe 687
Carpomya vesuviana Costa 522
Carposina niponensis Walsingham 598
Carposina sasakii Matsumura 598
Carsidara limbata (Enderlein) 137
Casmara patrona Meyrick 579
Celerio hippophaes (Esper) 879
Celypha pseudolarixicola Liu 659
Cephalcia flagellicornis (Smith) 1049
Cephalcia kunyushanica Xiao 1051
Cephalcia tianmua Maa, Xiao et Chen 1045
Cephalcia tienmua Maa 1045
Cephalcia yanqingensis Xiao 1047
Cerace stipatana Walker 644
Ceracris kiangsu Tsai 59
Ceracris nigricornis Walker 61
Cerambyx bungii Faldermann 340
Cerambyx populneus L. 370
Cerococcus muratae Kuwana 223
Ceroplastes japonicus Green 205
Ceroplastes rubens Maskell 208
Cerostoma sasahii Matsumura 657
Cerura menciana Moore 900
Cerura tattakana Matsumura 917
Chaitophorus albus Mordvilko 148

Chaitophorus hickelianae Mimeur 148
Chaitophorus inconspicuous Theobald 148
Chaitophorus populialbae (Boyer de Fonscolombe) 148
Chaitophorus tremulinus Mamontova 148
Chalcophora yunnana Fairmaire 312
Chalioides kondonis Matsumura 540
Chiasmia cinerearia (Bremer et Grey) 788
Chihuo zao Yang 764
Chilasa epycides epycides (Hewitson) 1030
Chilasa slateri Hewitson 1032
Chinolyda flagellicornis (Smith) 1049
Chionaspis euonymi Comstock 248
Chondracris rosea (De Geer) 65
Chondracris rosea brunneri Uvarov 65
Chondracris rosea rosea (De Geer) 65
Choristoneura lafauryana (Ragonot) 645
Chrysochroa fulminans F. 308
Chrysomela adamsi ornaticollis Chen 393
Chrysomela populi L. 394
Chyliza bambusae Yang et Wang 515
Cicadella viridis (L.) 120
Cifuna locuples Walker 1019
Cimbex connatus taukushi Marlatt 1062
Cinara formosana (Takahashi) 152
Cinara pinitabulaeformis Zhang et Zhang 152
Cinara tujafilina (Del Guercio) 155
Circobotys aurealis (Leech) 729
Clania larminati (Heylaerts) 542
Clania minuscula Butler 543
Clania variegata (Snellen) 547
Clanis bilineata (Walker) 881
Clanis bilineata bilineata (Walker) 881
Clitea shirahatai (Chûjô) 408
Clostera anachoreta (Denis & Schiffermüller) 903
Clostera anastomosis (L.) 906
Clytus validus Fairmaire 355
Cnidocampa flavescens (Walker) 621
Coccyx aeriferanus Herrich-Schäffer 654
Coleophora dahurica Falkovitsh 575
Coleophora longisignella Moriuti 575
Coleophora obducta (Meyrick) 575
Coleophora sinensis Yang 577
Colophorina robinae (Shinji) 132
Comstockaspis perniciosa (Comstock) 242
Conogethes punctiferalis (Guenée) 743
Coptotermes communis Xia et He 92
Coptotermes eucalyptus Ping 92
Coptotermes formosanus Shiraki 92
Coptotermes hongkongensis Oshima 92
Coptotermes rectangularis Ping et Xu 92

Corythucha ciliata (Say) 251
Cosmotriche davisa Moore 846
Cosmotriche inexperta (Leech) 806
Cossus arenicola Staudinger 603
Cossus chinensis Rothschild 600
Cossus cossus changbaishanensis Hua et al. 601
Cossus cossus chinensis Rothschild 600
Cossus cossus orientalis Gaede 601
Cossus orientalis Gaede 601
Cossus vicarius Walker 613
Cresococcus candidus Wang 222
Crinocraspeda inexperta Leech 806
Crocidophora evenoralis Walker 731
Cryphalus tabulaeformis chienzhuangensis Tsai et Li 474
Cryphalus tabulaeformis Tsai et Li 475
Crypsiptya coclesalis (Walker) 727
Cryptoblabes lariciana Matsumura 703
Cryptophlebia carophaga Walsingham 661
Cryptophlebia ombrodelta (Lower) 661
Cryptorhynchus lapathi (L.) 430
Cryptorrhynchus frigidus (F.) 466
Cryptorrhynchus gravis (F.) 466
Cryptorrhynchus mangiferae F. 468
Cryptorrhynchus olivieri (Faust) 470
Cryptorrhynchus olivieri Faust 470
Cryptotermes buxtoni Hill 90
Cryptotermes campbelli Light 90
Cryptotermes dentatus Oshima 90
Cryptotermes domesticus (Haviland) 90
Cryptotermes formosae Holmgren 90
Cryptotermes hermsi Kirby 90
Cryptothelea minuscula Butler 543
Cryptothelea variegata Snellen 547
Cryptotympana atrata (F.) 111
Culcula panterinaria (Bremer et Grey) 766
Curcuijo gigas F. 447
Curculio chinensis Chevrolat 433
Curculio davidi Fairmaire 435
Curculio frigidus F. 466
Curculio hippophes Zhang, Chen et Dang 437
Curculio lapathi L. 428
Curculio longipes F. 419
Cyamophila willieti (Wu) 130
Cyclidia substigmaria (Hübner) 749
Cyclopelta parva Distant 270
Cyclophragma xichangensis (Tsai et Liu) 808
Cycnia cunea Drury et Hübner 967
Cydalima perspectalis (Walker) 740
Cydia coniferana (Saxesen) 676
Cydia pomonella (L.) 646

Cydia strobilella (L.) 684
Cydia trasias (Meyrick) 663
Cydia zebeana (Ratzeburg) 678
Cyllorhynchites ursulus (Roelofs) 415
Cymolomia hartigiana (Saxesen) 665
Cyrtacanthacris lutescens Walker 65
Cyrtotrachelus buquetii Guérin-Méneville 417
Cyrtotrachelus longimanus Gtyllenhaal 419
Cyrtotrachelus thompsoni Alonso-Zarazaga et Lyal 419

D

Daphnis nerii (L.) 890
Dappula tertia (Templeton) 545
Dasineura datifolia Jiang 508
Dasmithius camellia (Zhou et Huang) 1089
Dasychira axutha Collenette 972
Dasyses barbata (Christoph) 533
Demobotys pervulgalis (Hampson) 733
Dendroctonus armandi Tsai et Li 477
Dendroctonus micans (Kugelann) 480
Dendroctonus valens LeConte 481
Dendrolimus albolineatus Matsumura 827
Dendrolimus grisea (Moore) 810
Dendrolimus houi Lajonquiére 810
Dendrolimus jezoensis Matsumura 827
Dendrolimus kikuchii kikuchii Matsumura 813
Dendrolimus kikuchii Matsumura 813
Dendrolimus punctata Kirby 815
Dendrolimus punctata punctata (Walker) 815
Dendrolimus punctata tehchangensis Tsai et Liu 818
Dendrolimus punctata wenshanensis Tsai et Liu 820
Dendrolimus punctatus spectabilis Zhao et Wu 822
Dendrolimus punctatus tabulaeformis (Tsai et Liu) 829
Dendrolimus punctatus Walker 815
Dendrolimus spectabilis Butler 822
Dendrolimus stbirtcus Tschetvcrikov 827
Dendrolimus suffuscus Lajonquiere 825
Dendrolimus suffuscus suffuscus Lajonquiere 825
Dendrolimus superans (Butler) 827
Dendrolimus superans Tsai et Liu 825
Dendrolimus tabulaeformis Tsai et Liu 829
Dendrolimus xichangensis Tsai et Liu 808
Dendrophilia sophora Li et Zheng 592
Dermestes sexdentatus Borner 490
Deserticossus arenicola (Staudinger) 603
Deserticossus artemisiae (Chou et Hua) 607
Diacrisia phasma Leech 965
Diaphania perspectalis (Walker) 740
Diaspidiotus gigas (Ferris) 240
Diaspidiotus perciciosus (Comstock) 242

Diaspidiotus slavonicus (Green) 244
Diaspis pentagona Targioni-Tozzetti 238
Dichelia privatana Walker 642
Dichocrocis punctiferalis Guenée 743
Dichomeris bimaculatus Liu et Qian 596
Diclidophlebia excetrodendri (Li et Yang) 136
Dictyoploca japonica Moore 863
Dictyoploca kurimushi Voolshow 863
Dictyoploca regina Staudinger 863
Dilophodes elegans sinica Wehrli 769
Dioryctria laurata (Sic) 707
Dioryctria mendacella Staudinger 707
Dioryctria mongolicella Wang et Sung 705
Dioryctria pryeri Mutuura 707
Dioryctria pryeri Ragonot 707
Dioryctria rubella Hampson 710
Dioryctria splendidella Herrich-Schäffer 710
Dioryctria yiai Mutuura et Munroe 723
Diprion jingyuanensis Xiao et Zhang 1066
Diprion liuwanensis Huang et Xiao 1069
Diprion nanhuaensis Xiao 1071
Dixippus bobaiensis (Chen) 77
Dolbina curvata Matsumura 883
Dolbina leteralis Matsumura 883
Dolbina tancrei Staudinger 883
Drosicha corpulenta (Kuwana) 173
Dryocosmus kuriphilus Yasumatsu 1115
Dundubia hainanensis (Distant) 114
Dyscerus cribripennis Matsumura et Kono 438
Dyscerus juglans Chao 440
Dysphania militaris (L.) 772

E

Earias pudicana pupillana Staudinger 953
Earias pudicana Staudinger 953
Ectropis excellens (Butler) 776
Eligma narcissus (Cramer) 955
Enarmonia cristata Walsingham 685
Enchera substigmaria Hübner 749
Endoclita excrescens (Butler) 528
Endoclita nodus (Chu et Wang) 531
Endoclita signifier (Walker) 525
Endropia consociaria Christoph 794
Entoria bobaiensis Chen 77
Eogystia hippophaecola (Hua, Chou, Fang et Chen) 609
Eotrichia niponensis (Lewis) 297
Epicopeia mencia Moore 796
Epinotia rubiginosana (Herrich-Schäffer) 667
Epinotia rubiginosana koraiensis Falkovitch 667
Epiplema moza Butler 798

Episparis tortuosalis Moore 963
Erannis ankeraria (Staudinger) 774
Erannis defoliaria gigantea Inoue 774
Ericeia inangulata (Guenée) 962
Ericerus pela (Chavannes) 209
Eriococcus kaki Kuwana 199
Eriococus lagerstroemiae Kuwana 201
Eriocrania semipurpurella alpina Xu 523
Eriogyna pyretorum (Westwood) 866
Eriogyna tegusomushi Sasaki 866
Eriosoma dilanuginosum Zhang 162
Etiella zinckenella (Treitschke) 713
Eucryptorrhynchus brandti (Harold) 442
Eucryptorrhynchus chinensis (Olivier) 444
Eucryptorrhynchus scrobiculatus (Motschulsky) 444
Euhampsonia cristata (Butler) 910
Eulecanium diminutum Borchsenius 212
Eulecanium giganteum (Shinji) 212
Eulecanium kuwanai Kanda 214
Eumeta minuscula Butler 543
Eumeta variegata (Snellen) 547
Eumorphobotys obscuralis (Caradja) 735
Euphalerus robinae (Shinji) 132
Euproctis bipunctapex (Hampson) 975
Euproctis combinate Walker 1021
Euproctis flava (Bremer) 978
Euproctis pseudoconspersa (Strand) 980
Euproctis scintillans Walker 1012
Eupterote chinensis Leech 855
Eupterote sapivora Yang 857
Eurytoma laricis Yano 1123
Eurytoma maslovskii Nikolskaya 1125
Eurytoma plotnikovi Nikolskaya 1128
Eutomostethus deqingensis Xiao 1095
Eutomostethus longidentus Wei 1091
Eutomostethus nigritus Xiao 1093
Eutricha fentont Butler 827
Euxoa segetum Schiffermüller 938
Euzophera alpherakyella Ragonot 714
Euzophera batangensis Caradja 716

F

Fenusella taianensis (Xiao et Zhou) 1085
Fuscartona funeralis (Butler) 634

G

Gadirtha inexacta (Walker) 949
Galleruca aenescens Fairmaire 411
Gampsocoris pulchellus (Dallas) 265
Gastrolina depressa Baly 396

Gastrolina depressa depressa Baly 396
Gastrolina depressa thoracica Baly 398
Gastrolina pallipes Chen 399
Gastrolina thoracica Baly 398
Gastropacha angustipennis Walker 832
Gastropacha populifolia (Esper) 832
Gastropacha populifolia (Esper) Lajonquiere 832
Gastropacha populifolia f. *fumosa* Lajonquiere 832
Gastropacha populifolia f. *rubatrata* Lajonquiere 832
Gastropacha tsingtauica Grunberg 832
Gastropacha vishnou Lefebure 848
Gatesclakeana idia Diakonoff 669
Gibbovalva urbana (Meyrick) 554
Gilpinia massoniana Xiao 1073
Girpa inangulata Moore 962
Glyphipterix semiflavana Issiki 566
Gonocephalum bilineatum (Walker) 325
Gonocephalum orarium (Lewis) 325
Gonocephalum seriatum Boisduval 325
Grapholitha separatana Herrich-Schäffer 676
Gravitarmata margarotana (Heinemann) 671
Gryllotalpa manschurei Shiraki 56
Gryllotalpa orientalis Burmeister 54
Gryllotalpa unispina Saussure 56
Gryllus flavicornis F. 65
Gyllorhynchites cumulatus (Voss) 415
Gynaephora recens Hübner 996
Gypsonoma minutana (Hübner) 673

H

Halyomorpha halys Stål 273
Haplonchrois theae (Kusnezov) 582
Haplothrips chinensis Priesner 286
Hapsifera barbata (Christoph) 533
Hedya mormopa Diakonoff 674
Hegesidemus habrus Darke 253
Hemiberlesia pitysophila Takagi 230
Hemichroa crocea (Geoffroy) 1097
Hieroglyphus tonkinensis Bolívar 63
Hippotiscus dorsalis (Stål) 275
Histia rhodope (Cramer) 636
Hodotermopsis dimorphus Zhu et Huang 91
Hodotermopsis fanjingshanensis Zhu et Wang 91
Hodotermopsis japonicus Holmgren 91
Hodotermopsis lianshanensis Ping 91
Hodotermopsis orientalis Li et Ping 91
Hodotermopsis sjostedti Holmgren 91
Hodotermopsis yui Li et Ping 91
Holcocerus arenicola (Staudinger) 603
Holcocerus arenicola var. *insularis* Staudinger 611

Holcocerus artemisiae Chou et Hua 607
Holcocerus hippophaecolus Hua, Chou, Fang et Chen 609
Holcocerus insulans Staudinger 611
Holcocerus japonicus Gaede 613
Holotrichia diomphalia Bates 295
Holotrichia oblita (Faldermann) 295
Holotrichia titanis Reitter 297
Homalocephala homali Yang et Li 135
Homona issikii Yasuda 648
Hulodes inangulata Guenée 962
Hyblaea puera Cramer 948
Hyles hippophaes (Esper) 879
Hylobitelus xiaoi Zhang 446
Hyphantria cunea (Drury) 967
Hyphantria cunea Drury et Fitch 967
Hyphantria cunea Drury et Hampson 967
Hyphantria textor Harris 967
Hypocala aspersa Butler 960
Hypocala subsatura Guenée 960
Hypocala subsatura var. *limbata* Butler 960
Hypogymna dispar L. 984
Hypomeces squamosus (F.) 447
Hyponomeuta variabilis Zeller 571
Hyposipatus gigas F. 425
Hypsauchenia chinensis Chou 118
Hypsipyla pagodella Ragonot 718
Hypsipyla robusta (Moore) 718

I

Icerya purchasi Maskell 176
Illiberis ulmivora (Graeser) 639
Inope ulmivora (Jordan) 639
Ips acuminatus (Gyllenhal) 484
Ips duplicatus (Sahalberg) 485
Ips hauseri Reitter 486
Ips nitidus Eggers 488
Ips sexdentatus (Börner) 490
Ips subelongatus (Motschulsky) 492
Ips typographus (L.) 494
Ips typographus De Geer 490
Iragoides conjuncta (Walker) 624
Iscadia inexacta Walker 949
Ivela ochropoda (Eversmann) 982

J

Jankowskia fuscaria (Leech) 778

K

Kakuvoria flavofasciata Nagano 568
Kermes nawae Kuwana 204

Kerria **spp. 187**
Kophene snelleni Heylaerts 538
***Kumasia kumaso* (Sugi) 942**
***Kunugia xichangensis* (Tsai et Liu) 808**
***Kurisakia sinocaryae* Zhang 168**
***Kytorhinus immixtus* Motschulsky 384**

L

***Lachnus tropicalis* (van der Goot) 157**
Lacida postica Walker 998
Lampra adustella (Obenberger) 313
Lampra limbata Gebler 313
Lampra mongolica (Obenberger) 313
***Lamprodila limbata* (Gebler) 313**
Larie dispar L. 984
Lasiognatha mormopa (Meyrick) 674
Laspeyresia coniferana (Saxesen) 676
Laspeyresia trasias Meyrick 663
Laspeyresia zebeana (Ratzeburg) 678
Latoia consocia (Walker) 628
***Lawana imitata* (Melichar) 105**
Lebeda bimaculata Walker 846
***Lebeda nobilis sinina* Lajonquiere 835**
Lebeda nobilis Walker 835
Lebeda plagifera Walker 844
Lecanium gigantea Shinji 212
Lecanium kuwanai Takahashi 214
***Lemyra phasma* (Leech) 965**
***Lepidiota stigma* (F.) 299**
Lepidosaphes conchiformioides Borchsenius 235
***Lepidosaphes conchiformis* (Gmelin) 235**
***Lepidosaphes salicina* Borchsenius 233**
***Leptocybe invasa* Fisher et La Salle 1118**
***Lepyrus japonicus* Roelofs 449**
***Leucoma candida* (Staudinger) 1015**
***Leucoma salicis* (L.) 1016**
***Leucoptera sinuella* (Reutti) 558**
Leucoptera susinella (Reutti) 558
Leucotermes speratus Holmgren 96
Liparis Mathura Swinhoe 988
Liparis salicis Bremer 1015
Liparis salicis L. 1016
***Lobesia cunninghamiacola* (Liu et Pai) 681**
***Locastra muscosalis* (Walker) 720**
***Lonchodes bobaiensis* (Chen) 77**
Loudonta dispar (Kiriakoff) 912
***Lycorma delicatula* (White) 103**
Lyda flagellicornis Smith 1049
Lyda nemoralis sensu Thomson 1043
Lygaeonematus compressicornis (F.) 1113

Lymantria aurora Swinhoe 988
***Lymantria dispar* (L.) 984**
***Lymantria mathura* Moore 988**
***Lymantria monacha* (L.) 990**
***Lymantria xylina* Swinhoe 992**

M

***Macellina digitata* Chen et He 73**
***Macrobathra flavidus* Qian et Liu 581**
***Macrosemia pieli* (Kato) 116**
***Macrotermes barneyi* Light 97**
***Mahasena colona* Sonan 549**
***Malacosoma dentata* Mell 837**
Malacosoma neustria dentata Mell 837
***Malacosoma neustria testacea* (Motschulsky) 839**
***Malacosoma rectifascia* Lajonquière 842**
***Maladera orientalis* (Motschulsky) 300**
Mallambyx raddei (Blessig) 356
Massicus raddei (Blessig) 356
***Matapa aria* (Moore) 1036**
Matsucoccus liaoningiensis Tang 178
Matsucoccus massonianae Yang et Hu 178
***Matsucoccus matsumurae* (Kuwana) 178**
***Matsucoccus shennongjiaensis* Young and Liu 180**
***Matsucoccus sinensis* Chen 184**
***Matsucoccus yunnanensis* Ferris 182**
Mecorhis ursulus (Roelofs) 415
***Megabeleses liriodendrovorax* Xiao 1099**
Meganoton analis (Felder) 875
***Megastigmus cryptomeriae* Yano 1117**
Megopis sinica (White) 358
***Meichihuo cihuai* Yang 781**
***Melanaphis bambusae* (Fullaway) 146**
Melanauster glabripennis Matsumura 330
***Melanocercops fi cuvorella* (Yazaki) 556**
Melanophila decostigma F. 315
Melanophila picta Pallas 315
Melanophila picta Saunders 315
***Melanotus cribricollis* (Faldermann) 319**
***Melolontha hippocastani mongolica* Ménétriés 302**
Messa acericola Xiao 1084
***Metacanthus pulchellus* Dallas 265**
***Metaceronema japonica* (Maskell) 215**
***Metasalis populi* (Takeya) 253**
***Micadina zhejiangensis* Chen et He 75**
Micromelalopha populivona Yang et Lee 915
***Micromelalopha sieversi* (Staudinger) 915**
Micromelalopha troglodyte (Graeser) 915
***Milionia basalis pryeri* Druce 783**
Mimopydna insignis (Leech) 893

Mirina christophi (Staudinger) 800
Monema flavescens Walker 621
Monochamus alternatus Hope 360
Monochamus tesserula White 360
Monochamus urussovii (Fischer-Waldheim) 362
Monosteira unicostata (Mulsant et Rey) 256
Moricella rufonota Rohwer 1101
Mylabris immixta Baudi 384
Mytilaspis conchiformis (Gmelin) 235

N

Nasutitermes erectinasus Tsai et Chen 99
Nasutitermes parvonasutus (Nawa) 100
Necroscia flavescens (Chen et Wang) 79
Nematus prunivorous Xiao 1103
Neoasterodiaspis castaneae (Russell) 224
Neocerambyx raddei Blessig 356
Neocerura tattakana (Matsumura) 917
Neocerura wisei (Swinhoe) 917
Neoclostera insignior Kiriakoff 906
Neodiprion dailingensis Xiao et Zhou 1074
Neodiprion huizeensis Xiao et Zhou 1076
Neodiprion xiangyunicus Xiao et Zhou 1078
Neostauropus alternus Kiriakoff 928
Nesodiprion huanglongshanicus Xiao et Huang 1080
Nesodiprion zhejiangensis Zhou et Xiao 1082
Nesticoccus sinensis Tang 190
Nezara viridula (L.) 277
Nigrotrichia gebleri (Faldermann) 295
Niphades castanea Chao 451
Niphades verrucosus (Voss) 452
Noctua heteroclita Müller 990
Noctua suffusa (Denis et Schiffermüller) 936
Norraca retrofusca de Joannis 897
Notobitus meleagris (F.) 267
Notobitus montanus Hsiao 269
Notodonta albibasis Oberthür 921
Notolophus ericae Germar 994
Notonagemia analis (Felder) 875
Numenes interiorata Walker 957
Nygmia bipunctapex Hampson 975
Nygmia pseudoconspersa Strand 980
Nygmia scintillans Walker 1012
Nygmia subflava Bremer 978

O

Obolodiplosis robiniae (Haldemann) 510
Ochrostigma albibasis Chiang 921
Ocneria furva Leech 1007
Ocneria monacha L. 990

Odites issikii (Takahashi) 584
Odites perissopis Meyrick 584
Odites plocamopa Meyrick 584
Odonestis plagifera Walker 844
Odonestis superans Butler, 827
Odonestis urda Swinhoe 844
Odontotermes formosanus (Shiraki) 101
Odontotermes qianyagensis Lin 101
Oides leucomelaena Weise 400
Olethreutes mormopa Clarke 674
Oligia apameoides Draudt 940
Oligia vulgaris Butler 940
Omalophora costalis (Matsumura) 109
Omphisa plagialis Wileman 745
Ophiorrhabda mormopa (Meyrick) 674
Ophrida xanthospilota (Baly) 402
Opogona sacchari (Bojer) 535
Oracella acuta (Lobdell) 192
Orgyia antiguoides (Hübner) 994
Orgyia antiqua (L.) 996
Orgyia caliacrae Caradja 994
Orgyia ceylanica Nietner 998
Orgyia gonostigma (L.) 996
Orgyia postica (Walker) 998
Otidognathus davidis (Fairmaire) 421
Oxycephala longipennis Gestro 391

P

Pachynematus itoi Okutani 1104
Paliga machoeralis Walker 747
Palmar limbata Zykor 313
Pammene ginkgoicola Liu 680
Pandemis heparana (Denis et Schiffermüller) 651
Pantana phyllostachysae Chao 1000
Pantana sinica Moore 1004
Papilio elwesi Leech 1028
Papilio epycides epycides Hewitson 1030
Papilio machaon L. 1033
Papilio slateri Hewitson 1032
Papilio xuthus L. 1034
Paralebeda plagifera (Walker) 844
Paralebeda urda backi Lajonquière 844
Parametriotes theae Kusnezov 582
Paranthrene tabaniformis (Rottemburg) 560
Parasa consocia Walker 628
Parasaissetia nigra (Nietner) 217
Parnops glasunovi Jacobson 404
Parocneria furva (Leech) 1007
Parocneria orienta Chao 1010
Parthenolecanium corni (Bouché) 219

Pegomya phyllostachys (Fan) 517
Pemphigus immunis Buckton 164
Percnia giraffata (Guenée) 786
Periergos dispar (Kiriakoff) 912
Perina basalis Walker 1021
Perina nuda (F.) 1021
Periphyllus koelreuteriae (Takahashi) 150
Petrova cristata Obraztsov 685
Phalaena (Bombyx) narcissus Cramer 955
Phalaena (Tortirx) resinella L. 687
Phalaena (Tortrix) mitterbacheriana Denis & Schiffer
Phalaena (Tortrix) zebeana Ratzeburg 678
Phalaena atlas L. 873
Phalaena cunea Drury 967
Phalaena dispar L. 984
Phalaena hartigiana Saxesen 665
Phalaena lactinea Cramer 964
Phalaena padella L. 571
Phalaena salicis L. 1016
Phalaena strobilella L. 684
Phalera (Noctua) bucephala L. 919
Phalera bucephala (L.) 919
Phalera bucephala infulgens Graeser 919
Phalera flavescens (Bremer et Grey) 930
Phalera flavescens kuangtungensis Mell 930
Phalera takasagoensis Matsumura 932
Phalera takasagoensis matsumurae Okano 932
Phalera takasagoensis ulmovora Yang et Lee 932
Phalerodonta albibasis (Chiang) 921
Phalerodonta bombycina (Oberthür) 921
Phassus excrescens (Butler) 528
Phassus nodus Chu et Wang 531
Phenacoccus azaleae Kuwana 194
Phenacoccus fraxinus Tang 196
Phenacoccus solenopsis Tinsley 197
Philosamia cynthia Walker et Felder 870
Philus antennatus (Gyllenhal) 364
Phlaeoba angustidorsis Bolívar 66
Phloeosinus aubei (Perris) 496
Phloeosinus sinensis Schedl 499
Phlossa conjuncta (Walker) 624
Phulaena gonostigma L. 996
Phycita pryeri Leech 707
Phycita rubella South 710
Pieris crataegi L. 1025
Pimelocerus juglans (Chao) 440
Pirkimerus japonicus (Hidaka) 262
Pissodes punctatus Langor et Zhang 453
Pissodes validirostris (Sahlberg) 455
Pissodes yunnanensis Langor et Zhang 457

Pityogenes chalcographus (L.) 501
Plagiodera versicolora (Laicharting) 406
Plagiosterna adamsi (Baly) 393
Platylomia pieli Kato 116
Platypeplus mormopa Meyrick 674
Platysamia atlas Oliver 873
Pleonomus canaliculatus Faldermann 321
Podagricomela shirahatai (Chûjô) 408
Podagricomela shirahatai Chen 408
Podontia lutea (Oliver) 409
Poecilocoris latus Dallas 278
Poecilonota limbata Gebler 313
Poecilonota variolosa (Paykull) 317
Polychrosis cunninghamiacola Liu et Pai 681
Polydesma vulgaris Butler 945
Polylopha cassiicola Liu et Kawabe 653
Polyphagozerra coffeae (Nietner) 615
Polyura athamas (Drury) 1022
Polyura athamas athamas (Drury) 1022
Polyura narcaea (Hewitson) 1023
Polyura narcaea meghaduta (Fruhstorfer) 1023
Polyura narcaea narcaea (Hewitson) 1023
Pontia crataegi L. 1025
Porthesia scintillans Walker 1012
Prays alpha Moriuti 570
Pristiphora beijingensis Zhou et Zhang 1106
Pristiphora conjugata (Dahlbom) 1109
Pristiphora erichsonii (Hartig) 1111
Proagopertha lucidula (Faldermann) 293
Procris pekinensis Draeseke 639
Prodiaspis tamaricicola (Malenotti) 237
Propachys nigrivena Walker 725
Protocryptis obducta Meyrick 575
Pryeria sinica Moore 640
Pseudaulacaspis pentagona (Targioni-Tozzetti) 238
Pseudopiezotrachelus collaris (Schilsky) 427
Pseudotomoides strobilella L. 684
Psilogramma menephron (Cramer) 885
Pterochlorus tropicalis van der Goot 157
Pterostoma grisea Graeser 923
Pterostoma sinica Moore 923
Ptilinus fuscus (Geoffroy) 323
Ptycholomoides aeriferanus (Herrich-Schäffer) 654
Purpuricenus temminckii (Guérin-Méneville) 366
Pydna anaemica Kiriakoff 893
Pydna goddrica Schaus 895
Pygaera anachoreta pallida Staudinger 903
Pygaera flavescens Bremer et Grey 930
Pygaera mahatma Bryk 903
Pygaera sieversi Staudinger 915

Pyralis resinana F. 687
Pyrosis eximia Oberthür 802
Pyrosis idiota Graeser 804
Pyrrhalta aenescens (Fairmaire) 411

Q

Quadraspidiotus populi Bodenheimer 244
Quadrastichus erythrinae Kim 1120
Quadrialcarifera cyanea (Leech) 926
Quadrialcarifera fransciscana Kiriakoff 926

R

Rabdophaga salicis (Schrank) 513
Rammeacris kiangsu (Tsai) 59
Ramulus chongxinense (Chen et He) 67
Ramulus minutidentatus (Chen et He) 69
Ramulus pingliense (Chen et He) 71
Reticulitermes chinensis Snyder 94
Reticulitermes flaviceps (Oshima) 96
Reticulitermes labralis Hsia et Fan 94
Retinia cristata (Walsingham) 685
Retinia margarotana Heinemann 671
Retinia pinicolana Doubleday 692
Retinia resinella (L.) 687
Retinia retiferana Wocke 671
Rhodoneura sphoraria (Swinhoe) 698
Rhyacionia duplana (Hübner) 688
Rhyacionia duplana simulata Heinrich 688
Rhyacionia insulariana Liu 690
Rhyacionia pinicolana (Doubleday) 692
Rhynchaenus empopulifolis Chen 459
Rhynchaenus frigidus (F.) 466
Rhynchophorus ferrugineus (Olivier) 423
Rhynchophorus gigas Herbst 425
Ricania speculum (Walker) 124
Rondotia menciana Moore 852

S

Salebria laruata Heinrich 707
Salebria laruata Issiki et Mutuura 707
Salebria laruata Mutuura 707
Samia cynthia (Drury) 870
Saperda carcharias (L.) 368
Saperda populnea (L.) 370
Sapporia repetita (Butler) 944
Sasakiaspis pentagona Kuwana 238
Saturnia atlas Walker 873
Schlechtendalia chinensis (Bell) 158
Scirtothrips dorsalis Hood 284
Scolytus schevyrewi Semenov 503

Scoparia resinalis Guenée 687
Scopelodes testacea Butler 632
Scythropiodes issikii (Takahashi) 584
Scythropus yasumatsui Kono et Morinoto 461
Selepa celtis Moore 950
Semanotus bifasciatus (Motschulsky) 372
Semanotus sinoauster Gressitt 374
Semiothisa cinerearia (Bremer et Grey) 788
Serica orientalis Motschulsky 300
Sesia siningensis (Hsu) 562
Shansiaspis sinensis Tang 246
Shirahoshizo coniferae Chao 463
Shirahoshizo patruelis (Voss) 464
Sinibotys evenoralis (Walker) 731
Sinomphisa plagialis (Wileman) 745
Sinonasutitermes erectinasus (Tsai et Chen) 99
Sinophasma brevipenne Günther 81
Sinophasma crassum Chen et He 86
Sinophasma largum Chen et Chen 82
Sinophasma maculicruralis Chen 84
Sinophasma mirabile Günther 86
Sinophasma pseudomirabile Chen et Chen 88
Sinorsillus piliferus Usinger 264
Sinoxylon japonicum Lesne 324
Sinuonemopsylla excetrodendri Li et Yang 136
Sipalinus gigas (F.) 425
Sipalus gigas (F.) 425
Sipalus hypocrite Boeheman 425
Sipalus misumenus Boeheman 425
Sirex gigas L. 1057
Sirex rufi abdominis Xiao et Wu 1053
Smerinthus planus Walker 887
Smerinthus tatarinovi Bremer et Grey 877
Smerithus planus planus Walker 887
Solenopsis invicta Buren 1133
Somena scintillans (Walker) 1012
Somera cyanea Leech 926
Sonsaucoccus sinensis (Chen) 184
Sphaleroptera diniana Guenée 694
Sphecia siningensis Hsu 562
Sphinx panopus Cramer 889
Spilosoma cunea Drury 967
Stathmopoda masinissa Meyrick 568
Stauronematus compressicornis (F.) 1113
Stauropus alternus Walker 928
Stegnoptycha rubiginosana Herrich-Schäffer 667
Stephanitis illicii Jing 258
Stephanitis laudata Drake et Poor 259
Stephanitis svensoni Drake 261
Sternochetus gravis (F.) 466

Sternochetus mangiferae (F.) 468
Sternochetus olivieri (Faust) 470
Sternuchopsis juglans (Chao) 428
Sternuchopsis sauteri Heller 429
Stilpnotia candida Staudinger 1015
Stilpnotia ochropoda Eversmann 982
Stilprotia salicis (L.) 1016
Streltzoviella insularis (Staudinger) 611
Strepsicrates coriariae Oku 693
Striglina bifi da Chu et Wang 700
Strobilomyia spp. 519
Suana ampla Walker 846
Suana cervina Moore 846
Suana concolor (Walker) 846
Subrita curviferella Walker 950
Sucra jujuba Chu 764
Sympiezomias lewisi Roelofs 472
Sympiezomias velatus (Chevrolat) 472
Synanthedon castanevora Yang et Wang 564
Syntomoza homali (Yang et Li) 135
Syntypistis cyanea (Leech) 926

T

Takecallis arundinariae (Essig) 166
Takecallis taiwana (Takahashi) 167
Tarbinskiellus portentosus (Lichtenstein) 52
Tenthredo erythrocephala L. 1038
Tenthredo stellata Christ 1043
Termes valgaris Shiraki 101
Tessaratoma papillosa (Drury) 281
Tetramesa phyllotachitis (Gahan) 1131
Tettigoniella viridis (L.) 120
Thalassodes quadraria Guenée 790
Thespea bicolor (Walker) 626
Thosea sinensis (Walker) 630
Thylactus simulans Gahan 376
Thyrgorina phasma Leech 965
Thysanogyna limbata Enderlein 137
Tilophora flavipes (Uhler) 110
Tomicus minor (Hartig) 505
Tomicus piniperda (L.) 506
Tomicus stenographus Duftschmidt 490
Tortrix coniferana Saxesen 676
Tortrix duplana Hübner 688
Tortrix griseana Hübner 694

Tortrix heparana Denis et Schiffermüller 651
Tortrix lafauryana Ragonot 645
Tortrix minutana Hübner 673
Tortrix retusana Haworth 655
Trabala cristata Butler 910
Trabala vishnou (Lefebure) 848
Trabala vishnou gigantina Yang 850
Trabala vishnou vishnou (Lefebure) 848
Trachea tokionis (Butler) 934
Trachypteris picta Marseul 315
Trachypteris picta picta (Pallas) 315
Tremex fuscicornis (F.) 1055
Trichoferus campestris (Faldermann) 377
Trioza camphorae Sasaki 139
Trioza magnisetosa Loginova 140
Trisula andreas Oberthür 930

U

Unaspis euonymi (Comstock) 248
Unaspis euonymi Takahashi et Kanda 248
Urocerus gigas taiganus Benson 1057
Urocerus gigas gigas (L.) 1057
Urocerus taiganus Benson 1057
Urostylis yangi Maa 283

X

Xerodes rufescentaria (Motschulsky) 794
Xyleutes persona (Le Guillou) 617
Xylococcus matsumurae Kuwana 178
Xylotrechus rusticus (L.) 378
Xystrocera globosa (Olivier) 381

Y

Yakudza vicarius (Walker) 613
Yemma signata (Hsiao) 266
Yponomeuta padella (L.) 571

Z

Zamacra excavate (Dyar) 792
Zeiraphera griseana (Hübner) 694
Zethenia rufescentaria chosenaria Bryk 794
Zethenia rufescentaria Motschulsky 794
Zeuzera coffeae Nietner 615
Zeuzera leuconotum Butler 617
Zeuzera multistrigata (Moore) 619

English Common Names Index

A

Ailanthus silkmoth 870
Ailanthus webworm 573
Alder sawfly 1097
Apple brown tortrix 651
Arborvitae aphid 155
Armand pine weevil 453
Army green moth 890
Asian longhorned beetle 330
Asian palm weevil 423
Asian swallowtail 1034
Asiatic rose scale 227
Atlas moth 873

B

Bamboo aphid 146
Bamboo cicada 116
Bamboo hemisphere scale 225
Bamboo nest scale 190
Bamboo shoot weevil 421
Bamboo weevil 417
Banana moth 535
Banded elm bark beetle 503
Barley wireworm 318
Black arched tussock moth 990
Black cutworm 936
Black locust aphid 144
Black locust gall midge 510
Black-spotted bamboo aphid 166
Black-veined white 1025
Blue gum chalcid 1118
Blue striped mime 1032
Box tree moth 740
Brindled shoot 673
Brown elm scale 219
Brown marmorated stink bug 273
Buff-tip moths 919

C

Camellia weevil 433
Camellia whitefly 142
Camphor psyllid 139
Camphor swallowtail 1030
Casuarina moth 992
Cherry ermine moth 571
Chestnut aphid 157
Chestnut gall wasp 1115
Chilli thrips 284
China fir borer 374
Chinese (oak) tussar moth 861
Chinese fir cone gelechiid 596
Chinese fir cone webworms 581
Chinese larch case bearer 577
Chinese moon moth 859
Chinese pine caterpillar 829
Chinese pistacia looper 766
Chinese processionary moth 855
Chinese sumac aphid 158
Chinese tasar moth 861
Chinese thrip 286
Chinese white pine beetle 477
Chinese yellow swallowtail 1034
Cocoa tussock moth 998
Coconut hispid 391
Coconut leaf beetle 391
Codling moth 646
Common redeye 1036
Common tiger moth 772
Cone flies 519
Cottony cushion scale 176
Crape myrtle bark scale 201
Cup moth 624
Cupreous chafer red-footed green beetle 291
Cutworm 938
Cypress bark beetle 496

D

Dadap shield scale 238
Dahurian larch case bearer 575
Dark fruit-tree tortrix 651
Dechang pine caterpillar 818
Deep mountain longhorn beetle 356
Dusky clearwing 560

E

Eight-spotted fulgorid 124
Elgin shoot moth 688
Elm carpenter moth 613
Elm caterpillar 796
Elm leaf worm 639
Emerald ash borer 304
Engraver beetle 484
Erythrina gall wasp 1120
Euonymus leaf notcher 640
Euonymus moth 762
Euonymus scale 248
European fruit lecanium scale 219
European spruce bark beetle 494

F

Fall webworm 967
False tiger moth 964
Fig leafminer 556
Fig scalem 235
Flattened eucleid caterpillar 630
Formosan subterranean termite 92
Fruit lecanium 219

G

Giant cutworm 934
Giant globular scale 212
Giant looper 755
Giant mealybug 173
Giant tea looper 757
Giant woodwasp 1057
Gold dust weevil 447
Golden rain tree aphid 150
Gray tiger longicorn beetle 378
Great spruce bark beetle 480
Green cochild 628
Green flower beetle 291
Green leafhopper 120
Grey double-bristled hawkmoth 875

Grooved click beetle 321
Gypsy moth 984

H
Hauser's engraver 486
Hevea tussock moth 998
Honey locust psyllid 132
Hornet moth 562

J
Japanese elm sawfly 1059
Japanese elm woolly aphid 162
Japanese giant silkworm 863
Japanese pagoda tree psyllid 130
Japanese pine bast scale 178
Japanese pine sawyer beetle 360
Japanese red pine caterpillar 822
Japanese swift moth 528
Jujube fruit fly 522
Juniper bark borer 372
Juniper tussock moth 1007

K
Kondo white psychid 540

L
Lac insects 187
Larch adelgid 170
Larch bark beetle 492
Larch bark moth 678
Larch caterpillar 827
Larch sawfly 1104, 1111
Larch tortrix 694
Larch wooly aphid 170
Large black-spotted geometrid 786
Large brown hawkmoth 885
Large green chafer beetle 291
Large pine weevil 425
Large poplar borer 368
Leaf beetle 388
Lesser pine shoot beetle 505
Litchi fruit moth 661
Litchi stink bug 281
Loblolly pine mealybug 192
Lobster caterpillar 928

M
Macadamia nut borer 661
Mango aphid 147
Mango hawk moth 889

Mango nut weevil 468
Mango pulp weevil 466
Mango seed weevil 468, 470
Mango stone weevil 468
Mango weevil 468
Masson pine caterpillar 815
Metallic-green beetle 288
Mitterbach's red roller moth 655
Mongolia mole cricket 56
Mulberry longhorn beetle 332
Mulberry psyllid 127

N
Narrow yellow-tipped prominent 932
Narrow-leaved oleaster psyllid 140
Narrow-necked click beetle 318
Nettle grub 630
Nigra scale 217
Northern bark beetle 485
Nun moth 990

O
Oleander hawk moth 890
Orange spotted shoot moth 692
Oriental brown chafer 302
Oriental cicada 111
Oriental eyed hawkmoth 887
Oriental leaf beetle 386
Oriental leopard moth 617
Oriental mole cricket 54

P
Palm heart leaf miner 391
Palm leaf beetle 391
Papaya scale 238
Parasol psyllid 137
Paulownia bagworm 547
Peach fruit moth (borer) 598
Peach pyralid 743
Peach twig borer 588
Pear oystershell scale 235
Pear white 1025
Persimmon fruit moth 568
Pine cone moth 671
Pine cone weevil 455
Pine false webworm 1038
Pine needle hemiberlesian scale 230
Pine resin-gall moth 687
Pine shoot beetle 506
Pine shoot moth 710

Pine spittle bug 110
Poplar and willow borer 428
Poplar and willow lace bug 253
Poplar bole clearwing moth 562
Poplar lace bug 256
Poplar lasiocampid 832
Poplar leaf aphid 148
Poplar leaf beetle 394
Poplar leaf roller weevil 414
Poplar prominent 900
Poplar sawfly 1106, 1109
Poplar stem gall aphid 164
Privet hawk moth 885
Pulse pod borer 713

R
Red costate tiger moth 964
Red imported fire ant 1133
Red palm weevil 423
Red turpentine beetle 481
Red wax scale 208
Red-necked longhorn beetle 340
Red-sided buprestid 313
Red-striped longhorn beetle 344

S
Sagebrush carpenter worm 607
Sago palm weevil 423
Salix gall midge 513
Sallow leafroller moth 586
San Jose scale 242
Sandthorn carpenter worm 609
Satin moth 1016
Scarce chocolate-tip 903
Seabuckthorn carpenter worm 609
Simao pine caterpillar 813
Six-toothed bark beetle 490
Sixtoothed spruce bark beetle 501
Slug moth 624
Small poplar borer 370
Small wrinkled stink bug 270
Southern green stink bug 277
Splendid knot-horn moth 707
Spotted lanternfly 103
Spotted locust 58
Spring cankerworm 751
Spruce seed moth 684
strawberry thrips 284
Striped alder sawfly 1097
Summer fruit tortrix moth 642

Summer shoot moth 688
Swift moth 528
Sycamore lace bug 251

T

Tallow aphid 147
Tallow tree brown tail moth 975
Tea bagworm 543
Tea moth 582
Tea seed bug 278
Tea tussock moth 980
Teak leaf skeletonizer 747
Teak moth 948
Temperate tussar moth 861
Ten-spotted buprestid 315
Tent caterpillar 839
Tent caterpillar 848
The common nawab 1022
the Xuthus swallowtail 1034
Thin-winged longicorn beetle 358
Top spotted tussock moth 996

Tortoise wax scale 205
Tremex wasp 1055
Tung oil tree geometrid 759
Turnip moth 938
Two-lined velvet hawkmoth 881

V

Vapourer moth 996

W

Wenshan pine caterpillar 820
Wheat wireworm 321
White mulberry scale 238
White peach scale 238
White satin moth 1016
White spotted longhorn beetle 351
White wax scale insect 209
White-striped longhorn beetle 348
Wide-tailed swallowtail 1028
Willow moth 1015
Willow oyster scale 233

Willow scale 240

X

Xiao pine weevil 444
Xichang pine caterpillar 808
Xingan larch case bearer 575

Y

Yellow margined buprestid 313
Yellow peach moth 743
Yellow swallowtail butterfly 1033
Yellow tail tussock moth 1012
Yellow tea thrips 284
Yellow-legged tussock moth 982
Yellow-spine bamboo locust 59
Yunnan pine caterpillar 810
Yunnan pine weevil 457

Z

Ziziphus leaf roller 657